Proceedings of the Second

Specialty Conference
on

Dynamic Response of Structures: Experimentation, Observation, Prediction and Control

Organized by the Engineering
Mechanics Division of the
American Society of Civil Engineers

CO-SPONSORED BY:
ASCE/EMD Technical Committee on Experimental
Analysis and Instrumentation
ASCE/EMD Technical Committee on Dynamics
ASCE/STD Technical Committee on Dynamic Effects
Earthquake Engineering Research Institute
National Science Foundation
Structural Engineers Association of California
Wind Engineering Research Council
Georgia Institute of Technology, School of Civil Engineering

Gary Hart, Editor

JANUARY 15-16, 1981
Sheraton, Atlanta
ATLANTA, GEORGIA

Published by the
American Society of Civil Engineers
345 East 47th Street
New York, New York 10017

Library of Congress Catalog Card No. 80-70135
ISBN 0-87262-261-4
Manufactured in the United States of America.

PREFACE

The Engineering Mechanics Division of the American Society of Civil Engineers has sponsored many valuable specialty conferences. On March 30 and 31, 1976, at the Universty of California, Los Angeles, it sponsored the specialty conference entitled "Dynamic Response of Structures: Instrumentation, Testing Methods and System Identification". That specialty conference was motivated by the belief of the Engineering Mechanics Division's Executive Committee that the general structural dynamics area of applied research would benefit from the bringing together of engineers interested in instrumentation, testing and system identification.

The Engineering Mechanics division selected Georgia Institute of Technology as host for the Second ASCE/EMD Specialty Conference on Dynamic Response of Structures. The Proceedings contained herein are for that conference and reflect new research in these areas of structural dynamics since the 1976 specialty conference.

The National Science Foundation provided financial support for the publication of these proceedings. I wish to especially acknowledge on behalf of many of the researchers involved in this specialty conference years of technical guidance and financial support by the National Science Foundation for research on the topics discussed herein. Financial support for this specialty conference was also provided by the American Society of Civil Engineers, Georgia Institute of Technology and the University of California at Los Angeles. On behalf of my conference co-chairmen, Professor Gerald A. Wempner and Professor Barry J. Goodno, I wish to express our appreciation to the conference steering committee for their direction and assistance.

Gary C. Hart
Conference Co-Chairman, Technical Program

CONFERENCE CO-CHAIRMEN

GERALD A. WEMPNER, GENERAL CHAIRMAN
School of Engineering Science and Mechanics
Georgia Institute of Technology

BARRY J. GOODNO, CO-CHAIRMAN, ARRANGEMENTS
School of Civil Engineering
Georgia Institute of Technology

GARY C. HART, CO-CHAIRMAN, TECHNICAL PROGRAM
School of Engineering and Applied Science
University of California, Los Angeles

CONFERENCE STEERING COMMITTEE

MELVIN L. BARON
WEIDLINGER ASSOCIATES
NEW YORK, NEW YORK

JACK E. CERMAK
DEPARTMENT OF CIVIL ENGINEERING
COLORADO STATE UNIVERSITY
FORT COLLINS, COLORADO

RAY W. CLOUGH
DEPARTMENT OF CIVIL ENGINEERING
UNIVERSITY OF CALIFORNIA
BERKELEY, CALIFORNIA

PAUL C. JENNINGS
DIVISION OF ENGINEERING
CALIFORNA INSTITUTE OF TECHNOLOGY
PASADENA, CALIFORNIA

ROBERT H. SCANLAN
DEPARTMENT OF CIVIL ENGINEERING
PRINCETON UNIVERSITY
PRINCETON, NEW JERSEY

HARESH C. SHAH
DEPARTMENT OF CIVIL ENGINEERING
STANFORD UNIVERSITY
STANFORD, CALIFORNIA

STUART E. SWARTZ
DEPARTMENT OF CIVIL ENGINEERING
KANSAS STATE UNIVERSITY
MANHATTAN, KANSAS

ANESTIS S. VELETSOS
DEPARTMENT OF CIVIL ENGINEERING
RICE UNIVERSITY
HOUSTON, TEXAS

RICHARD N. WHITE
SCHOOL OF CIVIL AND ENVIRONMENTAL ENGINEERING
CORNELL UNIVERSITY
ITHACA, NEW YORK

CONTENTS

ANALYSIS OF FLOATING BRIDGES: THE HOOD CANAL BRIDGE

By Rene W. Luft,[1] M. ASCE

INTRODUCTION

The floating bridge across the Hood Canal, completed over 15 years ago, is the last major floating bridge designed and built in the United States. The design of its continuous pontoon structure was based on analyses that gave vertical bending moments and torsional moments but no horizontal bending moments; however, it has been found that the horizontal bending moments (moments about a vertical axis) cause stresses of the same order of magnitude as the vertical bending moments. Vertical bending moments were computed first for the water in a static wave-like pattern along the length of the bridge and then increased due to dynamic amplification using Timoshenko's equation for vibrations of a beam on an elastic foundation (30). The design method did not account for the frequency dependent dynamic interaction between the waves and the bridge.

In the absence of a need for design methods for floating bridges, few researchers have studied their behavior. One major investigation, pursued at the University of Washington by Hartz and his students (6, 14, 15, 20), instrumented the Hood Canal bridge and then compared the measurements to theoretical predictions. These predictions were based on models of the bridge as beams on an elastic foundation (one each for heave, sway, and roll motion) with wave forces applied as load time histories.

On 13 February 1979, during a severe storm over the Olympic peninsula, the west span of the Hood Canal bridge failed and sank. As a consequence of that event, an investigation was performed to establish the probable causes of failure. The study included a complete analysis of the bridge for design and environmental loadings, and a scale model of the bridge to substantiate the analysis. During that study, methods of analysis of floating bridges were assembled and used.

The purposes of this paper are to present and discuss the steps involved in a floating bridge design and to discuss the model basin study as a calibration tool for the analysis. It contains an overview of the main aspects of floating bridge analysis and design, and it presents selected results of the review of the Hood Canal bridge. The following topics of this multidisciplinary design problem are briefly covered: wave forecasting and wave spectra, wind forecasting and turbulence spectra, hydrodynamics of floating bodies, and random vibration methods of solution. Because of the breadth of the presentation, no in-depth discussion of any single topic is attempted. A companion paper (3) gives the details of the model study performed by Alden Research Laboratory.

DESCRIPTION OF HOOD CANAL BRIDGE

The Hood Canal bridge is a floating concrete pontoon bridge in the Seattle, Washington, area. This bridge, which connects the Kitsap Peninsula to the east and the Olympic Peninsula to the west, is the only floating bridge in the United States to span a coastal inlet; it is also the longest with its 6,541 ft long floating

[1] Associate, Simpson Gumpertz & Heger Inc., Cambridge, Mass.

section. The total bridge length including fixed approach structures at each end is 7,861 ft.

The approach structures are of steel girder and reinforced concrete pier construction common to highway bridges of relatively short span in shallow water. The links between the ends of the floating structure and the approach structures are steel truss transition spans, 280 ft and 285 ft long. The transition spans provide longitudinal restraint for the floating portion, but they do not restrain the pontoons in the vertical direction during tidal fluctuations or in the horizontal direction transverse to the bridge. The transition spans are sufficiently high to allow small vessels to pass below.

The east and west spans of the floating portion of the bridge, which cross water up to 340 ft deep, were built using 10 and 13 individual pontoons, respectively. The pontoons are bolted and post-tensioned together to obtain continuous beams with full bending and shear continuity. The east span is 2,730 ft long; the west span is 3,810 ft long. A plan view of the floating pontoons and of the anchor cables is given in Figure 1.

The individual reinforced and post-tensioned concrete pontoons are typically 360 ft long, 50 ft wide, and 14 ft deep; their draft is about 8 ft. The pontoons are of cellular construction with bulkheads spaced at 12 ft-6 in. across the width and at 15 ft along the length. A two lane roadway is supported off the top deck of the pontoons by reinforced concrete columns. The roadway is 14 ft above the pontoons over most of the floating portion of the bridge, but it rises gradually near the ends to meet the transition spans at a maximum height of 53 ft-7 in. above the pontoons at the east end.

Over the deepest region of the channel, two drawspans are provided to create a 600 ft wide opening navigable by large vessels. The drawspans retract into lagoons created by flare and flanking pontoons resembling very large tuning forks in plan. The lines of traffic separate at the flare pontoons to travel around the lagoon; they then come together again on the draw pontoons.

Anchorage for the floating portion of the bridge is provided by a series of pretensioned anchor cables extending north and south from the bridge. Typically, one set of anchors goes from the midpoint of a pontoon to gravity anchors founded on the bottom of the canal. The slope of the chords of the cables is about 1 in 4; the cables thus provide effective lateral restraint to the bridge and a relatively compliant vertical restraint. The anchors are essentially solid blocks of concrete measuring 40 ft by 19 ft by 16 ft.

ENVIRONMENTAL CONDITIONS

Wind. Wind information is needed to estimate the wind forces on the bridge and to estimate the wave pattern which itself is needed to compute wave forces on the bridge.

The first design decision is the selection of a mean recurrence interval; for a bridge designed for a 50- to 75-year lifetime, a 100-year wind is often used. The probability of experiencing a 100-year storm is about 0.4 in 50 years and about 0.5 in 75 years; that is, there is a good chance of experiencing the design storm during the life of the structure.

For design one needs the mean wind velocity $U(y)$ at height y; the spectral density of the longitudinal component of wind velocity as a function of frequency, $S(y, f)$; and the cross-correlation between two points of coordinates x_1 and x_2 along the bridge, $S(y, x_1, x_2, f)$. Normally a map of basic design wind speeds (29) is adequate to obtain the mean or static wind force; however, if unusual conditions are expected, such as wind speed increase over a body of water (23), then a meteorological study is needed. For well-behaved climates, a probability

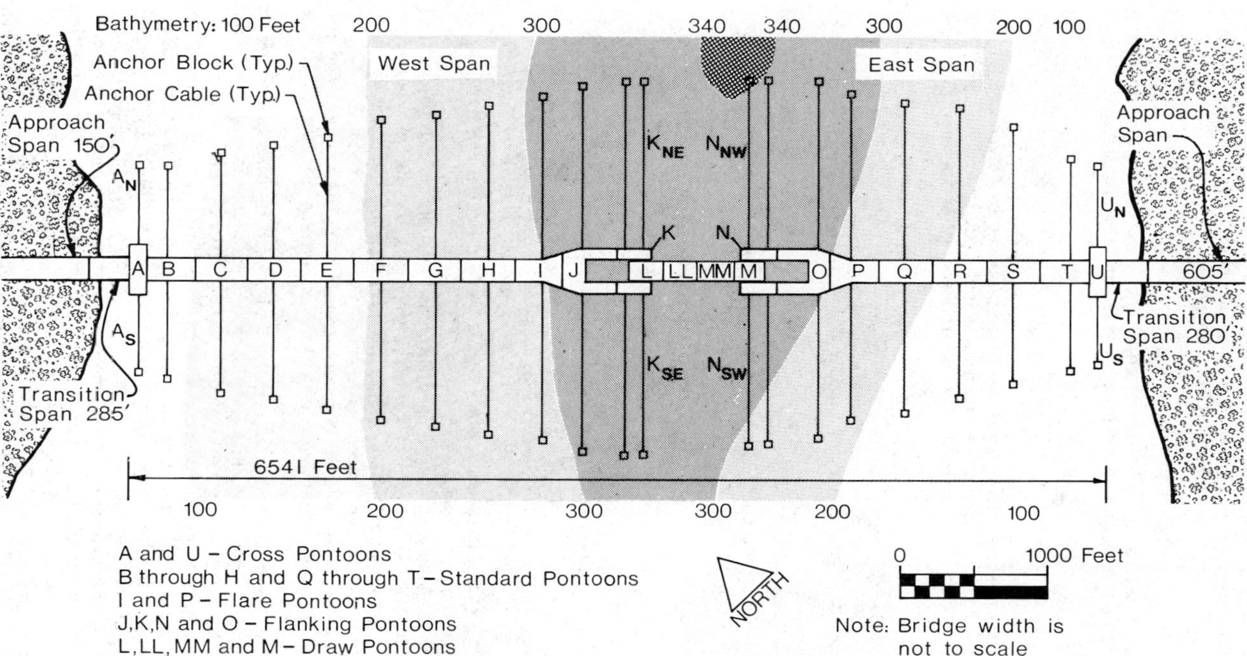

Figure 1 Plan View of Hood Canal Bridge

distribution function of Type I Extreme Value or Gumbel's Distribution is a good model for yearly maximum wind speeds; it is given by

$$F_I(x) = \exp\left(-\exp\left(-\frac{x-u}{\sigma}\right)\right) \quad (1)$$

The two parameters of the distribution, u and σ, are computed from the mean and variance of the measured maximum yearly wind speeds using

$$E(x) = u + 0.5772\,\sigma \quad (2)$$

$$V(x) = 1.645\,\sigma^2 \quad (3)$$

Spectral densities commonly used are those proposed by Davenport (9) and by Simiu and Scanlan (27). Davenport's spectrum is included in ANSI A58-1972 (1); Simiu and Scanlan's spectrum, defined only for frequencies larger than the modal frequency, is given by

$$S(y, f) = \frac{0.26\,u_*^2}{f}\left(\frac{f\,y}{U(y)}\right)^{-2/3} \quad (4)$$

where u_* is the friction velocity. The friction velocity is given by

$$u_*^2 = K\,U^2(10) \quad (5)$$

where K is the surface drag coefficient (about 0.005 to 0.006 for high winds over open sea) and U(10) is the mean wind velocity at the standard height of 10m (33 ft). The variance of the wind gust is approximately

$$V(u) = \int_0^{\infty} S(y, f)\,df = 6\,u_*^2 \quad (6)$$

The cross-correlation spectrum at height y, can be taken as (27)

$$S(y, x_1, x_2, f) = S(y, f)\,\exp\left(-\frac{C\,f\ \mathrm{abs}(x_1 - x_2)}{U(10)}\right) \quad (7)$$

where the constant C lies between 6 and 8 for winds in the 30 m/sec to 40 m/sec (67 mph to 89 mph) range.

The modal frequency of a wind spectral density occurs at a wavelength of about 600 m (2,000 ft); for a storm of 40 m/sec (89 mph), the 600 m wavelength corresponds to a period of about 16 sec. This period is of the same order of magnitude as that of a long continuous-pontoon bridge.

The wind velocity is used to compute drag forces, lift forces, and roll moments on the bridge. Roll moments are due to uneven lift forces on the roadway superstructure.

Waves. The greatest uncertainty in the design of a floating bridge lies in the definition of the wave pattern. The design could be based on directional energy spectra; but an accurate determination of directional spectra is still not possible. A practical approach is to define the wave conditions in three steps: (a) determine the design significant wave height and a range of significant periods from the wind speed, fetch, average depth of water, and width of the waterway using wave forecasting techniques; (b) select a spectral energy density, with significant wave height and period as parameters, that is a function of frequency only; and (c) estimate the directional content of the wave energy. Since floating bridges are sensitive to the frequency content of the waves, the analysis must be performed for a range of modal frequencies.

Widely used methods for wave forecasting include that by Sverdrup, Munk, and Bretschneider (SMB Method) described in the Shore Protection Manual (25),

Wilson's modified method (34), and the best fit equations developed during the Joint North Sea Wave Project, JONSWAP (2, 18). Several comparisons of the wave forecasting methods to actual sea conditions have been published (7, 10, 22). No consensus is yet available for the proper fetch for narrow waterways: the fetch may be reduced to account for the effect of the width (24, 25) or the actual fetch may be used (31).

The distribution of wave energy according to wave frequency is given by an energy spectrum (17). The two most widely accepted energy spectra are the Pierson-Moscowitz (P-M) spectrum for fully developed seas and the JONSWAP spectrum which is the P-M spectrum with a peak enhancement factor to account for fetch-limited sea conditions. The P-M spectrum, expressed as a function of frequency, is given by

$$S(f) = \frac{5 H_s^2 f_m^4}{16 f^5} \exp\left(-1.25 \frac{f_m^4}{f^4}\right) \quad (8)$$

where Hs is the significant wave height and fm is the modal frequency obtained from the significant period by the expression (see Ref. 26)

$$f_m = \frac{0.857}{T_s} \quad (9)$$

The JONSWAP spectrum depends on five parameters; they are Hs, fm, and three shape parameters. Ewing (11) proposes the following shape parameters: $\gamma = 3.3$, $\sigma = 0.07$ for $f \leq fm$, and $\sigma = 0.09$ for $f > fm$. The spectrum for these parameters is given by

$$S(f) = \frac{0.205 H_s^2 f_m^4}{f^5} \exp\left(-1.25 \frac{f_m^4}{f^4}\right) \gamma^{\exp\left(-\frac{(f-f_m)^2}{2 \sigma^2 f_m^2}\right)} \quad (10)$$

The P-M and JONSWAP spectra are one-sided, that is, defined only for positive values of frequency. The significant height is defined as

$$H_s = 4 \sqrt{m_o} \quad (11)$$

where m_o is the area under the spectrum. This definition is consistent with a significant height that is the average of the one-third highest waves.

Very limited information is available on the directional distribution of the wave energy. The difficulty lies in the measurements that require an array of gages at several adjacent points in the sea. Some directional wave information was obtained from pitch-roll buoys during the JONSWAP project (11). An alternative to directional energy spectra is a sea pattern of long-crested waves that travel normal to the bridge and that have a sinusoidally varying crest height parallel to the bridge. A commonly used crest length range is two to four times the wave length.

Tide and Current. Floating bridges over coastal sea waters are subject to tides. The design value of interest is the tide range, since it affects the loads in the anchor cables and hence the buoyancy loads on the bridge. Estimates of the 100-year peak high tide and 100-year peak low tide must be obtained; their difference is the design tide range.

At harbors, bays, estuary entrances, and coastal inlets the tidal fluctuations generate currents with velocities that are typically less than 2 knots (28), but currents may reach values as high as 6 knots. Measurements of tidal current velocity distribution in a vertical plane normal to the current (8) show that isotachs vary throughout the tidal cycle. Tidal velocity variations along the length of a pontoon bridge need normally not be accounted for in design. The variation of pressure along the length has little effect on the design because pressures from tidal currents are usually small compared to other loads on the bridge. Forces caused by tidal current can be safely taken as static.

The tidal currents affect the relative wind-to-water speed and, therefore, affect the wave heights. The actual fetch can be modified to account for the tidal current.

ANALYSIS OF FLOATING BRIDGES

Structural Modeling. Currently available methods of analysis of continuous-pontoon floating bridges are less precise than methods used to analyze other structures. Basic theory exists to develop a three-dimensional, fluid-structure interaction, finite element model; however, such an approach is not available today for practical applications. A simple yet reasonable approach is one that models the bridge as a linear beam element with six degrees of freedom per joint. The water is modeled as an elastic foundation for the beam; in addition, the water provides hydrodynamic mass and hydrodynamic damping to the beam.

Floating bridges are normally restrained by pretensioned anchor cables. These cables provide lateral and torsional restraint to the bridge, but they must also accommodate the change in elevation of the bridge with the tide. The pretension in the cables changes with elevation and with sway deflections of the bridge. Anchor cables are nonlinear structural elements; the degree of nonlinearity, however, depends on pretension and geometry (5).

The anchor cables can be modeled as pin-ended linear members; their cross sectional area, A^*, is proportional to the rate of change of the horizontal component of cable tension, H, with change in horizontal projection of cable length, L:

$$A^* = \frac{L}{E\cos^3\alpha}\,\frac{dH}{dL} \qquad (12)$$

where

α = angle of cable chord to horizontal
E = Young's modulus

The derivative dH/dL depends on the pretension of the cable at a given bridge elevation; it is computed from the condition that the unstretched cable length, S, is a constant (dS=0). This cable length is given by

$$S = L\left(\sec\alpha\left(1 + \frac{W^2}{24H^2\sec^4\alpha}\right) - \frac{H}{AE}\left(1 + \frac{W^2}{12H^2} + \tan^2\alpha\right)\right) \qquad (13)$$

where

W = total weight of cable
A = area of cable

The forces on the cable due to tidal currents may be included using the same formulation, but W is now the resultant of the weight and current forces. In computing dH/dL, the direction of this resultant W is taken as vertical and a line normal to it as horizontal.

The equivalent cable areas should be computed at low and high tide to estimate their effect on the dynamic response of the bridge; if necessary, two bridge models should be analyzed, one at low and one at high tide.

Since the water depth changes along the length of the bridge, if the slope of all cables is kept constant, then each anchor cable will be of a different length and will have a different force-deflection curve. In general, this leads to unbalanced anchor cable forces as the bridge goes from low to high tide and consequently to lateral movements and to horizontal bending moments in the bridge. Therefore, an objective of the design of anchor cables (selection of area, pretension, water elevation for pretensioning, and geometry) is to obtain a balanced set of cables, i.e., a set in which all cables have the same tension for every tide elevation. The balancing operation is a problem of optimization.

The modeling of the water and the water waves can be based solely on the hydrodynamic solutions developed for floating bodies (33), or it can be based on a combination of hydrodynamic solutions and experimental results simplified for design. Both approaches, however, are limited by the uncertainty in the wave parameters, especially the directional energy spectrum or, equivalently, the wave cross-correlation.

The hydrodynamic coefficients for mass, damping, and force are frequency dependent. The solution for the floating body problem using the Green's function method (12, 13, 16) gives fully coupled (heave, sway, and roll) frequency dependent mass and damping matrices; therefore, a harmonic analysis of the bridge requires at each forcing frequency the solution of a system of equations with complex coefficients. A simplification of the problem can be sought in two areas. First, the frequency dependence of the hydrodynamic mass is small in the dominant frequency range of the wave energy spectrum; therefore, a single constant hydrodynamic mass (or perhaps two values of hydrodynamic mass) can be chosen for the appropriate frequency range; and second, a modal damping value may be assigned to each heave, sway, and roll mode using the damping coefficient that corresponds to its frequency. This assignment of modal damping is possible in floating bridges because there is only minor cross coupling between heave, sway, and roll.

The floating body problem for a pontoon bridge is solved two-dimensionally in the vertical plane following the strip-theory of naval architecture (21). The velocity potential for the fluid motion is taken as the sum of five potentials: three correspond to the radiation problem for forced motion in heave, sway, and roll; one corresponds to the incident wave; and one is the scattering potential that results from diffraction of the incident wave. Each potential must satisfy Laplace's equation with a linearized boundary condition on the free surface; a boundary condition on the body expressing equal velocities for the fluid and the floating body; a boundary condition of zero velocity normal to the bottom; and a radiation condition at infinity that states that the free surface waves other than the incident wave are due to the floating body. A numerical solution of the five velocity potentials gives the added mass and the hydrodynamic damping. The exciting forces and moments for the bridge cross section are computed from the hydrodynamic damping using Haskind's relations (21). Alternatively, published results of theoretical or experimental investigations may be used to obtain all hydrodynamic coefficients. The most complete experimental study was performed by Vugts (32), whose results include a pontoon-like rectangular cross section.

The water provides heave and roll stiffness per unit length that is equivalent to an elastic foundation of subgrade modulus k given by

$$k = \gamma\, b \quad \text{for heave} \tag{14}$$

and by

$$k = \gamma\, b h \left(\frac{b^2}{12h} + \frac{h}{2} - y_g \right) \quad \text{for roll} \tag{15}$$

where

γ = unit weight of water
b = width of pontoon
h = draft or depth of pontoon below water
y_g = center of gravity of pontoon measured from pontoon bottom.

The stiffness matrix for the water is obtained from

$$k_{ij} = k \int_o^L \phi_i \phi_j \, dx = k \, m_{ij} \tag{16}$$

where

ϕ_i = interpolation function, quadratic or cubic for bending, linear for torsion

L = element length

m_{ij} = coefficients of water stiffness matrix for beam element (same as consistent mass matrix).

Methods of Solution. Static forces on a pontoon-type floating bridge include: (a) the load curves of dead and live load at several tide elevations (each load curve is obtained as the net vertical force resulting from weight, buoyancy, and anchor cable tension); (b) the tidal current pressure; (c) the mean wind force; and (d) the mean wave force. A solution for these static forces is obtained from a model of a beam on elastic foundation.

The dynamic forces on the bridge, due to wind and to water waves, are random in time and space. One may use a frequency domain solution or a time domain solution. The frequency domain solution or spectral analysis is based on a linear superposition of contributions from all frequencies; for this superposition to be valid the bridge must be linearly elastic, yet the properties of the bridge model may be frequency dependent. In the time domain solution, the properties of the bridge may be time-dependent to account for yielding, hysteretic behavior, or other nonlinearities; but the properties can not depend on frequency. A normal mode solution may be used in the frequency or time domain when the mass and damping are taken independent of frequency.

The solution steps for the wind vibration problem in the frequency domain are as follows: (a) obtain the spectral density functions of the external forces and moments along the length of the bridge, (b) compute the cross spectral densities of generalized forces, (c) compute the spectral density of response, (d) compute the root-mean-square (RMS) response, and (e) estimate the peak response. The spectral density of forces may be taken proportional to that of wind velocity since the gust velocity is small compared to the mean wind velocity.

When modal damping is used, the cross spectral densities of generalized forces, I_{jk} (f), for every pair of natural modes j and k are needed to obtain the response spectrum. With the wind spectrum given by Eq. 7, scaled by the appropriate drag factors, one uses (see Ref. 19)

$$I_{jk}(f) = \iint S(x_1, x_2, f) \, N_j(x_1) \, N_k(x_2) \, dx_1 \, dx_2 \tag{17}$$

$$S_r(x_1, x_2, f) = \sum_j \sum_k N_j(x_1) \, N_k(x_2) \, H_j(f) \, H_k(-f) \, I_{jk}(f) \tag{18}$$

where

$N_j(x_1)$ = shape of mode j

$H_j(f)$ = complex frequency response for mode j

The RMS response is obtained by numerical integration over frequency of $S_r(x_1, x_2, f)$ at $x_1 = x_2$.

The frequency domain solution for bridge vibrations due to waves differs from that due to wind because for waves no acceptable cross-spectral density of forces has yet been developed. One solution approach is to obtain the bridge response to a planar progressive wave of frequency f, wave number k, direction of travel θ, and amplitude $a(f,\theta)$. If the water height from mean water level at a point (x,z) is a linear superposition of plane progressive waves (21),

$$h(x,z,t) = \mathrm{Re}\int_0^{\infty}\int_0^{2\pi} a(f,\theta)\exp\left(-ik\,(x\cos\theta + z\sin\theta) + 2\pi i f t\right) df\, d\theta \quad (19)$$

then the bridge response is a linear superposition of the responses to the individual waves. The amplitude $a(f,\theta)$ in Eq. 19 is a random variable; the amplitudes $a(f,\theta)$ and their complex conjugates $a^*(f,\theta)$ are related to the RMS height and to the wave spectrum by

$$\overline{h^2} = 1/2\iint da\,(f,\theta)\, da^*\,(f,\theta) = \int_0^{\infty}\int_0^{2\pi} S\,(f,\theta)\, df\, d\theta \quad (20)$$

For each wave, the forces vary sinusoidally along the length of the bridge, with a peak amplitude obtained from the hydrodynamic solution; this force distribution, deterministic in space, reduces the randomness of the waves to the time variable only. (The implication is perfect correlation in space; such correlation is unlikely in reality. Determination of the proper space correlation is currently the most important unresolved theoretical question.) For constant mass and damping, the response for each incident wave is obtained by the use of the normal mode approach, and for a frequency-dependent model by the use of the equations with complex coefficients evaluated at the forcing frequency; the numerical effort required in the latter approach is much larger. The response for each wave is obtained by the use of an analysis program that accepts modal damping or a full damping matrix, as the case may be. A preprocessor is needed to obtain the joint forces from the floating body solution and a postprocessor is needed to accomplish the spectral superposition. This superposition will give the RMS values of response (displacements, moments, shears, etc.) from which the peaks are estimated by their multiplication by the probable ratio R of peak-to-RMS. The waves and wave forces follow a Rayleigh probability distribution (28); the response, being filtered through a multi-degree-of-freedom structure, is probably closer to the Gaussian probability distribution (4). The ratio R for a probability of 0.001 of being exceeded is 2.6 for the Rayleigh and 3.1 for the Gaussian distribution.

A time history solution may serve as a check method for the spectral analysis, and can be used to study bridge nonlinearities such as yielding of the bolts that connect pontoons. Time histories of wave forces are obtained from a superposition of the sinusoidal forces due to waves of frequency f and direction of travel θ. The superposition of waves must be consistent with the wave spectrum. (A time history of wave heights can not be used to obtain the forces, unless the Froude-Krylov assumption is made, because the wave history contains all frequencies and the forces are frequency-dependent). Wave force histories are computed for each structural joint; pairs of force histories must have the proper space correlation. The solution for the time histories of forces are obtained for a bridge model with constant masses and damping chosen for a representative frequency.

INVESTIGATION OF HOOD CANAL BRIDGE

Analytical Model. The methods of analysis described in the preceding section were used in the investigation of the Hood Canal bridge. Analytical models were developed for both the east span and the west span, and a hydroelastic model of the bridge was developed for the east span. A beam on elastic foundation model with six degrees of freedom per joint was selected. The hydrodynamic coefficients of mass and damping were obtained initially from results available in the literature; then, adjustments were made based on the results of the hydroelastic study. Modal damping was assigned to each mode based on the frequency of the mode and on the primary type of displacement (heave, sway, or roll) of the mode shape. A spectral analysis was performed for water waves consistent with the Bretschneider and with the JONSWAP sea spectra; ten frequencies and three directions per frequency were used for the spectral decomposition. A harmonic analysis of the bridge for each frequency/direction gave influence coefficients for use in the spectral summation; once these analyses were available, a parametric study of sea conditions was performed by changing the modal frequency of the sea spectra. In addition, as a check of the approach, a time history analysis was performed based on a Bretschneider wave spectrum with the best estimate significant height and period.

Hydroelastic Model Study. The purpose of the hydroelastic model was to serve as a calibration tool for the analysis. A 2,400 ft by 4,000 ft area of the Hood Canal that includes the eastern span of the floating bridge was simulated at a 1/60 geometric scale. The prototype bathymetry was simulated for depths of up to 100 ft because water waves of the frequencies of interest are unaffected by the bottom at greater depths. The model basin was capable of simulating flood or ebb tidal currents of up to 4.0 fps, and of simulating waves of periods ranging from 3 sec. to 8 sec. with approach angles from 0° to 45°. Waves were obtained from five independently driven, plunger-type, wave generators located on the south side of the bridge.

The model bridge was constructed of cellulose acetate butyrate (CAB) with a Young's modulus of 100 ksi. A typical pontoon was a geometrically scaled box beam with an external shell thickness of 0.06 in.; each pontoon cross section was modeled to have the properly scaled stiffnesses in sway, heave, and roll. The weight and the center of gravity of the model was adjusted by the addition of lead weights. The anchor cables were simulated by single braided 0.018-in.-diameter stainless steel cables that were run through a system of pulleys to an instrumented cantilever beam located at the edge of the model basin. The cantilevers provided most of the flexibility needed to model the stiffness of each cable; small adjustments in flexibility were achieved by changing the cantilever length.

The cable tensions and wave heights at several locations were measured. (Measurements of the freeboard on both sides of the bridge and a calibrated movie camera placed above the model to measure sway deflections were considered, but dropped because of time constraints.) Cable tensions were determined by measurement of the strain in the cantilevers by the use of foil type strain gages mounted in a full bridge arrangement. Wave height measurements were made by parallel-wire, resistance type, wave probes. Thirty-one channels of data were scanned serially in 3 milliseconds, multiplexed through an A/D converter, and stored in a LSI/1103 minicomputer. All data were automatically converted to forces or wave heights, and mean and RMS values were computed and displayed on a video terminal. The digitized data were then transferred for storage on disk memory of a PDP11-34A computer to enable further data processing.

Results of the Study. The analytical model was used to obtain results needed for comparison with the hydroelastic study and to verify the adequacy of the bridge under environmental loads. For calibration with the hydroelastic study, the following results were computed for the east span: (a) natural frequencies of vibration to check the hydrodynamic masses; and (b) average and peak forces in anchor cables due to sinusoidal waves, to check force coefficients. In addition, the hydrodynamic damping values measured were compared with the theoretical values. Thus, all hydrodynamic coefficients were checked by the model study.

Table 1 shows the natural periods of vibration and the damping coefficients for both the analytical model and the physical model of the east span of the Hood Canal bridge. The agreement between the periods obtained from the computer run and the periods measured in the model study is reasonably good; adjustments

TABLE 1 – Natural Periods and Damping Coefficients, east span of Hood Canal Bridge

Description of mode (1)	Natural period, in sec.		Modal damping, per cent of critical	
	Analytical model (2)	Physical model (3)	Analytical model (4)	Physical model (5)
Surge	15.1	14.2	10	5
First heave	7.3	8.0	30	35
Second heave	6.4	7.0	30	25
First sway	12.6	15.8	15	12
Second sway	9.2	9.8	15	13
First roll	7.8	6.7	20	13

TABLE 2 – Net anchor cable reactions, mean of north cable minus mean of south cable, kips

Anchor location (1)	Test 19		Test 20	
	Analytical model (2)	Physical model (3)	Analytical model (4)	Physical model (5)
O	85.5	94	84	109
P	85.5	66	84	64
Q	85.5	88	84	73
R	85.5	94	84	90
Average	85.5	85.5	84	84

TABLE 3 – Peak-to-peak anchor cable forces computed and measured in Test 19, kips

Anchor cable (1)	Peak-to-peak cable forces	
	Analytical model (2)	Physical model (3)
$N_{w,s}$	54.5	45.3
$N_{w,n}$	25.4	58.9
Q_s	29.4	44.5
Q_n	32.7	66.0
U_s	208.8	201.1
U_n	171.8	185.9

to the hydrodynamic masses used in the analysis could have been made to improve the agreement. The natural mode shapes for the first three heave and the first three sway modes for the east span of the bridge are shown in Figure 2.

Cable force measurements were used to check the applied wave forces and the structural response to these waves. Average forces on the bridge and peak dynamic forces in the cables confirmed that the results of the spectral analysis were of the proper order of magnitude. Table 2 gives, for tests 19 and 20, the net anchor cable reactions on the bridge measured in the physical model and computed analytically. In test 19 the average prototype wave height is 10.9 ft based on the readings of three wave channels (11.6 ft, 10.9 ft, and 10.1 ft); in test 20 the average wave height is 10.7 ft, also based on three channels (11.4 ft, 11.9 ft, and 8.9 ft). Because of the strip-theory assumption, the anchor force computed is the same for all anchors; the measured results differ among themselves, but their average corresponds to the computed value. Table 3 gives peak-to-peak anchor cable forces for three pairs of cables (see Figure 1). The comparisons of Table 3 are sufficient to serve the purpose of calibration. The differences between measured and computed results, however, are large near the middle of the bridge; the agreement is better for the two end regions. Further research with the model is needed to explain the differences; possible reasons include (a) difficulty of obtaining pure sinusoidal loading in the basin with five independent wavemakers, (b) reflections of waves at the basin edges, and (c) three-dimensional behavior rather than the two-dimensional behavior assumed.

After the analytical model was calibrated, it was available for use in the study of the bridge due to storm conditions. Of primary design interest are the vertical and horizontal bending moments in the pontoon structure. In Figure 3 are shown the shapes of three moment curves due to waves following a Bretschneider spectrum: a vertical moment curve, a horizontal moment curve for the dynamic load only, and a horizontal moment curve for the combined static and dynamic components of the load. The curves are shown through the flare pontoon; in the flanking pontoon, bending moments and axial forces for each arm were obtained. From the curves in Figure 3 one may observe that the highest bending moments occur close to the ends; in the central section the bending moment curves are reasonably flat.

CONCLUSION

Knowledge is available to design a floating bridge by use of analytical methods calibrated by a hydrodynamic model study. There still are, however, major elements of uncertainty; because of the sensitivity of floating bridges to the frequency content and spatial distribution of the applied forces, results that differ by factors of two or more may be obtained depending on the assumptions made. The uncertainties are in three areas: (a) structural model, including hydrodynamic mass and damping coefficients; (b) forces on the bridge for given wind and wave environments, especially the force distribution along the length of the bridge as a function of time or frequency; and (c) wind and wave environment. For a resolution of the first two areas of uncertainty, further basic research is needed, while for the third area, environmental measurements at the site of a particular bridge must be taken.

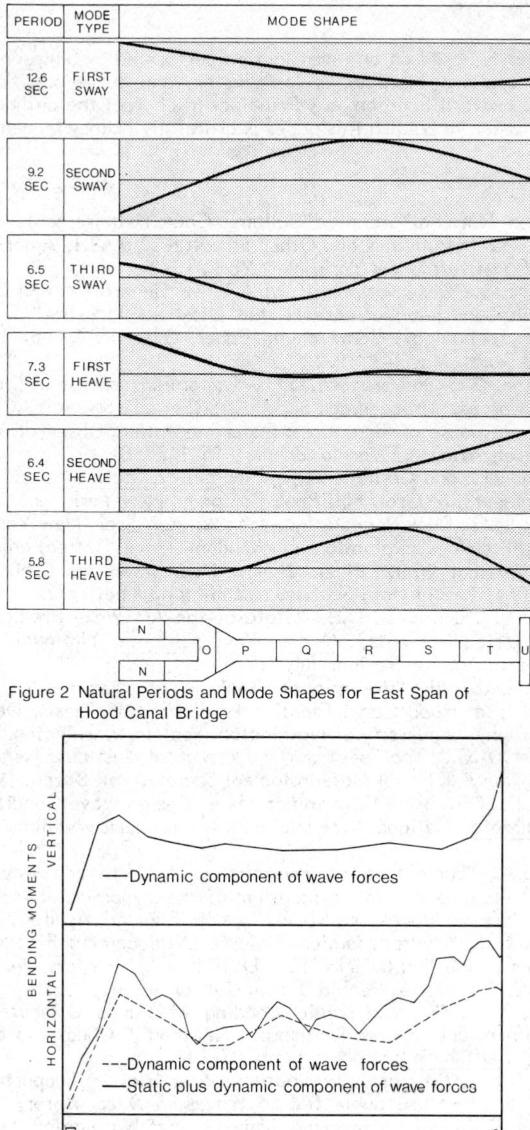

Figure 2 Natural Periods and Mode Shapes for East Span of Hood Canal Bridge

Figure 3 Bending Moments in Pontoons due to Random Wave Forces

ACKNOWLEDGMENTS

This paper is based on an investigation performed by Simpson Gumpertz & Heger Inc. for Toplis & Harding, Inc., Adjusters and Surveyors, San Francisco, California. Dr. Howard Simpson was Principal in charge; the author was Project Manager. Permission to publish this paper is gratefully acknowledged.

REFERENCES

1. "American National Standard Building Code Requirements for Minimum Design Loads in Buildings and Other Structures," A58.1, American National Standards Institute, New York, New York, 1972
2. Barnett, T.P., "Observations of Wind Wave Generation and Dissipation in the North Sea: Implications for the Offshore Industry," Fourth Annual Offshore Technology Conference, Paper OTC 1516, pp. 1-43 to 1-50, Houston, Texas, May 1-3, 1972
3. Beauchamp, C.H. and Brocard, D.N., "Dynamic Response of Floating Bridge to Wave Forces," Proceedings Second ASCE-EMD Specialty Conference on Dynamic Response of Structures: Experimentation, Observation, Prediction and Control, Atlanta, Georgia, January 15-16, 1981
4. Benjamin, J.R. and Cornell, C.A., Probability, Statistics, and Decision for Civil Engineers, McGraw Hill Book Company, New York, New York, 1970
5. Berteaux, H.O., Buoy Engineering, John Wiley & Sons, New York, New York
6. Burrows, F.G.A., "Evaluation of Random Wave Reflection Coefficients from a Vertical Wall," Ph.D. Thesis, Department of Civil Engineering, University of Washington, Seattle, Washington, June, 1972
7. Cardone, V.J. and Ross, D.B., "State-of-the-Art Wave Prediction Methods and Data Requirements," Ocean Wave Climate, National Oceanic and Atmospheric Administration, July, 1977
8. Creeden, J.J., III, "Preliminary Analysis of Currents of Hood Canal Obtained from Hood Canal Floating Bridge," S.M. Thesis, Department of Oceanography, University of Washington, Seattle, Washington, 1968
9. Davenport, A.G., "The Spectrum of Horizontal Gustiness Near the Ground in High Winds," J. Royal Meteorological Society, Vol. 87, pp. 194-211, 1961
10. Earle, M.D., "Practical Determinations of Design Wave Conditions," Ocean Wave Climate, National Oceanic and Atmospheric Administration, July, 1977
11. Ewing, J.A., "Some Results from the Joint North Sea Wave Project of Interest to Engineers," Int. Symposium of the Dynamics of Marine Vehicles and Structures in Waves, pp. 41-46, London, England, April, 1974
12. Garrison, C.J., "Hydrodynamics of Large Displacement Fixed and Floating Structures in the Sea (DYNRES6 & DYNRES6A)", Report No. 79-101, C.J. Garrison & Associates, Pebble Beach, California, May, 1979
13. Garrison, C.J., "Hydrodynamic Loading of Large Offshore Structures: Three-Dimensional Source Distribution Methods," Chapter 3 of Numerical Methods of Offshore Engineering, pp. 87-140
14. Gibson, G.E., "Pressure Distribution on a Rigidly Supported, Partially Submerged, Structure Subjected to Waves in Deep Water," S.M. Thesis, Department of Civil Engineering, University of Washington, Seattle, Washington, December, 1969
15. Hartz, B.J., "Summary Report on Structural Behavior of Floating Bridges," Final Report on Contract Y-909 with Washington State Department of Highways, Department of Civil Engineering, University of Washington, Seattle, Washington, June, 1972

16. Ho, R-T and Harten, A., "Green's Function Techniques for Solutions of Floating Body Problems," Civil Engineering in the Oceans/III, pp. 939-958, Specialty Conference, University of Delaware, Newark, Delaware, June 9-12, 1975
17. Hoffman, D., "Analysis of Measured and Calculated Spectra," International Symposium of the Dynamics of Marine Vehicles and Structures in Waves, pp. 8-18, London, England, April, 1974
18. Houmb, O.G. and Overik, T. "Parameterization of Wave Spectra and Long Term Joint Distribution of Wave Height and Period," Proceedings of the Conference on the Behavior of Offshore Structures (BOSS '76), pp. 144-169, The Norwegian Institute of Technology, Trondheim, Norway, 1976
19. Lin, Y.K., Probabilistic Theory of Structural Dynamics, McGraw-Hill Book Company, New York, New York, 1967
20. Mukherji, B., "Dynamic Behavior of a Continuous Floating Bridge," Ph.D Thesis, Department of Civil Engineering, University of Washington, Seattle, Washington, August, 1972
21. Newman, J.N., Marine Hydrodynamics, The MIT Press, Cambridge, Massachusetts, 1978
22. Resio, D.T. and Vincent, C.L., "Comparison of Various Numerical Wave Prediction Techniques," 11th Annual Offshore Technology Conference, Paper OTC 3642, pp. 2471-2478, Houston, Texas, April 30 - May 3, 1979
23. Resio, D.T. and Vincent, C.L., "Estimation of Winds over the Great Lakes," Journal of the Waterway, Port, Coastal and Ocean Division, ASCE, Vol. 103, No. WW2, Proc. Paper 12951, May, 1977, pp. 265-283
24. Seymour, R.J., "Estimating Wave Generation on Restricted Fetches," Journal of the Waterway, Port, Coastal and Ocean Division, ASCE, Vol. 103, No. WW2, Proc. Paper 12924, May, 1977 (See also Discussion in May, 1978, and Closure in February, 1979.)
25. "Shore Protection Manual," Volumes I, II, and III, U.S. Army Coastal Engineering Research Center, 1973
26. Silvester, R., Coastal Engineering, Volume I, Elsevier Publishing Company, Amsterdam, The Netherlands, 1967
27. Simiu, E. and Scanlan, R.H., Wind Effects on Structures: An Introduction to Wind Engineering, John Wiley & Sons, New York, New York, 1978
28. Sorensen, R.M., Basic Coastal Engineering, John Wiley & Sons, New York, New York
29. Thom, H.C.S., "New Distributions of Extreme Winds in the United States," Journal of the Structural Division, ASCE, Vol. 94, No. ST7, Proc. Paper 6838, July, 1968, pp. 1787-1801
30. Timoshenko, S., Young, D.H., and Weaver, W., Jr., Vibration Problems in Engineering, Fourth Edition, John Wiley & Sons, New York, New York, 1974
31. Vincent, C.L., "Wave Estimates Computed Using an Effective Fetch," Guidance Letter from Coastal Engineering Research Center, Document FGL-CO-2, 21 June 1979
32. Vugts, J.H., "The Hydrodynamic Coefficients for Swaying, Heaving and Rolling Cylinders in a Free Surface," International Shipbuilding Progress, Vol. 15, pp. 251-176, 1968
33. Wehausen, J.V., "The Motion of Floating Bodies," Annual Review of Fluid Mechanics, Volume 3, Annual Reviews Inc., Palo Alto, California, 1971
34. Wilson, B.W., "Numerical Prediction of Ocean Waves in the North Atlantic for December 1959," Deutsche Hydrographische Zeitschrift, Vol. 18, No. 3, pp. 114-130, 1966.

Dynamic Response of the Hood Canal Floating Bridge

by B. J. Hartz*, M. ASCE

Abstract

The calculation of the dynamic response of a long floating structure in which the length of the structure is long in comparison to the wave lengths of the waves is complicated by the necessity to consider the finite lengths of the wave crests. It is further complicated by the limited field data on which to base assumptions for dynamic response calculation. Extensive measurements were made on the Hood Canal Floating Bridge from 1966 to 1972. These field measurements are available for the "calibration" of model basin models and of mathematical/computer models of the behavior of floating bridges. In particular when put into the computer models developed during this project valuable information on the spatial correlation of the wave forces along the bridge was obtained. These computer models were subsequently able to furnish information on the dynamic behavior of this bridge and of the probable cause of sinking of the Western Section of the bridge in the storm of Feb. 13, 1979.

Introduction

The Hood Canal Floating Bridge, Fig. 1, was designed and built as a continuous concrete box floating structure, 6470 feet (1975m) long spanning a relatively deep natural water way a mile and a half wide in the State of Washington (1,2). The 50 foot wide by 14 foot deep pontoons (15m x 4.3m) support an elevated two lane roadway that carried about 6000 vehicles per day. Floating draw spans near the center allowed passage of large marine traffic. The bridge was completed at a cost of about 23 million dollars and opened for traffic in 1961. On Feb. 13, 1979 the 3775 foot (1150m) long Western Section of the bridge broke up and sank in a storm, Fig. 2.

From 1966 to 1972 the University of Washington had a research project on the determination of the Structural Behavior of Floating Bridges sponsored by the Federal Highway Administration and Washington State Department of Transportation (3). This project involved field instrumentation of the bridge to measure dynamic response including wind, waves, hydrodynamic pressures, accelerations, anchor cable forces and concrete strains. The results included field measurement of the actual dynamic response during periods of storm generated waves over a four year period.

---

* Professor of Civil Engineering, Univ. of Washington, Seattle, WA 98195

Fig. 1, Hood Canal Bridge - Before

Fig. 2, Hood Canal Bridge - After

The six year project also included development of mathematical models and computer programs for the prediction of the dynamic response of floating bridges and correlation of the analytical results with measured response from the field instrumentation (4). Supporting model studies were included in the area of determining hydrodynamic pressure distribution relationships (5,6,7,8) and in relating measured concrete strains to internal force resultants at a bridge cross-section (9). Some results of the field measurements and analytical calculations and their correlation have been previously published (10). Additional results are included in this paper.

Following the sinking of the Western Section of the bridge in the Feb. 13, 1979 storm additional studies were made for the purpose of trying to determine the cause of sinking (11,12,13,14). Results of some of these studies, correlation of these with the field measurements and some implications for the cause of failure of the bridge are also included in this paper.

Analytical Models for Dynamic Response of Floating Bridges

Analytical models for the dynamic response of long floating bridges or similar structures, such as continuous floating breakwaters, in which one dimension is long in comparison to the crest lengths of the existing waves have been developed (3,4,10,13,15,16,17,18,19). In general the analysis is either in the time domain using numerical simulation of the wave forces and integration of the equations of motion or in the frequency domain using either direct frequency response from cross-spectral densities of the wave excitation or superposition of responses at each frequency component with numerical simulation of the wave spectra with or without consideration of the directional spectra at each frequency. The advantage of frequency domain analysis is the ease with which the frequency dependence of hydrodynamic mass, damping and force coefficients can be included. However, studies have indicated that even for relatively broad spectra the error associated with using constant hydrodynamic mass, damping and force coefficients associated with the peak frequency in the spectra is very small. Langen (19) indicates less than 8% error for a broad wave spectra with a peakedness parameter $\gamma = 1$.

The advantages of time domain analysis are its much greater familiarity to engineers concerned with such structures, the availability of more structural dynamics computer programs that can handle such analyses and the possibility of including nonlinear response effects in the analysis. Some of the widely available computer programs that have been used for calculating the dynamic response of floating bridges or breakwaters are SAP IV, GTSTRUDL, STARDYN and NASTRAN. Some programs have both time domain and frequency domain capability.

In general the vertical or heave modes of a floating bridge are uncoupled from the lateral or sway and roll motions. The heave periods of vibration are determined primarily by the bouyancy stiffness and are closely spaced, characteristic of a beam on an elastic foundation with a high foundation modulus. For example, for the Hood Canal Floating Bridge, the "characteristic length" for the beam on an elastic foundation is about 200' (60m) and there are 11 vertical modes of vibration for the 3775' (1150m) long Western Section between the periods of about

6 seconds and 3 seconds. This must be considered in determing nodal spacings for a finite element representation of a continuous floating structure.

On the other hand, the sway and roll modes are normally coupled but this coupling is usually weak. The sway motion is affected primarily by the lateral anchoring system and the lowest sway modes can be expected to have relatively long periods of vibration. The lowest roll mode will most likely be rigid body roll with no deformation of the bridge unless forced by end constraints. In the case of the Hood Canal Bridge analyses in references 3,4,10 and 12, the sway and roll modes were assumed a priori to be uncoupled. The roll periods decreased rapidly from the essentially rigid body fundamental roll period and the response calculations indicated very small torsional deformation and associated torsional moments. These calculations were confirmed by the measured results (3,4). The negligible coupling between sway and roll was later independently confirmed by two different analyses of this same bridge (11,13) following the sinking of the Western Section.

With the sway, heave and roll motions uncoupled, and using constant hydrodynamic coefficients for the peak period in the wave spectra, the equations of motion for each for time domain analysis are of the form

$$\underline{M}\,\ddot{\underline{r}} + \underline{C}\,\dot{\underline{r}} + \underline{K}\,\underline{r} = \underline{R}(t) \qquad (1)$$

where the mass matrix, $\underline{M}$, includes the hydrodynamic added mass, the damping matrix, $\underline{C}$, is primarily hydrodynamic damping and the stiffness matrix, $\underline{K}$, includes the hydrostatic bouyancy stiffness in heave and roll and the lateral anchoring stiffness in sway. For the analysis results included here consistent mass and foundation stiffness matrices were generally used except for some results from one computer program in which it was necessary to lump the hydrostatic bouyancy stiffness at the node points. Static condensation was used to reduce out the rotational degrees of freedom associated with the polynomial beam functions. The bridge motions were assumed small compared to the wave particle motions and the displacement matrix, $\underline{r}(t)$, is the set of absolute displacements and the forcing function matrix, $\underline{R}(t)$, is the set of hydrodynamic wave pressure forces on a rigidly constrained body of waves. The hydrodynamic mass, damping and force coefficients are essentially those determined by Vugts (20,21) based on 2-dimensional theory. Although there is some discrepancy with some experimental results (22) and some additional analytical and test results are available (23,24), this is not a source of significant error in the analysis. Some correction to these coefficients for direction of wave travel relative to the normal can be incorporated for waves not normal to the structure or for directional wave components associated with a spreading function using results such as those in references 24 and 25. Such corrections or errors may be significant and the sensitivity of the response to variations in these coefficients can be found by a few additional relatively inexpensive computer runs.

The main difficulty in the above analysis is not in the mathematical modeling, the determination of hydrodynamic coefficients or in the computer solution, but in the spatial and temporal simulation of the wave environment. The question is not whether to use a JONSWAP, a Bretschneider, a Pierson-Moskowitz, or whatever spectrum (see for example ref. 26) what parameters to use in each, or precisely what the significant wave height, H_s, is or the peak period, T_p, for the spectrum, since, like the hydrodynamic coefficients, these generally do not cause significant differences for the sea states of concern, and again some inexpensive computer runs can determine the sensitivity of the responses to the shape and peak period of the spectra. For time domain analysis a digital time series representation of the spectra is available (27). The difficulty lies in determining the relationship between the wave heights, and consequently the hydrodynamic pressures and forces, along the length of the structures at each instant of time. The response is intimately related to the correlation or lack-of-correlation of these forces. A number of representations of the sea state are available (27,28) with the most common being the use of a vectorial or directional spectra obtained from one of the above scalar spectra through the use of an angular spreading function. One of the common forms (29) for the so-called spreading function is

$$f(\theta) = A(s) \cos^{2s}\theta \quad , \quad -\frac{\Pi}{2} \leq \theta \leq \frac{\Pi}{2} \tag{2}$$

Eq. (2) combined with the JONSWAP spectra has been used for calculation of responses of the Salhus floating bridge (17,18,19) and of the Hood Canal Floating Bridge (28). Monte Carlo simulations have also been used (30,31,16). It is still to be determined which, if any, of these approaches will give the real dynamic response of a real floating bridge in an actual short crested sea state such as will be found for the typical environment for a floating bridge. As indicated by Langen and Sigbjörnsson (18) the design procedures for offshore structures are not applicable and neither is the experience when the crest lengths become very short relative to a dimension of the structure.

Field Measurements on the Hood Canal Floating Bridge

In 1966-67 instrumentation was installed near the center of the 3775 ft. long (1150m) Western Section of the Hood Canal Floating Bridge to measure the following quantities (3). Wind speed and direction, wave heights at the side of the bridge and at a remote location, temperature, hydrodynamic pressures at 12 locations, 2 vertical and a horizontal acceleration at 3 cross-sections, 4 longitudinal and 4 rosette concrete strains and anchor cable strains on 3 sets of anchor cables. The data was digitized & recorded along with a clock reading on computer compatible digital tape at a rate of 120 readings/second. The system turned itself on every hour and recorded a set of 512 readings on each channel over about a 3 minute period during periods of Winter storms. Each computer tape recorded about one week of data. From 1967 to 1971 about 24 tapes of data during periods of storm activity on the Hood Canal were collected. A computer software package read the tapes, applied calibration factors for each transducer, made appropriate combinations of certain measured variables to convert them into more useful variables, such as converting the concrete strain measurements directly into internal force resultants of moments, shears, axial force and torque, extracted the mean value, maximum, minimum and standard deviation of each quantity, computed histograms and mean-square spectral densities and printed and plotted the results for each variable. A large number of these are contained in reference 3, particularly in Vol. II, and in reference 4.

Spatial Correlation Factor (SCF)

For some of the measurements on the Hood Canal Bridge 9 of the pressure transducers were aligned in a horizontal array along the windward vertical face of the bridge at a depth of about 5 ft. (1.5m) and the spacing was varied to as close as 60 ft. (18m). Calculation of the cross-correlation functions between adjacent gages from the recorded pressure time series indicated very little correlation even at 60 ft. spacing. Subsequent measurements on the Lake Washington Floating Bridge (32) with an array of pressure gages with a spacing 11.5' (3.5m) were used to calculate the value of the coherency function and the phase of the cross-spectral density at each of the spacings represented in the array for the frequencies in the incident waves. The coherency for these pressures was found to drop off rapidly at gage spacings of approximately 0.6 times the wave length of the waves. The phase generally became unstable at less than 0.5 times the wave length.

Based on the observed loss of coherency of hydrodynamic pressures at a distance of about 0.6 of the wave length of the significant waves in the incident sea state on each side of a node an estimate of the total hydrodynamic force to apply to any node can be made for various assumptions as to how the coherency drops off with distance from the node, providing the adjacent nodes for hydrodynamic force loading are sufficiently far away that the loadings at those nodes from the hydrodynamic force at the node under consideration are not significant. If the percentage of the tributary length at a given node that must be loaded with a uniform pressure to have the same "reaction" at the node as the reaction corresponding to the varying pressure drop off is defined as the SCF then an expression for the SCF can be derived based on the assumed distribution. For example, if the pressure is assumed to decrease linearly to zero in a distance of 0.6λ on each side of a node the SCF can be shown (33) to be

$$\begin{aligned} SCF &= \frac{0.6\lambda}{d}\left(1-\frac{1}{3}\frac{0.6\lambda}{d}\right) \\ &= \frac{0.6}{d/\lambda}\left(1-\frac{0.2}{d/\lambda}\right), \end{aligned} \tag{3}$$

where d/λ is the ratio of nodal spacing to wave length. The expression for a quadratic decrease of pressure in distance 0.6λ is

$$SCF = \frac{0.8}{d/\lambda}\left(1-\frac{0.225}{d/\lambda}\right) \tag{4}$$

For nodes with spacing, d, and a hydrodynamic force per unit length of $f_i(t)$ at node i the total hydrodynamic loading at i would then be

$$R_i(t) = SCF \cdot d \cdot f_i(t) \tag{5}$$

If the spacing of nodes is such that $d/\lambda \geq 1$ the hydrodynamic force at adjacent nodes associated with the pressure $f_i(t)$ at the ith node is negligible (33).

Empirical values of the SCF were obtained in references 3,4 and 10 based on correlating actual measured structural dynamic response of the Hood Canal Floating Bridge with calculated values. The calculated values were based on certain assumptions for hydrodynamic mass, damping and force coefficients. Generally they were also based on recorded time series of hydrodynamic pressures at certain nodes of the Bridge model and the rest of the nodes were loaded with these same time series with random shifting. However, the analysis was also done in the frequency domain with calculated wave spectra and specific assumptions of frequency dependence of the hydrodynamic quantities These results give another set of empirical SCF values by correlation of the variances of the measured and calculated responses.

Results for the Hood Canal Bridge were obtained using two different nodal spacings 360'(110m) and 180'(55m) for the computer model corresponding to two different d/λ ratios. For the wave length of about 90' (27m) corresponding to the peak of the wave spectra for storm conditions on the Hood Canal these give d/λ ratios of 4 and 2. For the draft 8'8" (2.6m) and width 50' (15m) of the Hood Canal Bridge and the peak period of the wave energy (4.2 sec.) on the Hood Canal the dynamic response computations generally used 0.15 damping ratio for both heave (vertical) and sway (horizontal) and force coefficients of $1.6^{K/ft/ft}$ ($77^{kN/m/m}$) wave amplitude and $0.54^{K/ft/ft}$ ($26^{kN/m/m}$) wave amplitude in heave and sway, respectively. Added mass coefficients were 2.0, vertical, and 0.11, sway. The SCF values obtained for these cases are shown in Figure 3 along with the curves corresponding to Eq. (3) and (4). The curve labeled "HCBR - Moment and Anchor Force Average" is based on averaging the results from vertical and horizontal moments and anchor cable forces for several storms. Individual results from the particular storm considered in Reference 4 and 10 are labeled separately for vertical moments, and anchor forces.

Because of the sensitivity of accelerations and shears to local hydrodynamic pressures the SCF values for these quantities are higher than for moments and anchor cable forces which depend upon displacement of the structure. A curve for average values of the SCF for these quantities obtained from the Hood Canal Bridge data is also shown in Figure 3. These are seen to be about twice the values for moments and anchor forces (also for displacements).

The differences between the SCF curves for lateral and vertical moment in Fig. 3 are due to the hydrodynamic force coefficients used. The value used for vertical force is believed to be too high. This is supported by data in references 22,23 and 24. In general the author believes that if the best current values for the hydrodynamic coefficients are used then the SCF values are best given by Eq. 3 for moments, anchor forces and displacements for a short-crested limited fetch sea state.

It should be noted that the wave directional effect appropriate to the Hood Canal is already incorporated into the SCF value. The Hood Canal Bridge is in a limited fetch basin with very short crested waves. The s-factor in the spreading function, Eq. 2 is probably only 1 or 2. This seems to be supported also by Langen and St. Denis in reference 19 and 29. For longer period waves the s-factor increases (29) and with less spreading the SCF value would increase somewhat since the wave forces would be more effective in exciting the structure. As the wave periods become longer the analysis approaches using a spreading function (for example 17,18,19,28) become more valid and the d/λ ratio is such

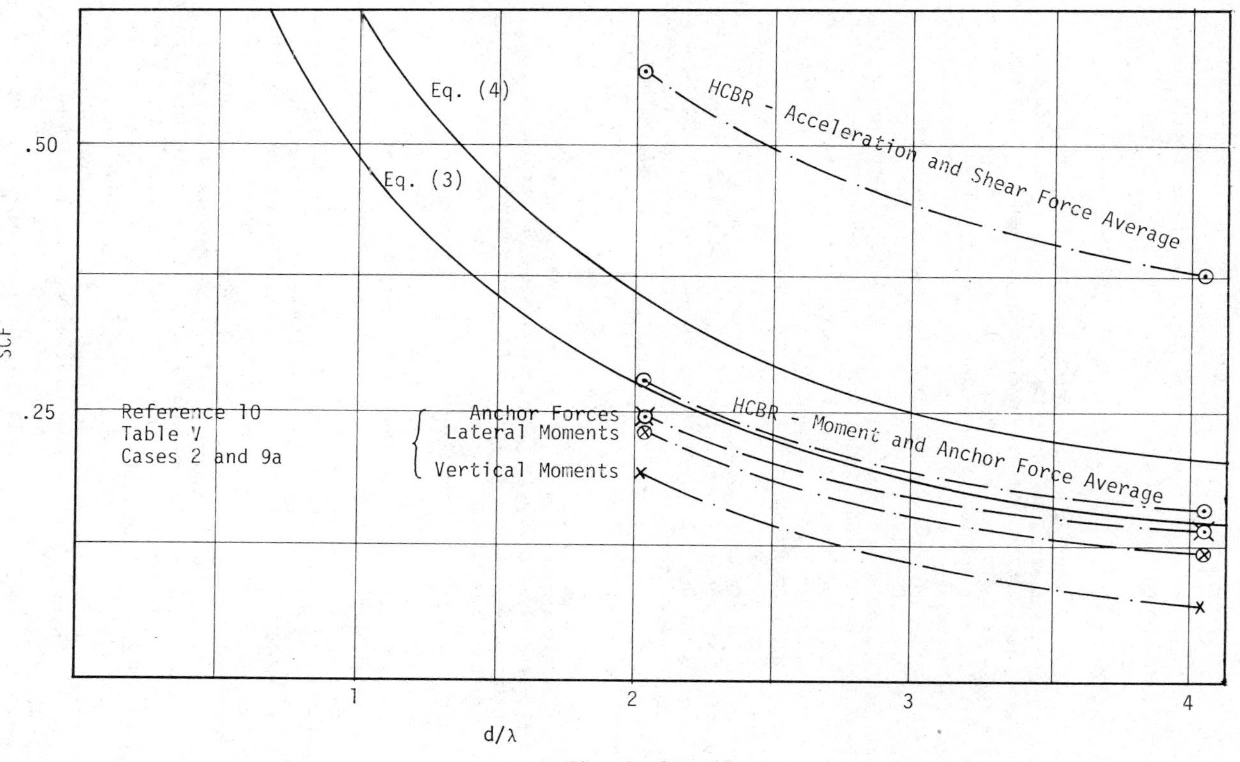

Fig. 3, SCF Values

that statistical independence of the hydrodynamic forces at adjacent nodes can not be assumed. It is expected that this approach can still be used by incorporating an appropriate force component at each node in-phase with the hydrodynamic forces at the adjacent nodes.

Application to the Hood Canal Bridge Sinking

Following the sinking of the Western Section of the Hood Canal Bridge in a violent storm on Feb. 13, 1979, hindcasts were made of the maximum sea states that might have existed at the peak of the storm. The spectra for 3 of these were used to generate time series of wave heights. These were randomly shifted for wave heights at nodes along the Western Section of the bridge and used to compute the dynamic response using the best estimate hydrodynamic coefficients for the peak period of each spectra and the SCF values from Fig. 3. The results for maximum dynamic lateral moment for each of the spectra with H_s and T_p shown are given in Figure 4 plotted along the section length from the West end. Runs were made both with SAP IV and with the UW program (3) and were found to be consistent. The bar near the center indicates the maximum measured during the 4 years of measurement and corresponds to an H_s of about 4 ft (1.2m) and T_p of 4.2 sec. The x's at this location are values from reference 11 for two damping values. Considering the interaction effect of simultaneous vertical moments the "functional dynamic capacity" (11) of pontoons B through H is in excess of 200000 ft-k and increases in the flared and flanking section I-K. The calculations here indicate that the maximum moments were well within the capacity for all possible wave loadings during the storm that sank the bridge.

Then why did the bridge sink? It is the author's firm conviction (12) that the pontoons flooded through open hatches on the deck from water from the waves trapped on the deck between the parapet walls. "Red" Taylor, the last person to leave the bridge watched the water pouring in two open hatches just before he left and the bridge sank.

Conclusion

The use of SCF values for the dynamic response of a long floating structure for which the crest lengths of the waves are short compared to the length of the structure normal to the waves appears to the author to be the best procedure currently available. Approaches using superposition of the responses to infinite crested waves at different angles of incidence representing the directional content of the waves appear to give excessively large bending moments and anchor forces, at least for the Hood Canal Bridge (28). This occurs because for each frequency of the wave spectra coinciding with one of the many natural frequencies of such a structure there is always one angle of incidence for which the wave lengths $\lambda/\sin\beta$ seen by the structure will coincide with modal wave lengths of the structure itself (λ = wave length of incident waves, β = angle of incidence). This gives excessively large amplification of the response. Such an approach was tried during the Hood Canal Bridge project and it was found that reasonable results could be obtained by filtering these wave frequencies. However, the filtering had to be empirically adjusted and this appeared less desirable and justifiable than the empirical adjustment of the SCF values. In addition the directional characterization of the waves in a fetch-limited basin must be known

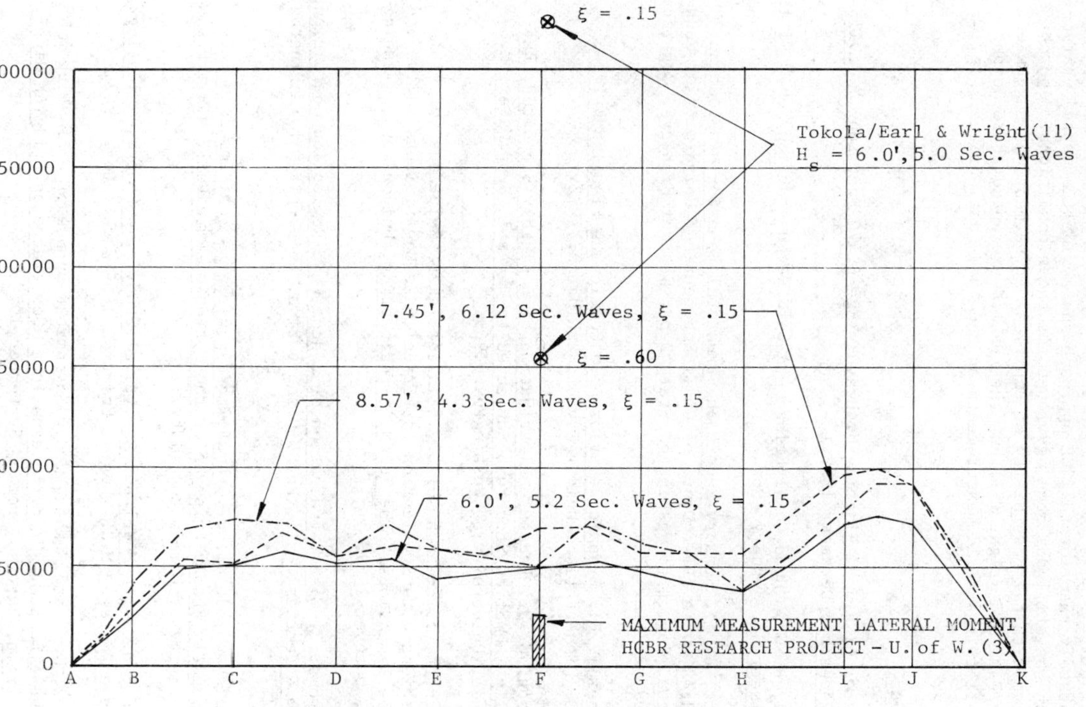

Fig. 4, Maximum Dynamic Lateral Bending Moment - Hood Canal Floating Bridge

more accurately than for the SCF approach and a widely accepted characterization has still not been developed for these conditions. Another alternative with the possibility of empirical adjustment was suggested by Tokola/Earl and Wright (11) but the results presented there do not relate well to the field measurements on the Hood Canal Bridge, Fig. 4.

References

1. Andrew, C. E., "A Floating Highway Bridge 6,470 Feet Long," Civil Engineering, December 1959.

2. Gaul, R. W., "Hood Canal Bridge - A Bold Reconstruction Job," Civil Engineering, January 1962.

3. Hartz, B. J., Summary Report on Structural Behavior of Floating Bridges, June 1972, Dept. of Civil Engineering, University of Washington, Report to Washington State Department of Highways, Contract Y-909, 2 Vols.

4. Mukherji, B., Dynamic Behavior of a Continuous Floating Bridge, Ph.D. Dissertation, University of Washington, 1972.

5. Gibson, G. E., Pressure Distribution on a Rigidly Supported, Partially Submerged Structure Subjected to Waves in Deep Water, M.S. Thesis, University of Washington, 1969.

6. Hartz, B. J. and Schmid, G., "Finite Element Solutions for Field Equations with Application to Virtual-mass Coefficients in Two and Three Dimensions," The TREND in Engineering, University of Washington, October 1969.

7. Burrows, F. G. A., Evaluation of Random Wave Reflection Coefficients from a Vertical Wall, Ph.D. Dissertation, University of Washington, 1972.

8. Sonnichsen, J. C., Jr., Dynamic Response of a Semi-Immersed Rectangularly Shaped Prism Under the Motion of Pure Roll, M.S. Thesis, University of Washington, 1967.

9. Dodds, B. J., A Model and Analytical Study of a Concrete Floating Bridge Pontoon, M.S. Thesis, University of Washington, 1967.

10. Hartz, B.J. and Mukherji, B., "Dynamic Response of a Floating Bridge to Wave Forces," Proc. International Conference on Bridging Rion-Antirrion, Patras, Greece, September 1977.

11. Tokola/Earl and Wright, Hood Canal Floating Bridge: Phase I Report: Determination of Cause of Failure, August 1979, to the State of Washington Department of Transportation. Contract Y-2000.

12. Hartz, B. J., The Hood Canal Bridge Failure: A Report of an Independent Investigation, September 1979 and November 1979, to the Washington State Transportation Commission.

13. Luft, Rene, W., "Analysis of Floating Bridges, The Hood Canal Bridge," to be presented at ASCE/EMD Specialty Conference on Dynamic Response of Structures, Atlanta, Ga., Jan. 15-16, 1981.

14. Beauchamp, Charles, "Dynamic Response of a Floating Bridge to Wave Forces," to be presented at ASCE/EMD Specialty Conference on Dynamic Response of Structures, Atlanta, Ga., Jan. 15-16, 1981.

15. Sigbjörnsson, R. and Langen, I., Wave-induced vibrations of a floating bridge: The Salhus Bridge, SINTEF Report STF71 A75018, Trondheim 1975.

16. Sigbjörnsson, R. and Langen, I., Wave-induced vibrations of a floating bridge: A Monte Carlo Approach, SINTEF Report STF71 A75039, Trondheim 1975.

17. Holand, I., Langen, I., and Sigbjörnsson, R., Dynamic Analysis of a Curved Floating Bridge, IABSE Proceedings P-5/77, 1977.

18. Langen, I. and Sigbjörnsson, R., "On Stochastic Dynamcis of Floating Bridges," Proc. International Conference on Environmental Forces on Engineering Structures, Editors: C. A. Brebbia, P. L. Gould and J. Munro, Pentech Press, London, 1979.
Also published in Eng. Struct., Vol. 2, 1980.

19. Langen, I., Frequency Domain Analysis of a Floating Bridge Exposed to Irregular Short-Crested Waves, SINTEF Report STF-71 A80008, Trondheim 1980.

20. Vugts, J. H., "The Hydrodynamic Coefficients for Swaying, Heaving and Rolling Cylinders in a Free Surface," Int. Shipbuilding Progr., Vol. 15, pp. 251-276, 1968.

21. Vugts, J. H., The Hydrodynamic Forces and Ship Motions in Waves, Dr. Thesis, Delft, 1970.

22. Holand, I. and Langen, I., Salhus floating bridge: Theory and Hydrodynamics Coefficients, SINTEF Report, Nov. 15th, 1972.

23. Webster, Wm. C., Hydrodynamic Tests of Four Cross-Sections for the Hood Canal Floating Bridge, UCB-Eng-7615, Berkeley, Jan. 1980.

24. Garrison, C. J., Interaction of Oblique Waves with an Infinite Cylinder, Report 80-101, January 1980, Pebble Beach, Ca.

25. Bolton, W. E. and Ursell, F., "The Wave Force on an Infinitely Long Circular Cylinder in an Oblique Sea," J. Fluid Mechanics, Vol. 57, 1968, pp. 241-256.

26. Handbook on Wave Analysis and Forecasting, World Meterological Organization, WMO - No. 446, Geneva 1976.

27. Borgman, L. E., "Ocean Wave Simulation for Engineering Design," Proc. ASCE, 95, WW4, 1969.

28. Hood Canal Floating Bridge Study, Structural Analysis of Existing East Half of Bridge, Tokola/Earl and Wright, prepared for the State of Washington, Department of Transportation, Mar. 1980, Agreement Y-2000, Supplement #3, Task #5.

29. St. Denis, M., "Some Cautions on Employment of the Spectral Technique to Describe the Waves of the Sea and the Response thereto of Oceanic Systems." Offshore Technology Conference OTC 1819, 1973.

30. Shinozuka, M., "Monte Carlo Solution of Structural Dynamics," Computers and Structures, 2, 1972.

31. Wen, Y-K. and Shinozuka, M., "Analysis of Floating Plate Under Ocean Waves," Proc. ASCE, 98, WW2, 1972.

32. Seltzer, G., "Wave Crests - How Long?," Dept. of Civil Engineering, University of Washington, October 1979, unpublished.

33. Hartz, B. J., Notes on the Spatial Correlation Factor, University of Washington, Seattle, Wa., June 1980, unpublished.

BRIDGE MODAL IDENTIFICATION PROBLEMS

Gerard C. Pardoen[1], M. ASCE
Athol J. Carr[2], M. ASCE
Peter J. Moss[2], M. ASCE

ABSTRACT

This note is intended to enunciate the difficulties encountered in identifying the modal characteristics of the Toe Toe Stream Bridge as a result of a recent experimental investigation. At first glance the relative structural simplicity and symmetry of the bridge would not suggest any unusual test procedures for modal identification. Retrospectively, this apparent structural simplicity and symmetry was misleading. The raked piers, which act as an elastic support for the bridge superstructure, the cross-fall and longitudinal slopes of the bridge deck, the bridge deck being removed from the neutral surface, etc. contributed to bridge vibration motion that was difficult to segregate into "clean" transverse, vertical, longitudinal or torsional modes with only 4 single axis instruments. As a result of the ambient and forced vibration tests the modal identification has been inconclusive and it has been suggested that too few instruments were used in the test procedure. This conclusion seems to be substantiated by the results of a subsequent analytical study and is offered so that others may profit by the authors' experience.

[1] Department of Civil Engineering, University of California, Irvine, Irvine, California.

[2] Department of Civil Engineering, University of Canterbury, Christchurch, New Zealand.

INTRODUCTION

This note is intended to enunciate the difficulties encountered in identifying the modal characteristics of the Toe Toe Stream Bridge as a result of a recent experimental investigation. At first glance the relative structural simplicity and symmetry of the bridge would not suggest any unusual test procedures for modal identification. However as a result of the ambient and forced vibration tests the modal identification has been inconclusive and it has been suggested that too few instruments were used in the test procedure. This conclusion seems to be substantiated by the results of a subsequent analytical study and is offered so that others may profit by the author's experience.

BRIDGE DESCRIPTION

The Toe Toe Stream Bridge, on the Mangaweka deviation just south of Taihape on State Highway 1 of New Zealand's North Island (see Figure 1), comprises a 3.5 m deep Ministry of Works and Development standard steel truss superstructure supported on inclined structural steel piers and on bearings at the abutments. With a two-lane reinforced concrete deck, the composite bridge has end spans of 15.743 m and 15.730 m and a central span of 36.643 m to give a total span of 73.823 m. Other details of the bridge are shown in Figure 1 and detailed below.

Piers: Three structural steel H-sections, 10 m in length and sloping at 45° to the bridge, with a transition length of 2.5 m to connect to hinges on the bottom chord of the four trusses.

Bearings: Instead of the typical Ministry of Works elastomeric bearings, lead-rubber dissipator devices were used. The lead-rubber device characteristics chosen were 280 x 230 x 127 mm elastomeric bearings with a central lead cylinder of 75 mm diameter to give a total dissipator force of 0.035 times the vertical force. The bearing was thicker than theoretically required to accommodate shear deformation; the additional compressive flexibility was desirable to delay yielding in tension of the bottom chord in the end span of the truss under the principal longitudinal sway mode.

Deck slab: 180 mm thick insitu concrete with local thickening over the trusses.

General views of the bridge are shown in Figure 2.

TEST PARTICIPANTS AND INSTRUMENTATION

The bridge testing was conducted as a cooperative venture between the Central Laboratories, Ministry of Works and Development and the Departments of Civil Engineering at the Universities of Auckland and Canterbury. The shaking machine used for the forced vibration of the bridge was one belonging to the Ministry of Works and Development Central Laboratories. For vibration measurements, displacement meters were used by Central Laboratories and Ranger seismometers by the Auckland and Canterbury Civil Engineering Departments. This note

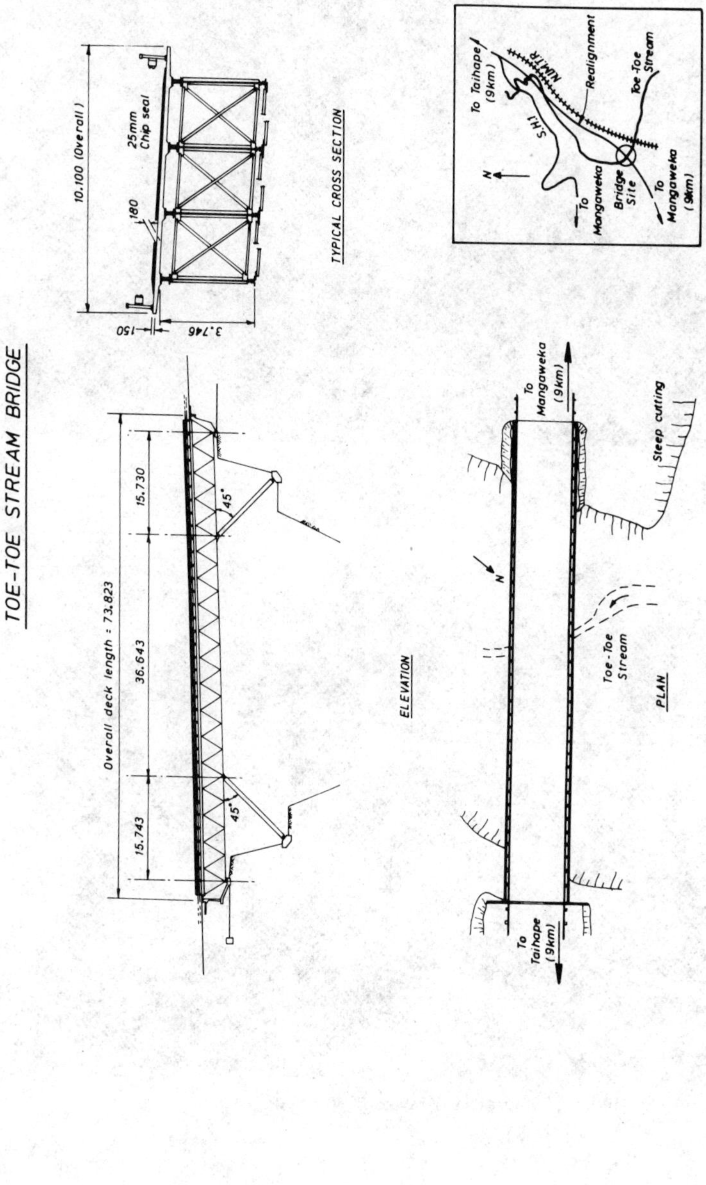

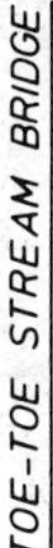

FIGURE 1. Plan and Elevation Views of Toe Toe Stream Bridge

FIGURE 2. General Views of Toe Toe Stream Bridge

describes only the tests carried out by the University of Canterbury and the subsequent data analysis of the ambient vibration data.

The bridge vibrations were measured using a set of four Ranger seismometers. The seismometers were each capable of measuring in a single direction which could be easily changed from vertical to horizontal and vice versa. The signals from the four seismometers were amplified and conditioned by a 4-channel signal conditioner which in turn provided the appropriate output to an FM analogue tape recorder. The signal conditioner could be set to give velocity, displacement and/or acceleration output signals from the velocity input.

Subsequent to the field recording of the bridge vibrations, the tape from the seismometers was replayed and each vibration record was processed by a Zonic DMS 5003 Data Memory System for frequency and spectral analysis. This digital signal processor consists of a two channel, high speed, fast fourier transform processor capable of sampling the data, averaging it as required, producing transfer functions and auto and cross-spectrum plots. The transfer function and spectrum plots are displayed on a graphics display terminal. This terminal is also used for all command control of the DMS 5003.

TEST AND ANALYSIS PROCEDURE

The seismometers were placed in selected patterns along either the center line of the bridge or the two edges in order to determine vertical, lateral, longitudinal and torsional frequencies of vibration for the bridge. The points used for the placement of the seismometers were: the center, third and quarter points of the central span, at the pier and at the center of the end span. By using the set of four seismometers in six different patterns, it was envisioned that the symmetrical modes of vibration would be identified. Approximately 40 minutes of data was recorded for each of the six ambient vibration instrument configurations.

By making use of the Fourier Transform procedures available on the Data Memory System, the vibration frequencies in the lateral, longitudinal and vertical were determined and are summarized in Table 1. The power spectrum results reflecting the test configurations used are shown in Figures 3→5. Each power spectrum curve represents the average power spectrum resulting from, in general, 25 "snapshots" of the time data. The letter-number code denoting the seismometer locations refers to the longitudinal and transverse positions respectively.

It should be noted that the data reduction of the recorded signal involves digitization at discrete time intervals. Discrete Fourier Transform procedures are then used to convert samples of the time data into the frequency domain in order to examine their frequency content. In the resulting Power Spectral Density plots such as Figures 3→5, the frequencies are not 'exact' but are more truly histograms giving values relating to a small band of frequencies. (In the reported analyses, each frequency band had a width of 0.049 Hz.)

One result of the digitization and analysis procedure is that it is

TABLE 1: Frequencies (Hz) Obtained from Ambient Vibration Tests Using Power Spectrum Density

Setup	Seismometer Direction	Seismometer Locations				Frequencies
		1	2	3	4	
1	Transverse	B-2	D-2	F-2	G-2	2.441, 4.883, 5.713
2	Transverse	B-2	E-2	I-2	L-2	1.807, 3.654, 5.566
3	Vertical	B-2	D-2	F-2	G-2	1.88, 3.223, 4.883
4	Vertical	B-2	E-2	I-2	L-2	1.807, 3.027, 4.883
5	Vertical	E-1	I-1	E-3	I-3	1.756, 2.979, 4.883, 5.615, 8.691, 12.16
6	Longitudinal	B-2	E-2	G-2	I-2	0.830, 3.076, 4.883, 8.057

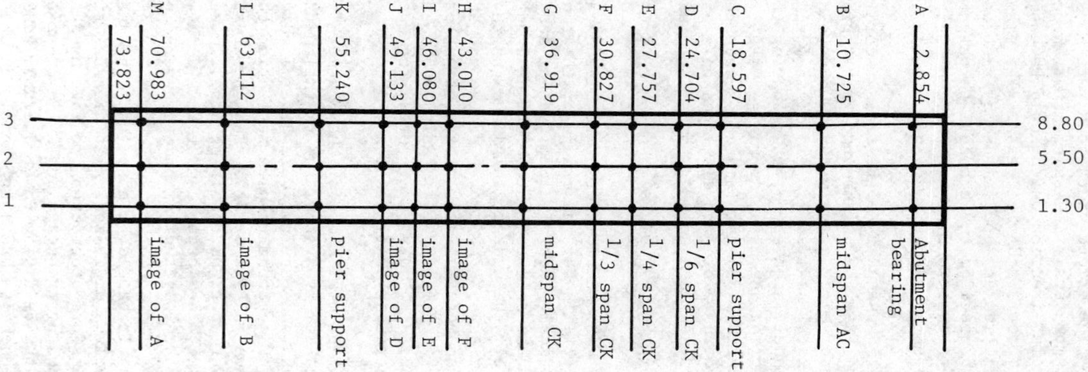

A 2.854 Abutment bearing
B 10.725 midspan AC
C 18.597 pier support
D 24.704 1/6 span CK
E 27.757 1/4 span CK
F 30.827 1/3 span CK
G 36.919 midspan CK
H 43.010 image of F
I 46.080 image of E
J 49.133 image of D
K 55.240 pier support
L 63.112 image of B
M 70.983 image of A
73.823

3 8.80
2 5.50
1 1.30

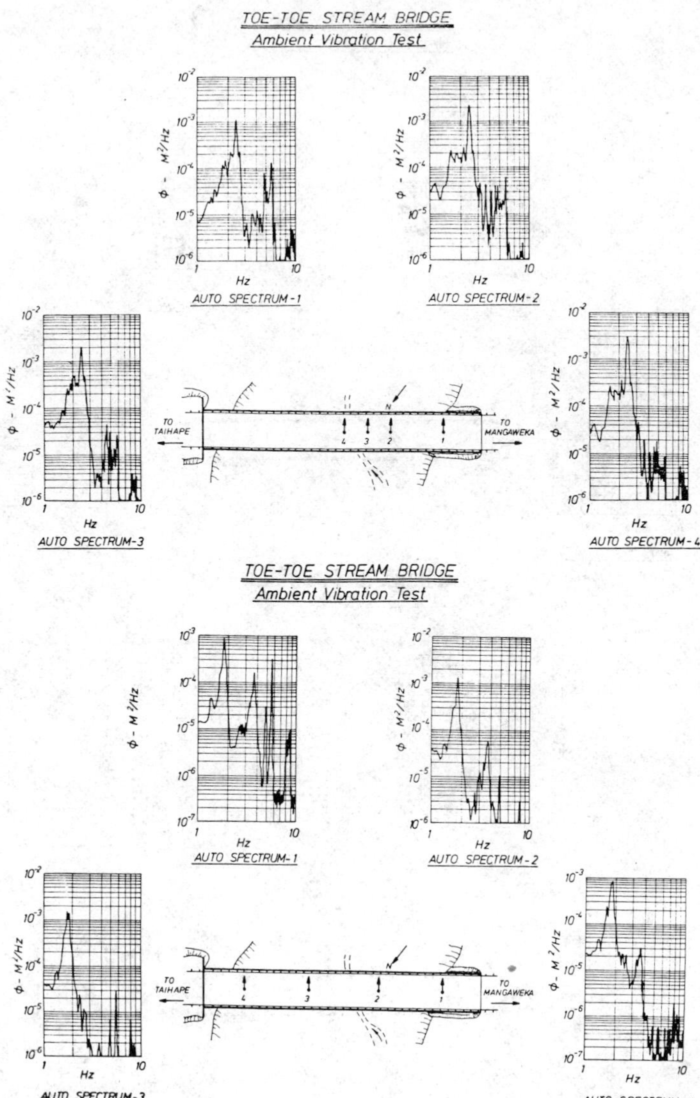

FIGURE 3. Ambient Vibration Results - Lateral Direction

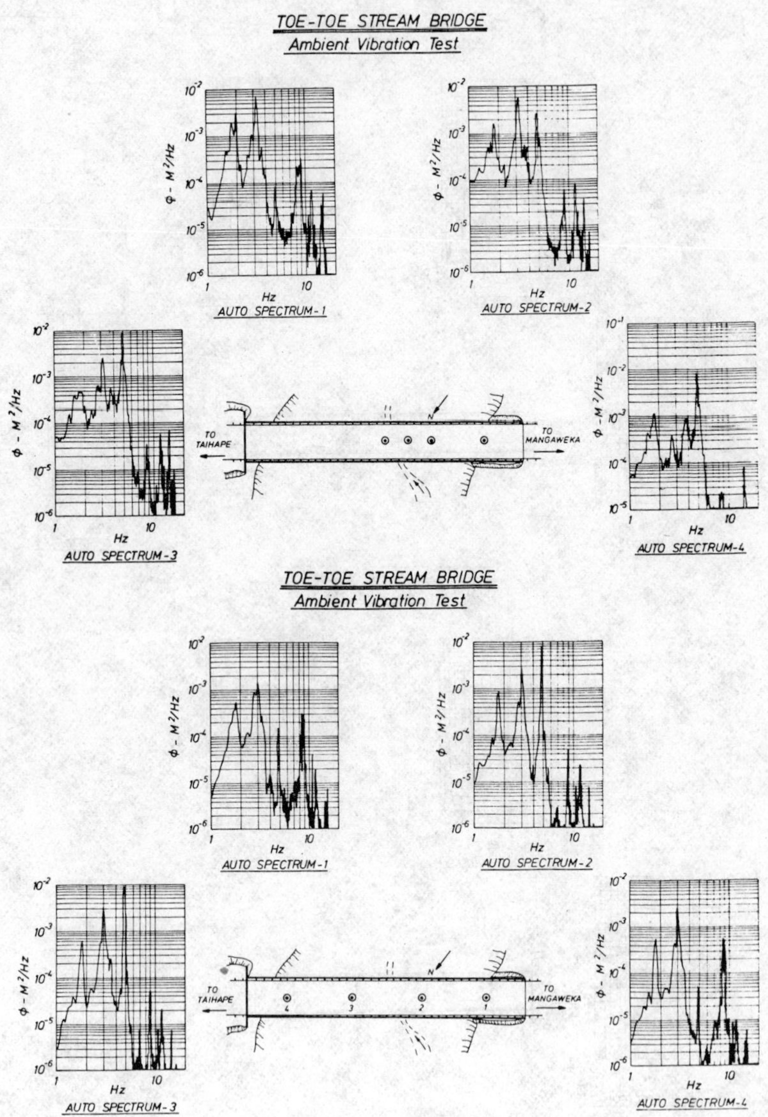

FIGURE 4. Ambient Vibration Results - Vertical Direction

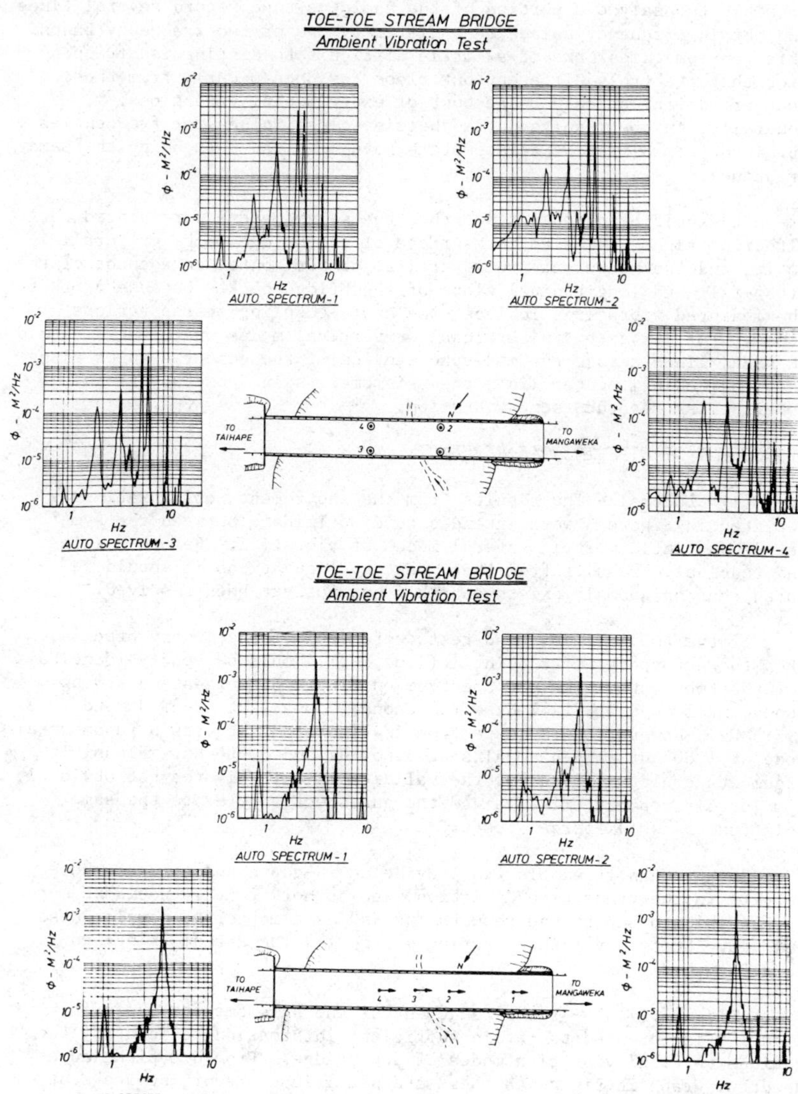

FIGURE 5. Ambient Vibration Results - Longitudinal and Torsion Directions

possible to analyze a portion of the analogue tape record several times and obtain frequency values that differ by one or two frequency bands. This can sometimes make it a little difficult to distinguish frequencies that are truly different but close (as does occur), from those that are different only on account of experimental variations. Generally, in the results quoted herein, close values for frequencies obtained from different seismometers have been taken as being the same frequency.

It should be noted further that the seismometers were placed either at right angles to the surface of the bridge deck, or parallel to it, and thus were not truly vertical or horizontal on account of the cross-fall and longitudinal slope of the bridge deck. Because of this, the measured vibrations include some cross-coupling of the various nodes. For instance, in torsional or vertical modes where there should be nodal lines across the mid-span section of the center span or along the longitudinal center line, the seismometers in those positions nevertheless recorded some vibration.

DISCUSSION OF EXPERIMENTAL RESULTS

(i) Lateral - The results from the instrument configurations depicted in Figure 3 were intended to discriminate between the symmetric and antisymmetric lateral modes of vibration. Before drawing any conclusive results from these test configurations it should be noted that an anomaly exists and this has not yet been resolved.

If the ambient vibration tests were truly a stationary process, then the autospectrum at point 1 (i.e. B-2) should be nearly identical for the two tests. The autospectrum on Figure 3a suggests a strong fundamental mode at 2.441 Hz with secondary peaks at 4.883 Hz and 5.713 Hz. However the autospectrum in Figure 3b suggests a fundamental mode at 1.807 Hz with harmonics at 3.654 Hz and 5.566 Hz. Plausible arguments could be given for the validity of both figures but obviously a major discrepancy exists since the autospectra refer to the same point under two different tests.

Rather than attempting to provide an in-depth analysis of these results in the transverse direction, the authors felt it would be more responsible to report the results "as is" as a matter of completeness and leave it to any future testing of the Toe Toe Stream Bridge to resolve the dilemma.

(ii) Vertical - The arrangement of the seismometers shown in Figure 4 is intended to obtain sufficient information to identify the various vertical vibration modes of the bridge. From a typical autospectrum graph in Figure 4a these are significant recurring peaks at 1.807, 3.223 and 4.883 Hz. From a study of the in-phase and out-of-phase relationships between the seismometers, as shown by graphs of the transfer function, it was possible to identify the following modes:

1.807 Hz - anti-symmetric vertical bending
3.223 Hz - anti-symmetrical vertical bending
4.883 Hz - symmetrical vertical bending.

Because of the limited number of seismometers available, it is impossible to make conclusive identification of these modes. This point will be amplified in a subsequent section. The second mode at 3.223 Hz seems to correspond with the longitudinal mode identified using test configuration 6 (see Table 1). The raked piers at each end of the bridge couple the longitudinal displacement of the deck with an antisymmetrical vertical displacement and, in hindsight, it would have been informative to have had a seismometer placed longitudinally so that the relative magnitudes of the vertical and longitudinal displacements could be measured in the first two modes above.

The first mode identified above appears to be the fundamental vertical bending mode shape of the bridge, this being an antisymmetric mode on account of the raked piers. This behavior may be regarded as analogous to that of a thin arch where the fundamental mode is antisymmetric.

The second mode as discussed above is probably the fundamental longitudinal mode which also produces an antisymmetric vertical bending component.

The third mode at 4.883 Hz appears to be the second vertical bending mode and may be regarded as analogous to the breathing mode in an arch structure.

(iii) Torsion - The seismometer configuration denoted in Figure 5a was intended to supplement the configurations of Figure 4 in order to segregate the torsional and vertical bending modes. From a typical autospectrum graph on Figure 5a there are significant recurring peaks at 1.758, 2.979, 4.883, 5.615, 8.691 and 12.16 Hz. From the in-phase and out-of-phase relationships between the seismometers (as determined from the transfer function graphs) and the results from Figure 4, it was possible to identify several torsional modes. Again with a limited number of seismometers it is difficult to identify conclusively the higher modes of vibration. However, if the fundamental torsional frequency is assumed to be 1.758 Hz, then the frequencies at 5.615, 8.691 and 12.16 Hz could represent higher torsional harmonics since they are in the approximate ratios of 3:1, 5:1 and 7:1 with the fundamental.

(iv) Longitudinal - The seismometer configuration shown in Figure 5b was intended to detect the longitudinal vibration behavior. The longitudinal ambient vibration test results show a predominant peak at 3.076 Hz with secondary peaks at 0.830, 4.883 and 8.057 Hz. However, inasmuch as these secondary peaks are approximately 2 orders of magnitude lower than that of the primary peak, they can be neglected as being meaningful. Thus the sharp peak at approximately 3.076 Hz represents the fundamental longitudinal frequency of the bridge.

ANALYTICAL MODEL

The general procedure in identifying the various modal parameters required the analysis of 2 channels of simultaneous vibration data. For a typical test configuration one channel was chosen as the reference station so that relative amplitudes as well as in-phase and out-of-phase relationships were with respect to the reference station. For instance, if channel 1 were chosen as the reference station, then 3 separate analyses were conducted to link the 4 data channels, namely, 2:1, 3:1, and 4:1. At times when the reference station was suspected to be at a nodal line for a particular frequency (e.g., reference channel at middle of bridge) another channel was chosen as the reference station and another series of 3 analyses were conducted. Thus the results reported in the previous section reflect the trends, in general, of several repetitive analyses of the same data. Despite this repetition of data reduction there appeared to be several discrepancies or inconclusive results in identifying the symmetric and anti-symmetric modes of lateral and vertical vibration.

In an endeavor to resolve the discrepancies, a computer model of the bridge was analyzed to determine its natural frequencies and mode shapes.

Rather than attempt a simplified analysis treating the deck and girder as an equivalent beam in order to reduce the dynamic problem, a full simulation was chosen so that the difficulties in determining the equivalent beam properties would not arise.

The four trusses were modelled with three dimensional truss members except for the two channels that represent the top chord which were modelled as continuous beam members.

The raked pier longitudinal members also had bending stiffnesses while the remainder of their members were treated as truss members as were all of the sway and wind bracing between the truss girders. The deck between the truss girders was represented by triangular high order finite shell elements with those parts of the deck cantilevered beyond the trusses being represented by three dimensional beam members with eccentric masses.

The lead-rubber bearings were represented by beams in shear and were assumed to be in their initial linear stiffness range which should be appropriate to the level of displacement exhibited during the experimental work. This model therefore had 186 nodal points, 108 triangular finite element and 598 beam and truss members. The displacement of all 4 nodes at every panel point along the trusses were slaved together in the transverse direction.

The mass of the bridge was assumed to be lumped at the deck level with five eighths of the mass of the raked piers was assumed to act with the deck, this being equally divided between the two panel points at the deck above the connection point of the piers to the trusses.

In order to further reduce the number of dynamic degrees of freedom the mass at each panel point along the bridge was lumped to the outer trusses and then to every second panel point along the bridge starting at the ends and at both panel points above the piers (see Figure 6). This gave 55 degrees of freedom compared with the 171 that would have resulted if every deck node had had its share of the mass.

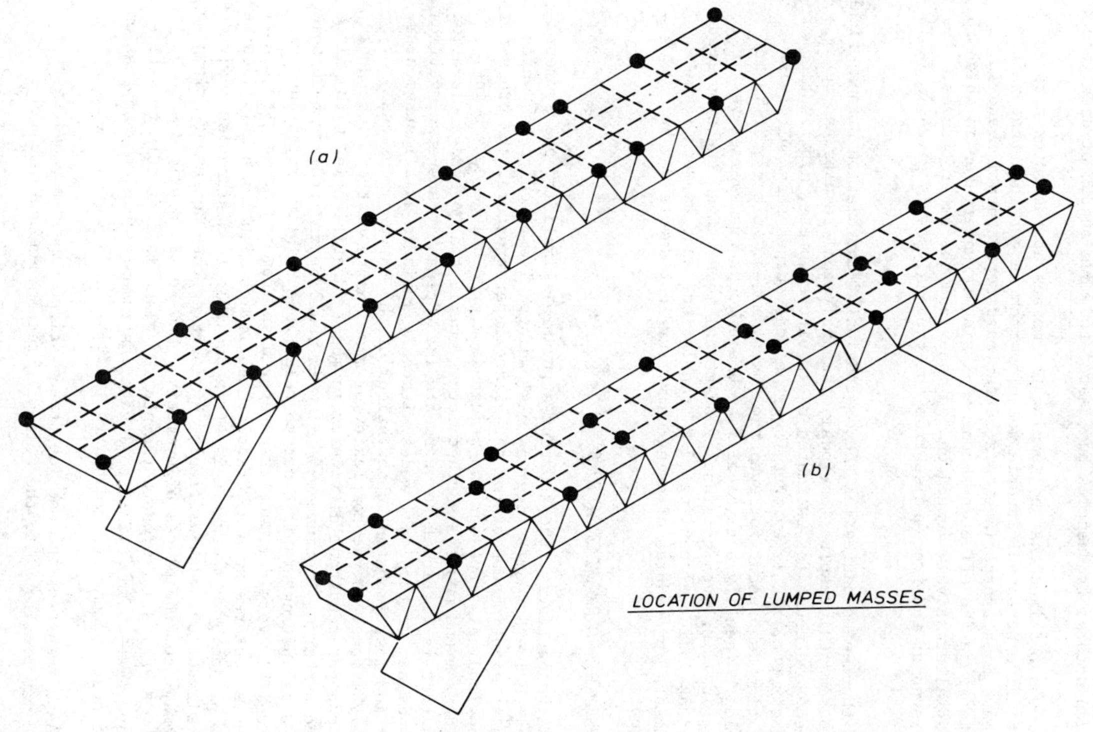

FIGURE 6. Lumped Mass Distributions of Bridge Deck

The results of this analysis are shown in Figure 7 which shows the first ten natural frequencies and associated mode shapes at deck level.

It was realized that by lumping the masses at the outer trusses the effective torsional interia would be much larger than that which would have been appropriate to a better mass representation and in order to gauge the effect a second analysis was undertaken in which the mass was alternatively lumped on the inner and the outer trusses. The comparisons of the frequencies from the two analyses is given in Table 2.

Without even addressing the discussion of correlation of analytical and experimental results (obviously the analytical results can be modified to account for estimates of in situ concrete, distribution of deck mass, etc.) it is evident from the mode shapes that multiaxis translation is present for even the lowest modes. Clearly it is not possible to identify analytically an unambiguous transverse, vertical, longitudinal, or even torsional fundamental mode. Even the simpliest vibrations indicate coupling of 2 translations and, as such, the experimental procedures of using only four instruments may be inadequate.

CONCLUSIONS AND RECOMMENDATIONS

Retrospectively, the apparent structural simplicity and symmetry of the Toe Toe Stream Bridge was misleading. The raked piers, which act as an elastic support for the bridge superstructure, the cross-fall and longitudinal slopes of the bridge deck, the bridge deck being removed from the neutral surface, etc. contributed to bridge vibration motion that is difficult to segregate into "clean" transverse, vertical, longitudinal, or torsional modes with only 4 single axis instruments. One of the most important observations of the three-dimensional analytical model is the complexity of mode shapes at the bridge deck level and the resultant difficulty in modal identification with only a limited number of instruments. Also it would be fruitful to have some multi-axis instruments in order to relate the relative magnitudes of longitudinal, vertical, and transverse displacements at any frequency. In some structures such as "regular" buildings with reasonably identifiable planes of symmetry it is often possible to distinguish individual modes without too much difficulty by locating the instruments along and then away from the planes of symmetry. However in a bridge such as the Toe Toe Stream Bridge, where the neutral plane is difficult to determine and next to impossible to locate instruments upon, the separation of the modes becomes exceedingly difficult.

As a result of the Toe Toe Stream Bridge experience and so that others may profit by it, the following recommendations are made:

1) Increase the number of instruments so that complex, multiaxis vibrations can be recorded simultaneously. For example, 8 single axis seismometers recorded on an 8-channel recorder should suffice for most applications whereas four channels may be insufficient.

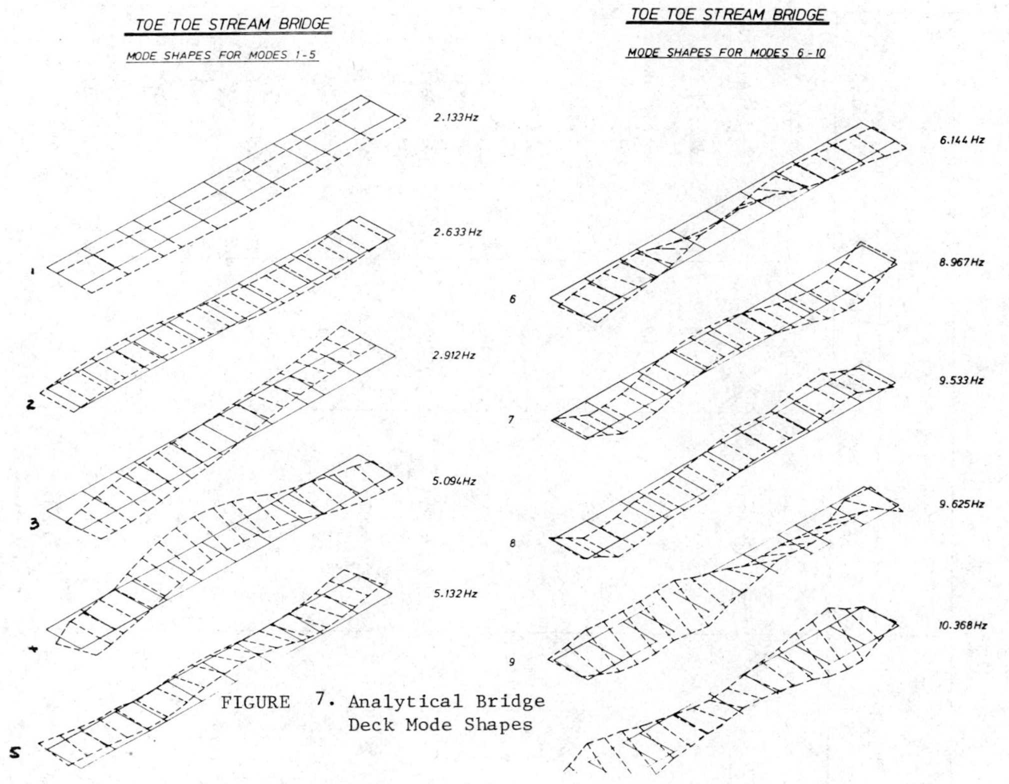

FIGURE 7. Analytical Bridge Deck Mode Shapes

Mode No.	Frequency (Hz) masses at		Mode Shape (see Figure 7 also)
	Outer Trusses	Alternate	
1	2.133	2.240	lateral
2	2.633	2.786	longitudinal
3	2.912	3.104	lateral
4	5.094	5.162	symmetric and vertical
5	5.132	5.366	symmetric and vertical
6	6.144	7.734	torsion
7	8.957	9.386	symmetric and vertical
8	9.533	10.389	anti-symmetric vertical
9	9.625	11.069	torsion
10	10.368	11.269	torsion
11	11.131	11.352	anti-symmetric lateral
12	11.263	12.058	anti-symmetric vertical
13	11.506	14.404	torsion
14	14.140	15.820	symmetric and vertical
15	14.516	15.912	torsion

TABLE 2: Analytically Predicted Bridge Frequencies

2) Record 2 and possibly 3 axes of data at key locations to obtain the relative degree of coordinate coupling.

3) Identify a sufficient number of instrument setups in order to define the significant vibration positions.

4) If at all possible determine the various modes of vibration to be expected of the structure from a reasonably well defined analytical model apriori to the testing.

ACKNOWLEDGMENT

This research was supported, in part, by the National Science Foundation under its U.S.-New Zealand Cooperative Science Program. The Support of the Foundation, under the direction of Dr. Alan Milsap, is gratefully acknowledged.

Additionally, the support of the Road Research Unit of the National Roads Board of New Zealand for providing financial assistance for the test participants is gratefully acknowledged.

DYNAMIC RESPONSE OF FLOATING BRIDGE TO WAVE FORCES

by C.H. Beauchamp[1], AM. ASCE and D.N. Brocard[2], AM. ASCE

Abstract

A hydroelastic model was constructed to investigate the dynamic response of the Hood Canal floating bridge and its mooring system to storm wave conditions. An automatic data acquisition system was used to simultaneously measure wave heights in the basin and tensions in the anchor cables. In a first series of tests, the bridge model was abruptly released from deflected states and allowed to oscillate freely in order to determine the periods and damping coefficients for the natural modes of vibration. The natural modes observed showed good agreement with those predicted by a numerical model of the bridge system. Storm wave conditions were then generated in the model basin and the anchor cable tensions were measured. Spectral analysis showed that although the waves excited several natural modes of vibration, the major energy associated with bridge motion was forced at the wave period. The hydroelastic model provided a reliable tool for investigating the complicated fluid/structure interactions of the bridge system.

Introduction

The Hood Canal floating bridge, which is located west of Seattle on Washington State Route 104, failed in a severe storm on February 13, 1979, and the western portion of the bridge sunk. This paper describes an investigation of the dynamic response of the Hood Canal bridge and its mooring system to wave forces under the storm conditions. For this investigation, a 1/60 scale hydroelastic model of a portion of the bridge and surrounding water body was constructed, instrumented, and tested. The results of this study were used to calibrate and verify a mathematical model of the bridge and mooring system which is discussed by Luft (1981).

---

[1] Research Engineer, Alden Research Laboratory, Worcester Polytechnic Institute, Holden, Massachusetts.

[2] Lead Research Engineer, Alden Research Laboratory, Worcester Polytechnic Institute, Holden, Massachusetts.

Description of the Prototype

The Hood Canal bridge was composed of two sections linked by a draw span to allow navigation passage. A schematic diagram of the modeled section of the bridge is shown in Figure 1. This section of the bridge was buoyed by a train of 9 rectangular pontoons which were typically 360 ft long, 50 ft wide, and 15 ft deep. These pontoons were constructed of reinforced concrete with a grid of internal ribs for flexural and torsional stiffening. The pontoons were fastened together by a system of pins and cable tendons which were pre-stressed such that the entire train behaved structurally as a homogeneous box beam. The mooring system consisted of nine pairs of opposing anchor lines which were prestressed with a load of 230 kips per anchor at high tide. The roadway was supported on a superstructure ranging in height from 15 to 60 ft above the surface of the pontoons. During severe storms, the draw span pontoon was drawn back in between the fork section (Figure 1) so that this end of the bridge was unconstrained except for the anchor lines. The floating portion of the bridge was connected to the shore via a transition span. The span was connected to the bridge superstructure in such a manner that it only constrained the bridge in the longitudinal direction; that is, it did not restrict vertical, horizontal, or rolling motions of the bridge.

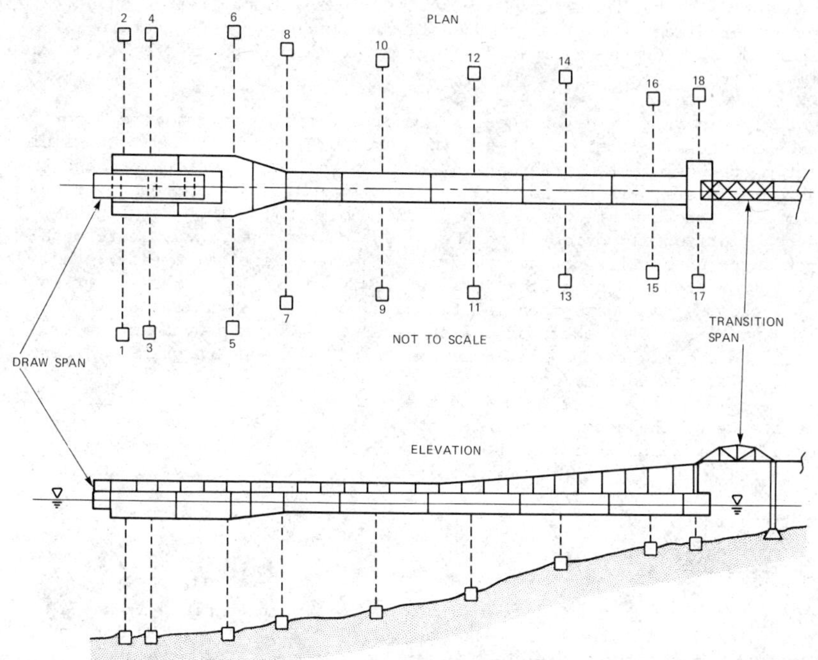

Figure 1 Schematic of the Modeled Portion of the Hood Canal Bridge

Model Similitude

Proper modeling of hydroelastic deformations and vibrations is dependent upon satisfying the condition of similitude as derived by non-dimensionalizing the fundamental laws governing hydrodynamics and structural elasticity. Dynamic similitude requires that the model to prototype scale ratios of forces must be equal for the dominant forces present in the system. A floating bridge is an elastic structure suspended at the free surface between air and water. The forces which are important in this system are inertia, gravity, hydrodynamic fluid pressure, and structural elasticity.

The presence of a free water surface in the modeled system required that the hydrodynamics be scaled according to Froudian similitude. That is, Froude numbers must be identical at corresponding points in the model and prototype. Using this criterion, the scale ratios of all the important hydrodynamic and structural properties can be derived from the geometric scale ratio (Haspra, 1978). Scale effects due to viscosity and surface tension were considered and the model scale was selected such that these scale effects were negligible (LeMehaute, 1976).

The important properties of the bridge were its geometric shape, elastic properties, total mass, and mass distribution. The external shape was important since it influences the fluid motion near the structure. Therefore, accurate scaling of all external dimensions of the bridge was imperative.

The elastic properties of the structure and mooring system were important in the interactions between the structure and the fluid. The fluid motion causes elastic deformations (vibrations) of the structure and the deformations simultaneously induce fluid motion. Four degrees of freedom were important in the motion of the Hood Canal bridge structure: surge (horizontal motion along the bridge centerline); sway (horizontal motion perpendicular to the centerline); heave (vertical motion); and roll (rotation around the centerline). Within the later three degrees of freedom, there was potential for many modes of vibration due to bending and torsional flexure of the structure. To obtain dynamic similarity, the elastic forces needed to be scaled as the hydrodynamic forces. Therefore, the spring constant of the mooring cables and the flexural and torsional stiffness moduli of the bridge structure were scaled as functions of the geometric scale ratio (Haspra, 1978). The stiffness modulus is defined here as the product, EI, where E is the modulus of elasticity and I is the moment of inertia. It is significant that in modeling the bridge structure, only the external geometry and the overall stiffness modulus needed to be simulated; the modulus of elasticity and the complex internal geometry of the bridge do not need to be scaled individually.

The total mass of the floating structure is important since it determines the draft and center of buoyancy, and the mass distribution is important since it determines the center of gravity and metacentric height. All of these properties control the righting moment of the structure, which is important in controlling the roll mode of motion (Berteaux, 1976).

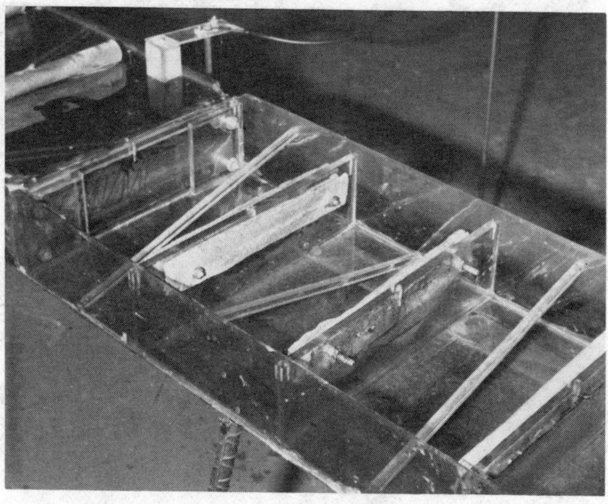

Figure 2 Internal Ribbing and Weights in Model Bridge

In the absence of damping, the natural modes of vibration and the frequencies of these modes are correctly scaled by the above similitude criterion. Damping forces in the floating bridge system include both structural and hydrodynamic damping. However, for both the prototype and model, the structural damping is an order of magnitude less than the hydrodynamic damping. For the important modes of vibration of the bridge, the generation of waves is the dominant cause of the hydrodynamic damping (Bhallachrayya, 1978), and since wave forces are gravitational, these damping forces are properly scaled by Froudian similitude.

Model Description

The model basin represented a 2400 ft by 4000 ft area of the Hood Canal estuary at a 1/60 geometric scale. For practical and economical reasons, the prototype bathymetry was only simulated to a maximum depth of 100 ft below MLLW. Areas with depths greater than 100 ft in the prototype were represented by a flat bottom at 100 ft in the model. This truncation of the bathymetry in the model was possible since, at this depth, the waves being tested were deep water waves and therefore unaffected by the bottom.

This basin was equipped with five individually driven plunger type wave generators. Each generator was capable of simulating waves with variable heights, periods, and approach angles to the bridge.

Figure 3 Instrumented Cantilevers Simulating Cable Elasticity

In modeling the bridge, it was assumed that the train of pontoons acted as a single homogeneous box beam. It was also assumed that the roadway superstructure had a negligible effect on the structural stiffness of the bridge and therefore the superstructure was not reproduced in the model except for its effect on the center of gravity.

To scale the elastic properties of the bridge, the model pontoons were constructed of cellulose acetate butyrate (CAB) which has a Young's Modulus of approximately 10^5 psi, as compared to reinforced concrete which has a modulus of 34×10^5 psi. It was computed that for the bridge pontoons, a geometrically scaled box beam of CAB with an external shell thickness of 0.060 inch and minimal internal ribbing had approximately the required stiffness moduli. Measurements on a section of the model indicated this design had adequate stiffness in sway and heave modes, but was too flexible in torsional mode. Internal diagonal beams were, therefore, added to the model pontoons to stiffen them in torsion without significantly affecting the flexural characteristics (see Figure 2). The measured moduli for this model were within ±15% of the values required to simulate the prototype. In view of the level of uncertainty in the calculated prototype values, further refinement was not pursued.

The specific gravity of CAB is only half that of reinforced concrete. Therefore, it was necessary to add lead weights to the model bridge to simulate the mass of the prototype. The lead weights were distributed in the model to correctly simulate the center of gravity and righting moment.

The bridge mooring system was simulated by a cable, pulley and cantilever system, as shown in Figures 3 and 4. In the model, anchor cables were run from the bridge through pulleys located on the basin floor to instrumented cantilevers. The first pulley on the basin floor was located so as to reproduce the angle of the prototype anchor line. The cable in the model was choosen so that it would be virtually non-elastic with model scale forces. The elasticity of the model anchor system was adjusted to simulate the prototype elasticity by adjusting the cantilever length.

Instrumentation

Cable tensions were determined by measuring the strain in the cantilever beams. Foil type strain gages, which were temperature compensated, were mounted in a full bridge arrangement on each cantilever. Wave height measurements were made by parallel wire resistance probes. These probes were compensated for the water conductivity and isolated to prevent varying ground potentials between probes. All the strain gage and wave probe signals were individually amplified and filtered prior to transmission to an automated data acquisition system.

The data acquisition system was controlled by a dedicated LSI/1103 computer which was accessed via a video terminal. All 31 channels of data (18 strains, 12 surface elevations, and 1 conductivity) were scanned serially, multiplexed through an A/D converter and stored digitally in the LSI/1103 in less than 3 milliseconds. This high scanning speed allowed a large number of scans during each vibration period. The scans were repeated automatically at a preselected time interval and for a preselected total number of scans. At the end of a test, all data were converted to forces or wave heights and transferred to a PDP11-34A computer for permanent storage and further analysis.

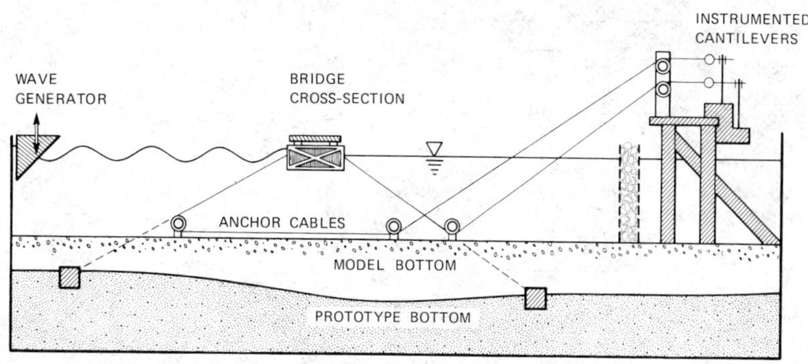

Figure 4 Schematic of Model Cross-Section

Testing for Natural Modes of Vibration

One objective of this model study was to determine the periods and damping coefficients for the natural modes of vibration; of particular interest was the measurement of hydrodynamic damping for which little information is available in the literature. This was accomplished by conducting a series of tests in which the model was manually deflected at one or more points and then abruptly released and allowed to oscillate freely in an otherwise still model basin. The locations which were selected for deflecting the bridge were those at which maximum deflections were found in the mode shape predicted by the numerical model. The resulting cable tension time series were used to determine the fundamental periods and corresponding logarithmic decrement damping coefficients.

The results of the natural mode tests are summarized in Table 1 and the observed periods are compared to the numerical model predictions of Luft (1981). The periods found for numerical and physical models agreed within ±20%. A large portion of this error can be attributed to the error in simulating the stiffness modulus of the bridge structure. It was found that due to relatively high torsional stiffness, all modes of torsional flexure vibrated at periods below the range of interest.

TABLE 1

Natural Modes of Vibration

Mode	Description*	Period (sec) Numerical Model (Luft)	Period (sec) Physical Model	Damping Coefficient Physical Model
Surge	Entire bridge in unison	15.1	14.0	0.05
First Sway		12.6	15.0	0.12
Second Sway		9.2	10.0	0.13
Third Sway	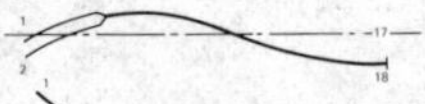	6.5	----	----
First Heave		7.3	8.0	0.35
Second Heave		6.4	7.0	0.25
Third Heave	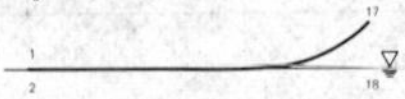	6.0	6.1	0.30
Roll	Entire bridge in unison	7.7	6.3	0.13

*Where applicable, schematic diagrams of the mode shape are shown with the location of some anchor cables indicated.

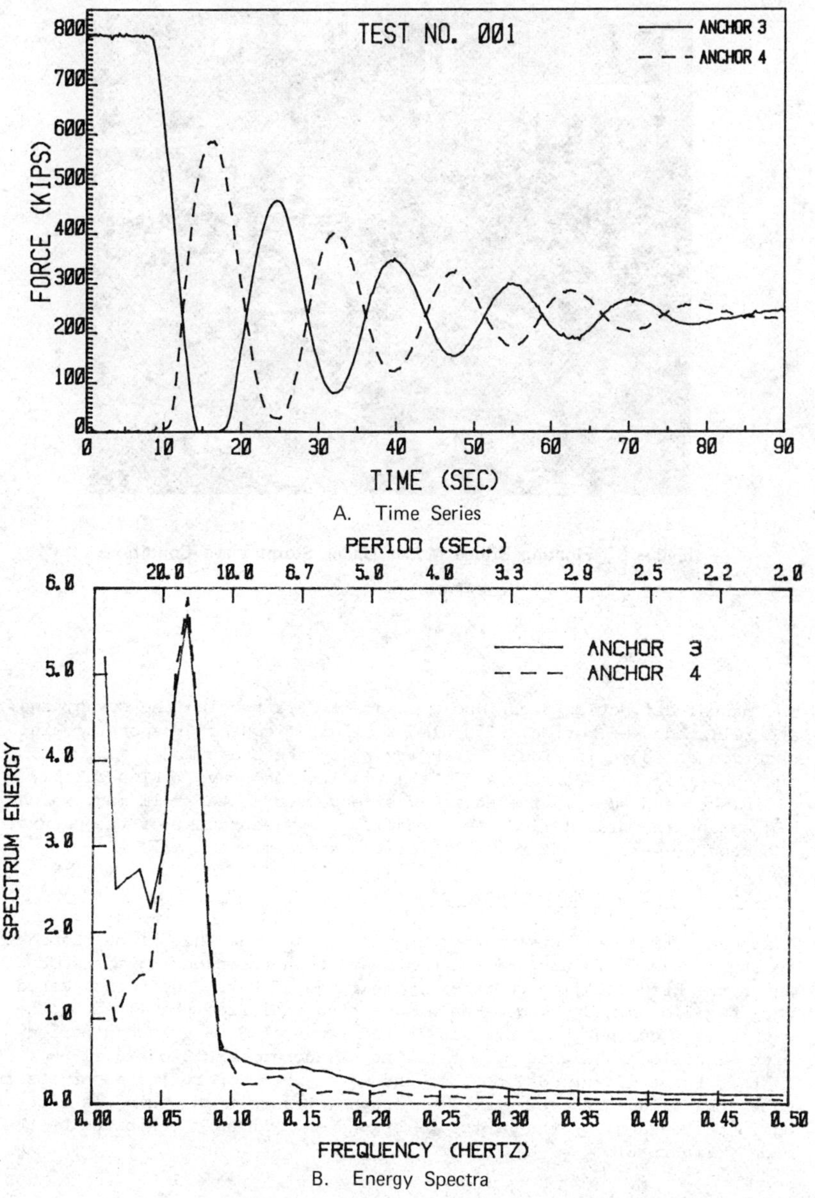

A. Time Series

B. Energy Spectra

Figure 5 Anchor Tension Responses of First Sway Vibrational Mode

Figure 6 Floating Bridge Model Under Storm Wave Conditions

The manual deflection technique worked well for testing the surge, sway, and roll modes of motion, with almost single frequency responses being observed as shown in Figure 5. However, in the heave mode, a high damping coefficient was observed. As a result, the heave motion quickly diminished and some of the energy was converted to the roll mode which had a close natural period. As a result, the measurement of the natural frequencies were not as reliable for the heave modes.

<u>Testing with Storm Wave Conditions</u>

A second series of tests was performed to determine the bridge response to storm waves. No wave measurements had been taken during the storm in which the bridge failed therefore, the range of wave conditions tested were based on hindcasting. The hindcasting indicated that most of the wave energy during the storm was in the range of 3 to 6 second periods with significant heights of 7 to 12 ft (Beauchamp and Brocard, 1980). In this range of periods, most of the wave energy was reflected or absorbed by the bridge and very little energy was transmitted. This was observed in photographs of the prototype bridge and was also exhibited by the model (Figure 6).

The model was tested with single frequency wave conditions. Time series and energy spectra for one test condition with a significant wave height of 7 ft and a mean period of 3.8 seconds are shown in Figure 7. At this period, the wave length (74 ft as computed by linear deep water wave theory) was of the same order as the width of the bridge (50 to 100 ft). As the wave length decreases to the width of a floating structure, the amplitude of the net vertical pressure fluctuation across the width of the structure becomes small resulting in small driving impulses in the heave and roll modes of vibration. Therefore, it was not expected that these modes of vibration would be induced in the bridge.

The bridge response to the 3.8 second, 7 ft waves is indicated by the anchor cable tensions time series and energy spectra, given in Figures 8 to 10. Inspection of the anchor tension time series showed that the mean deflection of the bridge was greatest in the center of the bridge (anchors 9 and 10). The difference between mean forces on windward and leeward anchors was 90 kips at the center, but only 20 to 30 kips at the ends. The mean deflection was primarily attributed to wave setup on the windward side of the bridge. This setup would tend to leak around the ends of the bridge, thus a larger setup and a larger mean deflection would be expected at the center.

The amplitude of the dynamic component of the forces was generally less than this mean difference. The rms amplitude was less than or equal to 8 kips for cables 1 to 14. However, the last two anchor pairs on the shore end of the bridge (anchors 15 to 18) had rms amplitudes between 13 and 24 kips.

Spectral analysis of the dynamic component showed four significant peaks at periods of 3.8 seconds, 5.1 seconds, 10 seconds, and 15 seconds. The phase information for all of these peaks showed that opposing anchor tensions vibrated approximately 180° out of phase. This would indicate that these vibrations were either roll or sway modes.

The peak at 3.8 second, which was the period of the driving wave, was generally the highest peak. The 15 second peak was most significant at anchors 1 and 2 (Figure 8), weak in the center of the bridge (Figure 9), and non-existent at anchors 17 and 18 (Figure 10). This period and mode shape are characteristic of the first sway mode of vibration (see Table 1). The peak of 10 second was only weakly observed near the center of the bridge. This peak was attributed to the second sway mode of vibration which showed a maximum deflection at anchors 11 and 12.

The observed peak at 5.1 second did not correspond to any of the natural modes tested on the physical model. It is probable that this peak resulted from the excitation of a higher order mode which was not tested. The numerical model of Luft predicted a third sway mode which showed a maximum deflection at anchor cables 17 and 18. The observed 5.1 second peak also showed maximum energy at anchor cables 17 and 18. The 5.1 second (0.20 hz) and 3.8 second (0.27 hz) periods had approximately equal energy in the spectra for anchors 17 and 18. The interaction of these two periods results in a beat pattern with a period of 15 seconds (0.07 hz) in the respective time series (see Figure 10). This measured beat frequency is equal to the difference between the two individual frequencies as predicted by linear superposition.

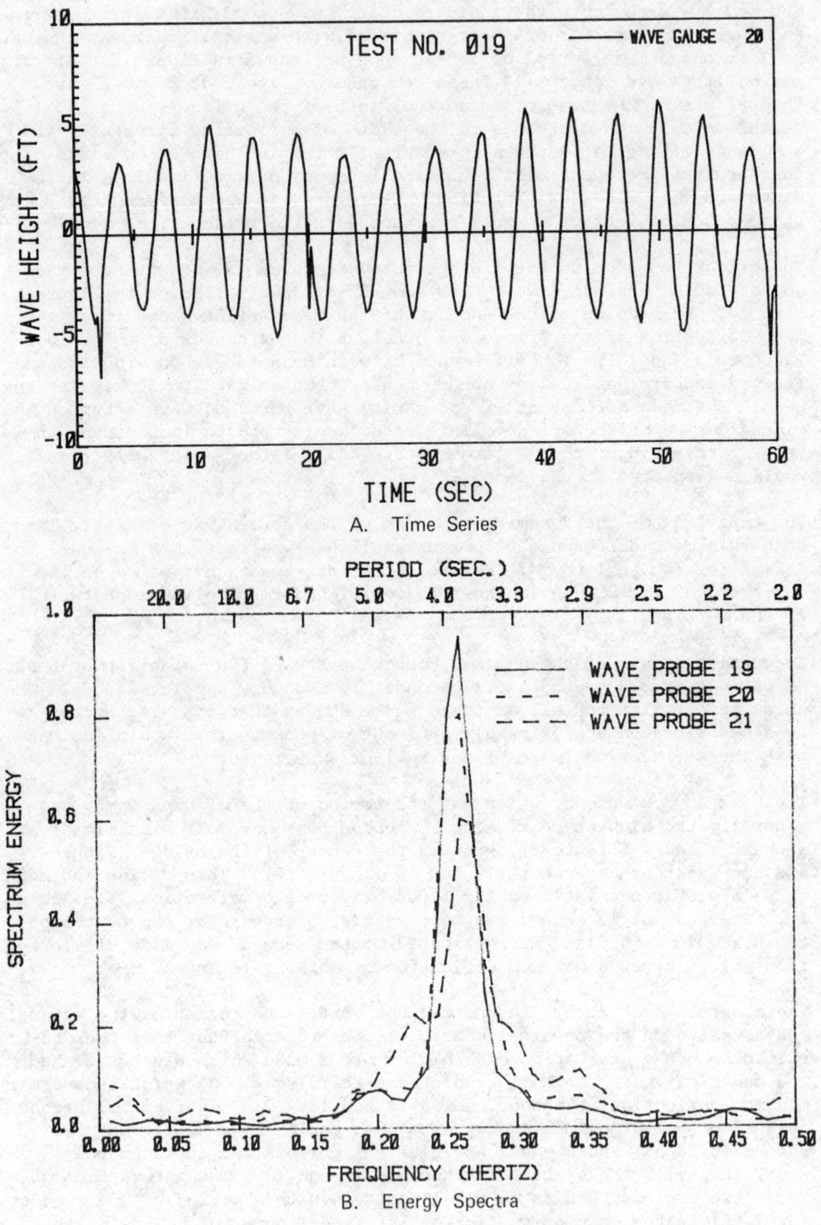

A. Time Series

B. Energy Spectra

Figure 7 Sample Test Wave Conditions

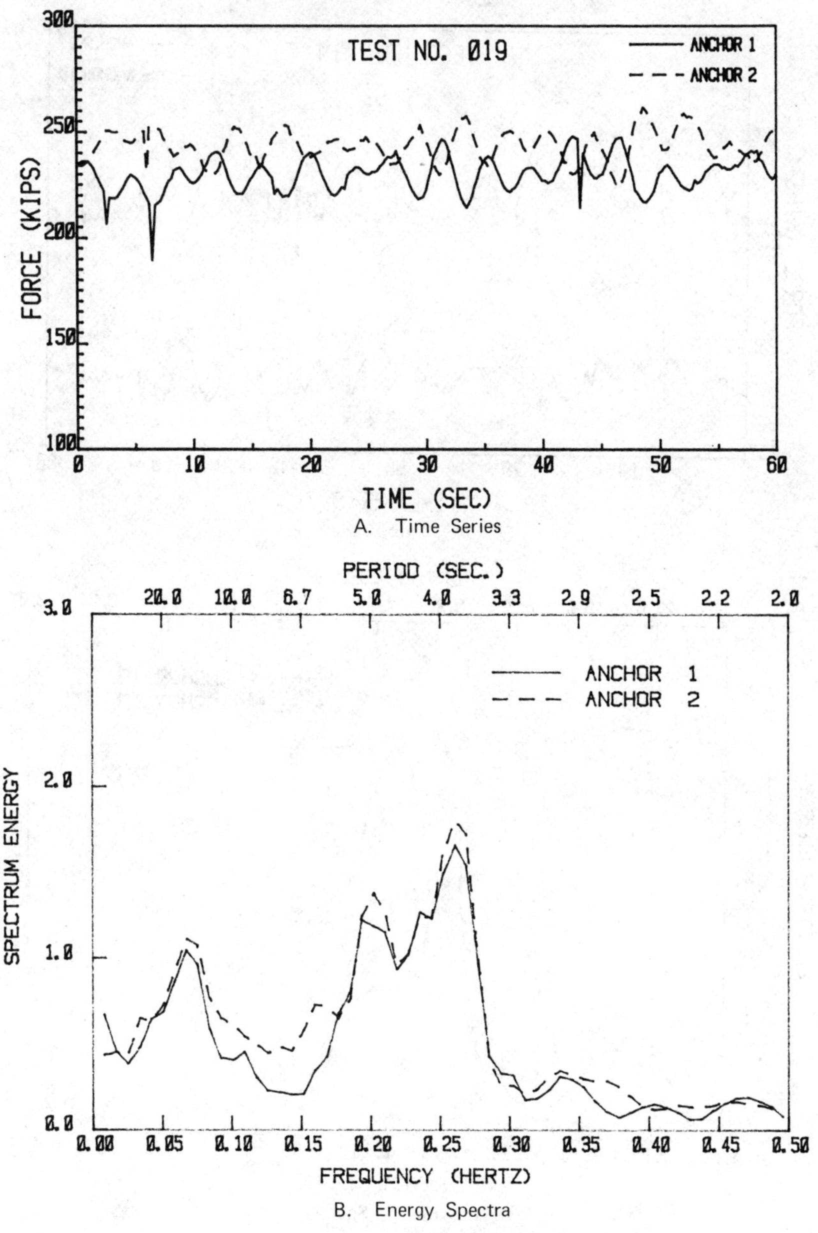

A. Time Series

B. Energy Spectra

Figure 8 Sample Anchor Tension Responses

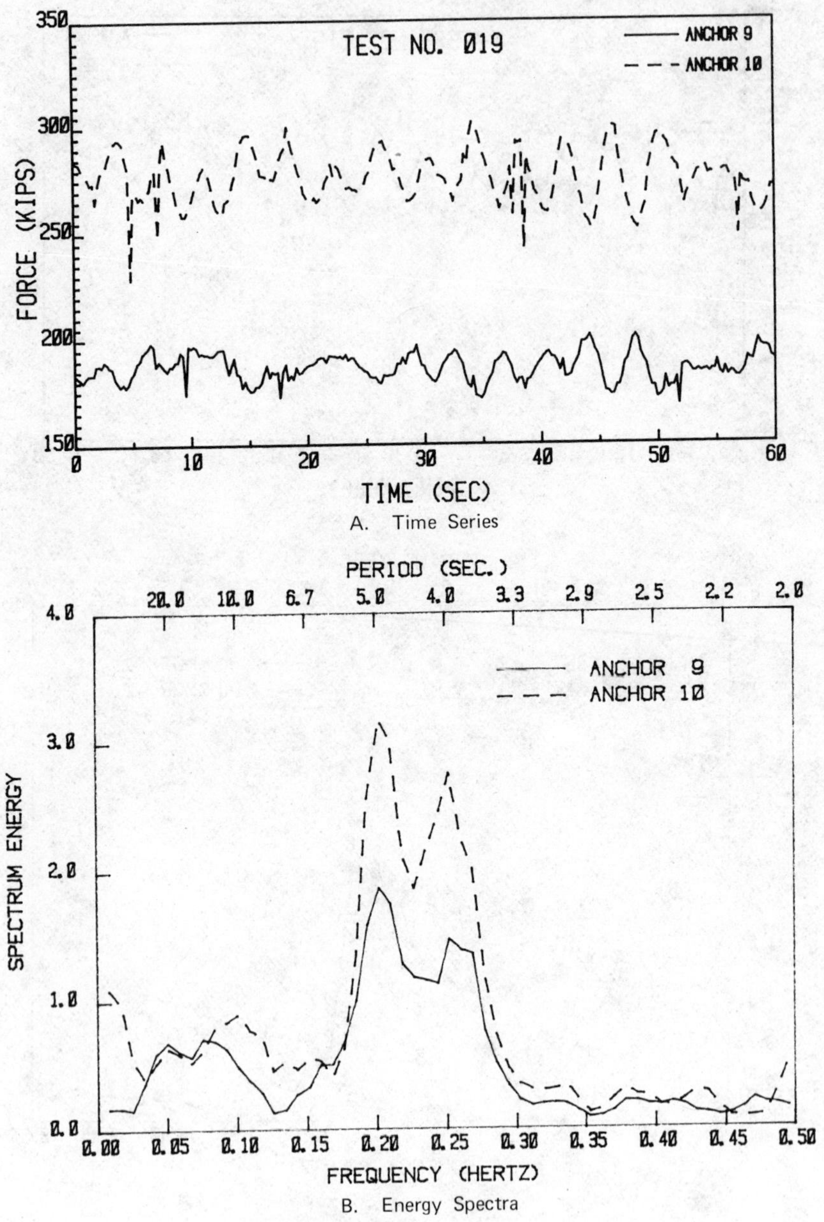

A. Time Series

B. Energy Spectra

Figure 9 Sample Anchor Tension Responses

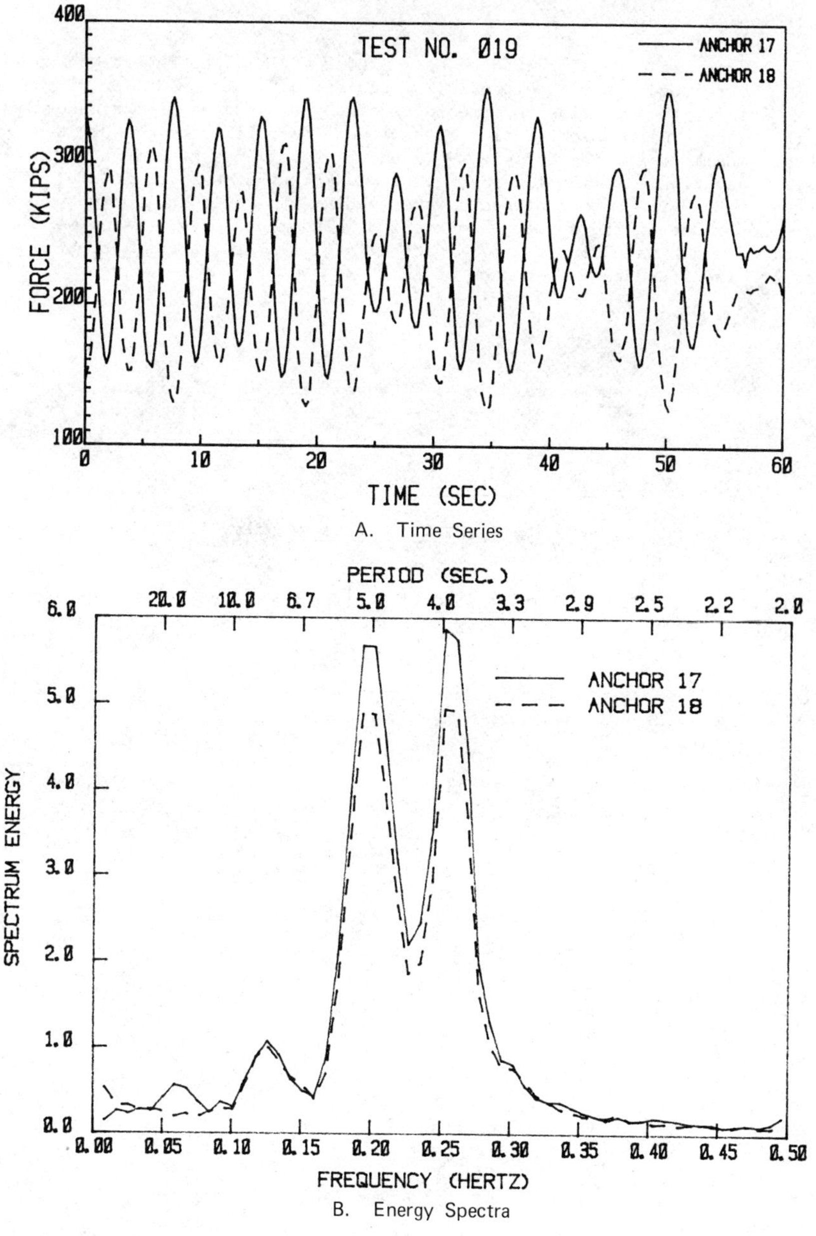

A. Time Series

B. Energy Spectra

Figure 10 Sample Anchor Tension Responses

Conclusion

The hydroelastic model proved to be a useful tool in the investigation of wave forces on the Hood Canal floating bridge. The model provides the ability to determine the frequencies and mode shapes of the natural vibrations. Of particular significance was the ability to measure the hydrodynamic damping coefficients in the various modes of motion which were unavailable in the literature. In addition, it was possible to subject the bridge to continuous wave loadings to determine the response of the bridge during a storm and to find which modes of vibration were excited under wave conditions. Spectral analysis of the response time series indicated the importance of the resonance energy relative to the energy at the driving period.

References

Beauchamp, C.H., and Brocard, D.N., "Hydroelastic Model Study of Wave and Current Forces on Hood Canal Floating Bridge," ARL Report in Progress, 1980.

Berteaux, H.O., Buoy Engineering, Wiley Interscience, 1976.

Bhattacharyya, R., Dynamics of Marine Vehicles, Wiley Interscience, 1978.

Haszpra, O., Modeling Hydroelastic Vibrations, Research Center for Water Resources Development (VITUKI), Akademiai Kiado, Budapest, Hungary, 1978.

LeMehaute, B., "Similitude in Coastal Engineering," ASCE Journal of Waterways, Harbors and Coastal Engr. Div., Vol. 102, WW3, p. 317, August 1976.

Luft, R.W., "Analysis of a Floating Bridge: The Hood Canal Bridge," Proc. of ASCE/EMD Spec. Conference on Dynamic Response of Structure, January 1981.

VIBRATION OF A STEEL TRUSS HIGHWAY BRIDGE

by

Celal N. Kostem,[1] M.ASCE, John W. Fisher,[2] F.ASCE,
Robert T. Reese,[3] T. Robert Kealey,[4] F.ASCE

Abstract

The results of the full scale testing and the computer simulation of a major steel truss highway bridge to determine its vehicle induced dynamic response are presented. The total length of the bridge is 2058 ft., and consists of two approach spans, two anchor spans and center span. The main objective of the research was to determine the dynamic behavior and fatigue life of twin bents that support an approach span and an anchor span. Several observations drawn from the study, that are applicable to similar bridges, are presented.

Introduction

The paper presents the results of a research program conducted on the vehicle induced vibration of a multi-span steel truss bridge. The research included the field testing of the bridge superstructure and pertinent parts of the support system, analytical simulation of the structural system, a limited parametric study on the effects of certain structural details on the overall dynamic behavior of the bridge, and the evaluation of the effects of the vibration on the fatigue life of key bridge components. Due to the multi-faceted nature of the research program the paper includes only the portions that could be applied to other similar bridge systems.

Problem Statement

During the inspection of the bridge, and according to reported

[1]Professor of Civil Engineering, Fritz Engineering Laboratory, Lehigh University, Bethlehem, PA.
[2]Professor of Civil Engineering and Associate Director, Fritz Engineering Laboratory, Lehigh University, Bethlehem, PA.
[3]Sandia Corporation
[4]Partner, Modjeski and Masters, Consulting Engineers, Harrisburg, PA.

field observations, it was noted that twin steel bents supporting the anchor and deck truss spans of the overall bridge system vibrate excessively with large amplitudes (Ref. 1). Even though the vibration was not perceptible to motorists, it was deemed necessary to identify the dynamic behavior of the bents, and to assess its significance on the fatigue life of the bents in light of the average daily truck traffic using the bridge.

The bridge carried four traffic lanes (two lanes in each direction) and consisted of nine spans, five of which were steel trusses (Fig. 1). The two deck trusses were simply supported. There was no noticeable static and dynamic interaction between the anchor spans and the deck truss spans. In the course of the field investigations it was noted that the pins at the end of the cantilevers, which were supposed to move freely, were frozen; resulting in a fully monolithic interaction between the suspended and cantilever spans. This observation required the determination of the effects of the frozen pins on the vibrational behavior of the twin bents.

The focus of the research was on the 78 ft. high twin bents which had been observed to vibrate. However, since the truck loads were transmitted through the deck trusses, and the 1204 ft. long main truss spans, it was essential to include the superstructure in the analysis of the structure. The experimental program was confined to the bents, whereas the analytical modeling considered each member of the bents and a simplified equivalent structural system for the main trusses.

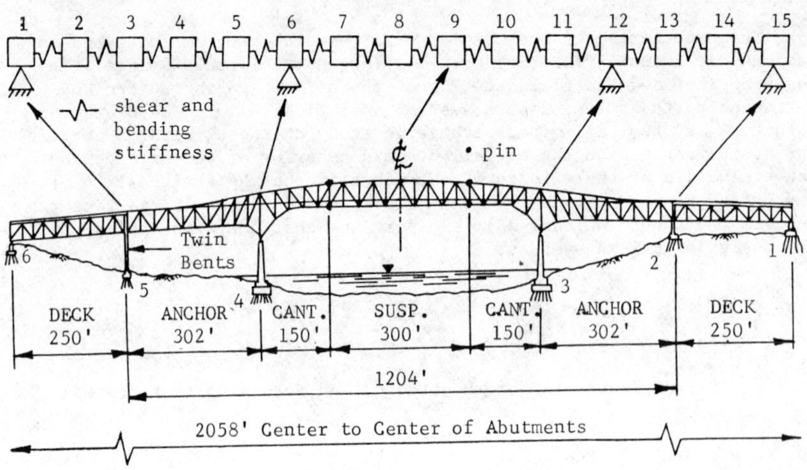

Fig. 1 Elevation and Lumped-Mass Model of the Bridge

The Experimental Research

In the experimental phase the twin bents were instrumented through the use of 10 deflection gages and 13 strain gages. The location of the instruments permitted the recording of in-plane and out-of-plane response, i.e. perpendicular and parallel to the bridge axis respectively.

The initial plans for the test loading consisted of one or two simulated AASHTO HS20 trucks traversing the bridge. The following test conditions were assessed (Ref. 4):

Trucks(s)	Condition
HS20	Outside lane each direction
HS20	Inside lane each direction
2-HS20 (1 lane)	Outside lane each direction
2-HS20 (1 lane)	Inside lane each direction
2-HS20 (adjacent lanes)	Both lanes each direction
2-HS20 (opposite)	Each lane

It was also desired to evaluate these loadings at crawl speed, and at speed runs of 5, 30, 45 and 60 mph. During the conduct of the tests it became apparent that fully loaded HS20 trucks were unable to exceed speeds in excess of 33 to 38 mph when traversing the bridge as a result of approach grades. Thus, all speed runs were conducted at 5 and 30 mph. The field observations and preliminary analytical studies also indicated that the worst loading for the support system occurred when two trucks crossed the structure side by side in a given direction. The detailed analytical study was carried out for this loading arrangement. The experimental research included the above defined load placements, and resulted in 100 separate loadings. Mean weights of the front, drive and rear axles of the trucks employed were 8,900, 31,135 and 32,435 pounds respectively. These weights closely simulated AASHTO HS20 Design Truck with corresponding axle weights of 8,000, 32,000 and 32,000 pounds (Ref. 4). The mean distance between the front and drive, and drive and rear axles of the test trucks was 15'-11" and 32'-3" respectively. The data, either raw or smoothened, is too extensive to be included in the paper. However, the strains and displacements of select recordings of the anchor span bents are shown in Figures 2 and 3. As can be noted, the static response of the instrumented points essentially becomes a "carrier wave" for the dynamic response. Furthermore, the amplitude of the dynamic response is fairly constant, with the exception of when the load is in the immediate vicinity of the instrumented member. Inspection of other recordings made at other points reveal similar characteristics.

Analytical Study

The main truss of the bridge superstructure was simulated by 15 masses, having vertical, translational and rotational degrees of freedom. These masses were interconnected, as shown in Fig. 1, by springs

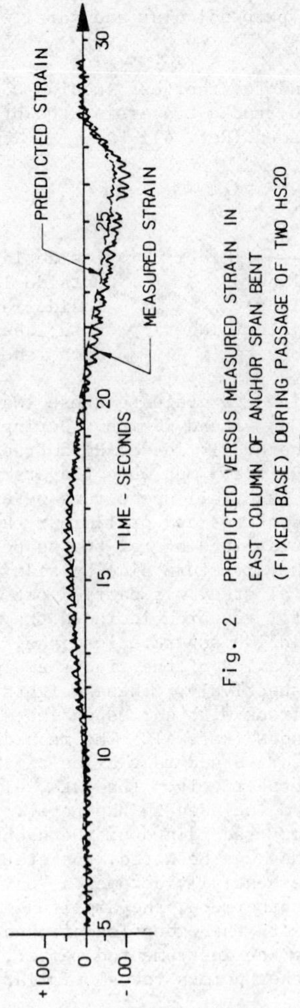

Fig. 2 PREDICTED VERSUS MEASURED STRAIN IN EAST COLUMN OF ANCHOR SPAN BENT (FIXED BASE) DURING PASSAGE OF TWO HS20 TRUCKS, RUN 70

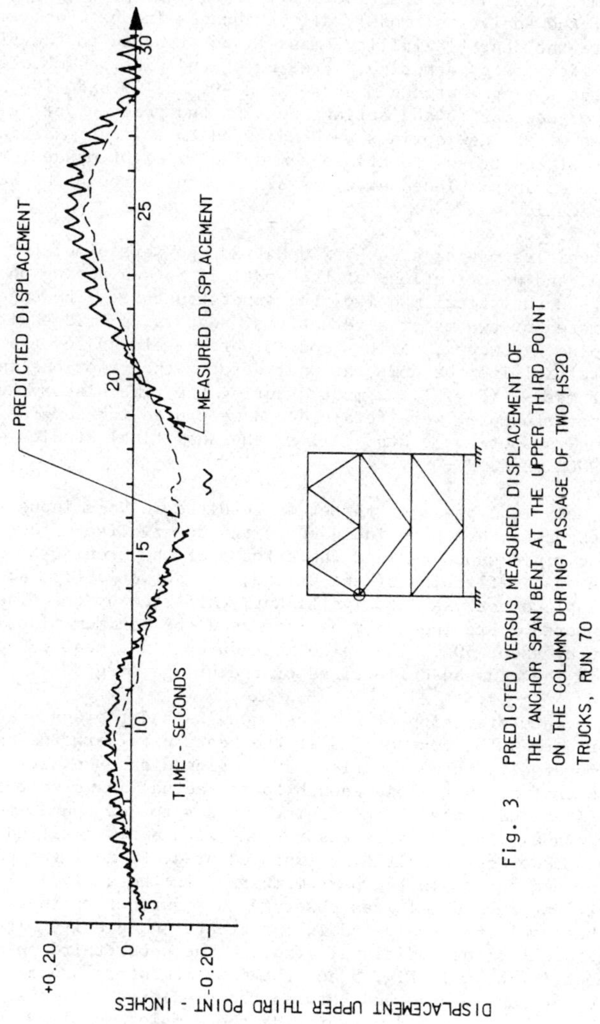

Fig. 3 PREDICTED VERSUS MEASURED DISPLACEMENT OF THE ANCHOR SPAN BENT AT THE UPPER THIRD POINT ON THE COLUMN DURING PASSAGE OF TWO HS20 TRUCKS, RUN 70

having the necessary stiffnesses. In the assignment of the inertial and stiffness properties results of a detailed static analysis were employed. This permitted accurate representation of the superstructure with minimal total degrees of freedom. Depending upon the specific purpose of the analytical model, slight changes in the model were made. To simulate the bent flexibility, mass-1 was attached to the fixed frame of reference by a spring. Freeing of the pins was modeled by removing the springs between masses-8 and -9, and -9 and -10, and insertion of new masses and springs between the removed springs. The stiffnesses of the new springs were adjusted to reflect the pinned connection of the truss. Bents were modeled by eight masses, and appropriate springs, placed at every sixth point through the height and at the foundation.

Two vehicles moving side by side at 30 mph were simulated through activation, and deactivation, of the vertical forces acting on the mode points of the analytical model of the superstructure. The superstructure consisted of two major alternatives, that is, a bridge with frozen or freed pins. However, various models were developed for the bents. In order to minimize the computational effort, the reactions and deformations of mass-1 (Fig. 1), computed during the time history analysis of the superstructure, were retained and were used for several time history analyses of the bents. All of the analytical studies employed program SHOCK (Ref. 2).

The analytical phase of the study resulted in voluminous data. Typical responses have been included in the paper. The effects of elastic and rigid supports, and the effects of the freeing of the pins on the mid-span deflection of the bridge, are shown in Fig. 4. The midspan displacement was not significantly affected by the elasticity of the supports. Freeing the pins increased the maximum midspan deflection by about 50%. This also introduced a noticeable oscillatory behavior in the midspan deflections of the bridge structure.

In the interpretation of the test results and the analytical work it was noted that the foundations of the bents were far from being fixed. Conversely, the assumption of the pinned base condition leads to results that are not close enough to the actual measured response. However, fixed and pinned base assumptions do provide bounds to the actual response. The analysis was re-done with a new modified stiffness. This model has resulted in improved prediction of the peak deflections and forces in the bent members. In the analytical prediction of bent response it was observed that further refinements were helpful, but would have resulted in far greater complexity with insufficient return for the additional effort. The upper third point bent displacement is shown in Fig. 5 and the base moment is shown in Fig. 6. Good agreement can be observed between the predicted and measured response. The analysis did not provide the complete duplication of the test results. The simulation of the superstructure with a limited number of degrees of freedom was more successful than the simulation of the bents. The dynamic response of the bents is influenced by the spring action provided by the soil-foundation interaction.

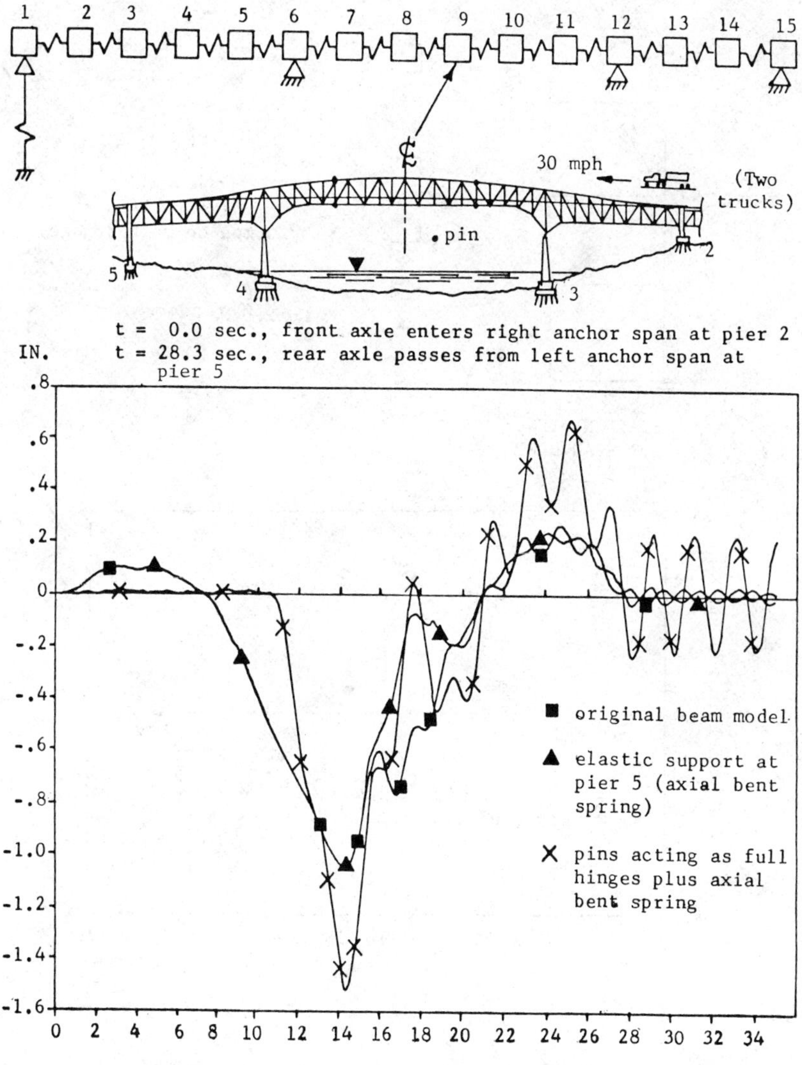

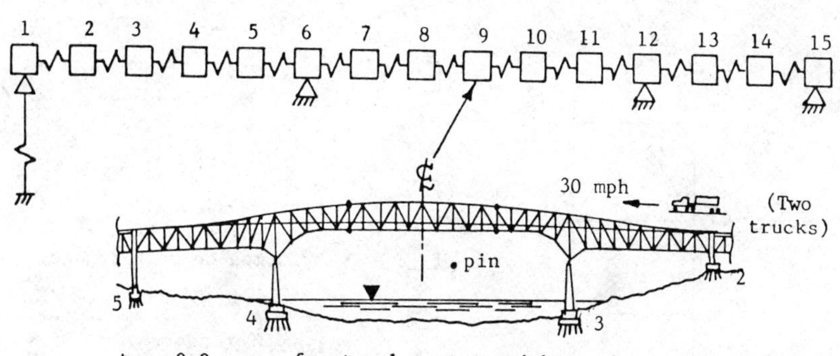

t = 0.0 sec., front axle enters right anchor span at pier 2

t = 28.3 sec., rear axle passes from left anchor span at pier 5

IN.

.8 .6 .4 .2 0 -.2 -.4 -.6 -.8 -1.0 -1.2 -1.4 -1.6

0 2 4 6 8 10 12 14 16 18 20 22 24 26 28 30 32 34

■ original beam model

▲ elastic support at pier 5 (axial bent spring)

X pins acting as full hinges plus axial bent spring

Fig. 4 Plot of Vertical Displacement of Mass vs. Time

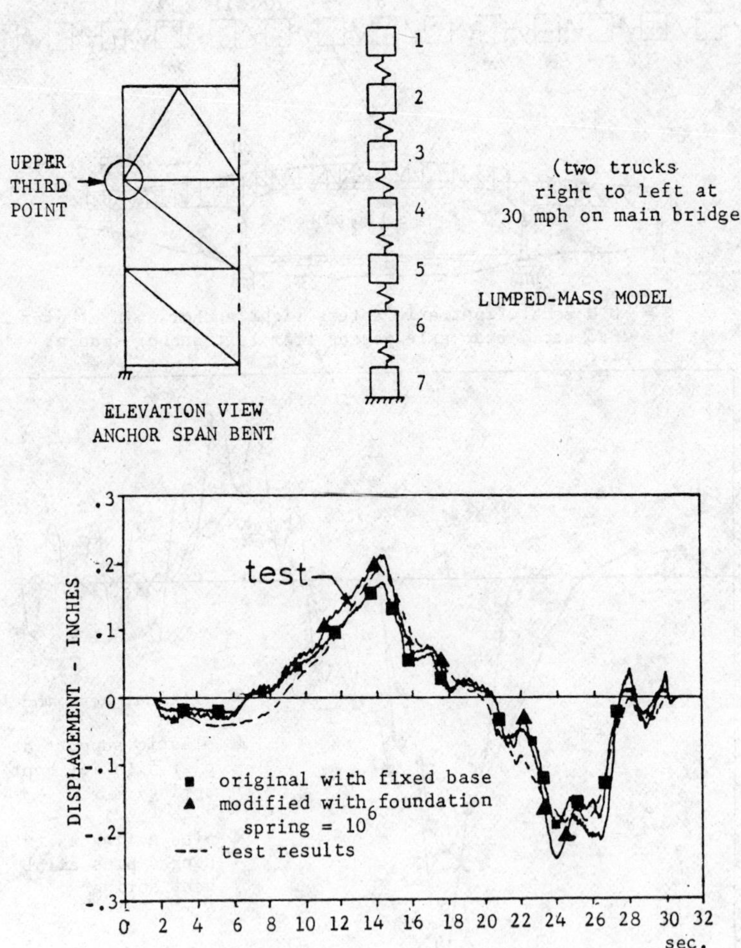

Fig. 5 Displacement vs. Time for Upper Third Point (Mass 3) for Anchor Span Bent

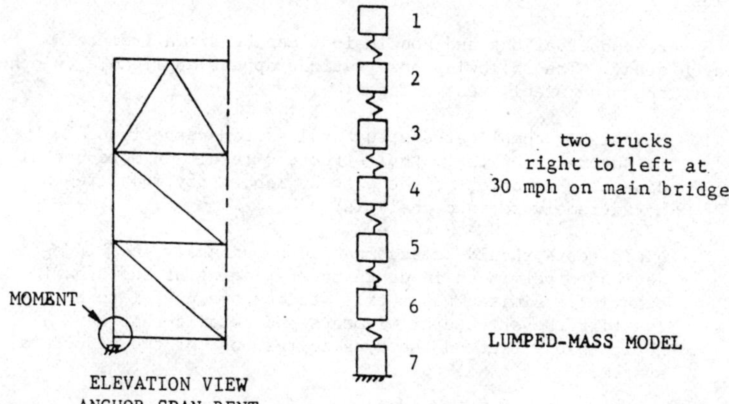

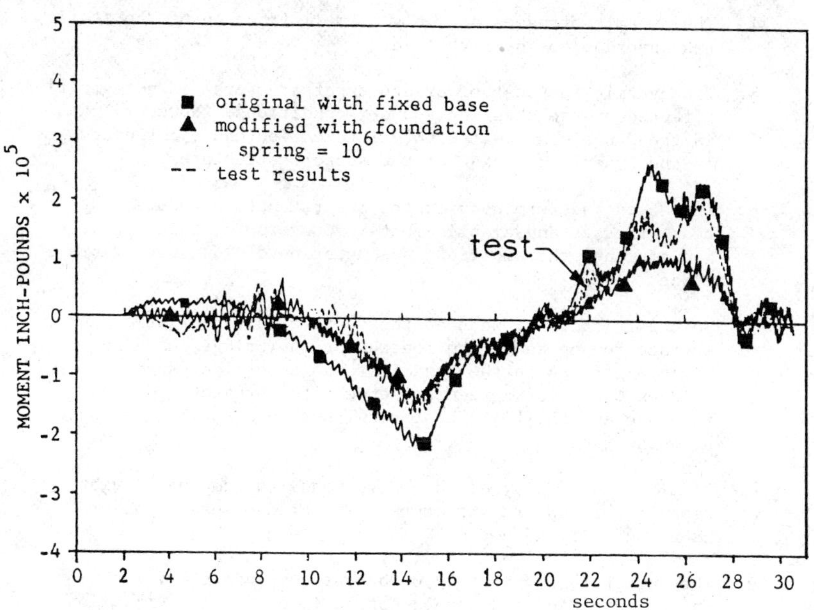

Fig. 6 Moment vs. Time for Base of Anchor Span Bent (Spring 6-7)

Conclusions

Several observations and conclusions can be drawn from this research study. The following observations appear applicable to other similar long span steel truss bridges:

1. For highly complicated structural systems experimental measurements on the actual structure result in more useful data at a fraction of the cost of laboratory model tests or extremely refined analysis.

2. Field tests should be accompanied by an analytical investigation to permit an in-depth understanding of the dynamic behavior. Limited parametric studies are useful to help identify the structural elements and details that have a noticeable effect on the dynamic response of the structural system.

3. Regardless of their complexity, long span truss bridges' and slender support bents' dynamic response are dominated by their first modes of vibration.

4. The dynamic stresses are only a small fraction of the static response due to the live load.

5. The dynamic response of bridge superstructures is not greatly affected by the changing support conditions. However, changes in the dynamic soil-structure interaction cause large changes in the dynamic response of the support structures.

6. Increased indeterminacy in the superstructure reduces the peak stresses and displacements. Conversely, this amplifies and prolongs the amplitude and duration of low level vibrations.

7. Secondary factors, such as the vehicle characteristics and surface roughness of the roadway of the bridge, have negligible effects on the dynamic response of long span bridges. Lack of agreement between the predicted dynamic response and the field test results should not be attributed to these factors.

8. Partial base fixity of the bents tends to dampen the vibration of the support structure. The bridge superstructure is not greatly affected.

9. The perceptible vibration of bridge components may lead to fatigue cracking and should not be ignored. The vibration should be investigated in order to establish whether or not fatigue cracking is likely. In this study the bents were not susceptible to fatigue damage because the Average Daily Truck Traffic (ADTT) was low. This may be a critical issue for some other bridges, depending upon the age of the bridge and the ADTT.

Acknowledgments

The reported research was directed by J. W. Fisher. The analytical study was conducted by C. N. Kostem and R. T. Reese. The study was part of an evaluation of the bridge by Modjeski and Masters, Consulting Engineers for the Corps of Engineers. The authors gratefully acknowledge the support of Modjeski and Masters in the conduct of the research program.

References

1. Modjeski & Masters, Consulting Engineers, "Summit Bridge Over Chesapeake and Delaware Canal, Vibration Study of Twin Bents at Pier 5," for Department of the Army, Philadelphia District, Corps of Engineers, Harrisburg, Pennsylvania, 1972.

2. Reese, R. T., "Multiple Configuration Analysis of Lumped Mass Dynamic Systems," Fritz Engineering Laboratory Report No. 400.8, Lehigh University, Bethlehem, Pennsylvania, 1972.

3. Reese, R. T. and Kostem, C. N., "A Method to Determine the Sensitivity of Mathematical Models in Deterministic Structural Dynamics," Fritz Engineering Laboratory Report No. 400.9, Lehigh University, Bethlehem, Pennsylvania, 1973.

4. The American Association of State Highway and Transportation Officials (AASHTO), "Standard Specifications for Highway Bridges," Twelfth Edition, Washington, D.C., 1977.

DEVELOPMENT AND USE OF FORCE PULSE TRAIN GENERATORS

By

Frederick B. Safford[1] and Sami F. Masri[2], M.ASCE

ABSTRACT

The study and investigation of the dynamic response of large structures to levels induced by natural or man-made events require excitation sources which possess large amounts of energy. Additionally, these energy sources must provide: Control of the excitation; portability to the site; ease of attachment to the structure; multiaxial excitation of the structure; and the ability to excite structures from simple harmonic motions to expected multifrequency response-time history motions. The development of force pulse train generators exhibits considerable promise to meet the foregoing for the study of linear and nonlinear dynamic responses of large structures.

INTRODUCTION

Recent analytical and experimental studies[1] indicate that a rudimentary series of rectangular or other simple pulses could be convolved with the impulse functions of a structure to induce motions closely approximating those caused by natural and man-made events. This result greatly simplified the control of high energy devices as the problem is reduced to three functions of on, off, and amplitude control. It was further determined that the excitation could also be placed directly on structures at one location or at multiple locations of test convenience and in single or multiple axes. When the structural excitation is caused by base motion, pulse simulation with generators attached to the same structure can duplicate the natural or man-made event with the exception of the rigid body modes.

Several large energy devices may be adapted for pulse train excitation. These include point source explosives, chemical rockets, metal cutting, and reactance by gas, water, or projectiles. The development and applications of several metal cutting systems are discussed together with a pulse modulated gas reactance system.

Interesting observations can be made through experiments and examinations of pulse trains in both the time and frequency domains. A single pulse is displayed in Figure 1 for both domains. However, a

[1]Associate, Agbabian Associates, El Segundo, California 90245
[2]Professor, School of Engineering, University of Southern California, Los Angeles, California 90007

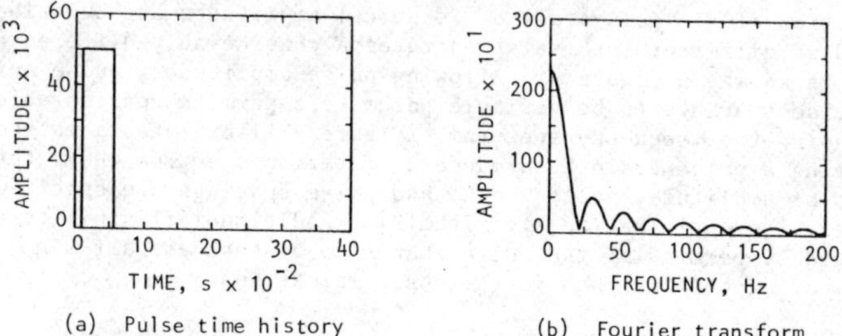

FIGURE 1. SINGLE PULSE TIME HISTORY WITH ASSOCIATED FOURIER TRANSFORM

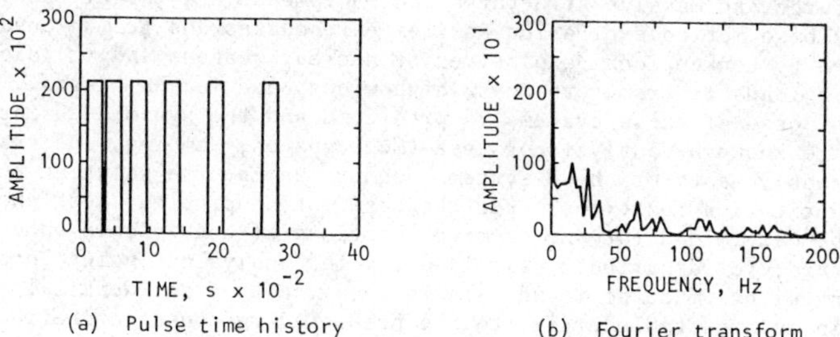

FIGURE 2. MULTIPULSE TIME HISTORY AND ASSOCIATED FOURIER TRANSFORM ROUGHLY DESIGNED TO POPULATE SPECTRAL ENERGY FROM 0 TO 40 Hz

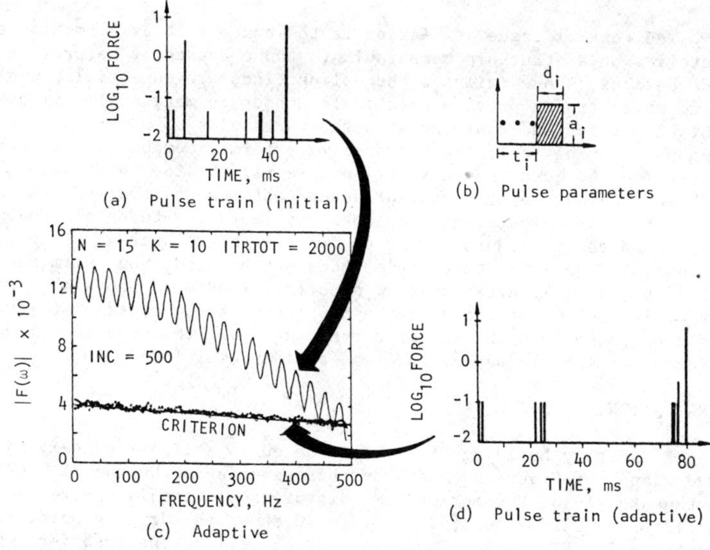

FIGURE 3. GENERATION OF PULSE TRAIN TO MATCH CRITERION FREQUENCY SPECTRUM, ADAPTIVE RANDOM SEARCH

collection of six pulses all of identical amplitudes and durations but spaced at different intervals generates a considerably different spectrum, as shown in Figure 2. Allowing pulse amplitudes, pulse durations, and pulse spacings to be variable parameters permits considerable leeway in shaping the frequency spectrum. Figure 3 illustrates a method of selecting a pulse train to produce a criterion spectrum where the parameters of amplitude, pulse width, and pulse spacings are specified by an iterative optimization algorithm[2]. Additional flexibility can be achieved by permitting the pulse shapes to be such as half-sine, saw tooth, or exponential and in various combinations.

MOTION SIMULATION OF STRUCTURES

Large and massive structures and equipments are largely subjected to (1) base motion excitation as from earthquakes and from ground shocks induced by conventional explosives or nuclear weapons and (2) distributed air loads as from tornadoes, high winds, and weapon blasts. Where the response of these systems is predicted and the dynamic characteristics are known by analysis or test (mode shapes, damping, and transfer functions), tests of these systems require an inverse solution to determine excitation functions. For the base motion problem, test excitation must be determined for one or more locations of convenience upon the structure; for distributed air loads, a few equivalent point locations of convenience must be found. It is required that an identical response (within an acceptable error) to the predicted response (criterion) caused by natural or man-made hazards be obtained. These new excitation functions can usually be found; however, the energy and waveforms required often make testing impractical due to the limitations of conventional shakers and/or costs.

The central issue in testing is to induce multiaxial motion-time histories on a structure comparable to those caused by natural or man-made hazards, since failures and malfunctions are essentially nonlinear. Force pulse trains provide practical methods in many applications to test structures and equipments to the thresholds of damage and malfunction by inducing realistic response wave forms. Computational methods to develop the required pulse trains are illustrated in Figures 4 and 5 with an application to a nuclear power containment structure shown in Figure 6. The computational methods are iterative to obtain the pulse train, and the procedure may be used for either linear[3] or nonlinear systems[4]. Various iteration methods may be used; the optimization algorithm[2] of Figure 5 using a spherical random search with a partially automatic control of the variance has proved to be quite efficient. It has been found that the variance can range over ten orders of magnitude, which permits rapid convergence and avoidance of local minima.

APPLICATIONS

A 25-story building (Fig. 7), modeled as a linear system, has the mode shapes of Figure 8, and under base excitation by the El Centro Earthquake yields the motion time histories plotted in Figure 11 for the 13th and 23rd floors[5]. Figure 10 shows the driving point impulse functions for Floors 8, 13, 18, and 23 as well as the transfer impulse functions between floors and each excitation location. Using the procedures of Figures 4, 5, and 6, the force pulse trains of Figure 9 were

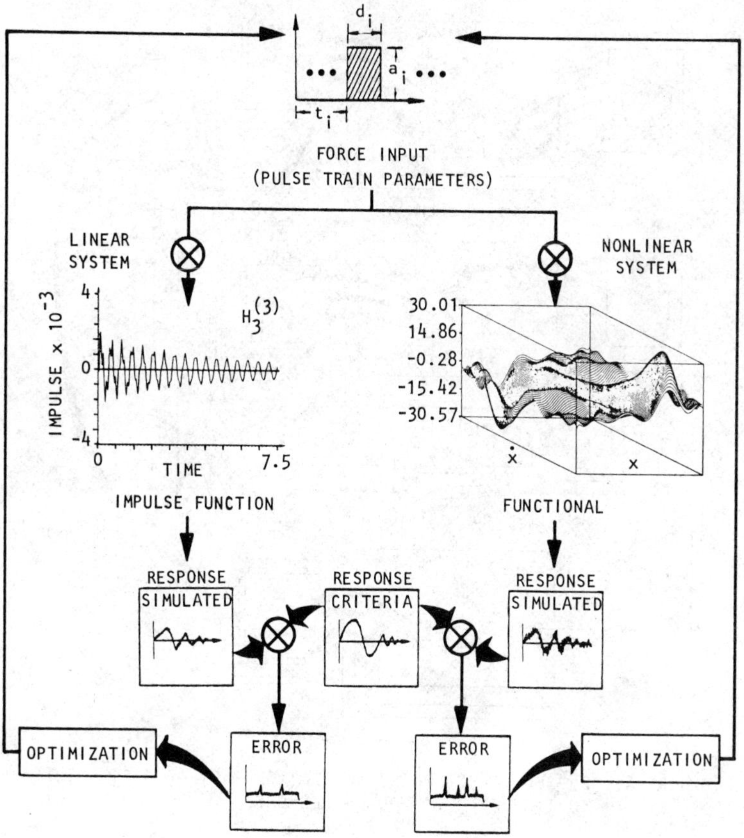

d_i

a_i

t_i

FORCE INPUT
(PULSE TRAIN PARAMETERS)

LINEAR SYSTEM

NONLINEAR SYSTEM

IMPULSE x 10^{-3}

$H_3^{(3)}$

4
0
-4
0
TIME
7.5

30.01
14.86
-0.28
-15.42
-30.57
$\dot{x}$
x

IMPULSE FUNCTION

FUNCTIONAL

RESPONSE SIMULATED

RESPONSE CRITERIA

RESPONSE SIMULATED

ERROR

ERROR

OPTIMIZATION

OPTIMIZATION

FIGURE 4. COMPUTATIONAL METHOD FOR PULSE TRAIN GENERATION

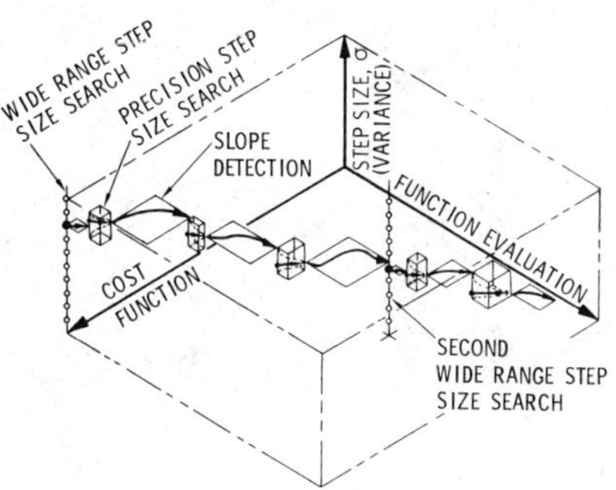

FIGURE 5. OPTIMIZATION ADAPTIVE RANDOM SEARCH ALGORITHM WITH RANGE AND PRECISION STEP SIZE

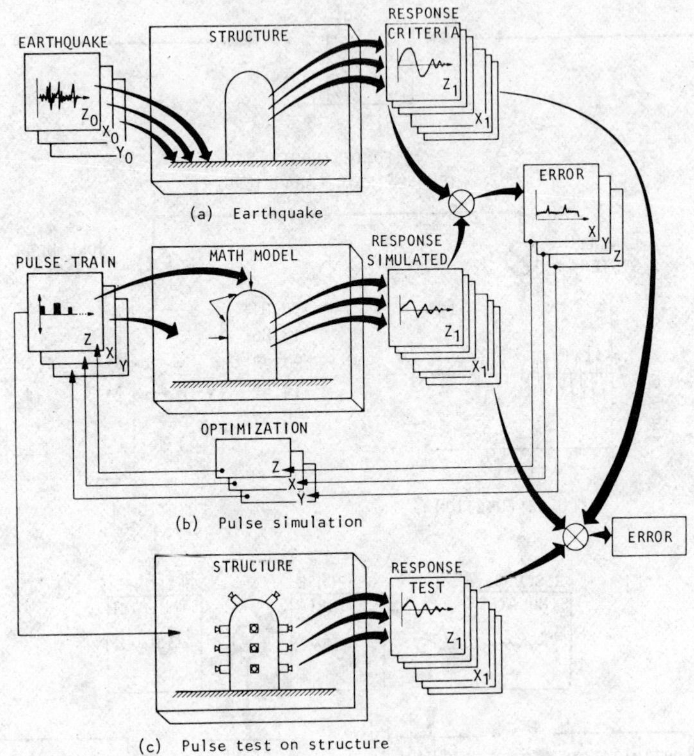

FIGURE 6. PROCEDURE FOR PULSE SIMULATION/TEST OF A STRUCTURE TO SIMULATE EARTHQUAKE RESPONSE

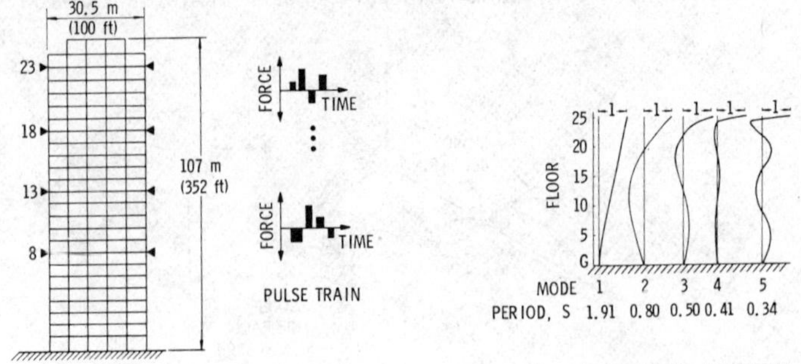

FIGURE 7. GAS PULSERS ARRAYED FOR EARTHQUAKE SIMULATION TEST OF A MULTISTORY BUILDING

FIGURE 8. BUILDING NATURAL MODES OF VIBRATION

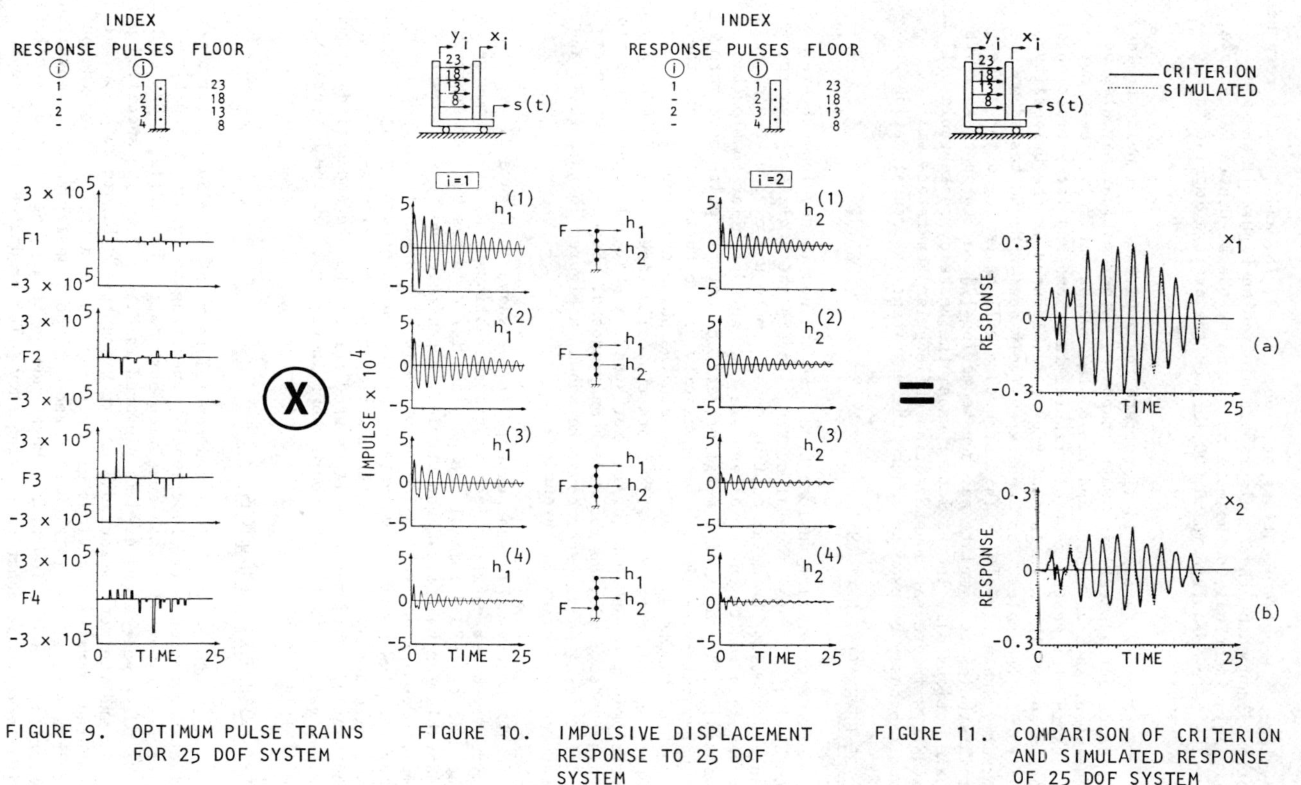

FIGURE 9. OPTIMUM PULSE TRAINS FOR 25 DOF SYSTEM

FIGURE 10. IMPULSIVE DISPLACEMENT RESPONSE TO 25 DOF SYSTEM

FIGURE 11. COMPARISON OF CRITERION AND SIMULATED RESPONSE OF 25 DOF SYSTEM

developed for pulse locations on Floors 8, 13, 18, and 23. The response motions induced by these pulsers are plotted on Figure 11 and are virtually identical to the motions caused by the El Centro Earthquake. Average force required is 50,000 lbf and the average impulse is 17 lbf-sec.

Figures 12 through 15 illustrate the application of pulse trains to induce a transient sine wave response. Other motions may also be generated, such as random and rapid sine sweeps (chirp). These motions are useful for measurement of transfer functions and for extraction of mode shapes. The structure of Figure 12, located at the University of California, Berkeley, has been extensively analyzed and tested on the Berkeley shake table[6]. This structure will be employed for pulse train studies to simulate earthquake motions[7] and to suppress earthquake motions[8]. In the latter case, the Berkeley shake table will be used to simulate earthquake motion while pulse generators will simultaneously be used to reduce the resultant structural motions.

PULSE GENERATING DEVICES

Pulse generating devices developed and now in development are composed of a class of metal cutting systems and a class of gas reactance systems. The metal cutting devices make use of the very high forces required to cut metal and are configured similar to broaching tools[9]. The shape of a metal projection controls pulse wave form (square, half-sine, etc.). The velocity of the cutting tool and the length of the metal projection govern the pulse duration[1]. For the metal removed, cutting coefficients typically range from 150,000 lbf/in^3 for aluminum to 300,000 lbf/in^3 for steel. Metal pulse generators producing up to 1,000,000 lbf have been proposed. Several varieties of metal cutting pulse generators are shown in Figures 16 through 19 and pictured in Figures 21 through 24.

The programmable gas pulse generator shown in Figure 20 is currently under construction; test stand calibrations will commence in the fall of 1980. Initial use provides for cold gas but the system is adaptable for both steam and chemically generated hot gas. Thrust amplitudes are controlled in the on-state by positioning the metering plug for flow control. Off-state for the pulse occurs by signaling the hydraulic actuator to move the metering plug to seal off gas flow at the nozzle. This device will be used for earthquake and anti-earthquake investigations on the structure shown in Figure 12 at the University of California, Berkeley.

TESTS WITH PULSE GENERATORS

A 20,000 lb shock isolated control room (50 ft x 50 ft) for a power plant was tested while functionally operating to induce motion-time histories expected from a specified base motion hazard[10]. One of the four pulse generators used is pictured in Figure 21. The specified pulse train and the four measured ones are presented in Figure 25. Typical measured transfer functions of the structure used to calculate the pulse train are presented in Figure 26 and include transfer function magnitude, phase, and impulse function. Magnitude units are the ratio of acceleration to force as function of frequency. The predicted control room

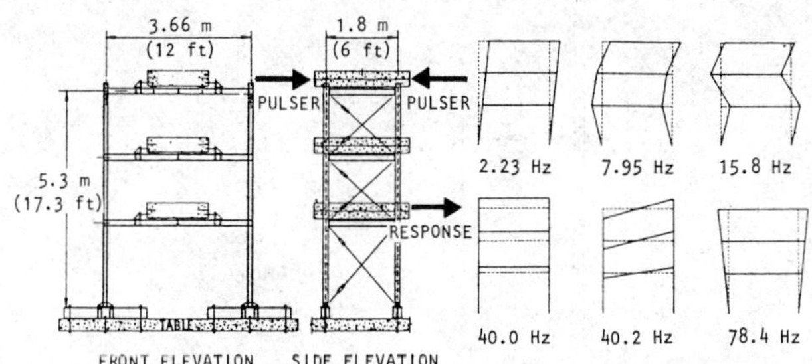

FIGURE 12. UCB TEST STRUCTURE

FIGURE 13. MODE SHAPES OF UCB FRAME DETERMINED BY SAP6

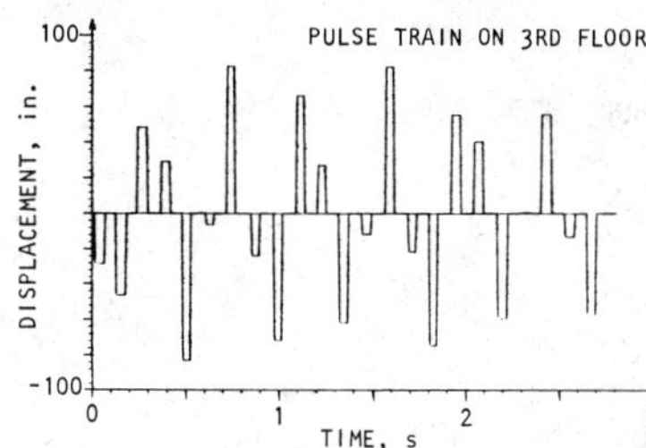

FIGURE 14. FORCE PULSE TRAIN LOCATED ON 3RD FLOOR OF UCB STRUCTURE TO INDUCE TRANSIENT SINUSOIDAL MOTION

1ST FLOOR

FORCE, lb

4

0

-4

0 1 2 3

TIME, s

FIGURE 15. TRANSIENT SINUSOIDAL MOTION ON FIRST FLOOR OF UCB STRUCTURE INDUCED BY PULSE TRAIN AT 2.23 Hz OF FIG. 13

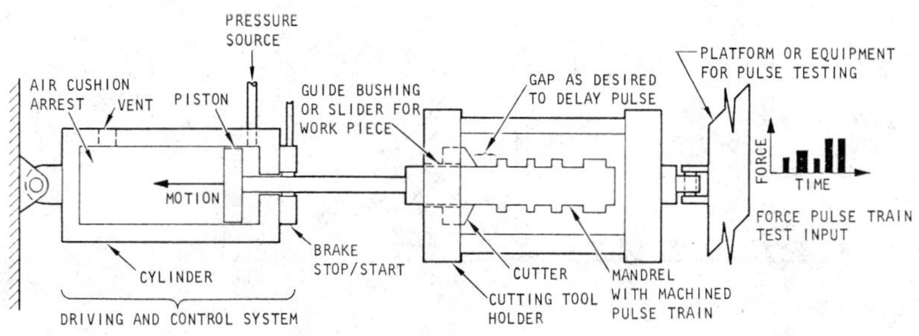

FIGURE 16. PULSE-FORMING DEVICE WITH DRIVING AND CONTROL SYSTEM MAXIMUM OUTPUT 8,000 lbf

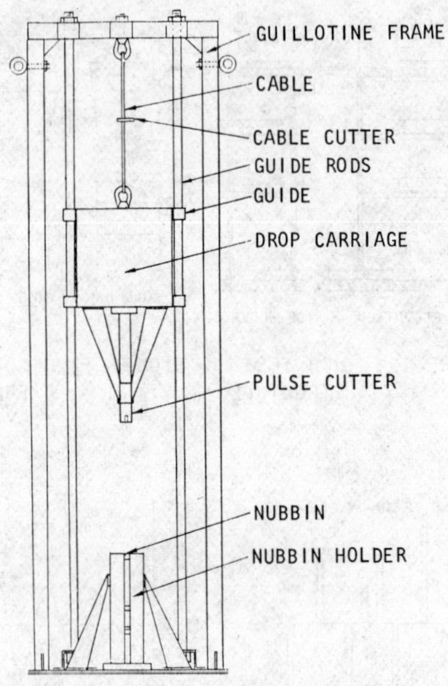

FIGURE 17. SKETCH OF MEDIUM-FORCE OUTPUT PULSE GENERATOR MAXIMUM OUTPUT 16,000 ft-lb AND 1,800 lbf-s

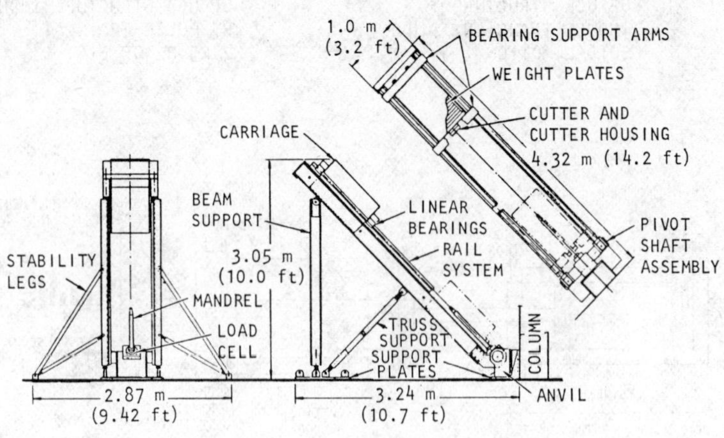

FIGURE 18. ASSEMBLY — IMPULSE TESTER MAXIMUM OUTPUT 54,000 ft-lb AND 4,700 lbf-s

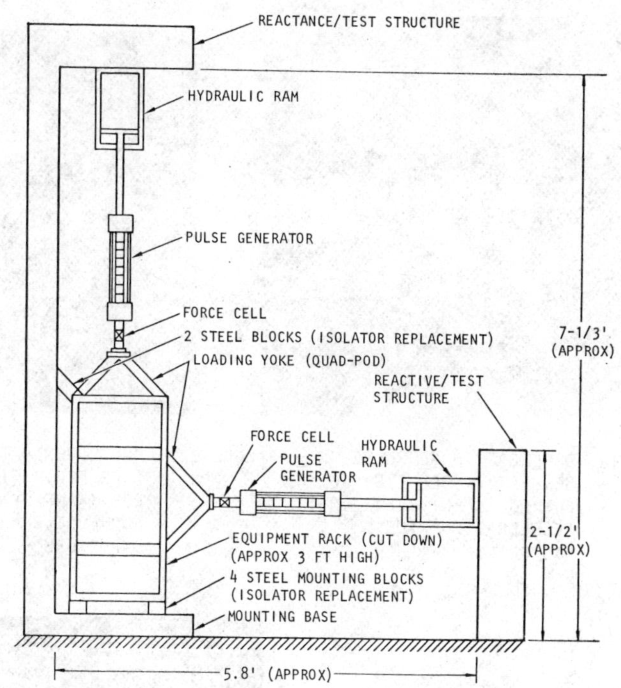

FIGURE 19. BIAXIAL TEST FACILITY CONFIGURATION MAXIMUM OUTPUT 15,000 lbf EACH AXIS

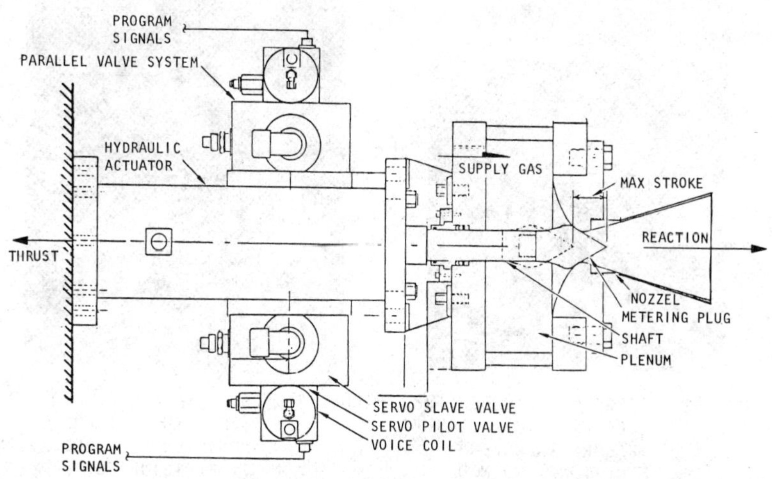

FIGURE 20. PROGRAMMABLE GAS PULSE GENERATOR COLD GAS SYSTEM — 500 TO 10,000 lbf

FIGURE 21. APPLICATION OF PULSE GENERATOR (SEE FIG. 16) FOR EQUIPMENT TEST ACCELERATION-TIME HISTORIES

FIGURE 22. PULSE GENERATOR (SEE FIG. 17) USED FOR TRANSFER FUNCTION AND MODE SHAPES OF STRUCTURE

FIGURE 23. LARGE PULSE GENERATOR (SEE FIG. 18) USED TO MEASURE TRANSFER FUNCTIONS AND MODE SHAPES OF LARGE STRUCTURES

FIGURE 24. BI-AXIAL PULSE GENERATOR (SEE FIG. 19) FOR EQUIPMENT TESTS ACCELERATION-TIME HISTORIES

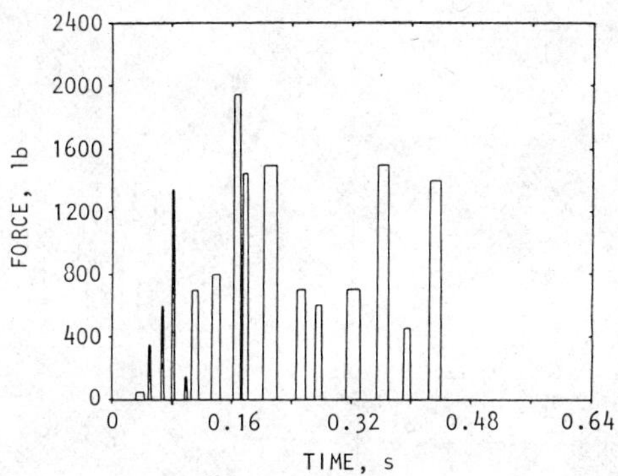

(a) Specified pulse train for each corner

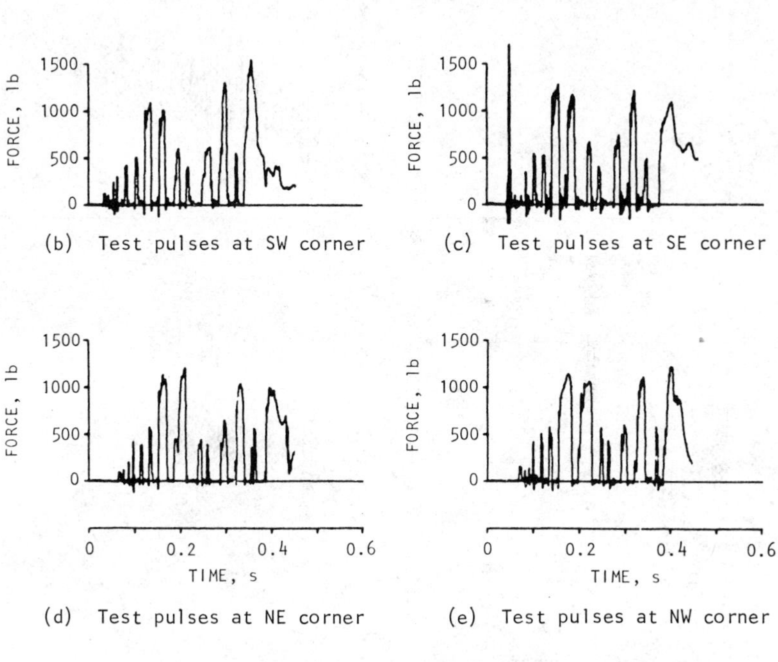

(b) Test pulses at SW corner

(c) Test pulses at SE corner

(d) Test pulses at NE corner

(e) Test pulses at NW corner

FIGURE 25. 200,000 LB SHOCK ISOLATED CONTROL ROOM PLATFORM: SPECIFIED AND ACTUAL TEST-INPUT PULSES USING FOUR PULSE GENERATORS SHOWN IN FIG. 21

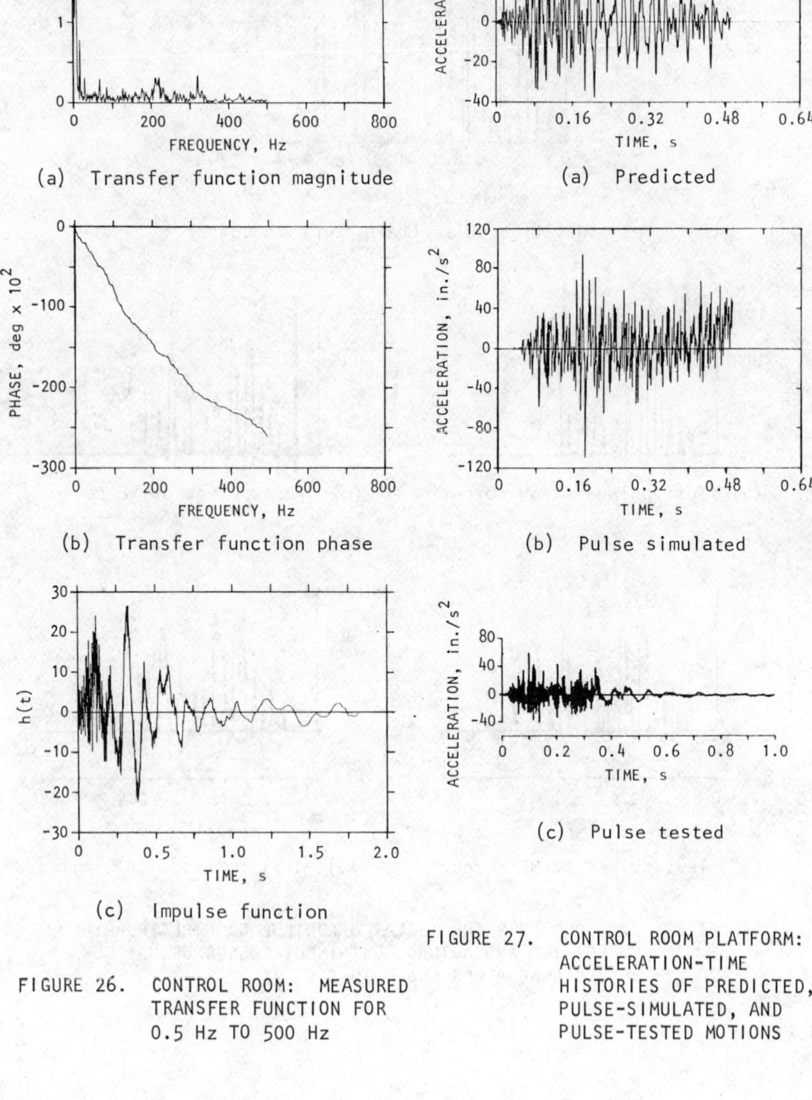

(a) Transfer function magnitude

(b) Transfer function phase

(c) Impulse function

FIGURE 26. CONTROL ROOM: MEASURED TRANSFER FUNCTION FOR 0.5 Hz TO 500 Hz

(a) Predicted

(b) Pulse simulated

(c) Pulse tested

FIGURE 27. CONTROL ROOM PLATFORM: ACCELERATION-TIME HISTORIES OF PREDICTED, PULSE-SIMULATED, AND PULSE-TESTED MOTIONS

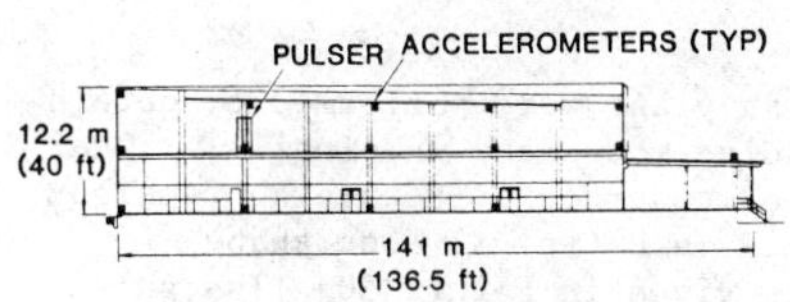

FIGURE 28. CROSS SECTION OF NUCLEAR PROCESSING PLANT SHOWING ACCELEROMETER AND PULSER LOCATIONS (PULSER EXCITATION NORMAL TO SECTION SHOWN)

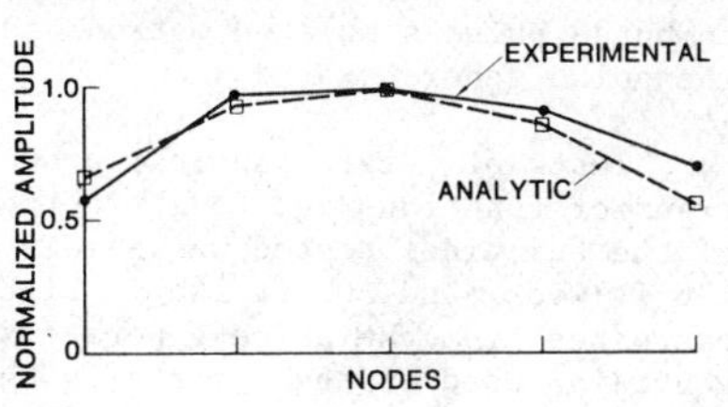

FIGURE 29. TYPICAL MODE SHAPE PLOT FROM EXPERIMENTAL AND ANALYTIC DATA (Frequency: 9.2 Hz)

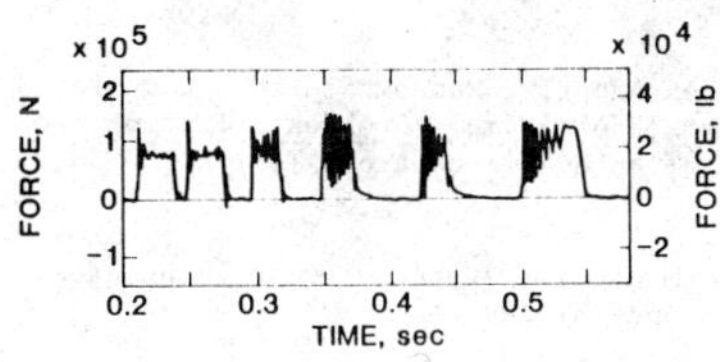

(a) Raw input force-time history

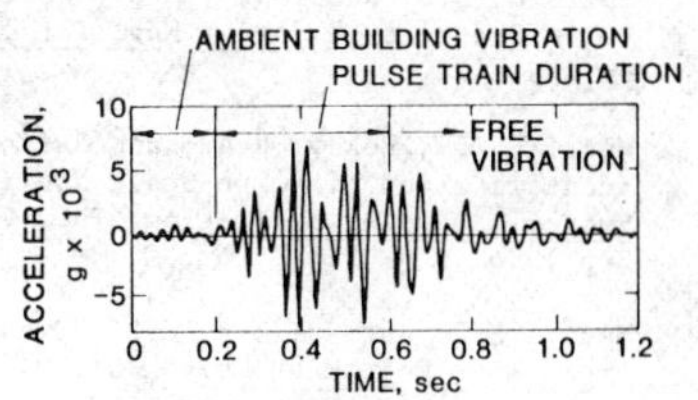

(c) Acceleration-time history (0 to 20 Hz)

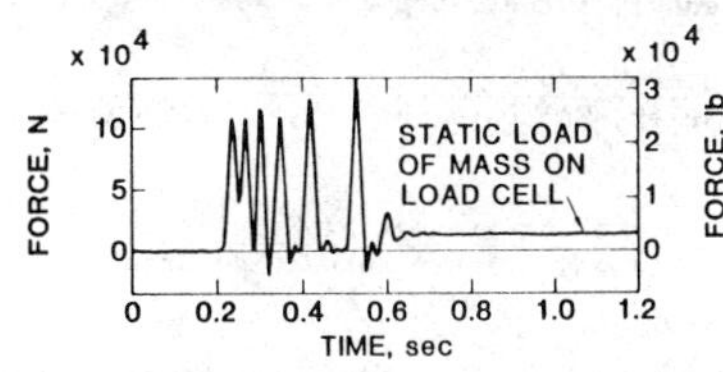

(b) Filtered input force-time history (0 to 20 Hz)

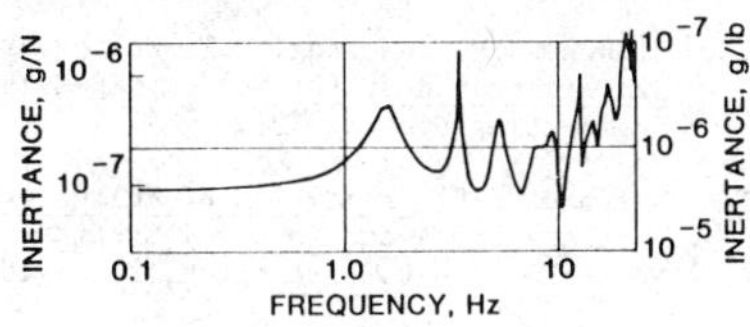

(d) Transfer function (acceleration divided by force)

FIGURE 30. TYPICAL DATA RECORDS FROM PULSE TRAIN TESTS AND RESULTING TRANSFER (INERTANCE) FUNCTIONS COMPUTED FROM DATA — PULSE GENERATOR SHOWN IN FIGS. 18 AND 23

motion due to base motion hazard is given in Figure 27 together with computed pulse simulated motion and motions measured during tests with the pulse generators.

Tests of a large nuclear processing plant were performed to obtain transfer functions and to extract building modes and damping[11]. One of the buildings tested was a reinforced concrete structure 40 ft high, 136 ft wide, and 300 ft long. Figure 29 is a typical mode shape extracted from the pulse generated data given in Figure 30. The pulse generator used in these tests is shown in Figures 18 and 23.

ACKNOWLEDGMENTS

This study was supported in part by the United States National Science Foundation under Grant No. PFR77-15010. The assistance and guidance provided by Dr. John B. Scalzi is greatly appreciated.

REFERENCES

1. Safford, F.B. and Masri, S.F. "Analytical and Experimental Studies of a Mechanical Pulse Generator," Jnl of Eng. for Industry, ASME, Series B, 96:2, May 1974, pp. 459-470.

2. Masri, S.F.; Bekey, G.A.; and Safford, F.B. "An Adaptive Random Search Method for Identification of Large Scale Nonlinear Systems," 4th Symp. for Identification and System Parameter Estimation, Int. Federation of Automatic Control, Tbilisi, USSR, Sep 1976.

3. Masri, S.F. and Safford, F.B. "Dynamic Environment Simulation by Pulse Techniques," Proc. ASCE Eng. Mech. Div. 101:EM1, Feb 1976, pp. 151-169.

4. Masri, S.F. and Caughey, T.N. "A Nonparametric Identification Technique for Nonlinear Dynamic Problems," Jnl of Appl. Mech. ASME, 46:2, Jun 1979.

5. Masri, S.F. and Safford, F.B. "Earthquake Environment Simulation by Pulse Generators," Proc. 7th World Conf. on Earthquake Eng. Istanbul, Turkey, Sep 8-13, 1980.

6. Clough, R.W. and Tang, D.T. Earthquake Simulator Study of a Steel Frame Structure, Experimental Results, Vol. 1, EERC 756. Berkeley, CA: Univ. of Calif. Earthquake Engineering Center, 1975.

7. Safford, F.B. Validation of Pulse Techniques for the Environmental Simulation of Earthquake Motions in Civil Structures, U.S. National Science Foundation Grant No. PFR77-15010, Washington, D.C., 1978.

8. Masri, S.F.; Bekey, G.A.; Safford, F.B.; and Dehghanyar, T.J. "Anti-Earthquake Application of Pulse Generators," Proc. Dynamic Response of Structures, Instrumentation, Testing Methods and System Identification, ASCE Specialty Conference, Atlanta, GA, Jan 1981.

9. Safford, F.B. "Mechanical Force Pulse Generator for Use in Structural Analysis," U.S. Patent Office No. 4,020,672, May 1977.

10. Safford, F.B. et al. "Air-Blast and Ground-Shock Simulation Testing of Massive Equipment by Pulse Techniques, 5th Int. Symp. on Military Application of Blast Simulation, Fortifikationsforvaltningen, Stockholm, Sweden, May 23-26, 1977.

11. Yates, D.G. and Safford, F.B. "Measurement of Dynamic Structural Characteristics of Massive Buildings by High-Level Multiple Techniques," Shock & Vibration Bull., 50, SVIC. Washington, DC: Naval Res. Lab, 1980.

ANTI-EARTHQUAKE APPLICATION OF PULSE GENERATORS

by

Sami F. Masri[1], George A. Bekey[2], Frederick B. Safford[3], Tejav J. Dehghanyar[4]

Abstract

This paper is concerned with a feasibility study into the use of servocontrolled gas pulse generators to mitigate the earthquake-induced motions of tall buildings. A simple yet reliable on-line active control algorithm is developed and applied, by means of numerical simulation studies, to a model of an existing steel frame structure. The control concept is shown to be effective in controlling the response of linear as well as nonlinear building systems under dynamic excitation.

Introduction

Analytical and experimental studies [1,2]* have shown that among the class of passive auxiliary mass dampers used for vibration control, the impact damper, which is a highly nonlinear version of such devices, has a superior performance record compared to the conventional dynamic vibration neutralizer when used under dynamic environments resembling earthquake excitations. However, due to the transient nature of earthquake ground motions, the impulsive forces imparted by the impact damper to the primary structure (see Fig. 1) do not always occur at the optimum time from the motion reduction point of view. Consequently, the efficiency of the impact damper, like that of other similar passive damping devices, is substantially reduced compared to its efficiency under periodic excitation.

A current research effort [3] is concerned with the validation of the concept of using pulse techniques [4] to simulate the response of structures to arbitrary dynamic environments [5]. In the course of this study, (1) an efficient algorithm has been devised for determining the optimum force pulse-train characteristics to be applied at different locations in the structure so as to match the criteria response [6], and (2) gas pulse generators, employing digital servocontrollers and hydraulic actuators in conjunction with a gas storage system and a nozzle with a metered flow to furnish the needed thrust, are being built.

---

[1]Professor, School of Engineering, University of Southern California

[2]Professor, School of Engineering, University of Southern California

[3]Associate, Agbabian Associates, El Segundo, California

[4]Graduate student, School of Engineering, University of Southern Calif.

*Numbers in brackets designate items in the reference list.

In view of the preceding discussion, this paper is concerned with a study of the feasibility of using servocontrolled gas pulse generators to mitigate the earthquake-induced oscillations of tall buildings or similar structures.

Control Algorithm

A recent study by the authors [7] presented a relatively simple on-line pulse control algorithm suitable for use with distributed parameter systems subjected to arbitrary nonstationary disturbances. The main idea of the algorithm is that resonance phenomena can be eliminated, or at least drastically reduced, by disorganizing the orderly and gradual buildup of the structural dynamic response by timed firing of a pulse of suitable magnitude applied in the proper direction. Furthermore, in order to minimize the amount of control energy utilized, the control force should be applied only when the structural response exceeds a certain threshold level related to the resistance of the structure.

Assume, as a first order approximation, that the structure to be controlled is modeled as an equivalent linear single-degree-of-freedom (SDOF) system as shown in Fig. 2. Suppose that at time $t = t_o$, the threshold barrier has been exceeded. Then the expected value of the response, under the assumption that the excitation is a zero-mean random process, is given by

$$E[y(t)] = y_o u(t - t_o) + \dot{y}_o v(t - t_o) + x_p(t) \tag{1}$$

where

$$u(t) = \exp(-\zeta\omega t)(\cos \omega_d t + \frac{\zeta\omega}{\omega_d} \sin \omega_d t) \tag{2}$$

$$v(t) = \frac{1}{\omega_d} \exp(-\zeta\omega t) \sin \omega_d t \tag{3}$$

$$x_p(t) = \int_{t_o}^{T_o} h(t - \tau)\, p(\tau) d\tau \tag{4}$$

$h(t)$ = Impulsive response of the system = $v(t)$ (5)

$p(t)$ = Impulsive control force of duration T_d and peak level P_o (6)

$$\omega_d = \omega\sqrt{1 - \zeta^2} \tag{7}$$

ω = Fundamental frequency of the system (8)

ζ = Ratio of critical damping of the system (9)

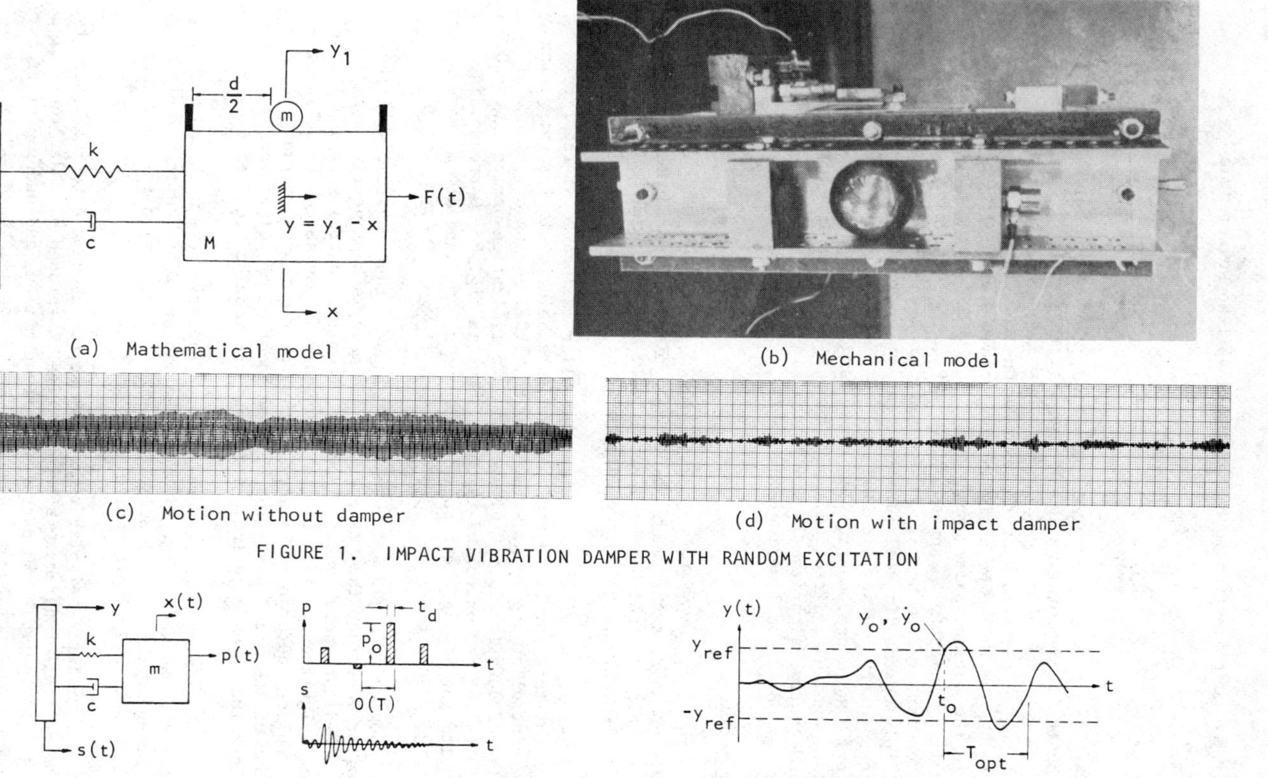

(a) Mathematical model

(b) Mechanical model

(c) Motion without damper

(d) Motion with impact damper

FIGURE 1. IMPACT VIBRATION DAMPER WITH RANDOM EXCITATION

FIGURE 2. EQUIVALENT SDOF SYSTEM TO BE CONTROLLED

The cost function to be minimized will be selected as

$$J(P_o) = \int_{t_o}^{t_o+T_{opt}} (E[y(t)])^2 dt \tag{10}$$

where T_{opt} is the optimization period, chosen to be O(T), with T being the fundamental period of the system.

Due to the nature of the expressions for u, v, and x_p appearing in Eq. 1, Eq. 10 can be analytically evaluated and differentiated to yield the optimum pulse amplitude P_{opt} which will minimize $J(P_o)$. The resulting analytical expression for P_{opt} will involve simple algebraic expressions (which need to be evaluated only once for a given system) multiplying the "initial" displacement and velocity y_o and $\dot{y}_o$. Thus, once a control pulse is called for, virtually negligible computational effort (and hence, control lag time) is needed to determine the optimum pulse magnitude, direction, and timing. This useful feature of the proposed control algorithm is significant in assessing the feasibility of the method for on-line implementation.

Applications

The utility of the proposed method will now be demonstrated by applying it to a three-degree-of-freedom model of a three-story steel frame (Fig. 3) that has been extensively studied, both analytically and experimentally [8,9] at the University of California, Berkeley (UCB).

Assume that a single pulse controller is located at the first story of this structure and that the threshold levels for triggering the controller are $\underset{\sim}{y}_{ref} = \{0.64, 0.88, 1.0\}$. Using a rectangular pulse of duration $T_d = 0.01$ sec and an optimization period $T_{opt} = 0.5$ sec results in the controlled response shown in Fig. 6. Corresponding results for a single controller location at m_2 and at m_3 are shown in Figs. 7 and 8.

Comparing Figs. 6, 7, and 8, it is seen that regardless of the controller location, the response of each mass is kept nearly bounded by the selected threshold levels. However, from the control energy point of view, significant reduction in the required impulse can be achieved if the pulse generator is located at the top floor, rather than the lower floor of the structure. The optimum location of pulse generators is being studied as a refinement to conserve impulse requirements.

The relative velocity response of the structure with and without control is shown in Fig. 9. Note that in spite of the transients corresponding to the control pulses, substantial reduction in the velocity is obtained with a single controller.

The relative displacement response of the same structure, with the same controller parameters used in Figs. 6 through 9, under the action of the El Centro, 1940 earthquake ground motion, and under a stationary random ground motion, is shown in Figs 10 and 11, respectively. As in the case of artificial earthquake D1, the results shown in Figs. 10 and 11 show that a single controller acting at the top floor can effectively control the system motion to within any reasonable threshold level.

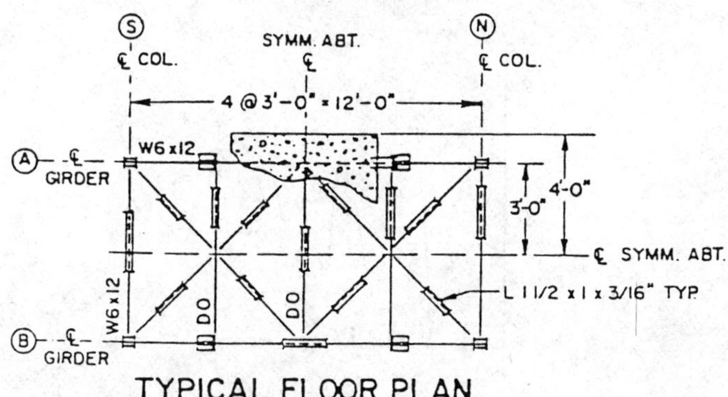

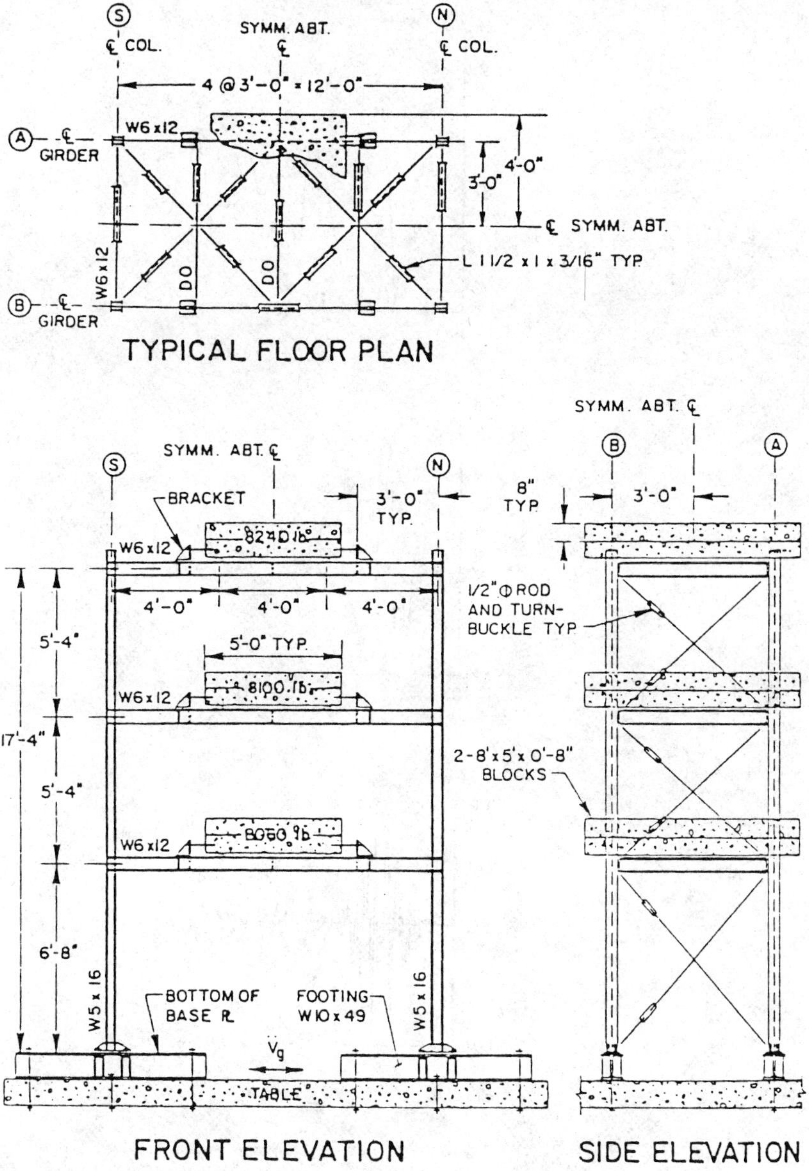

FIGURE 3. PLAN AND ELEVATIONS OF THE TEST STRUCTURE [8]

$\ddot{S}$ = E.Q. D1

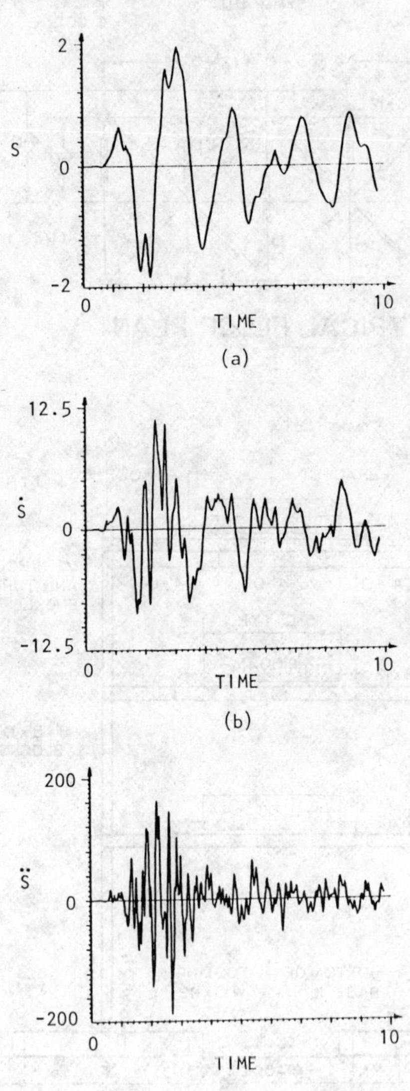

FIGURE 4. BASE MOTION CORRESPONDING TO E.Q. D1

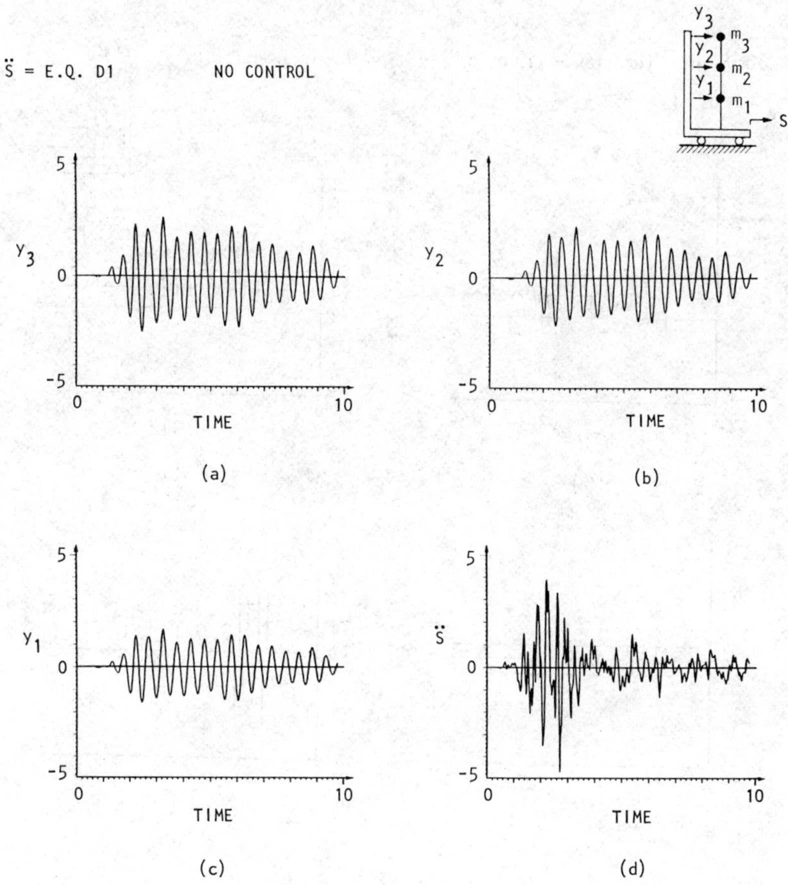

FIGURE 5. RESPONSE OF MODEL UCB-1 WITHOUT CONTROL UNDER EXCITATION E.Q. D1

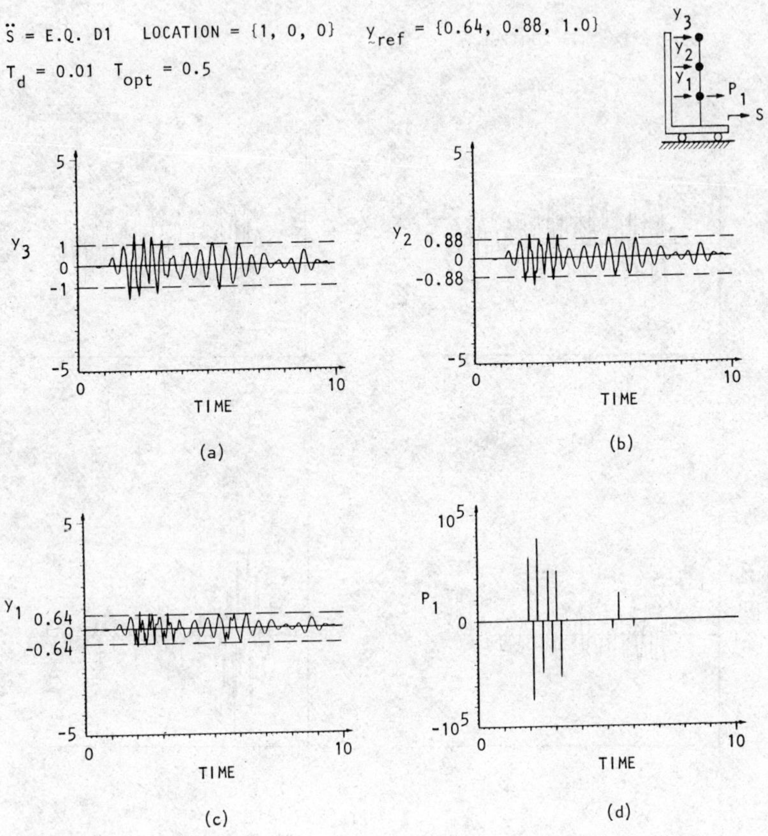

FIGURE 6. CONTROLLED RESPONSE OF MODEL UCB-1 UNDER EXCITATION E.Q. D1; CONTROLLER ACTING AT m_1

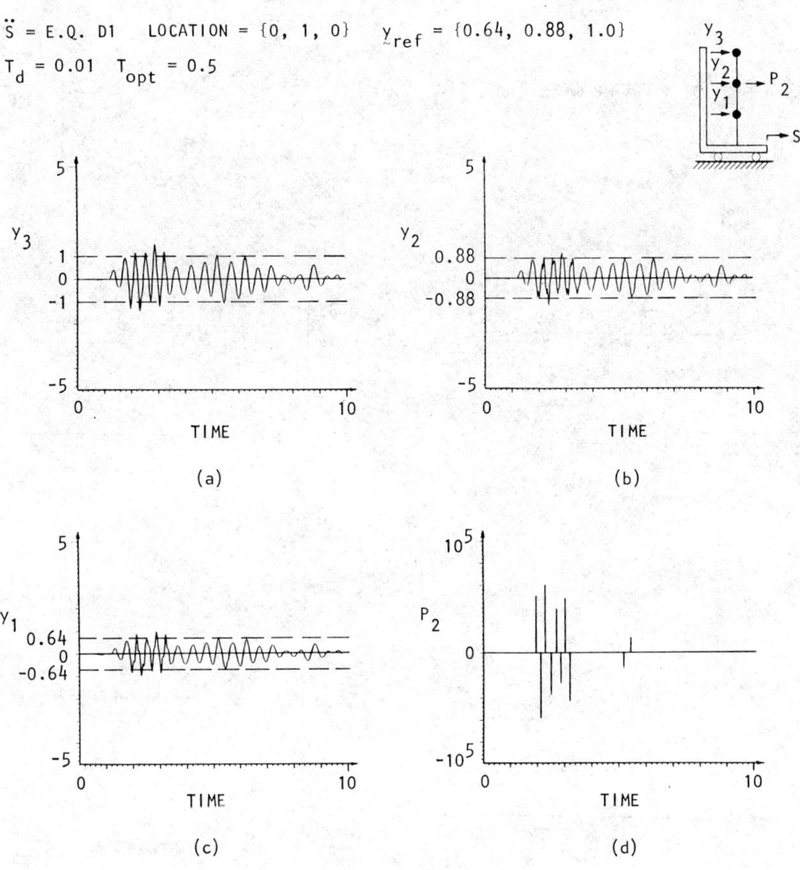

FIGURE 7. CONTROLLED RESPONSE OF MODEL UCB-1 UNDER EXCITATION E.Q. D1; CONTROLLER ACTING AT m_2

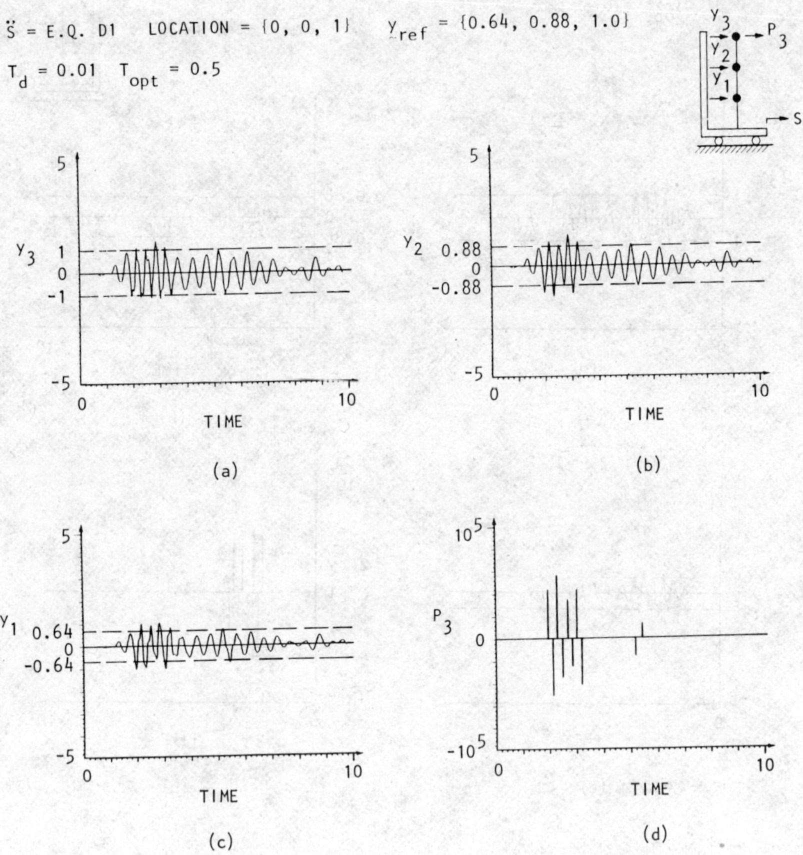

FIGURE 8. CONTROLLED RESPONSE OF MODEL UCB-1 UNDER EXCITATION E.Q. D1; CONTROLLER ACTING AT m_3

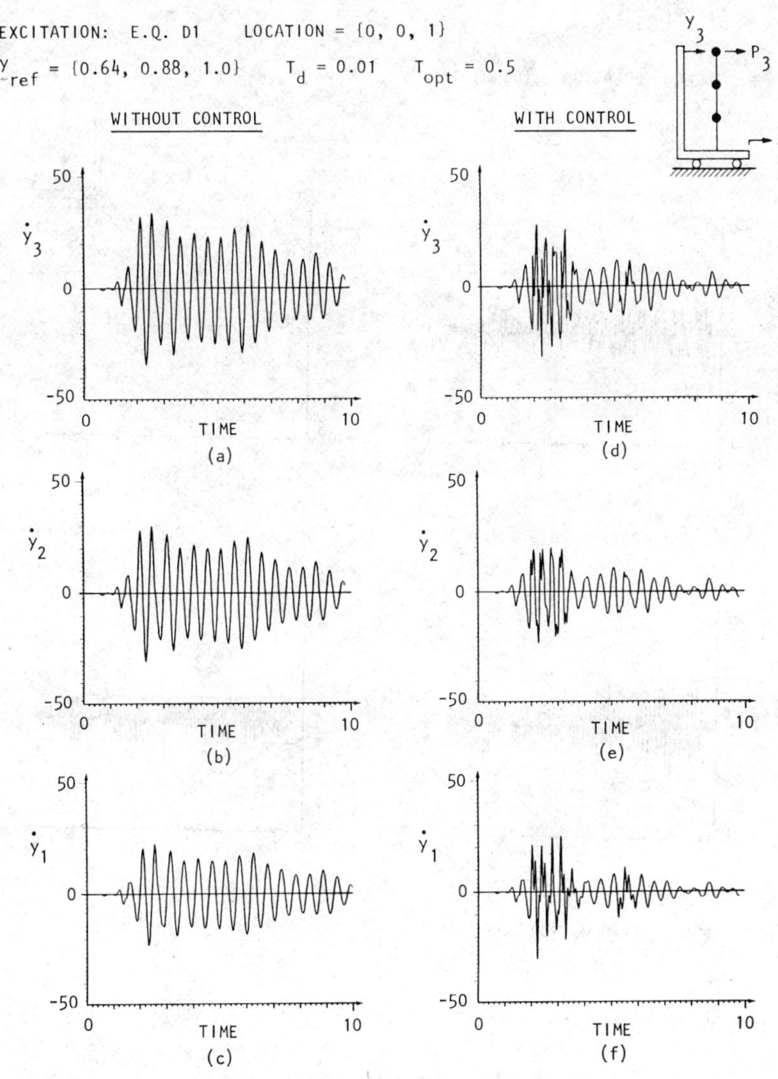

FIGURE 9. RELATIVE VELOCITY RESPONSE UNDER EXCITATION E.Q. D1; CONTROLLER AT m_3

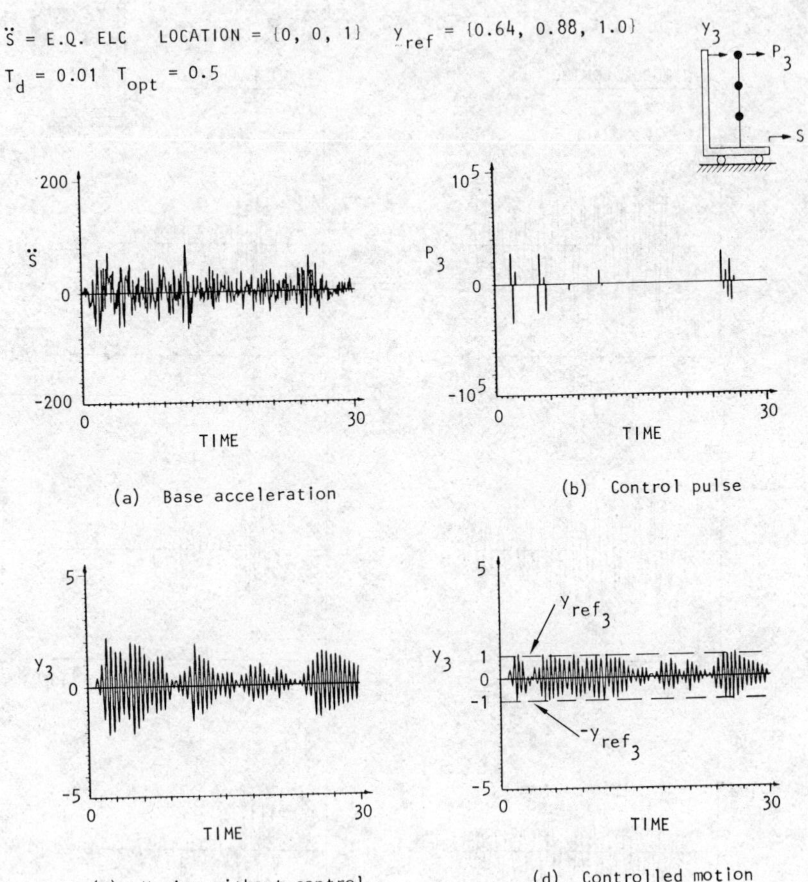

FIGURE 10. RELATIVE DISPLACEMENT RESPONSE UNDER EXCITATION E.Q. ELC; CONTROLLER AT m_3

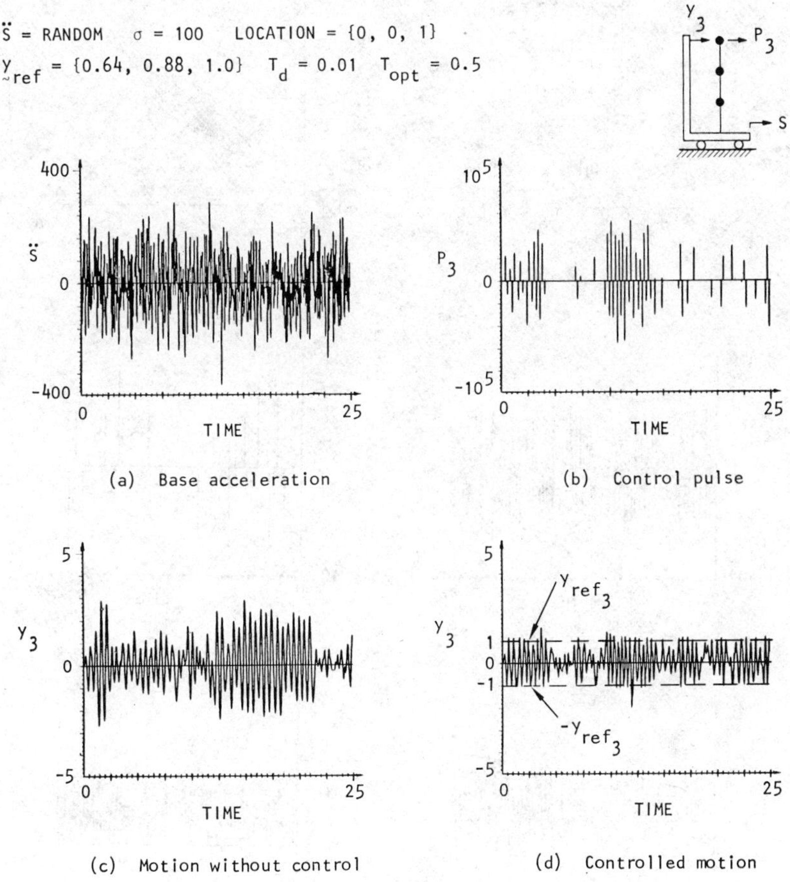

(a) Base acceleration

(b) Control pulse

(c) Motion without control

(d) Controlled motion

FIGURE 11. RESPONSE OF MODEL UCB-1 UNDER STATIONARY RANDOM EXCITATION; CONTROLLER AT m_3

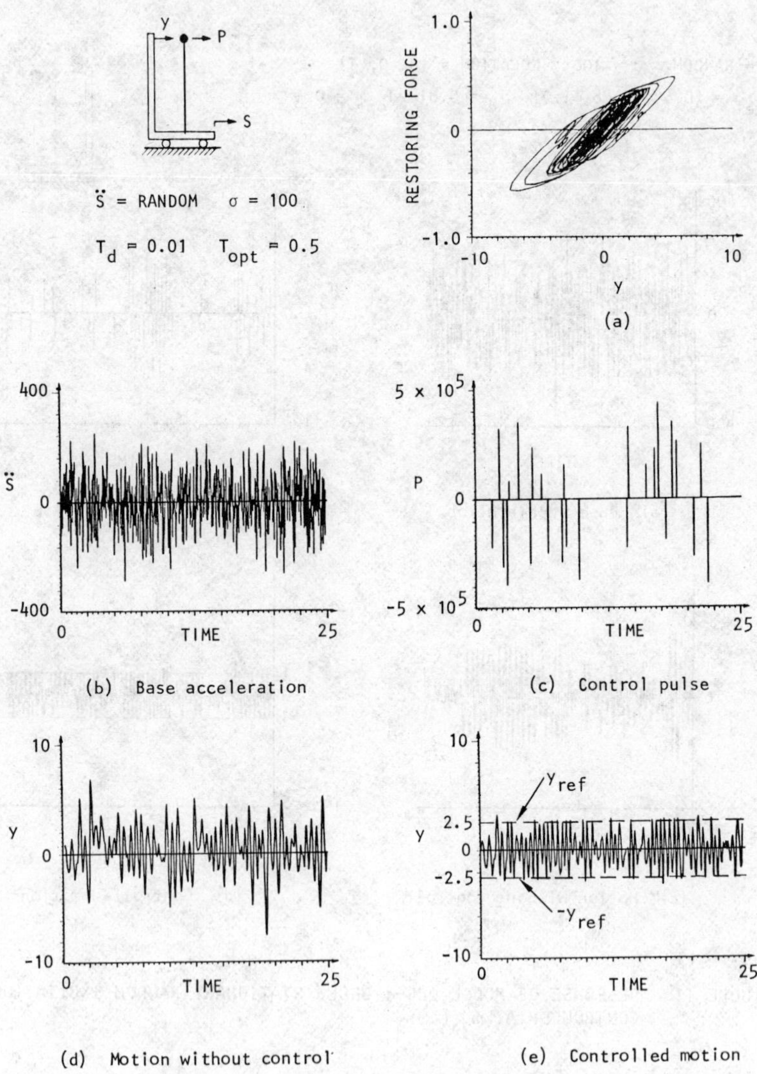

FIGURE 12. CONTROLLED RESPONSE OF A SDOF NONLINEAR SYSTEM UNDER STATIONARY RANDOM EXCITATION

To illustrate the performance of the proposed control algorithm when used in conjunction with nonlinear systems, consider a hypothetical SDOF system with the hysteretic characteristics shown in Fig. 12a. Under the action of stationary random excitation, the relative motion with and without control is shown in Fig. 12. As in the case of linear systems, it is clear that the control method is also successful in limiting the motion of this typical nonlinear system.

Conclusions

This paper shows the feasibility of using pulsed open-loop adaptive control for reducing the oscillations of tall buildings on similar distributed parameter systems subjected to strong ground shaking or arbitrary nonstationary disturbances.

The method is open-loop to reduce computing time, it is adaptive in order to take into account the varying nature of the system, and it uses pulse control to circumvent the problem of producing large control forces over sustained periods of time. Redundancy techniques using multiple microprocessors and motion sensors in parallel will ensure very high probability for amplitude and timing control of pulse generators.

Acknowledgment

This study was supported in part by a grant with the National Science Foundation.

References

1. Masri, S.F. "Steady-State Response of a Multidegree System with an Impact Damper," *J. Applied Mechanics*, Vol. 40, 1973, pp 127-132.

2. Masri, S.F. and Yang, L. "Earthquake Response Spectra of Systems Provided with Nonlinear Auxiliary Mass Dampers," *Proc. 5th World Conf. on Earthquake Engineering*, Rome, 1973.

3. National Science Foundation (NSF). *Feasibility of Force Pulse Generators for Earthquake Simulations*, NSF Grant PFR 77-15010. Washington, DC: NSF, 1979.

4. Safford, F.B. and Masri, S.F. "Analytical and Experimental Studies of a Mechanical Pulse Generator," *ASME J. Eng. Ind.*, Vol. 96, Series B, May 1974, pp 459-470.

5. Masri, S.F.; Bekey, G.A.; and Safford, F.B. "Optimum Response Simulation of Multidegree Systems by Pulse Excitation," *ASME J. Dynamic Systems, Measurement, and Control*, Vol. 97, Series G, No. 1, 1975, pp 46-53.

6. Masri, S.F. and Safford, F.B. "Earthquake Environment Simulation by Pulse Generators," *7th World Conf. on Earthquake Engineering*, Istanbul, 1980.

7. Masri, S.F.; Bekey, G.A.; and Udwadia, F.E. "On-Line Pulse Control of Tall Buildings," *Structural Control*, ed. H.H.E Leipholz. Amsterdam: North-Holland Publishing Co. and SM Publications, 1980.

8. Clough, R.W. and Tang, D.T. *Earthquake Simulated Study of a Steel Frame Structure, Vol. I: Experimental Results*, EERC 75-6, University of California, Berkeley, Apr 1975.

9. Tang, D.T. *Earthquake Simulated Study of a Steel Frame Structure, Vol. II: Analytical Results*, EERC 75-36, University of California, Berkeley, Oct 1975.

10. Jennings, P.C.; Housner, G.W; and Tsai, N.C. *Simulated Earthquake Motions*, California Institute of Technology Report, 1968.

A RECTILINEAR FORCE GENERATOR FOR FULL SCALE VIBRATION TESTING

J. I. Craig*
F. D. Lewis**

ABSTRACT

The design and construction of a unique force generator for use in full scale dynamic reponse testing of buildings is described. The design is rectilinear in concept and uses a servocontrolled hydraulic actuator coupled to a seismic mass as the driving element. The mass is supported on special trackless air bearings so that the force vector can easily be oriented in any horizontal direction. The control approach employed permits straightforward programming of the force level using a desktop computer and allows use of arbitrary periodic waveforms. Force generation can be synchonized with the acquisition of response data for minimum noise in the computed spectra.

INTRODUCTION

Full scale testing of tall buildings and other large civil engineering structures to determine their dynamic response characteristics typically requires the action of some forcing function. In systems terminology this is referred to as the input function and for practical, realizable systems must be specified in order to obtain a measurable output or response. The term forcing function is a generalized expression since the input can be a prescribed force (moment, pressure), acceleration, velocity, or a displacement. Forcing functions can be classified in a variety of ways, typically as either deterministic or nondeterministic, but in full scale testing it is more common to consider the input as due to either:

(a) ambient excitation, or
(b) prescribed excitation

In the first case, the forcing function is due to the presence and action of forces in the structure's environment, which can under certain circumstances include the action of internal machinery or building occupants. Typically, however these forces are due either to seismic events or to the action of wind or waves on the structure. The forces are inherently nondeterministic in nature and for the most part when utilized for testing purposes are very small in magnitude compared to static components. Forcing functions in the second case are all those due to deliberate action during the course of testing. While the magnitudes may be equal to or larger than those due to ambient forces, the prescribed forces are generally easily distinguished by being applied at essentially point locations as compared to the distributed nature of ambient forces. Prescribed forces can be either deterministic (sine, swept sine, impulse, psuedo random) or random in

* Associate Professor of Aerospace Engineering
Georgia Institute of Technology, Atlanta, GA
** Aircraft Development Engineering Specialist
Lockheed-Georgia Company, Marietta, GA

character, depending upon the objectives and methodology of the testing. Prescribed forces are generated by specific devices or pieces of equipment called shakers, exciters, drivers or vibration generators.

The present paper briefly presents the engineering considerations and design details of a unique rectilinear force generator for full scale vibration testing. This instrument was built specifically for use on an ongoing study of the interaction between cladding and the basic structure in tall buildings. The design uses a servo-controlled hydraulic actuator as the driving element which permits straightforward programming of the force level and allows use of arbitrary periodic forcing functions.

CONCEPT

Vibration or force generators are not new in concept. Small portable electrodynamic devices with peak outputs of from 1-50 lbf over a 20-5000 Hz bandwidth are commonly employed for model and small full scale testing. Similar but stationary units are available with outputs of up to several thousand pounds force. The major disadvantages of these devices for full scale testing of civil engineering structures are the very poor low frequency response and the low force output obtainable from portable-sized units. The frequency response problem is a fundamental limitation of the electronic amplifiers commonly used for this purpose while the force limitation is due largely to size and weight requirements for the field and armature assemblies.

Since the early sixties, electro-hydraulic equipment has found increasing use in the design of force generators. This equipment is commonly used in the aircraft industry and has become popular in many areas of mechanical and structural testing. In its basic form the key equipment consists of a continuously variable, electrically actuated, 4-way hydraulic spool valve. When properly connected to a double-acting hydraulic piston or actuator and furnished with high pressure (2000-3000 psi) hydraulic oil, the valve can be operated to produce advance or retraction of the actuator. The system becomes quite capable when closed loop or negative feedback control is used to operate the valve. In this approach, a stroke or a force transducer is used to measure the actuator position or force level and the signal is then used to control the valve in response to a command signal. The earliest testing applications were static in nature and used the electro-hydraulic equipment to generate prescribed static forces. Cyclic loading and arbitrary waveform dynamic loading applications quickly followed. Currently there are systems available for generating almost unlimited static forces and dynamic forces of several thousand pounds at up to 1000 Hz. Static limitations are due to actuator size while dynamic limitations are due to valve response and hydraulic flow rate limits as well as to problems with hydraulic resonance. The equipment need not be too costly but portability can be a problem due to the need for a high pressure hydraulic pump.

Finally, purely mechanical force generators have been used for many years to excite vibrations in structures. The simplest concept is an out-of-balance flywheel driven by a variable speed motor. A variety of rotating eccentric mass or reciprocating mass configurations are feasible. Perhaps the most popular and widely known technique for full scale vibration excitation in large civil engineering structures involves use of the counter-rotating eccentric mass vibration generators developed at Cal Tech (1). Their major advantages are simplicity and relatively small size, but the disadvantage is the inability to generate anything but sinusoidal waveforms.

A survey of full-scale testing techniques was presented at the preceeding conference in this series (2). Provide several papers a good description and discussion of various force generators that have been developed or adapted for this purpose.

The force generator described in this paper is basically an electro-hydraulic design. A small hydraulic actuator is used to move a seismic mass in a rectinlinear fashion in a horizontal plane. The generated force is thus the inertial reaction force, and consequently, its magnitude depends on the actuator displacement and square of the vibration frequency. This approach was taken for the following reasons:

(a) Electro-hydraulic actuator systems are mechanically simple and readily available. Conventional industrial pumps and actuators can be used and complete feedback control/servo amplifier electronics packages are commercially available.

(b) A rectilinear design generates the desired constant-direction vector force without the need for counter-rotating masses to cancel unwanted transverse components.

(c) The rectilinear electro-hydraulic approach permits straightforward programming of the force level and allows use of arbitrary periodic waveforms.

The major disadvantages to this approach are the somewhat larger size of the generator including the required portable hydraulic power supply and the greater complexity required to allow redirection of the force vector in a horizontal plane. This latter problem has, however, been eliminated by a unique design in the present case.

The rectilinear electro-hydraulic concept is not new, but no commercially available design is presently available. Several different custom designs have been used (3,4) and at least one approach involved vertical translation of the seismic mass for excitation of bridge structures (5). The present design was assembled in large part from conponents already available in the laboratory and current planning will allow the unit to remain intact as a versatile force generator for vibration testing.

DESIGN

The immediate application for the force generator is in vibration testing of tall buildings. Consequently, the baseline performance specifications were established as roughly equal to those of the Cal Tech/California design which is commercially available in similar form as the Kinemetrics VG-1 vibration generator (6). The VG-1 specifications and those of the present design are shown below.

Specification	Present	VG-1
Maximum force amplitued (lbs)	4,400	5,000
Maximum frequency (Hz)	15	9.7
Lowest frequency for max. force (Hz)	3.5	2.5
Force amplitude at 1 Hz (lbs)	1,200	805
Primary drive motor (Hp)	20	1.5

The next to last two specifications define the low frequency performance which is stroke limited in the present design and limited by the maximum eccentricity in the VG-1. Below about 1 Hz both designs roll off at the characteristic 24 dB per octave due to this limitation. The maximum force is limited in the present design by the maximum actuator force and is limited by strength considerations for the VG-1.

Mechanical Details

As noted in the previous section, a rectilinear design consists of a large seismic mass and a force actuator. In the present design the mass can be up to

3000 lbm and consists of a variable number of lead weights (4 x 4 x 8 in bricks) arranged on a "weight table". The table, in turn, is moved rectilinearily in a horizontal direction by a small (2 in. bore) hydraulic actuator. The weight table must be supported in such a manner that this motion is possible, and for versatility in testing, it must be possible to redirect this line of action with a minimum of effort.

The design approach at the outset was to use linear ball bushings riding on hardened, ground round rails. This eventually was ruled out due to a short projected bearing life at maximum design conditions (less than 100 hrs.) and due to the mechanical complexity required for easy realignment of the force vector. The solution to this problem represents one of the unique aspects of the present design. As shown schematically in Figure 1, the weight table is supported on four large air bearings which in turn ride on a large smooth and flat base plate. This

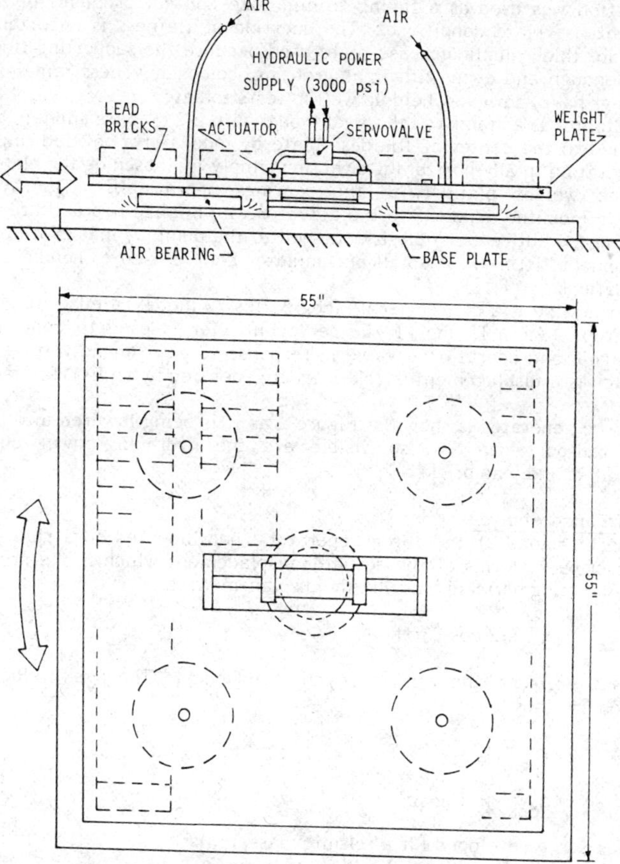

Fig. 1. Rectilinear Force Generator Schematic.

approach offers an almost unlimited bearing lifetime since there is nominally no bearing contact and the air flow tends to be self-cleaning. Since the bearings are trackless, the weight table can be aligned in any direction allowed by the actuator support design. The major disadvantage is the requirement for a low pressure (15 psi) air supply which in the present case is a portable compressor but which in future designs could be integrated into the hydraulic power supply. With the present bearing design, only 8 psi is required to lift the bearings but the flow rate can vary considerably depending on the flatness of the base plate.

Some difficulty was encountered in the design of the air bearings and several versions were tried. The major problem was an oscillatory instability or chatter in the bearings initially employed which had a roughly 1 in. wide by ¼ in. thick integral lip at the outer radius (this was to provide clearance in the event the base plate could not be maintained flat). The final solution was to employ a flat disk design which proved quite stable with the available air supply. An 0.03 in. layer of teflon was used as a facing to minimize contact problems or damage with uneven base plate conditions. The question of flatness is important even with the 1 in. thick aluminum base plate used because the supporting floors are never flat enough and even with light grouting, some uneveness remains. The teflon facings have, however, held up well in tests to date.

The actuator is a standard industrial model with a 2 in. bore and 8 in. stroke. It is attached to the center of the base plate by means of a beveled ring which allows orientation in any horizontal direction simply by loosening the clamps and rotating the weight plate (with the bearings operating!). The hydraulic servovalve is mounted adjacent to the actuator to minimize hydraulic resonance problems and the entire assembly is attached to the pump by means of hoses and quick-disconnect fittings. A small accumulator can be used to handle dynamic flow fluctuations.

The servovalve is a conventional Moog Series 73 model rated at 10 gpm at a 1000 psi drop. An MTS model 706 servocontroller is used to condition the feedback signal and operate the valve in response to a command voltage signal. The feedback is a displacement (stroke) signal generated by an LVDT attached to the actuator.

The force generator is shown in Figure 2 as it is being lowered into position from its transport cart. Also visible are the hydraulic power unit, the servocontroller, and lead bricks.

Performance Analysis

The performance of the force generator when mounted on a rigid base is easy to describe in terms of the actuator displacement which is the prescribed variable. Assuming harmonic motion the displacement is:

$$x = s/2 \cos 2\pi ft$$

where s is the stroke and f is the cyclic frequency. The peak velocity and acceleration are thus

$$v_{peak} = \pi sf$$

$$a_{peak} = 2\pi^2 sf^2$$

and the peak force developed for a seismic mass, m, is

$$F_{peak} = 2\pi^2 msf^2$$

Fig. 2. Complete System including Pump, Weights, and Controller being Installed.

The maximum force possible is limited by the actuator size and maximum hydraulic pressure:

$$F_{max} = p\,A$$

where A is the actuator piston area and p is the pressure. At high frequencies the maximum force limits the performance while at low frequencies the peak value is the limiting factor with the force level rolling off with the square of the frequency for a fixed stroke. The corner point occurs at

$$f_o = \pi(pA/2ms)^{1/2}$$

These conditions all apply provided the maximum hydraulic flow rate is not exceeded.

The maximun flow rate available from the power unit effectively imposes a constraint on the actuator velocity magnitude. The peak velocity is given above and occurs at two instants during each cycle. Typically, an accumulator is used to meet these peaks and the power unit must meet the average demand. Using this approach, the average velocity taken over the full stroke is:

$$v_{avg} = 2\,sf$$

and for a piston of area A with hydraulic flow rate Q, the piston velocity is Q/A. The average velocity increases with f for a given stroke and the limiting case occurs at a maximum frequency:

$$f_Q = Q/2As$$

Above this frequency, the maximum available force is limited to:

$$F_{peak} = \pi^2 \, mQf/A$$

and thus increases in proportion to f rather than f^2 for the stroke limited case.

These constraints are shown graphically in Figure 3. The operating region is to the right and below the curve. Clearly, the low frequency performance is stroke limited and rolls off sharply. The maximum force is limited by actuator size and hydraulic pressure. The knee between these extremes is defined by the flow rate limit. Finally, Figure 4 compares the performance envelop of the present force generator with the commercial VG-1 unit for reference purposes.

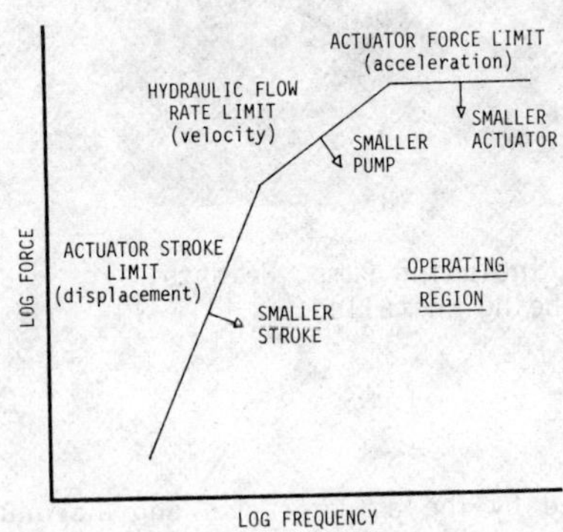

Fig. 3. Typical Performance Envelope for a Rectilinear Force Generator.

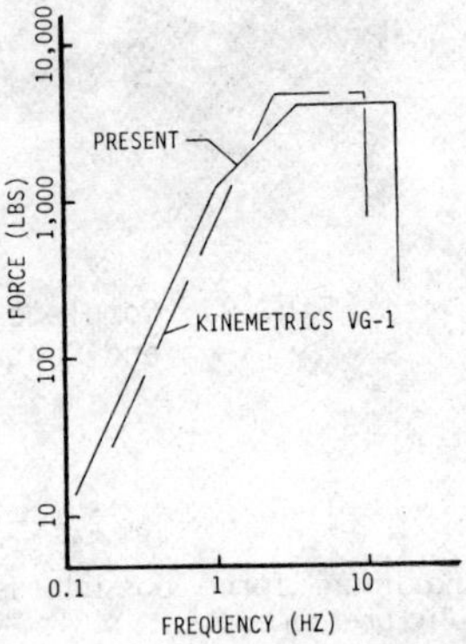

Fig. 4. Performance of the Present Unit with VG-1 as Reference.

Control Circuitry

In its basic configuration, the force generator produces a displacement of the seismic mass in response to a prescribed wavefrom supplied in analog form to the servocontroller. For sinusoidal, fixed frequency excitation it is only necessary to determine the appropriate stroke required to produce the desired force at the selected frequency, provided the operating point lies within the generator's capabilities.

The key advantage of a rectlinear generator in the present form is its ability to produce transient, nonsinusoidal forcing functions. However, in practice the desired force function is known but must be produced in the generator by means of a prescribed displacement. This transformation of the desired force function into a prescribed displacement can in principle be accomplished in analog terms by passing the given force function signal through an analog filter circuit which is the inverse of the function shown in Figure 4. Construction of such a circuit was not attempted in the present case, and instead, a digital approach was adopted that is more in keeping with the digital time series analysis techniques which are ultimately applied to the response signals.

In this approach, the force function is first constructed as a digital time series in the memory of a desktop computer. Then, the transformation to a prescribed actuator displacement function is carried out, and the result is checked to make sure performance limits will not be exceeded. Finally, this time series is transformed into an analog waveform which is used to command the servocontroller. This last stage is carried out by a digital-to-analog converter on a point-by-point basis at a specified clocking rate. This clock is also used to control the response sampling so that a minimum of noise is introduced into the response spectra ultimately computed. Further details of this aspect of the present application are beyond the scope of this paper. A schematic of the arrangement is, however, shown in Figure 5.

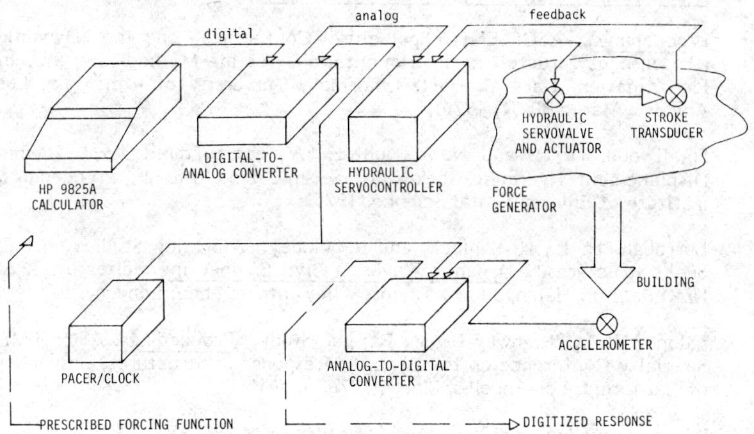

Fig. 5. Force Generator Control Schematic.

PERFORMANCE

The force generator described here has been built and is currently being used in full scale testing of several high rise buildings. In broad terms, the design has proven quite satisfactory, although as with all full scale work of this magnitude, the logistics of transporting 3000 lbs. of lead, a power unit, and the apparatus itself are not to be dismissed lightly. Several weight carts and a special fixture for storing the weight and base plates were fabricated in order to simplify transportation of the generator. The latter fixture can also be used to set the generator in place.

The most troubling aspect of the design was the air bearings themselves. As noted before, the initial approach was to use a cupped type of disk design for the bearings, but this was discarded because of a severe chatter. The cupped design was chosen to handle irregularities in the base plate, but apparently, insufficient air flow from the available air compressor was responsible for the chatter instability. This was overcome by changing to a flat disk design, and a teflon wear surface was added to protect the surfaces should contact occur.

At the present time, the unit is still undergoing performance testing and is being used on a trial basis to excite motion in a five story campus building. These tests involve integrated excitation and response measurements using transient excitation and time series analysis techniques. The unit has generally met all its requirements and specifications.

ACKNOWLEDGEMENT

The work reported here was sponsored in part by the National Science Foundation, Division of Problem-Focused Research Applications, under Grant ENV77-04269. Their support is gratefully acknowledged.

REFERENCES

1. Hudson, D. E., "Resonance Testing of Full-Scale Structures," ASCE Journal of the Engineering Mechanics Division, EM 3, June, 1964

2. Proceedings, ASCE/EMD Speciality Conference on the Dynamic Response of Structures: Instrumentation, Testing Methods and System Identification, Gary C. Hart, Chairman, University of California, Los Angeles, March 30-31, 1976.

3. Smallwood, D. O. and N. F. Hunter, "A Transportable 56kN, 200mm Displacement Hydraulic Shaker for Seismic Simulation," Proceedings, Institute of Environmental Science, 1975.

4. Burrough, H. L., H. Kinlock, and P. Winney, "Vibration Studies on Full Scale Structures," Dynamic Waves in Civil Engineering, edited by D. A. Howells, I. P. Haigh, and C. Taylor, Wiley-Interscience, London, 1971.

5. Salane, H. J., "Highway Bridge Fatigue Study," Proceedings, ASCE/EMD Specialty Conference on the Dynamic Response of Structures, University of California, Los Angeles, March 1976, p. 541.

6. Kinemetrics, Inc., 222 Vista Ave., Pasadena, CA 91107.

EARTHQUAKE SIMULATION USING CONTAINED EXPLOSIONS

John R. Bruce* and Herbert E. Lindberg†

Abstract

This paper describes a technique for using contained explosions to produce earthquake-like ground motion. The long range objective is in-situ testing of soil-structure interaction and of structures with complex internal equipment systems. The technique will be applicable to buildings, nuclear reactors, pipelines, power lines, dams, bridges, and tunnels.

The technique produces ground motion by simultaneous firing of a planar array of vertical line sources. The controlled release of high-pressure explosion products within each source allows controlled pressurization of the surrounding soil. In this way, both the amplitude and frequency content are controlled at levels suitable for testing with the array close to the test structure. This opens the possibility of in-situ testing at high levels of ground motion with a minimum of explosive and with little disturbance to the surroundings.

The results show that 160 pounds of explosive in an 80 by 40 foot array can produce accelerations of 0.6 g, velocities of 1 ft/s, and displacements of about 1 inch at frequencies of 3 to 5 Hz in a 30 by 30 foot test area. Earthquake-like motion lasting about 5 s can be produced by nine firings of a group of three parallel arrays. Each array is capable of three firings in a single experiment and can be reused indefinitely.

Introduction

The need for an in-situ test technique to aid in the design of earthquake-resistant structures has long been recognized. This need has become more acute with the development of nuclear reactors, greater population concentrations, and the more efficient designs that are made possible by computer technology. During the past three years, SRI International (formerly Stanford Research Institute) has been conducting a program funded by the National Science Foundation to develop an explosive method of testing in-situ structures at strong earthquake levels.

The objective of testing in-situ structures is to observe vibration modes and explore potential damage mechanisms in complete soil-structure and internal equipment systems. The technique will be applicable to buildings, nuclear reactors, pipelines, power lines, dams, bridges, and tunnels.

*Research Engineer, SRI International.

†Staff Scientist, SRI International.

The technique produces earthquake-like ground motion by simultaneous detonation of a planar array of vertical line sources placed in the soil near the test structure. The key feature of each line source is a cylindrical steel canister in which the explosive is detonated. Controlling the release of the high pressure detonation products from this canister allows controlled pressurization of the surrounding soil. In this way, both the amplitude and frequency content are controlled at levels suitable for testing with the array close to the test structure.

This technique opens the possibility of in-situ testing at high levels of earth motion with a minimum amount of explosive and with little disturbance to the surroundings. The duration of the simulated earthquake motion can be controlled by delayed multiple detonations within each line source and between groups of line sources.

Figure 1 shows a schematic of the array and a test structure. For testing a 30-foot-base structure, the array would measure about 80 feet long and 40 feet deep. It would consist of 10 line sources placed in vertical boreholes 40 feet deep and spaced on 9-foot centers, and it would be placed about 20 feet from the structure to be tested. Larger arrays can be built to test larger structures, or the larger structure can be built at an appropriate smaller scale. Results of the current program show that, in general, the array width should be two to three times the plan dimension of the test structure.

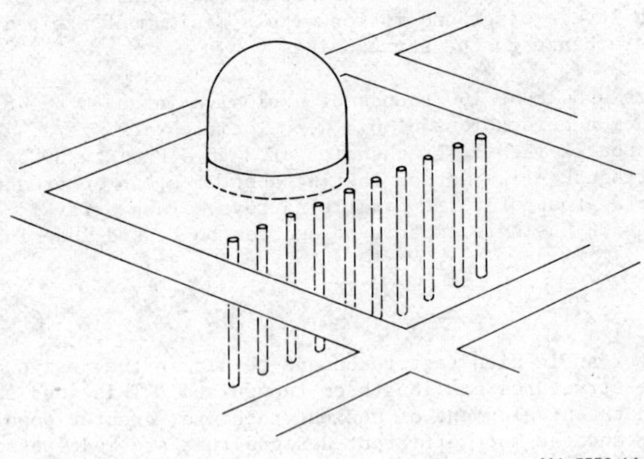

FIGURE 1 SCHEMATIC OF ARRAY AND TEST STRUCTURE

Line Source Design

Figure 2 shows the details of the line source design. The source shown is 12 inches in diameter and 39 feet in length. Approximately ten sources of this size are required to shake a 30 by 30 foot test area at full-scale earthquake levels.

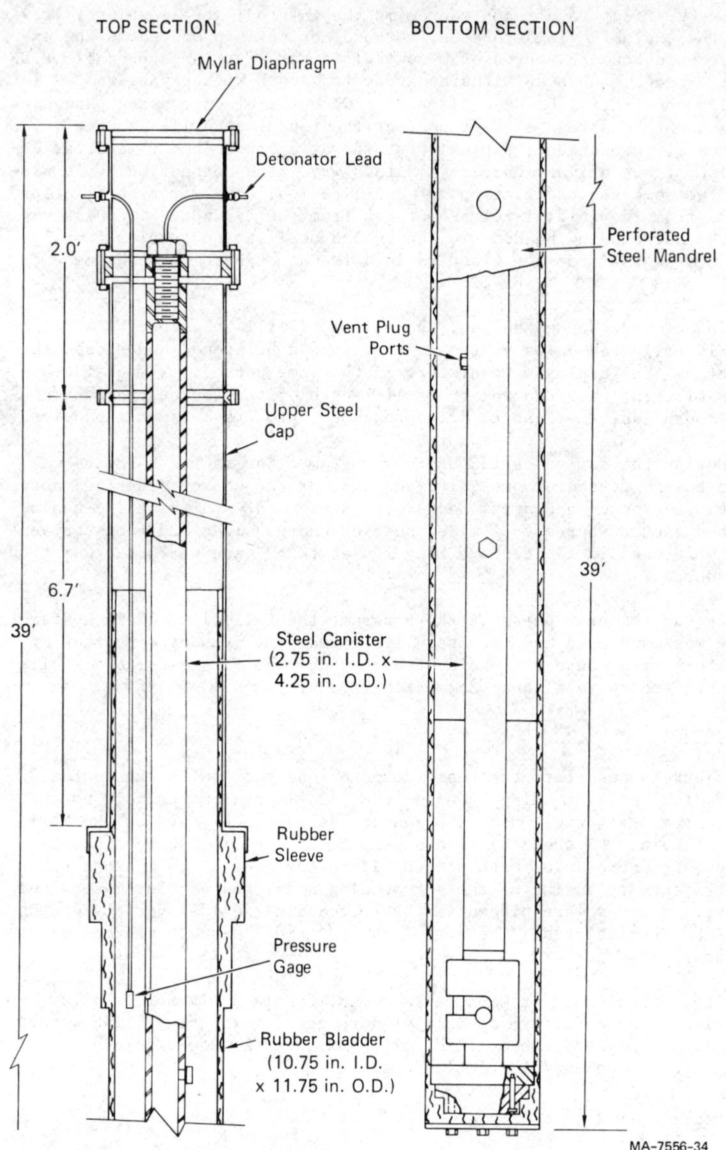

FIGURE 2 ASSEMBLY DRAWING OF THE 12-INCH DIAMETER SOURCE

The key feature of each source is the internal steel canister in which the explosive is detonated. Ground motion is produced as the explosive products are vented at a controlled rate into an expandable rubber bladder rugged enough to withstand repeated tests with expansions as large as twice the initial bladder diameter. Once the bladder has expanded, a Mylar diaphragm across an exit port at the top of each line source is cut by a small, independent, explosive charge at a prescribed time after detonation of the primary charge. In this way, each detonation cycle produces two complete oscillations of acceleration. The first oscillation results from the initial release of gas from the cylindrical canisters into the surrounding bladders. The second oscillation results from release of the gas from the bladders into the atmosphere through the exit ports.

In a future source design, the exit port will be opened by an electrically triggered valve system so that it can be opened and closed for each of several explosive canisters in the same source. Three side-by-side canisters, each running the full length of the source can each contain enough explosive gas to drive the source to its maximum expansion.

During the early portion of the program, line source hardware was developed and tested at one-third the size of the source described above. Both single source and array tests were performed with these 1/3-scale, 4-inch-diameter sources. A wide range of charge sizes and canister vent areas were used in the tests. Many repeat tests were performed using the same sources.

During the next phase of the program, the full-size, 12-inch-diameter source was developed and successfully tested. At present, an array of eight full-size sources is being fabricated. The first series of full-size array tests is planned for October 1980.

Test Setup

Figure 3 describes the test setup for one such full-size source. Figure 3(a) shows it being lowered into a 24-inch-diameter hole by the boom on the drill rig. Once the source was in place, sand was backfilled into the hole. Figure 3(b) is a schematic of the line source in the ground. Pressure gages were placed within the sources to measure the bladder pressure acting on the surrounding soil, and accelerometers and soil stress gages were placed in the surrounding soil at various depths and standoff distances. The test setup for the 1/3-scale sources was similar.

The test site is at Camp Parks in California's Livermore Valley, in a level area on a uniform deposit of dark grey, stiff clay. The water table is at 11 feet. Unconfined compression tests showed an average strength of 30 psi and a friction angle of 30°.

Test Results

In this paper, we will limit the discussion of the test results to the array test series performed with the one-third size sources. This series of seven tests was performed over a six-month time span using the same sources.

(a) PLACEMENT OF SOURCE

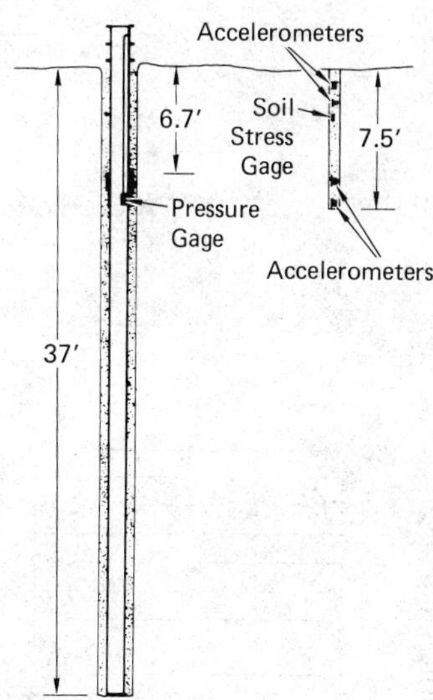

(b) SCHEMATIC OF SOURCE IN SOIL

MP-7556-36A

FIGURE 3 TEST SETUP FOR 12-INCH-DIAMETER SOURCE

Figure 4 shows the data from a typical array test using a total of 6.2 pounds of charge. The upper left-hand corner shows a plan view of the array layout of ten sources spaced on 3-foot centers and the associated instrumentation. Bladder pressure and expansion were measured in a central source and in an end source. The main ground motion instrumentation consisted of eight accelerometers in a T-pattern over an 10 by 10 foot region that would be occupied by a test structure.

Most of the accelerometers were at a 7.5-foot depth, the mid-depth of the sources, oriented to measure horizontal motion normal to the array. At the central location (center of test area), horizontal acceleration was also measured at a 2-foot depth, and vertical acceleration was measured at a 1-foot depth. Soil stress normal to the array was measured at a 2.5-foot depth at the central location and at a close-in location.

Time histories of the source pressure from one of the ten sources and of the soil stress at the central location are shown in the two blocks below the array layout in Figure 4. The bladder pressure rises linearly during the first 15 ms and then remains relatively constant until the Mylar diaphragm is ruptured and the pressure is vented to the atmosphere (at t = 60 ms for this test). The soil stress at the 10-foot standoff follows the general shape of the bladder pressure, but is delayed and spread out in time by the wave transit to the gage location.

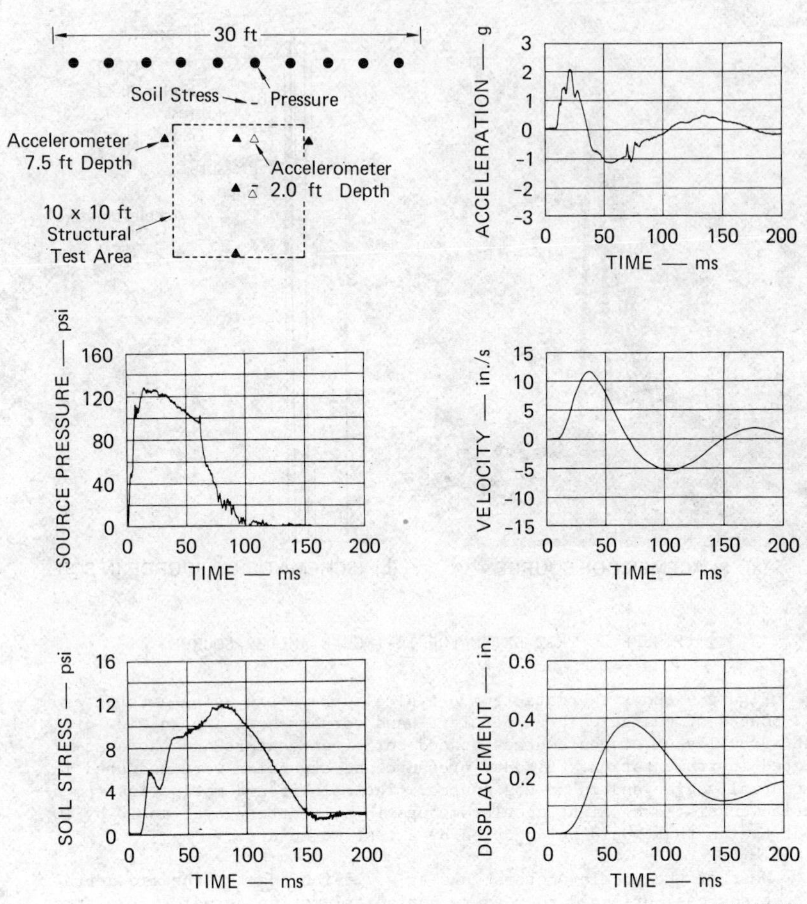

FIGURE 4 DATA FROM TYPICAL ARRAY TEST (Array of ten 4-inch-diameter sources using 6.2 pounds of propellant)

Ground acceleration, velocity, and displacement time histories at the central location of the instrumented area (from the 2-foot-depth accelerometer) are shown on the right-hand side of Figure 4. The fundamental period of motion is about 100 ms (10 Hz), which is consistent with the duration of the source pressure. The peak acceleration is 2 g, the peak velocity is 12 in./s, and the peak displacement is 0.4 inch. Use of geometric scaling laws to predict the corresponding near surface motions for a full-size array gives a frequency of 10 Hz/3 ≈ 3 Hz, a peak acceleration of 2 g/3 ≈ 0.7 g, a peak velocity of 12 in./s, and a peak displacement of 3 x 0.4 inch = 1.2 inches. The charge weight would be 3^3 x 6.2 pounds = 160 pounds.

Figure 5 shows the uniformity of ground motion across the length and width of the test area. These time histories were all recorded at a depth of 7.5 feet in an array test using 2.5 pounds of charge. The upper row in Figure 5 shows ground velocities at a 6-foot standoff from the array at locations 6 feet to the left, 6 feet to the right, and at the array centerline. They agree closely in shape and in magnitude.

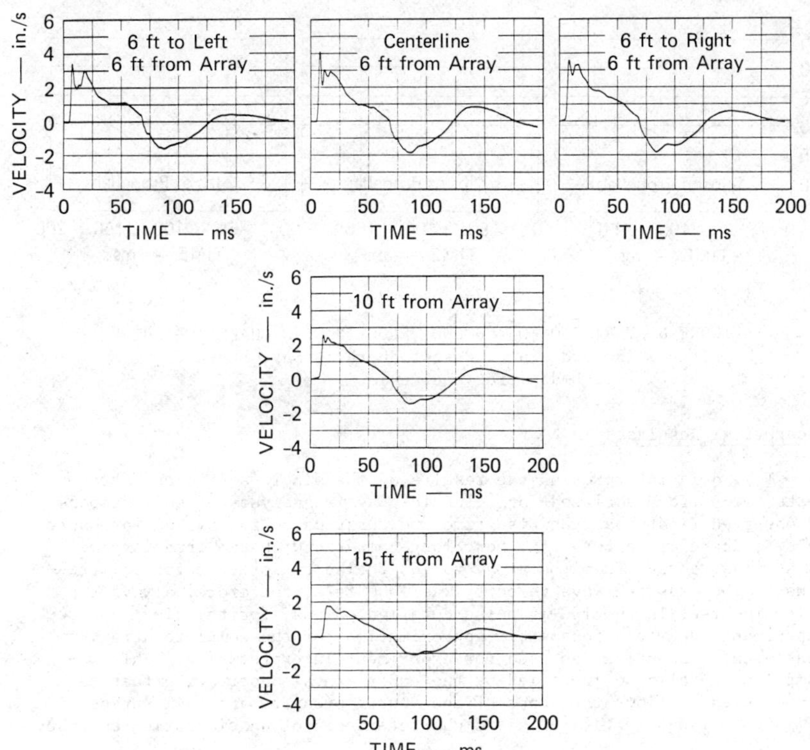

FIGURE 5 VARIATION OF GROUND VELOCITY OVER TEST AREA (Recorded at 7.5-foot-depth; array of ten 4-inch-diameter sources using 2.5 pounds of propellant)

The center column in Figure 5 shows the ground velocities along the array centerline at standoffs of 6, 10 and 15 feet. These agree closely in shape, but show a 30% decay in magnitude over the 11-foot distance. This spatial variation agrees with the results of an elastic analysis for the displacement field around the array, as is described in the next section. This decay can be reduced by several methods (longer array, increase in Poisson's ratio below water table, and a trench on the opposite side of the test area).

Figure 6 shows the increase in ground motion with increasing charge weight for three of seven tests using the same array. The shape and period of the ground velocity remain fairly constant, whereas the magnitude increases from 3.8 in./s with 2.5 pounds of charge to 12 in./s with 6.2 pounds of charge.

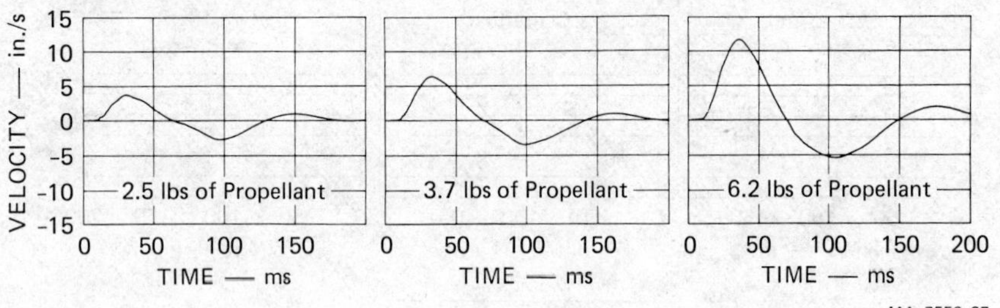

FIGURE 6 VARIATION OF GROUND VELOCITY WITH CHARGE WEIGHT (Recorded at 2.0-foot depth; array of ten 4-inch-diameter sources)

Theoretical Results

To aid in interpreting the results of the single-source and array tests, we performed closed-form elastic-plastic analyses of soil response. We observed first that the rise time and duration of the bladder pressure pulse (Figure 4, center left) were long compared with the transit time of an elastic wave from the array to the region of interest for soil response. The measured wave velocity was 1000 ft/s; therefore, a wave traveled 100 feet in the 100-ms pulse duration. Thus, for the single-source experiment, and to a reasonable approximation for the complete array, a quasi-static solution was used for experiment interpretation. This assumption can also be justified by the experimental observation that the displacement follows the shape of the source pressure and soil stress (Figure 4, lower left), the basic characteristic of quasi-static response.

An elastic-plastic plane strain analysis of a pressurized cylindrical hole showed that displacements from a single source will always be small. With properties measured for the Camp Parks soil (Young's Modulus 5000 psi, Poisson's ratio 0.25, and strength as given earlier), which are typical of those of many soils, the displacement at radius r from a 4-inch-diameter source at a source pressure of 75 psi is $w = 0.002\ R^2/r$, where

R is the radius of the elastic-plastic boundary around the source. To obtain elastic free-field response at the test structure, we must have $R < r$. Thus, $w < 0.002\ r$.

The advantage of the array is that, as the elastic-plastic boundary around each source grows, it approaches the boundaries around the adjacent sources. When the boundaries meet (more precisely, when they interact to form a continuous boundary) the length of the elastic-plastic boundary becomes equal to the array length.

Figure 7 shows a schematic of the developing plastic region around each source. An elastic plane strain analysis for the resulting elliptical boundary showed that the displacements for an array are two orders of magnitude greater than for a single source if the pressure acting on the elliptical array boundary is the same as the pressure acting on the circular, single source, elastic-plastic boundary.

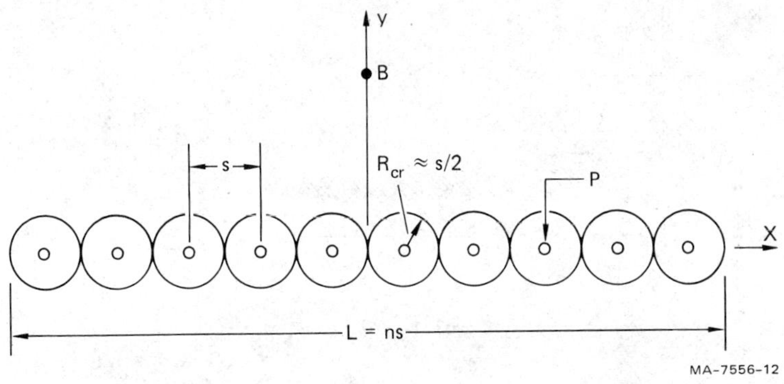

FIGURE 7 SCHEMATIC OF DEVELOPING PLASTIC REGION AROUND EACH SOURCE

Figure 8 shows the theoretical displacement field around an elliptical hole in an elastic material with Poisson's ratio ν equal to 0.25. The direction of each vector shows the direction of displacement, and the vector length shows its relative magnitude. The absolute magnitude is a function of the pressure on the elliptical boundary. The theory shows the same distribution of displacement that was observed in the tests (Figure 5); that is, the displacement is uniform across the length of the test area and decays by about 30% from the front to the back of the test area.

To use the elastic plane strain analysis to predict the magnitude of these ground displacements, it is necessary to determine the relationship between the source pressure and the pressure on the elliptical boundary. This was done with a finite element calculation using a Mohr-Coulomb elastic, perfectly plastic, material model to represent the soil. This accounted for the complex plastic interaction between the sources.

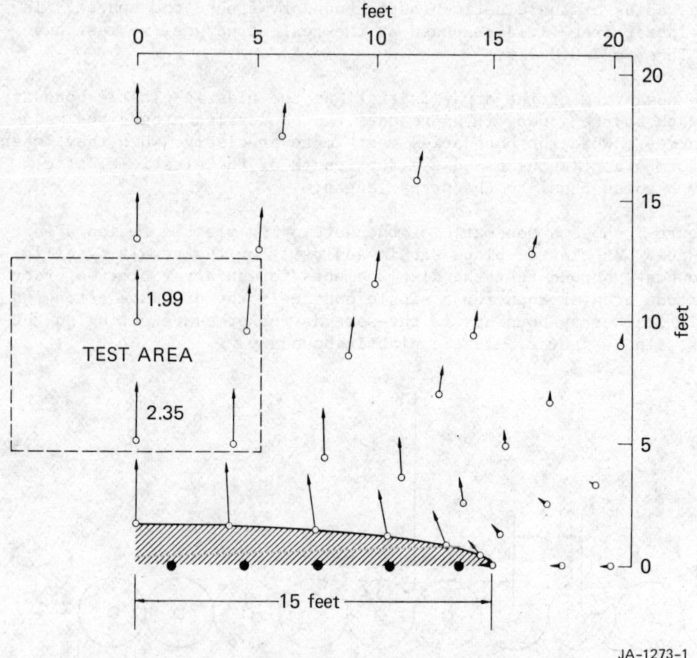

FIGURE 8 THEORETICAL DISPLACEMENT FIELD

Figure 9(a) shows the finite element grid used to model an individual source in an array. Symmetry boundaries were placed along the lines of symmetry of the source and along the line of symmetry between the sources. The elliptical boundary at the minor axis was represented by a series of elastic springs that give the pressure-displacement relation found from the elastic elliptical hole theory.

Calculations were performed with the finite element code NONSAP. Several parameters were varied from a set of baseline values. These parameters included the source radius a_s, the unconfined compressive strength of the soil σ_u, and the friction angle ϕ. The friction angle was varied at a constant cohesion.

Figure 9(b) shows the results of five of these code calculations. Boundary pressure P_e is shown as a function of source pressure P_s. Run 1 was performed for the baseline values of a_s = 3 inches, n = 10, s = 36 inches, E = 5000 psi, σ_u = 30 psi, and ϕ = 30°. (In the 1/3-scale array tests, the source radius a_s started at 2 inches and increased to about 4 inches, primarily because of local soil compressibility, a factor not accounted for in the Mohr-Comloub soil model.) In the other runs, one parameter at a time was varied from the baseline valves as indicated on Figure 9(b).

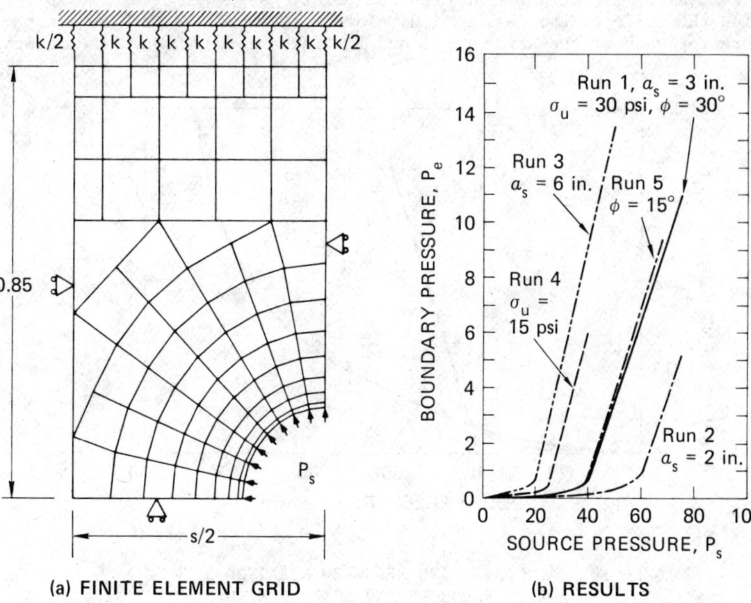

(a) FINITE ELEMENT GRID (b) RESULTS

FIGURE 9 CALCULATION OF PRESSURE ON THE IDEALIZED ELLIPTICAL BOUNDARY

In all cases, the ellipse boundary pressure increases very slowly until a critical source pressure is reached; it then increases rapidly. Examination of the results showed that, at these critical source pressures, the elastic-plastic radius is at about 12 to 14 inches. This fits with an intuitive observation that a critical source pressure is reached when the plastic interaction between the sources becomes strong. This occurs when the plastic radius approaches half the source spacing (s/2 = 18 inches in all cases). Until the source pressure reaches this critical value, the array is not being used to full advantage.

The variation of critical source pressure with changing parameters is shown in Figure 9(b). Increasing the source diameter decreases the critical source pressure (compare Runs 1, 2, and 3). Increasing the compressive strength of the soil increases the critical source pressure (compare Runs 1 and 4). Surprisingly, the friction angle of the soil appears to have little effect on critical source pressure (compare Runs 1 and 5).

The results of Run 1 were combined with the solution of an elliptical hole in a infinite elastic medium under internal pressure using E = 5000 psi and ν = 0.25 to determine the desired relation between source pressure and soil displacement. This calculated relation is compared in Figure 10 with the measured relation. The measured relationship was determined by plotting the value of the peak soil displacement versus the peak source pressure for each of the array tests and fitting a curve through the points.

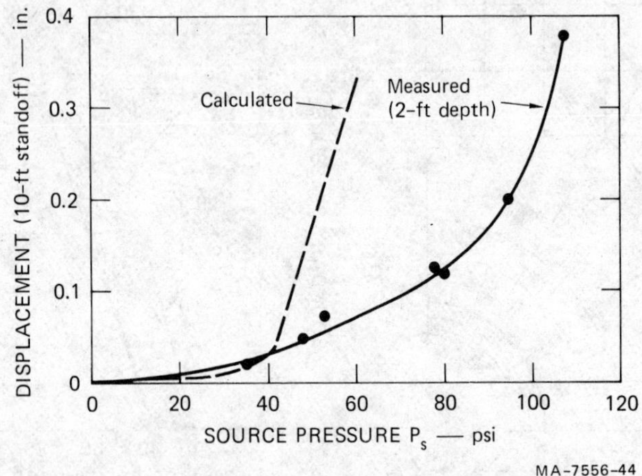

FIGURE 10 CALCULATED AND MEASURED RELATION BETWEEN SOURCE PRESSURE AND SOIL DISPLACEMENT

Qualitatively, the two curves show the same characteristic: the soil displacement increases slowly as source pressure is increased until a critical source pressure is reached; then the soil displacement increases at a much faster rate. The calculations indicated that this critical source pressure is reached when the plastic radii around each source in the array approach each other. At this point the array acts like a slit in the earth rather than as a series of individual single sources, and therefore an increase in source pressure contributes more directly to increasing the soil displacement rather than primarily increasing the elastic-plastic boundary radius around each source.

Response Spectra for Array-Produced Ground Motion

Figure 11 compares three response spectra: one for the estimated ground motion from a single detonation of an 80 by 40 foot array, one for the estimated ground motion from three such detonations spaced 1 s apart, and a design response spectrum* that envelops many spectra from statistically postulated future earthquakes. The ground motion for the 80 by 40 foot array was estimated by using the acceleration recorded at the

*N. Newmark, J. Blume, and K. Kapur, "Seismic Design Spectra for Nuclear Power Plants," ASCE J. Power Division (November 1973).

center of the structural test area in a 30 by 15 foot array test and applying geometric scaling laws with a scale factor of 3.

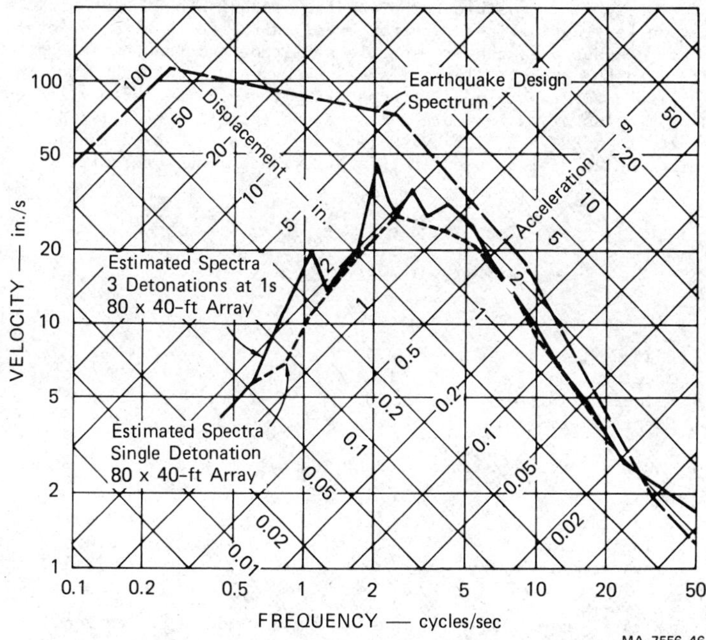

FIGURE 11 COMPARISON OF EARTHQUAKE DESIGN SPECTRUM AND RESPONSE SPECTRA FOR ESTIMATED GROUND MOTION FROM 80 by 40 FOOT ARRAY

Comparison of the two spectra for the 80 by 40 foot array with the Newmark, Blume, and Kapur spectrum shows good agreement in shape at frequencies higher than 2 Hz. At frequencies lower than 2 Hz, the array spectra fall off in amplitude from the Newmark, Blume, and Kapur spectrum. This falloff is a result of the 2-inch displacement limit for an 80 by 40 foot array operating at a 1-g level. Larger arrays will produce proportionately larger displacements. A 300-foot-long array is within conventional construction capability and would produce an amplitude of 7 inches at 1 g. This would match the design spectrum to frequencies at low as 1 Hz.

All three spectra are normalized to a peak ground acceleration of 1 g. Our experience with the 30 by 15 foot array and with scaling laws indicates that the 80 by 40 foot array can produce ground accelerations that exceed 0.5 g. This amplitude covers the range of expected accelerations from strong motion earthquakes.

Capabilities and Limitations

Results of the work to date demonstrate that the contained-explosion line source array is a feasible technique for producing earthquake-like ground motion for testing in-situ structures. Array tests with 1/3-scale sources have demonstrated that reasonable amplitude and frequencies can be coupled into the ground. Table 1 shows the ground acceleration, velocity, and displacement obtained in these array tests using three different charge sizes. Also given in Table 1 is the theoretical extrapolation of these results to a 80-foot-wide by 40-foot-deep array. The 12-inch-diameter source, which will be used in this array, has been successfully tested.

Table 1

SUMMARY OF ARRAY TESTS

30 by 15 foot Array (Test Results)			
Charge (lb)	2.5	3.7	6.2
Acceleration (g)	1.0	1.3	2.0
Velocity (in./s)	4.0	7.0	12.0
Displacement (in.)	0.13	0.25	0.4
80 by 40 foot Array (Extrapolated by Theory)			
Charge (lb)	65	100	160
Acceleration (g)	0.3	0.4	0.6
Velocity (in./s)	4.0	7.0	12.0
Displacement (in.)	0.4	0.8	1.2

The contained-explosion technique has two practical limitations. One is in producing large displacements. This leads to a limit on the earthquake motion that can be simulated at low frequencies. For the 80 by 40 foot array, simulation of the design response spectrum is limited to frequencies above 2 or 3 Hz. Larger arrays can be built to produce proportionally larger displacements and lower frequencies.

However, there is a practical limit to the depth of the array, which eventually limits the displacement. This depth limit is about 80 feet in soil similar to that at Camp Parks and is a result of the lateral earth pressure at the bottom of source reducing the effective source pressure to an inefficient level. Any increase in depth past this limit would be ineffective because high lateral earth pressure would prevent expansion in the lower portion of the source. In addition, we estimate that once the array length exceeds about four times its depth, the ground displacements are controlled primarily by the depth. Thus, any increase in array length past four times the depth is also ineffective in producing larger displacements. This gives 320 by 80 feet as the largest practical array size.

Until the 320 by 80 foot array size is reached, the displacement limitation is primarily a limitation of cost, an 80 by 40 foot array comprising 10 sources costs about $120,000 for construction and fielding and has a cost-per-test of about $8,000. A 160 by 40 foot array, comprising 20 of the same sources at the same depth above, costs about $240,000, and a 320 by 80 foot array, comprising 40 sources twice as deep as above, costs about $1 million.

A second limitation is in producing significant ground motion in very stiff soil or rock sites. The calculations discussed under theoretical results showed that test area displacements are proportional to the pressure on the elliptical array boundary and inversely proportional to the stiffness of the medium. Thus, bladder pressures must be higher for stiffer soils and are impractically high for rock, without some special provisions for enhancing displacements, such as, using trenches to provide a rock island.

Future Work

Tests with 12-inch-diameter sources in an 80 by 40 foot array are planned for October 1980. During 1981, a three-pulse-per-source capability will be developed and tested. This will provide the technological basis for the long-range program objective of designing groups of arrays, of this size and larger as needed, that can simulate motions lasting 5 to 10 s. For example, a group of three arrays of the size described above, with each array adjusted to produce a different pulse duration, could provide a sequence of 18 acceleration pulses (9 firings) and hence a simulated motion lasting 5 s and having a frequency content ranging from 2 to 10 Hz. We envision that such arrays would be built after completion of the above program as a cooperative effort among several universities or by industry for use in applied research and for immediate application to earthquake resistance research and certification of equipment.

Acknowledgments

The research was support by the National Science Foundation under grants No. PFR 78-00993 and No. PFR 79-20722, with Dr. M. P. Gaus as technical monitor. Use of reusable, pressurized bladders for explosive arrays was initially proposed by Dr. G. R. Abrahamson of SRI, who has also provided continuing technical guidance throughout the program. The NONSAP formulation and parameter study for elastic-plastic interaction between sources was performed by Dr. L. E. Schwer.

High Explosive Simulation of Earthquake-Like Ground Motions

Cornelius J. Higgins[1], Roy L. Johnson[2],
George E. Triandafilidis[2], Members, ASCE
and David W. Steedman[3], A.M., ASCE

ABSTRACT

This paper discusses several considerations in the design of earthquake simulation experiments including: simulation criteria, explosive ground motions phenomenology, ground motion prediction methods, enhancement methods, and some recently measured data. The development of simulation criteria involves a melding of an understanding of the engineering system and its important response mechanisms, on the one hand, and the practical capabilities of creating earthquake-like ground motions with high explosives, on the other. The area of ground motion phenomenology provides insight into the relation of source characteristics to the type of waves generated and their variation with range and depth. Prediction methods, based upon empirical analyses of explosive ground motion measurements and analytical calculations, are presented for dry soils. Some enhancements, which improve the fidelity of explosive events compared with earthquake motions, are described. Finally, some data from a recent series of simulation experiments are presented.

INTRODUCTION

Response measurements and post-event observations of actual earthquakes are a significant source of data on the behavior of prototype structures. Indeed, data obtained after every major earthquake have demonstrated serious deficiencies in our understanding of phenomena and associated design procedures. However, the available data from earthquakes are limited and will continue to be limited by uncertainties with regard to time and place of earthquake occurrence and by the limited amount of in-place instrumentation to record response. A more complete and adequate data base must come from simulations.

Simulation sources include field shaking machines, shake tables, snap-back methods, and explosions. This paper describes the current capabilities of high explosives for simulating earthquake ground motion effects. The method is especially well suited for investigating soil and soil-structure systems. In these cases, the response cannot be adequately evaluated independently of the medium through which the incoming waves propagate. In addition, high explosives can be used to excite prototype size structures to high levels of response.

---

1. Principal, Higgins, Auld & Associates, Inc., Albuquerque, NM
2. Professor of Civil Engineering, University of New Mexico, Albq., NM
3. Staff Engineer, Higgins, Auld & Associates, Inc., Albuquerque, NM

SIMULATION CRITERIA

The word simulation generally implies that the prototype environment cannot be generated at will and that some characteristics of the prototype environment will not be reproduced exactly in a simulation. In designing a simulation, it is necessary to determine those characteristics of the full-scale environment which are essential to adequately evaluate the system of interest. The system of interest and its dynamic response characteristics are major considerations. Certain features of earthquakes may be important for one structure but not for another. For example, aboveground structures founded on soil may be adequately tested by simulating certain kinematic features of earth motion (acceleration, frequency content, duration) while belowground structures may require both kinematic and dynamic simulation (i.e., both motions and stresses).

If the dynamic characteristics of the structure are such that maximum response will be achieved at, say, one-fourth the duration of the earthquake, then it may not be necessary to simulate the complete duration. If the structure is not acceleration sensitive but velocity or displacement sensitive, then certain acceleration amplitude features of the prototype earthquake may be compromised while still achieving an adequate simulation. The major point here is that the adequacy of a simulation should be judged by the degree to which the response of the system of interest matches or yields insight into prototype response.

No simulation will fully reproduce every feature of an earthquake and its interaction with an engineering system. At the very minimum, explosive simulation can produce environments equally as complex as earthquake environments and excite structures of prototype size. The most important value of such simulation is that both the excitation and the response can be used to judge the ability of current analytical techniques to predict response. Also, the experimental results provide a data base for developing new or improved techniques. If analytical techniques provide satisfactory prediction and/or understanding of a complex simulated dynamic environment and response, then the technical community will gain improved confidence that the same techniques can be adequately applied to prototype earthquake environments.

In developing simulation criteria, it is necessary to consider the two primary aspects of the prototype problem: (1) the characteristics of the prototype earthquake environment, and (2) the dynamic characteristics of the engineering system. Understanding of the prototype earthquake environment may be the weakest link in the process. Earthquake mechanisms and prediction are still incompletely understood. However, the area is developing and continually being updated. The recent engineering and geophysical literature contains summaries of prediction methods. The following paragraphs concentrate on the second consideration.

The important characteristics of earthquakes in terms of structural response and criteria for a good simulation are not universally agreed upon. Many American engineers and most U.S. design codes focus on peak acceleration. There are some applications where particle velocity is important. Frequency content and duration also play a role. The approach to simulation criteria here focuses on system response.

At one extreme, simulation criteria could require that the simulated environment contain a precise duplication of the prototype waves, and their stress and motion time histories. This would insure precise duplication of structure response for a full size prototype structure.

Such a severe criteria specification would be economically impractical and technically difficult to achieve. In fact, it is inconsistent with the state-of-knowledge of the prototype environment and understanding of the inelastic interactions which occur. A more realistic approach is to consider the type of structure, its dynamic characteristics, anticipated response in the prototype environment and the major uncertainties in the anticipated response. Simulation criteria should then be specified to insure similar response, especially excitation of the structure in such a way that the major uncertainties can be evaluated.

The criteria will probably vary from structure to structure and may include any or all of the following: (1) wave types (P, SV, SH or R), (2) stress-time history associated with the waves, (3) motion-time history at a point or points, (4) some level and type of response in the structure. The wave types and associated stresses might be important for underground or partially buried structures which are loaded directly by the waves. The stress fields associated with various wave types differ and the stress fields may be important for evaluation of the structural integrity of systems loaded by the stresses. It is possible that structure response may not be overly sensitive to the type of wave if the transit time across the structure is short compared to the structure response time. Analytical calculations (ref. 1) indicate that a relatively rigid (compared to soil) structure reflects incoming waves of all types and responds as if it were in a compression-dilatation field regardless of the type of incoming wave.

Another aspect of wave type relates to the propagation velocity of the different waves. Some aspects of the response of systems which are long (i.e., bridges, tunnels, dams) may be affected by the transit time of the wave across the structure and phase differences in the ground motion along the structure length.

For structures in which the stress system is not of major importance, i.e., where the structural strength is sufficient to withstand the incident stresses regardless of their distribution and type, then the specification of a motion time history at a point in the ground may be a sufficient criteria. This would be the case where the ground motion excites structure base motion which, in turn, excites motion and stresses in other parts of the structure not directly loaded by the incident waves (internal or aboveground components). This is a common problem in seismic design and is the one which most design codes treat for above-ground structures. If only motion time history at a point is of interest there is a wide range of wave types or combinations which can be used to produce it.

The least restrictive simulation criteria in terms of defining the waves, stresses, or ground motions is the use of one or more structural response parameters as measures of simulation. Normally, the behavior of various critical members or locations in a structure are of interest. The behavior, which determines whether the structure or some internal component is safe, may be defined by some level of stress, strain, displacement, velocity or acceleration at the location of interest. Relative displacement between an internal component and its attachment points may be of interest in a shock isolation problem. Acceleration may be a major concern in evaluating the behavior of electrical or electronic components. Whatever the parameter of interest, it is the level and type of response which is important rather than the specific details of the excitation causing the response. The response spectrum is a convenient tool for relating the characteristics of an input excitation to the response of a system.

Consider the idealized response spectra shown in figure 1. The spectra designated "prototype" are credible spectra for 1 g and 1/2 g magnitude 8 earthquakes. The specific spectral values are not especially important for illustrative purposes. The spectra designated "explosive ground motion" are credible spectra from hypothetical (but unspecified) ground motion experiments. The fact that the "explosive" spectra are shifted to the right relative to the earthquake spectra is a manifestation of the higher frequency content and higher accelerations of the explosive ground motion.

Suppose that a 2.0 Hz lightly damped system were of interest. Its peak responses to the 1 g and 1/2 g earthquakes are represented by the circled points on the "prototype" spectra. The spectrum for "explosive ground motion No. 1" passes nearly through the same point and, therefore, the peak response of the system would be about the same as in the 1 g earthquake in spite of the higher frequency content and higher acceleration of the high explosive experiment. The response of the system to "ground motion No. 2" would be about 65 percent that of the 1 g earthquake. Depending upon the objective of the experiment, these results might also be satisfactory for interpreting system response.

If the system of interest has multiple frequencies of importance to the overall response, then matching the prototype spectrum at a single point will probably not be satisfactory. The wider the range over which the system frequencies extend, the wider the range over which the "ground motion" spectrum would be required to approximate the prototype spectrum. This same comment applies if there is an uncertainty range in the fundamental frequency for a single degree of freedom system.

The use of models changes the required simulation requirements. Dimensional analysis and scaling must be introduced into the simulation design. For pure geometric scaling, accelerations must be multiplied by the scale factor and displacements divided by the scale factor while velocities remain unchanged.

In summary, simulation criteria should be specified in relation to the system of interest. If belowground stresses are important then simulations may be required to reproduce specific earthquake waves and their associated stress fields. If only the ground motion is important then there is flexibility in the methods that can be used to create the ground motion. In many instances, it appears that a good method for evaluating (and, inversely, designing) the ground motion is through the use of response spectra. If a ground motion produces system response at levels similar to those produced by an earthquake, then the specific details of the motion are important only to the extent that they influence the response spectrum. If modeling is used, appropriate scaling of both the system and the input environment is required.

EXPLOSIVE GROUND MOTION PHENOMENOLOGY

The basic high explosive charge configurations (figure 2) are spherical, cylindrical and planar. The selection of a charge configuration and appropriate enhancement technique will be dependent upon the simulation criteria for a particular engineering system of interest. Parameters which must be considered include motion amplitudes, frequency content and duration, perhaps as they influence response spectra, as well as the physical dimensions of the structure or desired test area.

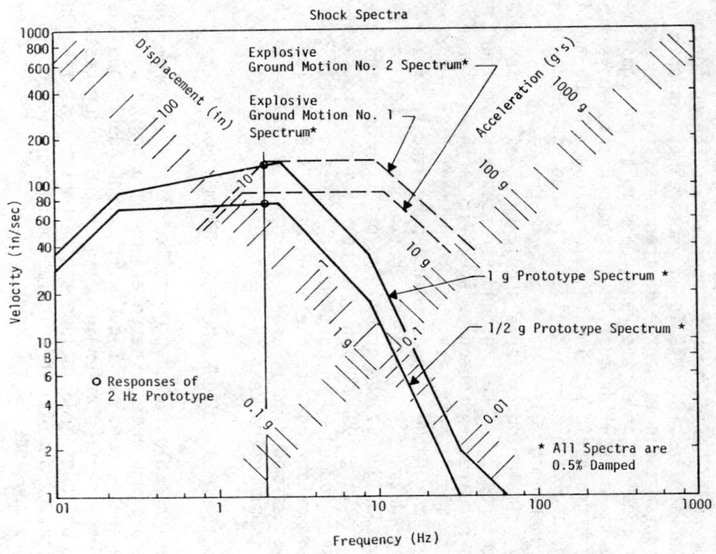

Figure 1. Illustration of the Use of Response Spectra for Developing Simulation Criteria

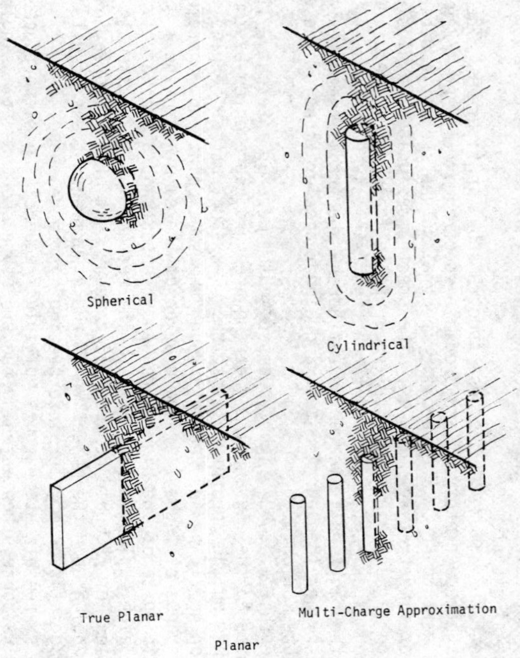

Figure 2. Three Classes of Explosive Charge Configuration

The planar array configuration was given the most attention for simulation applications because it provides relatively uniform motions across the width of the array and, in addition, allows a wider range of control over motion amplitudes and attenuation rates then is possible with the other charge configurations. Reference 2 provides an analysis of spherical, cylindrical and planar explosive data available prior to 1977, as well as analytical calculations to support the development of prediction relationships. The planar relationships have subsequently been updated based upon more recent data and further analysis. The updated results are provided here. First, it is important to examine the nature of the ground motions which propagate from a planar array.

It can be shown by dimensional analysis that the appropriate charge parameter for scaling ground motions from a planar array is the areal charge density of explosives in the array. For an infinite plane of explosives, the ground motions will be uniaxial. In any field problem, however, the array will always have finite dimensions. The finite dimensions will produce relief waves which originate at the edges of the array at the instant of detonation and propagate into the loaded region. Planar arrays are typically designed so that their length is longer than their height (usually 2 to 3 times). Hence, the major relief occurs from the top and bottom of the array. The loading waves and the relief or unloading waves from the top and bottom of a planar array believed to occur in elastic media are shown in figure 3. The major loading wave is a compression wave (P-wave) which causes motion away from the explosion and moves away from the array with velocity C_p. Along the length of the explosion it has a planar front in cross-section. At the instant of detonation, a compression wave and a shear wave are initiated at the ends of the explosion. In addition, a von Schmidt-like wave (a shear wave connected between the P-wave and S-wave fronts) is also expected to be initiated at the ends. In the region above and below the explosion these waves are loading waves, i.e., they are moving into unloaded material. In the region loaded by the incident P-wave, however, these waves are relief waves. Figure 3 shows that the direction of particle motion associated with the unloading waves tends to bring the material back to its initial condition, i.e., the major direction of motion is inward.

In inelastic soil media, the wave pattern will distort somewhat from that of figure 3, but the major loading and unloading waves will still exist. The P-relief wave travels faster than the loading wave. It will overtake and begin to erode the peak stress or velocity beyond some range. Using ray tracing and geometry, the range at which the P-relief wave overtakes the front of the P-loading wave can be estimated from wave speeds. Beyond this range, the attenuation of the peak stress, velocity and acceleration will approach cylindrical attenuation. The rate will probably not change instantaneously, but rather will begin to transition to a cylindrical condition.

Along the centerline, perpendicular to the array, attenuation of peak motion parameters is expected to occur as shown in figure 4. At ranges near the array, the relief effects do not affect peak parameters and the attenuation is planar and due only to inelastic effects. Beyond some range, relief waves from the array top and bottom begin to overtake the loading waves and cause the attenuation rates to approach cylindrical conditions. Finally, at some greater range, relief waves from the ends of the array overtake the loading waves and a transition to spherical attenuation might be expected.

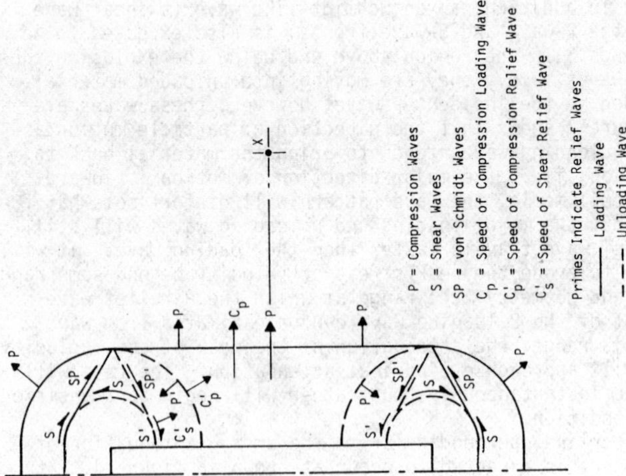

Figure 3. Major Wave Fronts From a Planar Explosion in Elastic Media

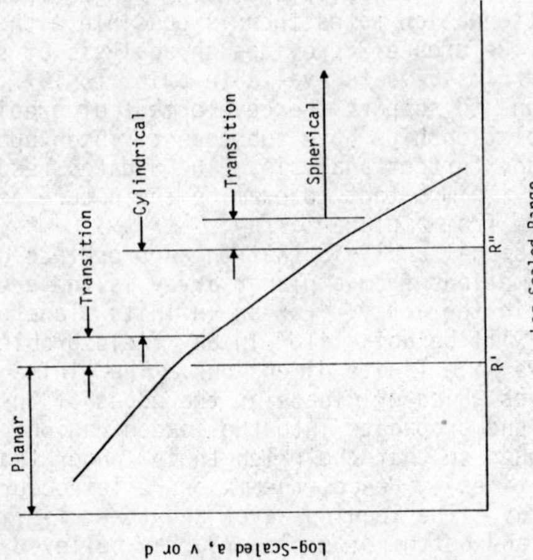

Figure 4. Effect of Finite Array Dimensions on Peak Ground Motion Attenuation

Most engineering systems for which ground motion simulation is required are located at or near the ground surface. Explosive arrays, due to the need for enhanced coupling, will be buried. The loading phenomena in the near-surface region is, therefore, very important. The loading waves above the explosive, upon encountering the free-surface, will reflect and cause enhancement in amplitudes near the surface compared with amplitudes near the center of the array, especially for vertical motions.

GROUND MOTION PREDICTION

Peak horizontal motion data at the array mid-depth for several planar events at McCormick Ranch, NM are given in figures 5, 6 and 7. Details on the specific events may be obtained from reference 2. The data clearly exhibit the expected transition from planar to cylindrical behavior. However, transition to spherical behavior is not evident. This may be due to reduced attenuation as the motions approach elastic levels compensating for the increased attenuations due to array end relief.

The range at which motions transition from planar to cylindrical attenuation is different for accelerations, velocities and displacements. Accelerations transition first (R/H = 0.52), velocities later (R/H = 1.15) and displacements last (R/H = 1.6). The data of figures 5, 6 and 7 are fit reasonably well by the following equations which can provide the basis for the design of planar simulation experiments in dry alluvial materials.

$$a\alpha = 1240\ (R/\alpha)^{-1.33} \qquad \text{for } R/H < 0.52 \qquad (1)$$

$$a\alpha = 503\ (H/\alpha)^{1.38}\ (R/\alpha)^{-2.71} \qquad \text{for } R/H > 0.52 \qquad (2)$$

$$v = 5.28\ (R/\alpha)^{-0.56} \qquad \text{for } R/H < 1.15 \qquad (3)$$

$$v = 6.54\ (H/\alpha)^{1.54}\ (R/\alpha)^{-2.1} \qquad \text{for } R/H > 1.15 \qquad (4)$$

$$d/\alpha = 0.0163\ (R/\alpha)^{-0.14} \qquad \text{for } R/H < 1.6 \qquad (5)$$

$$d/\alpha = 0.0414\ (H/\alpha)^{1.96}\ (R/\alpha)^{-2.1} \qquad \text{for } R/H > 1.6 \qquad (6)$$

where a = acceleration in g's; v = velocity in m/sec; d = displacement in m; α = areal charge density in kg/m^2 of equivalent TNT; H = array height in m. Reference 2 provides guidelines and procedures for predicting velocity-time histories.

In the near-surface region, outward motions are enhanced compared with center-line motions. Finite difference wave propagation calculations were performed to evaluate free-surface effects (ref. 2). Ratios of free surface amplitudes to centerline amplitudes, derived from the calculations, are given in figure 8.

Table 1 gives predicted horizontal motions and associated frequencies and durations at the 1 g acceleration level from a range of practical array heights and loading densities. It appears possible to achieve motions and frequency content which are within the ranges expected in strong earthquakes. Durations and number of motion cycles are limited for single arrays but these may be increased by using multiple, sequenced arrays.

Vertical motions are treated in some detail in reference 2. In general, vertical accelerations in the near-surface region can equal or slightly exceed the horizontal acceleration. This is because the initial

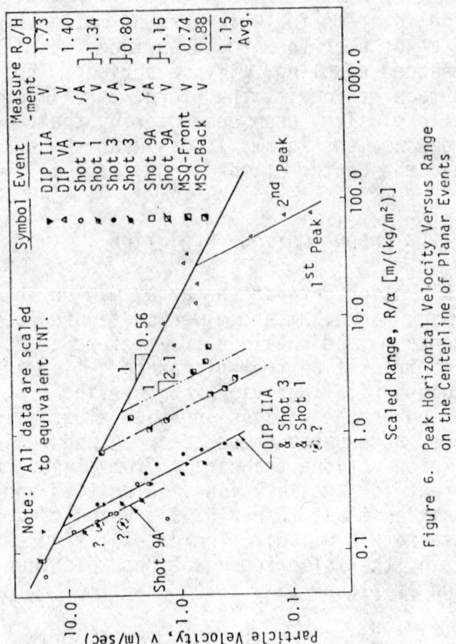

Figure 6. Peak Horizontal Velocity Versus Range on the Centerline of Planar Events

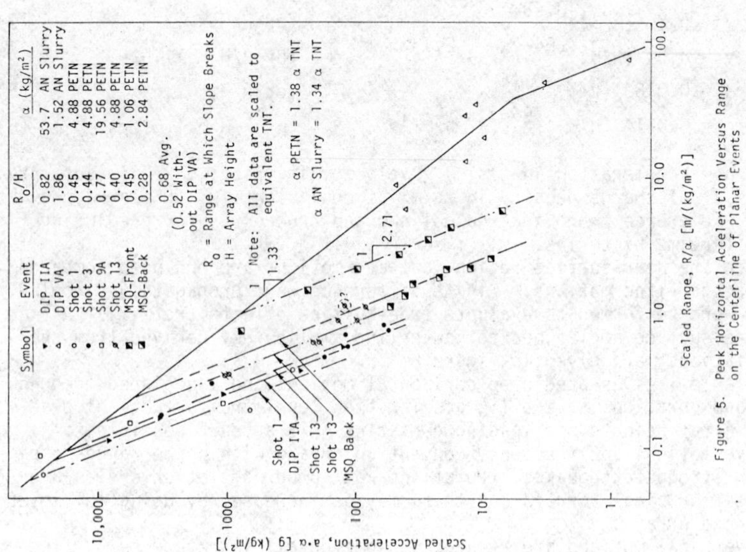

Figure 5. Peak Horizontal Acceleration Versus Range on the Centerline of Planar Events

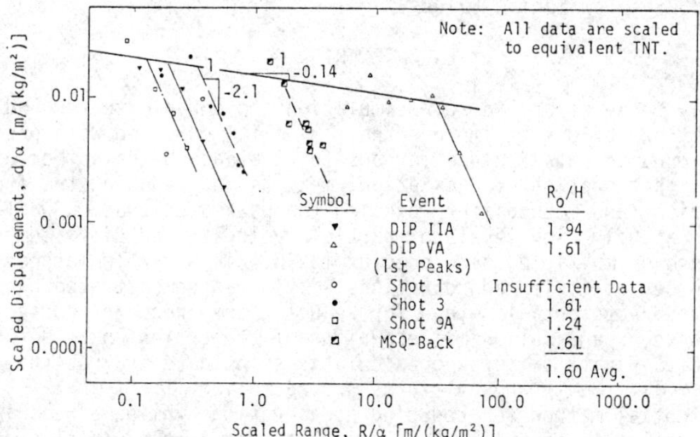

Figure 7. Peak Displacement Versus Range on the Centerline of Planar Events

TABLE 1

Estimated Motions at the 1 g Acceleration Level at the Ground Surface from Various Arrays

Array Height, ft (m)	Areal Charge Density, lb/ft² TNT (kg/m² TNT)	Range, ft (m)	Velocity, in/s (m/s)	Displacement, in (m)	Frequency, Hz	Duration s
25 (7.6)	2 (9.8)	121 (37)	12.2 (0.31)	0.8 (0.019)	7.1	0.12
	4 (19.6)	131 (40)	15.0 (0.38)	1.4 (0.035)	4.8	0.16
	8 (39.1)	141 (43)	17.7 (0.45)	2.5 (0.063)	3.0	0.24
50 (15.2)	2 (9.8)	171 (52)	20.5 (0.52)	1.4 (0.035)	3.7	0.16
	4 (19.6)	187 (57)	23.2 (0.59)	2.6 (0.065)	3.2	0.24
	8 (39.1)	203 (62)	27.6 (0.70)	4.6 (0.117)	2.2	0.32
75 (22.9)	2 (9.8)	210 (64)	33.1 (0.84)	2.2 (0.055)	4.4	0.20
	3.26 (15.9)	223 (68)	34.2 (0.87)	3.2 (0.081)	2.9	0.26
	4 (19.6)	230 (70)	35.0 (0.89)	3.8 (0.097)	2.4	0.28
	5.33 (26.1)	236 (72)	37.0 (0.94)	4.8 (0.122)	2.1	0.33
	8 (39.1)	249 (76)	37.8 (0.96)	6.6 (0.168)	1.8	0.41

wave is a P-wave and causes an upward Poisson's expansion at the free surface. The vertical velocity and displacement can be taken roughly as 2/3 and 1/3 the horizontal values.

ENHANCEMENTS

It may be necessary in some simulations to enhance the ground motions produced by explosions to improve their fidelity compared with earthquake motions. Enhancements include methods for increasing the number of motion cycles and the duration of excitation; methods for reducing the frequency of the motion; and methods for reducing the peak acceleration while maintaining relatively high levels of particle velocity and displacement.

The use of multiple, sequenced explosions is a viable method for enhancing number of cycles and durations. This was demonstrated by recent experiments. A potential method for reducing the frequency content of the explosive ground motion is decoupling of the explosion by detonating in cavities. Finite difference calculations indicate that as the cavity size increases, the frequency of the ground motion decreases.

A potential method for reducing accelerations while maintaining relatively high levels of velocity and displacement is the use of trench shields. Finite difference calculations suggest that accelerations can be reduced substantially with relatively modest trench depths. Velocities are also decreased, but not as much. Peak displacements do not change very much. Reference 2 provides more detail on these and other approaches.

SOME RECENT SIMULATION RESULTS

Early versions of the ground motion correlations were used to design the SIMQUAKE series of experiments (refs. 3 and 4). The series consisted of four events; Mini-SIMQUAKE, SIMQUAKE IA, SIMQUAKE IB and SIMQUAKE II. These were the first explosive experiments designed expressly for earthquake simulation. Mini-SIMQUAKE was a small feasibility experiment. The largest experiment was SIMQUAKE II, a double array test. The experiment involved two explosive arrays with approximately 30 tons and 40 tons of ANFO, respectively, in the front and back arrays. The back array was fired first and the front array was fired 1.2 seconds later. SIMQUAKE I was to be a sequenced experiment but only the back array fired due to a firing system malfunction. This experiment was designated SIMQUAKE IA. The front array was fired a few weeks later and designated SIMQUAKE IB. The arrays had the same design as SIMQUAKE II.

SIMQUAKE I and II horizontal accelerations, velocities and displacements in the near-surface region are compared with predictions in figures 9, 10 and 11. In general, the predictions compare reasonably well with the measurements. The major exceptions are the acceleration predictions for the front array which are low. The SIMQUAKE data are now being folded into the data base to enable improved predictions.

Figure 12 shows the horizontal accelerations and their integrations at the 107m range on SIMQUAKE II. The major frequency content of the motions is in the range 0.7 to 2 Hz and strong motion persists for about three seconds. Detailed information on the frequency content of both the vertical and horizontal motions in the experiments is exhibited directly in the response spectra shown in figure 13. These spectra are for measurements at the 61-m range.

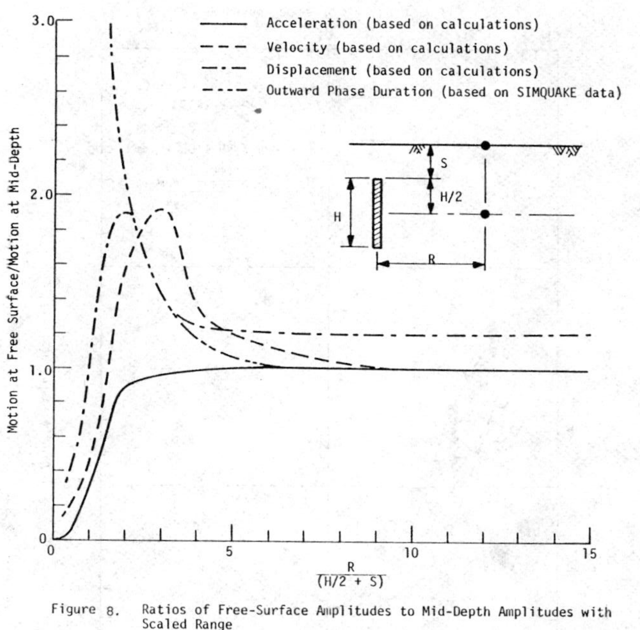

Figure 8. Ratios of Free-Surface Amplitudes to Mid-Depth Amplitudes with Scaled Range

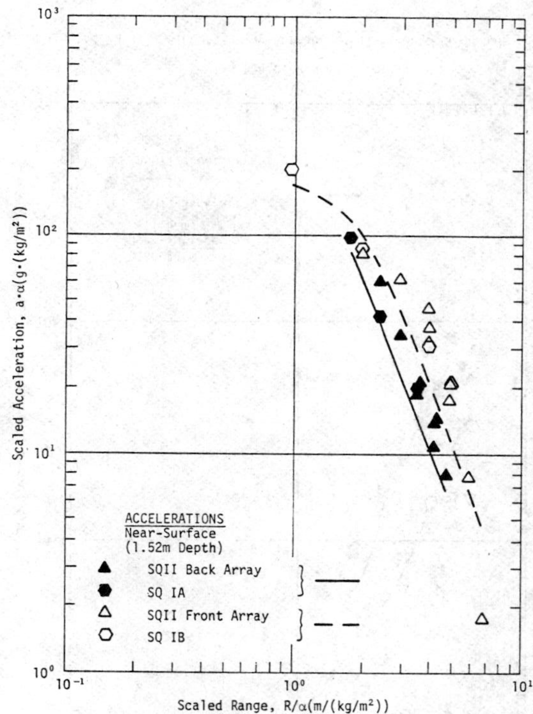

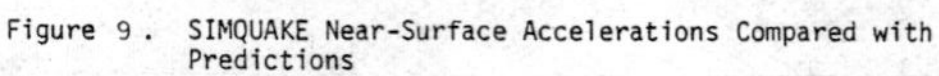
Figure 9. SIMQUAKE Near-Surface Accelerations Compared with Predictions

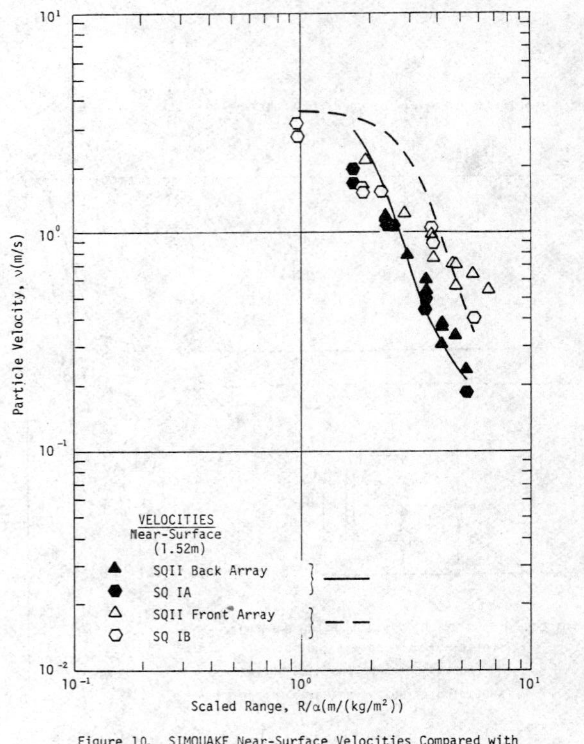

Figure 10. SIMQUAKE Near-Surface Velocities Compared with Predictions

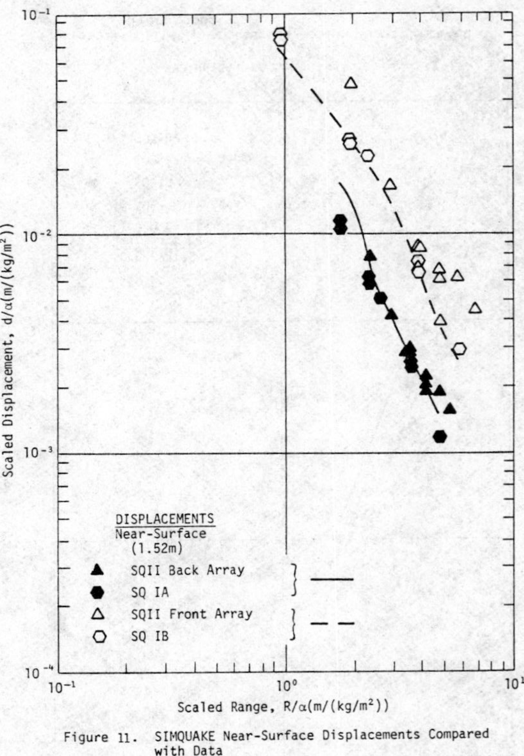

Figure 11. SIMQUAKE Near-Surface Displacements Compared with Data

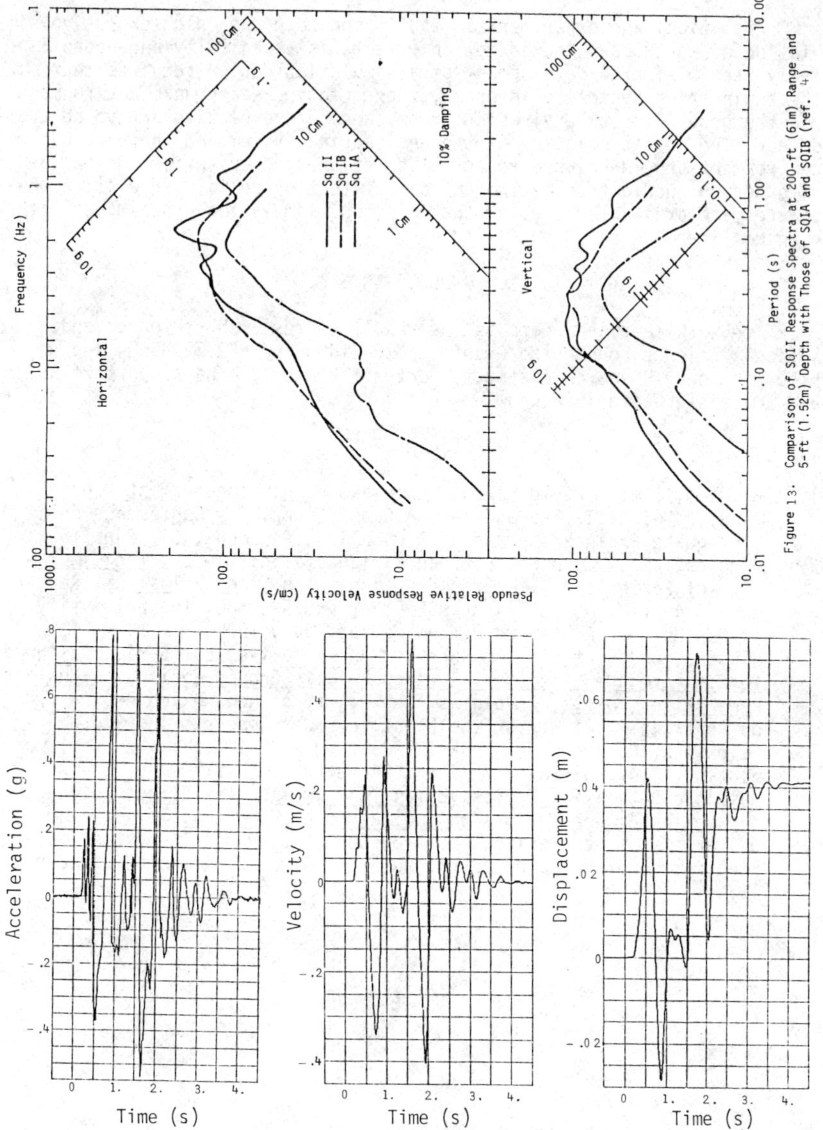

Figure 13. Comparison of SQII Response Spectra at 200-ft (61m) Range and 5-ft (1.52m) Depth with Those of SQIA and SQIB (ref. 4)

Figure 12. Horizontal Motions at 350 ft (106.7m) Range and 5-ft (1.52m) Depth on SIMQUAKE II (ref. 4)

CONCLUSIONS

Analytical and experimental results indicate that high explosive simulation of earthquake ground motion effects is technically and economically feasible. The development of a simulation design requires that the engineering system of interest, the prototype earthquake environment and the ground motion environment from candidate explosive arrays be evaluated. The characteristics of the engineering system and the earthquake environment should be used to establish simulation criteria. An explosive design should then be chosen to meet that criteria. Prediction procedures are available to allow the design of simulation experiments with high confidence.

ACKNOWLEDGEMENTS

The work described here is the result of research performed under National Science Foundation Grants ENG75-21580 and PFR-7923500, and Electric Power Research Institute Contract RP810-1. The support of NSF and EPRI is gratefully acknowledged.

REFERENCES

1. Higgins, C.J., "Field Measurements and Comparison Against Predictions," Vol. III, Proceedings, ASCE GED Specialty Conference on Earthquake Engineering and Soil Dynamics, Pasadena, CA, June 1978.
2. Higgins, C.J., Johnson, R.L. and Triandafilidis, G.E., The Simulation of Earthquake-like Ground Motions with High Explosives, Report No. CE-45 (78) NSF-507-1 on NSF Grant ENG75-21580, The University of New Mexico, Albuquerque, New Mexico, July 1978.
3. Higgins, C.J., et al, SIMQUAKE I - An Explosive Test Series Designed to Simulate the Effects of Earthquake-Like Ground Motions on Nuclear Power Plant Models, 7 Volumes, University of New Mexico Report on Contract RP 810-1 for the Electric Power Research Institute, Palo Alto, California, January 1979.
4. Higgins, C.J., Simmons, K.B. and Pickett, S.F., SIMQUAKE II - A Multiple Detonation Explosive Test to Simulate the Effects of Earthquake-like Ground Motions on Nuclear Power Plant Models, 8 Volumes, University of New Mexico Report on Contract RP 810-1 for the Electric Power Research Institute, Palo Alto, California, January 1979.

MODAL PARAMETER ESTIMATION FOR A HIGHRISE BUILDING USING AMBIENT RESPONSE DATA TAKEN DURING CONSTRUCTION

M. Meyyappa*
H. Palsson**
J. I. Craig***

ABSTRACT

Ambient response measurements were made on a highrise building during the construction phase. A series of measurements were made at different stages of construction so that any changes in the modal parameters with construction could be detected. A least-squares curve-fitting method is used to obtain these parameters from the output response spectrum. The results obtained are discussed and an attempt is made to determine the effects of cladding on frequencies and damping of different modes.

INTRODUCTION

Experimental determination of the modal parameters is of primary interest in the identification of dynamic structural systems. A structure is described by equations of motion involving mass, stiffness and damping matrices and its dynamic behavior is generally better understood in terms of these parameters. In many cases, some form of vibration testing is employed to obtain these parameters. Numerous discussions on various testing methods can be found, for example, in Ref. 1. For complex structures like buildings, forced vibration testing becomes a difficult task, even more so if the building to be tested is under construction. In such cases, a more practical approach is to use response data due to ambient excitation in the analysis.

This paper presents the results of a study carried out to determine the frequencies, damping factors and mode shapes of the first few modes of a highrise building using measured ambient data. Measurements were made over a period of several months while construction was in progress in order to trace the changes in the values of the parameters with construction. In particular, the measuring process began after erection of the steel structure when the installation of cladding had just started and continued at periodic intervals until the cladding was completely in place.

The most serious shortcoming in using ambient excitation is the fact that neither the spatial nor the temporal distribution of the input can be characterized. This eliminates the use of those methods that require explicit

*Graduate Research Assistant, Aerospace Engineering
**Graduate Research Assistant, Civil Engineering
***Associate Professor of Aerospace Engineering
Georgia Institute of Technology, Atlanta, Georgia

knowledge of the input in the parameter identification process. The most commonly used method utilizes the response power spectral density or PSD (2-7). Under the assumption that the input PSD is slowly-varying in the vicinity of each natural frequency, the output PSD can be taken as an approximation (to within a multiplicative constant) of the square of the magnitude of the frequency response function. The frequencies corresponding to the peaks in the output PSD are identified as the natural frequencies and the damping is usually determined by the half power bandwith method. Other methods of response analysis with random excitation input include the method of spectral moments (8, 9), random decrement method (10, 11), parametric curve-fitting (12) and auto-regressive moving average (ARMA) models (13, 14).

The method used here is basically parametric curve-fitting, but interaction between different modes is taken into account by fitting the magnitude of the frequency response function of a multi degree-of-freedom system to the response Fourier amplidute spectrum (or the square root of the output PSD) using the least squares criterion. The major emphasis in this paper is on the use of ambient response data in the identification of dynamic structural systems. In contrast to previous studies, the present work has succeeded in identifying up to the fifth bending mode by this manner, and it appears that adequate modal definition can be obtained for even higher modes, including torsion. The estimates of the modal parameters obtained using this method are given and the results are discussed following a description of the building and the measurements procedure.

STRUCTURE

The building selected for study is a 24-story steel frame office tower, 350 feet high with a lightweight partially opaque reflective glass facade (Fig. 1). The south wall of the building is inclined to increase shading of the face and reduce energy costs. The sloped face is obtained by incrementally increasing the width of floors 3 to 23 by 15 inches so that an overhang is formed at each level. The end bay on the south wall is therefore increased from 16.25 feet at the base to 43.5 feet at the roof. The increase in floor dimensions with height causes the center of gravity to be slightly eccentric, but the resulting overturning effect was accounted for in the design.

The building rests on 68 caissons with an average depth of about 65 feet and the caisson diameters range from 2.5 to 6 feet. Caissons which are supporting columns in potential uplift are drilled 8 to 10 feet into rock and are heavily reinforced. Grade beams are used to tie the caissons together around the tower perimeter.

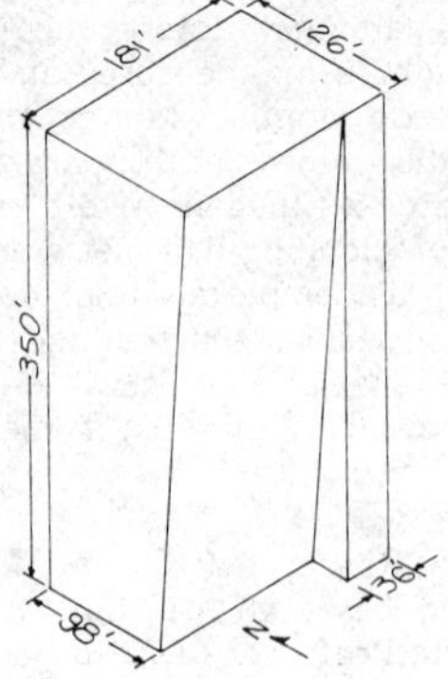

Fig. 1. Building Geometry

Steel stub-girder framing is used in the building floors in order to reduce the floor-to-floor height of each story. The slab is raised above the floor girders on short rolled stubs resting on and welded to the top flanges of the girders. Ductwork and other utilities pass between the stubs on top of the girders instead of below them. Shear connectors welded to the top of the stub girders make the slab composite with the steel structure.

MEASUREMENT

The ongoing construction activities precluded having any permanent fixtures and necessitated transportation of the measuring equipment to the building site every time a set of measurements was taken. Five low level force-balance accelerometers (Kinemetrics FBA-1) were used to measure the response simultaneously at different locations. The accelerometer signals, after amplification and filtering, were recorded on magnetic tape using an HP-3968 8 channel tape recorder (see Fig. 2).

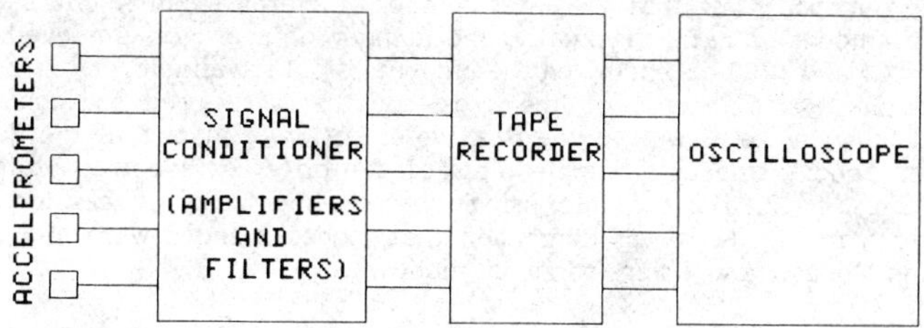

Fig. 2. Ambient Response Measurement Instrumentation

Eight sets of measurements were obtained in all, each of them lasting up to six hours and each made during the course of a particular day. Table 1 shows the level of cladding, i.e., the floor up to which cladding had been installed, on each of these days. On the first occasion (January 16, 1980) three accelerometers were used on the 22nd floor to identify the bending and torsional frequencies (Fig. 3). Accelerometer responses from locations 1 and 2 were used to detect the

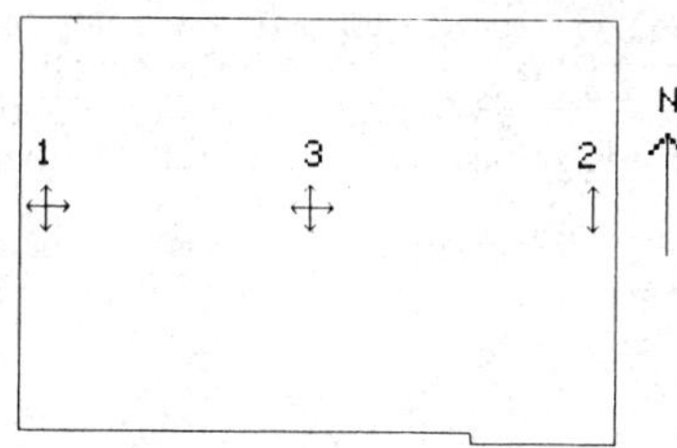

Fig. 3. Typical Accelerometer Layout for Bending and Torsion Response

TABLE 1- CLADDING LEVELS

DATE	FACES			
	EAST	WEST	NORTH	SOUTH
1/16	4	2	4	0
2/8	10	6	10	0
2/15	12	10	12	7
2/22	13	11	12	9
3/7	16	12	12	12
4/24	24	19	19	20
5/16	24	23	24	23
5/29	24	24	24	24

* AVERAGE HEIGHT OF CLADDING

torsional and N-S bending modes and the response at location 3 was used to detect the E-W bending modes. On all the other days except March 17, five accelerometers were placed on the 8th, 13th, 18th, 22nd and 24th floors. On one of these days (April 24), the 24th floor was inaccessible and the 20th floor was chosen instead. On the first three occasions the accelerometers were situated at location 3, twice in the N-S direction and once in the E-W direction and for the remaining three they were situated near the west face at location 1. It was decided to obtain data for all the modes on the last two days, so the accelerometers were rotated 90° midway through the measurement process, providing about 3 hours of data in each direction.

Data reduction was done using an HP 5420A Digital Signal Analyzer. The autospectrum for each output along with the cross-spectra between different outputs were obtained and stored for later use in modal parameter estimation. To reduce the data reduction time, the magnetic tape was played back 16 times faster, resulting in an amplification factor of 16 for the frequency as used in the analyzer. The analyzer uses ensemble averaging with overlapped processing to calculate the spectra so that a greater number of averages can be obtained using the same amount of data. Typically, the number of averages employed ranged between 400 and 1400 depending on the amount of data available.

The phase components of the cross-spectra were used to identify the torsional frequencies as well as the relative direction of motion of the different floors for bending modes. Sample plots of the different spectra obtained are given in Figs. 4 - 7. For example, using the cross-spectrum between locations 1 and 2 given in Fig. 6, it was concluded that modes 1 and 3 were N-S bending modes and Modes 2 and 4 were torsional modes.

ANALYSIS

For a linear, time-invariant system, the output PSD, $G_y(f)$, is related to the input PSD, $G_x(f)$, by

$$G_y(f) = |H(f)|^2 \, G_x(f)$$

where H(f) is the frequency response function of the system. For structural systems with low damping ($\rho < 0.1$) and a broadband input, $|H(f)|^2$ can be approximated by $G_y(f)$ (to within a multiplicative constant) provided $G_x(f)$ is devoid of any sharp variations around each natural frequency.

Assuming that the above conditions are satisfied, the modal parameter estimates are obtained by nonlinear least-squares curve fitting of the analytical form of $|H(f)|$ to $|Y(f)|$ (or $G_y(f)^{1/2}$) where Y(f) is the output Fourier spectrum. An interactive procedure is used starting with an initial guess for all the parameters to be estimated and adjusting these values for an acceptable fit with the data.

The frequency response function of an N degree-of-freedom system for an acceleration output measured at point i due to a force input-at-point j is given by

$$H(f) = \sum_{r=1}^{N} \frac{\phi_i^r \phi_j^r}{m_r} \frac{(f/f_r)^2}{1-(f/f_r)^2 + 2j\rho_r f/f_r} \qquad (1)$$

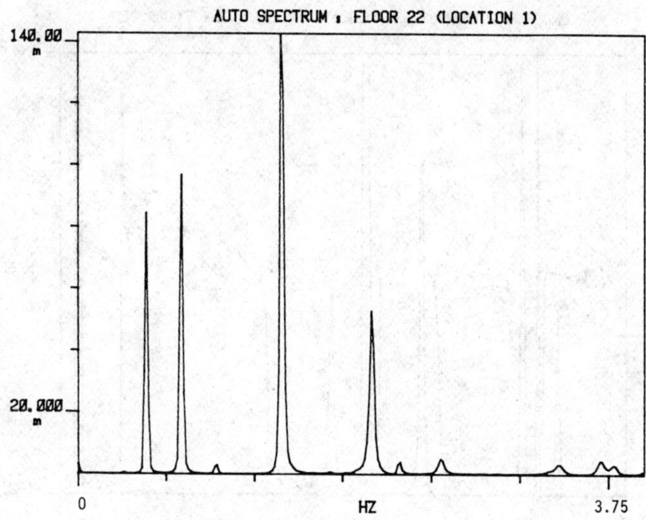

Fig. 4. N-S Response Spectrum at Location 1

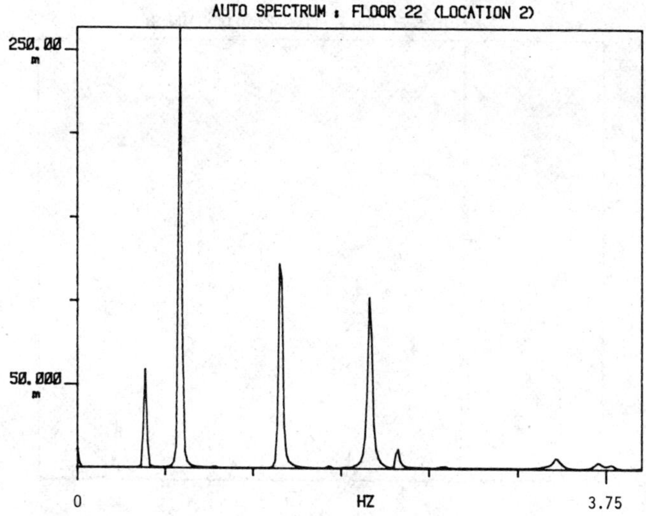

Fig. 5. N-S Response Spectrum at Location 2

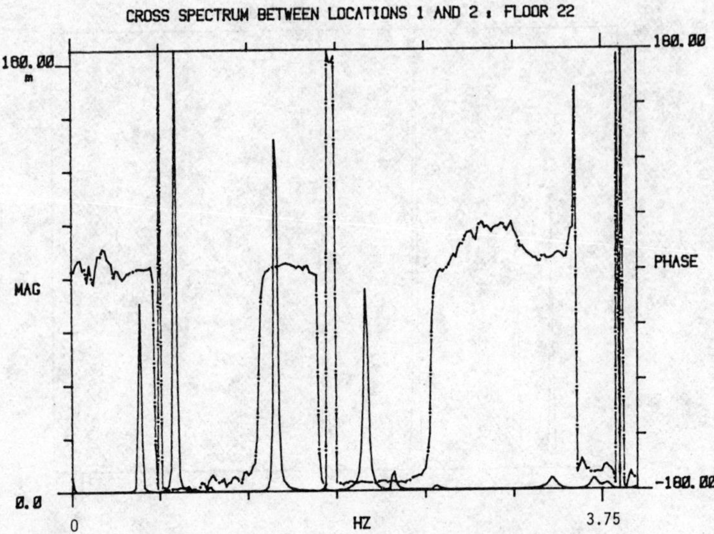

Fig. 6. Cross Power Spectrum Between Locations 1 and 2

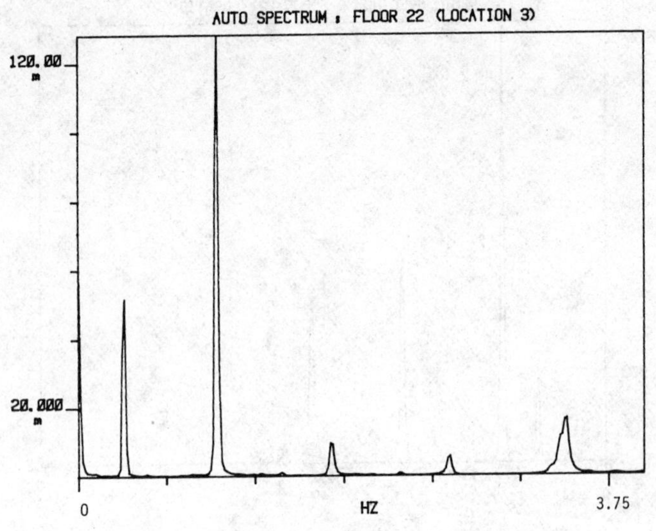

Fig. 7. E-W Response Spectrum at Location 3

where f_r = undamped natural frequency of mode
ρ_r = damping coefficient of mode
ϕ_i^r = r^{th} mode shape coefficient at the response point
ϕ_j^r = r^{th} mode shape coefficient at the input point
m_r = r^{th} modal mass

Defining the following quantities,

$$C_r = \frac{\phi_i^r \phi_j^r}{m_r} \qquad N_r = C_r (f/f_r)^2 \qquad \alpha_r = 1-(f/f_r)^2$$

$$\beta_r = \rho_r (f/f_r) \qquad D_r = \left[1-(f/f_r)^2\right]^2 + 4\rho_r^2 (f/f_r)^2$$

$|H(f)|$ can be written as

$$|H(f)| = \left[\sum_{i=1}^{N} \frac{N_i^2}{D_i} + \sum_{i=1}^{N-1} \sum_{j=i+1}^{N} \frac{2 C_i C_j (\alpha_i \alpha_j + 4\beta_i \beta_j)}{D_i D_j}\right]^{1/2}$$

The quantity C_r when measured at different output locations gives a direct estimate of the rth mode shape.

The error function to be minimized is

$$\psi = \sum_{K=1}^{P} \left| Y_K - |H(f)|_K \right|^2$$

where p is the number of points used in estimation and Y_k is the observed value of Y(f) at the kth point. The new estimates after (m-1) iterations are obtained as

$$\beta^{(m)} = \beta^{(m-1)} + \delta^{(m-1)}$$

where β is the parameter vector $\beta = \{f_1, \rho_1, C_1 \ldots f_N, \rho_N, C_N\}$ and δ is the increment vector.

The values for δ are determined using the Gauss-Newton Method with Levenberg-Marquardt modification (15, 16), i.e. δ is obtained by solving the following set of equations:

$$\left(A^{*(m-1)} + \lambda^{(m-1)} I\right) \delta^{*(m-1)} = g^{*(m-1)} \qquad (2)$$

where

$A^* = a_{ij}/(a_{ii} a_{ij})^{1/2}$

$g^* = g_j/(a_{jj})^{1/2}$

$A = J^T J$

$g = J^T Z$

λ = constant

and where J is the p x3N Jacobian matrix of partial derivatives of $|H(f)|$ with respect to the parameters to be estimated, evaluated at the p points and Z is the column vector of errors or residuals at these p points.

For sufficiently large values of λ, $(A^*+\lambda I)$ is positive-definite even for extremely ill-conditioned A^*, so that a solution always exists for Eqn. 2. Then, δ is obtained from δ^* as

$$\{\delta_j\} = \{\delta_j^* / (a_{jj})^{1/2}\}$$

Finally, λ is chosen according to the procedure given by Marquardt (16) with the modification suggested by Bard (17).

Before applying the above algorithm to the parameter estimation problem, the number of modes to be fitted (N), the specific points to be used around each frequency (and thus the total number of points, p), and the initial estimates for all the parameters f_i, ρ_i, C_i, i = 1, ...N must be determined. For the problem at hand, the peak frequencies obtained from the response spectrum were used as initial estimates for f_i, i = 1, ... N, and ρ_i for all i was initially 0.01. Estimates for all C_i were set equal to a fixed quantity that depended on the maximum amplitude of the spectrum. It should be noted that C_i can be either positive or negative. In most cases, the algorithm converges to a positive value of C_i if the initial estimate is positive and vice-versa. But a wrong choice of sign for one of the C_i results in a poorer fit. To determine the signs of the constants C_i in any given response spectrum, the following procedure was adopted. The dominant mode in the spectrum was taken with each of the other modes in turn and a two degree-of-freedom fit was carried out, first with the same signs for C_i of both the modes and then with opposite signs. The sign combination that gave a better fit was then tentatively chosen as the correct one.

In this way, the signs for the C_i of all the other modes with respect to that of the dominant mode were determined. Then a multi degree-of-freedom fit taking all the modes of interest into account was performed. If the fit was found to be poor, then the sign of one or more C_i was changed until a better fit was obtained. This procedure need be adopted for only one of the outputs since the phase component of the cross spectrum between this output and the others can be used to determine the signs for the C_i in the other outputs.

RESULTS

The results obtained by using the estimation algorithm described above with the ambient data from the building are discussed next. Fig. 8 shows a typical auto-spectrum and its curve-fit. It can be seen that the fit is very good around all the frequencies considered. Usually about 10 iterations were required to achieve the desired convergence, and the parameters were assumed to have converged if the relative changes in the estimates were less than 0.01%.

The average frequency estimates for the first three bending modes in both the directions and the first two torsional modes are given in Table 2 which also includes the different dates on which the corresponding measurements were made. Table 3 lists the average damping estimates for these modes. Estimates of the mode shapes were also obtained, but they will not be discussed here. Attention was directed in the present case at the first few modes only.

From Table 2 it is seen that except for the fundamental modes in torsion and N-S bending, the frequencies decrease initially and then show an increasing trend. For the fundamental modes, a decreasing trend is observed throughout. The initial decrease in the frequencies is largely due to the mass of cladding and other elements that were moved into the building and stored for later use in construction. During this period, the cladding levels increased by only a few

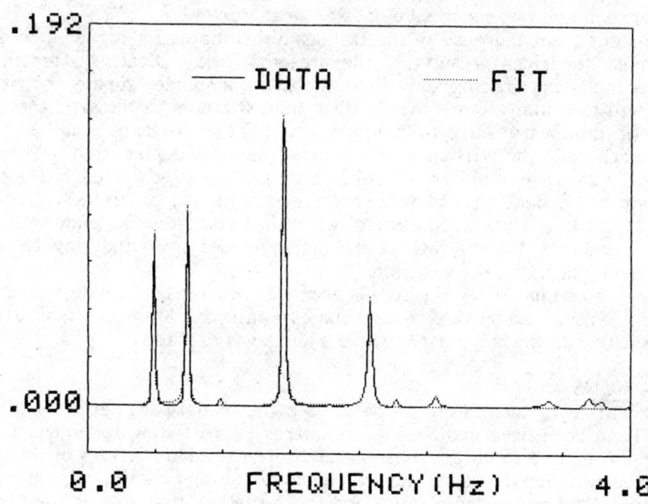

Fig. 8. Four Degree of Freedom Fit of the Spectrum Shown in Fig. 4

TABLE 2 - NATURAL FREQUENCIES

DIRECTION	MODE	TEST DATE 1/16	2/8	2/15	2/22	3/7	4/24	5/16	5/29
E-W	1	0.321			0.320	0.306		0.317	0.319
	2	0.977			0.955	0.908		0.962	0.980
	3	1.785			1.789	1.707		1.798	1.823
N-S	1	0.479	0.475	0.469		0.458	0.446	0.444	0.444
	2	1.444	1.424	1.412		1.367	1.365	1.368	1.375
	3		2.587	2.581		2.473	2.511	2.526	2.545
TORSION	1	0.732				0.702	0.694	0.692	0.692
	2	2.083				1.960	2.008	2.024	2.030

TABLE 3 - DAMPING FACTORS (% OF CRITICAL DAMPING)

DIRECTION	MODE	TEST DATE 1/16	2/8	2/15	2/22	3/7	4/24	5/16	5/29
E-W	1	0.44			0.56	0.53		3.15	2.32
	2	0.57			0.97	1.04		1.72	1.78
	3	1.03			0.87	0.62		1.76	1.49
N-S	1	0.91	0.61	1.92		1.02	0.88	0.82	0.71
	2	0.47	0.67	0.62		0.62	1.04	1.54	1.32
	3		0.66	0.92		1.10	1.56	1.30	1.23
TORSION	1	1.15				1.32	1.33	1.56	1.56
	2	0.71				0.79	1.11	1.56	1.47

floors between any two successive days of measurement.

The subsequent increase in the frequency estimates for higher modes is one of the most significant results of the present work. Clearly, after the main structure including bracing and floor slabs, is in place, further construction activity will tend more to add mass rather than stiffness to the structure, with a consequent downward shift in frequencies. The increasing trend suggests a general increase in the stiffness and the most plausible explanation for this is the continuing installation of the cladding previously stored on each floor. The fundamental N-S bending and torsional frequencies remain relatively constant over this period. This is consistent with the expected tendency of cladding stiffness to more noticeably affect the higher modes since they involve greater curvature and localized deformation.

Damping estimates for the torsional modes and the 2nd bending mode in the E-W direction show an upward trend. The estimates for all modes tend to be high on the last few days when most of the cladding was in place.

CONCLUSIONS

Ambient tests carried out to determine the modal parameters of a highrise building have been described. A least-squares curve-fitting technique has been successfully applied to obtain these parameters for all the modes of interest. It has been shown that by using several hours of ambient data, it is possible to obtain very good estimates of auto and cross-spectra that can be used in modal parameter estimation.

The results obtained indicate that cladding tends to increase the frequencies of the higher modes while the fundamental frequencies remain relatively unchanged. It has also been found that cladding has an increasing effect on the damping values in general, the increase being more pronounced in the case of torsional modes.

ACKNOWLEDGEMENT

The work reported on here was supported in part by the National Science Foundation, Division of Problem-Focused Research Applications under Grant ENV77-04269. Their support is gratefully acknowledged.

REFERENCES

1. Pilkey, W. D. and Cohen, R. (eds.), "System Identification of Vibrating Structures: Mathematical Models from Test Data", ASME New York, 1972.
2. Crawford, R. and Ward, H. S., "Determination of the Natural Periods of Buildings", Bull. Seism. Soc. Am., Volume 54, No. 6, 1743-1756, December, 1964.
3. Ward, H. S. and Crawford, R., "Wind Induced Vibrations and Building Modes", Bull. Seism. Soc. Am., Volume 56, No. 4, 763-813, August, 1966.
4. Trifunac, M. D., "Comparisons Between Ambient and Forced Vibration Experiments", Intl. J. Earthquake Eng. and Struct. Dyn., Volume 1, 133-150, 1972.
5. Udwadia, F. E. and Trifunac, M. D., "Time and Amplitude Dependent Response of Structures", Intl. J. Earthquake Eng. and Struct. Dyn., Volume 2, 359-378, 1974.
6. Trifunac, M. D., "Ambient Vibration Test of a Thirty-Nine Story Steel Frame Building", Earthquake Eng. Res. Lab., EERL 70-02, California Institute of Technology, Pasadena, 1970.
7. Bhat, W. V. and Wilby, J. E., "An Evaluation of Random Analysis Methods for the Determination of Panel Damping" NASA-CR-114423.

8. Vanmarcke, E. H., "Properties of Spectral Moments with Applications to Random Vibration", J. Eng. Mech. Div. ASCE, Volume 98, 425-446, 1972.
9. Vanmarcke, E. H., "Method of Spectral Moments to Estimate Structural Damping" in Stochastic Problem in Dynamics, (B. L. Clarkson, editor), Proc. Symp. held at Univ. of Southampton, 1977.
10. Cole, H. A., "On-the-line Analysis of Random Vibration", AIAA Paper No. 68-288, AIAA/ASME Ninth Structures, Structural Dynamics and Materials Conference, Palm Springs, California, 1968.
11. Cole, H. A., "On-line Failure Detection and Damping Measurement of Aerospace Structures by Random Decrement Signature", NASA-CR-2205, March 1973.
12. Schiff, A. J., "Identification of Large Structures Using Data from Ambient and Low Level Excitation", in System Identification of Vibrating Structures, (edited by W. D. Pilkey and R. Cohen), ASME, 1972.
13. Gersch, W., Nielsen, N. N. and Akaike, H., "Maximum Likelihood Estimation of Structural Parameters from Random Vibration Data", J. Sound and Vib., Volume 31, No. 3, 295-308, 1973.
14. Gersch, W. and Foutch, D. A., "Least-Squares Estimates of Structural System Parameters Using Covariance Function Data", IEEE Trans. on Autom. Control, Volume AC-19, No. 6, 1974.
15. Levenberg, K., "A Method for the Solution of Certain Nonlinear Problems in Least Squares", Quarterly Apppl. Math., Volume 2, 164-168, 1944.
16. Marguardt, D. W. "An Algorithm for Least Squares Estimation of Nonlinear Parameters", SIAM J., Volume 11, 431-441, 1963.
17. Bard, Y., "Nonlinear Parameter Estimation", Academic Press, New York, 1974.

STRUCTURAL CHARACTERIZATION OF CONCRETE CHIMNEYS

By Kalman L. Benuska[1], M. ASCE, Gerald Y. Matsummoto[2], Everett H. Killam[3], M. ASCE and John G. Diehl[4], A.M. ASCE

Abstract

Ambient wind vibrations were recorded on an 802 ft. chimney near Maysville, Kentucky. The test modal frequencies and shapes were compared to those calculated with a finite element model. The effect of structural taper, foundation stiffness, concrete modulus and steel reinforcement were investigated. Including the nonuniform structural taper and rotation at the chimney base gave good agreement between test and analysis. The measured damping was about one percent of critical for the ambient conditions.

Introduction

More stringent anti-pollution standards have placed new demands on the power generation industry to remove exhaust emissions from the ambient environment more effectively. These requirements often result in very tall power plant chimney stacks, many in the height range of 1000 feet. Design procedures specified by standard codes such as the American Concrete Institute do not completely address the added complexities associated with the new generation of large chimneys. Analytical methods currently used for predicting dynamic behavior, such as those described by Housner (2) and Rumman (4, 8), are based upon numerous simplifying assumptions; e.g., structural homogeneity with single elastic modulus, uniform vertical tapers, and no base rotations, which may or may not be sufficiently accurate. Empirical data derived from field measurements of dynamic chimney behavior are somewhat limited. In a study by Jeary (3) chimney frequencies and damping were measured but no attempt was made to correlate the data to analytical predictions.

In this investigation, the results of wind response measurements of the 802 ft. number two chimney at the Spurlock Station of the East Kentucky Power Corporation near Maysville, Kentucky, are compared to results based on a finite element analysis. Results from simpler structural models are also presented for comparison.

The Maysville chimney is a nonuniformly tapered concrete column having an outside diameter of 66.98 ft. (20.42 m) and thickness of 2.58 ft. (0.79 m) at the base and corresponding dimensions of 32.06 ft. (9.77 m) and 0.70 ft. (0.21 m) at the top (Figure 1). The structure is steel reinforced and constructed with three concrete corbels at 90, 525, and 780 ft. (27.43, 160.02, 237.75 m). The corbels, several construction openings, including one 25 ft. (7.62 m) wide by 75 ft. (22.86 m) high at

[1] Vice President, Kinemetrics Inc., Pasadena, CA.
[2] Technical Staff, TRW Defense and Space Systems Group, Redondo Beach, CA.
[3] Manager, Research and Development, Custodis Construction Co. Terre Haute, IN.
[4] Senior Engineer, Kinemetrics Inc., Pasadena, CA.

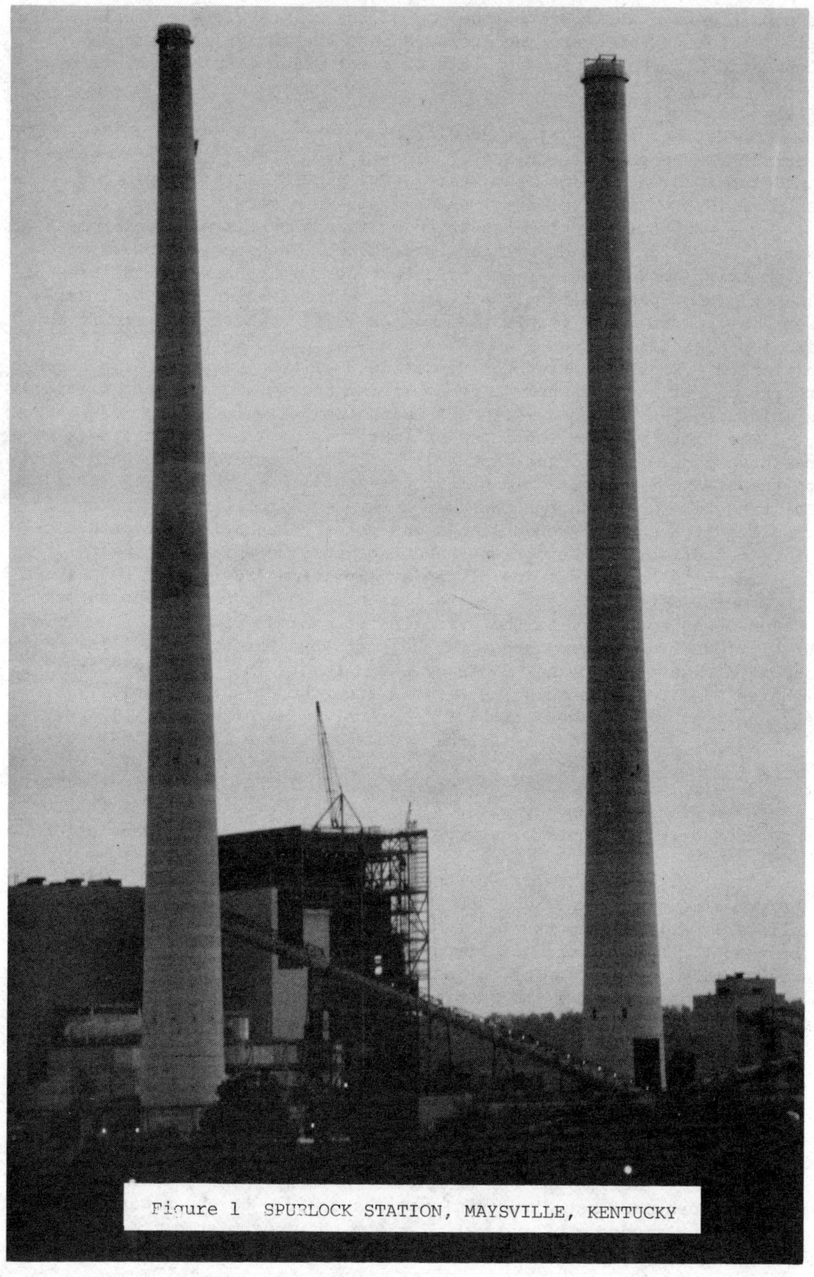

Figure 1 SPURLOCK STATION, MAYSVILLE, KENTUCKY

the base, and a work deck at the top of the chimney were present at the time of testing and were incorporated into the analytical model. The foundation of the chimney consists of steel piles with a concrete cap.

Field Program

On October 17, 1978, an acceleration monitoring system was installed in the concrete chimney. Wind-excited vibrations were recorded on October 18 and 19, and the system was removed on the evening of October 19. No wind velocity measurements were made.

The monitoring system consisted of five channels of accelerometers, signal conditioning and FM analog recording. The accelerometers were Kinemetrics Model FBA-1 force-balance types with 50 Hz natural frequency. High-pass filters, each with a corner frequency of 0.33 Hz, were introduced to AC-couple the signals. The signals were amplified and low-pass filtered at 5 Hz using Kinemetrics Model SC-1 signal conditioning. Figure 2 shows the resulting amplitude vs frequency characteristics for each channel. The amplified analog signals were recorded simultaneously on a Honeywell instrumentation tape recorder, Model 5600C.

The field program consisted of four runs. First, the channels were calibrated relative to one another. All five accelerometers were placed on top of the footing and identical horizontal base motion was recorded. The transfer functions for Channels 2 through 5 relative to Channel 1 resulted in the relative amplitude and phase calibration for each complete sensor, signal conditioner and tape recorder/playback channel.

The accelerometers were placed at elevation 538 ft. for Runs 2 and 3. Accelerometer orientations are presented in Figure 3. The purpose of Run 2 was to determine the direction of maximum response. The Ohio River flows from the southeast to the northwest through Maysville. As shown in Figure 3, Channel 1 was oriented toward the Ohio River and designated 0° (i.e., the Ohio River flows from the 90° direction). The strongest vibrations were at 0°, the cross-valley direction. The purpose of Run 3 was to determine the frequency of the lowest torsional mode. Channel 1 was oriented 0° and the phase of Channel 3 was compared to Channel 2. Channels 4 and 5 were compared in the same manner. The lowest torsional frequency was 3.5 Hz, which later analysis of the test data showed to be greater than the frequency of the third translational mode.

Figure 3 also illustrates the accelerometer locations for Run 4, the principal data run. The accelerometers were oriented in the cross-valley direction (Azimuth = 0°). A single horizontal accelerometer was attached to the platforms at elevations 278 ft., 538 ft. and 792 ft. Accelerations at the top of the chimney were about ±0.005g corresponding to approximately ±0.5 in. of dynamic displacement at the chimney top in the lowest mode. Two vertical accelerometers were attached to the top of the footing in the vertical plane containing the horizontal accelerometers. The vertical signals were subtracted electronically to produce a single output signal proportional to footing rotation. This sensor configuration contained the minimum number of sensors required 1) to measure the modal frequencies and the fundamental mode shape, 2) to indicate the shape of the second and third modes, and 3) to determine the response contribution due to rocking at the base. The experimental technique assumes that structural motions are due to the superposition of individual modes vibrating at discrete charactertistic frequencies. Relative amplitude and phase, the two quantities needed to determine

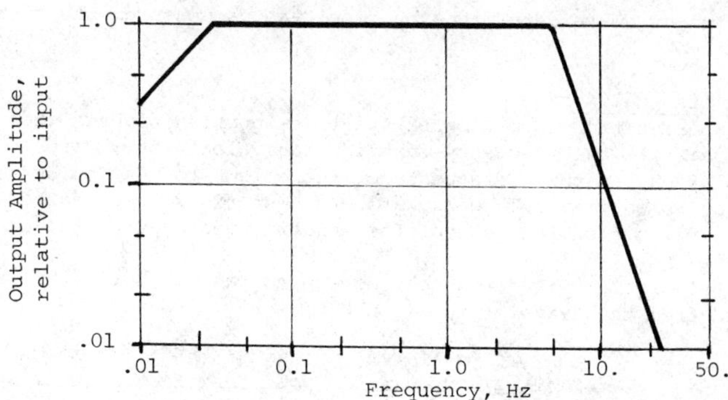

Figure 2 INSTRUMENTATION AMPLITUDE VS FREQUENCY CHARACTERISTIC

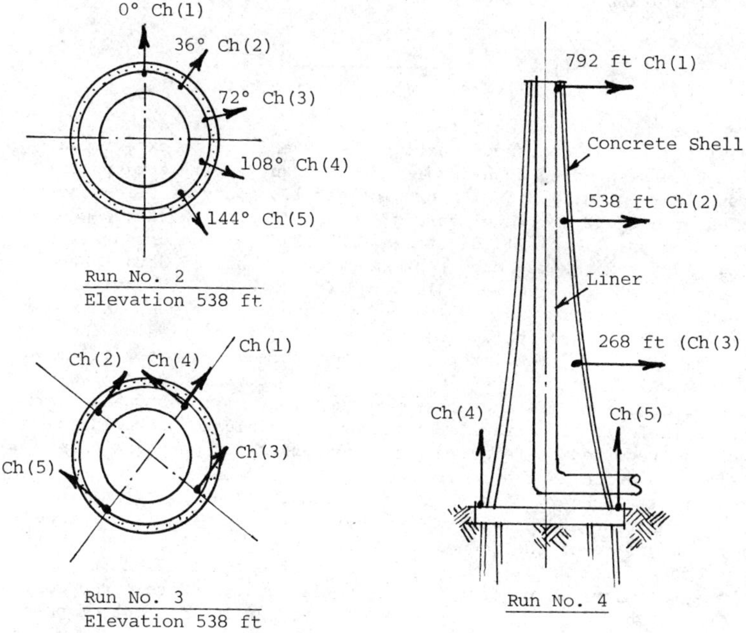

Figure 3 ACCELEROMETER LOCATIONS FOR DATA RUNS 2, 3 AND 4

mode shapes, were determined by simultaneous measurements at the five locations.

Test Data Analysis

Fourier transformation of the time-function signals into linear spectra X(f), the frequency-function equivalent form, gives spectrum peaks at each natural frequency of the column-foundation system. Modal damping is estimated by calculating a best fit single-degree-of-freedom transfer function for each of the spectrum peaks at the natural frequencies.

Relative amplitude and phase between two accelerometers are obtained from the transfer function H(f)

$$H_{ij}(f) = \frac{X_j(f)\ X_i^*(f)}{X_i(f)\ X_i^*(f)} = \frac{\text{cross power spectrum}}{\text{auto power spectrum}}$$

where the "jth" channel is compared to the "reference Channel i and X^* (f) is the complex conjugate of X(f). The correlation at each frequency is calculated by means of the coherence function $\gamma_{ij}{}^2(f)$, which is the squared correlation coefficient of the two spectra $X_i(f)$ and $X_j(f)$ (1).

$$\gamma_{ij}{}^2(f) = \frac{|X_j(f)\ X_i(f)|^2}{|X_i(f)\ X_i^*(f)|\ |X_j(f)\ X_j^*(f)|}$$

The coherence function varies between 1.0 (complete correlation) and 0 (no correlation). A coherence function value near 1.0 is expected at frequencies corresponding to natural modes. Figure 4 illustrates the linear spectrum calculated from approximately 1.8 hours of time-function data taken at the 792 ft. level during Run 4. The data were low-pass filtered (3.75 Hz, 11 poles), digitized at 10.24 sps, and divided into sixty-four 1024-point sets, each representing 100 sec. of data. After a 512-point linear spectrum was calculated for each set, the Kaiser-Bessel weighting function was applied to the spectrum values, and the ensemble of spectra were averaged. The modes are identified by the peaks in the spectrum. The first, second and third horizontal column-foundation modal frequencies are 0.31, 1.13, and 2.51 Hz respectively. The transfer function H(f) and coherence function $\gamma^2(f)$ test values are reported in Table 1. The transfer function was evaluated in polar coordinates and reported as relative amplitude and relative phase using the 792 ft. elevation as the reference.

In a similar manner, base rotation is reported in Table 1 as the horizontal amplitude at the 792 ft. elevation of a vertical line perpendicular to the plane of the foundation. The interpretation of Mode 1 is that 8 percent of the total horizontal motion at elevation 792 ft. was due to base rotation; 92 percent was due to deformation within the concrete column. In Mode 2, the base rotation was nearly 180° out of phase with the top of the column. Base rotation represents a normalized value of -41 percent when projected to elevation 792 ft. (see Figure 4 for a schematic representation). Base rotation in Mode 3 was 96° out of phase with the column deformation.

Damping was estimated from the spectra by calculating a best fit single-degree-of-freedom (SDOF) transfer function for the spectrum values at each peak. The data of Figure 5 contains fourteen spectral

TABLE 1 Transfer Function Values of Relative Amplitude, Relative Phase and Coherence Function in the Lowest Modes, Test Data

Mode	Location	Relative Amplitude	Relative Phase	Coherence
0.310 Hz	792'	1.00	0.0°	1.000
	538'	0.37	0.4	0.984
	268'	0.10	3.3	0.992
	Rotation	0.08	5.5	0.890
1.130 Hz	792'	1.00	0.0°	1.000
	538'	0.35	174.2	0.967
	268'	0.35	170.8	0.987
	Rotation	0.41	-145.7	0.732
2.510 Hz	792'	1.00	0.0°	1.000
	538'	0.41	158.0	0.985
	268'	0.27	-24.2	0.988
	Rotation	0.51	95.8	0.922

TABLE 2 Modal Frequencies for the Reinforced Concrete Chimney

	Frequency, Hertz		
	f_1	f_2	f_3
Test Data	0.310 (1.00)*	1.130 (1.00)	2.510 (1.00)
Tapered Column Analysis	0.387 (1.25)	1.440 (1.27)	3.473 (1.38)
Finite Element Analysis Fixed Base	0.322 (1.04)	1.246 (1.10)	2.956 (1.18)
Finite Element Analysis Allowing Base Rotation Including Steel Reinforcement	0.305 (0.98)	1.138 (1.01)	2.700 (1.08)
Finite Element Analysis Allowing Base Rotation Excluding Steel Reinforcement	0.299 (0.96)	1.103 (0.98)	2.624 (1.05)

* Number in parenthesis is ratio of analysis frequency to test frequency for each mode.

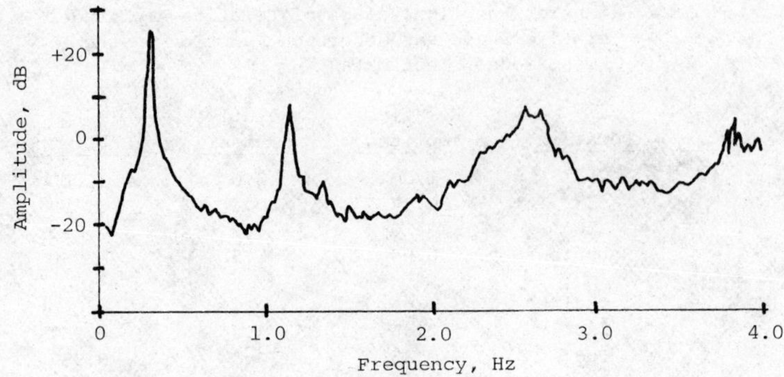

Figure 4 AMPLITUDE SPECTRUM, RUN 4, CH. 1, 792 FT. LEVEL

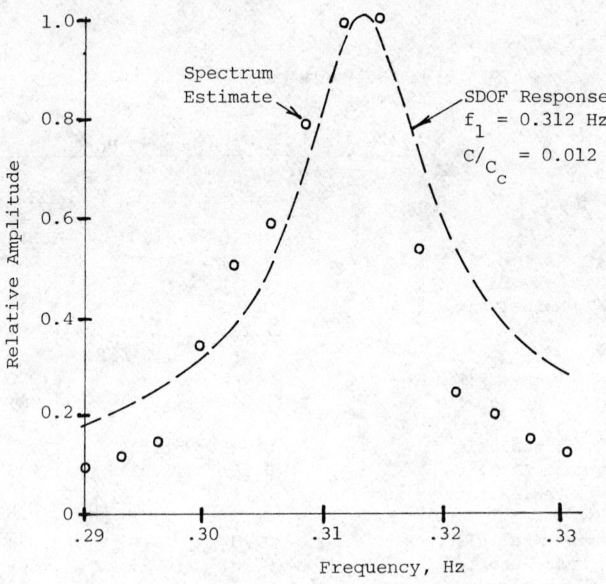

Figure 5 FIRST MODE DAMPING ESTIMATE FOR RUN 4, CH. 1, 792 FT. LEVEL

estimates in the first mode region of 0.29 Hz to 0.33 Hz. The dashed line represents the least-square fit of a SDOF transfer function with variable natural frequency and damping. Minimizing the square of the deviations provided a "best fit" solution of 0.31 Hz and 0.012 respectively for natural frequency and damping ratio. Damping ratios of 0.008 and 0.014 were calculated for the second and third modes respectively, applying the SDOF "best fit" to each spectrum peak.

It was not practical to test the stationarity of the wind loading on the column. Because the lightly damped structure had a highly amplified response at each of its normal modes, the spectral analyses accurately estimated the mode frequencies and mode shapes. The damping estimates are less reliable. Where modes were not closely spaced, e.g., at 0.31 Hz, the damping estimates probably understate the energy losses associated with brief periods of peak structural response because hysteresis damping tends to be amplitude-dependent. Where the third mode, 2.51 Hz, approaches the region of the first torsional mode, 3.5 Hz, the energy losses may be overstated because of overlap in the amplified response region.

Finite Element Analysis

The mesh geometry for the finite element model of the Maysville chimney is shown in Figure 6. To represent lateral bending stiffness, vertical pipe elements were located between 32 nodes which were placed to approximate nonuniformities in the vertical taper and changes in wall thickness. Each pipe element consisted of concentric concrete and steel cylinders. The steel cylinders were located at two-thirds the wall thickness, measured from the inside concrete surface, and represent reinforcement rods embedded in the concrete. They were sized from as-built steel-to-concrete area ratios along the vertical span.

Static compression tests were conducted on cylindrical samples during construction. The chimney concrete samples were taken at 7.5 ft. (2.286 m) height intervals. The analytical model used a modulus of elasticity derived from the sample strengths, f'_c, using the ACI formula $E_c = 57{,}000\sqrt{f'_c}$. However, they were increased 20 percent to account for loading rate effects which are described in experimental studies reported by McHenry and Shideler (5) and Popovics (7).

Allowances for the additional weight of a work deck at the top and several intermediate platforms were made by adding their mass to the appropriate nodal masses. A base coupling stiffness was selected by matching analytically determined base rotations to those found experimentally. All data were processed using the MRI/STARDYNE finite element program on a CDC 6600 computer.

Discussion of Results

A tabulation of measured and analytically derived natural frequencies are presented in Table 2. Calculations are included for a uniformly tapered, isotropic, homogeneous concrete column based on an average value for the "dynamic" modulus of elasticity. The steel reinforcements were not taken into account. The resulting first three nautral frequencies of 0.387, 1.440, and 3.473 Hz exceed the measured frequencies by 25, 27 and 38 percent, respectively. Because these analytical frequencies are directly proportional to $\sqrt{E_c}$, it is easily shown that reduction of E_c by 36 percent brings the first analytical

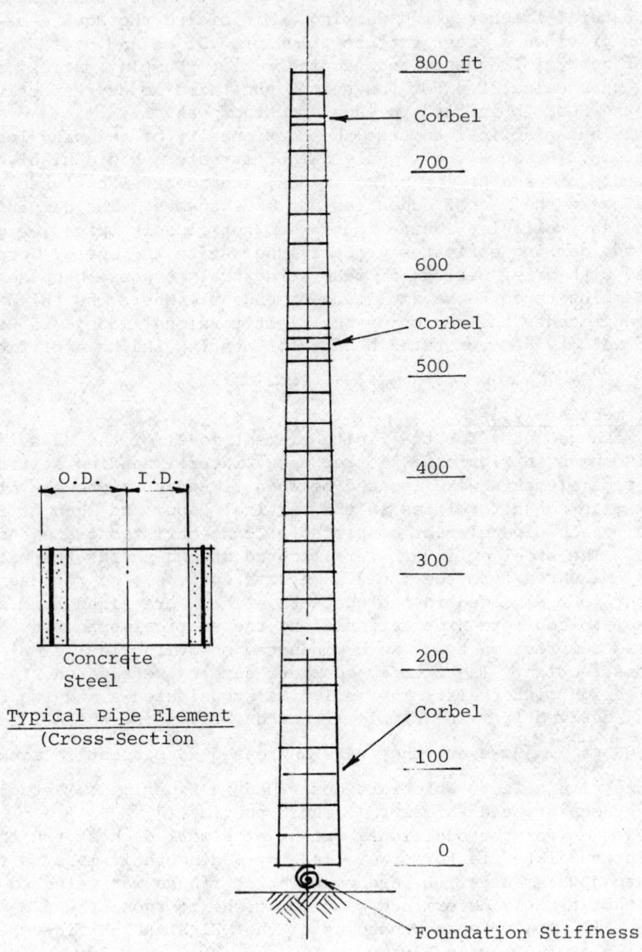

Figure 6 FINITE ELEMENT REPRESENTATION OF THE CHIMNEY

frequency in coincidence with the first measured frequency. It might therefore be implied that given an appropriate selection of E_c, the uniform tapered column approximation can adequately represent the chimney structure. However, this is not the case. The required value of the elastic modulus falls far below what has been found under laboratory test conditions. Thus the chimney designer, without prior knowledge of the frequencies, would significantly overestimate them, using standard accepted concrete properties.

Structural properties are more precisely simulated by the fixed-base finite element model which predicts the first three modal frequencies to within 4, 10 and 18 percent, respectively. Structural characterization is further improved by adding a base rotation spring to the model. Foundation stiffnesses are commonly ignored by the designer because of the difficulty in estimating their values, and, also, because of the belief that they have limited significance when dealing with low frequency structures. However, for the Maysville chimney, test base motion measurements indicated a noticeable contribution to the vibration mode shapes. A value for the chimney base stiffness of 3×10^{13} in.-lbs./rad. was assigned by running successive finite element analysis runs until the resulting contributions to the first and second mode chimney top vibrations reached 8.1 and 38 percent, respectively. See Figure 7. These compare well with the measured values. In addition, the first three computed natural frequencies, when including the base stiffness, are reduced to 0.305, 1.138, and 2.700 Hz, which vary by -1.6, +0.7, and 7.6 percent from the test values. Thus the first two mode shapes and frequencies are accurately predicted simultaneously. For the third mode the comparisons are less favorable, pointing to the inadequacy of characterizing soil-structure interaction by a single linear rotation spring.

Because the test data were collected under low wind conditions, the stiffness loss of concrete for high strain levels was not considered in this study. Including the vertical steel reinforcing bars embedded within the concrete results in about a 2 percent first mode frequency increase. The 20 percent increase applied in this study to the statically derived E values is in agreement with McHenry and Shideler (5) and Popovics' (7) contention that concretes modulus of elasticity is elevated during dynamic loading. Omission of this increase leads to a first mode frequency about 9 percent low.

The first mode damping was estimated at 1.2 percent of critical from the test data. This value is about double that of Jeary (3), although his measurements were made on chimneys containing multiple flues. The second and third mode damping were 0.8 and 1.4 percent of critical, respectively.

The modal frequencies and corresponding mode shapes calculated from the finite-element model are the undamped free vibration characteristic of the idealized structure and are uncoupled (i.e., independent of each other). Within a mode, all points move in-phase or out-of-phase; that is, with relative phase values of 0° or 180°. The vibration modes in a real structure may not be completely uncoupled because of damping and nonlinear materials. The phase values from the test reported in Table 1 are within 10° of in-phase or out-of-phase except the third mode and the base rotation within the second mode. Because the structure is damped lightly, the phase differences may indicate nonlinear soil-structure interaction at the foundation.

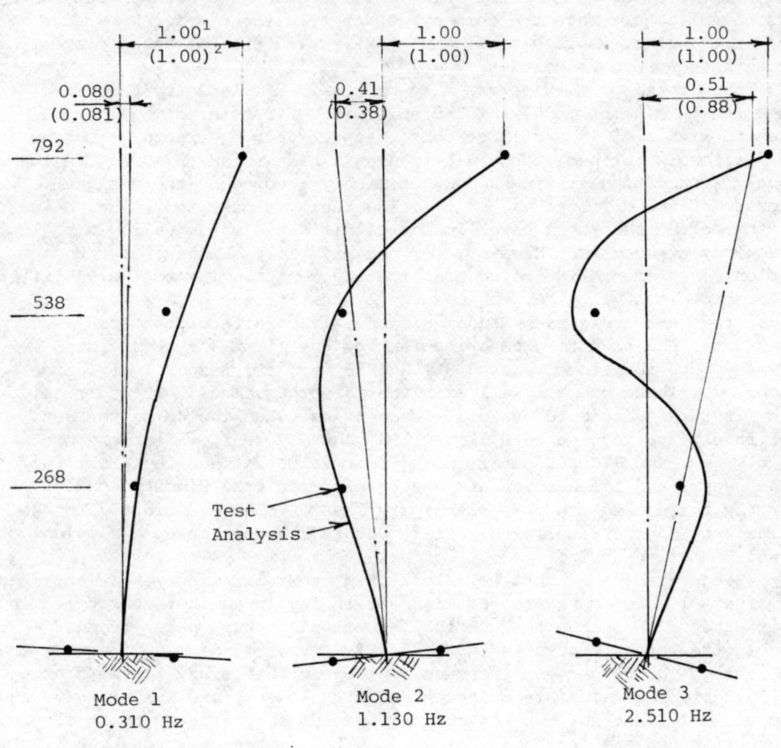

Notes:
[1] Mode values for test data
[2] Mode values for analysis are in brackets ( ).

Figure 7 COMPARISON OF TEST AND ANALYTICAL MODE SHAPES

Conclusions

The measured frequencies of vibration for the first three modes were 0.31, 1.13 and 2.51 Hz respectively. Overturning flexibility in the foundation contributed 8 percent to the first modes' horizontal displacement. Damping in the first mode was 1.2 percent of critical.

The simplest analytical model investigated was a uniformly tapered, homogeneous concrete column with its modulus of elasticity taken from the ACI formula, $E_c = 57{,}000 \sqrt{f'_c}$, and then increased by 20 percent to account for the dynamic loading rate. The resulting first three frequencies exceeded the measured by 25, 27 and 38 percent, respectively. Calculated modal frequencies were too high and would require an unrealistic low value for E_c to lower them to the measured values. E_c must be decreased by 56 percent to lower the fundamental frequency by 25 percent.

A finite element model was then used to represent the actual non-uniform stiffness and weight distribution. The first solution assumed a fixed base. Features were then added to incorporate overturning flexibility at the base and the stiffness contribution of steel reinforcement. In all cases, E_c was taken as 120 percent of the ACI formula.

The fixed base finite element model predicted the first three frequencies to within, respectively, 4, 10 and 18 percent of the measured values. Adding the foundation flexibility and steel reinforcement stiffness contributions to the finite element model gave the best fit. The first three frequencies were predicted to within, respectively, -1.6, +0.7 and 7.6 percent of the measured values.

The flexible base finite element model produced mode shapes which compared well to the measured modal values.

Of the parameters which affect the finite element model, flexibility of the chimney foundation appears to be most important. The omission of the flexibility leads to frequency overestimates which increase with the higher nodes. Additional factors which have an influence on estimated frequencies are assumed concrete modulus, steel reinforcements in the concrete, and the several concentrated masses within the chimney.

References

1. Bendat, J. S., & Piersol, A. G., Random Data: Analysis and Measurement Procedures, Wiley-Interscience, 1971, Chap. 5.
2. Housner, G. W., & Keightly, W. O., "Vibrations of Linearly Tapered Cantilever Beams," Proc. ASCE Jour. Eng. Mech. Div., Apr. 1962, pp. 95-123.
3. Jeary, A. P., "Damping Measurements from the Dynamic Behavior of Several Large Multi-flue Chimneys," Proc. Inst. Civil Eng., London, 1974.
4. Maugh, L. C., & Rumman, W. S., "Dynamic Design of Reinforced Concrete Chimneys," Jour. ACI, Sept. 1967, pp. 558-567.
5. McHenry, D., & Shideler, J. J., "Review of Data on Effect of Speed in Mechanical Testing of Concrete," P.C.A. Research Bull. D9, 1958.
6. Newmark, N. M., & Rosenblueth, E., Fundamentals of Earthquake Engineering, Prentice Hall, Inc., 1971, p. 35.
7. Popovics, S., "Method of Developing Relationships Between Mechanical Properties of Hardened Concrete," Jour. ACI, Dec. 1973, pp. 795-798.
8. Rumman, W. S., "Basic Structural Design of the Concrete Chimneys," Proc. ASCE Jour. Power Div., Jn. 1970, pp. 309-318.

Earthquake and Ambient Response of El Centro County Services Building

by

Gerard C. Pardoen*, M.ASCE

Gary C. Hart**, M.ASCE

Barrett T. Bunce***

Introduction

The full scale ambient and forced structural dynamic measurement of buildings has been an ongoing research activity for the past two decades, whereas the last decade has seen the emergence of strong motion data recorded by a number of building's seismographs due to significant seismic events. Recently the response of the Imperial County Services Building, due to the 15 October 1979 Imperial Valley earthquake, has stimulated interest within the earthquake engineering community because the structures response represents the first response measured in a building that has received major structural damage. This paper is devoted to assessing the column damage received by the Imperial County Services Building during the '79 earthquake and relating this damage to its strong motion data.

Building Description

The Imperial County Services Building (ICSB) serves as an office building for Imperial County. It was designed in 1968 (using the 1967 edition of the Uniform Building Code) and was completed in 1971 at a construction cost of $1.87 million. The building is 136 feet 10 inches by 85 feet 4 inches in plan and is founded on a Raymond step-taper concrete pile foundation. The piles are interconnected with reinforced-concrete link beams; they extend 45 feet to 60 feet into the alluvium foundation material composed primarily of sand with interbeds of clay to 60 feet (based on logs from 4 soil borings at the site).

Vertical loads are carried by reinforced-concrete floor slabs (5 inches thick at the second floor and 3 inches thick at the upper floors) supported by reinforced-concrete 5 1/2 inch-wide by 14 inch-deep pan joists spanning in the north-south (transverse) direction; the joists are supported by four longitudinal reinforced-concrete frames at 25 feet on center. The frame columns are typically 2 feet square, and the beams vary in size. Beams in the two interior frames are 2 feet wide by 2 feet 6 inches deep at all levels; those in the two exterior frames are 2 feet by 2 feet 5 inches at the second-floor level, 10 inches by 4 feet 5 inches

* Assistant Professor, University of California, Irvine, CA.
** Professor, University of California, Los Angeles, and Principal, Ruthroff, Englekirk & Hart, Inc., Los Angeles.
*** Senior Associate, Ruthroff, Englekirk & Hart, Los Angeles.

at the third-floor through sixth-floor levels, and 10 inches by 4 feet 2 inches at the roof level.

Lateral loads are resisted by the four reinforced-concrete frames in the east-west direction and reinforced-concrete shear walls in the north-south direction. The shear walls are discontinuous at the second-floor level. Below the second floor are three interior and one exterior 1-foot thick shear walls, and above the second floor, shear walls exist only at the east and west ends. Between the second and third floors, the walls are 7 1/2 inches thick, and above the third floor, they are 7 inches thick. According to the design calculations, the design "K" factor was 1.33 for the north-south shear walls, 0.67 for the east-west interior frames, and 1.0 for the east-west interior frames.

Instrumentation

This particular 6-story building has been instrumented by the California Division of Mines and Geology in accordance with its building instrumentation criteria for the California Strong Motion Instrumentation Program. In all the building is instrumented with 13 accelerometers as well as having a triaxial "free field" accelerograph located approximately 100 meters east of the building.

The 13-channel CRA-1 system in the building consists of 9 FBA-1 single-axis force-balance accelerometers located at various locations throughout the upper stories of the structure, an east-west oriented HS-0 horizontal starter at roof level, and one FBA-3 triaxial force-balance accelerometer package, one FBA-1 accelerometer, one 13-channel central recording unit, and a VS-1 vertical starter at ground level. The FBA accelerometers have a natural frequency of approximately 50 Hz and are connected by low-voltage data cable to the central recording unit. The recording unit is battery powered, is triggered by horizontal or vertical motion that equals or exceeds .01 g, and records on 7-inch (178-mm) light-sensitive film. The system is designed to record acceleration with frequency components in the 0 to 50 Hz range and with maximum amplitudes of 1 g (980 cm/sec^2). Real time is provided by a WWVB receiver and time-tick generator system; the recorder is not connected with the SMA-1 accelerograph located east of the building.

The FBA accelerometer location (Figure 1) were selected in order to provide information on overall building response as well as ground input motion. The primary purpose of the three north-south oriented accelerometers at the roof and second floor levels (accelerometers 1, 2, 3, 7, 8 and 9) is to obtain and isolate north-south translational, torsional, and in-plane floor bending response. In conjunction with the north-south oriented accelerometers at ground level (accelerometers 10 and 11), these accelerometers provide translational and torsional response, mode shape, and ground to second floor inter-story motion information. Similarly, the accelerometers at the ground floor, second floor, fourth floor, and roof levels in the more flexible east-west (frame) direction (accelerometers 4, 5, 6, and 13) provide east-west translational response, mode shape, and inter-story motion information. The two north-south-oriented accelerometers at ground level (accelerometers 10 and 11) are intended

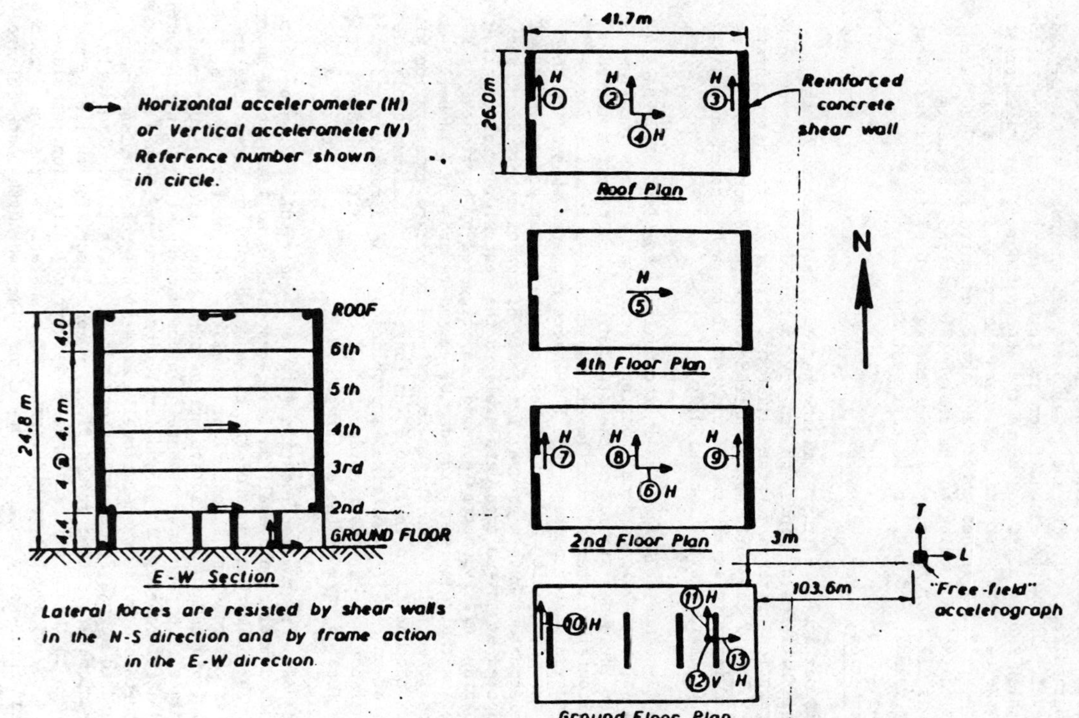

Figure 1 IMPERIAL COUNTY SERVICES BUILDING - STRONG-MOTION INSTRUMENTATION

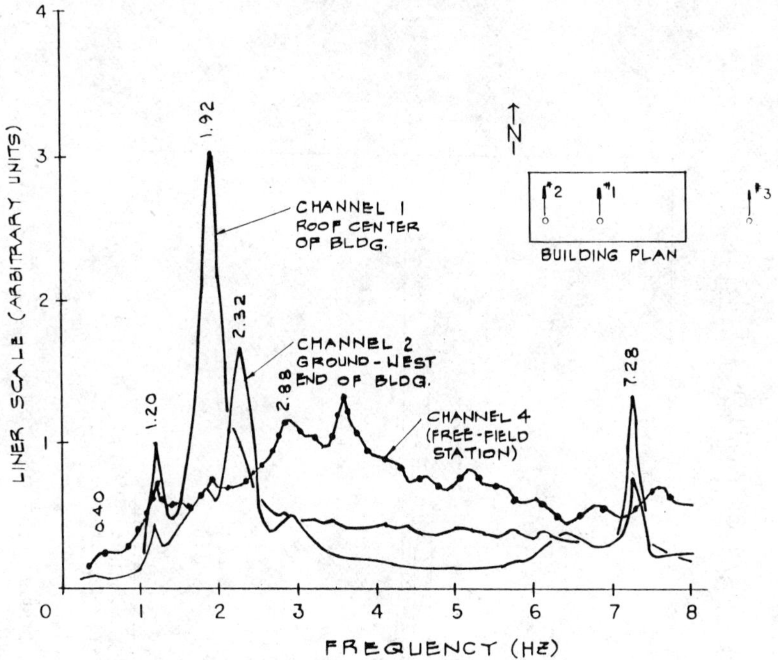

FIGURE 2 POST EARTHQUAKE AMBIENT GROUND LEVEL MOTIONS

Table 1 PRE AND POST EARTHQUAKE AMBIENT VIBRATION PERIODS

Date of Test	Frequency	Period (sec)	Direction
February 1979	1.55	.65	E
February 1979	2.24	.45	N
February 1979	2.81	.36	Torsion
March 1980	1.20	.83	E
March 1980	1.92	.52	N
March 1980	2.32	.43	Torsion

Note: Building was shored up when ambient natural frequencies were measured in March 1980

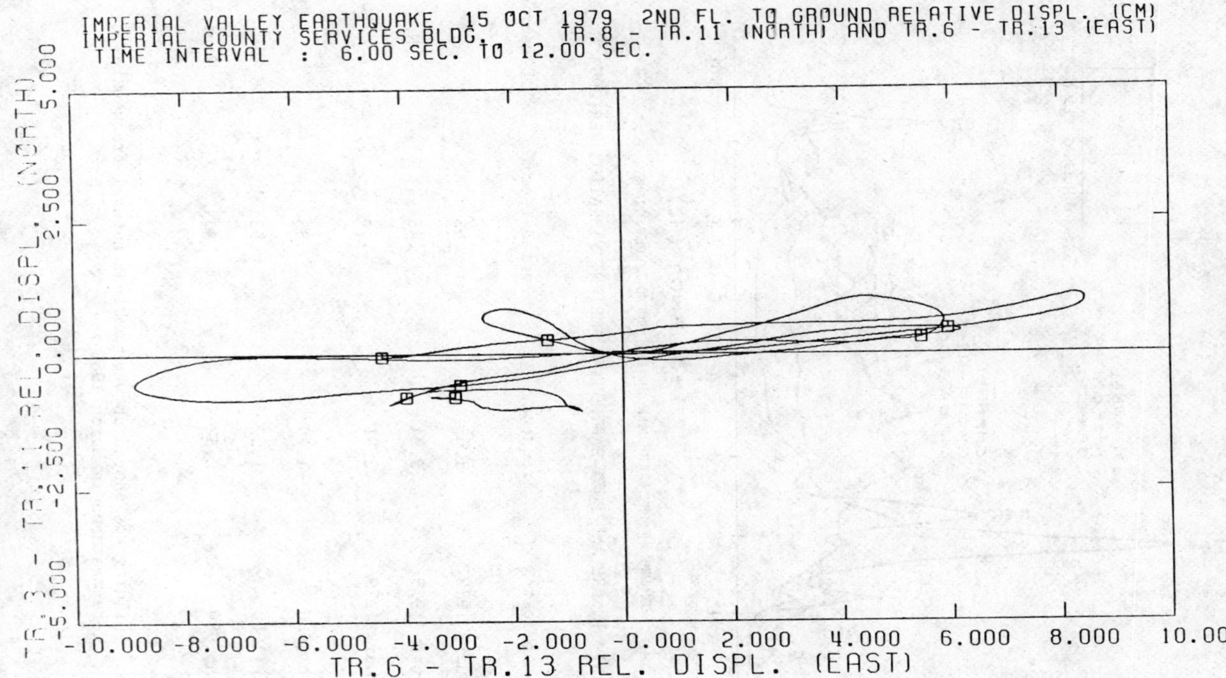

FIGURE 3 SECOND FLOOR TO GROUND RELATIVE DISPLACEMENT TIME HISTORY

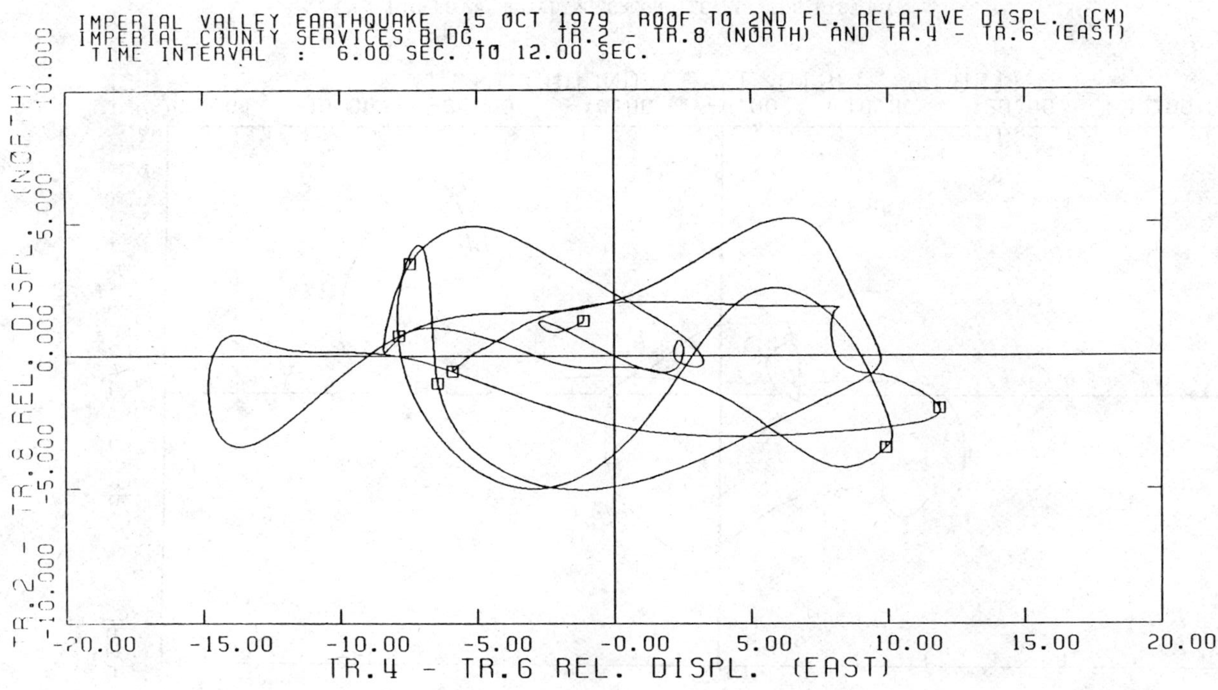

FIGURE 4 ROOF TO SECOND FLOOR RELATIVE DISPLACEMENT TIME HISTORY

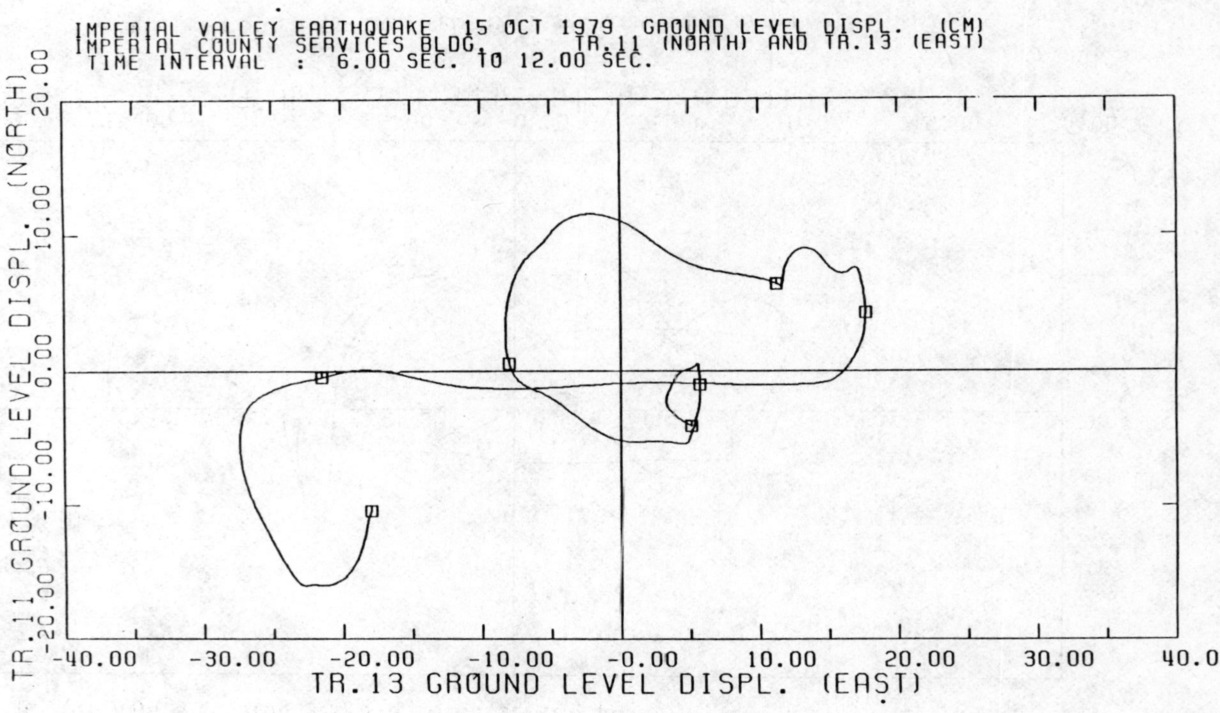

FIGURE 5 BUILDING GROUND LEVEL TIME HISTORY

Table 2 INTERSTORY DISPLACEMENTS (CM) AT MAXIMA TIME: (2ND FLOOR TO GROUND)

Time	Maximum (CM)	TR7 - TR10 (CM)	TR8 - TR11 (CM)	TR9 - TR11 (CM)	TR6 - TR13 (CM)
----	All Absolute	-2.648 (@6.58 sec)	-1.071 (@6.58 sec)	-3.442 (@9.24 sec)	-8.904 (@10.34 sec)
6.28	TR9=14.50	-2.40	-0.98	1.49	-2.57
6.37	TR8=-17.06 TR11=-16.04	-2.45	-1.02	1.63	-2.34
6.38	TR7=-16.98 TR10=-14.52	-2.46	-1.03	1.64	-2.30
6.58	(TR7-TR10)=-2.65 (TR8-TR11)=-1.07	-2.65	-1.07	1.63	-0.76
6.71	TR13=-27.41	-2.34	-0.92	2.18	-1.06
6.72	TR6=-28.68	-2.29	-0.90	2.25	-1.07
9.24	(TR9-TR11)=-3.44	0.20	-0.02	-3.44	-2.24
10.34	(TR6-TR13)=-8.90	-0.28	-0.54	0.15	-8.90

Table 3 INTERSTORY DISPLACEMENTS (CM) AT MAXIMA TIME (ROOF TO 2ND FLOOR)

Time (Sec)	Maximum (CM)	TR1 - TR7 (CM)	TR2 - TR8 (CM)	TR3 - TR9 (CM)	TR4 - TR6 (CM)
----	All Absolute	4.59 (@6.89 sec)	5.16 (@9.68 sec)	6.74 (@10.18 sec)	14.80 (@10.28 sec)
6.28	TR9=-14.50	-0.24	-0.22	-4.17	0.42
6.32	TR1=-17.36 TR3=18.92	-0.59	-0.63	-4.46	1.63
6.33	TR2=-17.58	-0.56	-0.63	-4.47	1.87
6.37	TR8=-17.06	0.01	-0.18	-4.21	2.46
6.38	TR7=-16.98	0.19	-0.01	-4.10	2.51
6.72	TR6=-28.48	1.65	1.51	-3.52	1.35
6.89	(TR1-TR7)=4.59	4.59	4.87	-1.26	-5.05
8.05	TR4=33.46	-1.82	-3.01	2.93	10.15
9.68	(TR2-TR8)=5.16	2.50	5.16	5.30	6.32
10.18	(TR3-TR9)=6.74	-2.95	-3.43	6.74	-13.77
10.28	(TR4-TR6)=-14.80	-1.36	-1.34	-5.19	-14.80

to identify collectively the extent to which differential horizontal ground motion has occurred, and the vertical accelerometer at ground level (accelerometer 12) provides information on vertical motion at this location. There are no vertically oriented accelerometers above ground level.

Pre- and Post-Earthquake Ambient Vibration Tests

Ambient vibration tests were performed on the building prior to and after the 15 October 1979 earthquake. The results of these tests, which were conducted as part of a cooperative effort of the Los Angeles and Irvine campuses of the University of California, are depicted in Fig. 2, tabularized in Table 1. It should be noted that the post earthquake results reflect the building on its shored up configuration.

Analysis of Earthquake Records

By manipulating the appropriate traces of the strong motion records, one can obtain an approximate horizontal plane movement time history such as those depicted in Figures 3, 4, and 5. Figure 3, for instance, represents the second floor to ground relative displacement time history during the 6-12 second duration of the earthquake. The time history curve in Figure 3 was obtained by plotting the E-W motion (the difference of traces 6 and 13) and the N-S motion (the difference of traces 8 and 11) at each time interval of the digitized earthquake response records. These figures are of particular interest when one considers the relative motion of the ground level to second floor and its effect on column failure.

Consider, for example, the effect of this interstory drift on the flexural behavior of the columns. Using standard structural engineering code calculations it can be shown that the ductile frames experienced E-W deflections of approximately 7 times that of the code allowables and yet these columns performed without flexural collapse. Frame flexural failure of the 4 easternmost columns must be ruled out since a flexural failure due to the interstory drift would have caused most, if not all, columns to fail. The lack of apparent frame flexural failure suggests that there was sufficient frame ductility despite the fact that comparing the moment developed by such interstory drift with column interaction diagrams was much greater than that needed to produce collapse.

A more probable, but tentative, explanation subject to a detailed nonlinear dynamic analysis of the four column's failure is due to overturning since the outer column pair showed more distress than the inner pair while the east end shear wall, the floor diaphragm, and the inner shear wall showed no signs of distress. The major difference in damage between the east and west ends of the building is due to the shear load and overturning moment resistance. The west end ground level shear wall beneath the upper wall prevents large axial forces from developing on the columns. However, the east end ground level shear wall is offset by some 30 feet from the upper wall requiring that the four slightly offset columns must resist the vertical and overturning loads.

DYNAMIC LONGITUDINAL RESPONSE OF AN EARTH DAM

Ahmed M. Abdel-Ghaffar,* A.M. ASCE

ABSTRACT

This paper concerns observations of the longitudinal vibrational behavior of a modern earth dam, including recorded earthquake responses, forced vibration tests, and ambient (unique in earth dam research) vibration measurements. Major dynamic characteristics of the dam to various forces are identified and new analytical elastic models are thereby developed to assess the vibrational behavior of the dam in a direction parallel to its axis. It was found that models in which the elastic moduli of the dam material vary along the depth are the most appropriate representations for predicting the dynamic characteristics of the dam. Finally, the agreement between the experimental and earthquake data and the theoretical results from some of the models is reasonably good.

INTRODUCTION

Transverse cracking of an earth dam can result from strains and stresses induced by earthquake longitudinal motion (as well as from differential settlements). Such cracks are of concern because they may later enlarge and produce catastrophic conditions under further earthquake impulses and other stresses. The 1971 San Fernando earthquake ($M_L = 6.3$) caused a transverse crack near the crest of the Santa Felicia Dam, a well constructed modern earth dam (2) that was shaken by a smaller earthquake ($M_L = 4.7$) in 1976 and was under comprehensive full-scale dynamic testing in 1978 (3). The unusual set of data accumulated on the dam during the tests and the earthquakes promotes development of analytical models for evaluating dynamic characteristics of earth dams in a direction parallel to the dam axis (1). This paper summarizes the earthquake observations, the full-scale experimentation's findings and the results of the analytical models, all relevant to the dynamic longitudinal response of the dam.

DESCRIPTION OF THE DAM

Santa Felicia Dam (Figs. 1 and 2) is a modern rolled-fill embankment constructed from well graded alluvial materials consisting of clay, sands, gravel, and boulders. The dam, which is 450 ft (137 m) long across the valley at the base and 1,275 ft (389 m) long at the crest, is equipped with two accelerographs that yielded data on how it responded to the two earthquakes (2).

EARTHQUAKE RESPONSE OBSERVATIONS: LONGITUDINAL COMPONENT

The San Fernando earthquake of February 9, 1971 ($M_L = 6.3$) caused a transverse crack on the dam crest at the east abutment (Fig. 3). The depth of the crack, approximately one-sixteenth of an inch in width, is not known. Investigation has implied that this narrow crack was caused by the dynamic strains induced by the longitudinal vibration resulting from the earthquake and not by any settlement. Fortunately, the crack does not seem to be structurally significant.

*Asst. Professor, Civil Engrg. Dept., Princeton University, Princeton, NJ.

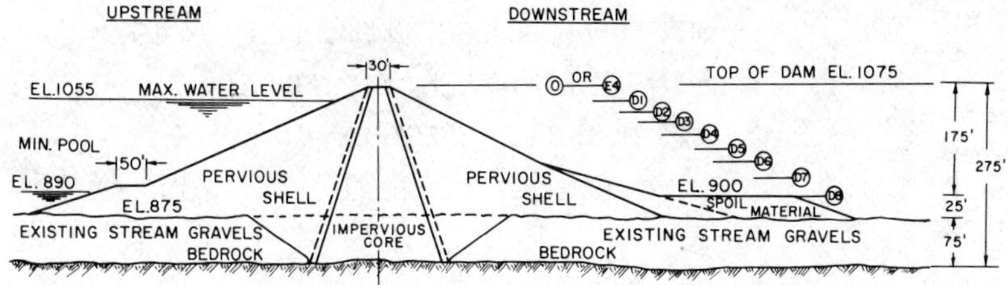

Fig. 1 Cross section of Santa Felicia Dam showing the measurement stations on the downstream face. (1 ft = 0.305 m)

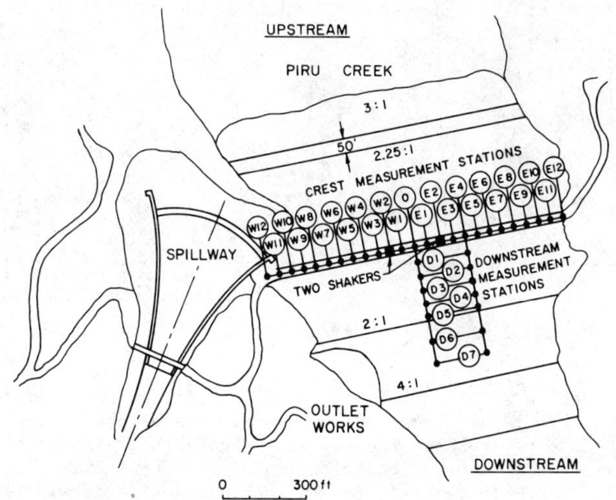

Fig. 2 Plan view showing measurement stations and location of two shakers. (1 ft = 0.305 m)

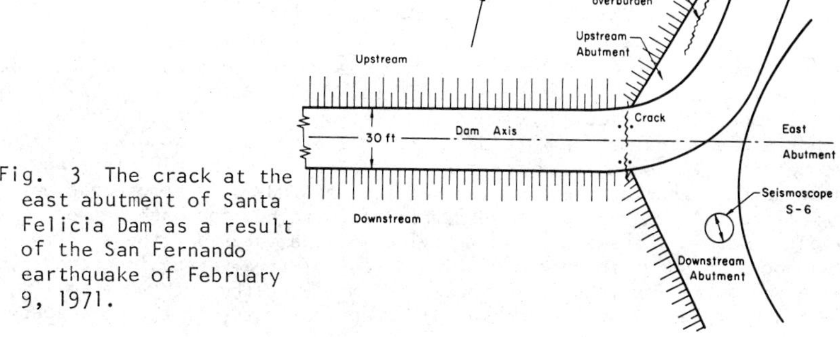

Fig. 3 The crack at the east abutment of Santa Felicia Dam as a result of the San Fernando earthquake of February 9, 1971.

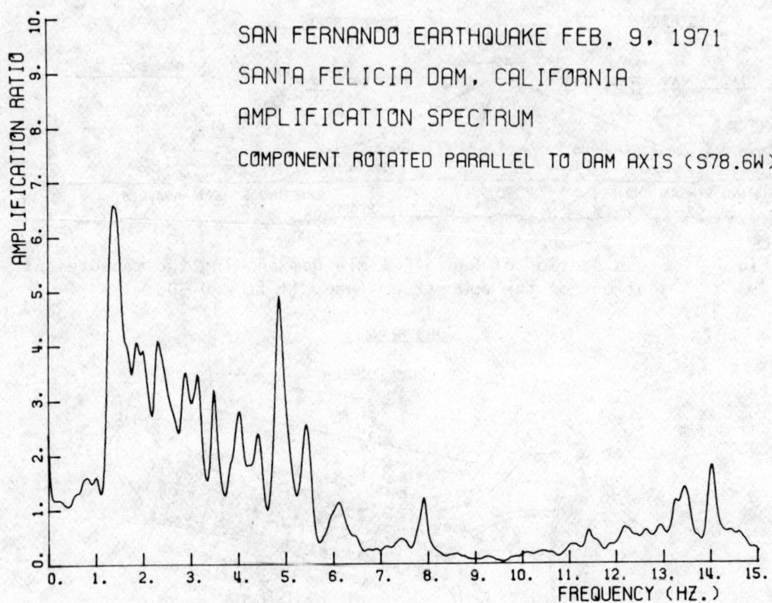

Fig. 4 Amplification spectrum of the 1971 earthquake.

Amplification spectra (Figs. 4 and 5) of the dam's earthquake records were computed by dividing Fourier amplitudes of acceleration of the crest records by those of the base records to indicate the resonant frequencies and to estimate the relative contribution of different modes in the longitudinal direction. In general, the two earthquakes revealed that the values of the resonant frequencies vary slightly from one earthquake to the other. In addition, amplification spectra of the upstream-downstream direction (2) showed that the dam responded primarily in its fundamental mode in that direction, but Figs. 4 and 5 of the longitudinal component are lacking pronounced single peaks.

FULL-SCALE EXPERIMENTAL INVESTIGATION

Forced Vibration Tests: The dam was excited into resonance (in the frequency range 1.0 to 6.0 Hz) in various modes in the longitudinal direction using only one shaker at Station E2 of Fig. 2 (See Ref.3). Eight seismometers, located at selected measurement stations, were used to pick up the steady state vibrations of the dam. Figures 6, 7, and 8 show some of the resonance curves determined for three selected stations. In plotting these curves, the magnitude of the double-amplitude of the response was normalized (divided by the largest response obtained for each set of weights used in the shaker), then divided by the square of the exciting frequency, as shown in Part b of Fig. 6, and plotted versus frequency. Because the force levels used to shake this massive structure were small, the peaks are not sharply defined. The best estimates of the mode shapes

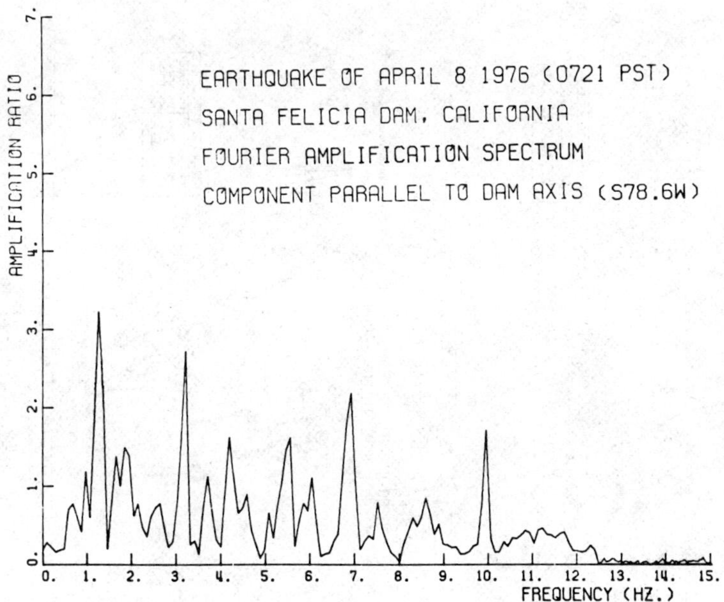

Fig. 5 Amplification spectrum of the 1976 earthquake.

of the crest vibrations at peak amplitudes obtained from the seismometer records are shown in Fig. 9; the local magnification effect of the soil surrounding the shaker block is also shown. It was found that some resonant longitudinal frequencies are very close (even identical) to some of the upstream-downstream frequencies. This proximity may suggest a strong coupling between these two horizontal directions, or it may suggest that due to both the eccentricity of the single shaker (it was not located on the longitudinal axis of the dam) and the fact that the dam is not symmetrical, the upstream-downstream modes containing significant longitudinal motions were excited.

Ambient Vibration Tests: The naturally-occurring vibrations of the dam caused by strong winds and the spilling of the reservoir were measured during the ambient vibration tests. Samples of the Fourier amplitude spectra of the recorded motions (at different stations and on different days) are shown in Fig. 10. The results indicated that there are many closely spaced frequencies not only in the two horizontal directions but also in the vertical direction and that there is modal coupling and interference between the three orthogonal directions; there were many frequencies with identical values measured in the three directions. Also, the correspondence between resonant frequencies from the tests and those estimated from earthquake records is reasonably good over the first few frequencies, but higher modes could not be reliably matched.

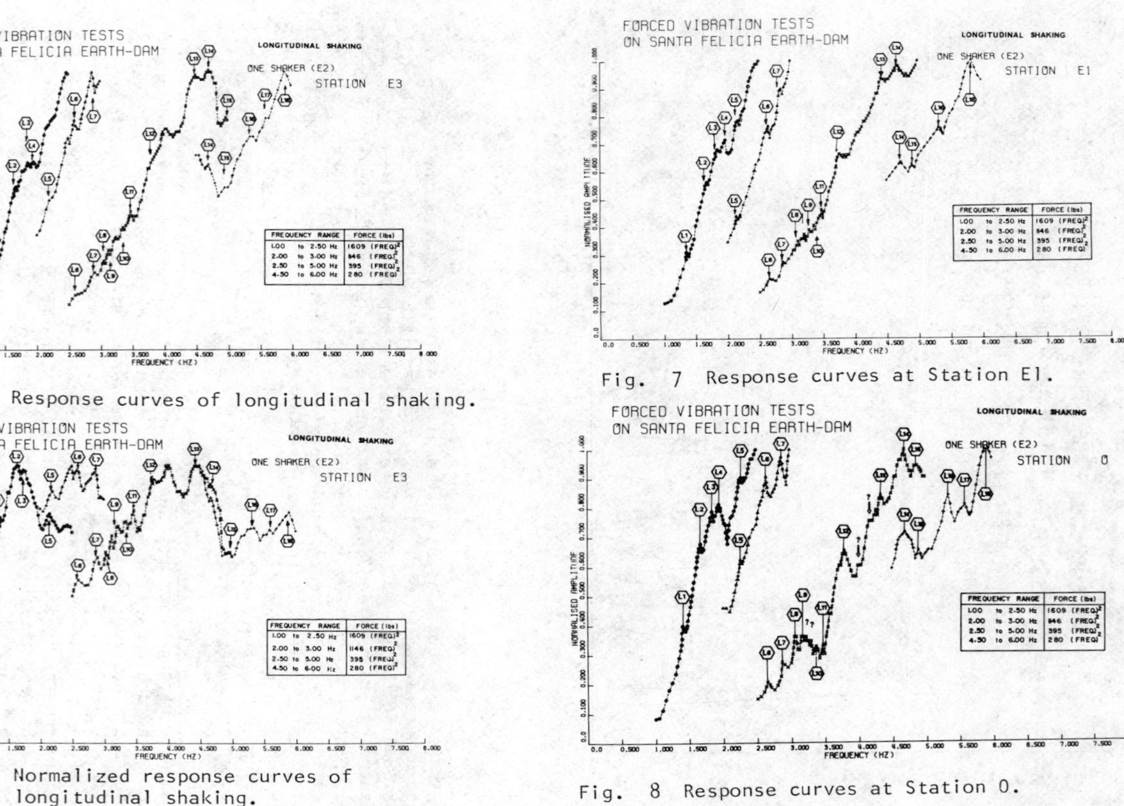

Fig. 6a Response curves of longitudinal shaking.

Fig. 6b Normalized response curves of longitudinal shaking.

Fig. 7 Response curves at Station El.

Fig. 8 Response curves at Station 0.

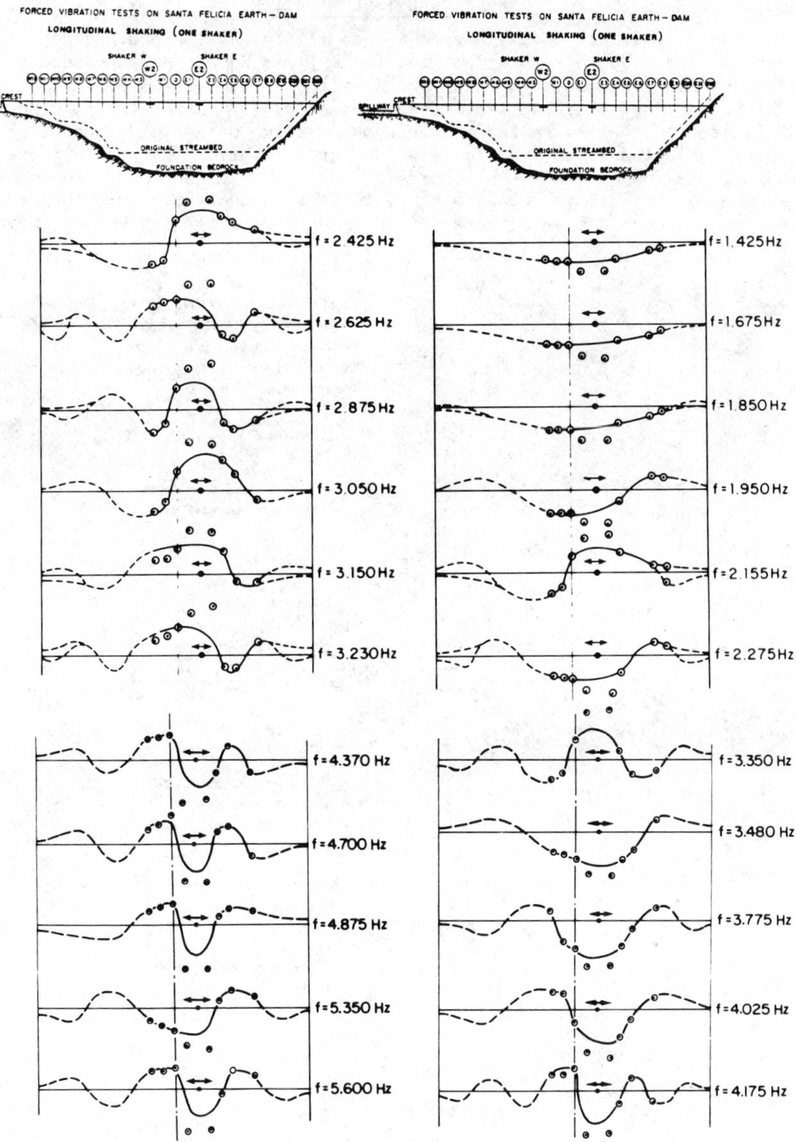

Fig. 9 Estimation of the resonating modes obtained during the longitudinal frequency sweeps.

ANALYTICAL INVESTIGATION

Two-dimensional models are developed (1) for evaluating natural frequencies and modes of vibration, and for estimating earthquake-induced stresses and strains (both shear and axial) of a wide class of earth dams in the longitudinal direction. The model (Fig. 11) is an elastic wedge of finite length in a rectangular canyon, resting on a rigid foundation and is subjected to uniform longitudinal ground motion $\ddot{w}_g(t)$. In these models, the nonhomogeneity of the dam materials is taken into account by assuming a specific variation of the stiffness properties along the depth (due to the continuous increase in confining pressure). Values of shear

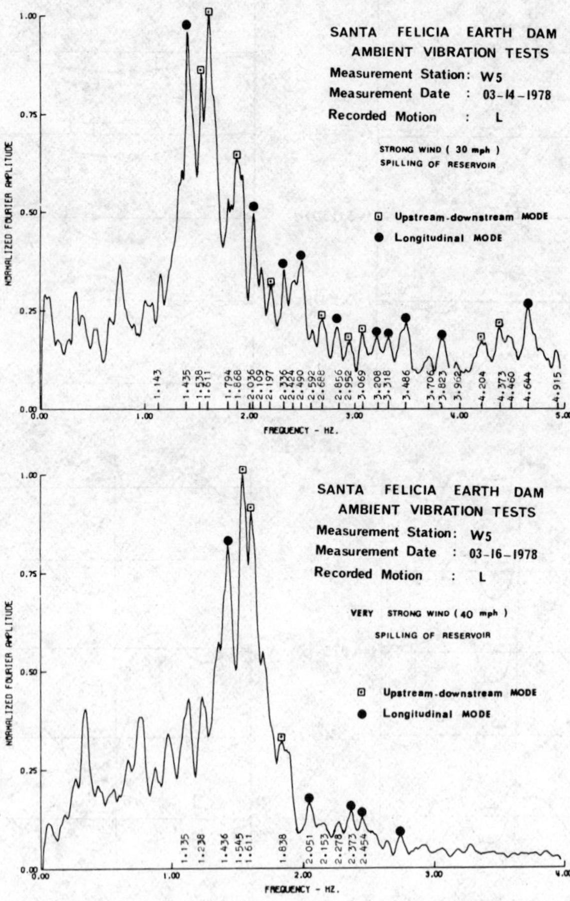

Fig. 10-a Fourier amplitude spectra of the longitudinal motion recorded at same stations on different days (the frequency resolution is 0.0073 Hz).

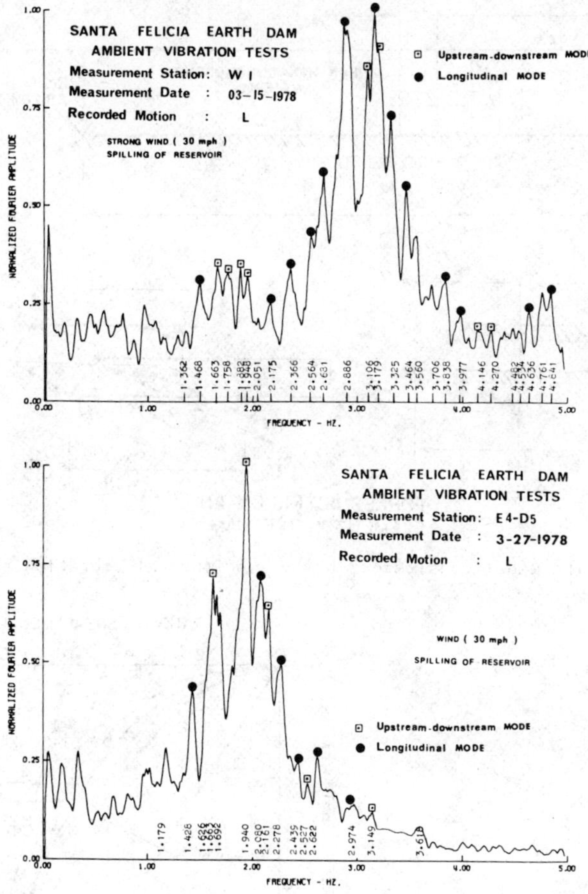

Fig. 10-b Fourier amplitude spectra of the longitudinal motion recorded at different stations on different days (the frequency resolution is 0.0073 Hz).

moduli (evaluated from in-situ wave-velocity measurements on Santa Felicia Dam, Fig. 12 and Ref. 2) suggest that the continuous variation of soil stiffness may be represented by the following relationships for both the shear modulus G(y) and the elastic modulus E(y) (See Fig. 12):

$$G(y) = G_0\left(\frac{y}{h}\right)^{\ell/m} , \qquad \left(\frac{\ell}{m} = 0,\ 1,\ \frac{1}{2},\ \frac{2}{5},\ \frac{1}{3}\right) , \qquad (1)$$

$$G(y) = G_0\left[(1-\varepsilon)\frac{y}{h} + \varepsilon\right] , \qquad \varepsilon = \frac{G_1}{G_0} , \quad (\varepsilon = 0.5 \text{ and } 0.6) , \qquad (2)$$

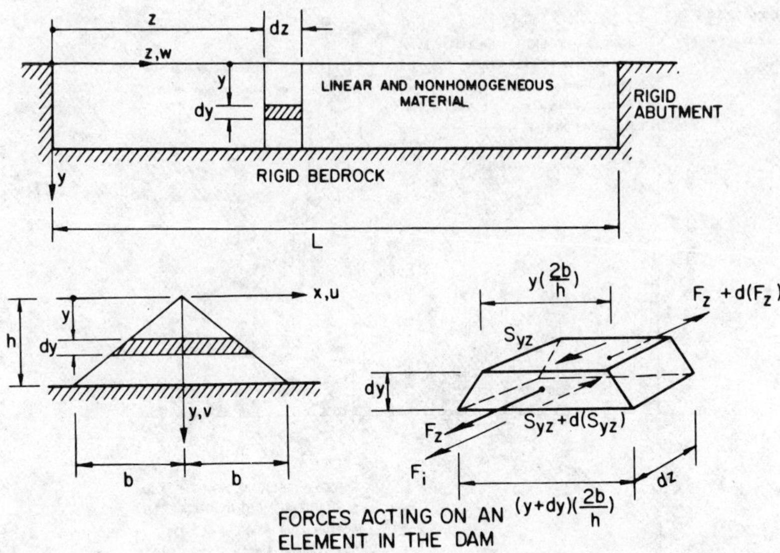

Fig. 11 The model considered in the longitudinal vibration analysis.

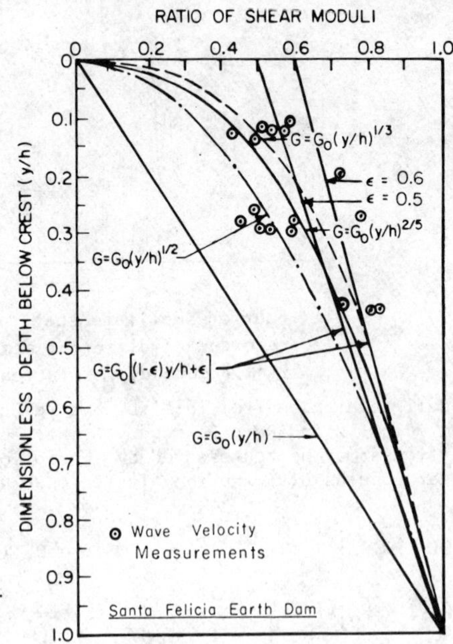

Fig. 12 Variation of stiffness properties along the depth of the dam and the results of in-situ wave-velocity measurements on Santa Felicia Dam.

where G_0 and G_1 are the shear moduli of the dam material at the base and at the crest, respectively; h is the height of the dam.

By considering the shear and axial forces acting in the longitudinal direction (Fig. 11), the equation of motion of the dam can be written as

$$\rho \frac{\partial^2 w}{\partial t^2} + c \frac{\partial w}{\partial t} - \frac{1}{y}\frac{\partial}{\partial y}\left[G(y)\frac{\partial w}{\partial y} y\right] - \frac{1}{y}\frac{\partial}{\partial z}\left[\eta G(y)\frac{\partial w}{\partial z} y\right] = -\rho \ddot{w}_g(t) \quad , \quad (3)$$

where ρ is the uniform mass density of the dam material, $w(y,z,t)$ is the longitudinal vibrational displacement, c is the damping coefficient, and $\eta[=E(y)/G(y) = 2(1 + \nu)]$ is an elastic constant (where ν is the Poisson's ratio of the dam material.)

For the undamped free vibration ($c = \ddot{w}_g(t) = 0$) the mode shapes of longitudinal vibration in the y and z directions can be given by (see Ref. 1)

$$Y_n(y) = \sum_{k=0}^{\infty} a_k y^{k+s} \quad , \text{ and } \quad Z_r(z) = \sin \frac{r\pi}{L} z \ , \quad n,r = 1,2,3,\ldots \ , \quad (4)$$

where the method of Frobenius is utilized to obtain the modes $Y_n(y)$ (since there is no closed form solution except for the case where $\ell/m = 0$); the number s and the coefficients $a_0, a_1, a_2, \ldots$ can be evaluated from the ordinary differential equation resulting from the separation of variables. In the above equation L is equivalent (average) length of the crest.

The frequency equation is obtained by satisfying the boundary conditions of zero displacements at the two end abutments and the base and of

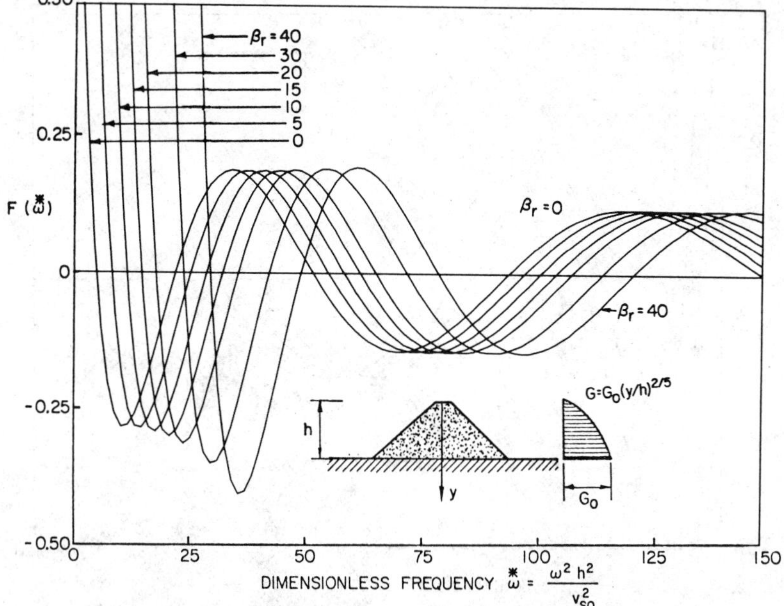

Fig. 13-a Plots of the frequency equation for the case of $\ell/m = 2/5$.

zero shear stress on the crest; the natural frequencies are defined through roots of

$$F\left(\frac{\omega^2 h^2}{v_{s0}^2}, \eta \frac{\pi^2 r^2 h^2}{L^2}\right) = 0 \qquad \text{or} \qquad F(\overset{*}{\omega}, \beta_r) = 0 \quad , \qquad (5)$$

where the dimensionless frequency $\overset{*}{\omega}$ equals $(\omega h/v_{s0})^2$ and the coefficient β_r is defined as $\eta(\pi rh/L)^2$; $\omega = \omega_{n,r}$ is the (n,r)th frequency, and $v_{s0} = \sqrt{G_0/\rho}$ is the shear wave velocity at the base of the dam material. For Santa Felicia Dam $\nu = 0.45$, $h/L = 236.5/912.5 = 0.26$; this gives $\beta_r = 1.92\ r^2$, $r = 1,2,3,\ldots$. The roots $\overset{*}{\omega}$ for different values of β_r and various stiffness variations are determined from plots (1) of the frequency equation (Eq. 5). Figure 13 shows the roots for the cases $\ell/m = 2/5$ and $\varepsilon = 0.5$; each β_r curve intersects the frequency axis at modes (1,r), (2,r), (3,r),... etc. The analytical models were used to establish, from the observed fundamental frequency $\omega_{1,1}$, of both earthquake records and full-scale results, a value for the shear wave velocity, v_{s0}, which was then used to calculate other frequencies higher than the fundamental. Tables 1 and 2 show a comparison between the observed resonant frequencies and estimated values computed from the various models; they also show the estimated shear-wave velocities from the observed data. The agreement between the experimental and earthquake data and the theoretical results is reasonably good.

Two modes, (1,1) and (2,1), of longitudinal vibration (along the depth of Santa Felicia Dam) resulting from various stiffness variations

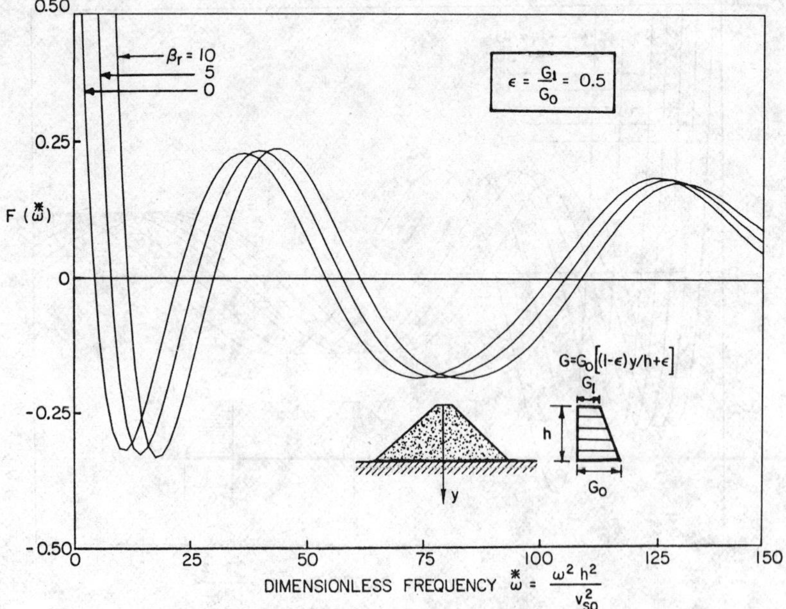

Fig. 13-b Plots of the frequency equation for the case of $\varepsilon = 0.5$.

are depicted in Fig. 14; modal configurations estimated from the ambient vibration measurements are also shown in the same figure. It is seen that models with $\ell/m = 1/2, 1/3$, or $2/5$ or the truncated cases ($\varepsilon = 0.5$ and 0.6) are the most appropriate representations for predicting the dynamic characteristics in the longitudinal direction. Furthermore, the ambient-measurement results in Fig. 15 confirm the prediction by the analytical models that for $n = 1$, the lower modes along the crest (low values of r, e.g., $r = 1,2$) are associated with shear-type modal configuration along the depth, while the higher modes (large values of r, e.g., $r \geq 3$) are associated with bending-type modal configurations.

Finally, the magnitude and distribution of shear stresses and normal (or axial) stresses in the $(n,r)^{th}$ mode are given by (see Ref. 1)

$$\tau_{n,r}(y,z,t) = \frac{4G_0}{r\pi h} \frac{\overset{*}{P}_{n,r}}{\omega_{n,r}\sqrt{1-\zeta_{n,r}^2}} \Phi_{n,r}\left(\frac{y}{h}\right) \sin\left(\frac{r\pi z}{L}\right) V_{n,r}(t) \quad , \qquad (6)$$

$$\sigma_{n,r}(y,z,t) = \frac{4\eta G_0}{L} \frac{\overset{*}{P}_{n,r}}{\omega_{n,r}\sqrt{1-\zeta_{n,r}^2}} \Gamma_{n,r}\left(\frac{y}{h}\right) \cos\left(\frac{r\pi z}{L}\right) V_{n,r}(t) \qquad (7)$$

where $\overset{*}{P}_{n,r}$ is the participation factor resulting from the modal configuration along the depth; for Santa Felicia Dam, values of $\overset{*}{P}_{1,1}$ are given by

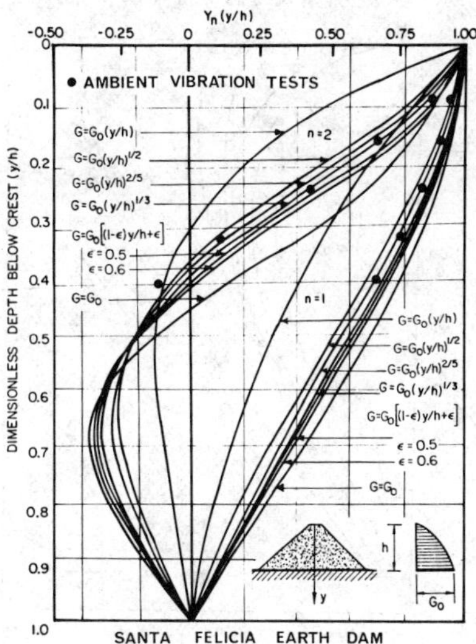

Fig. 14 Modes (1,1) and (2,1) resulting from various stiffness variations.

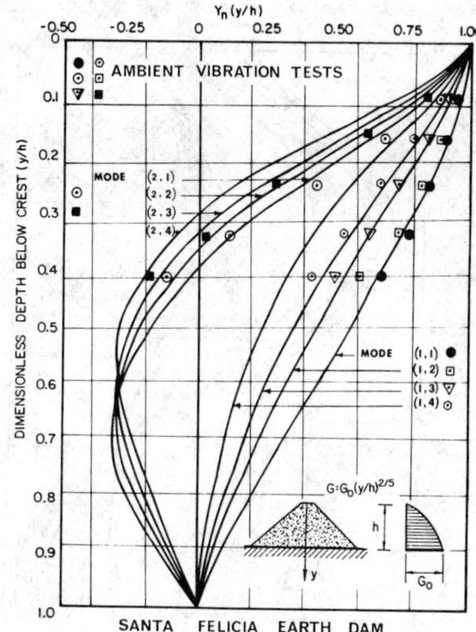

Fig. 15 Comparison between the results of ambient tests and those of the model $G = G_0(y/h)^{2/5}$.

Case:	$\ell/m=0$	$=1$	$=\frac{1}{2}$	$=\frac{2}{5}$	$=\frac{1}{3}$	$\varepsilon=0.5$	$=0.6$
$\overset{*}{P}_{1,1}$	1.818	2.575	1.959	1.891	1.802	1.749	1.705;

$\zeta_{n,r}$ is the modal damping factor; $V_{n,r}(t)$ is the convolution (or Duhamel) integral of the ground acceleration $\ddot{w}_g(t)$; $\Phi_{n,r}$ (or Φ_n) is the shear stress modal participation factor and $\Gamma_{n,r}$ (or Γ_n) is the normal stress modal participation factor along the depth of the dam. Figures 16 and 17 show Φ_n (n = 1 and 2) and Γ_n (n = 1) calculated for Santa Felicia Dam for the various stiffness variations. In general, it was found that the higher modes make a considerable contribution to the overall earthquake response of the dam (displacements, stresses, etc.); this is consistent with the earthquake amplification spectra of Figs. 4 and 5. Furthermore, the maximum tensile or compressive stresses (and strains) occur in the top region near the crest at the end abutments (where $\cos(r\pi z/L) \simeq 1$); this may explain the Santa Felicia Dam crack mentioned previously.

ACKNOWLEDGMENTS

This research was supported by a grant (PFR78-22865) from the National Science Foundation with Dr. William W. Hakala as the Program Manager.

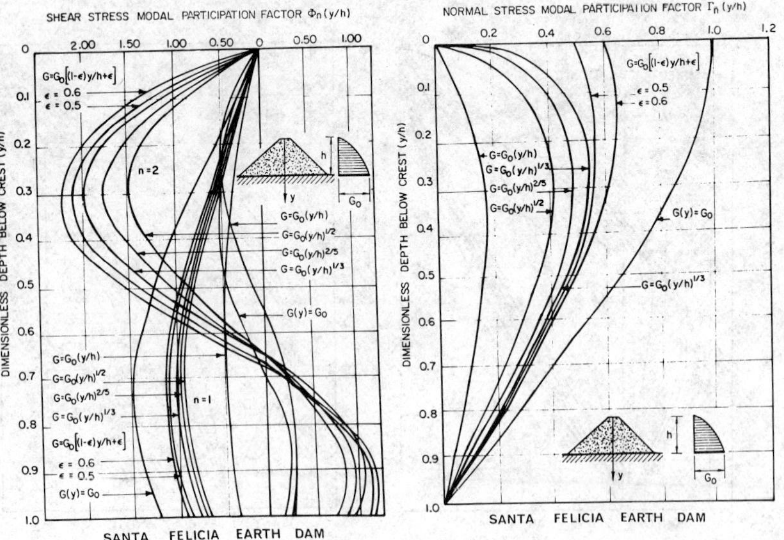

Fig. 16 Shear stress modal participation factor resulting from various stiffness variations.

Fig. 17 Normal stress modal participation factor resulting from various stiffness variations.

References

1. Abdel-Ghaffar, Ahmed M., "Earthquake Induced Longitudinal Vibration of earth dams," Report No. 80-SM-14, Civil Engineering Department, Princeton University, Princeton, NJ, August 1980.

2. Abdel-Ghaffar, A. M., and Scott, R. F., "An Investigation of the Dynamic Characteristics of an Earth Dam," EERL-78-02, Earthquake Engineering Research Laboratory, California Institute of Technology, Pasadena, CA, August 1978.

3. Abdel-Ghaffar, A. M., Scott, R. F. and Criag, M. M., "Full-Scale Experimental Investigation of a Modern Earth Dam," EERL-80-02, Earthquake Engineering Research Laboratory, California Institute of Technology, Pasadena, CA, February 1980.

Table 1 SANTA FELICIA EARTH DAM

Comparison Between Observed Resonant Frequencies (in Hz) During Two Earthquakes and Those Computed by the Proposed Analytical Models Longitudinal Direction

1971 Earthquake (M_L=6.3)		Mode	$G = G_0(y/h)^{\ell/m}$					$G = G_0\left[(1-\varepsilon)\frac{y}{h}+\varepsilon\right]$	
Freq.	Part. Fact.[a]	Order (n,r)	$\frac{\ell}{m}=0$	$\frac{\ell}{m}=1$	$\frac{\ell}{m}=\frac{1}{2}$	$\frac{\ell}{m}=\frac{2}{5}$	$\frac{\ell}{m}=\frac{1}{3}$	ε=0.5	ε=0.6
1.35	1.00	1,1	1.35	1.35	1.35	1.35	1.35	1.35	1.35
1.70	0.61	1,2	1.79	1.60	1.69	1.71	1.72	1.73	1.74
1.86	0.62	1,3	2.34	1.89	2.13	2.17	2.20	2.22	2.25
2.15	0.44	2,1	2.77	2.17	2.58	2.63	2.65	2.67	2.70
2.32	0.62	1,4	2.94	2.34	2.59	2.67	2.72	2.74	2.76
2.91	0.53	2,2	3.00	2.43	2.79	2.84	2.87	2.89	2.92
3.15	0.53	2,3	3.36	2.51	3.05	3.15	3.19	3.22	3.26
3.49	0.49	1,5	3.57	2.78	3.10	3.17	3.24	3.27	3.31
		2,4	3.81	3.10	3.49	3.55	3.59	3.62	3.66
3.85	0.34	1,6	4.21	3.20	3.50	3.66	3.77	3.82	3.87
		3,1	4.26	3.35	3.87	3.96	4.02	4.07	4.12
4.03	0.43	2,5	4.31	3.44	3.94	4.02	4.06	4.11	4.15
		3,2	4.42	3.48	4.01	4.10	4.16	4.22	4.27
4.42	0.36	3,3	4.63	3.68	4.23	4.33	4.39	4.56	4.51
		2,6	4.86	3.76	4.43	4.52	4.58	4.62	4.68
4.88	0.75	3,4	5.00	3.94	4.52	4.62	4.69	4.73	4.79
v_{s0} (in ft/sec.) =			722.8	967.8	827.0	804.0	789.5	801.7	782.7

1976 Earthquake (M_L=4.7)		Mode	$G = G_0(y/h)^{\ell/m}$					$G = G_0\left[(1-\varepsilon)\frac{y}{h}+\varepsilon\right]$	
Freq.	Part. Fact.[a]	Order (n,r)	$\frac{\ell}{m}=0$	$\frac{\ell}{m}=1$	$\frac{\ell}{m}=\frac{1}{2}$	$\frac{\ell}{m}=\frac{2}{5}$	$\frac{\ell}{m}=\frac{1}{3}$	ε=0.5	ε=0.6
1.27	1.00	1,1	1.27	1.27	1.27	1.27	1.27	1.27	1.27
1.66	0.67	1,2	1.68	1.50	1.59	1.61	1.62	1.63	1.64
1.86	0.73	1,3	2.20	1.78	2.00	2.04	2.07	2.08	2.11
2.15	0.38	2,1	2.60	2.20	2.43	2.47	2.50	2.51	2.54
2.64	0.38	1,4	2.77	2.04	2.44	2.51	2.57	2.59	2.61
		2,2	2.83	2.28	2.62	2.67	2.70	2.72	2.75
		2,3	3.16	2.37	2.87	2.97	3.00	3.03	3.07
3.22	0.91	1,5	3.36	2.52	2.91	2.98	3.05	3.08	3.13
		2,4	3.58	2.91	3.28	3.34	3.38	3.40	3.45
		1,6	3.96	2.56	3.29	3.44	3.54	3.57	3.61
3.71	0.55	3,1	4.01	3.16	3.65	3.73	3.78	3.83	3.88
		2,5	4.05	3.23	3.71	3.78	3.82	3.88	3.92
		3,2	4.16	3.27	3.77	3.86	3.91	3.97	4.02
4.20	0.79	3,3	4.39	3.46	3.98	4.07	4.13	4.19	4.25
		2,6	4.57	3.53	4.17	4.25	4.30	4.31	4.37
4.59	0.44	3,4	4.70	3.71	4.25	4.35	4.41	4.45	4.48
v_{s0} (in ft/sec.) =			680.0	910.5	778.0	756.4	742.7	754.2	736.4

[a]Participation factor was obtained by dividing the value of the amplitude corresponding to a given resonant frequency by the largest amplitude that corresponds to the fundamental frequency.

NOTE: Based on the in-situ wave-velocity measurements, the Poisson's ratio was taken to be 0.45.

Table 2

SANTA FELICIA EARTH DAM
Comparison Between Resonant Frequencies (in Hz) From Full-Scale Dynamic Tests and Those Computed by the Proposed Analytical Models
Longitudinal Direction

Forced Vibration Tests		Ambient Vibration Tests	Mode	$G = G_0(y/h)^{\ell/m}$					$G = G_0\left[(1-\varepsilon)\frac{y}{h}+\varepsilon\right]$	
Measured Frequency	Mode Designation and Remarks	Average Measured Frequency	Order n,r	$\frac{\ell}{m}=0$	$\frac{\ell}{m}=1$	$\frac{\ell}{m}=\frac{1}{2}$	$\frac{\ell}{m}=\frac{2}{5}$	$\frac{\ell}{m}=\frac{1}{3}$	ε=0.5	ε=0.6
1.425	L1	1.446	1,1	1.446	1.446	1.446	1.446	1.446	1.446	1.446
1.675	L2: Identical to U-D[a] Mode S1.[b]	1.638	-	-	-	-	-	-	-	-
1.850	L3: " " " " R1.	1.849	-	-	-	-	-	-	-	-
1.950	L4: " " " " S2.	1.986	-	-	-	-	-	-	-	-
2.155	L5	2.175	1,2	1.912	1.711	1.813	1.833	1.846	1.851	1.866
2.275	L	2.317	1,3	2.502	2.025	2.278	2.326	2.357	2.373	2.406
2.425	L	2.452	2,2	3.218	2.694	2.987	3.037	3.069	3.096	3.127
2.625	L6: " " " " R3.	2.638	-	-	-	-	-	-	-	-
2.875	L7	2.880	1,4	3.148	2.325	2.772	2.857	2.911	3.105	3.123
3.050	L8: " " " " AS3.	3.033	-	-	-	-	-	-	-	-
3.150	L9: " " " " R4.	3.150	-	-	-	-	-	-	-	-
3.230	L	3.229	1,5	3.821	2.598	3.263	3.392	3.476	3.498	3.513
3.350	L10	3.323	1,6	4.509	2.846	3.737	3.918	4.036	4.049	4.064
3.480	L11	3.487	2,3	3.600	2.978	3.319	3.377	3.416	3.453	3.490
3.775	L12	3.775	2,5	4.615	3.681	4.224	4.303	4.353	4.383	4.398
4.025	L	4.051	2,4	4.076	3.323	3.740	3.807	3.852	4.884	4.900
4.175	L : " " " " AS5.	4.190	-	-	-	-	-	-	-	-
4.370	L13: " " " " AR3.	4.358	-	-	-	-	-	-	-	-
4.700	L14	4.667	2,6	5.199	4.026	4.747	4.843	4.901	4.927	4.943
4.875	L15	4.876	3,4	5.354	4.225	4.843	4.952	5.023	5.052	5.078
5.350	L	-	3,5	5.776	4.572	5.223	5.337	5.413	5.438	5.454
5.600	L	-	3,6	5.920	4.949	5.660	5.778	5.857	5.880	5.896
		v_{s0} (in ft/sec)=		774.2	1037.0	885.8	861.2	845.6	858.7	838.4

[a] Upstream-Downstream direction.
[b] S and AS correspond to symmetric and antisymmetric shear modes, while R and AR correspond to symmetric and antisymmetric rocking modes (see Ref. 3).

A STUDY OF THE USE OF AMBIENT VIBRATION MEASUREMENTS TO DETECT CHANGES IN THE STRUCTURAL CHARACTERISTICS OF A BUILDING

by

P. R. Sparks[(1)], A. P. Jeary[(2)] and V. C. M. de Souza[(3)].

SUMMARY

Tests are described in which the lateral and torsional stiffnesses of a 1/4-scale model of an 18 storey, precast-concrete, building were studied as the structure was subjected to impact and explosion damage. The initial stiffness characteristics of the building were determined both dynamically, using an eccentric-mass vibrator, and statically using horizontally applied loads. Changes in these stiffnesses, due to damage, were then deduced from measurements of the fundamental natural frequencies, obtained from acceleration spectra of the response of the building to turbulence, generated by a laboratory air conditioning system. This procedure proved particularly useful in detecting structural changes which were not apparent from a visual inspection.

INTRODUCTION

There has been considerable interest in recent years in the use of ambient vibration measurements to detect changes in the structural characteristics of buildings subjected to damage. One of the problems in testing the validity of this technique is the difficulty of knowing exactly what structural damage has occurred and knowing what the initial characteristics of the structure had been.

Recently a 1/4-scale, micro-concrete, model of an 18 storey precast-concrete building, was constructed at the Building Research Establishment, U.K. The main purpose of this model was to test the behaviour of such a structure under impact loads and explosions but the opportunity was also taken to monitor the response of the structure to ambient vibrations as

---

(1) Department of Engineering Science and Mechanics, Virginia Polytechnic Institute and State University, Blacksburg, VA 24061 U.S.A.

(2) Building Research Establishment, Garston, Watford, WD2 7JR, U.K.

(3) Proconsulte Construcões Ltd. Rio de Janeiro, Brazil and University College, London, U.K.

the amount of damage increased. This paper describes these tests in which changes in the lateral and torsional stiffnesses of the building were deduced from measurements of its fundamental natural frequencies. These were obtained from acceleration spectra cf the response of the top of the structure to turbulence generated by an air-conditioning system.

The model building which was 11.8m high, 7.7m wide and 4.6m deep was made of 1500 precast concrete panels. Only the structural elements were modelled i.e. 6 cross-walls in the north-south direction, one spine wall in the east-west direction, a central core and all the floors, Figure 1.

The procedure adopted was to determine the initial stiffness characteristics of the structure dynamically and then to compare these stiffnesses with those obtained from static loading tests. Ambient vibrations were then analyzed to determine fundamental natural frequencies of the structure as damage progressed.

THE INITIAL CHARACTERISTICS OF THE BUILDING

The initial dynamic characteristics of the building were determined using an eccentric-mass vibrator. From these the translational and torsional stiffnesses were deduced. These were then used to predict the deflection of the structure under various types of statically applied loading and the results compared with the deflections observed in some preliminary loading tests on the structure.

The Vibrator Tests

The vibrator was mounted on the top of the core of the building and four accelerometers were placed around the edge of the roof with their sensitive axes parallel to the faces, Figure 1. The frequency of the force generated by the vibrator was slowly increased until resonance conditions were observed, indicated by a peak response from an accelerometer. For uncoupled modes one would expect to see similar, in phase, peak responses from the north and south accelerometers when the vibrator was operating at a natural frequency in the east-west direction. Similarly, the east and west accelerometers ought to have shown peak, in phase, responses at a natural frequency in the north-south direction.

At 4.67 Hz the accelerometers in the north and south showed peak signals which, although in phase, were of different amplitudes, indicating the existance of some torsion in a predominantly translational mode.

In the north-south direction no peak could be found in which the signals from the accelerometers in the east and west were in phase. One peak however did exist at 3.85 Hz in which the signals were in antiphase, the signal from the west accelerometer being somewhat greater than that from the east. It therefore appeared that a completely coupled torsional and north-south translational mode existed.

The ambient vibrations generated by the turbulence of the laboratory air-conditioning system were used to check this conclusion. The

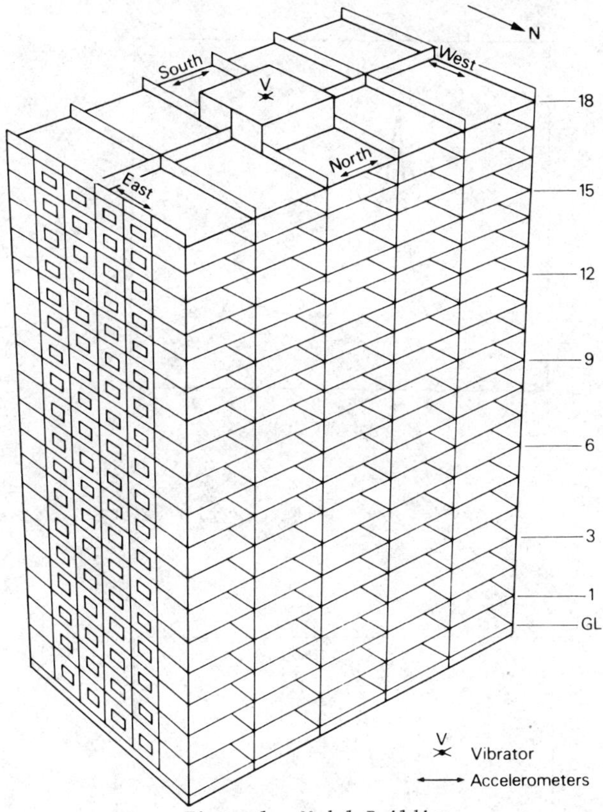

Figure 1. Model Building

accelerometer arrangement was the same as before. The signals from these were, after amplification and filtering, recorded on F. M. magnetic tape recorder.

Spectral analysis of the signals showed that the level of vibration was low and of the same order as peaks created by noise in the recording system. These noise peaks were also present on recordings made when no transducer was connected and a technique developed by Jeary[1] was used to identify those peaks in the spectra which were due to structural responses. This method rejects signals which are highly correlated between a transducer recording and a noise recording. Figures 2 and 3 show typical cleaned spectra obtained at this time.

In Figure 2 it can be seen that at the lower end of the spectrum obtained from the west accelerometer, a major peak exists at the frequency of the coupled torsional and north-south translational mode, $f_{(N-S, \theta)}$. A small peak also exists at the frequency of the predominantly translational east-west mode, $f_{(E-W)}$. This latter mode does contain some

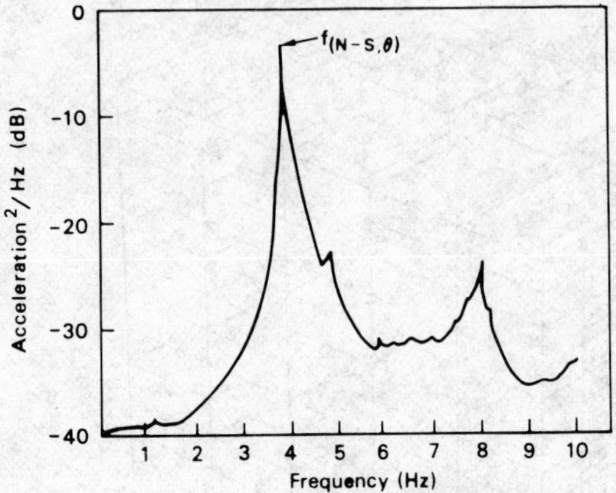

Figure 2. Initial Spectrum West Accelerometer

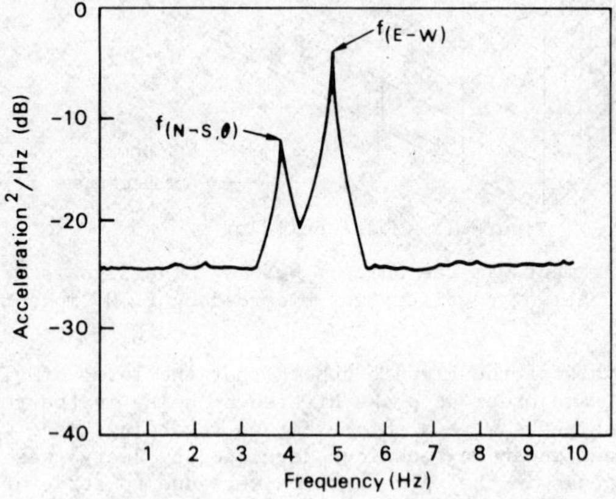

Figure 3. Initial Spectrum North Accelerometer

torsion and it is this component that is detected by the west accelerometer.

Figure 3 shows the spectrum from the north accelerometer. It contains two peaks, one associated with the torsional component of the (N-S,θ) mode and one with the predominantly translational component of the (E-W) mode, although this also contains an element of torsion. Neither spectrum, however, contains a peak at a third fundamental frequency.

Satisfied that only two fundamental frequencies existed, the vibrator tests were continued. Damping was determined by setting up each mode in turn at each of the two fundamental natural frequencies, suddenly stopping the vibrator and recording the decay of the vibrations as measured by each of the accelerometers. Mode shapes were determined by first setting up a mode at one of the fundamental natural frequencies and observing the levels of vibration at each of the floors in the centre of the south face of the building, in directions parallel to and at right angles to the face. This was then repeated at the second fundamental natural frequency. Figure 4 shows the translational mode shapes for the (E-W) and (N-S, θ) modes and the torsional mode shape for the (N-S,θ) mode.

When modes are uncoupled the stiffness of a particular mode of a structure can be obtained directly from a knowledge of the force generated by the vibrator, the response of the structure, the mode shape and the damping. In this case the modes were not pure and it was difficult to obtain accurate damping measurements or to say precisely what the response of the structure was in a particular mode.

Fortunately being an experimental building its mass was known accurately and, although the modes were complex, it was assumed that, for the purposes of comparison, the translational mode shapes, as measured on the south face, could be used to compute the translational modal mass. The modal mass of the rth mode, M_r, is given by

$$M_r = \int_0^H m(z)\phi_r^2(z)dz$$

where $m(z)$ is the calculated mass per unit height,

$\phi_r(z)$ is the translational mode shape and H is the height of the building.

The modal stiffness, K_r, is then determined from

$$K_r = 4\pi^2 f_r^{\,2} M_r$$

where f_r is the appropriate natural frequency. The corresponding equations for a torsional mode are

$$J_r = \int_0^H j(z)\ \phi_r^2(z)\ dz \text{ and } K_r = 4\pi^2 f_r^{\,2} J_r$$

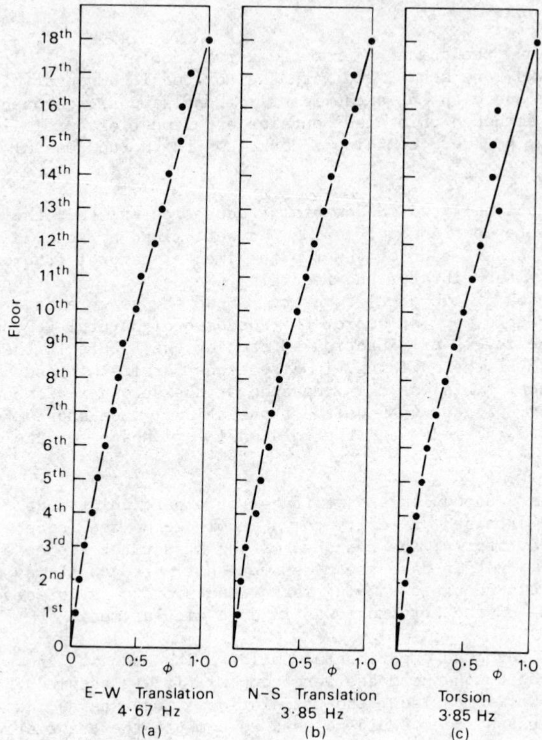

Figure 4. Normalized Mode Shapes

where J_r is the modal polar moment of inertia about an axis assumed to pass through the geometric centre of each floor of the building and j(z) is the corresponding value of the polar moment of inertia per unit height.

Values of stiffness, modal mass and frequency obtained in this way are given in Table 1 under state 1.

Static Load Tests

As part of the preliminary testing of the building a number of lateral load tests were carried out. Although not ideal for testing the stiffness values obtained from the vibrator tests they could be used as a check on the assumptions made in artifically decoupling the modes.

STATE	NATURAL FREQUENCY			MODAL MASS OR INERTIA			MODAL STIFFNESS		
	E-W (Hz)	N-S (Hz)	θ (Hz)	E-W (kg)	N-S (kg)	θ (kgm^2)	E-W (N/m)	N-S (N/m)	θ (Nm/rad.)
1	4.67	3.85	3.85	3.88×10^4	3.76×10^4	2.24×10^5	3.34×10^7	2.21×10^7	1.31×10^8
2	4.63	3.64	3.15	3.88×10^4	3.76×10^4	2.24×10^5	3.28×10^7	1.98×10^7	8.85×10^7
3	4.40	3.53	3.12	3.88×10^4	3.76×10^4	2.24×10^5	2.97×10^7	1.86×10^7	8.61×10^7
4	4.26	3.54	3.13	3.59×10^4	3.48×10^4	1.97×10^5	2.57×10^7	1.72×10^7	7.62×10^7
5	4.00	3.50	3.10	3.28×10^4	3.17×10^4	1.66×10^5	2.07×10^7	1.53×10^7	6.29×10^7

Table 1. Changes in Structural Properties

Three types of loading were used. Type A applied horizontal forces at the end of the spine wall in the direction from east to west. Table 2 shows the forces applied at each floor. Type B applied horizontal forces at the ends of each of the six cross-walls in the direction from north to south. Table 2 lists the forces applied at each cross-wall at a given floor level.

Type C applied horizontal forces at the end of the west cross-wall in the direction from north to south. Table 2 again shows the distribution.

If the assumption is made that the deflection of the structure at zero frequency (i.e. the static deflection) consists only of contributions from the fundamental modes, then the deflection at the top of the building, where the mode shape is equal to 1, should be given by F_r/K_r. F_r is the modal force defined by

$$F_r = \Sigma F'_n \phi_n$$

where F'_n is the force applied at the nth floor where the mode shape value is ϕ_n. The torsional rotation, θ, is given by

$$\theta = \frac{T_r}{K_r}$$

where the modal torque, T_r, is given by

$$T_r = \Sigma T'_n \phi_n$$

T'_n and ϕ_n being defined in a similar manner as for translation.

FLOOR	FORCE (kN)		
	Type A	Type B	Type C
18	2.033	0.454	2.724
15	4.218	0.908	5.448
12	3.534	0.794	4.764
9	3.534	0.794	4.764
6	2.736	0.567	3.402
3	2.280	0.454	2.724
1	1.140	0.227	1.362

Table 2. Force Distribution in Lateral Load Tests

On this basis one would expect from the estimates of stiffness made from the dynamic tests that the translation in the east-west direction under load type A would be 0.29 mm. The mean measured deflection was 0.28 mm. The computed deflection under type B in the north-south direction was 0.58 mm. In the load test there was some distortion of the structure as one might expect from the vibrator test results. The mean deflection of the centre of the building was judged to be approximately 0.63 mm. Type C loading provides a crude check on the torsional stiffness. Assuming the centre of rotation of the building to be at its geometrical centre, the torsional modal stiffness predicted a rotation of 0.00039 radians The measured rotation was of the order of 0.00043 radians.

STUDIES OF DAMAGE DUE TO IMPACT AND EXPLOSION

The comparisons between stiffnesses obtained from static and dynamic tests showed that reasonable estimates of translational and torsional stiffnesses could be obtained from measured modal frequencies and computed modal masses. Thus, provided that the assumption is made that after damage the mode shapes of the structure will not significantly change and that any significant changes in mass distribution can be observed, then only a measurement of frequency is required to provide an estimate of the stiffness of a given mode. There had been good agreement in the preliminary tests between natural frequencies obtained using the vibrator and spectral analysis of the response of the structure to ambient vibration. For safety reasons, therefore, it was decided to use the latter technique to determine the fundamental natural frequencies of the building as it received more and more damage.

The observed changes in frequency, modal mass and stiffness are given in Table 1. The nature of the impact or explosion and the ensuing damage, are given below.

(a) Between States 1 and 2. A swinging ball hit the building firstly at the north end of the west cross-wall and then at the south end of the west cross-wall, both near the bottom of the building. Only local damage to the lower flank wall panels appeared to result.

(b) Between States 2 and 3. One simulated gas explosion took place at the south-east corner bay of the 11th storey i.e. between floors 10 and 11 and caused the 10th, 11th and 12th floors to collapse onto the 9th floor which cracked and deflected severely. A block of 9 wall panels on the 10th, 11th and 12th storeys bulged outwards but remained in place, although they were apparently unable to carry any load. The spine wall moved back 20-30 mm. A second explosion at the south-west corner bay of the 13th storey caused similar damage with the collapse of the 12th, 13th and 14th floors. The spine wall appeared to move back about 100 mm.

(c) Between States 3 and 4. One simulated gas explosion took place in the north-west corner bay of the 15th storey. A progressive collapse occurred, half of the 16th floor and all of the 15th and lower floors of the bay ending up on the ground. The outer cross-wall remained in place, although severely distorted. The spine wall was again displaced in the region of the explosion.

A second explosion in the north-east corner of the 17th storey did not produce a progressive collapse but caused similar damage to that caused by the first two gas explosions.

(d) Between States 4 and 5. A simulated blast from a high explosive device took place in the centre-bay of the north face at ground level. All wall panels adjacent to the bay were blown out and the floor panels of the bay above and adjacent to the centre bay collapsed.

A second blast took place in the north-east corner bay of the ground floor. All six cross-wall panels were blown out as was the lowest element of the spire wall. After a delay of about 20 minutes the second cross-wall suddenly sheared at a line of door openings and with the exception of the 18th, all the floors of the north-east corner bay collapsed onto the ground. The end cross-wall remained in tact.

INTERPRETATION OF THE RESULTS

Before considering the damage detected by the vibration tests it is worth considering the way in which the structure was intended to resist lateral and torsional loads.

In the east-west direction the spine wall and the core provided the main resistance although it will be shown later that the floors and cross-walls acting as boxes probably provided some restraint.

The lateral resistance in the north-south direction was provided mainly by the cross-walls and the core. Torsional resistance was provided mainly by the outer cross-walls.

Changes Between States 1 and 2

The only oberved difference between these states was some minor damage to the north-west and south-west corners of the building. These corner walls apparently provided a significant part of the torsional restraint and the torsional stiffness dropped 34%. The north-south stiffness only dropped 10% and so for the first time separate (N-S) and θ modes could be detected. Loss of two walls perpendicular to the east-west direction had little effect on the east-west stiffness.

Change Between State 2 and 3

After the two gas explosions which occurred between states 2 and 3 some small lateral distortion of the spine wall was observed but it seems hardly enough damage to justify a 10% reduction in east-west stiffness. However, a number of floors collapsed, destroying about 5% of the boxes formed by the cross-walls and floors.

Apparently little damage occurred to the east-west lateral or torsional resistance of the structure.

Changes Between States 3 and 4

The progressive collapse that took place between states 2 and 3 removed the contribution of half of the end cross-wall i.e. about 8% of the lateral shearing elements. As a consequence of this the north-south stiffness dropped 8%. The outer cross-wall makes a greater contribution to the torsional stiffness and the torsional stiffness was seen to drop 12%.

It is interesting to note that the frequencies in the (N-S) and θ modes actually rose slightly but the progressive collapse brought about a significant redistribution of mass and when this was taken into account the stiffnesses were seen to drop. Although there appeared to be little damage to the spine wall, a further 10% of the floor-wall boxes were removed and the east-west stiffness dropped a further 18%.

Changes Between States 4 and 5

The progress collapse that occurred between these states removed about 20% of the remaining wall-floor boxes and removed a lower panel from the spine wall. As a consequence of this the east-west stiffness dropped a further 19%.

The north-south stiffness, losing the cross-wall contribution, dropped 11% and the torsional stiffness dropped 17% for similar reasons. Again changes in mass distribution had to be taken into account.

At this stage the east-west stiffness was only 63% of its initial value despite the fact that its supposedly main stiffening element, the spine wall, was largely in tact. The structure had however lost about 30% of the floor-wall boxes. In a real building the effect is likely to be more pronounced because partitions not included in this model would initially provide part of the stiffness. The north-south stiffness was 69% of its initial value but this seems reasonable because by this stage the equivalent of about 3 of the cross-walls were missing.

The torsional stiffness was only 48% of its initial value but one would expect the effect of the loss of cross-walls to be more pronounced in the torsional stiffness than in the north-south translational stiffness because the outer walls, those that provided the greatest torsional restraint, had been destroyed.

CONCLUSIONS

These tests have shown that provided good estimates of mass distribution and mode shapes are available, acceleration spectra of ambient vibrations can be used to detect changes in lateral and torsional stiffnesses of a building.

Ideally the structure's initial characteristics should be obtained using a vibrator. This form of controlled excitation enables mode shapes and damping to be obtained more readily, although in theory, with a sufficient number of accelerometers, ambient vibrations could be used to obtain these parameters.

Care should be taken in the interpretation of changes in natural frequency. Erroneous conclusions may be drawn concerning structural stiffnesses if significant redistribution of mass is not taken into account.

The method appears to be particularly useful in detecting changes which would not be apparent from a visual inspection. For example the detection of the large change in torsional resistance which resulted from apparently minor local damage and the discovery of the significant contribution of the boxes, formed by the floors and walls, in providing lateral restraint in the east-west direction. The method however does not give any indication of changes in the vertical load carrying capability of the structure.

ACKNOWLEDGEMENTS

The authors wish to acknowledge the contribution of G. S. T. Armer and his staff who conducted the main series of experiments and who were responsible for the design, construction and destruction of the model building.

The tests described in this paper formed part of the research programme of the Building Research Establishment and the results are published by permission of the Director.

The contribution of V. C. M. de Souza was made possible by a grant provided by the National Council for Scientific and Technological Development (C. N. P. q.), Brazil.

REFERENCE

1. Jeary, A. P., "The Estimation of Reliable Spectral Information When Recording Low Intensity Data," Journal of Sound and Vibration, Vol. 60, No. 3; 1978, pp. 401-409.

HYBRID MODELLING OF SEISMIC SOIL-STRUCTURE INTERACTION

by

Sunil Gupta[1] and Joseph Penzien[2], M. ASCE

ABSTRACT

A three-dimensional hybrid model for the analysis of soil-structure interaction under dynamic conditions is developed which takes advantage of the desirable features of the finite element and substructure methods and which minimizes their undesirable features. The modelling is achieved by partitioning the total soil-structure system into a near-field and a far-field with a hemispherical interface. The near-field, which consists of the structure to be analyzed and a finite region of soil around it, is modelled by finite elements. The semi-infinite far-field is modelled by distributed impedance functions at the interface which are determined by system identification methods. Numerical results indicate that the proposed model makes possible realistic and economical assessments of three-dimensional soil-structure interaction for both surface and embedded structures.

INTRODUCTION

Soil-structure interaction has considerable influence on the dynamic response of massive structures such as nuclear power plant buildings and offshore gravity towers supported on medium to soft soil conditions. While considerable effort has been made in the past to develop an understanding of this phenomenon, conceptual and computational difficulties still remain, primarily, due to the three-dimensional, semi-infinite nature of the soil medium, and the complex geometries caused by embedment of real structures. Non-homogeneity and strain dependency of soil properties and uncertainties associated with seismic input motions are factors which further complicate the modelling process.

The two basic methods currently available for the analysis of soil-structure interaction are the substructure method and the finite element method [1,2]. The substructure (or continuum) approach which accounts for radiation damping in the semi-infinite soil medium, provides a simple and economical three-dimensional model for a large class of practical soil-structure interaction problems. However, it is restricted to rigid foundations with simple geometries. Structures having embedded foundations and/or nonlinear soil behavior adjacent to its support cannot be modelled realistically. The finite element method, on the

---

[1] Assistant Research Engineer, Earthquake Engineering Research Center, University of California, Berkeley, Ca., U.S.A.

[2] Professor of Structural Engineering, University of California, Berkeley, Ca., U.S.A.

other hand, has the advantage that flexible, embedded structures with complex geometries can be modelled easily. However, a major disadvantage of this method is that the soil, essentially semi-infinite in nature, is normally modelled by a two-dimensional, finite-size system with a rigid lower boundary. Thus, radiation damping which accounts for the loss of energy in the form of waves travelling away from the foundation, cannot be modelled accurately.

Noting that generally the advantages of one approach are the disadvantages of the other, a hybrid model is developed in this investigation which takes advantage of the good features of each method while at the same time minimizing their undesirable features. This model provides a simple, rational, and economical method of treating soil-structure interaction in three-dimensional form.

HYBRID MODEL

For the hybrid model proposed, the total soil-structure system is partitioned into a near-field and a far-field as shown in Fig. 1. The near-field consists of the structure to be analyzed under prescribed loading conditions, and a finite portion of soil immediately surrounding its base which may encompass irregular geometries such as those produced by embedment. The entire near-field may be modelled in three-dimensional form using the finite element method which makes it possible to realistically represent the complex geometrical shapes associated with real structures. Spatial variations of soil properties within the near-field can also be taken into account by assigning appropriate material properties to the affected finite elements.

The far-field is a semi-infinite half-space which shares a common interface with the near-field. This interface has been judiciously chosen hemispherical in form so that it provides a smooth surface along which mathematical boundary conditions can be easily satisfied. Since the far-field radiates energy in the form of waves travelling away from the foundation, a realistic model must reflect this behavior. In the investigation reported herein, a far-field impedance matrix which relates far-field forces to far-field displacements at the interface nodal points is developed by methods of system identification. This matrix when combined with the near-field equations of motion effectively and efficiently simulates the total soil-structure system.

Equations of Motion

The equations of motion of the isolated near-field subjected to uniform ground motion along the interface can be written as

$$\underline{m}\,\underline{\ddot{u}} + \underline{c}\,\underline{\dot{u}} + \underline{k}\,\underline{u} = \underline{p}(t) + \underline{f}(t) \tag{1}$$

in which $\underline{u}(t)$ is the vector of nodal point displacements (including the interface nodes) relative to the motion of the boundary. Vectors $\underline{\dot{u}}(t)$ and $\underline{\ddot{u}}(t)$ are the corresponding velocities and accelerations. Vector $\underline{p}(t)$ contains the components of effective inertia loading on the system due to earthquake ground motion, and vector $\underline{f}(t)$ the interaction forces corresponding to the interface degrees of freedom. Matrices $\underline{m}$, $\underline{c}$ and $\underline{k}$ represent the mass, damping and stiffness, respectively, of the near-field finite element system.

The equations of motion for the near-field can be transformed into the frequency domain giving

$$(-\omega^2\underline{m} + i\omega\underline{c} + \underline{k})\ \underline{U}(\omega) = \underline{P}(\omega) + \underline{F}(\omega) \tag{2}$$

or

$$\underline{S}(\omega)\underline{U}(\omega) = \underline{P}(\omega) + \underline{F}(\omega) \tag{3}$$

where

$$\underline{S}(\omega) = -\omega^2\underline{m} + i\omega\underline{c} + \underline{k} \tag{4}$$

is the frequency dependent, complex valued impedance matrix characterizing the mass, damping, and the stiffness properties of the near-field. Vectors $\underline{P}(\omega)$, and $\underline{U}(\omega)$ are the Fourier transforms of load vector $\underline{p}(t)$ and displacment vector $\underline{u}(t)$, respectively. $\underline{F}(\omega)$is the Fourier transform of the interaction force vector $\underline{f}(t)$ and ω is the excitation frequency.

The vector $\underline{u}$ of nodal point displacements can be partitioned into two parts: $\underline{u}_b$ corresponding to the nodal displacements at the boundary common to the near- and far-fields, and $\underline{u}_s$ corresponding to the remaining nodal displacements of the near-field. Equation 3 can then be written in the partitioned form

$$\begin{bmatrix} \underline{S}_{ss} & \underline{S}_{sb} \\ \underline{S}^T_{sb} & \underline{S}_{bb} \end{bmatrix} \begin{Bmatrix} \underline{U}_s \\ \underline{U}_b \end{Bmatrix} = \begin{Bmatrix} \underline{P}_s \\ \underline{P}_b \end{Bmatrix} + \begin{Bmatrix} \underline{0} \\ \underline{F}_b \end{Bmatrix} \tag{5}$$

in which vector $\underline{F}_b$ represents the interaction forces at the interface between the near- and the far-fields.

For the isolated far-field, the interface dynamic force-deflection relationship is

$$\underline{S}_f(\omega)\underline{U}_f(\omega) = \underline{F}_f(\omega) \tag{6}$$

where $\underline{S}_f(\omega)$ is the far-field impedance matrix which has to be determined by a separate analysis. In rigorous form, it is a full matrix the elements of which characterize the mass, damping, and stiffness characteristics of the far-field. It is complex valued and frequency dependent.

The equations of motion for the far-field are incorporated into the frequency domain near-field equations by invoking the conditions of compatibility and equilibrium at the interface, i.e.,

$$\underline{U}_f = \underline{U}_b \tag{7}$$

and

$$\underline{F}_f + \underline{F}_b = \underline{0} \tag{8}$$

Substitution of Eqs. 6, 7 and 8 into Eq. 5, leads to the following equations of motion for the hybrid model of the entire soil-structure system in the frequency domain:

$$\begin{bmatrix} \underline{S}_{ss} & \underline{S}_{sb} \\ \underline{S}^T_{sb} & \underline{S}_{bb}+\underline{S}_f \end{bmatrix} \begin{Bmatrix} \underline{U}_s \\ \underline{U}_b \end{Bmatrix} = \begin{Bmatrix} \underline{P}_s \\ \underline{P}_b \end{Bmatrix} \tag{9}$$

or,

$$\hat{\underline{S}}(\omega)\underline{U}(\omega) = \underline{P}(\omega) \tag{10}$$

where $\hat{\underline{S}}(\omega)$ is the impedance matrix of the hybrid system including near- and far-fields.

Earthquake Input Motion

Definition of a realistic input motion is important for the earthquake response analysis of soil-structure systems. The seismic energy arriving at a particular site will depend upon several factors such as the fault rupture mechanism, the location and distance of the site relative to the earthquake epicenter, the intervening and local soil conditions, and the presence of topographical features such as mountains and canyons. A complete characterization of the earthquake ground motion unique to a particular site is, therefore, not possible within the present state of the art. In light of such uncertainties, it is reasonable and prudent to specify a site-dependent response spectrum based on the available seismological information and strong motion records consistent with the local soil conditions. Time-histories of motion, conforming to this response spectrum and having a designated peak acceleration, can then be generated for use as input to the soil-structure system.

In the hybrid model the seismic input is applied to the boundary nodes at the interface between the near- and the far-fields. Since the question concerning spatial variations of ground motion is still unresolved, the same free-field motion may often be applied along the entire boundary. Spatially varying ground motions, if known, can be incorporated into the solution using the hybrid model. This is accomplished by solving a modified set of dynamic equilibrium equations and combining the resulting nodal displacements with the corresponding quasi-static displacements produced by the prescribed interface displacements[3]. The lack of knowledge of appropriate input motions reflects a deficiency in the present state of the art and the need for additional research. It also reflects the need for sound judgement on the part of the analyst in selecting suitable input motions for soil-structure interaction studies.

Dynamic Response

Once the input motion has been defined, the Fourier amplitude $\underline{P}(\omega)$ of the resulting load vector, $\underline{p}(t)$, can be obtained from

$$\underline{P}(\omega) = \int_0^{T_d} \underline{p}(t)e^{-i\omega t}\, dt \tag{11}$$

where T_d is the time duration of excitation. The solution $\underline{U}(\omega)$ of Eq. 10 for discrete values of the excitation frequency completely characterizes response in the frequency domain. The time histories of response can then be obtained by the Fourier transformation of the complex frequency response into time domain using

$$\underline{u}(t) = \frac{1}{2\pi} \int_{-\infty}^{\infty} \underline{U}(\omega)e^{i\omega t}\, d\omega \tag{12}$$

Very efficient and economical Fast Fourier Transform (FFT) techniques are available[4] to carry out the transforms in Eqs. 11 and 12.

FAR-FIELD IMPEDANCE FUNCTIONS

Mathematical Modelling

An accurate representation of the far-field which accounts for radiation damping in the semi-infinite soil medium is central to the concept of hybrid modelling. The development of a far-field impedance matrix, $\underline{S}_f(\omega)$, as needed by the hybrid model, requires the solution of a set of partial differential equations with prescribed boundary conditions at the interface. For the case of torsional loading, it is possible to carry out this rigorous analysis and develop the corresponding far-field impedance matrix[5]. However, for general loading conditions, such a solution appears mathematically intractable at present. A semi-analytical approach is, therefore, adopted here in which the far-field may be thought of as being composed of infinitesimally thin soil elements, extending to infinity in the direction normal to the hemispherical surface cavity, which act independently of each other. This is a realistic assumption if the deformations are smooth and slowly varying functions over the interface, which can be assured by placing the interface at a reasonable distance from the structure. The dynamic load-deflection relationship of each of these infinitesimal soil elements can be characterized by impedance elements, the real part of which represent stiffness and the imaginary part radiation damping. Since there is an innumerable number of such closely spaced infinitesimal soil elements, the far-field, in the limit, may be replaced by continuous impedance functions placed in the three coordinate directions on the interface between the near- and far-fields, as shown in Fig. 2. Conceptually, this is the dynamic equivalent of the Winkler assumption made for the case of static loading of beams on elastic foundations. These continuous impedance functions can then be discretized at the interface nodal points to obtain the far-field impedance matrix.

An arbitrary variation of these far-field impedance functions can be represented in terms of a double Fourier series in θ and ϕ. However, for uniform or horizontally layered halfspaces, the far-field possesses material and geometric axisymmetry about the vertical axis; thus, the impedance functions must be independent of θ and be symmetric in ϕ, giving

$$\begin{aligned} S_R(R,\phi,b_o) &= \sum_{n=0}^{\infty} S_{Rn}(R,b_o)\cos n\phi \\ S_\phi(R,\phi,b_o) &= \sum_{n=0}^{\infty} S_{\phi n}(R,b_o)\cos n\phi \\ S_\theta(R,\phi,b_o) &= \sum_{n=0}^{\infty} S_{\theta n}(R,b_o)\cos n\phi \end{aligned} \tag{13}$$

where S_R, S_ϕ, and S_θ are the far-field impedance functions per unit interface area in the normal, tangential, and circumferential directions to the hemispherical interface, respectively, as shown in Fig. 2 . Coefficients S_{Rn}, $S_{\phi n}$, and $S_{\theta n}$ are complex valued functions of the interface radius R, the shear modulus G, and the non-dimensional frequency parameter b_o, defined by $b_o = \omega R/C_s$. In this expression, ω is the excitation frequency, $C_s = \sqrt{G/\rho}$ is the shear wave velocity, and

ρ is the mass density of the far-field material. The number of terms required in Eqs. 13 to properly represent the far-field depends upon the complexity of layering.

The discrete far-field impedances for node i on the interface are obtained by integrating the continuous boundary impedances over the tributary area A_i which extends halfway to the nodes adjacent to node i as shown in Fig. 2. Therefore, the discrete impedances at node i in the normal, tangential and circumferential directions, respectively, are

$$\left.\begin{aligned} S_R^i &= R^2 \int_{\phi_1^i}^{\phi_2^i} \int_{\theta_1^i}^{\theta_2^i} S_R(R,\phi,b_o)\sin\phi \, d\phi \, d\theta \\ S_\phi^i &= R^2 \int_{\phi_1^i}^{\phi_2^i} \int_{\theta_1^i}^{\theta_2^i} S_\phi(R,\phi,b_o)\sin\phi \, d\phi \, d\theta \\ S_\theta^i &= R^2 \int_{\phi_1^i}^{\phi_2^i} \int_{\theta_1^i}^{\theta_2^i} S_\theta(R,\phi,b_o)\sin\phi \, d\phi \, d\theta \end{aligned}\right\} \tag{14}$$

Thus, for each node on the interface the following 3 x 3 impedance matrix in spherical coordinates is obtained

$$\underline{S}^i = \begin{bmatrix} S_R^i & 0 & 0 \\ 0 & S_\phi^i & 0 \\ 0 & 0 & S_\theta^i \end{bmatrix} \tag{15}$$

leading in general, to a $3N_b x 3N_b$ far-field impedance matrix for the entire interface; where N_b is the number of nodes on the interface.

In the present investigation, the far-field considered is homogeneous, isotropic, and elastic leading to infinitesimal soil elements around the interface with the same properties giving rise to uniformly distributed impedance functions. Therefore, only the constant terms in Eqs. 13 need to be retained, giving

$$\left.\begin{aligned} S_R(R,\phi,b_o) &= S_{RO}(R,b_o) = \eta_R + i\xi_R \\ S_\phi(R,\phi,b_o) &= S_{\phi 0}(R,b_o) = \eta_\phi + i\xi_\phi \\ S_\theta(R,\phi,b_o) &= S_{\theta 0}(R,b_o) = \eta_\theta + i\xi_\theta \end{aligned}\right\} \tag{16}$$

where the η's and ξ's are the real and imaginary parts, respectively, of the unknown far-field impedance functions. These, as yet unknown, impedance functions are determined using system identification methods such that the resulting hybrid model provides a best-fit to the entire set of known response functions for a rigid massless circular plate on a uniform elastic halfspace. Because of axisymmetry of the system under consideration, both the near-field and the far-field can be formulated in an axisymmetric form. It is, therefore, necessary to consider only one quadrant in the $\theta = 0$ plane as shown in Fig. 3 where the nodal points actually describe nodal circles. Discrete far-field impedances as given by Eqs. 14 can be modified accordingly by extending the limits of integration in θ-direction from 0 to 2π instead of from θ_1^i to θ_2^i. The far-field impedance matrices so obtained can be combined with the finite element idealization of the near-field to yield the equation of motion for the hybrid model (Eq. 10) of the rigid massless circular plate in the torsional, vertical, and coupled translation and rocking modes of vibration.

Parameter Evaluation

For a prescribed value of excitation frequency ω and for assumed values of the far-field impedances, matrix Eq. 10 can be solved for the complex harmonic displacement vector $\underline{U}(\omega)$which includes the displacement amplitudes (compliances) of the rigid massless circular plate. Since the resulting compliances depend upon the assumed far-field impedances, errors are involved in their evaluation. To minimize these errors, an error function containing the sum of the squared errors of all the plate compliances is formed giving,

$$J(\underline{\beta},\omega) = \sum_{i=1}^{NC} \left[\mathrm{Re}(U_i) - \mathrm{Re}(C_i) \right]^2 + \sum_{i=1}^{NC} \left[\mathrm{Im}(U_i) - \mathrm{Im}(C_i) \right]^2 \qquad (17)$$

where, $\underline{\beta}$ is an n-dimensional vector containing all of the far-field impedance coefficients, $U_i = U_i(\underline{\beta},\omega)$ are the compliances of the plate as generated from Eq. 10 for the hybrid model, $C_i = C_i(\omega)$ are known plate compliances generated from analytical elasticity solutions, and NC is the number of plate compliances considered in the solution. The analytical solutions for the rigid massless circular plate in the torsional, vertical, and coupled translational and rocking modes of vibration have been presented in Refs. 6, 7 and 8.

Methods of system identification are used to systematically adjust the originally assumed values of the far-field impedance coefficients so that the error function $J(\underline{\beta},\omega)$ is minimized for discrete values of ω, thus, giving the far-field impedance vector $\underline{\beta}$ over the frequency range of interest. Most of these methods use the so-called gradient techniques in which new values for the components in vector $\underline{\beta}$ are obtained by following in the direction of the negative gradient of the error function in the n-dimensional parameter space. The method selected for the present study is the modified Gauss-Newton method which makes use of information on second derivatives; thus, resulting in an improved convergence rate [9]. It is defined by the following equation:

$$\underline{\beta}_i = \underline{\beta}_{i-1} - \lambda\, \underline{h}^{-1}(\underline{\beta}_{i-1},\omega)\, \underline{g}(\underline{\beta}_{i-1},\omega) \qquad (18)$$

where $\underline{\beta}_{i-1}$ and $\underline{\beta}_i$ are the parameter vectors at iterative steps i-1 and i, respectively,

$$\underline{g}(\underline{\beta}_{i-1},\omega) = \left\langle \frac{\partial J}{\partial \beta_1}, \frac{\partial J}{\partial \beta_2}, \ldots, \frac{\partial J}{\partial \beta_n} \right\rangle \qquad (19)$$

is the gradient vector, and

$$\underline{h}(\underline{\beta}_{i-1},\omega) = \begin{bmatrix} \frac{\partial^2 J}{\partial \beta_1^{\,2}} & \cdots & \frac{\partial^2 J}{\partial \beta_1 \partial \beta_n} \\ \cdot & & \cdot \\ \cdot & & \cdot \\ \frac{\partial^2 J}{\partial \beta_n \partial \beta_1} & \cdots & \frac{\partial^2 J}{\partial \beta_n^{\,2}} \end{bmatrix} \tag{20}$$

is the nxn Hessian matrix the inverse of which modifies both the magnitude and the direction of the steepest descent given by the negative gradient. Scalar λ is a positive parameter selected to ensure a decrease in error within each iteration cycle. The error function $J(\underline{\beta},\omega)$ defines an n-dimensional surface which in two dimensions is easy to visualize as shown in Fig. 4. The modified Gauss-Newton method is an interative process in which the error is minimized by obtaining successively better estimates of the far-field impedance vector $\underline{\beta}$ until a point $\underline{\beta}^*$ is located where the slope of the error surface approaches zero. For a detailed discussion of the method see the report by Gupta et al.[5].

NUMERICAL RESULTS

The near-field of the rigid circular plate on an elastic halfspace is modelled by a variable 3 to 9 node isoparametric element especially developed for this study[5]. The finite element mesh used is shown in Fig. 5. The largest dimension of elements anywhere in the mesh is approximately 1/4th of the wave length corresponding to a non-dimensional frequency, b_o, of 9.0. It is anticipated that errors in the displacement field for frequencies below this value will not be greater than 5%. Such a refined mesh was deemed necessary in this investigation to ensure that the far-field impedances obtained would not be unduly influenced by the near-field discretization.

The parameter adjustment algorithm described in the previous section was incorporated into a FORTRAN computer program. The far-field impedance functions, $\eta_R + i\xi_R$ in the normal direction, $\eta_\phi + i\xi_\phi$ in the tangential direction, and $\eta_\theta + i\xi_\theta$ in the circumferential direction were determined by minimizing the error function (Eq. 17) which is formed by considering the response of the plate in all the three modes of vibrations, namely -- torsional, vertical, and coupled translation and rocking. Far-field impedances so obtained are presented in Figs. 6-8 as a function of the non-dimensional frequency, b_o. For any particular frequency, these uniformly distributed far-field impedances are directly proportional to the shear modulus G, and inversely proportional to the interface radius R. The far-field impedance functions presented are for a Poisson's ratio of 1/3, a value which is fairly representative for soils.

The dynamic response of the rigid circular plate, using these far-field impedances are compared with the available solutions in Fig. 9. It is apparent from this figure that the proposed hybrid model is very effective in reproducing theoretical solutions. The discrepancies observed in the coupling compliances, C_{MH}, which are plotted in Fig. 13,

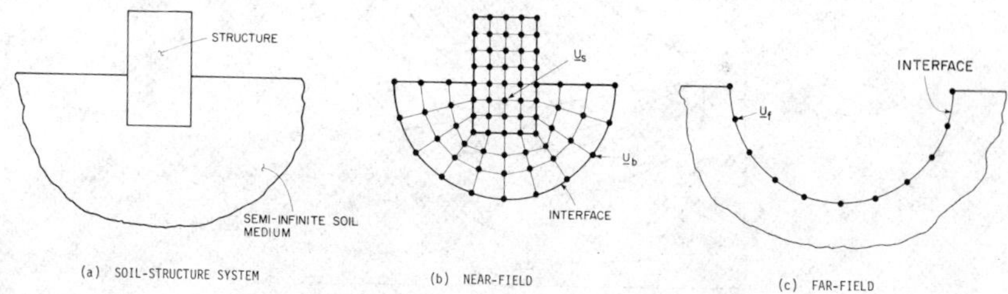

FIG. 1 HYBRID MODELLING OF SOIL-STRUCTURE INTERACTION

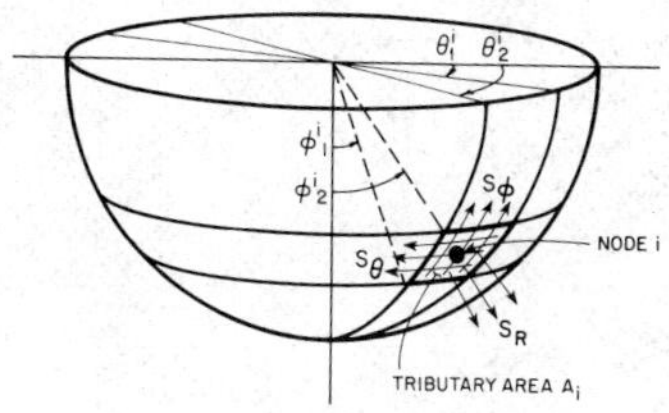

FIG. 2 FAR-FIELD IMPEDANCES IN SPHERICAL COORDINATES

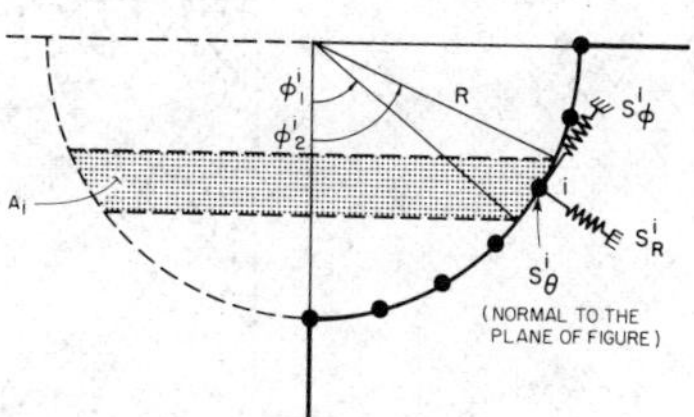

FIG. 3 DISCRETIZED IMPEDANCES IN AXISYMMETRIC FORMULATION

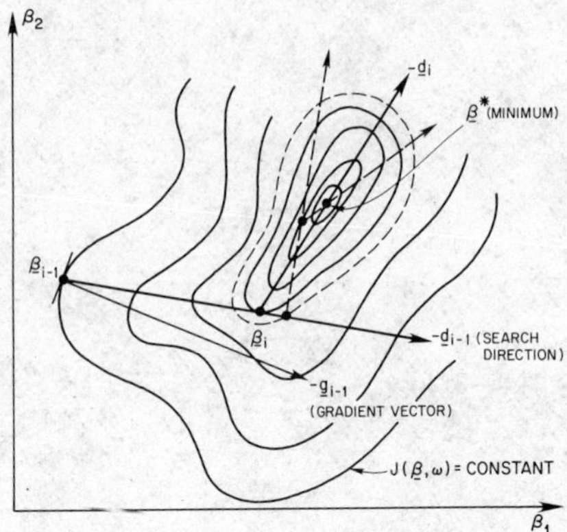

FIG. 4 ERROR SURFACE FOR TWO PARAMETERS

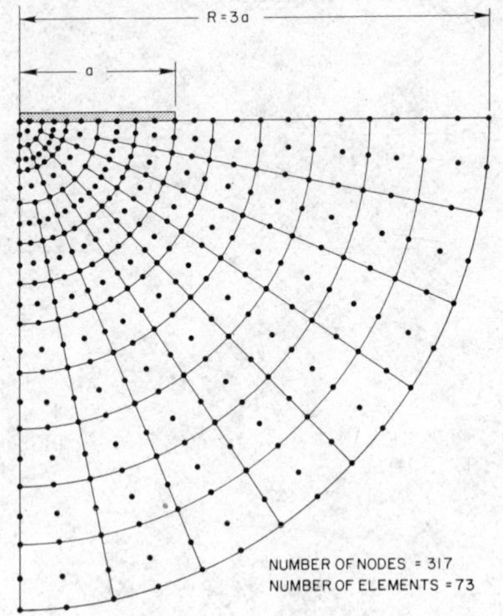

FIG. 5 NEAR-FIELD FINITE ELEMENT MESH FOR RIGID CIRCULAR PLATE

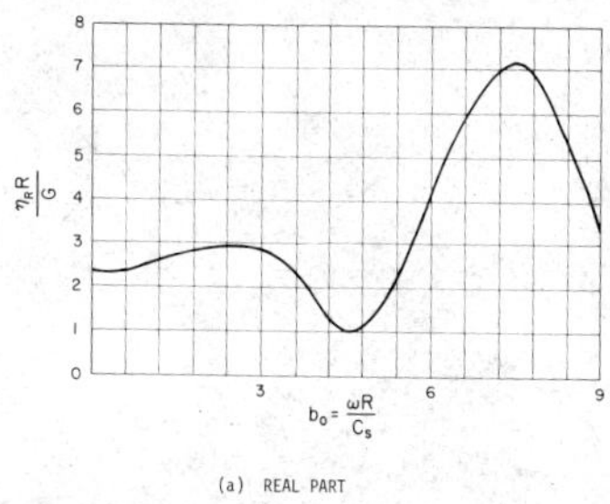

(a) REAL PART

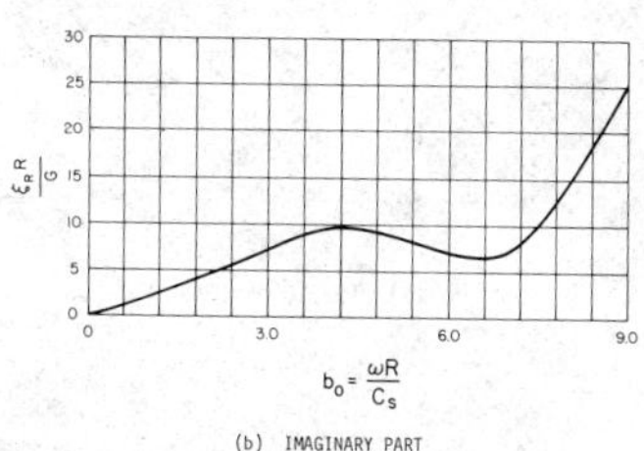

(b) IMAGINARY PART

FIG. 6 FAR-FIELD IMPEDANCE FUNCTIONS - NORMAL COMPONENT

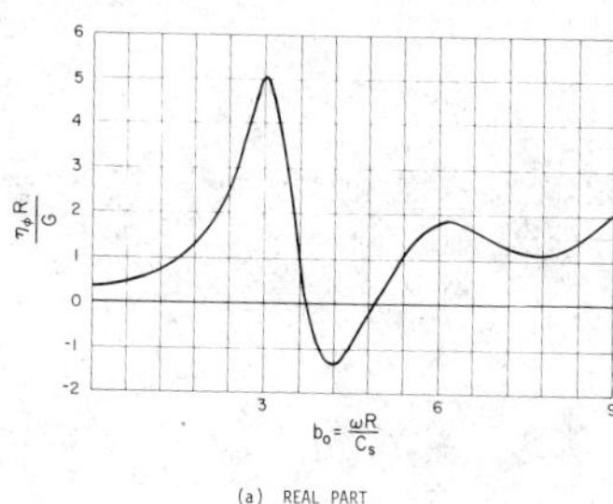

(a) REAL PART

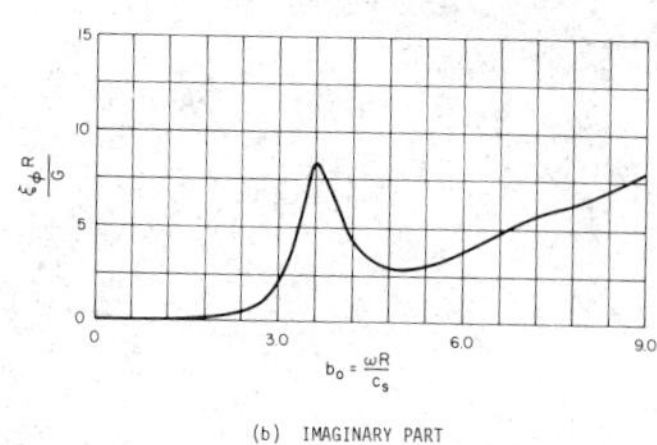

(b) IMAGINARY PART

FIG. 7 FAR-FIELD IMPEDANCE FUNCTIONS - TRANGENTIAL COMPONENT

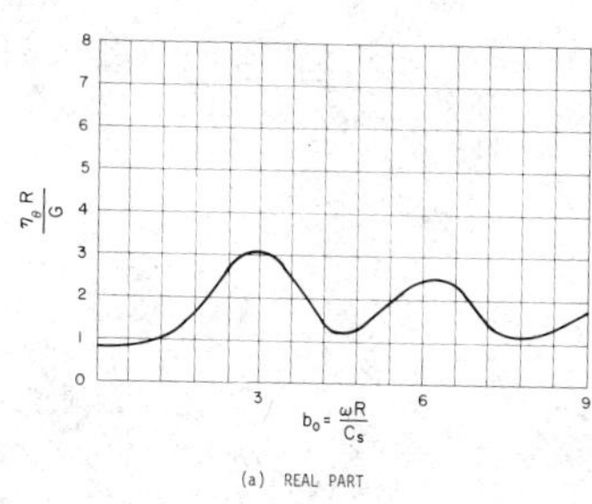

(a) REAL PART

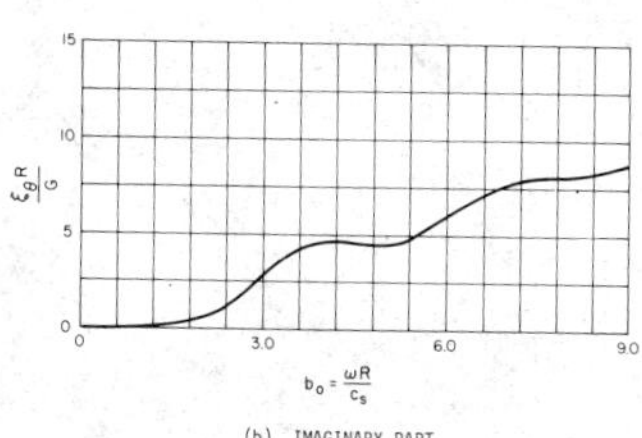

(b) IMAGINARY PART

FIG. 8 FAR-FIELD IMPEDANCE FUNCTIONS - CIRCUMFERENTIAL COMPONENT

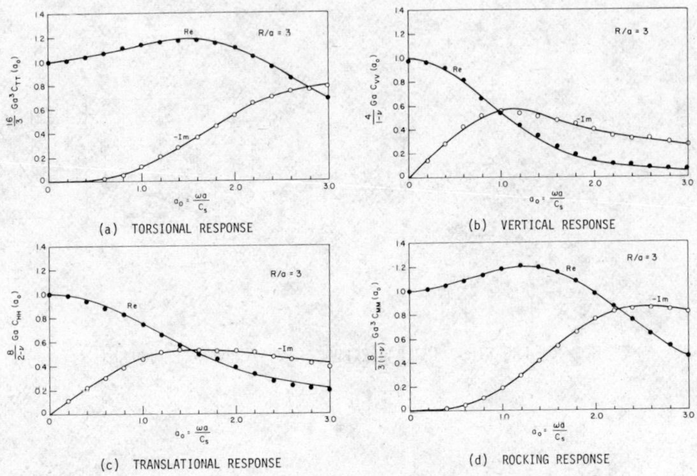

FIG. 9 COMPARISON OF PLATE COMPLIANCES USING IDENTIFIED IMPEDANCE FUNCTIONS; R/a = 3.0

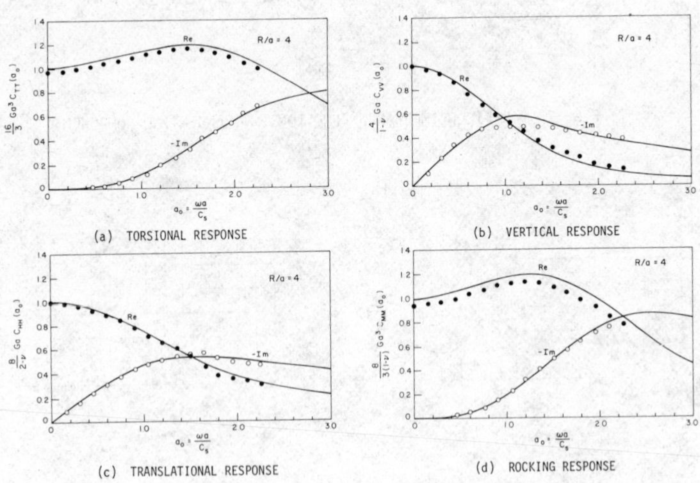

FIG. 10 COMPARIOSN OF PLATE COMPLIANCES USING IDENTIFIED IMPEDANCE FUNCTIONS; R/a = 4.0

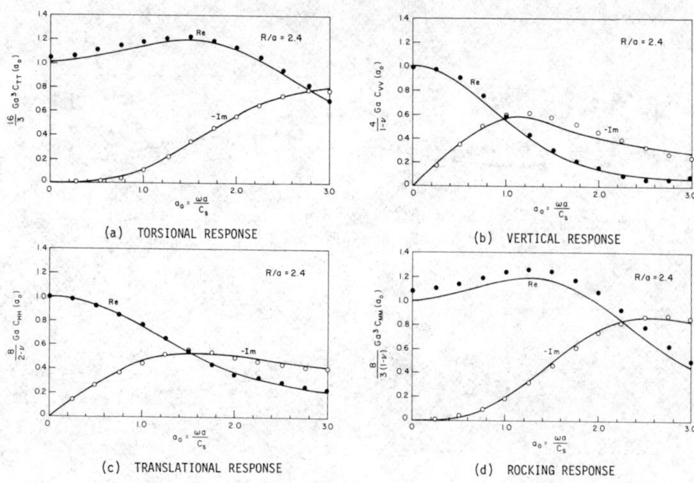

(a) TORSIONAL RESPONSE

(b) VERTICAL RESPONSE

(c) TRANSLATIONAL RESPONSE

(d) ROCKING RESPONSE

FIG. 11 COMPARISON OF PLATE COMPLIANCES USING IDENTIFIED IMPEDANCE FUNCTIONS; R/a = 2.4

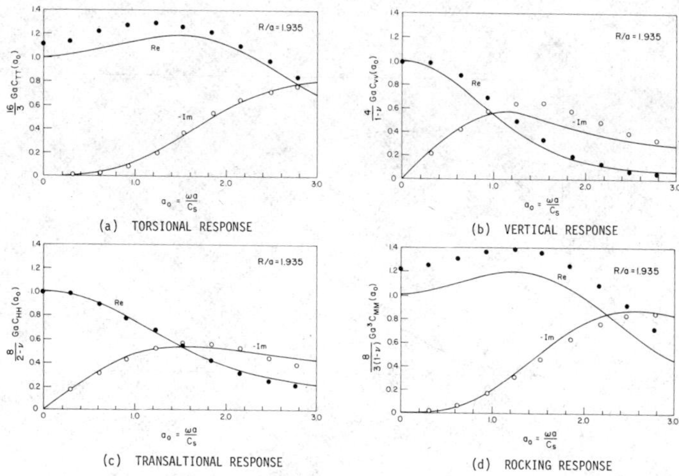

(a) TORSIONAL RESPONSE

(b) VERTICAL RESPONSE

(c) TRANSALTIONAL RESPONSE

(d) ROCKING RESPONSE

FIG. 12 COMPARISON OF PLATE COMPLIANCES USING IDENTIFIED IMPEDANCE FUNCTIONS; R/a = 1.935

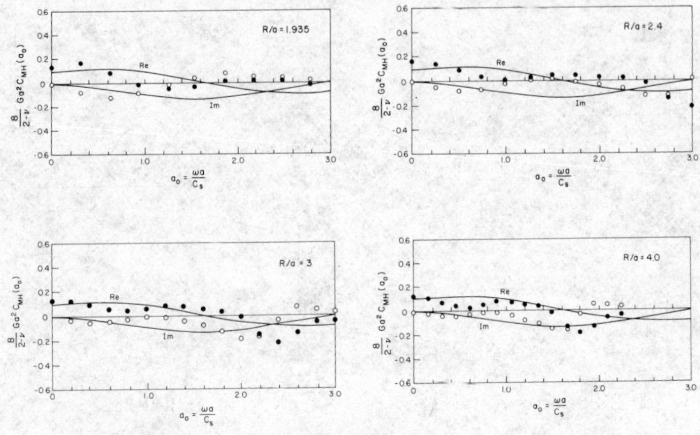

FIG. 13 COUPLING COMPLIANCES USING THE IDENTIFIED IMPEDANCE FUNCTIONS

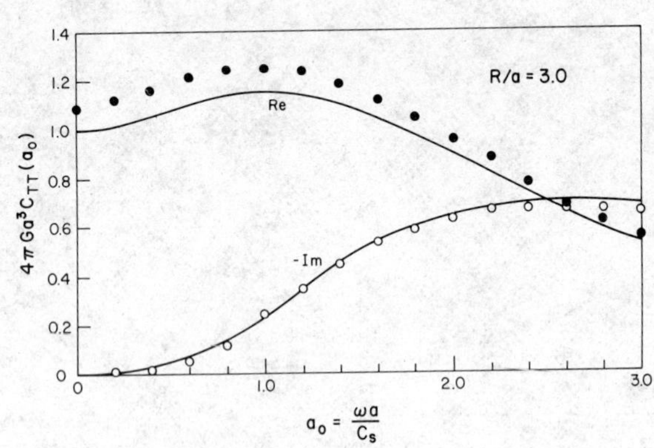

FIG. 14 TORSIONAL RESPONSE OF HEMISPHERICAL FOUNDATION USING IDENTIFIED IMPEDANCE FUNCTIONS

may be due to the assumption of relaxed boundary conditions in the theoretical solutions.

Far-field impedances must, in principle, be related only to the interface radius, R. However, due to the approximate and numerical nature of the modelling, these impedances may be expected to be influenced by the near-field model used in the system identification process. To investigate their range of applicability, these impedances are employed to calculate the compliances of the rigid plate for three other R/a ratios, namely 4.0, 2.4 and 1.935. In changing these ratios, R is held to a fixed value; thus, only the radius of the plate is changed. The resulting compliances are compared with their values obtained from the closed-form solutions in Figs. 10-12. These solutions agree very well for R/a ratios of 4.0 and 2.4 with the errors being of the order of 5 to 10%. The solutions for R/a = 1.935, which represents an extreme case of the far-field being placed at a distance less than two times the radius of plate, agree reasonably well except for the real part of the rocking compliance, C_{MM}, which has errors of the order of 20%. For all the results presented, the errors in imaginary parts are generally much smaller than in the real parts.

Finally, the identified impedances are used in the hybrid modelling of a rigid embedded hemispherical foundation whose response to torsional excitation has been evaluated analytically[10]. The finite element mesh used was obtained by modifying the mesh of Fig. 5. The results are presented in Fig. 14 which indicates that the numerical solutions are within about 10% of the analytical solutions.

CONCLUDING STATEMENT

The results presented here for uniform, elastic soil medium demonstrate the concept and the applicability of modelling the far-field through continuously distributed impedance functions. These functions, as required by the hybrid model, can also be generated for viscoelastic, and layered foundations. Non-linear soil behavior due to large strains in the immediate vicinity of structure's base can be modelled by incorporating this behavior into the affected finite elements in the near-field. This will, of course, require a time domain analysis which can be accomplished by averaging the impedance functions over the predominant frequency band of structural response. Further developments, including a general purpose computer program, are planned by the authors so as to make the method practical and reliable.

REFERENCES

1. Hadjian, A. H., Luco, J.E. and Tsai, N.C., "Soil-Structure Interaction: Continuum of Finite Element?", Nuclear Engineering and Design, Vol. 31, 151-167, 1974.

2. Analyses of Soil-Structure Interaction Effects for Nuclear Power Plants", Report by the Ad Hoc Group on the Soil Structure Interaction of the Committee on Nuclear Structures and Materials of the Structural Division of ASCE, 1979.

3. Clough, R. W. and Penzien, J., Dynamics of Structures, McGraw Hill, 1975.

4. Cooley, J. W. and Tuckey, J.W., "An Algorithm for the Machine Calculation of Complex Fourier Series", Mathematics of Computations, Vol. 19, 297-301, 1965.

5. Gupta, S. et al., "Hybrid Modelling of Soil-Structure Interaction", Report No. UCB/EERC 80-09, Earthquake Engineering Research Center, Berkeley, California, May 1980.

6. Lysmer, J. and Richart, F.E., Jr., "Dynamic Response of Footings to Vertical Loadings", Journal of the Soil Mechanics and Foundations Division, ASCE, Vol. 92, No. SM1, 65-91, Janaury 1966.

7. Veletsos, A.S. and Wei, Y.T., "Lateral and Rocking Vibrations of Footings", Journal of the Soil Mechanics and Foundations Division, ASCE, Vol. 97, No. SM9, 1227-1248, September 1971.

8. Luco, J. E. and Westmann, R. A., "Dynamic Response of Circular Footings", Journal of the Engineering Mechanics Division, ASCE, Vol. 97, No. EM5, 1381-1395, October 1971.

9. Matzen, V. C. and McNiven, H. D., "Investigation of the Inelastic Characteristics of a Single Story Steel Structure Using System Identification and Shaking Table Experiments", Report No. UCB/EERC 76-20, Earthquake Engineering Research Center, University of California, Berkeley, California, 1976.

10. Luco, J.E., "Torsional Response of Structures for SH Waves: The Case of Hemisperhical Foundations", Bulletin of the Seismological Society of America, Vol. 66, 109-123, 1976.

ON THE REDUCTION OF 3D SOIL-STRUCTURE INTERACTION PROBLEMS TO 2D MODELS

A. H. Hadjian[I], M. ASCE, J. E. Luco[II], and H. L. Wong[III]

ABSTRACT

The feasibility of modeling three dimensional soil-structure systems by two dimensional models is explored considering the three dimensional character of the superstructure, a single isolated foundation, a complex of closely spaced structures and structure-structure interaction effects. Guidelines are provided to minimize the impact of the two-dimensional simplification despite the fact that it is not a simple problem to obtain the response of three-dimensional structures by the use of two-dimensional models.

INTRODUCTION

A three-dimensional analysis of soil-structure interaction problems using the finite element method is, in principle, possible. However, at the present time, finite element analyses are usually carried out by taking two dimensional slices of sections through structures (5, 7, 13, 14). There are, however, several questions that must be adequately answered before an intrinsically three-dimensional problem could be reduced to a two-dimensional model. These questions are the subject of this paper. Specifically, the following issues are discussed:

a. Whether it is possible to take a single non-symmetric structure and model it as a plane structure.
b. Whether there are strip (two-dimensional) foundation models that would represent, with a reasonable approximation, isolated circular, rectangular and irregularly shaped foundations.
c. Whether it is possible to consider a slice or section through a power plant complex as representative of the structures at the plant.
d. Whether the interaction of two or more foundations can be represented by infinitely long strips.

The discussions presented in this paper are primarily intended to serve two purposes: a) to help obtain reasonable and acceptable solutions of three-dimensional problems when the only computational tool available is two-dimensional and b) to warn against the indiscriminate use of two-dimensional programs that could lead to unacceptable solutions even though the results of the computations are arithmetically correct.

[I] Principal Engineer, Bechtel Power Corporation, Los Angeles Power Division, Norwalk, California 90650.

[II] Associate Professor, Department of Applied Mechanics and Engineering Sciences, University of California, San Diego, La Jolla, California 92037.

[III] Research Assistant Professor, University of Southern California, Los Angeles, California 90007.

To avoid any of the arguments regarding the adequacy of the finite element method for solving the soil-structure interaction problems (1, 4, 12), all the results in this paper are obtained by non-finite element techniques. The examples are taken from the analysis of nuclear power plant structures. Considering the fact that four separate issues are being reviewed, conclusions and recommendations will be made as each topic is addressed.

STRUCTURAL MODELING

Fig. 1 shows a typical plan of a nuclear power plant structure. The structure is inherently three-dimensional in character, and hence torsional response is possible and may not be eliminated from the analysis. Additional attention in modeling of such a structure should include the proper consideration of the effects of shear areas, shear lag effects and mass moments of inertia (2, 8). It is clear then that the structure should be modeled in all its generality as a three-dimensional system. Any simplification to a two-dimensional model must be justified. One direct approach to show that such a simplification could be made is to first solve the correct problem in three dimensions and then to show that the results obtained can also be reasonably approximated by a two-dimensional model. Such an approach has been adopted herein for a simplified lumped mass model consisting of only two stories of the structure shown in Fig. 1. Discrete soil springs and dampers have been added to the base of the structure as shown in Fig. 2 to provide a convenient model that can be easily modified to two dimensions as will be described below. The properties of the complete model are given in Fig. 2. At first, the structure is analyzed as a three-dimensional model subjected to an earthquake type motion normalized to 1.0g peak acceleration, and acting along the y-direction. The damping of the superstructure is assumed to be 4% of critical for all fixed base modes, and the analysis is completed without introducing any approximations in the system damping characteristics, by first assembling the proper damping matrix (6) and then directly integrating the equations of motion.

Three-dimensional structures can be simplified to two-dimensional structures in a number of ways. Two of these possibilities are: a) the structure is constrained to respond only in the direction of excitation without regard to any change of frequencies of the constrained 2-D model, and b) the structure is constrained to respond only in the direction of excitation, but the stiffness properties are modified such that the fundamental frequency of the 2-D model is made equal to the fundamental frequency of the three-dimensional structure. The differences, for the present example problem, between the frequencies of the 3-D model and the constrained 2-D model are sufficiently small to justify the use of the first approach. Thus, those "soil springs" and "dashpots" that are no longer required in the 2-D model are simply eliminated to obtain the two-dimensional model. The "soil springs," for the purposes of this exercise, can be looked upon as a set of uncoupled stiffness elements of another floor. The uncoupled property of these elements is convenient when an input motion is to be defined at node 3 (Fig. 2) for subsequent analysis of the superstructure. This uncoupled property is also necessary for the proper definition of damping in the superstructure.

It has been suggested (5) that the proper response of the original three-dimensional structure can be restored if the calculated translational and rotational motions of a point in the two-dimensional model are used as base excitations to the original three-dimensional model above that point. This type of analysis is also performed herein by prescribing the translational and rocking motions of node 3 (Fig. 2) obtained from the 2-D model, as base input motions to the original three-dimensional model without the "soil springs." In this analysis, the superstructure damping of 4% for all modes is maintained.

The results from the three analyses prescribed above are shown in Fig. 3 as absolute acceleration response spectra at the top floor of the structure. Three spectra at each of five locations on the floor are shown: Response Curve 1 corresponds to the complete three-dimensional solution. This result is used as the criterion by which the other approximations are evaluated. Response Curve 2 corresponds to the constrained two-dimensional solution which gives identically zero response in the direction normal to the direction of excitation. This latter response is identified as Curve 2 = 0 in Figs. 3 and 4. The results presented as Response Curve 3 are obtained by applying the calculated translational and rotational motions at node 3 of the constrained 2-D model as input to the original three-dimensional superstructure. This is, of course, an attempt at recovering the lost response in the direction normal to the direction of excitation. The five locations on the top floor at which response spectra are compared correspond to points A, B, C, D and to the center of mass (circle with a cross) shown in Fig. 2. The accelerations shown at the high frequency end of the response spectra correspond to the peak floor accelerations at the respective locations.

As expected, the response in the direction of excitation (Fig. 3a, 3c and 3d) is the dominant response. Although, for the center of mass, Curves 2 and 3 are conservative approximations for Curve 1, this trend is lost when the response at the edges of the structure are studied. Thus, for example, in Fig. 3d Curve 2 underestimates and Curve 3 overestimates the correct floor peak acceleration at point D. With regard to the spectra amplitudes in the amplified region, significant differences may be observed.

As shown in Fig. 3b, the out-of-plane response is not significant for the center of mass; however, as expected, out-of-plane response becomes important at the edges of the floor (Figs. 3e and 3f). In the latter figures Curve 2 is, of course, zero, and Curve 3 does not properly represent the amplified regions of the spectra. The spectral peaks are shifted in frequency, even though their magnitude is about the same as for Curve 1. For this example, the out-of-plane peak floor accelerations are underestimated by about 30%.

To explore the problem further, the mass eccentricities of the original model shown in Fig. 2 have been doubled and the analyses repeated. The results are shown in Fig. 4. Additionally, in Fig. 4 a fourth response curve is added. Response Curve 4 is obtained by retaining the torsional "soil spring" when the response of the superstructure is calculated using as input the response at node 3 of the two-dimensional model. Convincing arguments may be made why torsional "soil springs" would be necessary to restore the three-dimensional response of the correct problem. However, referring to Fig. 4, one salient point immediately becomes obvious: Curve 4 is the worst solution, and, therefore, this particular attempt to "fix" the results should be discontinued. The following discussions are therefore confined to Curves 1, 2 and 3.

The differences between Curves 1 and 3 become important for the center of mass in both directions of response (Fig. 4a and 4b). This, of course, is expected as the eccentricities have been doubled. In general, the deviations of the approximate results (Curve 3) from the current results (Curve 1) do not follow a predictable pattern. For example, the ratios of approximate to correct peak floor accelerations range, depending on location, from 1.05 to 1.41 in the direction of excitation and from 1.95 to 4.6 normal to the direction of excitation. In addition to the above, all the comments made with regard to Fig. 3 apply also to Fig. 4.

Comparing the results of Figs. 3 and 4, it is found that for motion in the direction of the excitation as the eccentricity increases, the deviations of the approximate results from the correct three-dimensional solutions also increase. This observation is important in that the procedures recommended for restoring lost information produce worse results as the need for restoring the three-dimensional effects increases.

The response in the direction normal to the direction of excitation exhibits characteristics similar to those shown in Figs. 3e and 3f, i.e. the approximate results (Curve 3) have similar peak spectral amplitudes as the correct results (Curve 1) but the peaks occur at different frequencies.

The results shown in Figs. 3 and 4 strongly suggest that it is not a simple problem to first eliminate the three dimensionality of a structure and then restore the proper results by the use of two-dimensional approximations. Reported results that such approximations can be made are possibly due to the fact that the three-dimensional character of the problem was not significant to start with. An alternative treatment of the problem is to maintain the three-dimensional character of the superstructure while employing a dimensional soil model in a one-pass solution (3).

MODELING OF ISOLATED FOUNDATIONS

In the previous section, modeling of the foundation was intentionally omitted so that the parameter under consideration could be isolated and that potential compensating errors would not be introduced. The present discussion will emphasize the modeling of the foundation. The question that needs to be answered then is whether three-dimensional soil-structure interaction effects can be approximated by the use of strip (two dimensional) foundations. This can be studied exhaustively by examining the impedance functions. Since impedance functions are frequency dependent, there is no unique way that one may replace a foundation by an equivalent strip foundation. The problem is posed graphically in Fig. 5 for both circular and rectangular foundations. For the translational-rocking motion, a simplified impedance matrix (coupling terms being neglected) would involve two complex coefficients (four quantities, two real parts and two imaginary parts) at each frequency. Since there are only two unknowns (width and depth of the strip foundation), there is no possibility of obtaining a unique and complete equivalence between the real foundation and its two-dimensional substitute. If the impedances, as a function of frequency, are smooth, then there is some expectation that a reasonable equivalence can be obtained. However, if the soil medium has significant layering, there seems to be no basis for a rational equivalence since the impedance functions become severely oscillatory (9, 11). The question of equivalence will therefore be pursued further assuming no layering in the half-space. Even if an equivalence between a three-dimensional and two-dimensional foundation can be achieved, another important question is the loss of the potential for torsional foundation response due to non-vertically propagating waves or due to structure-structure interaction.

Circular Foundations

An earlier paper (10) explored the feasibility of this reduction for a circular foundation. Table 1 is reproduced from that paper. In Table 1, K_H and K_M represent the horizontal and rocking stiffness coefficients while C_H and C_M represent the corresponding radiation damping coefficients for the foundations. The superscripts s and d denote the strip and disc foundations, respectively. For nuclear power plant structures where structure-soil system frequencies are important, Model 3 of Table 1 is the most desirable, since stiffness ratios, both in rocking and translation, are kept close to unity for a wide range of frequencies. Fig. 6 shows the "best" relationship of the circular to strip foundations and Fig. 7 shows the comparison of the corresponding impedances as a function of the dimensionless frequency a_o. As expected, having matched the frequencies (i.e., the stiffness coefficients), the damping for the equivalent strip foundation is larger than that of the circular foundation for all frequencies. An example of the impact of this difference on the in-structure response spectra is shown in Fig. 8 (10). At the resonant frequency, the two-dimensional model underestimates the three-dimensional response. For a single structure, therefore, Model 3 may be used with the provision that an adjustment to the peak response is made to overcome this inherent problem of damping. It should be noted that the results illustrated in Fig. 8 were obtained on the assumption of nondissipating soil models. If material damping in the soil is considered, then the differences between the results of two and three-dimensional models may not be as pronounced as those shown in Fig. 8.

Rectangular Foundations

A similar study is made to obtain a strip foundation as a substitute for a rectangular foundation. Fig. 5b poses this problem schematically. The same considerations as for the circular foundation are employed in this case too; namely, to match the stiffnesses in translation and rocking. (For details of the calculations, refer to Appendix A.) Fig. 9 gives the equivalent width and length (depth of the two-dimensional model) of a strip foundation as a function of the dimensionless frequency a_o. Unlike the circular foundation, specific aspect ratios for the rectangular foundation must be studied. Five cases are shown that should provide, by interpolation, sufficient information for most structures. It is instructive to compare the dimensions of a rectangular foundation and its equivalent strip foundation for a given dimensionless ratio. Thus, from Fig. 9 and for $a_o = 1.0$, Fig. 10 is derived. As expected, it can be seen that, if a rectangular foundation is treated simply as a strip foundation without any adjustments, then the error introduced in the impedance functions would increase as the ratio of B/L increases.

Since structure-soil system frequencies are not known before an analysis is completed, one may guess at the possible frequency of the system and then compare the impedances at other frequencies to decide if an iterative solution is necessary. As an example, assume that B/L = 2.0 and that a_o is estimated as being equal to 1.15. From this information, it is possible (Appendix A) to derive the impedances for other frequencies. The results are shwon in Fig. 11. As expected, the stiffness coefficients between the strip and rectangular foundations match at $a_o = 1.15$. The divergence for other frequencies is minor for rocking motion throughout the frequency spectrum. For horizontal motion, the divergence is minor for $a_o > 1.15$; but for very low frequencies, singificant errors are possible indicating that an iterative approach may be necessary. With regard to damping, similar to the circular foundation, the equivalent strip overestimates the damping. Thus, at and about $a_o = 1.15$, the rocking damping coefficient is doubled and the translational damping is about sevenfold (Fig. 11b). The impact of these increased damping effects on the structural response can be studied as was reported in Ref. 10.

Arbitrary Shaped Foundations

Strip foundations equivalent to circular and rectangular foundations may be cautiously used if one recognizes the issue of the increase in damping and the fact that significant layering effects may invalidate the results obtained above. For foundations similar to the one shown in Fig. 1, a two level approximation could be attempted: first, to reduce the foundation to an equivalent circle or rectangle by judgement and then use the models developed in this paper to obtain the equivalent strip. Alternatively, one may generate the impedance functions for the arbitrary shaped foundation (15) and then, by trial and error, match it, at about the frequency of interest, to the strip foundation impedances.

MODELING OF THE PLANT

It is common practice with the two-dimensional approach to soil-structure interaction analysis to take a slice through the plant and analyze the resulting plane strain model. Two decisions are involved in this step of the analysis: a) which slice is the more appropriate, and b) what does a slice represent.

Since a proper modeling of single structures requires that the complete structure be considered, referring to Fig. 12, it becomes obvious that there is no unique way a strip or a slice can be taken through the plant. However, some liberties may be taken, such as shifting structures to line them up or modifying their base dimensions, to justify taking a strip. It should become obvious, though, that such a rearrangement may distort the effects of structure-structure interaction. Difficulties arise even in an extremely simplified idealization of structures arranged as shown in Fig. 13a. The analysis can be carried out by many possible re-arrangements of the structures. Two possibilities are shown in Figs. 13b and 13c.

The next question that needs to be answered relates to the fact that a proper solution of a complex plant layout, such as shown in Fig. 12, would require two or three slices in each direction. What this does is introduce slits that go down to the base of the finite-element model and that separate and independent soil slices would be excited by the same motion applied at the base of the models. The compatability of the motions at the interfaces between the slices needs to be evaluated.

In the previous section, the equivalent strip foundation models were discussed both for circular and rectangular foundations. Considering an isolated foundation, it was decided that there is, among many alternatives, a "best" model given some criterion of selection. Referring to Fig. 14a, assume that structures I and II have relative positions as shown. If the "best" models for the isolated foundations are used and the centers of mass for both foundations are kept fixed, the resulting gap between the structures tends to increase as shown in Fig. 14b and thus, possibly, alter the structure-structure interaction effects. An alternative is to maintain the actual widths of the structures (Fig. 14c). However, another equivalent strip must then be selected given this additional constraint. Recalling the problems discussed in a previous section, the resulting strip foundations would be less desirable than those shown in Fig. 14b. This issue is further explored as depicted in Fig. 15 for circular foundations and Fig. 16 for rectangular foundations. With the circular foundation, some "rational" approximations are possible as shown in Figs. 15b, 15c and 15d. These models are compared in Table 1 as Models 7, 8 and 9, respectively. As expected, it is to be noted that the impedance ratios decrease from Model 7 to Model 9. More importantly, a frequency match would be impossible and, depending on the model chosen, the damping would be larger or smaller than the three-dimensional model.

The impedance functions of the circular and equivalent strip of equal width and area, Model 8 (Fig. 15c), are compared in Fig. 17. The results over a wide range of dimensionless frequencies agree reasonably well with the expectations from Table 1, except of course for the horizontal impedances at the very low frequencies. If a model has to be chosen, it seems that a model between those of Models 7 and 8 would be the most appropriate. The half-length of this model would be equal to 0.85r, obtained by a simple ratio that makes $K_M^s/K_M^d = 1.0$. The choice of this model is based on the observation that the significant interaction mode is usually related to rocking motions. The attendant increase in damping must be remembered and the final results so adjusted.

The decision regarding the rectangular footing is more complicated since the aspect ratio enters into the problem. The equivalent length L_e of a strip foundation of width $B_e = B$ obtained by matching the real parts of the rocking impedance function with a rectangular foundation of width to length ratio B/L is plotted in Fig. 18a versus the dimensionless frequency $a_o = \omega B/V_s$ for different values of B/L. Although this equivalent strip foundation matches the rocking stiffness coefficients of the rectangular foundation, errors in modeling the radiation damping for the rectangular foundation are obtained as shown in Fig. 18b. In this figure, the ratio of the radiation damping of the equivalent strip to the radiation damping for the rectangular foundation are plotted versus a_o for different values of the aspect ratio B/L. Again, the error in modeling the damping terms will result in differences in response amplitudes.

Finally, it must be noted that the equivalent depths for the two-dimensional models of structures I and II shown in Fig. 14 are, in general, different. Since a common depth is required by the numerical two-dimensional solution, then, the masses of structures I and II must be distributed in such a way that the masses per unit depth correspond to M_I/c and M_{II}/L_e, respectively. The corresponding structural stiffnesses must be distributed in the same fashion. This procedure ensures that the interaction of each structure and the soil is not altered beyond the limitations already discussed for an isolated structure. On the other hand, this necessary procedure may introduce an additional distortion on the interaction between structures.

STRUCTURE-STRUCTURE INTERACTION

As depicted schematically in Fig. 19, the two-dimensional model will, relative to the three-dimensional model, overemphasize structure-structure interaction effects. Additionally, a study of wave propagation from point (3D) and line (2D) sources indicates that wave amplitudes in a three-dimensional model decay faster than in a two-dimensional model. In a 3D model, the Rayleigh wave decays with distance to the source as $r^{\frac{1}{2}}$, while, for a 2D model the amplitude of the Rayleigh wave remains constant independent of distance (in absence of material attenuation). Similarly, the far field body waves decay as r^{-2} in a 3D model while the decay is of the type $r^{-3/2}$ for a 2D model.

Another shortcoming of the two-dimensional model in the study of structure-structure interaction stems from the fact that, for a truly three-dimensional model of two foundations, the impedance matrix is coupled in all six degrees of freedom of the foundations. Thus, for example, twisting motions would be excited as a function of the parameters x and β of Fig. 20. In a two-dimensional analysis, although rocking and vertical response would be overemphasized as discussed above, twisting, translation normal to the direction of excitation and rocking about an axis parallel to the direction of excitation would be absent.

Additional evidence as to the distortions introduced by two-dimensional models of the structure to structure interaction phenomenon is presented in Fig. 21. In this figure, some of the static stiffness for two parallel rectangular foundations with length to width ratio (L/W) = 1, 2, 4 are plotted versus the separation ratio S/W. It may be seen in the figure that, for the horizontal stiffness K_{11}, the dependence on the separation ratio S/W is highly dependent on the length to width ratio L/W. In addition, as the length to width ratio increases, the amplitude of the coupling stiffness coefficient K_{14} becomes more important when compared with the stiffness coefficient K_{11}. In general, the longer the foundations, the stronger the foundation to foundation interaction effects. The results presented in Fig. 21 indicate that the interaction effects between two infinite strips ($L/W = \infty$) will be significantly different from those between foundations with finite length to width ratios.

The discussions and results presented in this and the previous section imply that, irrespective of whether structure-structure interaction is a significant effect, two-dimensional models are inadequate to provide useful insight into the structure-structure interaction phenomenon.

REFERENCES

1. Atalik, T. S., Hadjian, A. H. and Luco, J. E., "The Dynamic Modeling of the Visco-elastic Half Space by Finite Elements," Proc. 2nd ASCE Specialty Conf. on Structural Design of Nuclear Power Plant Facilities, New Orleans, Dec. 1975, pp. 763-785.

2. Hadjian, A. H. and Atalik, T. S., "Discrete Modeling of Symmetric Box Type Structures," Proc. Int'l Symp. of Earthquake Structural Eng'g, St. Louis, MO, Aug. 1976.
3. Hadjian, A. H. and Lin, S. T., "Three-Dimensional Response Calculations Using Two-Dimensional Codes," In preparation.
4. Hadjian, A. H., Luco, J. E., and Tsai, N. C., "Soil-Structure Interaction: Continuum or Finite Element?" Nuclear Engineering and Design, 31 (1974), pp. 184-194.
5. Hwang, R., Lysmer, J. and Berger E., "A Simplified Three-Dimensional Soil-Structure Interaction Study," Proc. 2nd ASCE Specialty Conf. on Structural Design of Nuclear Plant Facilities, New Orleans, Dec. 1975, pp. 786.
6. Ibrahim, A. M. and Hadjian, A. H., "Generalized Composite Damping Matrices for Three-Dimensional Structure-Foundation Systems," Proc. 2nd ASCE Specialty Conf. on Structural Design of Nuclear Plant Facilities, New Orleans, Dec. 8-10, 1975, pp. 932-960.
7. Isenberg. J. and Adham. S. A., "Interaction of Soil and Power Plants in Earthquakes," J. Power Division, ASCE, Oct. 1972, pp. 273-291.
8. Lin. Y. J. and Hadjian, A. H., "Discrete Modeling of Containment Structures," Proc. Int'l Symp. on Earthquake Structural Eng'g, St. Louis, MO, Aug. 1976.
9. Luco, J. E., "Vibrations of a Rigid Disc on a Layered Viscoelastic Medium," Nuclear Engineering and Design, 36 (1976), 325-240.
10. Luco, J. E., Hadjian, A. H., "Two-Dimensional Approximations to the Three-Dimensional Soil-Structure Interaction Problem," Nuclear Engineering and Design, 31 (1974), 195-203.
11. Luco, J. E. and Hadjian, A. H., "On the Importance of Layering on the Impedance Functions," Proc. 6th World Conf. on Earthquake Eng'g, New Delhi, India, Jan. 1976.
12. Luco, J. E., Hadjian, A. H., and Bos, H. D., "The Dynamic Modeling of Half-Plane by Finite Elements," Nuclear Engineering and Design, 31 (1974), pp. 184-194.
13. Lysmer, J. et al, "LUSH: A Computer Program for Complex Response Analysis of Soil-Structure Systems," Report EERC 74-4, Earthquake Eng'g Research Center, Univ. of Calif. Berkeley, Apr. 1974.
14. Seed, H. B. and Idriss, I. M., "Soil-Structure Interaction of Massive Embedded Structures During Earthquakes," Proc. 5th World Conf. on Earthquake Eng'g, Rome, Italy, 1973.
15. Wong, H. L., "Dynamic Soil-Structure Interaction," EERL 75-01 Report, Calif. Institute of Technology, Pasadena, CA, May 1975.

APPENDIX A. EQUIVALENT 2D MODEL FOR A RECTANGULAR FOUNDATION

A possible criterion to represent a three-dimensional rectangular foundation of total width B (in the direction of the excitation) and total length L by a two-dimensional strip foundation of total width B_e over a length L_e is to require that, for a particular frequency, the dynamic stiffness coefficients (real part of the impedance functions) be equal in both foundations for horizontal and rocking vibrations. The geometry of the problem is described in Fig. 5.b.

The horizontal K_H^R and rocking K_M^R stiffness coefficients for the rectangular foundation may be written in the form

$$K_H^R = B\ G\ k_H^R\ (a_o, \nu, B/L) \quad (1) \qquad K_M^R = B^2\ LG\ k_M^R\ (a_o, \nu, B/L) \quad (2)$$

where G is the shear modulus for the soil, $a_o = \frac{\omega B}{V_s}$ is a dimensionless frequency, ν denotes Poisson's ratio and B/L is the aspect ratio of the rectangular foundation. Numerical values for the functions k_H^R and k_M^R have been obtained by Wong and Luco (A.1) for the frequency range of interest and for different aspect ratios.

Similarly, the horizontal K_H^S and rocking K_M^S stiffness coefficients for a strip foundation of total width B_e over a length L_e can be expressed by

$$K_H^S = L_e G\, k_H^S\,(a_o',\nu) \quad (3) \qquad\qquad K_M^S = L_e G\, B_e^2\, k_M^S\,(a_o',\nu) \quad (4)$$

where a_o' denotes the dimensionless frequency $a_o' = \frac{\omega B_e}{V_s}$. Numerical values for the functions k_H^S and k_M^S have been presented by Karasudhi, Keer and Lee (A.2) and by Luco and Westmann (A.3).

The conditions $K_H^R = K_H^S$ and $K_M^R = K_M^S$, lead to

$$\left(\frac{B}{L}\right) k_H^R\,(a_o,\,\nu,\,B/L) = \left(\frac{L_e}{L}\right) k_H^S\,(a_o\,\frac{B_e}{B},\,\nu) \qquad (5)$$

$$k_M^R\,(a_o,\,\nu,\,B/L) = \left(\frac{L_e}{L}\right)\left(\frac{B_e}{B}\right)^2 k_M^S\,(a_o\,\frac{B_e}{B},\,\nu) \qquad (6)$$

Eliminating (L_e/L) from Eqs. (5) and (6) it is found that

$$\left(\frac{B_e}{B}\right)^2 = \frac{k_H^S\,(a_o\,\frac{B_e}{B},\,\nu)}{k_M^S\,(a_o\,\frac{B_e}{B},\,\nu)} \cdot \frac{k_M^R\,(a_o,\,\nu,\,B/L)}{\left(\frac{B}{L}\right) k_H^R\,(a_o,\,\nu,\,B/L)} \qquad (7)$$

For given values of a_o, ν and B/L, Eq. (7) may be solved by iteration to obtain the ratio B_e/B of the equivalent width to the width of the rectangular foundation. Having obtained the equivalent width, the equivalent length may be found by use of Eqs. (5) or (6):

$$\frac{L_e}{L} = \left(\frac{B}{L}\right) \frac{k_H^R\,(a_o,\,\nu,\,B/L)}{k_H^S\,(a_o,\frac{B_e}{B},\,\nu)} \qquad (8)$$

The ratios (B_e/B) and (L_e/L) are presented in Figs. 9a and 9b versus the dimensionless frequency $a_o = \frac{\omega B}{V_s}$ for rectangular foundations with aspect ratios B/L = ¼, ½, 1, 2 and 4 for a value of Poisson's ratio for the soil $\nu = 1/3$. It must be noticed that since $k_H^S\,(a_o',\,\nu)$ tends to zero as a_o' tends to zero, then $B_e/B \to 0$ and $L_e/L \to \infty$ as $a_o \to 0$.

A.1 Wong, H. L. and Luco, J. E., "Dynamic Response of Rigid Foundations of Arbitrary Shape," Earthquake Engineering and Structural Dynamics, Vol. 4, 1976, pp. 579-587.

A.2 Karasudhi, P., Keer, L. M. and Lee, S. L., "Vibratory Motion of a Body on an Elastic Half Plane," Journal of Applied Mechanics, Vol. 35, Trans. ASME, Vol. 90, Series E, Dec. 1968, pp. 1-9.

A.3 Luco, J. E. and Westmann, R. A., "Dynamic Response of a Rigid Footing Bonded to an Elastic Half Space," Journal of Applied Mechanics, Vol. 39, No. 2, June 1972, pp. 527-534.

Table 1. STIFFNESS AND DAMPING RATIOS FOR DIFFERENT FOUNDATION MODELS

MODEL	b/a	c/2a	K_H^s / K_H^d (AVERAGE)	C_H^s / C_H^d (HIGH-FREQ)	K_M^s / K_M^d (STATIC)	C_M^s / C_M^d (HIGH-FREQ)
1	0.618	1.27	1.	1.	0.57	0.51
2	0.695	1.76	1.38	1.56	1.	1.
3	0.816	1.27	1.	1.32	1.	1.18
4	0.866	0.906	0.71	1.	0.80	1.
5	0.866	1.13	0.89	1.25	1.	1.25
6	0.876	0.876	0.69	0.98	0.78	1.
7	1.0	1.0	0.79	1.27	1.18	1.7
8	1.0	0.785	0.62	1.0	0.925	1.33
9	1.0	0.59	0.46	0.75	0.69	1.00

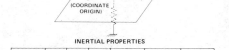

a
c
2b
EXCITATION

Figure 1. PLAN OF A TYPICAL AUXILIARY BUILDING IN A NUCLEAR POWER PLANT

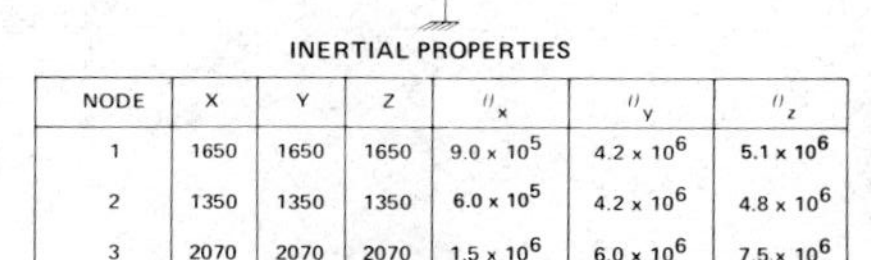

INERTIAL PROPERTIES

NODE	X	Y	Z	θ_x	θ_y	θ_z
1	1650	1650	1650	9.0×10^5	4.2×10^6	5.1×10^6
2	1350	1350	1350	6.0×10^5	4.2×10^6	4.8×10^6
3	2070	2070	2070	1.5×10^6	6.0×10^6	7.5×10^6

STIFFNESS PROPERTIES

ELEMENT	I_x	I_y	I_z	A_x	A_y	A_z
1-2	9.7×10^5	5.7×10^5	2.4×10^5	760	260	370
2-3	9.7×10^5	5.7×10^5	2.9×10^5	1080	460	460

SOIL STIFFNESS COEFFICIENTS

K_x	K_y	K_z	$K_{\theta x}$	$K_{\theta y}$	$K_{\theta z}$
5.3×10^5	5.5×10^5	6.7×10^5	2×10^9	5.3×10^9	5×10^9

SOIL DAMPING COEFFICIENTS

C_x	C_y	C_z	$C_{\theta x}$	$C_{\theta y}$	$C_{\theta z}$
3.6×10^4	3.7×10^4	6.7×10^4	6.0×10^7	1.3×10^8	6.0×10^7

STRUCTURE MODAL DAMPING – ALL MODES 4% CRITICAL

Figure 2. SYSTEM PARAMETERS IN KIP-FT-SEC UNITS

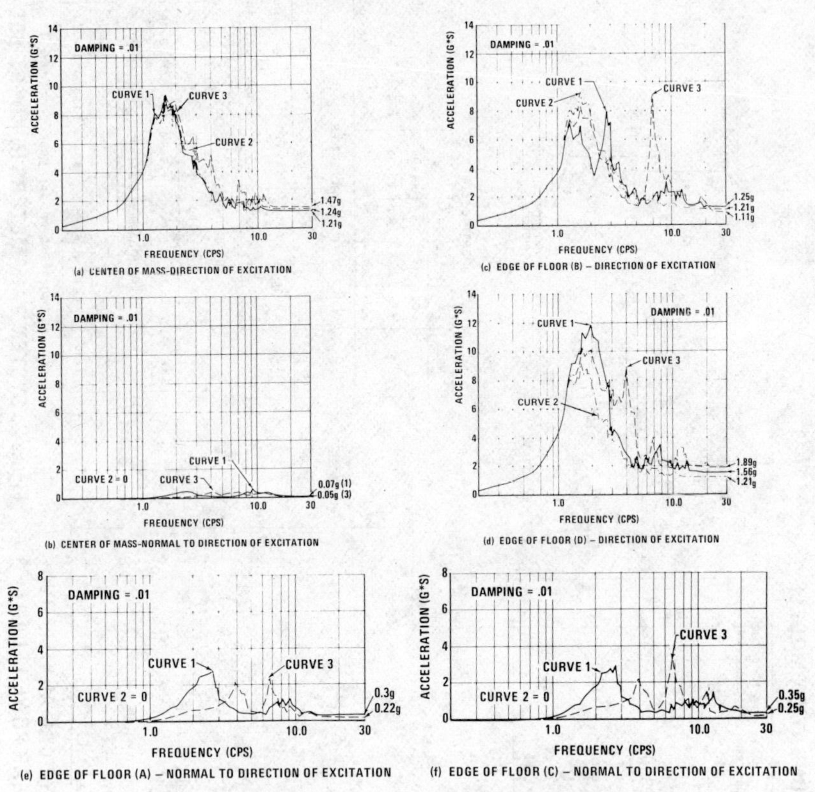

Figure 3. COMPARISON OF ACCELERATION FLOOR SPECTRA AT TOP OF STRUCTURE OF FIGURE 2

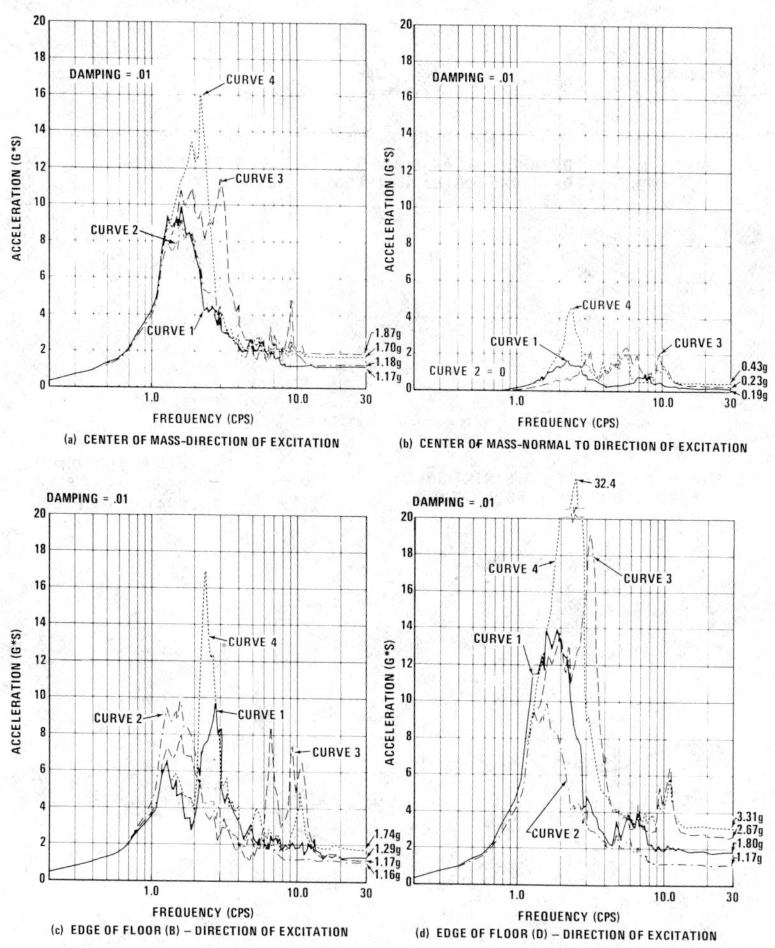

Figure 4. COMPARISON OF ACCELERATION FLOOR SPECTRA AT TOP OF STRUCTURE OF FIGURE 2 WITH MASS ECCENTRICITIES DOUBLED

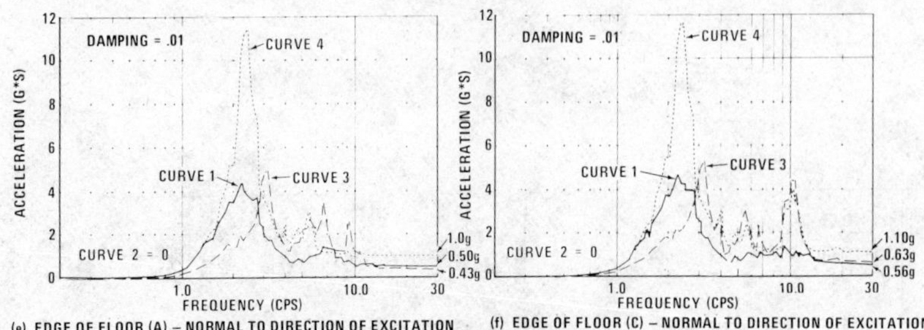

(e) EDGE OF FLOOR (A) – NORMAL TO DIRECTION OF EXCITATION (f) EDGE OF FLOOR (C) – NORMAL TO DIRECTION OF EXCITATION

Figure 4. (Continued) COMPARISON OF ACCELERATION FLOOR SPECTRA AT TOP OF STRUCTURE OF FIGURE 2 WITH MASS ECCENTRICITIES DOUBLED

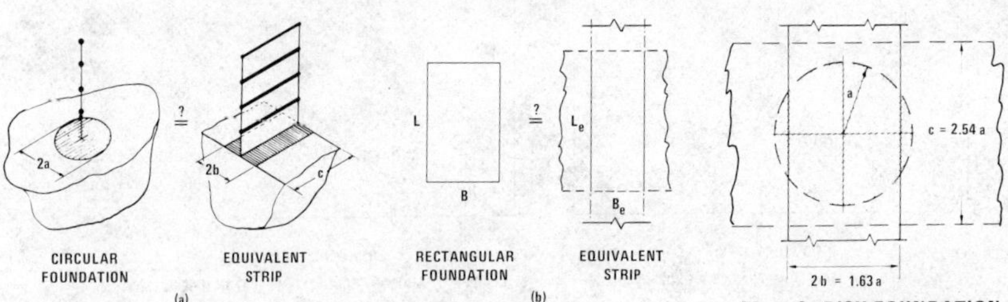

Figure 5. SCHEMATIC REPRESENTATIONS OF THE EQUIVALENCE OF FOUNDATIONS

Figure 6. DISK FOUNDATION AND 'EQUIVALENT' STRIP FOUNDATION (MODEL 3)

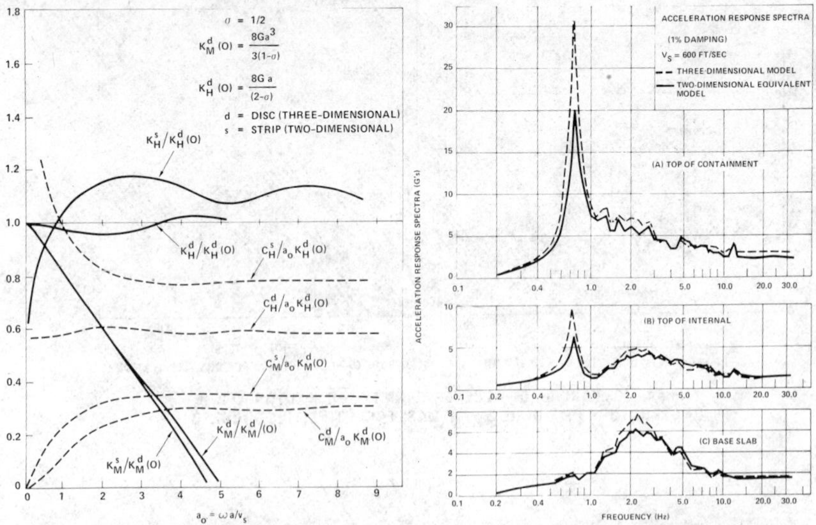

Figure 7. STIFFNESS AND DAMPING COEFFICIENTS FOR DISC AND 'EQUIVALENT' STRIP FOUNDATION (MODEL 3)

Figure 8. COMPARISON OF ACCELERATION RESPONSE SPECTRA FOR TWO- AND THREE-DIMENSIONAL MODELS (V_S = 600 FT/SEC)

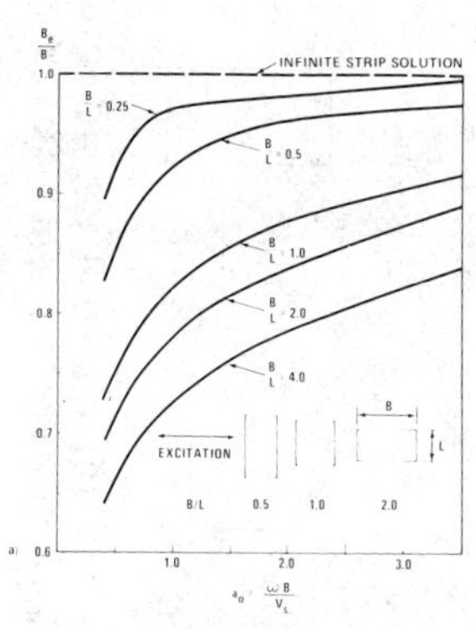

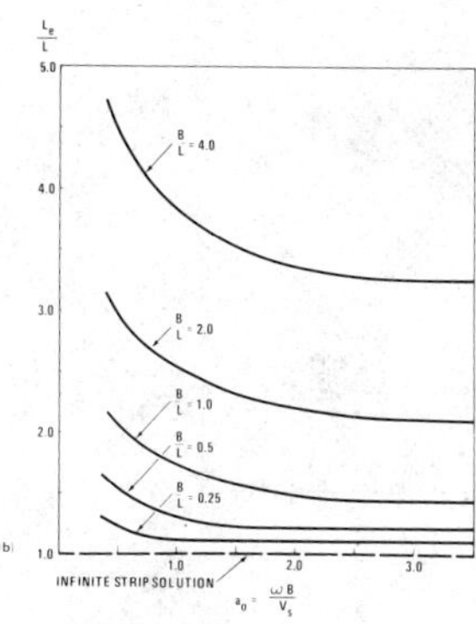

Figure 9. EQUIVALENT STRIP WIDTH AND LENGTH FOR RECTANGULAR FOUNDATIONS

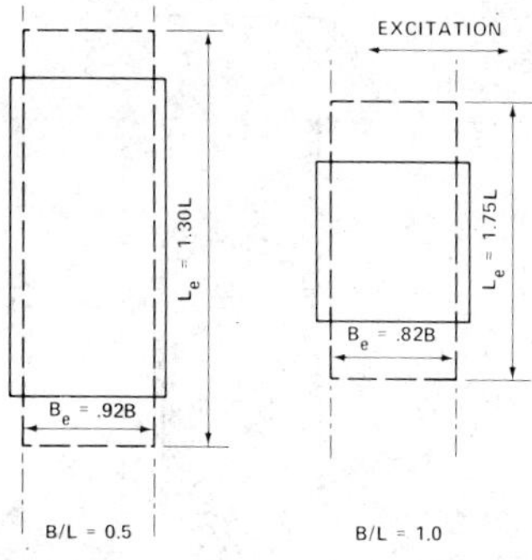

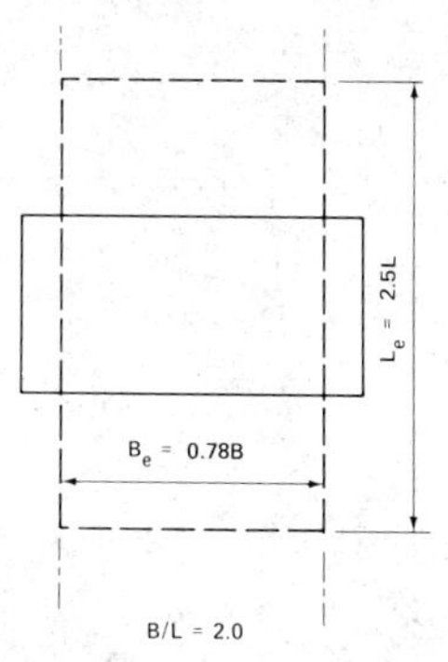

Figure 10. EQUIVALENT STRIPS FOR $a_0 = 1.0$

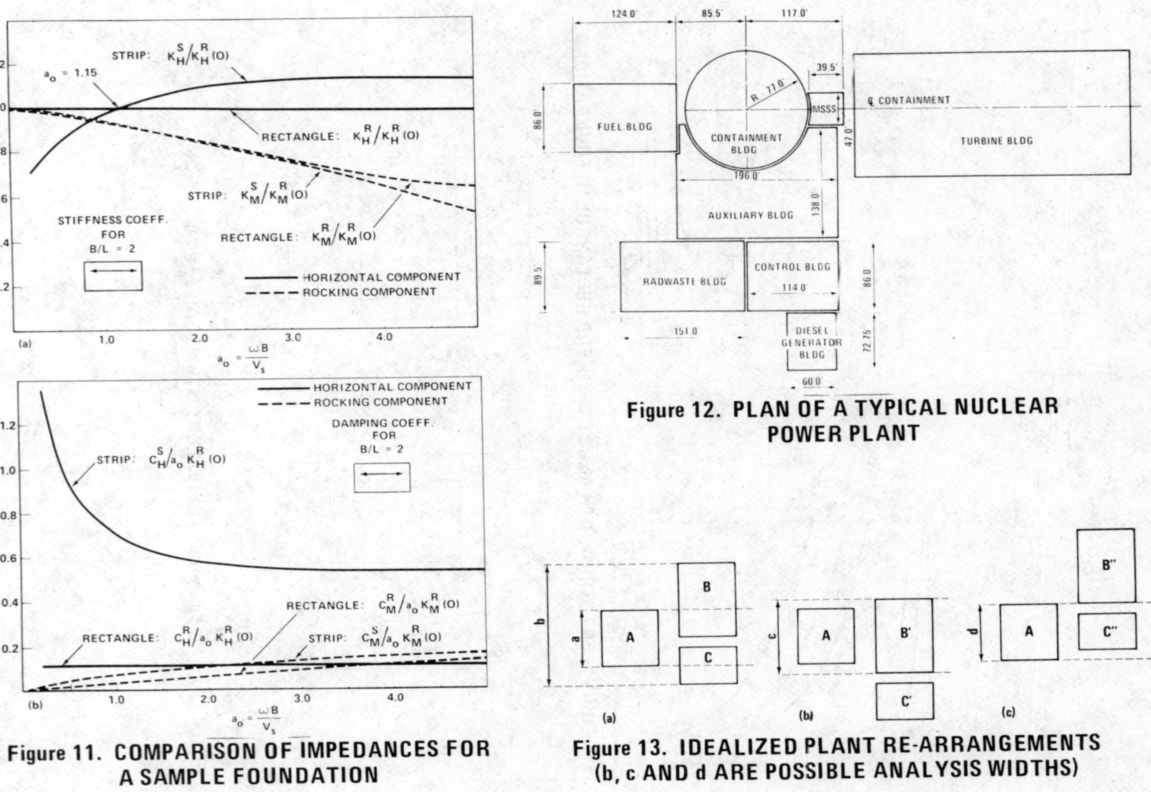

Figure 11. COMPARISON OF IMPEDANCES FOR A SAMPLE FOUNDATION

Figure 12. PLAN OF A TYPICAL NUCLEAR POWER PLANT

Figure 13. IDEALIZED PLANT RE-ARRANGEMENTS (b, c AND d ARE POSSIBLE ANALYSIS WIDTHS)

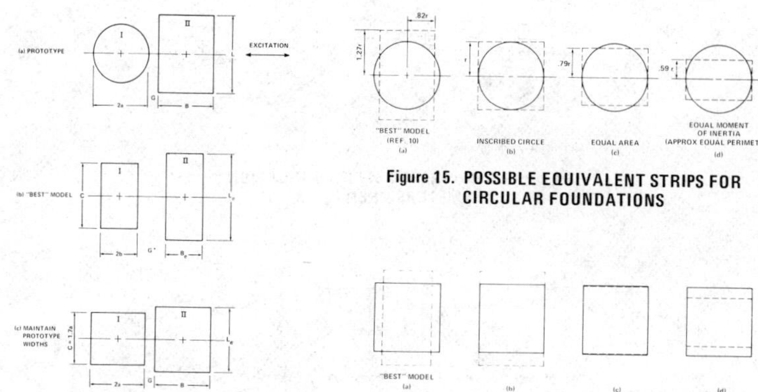

Figure 14. SCHEMATIC REPRESENTATION OF THE PROBLEM RELATED TO SEPARATION OF STRUCTURES

Figure 15. POSSIBLE EQUIVALENT STRIPS FOR CIRCULAR FOUNDATIONS

Figure 16. POSSIBLE EQUIVALENT STRIPS FOR RECTANGULAR FOUNDATIONS

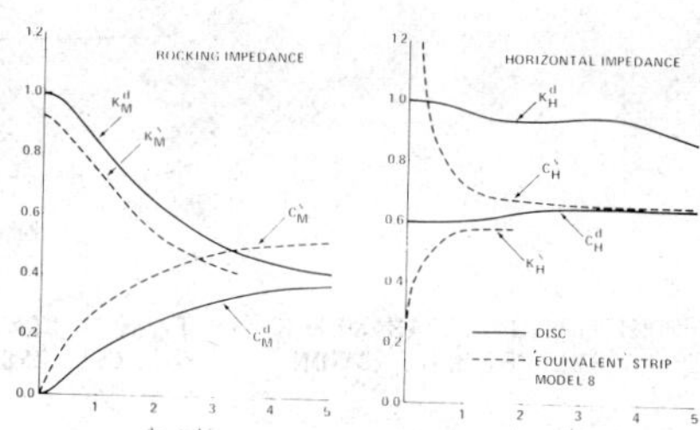

Figure 17. STIFFNESS AND DAMPING COEFFICIENTS FOR DISC AND EQUIVALENT STRIP FOUNDATION (MODEL 8)

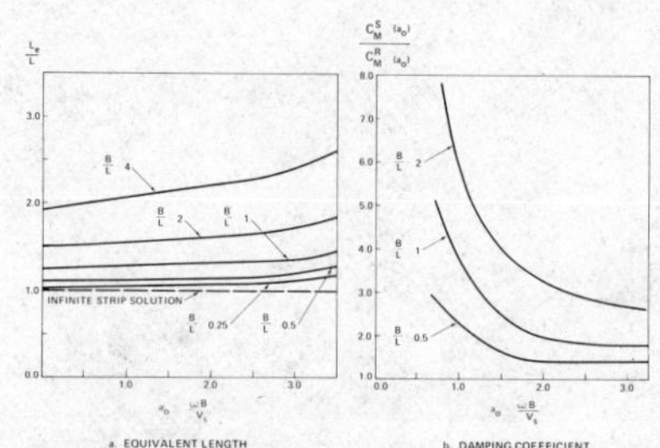

Figure 18. EQUIVALENT STRIP FOUNDATION PARAMETERS WHEN B_e = B

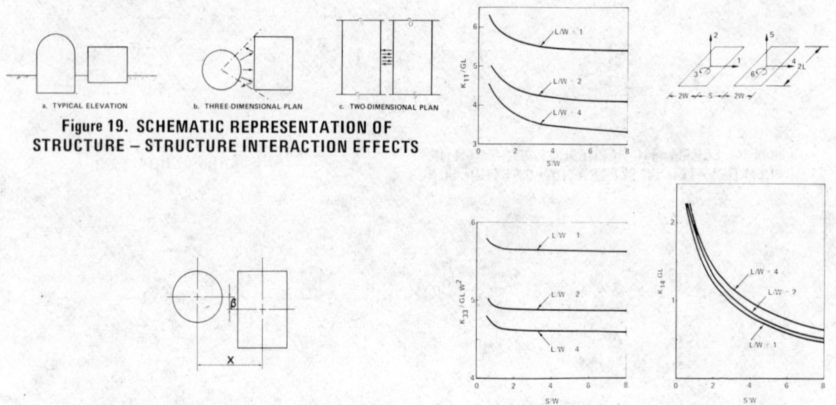

Figure 19. SCHEMATIC REPRESENTATION OF STRUCTURE – STRUCTURE INTERACTION EFFECTS

Figure 20. SCHEMATIC REPRESENTATION OF PARAMETERS AFFECTING STRUCTURE – STRUCTURE INTERACTION

Figure 21. STATIC STIFFNESSES FOR TWO PARALLEL RECTANGULAR FOUNDATIONS

A FAULT-RUPTURE PROPAGATION MODEL FOR GROUND MOTION SIMULATION

by

Yoshikazu Yamada[1)] and Shigeru Noda[2)]

ABSTRACT

Expected shear wave ground motion produced by finite, kinematic strike-slip fault surfaces is generated by Monte Carlo studies of dislocation models of nonuniform rupture. The particular earthquake for strong motion prediction in our proposed simulation procedure is a triggering model of completely random subfaults, which is described by eight source parameters. Superimposing the propagation effect of each mesh source, distributed on a fault-plane considering statistical uncertainty of the model parameters, realistic ground motion is synthesized and can be used as a rational basis for assigning future seismic risk.

INTRODUCTION

Stochastic methods for generating artificial strong motion for use in engineering design have been widely used [for example, Housner(1955), Shinozuka and Sato(1967), Amin and Ang(1968), Iyengar and Iyengar(1969), Saragoni and Hart(1974) et al. and Refs. 11,15,16,17]. Wong and Trifunac (Ref.17) gave a historical review for the generation of ground motion since 1955, and found that simulation models fall into two categories : 1)methods that utilize random functions, and 2)methods that involve source mechanism and wave propagation models.

Recently, in the former method, Kameda and Sugito et al.(Ref.11) proposed a prediction model of nonstationary earthquake motion scaled for magnitude, distance and local site conditions (N-value profile or transfer function of surface layers at the site). Otherwise, Wong and Trifunac (Ref.17) used the dispersion characteristics of the ground as the site parameter describing the geological conditions.

Most of the previous studies have been confined to the development of simulation models that predict earthquake motions through random functions of amplitude and phase or inclusion of regression equations of the scaling model parameters, considering their statistical uncertainties. However, these prediction models are based on empirical and/or statistical attenuation relations, rarely allowing for geophysical input and earthquake source mechanism. Rascón and Cornell(Ref.15), with the idea of a double couple source mechanism, developed a physically-based model to simulate strong ground motion, assuming earthquake generation along the fault line as a series of small and uniformly spaced foci. Then random P and S waves were propagated to the site.

Other modeling of the earthquake motion as the nonstationary second order autoregressive moving average (AR-MA) process has been also used to synthesize accelerograms [for example Toki and Sato(Ref.16)]. This result will allow the construction of artificial accelerograms for

---

1)Professor and 2)Research Associate of Civil Engineering, Faculty of Engineering, Kyoto University, Sakyo-ku, Kyoto 606 JAPAN

arbitrary epicentral distances and magnitudes.

In this paper our increasing interest in the engineering aspects of propagating ruptures over large fault planes lead use to study the effects of earthquake source parameters on expected strong motion. We propose a new simulation method for generating artificial strong motion accelerograms for use in engineering design, and present a statistical prediction model to determine future ground accelerations due to propagation of two-dimensional shear cracks. Therefore our method corresponds to the previous 2nd category, including scaling predictions dependent on more realistic and synthetic dynamic dislocation faults.

MATHEMATICAL FORMULATION OF DISLOCATION MODEL : REVIEW

Rupture propagating on a finite fault plane in an elastic medium represents an impotant aspect of seismic source mechanisms of earthquakes. The works of Maruyama(Ref.12), Knopoff(Ref.4) and Haskell(Refs.8,9) et al. are used in a dynamic dislocation model which mathematically represents the elastodynamic radiation resulting from the sudden occurrence of an earthquake due to faulting. The object of our approach is to predict strong motions using information on the techtonic stress and physical properties of the earthquake fault. We start from the representation theorem as follows.

The displacements of the surface of <u>a half space</u> are calculated by applying the theorem by Burridge and Knopoff(Ref.4):

$$U_i(x,t)=\int_{-\infty}^{\infty}dt'\iint_S \mu n_k(\xi)[u_j(\xi,t')](g_{ij,k'}(X,t-t':\xi,0) + g_{ik,j'}(X,t-t':\xi,0))d\xi_1 d\xi_3 \qquad \cdots\cdots\cdots\cdots (1)$$

The motions of the surface of a half space were calculated by a three dimensional Green's function for point dislocation which is a solution to Lamb's problem. Its complete solution are given by Johnson(Ref.10), who programmed it for evaluation of synthetic seismograms.

An alternative form of the theorem appropriate for the representation of faulting in <u>an infinite</u> homogeneous medium has been given by Haskell (Refs.8,9) as follows:

$$U_i(X,t)= -\iint_S\{\rho(\alpha^2-2\beta^2)n_j M_{iq,q}[D_j]+\rho\beta^2(n_q M_{ij,q}[D_j]+n_p M_{ip,q}[D_q])\}dS \qquad \cdots\cdots\cdots\cdots (2)$$

An operator $M_{ij,q}[\ .\ ]$ is as follows:

$$\begin{aligned} M_{ij,q}[\phi(\xi,t)]=(4\pi\rho)^{-1}\{&[15\gamma_i\gamma_j\gamma_q-3(\delta_{ij}\gamma_q+\delta_{iq}\gamma_j+\delta_{jq}\gamma_i)].r^{-4} \qquad \cdots\cdots (3)\\ &.\int_{r/\alpha}^{r/\beta}\phi(\xi,t-t')t'dt'+[6\gamma_i\gamma_j\gamma_q-(\delta_{ij}\gamma_q+\delta_{iq}\gamma_j+\delta_{jq}\gamma_i)](\alpha r)^{-2}\\ &.\phi(\xi,t-r/\alpha)-[6\gamma_i\gamma_j\gamma_q-(2\delta_{ij}\gamma_q+\delta_{iq}\gamma_j+\delta_{jq}\gamma_i)](\beta r)^{-2}\phi(\xi,t-r/\beta)\\ &+\gamma_i\gamma_j\gamma_q(\alpha^3 r)^{-1}\dot{\phi}(\xi,t-r/\alpha)-[\gamma_i\gamma_j\gamma_q-\delta_{ij}\gamma_q](\beta^3 r)^{-1}\dot{\phi}(\xi,t-r/\beta)\}\end{aligned}$$

In the above equations, $U=(u_i,i=1,2,3)$=cartesian components of displacement, X=cartesian coordinates, $\xi=(\xi_i,i=1,2,3)$=cartesian coordinates of point of integration on S, t & t'=time parameters, g_{ij}= displacement in the i-direction at location X, time t due to an impulse force in the j-direction at location ξ, time t', $g_{ij,k}$=spacial derivative of the Green's function, $g_{ij,k'}$with respect to the source coordinates $\xi_{k'}$, r=modulus of the distance of the point of rupture to site,

$\gamma_i=(x_i-\xi_i)/r$, S=fault plane area, ρ=density, α=P-wave velocity, β=S-wave velocity, μ=earth's crust rigidity, $D=(D_i)$=displacement discontinuity on S, $n=(n_i)$=unit normal on S, and δ_{ij}=Kronecker delta notation.

In the following section we discuss the statistical model of earthquake modeling a propagating rupture by summation of far-field approximate functions from Eq.(2) using the fumulae of Haskell.

GROUND MOTION FROM STOCHASTIC MODEL

The geometrical setting for an elementary rectangular fault model of length l_{ij} and width w_{ij} is as represented in Fig.1. Index ij indicates numbers of small finite elements of the fault. The fault element lies in a plane striking $E\theta_i^\circ S$ and dipping $\psi_1^\circ N\theta_i^\circ E$. The direction X_1 of the propagating plane dislocation front, traveling at the rupture velocity $V_{R_{ij}}$, is given by ϕ_1 while the final slip direction depends on the amplitude D_{ij}. Fig.1 includes also the coordinate system and faulting parameters. Fig.2 shows details of coordinates and system geometry corresponding to fault area, rupture area and small patch element. The coordinates of the site in a two dimensional reference system of the fault trace are transformed into the system of axis associated with the fault rupture plane.

We assume now a finite moving dislocation with unilateral type of faulting over a rectangular source and compute the radiational component of the acceleration field at the site for source geometry in Fig.2 of moving edge dislocation. Neglecting the terms with orders higher than r^{-2} in Eq.(2), an expression for mean Fourier spectral amplitude of earthquake acceleration at the site is obtained by summing source contribution of each element, ij:

$$\overline{F}_s(\omega) = 2\sum_{i=1}^{N}\sum_{j=1}^{M(i)} \overline{F}_{ij}(\omega).\overline{D}_{S_{ij}}(\omega).\overline{A}_{N_{ij}}(\omega) \quad \cdots\cdots\cdots\cdots\cdots (4)$$

The concept of continuously superposing each element's response is shown in Fig.3. From comparing Eq.(1) with Eq.(2) for evaluation of effects due to the free surface(Ref.2), this result is reasonably obtained by doubling the amplitude of motions in an infinite space for SH waves. In addition when the fault plane is divided up into small elements to fulfill the finiteness condition, this far-field condition is an effective approximation.

In Eq.(4),

$$\overline{F}_{ij}(\omega) = \frac{R_{ij}(\theta,\phi)}{4\pi\rho\beta^3}.\frac{Mo_{ij}}{r_{o_{ij}}}.\omega^2.\frac{\sin(\omega\chi_{W_{ij}})}{\omega\chi_{W_{ij}}}.\frac{\sin(\omega\chi_{L_{ij}})}{\omega\chi_{L_{ij}}}$$

$$\bullet \exp\left[-i.\omega\left(\frac{r_{o_{ij}}}{\alpha} + \chi_{L_{ij}} - \chi_{W_{ij}} + t_{ij}\right)\right] \quad \cdots\cdots (5)$$

where, $R_{ij}(\theta,\phi)$=radiation pattern correlation coefficient of strike-slip motion, $Mo_{ij}=Mo.(l_{ij}w_{ij})/LW$, Mo=sample seismic moment of fault plane, $r_{o_{ij}}$=distance from the origin of the ij-th element to the site(see Fig.2) and L,W=dimensions of total fault. θ and ϕ refer to spherical coordinate system in Fig.2. $\chi_{L_{ij}}$ and $\chi_{W_{ij}}$ are duration times associated with length and width of ij patch, respectively, and determined by element geometry, position of the site and rupture velocity $V_{R_{ij}}$ as follows:

$$\chi_{L_{ij}} = \frac{l_{ij}}{2}\left(\frac{1}{V_{R_{ij}}} - \frac{\sin\theta_{ij}\cos\phi_{ij}}{\alpha}\right), \quad \chi_{W_{ij}} = \frac{w_{ij}}{2}.\frac{\cos\theta_{ij}}{\alpha} \quad \cdots\cdots (6)$$

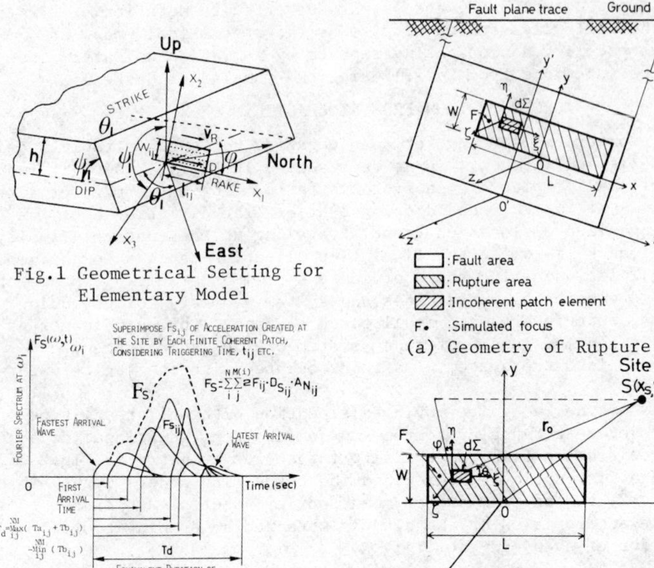

Fig.1 Geometrical Setting for Elementary Model

(a) Geometry of Rupture Plane

Fig.3 Conceptual Figure for Superimposing Each Element's FSP

(b) Relation between Site and Fault Rupture Plane

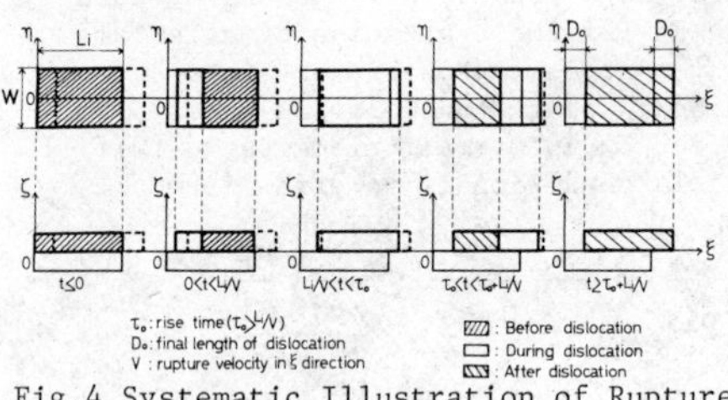

Fig.4 Systematic Illustration of Rupture Propagation

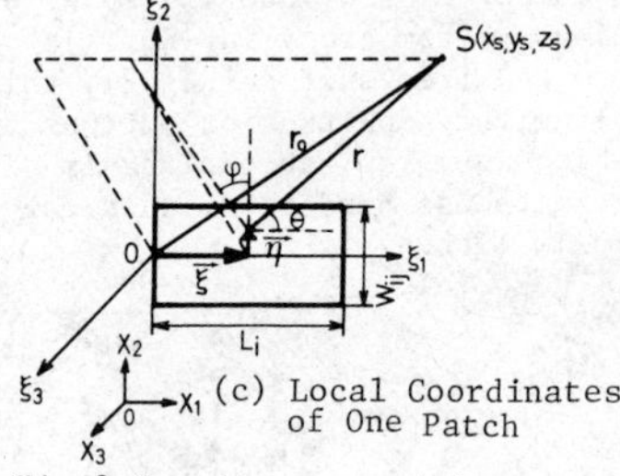

(c) Local Coordinates of One Patch

Fig.2 Coordinates and System Geometries

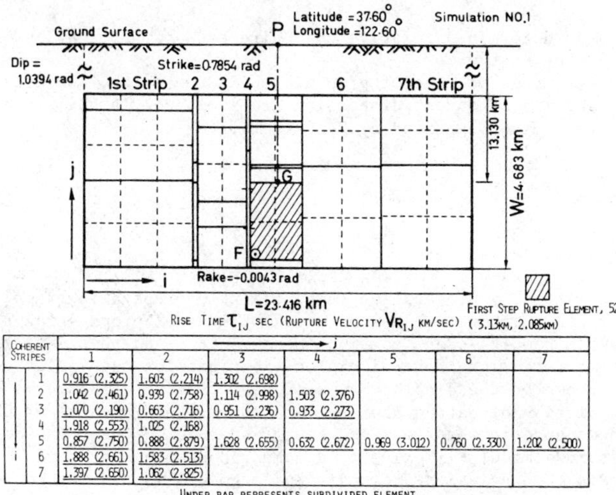

COHERENT STRIPES (i ↓ / j →)	1	2	3	4	5	6	7
1	0.916 (2.325)	1.603 (2.214)	1.302 (2.698)				
2	1.042 (2.461)	0.939 (2.758)	1.114 (2.998)	1.503 (2.376)			
3	1.070 (2.190)	0.663 (2.716)	0.951 (2.236)	0.933 (2.273)			
4	1.918 (2.553)	1.025 (2.168)					
5	0.857 (2.750)	0.888 (2.879)	1.628 (2.655)	0.632 (2.672)	0.969 (3.012)	0.760 (2.330)	1.202 (2.500)
6	1.888 (2.661)	1.583 (2.513)					
7	1.397 (2.650)	1.062 (2.825)					

UNDER BAR REPRESENTS SUBDIVIDED ELEMENT.

Fig.5 Typical Example of Each Rise Time, Rupture Velocity and Coherent Patch generated by Simulation No.1

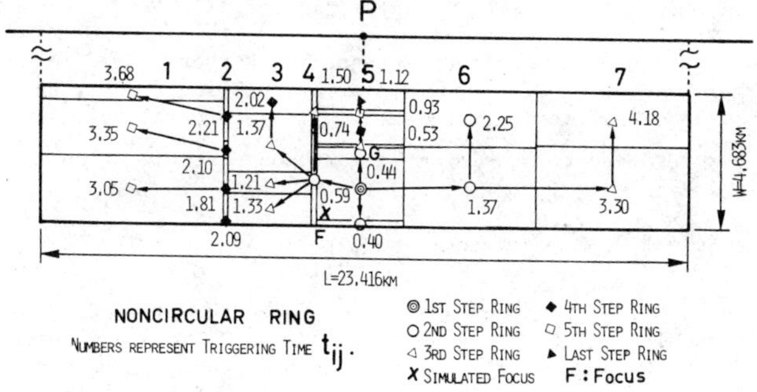

Fig.6 Triggering and Rupture Propagation Schemes of Simulation No.1

in which, t_{ij}=the triggering time of the element ij rupture as presented in the following section.
$\overline{F}_{ij}(\omega)$ are also determined relative to a direction of propagation, such as bilateral type motion.

To represent the effects of propagation in the earth, anelastic attenuation and the attendant dispersion are included via the operator using the formulation suggested by Stick(Ref.5).

$$\overline{A}_{N_{ij}}(\omega) = \exp\left[-\frac{r_{o_{ij}}\omega}{2Q\beta}\left\{1 + \frac{2.i}{\pi}\, ln\left(\frac{Q}{2000\pi}\right)\right\}\right] \qquad (7)$$

where Q is a quality factor.

$\overline{D}s_{ij}(\omega)$ is defined corresponding to the source dislocation function type — unilaterally propagating ramp, elliptic propagating ramp, self-similar crack, pinned annealing crack, bilateral propagating ramp, Ohnaka's ramp and multi-stage combined model, etc. Anderson(Ref.1) concluded that details of several dislocation function have relatively little infuluence on the displacements at near-field stations. We believe that the prediction of earthquake acceleration differs because of the change of dominance in the frequency energy spectrum. Our proposed source subdivided model will give good comparison with synthetic seismograms as the radiation from adjacent segments is assumed to be statistically independent or incoherent. But it is preferable to consider a simple source function than complicated models. If we assume exponential function $\overline{D}s_{ij}(t)=D_{ij}(1-\exp(-t/\tau_{ij})).H(t)$;$H(t)$=Heaviside step function,

$$\overline{D}s_{ij}(\omega) = (1+\omega^2\tau_{ij}^2)^{-1/2}\exp[-i.\tan^{-1}(\omega\tau_{ij})] \qquad (8)$$

Fig.4 shows a growth of a smoothed propagation ramp crack model on rectangular element in time. Shaded parts are healing fronts, separating areas which have ceased motion from areas which are still moving.

Finally we propose to model sample earthquake acceleration in the following way.

$$\ddot{X}_s(t)= \sum_{i=1}^{m} 2\sqrt{G_k(\omega)\,\Delta\omega}\,.\cos(\omega_k t + \phi_k(\omega)) \qquad (9)$$

in which m=number of superposed harmonic components, $G_k(\omega)$and $\phi_k(\omega)$ are respectively sample amplitude and phase functions. These are dependent on 1)radiation pattern geometry, 2)seismic moment energy level, 3)source mechanism time dependency, and 3)transmission path effects, and are determined from the statistics of Power Spectral Density (PSD) at the site generated in the simulation process along with the stochastic model of fault source. This is achieved by generating as many sample spectra in Eq.(4) as possible, considering random source parameters. Note that $\phi_k(\omega)$ were not generated by the pseudo random numbers distributed uniformly between $-\pi$ and π. Therefore Eq.(9) can lead to a realistic sample motion through Monte Carlo procedure.

TRIGGERING EFFECTS OF RUPTURE ACTION, REGRESSION ESTIMATION OF SEISMIC MOMENT AND DURATION

Random Propagation of the Rupture

We assume that the model of seismicity in this study is a stochastic process of trigger type.

For an explanation of the latter case, a typical example of gravity center,G, and dimensions of sample fault rupture with projection point, P, which defines the location of the sample fault plane is shown in Fig.5. Fig.5 shows also a typical example of coherent patch elements and focus location generated by simulation process. The patches are small enough to directly apply the wave equation. Otherwise they must be discretized in small finite elements. A patch which is too close to the site is further subdivided in order to fulfill the finite condition. These situations are shown in Fig.5 with rupture velocity of each element.

The focus point,F, is generated by simulation, assuming that its location can occure uniformly at any one of the patches. We assume that the propagation of the rupture in the fault plane is completely random and the rupture propagate in all directions in the fault plane from the rupture, ie, focal point. With that assumption, the occurence of transitions from one segment patch to next is governed by a Poisson distribution, and its process is highly irregular. Therefore we can consider this triggering action is a two-dimensional random walk from Ref.6. For each sample on this basis, the rupture front grows noncircularly outward at a constant simulated velocity V_R with each element from a prescribed point of initiation, until it reaches the boundaries of the fault rupture plane. The propagation ring by generating the next outer element of the ring enveloping using the former propagating ring corresponding to discrete sample fault model used in Fig.5 is shown in Fig.6.

From the choice of each element in the propagating ring, then, the triggering time, t_{ij}, of connected elements can be computed with the coordinates of the center, dimensions, direction of propagation, and rupture velocity of the patch, ij. The triggering time of a subdivided patch is also obtained by using the coordinates of 1)the patch which is triggered, 2)the patch which triggered, 3)the point on the border of former where that patch was triggered, the time at which the former patch was triggered at border point of both patches, and propagation velocity of both patches. Fig.6 includes rupture propagation schemes and each element triggering time.

Statistical Prediction of Seismic Moment and Duration of Strong Motion

The total duration of earthquake motion,T_d, is the sum of three items, depending on the earthquake mechanism and on the wave path.

$$T_d \simeq d_{source} + d_{\Delta} + d_{site} \qquad \text{.....................} (10)$$

where, d_{source}, d_{Δ} and d_{site} respectively represent, 1)the duration of source dislocation motion of fault plane, 2)the duration resulting from the dispersion of the seismic waves, and 3)the duration from the site geology on the depth of source layers. In the present study, three parameters are explicitly considered.

Time duration of simulated S wave, T_d, from propagating fault is determined from the viewpoint of superposition (for example Ref.14). In this study if source-to-site distance of element ij beyond which there are no significant motion or effective spectral value at the site contributes to total duration, T_d will be derived from:

$$T_d = \underset{ij}{\overset{NM}{\mathrm{Max}}}(Ta_{ij} + Tb_{ij}) - \underset{ij}{\overset{NM}{\mathrm{Min}}}(Tb_{ij}) \quad \cdots\cdots\cdots (11)$$

where,

$$Ta_{ij} = \frac{w_{ij}\cos\tilde{\theta}_{ij}}{\alpha} + l_{ij}\left(\frac{1}{V_{R_{ij}}} - \frac{\sin\tilde{\theta}_{ij}\cos\tilde{\phi}_{ij}}{\alpha}\right) + \tau_{ij},\quad Tb_{ij} = t_{ij} + \frac{\tilde{r}_{ij}}{\beta} \quad \cdots\cdots\cdots (12)$$

Tilda notation relates to the coordinates of the center of the rupturing element.

When sample ground motions are generated, regressions with constant variance are found(Refs.3,13). In order to predict the duration of earthquakes using that (Total Power)=(RMS)x(Duration), the following regression equation was used.

$$\ln T_{td} = d_1 + d_2 . \ln[\sum_{i=1}^{n} P_s(f_i) . \Delta f] \quad \cdots\cdots\cdots (13)$$

where, $P_s(f)$=Power spectral density of ground motion at a specified site. The coefficients, d_1 and d_2, of the expression of expected duration versus RMS are obtained from the method of least squares.

Similarly, the seismic moment parameter was scaled for duration parameter using the regression analysis. Although it is well known that RMS value of acceleration, a_{RMS}, is very weakly dependent on M_O($a_{RMS} \propto M_O^{1/6}$) but is directly proportional to stress drop, it is important for engineering purposes to represent expected seismic moment versus duration in the following form.

$$\begin{aligned} \ln M_0 &= m_1 + m_2 . \ln T_{td} \quad \cdots\cdots\cdots (14) \\ &= (m_1 + m_2 d_1) + m_2 d_2 . \ln[\sum_{i=1}^{n} P_s(f_i) . \Delta f] \end{aligned}$$

SIMULATION OF RANDOM SOURCE PARAMETERS

In this section the model of simulations is described in order to obtain fluctuation of $\overline{F}_s(\omega)$ in Eq.(4) along the frequency axis. Then $G_k(\omega)$ and $\phi_k(\omega)$ in Eq.(9) are calculated on this basis.

Seismic Moment, M_O

We assume that a suitable parameter of earthquake size is seismic moment, rather than magnitude M_L. The seismic moment is given as a truncated Weibull distribution to account for the strain energy buid-up proportional to time and for a statistics of shear crack's fracture based on the experimental results. In the dislocation source model, it seems appropriate bacause the use of Weibull distribution is consistent with the idea of failure of a mechanical component. Using the cumulative distribution function (CDF) with the parameters μ and γ of the inter-arrival time of radiated seismic energy, the CDF of seismic moment follows:

$$F_{M_0}(M_0) = N_{M_0}\left[1 - \exp\left\{-\mu\left(\frac{M_o - M_{o1}}{\underline{a}\, M_s}\right)^{\gamma}\right\}\right] \quad \cdots\cdots\cdots (15)$$

where, N_{M_0}=normalizing correlation factor

$$= \left[1 - \exp\left\{-\mu\left(\frac{M_{o2} - M_{o1}}{\underline{a}\, Ms}\right)^{\gamma}\right\}\right]^{-1}$$

$$M_s = \frac{\mu}{\eta \overline{\sigma}}$$

M_o is bounded by a lower bound M_{o1} and an upper bound M_{o2} corresponding to the earthquake size in a region considered. η_s is the seismic efficiency, $\overline{\sigma}$ is an average stress, a is a propagating constant. From Eq. (15) the simulated seismic moments are generated using standard random numbers(RN) in accounting for model parameters.

$$M_0 = M_{0_1} + \underline{a}M_s\left[\frac{1}{\mu} \cdot ln\left(\frac{N_{M_o}}{N_{M_o} - RN}\right)\right]^{1/\gamma} \quad \text{.................} \quad (16)$$

Geometry of the Fault Plane ——Location, Strike, Dip and Rake(Figs.1 and 2)

The uncertainty in the geometry (location and orientation) of the fault plane being not perfectly known except for very high seismic regions such as San Andreas fault is recognized. These parameters must therefore be developed through random variables with consideration of their uncertainties.

When the earthquakes are known to occur anywhere within a fault area, the possible epicentral point can be incorporated to locate uniformly between longitude and latitude of two points on fault trace which describe the orientation of the fault plane, and also to give a distance uncertainty on the longitude. Other parameters——strike(ie, spatial orientation of the fault), dip, and rake(angle which give the general direction of rupture propagation within the fault plane) may be more uncertain, and may take any value within the total range of variation defined by the lower and upper bounds around their mean values. For a narrow shallow fault, however, the rake may be taken as zero. Because of the nature of the uncertainty, the Monte Carlo procedure is used. The above five spatial parameters will each independently be generated by using a uniform distribution.

Focus Location

From determination of Point P in Fig.5 on this basis, the coordinates of the focus in the fault system is obtained by giving ranges of focus depth measured as the shortest distance from the surface of the earth in the fault plane. The hypocenter (ie, center of the rupture) will be probabilistically located within these ranges, using a similar uniform distribution(see Figs.5 and 6)

Outcropping of Fault Plane

The fault trace is well known in larger seismic regions, but its accurate sketch relative to the surface of the earth may be uncertain. This information will give the dispersion characteristics of ground trace for the basis of a perfectly unknown fault surface within the earth's crust obtained using previous steps. A remarkable effect of this phenomena is found in noting that a sample fault rupture may possibly not reach the surface of the earth.

The case when the fault breaks the surface of the earth for every release of energy is outcropping. Other case is that the fault may or may not outcrop depending on comparison between its dimensions and the depth of the focus. For $i \in n$ (sample number of random fault), let Y_i be a binary random variable such that

$$Y_i = \begin{cases} 1, & \text{if the former case} \\ 0, & \text{if the latter case} \end{cases}$$

If there is no information on the average depth of focus, the sample rupture area is assumed to reach the ground surface in all simulation processes. Although in this study Y_i is actually deterministic, note that a sample fault rupture may possibly reach the ground surface.

Ratio(Ra) of the Width(W) to the Length(L) of the Entire Fault Area

It has been shown that for a known fault in the world its width is strongly correlated with the length as follows:

$$W = R_a . L \qquad (17)$$

In the range of interest, then, statistical uncertainty of the parameter, Ra, is incorporated to give a rational basis for engineering decisions The lower and upper bounds, Ra_1, Ra_2 is taken here as 0.5, 2.0. Geller (Ref.7) showed that the average ratio Ra is 0.5, and found as first approximation a good regression relation between the fault area,S, and seismic moment, M_0:

$$\mathit{log}\, S = \frac{2}{3} \mathit{log}\, M_0 + a_2 (\Delta\sigma) \qquad (18)$$

where, a_2 is a parameter dependent on a constant stress drop, $\Delta\sigma$. Therefore once the seismic moment in first step is generated, using Eqs. (17) and (18) gives the dimensions of rectangular and planner fault surface. The length of the simulated fault rupture is known as follows.

$$L = \sqrt{\frac{a_3}{R_a} M_0^{1/3}} \qquad (19)$$

where, a_3 is a constant. Given the length of the fault, its width can be obtained by using the empirical Eq.(17) and uniform distribution of parameter Ra, lacking information to the contrary.

Subdivided Fault Segment Length and Width, l_{ij}, w_{ij}

When a sample seismic moment and shape of the rupture surface in the fault plane are generated by previous simulations, the fault plane in highly irregularly broken into NM subelements, as is shown in Fig.5. For simplicity it is assumed that Eq.(17) is valid for the average dimensions of each small element.

The probability function for these segment dimensions was assumed to be a negative exponential distribution, with mean and maximum values. The CDF of the coherent length is then:

$$F_{\tilde{l}}(l) = (1 - e^{-l/a_4}) \Big/ (1 - e^{-l_2/a_4}) \qquad \text{for } l_1 = 0 \le l \le l_2 \qquad (20)$$

when, $\overline{l}$ is a mean length, l_2 is greater than $(2\overline{l})$ and parameter a_4 which fulfills the following equation.

$$\overline{l}\Big/l_2 = a_4\Big/l_2 - (e^{l_2/a_4} - 1)^{-1} \qquad (21)$$

The ij-th length of the strip in Fig.5 is generated by a Monte Carlo procedure as follows:

$$l_{ij} = -a_4 . \mathit{ln} \{ 1 - RN. (1 - e^{-l_2/a_4}) \} \qquad (22)$$

Similarly the width of each element within a strip, w_{ij}, is simulated by using $R_b.l(R_b$: ratio of length and width in each element) instead of l in Eq.(20).

Let the dimensions of each coherent patch be small enough that the radiation in the frequency range of interest be equivalent to that of a point source. If the patch is not small enough to apply directly the wave equation,

$$l_{ij} \leq (l_{ij})_{max} = \frac{1}{\sin\phi_{ij}} \sqrt{\frac{r_{0ij}}{2} \frac{\beta}{f_{max}}} \qquad (23)$$

it must be further subdivided in smaller finite elements to fulfill the finiteness condition. Eq.(23) is described from the far-field approximation. Otherwise a patch which is too close to the site is also subdivided, using the dimensions of finite fault elements obtained by subdivision of its patch. These procedure are the same as the above process of discretization. Typical simulation result is Fig.5 given in the previous section.

Rise Time, τ_{ij}

In an actual earthquake, dislocation characteristics inevitably vary in some complex manner over the fault surface, but it may be evaluated rationally that the same slip function is used for each subdivided element on the rupture surface. The rise time parameter τ_{ij} of each element ij depends on spatial variations, and is strongly correlated to the rupture area S_{ij}. It is also controlled by the time it takes information for non-sliding portions on the fault surface to propagate to points where sliding is occuring. This choice of our source model is realistic because of the presence of stress singularities.

If the effective dynamic stress is equal to the static stress drop, on the average, the rise time τ_{ij} can be estimated as follows(Ref.7):

$$\overline{\tau}_{ij} = 0.41 l_{ij} \sqrt{R_b} / \beta \qquad (24)$$

It is assumed here that the fracture spreads radially over the fault surface from $\tau_1 = 0.5\overline{\tau}_{ij}$ to $\tau_2 = 2\overline{\tau}_{ij}$ by using the β distribution(Ref.3).

$$f_\tau(\tau) = \frac{1}{B(q,r)} \frac{(\tau-\tau_1)^{q-1}(\tau_2-\tau)^{r-1}}{(\tau_2-\tau_1)^{q+r-1}} \qquad (25)$$

where, B(q,r) is the Beta function. Parameter q and r can easily be solved with mean and variance from,

$$\left.\begin{aligned} \overline{\tau} &= \tau_1 + \frac{q}{q+r}(\tau_2 - \tau_1) \\ \sigma_\tau^2 &= \frac{qr}{(q+r)^2(q+r+1)}(\tau_2 - \tau_1)^2 \end{aligned}\right\} \qquad (26)$$

Rupture Velocity, $V_{R_{ij}}$

The rupture indicates at a focus on the fault surface. In this study it is assumed that a strike slip rupture spreads at a velocity varying from 50 percent to 91.94 percent of the local shear wave velocity β, the upper limit being set at the Rayleigh wave velocity according to fracture mechanisms. An average of rupture propagation velocity as shown by Geller(Ref.7) is $\overline{V}_{R_{ij}}$=0.72β. Sample values in element ij corresponding to spatial variations are generated as a random variable with the β distribution type the same as the rise time.

NUMERICAL RESULTS

In this section, we use the method presented above to simulate bilateral strike-slip earthquake ground motions for the parameters given in Table.1. Table 1 gives a summary of data re geometry of fault and each element patch, earthquake size and occurence parameters, source and propagation characteristics, and other dynamic parameters, including the uncertainty of the model parameters. The number of Monte Carlo simulations developed on this basis is 50, by varying the magnitude M_L between 5.6 and 7.4, on which focus take place from a very large hypocenter to relatively shallow point. In all of the following eaxmples one exponential source time function being the same over the fault surface is used, except for comparison of ground motion due to source functions. Our numerical results indicate that the rupture model gives a realistic of the energy in time, and that the directivity effects in a three dimensional field are accounted for.

Fig.7 shows a typical example of the geometry of this system on the fault trace at the surface of the earth, indicating a medium size earthquake on a fault approximately 57 km from the site. This illustration corresponds to the simulation No.1 shown in Figs. 5 and 6, indicating that total fault dimensions=(L,W)=(23.42km,4.68km), focal coordinates in rupture plane system=(F_X,F_Y)=(-1.34km,-1.96km), site coordinates in above system=(Xs,Ys,Zs)=(-6.59km,-63.96km,28.52km), seismic moment=$1.116\text{x}10^{20}$(m.N)(ie,M_L=7.33), duration of SH wave=7.04sec, and total number of finite elements=65, etc.

Fig.8 shows the scatter sample values of M_0, T_d and estimated a_{RMS} using the approximate relation $a_{RMS} \simeq \Sigma P_S(f_i) \cdot \Delta f$ about regression lines, where 40 data selected out of the total 50 are drawn. The values of the regression coefficients in Eqs.(13) and (14) are d_1=1.766, d_2=-0.086, m_1=45.16 and m_2=0.42. This result can be used for predicting earthquake motions for given values of seismic moment and duration at a specified site. This is achieved by generating sample earthquake motions using Eq.(9) in which the $G_k(\omega)$ and $\phi_k(\omega)$ are determined from Eq.(4), leading to the consistent ground motion.

Fig.9 shows typical examples of sample earthquake motions, where arbitrary two sets computed through 50 simulations were chosen. These consistent waves have a moment of $0.9843\text{x}10^{20}$(m.N), which corresponds to about magnitude 7.3. From Fig.8, overall expected duration of directed S-waves is 8.011sec. Then, the peak value in Fig.9(a) was 0.814m/sec^2 at 15.13sec after the first S wave energy radiation at the source, and similarly in Fig.9(b) 0.688m/sec^2 at 18.30sec. The spreading of rupture front on fault plane corresponding to seismic wave in Fig.9(a) is shown in Figs.5 and 6. At the top in each figure, bars labeled "⟷" given the time interval direct S waves are arriving from some part of the assumed fault plane, although observed records are actually accompanied by directed P waves and Surface waves not considered in this paper. In addition, the displacement in Fig.9(c) was derived from the corresponding consistent acceleration in Fig.9(b). The displacement and acceleration at the site caused by a sample fault rupture is obtained by superimposing the effect of each rupture element with some triggering phase decay in time. A displacement spectrum leads to apparant high-frequency decay, although this is observed from the corresponding RMS and Peak response spectra.

To examine nonstationary frequency characteristics of simulated waves, sample wave propagation characteristics within the earth are utilized in studying both scattering and attenuation of high-frequency waves by drawing arrival time curves. This aspect is discussed using the response envelope method. The digital plot curve of Fig.10 determined from Fig.9(b) shows the frequency content and the time dependent envelopes, normalized so that the largest amplitude is equal to 10. This result agrees with seismological knowledge.

In order to check the validity of the propagation, an additional result of the generation of the coherent patches by simulation, and triggering and rupture propagation schemes is also shown in Fig.11, against the subdivision model on the fault plane in Figs.5 and 6. Herewith, the total of finite fault elements required for the estimation of ground motion at the site was 65,41,61,36,51,45 and 70 etc, corresponding to each simulation ranging from Max. 80 to Min. 31. The number of simulations such that the number of subdivided elements is zero is 3 times within 50 times, corresponding to a total number of finite elements of 36,47 and 35. Although it is concluded that all the results are greatly influenced by the uncertainty of the parameters selected, it must be noted that for engineering purposes the choice of rupture velocities is important for the determination and characteristics of the frequency content as shown in Fig.10. Among the model parameters appearing in Figs.5 and 6, in general, rise time much greater than $w_{ij}/0.9194\beta$ would correspond to afterslip over the established fault plane, and the other way might correspond to a fault which seals itself behind the advancing rupture front.

The combination of randomness in the rupture propagation and source time functions leads to sample earthquake motions shown in Figs.12 and 13 with Fig.9. Figs.12 and 13 result from ramp and sinusoidal source functions against exponential function in Fig.9. Here, the conditions other than source time functions are all the same between each (a) or (b) in Figs.9,12 and 13. For example subdivision result on fault plane, triggering and rupture propagation algorithm by simulation appearing in Figs.5 and 6 leads to ground motion in Figs.9(a),12(a) and 13(a). Expected S wave duration is 8.009sec in Fig.12, and 8.242sec in Fig.13.

From comparison of corresponding simulated acceleration, the peak values using sinusoidal function as shown in Fig.13 are larger than the other cases in Figs.9 and 12, because of being affected strongly by the change of different power distribution dependent on the type of source functions. This result is also explained from RMS response spectra shown in Fig.14. A frequency where the mode of the spectrum is located is about 1Hz in Fig.14(c), on the contrary, showing flat spectra in 0.6 to 2.2Hz range as shown in Figs.14(a) and (b). This way can be justified by considering that a sinusoidal function's spectrum $\overline{D_S}(\omega)$ in Eq.(4) differs from the corresponding equation for the ramp function only by the response of the factor $[1-(\omega\tau/2\pi n)^2]$ (here, n=3) in the dominator, and causes a significant relative increase in the high frequency part on the energy spectrum

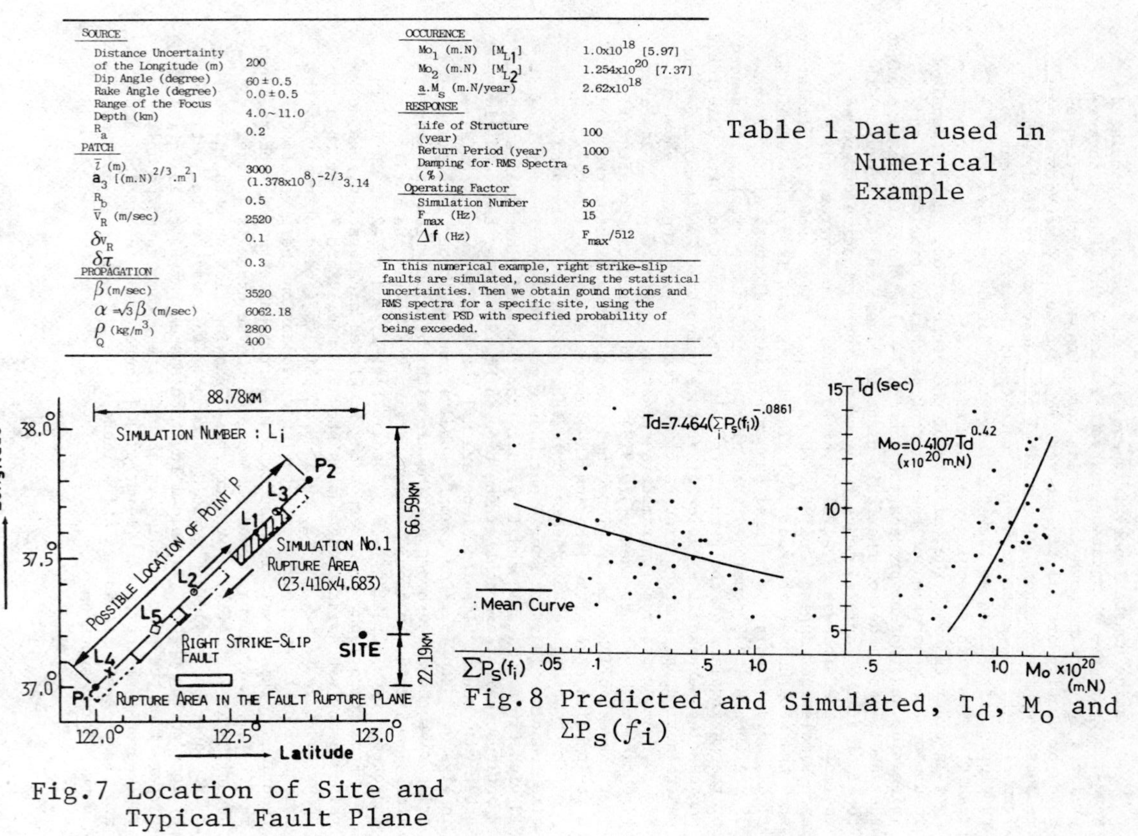

SOURCE	
Distance Uncertainty of the Longitude (m)	200
Dip Angle (degree)	60 ± 0.5
Rake Angle (degree)	0.0 ± 0.5
Range of the Focus Depth (km)	4.0 ~ 11.0
R_a	0.2
PATCH	
$\bar{l}$ (m)	3000
a_3 [(m.N)$^{2/3}$.m^2]	(1.378x10^8)$^{-2/3}$3.14
R_b	0.5
$\bar{V}_R$ (m/sec)	2520
δv_R	0.1
$\delta\tau$	0.3
PROPAGATION	
β (m/sec)	3520
$\alpha = \sqrt{3}\beta$ (m/sec)	6062.18
ρ (kg/m^3)	2800
Q	400

OCCURENCE	
Mo_1 (m.N) [M_{L1}]	1.0x10^{18} [5.97]
Mo_2 (m.N) [M_{L2}]	1.254x10^{20} [7.37]
$\underline{a}.M_s$ (m.N/year)	2.62x10^{18}
RESPONSE	
Life of Structure (year)	100
Return Period (year)	1000
Damping for RMS Spectra (%)	5
Operating Factor	
Simulation Number	50
F_{max} (Hz)	15
Δf (Hz)	F_{max}/512

In this numerical example, right strike-slip faults are simulated, considering the statistical uncertainties. Then we obtain gound motions and RMS spectra for a specific site, using the consistent PSD with specified probability of being exceeded.

Table 1 Data used in Numerical Example

Fig.7 Location of Site and Typical Fault Plane

Fig.8 Predicted and Simulated, T_d, M_o and $\Sigma P_s(f_i)$

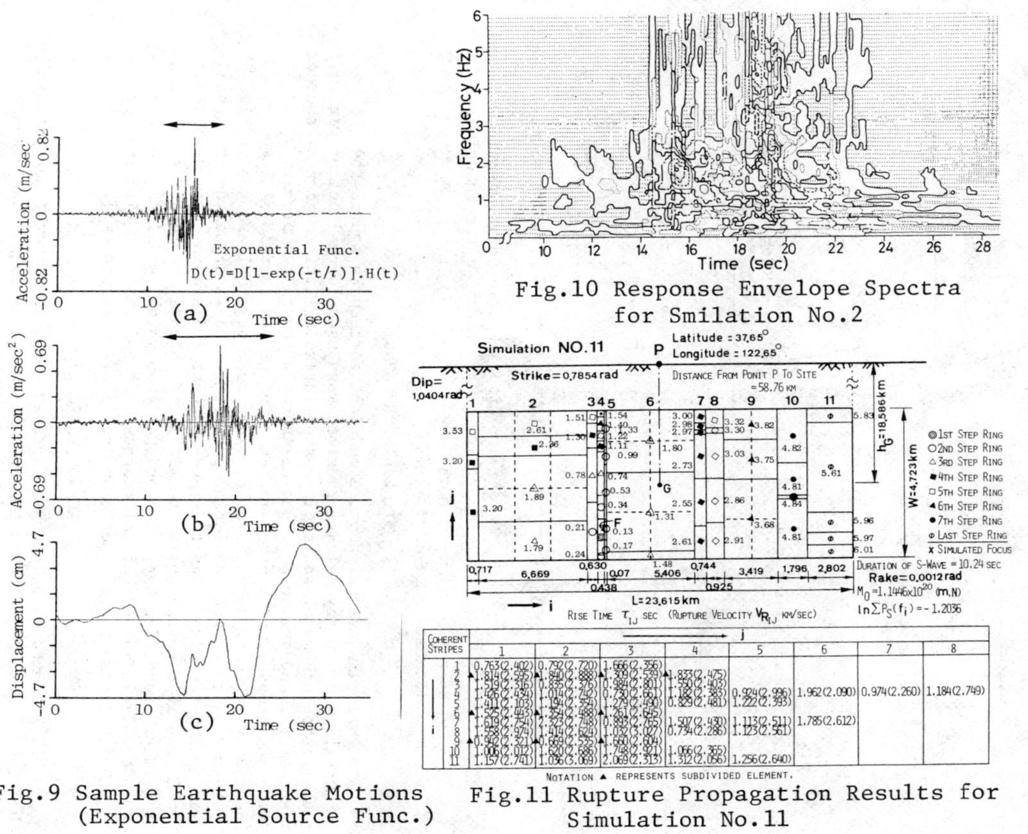

Coherent Stripes (i ↓)	j = 1	2	3	4	5	6	7	8
1	0.763(2.402)	0.792(2.720)	1.666(2.356)					
2	▲1.814(2.595)	▲1.840(2.888)	▲1.309(2.539)	▲1.833(2.475)				
3	1.519(2.316)	0.835(2.376)	0.984(2.801)	1.550(2.403)				
4	1.426(2.434)	1.014(2.742)	0.730(2.661)	1.182(2.383)	0.924(2.996)	1.962(2.090)	0.974(2.260)	1.184(2.749)
5	1.411(2.103)	1.194(2.354)	1.279(2.490)	0.829(2.481)	1.222(2.393)			
6	▲1.525(2.443)	▲1.354(2.488)	▲1.261(2.645)					
7	1.619(2.754)	2.323(2.748)	0.893(2.765)	1.507(2.430)	1.113(2.511)	1.785(2.612)		
8	1.558(2.974)	1.414(2.624)	1.032(3.027)	0.734(2.286)	1.123(2.561)			
9	▲0.942(2.321)	▲0.669(2.535)	▲1.660(2.604)					
10	1.006(2.012)	1.620(2.686)	1.748(2.921)	1.066(2.365)				
11	1.157(2.741)	1.036(3.069)	2.069(2.313)	1.312(2.056)	1.256(2.640)			

Notation ▲ represents subdivided element.

Fig.9 Sample Earthquake Motions (Exponential Source Func.)

Fig.10 Response Envelope Spectra for Smilation No.2

Fig.11 Rupture Propagation Results for Simulation No.11

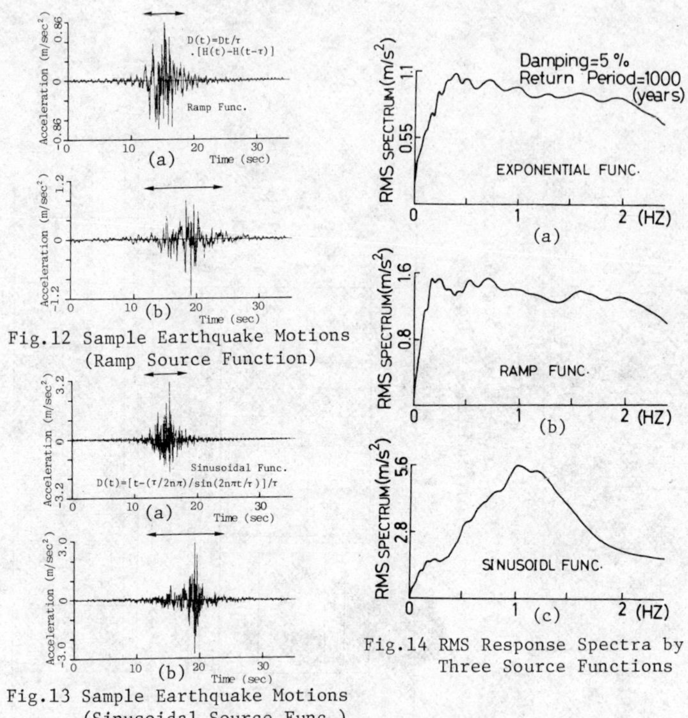

Fig.12 Sample Earthquake Motions (Ramp Source Function)

Fig.13 Sample Earthquake Motions (Sinusoidal Source Func.)

Fig.14 RMS Response Spectra by Three Source Functions

CONCLUDING REMARKS

A three dimensional rupture model of earthquake source and propagation of energy release in time, when used in conjunction with a subdivision scheme of fault plane and statistical information of source parameters, appears to be a useful approach toward the problem of modeling propagating rupture motion. Our proposed artificial model, dependent on focal mechanisms, provides more physical and qualitative information than site-consistent prediction models of earthquake ground simulation for given magnitude and epicentral distance developed so far. It is possible to synthesize realistic earthquake records by simulating the incoherence of a three dimensional geophysical model for engineering purposes including structural response, ground motion estimation for long-period structures, (for example, lifeline structures).

ACKNOWLEDGEMENT

Mr. Hakamata (Now, Hanshin Expressway Public Corporation) provided helpful suggestions both in the writing and debugging of the computer program.

REFERENCES

1) Anderson,J.G. and Richards,P.G., "Comparison of Strong Ground Motion from Several Dislocation Models," Geophys. J. R. astr. Soc., Vol.42, pp.347-373, 1975.
2) Anderson,J.G., "Motions Near a Shallow Rupturing Fault : Evaluation of Effects due to the Free Surface," Geohys. J. R. astr. Soc., Vol.46, pp.575-593, 1976.
3) Ang,A.H-S., and Tang,W.H., "Probability Concepts in Engineering Planning and Design, Vol.I : Basic Principles," John Wily & Sons,Inc., New York 1975.
4) Burridge,R. and Knopoff,L., "Body Force Equivalents for Seismic Dislocations," BSSA, Vol.54, pp.1875-1888, 1964.
5) Bache,T.C. and Barker,T.G., "The San Francisco Earthquake - A Model Consistent with Near-Field and Far-Field Observations at Long and Short Periods," Systems, Science and Software, Final Technical Report 78-3552, January 1978.
6) Boore,D.M. and Joyner,W.B., "The Influence of Rupture Incoherence on Seismic Directivity," BSSA, Vol.68, No.2, pp.283-300, April 1978.
7) Geller,R.J., "Scaling Relations for Earthquake Source Parameters and Magnitude," BSSA, Vol.66, No.5, pp.1501-1523, 1976.
8) Haskell,N.A., "Total Energy and Energy Spectral Density of Elastic Wave Radiation from Propagating Faults," BSSA, Vol.54, No.6, pp.1811 -1841, Dec., 1964.
9) Haskell,N.A., "Elastic Displacements in the Near-Field of a Propagating Fault," BSSA, Vol.59, No.2, pp.865-908, April 1969.
10) Johnson,L.R., "Green's Function for Lamb's Problem," Geophys. J., Vol.37, pp.99-131, 1974.
11) Kameda,H., Sugito,M. and Asamura,T., "Simulated Earthquake Motions Scaled for Magnitude, Distance, and Local Soil Conditions," Proc. 7WCEE, Istanbul, Sept. 1980.
12) Maruyama,T., "On the Force Equivalents of Dynamic Elastic Dislocations with reference to the Earthquake Mechanism," Bull. Earthq. Res. Inst., Vol.41, pp.467-486, 1963.
13) McGuire,R.K., "Seismic Structural Response Risk Analysis, Including Peak Response Regressions on Earthquake Magnitude and Distance," MIT, Dept. of Civ. Eng., Research Report R74-51, Aug., 1974.
14) Okada,Y., "The Difference in the Durations of P and S waves from Propagating Faults," Journal of Physics of the Earth, Vol.18, No.2, pp.193-202, 1970.
15) Rascón,O.A. and Cornell,C.A., "Strong Motion Earthquake Simulation," MIT, Dept. of Civ. Eng., Research Report R68-15, April 1968.
16) Toki,K. and Sato,T., "Detections of Nonstationary Filtering properties of Strong Motion Seismograms," Annals of the Disaster Prevention Research Institute, Kyoto University, No.22B-2, pp.25-36, April 1979 (in Japanese).
17) Wong,H.L. and Trifunac,M.D., "Synthesizing Realistic Ground Motion Accelerograms," USC, Dept. of Civ. Eng., Report No. CE78-07, Sept., 1978.

DYNAMIC RESPONSE WITH FOUNDATION INTERACTION

Timothy O. Weaver[1], Furman W. Barton[2], M.ASCE and Richard T. Eppink[3], M.ASCE

ABSTRACT

This paper describes a procedure for determining the earthquake response of building-foundation systems taking into account the frequency dependent characteristics of the soil. This procedure, which utilizes the Discrete Fourier Transform, initially solves only for the earthquake response of the foundation. Then, with this foundation response as excitation, the structural response is determined. This sequence of analysis permits efficient solution by such schemes as modal analysis.

Results are presented for a 50-story building in which the excitation is a truncated version of the Parkfield earthquake. The frequency response function for the base rotation and the top-floor displacement are determined for cases in which different numbers of modes are employed in the structural model. The effect of foundation interaction is evaluated by comparing the structural response of the fixed base system to that of the soil-structure system.

INTRODUCTION

The earthquake response of a structure can be significantly affected by its dynamic interaction with the foundation. However, a dynamic analysis which includes the interaction between the soil and the structure is a far more difficult task than solving for the response of a fixed-base system. In order to adequately represent the soil-structure interaction, it is necessary to take into account the frequency dependent characteristics of the soil. This frequency dependence in the resulting equations of motion for the system renders traditional solution procedures such

---

[1]Formerly Graduate Student, University of Virginia, currently with Exxon Production Research, Houston, TX.

[2]Professor of Civil Engineering, University of Virginia, Charlottesville, VA.

[3]Associate Professor of Civil Engineering, University of Virginia, Charlottesville, VA.

as the normal mode method or direct numerical integration unsuitable.

Rather than trying to analyze the entire soil-structure system at once, the standard procedure is to separate the structure from its base and determine the dynamic response of only the foundation when subjected to a harmonic load. Using this substructuring approach, the dynamic behavior of the foundation-soil subsystem may then be summarized in the form of frequency response curves for the foundation.

The most common procedure for determining structural response including foundation interaction is to transform the equations of motion into the frequency domain using Fourier transforms. The foundation dynamic characteristics which appear in the equations of motion, are determined separately by a steady-state analysis of the foundation subjected to harmonic excitation. The structural response is then obtained by solving the equations of motion in the frequency domain. This requires that the frequency domain response be determined over a range of frequencies for which the ground and structural response have significant components, which in turn requires the solution of the full set of equations for each frequency. This is equivalent to determining the complex frequency response function for the system. For systems having a large number of degrees of freedom, this analysis procedure can be prohibitively expensive. A number of modifications have been suggested to reduce the number of equations that have to be solved but the procedure is still costly.

This paper describes an alternate procedure for determining the earthquake responses of building-foundation systems taking into account the frequency dependent characteristics of the soil. This procedure, which utilizes the Discrete Fourier Transform, initially solves only for the earthquake response of the foundation. Then, using the foundation response as excitation, the structural response is determined by modal analysis. The efficiency of the normal mode method of analysis is enhanced since commonly only a small percentage of the total number of modes need be considered for satisfactory accuracy.

THEORY

Consider a multidegree-of-freedom model represented by the N-story shear building shown in Fig. 1. In a shear building the floor girders are assumed to be infinitely rigid and structural mass is concentrated at the floor levels. The effects of foundation interaction are modelled by providing a frequency dependent resistance to rotation at the base.

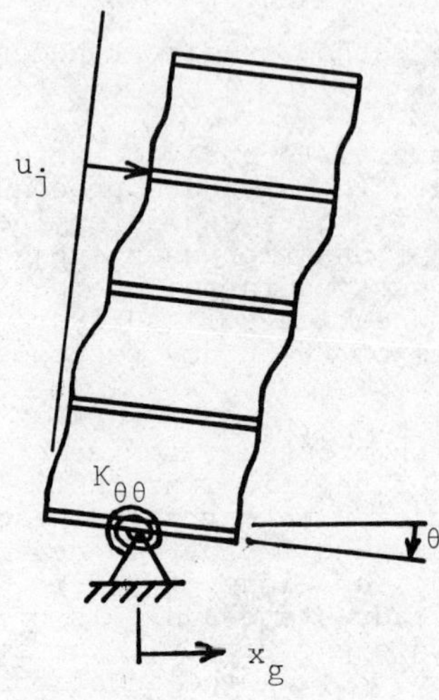

FIGURE 1: BUILDING SOIL-STRUCTURE MODEL

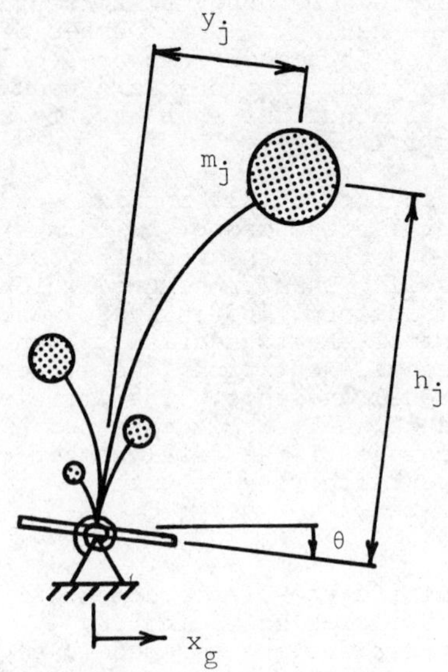

FIGURE 2: MODAL SOIL-STRUCTURE MODEL

The equations of motion for this system subjected to a ground acceleration $\ddot{x}_g$ are as follows:

$$[M]\{\ddot{u}\} + [C]\{\dot{u}\} + [K]\{u\} = \{m\}\ddot{x}_g \qquad (1)$$

where [M], [C], and [K] are, respectively, the mass, damping, and stiffness matrices of the system. The vectors $\{u\}$, $\{\dot{u}\}$, and $\{\ddot{u}\}$ represent, respectively, the displacement, velocity, and acceleration of the mass points. The vector {m} describes the distribution of inertial loads. It should be noted that frequency dependent terms appear in the damping and stiffness matrices of Eq. (1), corresponding to the foundation degrees of freedom, and, thus, these equations have no real meaning when viewed in the time domain.

A direct procedure for affecting the solution of Eq. (1) may be described as follows. The Fourier Transform of Eq. (1) is first obtained as

$$\{-4\pi^2f^2[M] + i2\pi f[C] + [K]\}\{U\} = -4\pi^2f^2\{m\}X \qquad (2)$$

where U and X are the Fourier Transform of u and x respectively. If the following definitions are made

$$[A] = [K] - 4\pi^2f^2[M]$$

$$[B] = 2\pi f[C]$$

then Eq. (1) may be rewritten as

$$\{[A] + i[B]\}\ \{U\} = -4\pi^2f^2\{m\}X \qquad (3)$$

where, again [A] and [B] are functions of the frequency f. The solution of Eq. (3) basically requires the inversion of the coefficient matrix {[A] + i[B]}. More importantly, however, the complex definition of the transformed responce {U} requires the complete solution of Eq. (3) at each frequency increment of interest. This approach is computationally expensive.

For a structure of the type described by Eq. (1), modal analysis is not an alternative because the equation cannot be uncoupled. However, the displacements of the base within the soil are due to the combined effect of forces imposed by the structure and resistance developed by the soil. If the base response were known, then the structural response could be determined by an analysis of the structure subjected to a known base excitation. Since the soil characteristics are included in the determination of base excitation, modal analysis can be used to calculate the response.

In order to solve for the base response of the soil-structure system, it is necessary to calculate the forces exerted by the building on its foundation. It is convenient to describe the vibration characteristics of the fixed-base structure in terms of a modal model. By doing so, it is possible to determine the contribution of each mode on the total force exerted by the structure on the foundation.

Consider the two systems shown in Figs. 1 and 2. Fig. 1 shows a model of an N-story building. The N natural frequencies of the fixed-base structure can be calculated by standard procedures. Each mode is assigned a damping ratio ξ, based on dynamic tests of similar buildings. Fig. 2 shows an equivalent modal model consisting of N discrete mass single-degree-of-freedom systems. The parameters of each single-degree-of-freedom system are chosen so that it is dynamically equivalent to a corresponding mode of the N-story building.

To solve for the physical quantities, m_j, h_j, and y_j of the modal model shown in Fig. 2, three conditions must be specified. These conditions are defined so that both systems will have the same base shear, base moment and kinetic energy.

When the multistory building is vibrating freely in the jth mode with mode shape, $\{\phi\}$, and circular natural frequency, p_j, the response is given by

$$\{u\} = \{\phi_j\}q_j \tag{4}$$

where q_j is the jth generalized modal coordinate. While vibrating in this mode, the base shear is given by

$$\{1\}^T[M]\{\phi_j\}\ddot{q}_j$$

the base moment by

$$\{h\}^T[M]\{\phi_j\}\ddot{q}_j$$

and the kinetic energy by

$$\frac{1}{2}\{\phi_j\}^T[M]\{\phi_j\}\dot{q}_j^2$$

where $\{h\}$ = vector of story heights of the building.

For the corresponding single-degree-of-freedom system of the modal model, the base shear is given by

$$m_j \ddot{y}_j$$

the base moment by

$$h_j m_j \ddot{y}_j$$

and the kinetic energy by

$$\frac{1}{2} m_j \dot{y}_j^{\,2}$$

Equating these quantities between the two models leads to the following relationship

$$m_j = \frac{(\{1\}^T[M]\{\phi_j\})^2}{\{\phi_j\}^T[M]\{\phi_j\}} \qquad (5)$$

$$h_j = \frac{\{h\}^T[M]\{\phi_j\}}{\{1\}^T[M]\{\phi_j\}} \qquad (6)$$

$$y_j = \frac{\{\phi_j\}^T[M]\{\phi_j\}}{\{1\}^T[M]\{\phi_j\}} \qquad (7)$$

If the modal model is excited by a base acceleration, $\ddot{x}_g$, the response of each single degree-of-freedom system will be governed by the differential equation

$$\ddot{y}_j + 2\xi_j p_j \dot{y}_j + p_j^{\,2} y_j = -\ddot{x}_g \qquad (8)$$

where the damping ratio, ξ_j, is the same as that assigned to the jth mode of the building model.

For a multidegree of freedom building system such as that shown in Fig. 1 which is subjected to a ground acceleration, the equations of motion for the actual structure may be written as follows:

$$[M]\{\ddot{u}\} + [C]\{\dot{u}\} + [K]\{u\} = -[M](\{1\}\ddot{x}_g + \{h\}\ddot{\theta}) \qquad (9)$$

For the modal model the equivalent relationship for each mode is

$$\ddot{y}_j + 2\xi_j p_j \dot{y}_j + p_j^{\,2} y_j = -(\ddot{x}_g + h_j \ddot{\theta}) \qquad (10)$$

If the base rotation θ were known, then Eq. 10 could be interpreted as the equation of motion of an ordinary base excited single-degree-of-freedom system subjected to a modified excitation. The solution to this problem is straightforward.

In order to determine θ, first express the base moment exerted on the foundation by the structure. The Fourier Transform of this expression is

$$\bar{M} = - \sum_{j=1}^{N} m_j h_j (\ddot{X}_g + \ddot{Y}_j + h_j \ddot{\Theta}) - I_0 \ddot{\Theta} \tag{11}$$

The Fourier Transform of Eq. 10 can be used to obtain an expression for $\ddot{Y}_j$. Then $\ddot{Y}_j$ can be eliminated from Eq. 11. The resulting expression for $\bar{M}$ is

$$\bar{M}(\omega) = (\sum_{j=1}^{N} m_j h_j H_j) X_g + (\omega^2 I_0 + \sum_{j=1}^{N} m_j h_j{}^2 H_j)\Theta \tag{12}$$

Eq. 12 has the form

$$\bar{M} = (K_{\theta x}^{Bldg})X + (K_{\theta\theta}^{Bldg})\Theta \tag{13}$$

where $K_{\theta x}^{Bldg}$ is a frequency dependent stiffness coefficient indicating the base moment exerted by the superstructure due only to harmonic base translation and $K_{\theta\theta}^{Bldg}$ defines the base moment due to a harmonic base rotation.

In a similar fashion, it is possible to represent the moment exerted by the soil on the base due to harmonic base rotation as

$$\bar{M} = (K_{\theta\theta}^{soil})\Theta \tag{14}$$

where $K_{\theta\theta}^{soil}$ is available from a separate analysis.

Equating the moments and solving for Θ yields

$$\Theta = \frac{K_{\theta x}^{Bldg}}{K_{\theta\theta}^{Soil} - K_{\theta\theta}^{Bldg}} X \tag{15}$$

This expression provides the steady-state base response of the soil-structure system to a harmonic base translation.

The rotational base response of the system in the time domain can be found by taking the inverse Fourier Transform of Θ. Once the time history is available, the structural response to the combined translational and rotational base excitation can be obtained by standard modal analysis.

EXAMPLE SOLUTION

The discrete model adopted as the example problem in this section represents a 50-story building of height H. The total building mass M is equally distributed and lumped at each floor level. The lateral stiffness is assumed to be the same for each floor. The cross-section of the building and its foundation is assumed to be circular with a radius r, and each of the fixed-base natural modes will be assumed to have a damping ratio of 0.05. The lowest natural frequency of the fixed-base structure is assumed to have a value $f_1 = p_1/2\pi = 0.2$Hz.

For illustrative purposes, the base excitation is taken to be a truncated version of the Parkfield earthquake shown in Fig. 3.

The dynamic characteristics of the foundation are modeled using the expressions derived by Meek and Veletsos (1) and are shown in Fig. 4. The foundation resistance is of the form,

$$K_{\theta\theta}^{Soil} = K_\theta(k_\theta + ia_0 c_\theta) \tag{16}$$

where K_θ is the static rotational stiffness of the foundation, a_o is a dimensionless frequency parameter given by

$$a_o = \frac{2\pi(f/f_1)}{\dfrac{c_s}{f_1 H}\ \dfrac{H}{r}} = \frac{\omega r}{c_s} \tag{17}$$

The quantities k_θ and c_θ are dimensionless stiffness and damping coefficients, respectively, and c_s is the shear wave velocity of the soil. For this example, it is assumed that $H/r = 5$ and $c_s/f_1H = 3$. These values describe a tall building on a relatively soft soil.

The first step in solving for the base rotation is to calculate the steady-state response or transfer function for the base rotation. This steady-state response expressed nondimensionally as $H\ddot{\Theta}/\ddot{X}_g$, is

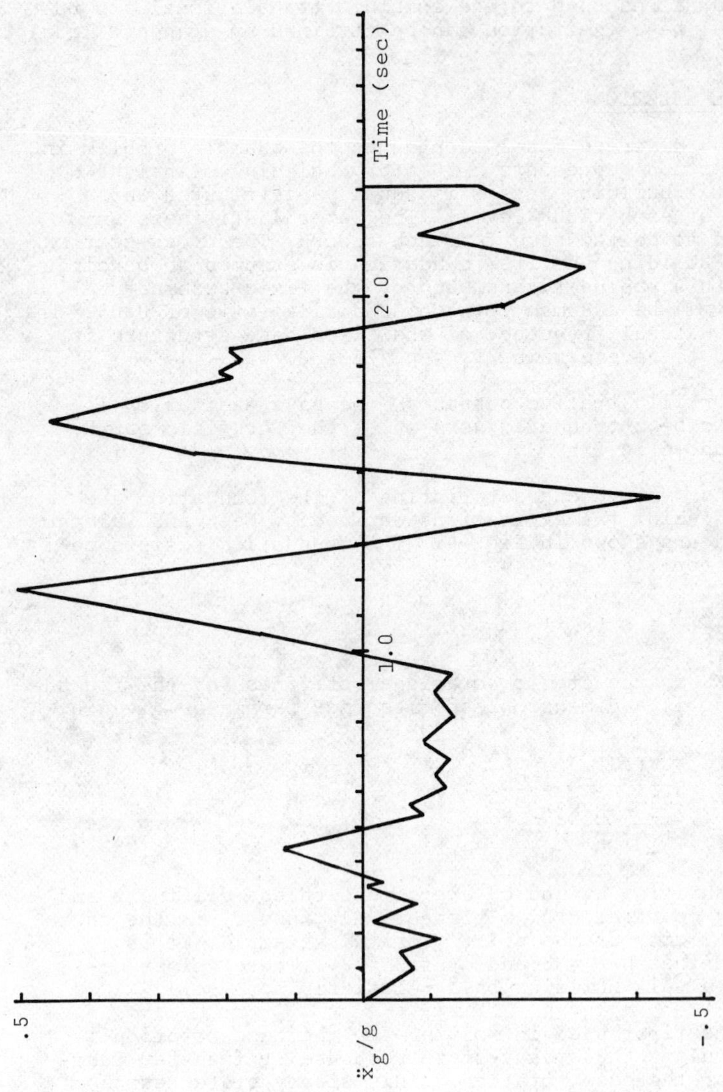

FIGURE 3: TRUNCATED VERSION OF PARKFIELD EARTHQUAKE

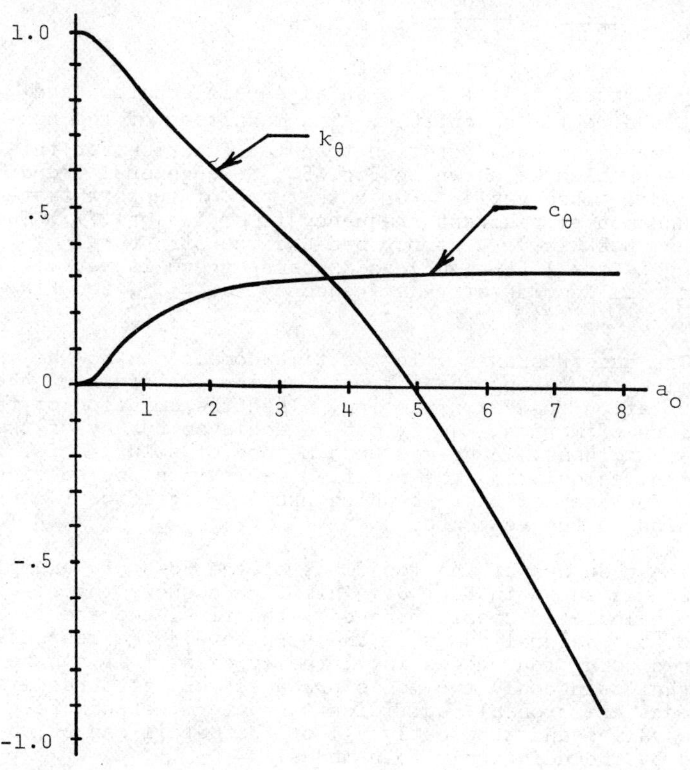

$$M = K_\theta(k_\theta + i a_o c_\theta)\theta$$

$$a_o = \frac{2\pi(f/f_1)}{\frac{c_s}{f_1 H} \quad \frac{H}{r}}$$

FIGURE 4: FOUNDATION DYNAMIC STIFFNESS COEFFICIENTS

$$\frac{H\Theta}{\ddot{X}_g} = \frac{\sum m_j h_j H_j}{\dfrac{K_{\theta\theta}^{Soil}}{p_1^2 MH^2} - \dfrac{(\omega/p_1)^2}{4(H/r)^2} + \sum m_j h_j^2 H_j} \quad (18)$$

It may be noted that this is an algebraic equation and $H\ddot{\Theta}/X_g$ is directly calculable. The magnitude of the non-dimensionalized base rotation transfer function for this example problem is shown in Fig. 5. Some general trends are readily observable. For both types of support systems, the fundamental resonant frequency is at $f/f_1 = .77$. The response peak is very narrow because the structural damping is small ($\xi = .05$) and because there is very little soil damping at this frequency ($a_o \simeq .3$, see Fig. 4).

The base response θ in the time domain can now be determined by taking the inverse Fourier Transform of the base rotation transfer function. Then the solution of the equations of motion, Eq. 9, can be achieved by any of the standard methods. A normal mode method of solution is completed by calculating the modified excitation and solving Eq. 10 for each of the j's which provide significant contribution to the response.

The response of the top floor of the 50-story example solution is shown in Fig. 6. This figure shows the response calculated considering only the first two and first four of the natural modes of the structure. The response obtained using four modes is virtually exact. It can be seen that using only two modes produces very good results. Two modes are probably sufficient to determine peak response values and apparently all of the detail can be obtained by including only four modes.

The structural response can also be continued in the frequency domain and the transfer function for the base response can be used as a transformed excitation to find the transfer function of the structural response. The frequency response of the top floor is shown in Fig. 7 for the building with a fixed base and also for a building in which the foundation interaction is included. For this particular system, the characteristics are such that the response of the flexible base system is larger at the top than that of the fixed base system at the fundamental frequency which is $f/f_1 = 0.77$ for the flexible-base system. The larger response occurs because at the low natural frequency of the structure, there is very little soil damping taking place. Thus, the flexibility of the system has been increased but with almost no change in

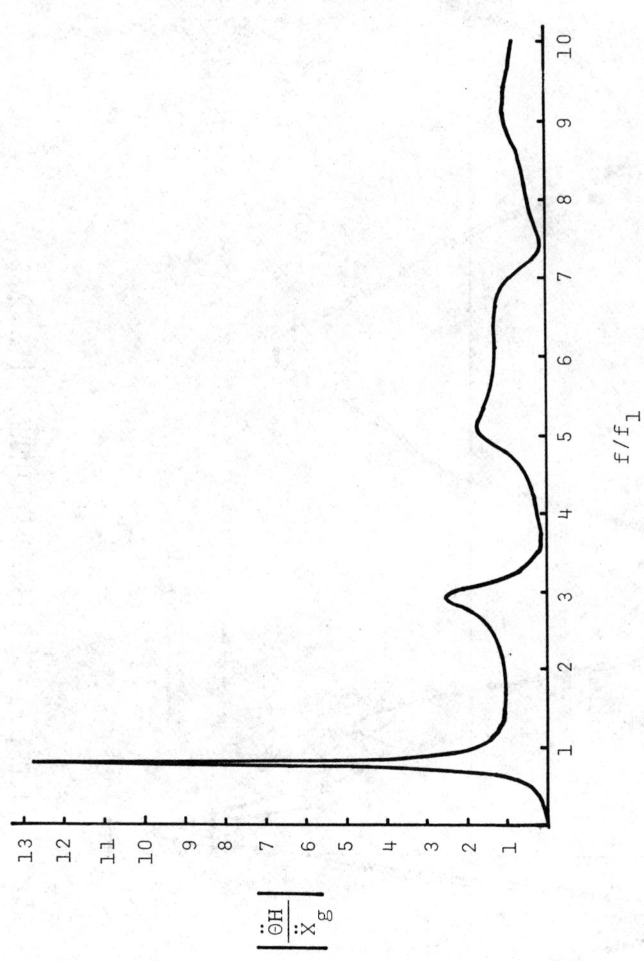

FIGURE 5: BASE ROTATION TRANSFER FUNCTION MAGNITUDE

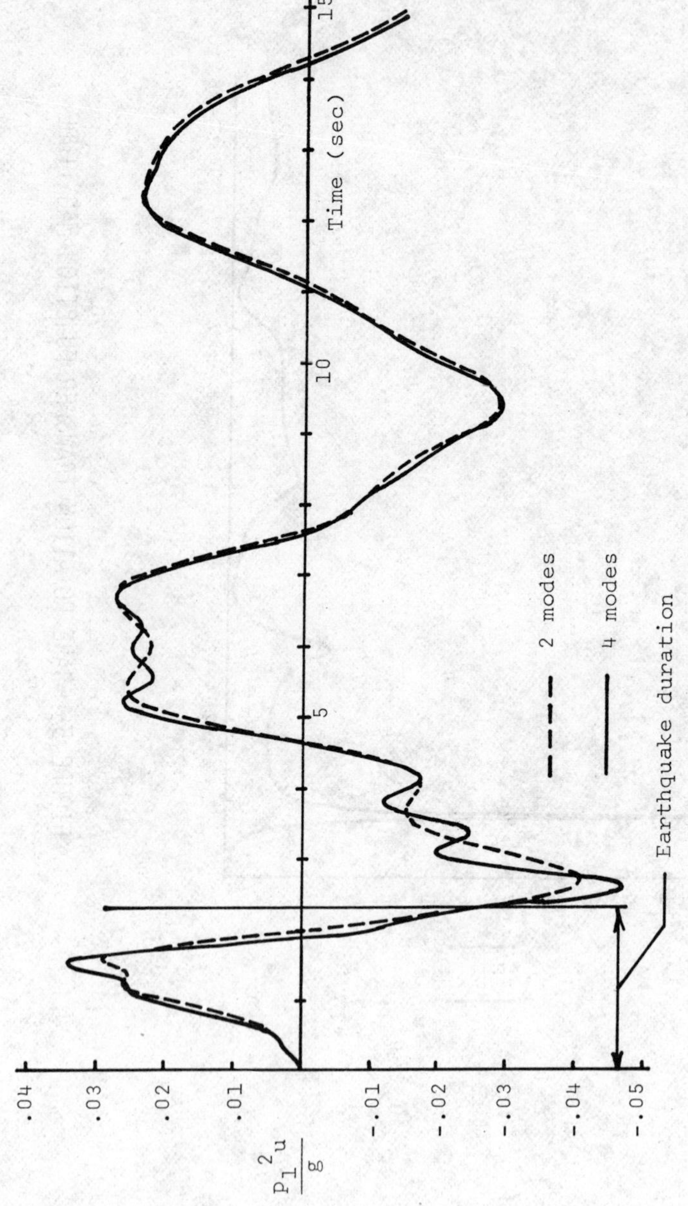

FIGURE 6: TOP FLOOR EARTHQUAKE RESPONSE

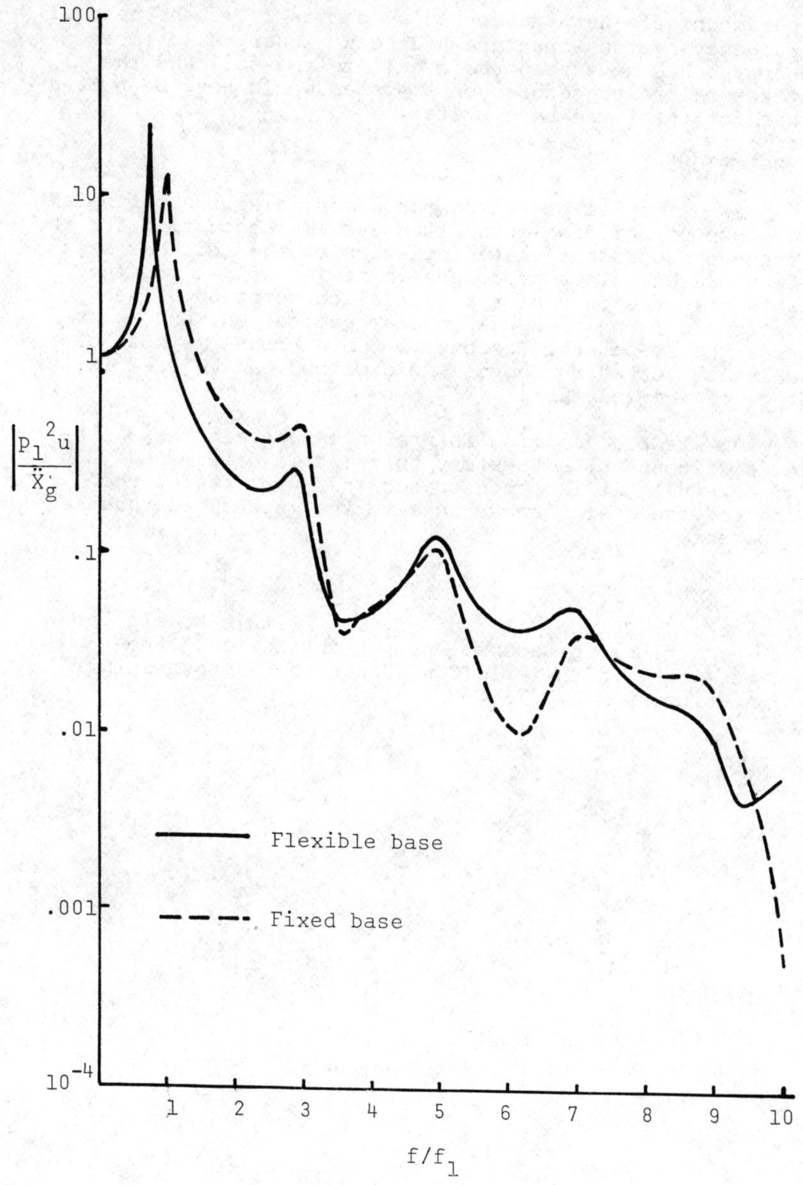

FIGURE 7: TOP FLOOR TRANSFER FUNCTION MAGNITUDE

the amount of energy absorbed by damping. If the natural frequency of the structure had been higher, a lot more energy could have been radiated into the soil and the structural response of such a system would have been less than for the fixed base system.

CONCLUSIONS

An alternate procedure for determining the earthquake response of building-foundation systems accounting for the frequency dependent characteristics of the soil has been demonstrated. This procedure initially solves for the earthquake response of the foundation rotation. Then, using this foundation response as excitation, one determines the structural response. Conventional methods of analysis such as the normal mode method can be used in this stage of the solution.

An example solution is presented wherein the earthquake response of a 50-story building is determined. In this example, it is demonstrated that very few of the modes of the system contribute to the general earthquake response.

REFERENCE

1. Meek, J. W. and Veletsos, A. S., "Simple Models for Foundations in Lateral and Rocking Motion," *Proceedings*, Fifth World Conference on Earthquake Engineering, Rome, 1973.

MEASUREMENTS OF ON-SITE DYNAMIC PARAMETERS FOR SEISMIC EVALUATIONS

By

Roger M. Zimmerman[1], M. ASCE

Kenneth R. White[2], M. ASCE

Abstract

The non-contact electromagnetic device developed in this study is capable of measuring bridge vibration and damping properties. The paper describes the development and field demonstration of the equipment. The tri-coil sensor proved to be an economical and relatively simple device to operate. A mobile unit could be developed using this equipment to excite bridges at low level vibrations and measure the desired dynamic properties.

I. INTRODUCTION

The purpose of this study was to develop and field test a new method to measure the resonant frequencies, modal shapes, and damping ratios for in-service bridges. These measurements are to be made in an effort to develop a low cost and relatively simple method to determine the sensitivity of existing bridges to earthquake. This information can serve two purposes. First the small data base used for AASHTO code formulation could be expanded and second bridges can be modified or retrofitted if they show particular earthquake sensitivity. In particular damping is a property that can only be determined through field data while vibrational modes can be modeled.

A relatively limited number of field tests have been conducted on bridges [1-19]. In some cases the bridges were carefully instrumented and tested to failure. Efforts have not been directed towards developing a mobile nondestructive forced vibration testing technique, although such methods are needed. In particular, there is a need to be able to induce controlled vibration energy such that amplitudes can be varied and the resulting vibrational responses studied. Such field data is extremely useful because it has been shown that damping varies with the magnitude of the amplitude [15]. More damping data can be used in improving modelling efforts.

[1] Member, Technical Staff, Sandia Nat. Lab., Albq., NM (formerly with NMSU)

[2] Professor of Civil Engineering, New Mexico State University, Las Cruces, New Mexico

Vibration sensing equipment has been developed around conventional force, acceleration, velocity, or displacement measuring devices. Each have their advantages and disadvantages. Accelerometers are probably the most popular, but they are expensive. A three-dimensional accelerometer can cost over $300. If six of these were placed on a bridge, the resulting cost would be significant. Economics suggested a study to develop a cheaper and simpler method to measure the key dynamic properties of vibrational magnitude, modes, and damping ratios.

It is known that coils moving in a static magnetic field induce currents. It was proposed that three small coils could be oriented orthogonal to each other and attached to a bridge and, if a static magnetic field were brought in proximity, three dimensional motion could be determined. This led to the proposal that was submitted and later funded under the NSF program titled, "Research Initiation in Earthquake Hazards Mitigation." This paper covers the study conducted under contract to New Mexico State University. The study objectives were:

1. Develop non-contact electromagnetic induction device to measure bridge vibration and damping properties.

2. Field test the device using forced vibrations on an existing bridge.

II. PROGRAM REQUIREMENTS

A. Experimental Developments

The program required the development of a shaker to induce the forced vibrations, a sensor to monitor the vibrations and an external magnet to activate the sensor. The program was funded as a research initiation effort and activities were directed solely towards fabricating equipment and conducting sufficient field studies to demonstrate the validity of the methods and not to perform a documented research study on bridge vibrations.

Bridge shakers have been developed around various principles. The double rotating mass and electrohydraulic oscillating mass shakers have been the most popular outside the United States [8, 9, 12-15, 19]. Most U.S. researchers have used traffic actuated vibrations using tractor-trailer combinations. It was felt that a precisely refined force generator was not necessary as the main requirement was to induce a vibration with controlled frequencies and amplitudes. This suggested the use of a single eccentric mass being driven by a variable speed motor. Older model pecan tree shakers operate on the principle that a cable is attached to a tree and to an eye bar that is connected to the housing enclosing a near spherical shaped ball bearing, approximately 6½ inches in diameter, which is eccentrically connected to a 1½ inch diameter shaft. The shaft can be connected to a power takeoff on a tractor with a belt and the frequency of shaking controlled by the tractor operator. This eccentric bearing unit was removed from its tree shaking apparatus and became the basic fixture which the forced vibration shaker was developed. The bearing provided a radial eccentric distance of 13/16 inch. The pecan shaker shaft was fixed to the top of twin channel beams which had been sectioned from a school bus used in an earlier study. The shaker unit was energized by placing a three-horsepower Century D. C. Motor on the channel beams. The motor operated at 120V and drew 22.6 Amp from a AC power generator that was rectified to DC voltage and oper, etc. The motor

Figure 1. Shaker on Big Dry Creek Bridge

Figure 2. Sensors D & E on Girder Ends with Electromagnet

had compound windings to give a variable speed control of ±10% for any pulley combination. Different size pulleys were coupled with the pecan shaker to give a range of speeds. For example a 3.5 inch pulley produced a speed of 810 RPM (13.5 Hz) and a 6.5 inch pulley produced a speed of 1520 RPM (25.3 Hz).

The overall shaker was designed such that the maximum vertical force due to the eccentric mass was less than that of the dead weight of the shaker itself. The addition of lead to the bus frame caused the total weight of the shaker unit to be approximately 1250 lbs.

A simple spring mass system analysis shows that the inertial force generated by eccentric masses is linear with the rotating mass and varies with the square of the frequency. This meant that the dead weight limit required matching of the frequency and weight. The eccentric bearing was connected to a 26 lb. steel basket that was guided to reciprocate in a vertical direction. This weight was adequate for high frequencies, and up to 150 lbs. of lead could be added to the basket for the lower frequencies. Thus the shaker was formed around a rigid frame containing a mass vibrating along a single axis. It was necessary to transport the shaker nearly 180 miles for a bridge test. The frame was attached to a trailer axle for this transporting. Upon arrival at the test site, the axle and wheel were removed and the entire unit lowered on two 4" x 40" channels which served as footings. One inch expanded PCV padding was placed under the footings for a cushioning effect. Figure 1 shows the shaker in place on the highway bridge.

The sensor, which was the central new feature to be incorporated in the study, was proposed in an effort to develop a simple and economical method that could be used to detect and measure the three desired properties. Figure 2 shows two sensors in place with the accompanying electromagnet. Three coils ¼ inches in diameter and 7/8 inches long manufactured as Miniature Type Magnetic Pickups by Power Instruments Inc., were rigidly attached to an aluminum block with nominal dimensions of 1" x 1" x 2". Aluminum was selected as it has approximately the same magnetic permeability as air. The aluminum coil unit was attached to a ¼ x 2" x 3" aluminum plate containing a 6-lead electrical plug and which could be bonded to a bridge with 3M contact cement. The sensor unit was inexpensive as each coil cost approximately $10 and the entire sensor could be fabricated for under $50.

The coils are oriented orthogonal to each other such that three dimensional vibrations can be monitored and the individual directional components identified. A schematic is shown in Figure 3. The induced signals are small and require amplification. Signal amplification consisted of a 48dB gain through the system. Calibration was with a sine wave at 100 Hz and signals were filtered above this frequency. Data was displayed on a Tekronix Type 564 B Storage Scope with a four trace Type 3A74 plug-in unit. Thus, four signals could be displayed and stored on the screen simultaneously. Permanent records of traces could be obtained with a polaroid camera.

The procedure was to activate the three coils of any one sensor by bringing a magnet in proximity and plugging in the leads from the oscilloscope and displaying the signal on the screen. The fourth challen was connected to a coil located on the shaker that recorded the RPM by sensing the passing of a magnet on the rotating eccentric shaft. Each time the rotating magnet passed the coil a sharp signal

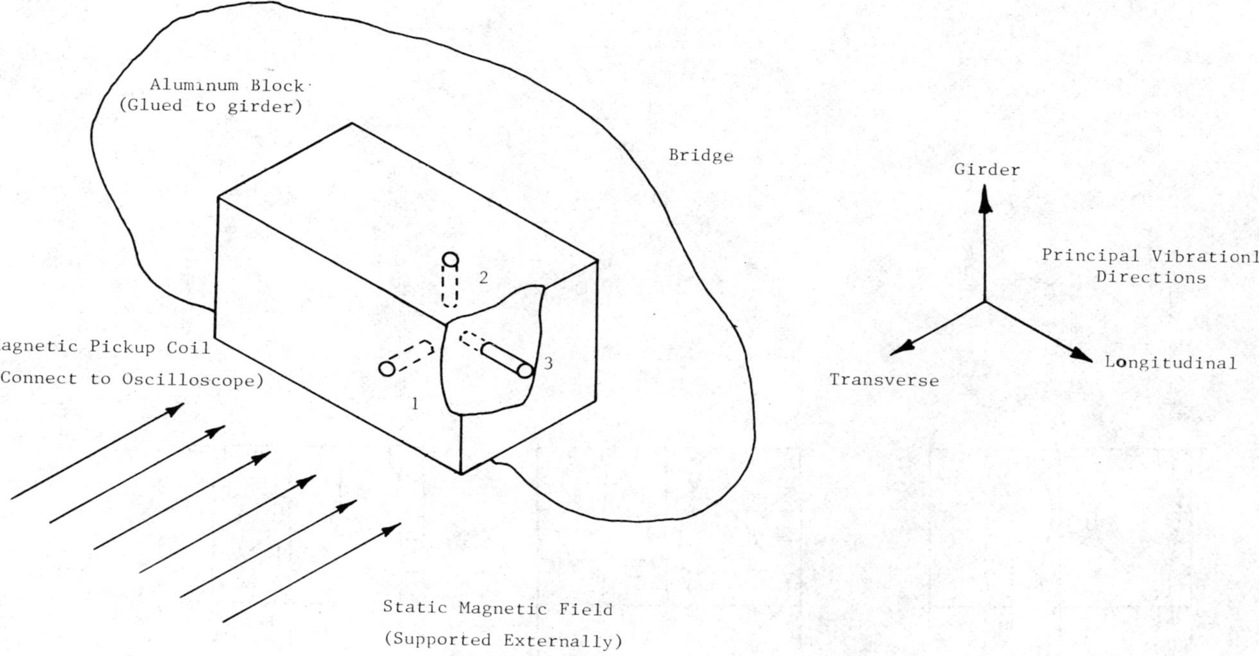

Figure 3. Tricoil Sensor Principle

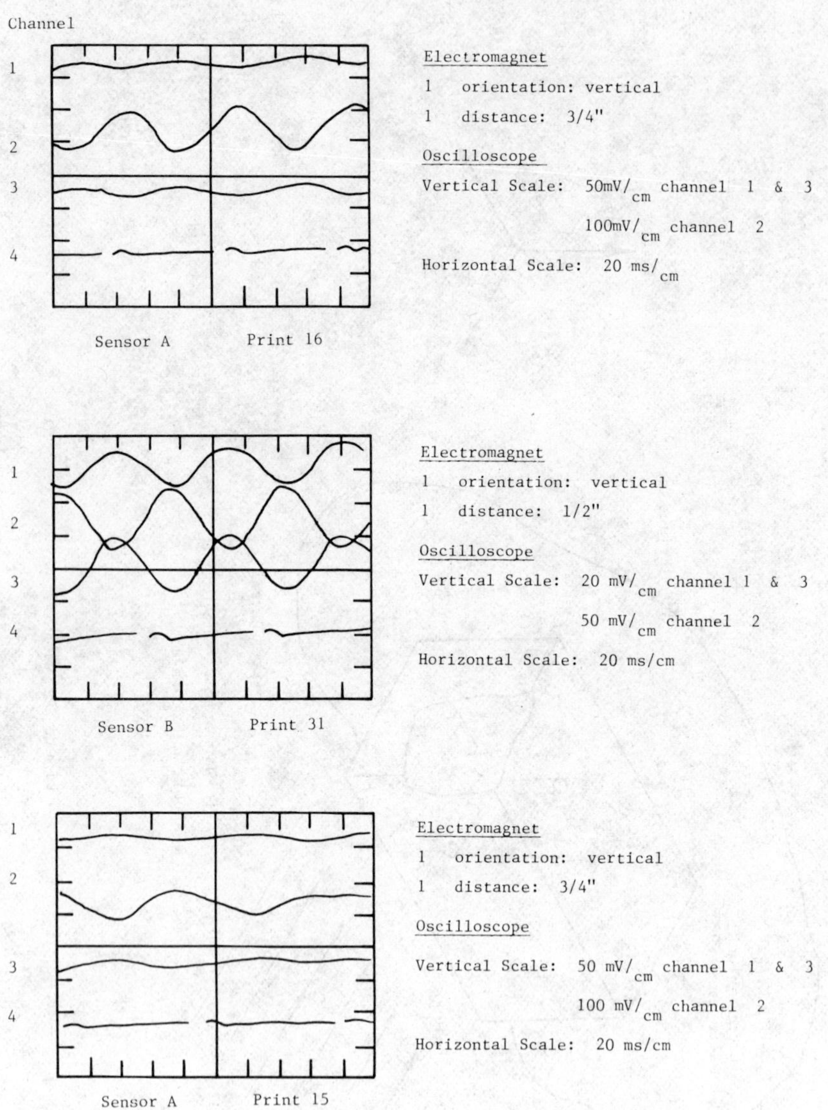

Figure 4. Typical Oscillograph Traces

was induced and the time required for one cycle could be accurately observed on the scope. Typical signals for the three coils and the RPM indicator are shown in Figure 4. The signals, as displayed in Figure 4, were manually triggered for storage on the scope by visually monitoring it during bridge vibration. A single electomagnet was used to induce the current in the coils in the figure, but the same result could have been obtained with a permanent magnet. A permanent magnet having a flux density of approximately 1100 gauss at the surface could energize the coils up to a distance of one inch.

An electromagnet was fabricated around a 2 inch diameter cold rolled steel core. Thirty pounds of #18 gage enamel coated copper wire were wrapped around this core over a 7 inch length. This provided a coil having approximately 35 layers of wire. The electromagnet operated at 110 volts DC and a 1.75 amps with minor heat vise over long periods of time. Electrical energy was provided in the field by rectifying 110 AC from a portable generator to DC. This magnet had a field density of approximately 2500 gauss at the center of the core. The electromagnet could activate the sensors for distances up to 5 inches.

The tri-coil sensor has an advantage and disadvantage in this mode of operation. The advantage is that the desired vibrational properties can be found as long as a signal can be received on the oscilloscope. The disadvantage is that there is no calibration as to the magnitudes of the vibrations. There must be controlled in the design of the force generating system.

B. Theoretical Developments

The program was designed as an experimental field oriented effort. The bridges to be tested were amenable to analytical modeling and vibrational modes and magnitudes were computed in an effort to validate the field measurements. ICES - STRUDL - II computational capabilities were available and contained dynamic analysis features. This computer package contains three analytical techniques appropriate for bridge modeling. The first is the plane frame solution in which all loads and deformations are in the same plane. In this case a bridge can be modeled as an equivalent beam. The second method is the plane grid. Here the loads and deformations are orthogonal to the plane of the structure. Finally, there is the finite element approach in which the total structure is defined by combining subelements with definable force-deformation properties.

Each of these programs has provisions whereby the structural properties of the materials and the corresponding structure geometry can be inputted and a dynamic model analysis requested. The computed results are eigenvalues and eigenvectors describing the model behavior. It should be pointed out that the plane frame solution was limited to planar eigenvectors.

The first three modal shapes for a typical simply supported bridge are illustrated in Figure 5. The torsional aspects can cause the second mode to be very near the first in magnitude. The modal shapes show that the first and second modes can be distinguished by observing the relative motions of points at the center and on the outside edges.

III. Field Tests

A. Test Progression

The test plan involved a progressive development of the test equipment. The first requirement was to verify that the tri-coil

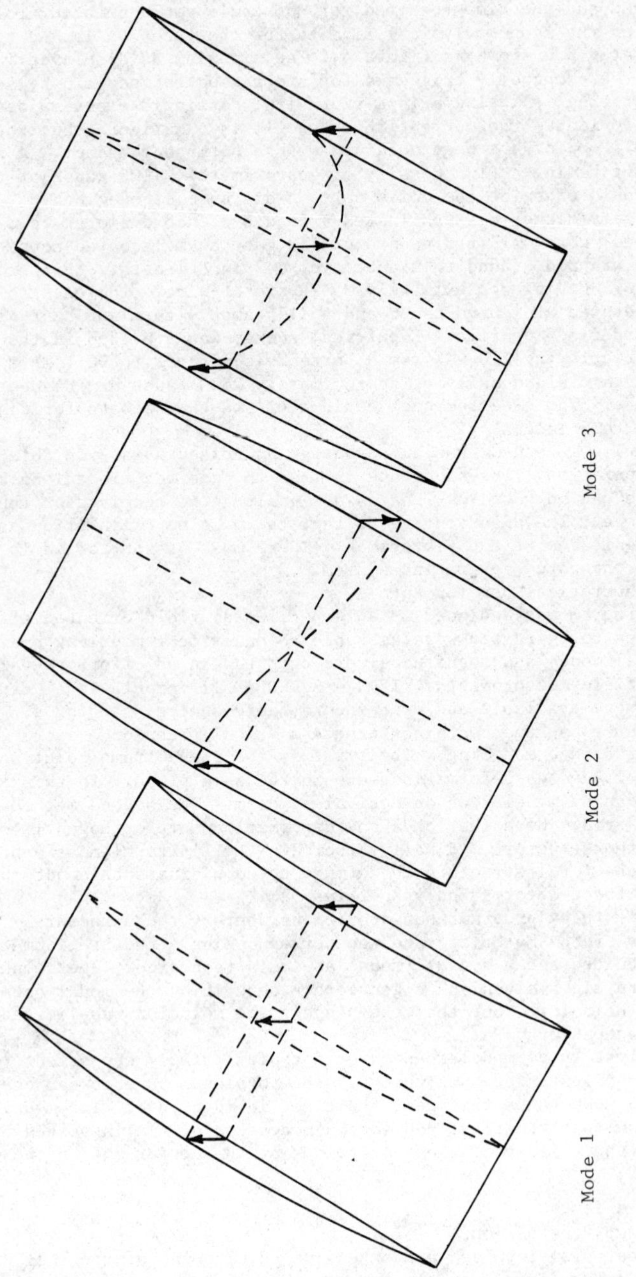

Figure 5. Typical Mode Shapes

sensor principle would work on a bridge in the field. A bridge on Interstate 25 over Missouri Street in Las Cruces, New Mexico was selected for the first test. In this case a single tri-coil sensor was glued to the center of an end span and a permanent magnet brought in proximity by supporting it on the slope of the abutment fill under the end support. Vibrations were actuated by traffic and the tri-coil sensor displayed adequate signal strength. No quantitative data was recorded, electrical grounding of the sensor became the only new variable to consider.

Two bridges were selected for analysis and field testing. These bridges were made available for field testing by the Elephant Butte Irrigation District and the New Mexico State Highway Department. The bridges were selected because of their accessibility and minimum inconvenience to users for the periods of testing. The general properties are given in Table 1. Both bridges were simply supported.

Table 1

Rio Grande Canal Bridge, Fairacres, New Mexico
- Length: 20' 0" 1 span
- Width: 16' 5"
- Construction: 6 prestressed twin tee sections with no deck surface
- Girder Depth: 15"
- Support: Simple
- Assume 28 day concrete strength of 4500 psi.

Big Dry Creek Bridge, Glenwood, New Mexico, US 180
- Length: 33' 7" each of 3 spans
- Width: 20' 0" at roadway
- Construction: Monilithic deck and 5 girders - reinforced concrete
- Girder Depth: 28"
- Supports: Abutments - with 2 intermediate reinforced concrete piers
- Assume 28 day concrete strength of 300 psi.

The next phase involved the testing of the shaker and verification that the basic system worked. The bridge selected was the short span over the Rio Grande Canal. The test involved the first field operation of the shaker and most efforts were directed towards matching the shaker to the relatively light bridge. The shaker was operated without the basket and the only eccentric weight was that of the eccentric bearing and housing. The housing of the bearing was connected to a short piece of pipe to keep it from rotating. The sensor was glued to the center of an outside Tee leg, and the magnet was attached to an external beam simply supported on the abutments. The magnetic support beam was relatively light and had a much lower fundamental frequency than the bridge. Therefore, the measured vibrations were that of the bridge relative to the magnet--both being supported on the same abutments. The bridge has been modeled as a plane frame and the first mode vibration predicted. Field measurements included measuring the periodic motion of the sensor by taking measurements on the memory oscilloscope. The theoretical resonant frequency was 21 Hz and the measured was 20 Hz. It was estimated that the deviator force caused by the shaker was $\pm$ 600 lbs. The preliminary tests on the Rio Grande Canal bridge verified that the forced vibration driver and sensor system would work. The passing of this stage led to the tests on the highway bridge where higher order vibrational modes and damping were to be measured.

The testing on the Big Dry Creek bridge involved some logistics. The bridge was located approximately 180 miles from Las Cruces which meant that all equipment had to be transported there. It was planned that the sensor system would be installed, and the electrical sensor and recording system would be checked out using traffic induced vibrations within one day. This method required the development of a procedure such that the triggering of the oscilloscope would coincide with the vibration induced by the passing vehicles. Many trials were necessary to obtain ten vibration traces. Of these ten, eight were taken with the permanent magnet and two with the electromagnet. No damping was observed. The traffic actuated measurements could be conducted without impeding normal traffic flow. Arrangements had been made with the New Mexico State Highway Department to close one lane of traffic for the second day of testing. The eastbound lane was closed and the shaker placed in the center of the lane on the end span as shown in Figure 1. This span was selected to facilitate the placement of the magnets under the sensors.

Six sensors were attached to the bridge. Sensor B was placed on the bottom of the center girder at midspan. Sensor F was placed on the lower side of this girder, again at midspan. These two sensors were placed near each other to verify the directional properties of the two sensors because they were orthogonal to each other in this placement. Sensor A was located on the bottom of the east end girder at midspan in order that distinction could be made between the first and second vibrational modes. Sensors D and E were located a mid depth on the ends of the end and center girders over the pier. These two are shown in Figure 2. This was done to detect relative longitudinal motion of the two girders, if any. Sensor C was placed on the end of the pier cap under the ends of the girders containing the other two sensors.

The test procedure involving the shaker and sensors consisted of locating the electromagnet near the sensor in either a vertical or horizonatal alignment. The magnet would be activated and the shaker started. The vibration would be visually monitored on the oscilloscope and then the frequency manually adjusted to a resonance. When an acceptable wave form was formed, it was stored on the scope and the shaker turned off. A polaroid picture was made of the oscilloscope trace in order that numerical data could later be obtained. In the case of measuring the damping ratio, the scope image was stored after the shaker had been shut off and the signal was decaying. The shaker stopped motion in just a few cycles and the bridge continued to vibrate for a few cycles longer. A total of 30 records were made using the shaker.

The damping ratio was caluclated using the logarithmic decrement of succeeding cycles. Data from as many cycles as possible were calculated from any one print and the results averaged for that particular run. A number of shaker runs were made to see if the damping ratio could be calculated by measuring the amplitudes of the received signals on both sides of resonance, and plotting these in order that the frequency equation could be used to calculate this quantity. This method did not work and all damping ratios were calculated using the logarithmic decrement method.

B. Test Results

The field tests on the bridge over Big Dry Creek exercized the entire system and the results are summarized in this section. Table III summarizes the first mode test data that was reduced to find the vibrational frequencies, modes, and damping ratios. The data is organized by sensor to show variation within the data and between the sensors. Samping ratio data is separated for clarity. The first set of data pertains to the fundamental resonant frequencies. The table shows that the 18 measurements taken with the shaker system were reasonably consistent with an average of 13.75 Hz. The two clean readings obtained with traffic actuations show more scatter and an average of 13.52 Hz. Multiple wheel effects caused a number of the traffic actuated prints to be voided. Although this data is easier and cheaper to obtain, it was found that the problems in triggering the system and the difficulties in making proper frequency identification make this a much less desirable method.

Buried in Table II are data from the ends of the girder and the pier cap. Small vibrations in the longitudinal direction, parallel to the girder, were observed for the pier cap. Even smaller vibrations in this same direction were noted for the end of the girder. This suggests some longitudinal freedom of motion of the pier cap. This observation was supported by a review of the prints of the pier cap motion under traffic actuated conditions. In one case a truck passed over the pier cap and a highly damped vertical vibration of the pier cap at a frequency of 20 Hz was observed. A review of the physical restraints affecting the pier cap suggests that these were the dominant conditions. The channel under the bridge has been cleaned out to a level near the footing of the pier, which in turn was resting on piles. The lack of fill material around the pier and the inherent longitudinal freedom found in simply supported spans sppears to allow the pier cap to vibrate vertically and longitudinally.

The damping ratio could be clearly measured on three prints. Ratios of all succeeding amplitudes were taken and an average established for a particular print. The table shows that there is reasonable consistency between the shaker and traffic actuated values. It was felt that the shaker method gave the best results and would be more consistent.

Table II

First Mode Resonance Measurements

Sensor	Print	Location	Frequency	Actuation	Magnet
A	16	Outside Girder	13.94 Hz	Shaker	Electromagnet
	37	"	14.02	"	"
	38	"	14.08	"	"
	39	"	13.30	"	"
	14	"	13.79	"	"
B	11	Center Girder	13.44	"	"
	25	"	13.37	"	"
	26	"	13.39	"	"
	27	"	13.70	"	"
	28	"	13.61	"	"
	31	"	13.92	"	"
C	22	Pier Cap	13.42	"	"
	23	"	13.70	"	"
E	19	Girder End	13.57	"	"

F	32	Center Girder	14.29	"	"
	33	"	14.13	"	"
	34	"	13.89	"	"
	35	"	13.94	"	"
		Ave	13.75		
B	2	Center Girder	11.90	Traffic	Permanent
B	10	"	15.15		Electromagnet
		Ave	13.52		Electromagnet

Damping Ratio

Sensor	Print	Location	Damping Ratio	Actuation	Magnet
B	29	Center Girder	0.06	Shaker	Electromagnet
A	15	Outside Girder	0.07	"	"
B	3	Center Girder	0.07	Traffic	Permanent

Table III summarizes the modal data for the three theoretical methods and the average field data. The table shows that the field data is approximately 9 percent higher for the first mode and 10 percent higher for the second mode. Third mode vibrations were not measured in the field. Several factors may explain the fact that the field measurements were higher. All properties of the bridge were assumed and only physical dimensions were measured. The bridge railing was omitted in the calculations. The bridge wearing surface was added as mass only. It was noted that sand had accumulated in the gap between the girders and the abutment and center span girders and the possibility is that this stiffened the net structural system. The net conclusion is that the different modeling methods compared well with each other that the field tests verified the models adequately.

Table III
Computer Method

Mode	Plane Frame	Plane Grid	Finite Element	Field Meas.
1	12.9 Hz	12.9 Hz	12.7 Hz	13.8 Hz
2	-	14.3	14.7	15.7
3	-	19.2	20.1	-

IV. SUMMARY AND CONCLUSIONS

The objectives of this study were to develop a non-contact electromagnetic induction device that could be used to measure bridge vibration and damping properties and to demonstrate this device in the field using forced vibration. These objectives were fulfilled. The tri-coil sensor proved to be an economical and relatively simple device to operate. A single bridge was tested in one day illustrating the relative ease of operation. The major disadvantages of the device is that the bridge vibrations are qualitative in nature and all data is reduced from relative vibrational magnitudes. Thus, displacements and stresses cannot be measured.

This pilot test demonstrated the feasibility of the tri-coil sensor and of the relative ease in performing non-destructive forced vibration tests. This leads to two possible future efforts. The first is to improve the sensor to be able to measure displacements accurately. The second is to develop a mobile unit that can be used to excite bridges to low level vibrations and to be able to measure the desired dynamic properties.

REFERENCES

1. Biggs, M. J. and H. S. Sver, "Vibration Measurements on Simple Span Bridges," Highway Research Board, Bulletin 124, pp. 1-16, 1956.
2. Foster, G. M. and L. T. Ohler, "Vibration and Deflection of Rolled Beam and Plate Girder Bridges," Highway Research Board, Bulletin 124, pp. 79-110, 1956.
3. Roesli, A., A. Smislova, C. E. Ekberg, and W. J. Eney, "Field Tests on a Prestressed Concrete Multi-Beam Bridges," Highway Research Board Proceedings, Vol. 35, pp. 152-171, 1956.
4. Hulsbos, C. L. and D. A. Linger, "Dynamic Test of a Three-Span Continuous I-Beam Highway Bridge," Highway Research Board, Bulletin 279, pp. 18-46, 1960.
5. Linger, D. A. and C. L. Hulsbos, "Forced Vibration of Continuous Highway Bridges," Highway Research Board, Bulletin 339, pp. 1-22, 1962.
6. Reilly, R. J. and C. T. Looney, "Dynamic Behavior of Highway Bridges," University of Maryland, Civil Engineering Department, 1966.
7. Novak, M. E., C. P. Heins, and C. T. Looney, "Induced Dynamic Strains in Bridge Structures due to Random Truck Loadings," University of Maryland, Civil Engineering Department, February 1968.
8. Kuribayashi, E. and T. Iwasaki, "Experimental Studies on Vibrational Damping of Bridges," Report of PWRI, No. 139, pp. 62-165, May 1970.
9. Charleson, A. W., "The Dynamic Behavior of Bridge Substructures," Report to the National Roads Board, University of Canterbury, New Zealand, 1970.
10 . Chen, C. H. and D. A. Van Horn, "Static and Dynamic Flexural Behavior of a Prestressed Concrete Bridge," Fritz Engineering Laboratory, Report 349.2, January 1971.
11. Chen C. H. and D. A. Van Horn, "Slab Behavior of a Prestressed Concrete I-Beam Bridge - Leighton Bridge," Firtz Engineering Laboratory, Report 349.5, July 1971.
12. Shepard, R. and A. W. Charleston, "Experimental Determination of the Dynamic Properties of a Bridge Substructure," Bulletin of the Seismological Socity of American, Volume 61, No. 6, pp. 1529-48, December 1971.
13. Kuribayashii, E. and T. Iwasakii, "Dynamic Properties of Highway Bridges," 5th World Conference on Earthquake Engineering.
14. Shepard, R. and G. K. Sidwell, "Investigations of the Dynamic Properties of Five Concrete Bridges," Fourth Australian Conference on the Mechanics of Structure and Materials, Department of Civil Engineering, University of Queensland, Brisbane Aust., August 1973.
15. Tilley, G. P., "Dynamic Response of Bridges," Transportation Research Record 665, Bridge Engineering, Volume 2, 1978.
16. King, J. P. C., M. Holowka, R. A. Dorton, A. C. Agarural, "Test Results from the Conestoga River Bridge," Transportation Research Record 665, Bridge Engineering, Volume 2, 1978.
17. Abdel, G. A. M., "Vibration Studies and Tests of a Suspension Bridge," California Instutute of Technology Earthquake Engineering Structural Dynamics, Volume 6, No. 5, 1978.

18. Baldwin, J. W., H. J. Salane, R. C. Daffield, "Fatigue Test of a Three-Span Composite Highway Bridge - Study 73-1," Missouri Cooperative Highway Research Program, Final Report 73-1, 1978.
19. Rainer, J. H., "Dynamic Testing of Civil Engineering Structures," Proceedings, 3rd Canadian Conference on Earthquake Engineering, Volume 1, pp. 551-574, June 1979.

ACKNOWLEDGEMENTS

The authors wish to thank the National Science Foundation for sponsoring this effort under Grant Number PRF-7823083 and the Engineering Experiment Station at New Mexico State University for their added support and services. The Elephant Butte Irrigation District and the New Mexico State Highway Department are thanked for making bridges available.

There are many unnamed professionals at NMSU that are thanked for their diligence and good work. A large part of the effort involved students and they should be thanked. In particular, Charles Cave, Alberto Arroyo, Gilbert Sollars, Jack Ayers and Mike Zimmerman are recognized for their contributions.

VIBRATION TESTS OF STRUCTURES AT VARIED AMPLITUDES

by Alan P Jeary, CEng FRMetS MIERE and Brian R Ellis BSc*

ABSTRACT

A new, four-unit, synchronised vibrator system has been used recently to calibrate the dynamic response of six tall buildings. The results of tests at various amplitudes are presented in this paper and these have been used to assess existing design guides. Comparison between predictions and measurements of response to wind are presented.

1 INTRODUCTION

During the last six years the Building Research Establishment(BRE) has been conducting vibration tests on tall buildings, one object of which has been to provide data for use with methods of predicting overall structural dynamic behaviour. The results obtained from tests on six buildings have been presented previously(1).

After the initial stages of the research programme into the dynamics of tall structures, the need for an improved vibration generator was identified. Accordingly, a four-unit synchronised exciter system was constructed, having one-thousandth of a hertz frequency stability and the capability of producing several different amplitudes of force or torque at any frequency. This system has been used to test a further six buildings and the results from the study of the dynamic characteristics of the structures are presented here.

As the results from this series of tests are expected to form some of the basic data for a revision of the UK wind loading code of practice (CP 3, Chapter 5, Part 2) emphasis has been placed on using the tested buildings to calibrate existing codes and design guides. Some comparisons between predictions and measurements of the response of various buildings to wind excitation are presented here.

2 THE VIBRATOR SYSTEM

Previous experience with vibration tests on tall buildings led to an identification of the need for a purpose built eccentric mass vibrator system, and accordingly, in 1976, the BRE let a contract to Bristol University to design and build a system which would meet a stringent specification. The prototype system was completed in December 1977 and the full system was operational by August 1978. Since that time the system has been used to test six tall buildings and three dams. The results from the work on dams have been presented elsewhere(2).

The vibrator system comprises four exciters, each of which has its own 'slave' control unit, driven by a master unit which controls the whole system. The frequency required is clocked from a crystal oscillator

*Building Research Establishment, Garston, Watford, Herts, UK.

and is accurate to one part in 10^6. The use of servo control techniques allows the frequency of the exciters to be controlled to a 0·001 Hz accuracy. Operation is possible up to a maximum frequency of 10 Hz.

The exciters each consist essentially of two motor driven sets of contra rotating weights to which may be added further weights allowing the force to be varied. The force produced in this way has been calibrated using a load cell and is well within the 3% accuracy of the specification. With the maximum number of weights attached a force of approximately one tonne (at 1 Hz) peak-peak is produced by the system. The servo control system allows each unit to be run either at 0° or at 180° relative phase and is accurate to $0{\cdot}6^o$. The exciters are mounted on rings of steel designed so that each unit may be rotated to any desired orientation in the horizontal plane.

Certain safety features have been incorporated into the design. For example, to avoid accidental over-excitation a conscious effort, on the part of the operator, is needed to exceed certain frequency limits.

3 THE RATIONALE FOR THE TESTING

We have assumed that the dynamic response of a structure can be described by the superposition of a series of similar equations, each of which describes the behaviour of a single mode of vibration. This technique is currently popular and forms the basis for most dynamic design methods. However, it will be shown later that the model is inadequate in certain cases.

The modal superposition method describes the modal response X_r for mode 'r' in the following way:

$$\ddot{X}_r + 4\pi f_r \zeta_r \dot{X}_r + 4\pi^2 f_r{}^2 X_r = \frac{F_r}{M_r} \qquad (1)$$

When f_r is the natural frequency, ζ_r is the modal damping ratio, M_r is the modal mass and F_r is the modal force.

If the mode shape $(\phi(x, y, z))$ is normalised to a value of 1 at the largest displacement then Lagrange's equation can be used to show that

$$M_r = \int_o^L \int_o^B \int_o^H m(x,y,z)\, \phi^2\, (x,y,z)\, dx\, dy\, dz \qquad (2)$$

where L, B and H are the length, breadth and height of the building and m(x,y,z) is the mass of the volume defined by dx, dy, dz. It is often assumed that L and B are constant. Measurements of M_r (from equation 1) and of the mode shape (ϕ) then allow the total mass of the structure to be calculated.

Thus the measurement of natural frequency, damping ratio, and mode shape defines all the necessary dynamic parameters, based on a linear viscoelastic theory, since the modal mass can be calculated using equation 1, once f_r, ζ_r and X_r have been measured for a particular value of F_r. The induced vibration testing undertaken by BRE seeks to define these parameters by selecting the following testing sequence:

1 Arbitrarily orientate the vibrator, increment the frequency sequentially and produce a response/frequency curve.

2 Select a frequency of resonance (defined by a peak response in (1)) and set the vibrator to operate at this frequency.

3 At this resonance frequency, monitor the directional response (for translational modes), by incrementally rotating an accelerometer. Establish the direction of the minimum response.

4 Set the vibrator to a direction which is orthogonal to this minimum response and incrementally increase the frequency of the vibrators through the range 0·1-10·0 Hz, taking measurements of vibration amplitudes at each frequency for one reference location in the building. In the jargon of vibration testing this operation is termed 'frequency sweeping'.

5 Repeat this operation for the orthogonal set of modes.

6 Repeat this operation for the torsional set of modes. This may involve setting the vibrators on opposite sides of the building, to operate in anti-phase.

7 Set the vibrators to a frequency of resonance, take an accelerometer to various locations throughout the building in turn, and measure the vibration amplitude at each position.

8 Repeat (7) for each mode of vibration in turn.

9 Tune the vibrators to a frequency of resonance, and monitor the decay of oscillation resulting when the vibrators are suddenly switched off. Modal damping values can be calculated from the decay.

10 Repeat (9) for each mode of vibration in turn.

11 Repeat (10) at various amplitudes of vibration of interest.

Wherever possible, recordings of the response of the building to wind excitation are also made.

In order to obtain acceptable statistical estimates of natural frequency and damping ratio it is necessary to use stationary samples of data. It has been found that for buildings with longer time constants ($\tau = 2\pi f_r \zeta_r$) it is uncommon to obtain stationary samples from sequential data and ensemble averaging techniques have to be used. The methodology for this type of analysis has been detailed previously(3,4).

4 RESULTS FROM TESTS OF BUILDINGS

The results of tests on six tall reinforced concrete buildings are presented, together with a short description of each building. The buildings are considered in the chronological order of testing.

4.1 National Westminster Tower, London (Fig 1)

This building is a 190 m tall, 46 storey, reinforced concrete office block. In plan the building resembles a 3 leaf clover pattern consisting of a massive heavily reinforced core, with the 3 'leaves' attached and supported by large concrete corbels at their bases. Service ducts and lifts are provided inside the core, and the leaves are mainly for office usage. The 3 leaves are each of different height. Leaf 'A'

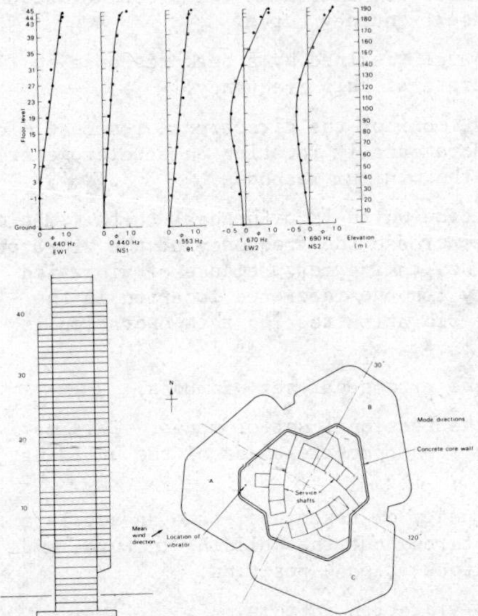

Figure 1

National Westminster Tower - mode shapes, plan and elevation

extends from the 1st to the 43rd levels, leaf 'B' from the 3rd to the 41st, and leaf 'C' from the 5th to the 39th. The exterior of the leaves consists of steel columns, to which the cladding panels are attached. The concrete floors are cantilevered from the core, and are attached to the steel columns at the periphery. Expansion joints are provided at the interfaces between leaves.

The top of the core is approximately 187 m above ground level and at the time of the forced oscillation tests the building was structurally complete, with the sole exception of the 46th floor. Additionally, the upper three floors were open to the atmosphere, and several minor items at or near ground level were not complete. Results of tests on this building are given in Table 1.

TABLE 1 NATIONAL WESTMINSTER TOWER - GENERAL SUMMARY

Mode	Torque (Nm)p-p	Force (N)p-p	Freq (Hz)	Angular Disp (rads)	Amplitude (mm)p-p	Damping %	Modal Mass ($x 10^6$ kg)	Modal Inertia (kg m²/rad)
EW1	-	201.0	0.440	-	0.10	0.5	24.9	-
NS1	-	201.0	0.440	-	0.16	0.5	16.0	-
θ1	26 553	-	1.553	$1.99.10^{-6}$	-	0.7	-	$1.00.10^{10}$
EW2	-	2 985.0	1.670	-	0.0026	2.8	-	-
NS2	-	3 056.9	1.690	-	0.0029	2.1	-	-

4.2 Sutherland House, Sutton, Surrey (Fig 2)

Sutherland House is a 10 storey, 35 m tall reinforced concrete building, 43 m x 12 m in plan shape. It has 21 columns along the periphery of

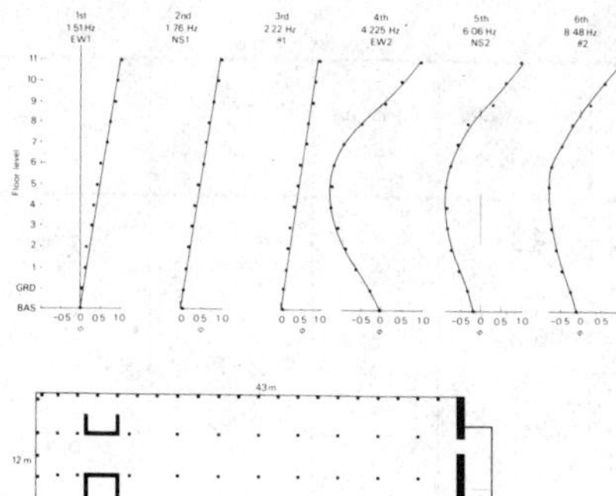

Figure 2

Sutherland House - mode shapes and floor plan

both longer edges. At one of the shorter edges there are 6 perimeter columns, whilst at the other end, shear walls are provided at both ends of the building and are integrated into a stairwell at one end. There are two rows of 10 columns inside the building and lightweight partitions are attached to these.

The vibrator system was fully operational for the first time for these tests, although only two exciters were placed in the building (at opposite ends). The forces generated by the vibrator system were sufficient to overcome the effects caused by the wind. This wind was, however, sufficient to allow an overnight recording of ambient-induced vibration to be made.

During the initial testing the force developed by the vibrator was insufficient to overcome the effects of wind activity in the fundamental modes and induced vibration testing was limited to higher frequency modes where the force developed by the vibrator was larger. A record of the response of the tower to a strong wind was made at this time. Shortly afterwards the tests were completed during a calm period and the calibrated first mode response was studied. The mass of the building was calculated from the dynamic tests and was close to that estimated from design drawings.

Table 2 summarises the major results found, and Fig 2 depicts the building together with the mode shapes of its resonances.

The mass of the building was calculated from the results of the induced vibration tests at various amplitudes and the results were consistent throughout. No design drawings were available which would have allowed a comparative check to be made.

4.3 Arts Tower, University of Sheffield, Yorkshire (Fig 3)

This 80 m high tower block consists of a series of cast in-situ reinforced concrete cores with deep floor slabs spanning between the core and the external reinforced concrete columns. It is 40 m x 20 m in plan. Above the 1st floor there are 88 columns each 203 mm x 406 mm, evenly spaced around the periphery. Below the 1st floor level this is reduced to 16 each 965 mm x 965 mm. The building which is clad in glass and lightweight panelling, is linked to an adjacent building at

TABLE 2 SUTHERLAND HOUSE GENERAL SUMMARY

Mode	Torque (Nm)p-p	Force (N)p-p	Freq (Hz)	Angular Disp (Rads)p-p	Amplitude (mm)p-p	Damping (%)	Modal Mass ($x\ 10^6$ kg)	Modal Inertia (kg m^2/rad)
EW1	-	5 600	1.46	-	0.41	2.77	2.930	-
	-	3 890	1.47	-	0.29	2.44	3.222	-
	-	2 220	1.51	-	0.16	2.11	3.653	-
	-	1 110	1.51	-	0.10	1.72	3.583	-
NS1		7 863	1.73	-	0.72	1.49	3.102	-
		5 450	1.74	-	0.48	1.32	3.598	-
		2 999	1.755	-	0.28	1.32	3.337	-
		460	1.78	-	0.06	1.29	3.065	-
θ1	183 360	-	2.17	$1.7.10^{-5}$	-	1.34	-	$21.6.10^{10}$
	28 880	-	2.22	$3.7.10^{-6}$	-	-	-	-
	28 400	-	2.25	$2.3.10^{-6}$	-	-	-	-
EW2	-	17 379	4.225	-	0.05	2.07	-	-
NS2	-	35 754	6.06	-	0.02	-	-	-
θ2	422 240	-	8.48	$1.4.10^{-7}$	-	-	-	-

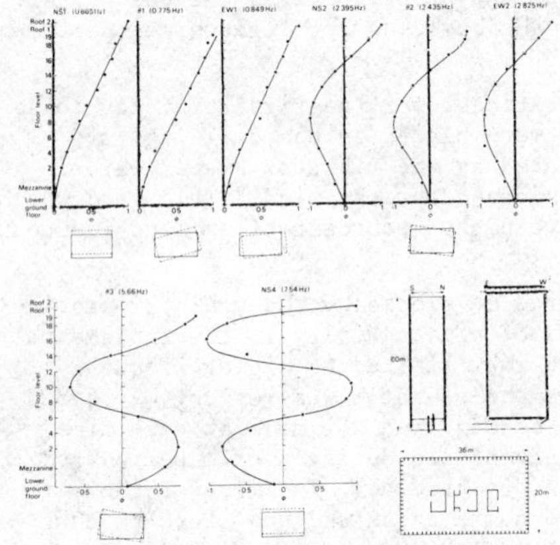

Figure 3

Arts Tower, Sheffield - mode shapes, plan and elevation

a mezzanine level, between the ground and 1st floors. There is a deep basement and the building is founded on piles driven into shale.

Two series of induced vibration tests have been conducted on this building. In the first series (made in 1976 in collaboration with CEBTP, Paris) a single unidirectional vibrator was used and some coupling of modes was found. In the second series, conducted in 1979, the four units of the BRE system were used to decouple the modes and eleven modes were identified (as opposed to three using a single vibrator). In the interim the usage of the building had changed, and a number of partitions had been removed from the building at 10th floor level.

Because of the willingness of the staff to co-operate in the tests the building was selected for a detailed study of response to wind excitation(5).

The mass of the building was calculated from the first vibrator test and from design figures, a very good agreement being obtained. No assessment of modal mass was possible from the second series of tests, because a fault in the phase control of the exciters was subsequently discovered.

Table 3 summarises the results obtained from the tests. It should be noted that the damping results for the EW2 modes are unreliable because there was significant interference from a torsional mode during the decay of oscillation. The values obtained for θ_1 suffered from similar interference.

TABLE 3 - SHEFFIELD GENERAL SUMMARY

Mode	Torque (Nm)p-p	Force (N)p-p	Freq (Hz)	Angular Disp (Rads)p-p	Amplitude (mm)p-p	Damping (%)	Modal Mass (x 10^6kg)	Modal Inertia (kg m²/rad)
NS1	-	-	0.665	-	1.59	1.18	-	-
	-	-	0.670	-	1.12	1.16	-	-
	-	-	0.672	-	0.63	0.99	-	-
	* -	538.7	0.678	-	0.23	0.8	6.35	-
θ1	-	-	0.775	$2.6\text{-}10^{-5}$	-	1.05	-	-
	-	-	0.779	$1.45.10^{-5}$	-	1.34	-	-
	*9908.5	-	0.787	$2.02.10^{-5}$	-	0.95	-	$10.6.10^{10}$
EW1	-	-	0.849	-	0.44	1.43	-	-
	-	-	0.853	-	0.24	0.98	-	-
	-	-	0.856	-	0.21	1.16	-	-
	-	-	0.859	-	0.12	1.27	-	-
	* -	868.8	0.861	-	0.24	0.8	5.84	-
NS2	-	-	2.380	-	0.51	1.48	-	-
	-	-	2.380	-	0.34	1.15	-	-
	-	-	2.385	-	0.27	1.01	-	-
	-	-	2.395	-	0.19	0.98	-	-
θ2	-	-	2.410	$2.2.10^{-5}$	-	1.67	-	-
	-	-	2.415	$1.85.10^{-5}$	-	1.63	-	-
	-	-	2.425	$1.45.10^{-5}$	-	1.41	-	-
	-	-	2.435	$1.0.10^{-5}$	-	1.41	-	-
	-	-	2.450	$6.5.10^{-6}$	-	1.29	-	-
EW2	-	-	2.825	-	0.14	1.14	-	-
	-	-	2.830	-	0.08	1.03	-	-
	-	-	2.840	-	0.05	0.93	-	-
	-	-	2.860	-	0.02	0.92	-	-
NS3	-	-	5.61	-	0.05	-	-	-
θ3	-	-	5.66	-	0.03	-	-	-
EW3	-	-	6.6	-	0.003	-	-	-
NS4	-	-	7.54	-	0.02	-	-	-

* Taken from 1976 tests. All others 1979

4.4 Dunstan Flour Mill (Fig 4)

The silo tower of Dunstan flour mill was built in 1938-9 on the south bank of the river Tyne. It has not been used since early 1978, and all equipment and fittings, apart from the steel stairs, had been removed from the tower before the tests were conducted.

The main part of the tower consisted of twelve storage shafts, 4.6 m x 2.7 m in section and 36.6 m tall (see Fig 4). At the top of these shafts was an unobstructed solid reinforced concrete distribution floor (the 7th floor). At the lower ends (7 m above the ground) the shafts were supported by the external walls and concrete columns (approx 0.5 m square) at the intersection of the shaft walls. The stairwell was housed in an extension to the storage silos (approx 49 m x 12.2 m x 6.7 m) and consisted of eight 5.2 m storeys and a basement. The structure was constructed from in-situ reinforced concrete, and stood over a 16 m thick deposit of alluvium (sand, gravel, clayey sand and silt). The whole structure was supported on concrete piles which passed through the alluvium into a layer of boulder clay.

Because the structure was to be demolished shortly after the testing, it was possible to vibrate it up to a large amplitude. The results of

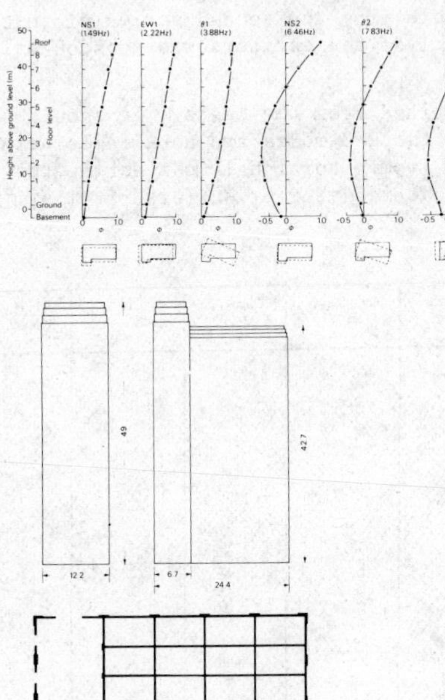

Figure 4

Dunstan Flour Mill - mode shapes, plan and elevation

the tests are given in Table 4. Estimates of modal mass calculated from these tests, were consistent over a range of amplitudes.

TABLE 4 DUNSTAN FLOUR MILL - GENERAL SUMMARY

Mode	Torque (Nm)p-p	Force (N)p-p	Freq (Hz)	Angular Disp (rads)p-p (x10^{-6})	Amplitude (mm)p-p	Damping % critical	Modal Mass (x 10^6 kg)	Model Inertia (x 10^9 kg m^2/rad)
NS1	-	18 000	1.45	-	2.75	2.60	0.875	-
	-	13 700	1.46	-	2.09	2.43	0.924	-
	-	9 300	1.47	-	1.42	2.42	0.914	-
	-	4 700	1.48	-	0.75	2.29	0.916	-
	-	2 900	1.48	-	0.47	2.18	0.938	-
	-	2 000	1.49	-	0.33	2.11	0.942	-
	-	1 100	1.49	-	0.18	2.02	1,000	-
EW1	-	41 800	2.21	-	1.73	3.24	1.209	-
	-	31 700	2.22	-	1.32	3.16	1.218	-
	-	21 800	2.23	-	0.90	3.00	1.259	-
	-	10 800	2.25	-	0.46	2.94	1.236	-
	-	6 700	2.26	-	0.29	2.85	1.264	-
	-	4 600	2.27	-	0.20	2.69	1.320	-
	-	2 500	2.28	-	0.11	2.57	1.278	-
θ1	272 800	-	3.90	31.7	-	3.07	-	0.157
	176 400	-	3.92	21.7	-	2.79	-	0.161
	96 400	-	3.94	13.1	-	2.44	-	0.165
NS2	-	12 300	6.46	-	0.012	3.08	-	-
θ2	210 000	-	7.83	-	-	1.20	-	-
EW2		20 700	8.4	-	$5.4.10^{-3}$	-	-	-

4.5 Exeter 6th Form College, Exeter, Devon (Fig 5)

This 10 storey, 27 m tall reinforced concrete building is 50 m x 13 m in plan and consists of columns and floor slabs, with lightweight concrete and glass cladding panels. Shear walls surround two stairwells located inside the building near the one third and two third positions at opposite sides of the longer plan dimension. There are 12 columns along the periphery of each long side and 7 larger columns along the spine of the building. The floors are of hollow-pot construction. Whilst a casual inspection of the building seems to show that it is of a fairly standard construction, it can be seen from the plan view in Fig 5 that not much structural stiffening is provided. Indeed there are numerous cracks throughout the internal blockwork partitions, and in our view these cracks are attributable to wind induced motion. At ground floor level single storey peripheral buildings are attached. Two of the BRE vibrators were used in the tests and modal masses were estimated from the tests alone.

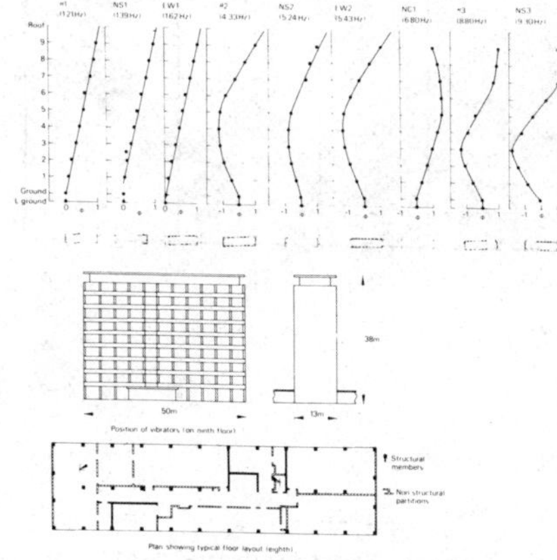

Figure 5

Exeter Sixth Form College Tower Block - mode shapes, plan and elevation

Table 5 gives a general summary of the results obtained and Fig 5 gives details of the building and its resonances.

4.6 Leicester University Engineering Tower, Leicester (Fig 6)

This tower is of unusual shape and is most easily described in terms of its East and West sides, which form two separate though interconnected buildings.

The structure on the East side is smaller and is connected to the main tower (West side) up to the 6th floor by solid reinforced flat slab concrete floors which also form the area between the two parts of the structure. In the interconnecting area are two staircases and a lift. The lift and stairs are carried in two separate in-situ concrete towers of approximately 2 m quasi-square shape with 150 mm walls. The West side is supported on 4 in-situ concrete columns, which are free standing above 1st floor level and carry a large lecture theatre which is cantilevered out 6 m. This theatre is a reinforced concrete base with 220 mm thick walls and slabs. At 4th floor level the 4 reinforced concrete columns project above the theatre roof to support the main tower which starts at 5th floor level.

TABLE 5 – EXETER: GENERAL SUMMARY

Mode	Torque (Nm)p-p	Force (N)p-p	Freq (Hz)	Angular Disp Rads)p-p	Amplitude (mm)p-p	Damping %	Modal mass (x 10^6 kg)	Modal inertia (kg m^2/rad)
θ1	80376		1.206	$3.3.10^{-5}$	–	2.0	–	$6.14.10^{10}$
	64962		1.207	$2.8.10^{-5}$	–	2.0	–	$5.81.10^{10}$
	49650		1.210	$2.0.10^{-5}$	–	2.0	–	$6.13.10^{10}$
	34134		1.212	$1.1.10^{-5}$	–	2.0	–	$7.83.10^{10}$
	18552		1.215	$0.4.10^{-5}$	–	2.0	–	$10.37.10^{10}$
	2798		1.215	$0.7.10^{-6}$	–	2.1	–	$9.67.10^{10}$
NS1	–	4153.7	1.393	–	0.432	2.3	4.05	–
	–	3346.7	1.392	–	0.357	2.3	3.96	–
	–	2545.3	1.392	–	0.250	2.3	4.32	–
	–	1751.8	1.395	–	0.141	1.5	8.10	–
	–	948.7	1.396	–	0.046	1.5	13.38	–
	–	143.0	1.396	–	0.008	1.5	12.09	–
EW1	–	5806.6	1.647	–	0.512	3.4	6.18	–
	–	4690.9	1.648	–	0.417	3.2	6.48	–
	–	3576.3	1.650	–	0.325	1.6	9.60	–
	–	2450.8	1.650	–	0.162	1.4	15.03	–
	–	1326.9	1.651	–	0.062	1.4	21.30	–
	–	200.0	1.651	–	0.012	1.4	16.80	–
θ2	17752	–	4.33	$1.1.10^{-6}$	–	1.5	–	–
NS2	–	2015.4	5.24	–	0.009	–	–	–
EW2	–	2164.2	5.43	–	0.011	1.6	–	–
NC1	–	3394.0	6.80	–	0.006	–	–	–
θ3	73325	–	8.80	$2.1.10^{-7}$	–	–	–	–
NS3	–	6348.4	9.30	–	0.002	–	–	–

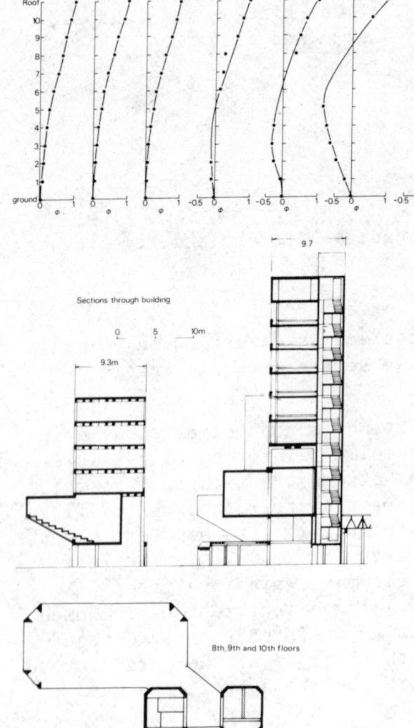

Figure 6

Leicester University Engineering Tower – mode shapes, plan and elevation

The main tower dimensions are 12 m x 6 m and it rises to 12th floor level at 37 m above ground. At 5th floor level a 200 mm in-situ concrete slab is supported by a central spine beam with bifurcations at both ends (in the form of a Y) so as to support the four corner columns. Between 5th and 6th floor levels these columns gradually change section until above 6th floor level they become twin triangular columns. Above this level one side is supported by a staircase tower wall whilst on the other three corners the triangular columns continue upwards. The floors from 6th to 10th levels are 200 mm hollow tile slabs which are carried by upstand edge beams. Above the 11th floor is a tank room which has solid concrete walls 180 mm thick. Lightweight aluminium and glass cladding is provided.

Two vibrators were installed at 10th floor level for these tests which were conducted during a single weekend. The complexity of the structure led to a number of rather complex mode shapes, but the modal mass, calculated from the fundamental translational modes gave consistent estimates at various amplitudes. The investigation of dynamic behaviour was limited to the main tower and staircases, although the effects of the East side on the dynamic behaviour of the West side were plain to see. The co-operation of the staff at Leicester University allowed tests to be performed with the water tank on the 11th floor full and with it empty.

Table 6 gives a general summary of the results obtained from this test and Fig 6 gives details of the more important responses and a plan of the building.

TABLE 6 - LEICESTER UNIVERSITY ENGINEERING TOWER - SUMMARY

Mode	Torque (Nm)p-p	Force (N)p-p	Freq Hz	Ang Disp (Rads)p-p	Amplitude (mm)p-p	Damping % critical	Modal Mass $x10^6$Kg	Modal Inertia x10 Kg m^2/rad
NS1	-	350.7	1.28	-	1.738	2.31	0.675	-
	-	2874.2	1.29	-	1.382	2.14	0.740	-
	-	2186.0	1.29	-	1.053	2.11	0.749	-
	-	1521.3	1.30	-	0.745	2.22	0.089	-
	-	848.2	1.32	-	0.407	1.74	0.867	-
EW1	-	4752.3	1.49	-	1.240	2.74	0.798	-
	-	3886.2	1.50	-	0.998	2.71	0.810	-
	-	2995.1	1.51	-	0.730	2.54	0.897	-
	-	2107.3	1.53	-	0.496	1.54	0.905	-
	-	1169.5	1.55	-	0.269	2.63	0.872	-
θ_1	52 030	-	1.98	$18.47.10^{-5}$	-	-	-	0.034*
	42 407	-	1.99	$14.78.10^{-5}$	-	-	-	-
	32 904	-	2.01	$11.30.10^{-5}$	-	-	-	-
	23 000	-	2.03	$8.04.10^{-5}$	-	-	-	-
	12 684	-	2.05	$4.35.10^{-5}$	-	-	-	-
θ_2	47 907	-	2.69	$2.59.10^{-5}$	-	-	-	-
	33 320	-	2.71	$4.82.10^{-5}$	-	-	-	-
	18 689	-	2.76	$6.91.10^{-5}$	-	-	-	-
NS2		6449.9	3.64	-	0.020	-	-	-
EW2		10033.7	4.54	-	0.008	-	-	-

* Modal Inertia estimated using a value of 2.5% for damping of 01 mode.
This value was derived from a half power bandwidth measurement from a wind response record.

DISCUSSION

It is impossible in one short paper to cover, in detail, all the aspects of work on dynamic behaviour of structures covered by our present research programme, and so only those aspects of major importance are considered here.

(i) Predictions of response caused by wind loading

One object of the programme of tests is to provide basic input data for the revision of the UK wind loading code of practice (CP 3 Chapter 5 Part 2) and accordingly emphasis has been placed on using the tested buildings to calibrate existing codes and design guides as well as to assess the likely values of parameters such as frequency and damping.

Table 7 summarises comparisons between measured responses to wind excitation on 5 of the recently tested buildings and predictions of the mode using the Engineering Sciences Data Unit item 76001(6). The calculated method is based on the Davenport spectral approach(7) and has been shown to give a similar result to the Canadian code prediction in one case. The estimates given in the table are based on the *measured* values of frequency and damping. At the design stage these values will have to be *estimated* and extra errors may result.

TABLE 7

Building	'50 year wind' (V_h) m/s	ESDU 76001 1/50 year prediction $\ddot{x}$	Actual windspeed (V_h) m/s	Measured response $\ddot{x}$	ESDU 76001 Prediction for actual wind $\ddot{x}$
Leicester	41.8	$4.71.10^{-2}$	10.8	$2.79.10^{-4}$	$6.11.10^{-3}$
Exeter	42.5	$1.58.10^{-2}$	8.4	$2.11.10^{-5}$	$3.29.10^{-4}$
Sutherland	35.9	$8.82.10^{-3}$	12.4	$4.46.10^{-4}$	$4.09.10^{-4}$
Sheffield	51.6	$3.60.10^{-2}$	21.3	$1.31.10^{-4}$	$3.83.10^{-3}$
Nat West	30.5	$6.24.10^{-3}$	11.0	$3.07.10^{-5}$	$1.85.10^{-4}$

It can be seen from these results that ESDU 76001 often overestimates the response, and this overestimate can be by more than an order of magnitude. The smaller buildings tend to lead to a smaller overestimate with the notable exception of the Exeter building. However, as can be seen from Table 5, this building has a torsional mode as the lowest frequency of resonance, and spectra show that this is the largest response. Deaves and Harris(9) have suggested that because turbulence decreases with height there is bound to be an overprediction unless this is taken into account. These results lend weight to this argument, although in three cases the wind data have been estimated from readings taken at some distance from the tested buildings. In the cases of the Sheffield Arts Tower and the Natwest tower anemometers on site were used. (In the case of the Arts tower the building was used as a worked example in the ESDU 76001 document.

(ii) The Exeter 6th Form College

Calculations of modal mass of the Exeter 6th Form College show there to be variations with amplitude indicating an increase of modal mass for smaller amplitudes. This is the only building tested which shows this effect and the only explanation that can be offered at present is that equations 1 and 2 do not represent the behaviour of this building very well.

There was a random and unpredictable variation of natural frequency (of the order of 0·06 Hz in the fundamental modes) which could not be correlated with any readily identifiable variables. This peculiarity may be caused by overall temperature effects.

(iii) Non-linear behaviour

The testing of buildings at various different amplitudes has shown the non-linearities which are to be found in the frequency and damping characteristics of any mode. From these tests the indication is that for any particular mode the smallest figures of damping and the highest natural frequencies should be used for design to withstand wind forces, because these correspond with the range of amplitudes shown to be likely for the '50 year wind'. The non-linearities were most apparent in the tests at Dunstan flour mill where, for a large range of amplitudes there was a 3% change in frequency and a 30% change in damping. This is evidence of the limitations of a linear model of dynamic behaviour over a range of amplitudes.

The mechanisms causing damping in a building are poorly understood at present and various studies are being conducted in order to gain more understanding. The data presented here (Tables 1-6) show that there is a reasonable correlation of the value of damping with natural frequency if damping values extrapolated to zero amplitude are taken. This adds weight to arguments previously promulgated(10,11).

CONCLUSIONS

The work reported here has shown that:

(i) the mathematical model often used to describe the overall dynamic behaviour of buildings is not always appropriate. For the case of one of the tested buildings this model has been shown to be unrepresentative of the real behaviour;

(ii) the prediction of the response of structures to wind loading is subject to large confidence limits, even when measured dynamic parameters are used;

(iii) there is non-linear variation of the dynamic parameters with amplitude, for six buildings.

These aspects need further research if improved methods of design are to be formulated.

The accumulation of data on the natural frequency and damping of real structures is also vital to good predictive methods. Other researchers are urged to inspect their data to look for a correlation between low amplitude damping and natural frequency.

It is intended that the continuing BRE research programme will concentrate less on ad hoc testing and more on a systematic study of the behaviour of building types which may lead to a better mathematical modelling of structural dynamic behaviour.

ACKNOWLEDGEMENTS

The authors are indebted to a large number of people who have made possible the tests reported here, but special thanks are due to G W Dupuch, K R Fry, J D Littler and M Beak who all helped with the operation of the tests, and to the owners of the buildings without whose co-operation the studies reported would have been impossible.

The work described has been carried out as part of the research programme of the Building Research Establishment of the Department of the Environment and this paper is published by permission of the Director.

REFERENCES

1 Jeary, A P and Sparks, P R. Some observations on the dynamic sway characteristics of concrete structures. ACI Symposium on Vibrations in Concrete Structures, New Orleans, October 1977

2 Severn, R T, Jeary, A P and Ellis, B R. Forced vibration tests and theoretical studies on dams. Proc Inst Civil Engineers, Pt 2, September 1980

3 Jeary, A P. Damping measurements from the dynamic behaviour of several large multi-flue chimneys. Proc Inst of Civil Engineers, 57, p 321-329, June 1974

4 Ellis, B R, Jeary, A P and Vasconcellos, R. The practical vibration analysis of structures with particular reference to offshore structures. Brasil offshore 79, Rio de Janeiro, October 1979

5 Jeary, A P. The dynamic behaviour of the Arts Tower, University of Sheffield, and its implications to wind loading and occupant reaction. BRE CP 48/78

6 Engineering Sciences Data Unit. Item 76001, ESDU 251-259, Regent Street, London, September 1976, Amended June 1977

7 Davenport, A G. The dependence of wind loads on meteorological parameters. Proc Int Conference on Wind Effects on Buildings and Structures, Ottawa, September 1967, Vol 1, pp 19-82

8 Jeary, A P and Ellis, B R. The response of a 190 m tall building and the ramifications for the prediction of behaviour caused by wind loading. Paper X91, 5th International Wind Conference. Colorado State University, Fort Collins, July 1979

9 Deaves, D M and Harris, R J. A mathematical model of the structure of strong winds. CIRIA, Report 76, May 1978

10 Kawasumi, H and Kanai, K. Vibrations of buildings in Japan. Proc 1st World Conference on Earthquake Engineering, Berkeley, California, 1956

11 Kobayashi, H and Sugiyama, N. Viscous damping of structures related to foundation conditions. Proc of 6th World Conference on Earthquake Engineering, New Delhi, January 1977

CALCULATED EXAMPLES OF LONG-SPAN BRIDGE RESPONSE TO WIND

by

Lawrence Rubin*
and
Robert H. Scanlan**
Dept. of Civil Engineering
Princeton University

Abstract

The response of long-span flexible bridges under wind excitation has been studied in the past; however, attacks on these problems continue to provide useful methods to aid in design. Prominent among the wind-induced phenomena are flutter, vortex-induced oscillation and buffeting. This paper addresses itself again to these wind-induced problems in an effort to provide useful procedures for design. Several example calculations will be included.

I. The Flutter Problem: Analytical Formulation

References [1], [2] treat this problem in detail. To analyze the motion of the full-span bridge, let $h(x,t)$ and $\alpha(x,t)$ represent respectively the bending and torsional displacements of the bridge as a function of spanwise position x and time t. Assuming that bending and torsion modes of the bridge are uncoupled from each other, let $h(x)$ be the model deflection form in the lowest bending mode and $\alpha(x)$ be the modal deflection form in the lowest torsion mode. Then,

$$h(x,t) = h(x)\, p(t) \tag{1}$$

$$\alpha(x,t) = \alpha(x)\, q(t) \tag{2}$$

where p and q are the generalized coordinates. The equations of motion for a bridge deck balanced about its c.g. can then be formulated as:

$$M_1[\ddot{p} + 2\,\zeta_h\omega_h\,\dot{p} + \omega_h^2 p] = \frac{1}{2}\rho U^2(2B)[KC_{11}H_1^*\frac{\dot{p}}{U} + KC_{12}H_2^*\frac{B\dot{q}}{U} + K^2C_{12}H_3^*q] \tag{3}$$

$$I_1[\ddot{q} + 2\,\zeta_\alpha\omega_\alpha\,\dot{q} + \omega_\alpha^2\,q] = \frac{1}{2}\rho U^2(2B^2)[KC_{12}A_1^*\frac{\dot{p}}{U} + KC_{22}A_2^*\frac{Bq}{U} + K^2C_{22}A_3^*q] \tag{4}$$

where:

---

* Graduate Student
** Professor

$M_1 = \int_{span} m(x)h^2(x)dx$ = the generalized mass

$I_1 = \int_{span} I(x)\alpha^2(x)dx$ = the generalized inertia

$$\left.\begin{aligned} C_{11} &= \int_{span} h^2(x)dx \\ C_{12} &= \int_{span} h(x)\alpha(x)dx \\ C_{22} &= \int_{span} \alpha^2(x)dx \end{aligned}\right\} = \text{modal factors}$$

ζ_h, ζ_α = damping ratios to critical in bending and torsion

ω_h, ω_α = natural circular frequencies in bending and torsion

U = cross-wind velocity

K = reduced frequency = $\frac{B\omega}{U}$

B = deck width

ω = circular frequency of flutter oscillation

ρ = air density $\begin{cases} \rho = 0.002378 \text{ slug/ft}^3 \\ \rho = 1.228 \text{ kg/m}^3 \end{cases}$

$\left.\begin{matrix} H_i^* \\ A_i^* \end{matrix}\right\}$ = self-excited aerodynamic coefficients derived from experiment [1], [2] (see Figure 1)

II. Solutions of the Flutter Equations

a) *Simplest Case: Pure Torsional Flutter*

A sure indication of flutter instability occurs when the torsional coefficient A_2^* *reverses sign* with increased value of reduced velocity $(\frac{U}{NB})$, where N is the natural torsional frequency of the bridge. This kind of instability occurs often in practice. To find the critical flutter velocity, it is first assumed (from wind-tunnel experience) that flutter occurs practically at the same frequency as the first mode, i.e. $\omega \approx \omega_\alpha$. Then, equating the net damping in equation (4) to zero, yields the critical value of A_2^*:

$$(5) \qquad (A_2^*)_{cr} = \frac{2I_1\zeta_\alpha}{\rho B^4 C_{22}}$$

The corresponding value $(U/NB)_{cr}$ from the plot (see Fig. 1) of A_2^* vs. U/NB is the critical one, and the critical wind velocity is:

$$(6) \qquad U_{cr} = NB(\frac{U}{NB})_{cr}$$

Example: Torsional Flutter

Consider the case where coefficient A_2^* of Figure 1 (deck section 4) is the only important torsional aerodynamic coefficient. Let $I = 857000$ lb. sec^2 with $B = 100$ ft. For a uniform deck, $I_1 = C_{22}\, I$, and taking $\zeta_\alpha = 0.01$,

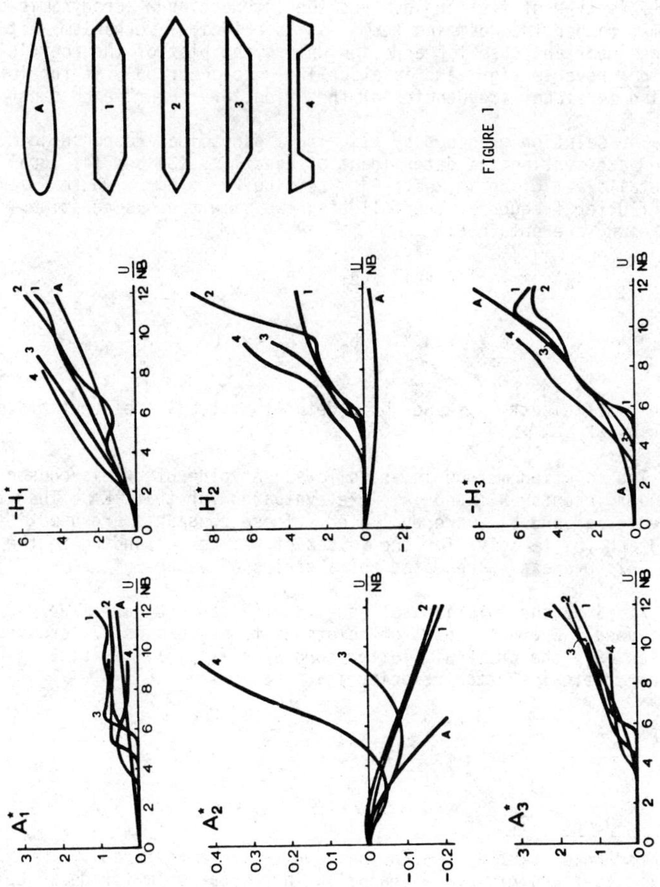

FIGURE 1

$$(A_2^*)_{cr} = \frac{2I\zeta_\alpha}{\rho B^4} = \frac{2(857000)(0.01)}{0.002378\ (100)^4} = 0.072$$

From Fig. 1, this corresponds to a $(U/NB)_{cr}$ value of 5.8 which, for a natural frequency of $N = 0.2$Hz corresponds to a flutter velocity of 116 ft/sec = 79.1 mph.

b) *Two-Degree-of-Freedom Flutter*

This type of flutter involves the interaction of aerodynamic stiffness terms, rather than damping terms as its principal mechanism. It will not be the important case for a design unless the plot of the coefficient A_2^* does not reverse sign, and is also likely to occur only if the lowest bending and torsional frequencies of the bridge are "near" each other.

The solution proceeds by assuming a sinusoidal response jointly in p and q and setting the determinant of equations (3) and (4) equal to zero. Details are given in Ref. [3]. Letting $X = \omega/\omega_h$, where ω is the circular flutter frequency, the following two simultaneous equations, with X as unknown, are obtained:

(7) $$a_4X^4 + a_3X^3 + a_2X^2 + a_1X + a_o = 0$$

(8) $$b_3X^3 + b_2X^2 + b_1X + b_o = 0$$

where the coefficients a_i and b_i (details omitted) are constants or functions of K.

The solution method is as follows. A value of K is chosen and all the coefficients H_i^* and A_i^* are evaluated for that K. Then the coefficients a_i and b_i are evaluated. These constants are used in equations (7) and (8) to solve for the values of X corresponding to the K chosen. This process is repeated for a series of values of K.

Plots of the solutions of equations (7) and (8) are made vs. the K values used. Where the plot of solutions from equation (7) crosses those of equation (8), the critical flutter condition (X_c, K_c) exists. The corresponding critical flutter velocity is:

(9) $$U_{cr} = \frac{B\omega_h X_c}{K_c}$$

Example: 2-Degree Flutter

For this example, the following data are considered: $\zeta_\alpha = \zeta_h = 0.01$. Aerodynamic coefficients are given below in tabular form for deck section 2 shown in Fig. 1. B = 100 feet, L = 4000 ft. The vertical and torsional modes are assumed as half sine waves. $I = 857000$ lb. sec^2, $M = 711.8$ lb. sec^2/ft^2. It will be assumed that $\omega_\alpha = 2\omega_h$, and that $\omega_h = 2\pi$ (0.1 Hz).

U/NB	A_1^*	A_2^*	A_3^*	H_1^*	H_2^*	H_3^*
2	0	0	0	-0.67	0	0
4	0	-0.03	0	-1.50	0	-0.05
6	0.75	-0.05	0.50	-2.05	0.70	-1.25
8	0.70	-0.10	1.00	-3.25	2.25	-3.35
10	0.68	-0.14	1.46	-4.25	4.25	-4.00
12	0.70	-0.16	1.69	-5.50	8.90	-5.00

When the above data are used in equations (7) and (8), the cubic equation possesses only one positive real root while the quartic equation has two, for $U/NB = 2,4,6,\ldots,12$.

Plots of the cubic equation's solution cross those of the quartic equation at $X = 1.62$ and $K = .535$ ($U/NB = 11.7$). Therefore,

$$U_{cr} = \frac{100\,[2\pi(.1)]\,1.62}{.535} = 190.3 \text{ ft/sec.}$$
$$= 129.7 \text{ mph.}$$

III. Vortex-Induced Oscillation

A long-span flexible bridge may, at a relatively low cross-wind velocity, exhibit oscillations of limiting amplitude due to the alternating fluid pressures accompanying vortex shedding. In this context, a self-limiting nonlinear Van Der Pol oscillator will be the model proposed in this report. Means for determining its two aerodynamic parameters will be described, and the extrapolation from tests of a bridge deck section model to prototype full-span bridge will be outlined and exemplified.

The equation of motion proposed for the vortex-induced vertical motion of a deck section model is as follows:

$$(10) \qquad m[\ddot{h} + 2\zeta_h\omega_h\dot{h} + \omega_h^2 h] = \frac{1}{2}\rho U^2(2B)H_0^*\left(1 - \varepsilon^2\frac{h^2}{B^2}\right)\frac{\dot{h}}{U}$$

where: ζ_h = damping ratio in h-motion; ω_h = natural circular frequency in h-motion; ρ = air density; H_0^* = aerodynamic damping coefficient; ε^2 = nonlinear aerodynamic response coefficient; U = steady cross-wind velocity; m = mass of deck section per unit span.

The model proposed is nonlinear, but based on some laboratory observation, the response will be assumed, approximately, to take the same form as that of a linear oscillator:

$$h = h_o \cos \omega t \tag{11}$$

where present interest will be focused only upon the vortex lock-in region, where:

$$\omega \simeq \omega_h \tag{12}$$

$$\frac{\omega A}{U} = 2\pi S \tag{13}$$

A being the projected (windward) area of the deck section per unit span, S being the Strouhal number governing vortex shedding, and ω = the circular frequency of vortex shedding.

For a steady oscillation to be maintained, the average damping energy per cycle (energy lost plus energy gained) will be zero. This fact leads to the assessment of amplitude as

$$h_o = \frac{2B}{\varepsilon}\left[1 - \frac{R\zeta_h}{H_o^*}\right] \tag{14}$$

where:

$$R = \frac{4\pi m S}{\rho A B} \tag{15}$$

Equation (14) contains two unknown constants, ε and H_o^*, that must be evaluated experimentally. Because the mechanical damping influences the response amplitude, a pair of experiments, each with different mechanical damping, will suffice to determine ε and H_o^*. Let (h_{o1}, ζ_{h1}) and (h_{o2}, ζ_{h2}) be the pairs of amplitude and corresponding damping ratios used in two successive deck model "resonance" experiments at the lock-in Strouhal frequency. Then:

$$H_o^* = \frac{R\left[\zeta_{h1} - \frac{h_{01}^2}{h_{02}^2}\zeta_{h2}\right]}{1 - \frac{h_{01}^2}{h_{02}^2}} \tag{16}$$

and

$$\varepsilon^2 = \frac{4B^2}{h_o^2}\left[1 - R\frac{\zeta_h}{H_o^*}\right] \tag{17}$$

where the value of h_o in equation (17) is either h_{o1} or h_{o2}, with ζ_h corresponding and H_o^* coming from (16).

To extrapolate to the full-span bridge; letting x be the spanwise deck coordinate, the vertical motion h at x can be expressed as:

$$h(x,t) = \phi(x)\,\xi(t)\,B \tag{18}$$

where $\phi(x)$ is the modal shape and $\xi(t)$ is the generalized coordinate. This leads to the result, analogous to (14), that

$$\xi_o = \frac{2}{\varepsilon}\left[\frac{\Phi_2}{\Phi_4} - \frac{4\pi M \zeta_h S}{\rho ABL\, \Phi_4\, H_o^*}\right] \tag{19}$$

where:

$$M = \int_0^L m(x)\,\phi^2(x)dx = \text{the generalized mass}$$

$$\left.\begin{aligned}\Phi_2 &= \int_0^L \phi^2(x)\,\frac{dx}{L}\\ \Phi_4 &= \int_0^L \phi^4(x)\,\frac{dx}{L}\end{aligned}\right\}\ \text{modal factors}$$

The spanwise distribution of maximum displacement is then given from (18) as:

$$h(x) = B\,\xi_o\,\phi(x) \tag{20}$$

Example: Vortex-Induced Oscillation

Consider a bridge of span $L = 600$ ft. with a running weight per foot of span of 4,500 lb. Assume a Strouhal number of $S = 0.15$ at a frequency of $n = 0.5$ Hz, with a corresponding bridge vertical mode that is a half sine wave over the span. Assume $B = 30$ ft. and $A = 5$ ft.

Assume that a 1/30 scale section model, 5 ft. long, of this bridge has been tested at vortex-shedding lock-in with the following results:

Damping Ratio	Amplitude No.
0.005	0.25 inch
0.035	0.125 inch

From (15) (either for model or prototype, assuming the model is correctly scaled):

$$R = \frac{4\pi mS}{AB} = \frac{4\pi\,\frac{4500}{32.2}\,(0.15)}{(0.002378)(5)(30)} = 738.5$$

from (16):

$$H_o^* = \frac{738.5[0.005 - 4(0.035)]}{1 - 4} = 33.233$$

from (17):

$$\varepsilon^2 = \frac{4 \times 12^2}{(0.25)^2}\left[1 - 738.5\,\frac{0.005}{33.233}\right] = 8192$$

so that

$$\varepsilon = 90.51$$

For 1/2 sine waves, $\Phi_2 = 1/2$ and $\Phi_4 = 3/8$. Then from (19) for an assumed value $\zeta_h = 0.01$:

$$\xi_o = \frac{2}{90.51}\left[\frac{0.5}{0.375} - 738.5\left(\frac{0.5}{0.375}\right)\frac{0.01}{33.233}\right]^{\frac{1}{2}} = 0.0225$$

From (20), the maximum deflection amplitude $h(x)$ (at center span, $x = 42$), is then

$$h(x) = 30 \times 0.0225 = 0.675 \text{ ft.}$$

IV. Buffeting Response of Long-Span Bridges

The present discussion returns to Ref. [4], reexamining and slightly modifying the theory relative to buffeting, first of single torsional modes, and second of single vertical modes. The extrapolation of wind tunnel model results to the full-span bridge for the purpose of calculating deflections due to buffeting will be presented here. No intermodal coupling, aerodynamic or otherwise, is considered in the present study.

a) Theory for Torsion

In the analysis of torsional buffeting response, the rth single spanwise torsional mode is assumed to be responding alone, and its response given by:

$$\alpha_r(x,t) = \alpha_r(x)\eta_r(t) \tag{21}$$

where η_r is the generalized coordinate.

By defining a net damping ratio γ_r as:

$$\gamma_r = \zeta_r - \frac{\rho B^4 L}{2I_r} A_2^* G_r \tag{22}$$

The equation of motion in torsion for the full-span bridge can be formulated as:

$$I_r[\ddot{\eta}_r + 2\gamma_r\omega_r\dot{\eta}_r + \omega_r^2\eta_r] = \int_0^L M_\alpha(x,t)\,\alpha_r(x)dx \tag{23}$$

where

$$I_r = \int_0^L I(x)\,\alpha_r^2(x)dx = \text{the } r^{th} \text{ generalized inertia}$$

$$G_r = \int_0^L \alpha_r^2(x)\,\frac{dx}{L} = \text{the } r^{th} \text{ modal factor}$$

$$\bar{M}_{\alpha,r(t)} = \int_0^L M_\alpha(x,t)\,\alpha_r(x)dx = \text{the generalized aerodynamic moment}$$

ζ_α = mechanical damping ratio in torsion

ω_α = natural circular frequency in torsion

ρ = air density

$\bar{U}$ = mean wind velocity normal to the span

B = deck width

$K = \frac{B\omega}{\bar{U}}$ = reduced frequency of torsional motion

A_2^* = flutter coefficient associated with torsional aerodynamic damping

$M_\alpha(x,t)$ = aerodynamic buffeting moment per unit span on section x ;
$M_\alpha(x,t) = \bar{U}B^2[C_M u(x,t) + \frac{1}{2} C_m' w(x,t)]$

C_M = moment coefficient of the bridge section at the equilibrium position under the steady wind

C_M' = $dC_M/d\alpha$ defined for this same position

$u(x,t)$ = horizontal wind gust velocity component

$w(x,t)$ = vertical wind gust velocity component.

Letting $S_u(n)$ and $S_w(n)$ be the single-point wind-gust spectra of u and w, assumed to be uncorrelated so that $S_{uw} = S_{wu} = 0$; and assuming the point-to-point lateral cross-spectra $S_u(X_A,X_B,n)$ and $S_w(X_A,X_B,n)$ between spanwise points X_A and X_B to exhibit coherence that falls off as:

$$e^{\frac{-C|X_A-X_B|}{L}}$$

where conservatively:

$$C = \frac{7nL}{\bar{U}}$$

where:

n = frequency of torsional motion
L = span length of bridge
$\bar{U}$ = mean cross-wind velocity normal to the span.

Then the power spectral density of $\alpha_r(x)$ can be expressed as:

$$S_{\alpha_r}(x,n) = \alpha_r^2(x)|A_r(n)|^2[\frac{\rho\bar{U}B^2}{I_r}]^2 \int_0^L\int_0^L \alpha_r(x_A)\alpha_r(x_B)[C_M^2 S_u(n) + \frac{1}{4} C_M'^2 S_w(n)]\, e^{-C|x_A-x_B|/L}\, dx_A\, dx_B$$

$$\simeq \alpha_r^2(x)|A_r(n)|^2[\frac{\rho\bar{U}B^2}{I_r}]\, L^2\, Gr[\frac{2(C-1)}{C^2}]\, [C_M^2 S_u(n) + \frac{1}{4} C_M'^2 S_w(n)] \qquad (24)$$

where:

$$|A_r(n)|^2 = \left\{[1 - (\frac{n}{n_r})^2 + [2\,\gamma_r \frac{n}{n_r}]^2\right\}^{-1} (2\pi n_r)^{-4} \tag{25}$$

n being the torsional frequency of oscillation while n_r is the natural torsional frequency of the rth mode.

The mean square deflection of $\alpha_r(x)$ about equilibrium is given by:

$$\sigma_{\alpha_r}^2(x) = \int_0^\infty S_{\alpha_r}(x,n)\,dn \tag{26}$$

where $S_\alpha(x,n)$ is given from (24).

The details of this integration are discussed in Reference [4]. With the forms for $S_u(n)$, $S_w(n)$ given as:

$$\frac{n\,S_u(n)}{u_*^2} = \frac{200\,f}{(1 + 50f)^{5/3}} \tag{27}$$

$$\frac{n\;S_w(n)}{u_*^2} = \frac{3.36f}{1 + 10f^{5/3}} \tag{28}$$

where: $f = nz/\bar{U}$, z being the height of the bridge, u_* the friction velocity, which depends upon $\bar{U}$ and the type of terrain over which the wind approaches the bridge:

$$\bar{U} = 2.5\,u_*\,\ell n\,\frac{z}{z_o} \tag{29}$$

where z_o is a fetch surface roughness length [3].

The mean square deflection can then be approximated by:

$$\sigma_\alpha^2 \simeq \frac{\alpha_r^2(x)}{K_r^4}\,[\frac{\rho B^4 L}{I_r}]^2\,\frac{G_r}{\bar{U}^2}\,[\frac{2(C-1)}{C^2}]\,F(\gamma_r, C_M, C_M', S_u, S_w) \tag{30}$$

where $K_r = 2\pi n_r B/\bar{U}$, and

$$F(\gamma_r,C_M,C_M',S_uS_w) = C_M^2[\frac{2\pi n_r S_u(n_r)}{8\,\gamma_r} + 6\,u_*^2] + \frac{C_M'^2}{4}\,[\frac{2\pi n_r\;S_w(n_r)}{8\,\gamma_r} + 1.70\,u_*^2] \tag{31}$$

In the case of a bridge with a constant section deck, $I_r = I\,G_r\,L$ and equation (30) becomes:

$$(32)\qquad \sigma_\alpha^2 = \frac{\alpha_r^2(x)}{K_r^4}\left[\frac{\rho B^4}{I}\right]^2 \frac{1}{G_r}\left[\frac{2(C-1)}{C^2}\right]\frac{F(\gamma_r, C_M, C_M', S_u S_w)}{\bar{U}^2}$$

Example: Buffeting in Torsion

A set of six torsional modes of sinusoidal character will be considered. The lowest will be a half sine wave; each subsequent mode will be assumed to possess one additional half sine wave. Arbitrary but reasonable frequencies $n = 0.1$ Hz, 0.2 Hz... 0.6 Hz will be assigned respectively to these modes. The response of each mode under naturally gusty winds with mean velocities of 30, 60 and 90 mph at an assumed bridge height $z = 200$ feet will be calculated. The net damping ratio will be calculated from equation (22) assuming a mechanical damping of $\zeta_r = 0.01$; the A_2^* will be taken from Ref. [4]. The deck width $B = 100$ ft., span length $L = 4000$ ft. The friction velocity when calculated from (29) for an assumed roughness length $z_0 = 0.0164$ feet (corresponding to an open sea wind fetch) results in:

$$u_* = \frac{\bar{U}}{23.52}$$

The coherence factor is C, while the spectral components of the wind velocity will be calculated from (27) and (28). The values of the deck section moment coefficient and its derivative will be taken as $C_M = -1$, $C_M' = 3.67$ and the moment of inertia per unit span $I = 857{,}000$ lb. sec^2.

A summary of the data used in the development of the mean square response σ_α^2 via equation (30) is exemplified in tabular form as follows for the case $\bar{U} = 60$ mph:

Data for $\bar{U} = 60$ mph = 88 ft/sec; $u_* = 2.551$ mph = 3.741 ft/sec

Mode	G_r	n_r(Hz)	C	$\bar{U}/n_rB$	A_2^*(Fig.1)	γ_r	f	S_u	S_w
1	0.5	0.1	31.82	8.80	0.15	Neg.	0.2273	96.2	57.9
2	0.5	0.2	63.64	4.40	-0.04	0.0155	0.4545	32.5	29.0
3	0.5	0.3	95.45	2.93	-0.07	0.0197	0.6818	16.9	17.0
4	0.5	0.4	127.27	2.20	-0.06	0.0183	0.9090	10.6	11.2
5	0.5	0.5	159.10	1.76	-0.05	0.0169	1.1364	7.4	8.0
6	0.5	0.6	190.91	1.47	-0.04	0.0155	1.3636	5.5	6.0

Equation (30) is then used to develop the results presented in the following table. The columns given represent σ_α^2 (in radians squared), σ_α (radians) and edge deflection $50\,\sigma_\alpha$ in feet (based upon a ½ deck width of 50 feet). On the assumption that the bridge random response will be normally distributed, maximum expected excursions could be taken as 3.5 to 4.0 times those presented below.

Mode	30 mph			60 mph			90 mph		
	σ_α^2	σ_α	Max.edge defl.ft 50 σ_α	σ_α^2	σ_α	50 σ_α^2	σ_α	σ_α	50 σ_α
1	2.20×10^{-4}	0.015	0.741		unstable			unstable	
2	4.64×10^{-4}	0.002	0.108	2.2×10^{-4}	0.0148	6.741		unstable	
3	5.82×10^{-7}	7.63×10^{-4}	0.038	2.1×10^{-6}	0.0046	0.229	2.4×10^{-4}	0.0155	0.775
4	1.29×10^{-7}	3.59×10^{-4}	0.018	4.6×10^{-6}	0.0021	0.107	3.8×10^{-5}	0.0062	0.308
5	4.08×10^{-8}	2.02×10^{-4}	0.010	1.5×10^{-6}	0.0012	0.061	1.2×10^{-5}	0.0034	0.170
6	1.60×10^{-8}	1.27×10^{-4}	0.006	5.7×10^{-7}	0.0008	0.038	4.5×10^{-6}	0.0021	0.106

It will be noted that for the bridge parameters chosen, the system is unstable at three calculated conditions affecting the first and second modes. This reflects the fact that the critical (zero damping) condition has been passed through and that flutter has occurred in the mode in question. It is further noted that buffeting response amplitude is particularly sensitive to the values of damping present as seen by equations (30) and (31).

Conclusions

The present paper first considers analytical models for three wind-response phenomena of flexible, long-span bridges, namely: flutter, vortex excitation, and buffeting. Based on minimal data acquired from section model tests of the deck sections of such bridges, examples are then presented that emphasize the practical application of the theories reviewed. The paper is intended to assist designers confronted with the problems in question.

The theory adduced relative to the vortex-response problem has not, to the authors' knowledge, been advanced heretofore in the same context.

REFERENCES

[1] Scanlan, R.H. and Tomko, J.J.: "Airfoil and Bridge Deck Flutter Derivatives," Jn. Eng. Mech. Div., Proc. Amer. Soc. Civ. Eng., Vol. 97, No. EM6, Dec. 1971, pp. 1717-1737.

[2] Scanlan, R.H.: "Recent Methods in the Application of Test Results to the Wind Design of Long, Suspended-Span Bridges," Report FHWA-RD-75-115, Federal Highway Administration, Office of Research and Development, Washington, DC, 1975.

[3] Simiu, E. and Scanlan, R.H.: Wind Effects on Structures, Wiley, New York, 1978.

[4] Scanlan, R.H. and Gade, R.H.: "Motion of Suspended Bridge Spans Under Gusty Wind," Jn. Struct. Div., ASCE, Vol. 103, No. ST9, Stpe. 1977, pp. 1867-1883.

DAMPING MEASUREMENTS OF TALL STRUCTURES

George T. Taoka*, Member, ASCE

ABSTRACT

The results of an investigation of damping measurements of five tall steel structures are presented. The structures include four tall steel buildings ranging in height from 103 to 170 meters, and a four-legged square steel tower of total height 333 meters.

For each structure, ambient vibration records were mechanically digitized and analyzed by three system identification methods; the correlation method, the spectral moments method, and the power spectral density method. A trapezoidal filter was used to isolate individual modes before the records were subjected to analysis.

In addition to estimates from random vibration data, damping estimates were also obtained under forced vibration rotating shaker tests. These estimates appeared consistent with those obtained from ambient data for fundamental modes of vibration. However, considerable variations were present when damping ratios for higher modes were compared.

I. THE CORRELATION METHOD

The theoretical basis for the use of the correlation function to estimate structural response parameters has been reviewed by Cherry and Brady [1]. A detailed discussion of the use of this method to estimate damping is available in a report by Taoka and Scanlan [10]. The following brief discussion of the general theory involved is taken from Cherry and Brady [1], and is included here for completeness.

If $h(\tau)$ is the system response function due to a unit impulse applied at $\tau = 0$, then the relation between an input function $x(t)$ and the resulting output response $y(t)$ is given by

$$y(t) = \int_0^\infty h(\tau)\, x(t-\tau)\, d\tau \tag{1}$$

For a lightly damped single degree of freedom oscillator having a natural circular frequency ω and a small critical damping ratio ζ the system response function is defined as

$$h(\tau) = \frac{e^{-\zeta\omega\tau}}{\omega\sqrt{1-\zeta^2}} \sin\left[\sqrt{1-\zeta^2}\,\omega\tau\right], \qquad \tau \geq 0 \tag{2}$$

It is assumed that $h(\tau) = 0$ for $\tau < 0$.

The autocovariance function for a function $y(t)$ is defined by

*Professor of Civil Engineering, University of Hawaii at Manoa

$$C_y(\tau) = \lim_{T\to\infty} \frac{1}{T} \int_{-\frac{T}{2}}^{\frac{T}{2}} y(t)\, y(t+\tau)\, dt \tag{3}$$

It can be shown that if the input function x(t) is composed of "white noise" with constant spectral density G_0, that the output autocovariance function is given by

$$C_y(\tau) = \frac{\Pi G_0}{2\zeta\omega^3} \left[e^{-\zeta\omega\tau} \left(\cos \sqrt{1-\zeta^2}\, \omega\tau + \frac{\zeta}{\sqrt{1-\zeta^2}} \sin \sqrt{1-\zeta^2}\, \omega\tau\right)\right] \tag{4}$$

Equation (4) represents a cosinusoidal function with exponential decay. The decay of the envelope of the autocovariance estimate of the output response can thus be used to estimate the critical damping ratio ζ of the system. The unbiased autocovariance estimate of a time series $\{y_n\}$ where n varies from 1 to N is given by

$$C_x(k) = \frac{1}{N-k} \sum_{n=1}^{N-k} y_n y_{n+k}\,, \qquad 0 \le k \le M \tag{5}$$

The corresponding autocorrelogram estimate is given by

$$R_y(k) = \frac{C_y(k)}{C_y(0)} \qquad 0 \le k \le M \tag{6}$$

In both Eqs. (5) and (6), the integer M, the maximum lag number, is kept small compared to N the number of points in the series.

The logarithmic decrement method [10] was used to estimate the critical damping ratio in each mode. The value is determined by

$$\zeta = \frac{1}{2\pi q} \ln\left[\frac{A_p}{A_{p+q}}\right] \tag{7}$$

where A_p is the peak amplitude at cycle p and A_{p+q} is the peak amplitude q cycles later.

II. SPECTRAL MOMENTS METHOD

In 1972, Vanmarcke [11] proposed a spectral moments method for estimating frequency and damping parameters of a randomly excited system. The zeroeth order moment gives the area under the power spectral density function. The first order moment is a function of its centroid, and the second order moment gives a measure of dispersion about a spectral peak indicating a central frequency.

If y(t) is a stationary random process with zero mean value, its autocorrelation function $R(\tau)$ is defined by

$$R(\tau) = E\,[y(t)\, y(t+\tau)]\ . \tag{8}$$

The "one-sided" Fourier Transform pair of formulas relating $G(\omega)$ and $R(\tau)$ is given by

$$G(\omega) = \frac{2}{\pi}\int_0^{\infty} R(\tau)\cos(\omega\tau)\,d\tau \tag{9}$$

$$R(\tau) = \int_0^{\infty} G(\omega)\cos(\omega\tau)\,d\omega \tag{10}$$

where $G(\omega)$ is the "one-sided" power spectral density function of the autocorrelation function of the zero-mean process $y(t)$. The mean square value $\langle y(t)^2 \rangle$ is obtained by setting $\tau = 0$, in Eq. (10), giving

$$R(0) = \int_0^{\infty} G(\omega)\,d\omega \tag{11}$$

The spectral moments are

$$\lambda_0 = \int_0^{\infty} \omega^0\,G(\omega)\,d\omega = \int_0^{\infty} G(\omega)\,d\omega \tag{12}$$

$$\lambda_1 = \int_0^{\infty} \omega^1\,G(\omega)\,d\omega \tag{13}$$

$$\lambda_2 = \int_0^{\infty} \omega^2\,G(\omega)\,d\omega \tag{14}$$

Vanmarcke has introduced the following quantities related to the spectral moments above.

$$\omega_1 = \frac{\lambda_1}{\lambda_2} \tag{15}$$

$$\omega_2 = \left[\frac{\lambda_2}{\lambda_0}\right]^{1/2} \tag{16}$$

Note that ω_1 and ω_2 as defined have dimensions of circular frequency. The parameter ω_2 will be directly related to ω_n, the natural circular frequency of vibration. The following dimensionless parameter

$$q = \left[1 - \frac{\lambda_1{}^2}{\lambda_0\lambda_2}\right]^{1/2} \tag{17}$$

will be directly related to the percentage of critical damping ζ of the vibrating system.

If $x(t)$ is the input and $y(t)$ the output of a linear system, the relationship between $x(t)$ and $y(t)$ is given by Eq.(1), where $h(\tau)$ is the impulse response function of the system. The system transfer function $H(\omega)$ is defined to be the Fourier transform of the impulse response function $h(\tau)$.

If $G_0(\omega)$ is the power spectral density function of the stationary "white noise" input $x(t)$, and $G(\omega)$ is the output spectral density function, the equation defining these relationship is

$$G(\omega) = |H(\omega)|^2 \, G_0(\omega) \, . \tag{18}$$

For a single degree of freedom system with natural circular frequency ω_n and damping ratio coefficient ζ, the square of the absolute value of the system transfer function is given by

$$|H(\omega)|^2 = \frac{1}{(\omega_n^2 - \omega^2) + 4\zeta^2 \omega_n^2 \omega^2} \tag{19}$$

It can then be shown that the following relationships exist:

$$\lambda_0 = \frac{\pi G_0}{4\zeta\omega_n^3} \tag{20}$$

$$\omega_2 = \omega_n \tag{21}$$

$$q^2 = \frac{4\zeta}{\pi} [1 - 1.1\zeta + 0(\zeta^2)] \, . \tag{22}$$

When $|\zeta| << 1$, it can be seen that

$$q^2 \cong \frac{4\zeta}{\pi} \tag{23}$$

for very light damping $[\zeta \leq 0.15]$. Thus the natural circular frequency and critical damping ratio is given by

$$\omega_n = \left[\frac{\lambda_2}{\lambda_0}\right]^{1/2} \tag{24}$$

$$\zeta = \frac{\pi}{4} q^2 = \frac{\pi}{4} \left(1 - \frac{\lambda_1^2}{\lambda_0 \lambda_2}\right) \, . \tag{25}$$

Since the output function $y(t)$ is filtered by a band-limited window function between circular frequencies ω_a and ω_b, the effect of excluding frequencies outside of (ω_a, ω_b) will now be discussed. It will be convenient to define dimensionless band limits

$$\Omega_a = \frac{\omega_a}{\omega_n} \, , \quad \Omega_b = \frac{\omega_n}{\omega_b} \, . \tag{26}$$

Vanmarcke has demonstrated that the natural frequency is still given by

$$\omega_n = \left[\frac{\lambda_2}{\lambda_0}\right]^{1/2} \tag{27}$$

However, the critical damping ratio is corrected according to the relationship

$$\zeta = \left[\frac{1+\Omega_a}{1-\Omega_b}\right] \frac{\pi}{4} \left[1 - \frac{\lambda_1^2}{\lambda_0 \lambda_2}\right] \tag{28}$$

for the special case when $\Omega_a = \Omega_b$. This condition was utilized in this investigation.

III. POWER SPECTRAL DENSITY METHOD

This method is based on analyzing the power spectral density function of the response spectra. In the output power spectrum, the spectral density becomes maximum at a frequency $\omega\sqrt{1-2\zeta^2}$, and the damping coefficient ζ can be calculated as follows, knowing the frequencies f_1 and f_2 where the spectral densities become $1/\lambda$ of its maximum.

$$\zeta \cong \frac{A}{2}\left(1 - \frac{3}{8}A^2\right) \tag{29}$$

where

$$A = \frac{f_2^2 - f_1^2}{f_2^2 + f_1^2} \cdot \frac{1}{\sqrt{\lambda - 1}} . \tag{30}$$

In practice, $\lambda = 2$ is usually used for simplicity.

Equation (30) is derived from the power spectral density function

$$S(\omega) = \frac{S_0}{(\omega_0^2 - \omega^2) + 4\zeta^2\omega_0^2\omega^2} . \tag{31}$$

This method has been used by Tanaka, Yoshizawa, Osawa, and Morishita [8] to estimate period and damping parameters of some buildings in Japan.

The power spectral density function is defined as the Fourier transform of the true autocorrelation function derived from a vibration record of infinite length of time. When the length of the record is finite, we cannot estimate the true autocorrelation function for arbitrarily long lags. In practical cases, therefore, we can only compute the so-called apparent autocorrelation function from a record of relatively short length. However, a good estimation of smoothened values of the true power spectrum can be obtained from the Fourier transform of a modified apparent autocorrelation function, which is the product of the apparent autocorrelation and a suitable even function called the lag window.

In the present investigation, the so-called 'hamming' type lag window expressed by Eq. (32) was used.

$$\begin{aligned} \omega(\tau) &= 0.54 + 0.46 \cos\frac{\pi\tau}{T_m} && \text{for } |\tau| < T_m \\ &= 0 && \text{for } |\tau| > T_m \end{aligned} \tag{32}$$

where, τ is a time lag and T_m is the maximum lag which we desire to use.

The corresponding frequency function is

$$W(\omega) = 0.54\,W_0(\omega) + 0.23\left[W_0\left(\omega + \frac{\pi}{T_m}\right) + W_0\left(\omega - \frac{\pi}{T_m}\right)\right] \tag{33}$$

$$W(\omega) = 2T_m\,\frac{\sin\omega T_m}{\omega T_m} \tag{34}$$

The smoothened spectral function is then expressed by

$$P(u) = K \int_{-\infty}^{\infty} \frac{1}{(1 - u'^2)^2 + 4\zeta^2 u'^2} [Y_0(u - u')$$

$$+ \frac{23}{54} \{Y_0(u - u' + \frac{1}{2f_0 T_m}) + Y(u - u' - \frac{1}{2f_0 T_m})\}]du' \qquad (35)$$

where,

$$K = \frac{1.08 T_m G_0}{\omega_n^4}, \quad Y_0(u-u') = \frac{\sin \omega_n T_m (u-u')}{\omega_n T_m (u-u')}, \quad u = \frac{\omega}{\omega_n}, \quad u' = \frac{\omega'}{\omega_n}.$$

This method has been used by Kobayashi and Sugiyama [3] to estimate the structural dynamic parameters of the five structures investigated in this report. The results of this analysis are presented later.

IV. RECORD DATA AND BUILDING DETAILS

Data Collection Procedure

The vibration sensors used in this investigation were three horizontal velocity meters manufactured by the Hosaka Instrument Company of Japan. The analog signals were subjected to low pass filtering effectively eliminating frequencies greater than 10 Hz. The resulting signals were then digitized at a constant inverval of 0.04 seconds, which gives a Nyquist frequency of 12.5 Hz. For the World Trade Center Building, a record length for each sensor of approximately 6,000 points, or four minutes, was recorded on each floor. For the other three buildings, as well as for the Tokyo Tower, 4,500 points, or three minutes of data, were continuously recorded.

The method for recording the ambient vibration data was the same for all the structures, so the method will only be described in detail for the World Trade Center Building. Data were collected on the 40th floor (1st mode), 28th floor (3rd mode), and 16th floor (2nd mode) for this building. On each floor level, the three vibration sensors, were placed as shown in Fig. 2. Sensor #1 was placed at the geometric center of the cross section, facing North-South. Sensor #2 was placed at the Northern tip of the cross section, facing East-West. Sensor #3 also faced East-West, but was located at the Southern tip of the cross section. Thus, if $y_1(t)$, $y_2(t)$, and $y_3(t)$ were the output from the three sensors, the North-South, East-West, and Torsional responses were given by

$$y_{N-S}(t) = y_1(t) \qquad (36)$$

$$y_{E-W}(t) = 1/2[y_2(t) + y_3(t)] \qquad (37)$$

$$y_T(t) = 1/2[y_2(t) - y_3(t)] \qquad (38)$$

The ambient responses for this building were subjected to a Trapezoidal Filter with passbands of 0.363 Hz, 0.394 Hz, and 0.607 Hz for the first three modes, rspectively. Figure 1 shows the details of the Trapezoidal Filter.

Tokyo World Trade Center Building (WTC)

This building rises 40 stories to a height of 152.2 meters at roof level. A small structure on the roof raises the overall height to 156.0 meters. The tower is a steel frame, almost symmetrical in plan view, being 51.4 m (E-W) by 48.8 m (N-S) in cross section. The area of a typical floor is 2,458 m^2. Details of the building are given in Figs. 2 and 3. Vibration tests on this structure have been reported by Muto [4].

International Tele-Communications Center (ITC)

This building is a steel frame which rises 32 stories to a height of 170 meters. It is almost square in plan view, with plan dimensions 51 m by 54 m. Its structural framing consists of closely spaced columns lying on the perimeter of the cross section. Its vibrational characteristics and design features have also been reported by Muto [5]. Records were taken from the 32nd floor for the fundamental modes and the 17th floor for the other modes. Its details are shown in Fig. 4.

Asahi Tokai Building (ATB)

The Asahi Tokai Building is a steel frame 30 story high-rise structure with a total height of 119 m. There is a concrete foundation of three floors below ground level. It is essentially square in cross section, with plan dimensions 36 m by 35 m. Each typical floor has a cross sectional area of 1,249 m^2. Its dynamic characteristics are described by Ichinose [2]. Its structural details are shown in Fig. 5. Data for this building were taken on the 22nd and 14th floors.

Yokohama Tenri Building (YTB)

The last building studied in this investigation is also a steel frame whose cross section is perfectly square, being 27 meters on each side. Its structural framing system consists of a "tube in tube" design. It consists of 27 stories and rises to a height of 103 m. There are also three basement floors below ground level. Its structural details are shown in Fig. 6. Vibration test for this building are described in a report by Tamano [7]. Ambient data were recorded on the 27th, 20th, and 13th floors.

Tokyo Tower (TT)

As can be seen in Fig. 7, the Tokyo Tower is an isolated free-standing steel-framed tower composed of two parts; a primary tower with a height of 253 m and another one of a height of 80 m being added on the top of the former. The latter tower may be divided into two parts, i.e., the super-gain-antenna of 60 m in height and the super-turn-antenna of 20 m in height. Thus, the total height of the tower is 333 m above ground level. The design of this tower, as well as its dynamic response to vibration test and under typhoon conditions, has been reported by Naito, Nasu, Takeuchi, and Kubota [6].

V. COMPARISON OF DAMPING RATIO ESTIMATES FROM DIFFERENT METHODS OF ANALYSES

In this section, parameter estimates for the four buildings obtained from the Correlation, Spectral moments, and Spectral Density methods of analysis will be compared. The estimates from the Spectral Density Method were calculated by Professor Hiroyoshi Kobayashi, with the aid of graduate student Nao Sugiyama [3]. It should be noted that each estimate was obtained from the same ambient vibration record, subjected to a Trapezoidal Filter. Thus any differences in these values would be solely due to the different method used to analyze the data. Forced vibration estimates from three structures are also included in the results.

World Trade Center (WTC)

The average values for the natural frequencies obtained from these analytical methods are 0.281, 0.861, and 1.60 Hz for the North-South direction, and 0.284, 0.870, and 1.61 Hz for the East-West direction. The first two Torsional frequencies are 0.350 and 0.977 Hz. The corresponding forced vibration natural frequency estimates for the translational modes are 0.318, 0.980 and 1.82 Hz for the North-South direction, and 0.315, 0.990, and 1.85 Hz for the East-West direction.

The critical damping ratio estimates are shown in Table 1. The last column in Tables 1 through 5, labeled S/N Ratio, are the signal-to-noise ratios present in the data record from which the modal damping estimates were calculated.

Table 1
CRITICAL DAMPING RATIO (ζ_n) FOR WTC BUILDING

Mode	Filtered Correlogram	Spectral Moments	Spectral Density	Forced Vibration	S/N Ratio
N-S First	0.0094	0.0070	0.008	0.007	32.8
N-S Second	0.0116	0.0072	0.008	0.013	17.3
N-S Third	0.0183	0.0152	0.016	0.014	6.9
E-W First	0.0100	0.0096	0.014	0.009	19.9
E-W Second	0.0096	0.0104	0.014	0.013	16.4
E-W Third	0.0108	0.0094	0.025	0.015	17.3
Torsion First	0.0128	0.0119	---	0.008	22.2
Torsion Second	0.0138	0.0116	---	---	14.5

The correlation curves for the first three North-South modes are shown in Fig. 11. The corresponding forced vibration frequency curves for the same three modes are shown in Figs. 8, 9, and 10. Further discussion of these curves are presented in an NSF report by Taoka [9].

International Telecommunications Center (ITC)

The system identification estimates are 0.324 and 0.955 Hz for North-South, and 0.314 and 0.929 Hz for the East-West directions, respectively. The first two torsional estimates are 0.413 and 1.05 Hz. The corresponding critical damping ratio estimates are listed in Table 2.

Table 2
CRITICAL DAMPING RATIO (ζ_n) OF ITC BUILDING

Mode	Filtered Correlogram	Spectral Moments	Spectral Density	S/N Ratio
N-S First	0.0050	0.0048	0.004	47.2
N-S Second	0.0070	0.0057	0.007	22.3
E-W First	0.0112	0.0115	0.009	23.0
E-W Second	0.0060	0.0060	0.006	22.8
Torsion First	0.0076	0.0091	0.005	26.3
Torsion Second	0.0144	0.0118	0.013	9.9

Asahi Tokai Building (ATB)

The system identification estimates are 0.380 and 1.14 Hz for the first two North-South directions. The fundamental East-West estimate is 0.387 Hz, and the fundamental Torsional estimate is 0.389 Hz. The corresponding forced vibration estimates are 0.434, 1.27, 0.436, and 0.562 Hz. The critical damping ratios for ATB Building are shown in Table 3. The signal-to-noise ratios are generally small, with the exception of the fundamental North-South mode.

Table 3
CRITICAL DAMPING RATIO (ζ_n) FOR ATB BUILDING

Mode	Filtered Correlogram	Spectral Moments	Spectral Density	Forced Vibration	S/N Ratio
N-S First	0.0052	0.0072	0.026	0.009	26.7
N-S Second	0.0209	0.0165	0.013	0.012	7.8
E-W First	0.0455	0.0287	0.024	0.009	7.8
Torsion	0.0473	0.0271	---	0.0073	7.7

Yokohama Tenri Building (YTB)

The system identification estimates are 0.460 and 1.32 Hz for North-South, 0.461 and 1.33 Hz for East-West, and 0.602 and 1.57 Hz for Torsional modes. The damping ratios are listed in Table 4.

Table 4
CRITICAL DAMPING RATIO (ζ_n) OF YTB BUILDING

Mode	Filtered Correlogram	Spectral Moments	Spectral Density	S/N Ratio
N-S First	0.0074	0.0075	0.005	24.5
N-S Second	0.0073	0.0059	0.005	14.5
E-W First	0.0129	0.0103	0.006	19.1
E-W Second	0.0091	0.0074	0.013	10.5
Torsion First	0.0125	0.0090	0.003	13.8
Torsion Second	0.0175	0.0173	0.014	6.5

The Tokyo Tower (TT)

Since the Tokyo Tower is symmetric in both North-South and East-West directions, only the North-South and Torsional modes were analyzed. The natural frequency estimates for the three North-South modes are 0.358, 0.594, and 1.25 Hz, with Torsional frequency estimates of 1.44 and 2.04 Hz. The forced vibration estimates for the translational modes were 0.377, 0.645, and 1.28 Hz. The damping estimates are shown in Table 5.

Table 5
CRITICAL DAMPING RATIO (ζ_n) OF TOKYO TOWER

Mode	Filtered Correlogram	Spectral Moments	S/N Ratio
N-S First	0.0078	0.0109	29.1
N-S Second	0.0082	0.0062	13.7
N-S Third	0.0010	0.0011	78.4
Torsion First	0.0025	0.0021	41.3
Torsion Second	0.0020	0.0019	29.1

VI. CONCLUSIONS

The signal-noise (S/N) ratio was found to have significant effect on the resulting accuracy of damping ratio estimates obtained in this study. Therefore, these ratios were listed in Tables 1 through 5. They measured the ratio of peak amplitude to the average noise amplitude in a region of 0.05 Hz on both sides of a natural frequency estimate. Generally speaking, damping estimates were accurate when the S/N ratio exceeded 15, but were unreliable when the S/N was below 10. The region $10<S/N<15$ was a "gray area" where no definitive statement could be made about damping accuracy.

For fundamental modes of vibration, all three methods of analysis gave reasonably close damping estimates, consistent with forced vibration measurements. For higher modes, however, considerable variations exist in the estimates obtained by different methods. With the exception of the ATB Building, whose records exhibited unfavorable signal-to-noise ratios, most damping estimates were in the range of about 0.5% to 1.5% of critical, under both ambient and forced vibration conditions.

VII. ACKNOWLEDGMENTS

The author wishes to acknowledge the support and cooperation of the Center for Engineering Research and the Computing Center of the University of Hawaii in the preparation of this paper. This research project was conducted under NSF Grant ENV-16926 of the National Science Foundation. The excellent consulting services of Professors Robert H. Scanlan (Princeton University) and Hiroyoshi Kobayashi (Tokyo Institute of Technology) are also gratefully appreciated.

VIII. REFERENCES

1. Cherry, S. and Brady, A.G., "Determination of the Structural Dynamic Properties by Statistical Analysis of Random Vibrations," Proceedings of Third World Conference on Earthquake Engineering, Vol. II, 1965.
2. Ichinose, K., Fujii, K., Ito, T., Hirose, M., and Yamahara, H., "Vibration Test of the Asahi Tokai Building," Proceedings of the Architectural Institute of Japan, November 1971.
3. Kobayashi, H. and Sugiyama, N., "Damping Characteristics of Building Structures by Measuring Oscillations due to Microtremors," Tokyo Institute of Technology Report, February 1975.

4. Muto, K., Ohta, T., Ashitate, T., and Uchiyama, M., "Vibration Test of WTC Building," Annual Proceedings of the Architectural Institute of Japan, September 1970.
5. Muto, K., Sato, K., Toyama, K., Uyeda, N., and Nagata, S., "High Rise Tube Design of the International Tele-Communications Center," Report of the Architectural Institute of Japan, 1974.
6. Naito, T., Nasu, N., Takeuchi, M., and Kubota, G., "Construction and Vibrational Characteristics of the Tokyo Tower," Bulletin of the Scientific and Engineering Laboratory, Waseda University, 1962.
7. Tamano, A., Asahi, K., and Abe, S., "Design of the Yokohama Tenrikyo Building," Report of the Architectural Institute of Japan, 1972.
8. Tanaka, T., Yoshizawa, S., Osawa, Y., and Morishita, T., "Period and Damping of Vibration in Actual Buildings During Earthquakes," Bulletin of Earthquake Research Institute, Vol. 47, 1969, Tokyo, Japan.
9. Taoka, G.T., "System Identification of Tall Vibrating Structures," Dept. of Civil Engineering, University of Hawaii, July 1979, NSF-ENV 75-16926.
10. Taoka, G. and Scanlan, R.H., "A Statistical Analysis of the Ambient Responses of Some Tall Buildings," Princeton University, April 1973.
11. Vanmarcke, E., "Properties of Spectral Moments with Applications to Random Vibrations," Journal of Engineering Mechanics Division, ASCE, Vol. 98, April 1972.

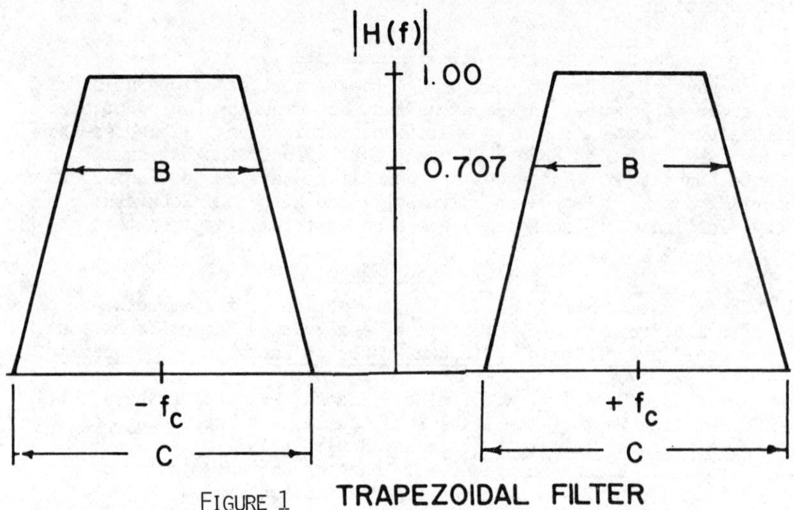

FIGURE 1 TRAPEZOIDAL FILTER

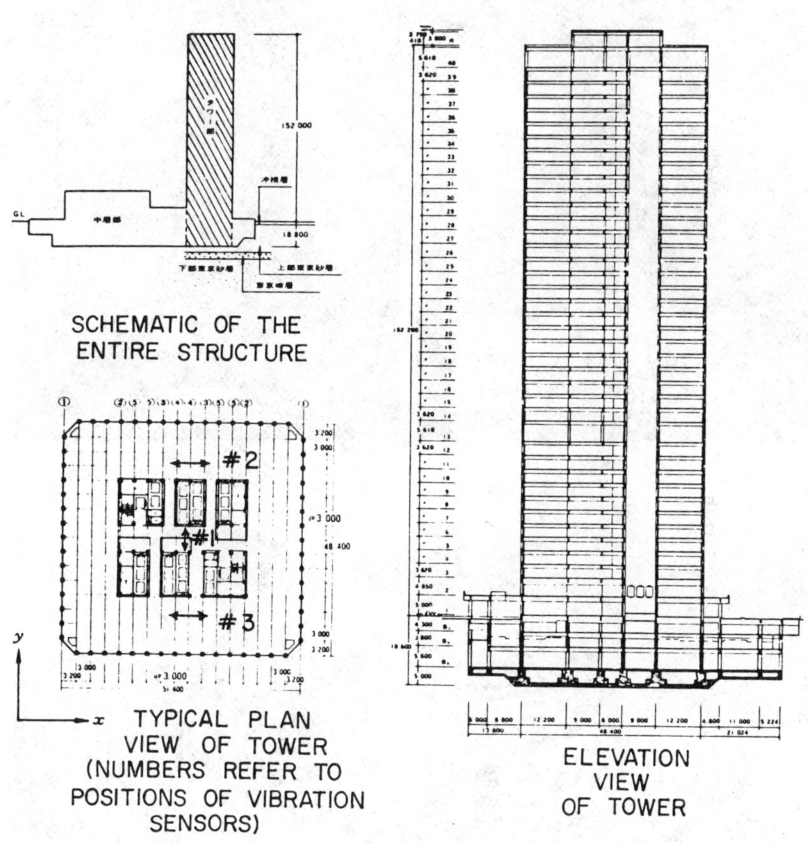

FIGURE 2
TOKYO WORLD TRADE CENTER BUILDING

FIGURE 3
TOKYO WORLD TRADE CENTER BUILDING

INTERNATIONAL TELE-COMMUNICATIONS CENTER

FIGURE 4

FIGURE 5 ASAHI TOKAI BUILDING

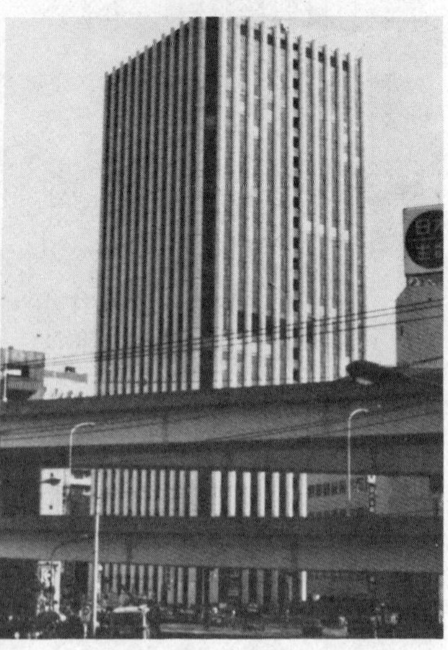

FIG. 6 YOKOHAMA TENRI BUILDING

FIGURE 7 TOKYO TOWER

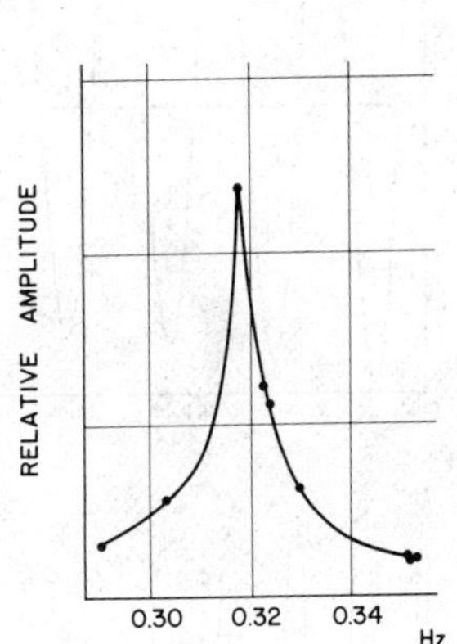

FIGURE 8.
FORCED VIBRATION AMPLITUDE
RESPONSE CURVE, WTC 1st MODE NS.

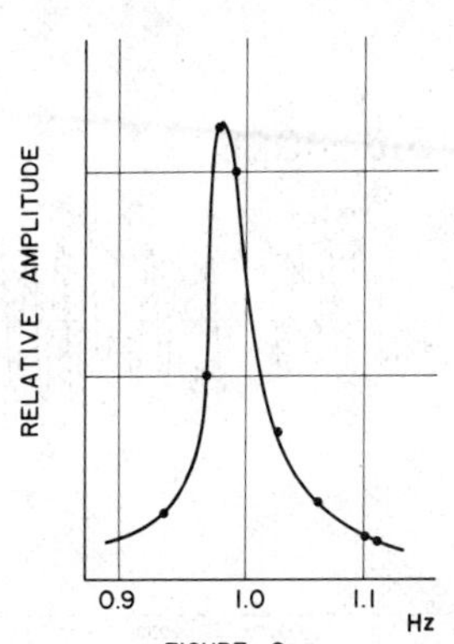

FIGURE 9.
FORCED VIBRATION AMPLITUDE
RESPONSE CURVE, WTC 2nd MODE NS.

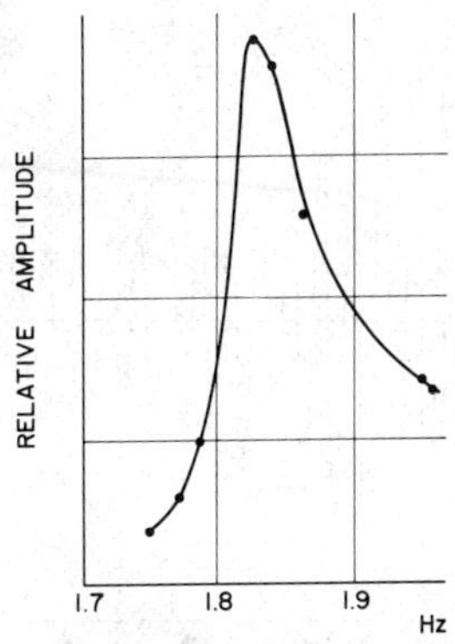

FIGURE 10.
FORCED VIBRATION AMPLITUDE
RESPONSE CURVE, WTC 3rd MODE NS.

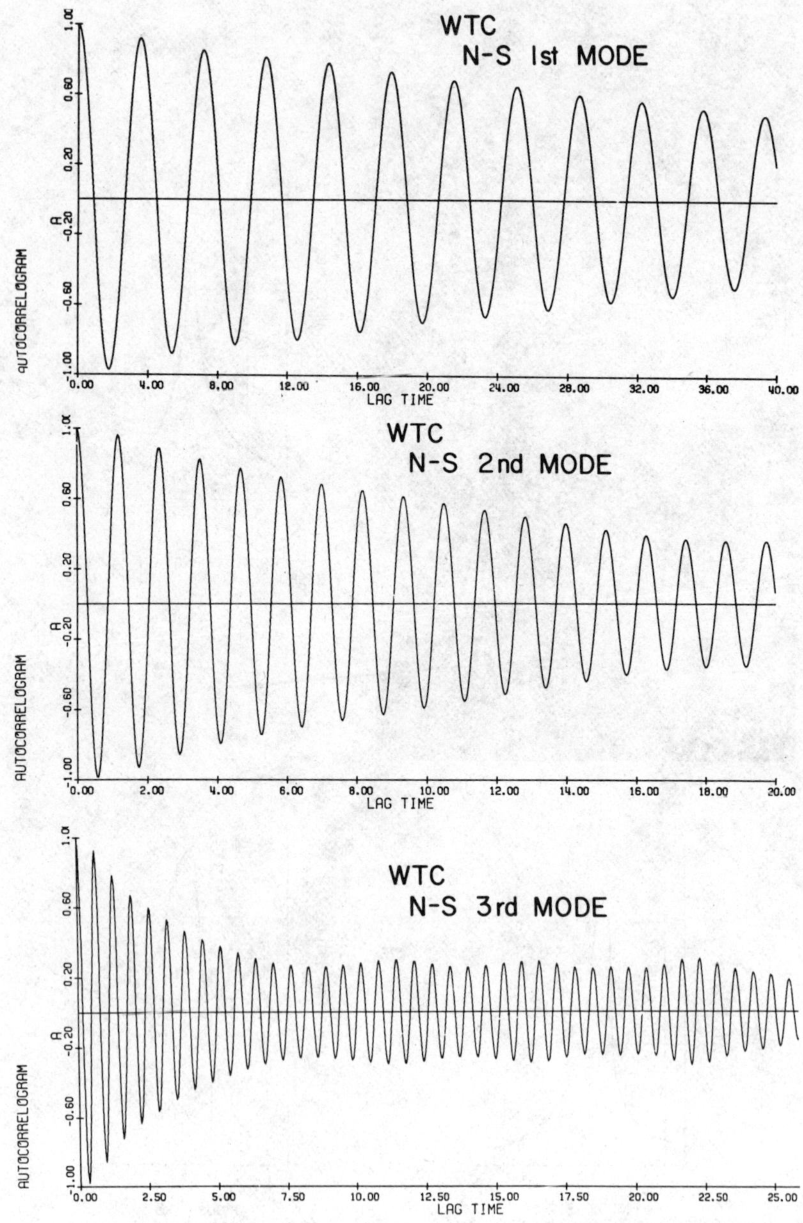

WTC
N-S Ist MODE
AUTOCORRELOGRAM
R
LAG TIME
WTC
N-S 2nd MODE
AUTOCORRELOGRAM
R
LAG TIME
WTC
N-S 3rd MODE
AUTOCORRELOGRAM
R
LAG TIME

FIGURE 11

WIND LOADING AND RESPONSE OF A HIGH-RISE BUILDING

by Russell S. Mills[1] and David Williams[2]

Abstract

An ongoing investigation concerned with the measurement and analysis of full-scale wind effects on high-rise buildings is presented. An instrumentation system suitable for the measurement of wind loading and structural response has been implemented in a 16-story Las Vegas building. The use of a displacement measuring system consisting of a vertically aligned laser and a light-sensitive diode permits evaluation of quasi-static as well as fluctuating response. Experimental results from several wind storms are compared to a recent method for analytically predicting alongwind structural response. The analytical method is found to underestimate the actual building response.

Introduction

Historically, the effects of wind on buildings have been idealized in design procedures by a deterministic, equivalent static load derived from mean wind speeds, drag coefficients, and gust factors. However, for flexible, lightly damped structures, such as modern high-rise buildings, consideration must be given to the dynamic characteristics of wind loading and structural response to permit a realistic estimate of building performance.

For very tall buildings, wind tunnel testing is commonly employed during design to provide greater understanding of the expected response to the wind environment; yet, for many buildings of intermediate height, wind tunnel testing may be prohibitively expensive. Since wind loading provisions may dictate the lateral force design of this class of structure, more accurate methods of wind analysis are clearly needed to avoid the uncertainties associated with current static procedures commonly employed in building codes.

Due to the random nature of the wind and the lack of knowledge concerning the effects of turbulent flow on bluff bodies, nondeterministic spectral analysis techniques are currently being developed as a promising methodology for predicting structural response to wind. In conjunction with this work, experimental measurements from wind tunnel and full-scale wind tests are continually needed to establish reliability in proposed procedures and to provide a basis for developing future

---

[1]Senior Research Engineer, URS/John A. Blume & Associates, Engineers, San Francisco

[2]Research Manager, URS/John A. Blume & Associates, Engineers, San Francisco

formulations. In this regard, results are presented from an ongoing full-scale experimental study of wind effects on a building of intermediate height. This effort forms part of the structural response investigations conducted by URS/John A. Blume & Associates, Engineers (URS/Blume) for the U.S. Department of Energy, Nevada Operations Office, in conjunction with the department's safety program for underground nuclear explosive (UNE) testing. The study was begun in recognition of the need for an evaluation of natural sources of structural loading in order to compare their effect with UNE-induced ground motion. Consequently, an instrumentation system suitable for measuring wind loading and structural response has been implemented in a 16-story, steel-frame office building located in Las Vegas, Nevada. Results obtained from three wind storms of moderate intensity are presented and a comparison is made to a spectral analysis procedure proposed by the National Bureau of Standards (10) and to observed amplitudes of structural response to UNE-induced ground motion.

Test Structure

The building chosen for study is a 16-story, steel-frame office building located in downtown Las Vegas. The building has a rectangular plan of 32 x 51 m (106 x 166 ft) and a height of 72 m (235 ft), including a 7 m (22 ft) high, precast roof parapet. A steel perimeter space frame is designed to carry the total building lateral forces while vertical loads are carried by both the perimeter frame and the interior bolted frames. An 8 in. thick concrete wall encloses one side of the building up to the 7.5 m (25 ft) high, second floor level and can be expected to contribute to the torsional response of the building due to the resulting eccentricity between the centers of mass and stiffness.

The structure is representative of typical modern design and construction practices and was designed according to the requirements of the 1970 edition of the *Uniform Building Code* (*UBC*). Consequently, the lateral design loads consisted of a basic wind pressure of 960 N/m^2 (20 psf) and a seismic base shear of $V = 0.015\ W$ based on the formula $V = ZKCW$ where $Z = 0.5$ (seismic Zone 2), $K = 0.67$, and $C = 0.044$. The structural members comprising the perimeter space frame were sized to restrict maximum story drift to less than 0.25%. Design base shears obtained from wind loading were larger than those obtained from code earthquake loading for both building axes. The structural foundation consists of reinforced concrete strip footings running along column lines in the longitudinal direction.

Although the building is located downtown, it has a relatively unobstructed wind exposure. Fig. 1 illustrates an estimate of the building exposure as a function of wind azimuth. The terrain 1 mile upwind of the structure was considered in these exposure estimates which are based on definitions suggested by Davenport (4).

Three possible sources of dynamic wind loading should be considered for any given situation (3). These are vortex shedding, galloping, and gust impingement (buffeting). Vortex shedding will become critical when the frequency of vortex shedding is tuned to a natural frequency of the structure. The shedding frequency of vortex trails can be estimated by

$$f_v = \frac{S\overline{U}}{B} \tag{1}$$

where:

$\overline{U}$ = mean wind speed

B = building width

S = Strouhal number $\cong$ 0.14

By setting the shedding frequency equal to the fundamental frequency of the study building, it can be seen that resonance will occur at a mean wind speed of 90 m/s (200 mph). Thus, this form of loading will not be of concern in this case, although it may be for exceptionally tall, slender buildings.

Galloping is generally not a predominant source of loading for buildings but is more common in very flexible, lightly damped structures such as bridges and towers. However, due to the low damping values associated with steel-frame buildings and to the slight eccentricity between the centers of mass and stiffness for the study building, galloping may make some contribution to the response of the structure, particularly when an additional eccentricity exists for the aerodynamic center.

However, buffeting can be expected to be the predominant source of loading for this building. Buffeting will be produced from both the natural gustiness of the wind and the turbulence generated by buildings located upwind of the study building. Due to the relatively open exposure of the study building, turbulence from other buildings should not be a primary component of the fluctuating wind except, perhaps, when winds originate from the northwest. Predominant strong winds for Las Vegas are from the northwest and southwest.

Instrumentation

The instrumentation employed at the study building is illustrated in Fig. 2 and consists of a wind sensor (anemometer/vane), pressure transducers, velocity meters, displacement measuring system, and an analog data acquisition system. Early experience with the instrumentation system has been reported by Williams (11).

Wind Sensor. A Gill 3-cup anemometer and microvane are mounted on a mast located near the center of the building, approximately 6.5 m (22 ft) above the top of the parapet wall. The anemometer has a distance constant of 2.4 m (7.9 ft) and, consequently, is capable of accurately measuring wind velocity fluctuations over the entire frequency range of interest.

It is likely that some distortion of wind velocity fluctuations is produced due to the relatively low height of the anemometer above the parapet wall. Ideally, the undisturbed wind flow should be measured at approximately 15 m (50 ft) above the rooftop, but building restrictions preclude raising the mast to this height. Recently, the mast has been extended to 10 m (32 ft) above the top of the parapet wall in an attempt to reduce possible distortions due to this effect in future analyses.

Pressure Transducers. A pressure measuring system utilizing SE 151 microbarographs, designed and supplied by Sandia Laboratories, provides an integrated measure of pressure at the exterior base of the parapet wall. Each set of 3 pressure ports along a building face are interconnected with 1.6 cm (5/8 in.), braided rubber hose and lead to a microbarograph. Two microbarographs are connected to opposite building sides and provide a measure of an effective pressure differential, while two additional pressure transducers measure pressures on the south and west faces relative to a building internal pressure. Consequently, effective differential pressure is obtained directly, while individual pressures on each building face can be determined from a simple numerical difference.

Due to the extensive length of hose between the pressure ports and the microbarographs, 46 m (150 ft) for all sets, pressure measurements above a cutoff frequency of approximately 1 Hz are likely to be distorted by air-column resonance in the pressure lines. This resonance phenomenon can be predicted from elementary fluid dynamics and has also been observed as a peak in the pressure power spectral density function for this particular case. However, since the significant natural frequencies of the study building are below 1 Hz, this limitation will not decrease the applicability of this measurement.

Velocity Meters. Two dual-channel orthogonal arrays composed of Model L-7 velocity meters are positioned on the north and south ends of the building's roof. This arrangement enables extraction of the two translational and the torsional components of the structural dynamic response. The L-7 system has a flat amplitude response from 0.1 to 34 Hz.

Displacement Measuring System. The displacement measuring system incorporates recent technological advances and has facilitated the acquisition of reliable data on total wind-induced structural response. This system, housed in a central elevator shaft, consists of a collimated light source provided by a low power (1 mw), alignment quality CW He Ne laser and a linearly responding, dual-axis, position-sensing photodiode with an active surface area of 3.5 cm square. The laser is mounted on the elevator pit floor and the beam is directed vertically toward a position sensor attached to a structural frame beam at roof level, 68 m (220 ft) above the laser. The displacement monitor compensates for variations in beam size and light intensity. Since only one system is utilized in the study building, total torsional response cannot be measured.

Through comparisons of the dynamic component of displacement response provided by this system and calculated through integration of the L-7 velocity response, the accuracy of the laser system has been thoroughly evaluated. Results indicate that displacement variations of the order of 0.01 cm (0.004 in.) can be accurately detected by the system. The displacement transducer is most applicable to measurement of the quasi-static component of structural response as well as to the displacement response contributed by the lower modes of the building, since the displacements associated with higher mode response are generally minute. Errors due to beam axis tilting produced by deformation of the building foundation have been negligible for the storms recorded to date.

The greatest hindrance to the successful application of the laser-based displacement transducer has been noise produced by thermal interference and by radio transmissions in downtown Las Vegas. The thermal noise, produced by heat generated by the laser, was minimized by shielding the lower 1 m of the laser beam with a cardboard tube. The RF interference was a more stubborn problem, requiring many months of shielding attempts and experimentation before its effect was reduced to a negligible quantity. Periodic upkeep of the system is necessary to prevent the recurrence of this interference.

Data Acquisition. Output from all sensors is recorded continuously on a 14-channel, FM magnetic tape system. Continuous recording is needed to obtain a low-wind record near each storm occurrence to serve as an instrument-zero reference for the pressure and displacement transducers. Absolute time is recorded on a tape channel to assist in the subsequent locating of segments desired for processing and analysis. The instrumentation system is inspected for malfunctions and the tape is changed every 5 days.

Data Analysis

Climatological records listing daily values for wind speed and direction are obtained monthly from McCarran International Airport, located approximately 10 km (6.2 miles) from downtown Las Vegas, and used to assist in the selection of storms suitable for detailed analysis. Three wind storms, two from the northwest and one from the southwest, were chosen for presentation in this report.

A standard data processing procedure has been established for wind storms analyzed in this study. The analog field tape containing a desired storm is scanned to locate the storm peak and to define the storm duration. At the same time, a calm period is found as close to the wind storm as is possible to serve as a zero reference for the displacement and pressure transducers. For the three storms to be discussed, calms were found within 1.5 hours prior to the storm's beginning in all cases.

The analog tapes are then played back at five times real-time while the selected calm and storm segments are digitized by a computer-based data acquisition system and stored on a computer-compatible, digital magnetic tape. Storm records of approximately 20 min duration are digitized at 0.05 sec intervals (real-time), producing a maximum frequency resolution of 10 Hz. Davenport (5) has reported that nontropical storms can be considered to have an approximately constant mean speed and direction for a 20 min time period. Frequency components above 10 Hz are filtered prior to digitization to prevent aliasing errors.

Data analysis consists of two forms: time history analysis and frequency domain spectral analysis. For time history representations, segments of 200 sec duration are located at the storm peak, digitally lowpass filtered at 4 Hz and decimated to 0.1 sec intervals. The time histories are then adjusted by the instrument-zero values obtained from the calm segment, resulting in final, corrected time histories of wind speed, direction and pressure, and building displacement and velocity response. Further processing is then undertaken to extract additional time domain parameters.

These include building translational and rotational acceleration, drag and lift pressures, and mean, root-mean-square (rms), and peak values of load and response. Fluctuating and quasi-static displacement components may be extracted, where the quasi-static component of response may be defined as the motion with frequency content below the lowest natural frequency of the building under study, taken as 0.0 - 0.3 Hz in this case. Also calculated are the vector amplitude of acceleration response at the building extremities and the area-averaged rms acceleration of the building's roof. The acceleration vector amplitude illustrates the largest magnitude of response experienced by the building and reflects the translational as well as the rotational response of the building. The average rms acceleration for a floor area of rectangular plan can be shown to be defined as,

$$\overline{\sigma} = \left[\sigma_{\ddot{x}}^2 + \sigma_{\ddot{y}}^2 + \frac{(x_0^2 + y_0^2)}{3} \sigma_{\ddot{\theta}}^2 \right]^{1/2} \qquad (2)$$

where:

$\overline{\sigma}$ = area averaged rms acceleration

$\sigma_{\ddot{x}}, \sigma_{\ddot{y}}, \sigma_{\ddot{\theta}}$ = rms acceleration in the x, y, and θ directions

x_0, y_0 = one-half the building's width and depth.

This measure is useful in human perception analysis, since it reflects the mean rms acceleration that would be felt over a given floor area and also incorporates the effect of rotational response.

Spectral analysis consists of obtaining estimates of auto- and cross-power spectral density functions through fast Fourier transforms and segment averaging techniques. For each power spectrum, 20 time segments of 1,024 points at 0.05 sec intervals are transformed and averaged, producing a spectrum with a maximum frequency of 10 Hz at an approximate resolution of 0.02 Hz. Twenty averages result in an estimate of power spectral density with a coefficient of variation of 22%.

Results obtained from analysis of the three wind storms are presented in Table 1. Mean and rms values were in most cases obtained from time segments of 20 min duration. The rms values are of the dynamic component (zero mean). Peak values were read from the 200 sec time histories taken at the region of maximum storm intensity. Fig. 3 illustrates several of these time histories of wind loading and building response as well as frequency domain power spectra of wind speed, pressure fluctuations, and building acceleration response. Positive wind azimuth and building rotation are taken as clockwise, positive building translation is to the north and east, while positive pressures imply a net resultant force to the north and east. Natural frequencies for N-S and E-W translational and the rotational modes are approximately 0.40, 0.44, and 0.57 Hz, respectively. Research has shown that a threshold value of human perception for linear acceleration may be taken as 0.005 to 0.015g (9). For storm 3, the largest storm analyzed to date, peak accelerations of 0.0016g occurred at the corners of the building's roof in an

11.6 m/sec (26 mph) mean wind. For a moderate intensity storm, then, human perception is not likely for this building.

Time history response of displacement (Fig. 3c) illustrates the large amplitude of the quasi-static component relative to the fluctuating component. For this case, the quasi-static component can be seen to average as much as three times the peak-to-peak fluctuating response. The fluctuating response may be more predominant in a building of higher flexibility or in one subjected to more turbulent wind conditions.

The power spectral density functions of building acceleration indicate that the building dynamic response is composed primarily of the two fundamental translational modes and the first rotational mode. The rotational mode is illustrated by the east-west building acceleration record because the recording instrument is eccentric to the building's east-west axis. The rotational mode can be seen to produce similar power spectrum amplitude as the east-west translational mode at the measurement location, indicating that torsional response is an important aspect of structural response to wind.

The wind speed and pressure spectra are shown in a normalized fashion (Figs. 3b and 3d). The spectral amplitude is normalized by the mean-square value of the measured parameter, producing a spectrum with unit area. The abscissa is wave number, obtained by dividing frequency by the mean wind speed, and indicates the average number of wave cycles per unit length. The leveling-off of the north-south pressure spectrum observable in Fig. 3d can be attributed to instrument noise.

Superimposed with the measured wind spectrum are two empirical representations of wind spectra as proposed by Davenport (4) and Simiu (10). The Davenport spectrum is currently utilized in building code requirements for wind design (1,2) and in normalized form is defined as,

$$\frac{S_u(f)}{\sigma_u^{\,2}} = \frac{4800}{\beta\,\overline{U}_{10}}\;\frac{x}{(1+x^2)^{4/3}}\;; \qquad x = \frac{1200f}{\overline{U}_{10}} \tag{3}$$

where the mean square value of wind speed fluctuations is modeled as,

$$\sigma_u^{\,2} = \beta u_*^{\,2} = \beta K \overline{U}_{10}^{\,2} \tag{4}$$

where β has a value between 4.75 and 6, u_* is the friction velocity, K is the drag coefficient, $\overline{U}_{10}$ is the hourly mean wind speed at 10 m elevation, and f is frequency in Hz. This representation produces a spectrum that is independent of height.

Simiu's spectrum attempts to account for the effect of height on wind turbulence and also is claimed to more accurately reflect the higher frequency portion of the spectrum. This formulation is given by,

$$\frac{S_u(f,z)}{\sigma_u^{\,2}} = \frac{200(z-z_d)f}{\beta(1+50n)^{5/3}}\;; \qquad n = \frac{f(z-z_d)}{\overline{U}_z} \tag{5}$$

where z is the height above ground and z_d is the zero-plane displacement, which can be assumed as zero except for in large cities.

Both spectra were calculated using coefficients applicable to city wind exposures and can be seen to give similar values for wind turbulence in the frequency range characteristic of the fundamental frequencies of tall buildings. However, the analytical spectra in this range appear to seriously underestimate the gustiness of the wind, as indicated by the measured wind spectrum. Some of the high amplitude turbulence present in the measured spectrum may be attributed to the relatively low position of the anemometer above the building top. However, it will be shown that an analytical estimate of building response based on the Simiu spectrum also underestimates the fluctuating response of the building.

During data analysis, an attempt was made to calculate transfer functions between wind speed, pressure, and building response. However, these functions exhibited values of coherence less than 0.2, as has been indicated by other researchers (6,10), for all but very low frequencies. This would indicate that a linear relationship between these parameters is a poor assumption, but one that is often made in analytical formulations.

Comparison to Analytical Procedure

An analytical procedure for predicting the alongwind response of tall buildings has been developed by Simiu and Lozier (10). This nondeterministic spectral analysis procedure is based on frequency domain transformations from wind speed to pressure to building response. The procedure has the capability to model structures with unusual mode shapes or for which the contribution of higher modes is significant.

The initial basis of the procedure is an empirical model for the normalized wind spectrum of alongwind velocity fluctuations, as was discussed previously, and a logarithmic profile for mean wind speed. A spatially dependent numerical expression for the cross-power spectral density function of longitudinal fluctuating wind velocities is then calculated from the model for the normalized wind spectrum and reduction factors based on the empirical formula for alongwind and acrosswind cross-correlation functions. A linearization of Bernoulli's equation, as applied to unsteady flow, is then made to arrive at predictions for the cross-spectra of longitudinal pressure fluctuations. Once this result is obtained, the mechanical admittance function of structural response is applied to arrive at an estimate for the spectral density of structural response and corresponding estimates of mean, rms, and probable maxima of displacement and acceleration response.

This methodology for obtaining estimates of structural response to wind was applied to the study building for the three wind storms under consideration. Predictions were also made for storms having estimated return periods of 150 and 2.5 years. Storms in Las Vegas having these frequencies of occurrence were found by Raggett (8) to have 1 hour averaged wind speeds at 10 m elevation in open terrain of 27.0 m/sec (60 mph) and 18.0 m/sec (40 mph) for 150 and 2.5 yr return periods, respectively. Table 2 summarizes the results of this analysis.

For the response predictions, a roughness length of 0.6 m was assumed, corresponding to a fairly unobstructed city exposure classification. Recommended values for the coefficients C_z and C_y are 10 and 16 for the acrosswind velocity cross-correlation factors, but Simiu and Lozier warn that large variations in these parameters have been observed in experimental studies and smaller values, implying greater acrosswind correlation, may be needed to provide accurate estimates of structural response to wind storms of moderate intensity. Consequently, values of $C_z = 4$ and $C_y = 6.4$ were assumed for all storms. An analysis was also performed for the wind storm with the 150 yr return period with the recommended values of $C_z = 10$ and $C_y = 16$ to evaluate the sensitivity of the results to variations in these parameters.

By comparing the experimental observations of structural response presented in Table 1 to the analytical predictions in Table 2, the accuracy of this procedure can be evaluated. For low intensity wind storms, the analytical method can be seen to produce response estimates considerably lower than the observed amplitudes of response, even with the adjusted values for C_y and C_z. Agreement improves with increasing storm intensity; however, the fluctuating response component is still seriously underestimated. The source of this error may be that the cross-correlation function has higher values at the natural frequencies of the building than has been assumed or, as was shown in the previous section, the adopted formula for wind spectra may underestimate the gustiness of the wind by as much as 500% in the region of the building fundamental frequencies.

Predictions for the wind storm with the 2.5 yr return period indicate peak displacements and accelerations of 3.3 cm (1.3 in.) and 0.004g, respectively. With the contribution of the torsional component of response, which this procedure neglects, this storm can be expected to produce response above the lower limits of human perception, particularly since this method appears to underestimate the amplitude of structural response.

The analysis of a storm with a return period of 150 yr shows little variation in displacement response with variations in C_z and C_y. Maximum displacement is approximately 7 cm (2.7 in.) for either case. A greater variation is seen in the estimates of acceleration response, with values of 0.007g and 0.013g resulting from the two analyses.

Comparison to Seismically Induced Response

Since its completion in 1975, the structure has been subjected to a number of UNE-induced ground motions and good records of roof-level acceleration and displacement have been obtained for several high-yield events. Maximum recorded displacement from these events has been 3.3 cm (1.3 in.) in both the longitudinal and transverse directions while maximum accelerations were approximately 0.025g.

The structural response to wind is characterized by a higher ratio of peak displacement to acceleration due to the quasi-static component of response, which is not present for seismically induced response. Thus, wind-induced response can be expected to exceed 3.3 cm (1.3 in.) displacement at an annual return period of 2.5 yr, while maximum acceleration response to wind is not likely to exceed 0.025g, even when high intensity storms of low probability are considered. Consequently, the lateral load due to wind response is comparable to that observed to date from seismic response, yet wind response would be less perceptible to occupants than seismic response.

Conclusions

A laser-based displacement measuring system has been successfully applied to the measurement of total building response to wind and has shown that the quasi-static component of structural response is the primary component of displacement response to wind for a modern office building of intermediate height in a relatively open wind environment.

An analytical method, developed by Simiu and Lozier at the National Bureau of Standards, for predicting alongwind structural response has been evaluated and shown to underestimate the actual building response, particularly in the case of the fluctuating component. Experimental results indicate that this error may be due to unrealistically low analytical estimates of the wind velocity fluctuations in the frequency range common to tall buildings.

This investigation of wind effects is planned to continue for an additional year at the study building. Future work will be directed toward further comparisons between experimental results and other analytical procedures, particularly methods that consider the important torsional component of wind response (7). An attempt will be made to quantify the discrepancies observed between the analytical formulation reviewed in this report and the experimental results.

Acknowledgments

This research study has been supported by the U.S. Department of Energy, Nevada Operations Office, under Contract No. DE-AC08-76DP00099. In addition, the authors wish to recognize the contributions of B. R. Bradley of URS/Blume, J. R. Banister and D. M. Ellett of Sandia Laboratories, and other URS/Blume employees involved in this project. The cooperation of the building owner is also deeply appreciated.

Appendix 1.--References

1. *American National Standard Building Code Requirements for Minimum Design Loads in Buildings and Other Structures*, ANSI A58.1-1972, American National Standards Institute, New York, 1972.

2. *Canadian Structural Design Manual*, Supplement No. 4 to the National Building Code of Canada, National Research Council of Canada, 1970.

3. Cermak, J. E., "Wind Tunnel Testing of Structures," *Proceedings*, ASCE/EMD Specialty Conference on the Dynamic Response of Structures, University of California, Los Angeles, Mar. 1976.

4. Davenport, A. G., "The Application of Statistical Concepts to the Wind Loading of Structures," *Proceedings*, Institution of Civil Engineers, Vol. 19, Aug. 1961.

5. Davenport, A. G., "The Relationship of the Wind Structure to Wind Loading," *Proceedings*, Symposium on the Effects of Wind on Structures, Her Majesty's Stationary Office, London, June 1963.

6. Holmes, J. D., "Pressure Fluctuations on a Large Building and Along-wind Structural Loading," *Journal of Industrial Aerodynamics*, Vol. 1, 1975/76.

7. Patrickson, C. P., and Friedmann, P., "A Study of the Coupled Lateral and Torsional Response of Tall Buildings to Wind Loadings," *UCLA-ENG-76126*, University of California, Los Angeles, Dec. 1976.

8. Raggett, J. D., "A Preliminary Study of Building Exposure to Wind in Las Vegas," File No. 77-001, URS/John A. Blume & Associates, Engineers, San Francisco, Feb. 1977.

9. Reed, J. W., "Wind-Induced Motion and Human Discomfort in Tall Buildings," *No. R71-42*, Massachusetts Institute of Technology, Cambridge, Nov. 1971.

10. Simiu, E., and Lozier, D. W., "The Buffeting of Tall Structures by Strong Winds," *National Bureau of Standards Building Science Series 74*, Washington, D.C., Oct. 1975.

11. Williams, D., "Measurement of Wind Effects on Tall Buildings," presented at the 1978 Third U.S. National Conference on Wind Engineering Research, held at University of Florida, Gainsville.

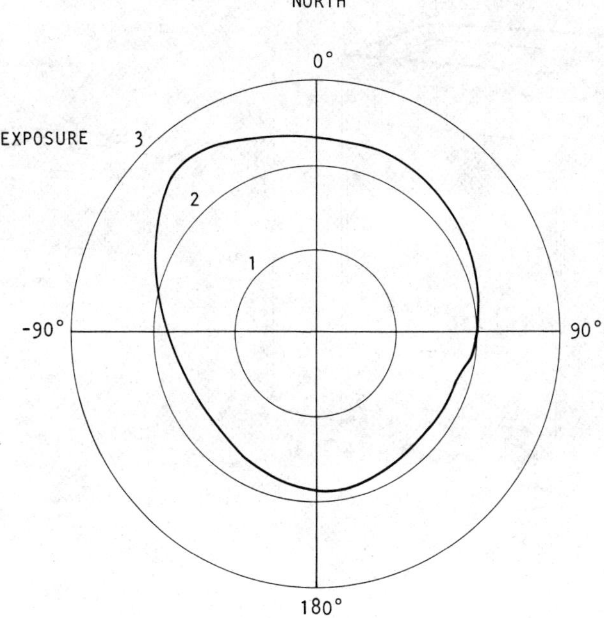

Exposure	Description
1	Open; few obstacles
2	Intermediate; obstacles 30-50 ft in height
3	City; large and irregular objects

FIG. 1--Wind Exposure of Study Building

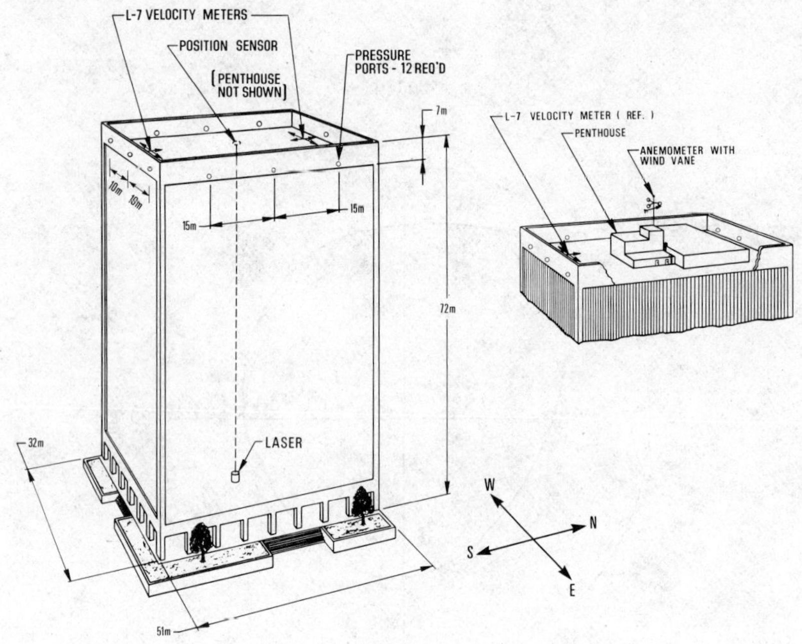

FIG. 2--Instrumentation System

(Fig. provided by Sandia Laboratories.)

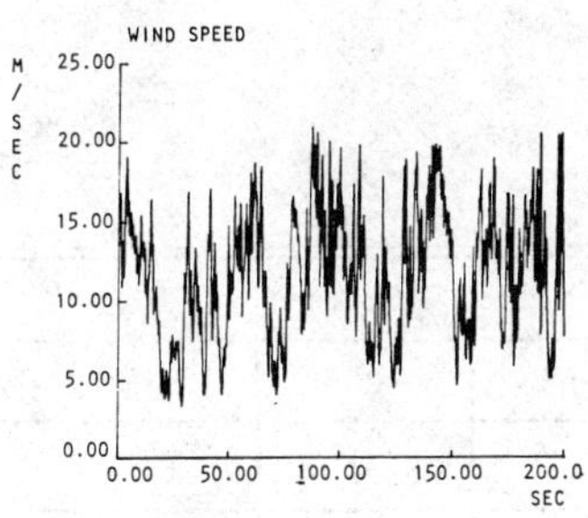

(a) Wind Speed

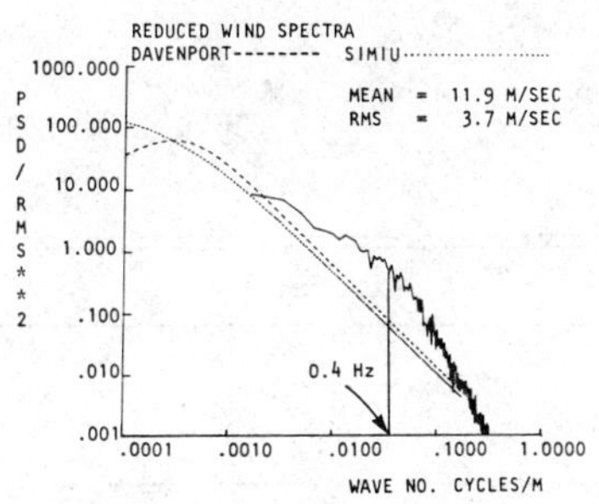

(b) Normalized Wind Spectra

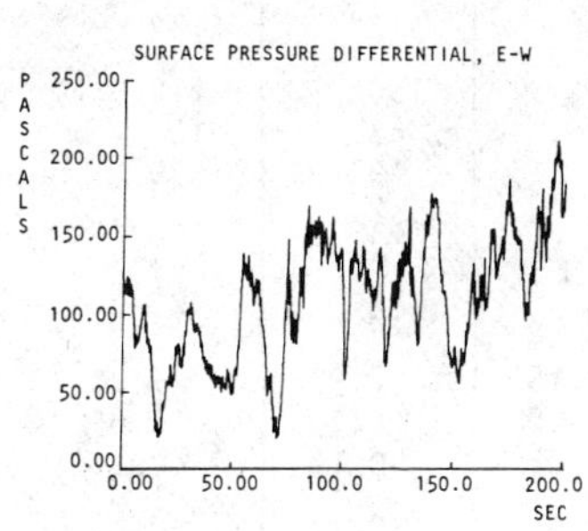

(c) Pressure Differential

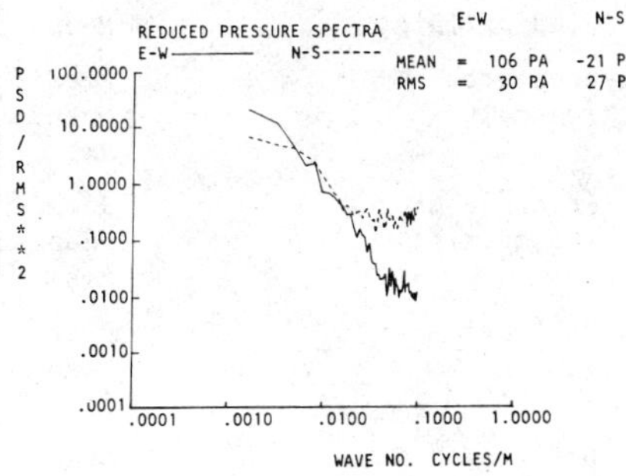

(d) Normalized Pressure Spectra

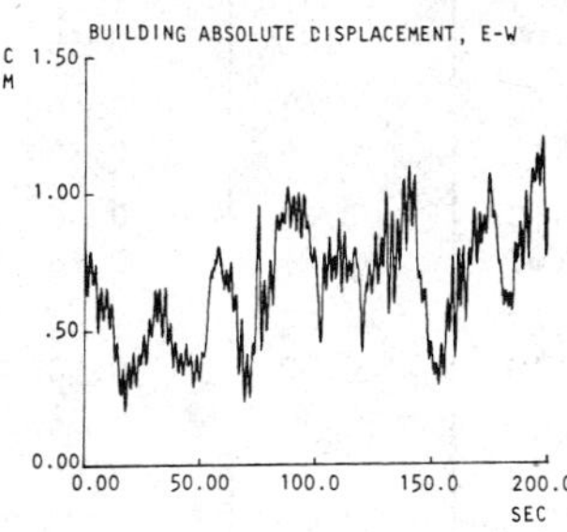

(e) Absolute Building Displacement, E-W

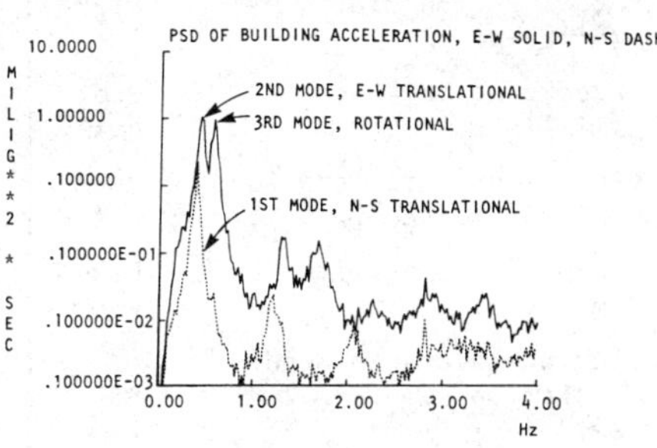

(f) Acceleration Power Spectra

FIG. 3--Wind Storm Number 3

TABLE 1.--Wind Storm Data

Parameter		Storm 1	Storm 2	Storm 3
(1)		(2)	(3)	(4)
Wind Speed: (m/sec)	Peak	12.6	19.3	20.9
	Mean	5.9	9.3	11.6
	Rms	1.9	2.9	3.7
Wind Azimuth:	Mean	215°	312°	317°
Pressure E-W: (Pascals)	Peak	30	150	210
	Mean	16	63	106
	Rms	7	19	30
Pressure N-S: (Pascals)	Peak	72	-103	-106
	Mean	41	-18	-21
	Rms	10	19	27
Deflection E-W: (cm)	Peak	0.15	0.97	1.20
	Mean	0.08	0.39	0.65
	Rms	0.03	0.14	0.21
Deflection N-S: (cm)	Peak	0.57	-0.17	-0.16
	Mean	0.27	-0.01	-0.03
	Rms	0.08	0.05	0.03
Acceleration E-W: (milli-g)	Peak	0.21	0.95	1.23
	Rms	0.08	0.30	0.35
Acceleration N-S: (milli-g)	Peak	0.60	0.41	0.40
	Rms	0.17	0.14	0.10
Rotational Acceleration: (milli-rad/sec^2)	Peak	0.10	0.23	0.40
	Rms	0.02	0.07	0.10
Corner Acceleration: (milli-g)	Peak	0.67	1.46	1.63
	Rms	0.19	0.42	0.50
Surface Averaged Accel-eration: (milli-g)	Rms	0.19	0.34	0.40

TABLE 2.--Predicted Alongwind Response From Analytical Method

Parameter (1)	Storm 1 E-W (2)	Storm 1 N-S (3)	Storm 2 E-W (4)	Storm 3 E-W (5)	RP = 150 years E-W (6)	RP = 150 years E-W (7)	RP = 2.5 years E-W (8)
z_0 (m)	0.6	0.6	0.6	0.6	0.6	0.6	0.6
C_z	4	4	4	4	4	10	4
C_y	6.4	6.4	6.4	6.4	6.4	16	6.4
X_{max} (cm)	0.24	0.15	0.62	0.97	7.60	6.91	3.26
$\overline{X}$ (cm)	0.12	0.08	0.31	0.48	3.47	3.47	1.54
σ_x (cm)	0.04	0.02	0.09	0.14	1.10	0.95	0.47
$\ddot{X}_{max}$ (milli-g)	0.09	0.07	0.35	0.70	13.11	6.95	3.98
$\sigma_{\ddot{x}}$ (milli-g)	0.02	0.02	0.09	0.17	3.20	1.69	0.97

z_0 = roughness length.

C_z, C_y = coefficients in acrosswind velocity cross-correlation function.

$\overline{X}$, σ_x, X_{max} = displacement mean, rms, and probable-maximum values.

$\sigma_{\ddot{x}}$, $\ddot{X}_{max}$ = acceleration rms and probable-maximum values.

milli-g = 10^{-3} g.

DYNAMIC RESPONSE OF INFLATABLE CABLE-REINFORCED STRUCTURES UNDER WIND LOADING*

By O. Vinogradov[†], D.J. Malcolm[††], AM.ASCE and P.G. Glockner[†††], M.ASCE

ABSTRACT

The influence of wind and internal pressure on the dynamic response of inflatable spherical caps is analyzed numerically using a finite element method and corresponding computer code capable of handling nonlinear static and small forced oscillation response of arbitrary cable networks. It is shown that horizontal and vertical fundamental frequencies of the cap are very sensitive to the wind velocity and when some "critical" velocity is exceeded the vertical stiffness of the structure may approach zero and the response becomes unstable. The relationship between the critical wind velocities and the values of internal pressure are presented for a spherical cap in graphical forms.

INTRODUCTION

The behaviour of a highly flexible air-supported structure is significantly different from that of a rigid structure [8]. It has been shown [4] that the initial stresses affect natural frequencies of these structures, and that the natural frequencies are approximately proportional to the square root of the internal pressure [10]. The influence of statically applied loads (drag force, dead weight, snow, etc.) on the dynamic response of an inflated spherical cap has been analyzed and it has been shown that this influence may be very significant [9,10]. An example of the dynamic response of an inflatable spherical cap, loaded by an idealized wind pressure (constant positive pressure on the leeward

---

* The results presented here were obtained in the course of research sponsored by the Natural Sciences and Engineering Research Council of Canada, Grant Nos. 69-1600 and 69-0743.

† Assistant Professor, Department of Mechanical Engineering, The University of Calgary, Calgary, Alberta, Canada, T2N 1N4.

†† Associate Professor, Department of Mechanical Engineering, The University of Calgary, Calgary, Alberta, Canada, T2N 1N4.

††† Professor and Head, Department of Mechanical Engineering, The University of Calgary, Calgary, Alberta, Canada, T2N 1N4.

side, constant negative on the windward side) has been considered [10] and it has been found that the response of the cap differs essentially from that without wind load, the reason being that the wind load changes the pressure distribution on the structure as a result of which some cable segments experience higher loads while on others the load is decreased. Since the stiffness of a cable segment is proportional to the pressure load [9], the stiffness of the overall structure varies with the wind load, underscoring the significance of the pressure distribution due to wind. The objective of this paper is to continue the analysis on the influence of wind on the dynamic response of a spherical cap, taking into account more realistically the wind pressure distribution.

The results presented are based on a numerical method with corresponding computer code developed for the nonlinear static and small forced oscillation analysis of arbitrary cable-reinforced air-supported structures [9]. The main assumptions are repeated here.

The inflatable cable-reinforced structure is modelled by a cable network with the action of the membrane neglected except for transmitting the pressure load to the cables. The pressure load is considered as distributed along cable elements while all other loads, including inertia forces in the dynamic problem, are concentrated at nodes. It is assumed that cables can be divided into segments in such a way that each is loaded by a uniform pressure line-load lying in one plane. For structures subjected to internal pressure plus a smoothly varying wind load, this decomposition can always be performed by increasing the number of elements sufficiently to approximate any pressure loading. The equilibrium of each segment under the uniform plane pressure line-load is described by a simple nonlinear algebraic equation.

The cables are considered as elastic elements undergoing large displacements. It is assumed that the change in shape of the structure does not influence the magnitude and distribution of the external pressure load. Under these assumptions the geometrically nonlinear static problem is reduced to the determination of the equilibrium shape of the structure under the given load. When the equilibrium shape of the structure, and the corresponding stiffness matrix are known, the small forced oscillations are investigated by transforming the static stiffness matrix into the dynamic complex stiffness matrix, including inertia terms from concentrated masses.

The method of assembling the overall structural stiffness matrix at each cycle is based on that used in the SAP IV computer program [1].

The numerical analysis has been applied to a spherical cap consisting of 14 reinforcing cables lying along geodesic lines with 147 degrees of freedom.

WIND PRESSURE

Bearing in mind that not the interaction between the wind and the structure but rather the influence of the wind on the dynamic response of the structure is the objective of this analysis, some simplifying assumptions concerning wind properties can be made. The wind is considered to have a steady mean velocity and high frequency fluctuations. The flow is regarded uniform, although it is well known [2] that due to surface friction, the wind velocity has a gradient. Even for such a simplified model of the wind, there are no experimental results for flow past a flexible structure. As a first approximation, the data for rigid bodies can be used. Experimental data from tests on rigid domes [5] has been used in [7] in simulating wind loading on inflatable domes. Of course, a check should be carried out after application of the wind to ensure that the shape of the structure does not change too much.

The assumption of uniform flow allows one to consider the flow past flexible spherical caps as part of the flow around a sphere. Experiments on spheres in steady flow show [3] that at large Reynolds numbers ($> 10^6$), corresponding to conditions existing for flow past large structures, the pressure distribution is close to that predicted by potential theory as

$$p = p_\infty + \frac{1}{2}\rho V^2 C_p \tag{1}$$

where

$$C_p = 1 - \frac{9}{4}\sin^2\theta \tag{2}$$

P_∞ = barometric pressure, V = wind velocity, ρ = air density, C_p = pressure coefficient as function of the angle θ. Eq. (2) is valid for $0 \leq \theta \leq (140°\text{-}150°)$. For $150° \leq \theta \leq 180°$, the pressure coefficient is equal to 0.15 [3]. For the spherical cap, the diagram of pressure coefficient as a function of θ is shown in Fig. 1. For comparison purposes, a simplified pressure coefficient variation, assumed in [10], is shown by dotted lines.

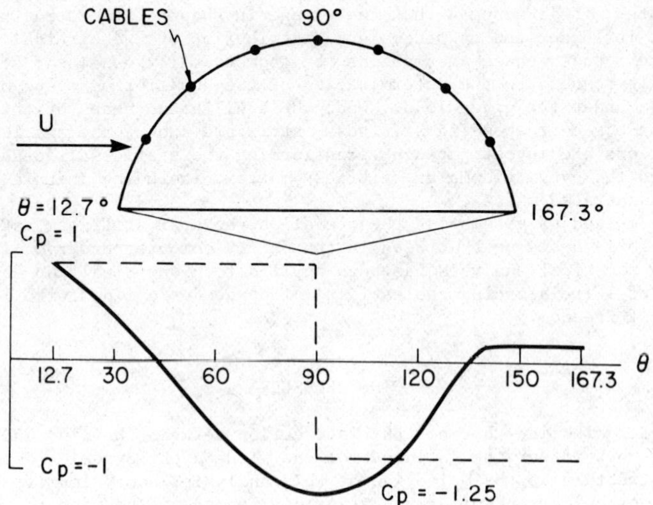

Fig. 1 Pressure Distribution Around the Spherical Cap

NUMERICAL RESULTS

The cable-reinforced spherical cap, shown in Fig. 2, has been used in [6] for the static analysis and in [10] for the dynamic analysis.

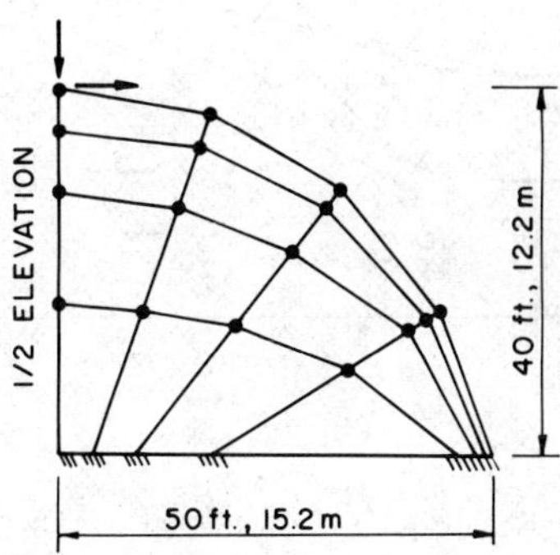

Fig. 2 Fourteen Cable Reinforcement of Spherical Cap (EA = 13.3 x 10^4 x N = 3 x 10^4 lbf)

The static load applied at nodes has been taken as the dead weight of the cables. The mass of the cables is also concentrated at nodes. Due to symmetry, all calculations are carried out for one half of the structure, consisting of 60 cable segments and 28 lamped masses. The dynamic characteristics of this structure were analyzed to demonstrate the influence of internal pressure and wind on the natural frequencies. A unit harmonic force is applied at the apex of the cap (see Fig. 2) in the vertical and horizontal directions, independently. The influence of the barometric pressure has been neglected in the analysis.

Fig. 3 shows diagrams of the vertical and horizontal fundamental frequencies vs. the wind velocity. For two different values of the internal pressure, the curves are similar. The horizontal frequency increases more or less gradually, with increasing wind velocity. The vertical frequency curve has three characteristic intervals. For comparatively low wind speeds, the fundamental frequency is almost constant. At some speed, this frequency starts to increase, almost proportionally with the wind, up to a point, after which it drops sharply. Thus, the first conclusion is that wind significantly changes the dynamic characteristics of inflatable structures. It increases the stiffness of the structure in the horizontal direction so that the structure has a stable horizontal response. The stiffness in the vertical direction undergoes more significant changes, and when the wind exceeds a definite velocity, which may be regarded as a "critical"

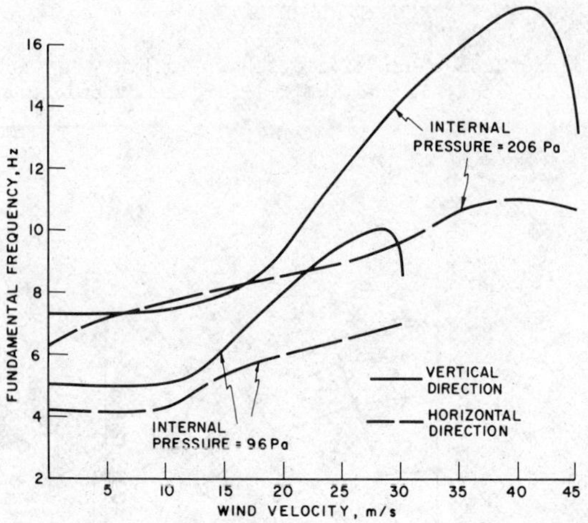

Fig. 3 Vertical and Horizontal Fundamental Frequencies vs. Wind Velocity

velocity, the stiffness characteristic becomes unstable in a sense that small fluctuations in wind velocity result in large changes in stiffness. Such behaviour is a distinctive characteristic of air-supported structures and differs from the rigid structure.

For every given internal pressure, there is a 'critical' wind velocity such that the larger the internal pressure, the larger this velocity. For the spherical cap under consideration, this relationship is shown in Fig. 4.

DISCUSSION OF RESULTS

The results presented are valid as long as the assumptions used are correct of which the interaction between wind and the structure is the least certain. Nonuniformity of wind flow, as opposed to the assumed 'uniform' flow, as well as change of shape of the structure, may be significant. However, we will discuss only the effects of change in shape of the structure on the pressure distribution. The deformed cross-sections of the cap are shown in Fig. 5 for two wind velocities and for an internal pressure of 206 Pa (20.6 mm water). As can be seen, even for the 'above-critical' wind speed (45 m/s) the deflected shapes are within 5% of the original geometry, thus displacements are obviously 'small'. But the main question was how such deflections may influence the critical wind speed, the speed above which the response of the structure becomes unstable due to a balancing of the positive wind pressure on the lower part of the structure by internal pressure. Thus, the upper part of the structure becomes more rigid due to a summation of

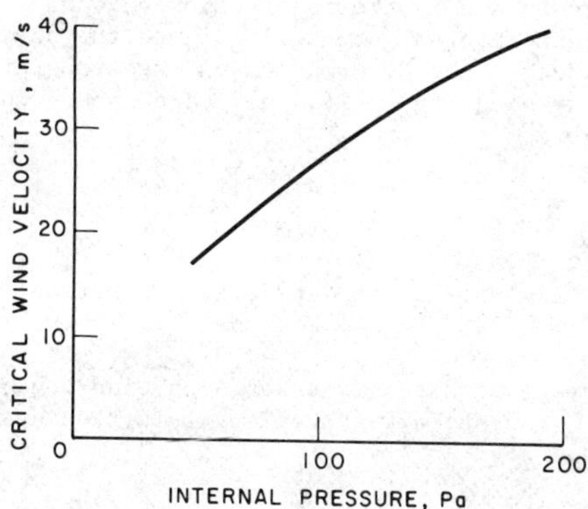

Fig. 4 Critical Wind Velocity vs. Internal Pressure

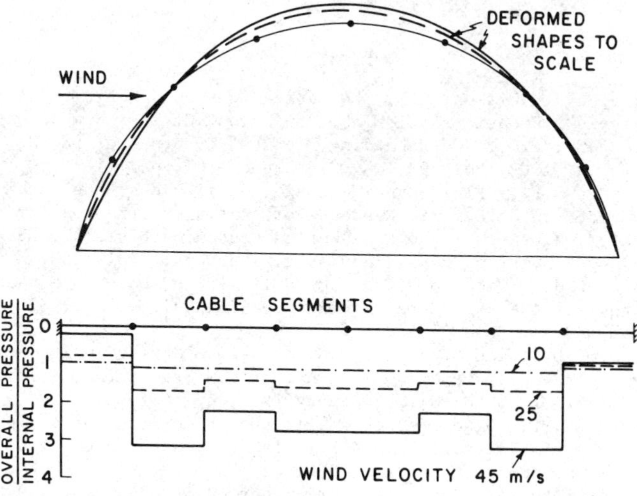

Fig. 5 Deformed Shapes and Pressure Distributions for Indicated Wind Velocities

wind and internal pressure, while the lower part becomes weaker due to a subtraction between wind pressure and internal pressure. This change in pressure distribution, for three wind velocities, is shown in Fig. 5. The instability occurs due to the interaction between alternating wind pressure and constant internal pressure. Since the positive wind pressure is increased due to the deformation of the structure, it follows that instability may occur actually under wind speeds which are lower than those calculated.

CONCLUSIONS

The numerical analysis of the inflatable spherical cap has shown that:

1. Its dynamic response strongly depends on wind conditions and when some critical wind speed is exceeded, this response becomes unstable;

2. Critical wind velocity increases almost proportionally with internal pressure.

The demonstrated sensitivity of the dynamic response of air-supported structures to wind load shows the need for further investigations in this field if inflatables are to be used under severe weather conditions.

REFERENCES

1. Bathe, K.J., Wilson, E.L., "Numerical Methods in Finite Element Analysis", Prentice-Hall, Inc., Englewood Cliffs, New Jersey, 1976.
2. Davenport, A.G., "Gust Loading Factors", Journal of the Structural Division, ASCE, Vol. 93, No. ST3, June 1967, pp. 11-34.
3. Hoerner, S.F., "Fluid-Dynamic Drag", published by the Author, 1965.
4. Leonard, J.W., "Dynamic Response of Initially-Stressed Membrane Shells", Journal of the Engineering Mechanics Division, ASCE, Vol. 95, No. EM5, Proc. Paper 6859, Oct. 1969, pp. 1231-1253.
5. Maher, F.J., "Wind Loads on Dome-Cylinder and Dome-Cone Shapes", Journal of the Structural Division, ASCE, Vol. 92, No. ST5, Proc. Paper 4933, October 1966, pp. 79-96.
6. Malcolm, D.J. and Glockner, P.G., "Optimum Cable Configuration for Air-Supported Structures", Journal of the Structural Division, ASCE, Vol. 105, No. ST2, February 1979, pp. 421-435.
7. Morris, N.F., "Wind Effects on Air-Supported Structures", presented at the April 25-29, 1977, ASCE Spring Convention and Exhibit, held at Dallas, Texas, (preprint 2860).
8. "State-of-the-Art Report on Air-Supported Structures", Task Committee on Air-Supported Structures of the Committee on Metals, of the Structural Division, ASCE, 1979.
9. Vinogradov, O.G., Malcolm, D.J., "Dynamic Response (Finite Element Approach) of Cable Reinforced Inflatables", presented at the April 14-18, 1980, ASCE Spring Convention and Exhibit held at Portland, Oregon, (preprint 80-101).

10. Vinogradov, O.G., Malcolm, D.J., Glockner, P.G., "Dynamic Response of Cable-Reinforced Inflatable Spherical Caps", Journal of the Structural Division, ASCE, in print.

DYNAMIC INTERACTION OF LIQUID STORAGE TANKS AND FOUNDATION SOIL

by

Medhat A. Haroun,[1] A.M. ASCE and George W. Housner,[2] M. ASCE

INTRODUCTION

Many studies have dealt with the dynamic interaction between different types of structures and the supporting soil during earthquakes. However, no attempt has been made so far to extend such analysis to the soil-tank system. A common approach in civil engineering practice is to consider the foundation soil to be rigid. The mechanical model derived by Housner [7] for rigid tanks can then be employed to estimate the maximum seismic response by means of a response spectrum.

As a natural extension of Housner's model, the effect of soil deformability on the seismic response of rigid tanks was investigated. A mechanical model was first derived to duplicate the lateral force and overturning moment exerted on the base of a rigid tank undergoing both translation and rotation. This model was then combined with another simplified model representing the flexibility of, and the damping in, the soil.

In recent years, much work involving the dynamic response of deformable containers has been made, but all of these studies have considered the foundation soil to be rigid. A complete analysis of the soil-tank system by the finite element method is both expensive and complicated. However, a simplified model of the soil can be employed with a finite element model of the shell to exhibit the fundamental dynamic characteristics of the overall system and to assess the significance of the interaction on the seismic response of deformable tanks. Since shell modes having more than one circumferential wave do not produce lateral force or moment, only the influence upon the $\cos\theta$-type modes should be investigated. Furthermore, rocking motion is most pronounced for "tall" tanks. The soil-tank interaction problem is therefore governed by a beam-type behavior rather than by a shell-type response. Consequently, the system has been modeled as a vertical cantilever beam supported by a spring-dashpot model.

Although these models represent a simplified version of the actual interaction problem, they provide insight into a complicated problem; and therefore, they are of a practical value.

(A). INTERACTION OF RIGID CYLINDRICAL TANKS AND SOIL

The formulation of the earthquake-response equations of the soil-

[1] Research Fellow, Civil Engineering Department, California Institute of Technology, Pasadena, California 91125.

[2] C F Braun Professor of Engineering, California Institute of Technology, Pasadena, California 91125.

tank system requires the evaluation of the lateral force and overturning moment at the base in terms of base motion. Therefore, it is advantageous to represent the liquid by an equivalent mechanical model capable of producing the same force and moment when subjected to the same base motion. Having obtained such a model, the equations of motion of the soil-tank system can be easily formulated. The foundation soil can be represented by a uniform elastic half space; however, because of the algebraic complexity associated with frequency-dependent impedance functions, the half space is further replaced by a discrete system of springs and dampers having constant values (frequency-independent).

(I). Liquid Oscillations in a Vibrating Rigid Cylindrical Tank

1. Statement of the Problem

Consider a rigid circular cylindrical tank of radius R, partly filled with a liquid of density ρ to an arbitrary depth H. The tank, which is initially at rest, is assumed to translate horizontally an arbitrary small displacement $x(t)$, and to rotate in a vertical plane an arbitrary small angle $\alpha(t)$ about a transverse axis through its base as shown in Fig.(1). Furthermore, the conventional assumptions regarding the liquid and its motion [1] are made.

The velocity potential function, $\phi(r,\theta,z,t)$, must satisfy the Laplace equation

$$\nabla^2\phi = 0 \quad ; \quad (0 \leqslant r \leqslant R \;,\; 0 \leqslant \theta \leqslant 2\pi \;,\; \text{and } 0 \leqslant z \leqslant H) \tag{1}$$

and the following boundary conditions:

(i) at the rigid tank wall

$$\frac{\partial\phi}{\partial r}(R,\theta,z,t) = \{\dot{x}(t) + z\dot{\alpha}(t)\}\cos(\theta) \tag{2}$$

(ii) at the rigid tank bottom

$$\frac{\partial\phi}{\partial z}(r,\theta,0,t) = -r\dot{\alpha}(t)\cos(\theta) \tag{3}$$

(iii) at the liquid free surface

$$\frac{\partial^2\phi}{\partial t^2}(r,\theta,H,t) + g\,\frac{\partial\phi}{\partial z}(r,\theta,H,t) = 0 \tag{4}$$

where g is the acceleration of gravity.

2. Transfer Functions of the Lateral Force and Overturning Moment

The analysis of liquid motion follows the method presented in [5] by employing Laplace transform to obtain the transfer functions of the lateral force and overturning moment exerted on the tank base. For the problem under consideration, it is convenient to write the velocity potential function as

$$\phi(r,\theta,z,t) = \phi_1(r,\theta,z,t) + \phi_2(r,\theta,z,t) \tag{5}$$

where ϕ_1 is the potential function associated with the motion of the wall which is given by

$$\phi_1(r,\theta,z,t) = \{\dot{x}(t) + z\dot{\alpha}(t)\}\, r\cos(\theta) \tag{6}$$

The velocity potential function $\phi_2(r,\theta,z,t)$ must therefore satisfy the homogeneous boundary condition

$$\frac{\partial\phi_2}{\partial r}(R,\theta,z,t) = 0 \tag{7}$$

and consequently, it can be written as

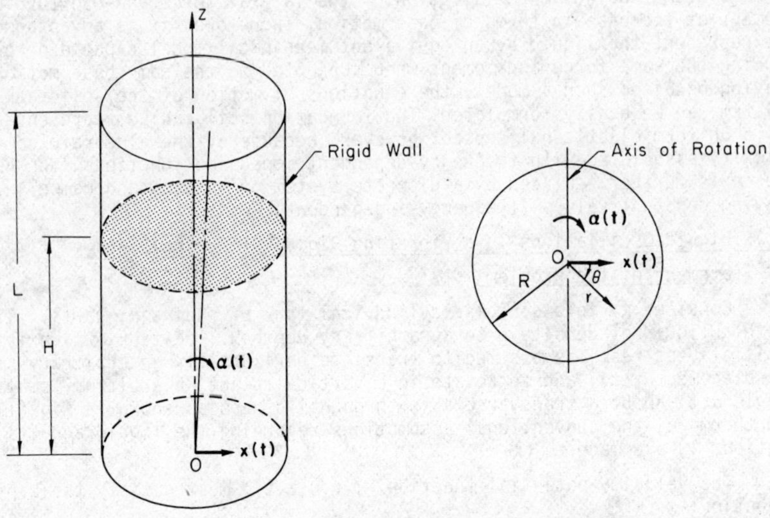

Fig.(1). Tank Geometry, Coordinate System and Degrees of Freedom.

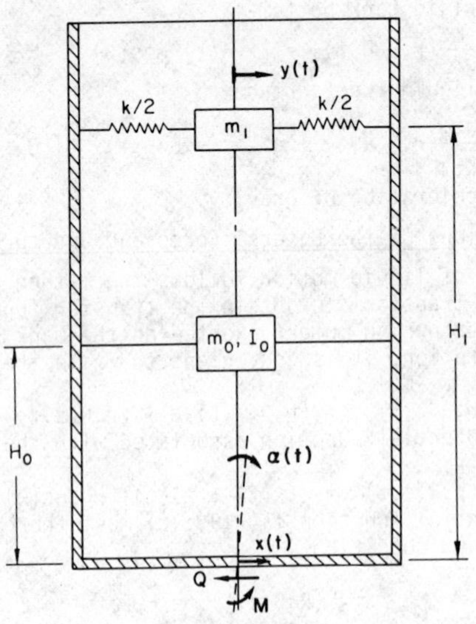

Fig.(2). Mechanical Analog.

$$\phi_2(r,\theta,z,t) = \left\{\sum_{i=1}^{\infty}\left[A_i(t)\cosh\left(\frac{\varepsilon_i z}{R}\right) + B_i(t)\sinh\left(\frac{\varepsilon_i(z-H)}{R}\right)\right]J_1\left(\frac{\varepsilon_i r}{R}\right)\right\}\cos(\theta) \tag{8}$$

where J_1 is Bessel function of the first kind of order 1; and ε_i are the zeros of $J'_1(\varepsilon_i) = 0$ (i=1,2,3,..). The arbitrary functions $A_i(t)$ and $B_i(t)$ can be determined by satisfying the remaining boundary conditions at z = 0 and z = H.

After the appropriate algebraic manipulations of the compatibility equation at the tank base and the use of the orthogonality relations of Bessel functions, the following expressions for $B_i(t)$ result

$$B_i(t) = -\frac{4R^2\dot{\alpha}(t)}{\varepsilon_i\left(\varepsilon_i^2-1\right)J_1(\varepsilon_i)\cosh\left(\frac{\varepsilon_i H}{R}\right)} \quad ; \quad i=1,2,... \tag{9}$$

To evaluate the functions $A_i(t)$, one can make use of Laplace transform. If $\bar{x}(s)$, $\bar{\alpha}(s)$, and $\bar{A}_i(s)$ denote the Laplace transform of $x(t)$, $\alpha(t)$, and $A_i(t)$, respectively, then the free surface boundary condition yields

$$\bar{A}_i(s) = \frac{-2Rs\left[s^2\bar{x} + H\bar{\alpha}(s^2+\lambda_i)\right]}{\left(s^2+\omega_i^2\right)\left(\varepsilon_i^2-1\right)J_1(\varepsilon_i)\cosh\left(\frac{\varepsilon_i H}{R}\right)} \quad ; \quad i=1,2,... \tag{10}$$

where λ_i are constants given by $\lambda_i = \frac{g}{H}\left[1 - 2/\cosh\left(\frac{\varepsilon_i H}{R}\right)\right]$ (11)
and

$$\omega_i^2 = \frac{g\varepsilon_i}{R}\tanh\left(\frac{\varepsilon_i H}{R}\right) \quad ; \tag{12}$$

ω_i being the natural frequencies of sloshing of a liquid of depth H contained in a stationary rigid cylindrical tank of radius R.

The hydrodynamic pressure is related to the velocity potential function by

$$p_d(r,\theta,z,t) = -\rho\frac{\partial\phi}{\partial t}(r,\theta,z,t) \tag{13}$$

and therefore,

$$\bar{p}_d(r,\theta,z,t) = -\rho s\bar{\phi}(r,\theta,z,s) \tag{14}$$

where $\bar{p}_d$ and $\bar{\phi}$ are the Laplace transform of p_d and ϕ, respectively.

By integrating the hydrodynamic pressure along the tank wall, the lateral dynamic force can be obtained

$$Q_d(t) = \int_0^H\int_0^{2\pi} p_d(R,\theta,z,t)\cos(\theta)R\,d\theta\,dz \tag{15}$$

Similarly, the moment exerted on the tank wall at its junction with the base can be computed from

$$M_d^*(t) = \int_0^H\int_0^{2\pi} p_d(R,\theta,z,t)\cos(\theta)\,z\,R\,d\theta\,dz \tag{16}$$

and therefore, the dynamic moment exerted on the base, including that which results from the pressure variation on the tank bottom, is given by

$$M_d(t) = M_d^*(t) + \int_0^R \int_0^{2\pi} p_d(r,\theta,z,t)\, r^2 \cos(\theta)\, d\theta\, dr \tag{17}$$

Equations (15) and (17) provide the lateral force and overturning moment due to tank acceleration; however, there is also an overturning moment directly proportional to "static" tank rotation. The "static" pressure distribution on the tank wall is given by

$$p_s(R,\theta,z,t) = \rho\, g\, .\, (H - z) + \rho\, g\, R\, \alpha(t) \cos(\theta) \tag{18}$$

and consequently,

$$Q_s(t) = 0 \quad ; \text{ and } \quad M_s(t) = mgH\left(\frac{1}{2} + \frac{R^2}{4H^2}\right)\alpha(t) \tag{19}$$

Upon using the Laplace transform function of the hydrodynamic pressure, the transfer functions of the lateral force and moment can be obtained. However, in an earthquake response analysis, it is usually satisfactory to retain only those terms due to the fundamental liquid sloshing mode and to neglect the effect of higher modes; i.e., to truncate the infinite series in Eq.(8) after the first term. The Laplace transform functions of the resultant force and moment ("static" plus dynamic) can now be written as:

$$\bar{Q}(s) = -s^2\bar{x}m - s^2\bar{\alpha}mH\left(\frac{1}{2} + \beta_1\right) + s^2 m\beta_2\left[\frac{s^2\bar{x} + H\bar{\alpha}\left(s^2 + \lambda_1\right)}{\left(s^2 + \omega_1^2\right)}\right] \tag{20}$$

and

$$\bar{M}(s) = -s^2\bar{x}mH\left(\frac{1}{2} + \frac{R^2}{4H^2}\right) - s^2\bar{\alpha}mH^2\left(\frac{1}{3} + \beta_3\right) + s^2 mH\beta_4\, .\left[\frac{s^2\bar{x} + H\bar{\alpha}\left(s^2 + \lambda_1\right)}{\left(s^2 + \omega_1^2\right)}\right] + mgH\left(\frac{1}{2} + \frac{R^2}{4H^2}\right)\bar{\alpha} \tag{21}$$

where

$$a = \left(\frac{\varepsilon_1 H}{R}\right) \; ; \quad \beta_1 = \frac{4}{a^2}\,\frac{[\cosh(a) - 1]}{\left(\varepsilon_1^2 - 1\right)\cosh(a)} \; ; \quad \beta_2 = \frac{2}{a}\,\frac{\tanh(a)}{\left(\varepsilon_1^2 - 1\right)} \; ;$$
$$\beta_3 = \frac{4[2\sinh(a) - a]}{a^3\left(\varepsilon_1^2 - 1\right)\cosh(a)} \; ; \quad \text{and} \quad \beta_4 = \frac{[2 - \cosh(a) + a\sinh(a)]}{a^2\left(\varepsilon_1^2 - 1\right)\cosh(a)} \tag{22}$$

3. Mechanical Model

Consider the mechanical model shown in Fig.(2). It consists of an oscillating mass m_1 located at a height H_1, together with a mass m_0, having a central moment of inertia I_0, rigidly attached to the container wall at a distance H_0 from the base. The model is assumed to perform small displacement and rotation, $x(t)$ and $\alpha(t)$, respectively.

The equations of motion of the system can be written as

$$Q = -m_0(\ddot{x} + H_0\ddot{\alpha}) - m_1(\ddot{x} + H_1\ddot{\alpha} + \ddot{y})$$
$$M = -m_0H_0(\ddot{x} + H_0\ddot{\alpha}) - m_1H_1(\ddot{x} + H_1\ddot{\alpha} + \ddot{y}) - I_0\ddot{\alpha} + \tag{23}$$

$$(m_0H_0 + m_1H_1)g\alpha + m_1gy \tag{23 cont.}$$

$$0 = m_1(\ddot{x} + H_1\ddot{\alpha} + \ddot{y}) + ky - m_1g\alpha$$

where $y(t)$ is the displacement of the oscillating mass m_1 relative to the wall of the tank.

Taking Laplace transform of all terms in Eq.(23) and eliminating $\bar{y}(s)$, one can obtain Laplace transform functions of the lateral force $\bar{Q}(s)$ and of the overturning moment $\bar{M}(s)$ in terms of the parameters of the mechanical model. A comparison between these functions and those given by Eqs. (20) and (21), respectively, yields ten equations relating the parameters of the mechanical system with the characteristics of the liquid and the dimensions of the tank. Six of these equations are independent and define the parameters m_0, H_0, m_1, H_1, k, and I_0(Table 1); and the remaining relations are identically satisfied. It should be noted that the mechanical model derived by Housner [7] considers only translational motion; and therefore, no value for I_0 was obtained.

(II). Formulation of the Soil-Tank Interaction Problem

Most seismic studies of liquid storage tanks consider the earthquake motion as a specified quantity at the tank support. In actual fact, however, the tank and the soil on which it is founded form a combined dynamic response mechanism. The soil-tank interaction problem can be broken down into three parts: the determination of the input motion to the foundation, the evaluation of the force-displacement relationship for the foundation, and the formulation of the equations of motion of the overall system.

1. Input Motion

For illustration purposes, the input ground motion (free field motion) is selected from available accelerograms. Two records are used in the analysis: (i) the N-S component of the 1940 El Centro earthquake, and (ii) the N-S component recorded at the Holiday Inn during the San Fernando earthquake in 1971.

2. Impedance Functions

Referring to the soil-tank system illustrated in Fig.(3), it is assumed that the tank is resting on a uniform elastic half-space which represents the foundation soil. The interaction force $Q(t)$ and moment $M(t)$ at the base can be expressed in terms of base translation $x(t)$ and rotation $\alpha(t)$ by complex frequency-dependent functions, the real part of which represents foundation stiffness, and the imaginary part damping [8]. However, it has been shown [6] that these functions can be reasonably approximated by constant values within the frequency range of interest. The constant parameter approximation as suggested in [3] takes the following form:

$$\begin{Bmatrix} Q(t) \\ M(t) \end{Bmatrix} = \begin{bmatrix} K_x & 0 \\ 0 & K_\alpha \end{bmatrix} \begin{Bmatrix} x(t) \\ \alpha(t) \end{Bmatrix} + \begin{bmatrix} C_x & 0 \\ 0 & C_\alpha \end{bmatrix} \begin{Bmatrix} \dot{x}(t) \\ \dot{\alpha}(t) \end{Bmatrix} \tag{24}$$

in which $K_x = \dfrac{32(1 - \nu_f)G_fR}{(7 - 8\nu_f)}$; $C_x = 0.576\, K_xR\sqrt{\rho_f/G_f}$;

$K_\alpha = \dfrac{8G_fR^3}{3(1 - \nu_f)}$; and $C_\alpha = \dfrac{0.3}{1 + [3(1 - \nu_f)I/(8\rho_fR^5)]} K_\alpha R\sqrt{\rho_f/G_f}$

Table (1)
Parameters of the Mechanical Model

m_0	$m(1 - \beta_2)$	ω_1^2	$\frac{1.84g}{R} \tanh\left(\frac{1.84H}{R}\right)$
H_0	$H\left(\frac{1}{2} + \frac{R^2}{4H^2} - \beta_4\right) \Big/ (1 - \beta_2)$	H_1	$H(\beta_4/\beta_2) - (g/\omega_1^2)$
m_1	$m\beta_2$	I_0	$mH^2(\frac{1}{3} + \beta_3 - \beta_4) - m_0H_0^2$
k	$m\beta_2\omega_1^2$	m is the total mass of the liquid	

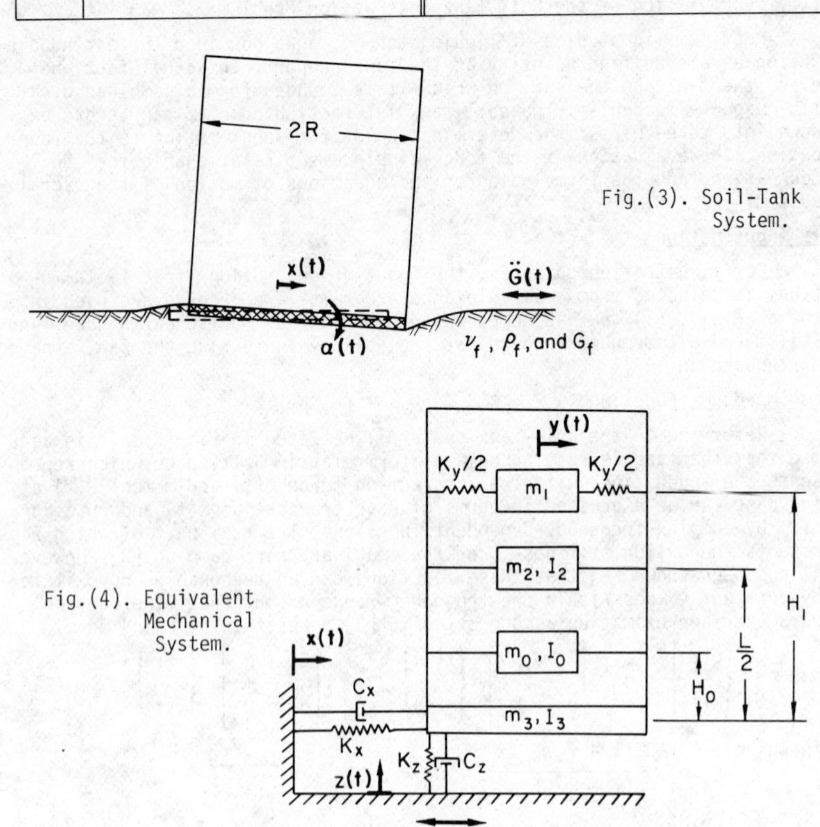

Fig.(3). Soil-Tank System.

Fig.(4). Equivalent Mechanical System.

where I is the total mass moment of inertia of the tank and the base about the rocking axis; and ν_f, ρ_f, and G_f are Poisson's ratio, mass density, and shear modulus of the foundation medium, respectively.

The most troublesome aspect of deriving a mechanical model for the foundation soil is defining the damping in a meaningful way. The assumption of elastic soil medium implies that no energy dissipating mechanism exists in the material itself; therefore, the dissipating terms of the impedance functions arise solely from the radiation of wave energy into the elastic half space. Some formulas have been suggested to account for the internal damping in the soil; these will be discussed in Sec.(III).

3. Equations of Motion of the Soil-Tank System

Consider the equivalent mechanical system shown in Fig.(4). The mechanical elements m_0, I_0, m_1, and K_y represent the contained liquid; m_2 and I_2 represent the rigid wall; m_3 and I_3 represent the tank base; and the springs K_x and K_z, and the dampers C_x and C_z represent the foundation soil. The motion of the system can be described by the base translation x(t) relative to the free field motion G(t); by the translation y(t) of the oscillating mass m_1 relative to the axis of the tank; and by the base rotation $\alpha(t)$.

Taking x(t), y(t), and $z(t)=R\alpha(t)$, as the generalized degrees of freedom, one can write the equations of motion of the damped soil-tank system in the following familiar form

$$[M]\{\ddot{q}\} + [C]\{\dot{q}\} + [K]\{q\} = -\ddot{G}(t)[M]\{r\} \tag{25}$$

where

$$\{q\} = \begin{Bmatrix} x(t) \\ y(t) \\ z(t) \end{Bmatrix} \quad ; \quad \text{and} \quad \{r\} = \begin{Bmatrix} 1 \\ 0 \\ 0 \end{Bmatrix}$$

(III). Earthquake Response of the Soil-Tank System

When analyzing the seismic response of the soil-tank model, one has to balance sophistication in the method of solution against the fact that there are many uncertainties in representing the actual foundation soil by a simple mechanical model. An exact solution for the mathematical model can be obtained by working in the frequency domain because the foundation impedances are frequency-dependent; however, by adopting the constant parameter approximation presented in Sec.(II-2), the earthquake response of the simplified model can be obtained by direct integration of the equations of motion and this has practical advantages.

In practice, the input motion is usually specified in the form of a response spectrum, and therefore, modal analysis becomes a direct and a logical procedure to be used. It should be mentioned, however, that modal analysis is not strictly applicable to a soil-structure system. The matrix equation cannot be rigorously solved by the normal mode method because the damping matrix [C] is not diagonalizable under the same transformation that diagonalizes both the mass and stiffness matrices. However, modal analysis can produce sufficiently good approximations from the engineering point of view provided that appropriate values of modal damping are used [6].

A digital computer program has been written to compute the seismic response of the soil-tank model. The program is first applied to compute the seismic response of a rigid tank founded on a rigid foundation, and the results are compared with those of Housner's model [7]. Such comparison helps in identifying the difference in response due to the difference in the values of the mechanical elements; the evaluation of that difference is important before any attempt can be made to interpret the effect of the foundation soil.

A rigid tank of a 30 ft (9.14 m) radius is considered, and the response is evaluated for different values of (H/R). It is assumed that the tank is subjected to the N-S component of the 1940 El Centro earthquake. The results of the calculations are presented in Fig.(5) in the form of a ratio of the maximum response of the present model to that of Housner's model. Three curves are given to illustrate the variation of the maximum base force, of the maximum overturning moment, and of the maximum wall moment with the height to radius ratio (H/R). It is clear that the two models give comparable values for the maximum base force Q and for the maximum overturning moment M, and that Housner's model underestimates the maximum wall moment M^*.

The program is then employed to investigate the influence of soil deformability on the response of the same tank. A direct integration technique is used to integrate the equations of motion. The analysis is carried out for two different values of (H/R) and for two different values of shear wave velocity. Inspection of Table (2) reveals that the deformability of the foundation soil can amplify the response considerably over that of a rigid tank on a rigid foundation. Such amplification arises mainly due to the rocking motion of the impulsive mass m_0. It should be noted, however, that large amplification occurs only for "tall" tanks and that "broad" tanks behave as if they are supported by rigid foundation, unless the foundation soil is very soft.

The influence of soil deformability on the dynamic response of the 30 ft radius tank is also evaluated using normal mode method. The response spectra of the 1940 El Centro earthquake and of the Holiday Inn record during the 1971 San Fernando earthquake are used as input. The analysis is carried out for 10 different values of shear wave velocity, and the results are displayed in Fig.(6). Inspection of this figure shows that the response of the soil-tank model to El Centro record is generally greater than its response to the Holiday Inn record. It is also seen that the influence of soil deformability on tank response to the Holiday Inn record becomes negligible for shear wave velocities higher than 2000 ft/sec. This is not true in the case of El Centro record which tends to amplify the response of the second mode of the system significantly, even for shear wave velocities higher than 2400 ft/sec. The validity of normal mode approximation is demonstrated in Table (2) where a comparison between the response obtained by modal analysis and that obtained by a direct integration technique is made.

It is important to note that damping in soil does not arise solely from radiation of wave energy into the half space; and therefore, one may consider additional damping to simulate the internal damping in the foundation. Two different approaches have been suggested [3,4] to evaluate such damping; these, when applied to the model under consideration, will result in an additional damping ratio 2 - 5% in the second mode. For illustration purposes, consider the case of H/R = 2.0, v_s = 1000 ft/sec, and assume that El Centro record is the input. The damping ratio (radiation only) is 14.2% which corresponds to a spectral acceleration of 0.524g. If internal damping of 3% is assumed, then the spectral acceleration becomes

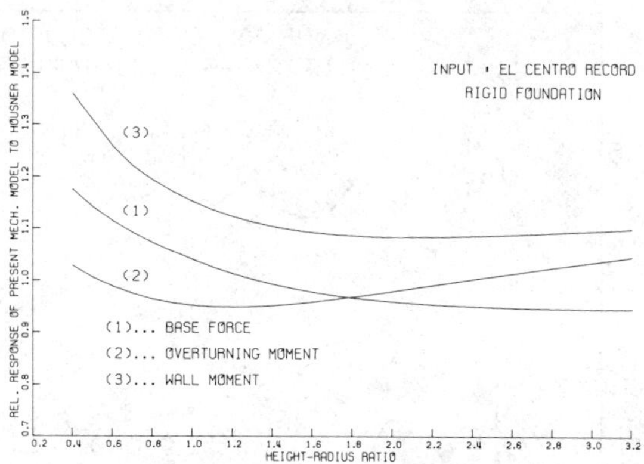

Fig.(5). Relative Response of Present Mechanical Model (on a rigid foundation) to the Response of Housner's Model.

0.495g; i.e., a reduction of 5.5%. However, this spectral value is still much higher than the maximum ground acceleration that would be used in case of a rigid foundation.

Based on the foregoing analysis, a set of design charts has been derived to provide practicing engineers with a direct method for estimating the maximum seismic response of rigid cylindrical tanks founded on a deformable foundation. The concept of equivalent simple oscillators has been used in the derivation of these curves which can be found in ref. [2].

Table (2)

Maximum Seismic Response of a 30 Feet Radius Tank*

Input : El Centro Record

	H/R	V_s (ft/sec)	Normalized Base Force (Q/mg)	Normalized Overturning Moment (M/mgH)	Normalized Wall Moment (M*/mgH)
Normal Mode Solution	2.0	∞ 1000 400	0.307 0.417 0.530	0.151 0.231 0.294	0.117 0.201 0.256
Direct Integration Solution	2.0	∞ 1000 400	0.310 0.412 0.511	0.153 0.222 0.276	0.119 0.192 0.239
	1.0	∞ 400	0.212 0.235	0.161 0.165	0.080 0.099

*Response on a rigid foundation was evaluated using Housner's model.

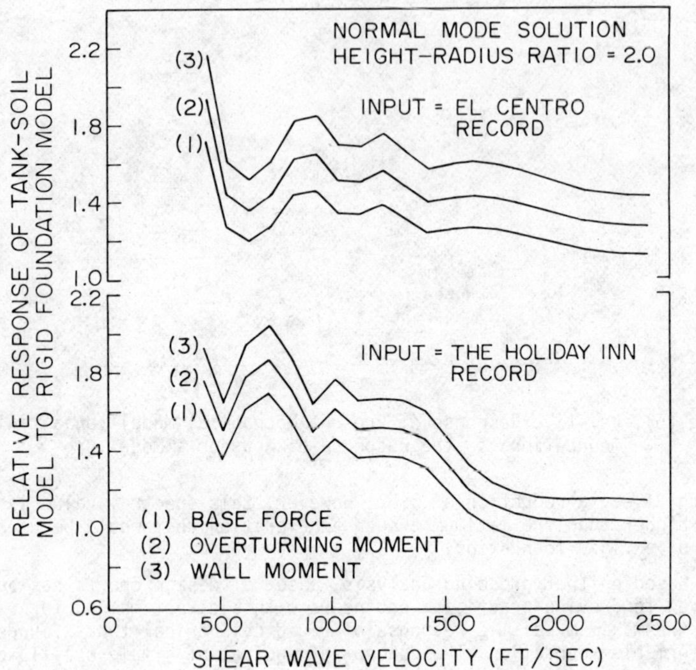

Fig.(6). Effect of Foundation Deformability on Seismic Response of Rigid Tanks

(B). INTERACTION OF FLEXIBLE CYLINDRICAL TANKS AND SOIL

The foregoing analysis is applicable only to tanks with rigid walls. In practice, many tanks are made of thin steel shells; and consequently, they are flexible. The vibrational motion of the shell involves both vertical and circumferential waves of deflection. However, soil interaction with shell modes having more than one circumferential wave can be neglected because these modes produce no lateral force or moment at the base. Furthermore, rocking motion is most pronounced for "tall" tanks. The soil-tank system is therefore modeled as a vertical cantilever beam supported by a spring-dashpot model.

(I). Hydrodynamic Pressure

In addition to the impulsive and convective pressures induced in a rigid cylindrical tank, there is also the short-period pressure component due to wall flexibility. In the following analysis, the effect of liquid sloshing is neglected because the interaction of sloshing modes with both shell deformation and foundation motion is weak. The remaining pressure components can be obtained from the appropriate expression of the velocity potential function which satisfies the Laplace equation and the following boundary conditions:

(i) at the deformable tank wall

Fig.(7). Hydrodynamic Forces.

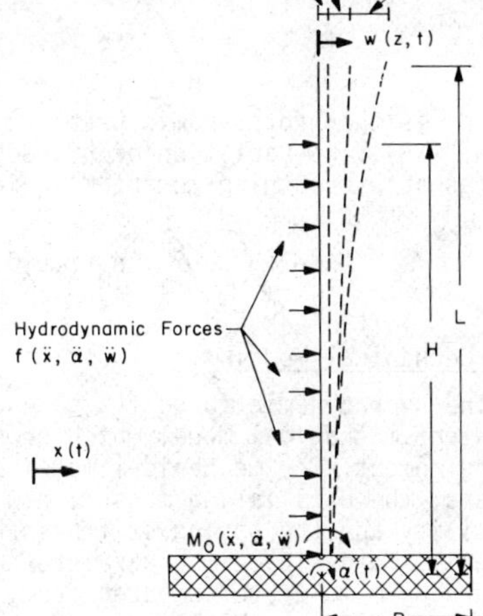

$$\frac{\partial \phi}{\partial r}(R,\theta,z,t) = \left\{ \dot{x}(t) + z\dot{\alpha}(t) + \dot{w}(z,t) \right\} \cos(\theta) \qquad (26)$$

where $w(z,t)$ is the deflection of the beam axis.

(ii) at the rigid tank bottom, the condition described by Eq.(3) must be satisfied.

(iii) at the liquid free surface

$$\frac{\partial \phi}{\partial t}(r,\theta,H,t) = 0 \qquad (27)$$

After the appropriate algebraic manipulations of Eqs. (3), (26), and (27), the following expression for $\phi(r,\theta,z,t)$ results

$$\phi(r,\theta,z,t) = \phi_1(r,\theta,z,t) + \phi_2(r,\theta,z,t) + \phi_3(r,\theta,z,t) \qquad (28)$$

where ϕ_1 is given by Eq.(6); and ϕ_2 takes the form given by Eq.(8) in which $A_i(t)$ are expressed as

$$A_i(t) = - \frac{2R(\dot{x} + H\dot{\alpha})}{\left(\varepsilon_i^2 - 1\right) J_1(\varepsilon_i) \cosh\left(\frac{\varepsilon_i H}{R}\right)} \; ; \quad i=1,2,... \qquad (29)$$

and $B_i(t)$ are given by Eq.(9). The velocity potential ϕ_3 is associated with wall deformation and can be expressed as

$$\phi_3(r,\theta,z,t) = \sum_{j=1}^{\infty} C_j \, I_1(\sigma_j r) \cos(\sigma_j z) \cos(\theta) \qquad (30)$$

where I_1 is the modified Bessel function of the first kind of order 1; $\sigma_j = (2j-1)\pi/2H$; and C_j are given by

$$C_j = \frac{2}{\sigma_j \, I_1'(\sigma_j R) \, H} \int_0^H \dot{w}(z,t) \cos(\sigma_j z) \, dz \qquad (31)$$

The lateral force exerted on the tank at any elevation z above the base can be computed from

$$f(z,t) = \int_0^{2\pi} p_d(R,\theta,z,t)\ R\cos(\theta)\ d\theta \tag{32}$$

where p_d is the hydrodynamic pressure which can be obtained from Eq.(13) and Eq.(28). The variation of hydrodynamic pressure acting on the base produces an overturning moment M_0 which can be calculated from

$$M_0 = \int_0^R \int_0^{2\pi} p_d(r,\theta,0,t)\ r^2 \cos(\theta)\ d\theta\ dr \tag{33}$$

(II). Mechanical Model

The hydrodynamic force f(z,t) and moment M_0 are first replaced by an equivalent mechanical model which produces the same base shear and overturning moment. The mechanical model shown in Fig.(2) is employed in the analysis; the oscillating mass m_1 now represents the influence of wall flexibility and the spring stiffness k represents the stiffness of the tank wall. To evaluate the parameters of the model, the base shear and overturning moment are computed from

$$Q(t) = \int_0^H f(z,t)\ dz \tag{34}$$

$$M(t) = \int_0^H f(z,t)\ z\ dz + M_0 \tag{35}$$

and these are compared with those obtained from the mechanical model which are given by

$$Q(t) = -\ m_0(\ddot{x} + H_0\ddot{\alpha}) - m_1(\ddot{x} + H_1\ddot{\alpha} + \ddot{y}) \tag{35}$$

$$M(t) = -\ m_0H_0(\ddot{x} + H_0\ddot{\alpha}) - m_1H_1(\ddot{x} + H_1\ddot{\alpha} + \ddot{y}) - I_0\ddot{\alpha} \tag{36}$$

Such a comparison yields five expressions for m_0, m_1, H_0, H_1, and I_0.

(III). Earthquake Response of the Soil-Flexible Tank System

Vibration tests on "tall" full-scale tanks [1] showed that the fundamental natural frequency of the wall is measurably lower than that computed assuming rigid foundation. In this section, the mechanical model derived in Sec.(II) is used with an assumed mode of tank deflection to investigate the effect of soil deformability on the natural frequency of completely filled tanks (L=H); a detailed analysis can be found in [2].

The analysis of liquid-shell interaction problem [1] revealed that the fundamental mode shape of moderately tall tanks can be approximated by a sine curve; therefore, one can assume that

$$\Psi(z) = \sin\left(\frac{\pi z}{2L}\right) \tag{37}$$

Taking y(t), x(t), and u(t)= $R\alpha(t)$ as the generalized degrees of freedom, one can write the matrix equation which governs the free vibration of the soil-flexible tank system as

$$[M]\{\ddot{q}\} + [K]\{q\} = \{0\} \quad ; \quad \{q(t)\} = \begin{Bmatrix} y(t) \\ x(t) \\ u(t) \end{Bmatrix} \tag{38}$$

Once the natural frequencies and the associated mode shapes are computed, the earthquake response of the soil-flexible tank system can be easily calculated [2].

(IV). Numerical Example

Consider a "tall" water storage tank, as shown in Fig.(8), which is anchored to a 2 feet thick R.C. slab on deep alluvium. The tank is 24 ft (7.32m) in radius, 71 ft (21.64m) in height, and consists of a thin steel shell of varying thickness; the maximum thickness at the bottom is 0.69 inch (1.75cm) and the minimum thickness at the top is 0.25 inch.

A series of ambient and forced vibration tests was conducted to determine the natural frequency of the tank [1]. Two seismometers were also placed at the ends of the principal diameter of the foundation slab to detect vertical motion, and thus to obtain a measurement of the amount of rocking at the base.

Figure (8) shows sample traces of the vertical motion at the foundation level during forced vibration tests. These show that the two seismometers (7) and (8) have the same amplitude and are 180 degrees out of phase. This rocking motion occurs at the fundamental natural frequency of the system which is clearly seen in the Fourier amplitude spectrum shown also in Fig.(8).

To correlate this result with the theoretical analysis, the natural frequency of the soil-tank system is computed for different values of

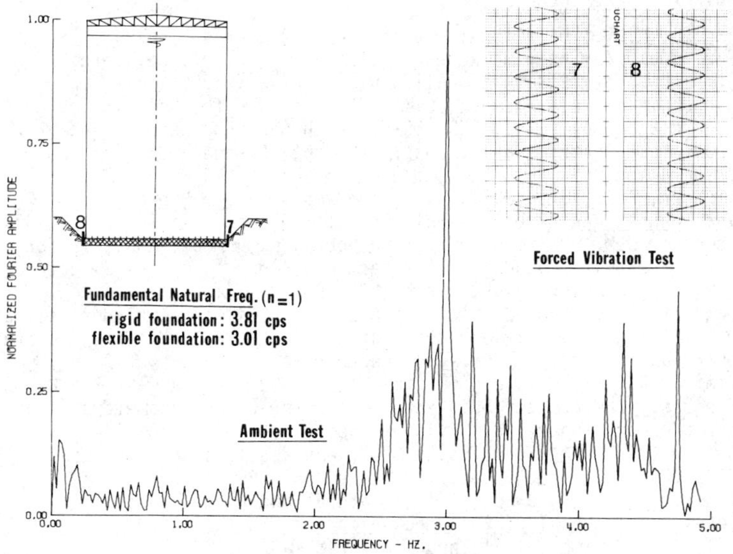

Fig.(8). Vibration Test of a Full-Scale Tank.

Table (3)

Fundamental Natural Frequency of the Soil-Tank System

V_S (ft/sec)	1000	1200	1400	1600	2000	2400	3600
f (cps)	2.79	3.03	3.21	3.34	3.49	3.59	3.72

shear wave velocity and the results are shown in Table (3). Inspection of this table indicates that the deformability of the foundation soil reduces the natural frequency of the tank, and as the shear wave velocity increases, the fundamental natural frequency approaches the value computed assuming rigid foundation. Once the parameters of the soil are specified, one can easily compute the seismic response of the soil-tank system. Due to space limitation, numerical examples of the earthquake response of the system will not be presented herein; these can be found in [2].

CONCLUSION

Two simplified methods are presented to compute the earthquake response of both soil-rigid tank system and soil-flexible tank system. For rigid tanks, soil deformability amplifies the response of "tall" tanks; however, "broad" tanks behave as if they are supported by a rigid foundation. The deformability of the foundation soil reduces the fundamental natural frequency of deformable containers; this was also observed in a vibration test of a full-scale tank.

APPENDIX (I) - REFERENCES

1. Haroun, M.A., "Dynamic Analyses of Liquid Storage Tanks," EERL 80-4, Earthquake Engineering Research Laboratory Report, California Inst. of Technology, Pasadena, California, February 1980.
2. Haroun, M.A., and Housner, G.W., "A Procedure for Seismic Design of Liquid Storage Tanks," EERL 80-5, Earthquake Engineering Research Laboratory Report, Calif. Inst. of Tech., Pasadena, Calif.(under preparation).
3. Richart, F.E., Jr., Hall, H.R., and Woods, R.D., Vibrations of Soils and Foundations, Prentice-Hall, Inc., Englewood Cliffs, N.J., 1970.
4. Roesset, J.M., Whitman, R.V., and Dobry, R., "Modal Analysis for Structures with Foundation Interaction," Journal of the Structural Division, ASCE, Vol. 99, March 1973, pp. 399-416.
5. Schmitt, A.F., "Forced Oscillations of a Fluid in a Cylindrical Tank Oscillating in a Carried Acceleration Field - A Correction," Convair Astronautics, Report ZU-7-074, February 1957.
6. Tsai, N.C., Niehoff, D., Swatta, M., and Hadjian, A.H., "The Use of Frequency-Independent Soil-Structure Interaction Parameters," Report of Power and Industrial Division, Bechtel Corporation, 1974.
7. U.S. Atomic Energy Commission, "Nuclear Reactors and Earthquakes," TID-7024, Washington, D.C., 1963, pp. 367-390.
8. Veletsos, A.S., and Wei, Y.T., "Lateral and Rocking Vibration of Footings," Journal of Soil Mechanics and Foundations Division, ASCE, Vol. 97, September 1971, pp. 1127-1248.

ENGINEERING APPROXIMATIONS FOR THE DYNAMIC BEHAVIOUR OF DEEPLY EMBEDDED FOUNDATIONS

A.P.S. Selvadurai[1], M. ASCE and M. Rahman[2]

ABSTRACT

The present paper provides certain engineering approximations for amplitude vs frequency response of foundations deeply embedded in isotropic elastic soil media. The approximate nature of the solutions stems from the introduction of certain simplifying assumptions. The contact tractions between the embedded foundation and the elastic medium are prescribed a priori and it is assumed that the depth of location of the foundation is such that the presence of a free ground surface has a negligible effect on the dynamic response of the foundation. Where relevant, the accuracy of the approximation is verified by appeal to available analytical results. Numerical solutions presented in the paper illustrate the correlation between the two solutions.

INTRODUCTION

The problem of the dynamic interaction between an embedded foundation and a soil medium is of considerable importance to the geotechnical design of deeply embedded footings, end-bearing screw piles and anchoring foundations which may be subjected to dynamic loads induced by earthquake and other wave phenomena (Fig. 1). In practice these isolated foundations are located at a finite depth below the ground surface. Also they possess finite relative rigidity characteristics consistent with their geometric and elastic properties. The analysis of a generalized three dimensional elastodynamic problem which takes into account not only the above factors but also the effects of complex constitutive relationships for soil media, the various interface conditions, arbitrary foundation geometries etc., is clearly a difficult task. To provide some insight into the problem of the dynamic behaviour of a foundation embedded within a soil medium it is instructive to consider a simplified idealization of the problem; namely, that of the dynamic response of a rigid flat foundation deeply embedded in an isotropic linear elastic solid. (The results for the surface foundation together with the results for the deeply embedded foundation would then provide two useful 'bounds' for the dynamic response of a foundation which is located at a finite depth.) In the conventional treatment of the above problem several simplifications are invoked to render the practical situation mathematically tractable. Firstly, the soil mass is idealized as an isotropic linear elastic medium. This excludes nonlinear and other dissipative phenomena which typify the mechanical response of real soils (Whitman [1]; Richart et al. [2]). Despite these limitations it is recognized that linear elasticity pro-

[1] Associate Professor, Dept. Civil Eng., Carleton University, Ottawa, Canada
[2] Research Associate, Dept. Civil Eng., Carleton University, Ottawa, Canada

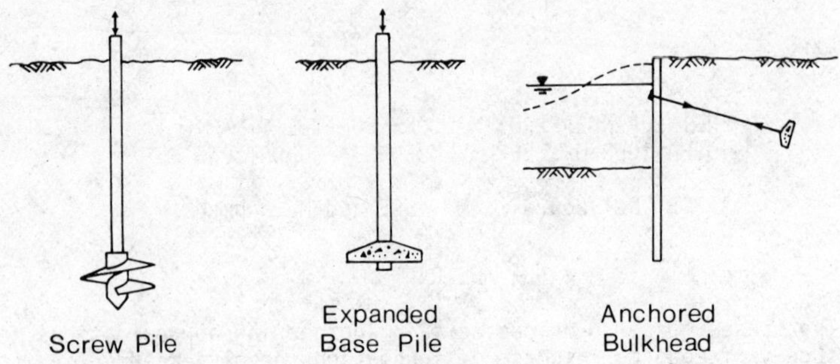

Fig. 1 Dynamic loading of deeply embedded foundations

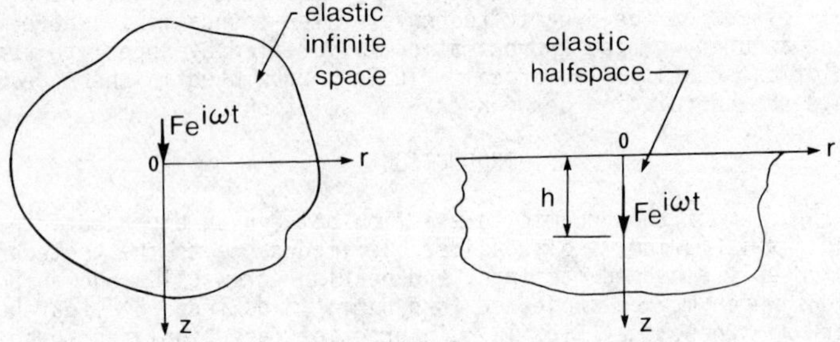

Fig. 2 Dynamic analogues of Kelvin and Mindlin problems

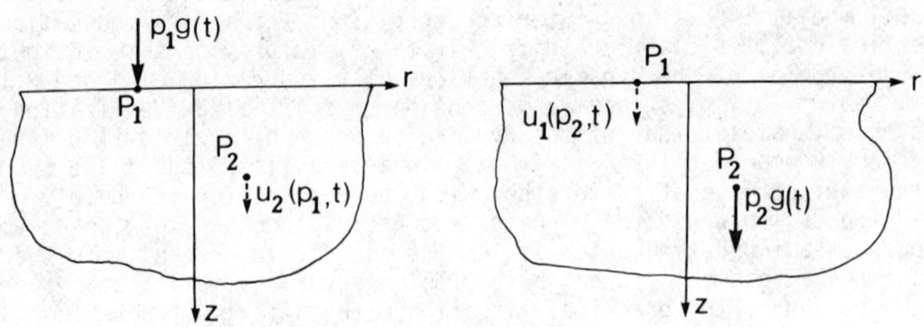

Fig. 3 Reciprocal states for the dynamic loading of a halfspace region

vides a useful first approximation for the modelling of the dynamic interaction problems associated with soil media. (For example, see Bycroft [3]; Awojobi and Grootenhuis [4]; Gladwell [5]; Luco and Westmann [6]; Novak and Sachs [7]; Hadjian et al. [8]; Wong and Trifunac [9]; Luco et al. [10]; Luco [11]; Selvadurai [12,13]. A comprehensive survey of these developments is given by Selvadurai [14].) Secondly, the soil region, which is finite in extent (for example a soil stratum), is represented by an elastic medium of infinite extent. Finally, it is assumed that the dimensions of the embedded foundation are such that the foundation region can be approximated by a flat disc-shaped rigid region (i.e. the flexible dynamic interaction between the foundation and the soil medium is assumed to be negligible). Despite these simplifications the analytical solution to the mixed boundary value problem associated with the dynamic behaviour of an embedded foundation can be attempted only in a limited number of situations. The dynamic interaction between an embedded disc inclusion and a normally incident plane harmonic compressional wave was examined by Mal et al. [15], Datta [16] and Mal [17]. Of related interest are the solutions for the steady oscillations of three dimensional objects such as spheres and ellipsoids embedded in elastic media developed by Chadwick and Trowbridge [18], Williams [19], Hill [20] and Datta and Kanwal [21]. The dynamic response of a rigid circular disc embedded in bonded contact with an isotropic elastic medium of infinite extent was examined by Selvadurai [12]. Kanwal [22] has presented an approximate solution to the embedded inclusion problem which is based on matched asymptotic expansion techniques. The class of problems which deals with foundations partially embedded in a halfspace region has been examined by Novak and Sachs [7], Luco [11] and others using both exact and approximate methods. In the above investigations either the foundation geometry or the mode of deformation is restricted to simplify the mathematical analysis.

The primary purpose of this paper is to investigate the possibility of developing certain approximate solutions to examine the dynamic response of foundations of arbitrary shapes deeply embedded in isotropic elastic soil media. To aid the development of such solutions it is necessary to introduce plausible assumptions concerning the distribution of contact stress at the soil - foundation interface. The a priori description of the contact stress distribution at the interface of the foundation in the analysis of dynamic soil - foundation interaction problems is not entirely novel; Reissner [23], Reissner and Sagoci [24], Arnold et al.[25], Bycroft [3], Warburton [26], Lysmer [27], Richart and Whitman [28], Hall [29] and others have used this type of approximation quite extensively in their analyses of the dynamic behaviour of rigid foundations resting on the surface of a halfspace region. In particular the contact stresses are approximated by the distributions in the associated static elasticity problem (Selvadurai [30]). (For example, in the axial vibration of a rigid circular foundation resting on an isotropic or transversely isotropic elastic halfspace the contact stress distribution is approximated by $\sigma_{zz}(r,0,t) = Pe^{i\omega t} / 2\pi a\sqrt{a^2-r^2}$, where a is the radius of the foundation.) In addition to the description of the contact stress distribution, it is assumed that the foundation is deeply embedded. According to this assumption the external boundaries have little influence on the dynamic response of the foundation. The analysis of the problem is approached by making use of the fundamental result developed by Eason et al. [31] for the harmonic loading of an elastic infinite space by a concentrated force. The justification for the use of the dynamic Kelvin source is invariably related to the depth of embeddment of the foundation. By considering the dynamic analogue of Mindlin's problem for a halfspace region (Selvadurai [14]) it can be shown that the effect of the free surface becomes negligible when the depth of location of the foundation exceeds approximately 10a, where a is the largest dimension of the embedded foundation. Furthermore, $a^2 = \mu/\gamma\omega^2$ (μ = linear elastic shear modulus; γ = mass density and ω = frequency of oscillation). The accuracy of this estimate is further supplemented by appeal to Graffi's [32] reciprocal theorem for dynamic systems and a dynamic analogue of St. Venant's principle. The solution for the periodic force embedded within an infinite space is used to generate approximate solutions for the amplitude frequency response of deeply embedded circular and rectangular foundations in which the contact stresses are prescribed a priori. The approximate solutions for the circular foundation are also compared with results derived for rigid foundations embedded in an elastic medium of infinite extent wherein the displacements within the foundation region are prescribed a priori.

DYNAMIC ANALOGUES OF KELVIN AND MINDLIN PROBLEMS

Kelvin's Problem: The axially symmetric problem related to the action of a periodic force of maximum magnitude F acting at the interior of an infinite space (Fig. 2) was examined by Eason et al. [31]. The components of the displacement vector $\underset{\sim}{u} = (u_r, 0, u_z)$ associated with the oscillatory load takes the form

$$u_r(r,z,t) = \frac{Fe^{i\omega t}a^2}{4\pi\mu k^2}\left(e^{-ikR/a}\left\{\frac{3rz}{R^5} + \frac{3ikrz}{aR^4} - \frac{k^2rz}{a^2R^3}\right\} + e^{-iksR/a}\left\{-\frac{3rz}{R^5} - \frac{3iksrz}{aR^4} + \frac{k^2s^2rz}{a^2R^3}\right\}\right) \tag{1}$$

$$u_z(r,z,t) = \frac{Fe^{i\omega t}a^2}{4\pi\mu k^2}\left(e^{-ikR/a}\left\{\frac{k^2}{a^2R} + \frac{(2z^2-r^2)}{R^5} + \frac{ik(2z^2-r^2)}{aR^4} - \frac{k^2z^2}{a^2R^3}\right\} + e^{-iksR/a}\left\{-\frac{(2z^2-r^2)}{R^5} - \frac{iks(2z^2-r^2)}{aR^4} + \frac{k^2s^2z^2}{a^2R^3}\right\}\right) \tag{2}$$

where

$$R^2 = r^2+z^2 \quad ; \quad k^2 = \frac{\omega^2a^2\gamma}{\mu} \quad ; \quad s^2 = \frac{\mu}{\lambda+2\mu} \tag{3}$$

λ,μ are the isentropic Lame constants, γ is the equilibrium density and a is a length parameter. The stress field associated with the harmonic Kelvin force can be evaluated by substituting (1) and (2) in the stress-strain relationships; i.e.

$$\{\sigma_{rr},\sigma_{\theta\theta},\sigma_{zz}\} = \lambda e + 2\mu\left\{\frac{\partial u_r}{\partial r}, \frac{u_r}{r}, \frac{\partial u_z}{\partial z}\right\}$$

$$\sigma_{rz} = \mu(\frac{\partial u_r}{\partial z} + \frac{\partial u_z}{\partial r}) \quad ; \quad e = \frac{\partial u_r}{\partial r} + \frac{u_r}{r} + \frac{\partial u_z}{\partial z} \tag{4}$$

Mindlin's Problem: We now turn our attention to the dynamical equivalent of Mindlin's problem. This concerns the evaluation of the stress and displacement fields associated with a harmonically oscillating force embedded at a finite distance from the free surface of an isotropic elastic halfspace. The formal procedures for the development of the solution to this problem are cited by Achenbach [33] and Ungar [34]. It may also be noted that the classical solutions of Cagniard [35], Pekeris [36,37], Pinney [38] and others for the response of an isotropic elastic halfspace due to a seismic pulse (of a Heaviside-type) located at the interior of the region can also be modified to cover the harmonic case. It is, however, relatively convenient to derive the solution to the dynamic Mindlin force problem by directly using the solution for the dynamic Kelvin force. Details of the procedure are given elsewhere (Selvadurai [14]). The result of primary importance to this paper concerns the surface settlement of the halfspace region due to the interior harmonic load. This result can be given in the integral form

$$u_z^m(r,0,t) = -\frac{k^2Fe^{i\omega t}}{2\pi a\mu}\left(\frac{h}{a}\right)\int_0^\infty \frac{\alpha\bar{f}(\xi)\xi J_0(\xi\rho)\,d\xi}{[(\beta^2+\xi^2)^2 - 4\xi^2\alpha\beta]} \tag{5}$$

where

$$\bar{f}(\xi) = \int_0^\infty \rho f(\rho)\, J_0(\xi\rho)\, d\rho \tag{6}$$

$$\begin{aligned} f(\rho) &= e^{-ik\Omega}\{\frac{\lambda}{\mu}(T_1+T_2+T_3) + 2T_3\} \\ &\quad + e^{-iks\Omega}\{\frac{\lambda}{\mu}(S_1+S_2+S_3) + 2S_3\} \end{aligned} \tag{7}$$

$$T_1 = \{-\frac{15\rho^2}{\Omega^7} + \frac{6k^2\rho^2}{\Omega^5} + \frac{3}{\Omega^5} - \frac{k^2}{\Omega^3}\} + i\{-\frac{15k\rho^2}{\Omega^6} + \frac{3k}{\Omega^4} + \frac{k^3\rho^2}{\Omega^4}\}$$

$$T_2 = \{\frac{3}{\Omega^5} - \frac{k^2}{\Omega^3}\} + i\{\frac{3k}{\Omega^4}\} \tag{8}$$

$$\begin{aligned} T_3 &= \{-\frac{5(2\zeta^2-\rho^2)}{\Omega^7} + \frac{3k^2\zeta^2}{\Omega^5} + \frac{4}{\Omega^5} + \frac{k^2(2\zeta^2-\rho^2)}{\Omega^5} - \frac{3k^2}{\Omega^3}\} \\ &\quad + i\{-\frac{5k(2\zeta^2-\rho^2)}{\Omega^6} + \frac{4k}{\Omega^4} + \frac{k^3\zeta^2}{\Omega^4} - \frac{k^3}{\Omega^2}\} \end{aligned}$$

$$S_1 = \{\frac{15\rho^2}{\Omega^7} - \frac{6k^2s^2\rho^2}{\Omega^5} - \frac{3}{\Omega^5} + \frac{k^2s^2}{\Omega^3}\} + i\{\frac{15k^2\rho^2}{\Omega^6} - \frac{3ks}{\Omega^4} - \frac{k^3s^3\rho^2}{\Omega^4}\}$$

$$S_2 = \{-\frac{3}{\Omega^5} + \frac{k^2s^2}{\Omega^3}\} + i\{-\frac{3k}{\Omega^4}\} \tag{9}$$

$$S_3 = \{\frac{5(2\zeta^2-\rho^2)}{\Omega^7} - \frac{3k^2s^2\zeta^2}{\Omega^5} - \frac{4}{\Omega^5} - \frac{k^2s^2(2\zeta^2-\rho^2)}{\Omega^5} + \frac{2k^2s^2}{\Omega^3}\}$$

$$+ i\{\frac{5k(2\zeta^2-\rho^2)}{\Omega^6} - \frac{4ks}{\Omega^4} - \frac{k^3s^3\zeta^2}{\Omega^4}\}$$

$$\Omega^2 = \zeta^2+\rho^2 \quad ; \quad \zeta = \frac{h}{a} \quad ; \quad \rho = r/a \tag{10}$$

$$\alpha = \begin{cases} (\xi^2-b^2)^{1/2} \ ; & \xi > b \\ -i(b^2-\xi^2)^{1/2} \ ; & 0 \le \xi \le b \end{cases} \tag{11}$$

$$\beta = \begin{cases} (\xi^2-k^2)^{1/2} & \xi > k \\ -i(k^2-\xi^2)^{1/2} & 0 \le \xi \le k \end{cases} \tag{12}$$

$$b^2 = \frac{\gamma\omega^2a^2}{(\lambda+2\mu)} \quad ; \quad k^2 = \frac{\gamma\omega^2a^2}{\mu} \quad ; \quad s^2 = \frac{1-2\nu}{2-2\nu} \tag{13}$$

Assuming that $u_z(0,0,t)$ can be represented in the form

$$u_z^m(0,0,t) = \bar{u}_m e^{i(\omega t+\varepsilon)} \tag{14}$$

where $\tan \varepsilon = m_2/m_1$, it can be shown that

$$\frac{\bar{u}_m}{(F/2\pi\mu a)} = k^2\frac{h}{a}\,[m_1^2+m_2^2]^{1/2} \tag{15}$$

The evaluation of the infinite integral in (5), which yields m_1 and m_2, is carried out by contour integration in the complex plane taking into account the appropriate branch cuts and Rayleigh poles (see, e.g., Eringen and Suhubi [39], Pilant [40]). Limitations of space precludes the inclusion of detailed results generated for the harmonic Mindlin force problem. It is sufficient to note that $\bar{u}_m/(F/2\pi\mu a)$ decreases as h/a increases.

Reciprocal Relationships: It is also worth noting that the result for the surface displacement of the halfspace due to the internally placed dynamic force can be inferred by appeal to the dynamic reciprocal theorem (see,e.g., Lamb [41], Rayleigh [42], Graffi [32], diMaggio and Bleich [43], Payton [44] and Gurtin [45]). Consider a concentrated force $p_1 g(t)$ applied at a point P_1 in an isotropic elastic body which is initially unstressed and at rest (Fig.3). This force produces a displacement $u_2(P_1,t)$ at the location P_2. Similarly, a concentrated force $p_2 g(t)$ applied at the point P_2 produces a displacement $u_1(P_2,t)$ at P_1. From the dynamic reciprocal theorem we have

$$p_1 u_2(p_1,t) = p_2 u_1(p_2,t) \tag{16}$$

and it is evident that under the same restrictions on forces and initial conditions, a relationship analogous to the Betti-Maxwell reciprocal theorem is valid. From this discussion we note that the surface displacement of the isotropic elastic halfspace region due to the embedded harmonic load can be directly inferred from the result for Lamb's [46] problem for a concentrated surface harmonic load. The integral expression for the displacement u_z^ℓ associated with the harmonic surface loading of a halfspace by a point source $Fe^{i\omega t}$ is given by

$$u_z^\ell(r,z,t) = \frac{Fe^{i\omega t}}{2\pi\mu a}\int_0^\infty \frac{\alpha\xi[2\xi^2 e^{-\beta z/a} - (\beta^2+\xi^2)e^{-\alpha z/a}]}{[(\beta^2+\xi^2)^2 - 4\alpha\beta\xi^2]} J_0(\xi\rho)d\xi \tag{17}$$

It is assumed that the displacement at the location (0,h) can be written as

$$u_z^\ell(0,h,t) = \bar{u}_\ell e^{i(\omega t+\chi)} \tag{18}$$

where $\tan\chi = \ell_2/\ell_1$ and ℓ_1 and ℓ_2 are determined from the integral expression

$$(\ell_1+i\ell_2) = \int_0^\infty \frac{\alpha\xi[2\xi^2 e^{-\beta h/a} - (\beta^2+\xi^2)e^{-\alpha h/a}]}{[(\beta^2+\xi^2)^2 - 4\alpha\beta\xi^2]} d\xi \tag{19}$$

Making use of (17) to (19) we have

$$\frac{\bar{u}_\ell}{F/2\pi\mu a} = [\ell_1^2+\ell_2^2]^{1/2} \tag{20}$$

Also as a consequence of the dynamic reciprocal theorem $\bar{u}_m = \bar{u}_\ell$, provided $h > 0$. The result (19) is, however, of a much simpler form than the auxiliary solution (15). Again, an evaluation of (19) shows that as h/a increases the displacement $u_z(0,h)$ decreases.

From these investigations it becomes evident that, provided the depth of location of the internal concentrated force exceeds approximately 10a (where a can be interpreted as the largest dimension of the embedded foundation) and provided $a^2 < \mu/\rho\omega^2$, the response of the halfspace due to the harmonic internal concentrated force can be approximated by the solution developed for the harmonic concentrated loading of an isotropic elastic infinite space. In this paper the latter solution is used to develop approximate solutions for the dynamic response of foundations deeply embedded in isotropic elastic soil media. In the problems examined here the foundation regions are of finite dimensions. It is therefore necessary to verify whether the approximate limits ($h > a$) concerning the depth of location of the concentrated harmonic load extend to other forms of distributed loadings. The most straightforward method would be to perform integrations of the dynamic Kelvin and Mindlin forces over the regions corresponding to the foundation areas, and the associated distributions, and examine the resulting correlations. The integrations involved in this procedure are somewhat cumbersome. Alternatively, a dynamic St. Venant principle could be invoked to reaffirm the conjecture that the approximate limits established for the depth of location of the harmonic concentrated force will also be applicable for the harmonic distributed loadings. As has been shown by Boley [47] and Gurtin [45], the classical St. Venant hypothesis in static elasticity can also be extended to the dynamic case provided the displacement and stress fields for the dynamic case can be expressed in the forms

$$u_i(x_i,t) = f(t)\, u_i^s(x_i) \quad ; \quad \sigma_{ij}(x_i,t) = f(t)\, \sigma_{ij}^s(x_i) \tag{21}$$

where the superscript s refers to the static solution. It is therefore to be expected that for low frequency oscillations (e.g. $k < 1$) the limits on h/a will extend to distributed internal loadings. For high frequency oscillations the limits of applicability of the dynamic Mindlin or Kelvin solution can be assessed only by recourse to the evaluation of the appropriate integral expressions.

APPROXIMATE SOLUTIONS FOR THE DYNAMIC BEHAVIOUR OF EMBEDDED FOUNDATIONS

In the preceding section we have established certain limits for the depth of location of a structural foundation, sufficient to ensure the applicability of the solution developed for a harmonic Kelvin force acting at the interior of an infinite space for the examination of dynamic problems associated with a halfspace region. We shall now utilize the harmonic Kelvin force solution to develop approximate solutions for the dynamic response of disc-shaped foundations of finite dimensions embedded in elastic media. The presence of the foundation is accounted for by assigning a priori typical variations for the normal contact stress distributions at the elastic medium - disc foundation interfaces. Since the foundation is assumed to occupy a disc-shaped region, these contact stresses are prescribed on the plane region $z = 0$. It should be noted that by virtue of the asymmetry associated with the internal loading of an infinite space region, the radial displacement u_r is zero on the plane $z = 0$ for all $t \geq 0$. The physical consequence of this result is that the foundation (semi-flexible) exhibits a bonded contact with the surrounding elastic medium. The specific forms of the normal contact stress distributions include the following:

(i) A uniform contact stress distribution over a circular region of radius a_0; i.e.

$$\sigma_{zz}(r,0,t) = \frac{Pe^{i\omega t}}{\pi a_0^2} \quad ; \quad 0 \leq r \leq a_0 \tag{22}$$

(ii) A parabolic contact stress distribution over a circular region of radius a_0; i.e.

$$\sigma_{zz}(r,0,t) = \frac{Pe^{i\omega t}}{2\pi a_0[a_0^2-r^2]^{1/2}} \quad ; \quad 0 \leq r \leq a_0 \tag{23}$$

(iii) A uniform contact stress distribution over a rectangular region of dimensions $2a_0 \times 2\phi a_0$; i.e.

$$\sigma_{zz}(x,y,0,t) = \frac{Pe^{i\omega t}}{4a_0^2\phi} \quad ; \quad |x| \leq a_0 \quad ; \quad |y| \leq \phi a_0 \tag{24}$$

(iv) A parabolic contact stress distribution over a rectangular region of dimensions $2a_0 \times 2\phi a_0$; i.e.

$$\sigma_{zz}(x,y,0,t) = \frac{Pe^{i\omega t}}{\pi^2[(a_0^2-x^2)(\phi^2 a_0^2-y^2)]^{1/2}} \quad ; \quad |x| \leq a_0 \; ; \; |y| \leq \phi a_0 \tag{25}$$

From (22) to (25) we note that the resultant contact force acting on the foundation area is $Pe^{i\omega t} = P(t)$. The dynamic response of the foundation is related to the weighted average of the displacements $u_z(r,0,t)$ in the foundation region. Also we assume that the weighted average of the displacements in the foundation region can be expressed in the form

$$\frac{\iint u_z(r,0,t)\, dA}{\iint dA} = w_a(t) = w_0 e^{i\omega t} \tag{26}$$

By making use of the Kelvin solution for $u_z(r,z,t)$ given by (2) and the contact stress distributions given by (22) to (25) it can be shown that the amplitude of the average displacement w_0 is related to the total applied force P(t) by the expression

$$P(t) = 4\pi\mu a_0 w_0 [g_1^2+g_2^2]^{1/2}\, e^{i(\omega t+\Psi)} \tag{27}$$

where $\tan\Psi = g_2/g_1$. The functions g_1 and g_2 are obtained from the following integral relationships:

For the circular foundation region:

$$(g_1+ig_2) = \left(\frac{2}{\pi k^2} \int_0^1 \rho \left\{ \int_0^{2\pi} \int_0^1 \sigma_i(\eta)\, [e^{-ik\zeta} \{\frac{k^2}{\zeta} - \frac{1}{\zeta^3} - \frac{ik}{\zeta^2}\} + e^{-iks\zeta} \{\frac{1}{\zeta^3} + \frac{iks}{\zeta^2}\}]\eta \, d\eta \, d\theta \right\} d\rho \right)^{-1} \tag{28}$$

where

$$\sigma_i(\eta) = 1 \quad ; \quad \text{uniform load}$$
$$\sigma_i(\eta) = \frac{1}{2[1-\eta^2]^{1/2}} \quad ; \quad \text{parabolic load} \tag{29}$$

$$\zeta^2 = \rho^2 + \eta^2 - 2\rho\eta\cos\theta$$

For the rectangular foundation region:

$$(g_1+ig_2) = \left(\frac{1}{\phi^2 k^2} \int_0^{X=1}\int_0^{Y=\phi} \left\{ \int_0^{\phi}\int_0^{1} \sigma_i(\xi,\eta) \{e^{-ik\delta}[\frac{k^2}{\delta} - \frac{1}{\delta^3} - \frac{ik}{\delta^2}] + e^{-iks\delta}[\frac{1}{\delta^3} + \frac{iks}{\delta^2}]\} \, d\xi \, d\eta \right\} dX \, dY \right)^{-1} \tag{30}$$

where

$$\sigma_i(\xi,\eta) = 1 \quad ; \quad \text{uniform load}$$
$$\sigma_i(\xi,\eta) = \frac{4\phi}{\pi^2[(1-\xi^2)(\phi^2-\eta^2)]^{1/2}} \quad ; \quad \text{parabolic load} \tag{31}$$

$$\delta^2 = (X-\xi)^2 + (Y-\eta)^2$$

MASS EFFECTS IN THE DYNAMIC BEHAVIOUR OF THE EMBEDDED FOUNDATION

The result (27) is essentially a relationship for the dynamic behaviour of an embedded massless foundation which is subjected to a harmonic load. The formal analysis can now be extended to include the effects of self weight of the deeply embedded foundation. We consider the problem where a foundation of mass M is subjected to a periodic force Q(t). The resulting weighted average of the displacements in the region corresponding to the foundation is denoted by $w_0 e^{i(\omega t+\phi)}$. The relationship between w_0 and the maximum amplitude Q* of the force Q(t) required to maintain the steady oscillation is given by

$$\frac{w_0 \mu a_0}{|Q^*|} = \frac{1}{4\pi\left[\{g_1 - \frac{k^2\Delta}{\pi(1-\nu)}\}^2 + g_2^2\right]^{1/2}} \tag{32}$$

where

$$\Delta = \frac{M(1-\nu)}{4\gamma a^3} = \text{mass ratio} \tag{33}$$

NUMERICAL RESULTS

The approximate solutions developed for the non-dimensional amplitude vs frequency response of embedded circular or rectangular foundations of mass M were numerically evaluated for a range of parameters of engineering interest. The evaluation of the integral expressions (28) and (30) is carried out by a repeated use of Gaussian quadrature techniques. The computer programme is so arranged that one can use 2,6,10, 20 or 40 point Gaussian nodes to check for convergence. A 20 point scheme was found to be satisfactory for the present purposes. The amplitude at $k = 0$ involves an arithmetic overflow; as such the limiting static value was ascertained by assigning $k = 0.0001$. In the case of the rectangular parabolic load $\sigma(\xi,\eta)$ becomes divergent when $\xi = X$ and $\eta = Y$. To carry out the numerical integration the ξ integration is carried out from 0 to 1 and the X integration is performed from ε to 1 (where $\varepsilon \ll 1$). The convergence of the integral is achieved with ε of the order of 0.001.

The Figures 4 and 5 show the amplitude vs frequency response of a deeply embedded circular foundation. The effects of both uniform and parabolic normal contact stress distributions are shown for purposes of comparison. Similar results obtained for the deeply embedded rectangular foundation are shown in Figures 6 and 7. It is instructive to assess the accuracy of the approximate solutions developed here by comparing these approximate results with numerical results derived from an analytical solution of a mixed boundary value problem. Such an analytical solution can of course be contemplated only in the case of the deeply embedded rigid circular foundation. Details of the analytical solution will not be pursued here (see, e.g., [12], [48]); the result equivalent to (32) takes the form

$$\frac{w_0 \mu a_0}{|Q^*|} = \frac{(3-4\nu)}{32(1-\nu)\left[\left\{p_1 - \frac{k^2\Delta(3-4\nu)}{8(1-\nu)^2}\right\}^2 + p_2^2\right]^{1/2}} \tag{34}$$

where

$$\begin{aligned} p_1 = 1 &+ k^2\left\{\frac{2I_2}{3\pi} - \frac{I_1^2}{\pi^2}\right\} + k^4\left\{-\frac{2I_4}{15\pi} + \frac{2I_1I_3}{3\pi^2} + \frac{7I_2^2}{15\pi^2} - \frac{2I_1^2I_2}{\pi^3} + \frac{I_1^4}{\pi^4}\right\} \\ &+ k^6\left\{\frac{4I_6}{315\pi} - \frac{4I_1I_5}{45\pi^2} - \frac{64I_2I_3}{315\pi^2} + \frac{2I_1^2I_4}{5\pi^3} - \frac{2I_3^2}{15\pi^2} + \frac{64I_1I_2I_3}{45\pi^3}\right. \\ &\left. - \frac{4I_1^3I_3}{3\pi^4} + \frac{233I_2^2}{315\pi^3} - \frac{123I_2^2I_1^2}{45\pi^4} + \frac{7I_1^4I_2}{2\pi^5} - \frac{I_1^6}{\pi^6}\right\} + O(k^8) \end{aligned} \tag{35}$$

$$\begin{aligned} p_2 = &\frac{kI_1}{\pi} + k^3\left\{-\frac{I_3}{3\pi} + \frac{4I_1I_2}{3\pi^2} - \frac{I_1^3}{\pi^3}\right\} + \\ &+ k^5\left\{\frac{2I_5}{45\pi} - \frac{4I_1I_2}{15\pi^2} - \frac{22I_2I_3}{45\pi^2} + \frac{I_1^2I_3}{\pi^3} + \frac{62I_1I_2^2}{45\pi^3} - \frac{8I_1^3I_2}{3\pi^4} + \frac{I_1^5}{\pi^5}\right\} + O(k^7) \end{aligned} \tag{36}$$

and

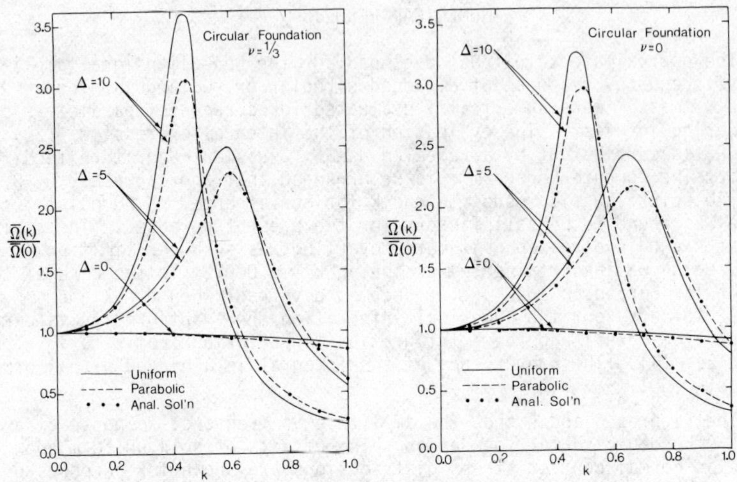

Figs. 4-5. Amplitude vs frequency response of embedded circular foundation [$\bar{\Omega}(k) = w_0\mu a_0/|Q^*|$; defined by (32)]
$\nu=0$; $\bar{\Omega}(0)=0.1027$(uniform); $\bar{\Omega}(0)=0.09213$(parabolic); $\bar{\Omega}(0)=0.09375$(analytical)
$\nu=1/3$; $\bar{\Omega}(0)=0.0856$(uniform); $\bar{\Omega}(0)=0.0768$(parabolic); $\bar{\Omega}(0)=0.07813$(analytical)

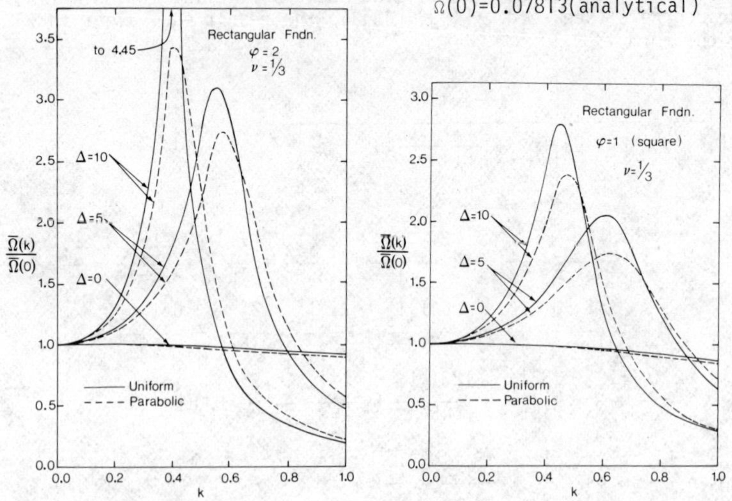

Figs. 6-7. Amplitude vs frequency response of embedded rectangular foundation [$\bar{\Omega}(k) = w_0\mu a_0/|Q^*|$; defined by (32)]
$\phi=1$; $\bar{\Omega}(0)=0.07353$(uniform); $\bar{\Omega}(0)=0.06294$(parabolic)
$\phi=2$; $\bar{\Omega}(0)=0.10095$(uniform); $\bar{\Omega}(0)=0.08649$(parabolic)

$$I_n = \frac{8(1-\nu)}{(3-4\nu)} \{(1-s^{n+2})\, L_{n+2} + s^{n+2} L_n\}$$

$$L_n = \frac{(n-1)}{n} \int_0^{\pi/2} \sin^{(n-2)}(\Omega)\, d\Omega \quad ; \quad n > 0 \qquad (37)$$

$$L_0 = \pi/2 \quad ; \quad L_1 = 1.$$

The numerical values of $w_0 \mu a_0/|Q^*|$ derived from (34) are shown in Figures 4 and 5.

CONCLUSIONS

This paper presents certain approximate techniques that may be used in the assessment of the dynamic response of deeply embedded foundations. The contact stress distribution at the interface is prescribed a priori. The numerical results developed for the circular foundation compare favourably with equivalent results derived for the square foundation. Furthermore, the analytical result for the low frequency approximation agrees, almost exactly, with the numerical result for parabolic loading. It may be noted that in the case of the parabolic approximation the fundamental solution is valid for all frequencies whereas the boundary conditions (for rigid foundation behaviour) are satisfied only approximately. In the case of the analytical solution the boundary conditions are satisfied exactly, whereas the frequency range is limited. From the degree of correlation which exists between the two solutions it is evident that at low frequencies (e.g. $k < 0.3$) the dynamic behaviour of the embedded foundation can be conveniently assessed by making use of the dynamic equivalent of the static contact stress distribution. This conclusion is strictly applicable to only circular or square-shaped rigid foundations. Whether such a correlation exists for rectangular foundations with arbitrary aspect ratios merits further investigation.

REFERENCES

[1] R.V. WHITMAN, Analysis of foundation vibrations. In: Vibrations in Civil Engineering (B.O. Skipp, Ed.), Butterworth. pp.159-179 (1966)

[2] F.E. RICHART, Jr., J.R. HALL, Jr. and R.D. WOODS, Vibrations of Soils and Foundations, Prentice-Hall, N.J. (1970)

[3] G.N. BYCROFT, Forced vibrations of a rigid circular plate on a semi-infinite elastic halfspace and on an elastic stratum. Phil. Trans. Roy. Soc. Ser.A, 248, 327-368 (1956)

[4] A.O. AWOJOBI and P. GROOTENHUIS, Vibration of rigid bodies on semi-infinite elastic media, Proc. Roy. Soc. Ser. A, 287, 27-63 (1965)

[5] G.M.L. GLADWELL, Forced tangential and rotary vibration of a rigid circular disc on a semi-infinite solid, Int. J. Eng. Sci., 6, 591-607 (1968)

[6] J.E. LUCO and R.A. WESTMANN, Dynamic response of circular footings, J. Eng. Mech. Div., Proc. ASCE, 97, 1381-1395 (1971)

[7] M. NOVAK and K. SACHS, Torsional and coupled vibrations of embedded footings, Earthquake Engineering and Structural Dynamics, 2, 11-33 (1973)

[8] A.H. HADJIAN, J.E. LUCO and N.C. TSAI, Soil structure interaction:

Continuum or finite element, Nuclear Engineering and Design, 31, 151-167 (1974)

[9] H.L. WONG and M.D. TRIFUNAC, Two dimensional antiplane, building-soil-building interaction for two or more buildings and for incident plane SH waves, Bull. Seis. Soc. Amer., 65, 1863-1885 (1975)

[10] J.E. LUCO, H.L. WONG and M.D. TRIFUNAC, A note on the dynamic response of rigid embedded foundations, Earthquake Engineering and Structural Dynamics, 4, 119-127 (1975)

[11] J.E. LUCO, Torsional response of structures for SH waves: The case of hemispherical foundations, Bull.Seis.Soc.Amer.,66,109-123 (1976)

[12] A.P.S. SELVADURAI, The dynamic response of a rigid circular foundation embedded in an isotropic medium of infinite extent, Proc.Int. Symp. of Soils Under Cyclic and Transient Loading (G.N.Pande and O.C. Zienkiewicz,Eds.), Balkema Press, 2, 597-608 (1980)

[13] A.P.S. SELVADURAI, Rotary oscillations of a rigid disc inclusion embedded in an isotropic elastic infinite space, Int.J. Solids and Structures (in press)

[14] A.P.S. SELVADURAI, Elastodynamic problems in soil-foundation interaction (unpublished work)

[15] A.K. MAL, D.D. ANG and L. KNOPOFF, Diffraction of elastic waves by a rigid circular disc, Proc.Camb.Phil.Soc.,64, 237-247 (1968)

[16] S.K. DATTA, The diffraction of a plane compressional elastic wave by a rigid circular disc, Quart. Appl. Math., 28, 1-14 (1970)

[17] A.K. MAL, Motion of a rigid disc in an elastic solid, Bull. Seis. Soc. Amer., 61, 1717-1729 (1971)

[18] P. CHADWICK and E.A. TROWBRIDGE, Oscillations of a rigid sphere embedded in an infinite elastic solid. II. Rectilinear oscillations. Proc. Camb. Phil. Soc., 63, 1207-1227 (1965)

[19] W.E. WILLIAMS, A note on the slow vibrations in an elastic medium, Quart. J. Mech. Appl. Math., 19, 413-416 (1966)

[20] J.L. HILL, Torsional wave propagation from a rigid sphere semi-embedded in an elastic halfspace, J. Acoust. Soc. Amer., 40, 376-379 (1966)

[21] S. DATTA and R.P. KANWAL, Rectilinear oscillations of a rigid spheroid in an elastic medium, Quart. Appl. Math., 37, 86-91 (1979)

[22] R.P. KANWAL, Dynamical displacements of an infinite elastic space and matched asymptotic expansions, J. Math.Phys.,44, 275-283 (1965)

[23] E. REISSNER, Stationare axialsymmetrische, durch eine schuttelnde masse erregte Schwingungen eine homogenen elastischen Halbraumes, Ing. Arch., 7, 381-396 (1936)

[24] E. REISSNER and H.F. SAGOCI, Forced torsional oscillations of an elastic halfspace, J. Appl. Phys., 15, 652-662 (1944)

[25] R.N. ARNOLD, G.N. BYCROFT and G.B. WARBURTON, Forced vibrations of a body on an elastic solid, J. Appl. Mech., 22, 391-401 (1955)

[26] G.B. WARBURTON, Forced vibrations of a body upon an elastic stratum, J. Appl. Mech., 24, 55-58 (1957)

[27] J. LYSMER, Vertical Motion of Rigid Footings, Dept. Civ. Eng., Univ. of Mich. Report 3-115 (1965)

[28] F.E. RICHART,Jr. and R.V. WHITMAN, Comparison of footing vibration tests with theory, J. Soil Mech.Fdn.Eng.Proc.ASCE,93,143-168 (1967)

[29] J.R. HALL,Jr., Coupled rocking and sliding oscillations of rigid circular footings, Proc. Int. Symp. Wave Propagation and Dynamic Properties of Earth Materials, Albuquerque, N.M. (1967)

[30] A.P.S. SELVADURAI, Elastic Analysis of Soil-Foundation Interaction, Developments in Geotechnical Engineering, Vol. 17, Elsevier,

Amsterdam (1979)
[31] G. EASON, J. FULTON and I.N. SNEDDON, The generation of waves in an infinite elastic solid by variable body forces, Phil. Trans. Roy. Soc. Ser. A, 248, 575-607 (1956)
[32] D. GRAFFI, Sui teoremi di reciprocita nei fenomeni dipendente dal tempo, Ann. Math., 18, 173-200 (1939)
[33] J.D. ACHENBACH, Wave Propagation in Elastic Solids, North Holland Publishing Co., Amsterdam (1973)
[34] A. UNGAR, The differential transform technique for solving problems of wave propagation. In: Modern Problems in Elastic Wave Propagation (J.Miklowitz and J.D.Achenbach,Eds.), John Wiley,N.Y.pp.83 - 102 , (1978)
[35] L. CAGNIARD, Reflexion et Refraction des Ondes Seismiques Progressives, Gauthier-Villars, Paris (1939)
[36] C.L. PEKERIS, The seismic surface pulse, Proc. Nat. Acad. Sci.,U.S. 41, 469-480 (1955)
[37] C.L. PEKERIS, The seismic buried pulse, Proc. Nat. Acad. Sci., U.S. 41 , 629-639 (1955)
[38] E. PINNEY, Surface motion due to a point source in a semi-infinite elastic medium, Bull. Seis. Soc. Amer., 44, 571-596 (1954)
[39] A.C. ERINGEN and E. SUHUBI, Elastodynamics, Vol.II. Linear Theory, Academic Press, New York (1975)
[40] W.L. PILANT, Elastic Waves in the Earth, Developments in Solid Earth Geophysics Vol. 11, Elsevier, Amsterdam (1979)
[41] H. LAMB, On reciprocal theorems in dynamics, Proc. Lond. Math. Soc., 19, 144-151 (1888)
[42] LORD RAYLEIGH, Some general theorems relating to vibration. Proc. Lond. Math. Soc., 4, 366-368 (1873)
[43] F.L. DiMAGGIO and H.H. BLEICH, An application of a dynamic reciprocal theorem, J. Appl. Mech., 26, 678-679 (1959)
[44] R.G. PAYTON, An application of the dynamic Betti-Rayleigh reciprocal theorem to moving-point load in elastic media, Quart. Appl. Math., 21, 299-313 (1964)
[45] M.E. GURTIN, The linear theory of elasticity. In: Handbuch der Physik, Band VIa/2 (C.Truesdell,Ed.), Springer-Verlag, Berlin, 1-295 (1972)
[46] H. LAMB, On the propagation of tremors over the surface of an elastic solid, Phil. Trans. Roy. Soc. Ser.A, 203, 1-42 (1904)
[47] B.A. BOLEY, On a dynamical Saint-Venant principle, J. Appl. Mech., 27, 74-78 (1960)
[48] A.P.S. SELVADURAI and M. RAHMAN, A note on the axisymmetric oscillations of a rigid disc inclusion embedded in an isotropic elastic medium (unpublished work)

LIQUEFACTION OF SOIL DURING EARTHQUAKES

S. Nemat-Nasser*, M. ASCE

ABSTRACT

Recent theoretical results on liquefaction of cohesionless saturated sand are summarized. Then some experimental observations of the effect of prior straining on the liquefaction resistance are mentioned.

1. INTRODUCTION

Liquefaction of saturated sand during earthquakes can lead to catastrophic failure. Therefore, considerable effort has been devoted to the understanding and quantification of the involved mechanisms. Most of the work in this area has been experimental in nature; see, for example, Seed and Lee (1966), Seed and Peacock (1971), Martin, Finn, and Seed (1975), and De Alba, Chan, and Seed (1975). Recently, Nemat-Nasser and Shokooh (1977, 1978, 1979) have proposed an energy-based simple model which seems to have good predictive qualities. The model, however, uses only one parameter to describe the microstructural changes that take place during the cyclic shearing of a sample, and therefore, its range of applicability is naturally limited. It has been shown experimentally by Finn, Bransby, and Pickering (1970) that prior large straining may reduce liquefaction resistance by an order of magnitude or even more. Our experiments (performed by Y. Tobita) on simple shearing of cohesionless sands, while confirming some of the results of Finn *et al.*, suggest that it is perhaps the change in fabric which may play a dominant role in the change of liquefaction resistance of the sample.

In this note the liquefaction model of Nemat-Nasser and Shokooh is first summarized. Then, a dilatancy equation which has been obtained by this writer, Nemat-Nasser (1979), on the basis of a microstructural consideration for a simple shearing of cohesionless sand, is discussed in an effort to explain the effect of prior straining on the liquefaction potential. Finally, some recent more fundamental work on fabric and its change during the course of deformation is mentioned.

2. AN ENERGY MODEL

Based on an energy consideration, a unified densification and liquefaction theory for cohesionless sand in cyclic shearing has been developed by Nemat-Nasser and Shokooh (1977, 1978, 1979). Here, a brief summary of the theory is presented. To this end the following physically obvious observations are made:

1. To change the void ratio from its current value e to $e + de$, an increment of energy dW is required for rearranging the sand grains (microstructural rearrangement);

---

* Professor of Civil Engineering and Applied Mathematics, Northwestern University, Evanston, IL.

2. This increment of energy dW increases as the void ratio approaches its minimum value e_m;
3. Since an increase in the excess pore water pressure, p_w, results in a decrease in the intergranular forces, the required incremental energy dW decreases with increasing p_w (here $p_w = \bar{p}_w/\sigma_c$, where $\bar{p}_w$ is the actual pore pressure, and σ_c is the applied confining pressure).

It then follows that

$$dW = -\tilde{\nu}\frac{de}{f(1+p_w)g(e-e_m)}, \tag{2.1}$$

where $\tilde{\nu}$ is a positive parameter, and the functions f and g are such that

$$f(1) = 1,\ f' \geq 0,\ g(0) = 0,\ g' \geq 0\ . \tag{2.2}$$

Since

$$de = -\frac{e\sigma_c}{\kappa_w}dp_w, \tag{2.3}$$

where κ_w is the bulk modulus of water, (2.1) is expressed as

$$dW = \frac{\nu}{\kappa_w}\frac{e\,dp_w}{f(1+p_w)g(e-e_m)}, \tag{2.4}$$

where $\nu = \tilde{\nu}\,\sigma_c = \nu(\sigma_c)$, i.e. a parameter which depends on the confining pressure.

For densification of drained sample $p_w = 0$, and (2.1) becomes

$$dW = -\tilde{\nu}\frac{de}{g(e-e_m)} \tag{2.5}$$

which upon integration yields

$$\Delta W = -\tilde{\nu}\int_{e_0}^{e}\frac{dx}{g(x-e_m)}, \tag{2.6}$$

where ΔW (measured per unit volume per unit confining pressure) is the total energy required to change the void ratio from e_0 to its current value e.

For liquefaction, integrate (2.3) to obtain

$$e = e_0 \exp\left\{-\frac{\sigma_c}{\kappa_w}p_w\right\}, \tag{2.7}$$

where initially $p_w = 0$. With the aid of (2.7), Eq. (2.4) is now integrated. But since σ_c/κ_w for pressures of several bars, is so small that $e \simeq e_0$ in (2.7), the calculations are considerably simplified (but without a loss in accuracy) by using e_0 instead of e in Eq. (2.4). Then upon integration, it follows that

$$\Delta W = \frac{\nu e_0}{g(e_0-e_m)}\int_0^{p_w}\frac{dx}{f(1+x)}\ . \tag{2.8}$$

To obtain explicit results, approximate expressions for functions g and f are used in such a manner that (2.2) is satisfied. Simplest functions of this kind are

$$g(e - e_m) \equiv (e - e_m)^n , \quad n > 1 ,$$
$$f(1 + p_w) \equiv (1 + p_w)^r , \quad r > 1 . \tag{2.9}$$

With these expressions, Eqs. (2.6) and (2.8) upon integration respectively yield

$$e = e_m + [(e_0 - e_m)^{1-n} + \bar{\nu} \Delta W]^{\frac{1}{1-n}}, \quad n > 1 , \tag{2.10}$$

and

$$\Delta W = \frac{\hat{\nu} e_0}{(e_0 - e_m)^n} [1 - (1 + p_w)^{1-r}] , \quad r > 1 , \tag{2.11}$$

where $\bar{\nu} = (n - 1)/\tilde{\nu}$, and $\hat{\nu} = \nu/(r - 1)\kappa_w$.

To complete the solution, the energy loss ΔW must be estimated. In a cyclic shearing, ΔW is related to the area of hysteretic loop and the number of cycles; Nemat-Nasser and Shokooh (1979). Here two cases are distinguished: (1) large amplitude shearing; and (2) small amplitude shearing, each either stress- or strain-controlled tests.

When the amplitude of shearing is large enough, all the particles are mobilized during each cycle, and hence the energy loss in each cycle is not very much dependent on the previous cycles. In this case it may be assumed that ΔW is proportional to the number of cycles, N. For very small amplitude shearing, on the other hand, the particles are only partially mobilized in each cycle, and hence the energy loss changes from cycle to cycle as the particles gradually take on new (more stable) positions relative to each other.

A series of experimental results on cyclic shearing of cohesionless sands, has been carefully examined by Nemat-Nasser and Shokooh (1979) and because of the observed symmetry of the hysteretic loop, it has been concluded that the area of hysteretic loop in the ith cycle may be approximated by $A_i = h_i \tau_0^{1+\alpha}$, where α must be an even positive integer (because of the symmetry), $\tau_0 = \hat{\tau}_0/\sigma_c$ is the shear stress amplitude normalized with respect to the confining pressure σ_c, and h_i is an increasing function of the number of cycles. Hence, for stress-controlled tests it follows that

$$\Delta W = \sum_{i=1}^{N} \lambda_i h_i \tau_0^{1+\alpha} = \tau_0^{1+\alpha} \sum_{i=1}^{N} \lambda_i h_i = \hat{h}(N) \tau_0^{1+\alpha} . \tag{2.12}$$

Moreover, for large stress amplitudes, it may be assumed that

$$\Delta W \simeq h N \tau_0^{1+\alpha} . \tag{2.13}$$

From (2.13) and (2.11) it now follows that

$$\tau_0^{1+\alpha} (e_0 - e_m)^n N = \hat{\hat{\nu}} e_0 [1 - (1 + p_w)^{1-r}] , \tag{2.14}$$

where $\hat{\hat{\nu}} = \hat{\nu}/h$. Since α is even, it can easily be fixed by inspection of experimental results. For cohesionless samples considered by De Alba et al. (1975), it is immediately seen that $\alpha = 4$. Moreover, this result is not very sensitive to small variations in the other parameters. For example for De Alba et al.'s (1975) experiments, n and r in (2.14) may be chosen in the ranges of 3 to 4 and 2 to 3, respectively, and hence $n = 3.5$ and $r = 2.5$ has been selected for comparison with the considered experimental results. The parameter $\hat{\hat{\nu}}$ then turns out to be a constant which does not vary much from test to test over a wide range of sample densities; see Nemat-Nasser and Shokooh (1979).

A loss of total bearing capacity occurs momentarily, when the pore pressure equals the confining pressure, so that $p_w = 1$ in (2.14). If the number of cycles to this liquefaction is denoted by N_ℓ, then (2.14) yields

$$\tau_0^{1+\alpha} = \frac{\eta\, e_0}{N_\ell (e_0 - e_m)^n}\,, \tag{2.15}$$

where η is a constant. This equation has been compared with experimental results, and it has been verified that over a wide range of densities, η indeed appears to be a constant which, of course, would depend on the material, the grain distribution, and the confining pressure. Figure 1 compares Eq. (2.15) with some experimental results; see Nemat-Nasser and Shokooh (1979). In Fig. 2 some recent liquefaction experiments performed at Northwestern University by the author's graduate student, Mr. Y. Tobita, on a simple shearing apparatus, are presented. As is seen, the gradient of the dimensionless shear stress amplitude plotted against the number of cycles to liquefaction on a log-log paper, is 5.0 for relative densities of 37, 53, and 61%, using Monterey No. 0 sand. This is an independent confirmation that $\alpha = 4$ is a reasonable approximation in (2.15).

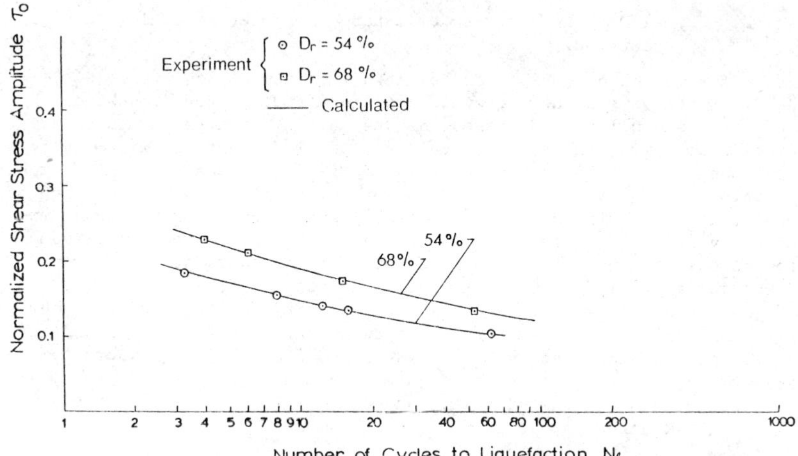

Fig. 1: Normalized shear stress amplitude vs number of cycles to liquefaction; solid curves display Eq. (2.15); data from De Alba, Chan, and Seed, 1975.

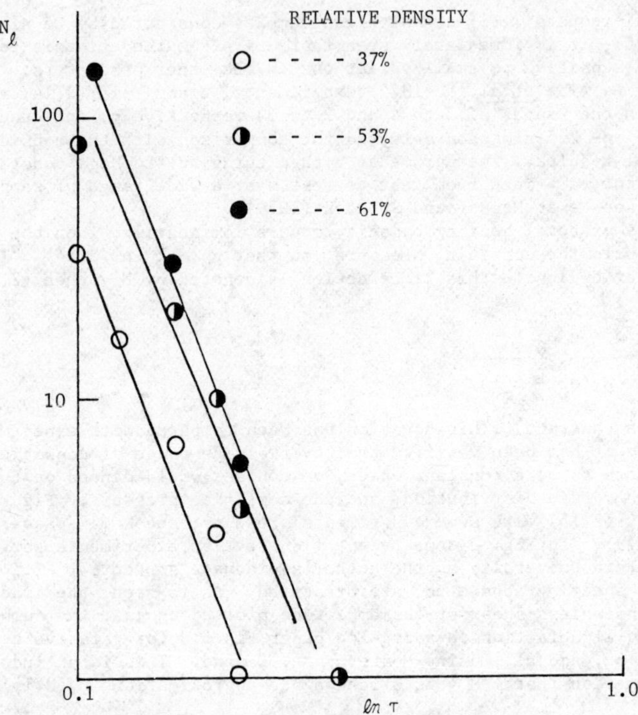

Fig. 2: Normalized shear stress amplitude vs number of cycles to liquefaction.

If α is taken equal 4, and since $n > 1$, it follows from (2.15) that, for the same number of cycles to liquefaction, the dimensionless shear stress amplitude increases, as the initial void ratio approaches its minimum value; in fact, for void ratios close to the minimum (very dense sand) a very large shear stress amplitude (approaching infinity as e_0 approaches e_m) is required. Moreover, since τ_0 is normalized with respect to the confining pressure, (2.15) shows that, for the same number of cycles to liquefaction, the dimensional shear stress amplitude is proportional to the confining pressure. These results are confirmed both experimentally and by field investigations.

It is possible to relate the shear strain amplitude to the number of cycles and the shear stress amplitude during liquefaction of undrained sand. This is discussed by Nemat-Nasser and Shokooh (1978) where comparisons with experimental results are also made.

For strain-controlled tests, and as a first-order-approximation, it may be adequate to take

$$\tau_0 = K \gamma_0^{1/\beta} , \tag{2.16}$$

where β must be an odd positive integer. This then gives

$$\Delta W \simeq \hat{k}(N)\, \gamma_0^{\frac{1+\alpha}{\beta}} \tag{2.17}$$

Extensive comparison with experimental results for densification has been made (Nemat-Nasser and Shokooh (1977, 1979)) employing (2.17) with $\hat{k}(N) = k_0 N$ for large strain amplitudes ($\gamma_0 > 0.1\%$), and $\hat{k}(N) = k_0\sqrt{N}$ for small strain amplitudes ($\gamma_0 < 0.1\%$), where n = 3.5, $\alpha = 4$, and $\beta = 5$ have been used. A typical result is shown in Fig. 3.

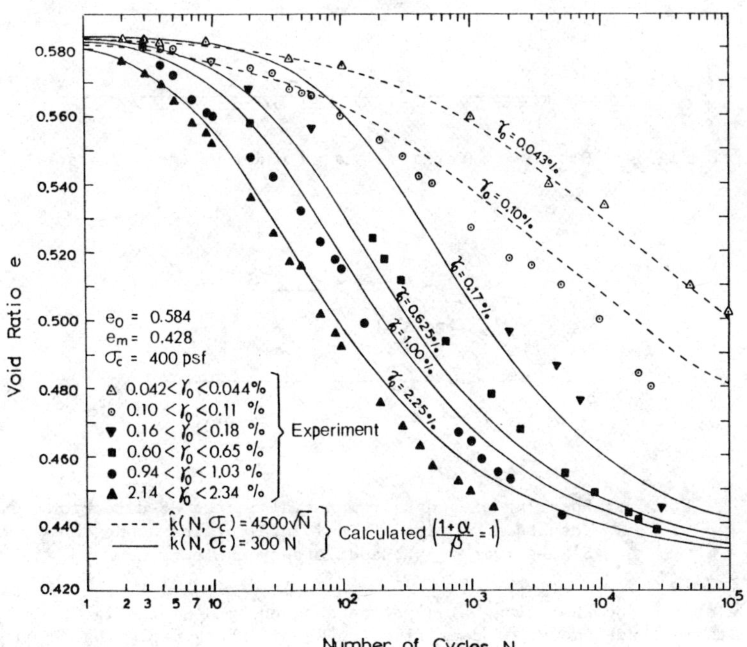

Fig. 3: Void ratio e vs number of cycles in cyclic shearing of dry sand (data from Youd, 1972); γ_0 = shear strain amplitude.

3. SIMPLE STATISTICAL MODEL

For a simple shearing of granular material a simple statistical model recently has been proposed by Nemat-Nasser (1979), which accounts for the observed initial densification, subsequent dilatancy when the material is densely packed, and a net amount of densification upon the completion of each cycle in cyclic shearing of drained samples. Here this model is briefly discussed.

When a granular material is sheared under normal pressure, as shown in Fig. 4, the individual grains do not move along the horizontal line, SS, which marks the macroscopic direction of shearing, rather they

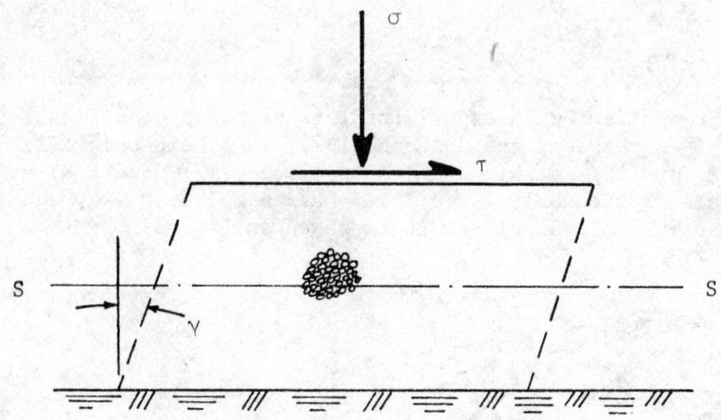

Fig. 4: Granular materials sheared under vertical pressure.

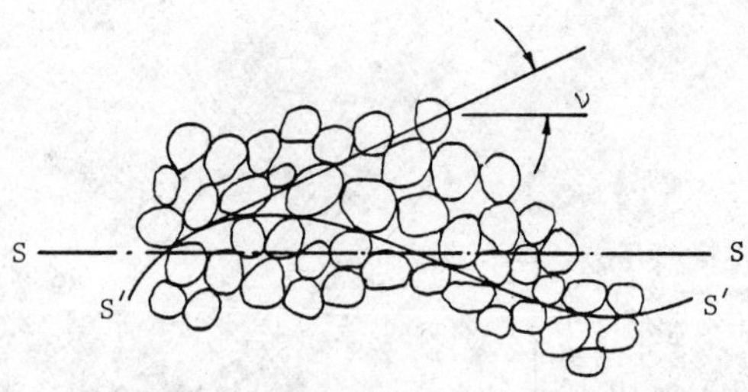

Fig. 5: Shearing of granular material in the SS-direction results in motion of individual grains along the S'S'-curve; ν is the dilatancy angle.

slide over each other along a wavy line denoted by S'S' in Fig. 5. This results in a net amount of densification or dilatancy depending on whether, on average, more particles move down or move up.

Consider a typical particle and let ν be the angle that the tangent to the S'S'-curve makes with the SS-direction. This angle is called the "dilatancy angle."

In a given sample which contains a very large number of particles, there exist many groups of particles, each with its own dilatancy angle at each instant. Let $p(\nu)d\nu$ be the volume fraction of particles whose dilatancy angle is between ν and $\nu + d\nu$. To calculate the overall rate of volume change, we obtain the rate of volume change associated with each group, and then integrate the results over the entire volume, using the weighting function $p(\nu)$.

To this end we calculate the rate of distortional work per unit volume associated with a group of particles whose dilatancy angle is ν. With ϕ_μ denoting the actual particle to particle angle of friction, this rate of work is given by, see Nemat-Nasser (1979),

$$\dot{w}' = \frac{\tau \cos \phi_\mu}{\cos(\phi_\mu + \nu)\sin \nu} \frac{\dot{v}}{v}, \tag{3.1}$$

where prime denotes the distortional part. Macroscopically, the rate of distortional work per unit volume is $\dot{w}' = \tau\dot{\gamma}$, and hence, (3.1) yields

$$\frac{1}{v}\frac{\dot{v}}{\dot{\gamma}} = \frac{\cos(\phi_\mu + \nu)\sin\nu}{\cos\phi_\mu} . \tag{3.2}$$

Now, if the dilatancy angle for a given macroscopic sample ranges over ν_0^- to ν_0^+, we obtain for the overall rate of dilatancy per unit rate of distortion,

$$\frac{1}{v}\frac{\dot{v}}{\dot{\gamma}} = \frac{1}{\cos\phi_\mu}\int_{\nu_0^-}^{\nu_0^+} p(\nu)\cos(\phi_\mu + \nu)\sin\nu\, d\nu, \tag{3.3}$$

where

$$\int_{\nu_0^-}^{\nu_0^+} p(\nu)\, d\nu = 1 . \tag{3.4}$$

Equation (3.3) has all the attributes necessary to yield the behavior of granular materials in simple shear.

To begin with, we observe that even if the density function $p(\nu)$ is symmetric about $\nu = 0$, the right-hand side of (3.3) becomes negative, i.e., densification. However, since the normal stress σ facilitates the movement of particles which are sliding down (negative ν) and hinders those which are sliding up (positive ν), we expect that, initially $p(\nu)$ should be biased toward negative ν's. This results in an initial densification, particularly for loosely packed granules in which the downward motion of particles is less apt to be hindered by neighboring particles.

As most particles become mobilized in a densely packed sample of granules, $p(\nu)$ tends to become biased toward positive ν's, and eventually leads to a positive value for the integral in Eq. (3.3). This results in dilatancy. It is easy to argue in this manner, that upon a load reversal the integral in (3.3) yields a net negative value, and hence a net amount of densification upon the completion of half a stress-cycle in cyclic shearing.

This statistical model permits a possible explanation for the observed reduction in resistance to liquefaction, that is induced by prior straining. The model requires that load-controlled and displacement-controlled prior straining be distinguished.

Consider a drained sample which is subjected to a half-cycle loading where the shear stress is increased from zero to a maximum value (associated with a large strain) and then decreased to zero. At this final state, according to the statistical model, the probability distribution $p(\nu)$ is highly biased toward the negative values of ν, which means that the continued stressing in the opposite direction leads to a large densification. Therefore, if this continued stressing is done under undrained conditions, liquefaction may occur within one cycle. In our laboratory at Northwestern University we have in fact observed this, which confirms the Finn *et al*. (1970) findings.

On the other hand, if a complete displacement reversal is implemented under drained conditions, then, in the subsequent undrained cyclic shearing with the same dimensionless shear stress amplitude, an increase in the liquefaction potential is observed. This is because a stress reversal with a finite value of the stress in the opposite direction is required in order to complete a half cycle shear strain loading. These and related results will be published elsewhere, together with experimental documentation.

4. CONCLUDING REMARKS

While the simple energy model presented in Section 2 provides a good first-order approximation for liquefaction prediction, it is certainly not adequate since only a one-parameter representation is used. The statistical model of Section 3 provides additional insight and suggests possible ways of improving liquefaction prediction. A more fundamental approach, however, is a careful microstructural development of rate constitutive relations.

If one assumes that individual grains are rigid and their mutual interaction involve only friction over the contacting regions, then it is reasonable to expect that the overall macroscopic quantities, such as stress, strain, and their rates should be expressible in terms of the relevant microscopic quantities by means of a systematic averaging process. A fundamental program of this kind has been recently initiated by the writer and his associates, where several preliminary but significant results have been obtained; see Christoffersen *et al*. (1980), Mehrabadi *et al*. (1980), Oda, Konishi, and Nemat-Nasser (1980), and Oda, Nemat-Nasser, and Mehrabadi (1980). Additional theoretical and experimental studies are in progress, and will be reported elsewhere.

ACKNOWLEDGMENT

This work has been supported in part by the U.S. Geological Survey under Contract No. 14-08-0001-17770, and in part by the U.S. Air Force Office of Scientific Research, Grant No. AFOSR-80-0017, to Northwestern University.

REFERENCES

Christoffersen, J., Mehrabadi, M.M., and Nemat-Nasser, S. (1980): "A micromechanical description of granular material behavior," Earthquake Research and Engineering Laboratory Tech. Rep. No. 80-1-22, Dept. Civil Engrg., Northwestern University, Evanston, Ill.; to appear in *J. Appl. Mech.*

De Alba, P.S., Chan, C.K., and Seed, H.B. (1975): "Determination of soil liquefaction characteristics by large-scale laboratory tests," NUREG-0027, NRC-6, Shannon and Wilson, Inc., and Agbabian Assoc., Seattle WA. Prepared for the U.S. Nuclear Regulatory Commission under Contract AT(04-3)-954. See also EERC Rep. 75-14, University of California Berkeley, Calif.

Finn, W.D.L., Bransby, P.L., and Pickering, D.J. (1970): "Effect of strain history on liquefaction of sand," *J. Soil Mech. Found. Div., ASCE*, *SM6*, 1917-1933.

Martin, G.R., Finn, W.D.L., and Seed, H.B. (1975): "Fundamentals of liquefaction under cyclic loading," *Journal of the Geotechnic Engineering Division, ASCE*, *101*, 423-438.

Mehrabadi, M.M., Nemat-Nasser, S., and Oda, M. (1980): "On statistical descriptions of stress and fabric in granular materials," Earthquake Research and Engineering Laboratory Tech. Rep. No. 80-4-29, Dept.

Civil Engrg., Northwestern University, Evanston, Ill.

Nemat-Nasser, S. (1979): "On behavior of granular materials in simple shear," Earthquake Research and Engineering Laboratory Tech. Rep. No. 79-6-19, Dept. Civil Engrg., Northwestern University, Evanston, Ill.; to appear in Soils and Foundations.

Nemat-Nasser, S. and Shokooh, A. (1977): "A unified approach to densification and liquefaction of cohesionless sand," Earthquake Research and Engineering Laboratory Tech. Rep. No. 77-10-3, Dept. Civil Engrg., Northwestern University, Evanston, Ill.

Nemat-Nasser, S. and Shokooh, A. (1978): "A new approach for the analysis of liquefaction of sand in cyclic shearing," Proc. Second Int. Conf. on Microzonation, 2, 957-969, San Francisco, Calif.

Nemat-Nasser, S. and Shokooh, A. (1979): "A unified approach to densification and liquefaction of cohesionless sand in cyclic shearing," Canadian Geotechnical J., 16, 659-678.

Oda, M., Konishi, J., and Nemat-Nasser, S. (1980): "Index measures for granular materials," Earthquake Research and Engineering Laboratory Tech. Rep. No. 80-3-26, Dept. Civil Engrg., Northwestern University, Evanston, Ill.

Oda, M., Nemat-Nasser, S., and Mehrabadi, M.M. (1980): "A statistical study of fabric in a random assembly of spherical granules," Earthquake Research and Engineering Laboratory Tech. Rep. No. 80-4-28, Dept. Civil Engrg., Northwestern University, Evanston, Ill.

Seed, H.B. and Lee, K.L. (1966): "Liquefaction of saturated sands during cyclic loading," J. Soil Mech. Found. Div., ASCE, 92, 105-134.

Seed, H.B. and Peacock, W.H. (1971): "Test procedures for measuring soil liquefaction characteristics," J. Soil Mech. Found. Div., ASCE, 97, 1099-1119.

Youd, T.L. (1972): "Compaction of sands by repeated shear straining," J. Soil Mech. Found. Div., ASCE, 98, 709-725.

DYNAMIC RESPONSE OF LATERALLY-LOADED PILES IN CENTRIFUGE

Jean H. Prevost[(1)], M. ASCE, James D. Romano[(2)], Ahmed M. Abdel-Ghaffar[(1)], A.M. ASCE, Rod Rowland[(3)]

ABSTRACT

The dynamic response of a laterally loaded pile embedded in sand is studied using the centrifugal modeling technique and the results are presented in this paper. The experimental procedure and limitations are presented.

Pile response to forced vibrations was found to depend strongly on the magnitude and frequency of loading as well as the density of the soil. Increased soil density was found to decrease both the displacement and damping of the pile while increased density had the opposite effects. Nonlinear behavior was evidenced by a decrease in the resonant frequency of the soil pile system as the loading was increased. Moment points were obtained from strain measurements and a curve was spline fitted. This curve was integrated and differentiated twice to obtain pile displacement and soil reaction, respectively.

Results of free vibration tests shows that the damping ratio of the soil-pile system increased with increasing load. Finally, the results of the tests show that experimental data obtained from centrifuge simulation accurately reproduce prototype results when scaled accordingly.

INTRODUCTION

Horizontal, time dependent loadings are caused by earthquakes, machine vibrations, and wind or wave action. The prediction of pile response to such loadings is one of the most challenging problems in foundation engineering. Reliable predictions are made difficult by the complex interaction between the elastic pile and the inelastic soil.

---

(1) Assistant Professor, Civil Engineering, Princeton University, N.J.

(2) Engineer, Schlumberger, Texas; formerly student at Princeton University.

(3) Research Engineer, Department of Civil Engineering, Princeton University, Princeton, New Jersey 08544.

Traditionally, designers have thus had to rely upon full-scale, in-situ tests as the principle source of information for design. These tests require much time and capital for successful completion. A centrifuge can simulate gravity-induced stresses in soil deposits at a reduced geometrical scale through centrifugal modeling. This modeling technique leads to a set of scaling relationships, or scaling laws, that affect time relationships, physical dimensions, and the many derivatives of these combinations such as velocity, acceleration, force, etc. These scaling relations have been discussed in the literature (see e.g. [4], and the relations between the quantities of interest here are shown in Table 1. The advantage of using a centrifuge to test soil models is that at homologous points the stresses in the soil model are the same as in the full-scale prototype situation.

TABLE 1

Scaling Relations

Quantity	Full Scale (Prototype)	Centrifugal Model at n g's
Linear Dimension, Displacement	1	1/n
Area	1	$1/n^2$
Volume	1	$1/n^3$
Stress	1	1
Strain	1	1
Force	1	$1/n^2$
Mass	1	$1/n^3$
Acceleration	1	n
Velocity	1	1
Time in Dynamic Terms	1	1/n
Frequency in Dynamic Problems	1	n

When the soil is the same in both model and prototype, then the stress-strain behavior is identical at homologous points. The technique promises to be an invaluable aid for studying a variety of complex geotechnical problems and in particular, for studying dynamic soil-structure interaction problems.

This investigation reports on the dynamic response of laterally-loaded piles embedded in sand using the centrifuge technique.

EQUIPMENT AND INSTRUMENTATION

(a) Centrifuge

The centrifuge used is a Model 1230-1 Genesco "G-accelerator" hydraulically driven and which consists of an 84-inch diameter aluminum arm which rotates in the horizontal plane and is rated at 15 g-tons payload capacity. At each end of the arm is located a swinging basket in which a $14 \times 14 \times 9.5$ inch bucket is attached to contain the experiment. For these experiments, a 150 lb. payload was accelerated to 100 g's.

Electrical power and signals to and from the rotating arm or basket are conducted through sliprings. The low voltage signals from the strain-gages, displacement transducers and the accelerometer were transmitted via 20 gold-plated low noise sliprings manufactured by Michigan Scientific.

(b) Pile and Deformation Sensors

The pile model used for these experiments was machined from .25 inches OD aluminum tubing to a diameter of .21 inches and wall thickness of .02 inches. The pile weighted 0.0075 lbs. was 9.9 inches long, and was embedded in the sand to a depth of 8.60 inches. The pile was gaged with 12 Micro-Measurements strain gages located at 0.13, 0.58, 0.91, 1.25, 1.93 and 2.93 inches below the soil surface. To increase the strain reading, two gages were glued at each depth on opposite sides of the pile and wired into adjacent positions in a common bridge. Two semi-conductor strain gages were located at 4.18 and 5.25 inches below the soil surface, but due to the extremely small strains in the pile at that depth, no meaningful reading was ever obtained from these gages. The entire pile was coated with Micro-Measurement "M-bound" protective coating, and had a final flexural rigidity, EI, of approximately 500 lbs.-in^2.

The pile was mounted with a cap which contained a coil and an accelerometer, and weighted 0.330 lbs., bringing the total mass of the pile and cap to 0.405 lbs.

Displacements of the pile head were obtained by integrating twice the reading of the Endevco 2221-A model accelerometer attached to the pile cap. A charge amplifier mounted closed to the axis of the centrifuge amplified the accelerometer signal before it reached the sliprings, thus greatly enhancing the signal to noise ratio.

(c) Signal Conditioning and Data Acquisition

Both the accelerometer and strain gage waveforms were recorded on a Tektronix 549-A dual-beam storage oscilloscope. Despite the use of high quality sliprings, enough noise was still produced which made accurate resolution of strain gage readings below 50 μ in/in impossible. The strain gage signals were first preamplified, using a Tektronix preamp and balanced, using a Tektronix 502-A bridge circuit. This output was then filtered and sent to the 549-A.

(d) Specialized Test Apparatuses

1. *Free vibration device:* Static monotonic loading and free vibration of the pile were achieved by a displacement-cam device similar to the one first designed by Scott [6]. The applied load was measured with a Celesco MB101-3 model load cell attached to the lever arm.

2. *Forced vibration device:* An electromagnetic "shaker" was designed to vibrate the pile over a wide range of frequencies and loadings. The device consists of two stationary electromagnets mounted on an aluminum frame and a smaller dumbell shaped coil coaxially located between the two larger electromagnets and mounted to the pile cap. Details about the design may be found in [3]. There are a number of advantages associated with the use of an electromagnetic shaker. The system is not mechanically operated and is independent of the high g's created in the centrifuge. Further, the system has high frequency capabilities (over 1 kHz), allowing investigation of higher mode response.

(e) Soil Tested

A uniform silica sand (Ottawa sand) with a mean grain diameter of 0.008 inches (0.2 mm) was used in the experiments. Various density and water content of the deposit were used:

1. *Loose, dry sand:* prepared by pouring the sand into the bucket and smoothing the surface.

2. *Dense, dry sand:* the top five inches of the deposit were removed from the bucket and replaced in 0.25 to 0.50 inch layers which were tamped with a 2 lb. weight.

3. *Dense, saturated sand:* sand was prepared as in (2) above, except that the water table was brought up to the current soil surface after each new layer had been tamped.

TEST RESULTS

(a) Prototype Simulation

The response characteristics of a pile measured in a centrifuge would be of little value if centrifugal modeling does not accurately simulate prototype behavior. In order to investigate the accuracy of the scaling equations, the pile and sand used in these experiments were modeled after full-scale tests [1,2]. A similar study has been previously attempted and is reported in [5].

In the Mustang Island full-scale tests [2], a 24-inch diameter pile, 3/8 inch thick, was embedded in a saturated dense sand deposit to a depth of 828 inches. The bending stiffness of the pile was 600×10^6 lbs. - in^2 (= 600 lbs. - in^2 in model dimension). The aluminum model pile was embedded to a depth of 8.65 inches, had the same cylindrical shape, but had a bending stiffness 516 lbs. in^2 in less than that required to exactly simulate the prototype pile. The load was applied at ground level.

Figure 1 shows the moment distribution along the pile for various loads. Note that the gage located at 0.91 inches below the surface was not operating in this test, and that the resolution in the gages below 1.93 inches was so poor that no signal could be detected above noise level, thus limiting the amount of information that could be gathered from this test. The moment curves were integrated, and Figures 2 and 3 show comparisons between measured model and prototype behavior. In spite of the above mentioned limitations, the agreement between the two sets of results is seen to be fairly good.

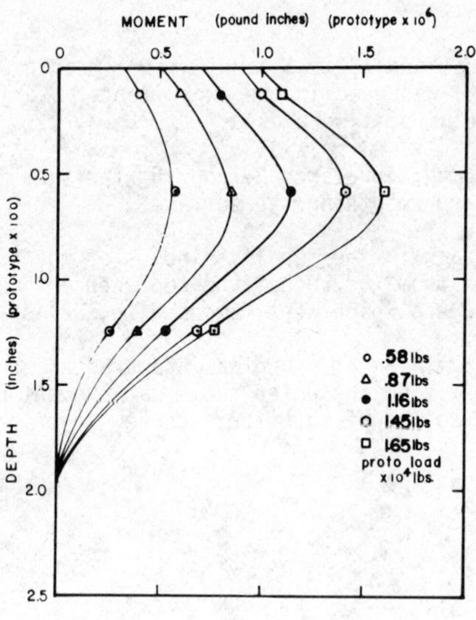

Figure 1

Moment Curves Attained During the Saturated Simulation of the Mustang Island Test [2]

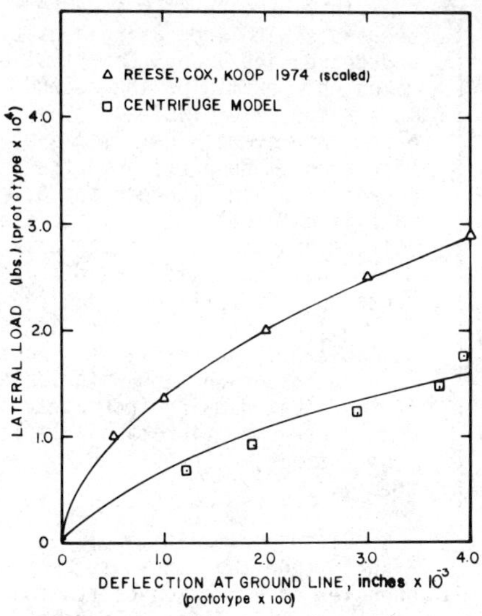

Figure 2

Comparison of Centrifuge Moment Values with Full Scale Values in Saturated Sand

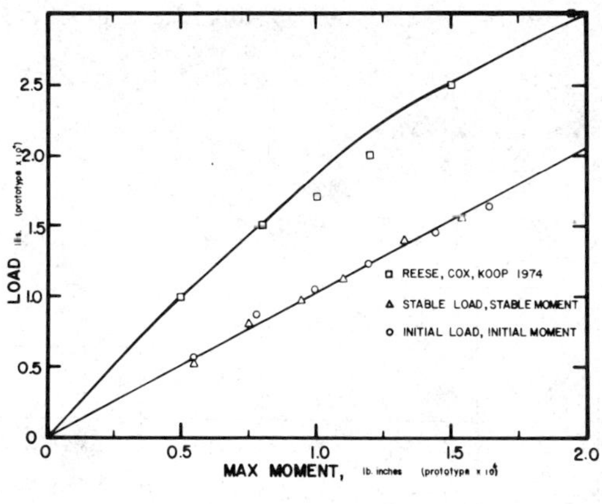

Figure 3

Pile Head Displacements Obtained in the Centrifuge Compared with Full Scale Results

(b) Free Vibration Test Results

The complete set of data generated in these tests is displayed in Table 2. The damping ratio's, measured for each of the loadings over the first cycle of vibration, are plotted in Fig. 4. From this figure, it is concluded that the damping ratio increases non-linearly with increased load. The damping ratio decreased as the amplitude of vibration decreased. While measurements made over the 8th cycle of vibration were not very accurate, they consistently showed a large variation from the damping ratio measured over the first cycle of vibration. Thus, it is concluded that damping ratio increases with increased soil strain. The second column of Table 2 shows that the first resonant frequency of vibration did not change with loading in the range of loads tested. The average fundamental frequency of the pile was 139 cps, the same resonant frequency observed for loading one in the forced-vibration tests.

(d) Forced Vibration Test Results

1. Response Curves

The response curves of the acceleration amplitudes (absolute values; peak-to-peak) versus the exciting frequencies, are shown in Figures 5 and 6. First and second mode resonsnaces are easily descernible. The high sharp peaks of first mode response, indicate that there is little damping associated with this mode. Damping results, calculated using the one-half-power points method, are presented in Tables 3 and 4. The data

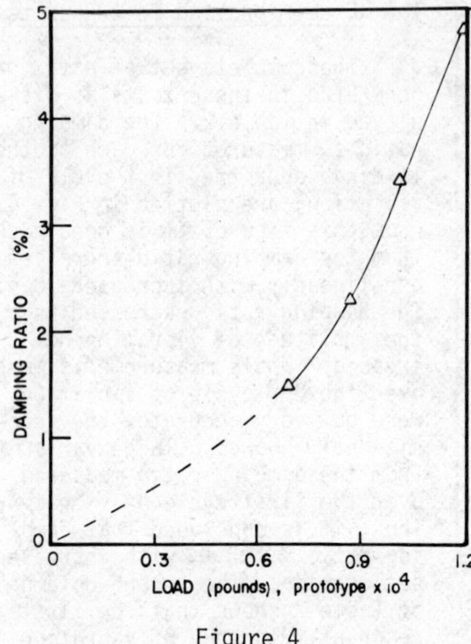

Figure 4

The Damping Ratio of a Function of a Pile Load

show that damping at resonance decreases with increasing density and decreasing load. Care must be taken when examining the second mode response of Table 4. The broad, non-symmetrical forms of the second mode portion on the curves shown in Figure 5 and 6, make it difficult to estimate the half points for the damping calculation. Peak pile head displacements for the different loading and soil conditions are presented in Table 5. The results of Table 3, used in conjunction with Table 5, show that damping increases with increasing vibrational amplitude (strain) as predicted by theory.

Closer examination of Figs. 5 and 6 show that the natural frequencies of vibration decreases with increasing load, which is a typical behavior of a softening dynamical system. This is the first evidence of soil non-linearity and it is caused by a decrease in soil stiffness with an increase in dynamic stress and strain.

The strain data measured along the length of the pile were converted to moment values and are plotted in Figs. 7 and 10. The high value of the moments at first resonance is a result of the position of load application.

TABLE 2

Damping Data for a Freely Vibrating Pile*

Point Number	Load (Pounds)	First Resonant Frequency (Hertz)	Acceleration at Pile Head (g's-peak)	Pile Head Displacement (inch-peak)	CYCLE 1 OR 2: Logarithmic Decrement $\delta = \ln \frac{x}{x+1}$	CYCLE 1 OR 2: Damping Ratio %	CYCLE 8: Logarithmic Decrement δ	CYCLE 8: Damping Ratio %
1	.656	138	1.86	$1.0\text{x}10^{-3}$	.100	1.5	0.051	0.8
2	.878	140	2.75	$1.5\text{x}10^{-3}$	.142	2.3	0.087	1.2
3	.969	140	3.50	$1.8\text{x}10^{-3}$	.215	3.4	0.074	1.4
4	1.14	138	3.75	$2.0\text{x}10^{-3}$	.300	4.8	0.095	1.5

*mean frequency is 139 Hz.

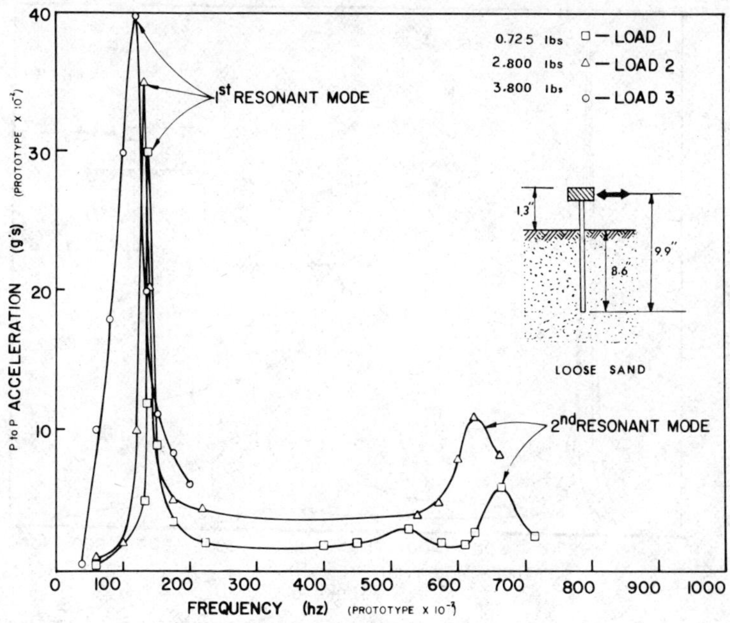

Figure 5

Pile Head Acceleration Plotted Against Frequency for 3 Loadings in Loose Sand

TABLE 3

Damping Ratio at Pile First Resonance

SOIL CONDITION	LOAD 1	LOAD 2	LOAD 3
Loose	2.5%	6.7%	10%
Dense	1.6%	3.4%	5.7%

TABLE 4

Damping Ratio at 2nd Resonance

SOIL CONDITION	LOAD 1	LOAD 2
Loose	4.7%	5.5%
Dense	4.6%	6.7%

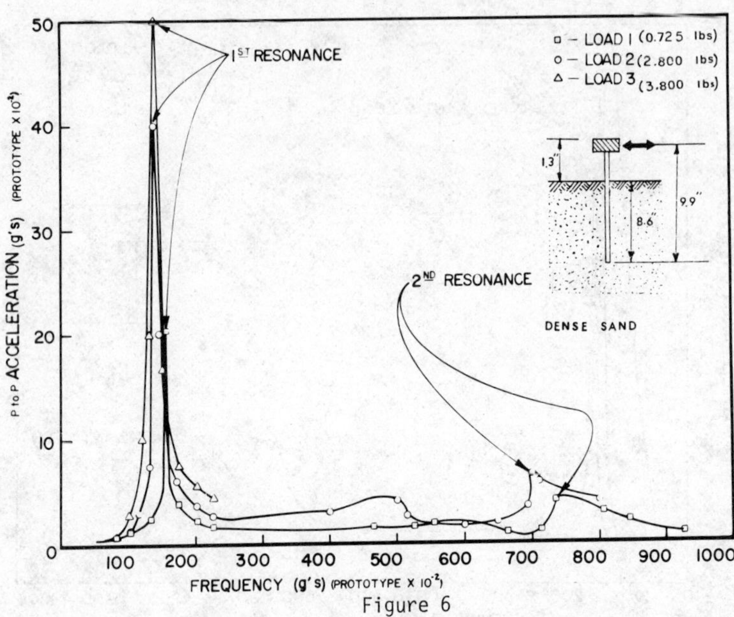

Figure 6

Pile Head Acceleration Plotted Against Frequency for 3 Loadings in Dense Sand

TABLE 5

Pile Cap Displacements in Inches x 10^{-3} at Pile First Resonances (prototype X100)

SOIL CONDITIONS	LOAD 1	LOAD 2	LOAD 3
Loose	7.5×10^{-3}	8.5×10^{-3}	14.0×10^{-3}
Dense	4.0×10^{-3}	9.0×10^{-3}	12.0×10^{-3}

The load for these tests was applied 1.25 inches (scaled 10.5 ft.) above the soil surface so the pile acted much like a cantilever.

2. Spline Fitting

A cubic spline was fitted to the 6 data points of the moment curves of Figs. 7 to 10; the fit requires the specification of the first and/or second derivatives of each point. The second derivative of the bottom data point (the soil "reaction") was set to zero. Due to the difficulties of specifying the derivative of the top data point, the spline fit

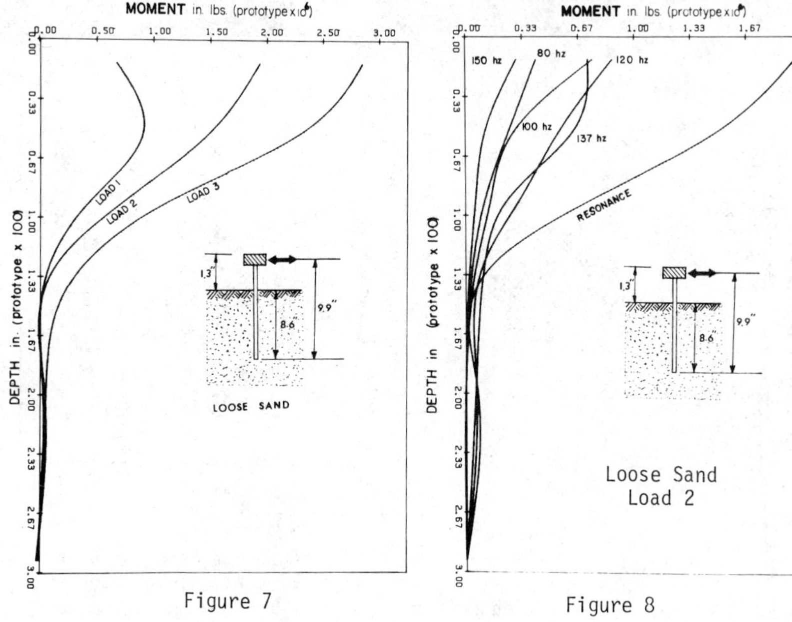

Figure 7

Bending Moment as a Function of Pile Length at 3 Loads in Loose Sand

Figure 8

The Bending Moment in a Pile as a Function of Depth for a Number of Exciting Frequencies

indices some inaccuracies into the moment curve. Since the gages on the pile provided only the absolute value of the strain, the negative values of the moment (Figure 10) are creations of the spline interpolation. The values of the second derivatives of the moment curves were calculated at each data point and were also fitted with the same spline interpolation (Figures 11, 12, and 13).

If a perfect bond between the soil and pile is assumed, the second derivative of the moment curves (of Figs. 11, 12 and 13), is a profile of the total soil reaction. However, the total soil reaction is composed of compression of the sand on one side of the pile and shear along the profiles of the pile. Thus, the term "soil reaction" should be applied to the second derivative figures with reservation.

3. Moment, Displacement and Reaction Curve

The moment diagram of Figure 7 shows that only the upper 3 inches of the pile are actively involved in bending; the lower seven inches, necessary for rigidity, are uninfluenced by the pile load. It is also obvious that the pile maximum bending moment increases with increasing load and moves toward the soil surface in response to the soil nonlinear

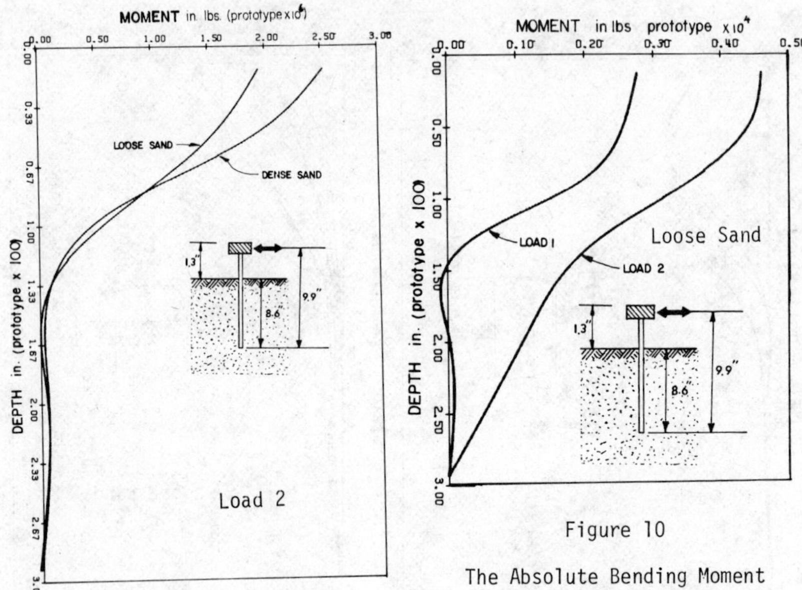

Figure 9
Bending Moment as a Function of Pile Length in Sand of Two Densities

Figure 10

The Absolute Bending Moment in the Pile at Second Resonance

behavior. Detailed examination of the displacement diagram (Fig. 14) shows that for load one, 75% of the total displacement occurs within the upper two-thirds of the active displacement length. When load 3 was applied, 75% of the bending now occurred in the upper third of the active length of the pile. The effect of the increased displacement was confined to the uppermost layers of the soil. Examination of the soil reaction in Figure 11 shows that while the soil reaction increases with increased applied load, the location of maximum reaction does not change. To explain the large increase in displacement with only a slight increase in load, one must assume that yielding is occuring.

From the variation in pile response with frequency, the soil reaction, the bending moment, and the pile displacement (Figs. 8, 12 and 15) are appreciably greater at resonance than at any other frequency. As the amplitude of vibration increases, the location of the maximum moment rises to the soil surface, again indicating that yielding is occurring. In this instance, the non-linear behavior of the sand is caused by increased soil strain. Pile response was greater for frequencies below the resonant point than for those above it.

The maximum moment of a pile embedded in sand increases with density (Figure 9), but this value quickly drops below the moment values measured for the pile in loose sand. The stiff, dense sand provides a rigid foundation for the pile, forcing it to behave as a cantilever. As a result,

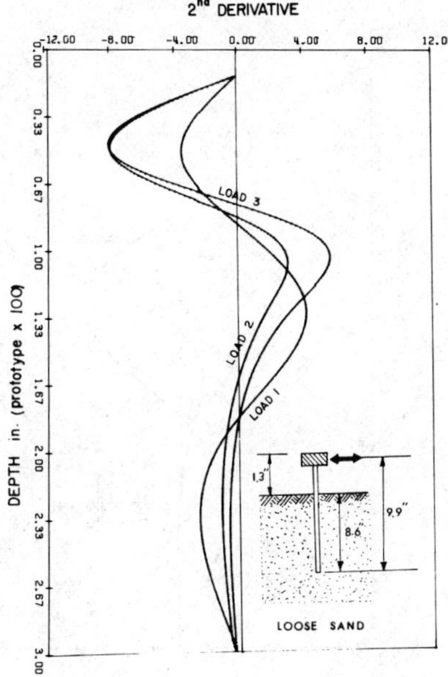

Figure 11

The "Soil Reaction as a Function of Depth for 3 Loads in Loose Soil

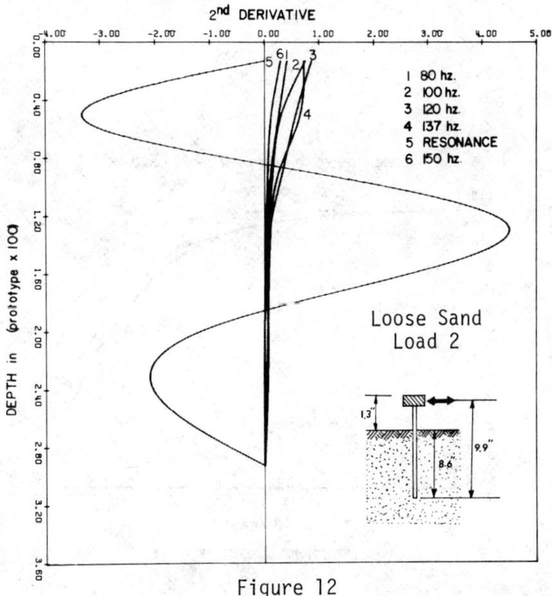

Figure 12

The "Soil Reaction" as a Function of Loading Frequency

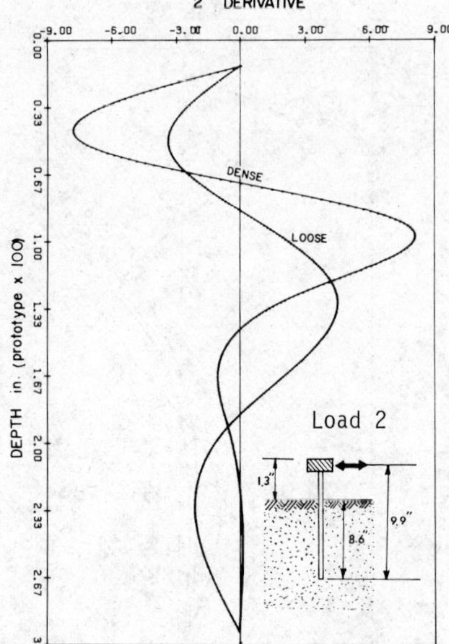

Figure 13

The "Soil Reaction" as a Function of Depth for Two Densities of Sand

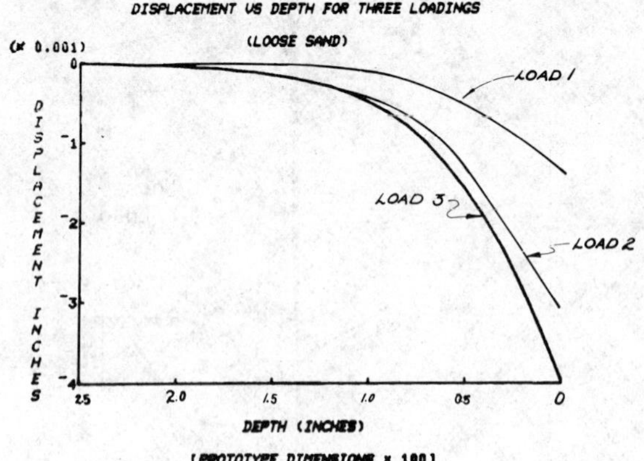

Figure 14

Displacement vs. Depth for Three Loadings

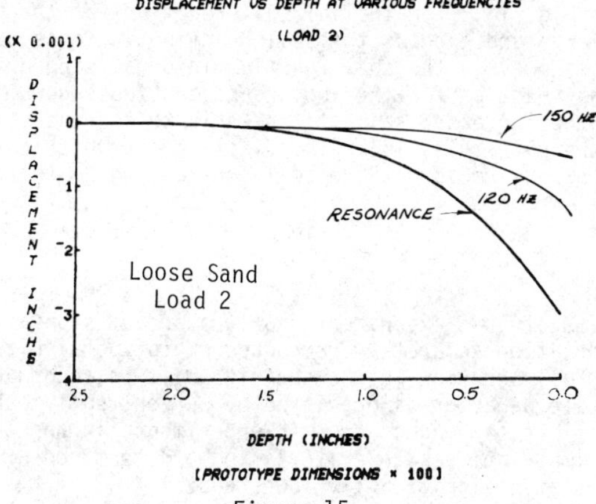

Figure 15

Displacement vs. Depth at Various Frequencies

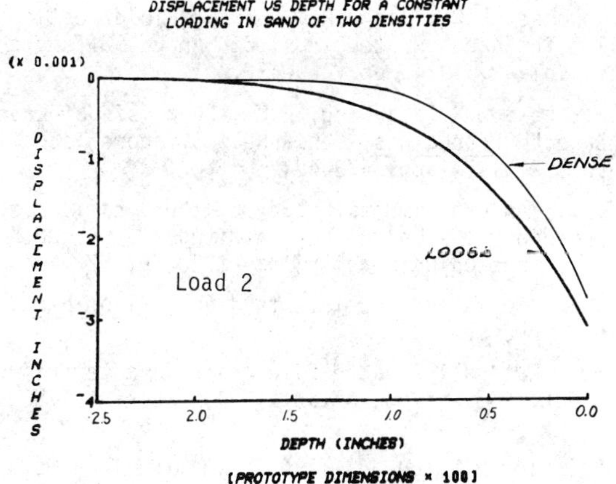

Figure 16

Displacement vs. Depth for a Constant Loading in Sand of Two Densities

high moment values are recorded near the soil surface. This effect can be observed in Fig. 13. The initial soil reaction for the pile in dense sand is very large and exceeds the soil reaction in loose sand by a significant amount. However, the soil reaction in dense sand quickly drops to zero, while the reaction for the loose sand continues down the pile length. The effect of the increased soil stiffness is to decrease the total displacement of the pile (Fig. 16). The displacement which does occur is confined to a narrow region at the soil surface, where the sand is the weakest.

Conclusions

The results of this experiment show that results obtained from centrifuge simulation accurately reproduce prototype results when scaled accordingly. Soil behavior in the centrifuge is in accordance with observed full scale behavior as predicted by current soil mechanics theory. It is also possible to conduct dynamic experiments in the centrifuge, allowing many important design criteria to be determined at a greater convenience and at a significantly lower cost.

Acknowledgements

The centrifuge facility is sponsored by grants from Princeton University Civil Engineering Department, Woodward-Clyde and Dames and Moore consulting firms, Exxon, Amoco and Shell Oil Companies. Their support is most gratefully acknowledged.

References

1. Cox, W.R., Reese, L.C., and B.R. Grubbs, "Field Testing of Laterally Loaded Piles in Sand," Proceedings, 6th Annual Offshore Technology Conference, Vol. 2, 1974, pp. 459-472.

2. Reese, L.C., Cox, W.R. and Koop, F.D., "Analysis of Laterally Loaded Piles in Sand," Proceedings, 6th Annual Offshore Technology Conference, Vol. 2, 1974, pp. 473-484.

3. Romano, J.D., "An Experimental Study of the Dynamic Response of Laterally Loaded Piles Using the Centrifuge Technique," Civil Eng. Department, Princeton University, May 1980, p. 82.

4. Roscoe, K.H., "Soils and Model Tests," Journal of Strain Analysis, Vol. 3, 1968, pp. 57-64.

5. Scott, R.F., "Laterally-Loaded Pile Tests in a Centrifuge," Proceedings, ASCE Nat. Convention, Atlanta, GA, Preprint 3786, October, 1979, p. 22.

6. Scott, R.F., Liu, H.P. and J. Ting, "Dynamic Pile Tests by Centrifugal Modeling," Proceedings, 6th World Conference on Earthquake Engineering, Paper 4-50, New Delhi, India, 1977.

TRAVELING WAVE ANALYSIS OF A LARGE FOUNDATION

Philip J. Richter[1] and Elwood A. Smietana[2]

ABSTRACT

A method is developed and utilized to perform seismic traveling wave analysis of a large foundation supporting important equipment. Results of these time-history analyses were used in the foundation design and to provide base motions for evaluation of equipment response.

INTRODUCTION

General Background

The Gas Centrifuge Enrichment Plant (GCEP) is currently under construction near Portsmouth, Ohio. The GCEP program is directed by the Deputy Manager for Enrichment Expansion Projects, U.S. Department of Energy, Oak Ridge, Tennessee, with technical oversight by Union Carbide Corporation-Nuclear Division. The GCEP is intended to produce uranium hexafluoride of nuclear fuel grade with a great reduction in the use of electrical energy over the existing gas diffusion process.

The heart of the GCEP will be large process buildings (seven acres each in plan area) with several centrifuge foundations approximately 90 feet by 280 feet by 2 feet thick. Each of these foundations supports a large number of centrifuges and a complex network of piping on service modules which transports the gaseous materials to and from the centrifuges. For a foundation of such large plan dimensions, the weight of equipment and piping supported is quite light.

Earthquake Engineering Program

The GCEP site is in a relatively quiescent area with regard to earthquake activity. However, because of the large capital investment and potential sensitivity of equipment, an extensive

[1] Principal Structural Engineer, Fluor Engineers and Constructors Inc., Irvine, California

[2] Structural Engineer, Fluor Engineers and Constructors, Inc., Irvine, California

seismic analysis and design effort is being carried out in conjunction with the project. A significant portion of the seismic analysis and design effort has been expended on the large centrifuge foundations (Reference 1).

Traveling Wave Consideration

During an earthquake, ground motions propagate outward from a seismic source (the epicenter). Thus, the point on a structure foundation nearest the epicenter will be subjected to ground motions slightly before other points along the foundation. For most common size structures, this difference in arrival times is so small that it is generally neglected and the structure is designed as if all points of the foundation are subjected to the ground motion simultaneously (in-phase). This is often referred to as the "fixed-base" assumption for analysis purposes. Earthquake damage experience indicates that for relatively long structures, such as bridges and some buildings, an in-phase assumption is generally not adequate, and out-of-phase ground motion effects must be considered. Figure 1 provides a simplified and schematic view of the traveling wave (TW) phenomena.

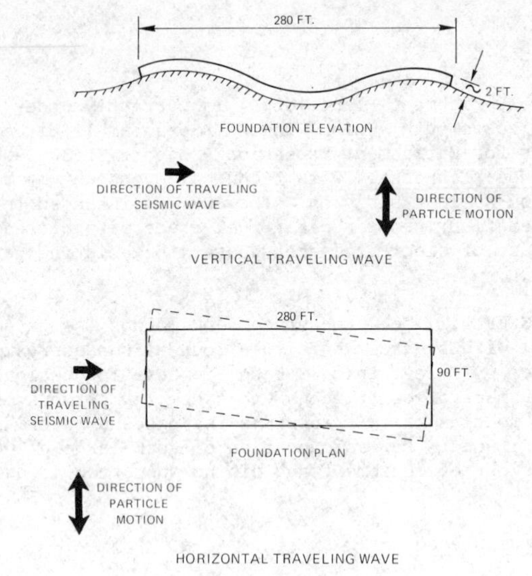

FIGURE 1
TRAVELING WAVE – SIMPLIFIED COMPONENTS

Because of lack of information and understanding of the traveling wave effect on structures, there is no established procedure for its consideration in seismic design. The purpose of this paper is to describe the approach developed for the GCEP to assess the traveling wave effect on the large centrifuge foundations. The motivation in development of the approach was to assess the possible impact of traveling wave effects on the foundation design and to provide some indication of the foundation seismic motions for purposes of bounding the supported equipment response.

Basic Premises

Investigators in the last several years have conducted a considerable amount of research in an effort to gain insight into the behavior of foundation systems responding to traveling seismic waves. The TW analyses described herein are intended to be a design-supportive effort rather than a research-oriented effort. The following basic premises, founded on experience and engineering judgement, served to reasonably limit the scope of the TW anaylses while focusing the results on the most pertinent factors:

a. Typical soil-structure interaction (SSI) effects are negligible. Two possible types of SSI may be considered in seismic design of structure-foundation systems. The more typical SSI phenomenon is a feedback or exchange of energy between foundations and adjacent soil during excitation which leads to modification of free-field ground motions in the vicinity of the foundations. Because of the relatively shallow layer (30 feet to 35 feet) of stiff clay overlying rock, and the relatively light foundation, the typical SSI in the vertical direction was considered to be small and was neglected for purposes of analysis and the results reported herein. This conclusion was based on review of the project soil report, standard references (see for example, Reference 2), and engineering judgement. Horizontal SSI was considered by inclusion of appropriate soil springs and damping in the analytical models. Though again, this effect is shown to be relatively small in the case of horizontal translational response.

b. A second type of SSI is site-structure interaction, that is the effect of local soil layers and geometry on the free-field ground motions. This was also considered to be negligible. The free-field motion is defined for the surface and near surface foundations by the GCEP criteria (Reference 3). It is judged that there is little difference between free-field and bedrock motions because of the stiff clay soil which occurs in a uniform, shallow layer (Reference 4).

c. It is recognized that propagating seismic waves are complex wave forms which may occur at oblique angles, horizontal and vertical, with the foundation and the surface. By utilizing the standard criteria free-field waves, assuming propagation in a horizontal longitudinal direction, it is expected that input and thus analytical response are near an actual upper bound. The standard input forms contain phases which may be considered as P and S waves as well as Rayleigh (surface) waves. See Figure 2.

d. Relating to item c, the effect of TW on the foundation was determined in two independent sets of analyses for longitudinal wave propagation. The first set utilized a model to consider transverse horizontal free-field particle motion. The second set utilized a model to consider vertical free-field particle motion. Thus the premise implied is that the complex response of the foundation can be separated into two relatively uncoupled components, one horizontal (transverse) and one vertical. This paper is limited to reporting on the more significant analyses considering horizontal particle motions only.

e. Several sets of earthquake records were developed to represent appropriate seismic ground motion for the project. Also, both Operating Earthquake (OE) and Maximum Earthquake (ME) levels were considered in incorporating a two-level design approach. The peak ground acceleration levels are .04 g and .15 g for the OE and ME, respectively. One set of five acceleration records of about 40 seconds duration, each covering a broad frequency band, were developed as the standard time-histories for the project. Each of these records has response spectra which essentially envelope project criteria design spectra (Reference 5). These records were artifically generated but were based on parameters selected from real records. One record entitled "GATH .02" was adopted for the purposes of the TW analyses as being representative of all five of the standard time histories. See Figure 2. The results discussed in this paper are limited to those based on the GATH .02 input scaled for either OE or ME level ground motions.

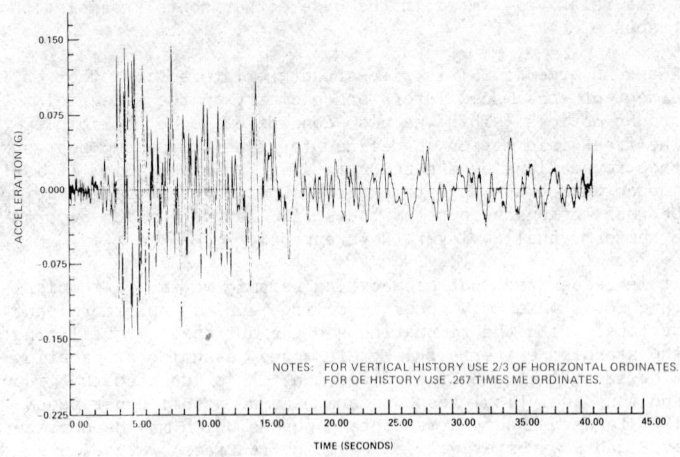

FIGURE 2
GCEP STANDARD FREE FIELD EARTHQUAKE ACCELERATION
TIME HISTORY ME – GATH .02 HORIZONTAL

ANALYTICAL METHOD

Assumptions

In addition to the basic general premises cited above, a number of more specific assumptions are fundamental to the development of the horizontal traveling wave analysis method. They are as follows:

a. The criteria free-field time history (Figure 2) is assumed to represent the transverse component of response of a horizontal shear wave propagating parallel to the longitudinal axis of the foundation.

b. Locally, each point on the ground's surface along a given wave path has the same ground motion, that being the criteria free-field motion. The only difference is arrival times.

c. Along a given wave path, the difference in arrival times is based on constant shear wave propagation velocity. The shear wave propagation velocity used in the GCEP analysis, based on the project foundation report, is 920 ft/second for the Operating Earthquake.

Description of Approach

The approach adopted for modeling traveling wave effects is to prescribe the criteria free-field motion time history at the base of the soil-foundation system by the use of "artificial ground nodes." The term "artificial" is derived from the fact that the base of the soil springs are not actually fixed to a global reference system, as in the typical case of uniform base excitation, but rather free to translate under the action of the input forcing function described below. As discussed above, the input motion is identical at each artificial ground node, and is phased according to the shear wave propagation velocity. Each artificial ground node is connected to a soil spring which is in turn connected to a node on the beam element model representing the structure foundation. Though a basic premise of the analyses is that translational SSI is negligible, soil springs are utilized to allow for such interaction effects as may occur as a result of the traveling wave phenomena including those related to torsional response. The soil springs were developed utilizing half-space theory with the method summarized in Reference 6. Briefly, the horizontal spring stiffness is calculated for the entire foundation area, then distributed to each node in proportion to the tributary length of adjoining beam elements.

A basic general purpose finite element computer code, SAP, (Reference 7) was used for the purpose of performing the time-history analyses. Key to the specific method for prescribing motion at the artificial ground nodes are the dynamic analysis capabilities of SAP. General structure excitation may be accomplished by uniform base excitation with acceleration time histories, or by multiple arbitrary force time histories with different arrival times and points of application. Utilization of a phased input approach to simulating traveling wave effects required the use of a force time-history applied at different times along the base of the soil-structure system.

In order to utilize the phased force input capability of SAP to obtain the desired response, a large (relative to the structure) spring and mass are attached to the artificial ground node. This spring-mass system is called the "artificial oscillator." See Figure 3. The artificial oscillator spring (an axial boundary element typical to SAP) provides for overall structure grounding to the global absolute system. A force time history, F(t), is applied to the ground node which results in a time-history of response at the node which is, essentially, the criteria free-field time history. To minimize modal coupling between the artificial oscillators and the soil-foundation system, the mass at each ground node is taken as at least 1000 times the mass of the entire structural system. In order to reproduce free-field motion at the artificial ground node, the input F(t) must be defined as follows:

$$F(t) = K\,X(t) + M\,\ddot{X}(t)$$

where

K = Artificial oscillator spring constant

M = Artificial oscillator mass

X(t) = Free-field displacement time-history

$\ddot{X}(t)$ = Free-field acceleration time-history

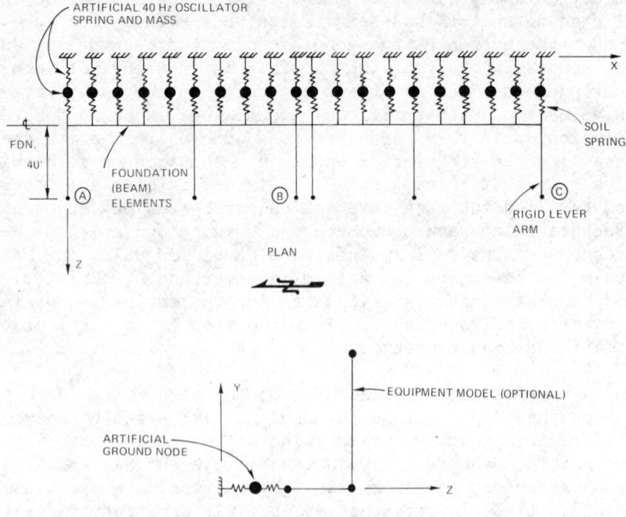

FIGURE 3
HORIZONTAL TRAVELING WAVE – FOUNDATION MODEL

It is of interest that for a very flexible spring, K becomes small relative to M, and F(t) can be represented by M $\ddot{X}(t)$ alone. Conversely for a rather stiff spring, such that the artificial oscillator frequency is beyond the ground motion amplification region, M $\ddot{X}(t)$ becomes small relative to K X(t), and F(t) can be represented by K X(t).

Studies were performed utilizing artificial oscillators with fundamental frequencies in the range of 10^{-4} to 10^{+4} Hz. It was found that when M $\ddot{X}(t)$ was applied to a single low frequency oscillator (10^{-4} to 10^{-1} Hz) the resulting response was identical to that of the input motion. However, when the low frequency oscillator was used in conjunction with a multi-degree-of-freedom system, instability problems arose resulting in non-convergence of the eigenvalue solution. As solution of the eigenvalue problem is fundamental to the mode super-position method, this approach was disgarded for use with SAP.

When K X(t) was applied to very high frequency oscillators, the response was that of high frequency oscillations centered about a waveform very similar in shape to that of the input waveform. The response of a marginally rigid oscillator (fundamental frequency in the vicinity of 30 Hz and less) when subjected to K X(t) tended to be amplified beyond that of the input motion. Thus an artificial oscillator was adopted with a frequency above the amplification-susceptible range (approximately 1 to 30 Hz), but below the very high frequency range (10^{+3} to 10^{+4} Hz).

Practical considerations precluded extensive research into the development of the ideal combination of artificial oscillator frequency and integration time step. Ultimately, a 40 Hz oscillator, in combination with an integration time step of 0.01 seconds, was selected for the phased force input approach with F(t) = K X(t). The 0.01 second integration time step provided for a reasonable reproduction of free-field motion at the artificial ground node, and a good determination of structure response for the below 20 Hz frequency range of interest. Studies showed that smaller time steps, on the order of 0.0025 seconds, provided a more accurate representation of free-field motion at the ground node. The use of such a small time step, however, resulted in a four-fold increase in computing time, and was therefore not used because of the many analyses performed with the 40 second duration record.

In the determination of foundation response, it was unnecessary to include the many individual artificial oscillator modes in the mode super-position analysis because foundation and oscillator responses are uncoupled. In order to obtain a reasonable approximation of soil stresses, however, it was necessary to include each oscillator mode in the solution. The soil stresses, which are defined as a function of the axial deformation of the soil springs, cannot be determined without including the total deformational response of the springs. The total deformation of an individual soil spring is the difference in absolute displacement between the foundation node at one end, and the artificial ground node at the other. Thus the modal responses of the oscillators are required in this case. Unique oscillator modes were achieved by slightly varying the magnitude of each artificial oscillator spring stiffness such that each oscillator frequency was different, but very nearly equaled 40 Hz.

Analytical Model

The analytical model of the foundation is shown in Figure 3. The soil springs are located at each node connecting the discrete 3-dimensional beam elements. The beam elements represent the slab and are located at the slab centerline. The beams represent horizontal in-plane slab behavior in response to horizontal traveling wave input and therefore vertical degrees of freedom are eliminated. The beam elements allowed consideration of shear deformation. Also, rotary inertia at each node was included in the model. The rotary inertia tributary to each node (not automaticaly calculated in SAP as is tributary translational mass) is calculated as that of the mass moment of inertia of a rectangular plate about a vertical axis through its center of gravity. The dimensions of the elemental plate are: slab width (88 feet) by beam element length (variable) by slab thickness (2 feet).

The rigid lever arms with off-center nodes are provided as a convenience so that motions away from the foundation center may be calculated. Assuming plane bending, the motions away from the centerline at each longitudinal location may be calculated considering rigid body rotations about the centerline point.

The mass of all equipment is distributed proportionately to the beam nodes and the effect of equipment and piping dynamic response, i.e. coupling, was considered to be negligible with regard to the foundation response.

Based on the premise of negligible SSI, the transverse rigid body translation of the foundation under global (uniform at all points) seismic shaking should very closely approximate the free-field motion. Note that only the rigid translational mode, with 100 percent of the total foundation mass participating, responds to uniform base excitation. As the rigid body translational frequency of the soil-foundation system is in the vicinity of 10 Hz, the system must be heavily damped for the slab to move with the ground. This heavy damping was verified utilizing simplified calculations consistent with the half-space approach used to calculate soil stiffness. A value of 60 percent damping yields slab response essentially equal to the ground motion for uniform base excitation of the soil-foundation system. This damping value is thus appropriate for the rigid translational mode. Simplified calculations for the torsional mode also indicated relatively high damping. However, in order to provide for a bounding result, and because of lack of prior knowledge with this type of analysis, an estimated lower bound value of 15 percent was selected for damping in the rigid body torsional mode. Damping for all structural modes including slab bending was taken to be 3 percent for OE level events. A study was made to assess the appropriate damping for the artificial oscillators. For damping levels between 0 and 7 percent, no appreciable difference in response at the ground node was observed. Thus undamped oscillators were selected for the determination of soil stresses wherein the inclusion of oscillator modes in the analysis is required. Table 1 illustrates the first few, and most important, soil-foundation system mode shapes and frequencies.

Alternate Method

As an additional check on the analyses described herein, a second method of traveling wave analysis was developed. This method consisted of a single rigid beam model (same length) with only two independent degrees of freedom, one for translation and one for rotation. Similarity in ground motion input and foundation behavior (except for beam-flexure modes) was accomplished by the use of a single artificial oscillator and soil spring to accommodate translational input/response, and likewise for torsional input/response. The translational and torsional input forcing functions were calculated as the summation over the length of the beam of discrete force time histories, taking into account the distance to the C.G. of the beam for the net torsional effect. The distribution of discrete input points for the above calculations was similar to that of the primary beam model of Figure 3. This second method of analysis is analogous to the method developed by Newmark and Hall (Reference 8). In comparing results of the two methods, both models predict some attenuation of free-field motion in the central portion of the foundation, and a significant amplification of motion at each end. The few isolated cases where foundation response in the two methods differed are probably attributable to the role that flexural and shear deformation play in the primary flexible beam model.

SUMMARY OF RESULTS

Method and Model

In addition to the numerical values for response outlined below, development of a satisfactory method and model for TW analysis of the large foundation must be considered as a result. A significant amount of effort provided the model and approach which in turn provided numerical results which are reasonable on the basis of analytical check points and engineering judgement. One check point is observed in that the artificial oscillator and soil spring system with the foundation beam model, when subjected to uniform seismic input, gives foundation responses which are essentially equal to the free-field input. This model has a 60 percent damping factor on the first translational (rigid body) foundation mode. Other check points, i.e. correlations by independent assessments and calculations, are discussed above under "Analytical Method," and below.

TW Analytical Results

Table 1 gives the first four horizontal in-plane mode shapes and frequencies for the foundation model shown in Figure 3. The first two modes are rigid body rotation and translation, and the second two modes are the first and second slab bending modes. The first two modes compare favorably with those based on simplified calculations using half-space theory (Reference 6). The first slab bending mode frequency compares closely to that calculated with a simplified approach for a beam on an elastic foundation.

TABLE 1
FOUNDATION MODES

MODE NO.	MODE SHAPE	FREQUENCY (Hz) OE	ME
1	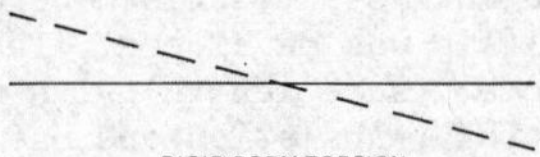 RIGID BODY TORSION	10.8	9.3
2	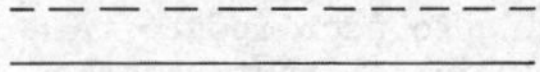 RIGID BODY TRANSLATION	11.1	9.5
3	 FIRST BEAM	13.9	12.8
4	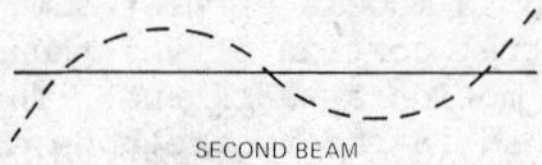 SECOND BEAM	22.4	21.7

NOTE: Frequencies differ between OE and ME cases because of differing soil stiffness.

Attention is called to the fact that modal participation factors calculated within SAP are applicable to uniform seismic shaking only and lose their direct significance for TW (phased force) input definitions. Thus, a zero participation factor calculated for the torsional foundation mode is not an indication of the effect this mode will have on foundation response under the action of a phased force input.

Table 2 summarizes in-plane foundation shears and moments resulting from various analysis cases. For the sake of gaining insight into the sensitivity to soil stiffness, an OE analysis series was made utilizing 70 percent of the normal stiffness. Results were close to being linearly proportional to the fraction of soil stiffness. The standard analysis adopted herein for obtaining TW response omits the artificial oscillator modes. Though values for moment are close, the values for shear vary significantly. The effect of varying Poisson's ratio for the beam material from 0.0 to 0.2 is seen to be relatively small.

Upon review and evaluation of the overall results it was concluded that chord steel should be included at each longitudinal edge of the slab. This steel was designed for the TW OE case. Design for ME moments and shears are not included as the analysis results at this level are much more uncertain because the response may occur in the nonlinear domain, and the GCEP criteria allows utilization of a significant amount of ductility in the ME case. Soil stresses were determined to be relatively low. The maximum soil shear stress under the OE event was 0.4 KSF. This value is of a realistic order and thus provides a checkpoint on the overall results.

TABLE 2
SHEARS, MOMENTS, SOIL STRESSES AND DISPLACEMENTS

EFFECT CONSIDERED	CASE DESCRIPTION	SHEAR (V) (KIPS)	MOMENT (M) (FT.–KIPS)
VARYING* SOIL STIFF-NESS (1)	OPER. E.Q. NORM. STIFFNESS	940.	55,000.
	OPER. E.Q. 70% NORM. STIFF.	710.	43,000.
ARTIFICIAL** OSCILLATOR MODES	OPER. E.Q. MODES NOT INCL.	840.	43,000.
	OPER. E.Q. MODES INCL.	1,420.	35,000.
*POISSON'S = 0.0 **POISSON'S = 0.2	MAX. E.Q. (2) NORM. STIFFNESS	2,860.	186,000.

(1) Artificial Oscillator Modes Included

(2) Artificial Oscillator Modes Not Included

CASE DESCRIPTION	SOIL STRESS (KSF)	Δz END-TO-END FDN. DISPL. (IN.)
OPER. E.Q.(1)**	0.4	0.8
MAX. E.Q. (1)**	1.2	3.4

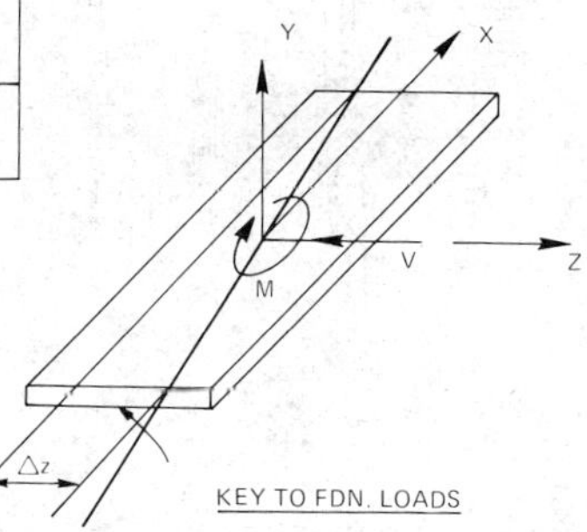

KEY TO FDN. LOADS

Absolute transverse foundation displacements calculated resemble free-field displacements both in waveform and amplitude. A displacement of interest is the maximum relative displacement from one end of the foundation to the other. This value is 0.8 inch for OE and 3.4 inches for ME. Refer to Table 2. These displacements are a measure of the amount of torsion to which the foundation is subjected. The maximum relative displacement between the foundation and ground (free-field) is of interest for piping connections from external locations. The peak values of 0.1 and 0.3 inches for OE and ME, respectively, were determined from the maximum soil spring stress.

TW acceleration time-history response of the foundation is shown in Figures 4 and 5. Comparison with the input free-field motion GATH .02, Figure 1, indicates that some of the higher frequency content of the free-field is filtered out in the foundation response. Table 3 summarizes maximum accelerations for points A, B, and C (Figure 3) for selected analysis cases. Generally, response at the middle point is somewhat attenuated from the free-field, and the end points have somewhat higher accelerations than the free-field. Also a significant longitudinal acceleration component, a major fraction of the free-field acceleration, is observed. It is noteworthy that attenuation in translational effects and amplification related to torsion are more pronounced for the ME case where the TW velocity is lower. This behavior may possibly be explained in that with lower wave velocity there are less full cycles of wave action interacting with the foundation and thus less of a cancelling effect between positive and negative particle motions, with torsional response more emphasized as a result. This behavior is consistant with findings in other analyses (see for example, Reference 8).

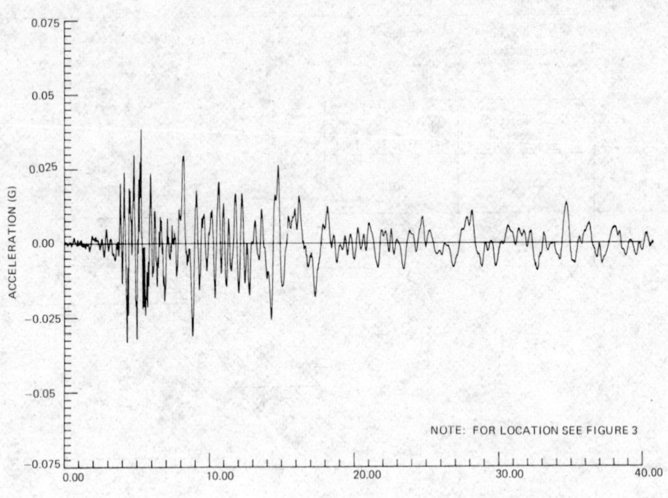

FIGURE 4
TRAVELING WAVE ANALYSIS
Z-COMPONENT ACCELERATION TIME HISTORY AT POINT Ⓑ OPERATING EARTHQUAKE

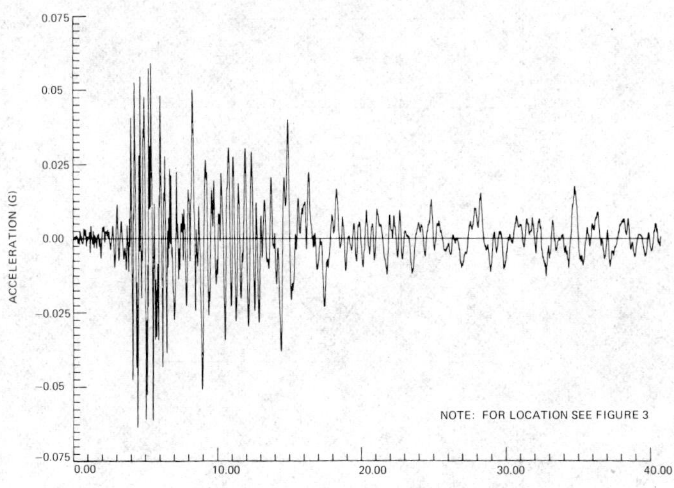

FIGURE 5
TRAVELING WAVE ANALYSIS
Z-COMPONENT ACCELERATION TIME HISTORY AT POINT Ⓒ OPERATING EARTHQUAKE

TABLE 3
SUMMARY OF MAXIMUM ACCELERATIONS

EFFECT CONSIDERED	CASE DESCRIPTION	MAXIMUM FOUNDATION ACCELERATIONS (2) (IN G's)			
		POINT Ⓐ Z COMP.	POINT Ⓑ Z COMP.	POINT Ⓒ Z COMP.	POINT Ⓒ X COMP
VARYING* SOIL STIFFNESS (1)	OPER. E.Q. NORM. STIFFNESS	0.05	0.04	0.06	0.03
	OPER. E. Q. 70% NORM. STIFF.	0.05	0.03	0.07	0.03
	MAX. E.Q.**(1) NORM. STIFFNESS	0.19	0.11	0.22	0.09
ARTIFICIAL** OSCILLATOR MODES	OPER. E.Q. MODES NOT INCL.	0.05	0.04	0.06	0.03
	OPER. E.Q. MODES INCL.	0.05	0.04	0.06	0.03

*POISSON'S = 0.0
**POISSON'S = 0.20

(1) Artificial Oscillator Modes Not Included.
(2) For Locations See Figure 2.

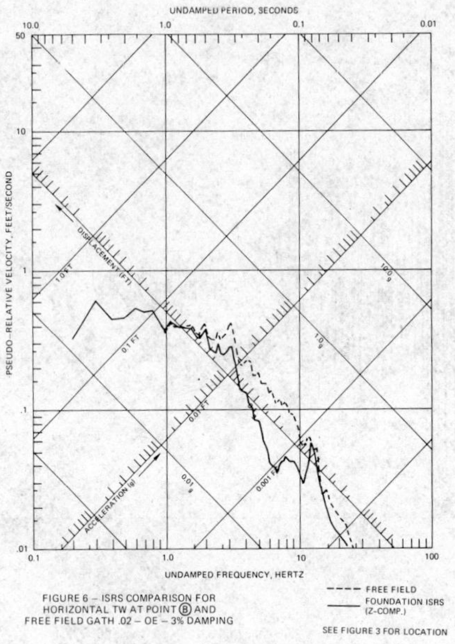

FIGURE 6 – ISRS COMPARISON FOR HORIZONTAL TW AT POINT Ⓑ AND FREE FIELD GATH .02 – OE – 3% DAMPING

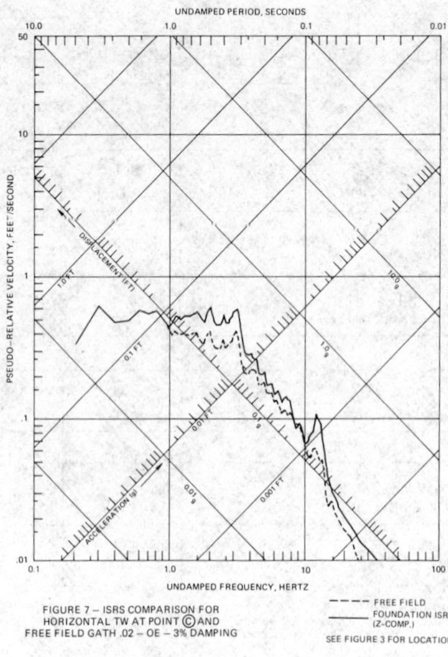

FIGURE 7 – ISRS COMPARISON FOR HORIZONTAL TW AT POINT Ⓒ AND FREE FIELD GATH .02 – OE – 3% DAMPING

A comparison of foundation response spectra for selected foundation points with the free-field response spectra is shown in Figures 6 and 7. For the central point (point B) transverse translational response is attenuated at frequencies above approximtely 1 Hz. Some amplification of response due to torsion is indicated for point C representing the end of the foundation.

In summary, the calculated TW motion responses are not grossly different from the free-field motions, but there is a significant torsional effect, and the translational motion is somewhat attenuated. The torsional effects are more pronounced for the lower wave propagation velocity associated with an ME. The results compare well quantitatively with those reported by other investigators for similar circumstances. Reference 8 relates to a rigid body model to assess torsional affects which give numerical results for response quite similar to the results developed herein. However, in the method developed for GCEP, an assessment of in-plane bending of the foundation can also be made because of the use of non-rigid elements in the model.

REFERENCES

1. Fluor Engineers and Constructors, Inc., Seismic Analysis of Service Module and Train Foundation, prepared for the U.S. Department of Energy, Oak Ridge Operations Office, December, 1979.

2. Newmark, N.M. and Rosenblueth, E., Fundamentals of Earthquake Engineering, Prentice-Hall, Inc., 1971.

3. Seismic Design Criteria for the Gas Centrifuge Enrichment Plant - GCEP, 1978, U.S. Department of Energy, Oak Ridge Operations Office (ORO-EP-120).

4. Seed, H.B. "Design Provisions for Assessing the Effects of Local Geology and Soil Conditions on Ground and Building Response During Earthquakes," presented at the Earthquake Symposium sponsored by the Structural Engineer's Association of Southern California, Los Angeles, California, June, 1975.

5. U.S. Atomic Energy Commission, "Design Response Spectra for Seismic Design of Nuclear Power Plants, Regulatory Guide 1.60, October, 1973.

6. Richart, F.E. Jr., Hall, J.R. Jr., and Woods, R.D., Vibrations of Soils and Foundations, Prentice-Hall, Inc., 1970.

7. Bathe, K.J., Wilson, E.L., and Peterson, F.E., "SAP IV-A Structural Analysis Program for Static and Dynamic Response of Linear Systems," EERC Report 73-11, College of Engineering, University of California, Berkeley, April, 1974.

8. Hall, W.J., Morgan, J.R. and N.M. Newmark, "Traveling Seismic Waves and Structural Response," Proceedings of the Second International Conference on Microzonation for Safe Construction--Research and Application, San Francisco, California, November, 1978.

INCREASED SEISMIC RESISTANCE OF HIGHWAY BRIDGES USING IMPROVED BEARING DESIGN CONCEPTS

BY

ROY A. IMBSEN[1], A.M. ASCE
AND
RICHARD V. NUTT[2], A.M. ASCE

ENGINEERING COMPUTER CORPORATION
Sacramento, California

ABSTRACT

The possibility of increasing the earthquake resistance of bridges through the use of improved bearing concepts is explored. An overview is given of the current use and future trends in bridge bearings. The past performance of different bearing types during earthquakes is evaluated. Several new design concepts for improved bridge bearings are presented. Both expansion and "fixed" bearing design concepts are included. The general conclusions of analytical case studies of bearing response are presented. In addition, the results of analytical studies into the sensitivity of bearing parameters are presented. Finally, a design framework for selecting the most cost-effective bearing design is suggested.

[1]Vice President and Principal Partner, Engineering Computer Corporation, Sacramento, California.

[2]Civil Engineer, Engineering Computer Corporation, Sacramento, California.

I. INTRODUCTION

Bridges are important links in our surface transportation network since they provide the means for overcoming both manmade and natural obstacles. It is crucial that they continue to function in this vital role following an earthquake when protection of lives and property depends on the efficient movement of emergency traffic.

The current seismic design philosophy for bridges accepts yielding in the supporting members as a method of preventing structure collapse during the maximum credible earthquake. Because yielding usually results in structural damage, expensive repairs may be incurred even after an earthquake of moderate intensity.

An alternate or supplemental means of earthquake resistance which will prevent or minimize yielding, is desirable from both a functional and economic point of view. One method of providing such earthquake resistance is through the imaginative design of bearings. A prerequisite to improved bearing design is a thorough understanding of the effect of a bearing system on the earthquake resistance of a bridge and the importance of bearing design in the overall seismic design process.

The primary objective of this paper is to explore cost effective means of increasing the earthquake resistance of bridges through the use of improved bearing systems.

II. BACKGROUND

Bridge bearings have traditionally been used at superstructure/substructure interfaces to provide for structural support, relative movements and incompatibility of materials. "Fixed" bearings must provide for the transmission of vertical and horizontal forces to the substructure while accommodating the non-compatible rotations of the superstructure with respect to the substructure. Expansion bearings on the other hand are required to accommodate large relative translational movements due to temperature, shrinkage, creep, prestressing, etc., while maintaining their vertical load carrying capabilities. Most bridges in existence today use either sliding plates, rolling devices, linkage or rocking devices, or elastomeric pads as bearings.

Prior to the San Fernando earthquake of 1971, resistance to lateral forces at bridge bearings was provided by shear keys, keepers, or other restraining devices of nominal design. Longitudinal force was transferred to the substructure through fixed bearings at one or more points on the superstructure. In most cases, adjacent superstructure sections were not tied together at the expansion joints which led to several bearing and structural failures during earthquakes.

Within the United States, earthquake resistance at bearings is still provided in the form of shear keys, keepers, cables, etc., which are designed to elastically resist the forces induced during an earthquake. Restraint of this type represents a cost effective means of mitigating catastrophic failures, but is not effective in absorbing the energy generated by an earthquake. Thus post-elastic behavior of the supporting members must still be relied on for this purpose.

Many researchers and innovative engineers have recently been considering a broader role for bearings in bridge earthquake resistance. Most of these efforts are directed toward the use of isolation and/or energy dissipation in the bearings.

Within the United States, the acceptance of bridge bearings as a means of improving earthquake resistance has been slow. In many cases, bearings are an "after thought", selected near the end of the design process when the earthquake resistant mechanism has already been established. One reason for this reluctance to use bearings for earthquake resistance is the lack of readily available earthquake resistant bearing systems of proven reliability. Another reason is the lack of reliable, cost-effective analytical tools with which the earthquake resistant qualities of bearings can be assessed. It is, therefore, necessary to overcome both of these deterrents if bearings are to play a larger role in the earthquake resistance of structures.

III. DEVELOPING RELIABLE, COST EFFECTIVE BEARINGS FOR EARTHQUAKE RESISTANCE

A. Evaluation of Existing Bearings

Before developing a more promising earthquake resistant bearing design concept, it is necessary to understand the relative merits of existing bearing types.

Rocker bearings have been used in bridges for several years. They are generally made of steel and provide for movement in the direction of expansion by rocking of a steel link which bears on plates imbedded in both the superstructure and the substructure.

During earthquakes, these bearings have not performed very well. They become unstable outside a relatively narrow range of movement and numerous instances of toppled rocker bearings have been reported. A common problem with these bearings during an earthquake is the failure of the transverse keeper plates. This type of failure may increase the energy dissipation potential of these bearings due to subsequent sliding, but it also increases the likelihood of the bearing becoming unstable.

Roller bearings are also generally made of steel, and provide for movement by rolling of a cylinder.

Under earthquake loading they are much more stable than rockers, since their movement range is generally much larger. Like the rockers, they are difficult to restrain transversely and possess little energy dissipation potential unless sliding occurs. They generally perform satisfactorily during an earthquake, but do not improve the overall dynamic response of the structure.

Sliding bearings range from simple asbestos sheet packing between the substructure and the superstructure for short span bridges to elaborate "pot" bearings with teflon and stainless steel sliding surfaces for structures with large movements.

Because of their nature, sliding devices are extremely stable during an earthquake. Although the sliding of two surfaces will dissipate energy, the low friction values used limit the energy dissipation in expansion bearings. Generally, sliding bearings have performed well during past earthquakes.

Elastomeric pads are widely used for bridges and have gained in popularity over the past several years. Most pads are constructed by bonding alternating layers of elastomer and reinforcement (steel or fabric) together. The elastomer consists primarily of either natural rubber or neoprene although other synthetic rubbers have a potential for bearing applications.

In general, these bearings have performed well during past earthquakes although they have been known to "walk-out" if not properly secured.

Although many bearings have suffered damage during earthquakes, it is important to remember that loss of vertical support has resulted in the most spectacular failures. Although qualities of energy dissipation, isolation, and restraint can reduce force levels and/or mitigate structure response, experience has shown that the primary performance characteristic of an earthquake resistant bearing is the maintenance of vertical support. Therefore, any bearing worthy of consideration must possess a high degree of reliability in this area.

B. Promising New Design Concepts

In exploring potential new bridge bearing design concepts for earthquake resistance, it is convenient to categorize bearings as either "fixed" or expansion.

There are several concepts for expansion bearings that show promise for improved earthquake resistance. One such concept employs a more or less conventional elastomeric bearing pad. During normal movement and moderate earthquakes, the pad would function exactly as present elastomeric pads. However, during a strong earthquake, one surface would be designed to slide. The coefficient of friction would be selected to preserve the stability and integrity of the pad itself while limiting the inertia force that could be transferred to the bridge supports. The remaining surface of the pad would be physically attached to the substructure to insure that all movement would take place on the sliding surface and to prevent pad "walk-out". A schematic drawing of this design concept is shown in Figure 1.

Such a system has the following advantages:

1. A high degree of reliability in maintaining vertical support.
2. Protection of the supports from loads which could cause excessive damage.
3. Potential for significant energy dissipation at high displacement levels.
4. A high degree of isolation which will tend to limit structural response.
5. Compatibility with current production capabilities of bearing manufacturers.
6. Designer acceptance of elastomeric bearing pads.

A somewhat different concept is to use viscous dampers at the expansion bearings. These devices have force-displacement characteristics that vary with the rate of displacement. At slow rates of displacement, such as temperature movement, very little force is generated. At high rates, such as in an earthquake, more force is required to deform the device. This allows the designer to make use of all supports to resist the earthquake load. In addition, energy is dissipated throughout the course of rapid movement. This tends to mitigate the resonant condition that can increase force levels so dramatically during an earthquake.

The Japanese have used this concept extensively for earthquake resistance. To date viscous dampers have been separate devices which are used in conjunction with conventional bearings. These devices have a history of maintenance difficulties due to leaking seals, etc.

To meet the requirements of a self-contained bearing which will be relatively maintenance free and stable under sustained load, a bearing such as the one shown in Figure 2a is proposed. This bearing would consist of a strong, reinforced flexible membrane which would contain an

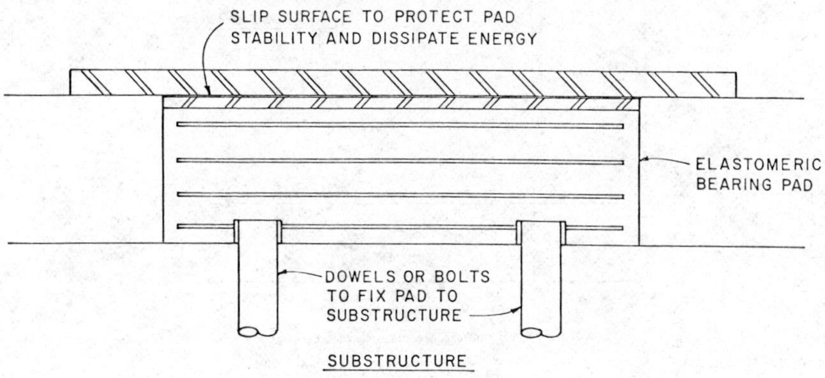

EARTHQUAKE RESISTANT BEARING CONCEPT IE

Elastomeric Pad with Slip Surface

Figure 1

incompressible viscoelastic material. Relative movement of the two bearing surfaces would cause the block of viscoelastic material to deform. During an earthquake, this deformation would require more force and therefore the device would behave like a viscous damper.

This type of bearing is susceptible to damage caused by vandalism, fire, or tearing due to excessive relative movement of the bearing surfaces. Any loss of viscoelastic material could result in collapse of the bearing. To prevent damage due to vandalism and fire, a shield could be constructed around the bearing. This shield would be used to induce sliding of the bearing which could prevent the membrane from being torn apart due to excessive movements. A schematic drawing is shown in Figure 2b.

Another solution is to use a normal elastomeric bearing pad to support the vertical load. The lateral sitffness of the elastomeric pad would induce sliding at excessive displacements. The pad would continue to support the vertical load even if the flexible membrane was punctured. This concept is shown schematically in Figure 2c.

A considerable amount of development is required before this flexible membrane concept can be used with

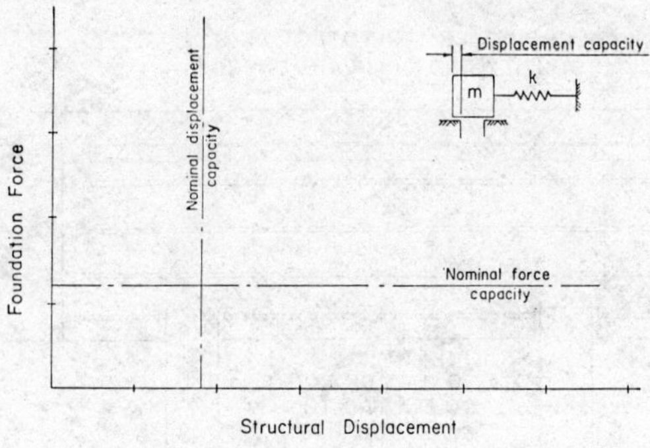

Maximum Structural Response—Non Seismic Loading

Figure 7a

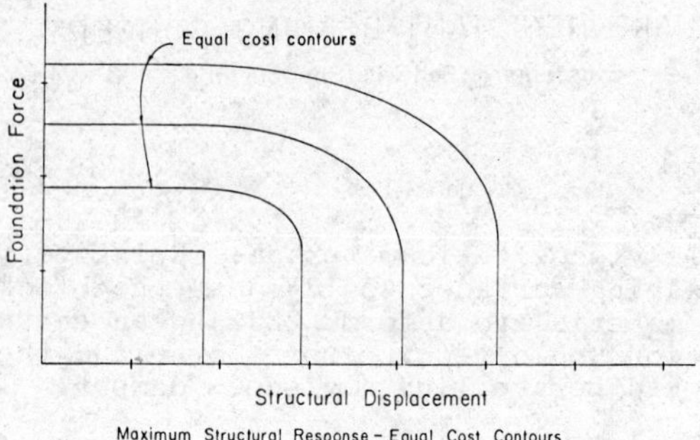

Maximum Structural Response - Equal Cost Contours

Figure 7b

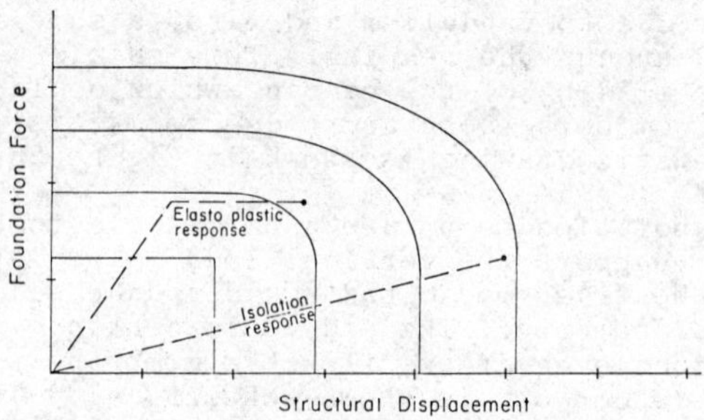

Maximum Structural Response - Sample Cost Comparison of Earthquake Resistant Mechanism

Figure 7c

confidence in a bridge structure. The membrane must possess unusually high strength, stability, and durability. The viscoelastic material must have suitable force-displacement characteristics over the necessary range of temperatures. These materials must be developed and tested. In addition, these materials must be made to work as a unit. Tests of the assembled bearing system should be conducted to determine performance characteristics during an earthquake. Although a considerable development effort is required, this concept may be the answer to providing reliable, cost-effective viscous damping at bridge bearings.

"Fixed" bearings are in many ways similar to bearings used for buildings or other similar structures. There are essentially two innovative approaches to improving the earthquake resistance of these bearings.

One method of improving earthquake resistance is to provide nominal fixity of a "free" or expansion bearing. Fixity is provided by a "fuse" which will resist lateral loads caused by wind or braking, but will fracture during a strong earthquake. Once the "fuse" has failed, the structure will be isolated from the earthquake ground motion. This will result in very small force levels in the supporting members, but larger relative displacements between the supports and the bridge superstructure. Generally, a restoring force should be provided to return the structure to a neutral position following the earthquake. Several individuals have proposed this type of bearing system.

The second approach to the design of earthquake resistant "fixed" bearings attempts to mitigate the effect of high relative displacements experienced in isolated structures. Although relative movement will be allowed at the bearings under severe lateral loading, the primary mechanism of earthquake resistance is energy dissipation as opposed to isolation. To provide for energy dissipation, several devices have been proposed. In mose cases these devices also serve as "fuses" to prevent movement at low force levels.

A currently popular method of providing energy dissipation is through the use of steel devices designed to yield under severe loading. The devices are relatively inexpensive, maintenance free, and if properly designed, reliable through several cycles of yielding. Some of the more popular devices relay on twisting of a torsion bar, bending of a tapered cantilever beam, and bending and rolling of a steel hoop. Another device, which has the advantage of utilizing "off-the-shelf" steel components, is the constant moment simple beam device shown in Figure 3. These devices are very adaptable to bridge use and many of them have been used successfully on bridge in

New Zealand.

Energy is also dissipated due to the sliding of two surfaces. By selecting the appropriate coefficient of friction, many sliding expansion bearings in use today could be made to function as "fixed" bearings. In this type of installation, attention should be given to preventing sliding of the surfaces caused by vibration of the superstructure under normal loading conditions. A suggested approach to this problem at a typical bridge abutment is shown in Figure 4. In this case, enclosed elastomeric bearings are sloped to keep the structure in a neutral position under normal conditions. During an earthquake, the superstructure would have to be lifted to cause sliding.

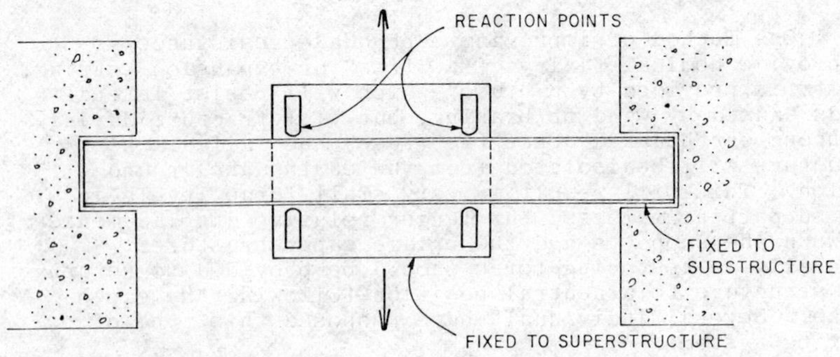

CONSTANT MOMENT SIMPLE BEAM DEVICE

Figure 3

There is not enough space to discuss in detail the many bearing concepts that have been developed to date. The innovative designer has several reliable bearing systems from which to choose. With continued research in this area, the number of possibilities can only increase.

IV. SEISMIC RESPONSE OF BRIDGE BEARINGS

A. Analytical Techniques

Faced with a wide choice of bearing types and functions, the designer must be able to select the bearing most appropriate for his bridge. Since many bearings have non-linear force/displacement characteristics, it is difficult for the designer to determine the overall effect

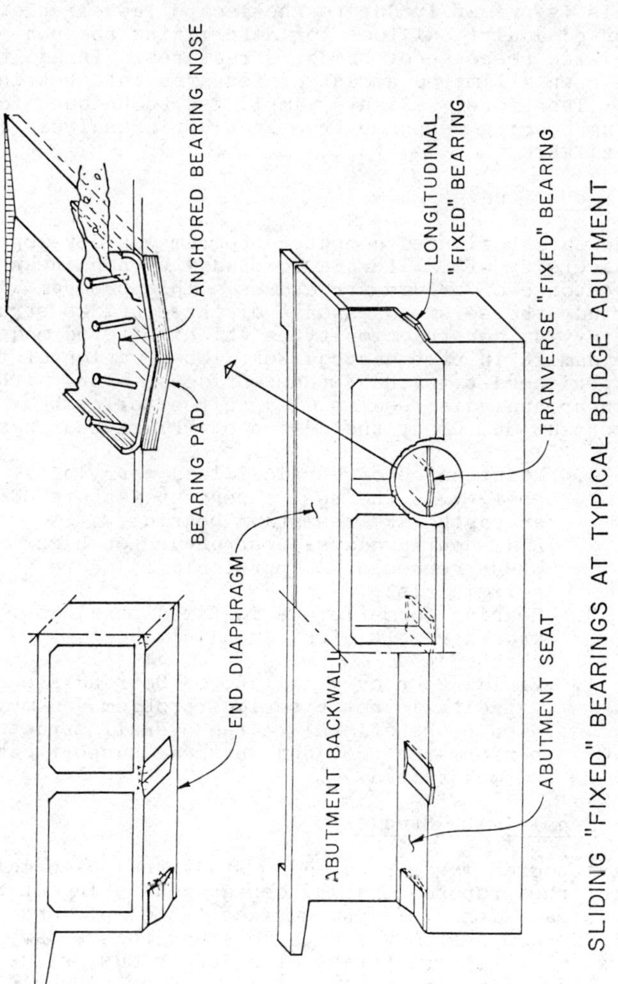
ANCHORED BEARING NOSE
BEARING PAD
LONGITUDINAL
"FIXED" BEARING
TRANVERSE "FIXED" BEARING
END DIAPHRAGM
ABUTMENT BACKWALL
ABUTMENT SEAT
SLIDING "FIXED" BEARINGS AT TYPICAL BRIDGE ABUTMENT

Figure 4

of bearing response on the seismic response of his structure. This is primarily due to the lack of a reliable, user-oriented analytical tool for determining the nonlinear seismic response of bridge structures. In addition, there has been a limited amount of research into bearing response. Therefore, reliable simplified techniques for determining bearing response from an elastic analysis are also unavailable.

B. Case Studies

A research oriented computer program was developed at the University of California to study the nonlinear seismic response of bridge structures. This program was used to study the seismic response of three bridge structures of varying construction types which suffered moderate to severe damage in past earthquakes. Observation of the damage experienced by each of these bridges, along with the nonlinear analysis results, led to the following general conclusions about the design of bridge bearings:

1. Maintenance of vertical load carrying ability is the most important feature of an earthquake resistant bearing.
2. A discontinuous superstructure on high piers represents a particularly acute seismic risk.
3. Bearing force levels in fixed bearings are extremely high, even in moderate earthquakes.
4. Yielding or even failure of bearing shear capacity is not a serious problem and may even be beneficial to the overall structure performance provided vertical support can be maintained.

C. Sensitivity Studies

This program was also used to study the potential behavior of the proposed bearing design consisting of an elastomeric pad with a sliding surface. A simple one degree of freedom system was used to study this behavior. Such a system is representative of a continuous bridge structure responding in the longitudinal direction. A schematic diagram of this behavior and the single degree of freedom idealization is shown in Figure 5.

For the purpose of this study, all bearings were assumed to reach the maximum force level simultaneously. In using this earthquake resisting mechanism, the designer is faced with selecting the proper longitudinal stiffness and maximum force level.

In order to investigate the sensitivity of the response to several of the parameters over which the

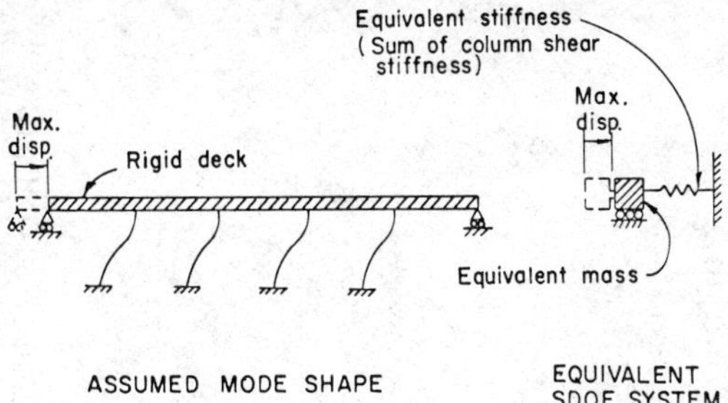

SINGLE DEGREE OF FREEDOM IDEALIZATION OF BRIDGE LONGITUDINAL MODE

Figure 5

designer has control, the simplified model was analyzed with different values for each of the parameters. A time history of the force and displacement response of each of the simplified models to two different earthquake motions was determined. The maximum responses for each analysis were recorded and plotted on a force vs. displacement graph. This graph is shown in Figure 6. Notice the characteristic shape that these graphs take when bearings of constant lateral stiffness have the coefficient of friction of the sliding surface varied. As can be seen, there is a coefficient of friction at which both displacement and force will be relatively low. This is the type of information the designer must have when selecting the best bearing design for his application.

V. DESIGNING COST EFFECTIVE EARTHQUAKE RESISTANT STRUCTURES USING BEARINGS

Once he is able to determine the maximum displacement and acceleration responses caused by a particular bearing system, the designer must select the most appropriate bearing for his bridge. Criteria has been suggested which seeks to minimize some aspect of structural response, such as acceleration. However, given equal reliability of bearing systems, the most meaningful approach is to

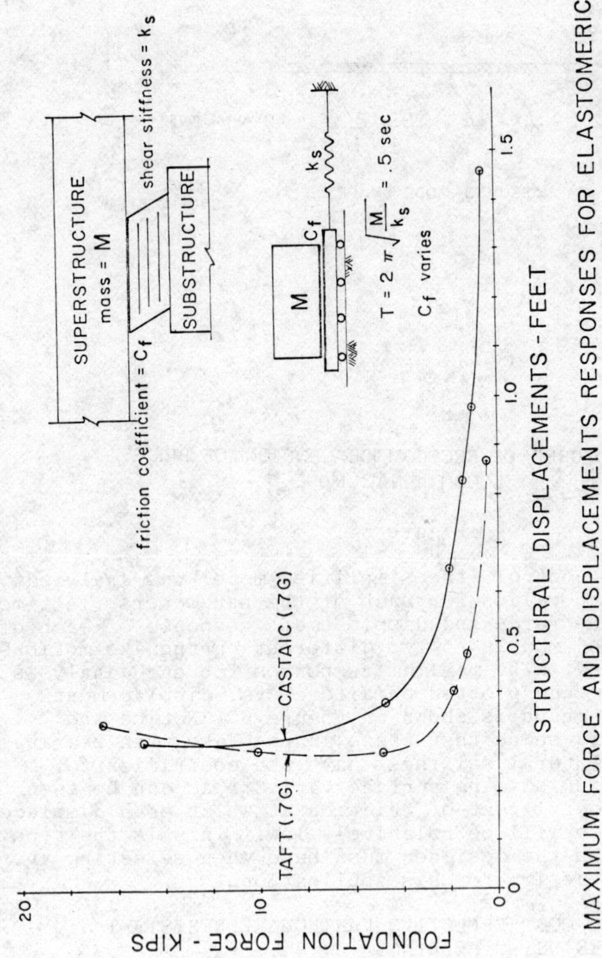

MAXIMUM FORCE AND DISPLACEMENTS RESPONSES FOR ELASTOMERIC BEARING PAD SYSTEMS OF PERIOD = .5 SEC & VARIABLE COEFFICIENTS OF SLIDING FRICTION

Figure 6

minimize cost. A suggested technique for investigating the relative cost of using various bearing systems is illustrated graphically in the following paragraphs.

Consider the maximum force vs. maximum displacement graphs developed in the previous section. Structural displacements of a bridge deck in the longitudinal direction are usually limited by the available seat length and bearing movement capacity at certain supports such as the abutments. When earthquake loading is not considered, these lengths will usually be governed by expansion and contraction of the superstructure and/or minimum detailing standards. In the case of foundation forces, lateral loads other than those created by seismic loadings will require design to a certain capacity. These nominal displacement and force values which do not include the effects of earthquake loading may be plotted on a maximum displacement versus maximum foundation force graph as shown in Figure 7a. Costs associated with increasing the displacement and force capacity are generally the most significant costs in any seismic design.

To assess the economic advantages of a particular design involving a simple structural system such as the one mentioned previously, it is necessary to evaluate the increased displacement or foundation force capacity that can be purchased for a given increase in total structure cost. By plotting the increase in these capacities for incremental increases in cost and by evaluating the various combinations of displacement and force capacities that can be purchased at each level of cost increase, contours of equal incremental cost increases can be plotted on our graph as shown in Figure 7b. With these contours, a means now exists for evaluating the economic merits of a given design approach.

Consider as a hypothetical example the selection of a bearing system for a simple highway bridge. The designer may first wish to consider isolating bearings in which a "fuse" is provided to resist normal wind and braking forces. Once the fuse has failed during an earthquake, the foundation will be isolated from the inertia forces in the superstructure, and foundation forces will remain at a low level. The resulting structural displacements will generally be large, however. The resulting response may be plotted on the graph as shown in Figure 7c.

As a second alternative, the designer might consider a bearing which uses an elastoplastic energy dissipation mechanism. In such a design, the displacement response would be reduced, but at the expense of increased forces. If plotted on the force-displacement graph as shown in Figure 7c, the relative cost can be compared with other alternatives. In this example, the second alternative is more cost effective than isolation. By varying the initial stiffness and yield level of the elasto-plastic energy dissipation mechanism, it may be possible to achieve even better economics. By comparing alternatives based on equal cost contours, the optimum design can be selected. This procedure may be extended to more complicated design problems provided the analytical tools are available to predict structure response.

VI. CONCLUSIONS AND RECOMMENDATIONS

The possibility of improved earthquake resistant bridge design through the effective use of bearings shows great promise. Before this can become a reality, the designer must be convinced that this approach will produce cost effective bridge designs. This requires that reliable earthquake resistant bearing systems be readily available or easily manufactured. To prove the reliability of new design concepts, laboratory testing is needed. This testing should produce information that is meaningful to the designer and can be used in designing similar bearings.

In addition, the designer must be provided with the analytical tools with which to assess bearing response. Not only should the designer have access to easy-to-use non-linear dynamic analysis computer programs, but he should also have guidelines for using normal elastic analysis techniques to predict bearing behavior.

Once reliable bearings are made available and the designer can analyze their behavior during an earthquake, bridge bearings will begin to achieve their full potential in improving the earthquake resistance of bridge structures.

SEISMIC STUDY OF IMPERIAL COUNTY SERVICES BUILDING

by

Mehdi Saiidi,[1] A.M. ASCE

ABSTRACT

The response of the Imperial County Services Building to the earthquake of October 15, 1979 is calculated using complex and simple nonlinear analytical models. The analytical results are compared with the measured responses and the correlations are discussed. Conclusions are reached regarding the assumptions and idealizations made in the analyses in relation to the actual behavior of the structure.

INTRODUCTION

Calculated responses of structures are usually influenced by the many assumptions made in the analysis. In relatively complex analytical models, the results are affected by several factors, many of which are not well established. To determine the validity of the assumptions made and the reliability of the analytical models, the calculated results need to be somehow verified.

The nonlinear analysis of reinforced concrete structures subjected to strong ground motions is among the complex problems with several parameters involved [4]. To verify the analytical results, model structures have been tested and the experimental results have been compared with those calculated [3,8]. The physical models have often provided valuable information about the behavior of structures. Nevertheless, there has been a considerable interest in the measured response of structures built in ordinary field condition, subjected to earthquakes strong enough to cause nonlinear deformations. Such responses are relatively rare.

During the earthquake of October 1979 in Southern California, the acceleration response of the Imperial County Services Building was measured [5]. The structure is a six-story reinforced concrete frame building with walls in one direction. This paper presents an analytical study of the behavior of the building, using the "multi-degree" and "single-degree" nonlinear models which were previously introduced and examined against the observed responses of eight small-scale structures tested at the University of Illinois [8]. It is not the intention of this paper to explain the causes of the local failures in the building. Rather, the paper presents a discussion regarding the correlation between the calculated and measured results.

[1]Assistant Professor of Civil Engineering, University of Nevada, Reno

DESCRIPTION OF THE STRUCTURE

The structure is a six-story reinforced concrete building with five bays in east-west and three bays in north-south directions. The overall plan view dimensions of the structure are 136'-10" in EW and 85'-4" in NS, and the overall height is 81'-8" above ground. The structural system consists of reinforced concrete columns, slabs, and joists supported by beams in the longitudinal direction (Fig. 1). There are four shear walls in the first story in the transverse direction. Parallel to these walls, two exterior walls extend from the second story to the top of the building. The story heights are given in Fig. 2. The specified ultimate compressive strength of concrete for walls, beams, joists, and slabs was 4000 psi and for columns was 5000 psi at 28 days. The yield stress of the reinforcing steel bars was 50,000 psi. Detailed information about element dimensions and reinforcement distribution is provided in Reference 5.

The building was instrumented by a 13-channel recording system. The accelerometers were located on the ground floor, second floor, fourth floor, and the roof of the building. The acceleration histories were measured in the longitudinal, transverse, and vertical directions [5].

THE ANALYTICAL MODELS

Two analytical models, one "multi-degree" and the other "single-degree" (called the Q-Model), were used in the present study. Both these models are for nonlinear analysis of reinforced concrete planar frames subjected to base accelerations. The models have been successful in simulating the response of several test structures [8].

The MDOF Model - This model idealizes the structure as an assemblage of line members coinciding with the axes of structural elements. The structure is assumed to be supported by a rigid foundation with no rocking. A structural element is assumed to consist of (a) two infinitely rigid end portions which represent the parts of the element in the adjacent joint cores, (b) a flexible elastic middle portion, and (c) two nonlinear springs connecting the middle portion to the rigid ends. The hysteretic behavior of the springs is governed by the rules of a number hysteresis models. For the present study the Takeda model was used [9]. In formulating member stiffnesses, axial deformations are ignored, and shear deformations are taken into account only for walls. Only flexural nonlinearity is considered. Masses are assumed to be lumped at floor levels. The differential equation of motion is numerically integrated, and displacement and acceleration response histories at different floors as well as base shear and base moment histories are recorded and plotted [8].

The Q-Model - This is a very simple and economical model which treats the structure as a nonlinear single-degree-of-freedom system. The structure is idealized by a lumped equivalent mass mounted on a massless rigid bar which in turn is connected to the ground by a hinged support and a nonlinear rotational spring [8]. To account for energy dissipation during elastic stage, a viscous damper is connected to the mass. The primary force-deformation characteristics of the spring are related

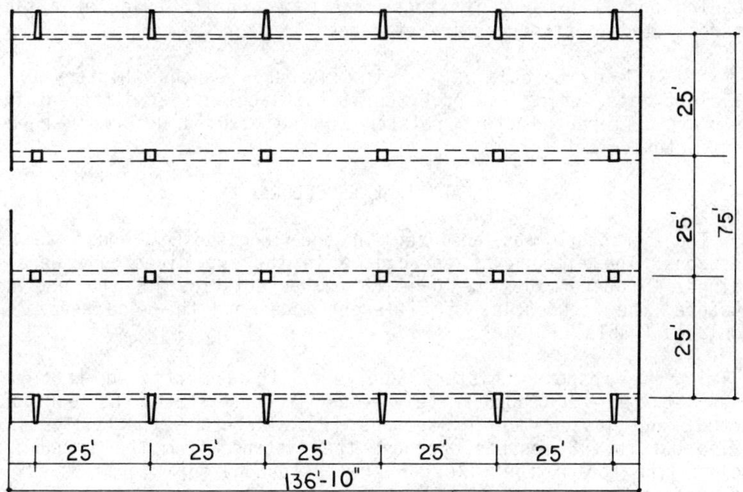

Fig. 1 Plan View of a Typical Floor

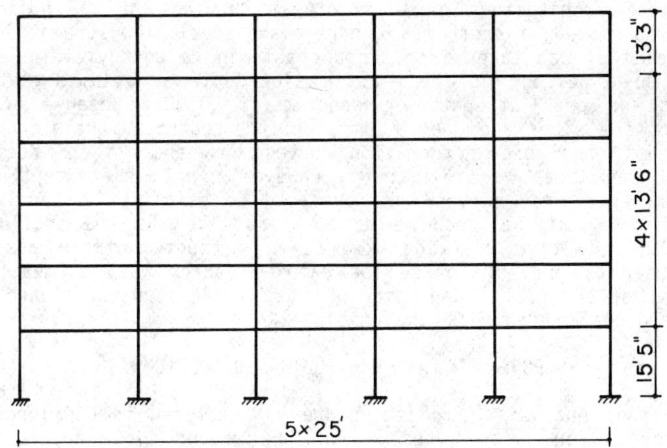

Fig. 2 Elevation of the Structure in the Longitudinal Direction

to the behavior of the multistory structure subjected to static lateral loads at floor levels. The stiffness properties of the spring at unloading and load-reversal stages are based on the rules of a simple hysteresis model introduced in Reference 8.

The Q-Model calculates the displacement response history of the equivalent mass, which is converted to displacements at different floor levels by different factors related to the assumed deformed shape of the structure.

ANALYTICAL PROCEDURE

The structure was analyzed in the longitudinal and transverse directions. The effects of earthquake in the two directions were not combined. To determine the dynamic characteristics of the uncracked structure, the frequencies of different modes and the mode shapes were calculated (Table 1).

In the response history analysis, it was noticed that major pulses started in approximately four seconds after the earthquake started, and lasted for 15 seconds [2]. Therefore the analysis was carried out for 15 seconds. Because the distance from the ground floor level to the top of the pile cap in the foundation is only 2'-2", the acceleration at the ground floor was assumed to represent the base acceleration.

The moment-curvature relationships for structural members were calculated based on the dimensions and reinforcement distribution provided in the architectural and structural drawings of the building, using the specified strength for concrete and steel. The strain hardening of the steel was taken into account assuming a post-yielding slope equal to three percent of the elastic slope. It was assumed that the strain at the peak strength of concrete is 0.002. Because of the presence of the slabs and due to different top and bottom reinforcement in the beams, the force-deformation curves for negative and positive moments of each beam were different. However, with the exception of cracking moments, the curves were averaged to simplify the analysis. The cracking moment for each member end was taken as the smaller of negative and positive cracking moments to account for the possible cracks caused by gravity loads. The damping factor was assumed to be two percent of the critical damping of the elastic structure (uncracked for MDOF and cracked for SDOF analysis).

RESPONSE IN THE LONGITUDINAL DIRECTION

The MDOF Model Analysis - The structure was analyzed in the longitudinal direction for fifteen seconds of the ground floor acceleration record. The time interval of numerical integration was 0.01 second. The acceleration and displacement response histories at different levels were calculated and plotted. The calculated responses were superimposed on the measured ones, where available (Fig. 3). The "measured" relative displacement responses are those integrated from the measured accelerations [2]. The acceleration response maxima are listed in Table 4.

Figure 3 represents the calculated and measured acceleration and

TABLE 1 - MODE SHAPES AND PERIODS IN THE LONGITUDINAL DIRECTION

Mode No. / Floor	1	2	3	4	5	6
Roof	1.0	1.0	1.0	1.0	1.0	1.0
6	0.94	0.49	-0.33	-1.41	-2.49	-3.30
5	0.83	-0.26	-1.38	-1.04	1.33	4.44
4	0.68	-0.91	-0.82	1.50	1.10	-4.44
3	0.50	-1.15	0.69	0.90	-2.47	3.29
2	0.27	-0.84	1.33	-1.54	1.46	-1.30
Period (Sec.)	0.62	0.20	0.12	0.08	0.07	0.06

TABLE 2 - ASSUMED DISPLACED SHAPE OF IN THE LONGITUDINAL DIRECTION

Floor	2	3	4	5	6	Roof
Displacement	0.35	0.58	0.76	0.90	0.97	1.0

TABLE 3 - PROPERTIES OF THE Q-MODEL IN THE LONGITUDINAL DIRECTION

Parameter	Quantity
Equivalent Height (in)	677.
Equivalent Mass (Kip - S^2/in)	11.8
Base Moment at the "yield" point (kip-in)	709100
Displacement at the "yield" point (in)	2.48
Base Moment at a point beyond the "yielding" (kip-in)	1174600
Displacement at a point beyond the "yielding" (in)	12.42

TABLE 4 - MAXIMUM ACCELERATION RESPONSES

Unit = G

Floor		2	3	4	5	6	Roof
Long.	Meas.	0.27	-	0.26	-	-	0.45
	Calc.	0.31	0.34	0.33	0.33	0.36	0.38
Trans.	Meas.	0.31	-	-	-	-	0.56
	Calc.	0.44	0.40	0.33	0.29	0.34	0.42

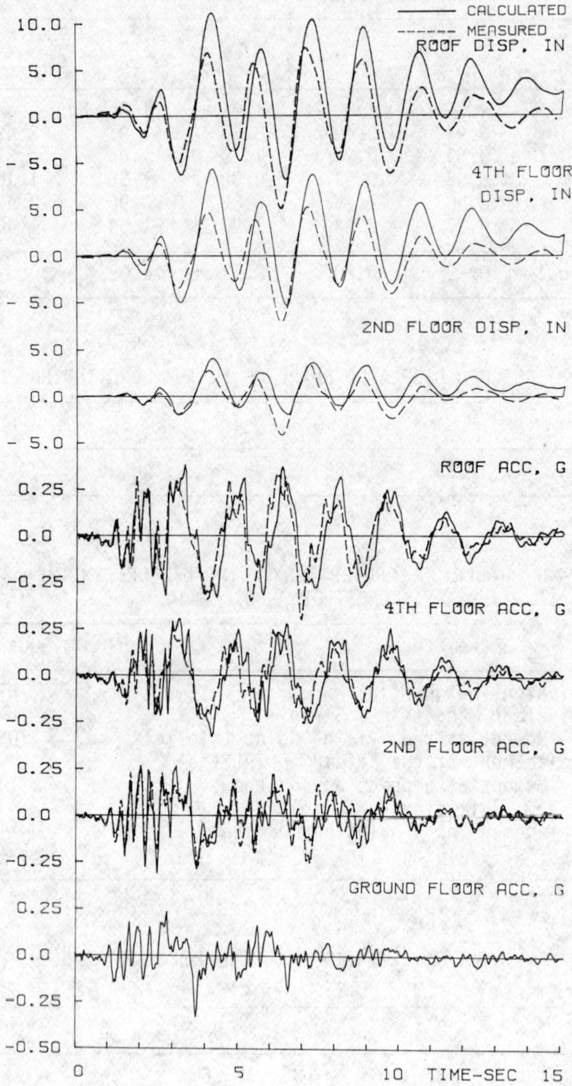

Fig. 3 Calculated and Measured Responses in the Longitudinal Direction

relative displacement responses. It can be seen that the calculated waveforms were generally similar to those measured. The fundamental frequencies of the responses were close and the peak values occurred simultaneously for most instances. However, differences in high-frequency amplitudes were noticeable. At the roof, the calculated acceleration was nineteen percent smaller than the measured value. The two peaks did not occur simultaneously. At the fourth floor, the calculated peak value was twenty-six percent greater than the measured maximum acceleration. During the first and last five seconds of the response at the second floor the calculated and measured responses were quite close. At the middle, the calculated peaks were generally lower than the measured values and the correlation was poor. However, the calculated and observed maximum responses occurred at the same time. The maximum response was overestimated by fifteen percent.

The broken curves in the top three responses in Fig. 3 represent the displacements, relative to the ground floor of the structure, integrated from the measured accelerations. Here, the curves are referred to as "measured" displacements. It can be seen that for the first three seconds there was a close agreement between the calculated and measured responses. Beyond this point, although the waveforms were generally similar, there was a period elongation between T=4 and T=5 seconds in the calculated results. The calculated curves were shifted toward the positive region. The measured and calculated double-amplitude displacements at the roof were about the same. During the last four seconds of the response, the measured and calculated curves experienced small amplitudes. The absolute calculated and measured maxima are shown in Fig. 5. The calculated and measured maxima did not occur at the same time. However, in terms of absolute values, the agreement was close.

In view of the simplifications made in the analysis and the uncertainties involved in the concrete and steel strength used in the analysis, the overall performance of the analytical model for the longitudinal direction is regarded as satisfactory.

<u>The Q-Model Analysis</u> - The displacement response history of the structure was calculated using the Q-Model. To determine the lateral force-deformation characteristics, the structure was analyzed for static lateral loads applied in small increments at floor levels. The loads had a triangular distribution with the maximum at the top. The resulting base moment top-level displacement curve was approximated by a bilinear relationship. The lateral displacements of the structure corresponding to the idealized "yield" point were normalized with respect to the top-level displacement and used as the deflected shape of the structure during the earthquake (Table 2).

The equivalent height and mass of the "single-degree" structure were found based on the methods described in Reference 8, and are listed in Table 3. The idealized curve was converted to represent the displacement at the equivalent height. The coordinates of the "yield" point and an arbitrary point beyond the yielding are presented in Table 3. The latter point is used to define the slope of the "post-yielding" branch.

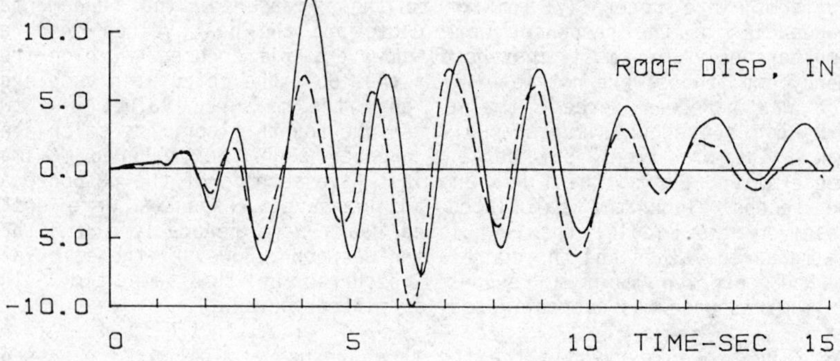

Fig. 4 Calculated (solid line) and Measured (broken line) Relative Displacement Response Based on the Q-Model

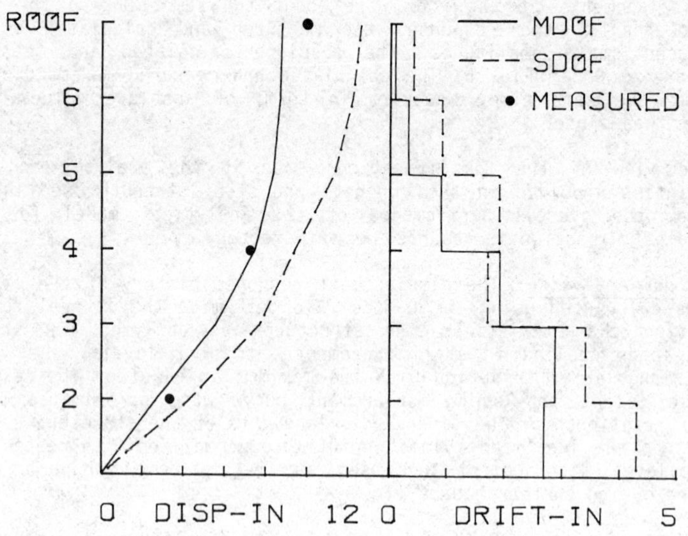

Fig. 5 Maximum Relative Displacements in the Longitudinal Direction

The structure was analyzed for fifteen seconds of ground floor acceleration record. The time interval for numerical integration was 0.07 seconds. The calculated roof displacement history superimposed on the measured response is shown in Fig. 4.

The Q-Model results were very similar to those calculated using the MDOF model. In relation to the measured response, the Q-Model produced a curve with reasonable agreement with the measured curve in frequency content and waveform. The calculated and measured maxima did not occur at the same time, and the absolute calculated maximum response was twenty-seven percent greater than the measured value.

The maximum floor displacements and story drifts calculated using the Q-Models are shown in Fig. 5. It can be seen that the maximum displacements were generally overestimated compared to the measured values and MDOF results. The same was true for the maximum story drifts.

The above observations show that the Q-Model resulted in a response similar to that obtained from a MDOF model, while the computational effort was considerably less than that required for the MDOF analysis. The performance of the Q-Model for the east-west direction of the structure studied here was satisfactory and promising.

RESPONSE IN THE TRANSVERSE DIRECTION

The lateral load resisting system of the structure in the transverse direction consists of four shear walls in the first story and two exterior walls from the second story to the top of the building. The floors consist of joists supported by girders in the longitudinal direction. There are no girders in the transverse direction.

In the analysis, it was assumed that frames consisted of the columns and beams representing the joist system bounded by panel center lines for the interior frames, and panel center line and the edge of the structure for the exterior frames. The six frames were combined into one three-bay six-story frame. In addition, the walls were combined into one six-story wall, with the stiffness at each level equivalent to the sum of the stiffnesses of the walls at that level of the structure. Therefore, the idealized structure consisted of a frame and a wall. By this idealization, in-plane deformation of floors were ignored, and all points at each floor were assumed to have equal displacements. This assumption was in clear contrast with the measured response of the structure. Nevertheless, the analysis was carried out. Because it was realized that the Q-Model is too simple to represent the structure, only the MDOF model was used for the analysis.

The calculated results were superimposed on the response of the structure measured at the middle of the second floor and the roof (Fig. 6). Excellent agreement was seen between the calculated and measured acceleration response at the second floor. The close agreement throughout the response was not surprising because the structure is relatively stiff and the acceleration at the second floor is similar to the ground floor acceleration which was the input to the analytical model. The correlation between the calculated response and the acceleration measured at the middle of the roof was poor, and the peak value was underesti-

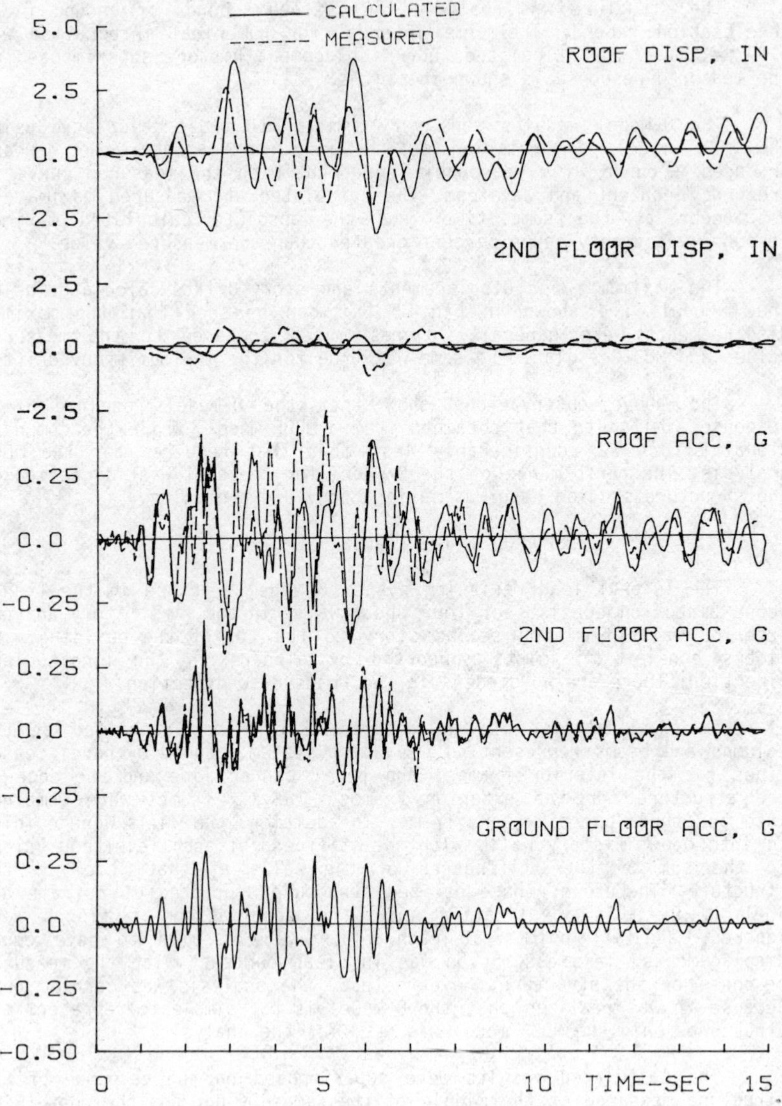

Fig. 6 Calculated and Measured Responses in the Transverse Direction

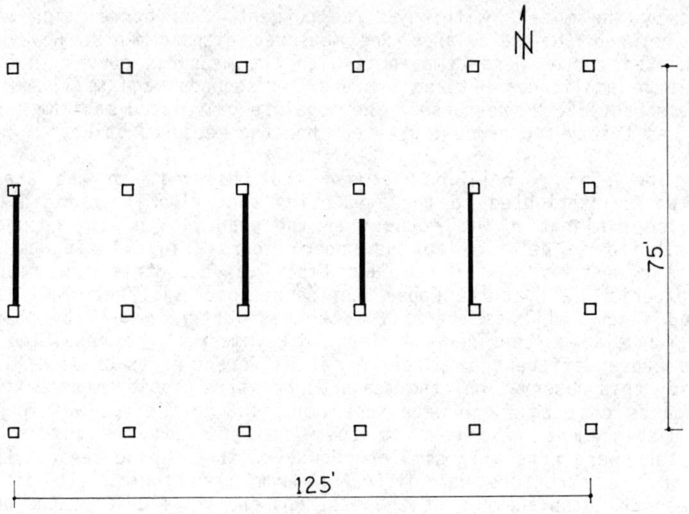

Fig. 7 Plan View of the Ground Floor

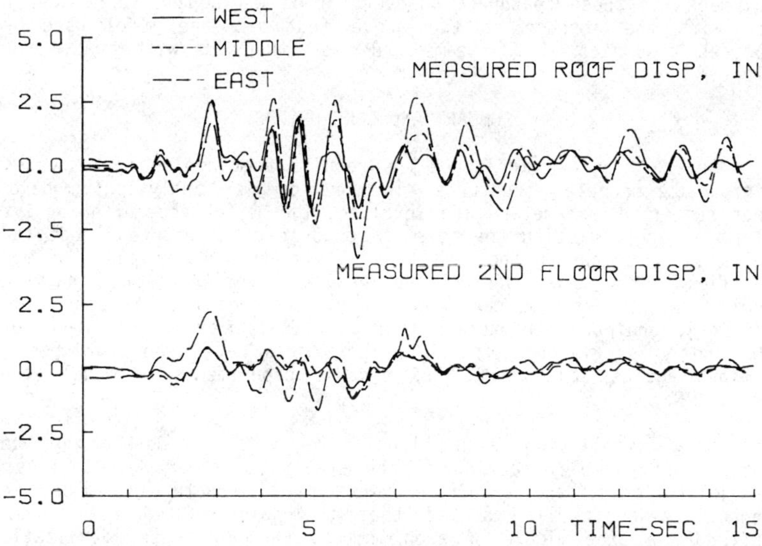

Fig. 8 Measured Relative Displacement Responses in the Transverse Direction

mated by the model by thirty-three percent. Poor correlation was also seen between the calculated and measured displacements. However, the calculated curves were comparable with the measured curves in exhibiting large amplitudes between T=2 and T=6 seconds, and small amplitudes for the rest of the responses. The absolute calculated maximum displacement was thirty-two percent greater than the measured value.

The poor correlation between the observed and the analytical results is attributed to the assumption that floors are rigid. A more detailed explanation was found when the plan view of the ground floor was studied in relation to the upper floors (Fig. 7). It was noticed that the west wall on the ground floor was eccentric with respect to the exterior wall at the upper stories by about six feet while the east ground floor wall was apart from the east exterior wall by thirty-one feet, causing a considerable lack of symmetry. The resulting torque would cause different displacement at different parts of each floor. To clarify this observation, the measured relative displacements at different parts of each floor were superimposed (Fig. 8). It was found that the displacement responses at the west end and the middle of the structure were generally similar. However, the displacement at the east end of the structure was different from the others. The difference between the displacement at the east end and the middle of the building was as great as 1.4 inches at T=3 seconds of the second floor respose, and 1.5 inches at T=6 seconds of the roof response. It was concluded that, to perform a successful analysis of the structure the eccentricity of the walls and the flexibility of the floors should have been included. The analytical model used in the present study is not equipped with these features. However, it is worthwhile to notice that even with the absence of the above features, the model used here produced responses of the same order of magnitude with the measured values.

SUMMARY AND CONCLUSIONS

Several factors influence the results from nonlinear analysis of reinforced concrete structures. In more complex analytical models a great number of parameters are involved. The performance of an analytical model in predicting the seismic response of structures needs to be evaluated by comparing the results with the observed response of model and field structures and/or the results from reliable analytical models. In the study presented here, the measured response of a full-scale ordinary structure was used to evaluate the performance of a MDOF and a "single-degree" model. The models have been developed and examined against the results of dynamic experiments on model structures.

Many simplifying assumptions were made in developing these models. For example, rocking of the structure, shear deformations of the joint cores, and axial deformation of structural members were ignored. Also it was assumed that the point of contraflexure is stationary at the middle of each member. Each of these idealizations may bring inaccuracy in the results. However, the approximations involved may not influence the results to any extent beyond the uncertainties associated with earthquake loads.

The results from the present study indicate that the assumptions made in the analytical models were accurate enough to simulate the overall longitudinal response of the building studied. It was further shown that the Q-Model led to results comparable to that of the MDOF model, while it required considerably less computation. In the transverse direction the assumptions made in the models were not representative of the actual behavior of the structure. It is concluded that, for structures with significant eccentricities, the model should allow for differential displacements at each level.

ACKNOWLEDGEMENTS

Special gratitude is due to Dr. Mete Sozen, Professor of Civil Engineering at the University of Illinois at Urbana, for his encouragement and advice. Thanks are also due to Dr. Bruce Douglas, the Chairman of the Civil Engineering Department at the University of Nevada, Reno, for his helpful discussions. John Ragsdale of the Office of Strong Motion Studies of the State of California and Mike Kreger, graduate research assistant at the University of Illinois are thanked for providing the information about the Imperial County Services Building and the measured data.

REFERENCES

1. Arnold, C., "Architectural Implications," Imperial County, California, Earthquake October 15, 1979, Earthquake Engineering Research Institute, February 1980, pp. 111-116.

2. Haroun, M., "Corrected Accelerograms and Response Spectra, Imperial Valley Earthquake of October 15, 1979," Earthquake Engineering Research Laboratory, California Institute of Technology, March 1980.

3. Otani, S. and M.A. Sozen, "Behavior of Multistory Reinforced Concrete Frames During Earthquakes," Civil Engineering Studies, Structural Research Series No. 392, University of Illinois, Urbana, November 1972.

4. Powell, G.H. and D.G. Grow, "Influence of Analysis and Design Assumptions on Computed Inelastic Response of Moderately Tall Frames," Report No. EERC 76-11, University of California, Berkeley, April 1976.

5. "Preliminary Data, Imperial Valley Earthquake of 15 October 1979, Imperial County Services Building," Department of Conservation, Division of Mines and Geology, Office of Strong Motion Studies.

6. Rojahn C. and J. T. Ragsdale, "Strong-Motion Records from the Imperial County Services Building, El Centro," Imperial County, California, Earthquake October 15, 1979, Earthquake Engineering Research Institute, February 1980, pp. 173-184.

7. Saiidi, M., "User's Manual for the LARZ Family, Computer Programs for Nonlinear Seismic Analysis of Reinforced Concrete Planar

Structures," Civil Engineering Studies, Structural Research Series No. 466, University of Illinois, Urbana, November 1979.

8. Saiidi, M. and M.A. Sozen, "Simple and Complex Models for Nonlinear Seismic Response of Reinforced Concrete Structures," Civil Engineering Studies, Structural Research Series No. 465, University of Illinois, Urbana, August 1979.

9. Takeda, T., M.A. Sozen, and N.N. Nielsen, "Reinforced Concrete Response to Simulated Earthquakes," Journal of Structural Division, ASCE, Vol. 96, No. ST12, December 1970, pp. 2557-2573.

10. Wosser, T.D. M. Fovinci, and W.H. Smith, "On the Earthquake-Induced Failure of the Imperial County Services Building," Imperial County, California, Earthquake October 15, 1979, Earthquake Engineering Research Institute, February 1980, pp. 159-172.

DETECTION OF FATIGUE CRACKS IN STRUCTURAL MEMBERS

by

E. Kummer[I], J. C. S. Yang[II], N. Dagalakis[III]

ABSTRACT

This study presents the application of the Random Decrement analysis technique for the detection of fatigue cracks on end-bolted cover plated beams. The specimens tested were 15 1/2 ft long W14 x 30 beams, reinforced by two cover plates at the top and bottom flange in the center region of the beam. The cover plates are attached by two 1/4 in welds on both sides and two rows of two bolts on each side.

The specimens were fatigue loaded on a 15 ft span with two concentrated and equal loads applied with a spreader beam symmetrically to the center line. The fatigue loading was interrupted from time to time to record the response of the structure to a random input excitation. The response signal was analyzed and its random decrement was correlated with the size of the fatigue cracks.

INTRODUCTION

The change in steel construction, especially for civil engineering types of structures, has been quite significant in the past thirty years. This applies particularly to the connectors. Where, in many cases, the weldment has been replaced by welded joints.

The welded connection, however, has some special features, which have important effects on the dynamic behavior of structures. The stiff weldment creates abrupt changes in the cross-sectional properties of structural members. This leads to large stress concentrations in the transition area [1]. This means for dynamically loaded structures fatigue has to be taken into consideration. A microscopic examination of the weld reveals that the tip contains a large number of microcracks formed during the cooling process, which serve as initiation of fatigue cracks.

As a consequence, the determining factors for the design of many structures have shifted from a statical control of the allowable stress

I Graduate Research Assistant, Mechanical Engineering Dept., University of Maryland, College Park, MD.
II Professor, Mechanical Engineering Dept., University of Maryland, College Park, MD.
III Assistant Professor, Mechanical Engineering Dept., University of Maryland, College Park, MD.

to fatigue verification. For example the large experimental research program by Fisher [2, 3, etc.] along with the analytical work by Irwin [4] and Paris [5] initiated a rapid development of knowledge of the fatigue phenomenon and provided the designing engineer with a relation between fatigue life and induced stresses in a structural member.

The results showed that the fatigue life of a structural member cannot readily be predicted by a simple number of load cycles for a certain stress range, because there are a large number of influence factors involved that cannot be included in a deterministic computation. Consequently, the fatigue life of a structural member is usually expressed as a probability of survival.

Probability, however, is an overall consideration of a large number of cases and does not necessarily give the needed information for the individual case. Only its combination with safety considerations can it provide the designing engineers with a more exact value of fatigue life permitted for a certain structure. But, since in many cases human lives are involved, safety factors have to be quite restrictive in order to cut down the risk of failure to a reasonable low rate. This means only a small part of the actual fatigue life can be used before replacement of the structure is required. This procedure is not very economical and the tendency is now to extend the fatigue life of structural elements by using crack detection techniques to locate developing fatigue cracks early enough so that they can be repaired. In this way the life of structures can be extended by a factor of two or more. In order to minimize the costs of crack detection, various scientists have put considerable efforts into developing new, more reliable and less time consuming detection methods. Special significance should be given to on-line techniques, since in some cases, an off-line examination of the structure could be very expensive and impractical (nuclear power plants, etc.).

In recent years a wide range of structural monitoring techniques have been developed such as ultrasonic, acoustic emission, vibration, etc. There are advantages and limitations to all these techniques. Some of the limitations are the inability of monitoring on line and on large systems, of detecting internal cracks and on the reliability due to the effect of input excitation.

Random Decrement analysis is a fast-converging method for extracting meaningful information from random data. When applied to the random output of a transducer placed on an object which is subjected to random excitation, a signature is obtained which can be used to measure damping or to detect incipient failures. The method is particularly useful in field measurements of structures and mechanical systems because excitation is provided naturally by such random inputs as acoustic noise, wind, seismic disturbances, traffic loads, etc.

The randomdec analysis was developed by Mr. Henry A. Cole about a decade ago. In his work for NASA on flutter problems associated with the Saturn launch vehicle he found the autocorrelation function of random vibrations to change not only with the development of a fatigue crack in a structure but also with variations in the random environment, thus giving false indications of failure. Theoretically, the problem of changes in vibrations which result from changes in the input environment could be overcome by measuring both the input forces and the output vibrations and then calculating cross-spectra or cross-correlations. But this is extremely difficult to do in practice because the input forces occur at so many points that quantification is almost impossible. The problems with spectral and correlation methods are further complicated if a structure has damping which is nonlinear with amplitude, which is often the case.

The Random Decrement method analyzes the measured output of a system subjected to some ambient random input. After analysis a signal results that is the free vibration response or signature of the mechanical structure. This signal is independent of the input and represents the particular structure tested. The ability to obtain response signatures for different modes (usually accomplished by filtering the output) enables one to detect early damage before overall structural integrity is affected. Local flaws, such as cracks, too small to affect the overall structural integrity, have a significant effect on the signatures of the higher modes. As a flaw grows, progressively lower modes are affected until overall failure occurs. Damage is detected by studying and comparing the signatures of the higher structural modes.

When a fatigue crack develops, it introduces additional degrees of freedom which are then excited by random forces. Small cracks show up as small blips in the "hashy" high modal density region of the response, and as these cracks grow, the failure mode frequencies decrease and approach the fundamentals where failure becomes imminent. The detection of flaws consists of intercepting failure modes at high enough frequencies such that corrective action can be taken. This procedure is accomplished by passing a random signal through a bandpass filter which is set at a high frequency band. If a failure develops it will have a dramatic effect on the signature because it will dynamically couple with structural modes within these bandpass frequencies. Figure 1 illustrates the acquisition of Randomdec Signatures, and hypothesis on the sensitivity of the signatures to flaws.

For the failure detector, once the standards have been established only parts of the signature at peaks need to be recalculated with warning devices sensitive to voltage changes in the peak values.

A procedure for failure detection is outlined on figure 2, which shows only a single peak for illustration. The standard signature region is first established to a confidence level consistent with percent of false alarms which could be tolerated. For the 95-percent confidence level shown, of course, false warnings would occur 5 percent of the time. Detection would be as shown on the figure. The check on standard deviation, $\bar{\sigma}$, is to prevent false indications due to extraneous input sources other than the normal random excitation, i.e., a sinusoidal force or signal in the electronics. For example, if a sinusoidal force was applied to the structure, the signature would become an undamped cosine wave and fall outside the standard region, but the standard deviation would fall to zero. In this case the amber light would go on.

Several publications [6-13] have already examined the theoretical background of the Random Decrement method and have pointed out its usefulness. A brief, rather intuitive explanation of the principles of Random Decrement Technique is given in Appendix A.

In this paper a laboratory study of fatigue cracks on end-bolted cover plated beams was conducted. The specimens used for the study are 15 1/2 ft long beams made of wide flange shapes W14 x 30, reinforced with two end-bolted coverplates attached to both flanges. These beams have been Randomdec analyzed at the beginning and during the fatigue testing until they failed.

EXPERIMENTAL SETUP

Introduction

It is a very common practice in bridge structures to reinforce the main girders with coverplates at locations of high bending moment. Coverplates, however, have a very low fatigue strength. As a consequence the coverplates have to be extended considerably until the end reaches an area where the stress range is relatively low. Several researchers have already tried to increase the fatigue strength by using new construction details [14]. The Maryland State Highway Administration has sponsored an extensive research program investigating the fatigue behavior of end-bolted coverplates. The fatigue tests are currently going on; a preliminary report is being published [15].

Specimens

The specimens tested in this study consist of a 15 1/2 ft long beam w14 x 30, reinforced with two coverplates 5 3/4 in x 1/2 in at the top and bottom flange in the center region of the beam. The coverplates are attached by two 1/4 in welds on both sides and two rows of bolts at the end of the coverplates, each row containing two 7/8 in diameter A325 (type 3) high-strength bolts, providing a no-slip, friction type connection (Fig. 3).

The plates and beams are fabricated from material satisfying the chemical and mechanical property requirements of ASTM A588 high-strength, low-alloy structural steel. The yield point of this material was found to be 71.8 psi and 55.1 psi for the plate and the beam respectively.

The specimens are tested on a 15 ft span with two concentrated loads of equal value, applied with a spreader beam symmetrically to the centerline of the specimen, spaced at 2 ft-6 in (Fig. 4). The tests are carried out by a load controlled MTS closed-loop system of 50 kips dynamic capacity. A constant amplitude, sinusoidal loading is used throughout the testing. The stress range σ_r is set to σ_r = 31 ksi (minimum stress σ_{min} = 1.5 ksi) at the location of the first row of bolts, computed with the section properties of the w14 x 30 alone. In order to minimize lateral movement wooden stiffeners are placed between the flanges at the reaction and the load points of all beams.

Random Excitation and Response Recording

In order to monitor the vibrational behavior of the beam the fatigue testing was been interrupted at regular intervals and the response to random excitation recorded.

The random input is produced by a Random Noise Generator (GENERAL RADIO COMPANY, 1390-B), connected to a power amplifier (ELECTRODYNE, N-100) whose signal was then input into a 25 lb shaker (MB ELECTRONICS, PM 50). This shaker is attached to the beam at 3 ft-6 in from the left end (Fig. 5).

The beam response is picked up by two accelerometers (COLUMBIA, 504-3), gluded to the inside surface of the tension flange. The output signal is amplified by charge amplifiers (KISTLER, 504) and then recorded on magnetic tape with an AC/DC tape recorder (B&K, 7003).

The recorded response signal is first passed through a band pass filter before it is fed into a programmable amplifier, which is tied to a minicomputer (CROMEMCO, 32-K bit memory). The computer program is set up such that it automatically amplifies the input signal to the full' range capacity of $\pm$ 8 volts of the following analog-to digital converter (8 bits). The digitized signal is then stored into the memory of the computer and Randomdec analyzed.

RESULTS

Although the length of the coverplate and the number of bolts vary for the three different types of specimens their vibrational behavior turned out to be very similar so that we can restrict ourselves to the presentation of the results of one beam.

The specimen k49 with a four bolt end connection will be discussed. This beam had a fatigue life of roughly one million cycles (σ_r = 31 ksi). The fatigue testing was interrupted seven times to take a response recording. The eighth recording was taken after the testing machine was stopped from a displacement switch and a significant crack had appeared.

Previous tests had shown that the cracks developed in almost all cases at the flange bolt hole, propagating in perpendicular direction to the axis of the beam. Furthermore, it was found that the failure usually occurred in the first row of bolts. In beam k49, however, cracks on two bolts were found, both at the north-west end of the coverplate. A major crack appeared at the second bolt from the end of the coverplate, which had extended 1 1/8 in towards the edge of the flange and 11/16 in towards the web. A second, minor crack had developed at the first bolt having reached a length of 13/16 in towards the web.

The eight response recordings were band pass filtered with the following filter limits:

- lower cutoff frequency: 5000 Hz
- higher cutoff frequency: 7000 Hz

The resulting Randomdec signatures are represented in fig. 6. They can be interpreted as follows:

- The Randomdec signature of the response of the undamaged beam contains basically only one natural frequency inside the filter range.
- The frequency of the signature dropped slightly between the first (taken after 273420 load cycles) and the second recording (taken after 292160 cycles). This can be related to the slip occurring between flange and coverplate. Strain gage measurements on the coverplate have also shown this slip, which happened during the first 200,000 - 300,000 load cycles.
- Signatures two, three and four are identical, indicating no fatigue damage in the beam.
- The fifth signature (796,490 cycles) shows a slightly decreased frequency. This is the first indication of beginning fatigue deterioration in the tested beam.
- The sixth signature is a superposition of two natural frequencies. In addition to the natural frequency previously encountered, the next higher one shifted down into the filter range.
- The seventh frequency, again, contains only one natural frequency. The natural frequency originally considered shifted further down and is now outside the filter range.
- The final signature is again a superposition of two frequencies, suggesting that a new natural frequency has shifted into the selected filter range.

The method described above with constant filter range is usually used for crack detection but gives no indication of the severity of the deterioration. A changed signature merely indicates the fact that a crack exists. If an estimation of the crack size is desired it seems more appropriate to focus on the frequency of the Randomdec signature. It was shown [16] that a relation between natural frequency and crack size exists.

It is clear, that a constant filter range cannot be used any more because the natural frequency considered would shift out of the filter band with increasing crack size. The filter range has therefore to follow the natural frequency. For this, however, it is essential to know the approximate value of the natural frequency in order to set the filter limits. This information can be obtained from the power spectral density function.

The results for these frequencies plotted as a function of the applied load cycles is displayed in fig 6. For convenience the frequencies are divided by the frequency of the first recording. The results are remarkable in two ways:

- The slip that occurred between the flange and the coverplate between the first and the second recording has also an effect of the signature frequency.
- For small crack sizes when the loading edge of the crack is still under the washer the signature shows only a small decrease in frequency. These results are in direct contrast to those findings on the completely welded specimens, where the signature

frequency showed a sharp decline even for small cracks [16]. This can be explained by the fact that the two crack surfaces are not free to move because they are held together by the pretension force in the bolt and the friction between washer and flange surface. Only when the crack emerges from under the washers the signature frequency drops significantly.

CONCLUSION

The Random Decrement Technique is a crack detection method, which analyzes the overall vibrational behavior of a system subjected to some unknown, random excitation. This concept features some significant advantages:

- The entire structure can be tested and analyzed with a very limited number of transducers.
- Points or components with access difficulty pose no problem.
- The structure does not need to be disassembled for the testing.
- The recording of the response signal requires only a very short time.
- No heavy equipment or instrumentation is needed.
- The structure can be tested in service, as long as the natural input is relatively random, which is the case for many facilities (bridges, piping systems, aircrafts, etc.).

The results obtained from the Randomdec analysis of a 15 1/2 ft beam w14 x 30, reinforced with two end-bolted coverplates points out some of the capabilities of this technique:

- The Random Decrement Technique is able to detect fatigue cracks in bolted joints before the crack emerges from under the washer.
- The change in the frequency can be related to the crack size. However, the frequency change is not proportional to the size of the fatigue crack as long as the crack front is under the washer. This drawback does not exists for welded connections [16].
- Randomdec can detect fatigue deterioration after roughly 80% of the total fatigue life, whereas the visual detection of the crack was at 98%.

FURTHER INVESTIGATION

A number of research projects is on going at the University of Maryland in an effort to further develop the Random Decrement Technique.

Random Decrement Technique

It is advantageous for any crack detection method to be able to locate the position of failure in addition to the detection of a crack. A first effort in this direction has been done in [16] and further investigations are needed to explore the capabilities of Randomdec to track down a fatigue deterioration.

Another point of interest is related to the sensibility

of Randomdec. Up to now very little research has been done to find out how large a crack has to be for a certain structure in order to be detected by Randomdec. Obviously this question cannot be answered in general because the size and complexity of a structure has to be taken into consideration.

In addition changes in the boundary conditions or other variations could also affect the signature without being related to a fatigue deterioration. Further research is needed to distinguish the effects of cracks from the effects of other changes which might lead to a signature change.

OUTLOOK

Today, there are more and more large systems subjected to fatigue loading whose failure might have catastrophic consequences. Furthermore, it is often very costly, to say the least, to shut them down for a certain time and inspect them for fatigue cracks. Recent failures in nuclear power plants, aircrafts and off-shore structures have emphasized the need for simple and reliable on-line testing techniques to ensure the structural integrity of a system. It might very well be that this is the major field of application for Randomdec in the future. Its relative simplicity and the fact that the natural excitation can be used as input make this method especially attractive to in service testing. The ability to continuously monitor the structural integrity of a system using the Randomdec computer is another important factor.

ACKNOWLEDGEMENTS

The results of this investigation were obtained in research contracts funded by the National Science Foundation under Contract No. ENG 77-12745 and Contract No. INT76-14751-A03.

Special thanks are given to partial supports from the Maryland State and Federal Highway Administrations.

REFERENCES

1. Hirt, M.A., Kummer, E., "Einfluss der Spannungskonzentrationauf die Ermüdungsfestigkeit geschweisster Konstruktionen," ICOM 044, Ecole Polytechnique Federale de Lausanne, 1978.

2. Fisher, J.W., Frank, K.H., Hirt, M.A. and McNamee, B.M., "Effect of Weldments on the Fatigue Strength of Steel Beams," NCHRP Report No. 102, Highway Research Board, Washington, DC., 1970.

3. Fisher, J.W., Albrecht, P.A., Yen, B.T., Klingeermann, B.J. and McNamee, B.M., "Fatigue Strength of Steel Beams with Transverse Stiffeners and Attachments," NCHRP Report No. 147, Highway Research Board, Washington, DC., 1974.

4. Irwin, G.R., "Analysis of Stresses and Strains Near the End of a Crack Traversing a Plate," Journal of Applied Mechanics, Vol. 24, 1957.

5. Paris, P.C., "The Fracture Mechanics Approach to Fatigue," Proc. Tenth Sagamore Army Materials Research Conference, Syracuse University Press, 1964.

6. Cole, H.A., "Method and Apparatus for Measuring the Damping Characteristics of a Structure," United States Patent No. 3,620,069, 1971.

7. Cole, H.A., "On-Line Failure Detection and Damping Measurement of Aerospace Structures by the Random Decrement Signatures," NASA CR-2205, 1973.

8. Reed, R.E., Cole, H.A., "Mathematical Background and Application to Detection of Structural Deterioration in Bridges," FHWA Report No. 181, Federal Highway Administration, Washington, DC, 1976.

9. Caldwell, D.W., "The Measurement of Damping and the Detection of Damage in Linear and Nonlinear Systems by the Random Decrement Technique," Ph.D. Thesis, University of Maryland, 1978.

10. Caldwell, D.W., "The Measurement of Damping and the Detection of Damage in Structures by the Random Decrement Technique," M.S. Thesis, University of Maryland, 1975.

11. Yang, J.C.S., Caldwell, D.W., "measurement of Damping and the Detection of Damages in Structures by the Random Decrement Technique," 46th Shock and Vibration Bulletin, 1976, pp. 129-136.

12. Yang, J.C.S., Caldwell, D.W., "A Method for Detecting Structural Deterioration in Piping Systems," ASME Probabilistic Analysis and Design of Nuclear Power Plant Structures Manual, PVB-PB-030, Dec. 1978, 97-117.

13. Yang, J.C.S., Dagalakis, N., Hirt, M., "Application of the Random Decrement Techniques in the Deteciton of Induced Crack on an Offshore Platform Model," to be published, ASME Winter Meeting, Nov. 17, 1980, Chicago, Ill.

14. Van der Schaaf, T., Jr., The Bridges of the Kreekrakdam on the new Scheldt-Rhine Canal (Netherlands), Acier-stahl-Steel, November 1974.

15. Worthington, C.G., "initial Fatigue Test Data for End-Bolted Cover-plated Beams," M.S. Thesis, University of Maryland, 1980.

16. Kummer, E., "The Detection of Fatigue Cracks in Structural Members by the Random Decrement Technique," M.S. Thesis, University of Maryland, 1979.

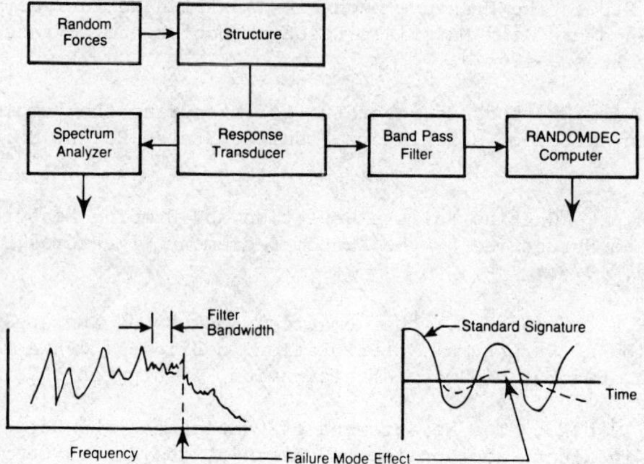

Random Decrement Average

Figure 1: Sensitivity of random decrement signature to flaws

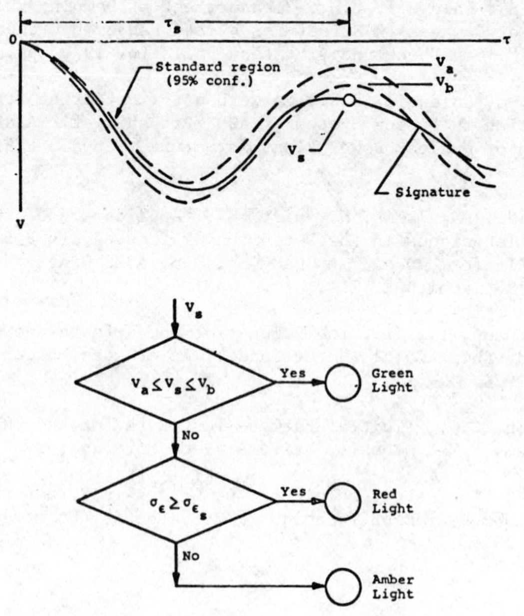

Figure 2: On-line failure detection at a single point on the signature

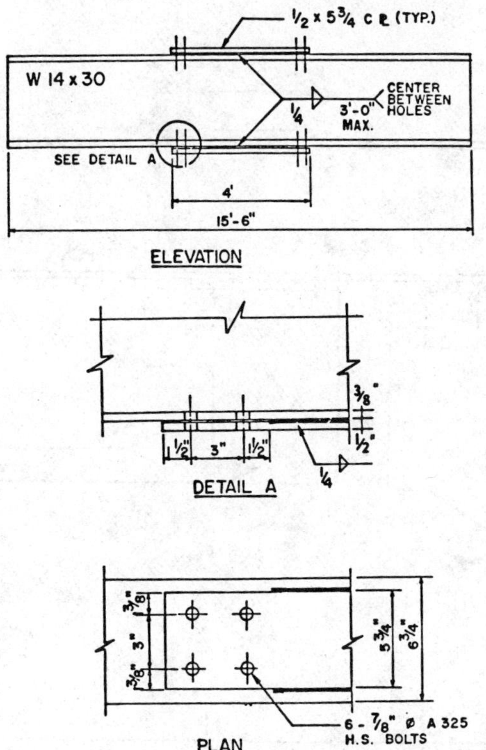

Figure 3 : Dimensions of specimens with four bolts at the cover plate end

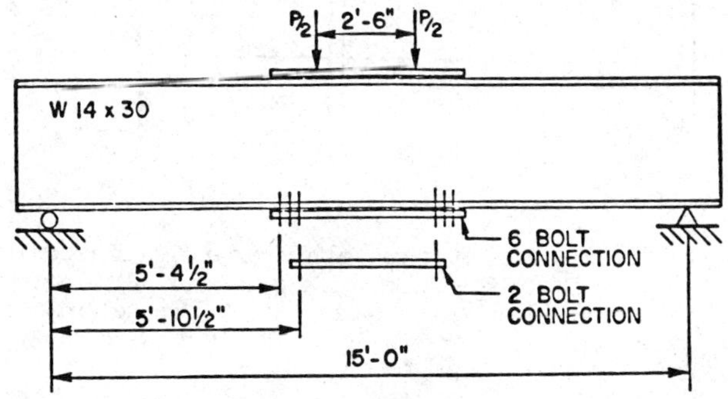

Figure 4 : Test setup

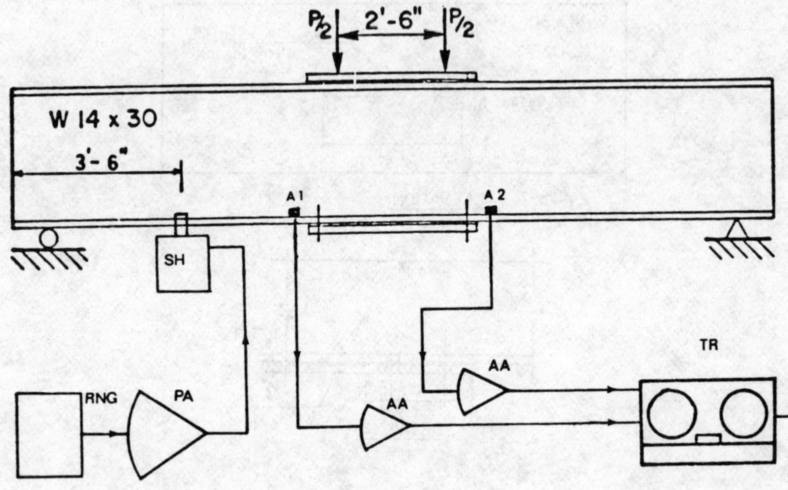

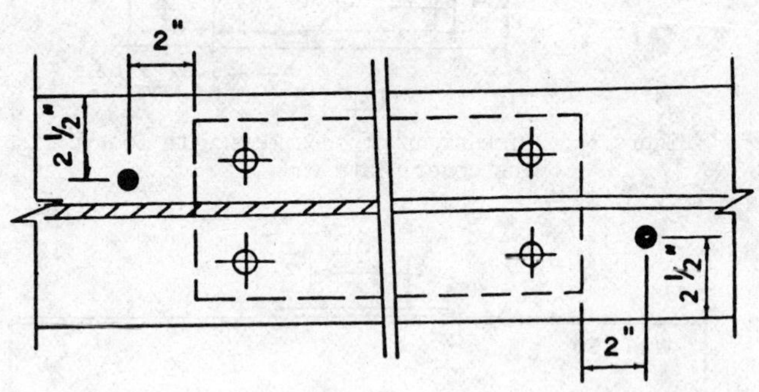

Figure 5 : Random excitation and response signal recording, with RNG: random noise generator, PA: power amplifier, SH: shaker, A1, A2: accelerometers, AA: charge amplifiers, TR: taperrecorder

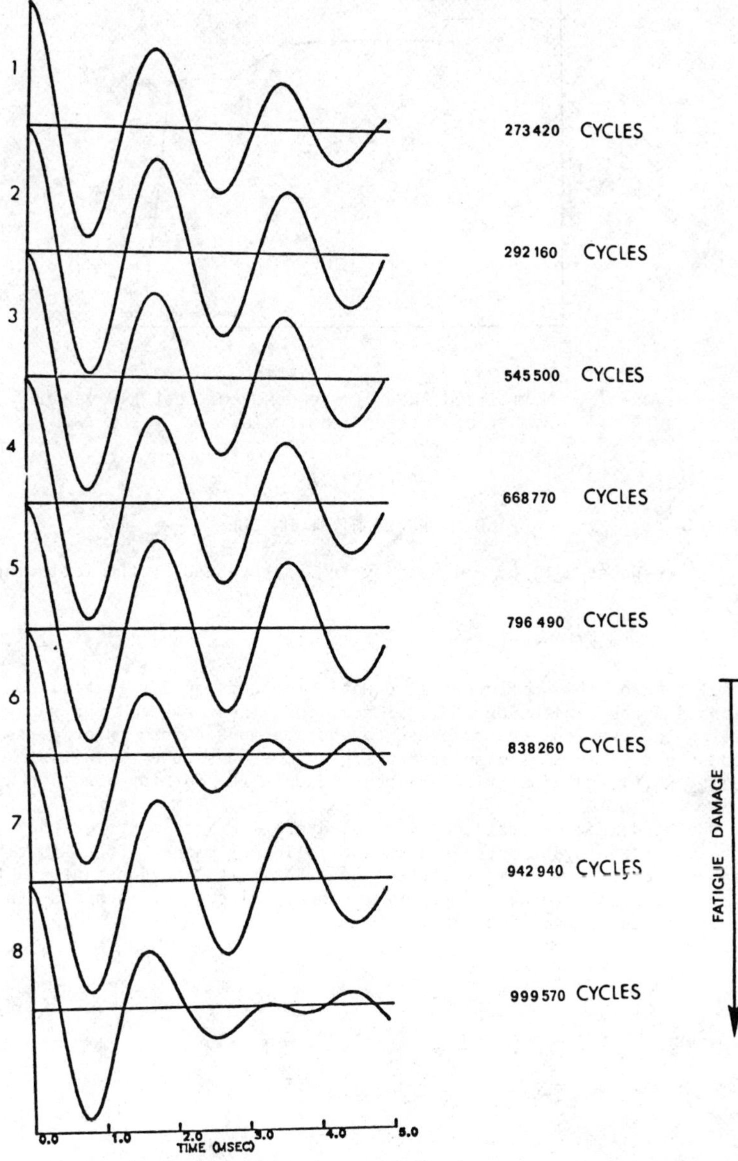

Figure 6 : Randomdec Signatures of test k49 using a constant filter range

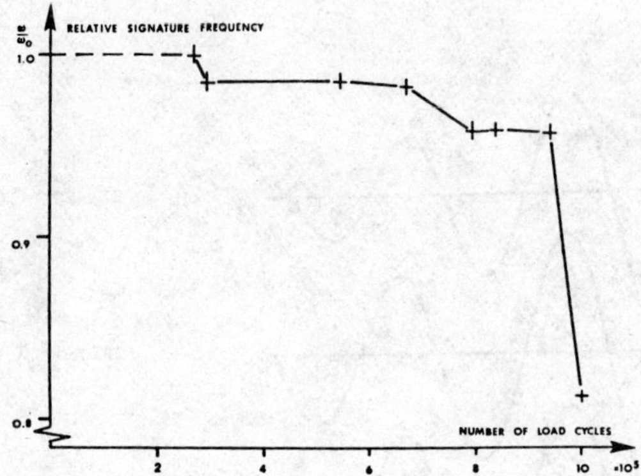

Figure 7 : Normalized natural frequency of test K49 versus number of fatigue load cycles

APPENDIX A

RANDOM DECREMENT TECHNIQUE

The response x(t) of a linear system is governed by the following basic equation:

$$m\,\ddot{x}(t) + c\,\dot{x}(t) + k\,x(t) = f(t) \qquad (1)$$

The solution of this differential equation depends on its initial conditions and the excitation f(t). Since, for linear systems the superposition law applies, the response can be decomposed into three parts: response due to initial displacement $x_d(t)$, response due to initial velocity $x_v(t)$ and finally the response due to the forcing function $x_f(t)$.

The Randomdec analysis consists of averaging N segments of the length τ_1 of the system response in the following manner: the starting time t_i of each segment is selected such that $x_i(t_i)=x_s$=constant and the slope $\dot{x}_i(t_i)$ is alternating positive and negative. This process can be represented in mathematical form:

$$\delta(\tau) = \frac{1}{N}\sum_{i=1} x_i(t_i+\tau) \qquad (2)$$

where $x_i(t_i) = x_s \qquad i = 1, 2, 3 \ldots.$

$\dot{x}_i(t_i) = \geq 0 \qquad i = 1, 3, 5 \ldots.$

$\dot{x}_i(t_i) = \leq 0 \qquad i = 2, 4, 6 \ldots.$

The function $\delta(\tau)$ is called the Randomdec signature and is only defined in the time interval $0 \leq \tau \leq \tau_1$. The meaning of the Randomdec signature can now be determined. If the parts due to initial velocity are averaged together, they cancel out because alternately parts with positive and negative initial slopes are taken and their distribution is random. Furthermore, if the parts due to the excitation are averaged they also vanish because, by definition, the excitation is random. Finally only the parts due to initial displacement are left and their average is the Randomdec signature representing the free vibration decay curve of the system due to an initial displacement, which corresponds to the bias level x_s. (Fig 1)

In reality the Randomdec computer converts each segment into digital form and adds it to the previous segments (Fig 2); the average is then stored in the memory and can be displayed on a screen. The number of segments to be averaged for the Randomdec signature depends on the signal shape, usually 400 to 500 averages are sufficient to produce a repeatable signature.

One particularly interesting characteristic of Randomdec technique should be mentioned: it requires no knowledge of the excitation f(t) as long as it is random. Neither the type nor the intensity of the input affect the signature.

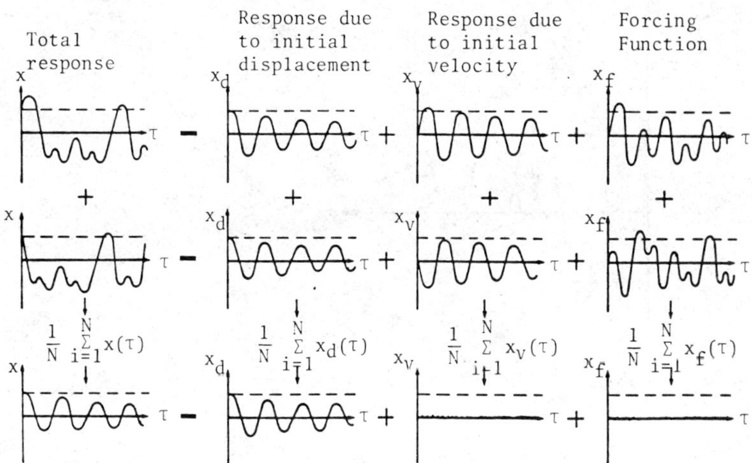

Fig. 1 Principles of Random Decrement Technique

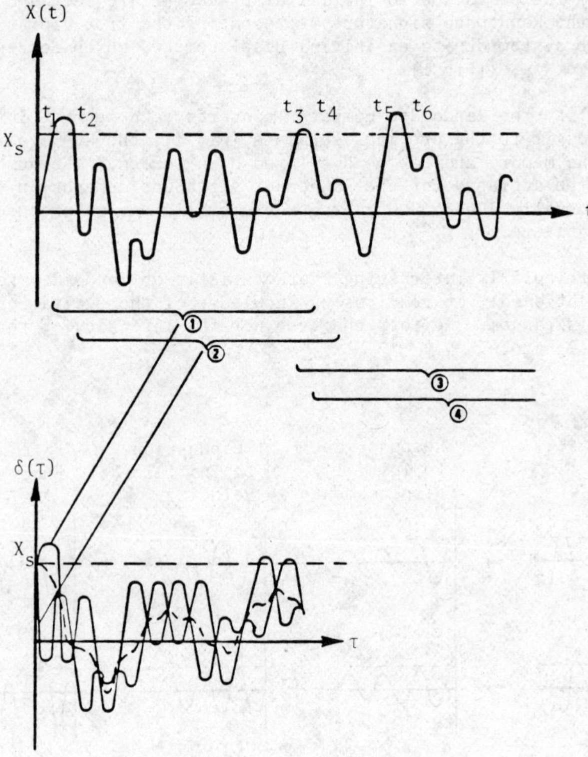

Fig. 2 Extraction of the Random Decrement Signature

BOUNDARY ELEMENT ANALYSIS OF RESERVOIR VIBRATION

By Jagmohan L. Humar[1], Youssef G. Hanna[2]

ABSTRACT

The boundary element method of analysis is applied to the calculation of hydrodynamic pressures in a reservoir impounded by a gravity dam and subjected to a hormonic ground motion. It is shown that the differential equation governing the small amplitude vibrations of the reservoir reduces to the familiar Helmholtz equation. This equation is converted to an integral equation which involves integrations only on the boundary of the reservoir. The integrals are evaluated numerically by using a series of straight line elements to represent the boundary. The resulting simultaneous equations are solved by standard techniques to obtain the pressures and pressure derivatives on the boundary. Analytical results are presented to show that the proposed method gives results that are in very good agreement with the known classical solutions. The method can be applied to the solution of vibration problems of reservoirs with irregular boundaries for which classical solutions do not exist. The method is also successful in dealing with cases of infinite reservoirs and in modelling the energy loss in the waves moving to the infinity.

INTRODUCTION

In the analysis of stresses and deformations in a dam subjected to earthquake ground motion, consideration must be given to the effect of hydrodynamic forces cuased by the impounded water. In 1933 Westergaard published a study on the hydrodynamic pressures on rigid dams (12). Since then, a number of analytical studies have been carried out on the dynamic interaction problem of dam reservoir systems. The earlier investigations all treated the dam as rigid (3,7). Later studies have been accounted for the flexibility of the dam (2,4,5,10). The important conclusions drawn from these studies were that the earthquakes can cause significant hydrodynamic pressures on the dam face, the compressibility of water cannot be neglected in the analysis and that the deformations of the dam may alter the pressures significantly.

A fundamental assumption made in all studies referred to above is to treat the problem as two dimensional. This assumption is reasonable for gravity dams. Two different approaches have been

---

1. Associate Professor, Dept. of Civil Engineering, Carleton University, Ottawa, Ontario, Canada
2. Graduate Student.

used in the solution of the simplified two dimensional problem.

In one approach, the dam is represented as an assemblage of finite elements and the reservoir is modelled as a continuum with simple boundaries for which classical solutions can be easily obtained. The two solutions are then coupled by accounting for the forces interacting between the reservoir and the face of the dam. A limitation of this approach is that classical solutions for hydrodynamic pressures in reservoirs are available only for simple boundaries. It is therefore not possible to treat cases in which the reservoir has a sloping bed or the dam face is inclined.

The second approach to the solution of the interaction problem is to model the complete system by finite elements considering displacements as the basic unknowns for the dam and the pressures as the unknowns for the reservoir. Reference 13 presents the basic formulation of the method and reference 10 describes its application to the dam-reservoir problem. Apart from the considerable computational effort required by the method, a major difficulty arises in representing the infinite boundary of the reservoir. Efforts to model the energy loss in the outgoing waves by including a radiation damping term in the finite element formulation are not very successful.

The method presented in this study uses the boundary element method to model the infinite or finite reservoir. The method is quite suitable for boundaries of irregular shape. In addition it can very efficiently model reservoirs extending to infinity. This paper presents the formulation of the method and its application to the calculation of hydrodynamic pressure on a rigid dam face whose motion is a harmonic function of time. Extension to the case of flexible dam in which the motion of the dam face is prescribed is straight forward. Once a method has been developed to calculate the hydrodynamic pressure for a prescribed motion of the dam face, the complete dam-reservoir problem can be solved by the method of substructures. The substructure analysis is however not presented here.

ASSUMPTIONS

The following assumptions are made in determining the hydrodynamic pressure in water reservoirs.

1. Water is assumed to be compressible but its internal viscosity is neglected.
2. The effect of waves at the free surface of the water is ignored; the associated errors have been found to be small by other investigators (3).
3. The motion of water is limited to small amplitudes.
4. The upstream face of the dam is assumed to undergo a prescribed motion which is a harmonic function of time. For a rigid dam this motion is equal to the horizontal motion of the ground.

5. The motion of the dam-reservoir system is considered as two dimensional, i.e. it is the same for any vertical plane perpendicular to the axis of the dam.

In the present study, no restrictions are imposed on the shape of either the upstream face of the dam or the bed of the reservoir.

FORMULATION OF THE BOUNDARY INTEGRAL EQUATIONS

With the assumptions described above, the motion of the water in a reservoir is governed by the following equations

$$\frac{\partial^2 u}{\partial t^2} = -\frac{1}{\rho}\frac{\partial p}{\partial x} \tag{1a}$$

$$\frac{\partial^2 v}{\partial t^2} = -\frac{1}{\rho}\frac{\partial p}{\partial y} \tag{1b}$$

$$\frac{\partial^2 p}{\partial x^2} + \frac{\partial^2 p}{\partial y^2} = \frac{1}{c^2}\frac{\partial^2 p}{\partial t^2} \tag{1c}$$

where x and y are the cartesian coordinates (Fig. 1), u and v are respectively the x and y components of the displacement of a particle of water, t is the time variable, ρ is the mass of water per unit volume, p is the hydrodynamic pressure (in excess of the hydrostatic pressure) and c is the velocity of sound in water.

If the hydrodynamic pressure is caused by the motion of the upstream face of the rigid dam moving horizontally with an acceleration $e^{i\omega t}$, ω being the exciting frequency, it is reasonable to assume that the pressure p will be of the form

$$p = p_o(x,y,\omega)\, e^{i\omega t} \tag{2}$$

Substitution of Eqn (2) into Eqn (1) gives the familiar Helmholtz equation

$$\frac{\partial^2 p_o}{\partial x^2} + \frac{\partial^2 p_o}{\partial y^2} + k^2\, p_o = 0 \tag{3}$$

where $k^2 = \frac{\omega^2}{c^2}$

On multiplying Eqn. (3) by p_o^*, an arbitrary function of x and y, and integrating over the domain under consideration (water reservoir) the following integral equation is obtained.

$$\int\!\!\int_A \left(\frac{\partial^2 p_o}{dx^2} + \frac{\partial^2 p_o}{\partial x^2} + k^2 p_o \right) p_o^*\, dxdy = 0.0 \tag{4}$$

where A represents the domain.

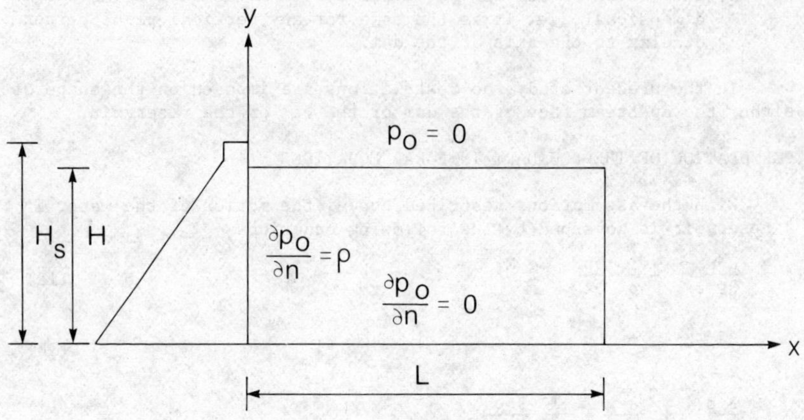

Fig. 1 Rigid Dam Storing Finite Water Reservoir.

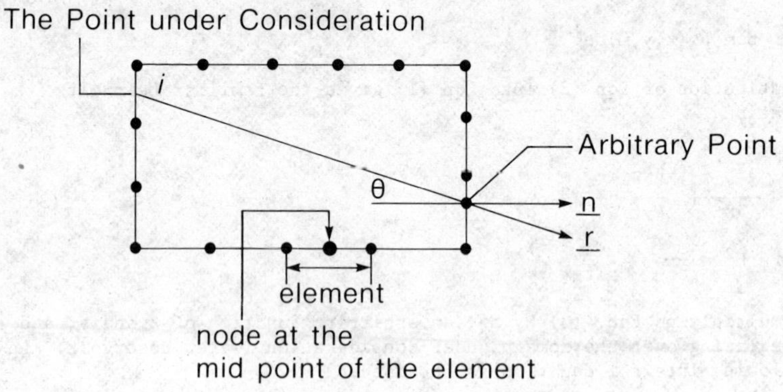

Fig. 2 Discretization of Reservoir Boundary

By the application of Green's theorem, Eqn. (4) can be reduced to the following form.

$$\iint_A \left(\frac{\partial^2 p_o^*}{\partial x^2} + \frac{\partial^2 p_o^*}{\partial y^2} + k^2 p_o^* \right) p_o \, dxdy = \int_C p_o \frac{\partial p_o^*}{\partial n} dC - \int_C p_o^* \frac{\partial p_o}{\partial n} dC \quad (5)$$

where C is the boundary of the water reservoir and n is the outward normal to the boundary.

If the as yet arbitrary function p_o^* is chosen so as to satisfy the following equation

$$\frac{\partial^2 p_o^*}{\partial x^2} + \frac{\partial p_o^*}{\partial y^2} + k^2 p_o^* = \Delta^i \quad (6)$$

where Δ^i is a Dirac Delta function centered at point 'i' having coordinates (x_i, y_i), then Eqn. (5) becomes:

$$\iint_A \Delta^i p_o \, dxdy = \int_C p_o \frac{\partial p_o^*}{\partial n} dc - \int_C p_o^* \frac{\partial p_o}{\partial n} dc \quad (7)$$

By using the properties of the Dirac Delta function, it can be shown that Eqn. (7) reduces to the following form (1)

$$c_i (p_o)_i = \int_C p_o \frac{\partial p_o^*}{\partial n} dc - \int_C p_o^* \frac{\partial p_o}{\partial n} dc \quad (8)$$

where c_i = 1 for a point inside the water reservoir

c_i = ½ for a point on the smooth boundary of the water reservoir

c_i = 0 for a point outside the water reservoir.

Eqn. (8) relates the value of the function p_o at any point i to its values on the boundary of the water reservoir. It should be noted that the integrations contained in Eqn. (8) are to be carried out only on the boundary.

On setting

$$\frac{\partial p_o^*}{\partial n} = q^* \text{ and } \frac{\partial p_o}{\partial n} = q$$

Eqn. (8) becomes:

$$c_i (p_o)_i = \int_C p_o q^* \, dc - \int_C p_o^* q \, dc \quad (9)$$

In order to evaluate the integrals contained in Eqn. (9), it is assumed that the boundary of the water reservoir is divided into N straight line boundary elements as shown in Fig. 2 and that the values

p_o and q are constant over each element and equal to their values at the mid-point of the element. Eqn. (9) for a given point 'i' on the boundary then takes the following discretized form

$$-\tfrac{1}{2}\,(p_o)_i + \sum_{j=1}^{N} (p_o)_j \int_{c_j} q^*\, dc = \sum_{j=1}^{N} q_j \int_{c_j} p_o^*\, dc \tag{10}$$

where $(p_o)_j$ and q_j are the values of these variables over the jth element and c_j represents the boundary of that element.

Setting

$$\int_{c_j} q^*\, dc = h_{ij}$$

$$\int_{c_j} p_o^*\, dc = G_{ij}$$

$H_{ij} = h_{ij}$ when $i \neq j$ and

$H_{ij} = h_{ij} - \tfrac{1}{2}$ when $i = j$

Eqn. (10) becomes:

$$\sum_{j=1}^{N} H_{ij}\,(p_o)_j = \sum_{j=1}^{N} G_{ij}\, q_j \tag{11}$$

The complete set of equations for the N nodes can now be expressed in matrix form as:

$$[H]\,\{p_o\} = [G]\,\{q\} \tag{12}$$

where

[H] is a matrix containing the elements H_{ij}

[G] is a matrix containing the elements G_{ij}

$\{p_o\}$ is a vector containing the known and unknown values of the pressure p_o

{q} is a vector containing the known and unknown values of the derivates of p_o

Now, if it is assumed that n_1 values of p_o and n_2 values of q are known so that $n_1 + n_2 = N$, there are N unknowns in Eqn. (12). Reordering the equations in such a way that all unknowns are transferred to the left hand side, one can write:

$$[B]\,\{X\} = \{F\} \tag{13}$$

where X is the vector of unknown p_os and qs.

Once the values of p_os and qs on the entire boundary are known, one can calculate the value of p_o at any interior point using Eqn. (8), which after discretization becomes

$$(p_o)_i = \sum_{j=1}^{N} (p_o)_j \int_{c_j} q^* \, dc - \sum_{j=1}^{N} q_j \int_{c_j} p_o^* \, dc \qquad (14)$$

The above formulation effectively reduces the dimensionality of the problem by one converting an integration over the domain to a set of integrals over the boundary. The success of the method depends on an appropriate choice for the function p_o^* which must satisfy Eqn. (6). The solution of that equation is called the fundamental solution (1). It can be shown that Eqn. (6) has two linearly independent solutions $J_o(kr)$ and $Y_o(kr)$, where $J_o(kr)$ is a Bessel function of zero order and first kind, $Y_o(kr)$ is a Bessel function of zero order and second kind and r is the radial distance variable given by

$$r = \sqrt{(x-x_i)^2 + (y-y_i)^2}$$

The choice of an appropriate solution for p_o^* is discussed below for both the finite and the infinite water reservoir.

Finite Water Reservoir

Of the two linearly independent solutions for Eqn. (6), the Bessel function of the first kind and order zero, $J_o(kr)$ will not be considered since it is regular (has a value of 1.0) at r = 0.0 and hence cannot lead to the desired delta function singularity at that point. Hence, the fundamental solution for the case of finite water reservoir will be taken as

$$p_o^* = D\, Y_o(kr)$$

in which D is a constant. It can easily be proved (6) that D is equal to 1/4; the fundamental solution therefore takes the form

$$p_o^* = \tfrac{1}{4}\, Y_o(kr) \qquad (15)$$

Using the fundamental solution given in Eqn. (15), the elements of the matrices [H] and [G] of Eqn. (12) can be evaluated numerically by using say a 4-point Gauss quadrature formula (11). Thus

$$H_{ij} = \frac{k\ell}{8} \sum_{m=1}^{4} Y_o'(kr)_m \,(\cos\theta)_m\, w_m \qquad (16)$$

$$G_{ij} = \frac{\ell}{8} \sum_{m=1}^{4} Y_o(kr)_m\, w_m \qquad (17)$$

Where w_m is the weighting factor used in the Gaussian quadrature and ℓ is the length of the element j. For the special case of i = j,

$$H_{ii} = -\tfrac{1}{2} \qquad (18)$$

and the term G_{ii} is calculated analytically using the approximation

$$Y_o(kr) \simeq \frac{2}{\pi} \ln (kr), \text{as } r \rightarrow 0.$$

which gives

$$G_{ii} = \frac{\ell}{2\pi} [\ln(\frac{k\ell}{2})-1] \tag{19}$$

After the elements of [G] and [H] matrices have been obtained, the following boundary conditions are applied.

1. The hydrodynamic pressure p should vanish at the free surface of the reservoir, i.e. $p_o(x,H) = 0$
2. For a horizontal ground acceleration, the vertical acceleration of the particles of water on the bed of the reservoir should vanish. This condition gives $\frac{\partial p_o}{\partial n} (x,o) = 0$
3. The horizontal acceleration of the particles of water of the upstream face of the dam should be equal to the horizontal ground acceleration. This condition gives $\frac{\partial p_o}{\partial n} (o,y) = \rho$
4. For the boundary condition at the far end of the reservoir, the following cases arise:
   a. The horizontal acceleration of the particles of water at the far end of the reservoir is assumed to be zero. In that case

   $$\frac{\partial p_o}{\partial n} (L,y) = 0.$$

   b. The horizontal acceleration of the particles of water at the far end is assumed to be equal to the horizontal ground acceleration. In that case $\frac{\partial p_o}{\partial n} (L,y) = -\rho$

<u>Infinite Water Reservoir</u>

The fundamental solution for p_o^* requires special consideration for the case of infinite water reservoir. From the known asymptotic behaviour of $Y_o(kr)$ (8), it is noted that:

$$p_o^* \; e^{i\omega t} = \tfrac{1}{4} Y_o(kr) \; e^{i\omega t}$$

$$\simeq \tfrac{1}{4} \sqrt{\frac{2}{\pi kr}} \; \frac{1}{2i} \; [e^{ik(r+ct-\pi/4k)} - e^{ik(r-ct-\pi/4k)}], \text{ as } r \rightarrow \infty \tag{20}$$

Now, the general form F(r-ct) represents a wave moving radially outwards at a constant speed c and the general form F(r+ct) represents a wave moving inwards at a constant speed c[5]. The first term in the last expression for $p_o^* e^{i\omega t}$ thus represents an inward moving wave and the second term an outward moving wave. The solution given by Eqn. (15) is not suitable for the case of an infinite water reservoir because the presence of waves coming in from infinity is unacceptable from a physical point of view.

It is known that Hankel functions of zero order and the first and second kind, are also solutions of Eqn. (6). Consider the Hankel function of zero order and second kind, $H_o^{(2)}(kr)$. From the asymptotic behaviour, it can be shown that:

$$H_o^{(2)}(kr)\, e^{i\omega t} \simeq \sqrt{\frac{2}{\pi kr}}\, e^{-ik(r-ct-\pi/4k)}, \text{ as } r \to \infty \tag{21}$$

The solution given by Eqn.(21) contains only outgoing waves and this satisfies the physical requirements. It may be noted that the waves have a diminishing amplitude because of the $r^{-\frac{1}{2}}$ factor.

The following fundamental solution is therefore adopted for the case of infinite water reservoir.

$$p_o^* = i/4\, H_o^{(2)}(kr) \tag{22}$$

Noting that

$H_o^{(2)}(kr) = J_o(kr) - i\, Y_o(kr)$, expressions can be derived for the elements of the [H] and [G] matrices by a procedure similar to the one described for a finite reservoir.

As in the case of a finite water reservoir, the complete system of equations can be written as

$$[H]\,\{p_o\} = [G]\,\{q_o\} \tag{23}$$

where H, G, p_o and q_o are now complex.

The boundary conditions are similar to those for the finite reservoir case except that the far end of the reservoir extends to infinity and the boundary condition at that point is given by

$$p_o\,(\infty,y,t) = 0$$

The above condition and the fact that the fundamental solution p_o^* tends to zero as kr tends to infinity implies that the integration on the right hand side of Eqn. (8) need not cover the boundary at infinity.

ANALYTICAL RESULTS

The analysis procedure described above is applied to the study of hydrodynamic pressures caused in both finite and infinite water reservoirs by a moving rigid dam whose motion is a harmonic function of time. In the studies presented here, the height of reservoir is taken as 100 m, the velocity of sound in water as 1438 m/s, and the amptitude of the horizontal acceleration of dam as 1 m/s^2.

The natural frequencies of a reservoir impounded by a dam with vertical upstream face are given by

$$\omega_n = \frac{(2n-1)\pi c}{2H}$$

For H = 100 m, the fundamental natural frequency of the reservoir works out to 22.59 rad/sec.

Finite Water Reservoir

A finite water reservoir of height 100 m and length L is analysed by the method described above. The boundary of the reservoir is divided into 35 constant boundary element. The hydrodynamic pressure distribution on the upstream face of the dam for excitation frequency equal to half of the first natural frequency is shown in Fig. 3. For the purpose of comparison the pressure distribution obtained from known classical solutions (9) is also shown. The agreement of the results is excellent.

It is known that when the exciting frequency is less than the fundamental natural frequency of the reservoir, the pressure is always in phase with the excitation irrespective of whether the reservoir is finite or infinite. In such a case the pressure in finite reservoir will approach that in an infinite reservoir as the reservoir length L is increased. It will be noted from Fig. 3 that when L = 2H, there is no difference between the classical solutions for the finite and the infinite water reservoirs.

The situation is different for an excitation frequency greater than ω_1. The pressure is still in phase with the excitation if the reservoir is finite, but a phase difference exists between the two for an infinite reservoir. The phase angle, in fact, varies along the height of the reservoir and the maximum pressures at different levels occur at different instants of time. The hydrodynamic pressures on a rigid dam for an excitation frequency greater than ω_1 are presented in Fig. 4. The pressures obtained from the boundary element method are again in very good agreement with the classical solution for a finite reservoir. However, the solution for an infinite reservoir is quite different.

Fig. 5, shows a plot of the total hydrodynamic pressures on the face of the dam for different values of ω/ω_1. Results obtained from the classical solution are also shown. As would be expected, when the exciting frequency is close to one of the natural frequencies of the reservoir, the response becomes very large. A problem associated with the use of the boundary element method in the present application is that the frequencies of the domain external to the one being considered may be intermixed with the latter. When the exciting frequency is close to one of these spurious frequencies, the response values obtained by the boundary element method becomes unreliable. The extra peak in the curve for boundary element method seen in Fig. 5, repre-

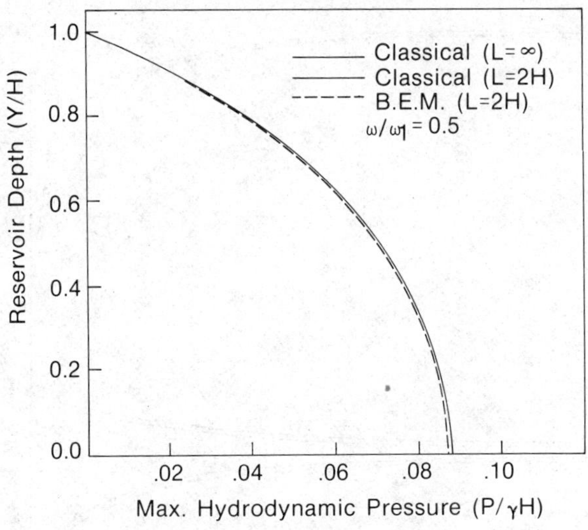

Fig. 3 Hydrodynamic Pressure in Finite Water Reservoir with Stationary Far End.

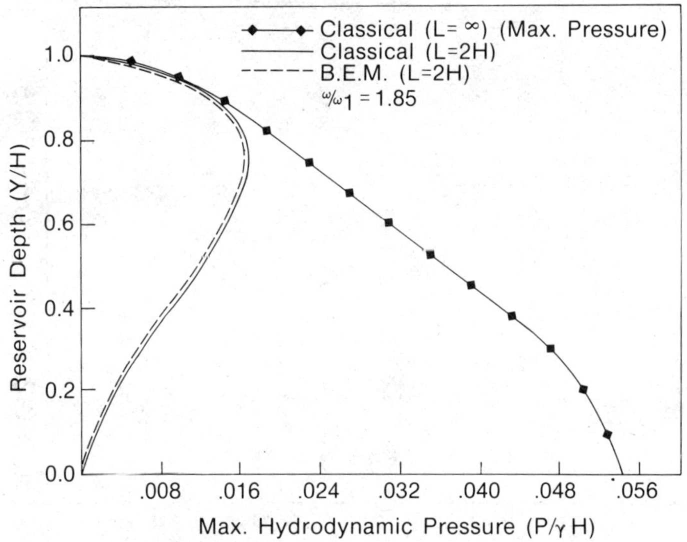

Fig. 4 Hydrodynamic Pressure in Finite Water Reservoir with Stationary Far End.

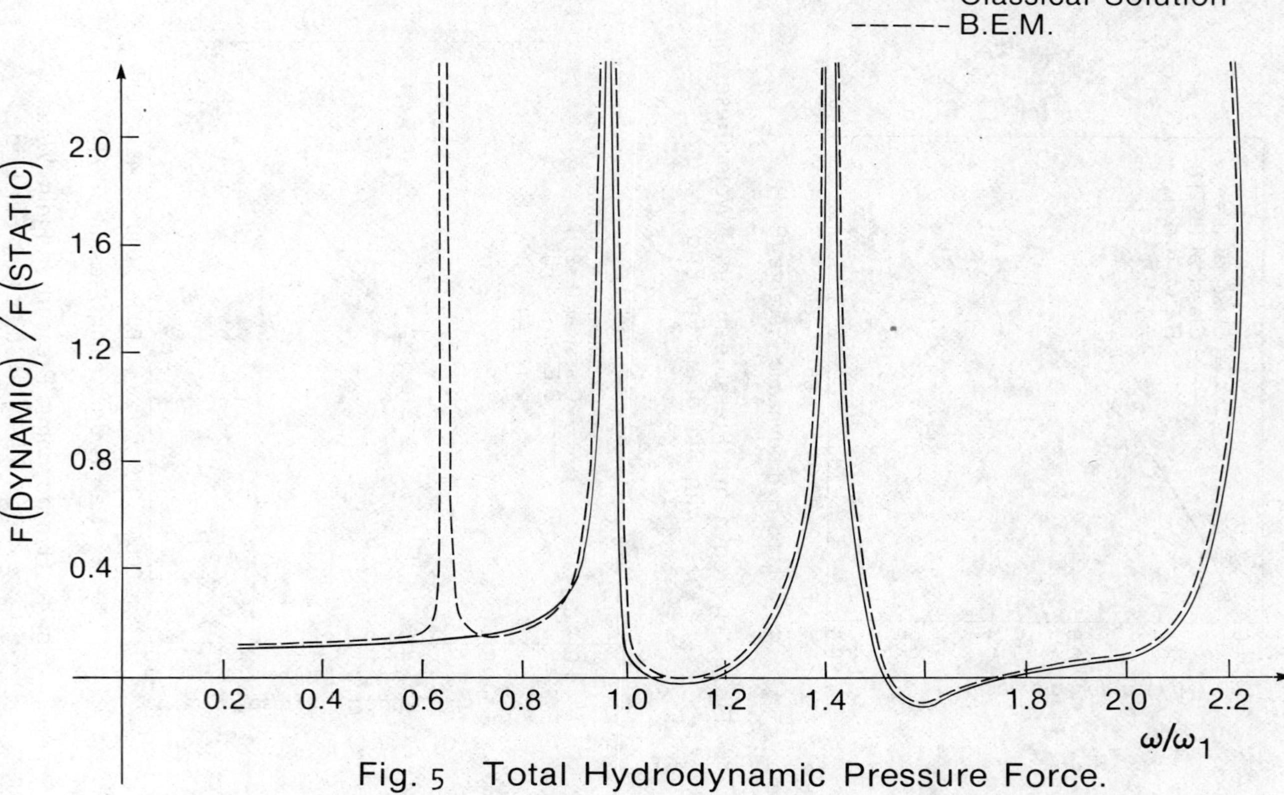

Fig. 5 Total Hydrodynamic Pressure Force.

sents one such spurious frequency.

The existance of spurious frequencies may be considered as a limitation of the boundary element method. As will be shown later, the problem does not occur in reservoirs of infinite extent.

Infinite Water Reservoir

The hydrodynamic pressure distribution on the upstream face of the dam impounding an infinite water reservoir is shown in Fig. 6 for $\omega/\omega_1 = 2.66$. As stated earlier, the pressure is not in phase with the exciting frequency in this case. If the hydrodynamic pressure is expressed as $p_o e^{i\omega t}$, p_o has a real part, $(p_o)_R$ and an imaginary part $(p_o)_I$ which are both shown in Fig. 6a. The amplitude of pressure is shown in Fig. 6b. It should be noted that the phase angle also varies along the height. The pressures obtained from the classical solutions are also presented in Fig. 6. The agreement between the two sets of results is satisfactory.

The total hydrodynamic pressure force obtained for different values of ω/ω_1 is compared in Fig. 7 with that obtained from the classical solution. The agreement of results is satisfactory for the entire range of frequencies studied. The problem of spurious frequencies pertaining to the exterior domain does not appear in this case.

ACKNOWLEDGEMENT

The work presented here was supported by a grant from the National Science and Engineering Research Council of Canada.

APPENDIX I - REFERENCES

1. Brebbia, C.A., "The Boundary Element Method for Engineers", Pentech Press, Devon, 1978.

2. Chakrabarti, P., Chopra A.K., "Earthquake Analysis of Gravity dams including Hydrodynamic Interaction", Earthquake Engineering and Structural Dynamics, Vol. 2, 1973, pp. 143-160.

3. Chopra, A.K. "Hydrodynamic Pressures in Dams during Earthquake", Journal of Engineering Mechanics Division, ASCE Vol. 93, No. EM6 1967, pp. 205-223.

4. Chopra, A.K. "Earthquake Behaviour of Reservoir-Dam Systems" J.Eng. Mech. Dn. ASCE Vol. 94, No. EM6 1968, pp. 1475-1500.

5. Chopra, A.K. "Earthquake Response of Concrete Gravity Dams", J.Eng. Mech. Dn. ASCE Vol. 96 No. EM4, 1980 pp. 443-454

6. Hanna, Y.G., "Application of Boundary Element Method to Certain Problems in Structural Dynamics", M.Eng. Thesis, Carleton University, Ottawa, 1980.

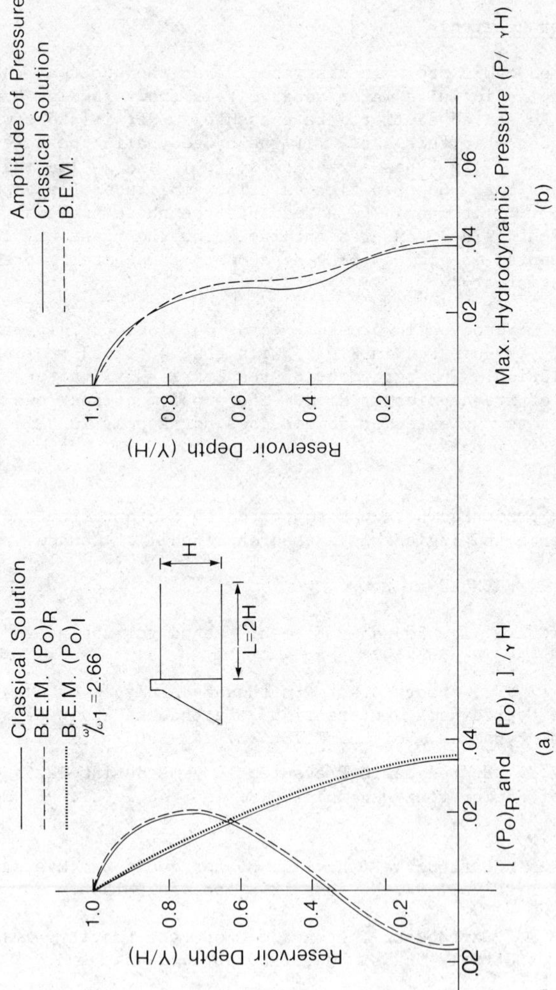

Fig. 6 Hydrodynamic Pressure in Infinite Water Reservoir

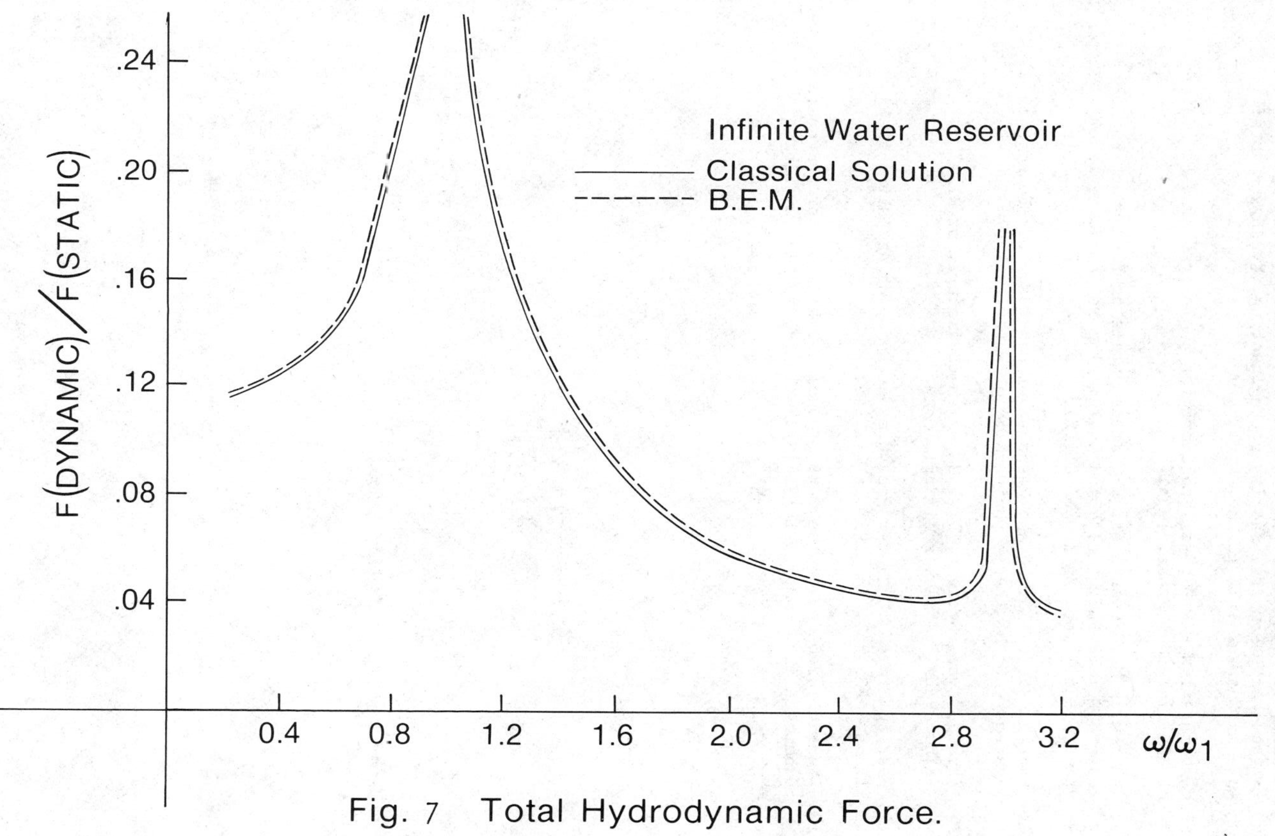

Fig. 7 Total Hydrodynamic Force.

7. Kotsubo, S. "Dynamic Water Pressures on Dam due to Irregular Earthquakes", Memoirs, Faculty of Engineering, Kyushy University, Iukuoka, Japan, Vol. 18, No. 4, 1959.

8. MacLachlan, N.W., "Bessel Functions for Engineers", 2nd Ed. Oxford at the Clarendon Press, 1955.

9. Newmark, N.M., Rosenblueth E. "Fundamentals of Earthquake Engineering", Civil Engineering and Engineering Mechanics Series. Prentice-Hall Inc. Englewood Cliffs, N.J.

10. Sharan, K.S., "Earthquake Response of Dam-Reservoir-Foundation Systems", Ph.D. Thesis, University of Waterloo (1978).

11. Stroud, A.H., Secrest, D., "Gaussian Quadrature Formulae" Prentice-Hall, New York, 1966.

12. Westergaard H.M., "Water Pressures on Dams during Earthquake" Trans. ASCE Vol. 98, 1933 pp. 415-433.

13. Zienkiewicz O.C., Newton R.E., "Coupled Vibrations of a Structure Submerged in a Compressible Fluid", Int. Symposium on Finite Element Techniques, Stuttgart, 1969.

DYNAMIC STABILITY OF STATICALLY PRE-LOADED SYSTEMS

G. J. Simitses[+], M. ASCE, and

C. A. Lazopoulos[++]

Abstract

The paper deals with systems, which are subject to either limit point instability or unstable bifurcational instability, when loaded quasi-statically. These systems are first loaded quasi-statically to a level below the static critical load and subsequently they are subjected to suddenly applied loads of constant magnitude and finite duration. Clear concepts of dynamic stability or instability are presented, including criteria and estimates for critical conditions. These are demonstrated through several one- and two-degree-of-freedom models. Moreover, the analysis encompasses the extreme cases of ideal impulse (duration time of the sudden load approaching zero) and suddenly applied loads of constant magnitude and infinite duration (duration time of the sudden load approaching infinity).

Introduction

The term "Dynamic Stability" encompasses several classes of problems and it has been used, by the various investigators, in connection with a particular study. Examples of these include problems of parametric excitation [1] "follower force" type of problems [2,3], problems of aeroelastic instability and others [4-8].

A large class of structural problems, that has received attention recently, is that of impulsively loaded systems as well as suddenly loaded systems with constant loads of infinite duration [9-20]. The systems considered in the above studies are subject to either limit point instability or unstable bifurcation instability, when loaded quasi-statically. Moreover, a few solutions have appeared in the literature, which deal with this type of systems, when loaded suddenly with a constant load of finite duration [21].

The paper deals with systems (of this latter category) which are first loaded quasi-statically to a level below the static critical load, and subsequently they are subjected to sudden loads of constant magnitude and finite duration. Clear concepts of dynamic stability or instability are presented including criteria and estimates for critical conditions. These are demonstrated through several one- and two-degree-of-freedom models.

Statement of the Problem

Consider a model at its stable equilibrium position $L_S^{P_o}$, when subjected to an initial static load P_o. At time $t = 0$ an additional constant load P is suddenly applied to the system and acts only for finite duration time, $t = T_o$. After the release of the force P, the system moves because of the acquired total energy during the action of the load

+ Professor of Engineering Science and Mechanics, Georgia Institute of Technology, Atlanta, Georgia.

++ Graduate Student.

P. The system will be called dynamically stable is its motion is "unbuckled", meaning that motion is bounded [12]. Since the systems under consideration exhibit limit or unstable bifurcation point instability, the system is stable if the energy, imparted through the action of the load P, is insufficient for the system to reach the unstable static equilibrium point on the "P_o-load" total potential of the system with zero velocity (zero kinetic energy). For each individual model, the criterion is invoked and estimates for critical conditions are found. The extreme cases of $T_o \rightarrow \infty$ (constant load of infinite duration) and $T_o \rightarrow 0$ (Ideal Impulse) are treated as special cases.

Concepts and General Procedure

The concept of dynamic stability and the general procedure are extensions of those developed for the case of suddenly loaded systems without preloading [22]. The static equilibrium positions of the preloaded system are given as solutions to

$$\partial U_T^{P_o}(L)/\partial L = 0 \tag{1}$$

where L is the position of the system. Thus, one may find all the "P_o-load" static equilibrium positions including the near stable position $L_S^{P_o}$ and the unstable position $L_u^{P_o}$, through which dynamic instability can be realized (see Fig. 1).

Keeping the same generalized coordinates for all models and the same expressions for the total potential and for the kinetic energy, one may apply the concepts already developed [22]. These are next explained through the use of Fig. 1, which holds for one-degree-of-freedom systems, but the explanation is applicable to all finte-degree-of-freedom systems.

The system is initially loaded quasi-statically by load p_o and it reaches point A ($L=L_S^{P_o}$; stable static equilibrium point). Then, a load p is applied suddenly for a time τ_o (finite duration). At τ_o the load p is removed.

A potential $\bar{U}_T^P$ is defined, such that $\bar{U}_T^P = \bar{U}_T^{P_o}$ at $\tau=0$ or $L=L_S^{P_o}$ (see Fig. 1).

$$\bar{U}_T^P = \bar{U}_T^{P_o+P} + \left[\bar{U}_T^{P_o}\left(L_S^{P_o}\right) - \bar{U}_T^{P_o+P}\left(L_S^{P_o}\right)\right] \tag{2}$$

Since the system is conservative, during the action of p, one may write

$$\bar{U}_T^P + \bar{T}^{P_o+P} = U_T^{P_o}\left(L_S^{P_o}\right) \; ; \; 0 \le \tau \le \tau_o \tag{3}$$

where $\bar{T}^{P_o+p}$ is the kinetic energy of the system. Making use of Eq. (2), Eq. (3) becomes

$$\bar{U}_T^{P_o+P} + \bar{T}^{P_o+P} = \bar{U}_T^{P_o+P}\left(L_S^{P_o}\right) \; ; \; 0 \le \tau \le \tau_o \tag{4}$$

For times greater than τ_o, the system is also conservative and conservation of energy yields

$$\bar{U}_T^{P_o} + \bar{T}^{P_o} = \bar{U}_T^{P_o}(\tau_o) + \bar{T}^{P_o}(\tau_o) \; ; \; \tau > \tau \tag{5}$$

If the force p has imparted sufficient energy into the systems, such that it can reach the unstable point B ($L = L_u^{P_o}$; see Fig. 1) on the "p_o-load" potential with zero kinetic energy (velocity), then "buckled" motion is possible, and the system becomes dynamically unstable.

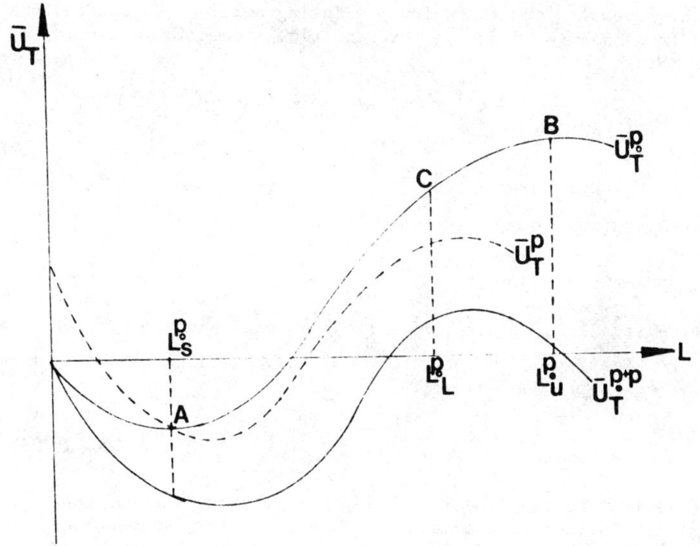

Fig. 1. Definitions of Total Potentials

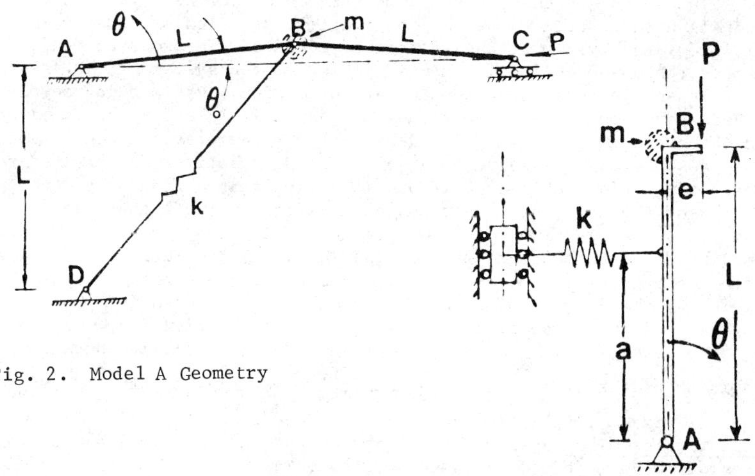

Fig. 2. Model A Geometry

Fig. 3. Model B Geometry

The governing equations for predicting critical conditions are obtained from Eqs. (4) and (5). By requiring kinematic continuity at $\tau = \tau_o$ one may write

$$\bar{T}^{P_o+P}(\tau_o) = \bar{T}^{P_o}(\tau_o) \tag{6}$$

From Eq. (4) one may write

$$\bar{T}^{P_o+P} = \bar{U}_T^{P_o+P}\left(L_S^{P_o}\right) - \bar{U}_T^{P_o+P}(\tau_o) \tag{7}$$

Substitution of Eqs. (6) and (7) into Eq. (5) yields

$$\bar{U}_T^{P_o} + \bar{T}^{P_o} = \bar{U}_T^{P_o}(\tau_o) - \bar{U}_T^{P_o+P}(\tau_o) + \bar{U}_T^{P_o+P}\left(L_S^{P_o}\right); \text{ for } \tau > \tau_o \tag{8}$$

A critical condition exists, if the system can reach point B (see Fig. 1) with zero kinetic energy $\left[\bar{T}^{P_o}\left(L_u^{P_o}\right) = 0\right]$; by denoting the critical condition by $(p, \tau_{o_{cr}})$, one may write

$$\bar{U}_T^{P_o}\left(L_u^{P_o}\right) = \bar{U}_T^{P_o}\left(\tau_{o_{cr}}\right) - \bar{U}_T^{P_o+P}\left(\tau_{o_{cr}}\right) + \bar{U}_T^{P_o+P}\left(L_S^{P_o}\right) \tag{9}$$

Note that Eq. (9) relates p, τ_{ocr} and the position of the system, L_{cr}, at the instant of the release of the force p. Also, observe, that the above described critical condition depends on the evaluation of two parameters, p and τ_o. One approach is to prescribe τ_o and find the corresponding p_{cr} and the other is to prescribe p and find the corresponding $\tau_{o_{cr}}$. The two are equivalent. Regardless of the approach, Eq. (9) relates three parameters p, $\tau_{o_{cr}}$ (or p_{cr}, τ_o) and L_{cr} (the position of the system taken as one parameter).

The second (needed) equation is obtained from Eq. (4). This equation is used to relate the load p, the time of release, τ_o, and the position of the system at the instant of release. In order to find, for a prescribed sudden load, p, the position of the system at the instant of release, one must specify the path of motion. For one-degree-of-freedom systems there is only one path and this is easily accomplished. On the other hand, for a multi-degree-of-freedom system there are numerous possible paths leading to a multitude of positions for a given release time. In such cases, if one is interested in finding a lower bound for the critical condition, he may find the path that yields the smallest possible time. This may be accomplished by solving the corresponding "brachistochrone" problem (see Ref. [23]). Thus, Eq. (4) along with the path of motion, relates the release time τ_o and the system position at the instant of release. These steps are clearly demonstrated for each of the three models, in the subsequent articles.

Moreover, if one is dealing with a deflection limited design (the position of the system cannot exceed $L_L^{P_o}$ - see Fig. 1), then $\bar{U}_T^{P_o}(L_u^{P_o})$ is replaced by $\bar{U}_T^{P_o}(L_L^{P_o})$, in the outlined procedure.

<u>Parenthesis</u>: The cases of ideal impulse and suddenly applied load of constant magnitude and infinite duration, may be obtained as special cases of the present procedure. However, critical conditions for these two load cases may also be obtained independently.

<u>For the ideal impulse case</u>, one may relate the impulse to an initial kinetic energy, and from the conservation of energy

$$\bar{U}_T^{P_o} + \bar{T}^{P_o} = \bar{U}_T^{P_o}(L_S^{P_o}) + \bar{T}_i^{P_o} \tag{10}$$

Then $\bar{T}_i^{P_o}$ is critical (related to the critical ideal impulse) if the system reaches position $L_u^{P_o}$ (pt. B of Fig. 1) with zero kinetic energy. Thus,

$$\bar{T}_{i_{cr}}^{P_o} = \bar{U}_T^{P_o}(L_S^{P_o}) - \bar{U}_T^{P_o}(L_u^{P_o}) \tag{11}$$

For the second extreme case ($\tau_o \to \infty$), p_{cr} may be obtained from Eq. (3) or (4), which, for this case, holds true for all τ $(0 \le \tau \le \infty)$. Thus, p_{cr} corresponds to the solution of

$$\bar{U}_T^{P_o+P}(L_u^{P_o+P}) = \bar{U}_T^{P_o+P}(L_S^{P_o}) \tag{12}$$

where $L_u^{P_o+P}$ denotes the unstable static equilibrium position of the system when the static load is equal to p_o+p.

Applications and Results

Three mechanical models are chosen in order to demonstrate the concepts, discussed in the previous article, and the solution technique, needed to obtain estimates of the critical conditions. These same models have been employed in [22,23]. In these references, the complete static and dynamic stability analysis of the models is presented, but without preloading for the dynamic case. The geometry and sign convention of the models is shown on Figs. 2, 3, and 4. Model A is characteristic of a geometrically imperfect system, Model B of an eccentrically loaded system, and Model C of a system that exhibits unstable bifurcational branching under static loading.

Model A

In evaluating the effect of static preloading for this model, three imperfection angles are chosen and three preloading levels, p_o, for each imperfection angle, θ_o. These are given below, along with the static critical load for each θ_o (see [22,23]).

$\theta_o = 0.005$; $p_o = 0.34, 0.38, 0.42$; $p_{cr_s} = 0.440$

$\theta_o = 0.010$; $p_o = 0.30, 0.36, 0.40$; $p_{cr_s} = 0.415$

$\theta_o = 0.020$; $p_o = 0.30, 0.34, 0.37$; $p_{cr_s} = 0.384$

The expressions for the total potential and kinetic energy, in terms of nondimensionalized parameters (see [23]) are given by

$$\bar{U}_T^P = \frac{U_T^P}{kL^2} = \left[\sqrt{1+\sin\theta} - \sqrt{1+\sin\theta_o}\right]^2 - p\,(\cos\theta_o - \cos\theta) \tag{13}$$

$$\bar{T} = \frac{T}{kL^2} = \frac{1}{2}\left(\frac{m}{k}\right)\left(\frac{d\theta}{dt}\right)^2 = \left(\frac{d\theta}{d\tau}\right)^2 \tag{14}$$

where $p = 2P/kL$, $\tau = t\,(2k/m)^{\frac{1}{2}}$, and m is the small concentrated mass (rigid bars assumed weightless).

The following steps are employed, in order to obtain estimates for the critical conditions, for the case of suddenly applied load of constant magnitude and finite duration, after static preloading.

(a) From the static stability analysis obtain $\theta_s^{P_o}$ and $\theta_u^{P_o}$ for each p_o.

(b) Use of Eq. (9) yields

$$p\left(\cos\theta_s^{P_o} - \cos\theta_{cr}\right) = \left[\sqrt{1+\sin\theta_u^{P_o}} - \sqrt{1+\sin\theta_o}\right]^2$$

$$- \left[\sqrt{1+\sin\theta_s^{P_o}} - \sqrt{1+\sin\theta_o}\right]^2 + p_o\left(\cos\theta_u^{P_o} - \cos\theta_s^{P_o}\right) \quad (15)$$

where θ_{cr} is the position θ at the instant of the release of the force p ($\tau = \tau_o$).

In Eq. (15), for a given, p_o, everything is known (p_o, θ_o, $\theta_s^{P_o}$, and $\theta_u^{P_o}$) except p and θ_{cr}. Therefore, Eq. (15) relates p and θ_{cr} for a critical condition to exist.

(c) Since $\bar{T}^{P_o+P} = \left(\frac{d\theta}{d\tau}\right)^2$, then from Eq. (4) one may write

$$d\tau = \left[U_T^{P_o+P}\left(\theta_s^{P_o}\right) - \bar{U}_T^{P_o+P}(\theta)\right]^{-\frac{1}{2}} d\theta \quad (16)$$

Integration from $\tau = 0$ to $\tau = \tau_o$ and use of the expression for the total potential [see Eq. (13)] yields

$$\tau_o = \int_{\theta_s^{P_o}}^{\theta_{cr}} \left\{\left[\sqrt{1+\sin\theta_s^{P_o}} - \sqrt{1+\sin\theta_o}\right]^2 - \left[\sqrt{1+\sin\theta} - \sqrt{1+\sin\theta_o}\right]^2\right.$$

$$\left. + \left(p_o + p\right)\left(\cos\theta_s^{P_o} - \cos\theta\right)\right\}^{-\frac{1}{2}} d\theta \quad (17)$$

Note that Eq. (17) also relates θ_{cr} to p.

A critical condition is characterized by (p, τ_o) that satisfies Eqs. (15) and (17). This means that for a given release time, τ_o, find p_{cr} or for a given p find $\tau_{o_{cr}}$. Computationally, though, it is easier to assign values of θ_{cr}, solve for p from Eq. (15) and then for the corresponding τ_o from Eq. (17).

A computer program has been written for these computations. Values of θ_{cr} are assigned, starting with $\theta_s^{P_o} + \delta\theta$, where $\delta\theta$ is very small, and computing the corresponding values of p and τ_o for each $\delta\theta$ ($\theta_{cr} = \theta_s^{P_o} + \delta\theta$).

Partial results are presented graphically on Fig. 5 for $\theta_o = 0.010$. Critical conditions appear as plots of p versus duration time, τ_o . Note that as $\tau_{o_{cr}}$ increases, the corresponding value of p approaches $p_{cr\infty}$ (see Table 2). The results for the two extreme cases, $\tau_o \to 0$ and $\tau_o \to \infty$, are presented in tabular form on Tables 1 and 2.

Table 1 - Critical Ideal Impulse, $(p\tau_o)_{cr}$, Model A

θ = 0.005		θ = 0.010		θ = 0.020	
p_o	$(p\tau_o)_{cr}$	p_o	$(p\tau_o)_{cr}$	p_o	$(p\tau_o)_{cr}$
0	162.000	0	81.00	0	40.00
0.340	8.246	0.300	6.980	0.300	2.900
0.380	3.498	0.360	2.247	0.340	1.207
0.420	0.758	0.400	0.486	0.370	0.338
0.440	0	0.416	0	0.384	0

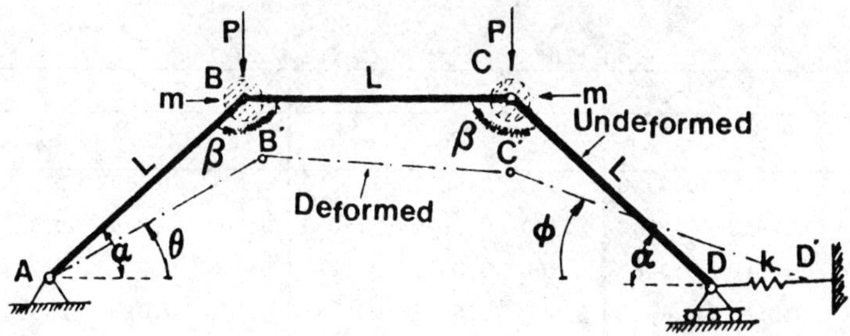

Fig. 4. Model C Geometry

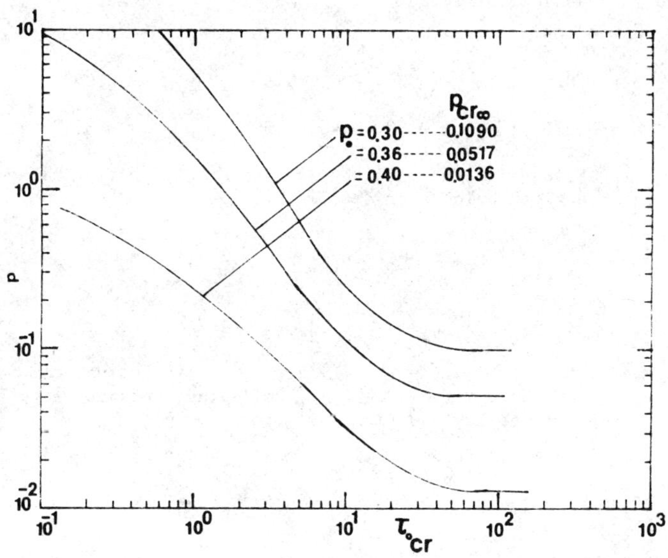

Fig. 5. Critical Conditions; Model A, θ_o = 0.01

Table 2 - Critical Dynamic Load, p_{cr_∞}, Model A

(Constant Load of Infinite Duration)

$\theta_o = 0.005$			$\theta_o = 0.010$			$\theta_o = 0.020$		
p_o	p_{cr_∞}	$p_o+p_{cr_\infty}$	p_o	p_{cr_∞}	$p_o+p_{cr_\infty}$	p_o	p_{cr_∞}	$p_o+p_{cr_\infty}$
0	0.4320	0.4320	0	0.4010	0.4010	0	0.3690	0.3690
0.340	0.0940	0.4340	0.300	0.1090	0.4090	0.300	0.0751	0.3751
0.380	0.0560	0.4360	0.360	0.0517	0.4117	0.340	0.0396	0.3796
0.420	0.0165	0.4365	0.400	0.0136	0.4136	0.370	0.0116	0.3816
0.440	0	0.4400	0.4170	0	0.4160	0.384	0	0.3840

Note that the total load, that the system can withstand, varies from p_{cr_∞}, when $p_o = 0$, to the static critical load, when $p_o = p_{cr_s}$.

Model B

In evaluating the effect of prestress for this model, three load eccentricities are chosen, and, for each eccentricity, three preloading levels, p_o. These numbers are given below along with the static critical load for each eccentricity.

$$\bar{e} = 0.005 \quad ; \quad p_o = 0.35,\ 0.40,\ 0.50 \quad ; \quad p_{cr_s} = 0.955$$

$$\bar{e} = 0.010 \quad ; \quad p_o = 0.30,\ 0.40,\ 0.50 \quad ; \quad p_{cr_s} = 0.932$$

$$\bar{e} = 0.020 \quad ; \quad p_o = 0.30,\ 0.40,\ 0.50 \quad ; \quad p_{cr_s} = 0.898$$

The expressions for the nondimensionalized total potential and kinetic energy are given below (for details see [23]).

$$\bar{U}_T^P = \frac{2U_T^P}{ka^2} = \sin^2\theta - 2p\,(1 - \cos\theta - \bar{e}\sin\theta) \qquad (18)$$

$$\bar{T} = \frac{2T}{ka^2} = \frac{I}{ka^2}\left(\frac{d\theta}{dt}\right)^2 = \left(\frac{d\theta}{d\tau}\right)^2 \qquad (19)$$

where $\bar{e} = e/L$, $p = PL/ka^2$, $\tau = (ka^2/I)^{\frac{1}{2}}\, t$, and I is the system moment of inertia about the hinge A (see Fig. 3). Following the same steps as those for Model A, the key equations are

$$2p\,(\cos\theta_s^{P_o} - \bar{e}\sin\theta_s^{P_o} - \cos\theta_{cr} + \bar{e}\sin\theta_{cr}) = \sin^2\theta_u^{P_o} - \sin^2\theta_s^{P_o}$$
$$- 2p_o\,(\cos\theta_s^{P_o} - \bar{e}\sin\theta_s^{P_o} - \cos\theta_s^{P_o} + \bar{e}\sin\theta_u^{P_o}) \qquad (20)$$

$$\tau_o = \int_{\theta_s^{P_o}}^{\theta_{cr}} \left\{\sin^2\theta_s^{P_o} - \sin^2\theta - 2(p_o+p)\,(\bar{e}\sin\theta_s^{P_o} - \cos\theta_s^{P_o} - \bar{e}\sin\theta + \cos\theta)\right\}^{-\frac{1}{2}} d\theta \qquad (21)$$

A critical condition is characterized by (p, τ_o) that satisfies Eqs. (20) and (21). This means that for a given release time, τ_o, find p_{cr} or for a given p find $\tau_{o_{cr}}$. Computationally, though, it is easier to assign values of θ_{cr}, solve for p from Eq. (20) and then for the corresponding τ_o from Eq. (21).

Partial results are presented graphically, Fig. 6, and in tabular form, Tables 3 and 4.

Table 3. Critical Ideal Impulse, $(p\tau_o)_{cr}$, Model B.

$\bar{e} = 0.005$		$\bar{e} = 0.01$		$\bar{e} = 0.02$	
p_o	$(p\tau_o)_{cr}$	p_o	$(p\tau_o)_{cr}$	p_o	$(p\tau_o)_{cr}$
0	200	0	100	0	50
0.35	192	0.30	88	0.30	42
0.40	164	0.40	76	0.40	36
0.50	139	0.50	62	0.50	29
0.955	0	0.932	0	0.898	0

Table 4. Critical Dynamic Load, p_{cr_∞}, Model B.

(Constant Load of Infinite Duration)

$\bar{e} = 0.0005$			$\bar{e} = 0.010$			$\bar{e} = 0.020$		
p_o	p_{cr_∞}	$p_o+p_{cr_\infty}$	p_o	p_{cr_∞}	$p_o+p_{cr_\infty}$	p_o	p_{cr_∞}	$p_o+p_{cr_\infty}$
0	0.948	0.948	0	0.912	0.912	0	0.85	0.850
0.35	0.600	0.950	0.300	0.621	0.921	0.300	0.57	0.870
0.40	0.552	0.952	0.400	0.522	0.922	0.400	0.48	0.880
0.50	0.453	0.953	0.500	0.422	0.922	0.500	0.38	0.880
0.955	0	0.955	0.932	0	0.932	0.898	0	0.898

Model C

For this particular model, three values of Λ (rise parameter) are chosen and three values of preloading, p_o, for each Λ-value. These are given below.

$$\Lambda = 5.0 \quad ; \quad p_o = 2.8,\ 3.4,\ 4.0 \quad ; \quad p_{cr_s} = 5.154$$

$$\Lambda = 6.0 \quad ; \quad p_o = 3.2,\ 3.8,\ 4.2 \quad ; \quad p_{cr_s} = 6.000$$

$$\Lambda = 8.0 \quad ; \quad p_o = 4.8,\ 5.4,\ 6.0 \quad ; \quad p_{cr_s} = 7.310$$

The expressions for the nondimensionalized total potential and kinetic energy are given below (for details see [23]).

$$\bar{U}_T^P = \frac{U_T}{\bar{\beta}^2 kL^2} = (r^2 + 9s^2 - 2\sqrt{\Lambda}r + \Lambda) + \frac{1}{2}(\Lambda - r^2 - 3s^2)^2 - 2p(\sqrt{\Lambda} - r) \tag{22}$$

where $r = (\theta + \varphi)/2\sqrt{\bar{\beta}}$; $s = (\varphi - \theta)/2\sqrt{\bar{\beta}}$;

$\bar{\beta} = \beta/kL^2$; $\Lambda = \alpha/\sqrt{\bar{\beta}}$; $p = P/(kL\bar{\beta}^{3/2})$

and

$$\bar{T} = \frac{T}{\bar{\beta}^2 kL^2} = \frac{1}{2}\left[\left(\frac{dr}{d\tau}\right)^2 + \left(\frac{ds}{d\tau}\right)^2\right], \quad \tau = t\,(\bar{\beta}k/2m)^{\frac{1}{2}} \tag{23}$$

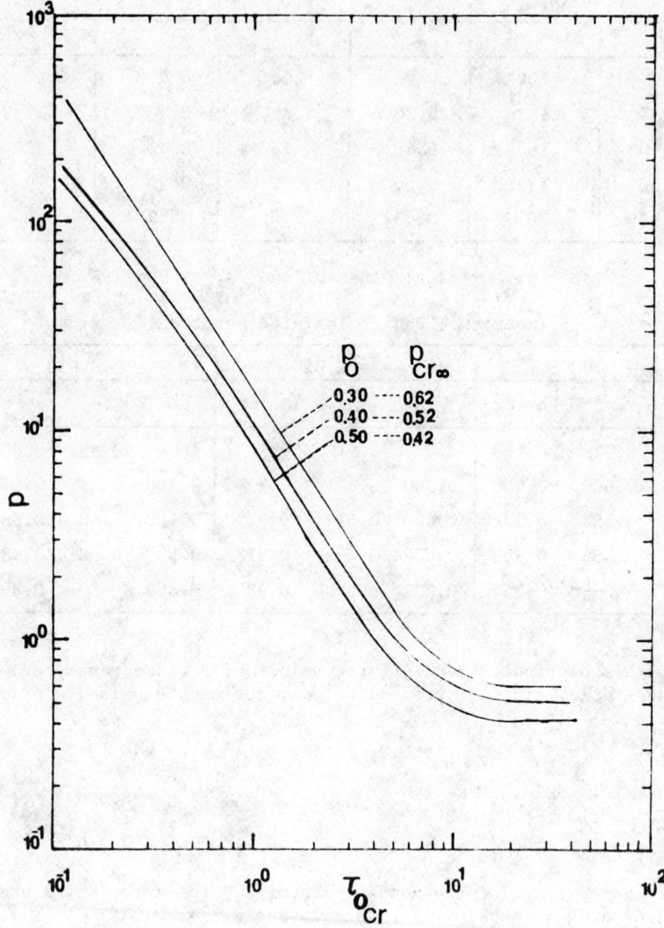

Fig. 6. Critical Conditions; Model B, $\bar{e}$ = 0.01

Note that the bars of this model are assumed weightless (see Fig. 4), and the mass of the system, 2m, is assumed as concentrated at points B and C of the model (m at B and m at C; not shown on Fig. 4).

The steps for this model are the same as before. The only difference is that, in finding the smallest time, $\tau_{o_{cr}}$, required for a given p to reach a critical condition, the path of the motion must be known. In order to find τ_o, the related "brachistochrone" problem is solved (reference [23]). The solution shows that the path is the symmetric one ($s \equiv 0$).

Thus, the two governing equations obtained from Eqs. (9) and (4), for this model are

$$2p\left(r_s^{P_o} - r_{cr}\right) = \left[\left(r_u^{P_o}\right)^2 + 9\left(s_u^{P_o}\right)^2 - 2\sqrt{\Lambda}\left(r_u^{P_o}\right) + \Lambda\right] +$$

$$\frac{1}{2}\left(\Lambda - r_u^{P_o}\right)^2 - \left(r_s^{P_o} - \sqrt{\Lambda}\right)^2 - \frac{1}{2}\left[\Lambda - \left(r_s^{P_o}\right)^2\right]^2 + 2p_o\left(r_u^{P_o} - r_s^{P_o}\right) \quad (24)$$

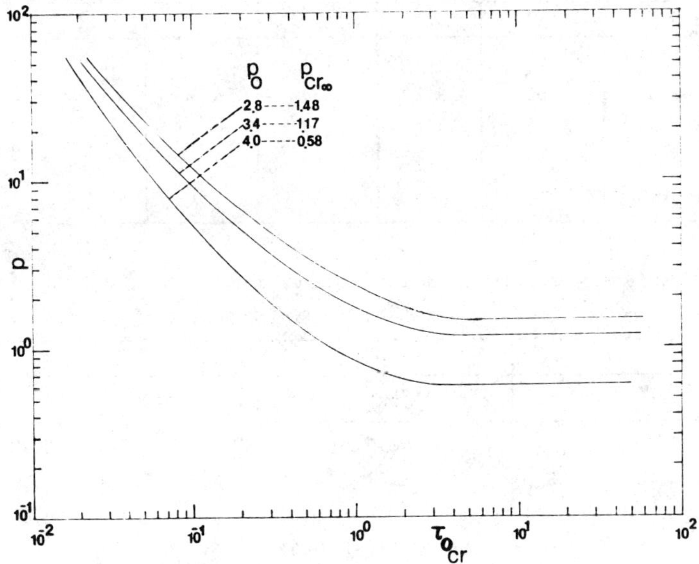

Fig. 7. Critical Conditions; Model C, $\Lambda = 5$

and

$$\tau_{o_{cr}} = \int_{r_s^{p_o}}^{r_{cr}} \left[\left(r_s^{p_o} - \sqrt{\Lambda}\right)^2 + \frac{1}{2}\left(\Lambda - r_s^{p_o}\right)^2 - \left(r - \sqrt{\Lambda}\right)^2 - \frac{1}{2}\left(\Lambda - r\right)^2 - 2\,(p + p_o)\,(r - r_s^{p_o}) \right]^{-\frac{1}{2}} dr \qquad (25)$$

where the position of the system at the instant of application of the sudden load, p, $(r_s^{p_o}, 0)$ and the position of the "p_o-loaded" saddle point, $(r_u^{p_o}, s_u^{p_o})$ are known, for each Λ and p_o, from the static analysis.

Note that a critical condition is characterized by finding $\tau_{o_{cr}}$ for a given p. Also note that Eqs. (24) and (25) relate $\tau_{o_{cr}}$ and r_{cr} (two equations in two unknowns) for a preassigned value of p and for known values of p_o, Λ, $r_s^{p_o}$, $r_u^{p_o}$, and $s_u^{p_o}$ (obtained from the static analysis).

Partial results are presented graphically, Fig. 7, and in tabular form, Tables 5 and 6.

Table 5. Critical Ideal Impulse, $(p\tau_o)_{cr}$, Model C.

$\Lambda = 5.0$		$\Lambda = 6.0$		$\Lambda = 8.0$	
p_o	$(p\tau_o)_{cr}$	p_o	$(p\tau_o)_{cr}$	p_o	$(p\tau_o)_{cr}$
0	6.00	0	6.71	0	7.93
2.8	3.06	3.2	3.39	4.80	3.04
3.4	2.32	3.8	2.67	5.40	2.31
4.0	1.32	4.2	1.67	6.00	1.44
5.145	0	6.0	0	7.31	0

Table 6. Critical Suddenly Applied Loads, p_{cr_∞}, Model C. (Constant Magnitude and Infinite Duration)

$\Lambda = 5.0$			$\Lambda = 6.0$			$\Lambda = 8.0$		
p_o	p_{cr_∞}	$p_o+p_{cr_\infty}$	p_o	p_{cr_∞}	$p_o+p_{cr_\infty}$	p_o	p_{cr_∞}	$p_o+p_{cr_\infty}$
0	3.70	3.70	0	4.35	4.35	0	5.70	5.70
2.8	1.48	4.28	3.20	1.68	4.88	4.8	1.59	6.39
3.4	1.17	4.57	3.80	1.32	5.12	5.4	1.21	6.61
4.0	0.58	4.58	4.20	0.75	4.95	6.0	0.68	6.68
5.154	0	5.154	6.00	0.00	6.00	7.31	0.00	7.31

Discussion

Concepts of dynamic instability of preloaded systems subjected to suddenly applied step-loads have been presented. Moreover, criteria of dynamic instability and estimates of critical conditions have been discussed. The above have been demonstrated through three simple mechanical models.

For the one-degree-of-freedom models, the estimates represent the true critical conditions for dynamic instability to occur. In other words, there are no upper and lower conditions. On the other hand, for the two-degree-of-freedom model the estimates, presented herein, represent a lower bound of critical conditions (MPCL according to [12] and sufficiency conditions for stability according to ([13-17]).

One important result, though, is that the effect of preloading is independent of the particular model. Specifically:

(1) For the extreme case of the ideal impulse (see Tables 1, 3, and 5), the critical impulse decreases continuously from the value corresponding to $p_o = 0$ down to zero, when $p_o = p_{cr_s}$.

(2) Similarly, for the case of suddenly applied constant loads of infinite duration, the total load, $p_o + p_{cr_\infty}$ (see Tables 2, 4 and 6), that the system can withstand, increases from the value of p_{cr_∞} corresponding to $p_o = 0$, up to the static critical load, p_{cr_s}.

Both of these observations are very significant in the design of dynamically loaded system, since the analyses corresponding to zero prestress are much simpler in execution. Moreover, the static analysis is comparatively simpler to perform.

Acknowledgement

This work was performed under Contract No. F 33615-79-C-3221 with the U.S.A.F., Aeronautical Systems Division (AFSC), Wright-Patterson Air Force Base. The financial support provided by the United States Air Force is gratefully acknowledged.

References

1. Bolotin, V. V., The Dyanmic Stability of Elastic Systems, (Translated by V. I. Weingarter, et al.) Holden Day, San Francisco, 1964.

2. Bolotin, V. V., Nonconservative Problems of the Theory of Elastic Stability, Moscow 1961; English Translation published by Pergamon Press, New York, 1963.

3. Herrmann, G., "Stability of Equilibrium of Elastic Systems Subjected to Nonconservative Forces," Applied Mechanics Reviews, Vol. 20, 1967, pp. 103-108.

4. Stoker, J. J., "On the Stability of Mechanical Systems," Communication on Pure and Applied Math., Vol. VIII, 1955, pp. 133-142.

5. Seckel, E., Stability and Control of Airplanes and Helicopters, Academic Press, New York, 1964.

6. Lefschetz, S., Stability of Nonlinear Control Systems, Academic Press, New York, 1965.

7. Paidoussis, M. P., "Dynamics of Tubular Catilevers Conveying Fluid," J. Mech. Eng. Science, Vol. 42, No. 2, 1970, pp. 85-103.

8. Bohn, M. P., and Herzmann, G., "Instabilities of a Spatial System of Articulated Pipes Conveying Fluid," J. Fluid Engineering, 1974, pp. 289-296.

9. Hoff, N. J., and Bruce, V. G., "Dynamic Analysis of the Buckling of Laterally Loaded Flat Arches," J. Math. and Phys., Vol. 32, 1954, pp. 276-388.

10. Budiansky, B., and Roth, R. S., "Axisymmetric Dynamic Buckling of Clamped Shallow Spherical Shells," Collected Papers on Instability of Shell Structures, NASA TN D-1510, 1962.

11. Budiansky, G., and Hutchinson, J. W., "Dynamic Buckling of Imperfection-Sensitive Structures," Proceedings of XI International Congress of Applied Mechanics, Munich, 1964.

12. Simitses, G. J., "Dynamic Snap-Through Buckling of Low Arches and Shallow Caps," Ph.D. Dissertation, Department of Aeronautics and Astronautics, Stanford University, June 1965.

13. Hsu, C. S., "On Dynamic Stability of Elastic Bodies with Prescribed Initial Conditions," International J. of Eng. Sciences, Vol. 4, 1966, pp. 1-21.

14. Hsu, C. S., "The Effects of Various Parameters on the Dynamic Stability of a Shallow Arch," J. Appl. Mech., Vol. 34, No. 2, 1967, pp. 349-356.

15. Hsu, C. S., "Stability of Shallow Arches Against Snap-Through Under Tunewise Step Loads," J. Appl. Mech., Vol. 35, No. 1, 1968, pp. 31-39.

16. Hsu, C. S., "Equilibrium Configurations of a Shallow Arch of Arbitrary Shape and their Dynamic Stability Character," International J. Nonlinear Mechanics, Vol. 3, June 1968, pp. 113-136.

17. Hsu, C. S., Kuo, C. T., and Lee, S. S., "On the Final States of Shallow Arches on Elastic Foundations Subjected to Dynamic Loads," J. Appl. Mech., Vol. 35, No. 4, 1968, pp. 713-723.

18. Budiansky, B., "Dynamic Buckling of Elastic Structures: Criterion and Estimates," Dynamic Stability of Structures (Edited by G. Herrmann), Pergamon Press, 1967.

19. Simitses, G. J., "On the Dynamic Buckling of Shallow Spherical Caps," J. Appl. Mech., Vol. 41, No. 1, 1974, pp. 299-300.

20. Tamura, Y. S., and Babcock, C. D., "Dynamic Stability of Cylindrical Shells Under Step Loading," J. Appl. Mech., Vol. 42, Series E, No. 1, March 1975, pp. 190-194.

21. Zimcik, D. G., and Tennyson, R. C., "Stability of Circular Cylindrical Shells Under Transient Axial Impulsive Loading," Proceedings AIAA/ASME/ASCE/AHS 20th Structures, Structural Dynamics and Materials Conference, St. Louis, Missouri, April 4-6, 1979.

22. Simitses, G. J., "Dynamic Stability of Structural Elements Subjected to Step-Loads," Proceedings, Army Symposium on Solid Mechanics, 1980; Designing for Extremes: Environment, Loading, and Structural Behavior, South Yarmouth, Cape Cod, Massachusetts, Sept. 30-Oct. 2, 1980.

23. Lazopoulos, C. A., "Dynamic Stability of Structural Elements Under Step-Loads," Ph.D. Dissertation, School of Engineering Science and Mechanics, Georgia Institute of Technology, Atlanta, Georgia, July 1980.

STATIONARY RESPONSE OF STRUCTURES WITH CLOSELY SPACED MODES

By

Armen Der Kiureghian[1], A.M. ASCE

ABSTRACT

Analytical solutions for the stationary response of linear, multi-degree-of-freedom structures having closely spaced modes and subjected to white-noise or filtered white-noise inputs are derived. Spectral moments of the response power spectral density are obtained as a superposition of the corresponding spectral moments of modal responses for which closed-form results are given. The superposition procedure includes the effect of correlation between modal responses which is significant for modes with closely spaced frequencies. Simplified approximations are derived for responses to wide-band inputs.

INTRODUCTION

Modes with closely spaced frequencies occur in many structural systems. Buildings with eccentric masses, equipment-structure or piping-structure systems, and certain shell and plate structures are common examples. It is known that the modal responses of structures with closely spaced frequencies and subjected to stochastic inputs are highly correlated. Because of this correlation, application of the mode-superposition technique for random vibration analysis of such structures requires special attention, since, as shown in this paper, cross-terms between modal responses should properly be included in the superposition process. In this paper, a mode-superposition procedure is formulated for the stationary response of linear structures including the effect of correlation between modal responses. Through this procedure, the moments of the response power spectral density about the frequency origin are obtained in terms of the corresponding moments of individual modal responses and the correlation coefficients between them. Closed-form solutions for the first three spectral moments of modal responses and the associated correlation coefficients are derived for the cases of white-noise and filtered white-noise input. It is well known that these moments sufficiently describe most statistical measures of the response that are of engineering interest (5).

Special attention is given to the response of structures with closely spaced frequencies that are subjected to wide-band inputs. For such inputs, simplified, approximate expressions for the correlation coefficients between modal responses are obtained. These expressions are examined by comparison with exact results for the response to filtered white-noise input. This analysis also demonstrates the significance of the correlation between responses in modes with closely spaced frequencies.

RESPONSE OF MDF SYSTEMS

Consider an n-degree-of-freedom linear structure with mass, damping, and stiffness matrices **M**, **C**, and **K**, respectively. Assume that the structure has classical modes with natural

[1]Department of Civil Engineering, University of California, Berkeley, California.

frequencies ω_i, modal damping coefficients ζ_i, and modal vectors $\mathbf{\Phi}_i = \left[\phi_{1i} \ \phi_{2i} \ \cdots \ \phi_{ni}\right]^T$, where $i = 1, 2, \ldots, n$, and a superposed T denotes a transpose. It is well known that any response of such a system to an input excitation, $F(t)$, can be obtained as a superposition of its modal responses as

$$R(t) = \sum_{i=1}^{n} R_i(t) \tag{1}$$

where $R_i(t)$ is the response to $F(t)$ in mode i. Each modal response is given by

$$R_i(t) = \Psi_i S_i(t) \tag{2}$$

where $S_i(t)$ is the i-th normal coordinate, representing the response to $F(t)$ of a linear, single-degree-of-freedom oscillator of frequency ω_i and damping coefficient ζ_i, and Ψ_i is the effective participation factor for mode i given by

$$\Psi_i = \Gamma_i \alpha_i = \frac{\mathbf{\Phi}_i^T \mathbf{M I}}{\mathbf{\Phi}_i^T \mathbf{M} \mathbf{\Phi}_i} \alpha_i \tag{3}$$

where Γ_i is the conventional participation factor for mode i (1), $\mathbf{I}$ is the influence vector coupling the input into the degrees of freedom of the structure, and α_i is the response quantity expressed in terms of the i-th modal vector. For example, for $F(t)$ applied along the k-th degree of freedom, $\mathbf{I}$ contains a 1 in the k-th row and zeros elsewhere. Also, the quantity α_i is generally given in terms of the properties of the structure and a linear combination of the elements of the i-th modal vector. For example, for displacement of the k-th degree of freedom, $\alpha_i = \phi_{ik}$, and for the force in a spring connecting the k-th and l-th degrees of freedom, $\alpha_i = (\phi_{ik} - \phi_{il}) k_{kl}$, where k_{kl} is the spring constant.

For a zero-mean stationary excitation $F(t)$, using Eq. 1, the power spectral density of the stationary response, $R(t)$, is

$$G_R(\omega) = \sum_{i=1}^{n} \sum_{j=1}^{n} \Psi_i \Psi_j G_F(\omega) H_i(\omega) H_j^*(\omega) \tag{4}$$

where $G_F(\omega)$ is the power spectral density of $F(t)$, and

$$H_i(\omega) = \frac{1}{\omega_i^2 - \omega^2 + 2i\zeta_i\omega_i\omega} \tag{5}$$

is the complex frequency response function for mode i, representing the steady-state solution of $S_i(t)$ for a forcing function of the form $\exp(i\omega t)$, and the asterisk denotes a complex conjugate. Observe that because of symmetry, complex terms in the summation in Eq. 4 cancel and, as a result, $G_R(\omega)$ is always real-valued.

Of primary interest from an engineering viewpoint are the moments of the power spectral density about the frequency origin. Using one-sided spectral densities, these moments for the response are defined as

$$\lambda_m = \int_0^\infty \omega^m G_R(\omega)\, d\omega; \quad m = 0, 1, 2, \ldots \tag{6}$$

In particular, it has been shown that most statistical measures of the response that are of engineering interest can be obtained in terms of the first three spectral moments, i.e., for $m = 0$, 1, and 2 (5). For example, it is well known that $\sigma_R = \sqrt{\lambda_0}$ and $\sigma_{\dot{R}} = \sqrt{\lambda_2}$ are, respectively, the root-mean-squares of $R(t)$ and its time rate, $\dot{R}(t)$, whereas λ_1 is shown to be related to the envelop process of $R(t)$ (5). Also, $\nu = \sqrt{\lambda_2/\lambda_0}/\pi$ denotes the mean zero-crossing rate of the response, and $q = \sqrt{1 - \lambda_1^2/\lambda_0\lambda_2}$ is a measure of dispersion of the power spectral density of the response. The latter two parameters together with σ_R are also sufficient to describe the cumulative distribution (6) and the mean and variance (2) of the peak value of the response over a specified duration.

Using Eq. 4 in Eq. 6, the spectral moments of the response can be expressed as

$$\lambda_m = \sum_{i=1}^{n}\sum_{j=1}^{n}\Psi_i\Psi_j\lambda_{m,ij}; \quad m=0,1,2,... \tag{7}$$

where

$$\lambda_{m,ij} = \text{Re}\left[\int_0^\infty \omega^m G_F(\omega)H_i(\omega)H_j^*(\omega)\,d\omega\right]; \quad m=0,1,2,... \tag{8}$$

are *cross-spectral moments* of normal coordinates $S_i(t)$ and $S_j(t)$ associated with modes i and j. For $i=j$, Eq. 8 reduces to

$$\lambda_{m,ii} = \int_0^\infty \omega^m G_F(\omega)|H_i(\omega)|^2 d\omega; \quad m=0,1,2,... \tag{9}$$

which describes the spectral moments for the i-th normal coordinate, $S_i(t)$. It is clear from the above definition that $\lambda_{0,ii}$ and $\lambda_{2,ii}$ represent, respectively, the mean-squares of $S_i(t)$ and $\dot{S}_i(t)$, whereas $\lambda_{0,ij}$ and $\lambda_{2,ij}$ represent the covariances between $S_i(t)$ and $S_j(t)$ and between $\dot{S}_i(t)$ and $\dot{S}_j(t)$, respectively. The first moments $\lambda_{1,ii}$, and $\lambda_{1,ij}$ are related to the envelop processes of $S_i(t)$ and $S_j(t)$. It is useful to introduce a set of dimensionless coefficients defined as

$$\rho_{m,ij} = \frac{\lambda_{m,ij}}{\sqrt{\lambda_{m,ii}\lambda_{m,jj}}}; \quad m=0,1,2,... \tag{10}$$

Clearly, $\rho_{0,ij}$ and $\rho_{2,ij}$ are correlation coefficients between $S_i(t)$ and $S_j(t)$ and between $\dot{S}_i(t)$ and $\dot{S}_j(t)$, respectively. These quantities behave like correlation coefficients for other values of m also, since $-1 \leqslant \rho_{m,ij} \leqslant 1$ and $\rho_{m,ii}=1$ for all m. Thus, the name *correlation coefficient* will be used to describe $\rho_{m,ij}$ for all m.

Introducing Eq. 10 in Eq. 7, an alternative superposition formula for the spectral moments of the response is found in terms of the correlation coefficients and the spectral moments of individual normal coordinates

$$\lambda_m = \sum_{i=1}^{n}\sum_{j=1}^{n}\Psi_i\Psi_j\rho_{m,ij}\sqrt{\lambda_{m,ii}\lambda_{m,jj}}; \quad m=0,1,2,... \tag{11}$$

Either of Eqs. 7 or 11 can be used to obtain the moments of the response power spectral density. However, Eq. 11 is more useful for responses to wide-band inputs since, as will subsequently be shown, for such inputs simple approximate expressions for $\rho_{m,ij}$ can be obtained, thus eliminating the need for the cross-spectral moments, $\lambda_{m,ij}$, required in Eq. 7, which are usually difficult to evaluate.

For structures with well spaced frequencies the cross-spectral moments $\lambda_{m,ij}$ are generally small in comparison to the terms $\lambda_{m,ii}$ and $\lambda_{m,jj}$. For such structures, therefore, Eq. 7 can be reduced to

$$\lambda_m = \sum_{i=1}^{n}\Psi_i^2\lambda_{m,ii}; \quad m=0,1,2,... \tag{12}$$

This form of superposition of spectral moments has been studied before (5). However, for structures with closely spaced frequencies this reduction is not valid since the cross-spectral moments for the closely spaced modes are generally of the same order as the individual mode moments and their elimination may lead to erroneous results. In the following, solutions of the spectral moments for individual modes, $\lambda_{m,ii}$, the cross-spectral moments, $\lambda_{m,ij}$, and the correlation coefficients, $\rho_{m,ij}$, for $m=0,1,2$ and for responses to white-noise and filtered white-noise inputs are summarized. Detailed discussions and derivations of these results are given in Ref. 2.

Response to White-Noise Input -- For a white-noise input

$$G_F(\omega) = G_0 \tag{13}$$

where G_0 is a constant scale factor. Substituting the above in Eq. 8 and evaluating the integral, one obtains

$$\lambda_{0,ij} = \frac{2\pi G_0}{K_{ij}}(\zeta_i\omega_i + \zeta_j\omega_j) \tag{14}$$

$$\lambda_{1,ij} = \frac{2\pi G_0}{K_{ij}}\left\{\frac{1-\frac{2}{\pi}\tan^{-1}(\zeta_i/\sqrt{1-\zeta_i^2})}{4\sqrt{1-\zeta_i^2}}\left[\omega_i(\zeta_i\omega_i+\zeta_j\omega_j) + \omega_j(\zeta_i\omega_j+\zeta_j\omega_i)\right]\right.$$
$$+\frac{1-\frac{2}{\pi}\tan^{-1}(\zeta_j/\sqrt{1-\zeta_j^2})}{4\sqrt{1-\zeta_j^2}}\left[\omega_i(\zeta_i\omega_j+\zeta_j\omega_i) + \omega_j(\zeta_i\omega_i+\zeta_j\omega_j)\right].$$
$$\left.-\frac{\ln\frac{\omega_i}{\omega_j}}{2\pi}(\omega_i^2 - \omega_j^2)\right\} \tag{15}$$

$$\lambda_{2,ij} = \frac{2\pi G_0}{K_{ij}}(\zeta_i\omega_j+\zeta_j\omega_i)\omega_i\omega_j \tag{16}$$

where

$$K_{ij} = (\omega_i^2-\omega_j^2)^2 + 4\zeta_i\zeta_j\omega_i\omega_j(\omega_i^2+\omega_j^2) + 4(\zeta_i^2+\zeta_j^2)\omega_i^2\omega_j^2 \tag{17}$$

For $i=j$, Eqs. 14-16 give the spectral moments for the individual normal coordinates as

$$\lambda_{0,ii} = \frac{\pi G_0}{4\zeta_i\omega_i^3} \tag{18}$$

$$\lambda_{1,ii} = \frac{\pi G_0}{4\zeta_i\omega_i^2}\,\frac{1-\frac{2}{\pi}\tan^{-1}(\zeta_i/\sqrt{1-\zeta_i^2})}{\sqrt{1-\zeta_i^2}} \tag{19}$$

$$\lambda_{2,ii} = \frac{\pi G_0}{4\zeta_i\omega_i} \tag{20}$$

The results in Eqs. 18 and 20 are well known (4), whereas that in Eq. 19 is somewhat simpler than a similar result given by Vanmarcke (5).

Using Eqs. 14-16 in Eq. 10, the correlation coefficients, $\rho_{m,ij}$, for response to white-noise input are obtained. For $m=0$ and 2 the results are

$$\rho_{0,ij} = \frac{8\sqrt{\zeta_i\zeta_j\omega_i\omega_j}(\zeta_i\omega_i + \zeta_j\omega_j)\omega_i\omega_j}{K_{ij}} \tag{21}$$

$$\rho_{2,ij} = \frac{8\sqrt{\zeta_i\zeta_j\omega_i\omega_j}(\zeta_i\omega_j + \zeta_j\omega_i)\omega_i\omega_j}{K_{ij}} \tag{22}$$

The corresponding expression for $\rho_{1,ij}$ is too complicated and for that reason has little practical value. For small damping and closely spaced modes, Eqs. 21 and 22, as well as the exact expression for $\rho_{1,ij}$, can be reduced to dominant terms to

$$\rho_{0,ij} \approx \frac{2\sqrt{\zeta_i\zeta_j}\left[(\omega_i+\omega_j)^2(\zeta_i+\zeta_j) + (\omega_i^2-\omega_j^2)(\zeta_i-\zeta_j)\right]}{4(\omega_i-\omega_j)^2 + (\omega_i+\omega_j)^2(\zeta_i+\zeta_j)^2} \tag{23}$$

$$\rho_{1,ij} \approx \frac{2\sqrt{\zeta_i\zeta_j}\left[(\omega_i+\omega_j)^2(\zeta_i+\zeta_j) - 4(\omega_i-\omega_j)^2/\pi\right]}{4(\omega_i-\omega_j)^2 + (\omega_i+\omega_j)^2(\zeta_i+\zeta_j)^2} \tag{24}$$

$$\rho_{2,ij} \approx \frac{2\sqrt{\zeta_i\zeta_j}\left[(\omega_i+\omega_j)^2(\zeta_i+\zeta_j)-(\omega_i^2-\omega_j^2)(\zeta_i-\zeta_j)\right]}{4(\omega_i-\omega_j)^2+(\omega_i+\omega_j)^2(\zeta_i+\zeta_j)^2} \tag{25}$$

The exact and approximate expressions of $\rho_{m,ij}$ for response to white-noise input are plotted in Fig. 1 against the ratio of frequencies, ω_i/ω_j, and for various values of the modal damping coefficients, ζ_i and ζ_j. Observe that the approximate expressions closely agree with the exact results even for damping ratios as high as 20 percent and frequency ratios of 0.5 or less. It is

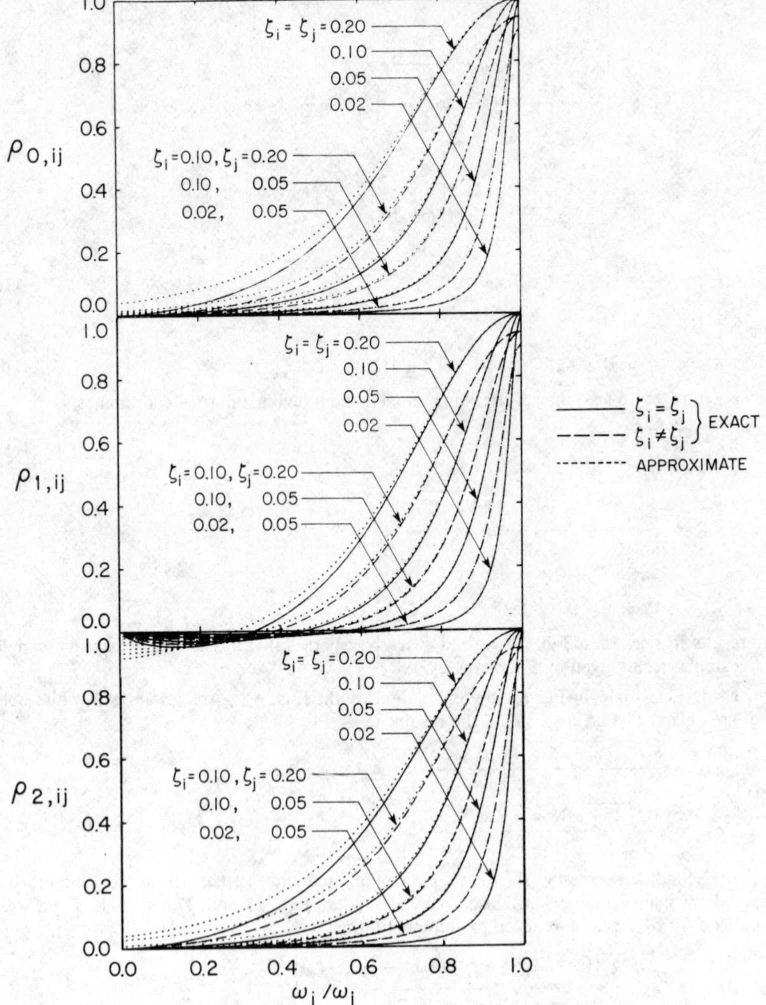

FIG. 1.- Correlation Coefficients for Response to White-Noise Input

important to note in this figure that the coefficients $\rho_{m,ij}$ rapidly diminish as the two modal frequencies become widely separated. The decay in these coefficients is especially rapid for small values of damping coefficients. Thus, it is evident from this figure that the cross-modal terms in Eqs. 7 or 11 are only significant for modes with closely spaced frequencies and that the reduction to Eq. 12 is only valid when the modal frequencies are well separated. As a simple rule, the cross terms in Eqs 7 or 11 may be considered as insignificant when

$$\frac{\omega_i}{\omega_j} \leqslant \frac{0.2}{\zeta_i+\zeta_j+0.2} \tag{26}$$

This approximately corresponds to $\rho_{m,ij} \leqslant 0.1$.

Response to Filtered White-Noise Input -- Formally, filtered white noise results as the response of an oscillator to a white-noise input. It is often used to represent the input into a structure supported by a single-degree-of-freedom primary system which itself is subjected to a white-noise excitation. A common example is the base input into a structure situated on a soil layer which is excited by earthquake motions. More generally, however, the filtered white noise may be used as a convenient model for a large class of excitations. Consider an input power spectral density of the form

$$G_F(\omega) = \frac{\omega_g^4 + 4\zeta_g^2\omega_g^2\omega^2}{(\omega_g^2-\omega^2) + 4\zeta_g^2\omega_g^2\omega^2}\, G_0 \tag{27}$$

where ω_g and ζ_g are constants. This may represent the power spectral density of the absolute acceleration response of a single-degree-of-freedom oscillator subjected to a white-noise base acceleration, where ω_g and ζ_g are the natural frequency and damping coefficient of the oscillator, respectively. (For example, Kanai (3) has suggested $\omega_g = 5\pi$ and $\zeta_g = 0.6$ for modeling the ground acceleration during earthquakes.) However, by proper selection of ω_g and ζ_g, Eq. 27 may be used to represent excitations with varying power spectral density shapes. In particular, the filter frequency, ω_g, determines the dominant range of input frequencies, whereas ζ_g determines the smoothness of the power spectral density shape. Fig. 2 illustrates the power spectral density in Eq. 27 for three selected values of ζ_g. Observe in this figure that the spectra become wide-banded for increasing values of ζ_g. In practice, a filtered white-noise process with $\zeta_g \geqslant 0.6$ is often considered to be a wide-band process.

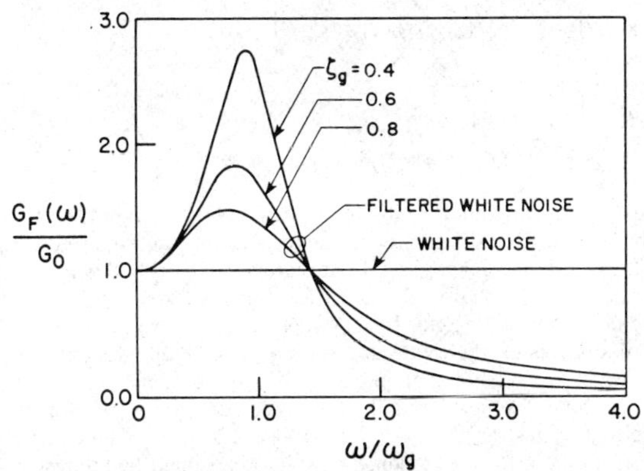

FIG. 2.- Power Spectral Density of Filtered White-Noise Process

Closed-form solutions for the cross-spectral moments of modal responses to a filtered white-noise input with the power spectral density in Eq. 27 are reported in Ref. 2. These are rather long expressions in terms of the modal properties, ω_i, ω_j, ζ_i, and ζ_j, the filter parameters, ω_g and ζ_g, and the scale factor, G_0. For the purpose of brevity, these solutions are not repeated here and the interested reader is referred to the original work. Setting $i=j$ in these solutions, the results for the spectral moments for individual normal coordinates are obtained as follows:

$$\lambda_{0,ii} = \frac{\pi G_0}{4\zeta_i\omega_i^3}\frac{A}{F} + \frac{\pi G_0}{4\zeta_g\omega_g^3}\frac{B}{F} \tag{28}$$

$$\begin{aligned}\lambda_{1,ii} = {} & \frac{\pi G_0}{4\zeta_i\omega_i^2}\frac{1-\frac{2}{\pi}\tan^{-1}(\zeta_i/\sqrt{1-\zeta_i^2})}{\sqrt{1-\zeta_i^2}}\frac{A'}{F} \\ & + \frac{\pi G_0}{4\zeta_g\omega_g^2}\frac{1-\frac{2}{\pi}\tan^{-1}(\zeta_g/\sqrt{1-\zeta_g^2})}{\sqrt{1-\zeta_g^2}}\frac{C}{F} - \frac{2G_0}{\omega_g^2}\frac{D}{F}\ln r\end{aligned} \tag{29}$$

$$\lambda_{2,ii} = \frac{\pi G_0}{4\zeta_i\omega_i}\frac{A''}{F} + \frac{\pi G_0}{4\zeta_g\omega_g}\frac{E}{F} \tag{30}$$

where $r=\omega_i/\omega_g$ and

$$A = A_0+4\zeta_i^2r^2\left[2-4\zeta_g^2-(3-4\zeta_i^2)r^2-4\zeta_g^2r^4\right]$$

$$A' = A_0+4\zeta_i^2r^2\left[1-4\zeta_g^2-2(1-\zeta_i^2)r^2-2\zeta_g^2r^4\right]$$

$$A'' = A_0+4\zeta_i^2r^2\left[-4\zeta_g^2-r^2\right]$$

$$B = 1-8\zeta_g^2-2(1-2\zeta_i^2)r^2+(1+4\zeta_g^2)r^4$$

$$C = 1-4\zeta_g^2-2(1-2\zeta_i^2+2\zeta_g^2-4\zeta_i^2\zeta_g^2)r^2+(1+4\zeta_g^2-8\zeta_g^4)r^4$$

$$D = 1-(1-2\zeta_i^2)r^2-2\zeta_g^2r^4$$

$$E = 1-2(1-2\zeta_i^2+4\zeta_g^2-8\zeta_i^2\zeta_g^2)r^2+(1+4\zeta_g^2-16\zeta_g^4)r^4$$

$$F = F_0+8\zeta_i^2r^2\left[1-2\zeta_g^2-2(1-\zeta_i^2)r^2+(1-2\zeta_g^2)r^4\right]$$

in which

$$A_0 = 1-2(1-4\zeta_g^2)r^2+(1+4\zeta_g^2)r^4$$

$$F_0 = 1-4(1-2\zeta_g^2)r^2+2(3-8\zeta_g^2+8\zeta_g^4)r^4-4(1-2\zeta_g^2)r^6+r^8$$

Several special cases of the solutions in Eqs. 28-30 are noteworthy. First, as ω_g approaches ∞, the filtered white-noise process approaches the white-noise process. In this case r approaches 0, A, A', A'', and F approach 1, and terms in Eqs. 28-30 with ω_g in their denominators vanish. The expressions for the spectral moments then reduce to Eqs. 14-16 for response to white-noise input. Second, for $\zeta_i \ll \zeta_g$, i.e., for a lightly damped oscillator and a wide-band input, first terms on the right-hand sides of Eqs. 28-30 are dominant, provided ω_i is not much greater than ω_g. (Note that for $\omega_i \gg \omega_g$, the oscillator frequency is beyond the significant range of input frequencies and, therefore, such a case is not of much practical interest.) Observe that in this case $A \approx A' \approx A'' \approx A_0$ and $F \approx F_0$. Thus, the spectral moments in this case differ from those for the response to white-noise input by the factor A_0/F_0, which is only a function of r and ζ_g. Next, consider the case where ζ_i and $\zeta_g \ll 1$, i.e., a lightly damped oscillator and a narrow-band excitation. Under the condition that r is not in the neighborhood of 1, Eqs. 28-30 can be reduced to

$$\lambda_{m,ii} \approx \omega_i^m A_{H_i} G_F(\omega_i) + \omega_g^m A_F |H_i(\omega_g)|^2; \quad m=0,1,2 \tag{31}$$

where $A_{H_i} = \pi/4\zeta_i\omega_i^3$ and $A_F = \pi G_0 \omega_g/4\zeta_g$ are areas under $|H_i(\omega)|^2$ and $G_F(\omega)$ diagrams, respectively. This simple relation can be used, for example, to approximate the response of a secondary system supported by a lightly damped primary system whose natural frequencies are well separated. Finally, in the case of resonance when $\omega_i = \omega_g$ and $r=1$, Eqs. 28-30 reduce to

$$\lambda_{0,ii} = \frac{\pi G_0}{4\zeta_i\omega_i^3} \frac{1+4\zeta_g(\zeta_g+\zeta_i)}{4\zeta_g(\zeta_g+\zeta_i)} \tag{32}$$

$$\lambda_{1,ii} = \frac{\pi G_0}{4\zeta_i\omega_i^2} \left[\frac{1-\frac{2}{\pi}\tan^{-1}(\zeta_i/\sqrt{1-\zeta_i^2})}{\sqrt{1-\zeta_i^2}} \frac{1+4\zeta_g^2-2\zeta_i^2}{4(\zeta_g^2-\zeta_i^2)} - \frac{1-\frac{2}{\pi}\tan^{-1}(\zeta_g/\sqrt{1-\zeta_g^2})}{\sqrt{1-\zeta_g^2}} \frac{\zeta_i(1+2\zeta_g^2)}{4\zeta_g(\zeta_g^2-\zeta_i^2)} \right] \tag{33}$$

$$\lambda_{2,ii} = \frac{\pi G_0}{4\zeta_i\omega_i} \frac{1+4\zeta_g^2}{4\zeta_g(\zeta_g+\zeta_i)} \tag{34}$$

It is observed that in this case the spectral moments are equal to those for the response to white-noise input as amplified by a factor in terms of ζ_g and ζ_i. Note that the amplification factor can be very large for small ζ_g, i.e., for narrow-band inputs.

Ratios of spectral moments for response to filtered white-noise input, Eqs. 28-30, to the corresponding values for response to white-noise input, Eqs. 14-16, are plotted in Fig. 3 against the ratio $r = \omega_i/\omega_g$ and for three values of ζ_g representing narrow-band to wide-band inputs. Results for $\zeta_i = 0.02$ and 0.10, representing lightly and moderately damped oscillators, respectively, are shown in this figure. The ratios A_0/F_0 are also shown for comparison. It is interesting to note that the ratios for the three spectral moments are nearly the same in all cases. Observe that the ratios are considerably different from unity even for the filtered white-noise input with $\zeta_g = 0.6$. This result indicates the sensitivity of the spectral moments to the shape of the response power spectral density. It should be clear from this observation that a white-noise approximation for a wide-band, filtered white-noise input with ζ_g as high as 0.6 would not be appropriate for computation of the spectral moments. It is also interesting to observe in Fig. 3 that the ratio A_0/F_0 closely agrees with the exact ratios of the spectral moments for small ζ_i/ζ_g, say when it is of order 0.1 or less. Thus, this ratio can be used together with the expressions of the spectral moments for response to white-noise input to approximate the spectral moments for response to filtered white-noise input when $\zeta_i/\zeta_g \leqslant 0.1$.

From the exact solutions of $\lambda_{m,ij}$ for response to filtered white-noise input reported in Ref. 2, the correlation coefficients $\rho_{m,ij}$, for $m=0,1,2$, have been computed and are shown in Fig. 4 for $\zeta_g = 0.6$ and for three different values of ω_g (solid curves). These are compared in the figure with the corresponding values of the correlation coefficients from Eqs. 23-25 for response to white-noise input (dashed curves). Observe that for the most part Eqs. 23-25 agree with the exact coefficients for response to filtered white-noise input with $\zeta_g = 0.6$. This is especially true for closely spaced frequencies, which is when the coefficients are of most interest, and for small damping. The biggest deviations occur when ω_i and ω_j are both much greater than ω_g (see Fig. 4 or $\omega_j = 2\omega_g$ and $\omega_i/\omega_j = 2$), i.e., when the modal frequencies are beyond the significant range of input frequencies. This case, of course, is not of much practical interest. Thus, it appears that the correlation coefficients based on white-noise input can be used for wide-band, filtered white-noise inputs as long as the frequencies of important modes are not far

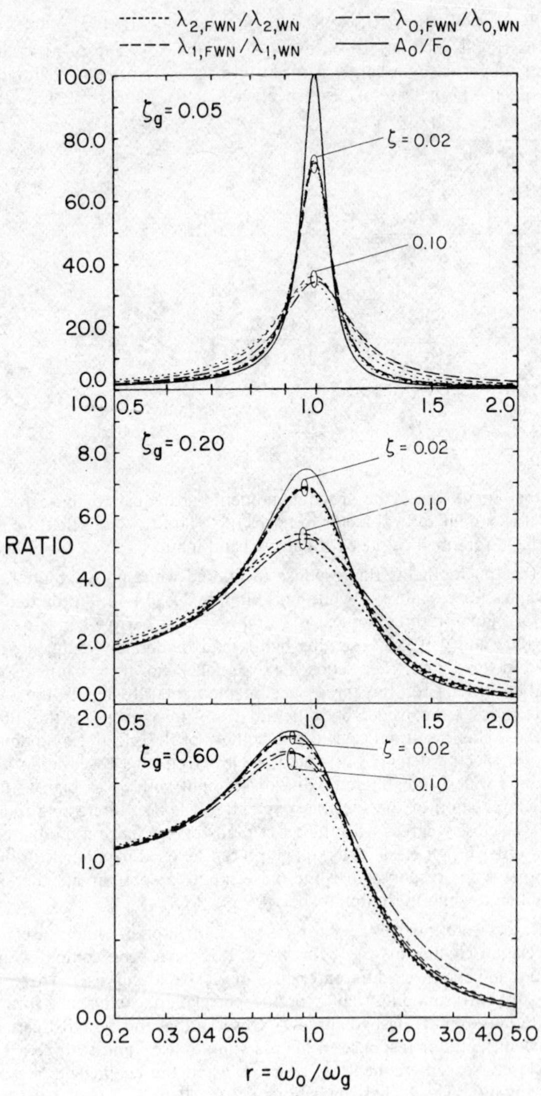

FIG. 3.- Ratios of Spectral Moments for Responses to Filtered White-Noise (FWN) and White-Noise (WN) Inputs

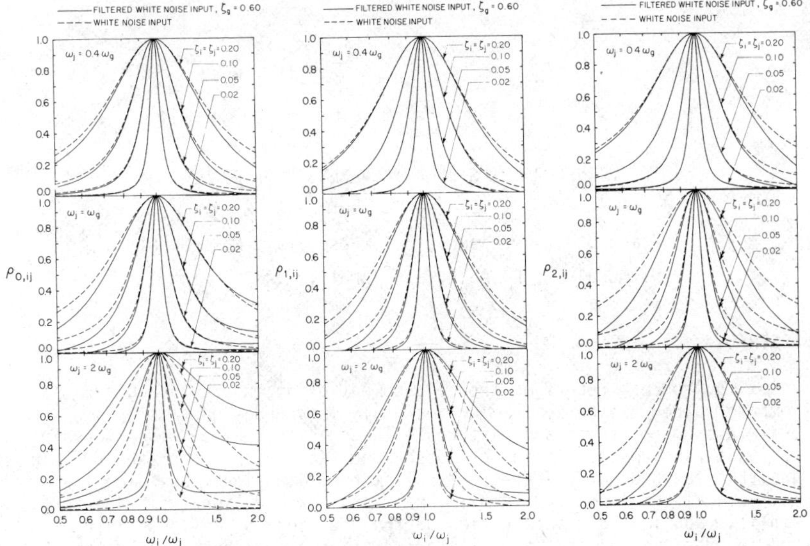

FIG. 4.- Correlation Coefficients for Response to Filtered White-Noise Input

beyond the significant range of input frequencies. With this approximation, Eqs. 23-25 together with the spectral moments for individual modes can be used in Eq. 11 to compute the spectral moments for structures with closely spaced frequencies which are subjected to wide-band, filtered white-noise inputs. It is intuitively obvious that the same approximation would also be valid for other wide-band excitations for which the spectral moments for individual modes are available.

ACKNOWLEDGEMENT

This research work was supported by the U.S. National Science Foundation under Grant No. ENG-7905906. This support is gratefully acknowledged.

REFERENCES

[1] Clough, R.W., and Penzien, J., *Dynamics of Structures*, McGraw-Hill, New York, NY., 1975.

[2] Der Kiureghian, A., "On Response of Structures to Stationary Excitation," *Report No. UCB/EERC-79/32*, Earthquake Engineering Research Center, University of California, Berkeley, CA., December, 1979.

[3] Kanai, K., "Semi-Empirical Formula for Seismic Characterization of the Ground," *Bulletin of Earthquake Research Institute*, University of Tokyo, Japan, Vol. 35, June, 1967.

[4] Lin, Y.K., *Probabilistic Theory of Structural Dynamics*, McGraw-Hill, New York, NY., 1967.

[5] Vanmarcke, E.H., "Properties of Spectral Moments with Application to random Vibration," *Journal of the Engineering Mechanics Division*, ASCE, Vol. 98, No. EM2, Proc. Paper 8822, April, 1972, pp. 425-446.

[6] Vanmarcke, E.H., "On the Distribution of the First-Passage Time for Normal Stationary Processes," *Journal of Applied Mechanics*, Vol. 42, March, 1975, pp. 215-220.

INTERACTIVE GRAPHICS DYNAMIC ANALYSIS OF FRAMES

By M. Gattass[1], J.G. Orbison[1], C.I. Pesquera[1], M.A. Schulman[1], W. McGuire[2], F. ASCE, J.F. Abel[3], M. ASCE

ABSTRACT

A progress report on the development of an interactive graphics computer aided design system for the analysis and design of steel buildings under static and earthquake loads is presented. Preprocessing, analysis, and postprocessing are described, with emphasis on the dynamic analysis features of the system. Full three-dimensional building frames may be analyzed linearly. Provision exists for modal analysis and for explicit and implicit integration of the equations of motion. Both problem description and result interpretation are graphical. Vector refresh graphics and color raster graphics are used for describing dynamic effects.

INTRODUCTION

Extensive research on the use of interactive computer graphics in structural engineering has been conducted by Cornell's Department of Structural Engineering and Program of Computer Graphics (Refs. 1, 2, 3, and 4). One part of this effort is directed toward the development of a system for the computer aided design of steel framed buildings subjected to static and earthquake loading. This is a progress report on the portions of this research that are particularly relevant to earthquake engineering and the dynamic analysis of three-dimensional systems. Other features of the work, such as steel design capability and developments in nonlinear analysis, are mentioned only briefly.

The Laboratory of Computer Graphics currently has the facilities to generate complex static and dynamic vector displays with graphic output devices ranging from storage tube technology to fast, vector-refresh displays. Color images can be generated on 24-bit frame buffers providing perceptually continuous tone images. Drawings, photographs, and slides can be digitized at video rates. All of the graphic input and output devices are supported by a virtual-memory, 32-bit mini-computer.

All of the research reported on has been conducted in this environment. A major aim has been to explore ways in which to apply

---

1. Graduate Res. Asst., Cornell Univ., Ithaca, NY
2. Prof. of Civ. Engng., Cornell Univ., Ithaca, NY
3. Assoc. Prof. of Civ. Engrg., Cornell Univ., Ithaca, NY

this type of equipment most effectively to the problems investigated. It is hoped to demonstrate the power, the economy, and the attractiveness of interactive graphics computer aided design. As an example of this effort, it may be noted that in all parts of the programs to be described, control may be exercised by the operator through a keyboard terminal or directly through a digitizing tablet placed in front of the picture screen. The position of a stylus as it is moved on this tablet is represented by a cursor displayed on the screen. Through the manual operation of the stylus, figures may be created, commands issued, and the flow of the program controlled. Maximum use is made of this user-convenient medium.

Three subdivisions may be defined in most computer aided design programs: 1) a preprocessor or problem description component, 2) a processor, in this case a body of earthquake analysis programs, and 3) a postprocessor or result interpretation component. These divisions will be used for convenience below, but it should be pointed out that they are not independent; the system is an integrated, modular one.

PREPROCESSOR

An interactive vector graphics preprocessor has been developed for the generation and specification of problem data and analysis control parameters. Use of the preprocessor simplifies and accelerates the input phase of the analysis/design sequence. Further, by providing a visual display of the modeled structure, the occurrence of input errors is largely eliminated. To assist the analyst, options have been included to allow rotation and translation in space, and to permit "zooming in" on a portion of the structure. These capabilities are particularly useful when generating large, complex structures. Through the use of interactive options and permanent data bases, the required entry of repetitive information has been minimized.

Geometry Definition. General space frames having six degrees of freedom at each node may be generated. Particular attention has been given to building frames. Many of the features to be described represent simplifications permissible in that type of structure. The geometry of the structure is created by assembling planar or space subframes. Minimal data entry is required for subframe generation: bay spacings, story heights, and number of bays and stories are sufficient for basic frame description. Modifications to the frame geometry may be performed on the subframe, or delayed until frame assembly has been completed.

To assemble the subframes in the global coordinate system, it is necessary to specify the location in space of 3 nodes or "keypoints" on the frame. One of the keypoints is automatically located at the origin of the subframe's local coordinate system; the remaining keypoints are specified interactively. Figure 1 illustrates a space frame with the keypoint nodes highlighted in a constituent planar subframe. During the frame assembly process, duplicate nodes and elements (typically arising at the intersection of subframes) are internally deleted. Node and element numbers, nodal coordinates, and member end incidences are generated internally. A bandwidth minimizer has been implemented, the node numbering sequence being dependent upon the relative number of bays (or stories) in each direction. A building frame generated in this manner is shown in Figure 2.

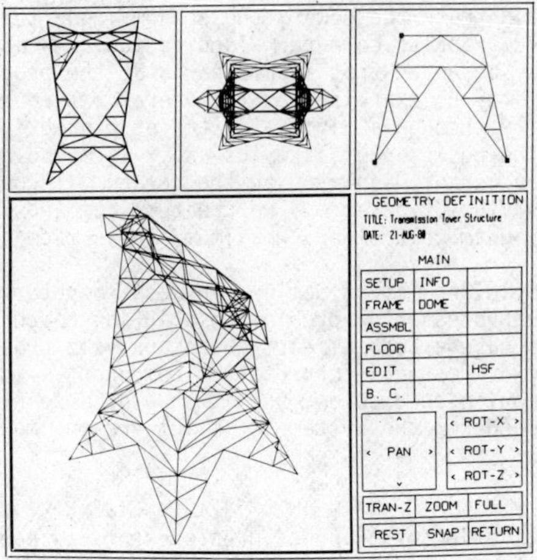

Figure 1. Transmission tower space frame. A planar subframe with highlighted keypoints is displayed in the upper right viewport.

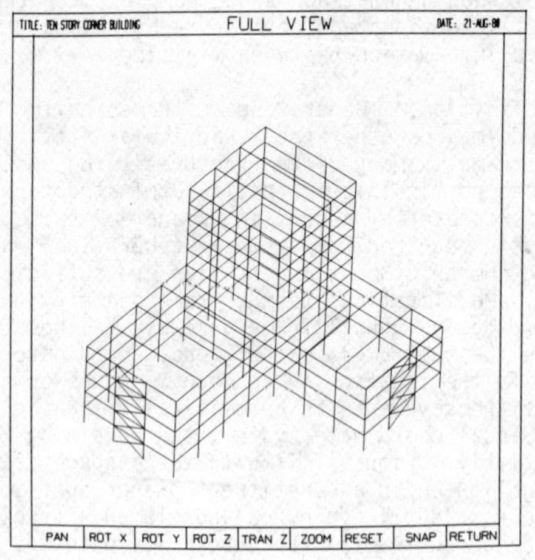

Figure 2. Full-screen view of a perspective display of an assembled building frame.

The present dynamic analysis program provides the option for specifying rigid floor diaphragms. Floors are described after assembly of the space structure. Any combination of closed polygons may be used to define the geometry of the floor; openings and cutouts are permitted. The floor weight per unit area is specified, and the center of mass is internally computed using line integration. The total translational and rotational masses are lumped at the center of mass, which is treated as a structural node.

Extensive frame editing capabilities have been incorporated. A node may be added by specifying its global coordinates; conversely, nodes and elements may be deleted by "pointing" to them with the stylus and cursor. Any two nodes may be used to define a new element spanning between them. An existing element may be subdivided into a specified number of equal-length segments as a simplified mesh refinement technique.

Several interactive options have been developed for the specification of nodal fixity conditions. The global coordinate system is used as reference, with no external restraint in any direction being the default condition. The fixity may be defined in a single step for a) one node alone, b) all nodes lying on a specified line, c) all nodes between two given elevations, and d) all nodes in a defined plane. The fixity condition is chosen from a displayed table, and assigned to the appropriate node(s) with the stylus and cursor.

Element Properties. At present, a three-dimensional beam-column element has been implemented. As steel frames are emphasized in this research, a permanent data base has been developed to assist in the assignment of element properties. The data base consists of relevant section properties for standard Wide Flange and S-Shapes. Sections are chosen from displays of available sections, as shown in Figure 3. Provision has been made to allow the specification of non-standard element properties. For three-dimensional frame structures, it is necessary to define the orientation of the element cross section in space. For the general case, this is accomplished by specifying the coordinates of a third node in the plane of the element web. To simplify this task, four common orientations have been identified: 1) beam with vertical web, 2) beam with horizontal web, 3) column with web parallel to the global X axis, and 4) column with web parallel to the global Y axis. The location of the third node is internally generated if one of these cases is selected.

In preparation for future nonlinear analysis capabilities, the yield strength of an element may be specified using a displayed table of steel types. Availability of the section in the selected steel type is verified internally with reference to ASTM Specification A6 (Ref. 5). Further, the correct yield strength is assigned according to the section's group number, stored in the permanent data base.

Static Analysis Information. To describe a loading condition for static analysis, different load types may be used (e.g. dead load, live load, etc.). Different combinations of load types may be specified to generate a load vector for analysis. All loads are considered as nodal loads, and may be specified in ways similar to the assignment of nodal fixity conditions. Figure 4 illustrates a structure after a loading condition for a number of nodes has been specified.

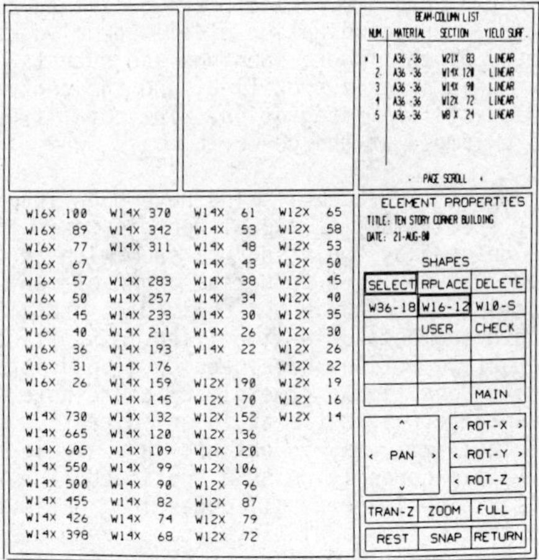

Figure 3. Display of a portion of the steel sections table used to select stored standard element properties.

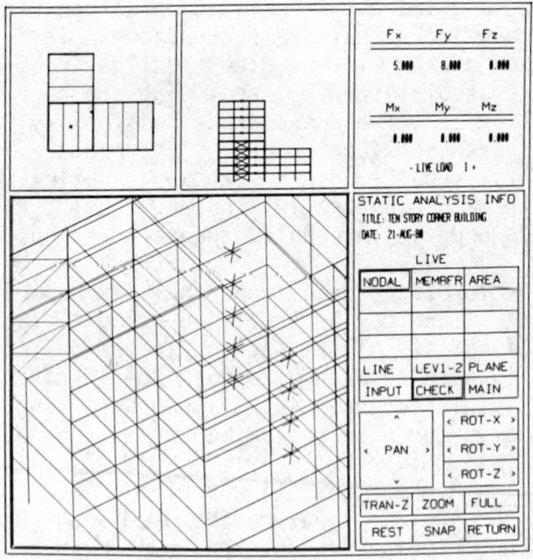

Figure 4. Frame with specified nodal loads (enlared partial perspective view in main viewport).

Earthquake Analysis Information. Ground motion may be defined either by an earthquake record or a design response spectrum, depending upon the type of analysis to be performed. An algorithm has been incorporated to generate response spectra from an accelogram (Ref. 6). Either form may be specified in any of the three global directions, and amplitude scaling factors may be entered for a particular direction. To assist the user, a permanent library of earthquake records and response spectra has been developed, and may be retrieved through the preprocessor. Stored information includes the title of the record, record type, and the appropriate record coordinates. The records are displayed in sequence, and selection of the desired earthquake is made from the display. Figure 5 shows the North-South component of the El Centro earthquake as retrieved from the library.

DYNAMIC ANALYSIS PROCEDURES

Modeling of the spatial coordinates of buildings has been the subject of considerable attention by many investigators. References 7, 8, 9, and 10 illustrate typical stages in the progress of these studies. Topics of particular concern have been the fidelity of the idealization, computational and modeling economy, user convenience, and the proper compromise between the sometimes conflicting demands of accuracy, economy, and convenience.

In most of the earlier work, the building was viewed as an assemblage of plane frames. Generally, such substructures were assembled in ways that did not satisfy all the requirements of compatibility. In addition, the frames were usually treated as systems having no out-of-plane rigidity. These procedures were justified by the observation that they yielded reasonably accurate results for certain limited classes of buildings.

Some of the advantages of these earlier procedures, such as simplified data preparation and output format, and the effective use of out-of-core memory, have declined in significance as computer technology has advanced. For example, the advent of computer graphics with three-dimensional capabilities has limited the desirability of relating all input-output data to planar systems. The virtual memory systems of the present generation of minicomputers has minimized the need to resort to less efficient out-of-core programming.

For these reasons, in the present work buildings have been modeled as three-dimensional structures with full displacement compatibility (frequently violated in previous work at column lines common to two or more planar frames) and correct stiffness modeling, particularly out-of-plane rigidity in subframes.

A particular problem in spatial representation is the modeling of building floors. In some earlier studies (Ref. 7), floors were modeled as rigid struts linking the planar frames. One disadvantage of this model for dynamic analysis is that the torsional modes of vibration are not represented. In agreement with later works (Refs. 9, 10), the floor is assumed to be infinitely rigid in its own plane, and infinitely flexible out of its plane. This option for modeling building structures appears to represent accurately the behavior of typical flooring systems, and significantly reduces the number of degrees of freedom in the idealization. Each floor is represented by a node located at the center of mass, and has three degrees of freedom associa-

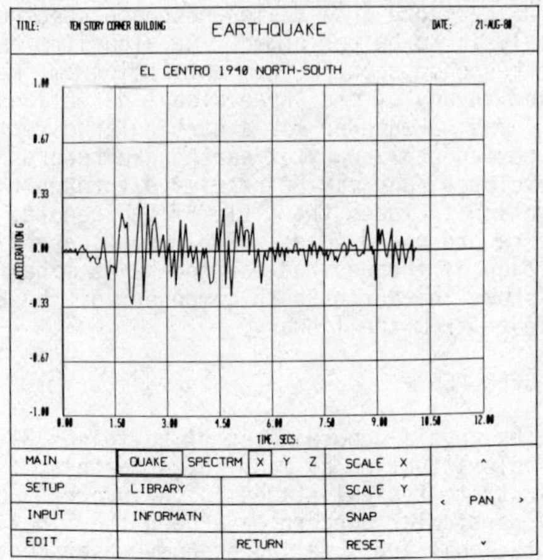

Figure 5. El Centro Earthquake accelogram, North-South Component, retrieved from permanent data base.

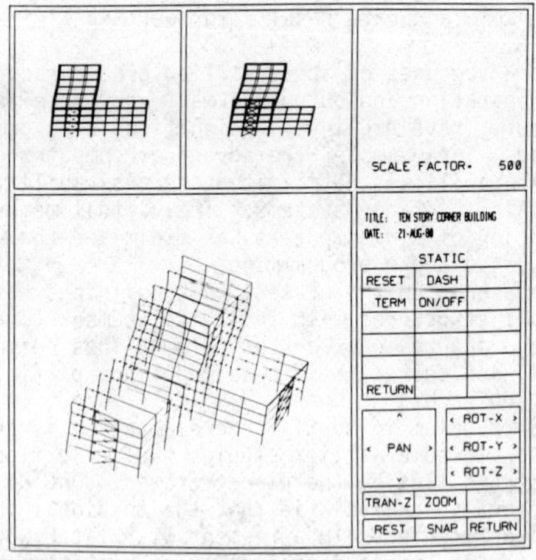

Figure 6. Deflected shape of building frame under quasistatic wind loading.

ted with it: two in-plane translations and an in-plane rotation. The floor behavior is modeled by introducing constraint equations over the horizontal displacement and torsional rotation of the nodes linked to the floor. These constraints are applied during the formulation of the element stiffness matrices in the global system. A common restriction that the columns framing into the floor must be vertical has thus been eliminated.

The analysis procedures utilize a lumped mass idealization, resulting in a diagonal mass matrix. An automatic procedure has been implemented in which any degree of freedom with an associated mass is considered a dynamic degree of freedom, and is present in the differential equations of equilibrium. Those degrees of freedom having no associated mass are condensed out by static condensation.

Time Integration in Dynamic Analysis. Given a spatial discretization, the equations of dynamic equilibrium can be written as the following system of second order ordinary differential equations:

$$[M]\{\ddot{U}\} + [C]\{\dot{U}\} + [K]\{U\} = -[M]\{\ddot{U}_g\} \qquad (1)$$

The dissipative forces are included in their linearized viscous form. The resulting damping matrix [C] is assumed to be a linear combination of the mass and stiffness matrices:

$$[C] = \alpha[M] + \beta[K] \qquad (2)$$

where the parameters α and β are specified by the analyst. For modal analysis, damping is specified as a fraction of critical damping; each mode may have a different damping value.

Several procedures exist to carry out the integration of these equations. The choice of the most appropriate method is dependent upon specific problem parameters such as the total number of degrees of freedom, the number of time steps, the length of the earthquake record, and whether or not linear behavior is being assumed. At present, three methods of integration have been implemented.

A Modal Analysis technique has been included, in which the integration is carried out over the uncoupled modal coordinates of the system. Integration is not performed numerically, but rather by evaluating the analytic response of a linearized forcing function (Ref. 9). Modal analysis is frequently the most efficient method for approximating the time history of linear elastic structures. Often only the first few modes of vibration significantly affect the structural response; the principal advantage of modal analysis is that the number of modes to be considered can be limited. However, in problems where a large number of modes must be modeled, or where the number of time steps is small, a direct integration scheme may provide superior efficiency.

The second method implemented is an explicit time integration algorithm. The Central Difference Method (Ref. 11) has been selected for its accuracy and computational efficiency, but a modification to this method has been incorporated. In the Central Difference Method, the acceleration is approximated by:

$$\{\ddot{U}_t\} = \frac{1}{(\Delta t)^2}[\{U_{t-\Delta t}\} - 2\{U_t\} + \{U_{t+\Delta t}\}] \tag{3}$$

The error of this expansion is of the order $(\Delta t)^2$, and to maintain the same accuracy in the velocities, they are approximated as:

$$\{\dot{U}_t\} = \frac{1}{2\Delta t}[\{U_{t+\Delta t}\} - \{U_{t-\Delta t}\}] \tag{4}$$

However, in the present work the following backward difference is used:

$$\{\dot{U}_t\} = \frac{1}{\Delta t}[\{U_t\} - \{U_{t-\Delta t}\}] \tag{5}$$

While the error of this approximation is greater, on the order of (Δt), it provides a significant increase in computational efficiency under certain damping conditions. In the classic Central Difference Method, the solution of the system of simultaneous equations is not required if the damping matrix is null or diagonal; rather, only matrix multiplication is required at each step. This significant advantage does not hold if the damping matrix contains non-zero off-diagonal terms. With the modified velocity assumption, the solution of the system of simultaneous equations is not required regardless of the form of the damping matrix, and the more accurate nondiagonal representation of the damping of eqn. (2) is permitted. Comparisons between the two procedures have indicated that the error incurred is minimal.

The Central Difference Method is conditionally stable, and it is this characteristic which frequently governs the efficient use of the algorithm. The stability criterion requires that:

$$\Delta t \leq T_n/\pi \tag{6}$$

where T_n is the smallest period of vibration of the idealized structure. Thus, the stable time increment is frequently quite small, and but rarely known beforehand. The small time step may result in great computational effort if the structural response is to be determined over a significant time span. Further, care must be taken to eliminate massless degrees of freedom, which would produce modes of vibration with zero periods. Judgment may be exercised during the mass lumping procedure to eliminate modes with unnecessarily small periods. Fortunately, the stable time step is typically small enough that the Central Difference Method will provide excellent accuracy.

In those cases where the required time step is small, requiring a large number of increments, it may be more economical to employ an unconditionally stable method. As none of the explicit schemes are unconditionally stable, the use of an implicit integration scheme is indicated. The Newmark Method (Ref. 11) of implicit integration was selected as the third method to be implemented because of its flexibility and superior accuracy. The velocities and displacements are approximated by:

$$\{\dot{U}_{t+\Delta t}\} = \{\dot{U}_t\} + [(1-\delta)\{\ddot{U}_t\} + \delta\{\ddot{U}_{t+\Delta t}\}](\Delta t) \tag{7}$$

$$\{U_{t+\Delta t}\} = \{U_t\} + \{\dot{U}_t\}\Delta t + [(1/2-\alpha)\{\ddot{U}_t\} + \{\ddot{U}_{t+\Delta t}\}](\Delta t)^2 \qquad (8)$$

This method is unconditionally stable for values of $\delta \geq 0.50$ and $\alpha \geq 0.25\ (0.50 + \delta)^2$. The unconditional stability allows the use of much larger time steps in comparison with the explicit methods. However, too large a time increment will result in inaccurate results. The primary disadvantage of this method (and all implicit methods) is that it is necessary to solve the system of simultaneous equations at each time step. Static condensation of the massless degrees of freedom, while not required for numerical stability, is performed for computational efficiency. Typically, output results are not required at each time step. Thus, it is economical to operate on the condensed system, recovering the full displacement vector only at those time steps designated for output.

As stated, no integration scheme can be considered superior for all structural models. The three methods implemented namely, modal analysis, explicit integration and implicit integration, provide excellent analytical flexibility, enabling the experienced analyst to perform efficient, economical, and accurate linear dynamic analyses.

Computer Implementation. The program package was designed for use on virtual memory mini-computers. As out-of-core file operations are quite slow compared with the virtual memory paging system, efficient use of the virtual memory was a prime consideration. Dynamic memory allocation, in its restricted Fortran form, was used throughout the package. To minimize the number of address computation operations, one dimensional arrays are used extensively. By using a character manipulation program developed as part of this research, a particularly efficient procedure for transforming the element stiffness matrix to the global coordinate system was implemented. The congruent transformations required are explicitly coded, fully accounting for matrix symmetry and sparcity.

In comparison with a previous procedure, which also recognized matrix sparcity but performed the two matrix multiplications, the transformations produced by the explicit code are accomplished in one-tenth the time.

POSTPROCESSOR

Clear, efficient interpretation of analysis results is vital to the analysis/design procedure. The degree of understanding of the structural behavior determines the effectiveness of the design process. A postprocessor that has been developed - and continues to be refined - allows the engineer to evaluate the analysis results in an interactive, graphical environment. The postprocessor is implemented on a vector refresh display device. Dynamic response may be examined, and translation, rotation, and zooming of the structure display are available for clarity.

Current capabilities of the postprocessor include the ability to examine the deflected shape of the structure, and to display member end forces, reactions, and deflections as a function of time or load intensity. Depending upon the method of analysis, static load results, dynamic mode shapes and modal analysis time histories, and direct transient results may be displayed.

Several options are available while displaying the deflected shape of the structure. The user may interactively rotate, translate, and scale the structure in the main "viewport" (display area). In addition, two orthogonal views of the structure are displayed in the upper viewports. Figure 6 illustrates the displaced shape of a three-dimensional structure. Deactivation of the upper views is optional, and decreases the number of lines to be drawn, thus increasing the speed of the display change. The analyst may also choose to display the undeflected shape as dashed lines, providing a comparison of the deflected structure with the initial geometry.

Static deflections of the structure are displayed by scaling the global displacements of the nodes, while members are displayed as a series of line segments depicting the cubic displacement function. The scaling factor may be interactively controlled to enhance display clarity.

Mode shape displays have also been implemented. The user enters the desired mode shape number by using a numerical "keypad" on the display screen, pointing to the desired number with the stylus and cursor. A precomputed scale factor is used, with an option to override. The appropriate mode shape is displayed in the three viewports, and the manipulation options described above are available. Mode shape time displacements are displayed by multiplying the eigenvector by a sine function. Figure 7 contains displays of three modes of a building frame.

Time histories of the resultant deflected shape may also be displayed. To enhance the speed of display change, members are displayed as single straight lines. The speed of the response may be controlled by the user.

Graphical plots may be generated depicting the response of a member or node as a function of time or load. After identifying a node or member to be examined, the user may then specify the type of information to be displayed. Any computed quantity may be displayed on any orthogonal axis. For example, with the horizontal axis defined as the time axis, the deflection of any degree of freedom may be assigned to the vertical axis. Figure 8 depicts a typical time-displacement plot for a structure. The structure is displayed in an upper view-port with the selected node or member numbered and highlighted. The analyst may display up to three plots simultaneously, using a constant scale or specific scales for each plot. The displayed plots may be enlarged to fill the screen for clarity. Reference grid lines may be activated and scaled.

In anticipation of nonlinear analysis capabilities, the post-processor can provide a display of a three-dimensional yield surface. This feature will allow the analyst to observe the proximity to yield of a member, assumed to be a function of axial force and biaxial bending. Figure 9 illustrates the yield surface. A series of points could be plotted interior to or on the surface to describe the member end-force history.

Color Displays. Color output can be displayed on a raster scan device. Color postprocessing has been incorporated extensively in earlier, related two-dimensional static analysis work at Cornell (Ref. 2). Research in three-dimensional and dynamic color is currently underway.

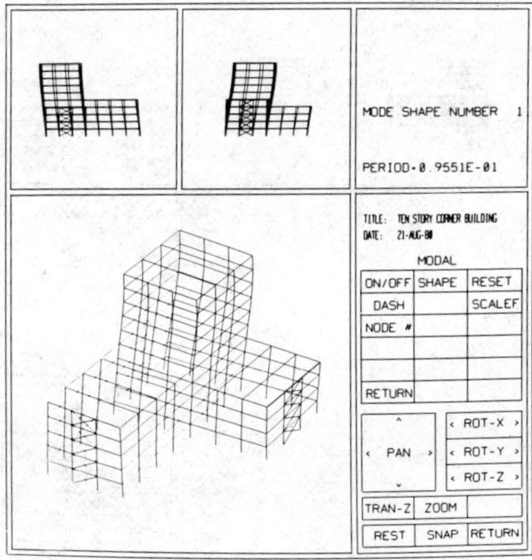

Figure 7(a). First free vibration mode of building (perspective view in main viewport).

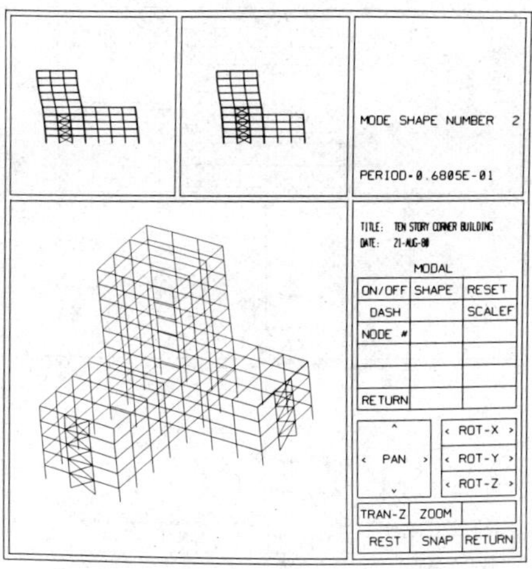

Figure 7(b). Second free vibration mode of building (perspective view in main viewport).

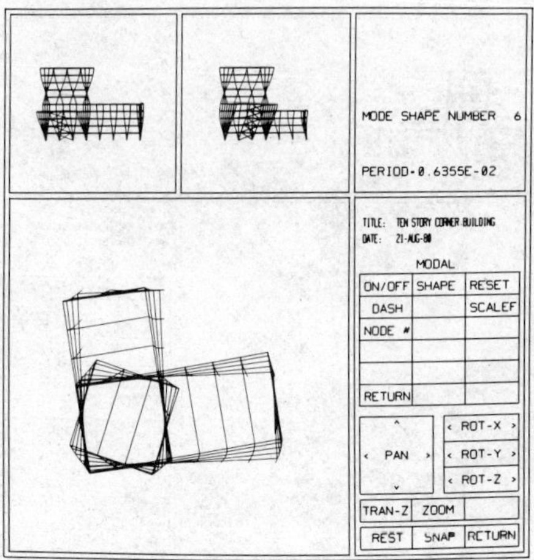

Figure 7(c). Sixth free vibration mode of building (plan view in main viewport).

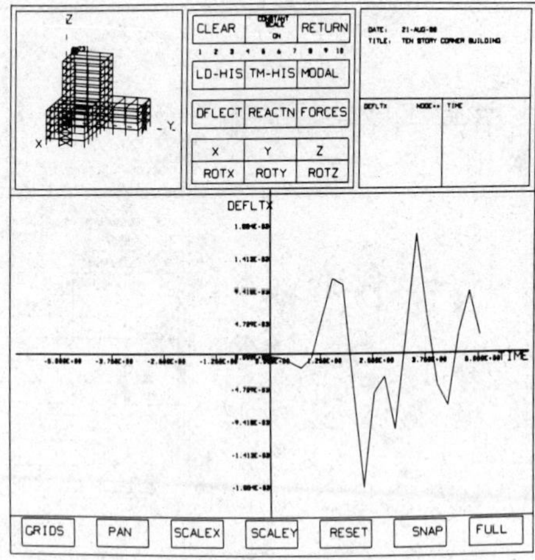

Figure 8. X-Displacement vs. time for a top-floor node highlighted in the upper left view.

One possible overall color display is the deflected shape of the structure with the element colors assigned as a function of their internal force level. Force histories would be observed by storing each time increment display on a video disk, thus permitting the high speed playback of the structural response. An alternative that has been implemented is one that specifies a direction normal to the element for the time scale, and displays the change in an element's stress resultant along this axis. Thus, each member response is depicted by a color rectangle with the length of the element along one axis, and time along the other.

SUMMARY

Progress in the development of an interactive computer graphics system for the analysis of frame structures has been presented. A comprehensive system for the linear analysis of three-dimensional building frames under earthquake loading has been developed. Research is continuing with emphasis on nonlinear, three-dimensional static and dynamic analysis and on the problems of the design and proportioning of steel frame structures.

ACKNOWLEDGEMENTS

This work was supported by the National Science Foundation under Grant No. PFR-7815357. The authors wish to thank Donald P. Greenberg, co-principal investigator with W. McGuire and J.F. Abel, for his contributions to the project as Director of the Laboratory of Computer Graphics.

REFERENCES

1. Abel, J.F., Greenberg, D.P., McGuire, W., and Gallagher, R.H., "Interactive Graphics for Finite Element Analysis", Proceedings of the Seventh Conference on Electronic Computation, ASCE, St. Louis, Mo., 1979.
2. Gross, J.L., Mutryn, T.A., and McGuire, W., "Computer Graphics in Nonlinear Design Problems", Canadian Journal of Civil Engineering, Vol. 6, 1979, pp. 97-111.
3. McGuire, W., "Interactive Computer Graphics and the Design of Steel Frames", presented at the First Symposium on Metal Structures, Mexical Society of Structural Engineers, Queretaro, Mexico, July 1978.
4. Gross, J.L., Mutryn, T.A., and McGuire, W., "Computer Graphics and Nonlinear Frame Analysis", Proceedings of the Seventh Conference on Electronic Computation, ASCE, St. Louis, Mo., 1979.
5. AISC, Manual of Steel Construction, Seventh Ed., 1970.
6. Nigam, N.C., and Jennings, P.C., "Digital Calculation of Response Spectra from Strong-Motion Earthquake Records", Research Report, Earthquake Engineering Research Laboratory, California Institute of Technology, Pasadena, California, June 1968.
7. Clough, Ray W., Wilson, Edward L., and King, Ian P., "Large Capacity Multistory Frame Analysis Programs", Journal of the Structural Div., ASCE, Vol. 89, No. ST4, Proc. Paper 3592, August, 1963, pp. 179-204.

8. Weaver, William Jr., and Nelson, Mark F., "Three-Dimensional Analysis of Tier Buildings", Journal of the Structural Div., ASCE, Vol. 92, No. ST6, Proc. Paper 5019, Dec. 1966, pp. 385-404.
9. Wilson, E.L., Hollings, J.P., and Dovey, H.H., "ETABS Three Dimensional Analysis of Building Systems (Extended Version)", Report No. EERC 75-13, College of Engineering, Univ. of Calif., Berkeley, Calif., April 1975.
10. Nair, R. Shankaran, "Linear Structural Analysis of Multistory Buildings", Journal of the Structural Div., ASCE, Vol. 101, No. ST3, March 1975, pp. 551-565.
11. Bathe, K.-J., and Wilson, E.L., Numerical Methods in Finite Element Analysis, Prentice-Hall, Inc., Englewood Cliffs, New Jersey, 1976.

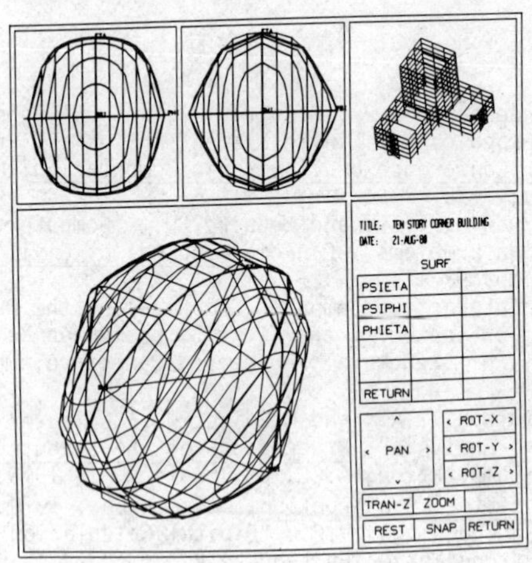

Figure 9. Wide flange section three-dimensional normalized yield surface relating bending moments about 2 orthogonal axes and axial force.

EARTHQUAKE BEHAVIOR OF ONE-STORY MASONRY HOUSES

by

R. W. Clough[I], P. Gülkan[II], and R. L. Mayes[III]

SUMMARY

Observations on the seismic behavior of simple masonry houses subjected to simulated earthquake motions are presented. The houses were constructed so that similar wall components were oriented both parallel and transverse to the shaking table motion. In general, a nominal amount of partial reinforcement appeared to have a very beneficial effect in improving the strength of the wall components and in preventing excessive cracking. Final recommendations will await the results of an additional test in which the walls of the house will be subjected to combined in-plane and out-of-plane loads.

INTRODUCTION

A major deterrent to a more widespread use of masonry construction in earthquake prone areas has been the unfavorable response of masonry buildings to earthquake motions observed in the past. For example, the Long Beach earthquake of 1933 provided the impetus for more stringent seismic requirements in California particularly in relation to masonry construction and the use of masonry for school buildings was no longer allowed. The absence of an accepted design basis for brick or concrete block unit masonry tends to discourage its use and supports the fallacious belief that this is an inherently inferior material which must be heavily reinforced if it is intended for earthquake resistant construction. Indeed, the unit masonry composite, brick or block and mortar, is a highly variable and complex material. Possible variations in construction are unlimited and strength parameters vary from one location to another.

Seismic design requirements specified by the U.S. Department of Housing and Urban Development (HUD) are referenced to "seismic risk zones" defined by the Uniform Building Code (UBC).[1] Partial reinforcement is required for all residential masonry construction in Seismic Zone 2, if compliance with HUD's Minimum Property Standards (MPS)[2] is to be maintained. The MPS requirements apply to all housing produced under HUD programs or covered by HUD mortgage insurance. They are considered as minimum requirements for the design and construction of adequate housing. Where local codes or standards permit lower requirements than the MPS, the latter shall apply. Local Acceptable Standards (LAS) intended to modify the MPS may be issued by HUD field offices with concurrence by HUD Headquarters.

I Professor of Civil Engineering and Assistant Director, Earthquake Engineering Research Center, University of California, Berkeley, U.S.A.

II Associate Professor of Civil Engineering, Middle East Technical University, Ankara, Turkey.

III Assistant Research Engineer, Earthquake Engineering Research Center, University of California, Berkeley, U.S.A.

In August 1974, the HUD office in Phoenix, Arizona, issued an LAS to cover the design and construction of masonry buildings. Specifically, unreinforced masonry was no longer permitted, and masonry walls of single-story residences located in Seismic Zone 2 (where Phoenix lay after the 1973 UBC seismic map changes) were required to contain at least the following amounts of vertical reinforcement:

(1) No. 4 (12.7 mm diameter) bars at all building corners, at wall ends, door and window openings, and in wall sections 2 ft (0.6 m) wide or less. Openings larger than 12 ft (3.6 m) in width require "special" analyses.

(2) No. 4 bars spaced at no more than 12 ft (3.6 m) on center.

(3) All reinforcement to be matched with dowels embedded in foundation walls or slabs.

The promulgation of this standard was questioned by representatives of the local housing industry on the grounds that compliance with it would lead to increased costs to the consumer as well as unnecessarily high factors of safety. To address the problem HUD contracted with the University of California, Berkeley, to determine the behavior and reinforcement requirements of single-story masonry structures under seismic loads by testing full scale specimens on the shaking table. This paper presents an overview of the experimental investigation designed to meet the following objectives:

(1) Determine the maximum earthquake intensity that could be resisted satisfactorily by an unreinforced house, and

(2) Evaluate the additional resistance that would be provided in the structure by partial reinforcement.

Details of the experimental phase of the study are given in References 3 and 4. Conclusions and recommendations will be made available in a separate publication in the same series.

EXPERIMENTAL PROCEDURES

House Models

A unique feature of the study was the testing of full scale components of typical masonry houses under simulated earthquake motions. The feasibility of one-half or one-third scale models of a full size house for the test specimens was studied but not adopted because there would have been problems associated with extrapolating the laboratory results to field conditions. Consequently, wall panels 8 ft-8 in. (2.64 m) high and up to 16 ft (4.9 m) long were constructed with commercially available concrete blocks or clay brick units. The wall panels were assembled to form 16 ft (4.9 m) square test specimens built upon reinforced concrete strip footings. The individual wall units were connected at the top by a timber roof structure bolted to the walls so that the house model would behave as a unit during applied base motions. Concrete slabs were attached to the top of the plywood sheathing to account for the reduction of mass resulting from having scaled down the plan dimensions. The total weight of these slabs was adjusted so that the ratio of total roof load to total peripheral wall length was similar to that for a

40x50 ft (12x15 m) prototype house with a specified roof load of 20 psf (1 kN/m^2).

In the reported phase of the work, four houses were tested. In Fig. 1 a three-dimensional view is given for the basic type of structure in the program; Table 1 lists material properties and other pertinent information. All houses were designed so that simultaneous out-of-plane and in-plane response of both partially reinforced and unreinforced panels could be measured. All partial reinforcement consisted of vertical bars.

For House 1 (concrete block) walls W1 and W2 were reinforced with one #4 (12.7 mm diameter) bar at each end cell; corners designated as C1 and C3 had three #4 bars. In House 2 (concrete block) and House 3 (clay brick) wall panels B and B1 were reinforced with #4 bars at each end; walls A and A1 were unreinforced. In House 4, no reinforcement was provided in any wall for the initial part of the test program. After two earthquakes with maximum accelerations in the 0.3 g range had been applied to the structure all wall panels were reinforced with two #3 (9.5 mm diameter) bars; in panels A1 and B these bars were dowelled whereas for walls A and B1 the reinforcement simply butted up against the top of the footings.

Earthquake Simulator Facility

The earthquake simulator facility at Berkeley consists of a 20x20 ft (6.1x6.1 m) shaking table, its hydraulic actuator and control systems, a data acquisition system, and the laboratory building that houses these units. The shaking table is a concrete slab weighing 90,000 lb (41,000 kg); the maximum weight of a test structure that can be mounted on the table is 110,000 lb (50,000 kg). The actuators can move the table in one horizontal and vertical direction. Further details of the earthquake simulator characteristics are available elsewhere[5].

The data acquisition system incorporates a high speed scanner under the control of a minicomputer. It is capable of sampling up to 128 channels of data at a rate of up to 100 readings/s/channel. The data are stored temporarily on disc and later transferred to magnetic tapes for subsequent analyses[3].

Instrumentation

The responses of the test structures were measured by means of an array of displacement or acceleration measuring devices. Displacements of the house specimens relative to the shaking table were monitored with either direct current differential transformers (DCDTs) or potentiometers, and accelerations with either strain gage or force balance type accelerometers. The ranges of the transducers were selected so as to avoid saturation at the maximum expected response peak. The numerals indicated in Fig. 1 denote instrument type and location.

Test Parameters

The principal test parameters considered in this investigation were as follows: simulated earthquake motions - three different recorded motions at different intensities, roof truss orientation, base fixity of walls, and local repair of cracked walls. No scaling of the time axis in the simulated motions was considered; the applied base motions lasted up to 25 seconds.

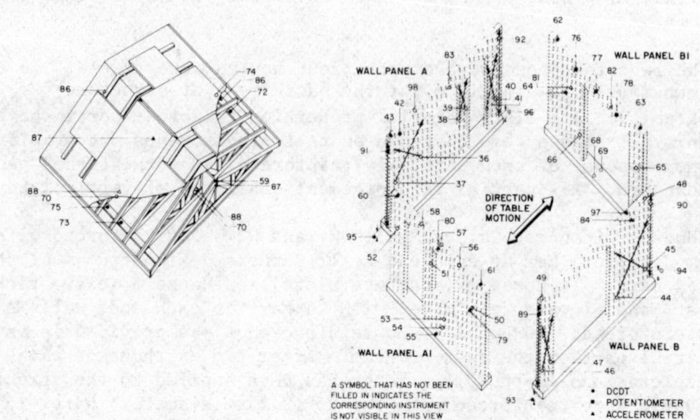

FIGURE 1 TYPICAL TEST SPECIMEN

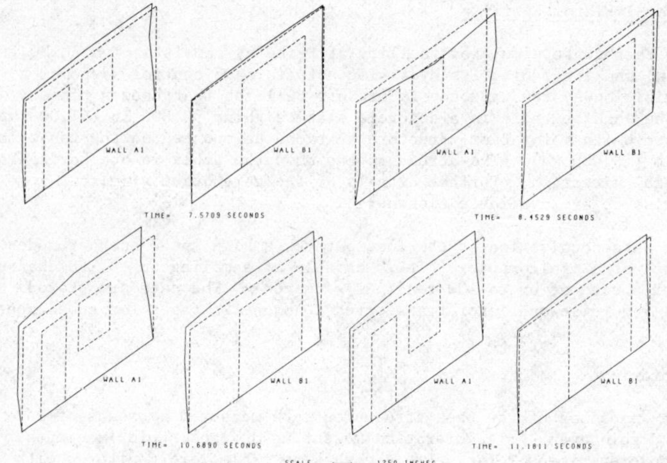

FIGURE 2 DEFLECTED SHAPES FOR OUT-OF-PLANE WALLS: HOUSE 3, TEST 17 (PACOIMA - 0.26 g)

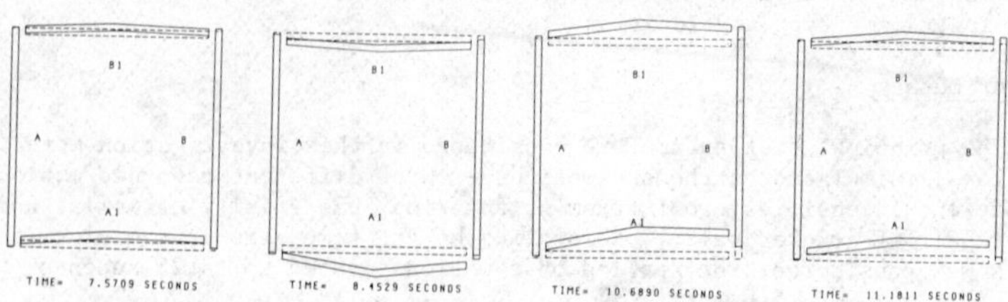

FIGURE 3 DEFORMED PLAN VIEWS; HOUSE 3, TEST 17 (PACOIMA - 0.26 g)

TABLE 1

MATERIALS FOR THE TEST STRUCTURES

House No.	Type 1 Masonry Unit	Masonry Strength psi (kN/m^2)	Mortar Strength psi (kN/m^2)	Grout Strength psi (kN/m^2)	f_y ksi (kN/m^2)	Reinforced Panels (Fig. 1)
1	6x4x16 in. Concrete Block	1,530	2,650	6,350	54.0	W1,W2,C1,C3
2	6x4x16 in. Concrete Block	1,983	4,730	3,360	59.3	B,B1
3	6x4x12 in. Concrete Block	4,970	2,336	2,536	59.3	B,B1
4	6x4x16 in. Concrete Block	2,778	1,944	2,324	54.0	See Note(3)

Notes: (1) Masonry strength is based on net area of units.
(2) Type S mortar of the UBC was specified for all houses.
(3) House 4 was tested twice with no reinforcement in any of the four walls before two #3 bars were grouted in each wall.

TABLE 2

EPA VALUES FOR NORMALIZED SHAKING TABLE MOTIONS

Earthquake	Taft	El Centro	Pacoima
Houses 1 and 2	0.90 g	0.82 g	0.79 g
Houses 3 and 4	0.91 g	0.72 g	0.78 g

SELECTED EXPERIMENTAL RESULTS

In designing the test structures and planning the outline of the test program it was recognized that the behavior of a particular group of masonry house models on the shaking table would not necessarily constitute the final arbiter in arriving at definitive conclusions regarding reinforcement requirements. The projection of laboratory data to field conditions requires an assessment to be made of the effects of parameters which could not be included or adequately represented in the program. These include: (1) foundation flexibility, (2) variations in workmanship, (3) pre-existing cracks in the walls, (4) seismic input, (5) roof load, (6) geometric effects, (7) roof diaphragm flexibility, (8) torsional response mechanisms, and (9) progressive damage. Each of these factors was evaluated in detail. The ultimate conclusion was that considered in its entirety, the behavior observed in the shaking table tests was quite similar to the performance expected of real masonry houses subjected to real earthquakes, provided an adjustment were made in the determination of the critical peak table acceleration to account for the coupling between the two horizontal components of an actual ground motion. The second horizontal component was expected to have a greater detrimental effect on unreinforced masonry walls, and this was judged to be equivalent to an increase in the intensity of single horizontal component motion by 50 percent. For partially reinforced walls a 20 percent increase was deemed appropriate. This consideration is expected to play a major role in the final recommendations.

Because this paper is intended to serve as a summary of References 3 and 4, detailed descriptions of the observed response will not be provided. Rather, a general overview of the test behavior will be presented and illustrated with data from all four house tests.

General Comments

The specimens used in this study were typical of "box" structures that derive their lateral force resistance from membrane action of the walls. The major part of the lateral force for the houses resulted from the concrete slabs on the roof. Resistance to this force was provided by a mechanism dependent on the relative in-plane shear rigidity of the roof and the wall components; the out-of-plane rigidity of the wall panels and the flexural stiffness of their connections to the roof were of negligible value in resisting the roof load. The roof structure simply provided the top support for the out-of-plane walls. From this description it is clear that the out-of-plane walls of a masonry house must have sufficient flexural strength to resist their own inertia forces when acting as vertical beams while the in-plane walls must have the capacity to resist the inertia forces on the entire roof system plus the top half of the walls.

In general the observed behavior was consistent with this description of box structures subjected to lateral forces. During the tests, roof displacement amplitudes were directly related to the behavior of the in-plane walls (designated as W1 and W3 in House 1 and A and B for the others; see Fig. 1). Differential displacements of the in-plane walls were accommodated by racking distortions of the roof, relatively small in-plane distortions were observed in the out-of-plane walls, so it may be concluded that the roof structure did not rotate as a rigid unit. This is consistent with the customary design assumption that plywood diaphragms are much more flexible in shear distortion that are masonry walls. The magnitude of differential displacements on two sides of the roof structure can be assessed from a simultaneous study of Figs. 2 and 3. Three dimensional deflected shapes of out-of-plane walls are plotted at four successive instants of time in the former. These shapes have been obtained by joining the readings of displacement gages; it is seen clearly that because of the unequal displacements permitted by the in-plane walls on either side of the out-of-plane walls, a complex pattern of displacements is imposed on them. Generally, there appears to be bending both parallel and perpendicular to the bed joints and the walls are subjected to twist as well. Figure 3 indicates the plan view distortions of the house at the same instants of time as in Fig. 2; the amount of racking distortions imposed on the roof structure is also illustrated in this succession of figures.

A significant observation made from these experiments was that typical single-story houses are so rigid that they do not develop complicated response mechanisms during an earthquake. Motions of the test structures followed the shaking table motions very closely, with distortions generally proportional to, and in phase with, the base accelerations. This is verified in Fig. 4 which shows a 6-second long part of the response of House 2. The peak input acceleration may therefore be cited as the dominant quantity controlling response. The most significant features of the observed response of the test structures considered as a whole may be summarized as follows.

Unreinforced Walls

No cracking was observed in any major unreinforced wall unit for tests with peak acceleration less than 0.2 g. The lowest intensity of shaking that caused cracking of a non-bearing in-plane wall occurred during tests with peak acceleration of 0.21 g (House 1); the minimum intensity to cause cracking of an out-of-plane wall was 0.25 g.

Unreinforced out-of-plane walls continued to perform satisfactorily after cracking during several tests of increased intensity, but the displacements of these walls generally became excessive during tests with accelerations greater than 0.4 g. These large displacements involved hinging at the horizontal crack line and exhibited potential instability.

Cracking of unreinforced in-plane walls was of two types; horizontal cracks in panels without openings (as in House 1) and a diagonal crack extending downward from the window corner in the wall units with window penetrations (as in Houses 2, 3 and 4). Permanent displacements generally were not associated with the horizontal cracks; however, the diagonal cracks led to permanent displacements which became unacceptably large with further testing. The typical pattern of damage shown in Fig. 5 was observed in all in-plane walls with window penetrations but no limit of peak acceleration could be established which would ensure acceptable performance of cracked window walls.

Partially Reinforced Walls

Nearly all partially reinforced wall units performed satisfactorily in all tests. None of the partially reinforced out-of-plane components developed any important cracks during any test, including several that were subjected to peak accelerations in excess of 0.5 g. In general, the partially

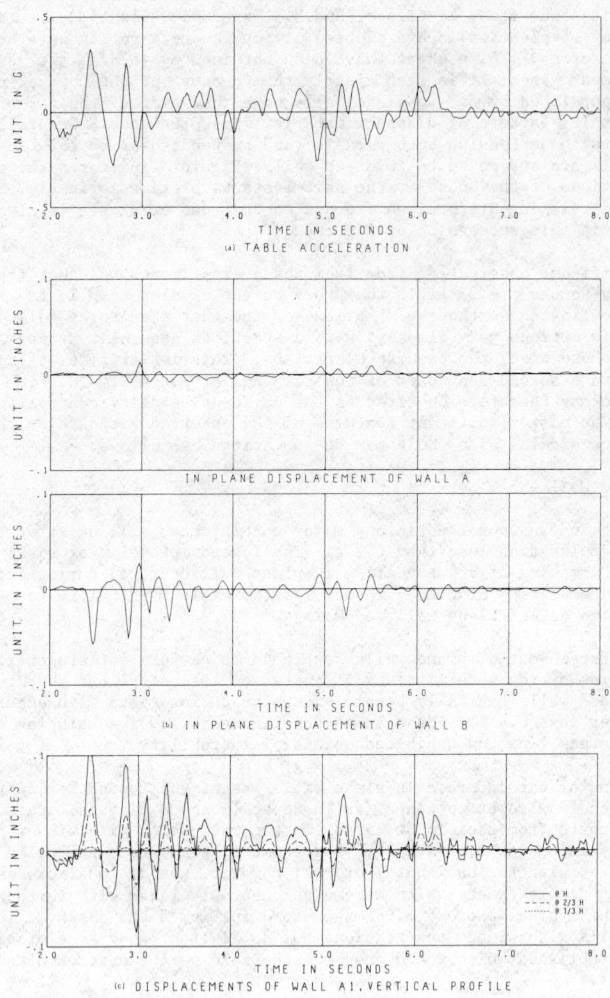

FIGURE 4 MEASURED RESPONSE OF HOUSE 2 DURING TEST 14 (EL CENTRO - 0.33 g)

FIGURE 5 DIAGONAL CRACKING IN UNREINFORCED IN-PLANE WALLS WITH WINDOW PENETRATIONS

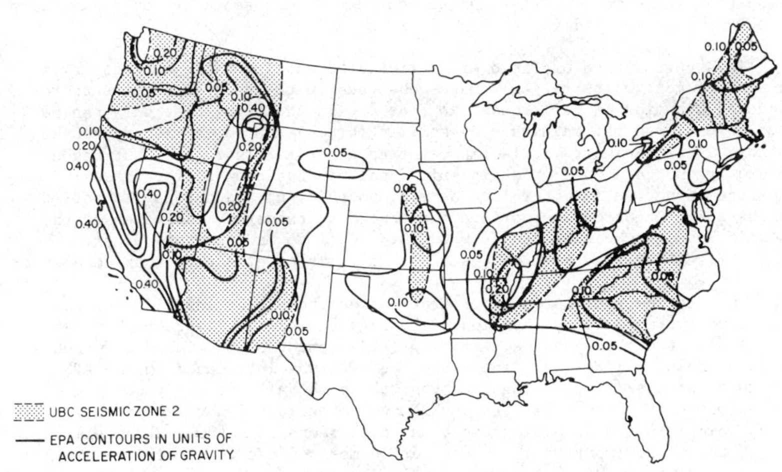

FIGURE 6 ATC AND UBC SEISMIC MAPS

reinforced in-plane walls also performed satisfactorily although they developed some cracks when peak accelerations were greater than 0.3 g. Cracking in the pier units without window openings was associated with rigid body rocking, and included a horizontal crack due to the uplift near the base of the wall as well as cracks at the ends of the door spandrels. Residual cracks after the tests were all narrow and easily repairable.

The only partially reinforced wall which exhibited unsatisfactory behavior was the window wall of House 4 (designated as A in Fig. 1). A typical diagonal crack extending from the window corner developed during the first phase of testing when this specimen was unreinforced. After repair and reinforcing the wall resisted a 0.32 g peak acceleration base motion without additional cracking. However, in subsequent tests with peak accelerations in the range of 0.47 g to 0.68 g in the horizontal direction, additional cracks developed in spite of the reinforcing bars. It may be significant to note that no dowels were provided with this reinforcement because the ultimate failure included displacements at the base of the wall that probably would have been constrained by dowel action.

Seismic Input for Zone 2

From the earliest stages of the experimental investigation, one of the most critical questions was the intensity of shaking table accelerations that should be used to represent the maximum earthquake motions expected in UBC Zone 2. No decision on this matter was required while the experimental studies were being performed because the test program was planned to determine the seismic capacity of both unreinforced and partially reinforced masonry housing. In other words, the test served to determine the peak shaking table accelerations at which cracking was initiated and unacceptable damage was observed. It is therefore necessary to establish the intensity of shaking table motions that can be considered representative of the maximum expected field conditions.

This correlation of field excitation with shaking table motions is required, of course, to relate damages observed in the test structures to the expected behavior of real houses in Zone 2. In addition, test data can be used to estimate the magnitude of seismic forces induced in masonry houses by Zone 2 earthquakes. It is important to note in this regard that masonry houses are not designed by engineering analysis based on design forces; their construction is controlled by minimum building standards. To assist in the formulation of appropriate construction standards, estimates of the seismic loads to be expected in UBC Zone 2 will be made in this paper. The level of seismic load will be described in terms of effective peak acceleration.

The best current estimate of expected earthquake intensities for the U.S. was developed by the Applied Technology Council[6]. Figure 6 shows the ATC map of effective peak acceleration (EPA) contours superimposed on the 1976 UBC Seismic Zone Map. The concept of EPA was introduced in the ATC document because it was recognized that the peak value of the ground acceleration may not relate well with the damage potential of a given earthquake. Sharp spikes in the accelerograms tend to overemphasize the peak acceleration value, while the importance of long period motions due to distant events may be underestimated. A critical transition needs to be made, therefore, regarding the interpretation of shaking table accelerations in terms of EPA.

The EPA of a given ground motion is defined by the ATC in terms of the motion evaluated for 5 percent damping. Specifically, a line of constant spectral acceleration, S_a, is drawn on the response spectrum which approximates the average spectral acceleration in the period range of 0.1 - 0.5 s. The EPA is given by this average S_a divided by 2.5, where the divisor is a typical response amplification factor. The EPA values of the shaking table motions were determined by applying this procedure to the response spectra of the motions recorded during the testing of each house. The EPA values are presented in Table 2. For the purpose of this discussion, the variations of EPA shown in Table 2 are not significant, and a single average value of 0.82 g may be adopted. This means that a table motion having a peak acceleration of 1.0 g is assumed to have an EPA of 0.82 g. It is therefore assumed that the maximum EPA of 0.2 g indicated in Fig. 6 is represented by a peak shaking table acceleration of 0.24 g.

Test Structure Response Acceleration

Although masonry houses are relatively rigid structures they do have some flexibility and therefore exhibit some vibratory response mechanisms. The significant aspect of these mechanisms is that applied motions are amplified by structural response so that peak accelerations recorded on the structure are greater than the peak input acceleration. This amplification effect is represented in the definition of EPA by the factor 2.5 by which spectral accelerations are divided. The ATC map has tacitly assumed an amplification factor of 2.5 to be appropriate for typical building structures.

The experimental data obtained in this investigation provide a direct measure of the amplification of seismic excitations for single-story masonry houses. During tests, accelerations were measured at various points on the structures, and ratios of the peak values of these local accelerations to the peak input acceleration indicated the local structural response amplification. A typical set of amplification ratios obtained from House 3 is illustrated in Fig. 7. These figures demonstrate that the magnification factors vary considerably, from point to point on the structure, and with varying test conditions. For example, amplification values at the roof level average about 1.8 when in-plane walls are load bearing, but reach values of over 3 for the more intense excitations when the in-plane walls carry no dead load from the roof.

Although it is evident from Fig. 7 that no single number can represent the response amplification for all parts of the test structure under all conditions, the ATC value of 2.5 appears to be reasonable. For the most part of the test data this figure is conservative so that seismic force estimates based on this factor will exceed the actual load. The number of cases where the observed amplification exceeds 2.5 are few, and apply to localized parts of the structure or to unusual test behavior. Therefore, the amplification factor of 2.5 is proposed for estimating the seismic forces induced in the test structures by any given peak table acceleration.

CONCLUSIONS

On the basis of the results obtained in this investigation a number of conclusions may be drawn concerning the earthquake behavior of simple masonry houses, although it is evident that certain aspects of such behavior are not fully understood.

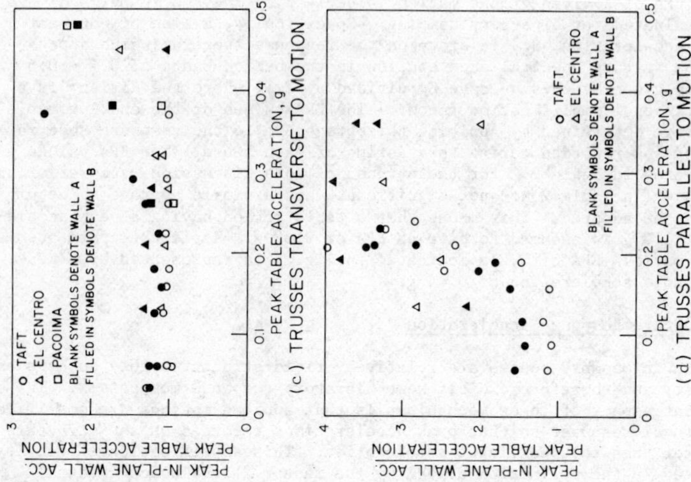

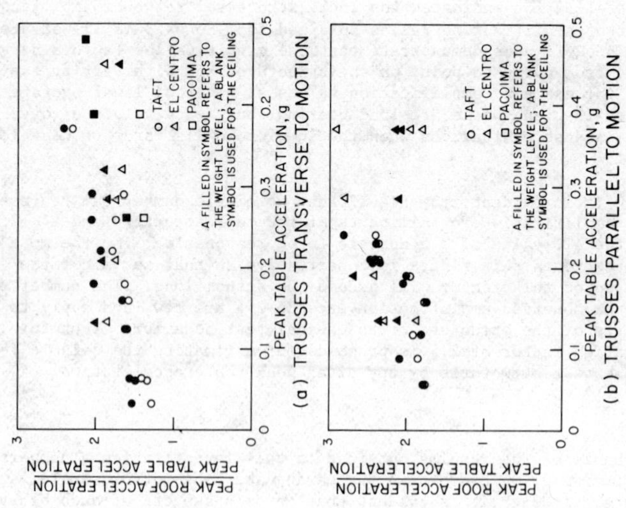

FIGURE 7 AMPLIFICATION OF BASE ACCELERATION: HOUSE 3

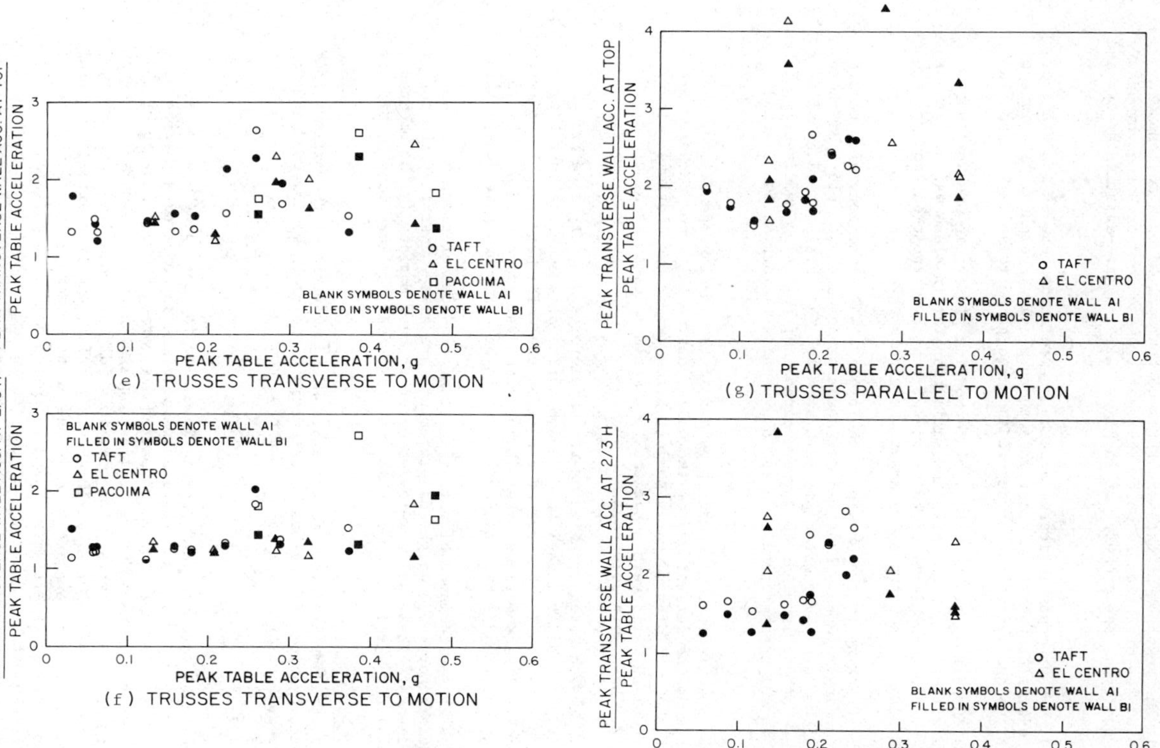

FIGURE 7 (CONT.) AMPLIFICATION OF BASE ACCELERATION: HOUSE 3

(1) The response of the structures to the applied base motions was affected by the direction in which the roof trusses were oriented, by base fixity conditions and geometry of in-plane walls, by pre-existing cracks and reinforcement.

(2) Measurements indicated a strong correlation between the peaks of the table accelerations and recorded structural displacements. The type of base motions applied to the structure appeared to be relatively unimportant.

(3) Repair of cracks with surface bonding plaster in Houses 1 and 3 was more effective for out-of-plane walls than for in-plane walls. Investigation of the feasibility of this type of repair for post-earthquake strengthening of similar structures appears to be promising.

(4) Cracking in the unreinforced in-plane walls was generally caused by overturning moments. Measured net shear stresses on the piers of these walls reached a maximum of 40 psi (270 kN/m^2). The unreinforced cantilever type piers of the in-plane walls were capable of resisting unidirectional base motions in the cracked condition and they tended to rock at the cracked section.

(5) Dowelled reinforcement provided in the in-plane cantilever type piers reduced the deflections of the walls considerably when compared with those of the unreinforced walls subjected to similar base motions. The reinforced in-plane walls were subjected to a maximum net shear stress of about 40 psi (270 kN/m^2) in both Houses 3 and 4.

(6) The observed cracking in the unreinforced out-of-plane walls generally occurred at midheight or above during peak accelerations between 0.2 g and 0.45 g. In the cracked condition the walls were capable of resisting unidirectional base motions; the hinging mechanism associated with large displacements at the 2/3 height occurred during base motions in excess of 0.45 g.

(7) The performance of partially reinforced out-of-plane walls was very satisfactory; 2/3 height displacements are always less than the top displacements. These walls were subjected to base motions (with combined horizontal and vertical components) including peak accelerations as high as 0.68 g.

(8) The amplification factor expressing the ratio of the maximum roof level acceleration to peak base acceleration appears to be independent of the type and intensity of the applied base motion.

POSTSCRIPT

The tests conducted during this investigation have provided for the first time quantitative and qualitative information concerning the response of typical single-story houses to earthquake motions. Most aspects of the behavior of real houses during real earthquakes have been simulated adequately by the shaking table tests. However, it should be emphasized that one important feature of the real earthquake situation, namely biaxial horizontal excitation, has not been included in the tests. Because this factor represents the major area of uncertainty, it is recommended that an additional investigation be undertaken to carry out the testing of an additional house oriented at a skew angle on the shaking table. The relatively limited test would greatly increase the value of the work already done and summarized in this paper.

ACKNOWLEDGEMENT

The financial support of HUD and the continuing technical guidance of the ATC-5 Advisory Panel are gratefully acknowledged.

REFERENCES

1. Uniform Building Code, 1976 Edition, International Conference of Building Officials, Whittier, California, 1976.

2. U.S. Department of Housing and Urban Development, "Minimum Property Standards, One- and Two-Family Dwellings", HUD 4900.1, 1973 Edition, Washington, D.C., 1973.

3. Gülkan, P., Mayes, R. L., and Clough, R. W., "Shaking Table Study of Single-Story Masonry Houses - Volume 1, Test Structures 1 and 2", Earthquake Engineering Research Center Report No. UCB/EERC-79/23, University of California, Berkeley, September, 1979.

4. Gülkan, P., Mayes, R. L., and Clough, R. W., "Shaking Table Study of Single-Story Masonry Houses - Volume 2, Test Structures 3 and 4", Earthquake Engineering Research Center Report No. UCB/EERC-79/24, University of California, Berkeley, September, 1979.

5. Rea, D., and Penzien, J., "Dynamic Response of a 20'x20' Shaking Table", Proceedings of the Fifth World Conference on Earthquake Engineering, Rome, 1973.

6. "Tentative Provisions for the Development of Seismic Regulations for Buildings", Applied Technology Council Publication ATC 3-06, (NSF Publication 78-8, NBS Special Publication 510), U.S. Government Printing Office, Washington, D.C., June, 1978.

Wind Damage in Hurricane Frederic

T. A. Reinhold[1] and K. C. Mehta[2], M.ASCE

ABSTRACT

Damage to buildings and other structures caused by Hurricane Frederic's winds is reviewed and classified by level of engineering design and by range of maximum wind speed. Observations concerning the performance of buildings and structures are grouped according to whether the buildings and structures are fully engineered, pre-engineered, marginally engineered, or non-engineered. A major emphasis is placed on estimating wind speeds, at sites of damage observations, from anemometer data. Wind speed data is converted to equivalent maximum fastest-mile wind speeds at elevations of 10 meters over open terrain. Performance observations are further correlated with fastest-mile wind speed ranges of 70-85 mph, 85-100 mph, and 100-110 mph. These observations indicate that significant damage did not occur until wind speeds reached or exceeded design values.

INTRODUCTION

On the evening of September 12, 1979, Hurricane Frederic struck the Gulf Coast of the United States just east of Pascagoula, Mississippi and hurricane winds impacted coastal areas of Alabama, Florida, and Mississippi. Preliminary reports of the resulting damage indicated that wind damage to buildings, power lines, and trees was widespread; but, that buildings were only partially demolished, except along the shoreline where the storm surge was heaviest. Consequently, it was felt that a survey of structural damage in the affected area could provide information which would be useful in evaluating current building codes and construction practice.

The Institute for Disaster Research (IDR) at Texas Tech University and the National Bureau of Standards cooperated to send a team of engineers to perform a survey of structural damage. Communities which were surveyed include Mobile, Gulf Shores, and Dauphin Island, Alabama; Pensacola, Florida; and Pascagoula, Biloxi, and Gulfport, Mississippi. Damage to buildings and structures is documented by approximately one thousand photographs (color slides and black and white prints). The photographs are supported by notes, transcripts of voice-recorded observations, and some measurements of building material dimensions.

---

[1]National Bureau of Standards, Washington, D.C.

[2]Institute for Disaster Research, Texas Tech University, Lubbock, TX.

In order for damage observations to provide information which is meaningful in evaluating building codes and construction practice, a knowledge of wind speeds encountered is critically needed. In previous studies of hurricanes which have caused significant damage, wind speed and wind direction data have generally been quite limited and consequently estimates have been based largely on "educated guesses" or on the damage itself. Fortunately, a large number of anemometers survived Hurricane Frederic and an extensive amount of data is available. Thus, it is possible to produce what is perhaps the most complete description available to date of maximum hurricane surface winds over land, that is based on actual wind measurements.

A FRAMEWORK FOR EVALUATING WIND DAMAGE

Upon viewing the often awesome damage caused by a natural disaster such as a hurricane, it can be difficult to place damage levels into perspective. One format for placing damage observations into an engineering perspective has been suggested by Minor and Mehta [1]. The format they suggest is to divide structures according to the level of engineering attention given to design of the structure. The following four levels are defined in reference 1 and are used in this paper.

1. Fully engineered - Buildings or structures which receive specific, individualized attention from professional architects and engineers. Usually the design is site specific. Examples include tall buildings and special purpose structures such as sports arenas and hospitals.

2. Pre-engineered - Buildings or structures which receive engineering attention during the design stages and which are manufactured and marketed as many similar units for a wide variety of uses in a variety of locations. Examples include mobile homes and metal buildings.

3. Marginally engineered - Buildings built with some combination of masonry, steel framing, open-web steel joists, wood framing, and wood rafters with incomplete engineering design. For example, in a masonry and steel frame warehouse the frame may have received engineering design while the masonry walls received little or no attention. Examples may include schools, motels, apartments, and commercial or light industrial buildings.

4. Non-engineered - Buildings which receive no specific engineering attention. Examples include most single and multifamily residences, many one- or two-story apartment units, and many small commercial buildings.

These four categories are used as part of the framework for presenting observations of damage to buildings and structures caused by Hurricane Frederic. The framework is completed by subdividing damage observations for each category according to whether the estimated maximum wind speeds are lower than, equal to, or greater than the design wind speeds for the area.

There are other parameters which clearly influence damage levels but are not included in this framework. Examples include adequacy of local building codes, level of code enforcement, and quality of workmanship. While these parameters are not considered explicitly in the categorization of damage observations, they are subsequently considered and their influence will be discussed.

FASTEST-MILE WIND SPEEDS BASED ON ANEMOMETER RECORDS

Wind data from Hurricane Frederic were obtained from ten sites following the storm. Four of the sites were near Mobile, Alabama and six were located along the coast between Gulfport, Mississippi and Fort Walton Beach, Florida. The locations of the sites are shown in figure 1 together with their relation to the estimated storm track. None of the stations were primary National Weather Service (NWS) stations with fastest-mile recorders. Furthermore, the form of the wind speed recordings varies from site to site as does the terrain exposure and elevation and type of anemometer used. Consequently, the raw wind speed data obtained from the various sites consisted of a nonhomogeneous set of recordings which could not be effectively used in an engineering assessment of the damage. The following subsections provide information on the sites where data was obtained and on procedures used to transform the data into estimates of fastest-mile wind speeds.

Anenometer Sites and Condition of Equipment

A brief description of the anemometer sites listing the type and elevation of the anemometers, type of data available, and assumptions about the most probable terrain type are listed in table 1. Most of the sites were visited in order to obtain the information directly and to make an estimate of the terrain exposure. A classification of candidate terrain types is given in table 2 where site characteristics are given in terms of the commonly used log-law and power-law. Sources of these characteristics will be described in a later section.

Three of the four airport stations listed in table 1 could be classified as having open exposure, type 2, conditions for the wind directions experienced during the passage of the hurricane. However, the Pensacola Naval Air Station (PNA) site was not as clean. At PNA, the anemometer was mounted on the control tower roof and experienced an open exposure for winds blowing from directions of 0 to 70 degrees and 185 to 360 degrees, measured clockwise from North. For wind directions between 75 and 180 degrees winds passed over trees which ranged from 20 to 40 feet (6-12 meters) in height and the terrain is assumed to be type 3 or similar to that over the outskirts of towns. It is noted that the coastal areas surrounding the airports were heavily forested with 7 to 10 meter tall pine trees. Consequently, the assumption of open exposure is made with some reservations despite the fact that the wind generally passed over large open areas immediately surrounding the airport runways.

Terrain exposure at several other sites could not be clearly classified as fitting a specific type without more serious reservations. A brief discussion of the sites follows:

a) Dauphin Island Bridge (DIB): The instrumentation is located on top of the drawbridge super structure approximately 60 feet (18 meters) above the water level. It is assumed that the exposure is type 1 (sea surface in strong winds), although Dauphin Island forms a barrier to the south and southeast of the bridge. It was not possible to visit the site because of damage to the causeway and bridge. However, because of the height of the instrumentation, the predominance of a sea exposure, particularly from the east, and the fact that the drawbridge portion is at some distance from the main body of the island or mainland, a sea exposure is considered appropriate.

b) Gulfport Civil Defense (GCD): The instrumentation is located on a tower mounted atop a 25 to 30 foot (8 to 9 meter) building near the center of Gulfport. The strongest winds (only the peak wind speed was recorded) came from the north or northwest across a large portion of the city. The city is not heavily built up with large buildings and, consequently, the terrain is assumed to correspond to the centers of towns, type 4.

c) Ingalls Shipbuilding (ISB): The exposure to the north and west is that of open terrain with some small bodies of water. However, there are some large warehouse-type buildings near the fire station where the anemometer was located. Winds from the south and southwest would have a sea exposure but the site was obstructed by cranes and buildings. The assumption that the terrain exposure is type 2 (open terrain) was considered to be reasonable, despite possible local effects due to neighboring buildings. During the storm, a portion of the tower above the anemometer collapsed, the direction vane was destroyed and the anemometer was tilted to the south at approximately 45 degrees. Fujita, et al. [2] also noted this damage and a picture of the anemometer is shown in their report. Wind tunnel tests on a similar anemometer tilted at 45 degrees indicated little change in the velocity measurements and consequently it is suggested that the data not be changed due to the tilt.

d) Mobile Civil Defense (MCD): The anemometer was located two-thirds of the way up a water tower and was shielded for a number of wind directions by the tower legs, struts, and center pipe. The tower was located on a rise west of the center of Mobile, about two miles into the suburbs. The city and the suburbs are heavily wooded with the trees being 30 to 50 feet tall (9 to 15 meters). Terrain type 3 or 4 was assumed.

e) Coast Guard Cutters: The "Salvia" was moored at the Alabama State Docks and was shielded by a warehouse and a larger boat. Consequently, its wind data is highly questionable. The "White Pine" was located midstream in the south channel of the Mobile River south of Twelve Mile Island. The river banks are well forested with pine trees having an average hight of 25 to 30 feet (7.5 to 9 meters). It was not possible to visit this site but a detailed report was obtained courtesy of the Quarter Master on the White Pine. This site was one of the most difficult to characterize. Information in table 2 would lead to an assumption of category 6, pine forest, but that assumption produced wind speeds which were not consistent with any of the other data. Based partly on the results of analyzing the data using different assumptions of terrain type, it was decided to use terrain types 2 or 3.

Most of the instrumentation at the various sites received regular service by either military or NWS personnel and it can be assumed that the data are fairly accurate, indeed as accurate as the data which forms the base for extreme wind speed estimates in other parts of the United States. There are several exceptions. The anemometer at the MCD site had not been serviced for a number of years but it was removed following the storm and calibrated in the NBS wind tunnel by personnel from the Fluid Mechanics Division. Subsequently, data were corrected using the calibration curve. Also, the anemometers on the Coast Guard cutters had reportedly not been serviced since installation.

Storm Track

An estimate of the storm track is shown in figure 1. The track was determined on the basis of radar pictures obtained from the Naval Weather Service at the Pensacola Naval Air Station. The track is similar to the one reported by Fujita, et al. [2]. It is recognized that the center of minimum pressure, the center of the region of minimum winds and the center of the rainless portion in the radar pictures may not coincide. However, in the absence of the detailed pressure and velocity information required, the center of the rainless portion of the radar picture provides a useful approximation of the location of the eye of the storm.

Transformation to Fastest-Mile Wind Speeds

In order to compare the wind speed data obtained from Hurricane Frederic with design wind speeds, it is necessary to express them in the form of fastest-mile wind speeds in a standard exposure. Normally, such design wind speeds are given for open exposure conditions and a height of 30 feet (10 meters) [3, 4]. Consequently, it is necessary to transform the data from the exposure and height corresponding to the actual environment of the instrument to standard exposure and height conditions, and from a mean or gust value to a fastest-mile value.

Procedures for transforming from one averaging time to another, from one terrain to another, and from one elevation to another are contained in reference 5. The transformation for different averaging times is based on work by Durst as modified by Simiu and Scanlan [5]. The equation is as follows:

$$U_t(z) = U_{3600}(z)\left(1 + \frac{0.98\ c(t)}{\ell n\ (z/z_o)}\right) \qquad (1)$$

where $U_t(z)$ = velocity for averaging time t in seconds at height z

$U_{3600}(z)$ = mean hourly velocity at height z

z_o = roughness parameter for site exposure

$c(t)$ = coefficient obtained from table 3, linear interpolation is allowed.

This procedure had not been verified for winds in hurricanes. Consequently, initial work was performed to check the procedure. The check was accomplished by determining peak to mean velocity ratios for 20 to 30 minute segments of the wind records from the seven anemometer sites with strip chart recordings. The peak to mean velocity ratios represent the effect of averaging time, between 2 second averages and 1200 second averages, on the peak velocity estimates. Probability distributions of the peak to mean ratios are plotted in figure 2 for the seven sites which had strip chart records. Corresponding ratios based on Durst's work as modified by Simiu and Scanlan are plotted at the 50% probability level. In general, the agreement is quite good except for the data from MCD where interference of the water tower may affect the data. Based on this comparison, it was determined that equation 1 and the values of c(t) in table 3 could be used with hurricane wind data.

Procedures for transforming wind speeds from one elevation to another and from one exposure to another are also contained in reference 5. The transformations are based either on the power-law or the log-law and are designed to be used with mean hourly wind speeds. The power-law transformation requires knowledge of the power-law component, α, and the boundary layer height, δ, for each terrain in question. The log-law transformation requires knowledge of the roughness length parameter for each terrain. Table 2 contains values of these parameters for the various types of exposure assumed. The values are obtained largely from tables 2.2.1 and 2.2.2 of reference 5 but also rely on input from references 6 and 7. Judgment was used in order to produce an even gradation of exposure types.

The log-law and power-law relationships may not apply to hurricanes, especially near the eye wall where winds are highest. Furthermore, even if they do apply there is no guarantee that the parameters developed from data for other types of storms are applicable to hurricane winds. However, the differences between the anemometer elevations and the standard reference elevation are generally small and the approximations involved are likely to be acceptable, hopefully, less than 5 miles per hour.

Available mean, one minute average and gust velocity data were transformed to equivalent mean, gust, and fastest-mile wind velocities for open terrain conditions at 10 meters. Initially, all data was transformed to equivalent mean hourly values. Next, the mean hourly values were transformed from the specific terrain and elevation of the instrumentation to the standard exposure conditions. Finally, the hourly values were transformed to gust and fastest-mile values for the standard exposure conditions. Consistency of the transfomrations was checked by recalculating the peak to mean velocity ratios for the transformed data. To obtain the peak to mean ratios, gust velocities for standard exposure conditions which were obtained from original gust velocity data were divided by corresponding mean velocity data for standard exposure conditions which were obtained from the original mean velocity readings.

Transformations of the one minute average readings involved an additional step because the data does not represent the maximum one minute average during a certain period of time. Rather, the one minute average reading is taken at an arbitrary time, usually every hour. Consequently, a transformation which uses equation 1 and a c(t) corresponding to 60 seconds would be misleading since equation 1 refers to maximum values. A relationship between the arbitrary one minute average reading and the fastest-mile reading during that hour has been suggested by Thom [8] and is expressed as follows:

$$U_{fm} = 9.55 + 0.999\ U_m \qquad (2)$$

where U_{fm} = fastest-mile wind speed during an hour

U_m = one minute average reading during the same hour.

Consequently, one minute average readings were first transformed to fastest mile wind speeds which were then transformed to mean hourly wind speeds using equation 1.

Maximum Fastest Mile Wind Speeds

Sample results of the wind speed data analysis are shown in figure 3. Maximum fastest-mile wind speeds for each hour interval during the passage of the storm are shown along with an approximate wind direction. The different symbols represent estimates of fastest-mile wind speeds at 10 meters elevation over open terrain which are obtained from raw data of the various forms indicated. Note that the fastest-mile wind speeds predicted from the different types of data, including one minute average readings, are quite consistent (figure 3).

Maximum fastest-mile wind speeds for each of the anemometer sites except MCD are shown in figure 4. It is possible to estimate maximum fastest-mile wind speeds at damage observation sites from the values given in figure 4. Contour maps of maximum wind speed are being developed using available data and mathematical models for the distribution of hurricane winds. However, the contour maps were not completed at press time.

The design wind speeds specified in reference [4] for the coastal area affected by Hurricane Frederic are fastest mile wind speeds, at a 10 meter elevation over open terrain, of between 90 and 100 miles per hour. These speeds are based on a 50-year mean recurrence interval. Results presented in reference [3] indicate a design wind speed of 91 miles per hour. Building officials in the area indicated that a design wind speed of 90 mph is used in Mobile. Consequently, damage observations are categorized according to wind speed using the following ranges: 1) Wind speeds lower than design values, 70 to 85 mph; 2) Wind speeds near design values, 85 to 100 mph; and 3) Wind speeds higher than design values, 100 to 110 mph.

DAMAGE OBSERVATIONS

It is possible to associate damage in the communities surveyed with the wind speed ranges chosen. Structures in Gulfport and Biloxi, Mississippi, and in Pensacola, Florida experienced wind speeds below design values. Structures in Mobile, Gulf Shores and Foley, Alabama experienced wind speeds near design values. Structures in Pascagoula, Mississippi and Dauphin Island, Alabama experienced wind speeds in excess of design values. Damage observations are presented in the following paragraphs, according to level of engineering design and range of maximum wind speed.

Fully Engineered

1. Wind speeds below design values - No damage was observed.

2. Wind speeds near design values - Some damage to cladding and building facades was observed. An example of facade damage to a building on the west side of Mobile, Alabama is shown in figure 5. Several cases were observed where windows on fully engineered buildings were shattered by wind blown pebbles and debris from the roof of an upstream building.

3. Wind speeds higher than design values - Damage was limited to cladding and facades. Broken windows constituted the most common type of damage.

Pre-Engineered

1. Wind speeds below design values - No damage to mobile homes or metal buildings was observed.

2. Wind speeds near design values - Numerous cases of damage to metal buildings and several cases of damage to mobile homes were documented. Figure 6 illustrates typical types of damage to metal buildings. In nearly all cases, at least one metal door in the building was also damaged or missing and generally, there were other metal buildings nearby which suffered no apparent damage. While it is difficult to reconstruct modes or sequences of failure with a great deal of certainty, it appears that the doors were the weak link and that large internal pressures occurring after failure of the door led to damage of other cladding elements. Apparently, if all doors remained intact, the building suffered no visible damage. Some of the damage to mobile homes may have been due to fatigue. The owner of one mobile home which lost its roof reported that the roof failed some time between 600 and 800 hours Greenwich Mean Time (GMT) which was well after the strongest winds occurred. Some mobile homes weathered the storm with little visible damage. Damage of a trailer park between Biloxi and Pascagoula, Mississippi has been carefully documented and will be presented in a complete report of damage observations.

3. Wind speeds higher than design values - Damage similar to that described for wind speeds near design values was observed.

Marginally Engineered

1. Wind speeds below design values - Broken plate glass windows of several stores were observed.

2. Wind speeds near design values - Numerous instances of damage were observed. Unreinforced or under-reinforced and unbraced masonry walls of several light commercial or warehouse type structures collapsed. A typical example is shown in figure 7. In most cases, failure of a metal overhead door was also discovered and it is likely that the

increased internal pressure, arising after the door failure, led to the collapse of the wall. Built up flat roofs are used extensively throughout the area and many suffered various levels of damage and most had developed leaks. Several motels built using common wood frame construction techniques suffered extensive damage. The most severe damage to these motels occurred to those with covered exterior walkways. The entire roof was blown away in a couple of instances. The indication is that the roof connections could not withstand the high wind loads associated with the overhang.

3. Wind speeds higher than design values - Damage similar to that described for wind speeds near design values was observed.

Non-Engineered

1. Wind speeds below design values - Little damage was observed except to roofing shingles or old dilapidated farm or storage buildings. Damage to plastic signs was fairly widespread.

2. Wind speeds near design values - Widespread damage to large elevated signs and their frames; particularly frames constructed using formed or rolled steel shapes. Frequent damage to roofing. Some extensive damage to roof systems, particularly to those with overhangs. One of the more dramatic examples of damage to roof systems is shown in figure 8. These buildings are part of an appartment complex just south of Mobile, Alabama. Nearly every building in the complex experienced similar roof damage where an overhanging portion blew away. The majority of the damage to homes was related to falling trees and in general wood frame homes stood up quite well.

3. Wind speed higher than design values - Damage similar to that observed for wind speeds near design values. However, in some instances houses were completely demolished. Other houses stood up very well to the high wind speeds.

OTHER PARAMETERS

It was clear from the damage observations that some of the more serious damage occurred to structures which exhibited one or more design or construction flaws. Note that most of the buildings in the various categories withstood wind speeds nearly equal to or even greater than the design wind speeds.

There are indications that, in some communities, diligent enforcement of building codes produced non-engineered structures which withstood wind speeds well above design values. For example, Federal Emergency Management Association personnel who conducted a workshop for building officials indicated that the inspector for Dauphin Island, Alabama was especially diligent. An indication that his efforts may have had an impact on building performance is obtained from the following comment in reference 2:

"Wind effects on the beach houses (on Dauphin Island) were minimal, and were rated from aerial surveys as being F0 or F1. If there were no recorded wind traces from the Dauphin Island Sea Lab and bridge, the estimated wind speeds based on damage would have been less than 100 mph." (reference 2 refers to peak wind speeds, a corresponding fastest-mile wind speed would be between 80 and 90 mph and is considerably less than the 106 mph value shown in figure 4).

SUMMARY AND CONCLUSIONS

A framework for evaluating wind damage observations from an engineering perspective has been presented. The framework allows consideration of both the maximum wind speed at a site and a definition of the level of engineering attention devoted to a structure. Note that similar damage level observations were made for the last three categories of structures at similar wind speeds. Consequently, it may be possible to define other categories which would provide greater insight into structural damage. Wind speeds at various damage sites are estimated from anemometer records and are expressed in terms of fastest-mile wind speeds which can be compared with design values.

Damage observations are summarized and indicate that signficant damage did not occur until wind speeds reached or exceeded design values. This was true for marginally engineered and non-engineered structures as well as full-engineered and pre-engineered structures. There is some indication that existence and enforcement of adequate building codes can produce marginally engineered and non-engineered buildings which perform adequately under design wind loads. These observations indicate that design procedures and existing building codes are producing structures which are in most cases performing up to desired levels.

REFERENCES

1. Minor, J.E., and Mehta, K.C., "Wind Damage Observations and Implications," Journal of the Structural Division, ASCE Vol. 105, No. ST11, November 1979, pp. 2279-2291.

2. Fujita, T., Wakimoto, R., and Stiegler, D., "Mesoscale Damage Patterns of Hurricane Frederic in Relation to Enhanced SMS Imagery," Proceedings of the 19th Conference on Radar Meteorology, Miami Beach, Florida, April 15-18, 1980.

3. Batts, M., Cordes, M., Russell, L., Shaver, J., and Simiu, E., Hurricane Wind Speeds in the United States, Building Science Series Report 124, National Bureau of Standards, Washington, DC, May 1980.

4. "Building Code Requirements for Minimum Design Loads in Buildings and Other Structures," (ANSI A58.1-1972), American National Standards Institute, New York, 1972.

5. Simiu, E., and Scanlan, R., Wind Effects on Structures, Wiley-Interscience, New York, NY, 1978.

6. Counihan, J., "Adiabatic Atmospheric Boundary Layers: A Review and Analysis of Data from the Period 1880-1972," Atmospheric Environment, Vol. 9, 1975, pp. 871-905.

7. Engineering Sciences Data Unit, "Characteristics of Wind Speed in the Lower Layers of the Atmosphere Near the Ground: Strong Winds (Neutral Atmosphere)," ESD Item No. 72026, London, England, November, 1972 (with Admendments A and B, October, 1974).

8. Thom, H.C.S., "Prediction of Design and Operating Velocities for Large Steerable Radio Antennas," Large Steerable Radio Antennas - Climatological and Aerodynamic Considerations, Annuals of the New York Academy of Sciences, Vol. 116, Art. 1, pp. 90-100, New York, NY, 1964.

Table 1. Characteristics of Anemometer Sites

Sites	Terrain Type(1)	Height of Anemometer	Type of Instrumentation	Types of Data Available		
				Strip Chart	Peak Observ.	1 Min. Avg. Readings
Dauphin Island Bridge (DIB)	1	∿18m	3-cup and vane, NWS type	Yes	No	No
Eglin Air Force Base (EAF) (airport)	2	4.6m	3 blade impeller and aerovane	Yes	Yes	Yes
Gulfport Civil Defense (GCD)	4	∿27m	3-cup and vane, NWS type	No	Yes	No
Ingalls Shipbuilding (ISB)	2	∿10m	3-cup and vane, NWS type	Yes	No	No
Mobile Civil Defense (MCD)	3 or 4	∿23m	3-cup and vane, NWS type	Yes	No	No
Mobile NWS, Bates Field (MOB) (airport)	2	6.7m	3-cup and vane, NWS type	Yes	Yes	Yes
Pensacola Naval Air Station (PNA) (airport)	2 and 3(2)	∿23m	impeller and aerovane	Yes	Yes	Yes
Pensacola Regional Airport (PRA)	2	6.7m	3-cup and vane, NWS type	Yes	Yes	Yes
Coast Guard Cutter White Pine (near Twelve Mile Island)	2 or 3	∿15m above water level	impeller and aerovane	No	Yes	Yes
Coast Guard Cutter Salvia (at Alabama State Docks)	shielded	∿15m above water level	impeller and aerovane	No	No	Yes

(1) See table 2 for description of terrain type.
(2) Type 2 for wind directions 0-70 and 185-360, type 3 for wind directions 75-180.

Table 2. Velocity Profile Characteristics

Exposure Type	Type of Surface	Log-Law Surface Roughness Parameters z (m)	Power-Law Exponent α	Nominal Boundary Layer Height for Use with Power-Law δ (m)
0	Sand	.0005	.10	225
1	Sea surface (high winds)	.005	.12	250
2	Grass (open)	.05	.15	275
3	Outskirts of towns	.20	.19	340
4	Centers of towns	.40	.22	400
5	Centers of large cities	.70	.26	430
6	Pine forest	1.00	.30	460
7	Extremely high roughness	2.00	.35	500

Table 3. Coefficient c(t)

t	1	10	20	30	50	100	200	300	600	1000	3600
c(t)	3.00	2.32	2.00	1.73	1.35	1.02	0.70	0.54	0.36	0.16	0.00

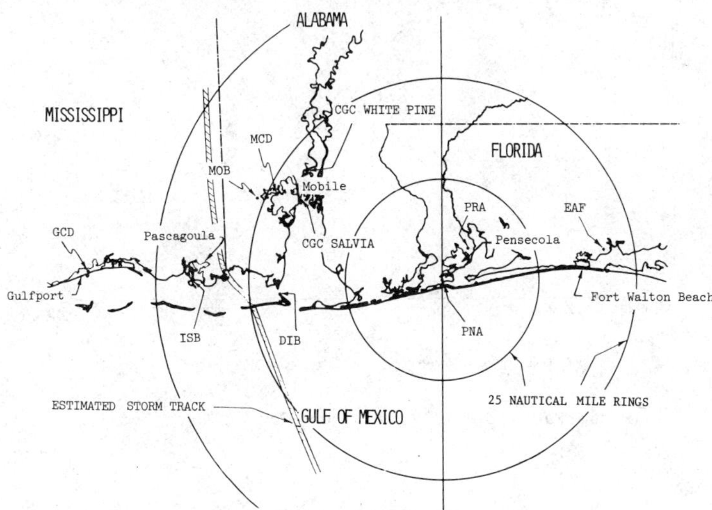

Fig. 1. Map of Gulf Coast Near Mobile,Alabama Showing Anemometer Sites.

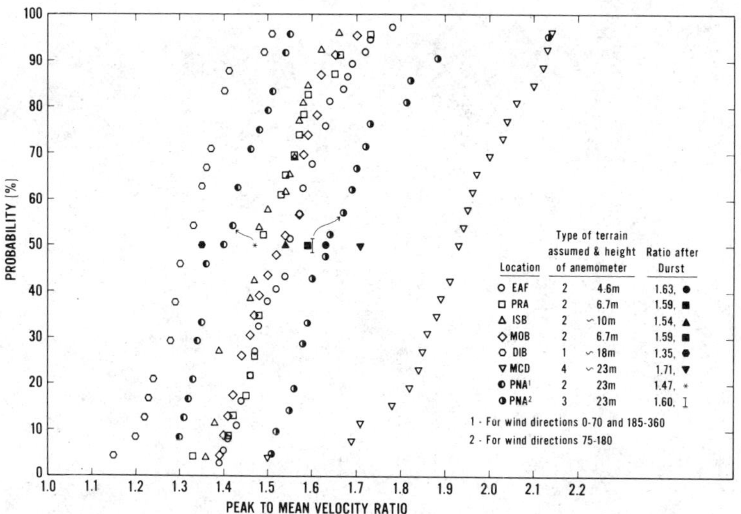

Fig. 2. Probability Distributions of Peak to Mean Velocity Ratios from Actual Data.

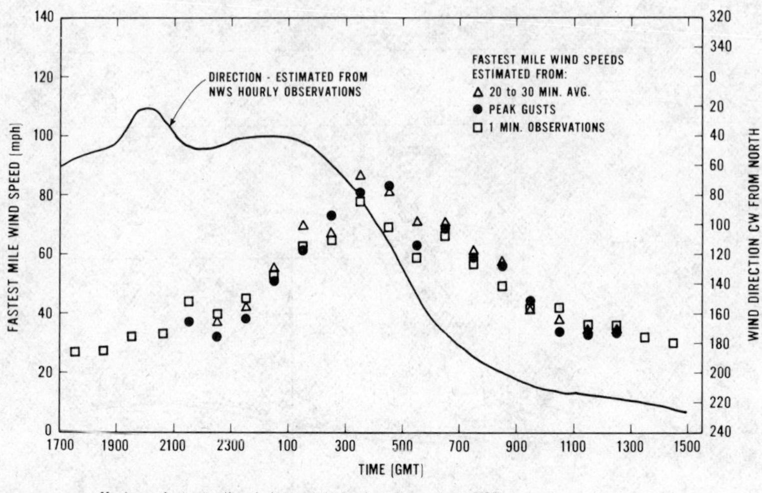

Fig. 3. Maximum Fastest-Mile Wind Speeds During Hour Intervals at MOB.

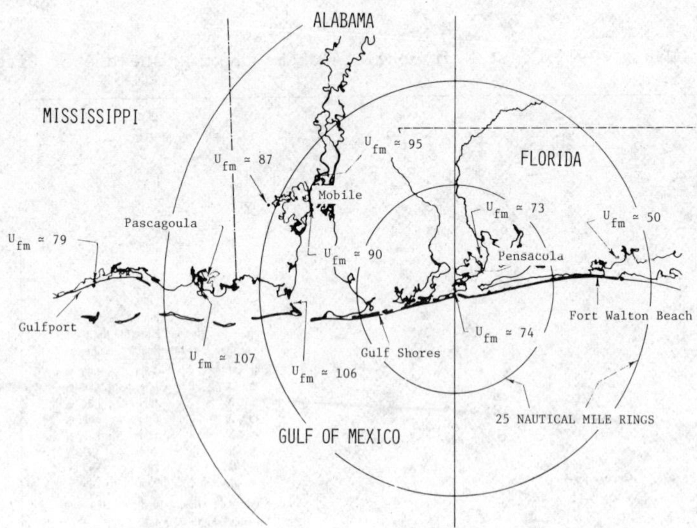

Fig. 4. Maximum Fastest-Mile Wind Speeds at Anemometer Sites, Data Converted to 10 m in Open Terrain.

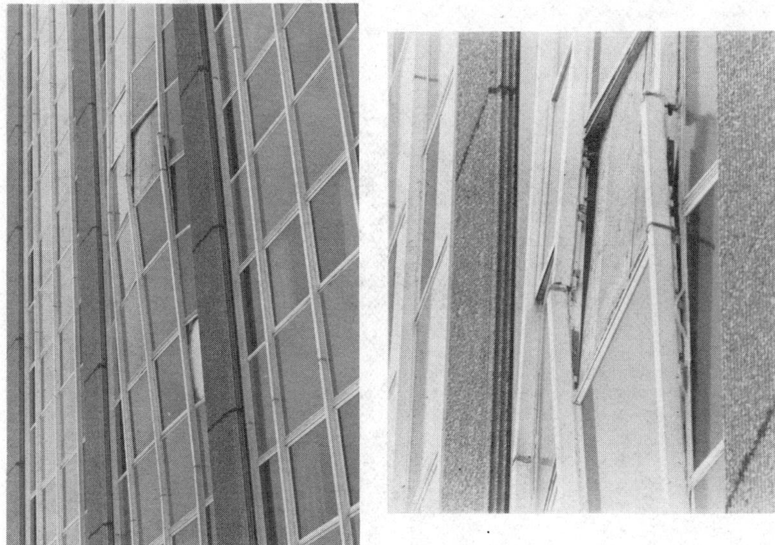

Fig. 5. Damage to Facade of Tall Building in Mobile, Alabama.

Fig. 6. Damage to Metal Building Located North of Gulfshores, Alabama.

Fig. 7. Collapsed Wall of Commercial Building in Mobile, Alabama.

Fig. 8. Roof Damage to Apartments in Complex South of Mobile, Alabama.

BUILDING RESPONSE TO EXTREME WINDS AND TORNADOES

by

James R. McDonald[1], M. ASCE
Kishor C. Mehta[1], M. ASCE

ABSTRACT

The response of buildings and their components to severe wind loads are systematically catagorized, based on wind storm damage investigations over a period of years. Building response, as defined herein, implies more than displacement or distortion of the structural system. It encompasses the performance of the entire building, including both the structural and architectural systems. Three types of building failures are identified. Five types of structural systems and seven components of the architectural system are evaluated.

INTRODUCTION

Each year, a wide variety of buildings and other structures are damaged by various types of wind storms, including winds from thunderstorms, downbursts, hurricanes and tornadoes. While it is not economically feasible to design all buildings to resist the worst of these winds, knowledge of how certain building systems and components respond to severe winds is beneficial to those involved in design, in standard writing, and in insurance coverage.

This paper summarizes the observed response of ordinary buildings and structures to severe winds. The data has been collected and summarized by various investigators over a period of years. Response, as used herein, implies more than displacement or distortion of the structural system under implied loads. Response encompasses the performance of the entire building, including both the structural and architectural systems.

The structural system involves those members and components that carry a calculable proportion of the wind loads. Beams, columns, frames and shear walls are typical members of the structural system. The architectural system consists of the nonelectrical, nonmechanical, and non-load carrying components of the building. The architectural

---

[1]Professor of Civil Engineering, Texas Tech University, Lubbock, Texas

system includes windows, exterior non-load-bearing walls, interior partitions, parapet walls, fascia, roofing system, canopies, walkways, and other appurtenances attached to the building. If the mechanical and electrical systems of a building are affected by the winds to the extent that the normal functions of a building cannot continue after the storm, then they too are a part of the "performance" of the building under wind load.

One of the first investigations and subsequent analyses of wind damage was performed by Segner [1] in his study of the Dallas tornado of 1957. Mehta, et al. [2] systematically studied more than 90 structures damaged by the Lubbock tornado of 1970. Much of the information gained from ten years of storm damage investigations by the Institute for Disaster Research at Texas Tech University was summarized by Minor, et al. [3]. McDonald and Lea [4] conducted a historical survey of building damage to medical facilities and systematically classified the types of damage to these and various other buildings. The above data, along with the damage investigation experiences of the authors, are used to systematically classify the response of buildings to severe wind storms.

An interesting and perhaps surprising fact about wind damage is that the appearance of damage is not unique to a certain type of wind storm. If the wall of a building collapses under wind load, the appearance of the damaged wall looks the same, whether it was caused by a tornado, hurricane or other severe wind. In subsequent discussions of damage, no distinction is made between the different types of wind storms.

TYPES OF FAILURES

The types of failure and extent of damage to a building depends to a great extent on the degree of engineering attention paid to the building prior to construction. Minor and Mehta [5] classified the degree of engineering attention to buildings as fully engineered, pre-engineered, marginally engineered, and nonengineered. Those buildings that receive the most engineering attention have been observed to perform substantially better in wind storms than those receiving a lesser amount. The types of failures observed in buildings fall into three basid categories:

1) overall structural failure
2) component failure
3) breach of building containment

Overall Structural Failure

This mode of failure includes structural collapse, overturning, sliding and structural response to impact from large objects propelled by the wind. Structural collapse is generally the most catastrophic type of failure. It can occur suddenly and without warning. The most common types of buildings that have undergone structural collapse are pre-engineered metal buildings and gymnasium-type structures. In nearly every case, the collapse occurs when the winds are blowing perpendicular to the main structural frames. The other type of collapse is due to the failure of loadbearing walls. The loadbearing walls collapse first and the roof comes down.

The most common examples of overturning occurs with unanchored mobile homes. This situation does not necessarily indicate a design deficiency, but a lack of knowledge or concern of the owner. Examples of sliding have been observed where residences are not anchored to basement walls. Although the idea of a house being unanchored to its foundation sounds preposterous, examples have been found in storm damage in Omaha, Nebraska, Cheyenne, Wyoming, and Birmingham, Alabama, to name a few.

Another type of overall structural response is the impact of a large object such as an automobile with a wall such that the wall undergoes excessive deflection or collapse. This type of potential tornado-generated missile is sometimes overlooked in favor of the type that only causes localized damage from penetration or perforation.

Component Failure

This type of failure occurs because of inadequate size and/or strength of the load carrying member. Also associated with component failure is the strength of the component connection or anchorage. Failure of an individual component in itself may not be serious. However, such a failure often leads to overload of other components and subsequently a progressive failure occurs. For example, the anchorage of roof joists sitting on top of a loadbearing wall fails due to uplift created by the wind forces. As the roof uplifts and is possibly removed, the top of the wall is left unsupported. The wall now acts as a cantilever, rather than a single beam, and fails at a smaller load than before. The progressive failure is initiated by the failure of a single component.

Breach of Building Containment

While overall collapse or failure of a component is of concern, breach of the building containment can occur without any other type of structural damage and still cause high dollar damage to a building. Whenever the containment of the building is breached, both wind and water that often accompany a severe wind storm get into the building and cause danger to occupants and damage to equipment and furnishings of the building. In the case of nuclear material handling facilities, breach of the containment allows radioactive materials to get out of the building and to be exposed to the environment.

PERFORMANCE OF STRUCTURAL SYSTEMS

McDonald and Lea [4] surveyed the performance of hospital buildings subjected to tornadoes, hurricanes or other extreme winds. Five different types of structural systems were identified. The systems included:

1) cast-in-place reinforced concrete frame,

2) steel frame with nonloadbearing walls,

3) steel frame with loadbearing walls,

4) precast concrete frame with loadbearing walls, and

5) timber construction.

Cast-in-Place Reinforced Concrete Frame

None of the six buildings that were constructed with reinforced concrete suffered any structural damage [4]. The inherent rigidity, which is produced by strong connections and diaphragm action of the roof and floor slabs limits lateral deflections due to wind loads.

Steel Frame with NonLoadbearing Walls

Typically, this type of structural system has been used for both low and high-rise buildings. Connections may be bolted, welded, or a combination of both. The frames may be braced or unbraced. Steel framed buildings, except for single-story warehouse types, do not usually undergo overall collapse.

The Great Plains Life Building (known today as Metro Tower) survived the Lubbock, Texas tornado of 1970, even though the average wind pressures were more than double the 20 psi uniform wind pressure used in the original design [6]. In this case, the building utilized additional stiffnesses from exterior masonry walls and relatively stiff interior partitions. Older buildings such as the Great Plains Life Building (GPL) have a greater inherent resistance than modern steel-framed buildings with glass or metal curtain walls and lightweight movable partitions.

Excessive wind drift causes damage to interior partitions and ceilings. It can result in permanent deformation of the structural frame. The GPL building frame suffered 10 in. of permanent deformation. Cracked plaster and other interior damage to door frames indicated that the actual deformations must have been two to three times this large. An eyewitness to the storm, who rode it out on the top floor of the building, described the movement as being similar to a ship rolling in a heavy sea. There was evidence of yielding in the form of yield lines that showed up in the shop paint on the girders. The riveted moment-resistant connections, however, did not fail and the building was able to utilize its reserve strength when loaded beyond that called for by the local building code.

Steel Frame With Loadbearing Walls

This type of construction is typical of that used in light commercial buildings. Two modes of failure have been observed with this type of structural system. In one case, unreinforced loadbearing walls constructed of masonry or concrete blocks fail in flexure under the wind loads. Subsequent to failure of the wall, the roof joists collapse downward, causing almost complete destruction of the building. The second mode of failure was described above where the roof uplifts, leaving the wall unsupported at the top, and collapse of the wall occurs.

Precast Concrete Frame With Loadbearing Walls

The same comments for loadbearing walls mentioned above hold true for this case. Precase concrete members provide considerably more dead weight than a roof consisting of steel joists and metal deck. Sometimes the precast members rock or slide off column corbles or a weld plate fails, causing collapse of the roof.

Timber Construction

Used typically in residential, apartment and motel applications, timber construction generally receives little or no engineering attention. Practices are dictated by tradition in the building trade, which strongly resists suggestion of change. Roof overhangs and weak roof-to-wall anchorages are the primary causes of failure in the buildings.

PERFORMANCE OF ARCHITECTURAL SUBSYSTEMS

In the historical survey performed by McDonald and Lea [4], seven elements were identified as a part of the architectural system as defined herein.

Windows

Window glass breakage is the major cause of damage to buildings in wind storms. Breach of the building containment allows wind and rain to get inside the building, causing large dollar values of damage. Window glass traditionally has been designed only to resist wind pressures, with little attention given to the impact of windborne debris [7]. The First National-Pioneer National Gas Building in Lubbock was located adjacent to buildings covered with roof gravel. The building suffered extensive damage to window glass from the flying roof gravel in the Lubbock tornado [2]. On the other hand, the Mississippi Power Company building in Biloxi, Mississippi had screens to protect it from small flying debris. Very few windows were broken in this building, even though wind pressures were at design values or above.

Exterior NonLoadbearing Walls

Both metal curtain walls and nonloadbearing masonry walls are included in this discussion. Stud walls with brick veneer are also included. Most metal curtain walls perform well in wind storms, because they must pass certain standard wind and water leakage tests. Problems generally occur because of weak anchorages to the main structural system.

Interior Partitions

Since they are nonloadbearing, interior partitions do not always receive the design attention needed. Because the wind can get inside a building, partitions in exterior rooms sometimes collapse. The cause is weak anchorage at floor and ceiling.

Parapet Walls

Parapet walls are damaged by wind forces and the impact of flying debris. Sometimes the coping material, which is typically stone or metal, is ripped off by the winds and becomes flying debris. The collapse of a parapet wall can cause damage to a lower roof level.

Canopies and Covered Walkways

The consequences of damage are marginal. Canopy material can become airborne and strike other areas.

Fascia

Fascia damage is usually of little consequence except it too can become airborne debris. If removal of fascia material allows wind to get underneath the roof deck, additional uplift pressures can cause the roof to pull up and the entire roof section may be ripped off.

Built-Up Tar and Gravel Roofs

The primary problem observed with roofing material is weak bonding to the roof deck. In many examples observed, there is little or no bond between the roof membrane and the deck or slab. Next to window glass breakage, loss of roofing material is the most common type of wind damage.

IMPLICATIONS OF BUILDING PERFORMANCE

Lessons learned from investigation of damages buildings are important to continued improvements in occupant safety and in the mitigation of property loss. Numerous conclusions have been drawn in earlier studies of storm damage. However, three observations stand out in the authors' minds at this point in time.

Appearance of Damage

Earlier discussions of component failure pointed out that a component failure can lead to progressive failure of the building. Whereas a weak component might initiate a progressive failure at relatively low wind speeds, the final appearance of damage might look extremely bad to the untrained eye. Appearance of damage is one of the indirect methods of estimating wind speeds in tornadoes. Use of the Fujita scale [8] allows one to estimate the relative severity of a tornado. Yet, this approach must be used with extreme caution, because appearance of damage can be very misleading.

Standards of Practice

Observation of building performance tends to substantiate the approaches to design for wind loads taken in building codes and standards of practice (e.g. ANSI A58.1-1972). While some debate lingers about pressure coefficients, gust response factors and the like, observed damage verifies the types of loading conditions postulated by the codes and standards. For example, failures often initiate at roof corners. While there may be some debate about magnitude of localized pressures, codes and standards today do take into account localized pressures at roof corners. The same thing can be said about failure initiating at ridges and eaves.

Existing Buildings

The knowledge gained from storm damage observations allows the prediction of potential wind damage to existing buildings. The approach was first applied to VA hospital facilities by McDonald and Lea [4], and has been extended to a more quantitative approach by Smith, et al. [9]. The same information can be used for identifying safe places in buildings for occupant protection. Safety in schools during the threat of severe weather has been tremendously inhanced by knowledge of building performance in wind storms.

ACKNOWLEDGEMENTS

This paper is an assembly of information from many sources. The contributions of many colleages and fellow researchers are gratefully acknowledged by the authors. Financial support, both directly and indirectly, were provided by the National Science Foundation, the Veterans Administration, the Nuclear Regulatory Commission, and Argonne National Laboratory.

LIST OF REFERENCES

1. Segner, E.P., "Estimate of Minimum Wind Forces Causing Structural Damage in the Tornadoes at Dallas, Texas, April 2, 1957," U.S. Weather Bureau Research Paper, No. 41, 1960, pp. 169-175.

2. Mehta, K.C., McDonald, J.R., Minor, J.E., and Sanger, A.J., "Response of Structural Systems to the Lubbock Storm, "Texas Tech University Storm Research Report SRR 03, October, 1971 (NTIS Accession No. PB-204-983).

3. Minor, J.E., McDonald, J.R., and Mehta, K.C., "The Tornado: An Engineering-Oriented Perspective," NOAA Technical Memorandum ERL NSSL-82, National Severe Storms Laboratory, Norman, Oklahoma, December, 1977.

4. McDonald, J.R., and Lea, P.A., "A Study of Building Damage Caused by Wind Forces," Technical Report prepared for Veterans Administration, Office of Construction, Washington, D.C., January, 1978.

5. Minor, J.E., and Mehta, K.C., "Wind Damage Observations and Implications," Journal of the Structural Division, ASCE, Vol. 105, No. ST11, Proc. Paper 14980, November, 1979, pp. 2279-2291.

6. McDonald, J.R., "Structural Response of a Twenty-Story Building to the Lubbock Tornado," TTU SRR 01, Texas Tech University, Lubbock, Texas, October, 1970.

7. Minor, J.E., and Beason, W.L., "Window Glass Failures in Windstorms," Journal of the Structural Division, ASCE, Vol. 102, No. ST1, Proc. Paper 11834, January, 1976, pp. 147-160.

8. Fujita, T.T., "Proposed Characterization of Tornadoes and Hurricanes by Area and Intensity," SMRP No. 91, Satellite and Mesometeorology Research Project, the University of Chicago, Chicago, Illinois, 1971.

9. Smith, D.A., Mehta, K.C., and McDonald, J.R., "Procedure for Predicting Wind Damage to Buildings," IDR-45D, Institute for Disaster Research, Texas Tech University, Lubbock, Texas, September, 1979.

TORNADO WIND EFFECTS ON THE GRAND GULF COOLING TOWER

by

T. Theodore Fujita[1)]

A B S T R A C T

The cooling tower of the Grand Gulf Nuclear Power Plant was damaged by the April 17, 1978 tornado. The damage, visible from the air, was caused by the crane standing at the center of the tower. Although no wind-effect damage was detected from the air, an attempt was made to estimate the differential pressure induced by the tornado. The computed pressures include both aerodynamic and vortex pressures.

INTRODUCTION

The Grand Gulf, Mississippi Nuclear Power Plant was under construction when a tornado moved through the plant site shortly before midnight of April 17, 1978. It was the first of the two tornadoes, each rated as an F3 tornado.

Similar to the Richter-scale earthquake magnitude, the Fujita-scale tornado intensity refers to the highest F rating inside a specific tornado. The F-scale damage at various parts of a tornado area varies, like the Mercalli earthquake scale, as a function of the location. For the original paper defining the F-scale, refer to Fujita (1971).

Scale	F0	F1	F2	F3	F4	F5
Windspeeds	40-72	73-112	113-157	158-206	207-260	261-318

Table 1. Fujita-scale windspeeds in mph

Windspeed estimates at the Grand Gulf Power Plant site were made independently by three authors.

[1)]Professor of Meteorology, Department of Geophysical Sciences, The University of Chicago, Chicago, Illinois, 60637.

References	Windspeeds
Fujita (1978)	113 - 157 mph (F2 range)
Rotz (1978)	120 - 140 mph
McDonald (1978)	125 - 150 mph

Table 2. Estimated windspeeds at the Grand Gulf plant site.

PATH OF TORNADO RELATIVE TO COOLING TOWER

Fig. 1 shows a distant view, looking northeast, of the cooling tower No. 1 which was being constructed toward the final height of 520 ft AGL. The tower height at the time of the tornado was 461 ft topped with 30 segments of the scaffold.

FIG. 1. Aerial view (looking northeast) of the Grand Gulf Power Plant, Mississippi, affected by the April 17, 1978 tornado (Photo by Fujita on April 19).

FIG. 2. A high-oblique view of the cooling tower (Photo by Fujita on April 19).

A close-up view of the tower in Fig. 2 shows that the top wall was damaged. The tornado moved near the right edge of the tower (See Fig. 3). A tower crane, standing at the center, with the horizontal boom set free, fell toward the east-southeast.

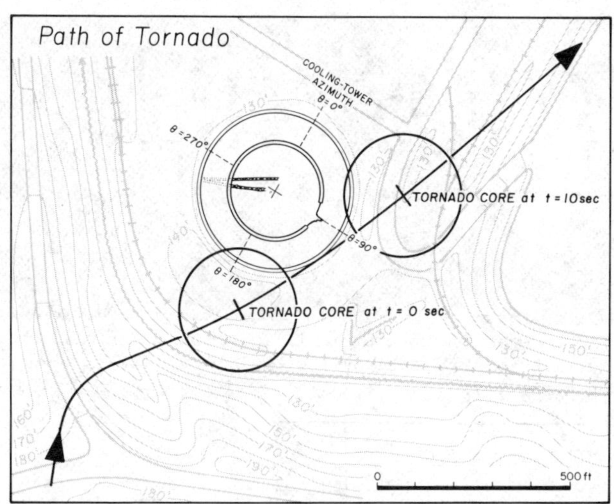

FIG. 3. Path of the tornado in relation to the cooling tower. The azimuth angle assigned to the tower is 30° less than the true azimuth.

FIG. 4. A close-up view of the scaffold/wall damage caused by a tower crane which fell on the eastern wall (Photo by Fujita on April 19).

Fig. 4 taken from 700' shows the feature of a deep cut on the shell. Fig. 5 shows the landed concrete and crane. Estimated trajectories of the shell concrete in Fig. 6 reveal that not all concrete fell straight downward. The cause of the non-vertical fall could be the wind or the momentum given by the impact of the crane.

FIG. 5. Wreckage of the concrete wall and crane which slid down the shell surface and landed along the base of the tower. (Photo by Fujita on April 19).

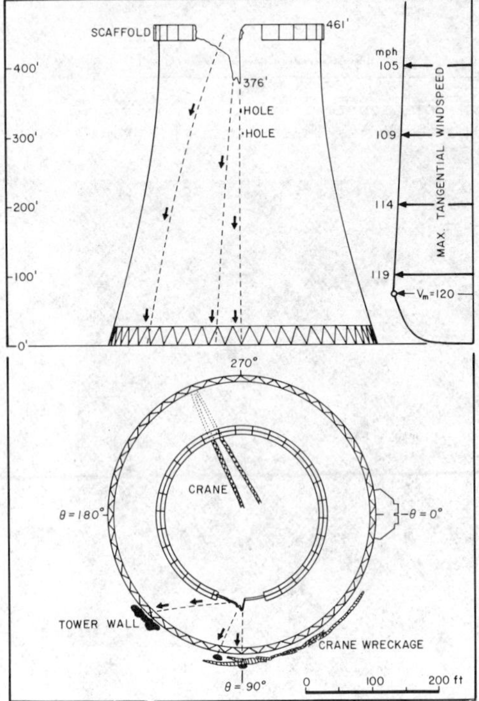

FIG. 6. Vertical and plan views of the cooling tower. The vertical stem of the crane, folded into two sections, landed inside the tower. The maximum tangential windspeeds of the tornado core are shown on the right side.

COMPUTATION OF PRESSURE COEFFICIENT

As shown in Fig. 7, θ_s denotes the center angle of stagnation point where $\beta = 0°$. Likewise θ_w is located at the center of the wake where $\beta = 180°$.

There are a number of measurements of pressure coefficients around the wall of cooling towers. Used for this calculation are Niemann's (1977) results which vary with meridional wind ribs characterized by height, k, and the distance, a, of the ribs. In this paper, it was assumed that the pressure coefficients, measured as a function of β, vary from +1.0 at the stagnation point to -0.5 at the wake center.

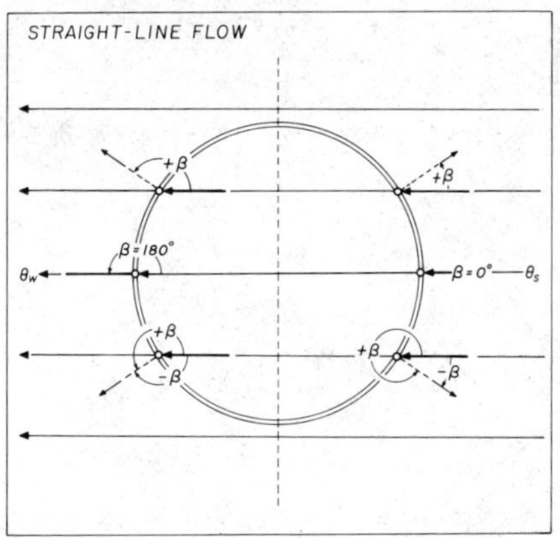

FIG. 7. Definition of flow angle, β, when the circular shell of a cooling tower is affected by a straight-line flow. θ_s and θ_w denote the center of stagnation and the wake, respectively.

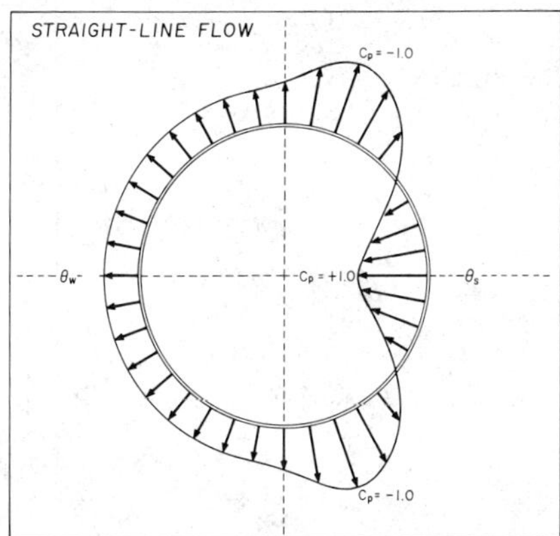

FIG. 8. Pressure coefficient, C_p, which varies with β.

The distribution function of C_p in Fig. 8 is expressed by

$$C_p = f(\beta) \tag{1}$$

This is the case of a straight-line flow impinging upon the shell surface.

For a curved flow, however, measurements become very difficult especially when one wishes to obtain results with a stable flow.

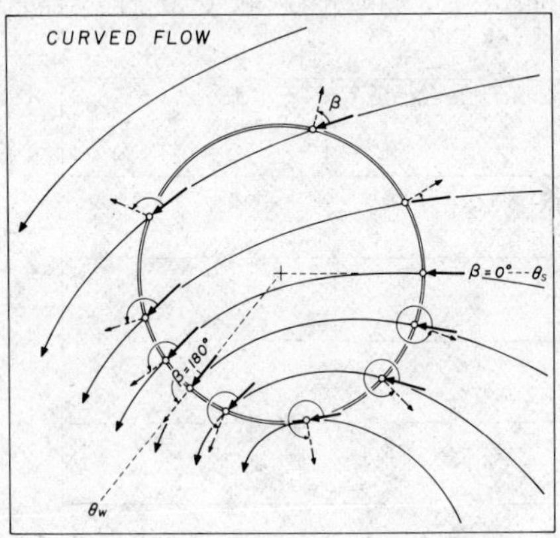

FIG. 9. Definition of flow angle, β, when a curved flow interacts with a cooling tower. For curved flow, the angular distance between θ_s and θ_w is not 180°.

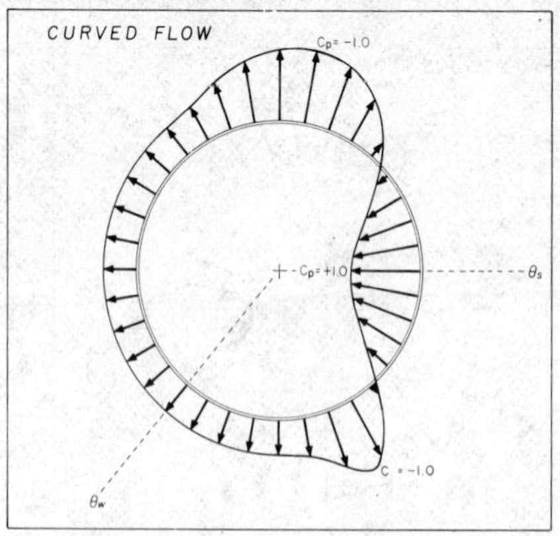

FIG. 10. Distribution of pressure coefficient, C_p, around the shell of a cooling tower affected by a curved airflow.

In this paper, the flow angle, β, of a curved flow was measured as angles given in Fig. 9. Pressure coefficients in Fig. 10 were computed from Eq. (1) which is the result obtained from straight-line flow.

TORNADO PARAMETERS

The basic parameters of the tornado at t = 0 were estimated from detailed aerial photography and mapping of the damage in and around the plant site. These parameters are

Radius of outer core, R_o = 150 ft
Top of inflow height, H_i = 72 ft
Maximum tangential velocity, V_m = 120 mph
Translational velocity, T = 34 mph

The model of tornado used in computing the three-dimensional airflow is DBT-77 described in the Workbook of Tornadoes and High Winds by Fujita (1978).

The maximum tangential windspeed, V_m = 120 mph of this model tornado occurs at 72 ft AGL. The maximum values decrease to 119 mph at 100 ft, 109 mph at 300 ft, etc. It decreases to zero at the ground (See Fig. 6).

OUTSIDE PRESSURE, P_{out}

When wind is calm, the difference in the pressure on the inner and outer walls is zero. During a tornado, however, the outer wall receives two types of pressure

P_a (Aerodynamic Pressure): Pressure exerted upon a surface by the air moving around a body; this pressure decreases to zero when wind dies down.

P_t (Tornado-vortex Pressure): Negative pressure inside a tornado vortex; the lowest pressure is located at the vortex center.

The outside pressure, P_{out}, is defined as

$$P_{out} = P_a + P_t \,. \tag{2}$$

P_t is computed from the pressure equation of DBT-77 while P_a is computed from

$$P_a = C_p \times \frac{1}{2} \rho V^2$$
$$= f(\beta,\ V) \tag{3}$$

where C_p is computed from Eq. (1) and V, from 3D windfield of DBT-77.

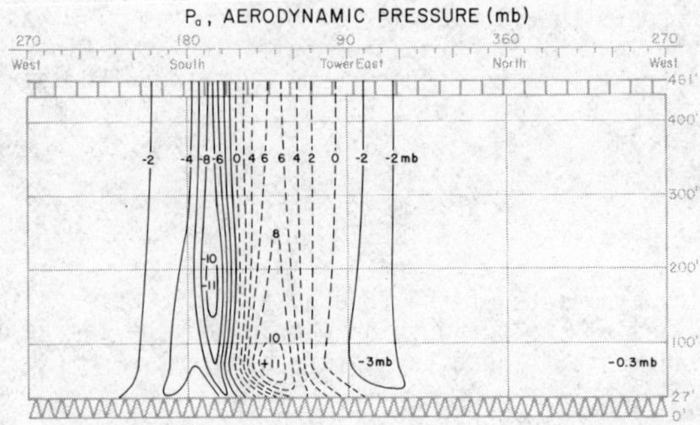

FIG. 11. Aerodynamic pressure on the outer surface of the cooling tower when the tornado was located at $t = 0$ position in Fig. 3.

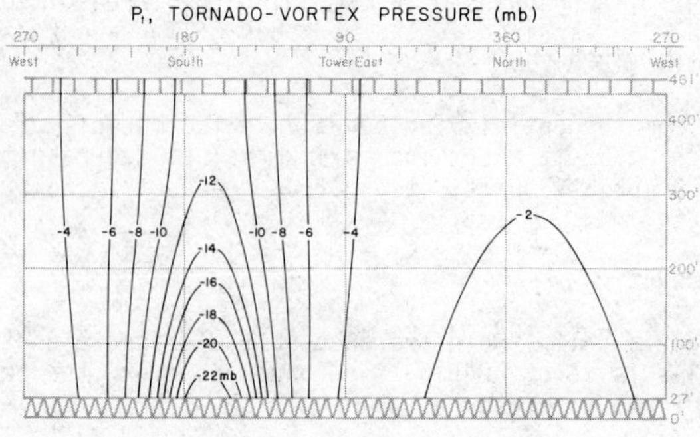

FIG. 12. Tornado-vortex pressure on the outer surface of the cooling tower at $t = 0$.

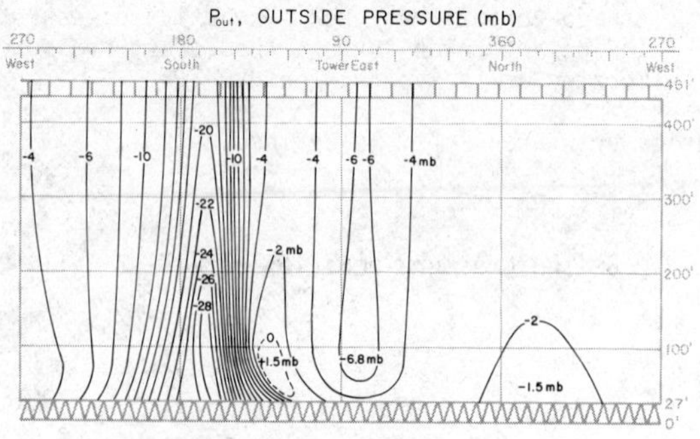

FIG. 13. Outside pressure $P_{out} = P_a + P_t$ obtained by adding pressure fields in Figs. 11 and 12.

Fig. 11 shows the distribution of aerodynamic pressure at t = 0. Due to a strong flow impinging upon the shell wall from the southeast, stagnation pressure of +11 mb is seen at 70 ft AGL where θ = 130°. The maximum suction is seen at 170 ft and 170°.

Pressure in mb	0	10	20	30	40
Pressure in psi	0	0.14	0.29	0.44	0.58

Table 3. Conversion of millibar (mb) into psi

Vortex pressure in Fig. 12 is minimum at the base of the shell, which is closest to the tornado at t = 0. Negative pressure of 22 mb is lower than the aerodynamic suction pressure.

The outside pressure, $P_{out} = P_a + P_t$, in Fig. 13 is the lowest along the tower azimuth, θ = 160°. There is a small area of positive pressure at 70 ft and 130° where the tornado wind induces the stagnation point.

INSIDE PRESSURE, P_{in}

If we assume that there is no air motion inside the tower, the inside pressure is uniform. This uniform pressure or the mean inside pressure is computed as

$$\overline{P}_{in} = \frac{1}{2}(\overline{P}_{top} + \overline{P}_{base}). \qquad (4)$$

Fig. 14 shows the variation of outside pressure along the edges of the tower, at its top and at its base.

Theoretically, the pressure gradient inside the atmosphere cannot be infinity. Should it happen, a shock wave will form and propagate out immediately. To avoid the finite pressure difference on both sides of the shell wall, assumed paper thin, situation 1 in Fig. 15 changes into situation 2 so that the outside pressure decreases by B and the inside pressure increases by A.

Since the net effect of the pressure upon the wall is the differential pressure, ΔP, we write

$$\begin{aligned} \Delta P &= P_{out} - P_{in} \qquad (5) \\ &= (P_{out} - B) - (\overline{P}_{in} + A) \\ &= P_{out} - (\overline{P}_{in} + A + B) \end{aligned}$$

to obtain

$$P_{in} = \overline{P}_{in} + (A + B). \qquad (6)$$

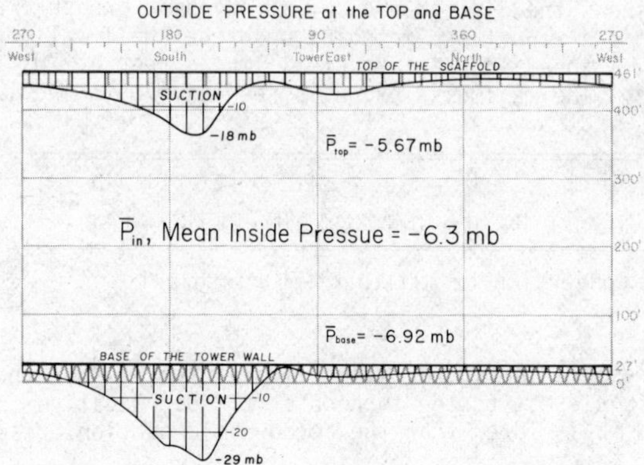

FIG. 14. Outside pressure along the top and base edges of the cooling tower at $t = 0$. The mean inside pressure is assumed to be the mean values of the pressures at the top and at the base.

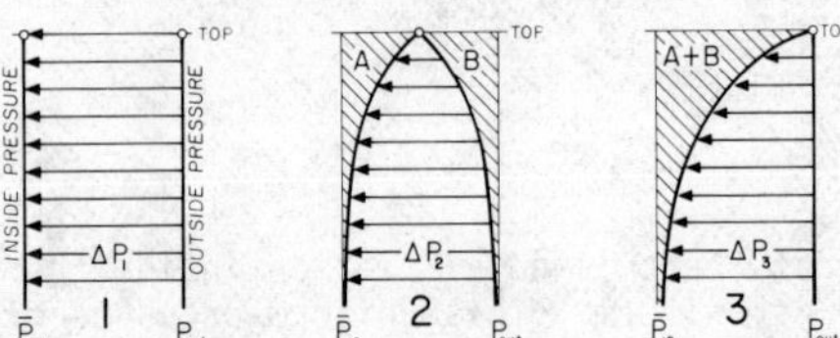

FIG. 15. Equalization of the pressure at the edge of the tower wall, which takes place in order to eliminate a "pressure jump".

In computing inside pressure, P_{in} , in Fig. 16, an exponential function of A + B was assumed. The function is

$$A + B = (P_{out} - \overline{P}_{in})\, e^{-0.05\Delta H} \qquad (7)$$

where ΔH denotes the distance (ft) from the edge to a given point on the inside wall.

The distribution of inside pressure computed from Eqs. (6) and (7) is presented in Fig. 16.

Differential pressure, ΔP, which is the difference between P_{out} and P_{in}, is the pressure which causes the compression and the suction of the tower wall. Its pattern in Fig. 17 reveals a strong suction at 110 ft and 165° where the tornado airflow is tangent to the wall surface and the distance to the tornado center is closest.

It should be noted that the pressure pattern shifts along the tower surface rapidly as the tornado passes by the tower, taking only a few seconds.

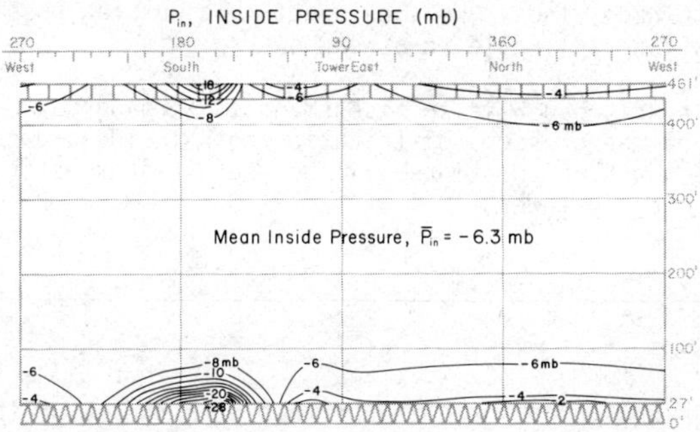

FIG. 16. Pressure inside the cooling tower computed as the sum of (A + B) in Fig. 15 and the mean inside pressure in Fig. 14.

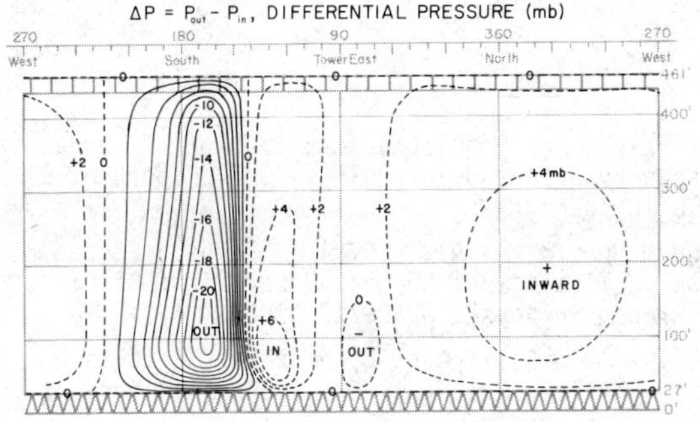

FIG. 17. Differential pressure obtained by subtracting the inside pressure in Fig. 16 from the outside pressure in Fig. 14. This is the net pressure acting upon various sections of the tower wall.

CONCLUSIONS

It is rare that a tornado passes within close proximity to a cooling tower. Nevertheless, the Grand Gulf cooling tower was affected by F2 tornado winds. We should not rule out the possibility of a worse situation applicable to other power plants.

Presented in this paper is an assessment of tornado parameters and computation of the pressure distribution both inside and outside the tower shell. Such computations will be useful in assessing wind effects on cooling towers.

Acknowledgements

The research presented in this paper has been sponsored by NRC under contract 04-74-239.

References

Fujita, T. T., (1971): "Proposed Characterization of Tornadoes and Hurricanes by Area and Intensity", SMRP Paper No. 91, University of Chicago.

Fujita, T. T. and McDonald, J. R., (1978): "Tornado Damage at the Grand Gulf, Mississippi Nuclear Power Plant Site: Aerial and Ground Surveys", NUREG/CR-0383.

Niemann, H. J., (1977): "Static and Dynamic Wind Effects on Cooling Tower Shells", Preprint 3031, ASCE Fall Convention, San Francisco, California.

Rotz, J. V., (1978): "Tornado Damage Report, Grand Gulf Generating Station, Port Gibson, Mississippi," Internal Report, Bechtel Power Corporation, San Francisco, California.

THE KALAMAZOO TORNADO: BUILDING DAMAGE IN THE CENTER CITY

by

James R. McDonald[1], M. ASCE
James J. Abernethy[2]

ABSTRACT

On May 13, 1980 a tornado struck the city of Kalamazoo, Michigan, passing through the heart of the downtown business district. Although relatively weak compared to other "maxi tornadoes," the damage in the downtown area was of particular interest to engineers and preparedness officials. This paper reviews the performance of selected buildings and discusses the ramifications of other cities being struck by tornadoes.

INTRODUCTION

On the afternoon of May 13, 1980, a tornado touched down at 3:00 p.m. (EST) at a point eight miles west of the city of Kalamazoo, Michigan. Upon entering the city, it traveled eastward through residential areas until it reached the downtown business district. There, it did extensive damage to engineered structures, including several multi-story buildings. Five persons were killed in the Kalamazoo tornado and 79 were injured.

Because of the damage to the engineered structures, the extent of damage and the potential for obtaining useful information, a storm study team from the Institute for Disaster Research was sent to the scene to evaluate and document the damage caused by the storm.

The basic objectives of the storm study effort were to

1) Document the damage by means of ground level and aerial photographs,

---

[1]Professor of Civil Engineering, Texas Tech University, Lubbock, Texas

[2]Professor of Architecture, Lawrence Institute of Technology, Detroit, Michigan

2) study the performance of engineered structures,

3) document patterns of debris and types of damage for possible correleation with concepts of near-ground wind fields, and

4) provide input for the F-scale classification of the tornado.

The uniqueness of the event is the fact that the tornado passed through the heart of the downtown business district. In this report, the performance of selected engineered structures are described. In addition, the ramifications of the center city being hit be a tornado are described. Time and space do not allow consideration of all of the objectives set forth by the storm study team.

THE KALAMAZOO TORNADO

The Kalamazoo tornado first touched down in relatively open country eight miles west of the city. Only minor damage to trees, fences, and power poles could be observed until the storm entered the city limits at the intersection of Drake Road and Croyden Avenue. At this point, the roof of the Church of Jesus Christ of the Latter Day Saints was partially removed and there was damage to the educational wings of the facility. Figure 1 shows the path of the tornado as it translated through three miles of residential area before it passed through the heart of the downtown business district.

After crossing Mountain Home Cemetary, where there was extensive damage to trees and monuments, the tornado crossed Michikal Avenue and struck the St. Augustine Catholic Church and school facility. Beyond St. Augustine, five multi-story buildings were located in the path. They ranged in height from seven to fifteen stories. Other commercial and industrial buildings were damaged by the storm as it moved through the east side of town. Figure 2 shows details of damage to the downtown area. The storm apparently lifted near Comstock east of the city limits, and dissipated into the clouds. Table 1 contains a summary of information on the tornado characteristics and statistics.

PERFORMANCE OF ENGINEERED STRUCTURES

Performance of selected engineered structures are reviewed briefly in this section. The objective of the prelimineary analysis is to use the structure as an indicator of tornadic wind speed. The following structures are reviewed:

1) St. Augustine School

2) U.S. Post Office

3) ISB Building

4) Hilton Hotel

5) Gilmore Department Store

Figure 2 indicates damage to a number of different types of structures besides the ones mentioned above. These other structures are of

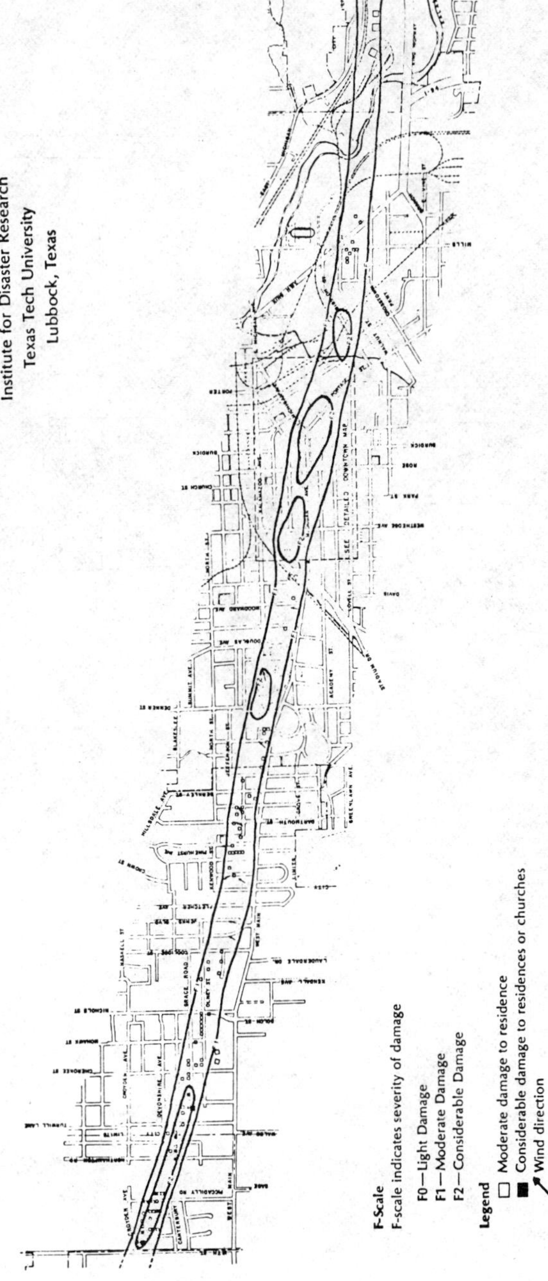

FIGURE 1. DAMAGE PATH OF KALAMAZOO TORNADO

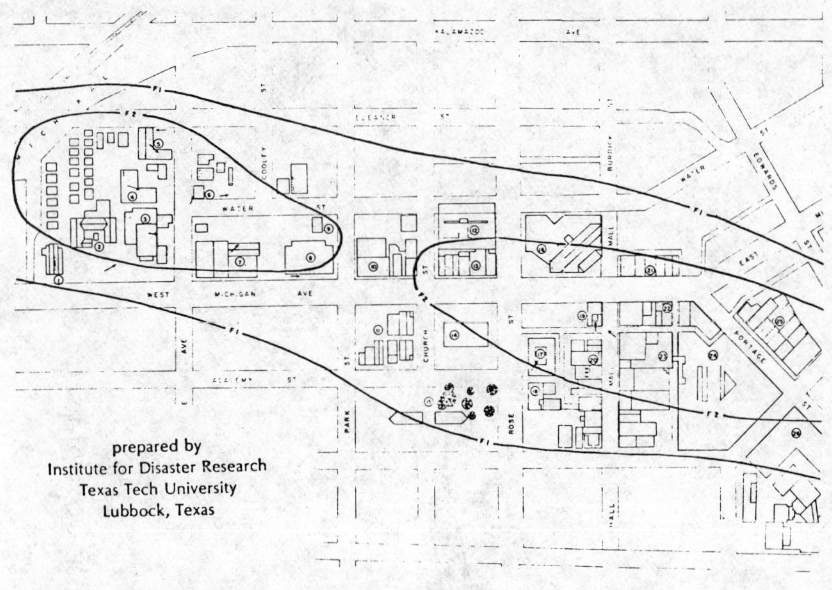

Building Key

1. St. Augustine Cathedral
2. Rectory
3. Parochial Building (gym)
4. Elementary School
5. Kelly's Auto Glass
6. "I Am" Ministry
7. Stuart-Clark Furniture
8. Post Office
9. Law Office
10. First Federal Savings and Loan
11. First Baptist Church
12. Parking Garage
13. Small Businesses
14. Courthouse
15. Bronson Park
16. Hilton Inn
17. ISB Building
18. YMCA Building
19. Kalamazoo Building
20. Shops facing Mall
21. Small Businesses
22. American National Bank Building
23. Gilmore Department Store
24. Parking Garage
25. Small Businesses
26. Upjohn Building
27. Small Businesses
28. Power Plant
29. Auto Dealership
30. European Motors

FIGURE 2. DOWNTOWN AREA

TABLE 1

KALAMAZOO TORNADO CHARACTERISTICS AND STATISTICS

PATH LENGTH		11 miles
PATH WIDTH		100 yards (average)
ESTIMATED WIND SPEED		150 mph
TIME OF INITIAL TOUCHDOWN		3:00 p.m. (EST)
TIME OF LIFT UP		3:25 p.m. (EST)
AVERAGE TRANSLATIONAL SPEED		30 mph
GRADATIONS OF DAMAGE		11 miles
Length of F0 Damage	5 mi	
Length of F1 Damage	4 mi	
Length of F2 Damage	2 mi	
SENSE OF ROTATION		Counterclockwise
DIRECTION OF TRAVEL		ESE
NUMBER OF DEATHS		5
NUMBER OF INJURIES		79
ESTIMATED TOTAL PROPERTY DAMAGE*		$50 million
RESIDENTIAL UNITS		407
Destroyed	23	
Severe Damage	84	
Light to Moderate Damage	300	
COMMERCIAL STRUCTURES		150
Destroyed	22	
Severe Damage	57	
Moderate Damage	71	
INDUSTRIAL STRUCTURES		8
Severe Damage	7	
Moderate Damage	7	
PUBLIC STRUCTURES*		4
Moderate Damage	4	

*Kalamazoo Gazette

no less importance in the overall assessment of damage. However, the ones discussed below play a primary role in estimating maximum wind speeds of the tornado.

St. Augustine School

The St. Augustine School facility consisted of the parochial building and the elementary school building (buildings 3 and 4 in Figure 2). The parochial building contains classrooms, offices and a gym. Damage to the St. Augustine gym is shown in Figure 3. Steel trusses span 80 ft in the east-west direction. They are supported on a loadbearing wall constructed of 12-in. concrete masonry and 4-in. face brick. The roof was constructed of metal deck with insulation board and a built-up tar and gravel roof. The loadbearing walls collapsed toward the east, causing the entire roof to shift eastward about 15 ft. The loadbearing walls were not vertically reinforced. There was horizontal truss-type reinforcing at every third course (24 in.). The walls were 18-20 ft high. Evidentally, the loadbearing walls were forced to resist the lateral loads with little help from shear wall action. Thus, the wall failure and subsequent roof collapse could have occurred at wind speeds of 150 mph or less.

In the elementary school building, the greatest wind damage was experienced by the windward (primarily west) and south second floor exterior walls. Serious injuries would have been sustained by occupants located in rooms enclosed by these walls. However, since the building was on the "left" side of the tornado path, the north walls also were hit with high winds, sufficient to cause extensive window breakage and some inblown debris. Incoming winds buckled a second floor interior loadbearing wall, causing roof collapse. The flat built-up roof with its lightweight steel deck was uplifted and removed in numerous locations along the windward edges. This damage is shown in Figure 4. Inblown missiles (primarily gravel and roofing from nearby buildings) penetrated the windward facing spaces, impacting on the first interior vertical surface. The interior walls, doors and wired-glass windows stopped further penetration. Door hinges were strained, but did not fail. Closed interior doors did not open. As a result, interior locations on the first floor would have offered protection to occupants of the building. The lower level was clearly the safest. Completely interior spaces, especially those with short spans, offered high degrees of occupant protection. There was no evidence of any failure of the floor system.

U.S. Post Office Building

The only damage suffered by this well-engineered building was several broken windows on the west side and a missile impact from a piece of concrete block that passed through a metal canopy on the north side and continued through a window into the ladies' restroom. Figure 5 shows a view looking south at this building. A law office and a furniture store next to the Post Office were heavily damaged (see Fig. 2, Structure No. 9). The tendency is to say that the tornado "skipped" over the Post Office building because there is no damage. The skipping does not appear to be true in this case. It is simply a matter of a well-engineered building standing up to an overload.

FIGURE 3. COLLAPSED ROOF OF ST. AUGUSTINE GYM

FIGURE 4. CLASSROOM WING OF ST. AUGUSTINE ELEMENTARY SCHOOL

FIGURE 5. WEST END OF U.S.
POST OFFICE BUILDING

FIGURE 6. VIEW OF ISB BUILDING
LOOKING SOUTHEAST

ISB Building

The Intrastate Bank and Trust (ISB) building is an example of a modern type of building, typical of what can be found in almost any city the size of Kalamazoo. The building has nine stories, a small aspect ratio (width/height or length/height), and is completely glass enclosed (See Fig. 2, Structure No. 17). The basic structural system consisted of a steel framework with the floor constructed of metal deck and a topping of concrete. Moment resisting frames coupled with a central elevator and stair core adequately resisted the lateral wind loads. There was no damage to the structural system. However, damage to the building itself was both extensive and expensive. As shown in Figure 6, the exterior cladding was damaged on all sides. Much of the glass breakage could be attributed to impact from flying debris, mostly roof gravel. Broken glass was found inside the offices on the west side of the building, which would have been a windward wall. Whether the glass was broken due to outward acting pressure or due to missile impact on the other three sides was inconclusive. Interior partitions were collapsed due to air circulating through the building. Suspended ceiling panels also were severely damaged. Wind and water caused extensive damage to office furnishings, equipment and other contents of the building. The above-design-code wind loads coupled with impacts from flying debris contributed to the extensive damage to the cladding. Wind speeds in excess of 150 mph are not required to explain the extensive damage to this building.

Hilton Hotel

Shown in Figure 6 (Structure No. 16 in Fig. 2), damage to the Hilton Hotel is very similar to that of the ISB Building. The Hilton is located on the opposite side of the tornado from the ISB building. There was no damage to the structural system. All other damage was restricted to cladding and contents of the building. Window glass breakage was primarily due to impact of small missiles such as roof gravel.

Gilmore Department Store

The Gilmore Department Store building is located on the mall in downtown Kalamazoo (see Fig. 2, Structure No. 23). The heavily damaged part of this building is six stories high. The structure has loadbearing masonry walls, and a wooden floor structure supported on interior walls and pipe columns. The east wall collapsed outward, sending piles of debris to the street below. Two persons were killed by the falling debris. The floors had enough support from the interior walls and columns that they did not collapse subsequent to failure of the loadbearing wall. The wall failure was induced by outward acting external wind pressures, possibly in combination with outward acting internal pressures. Lack of a substantial tie of the wall to the floor diaphragm suggests that extreme wind speeds are not required to explain the rather dramatic failure. Figure 7 shows the damage to the east side of the building.

FIGURE 7. VIEW LOOKING NORTH SHOWS CLADDING DAMAGE TO HILTON HOTEL

FIGURE 8. EAST WALL OF GILMORE DEPARTMENT STORE COLLAPSED OUTWARD

F-SCALE RATING OF KALAMAZOO TORNADO

The Fujita Scale (F-Scale) is used to classify the relative intensity of tornadoes (Fujita, 1971). The F-scale rating assigned to a tornado is based on the appearance of the worst damage observed anywhere along the damage path. The F-scale classification is thus assigned by comparing word descriptions of each F-scale classification with observed damage. As an additional aid, photographs representing typical F-scale damage are furnished by Dr. Fujita.

While there are certain inherent fallacies with the concept of the Fujita scale, it has become firmly entrenched within the profession as a system for rating tornadoes. When used with data needed for tornado hazard probability assessment, the F-scale rating system is workable and useful. The major problem with the F-scale rating system is the somewhat ambiguous manner in which the extent of damage is described. To the untrained eye, extensive damage to a weak structure may be thought to be done by high winds, whereas little damage to a well constructed building is wrongly interpreted as low wind speeds.

In identifying the worst tornado damage within the path, which is the basis for the F-scale rating, some judgment is required with respect to the strength of the structures involved. In looking at the Kalamazoo damage path, the worse damage could be at either the St. Augustine gym, the ISB building, or the Gilmore Department Store. Each of these buildings had a specific feature that caused severe looking damage that could be caused by relatively low wind speeds. In the case of the St. Augustine gym, there was little resistance to lateral loads from shear walls or other bracing systems. The rather tall unreinforced (vertically) walls had to carry the lateral loads themselves. The ISB building suffered no structural damage. However, the damage to glass cladding and to interior contents of the building made the damage look severe, even though it was caused by relatively moderate wind speeds. The dramatic failure of the east exterior wall of the Gilmore Department Store was the result of an unexpected weak link in the construction. Based on the above discussion, it is the authors' opinion that the observed damage in the Kalamazoo tornado could have reasonably been caused by wind speeds less than 150 mph (F2 F-scale range).

POTENTIAL TORNADO DAMAGE TO THE CENTER CITY

The Kalamazoo tornado was relatively weak when compared with the "maxi-tornadoes" that struck Xenia, Ohio (1974), Birmingham, Alabama (1977), and Wichita Falls, Texas (1979). Based on statistics from the data tape assembled at the University of Chicago (Tecson, et al., 1979), approximately 85 percent of all reported tornadoes have maximum wind speeds less than or equal to 150 mph. If a very severe tornado should hit a center city, the damage in terms of dollar loss would be immense. If given sufficient warning, occupants should be able to find places of safety within a building. These places should be identified and marked as a part of the preplanning for such an emergency. Occupants of the ISB building could have taken interior hallways to the inner core stairway and could have made their way to the basement in relatively safety. The likelihood of collapse of a multi-story building appears to be remote. Two high-rise buildings in Lubbock, Texas survived wind speeds in excess of 175 mph at roof level (Mehta, et al., 1971).

Low-rise masonry structures are likely to suffer collapse of their loadbearing walls and subsequent collapse of roof and some floors. We hasten to point out that design criteria for masonry construction according to recent codes (ACI 53.1-76) should perform very well in severe wind situations--provided the criteria are followed in actual construction. The reinforcement requirements and the provision for connections and anchorages should provide the needed ductility to permit a structure to undergo overload without collapse.

CONCLUSION

Although relatively weak, the Kalamazoo tornado of May 13, 1980 was a very interesting one from the perspective of the researcher and the engineer. The storm marked one of the few times when a tornado struck the heart of the downtown business section of a fairly large city. A number of "engineered" structures were located in the path, including several multi-story buildings. None of the high-rise buildings suffered damage to their structural systems, even though winds exceeded design values in some cases.

The tornado was a classic type, typical of those that occur each year in the Midwest. Tempered with experience and engineering judgment, the maximum wind speeds appear to be less than 150 mph, which puts it in the F2 F-scale range. When the appearance of damage is tempered with engineering judgment, as discussed in this report, an F3 rating does not appear to be justified.

It is concluded, from this and other studies of tornadoes, that if a center city is hit by a severe tornado, the damage will be immense, but high-rise buildings are not likely to collapse. With proper pre-planning, occupants of the buildings should be able to find areas of safety within the buildings.

ACKNOWLEDGEMENTS

Support for this research was provided by the Nuclear Regulatory Commission, Office of Nuclear Regulatory Research. The guidance and encouragement of Robert F. Abbey, Jr. of NRC is gratefully acknowledged.

REFERENCES

Fujita, T. T., 1971: "Proposed Characteristics of Tornadoes and Hurricanes by Area and Intensity," SMRP Research Report No. 91, Department of Geophysical Sciences, University of Chicago, Chicago, Illinois.

Mehta, K. C., McDonald, J. R., Minor, J. E., and Sanger, A. J., 1971: "Response of Structural Systems to the Lubbock Storm," Texas Tech University Storm Research Report SRR 03, (NTIS Accession No. PB-204-983), Lubbock, Texas.

Tecson, J. J., and Fujita, T. T., 1979: "Statistics of U.S. Tornadoes Based on the DAPPLE Tornado Tape," Preprints, 11th Conference on Severe Local Storms, October 2-5, 1979, Kansas City, MO, published by American Meteorological Society, Boston, MA.

EVALUATION OF SOIL-STRUCTURE INTERACTION METHODS

by

Anand K. Singh,[1] Tzu-I. Hsu,[2]

Tara P. Khatua,[3] and Shih-Lung Chu[4]

ABSTRACT

In the last several years, two methods of analysis have evolved for the evaluation of the effects of seismic soil-structure interaction for nuclear power plants: the impedance function approach and the finite element method.

In this paper, the various methods are briefly reviewed and their relative merits evaluated. Results for a typical power plant are presented using the elastic half-space approach, the viscoelastic layered half-space approach, and the finite element method using two- and three-dimensional models. It is concluded that the various approaches lead to different responses due to inconsistencies in the basic assumptions required for analysis. It is felt that further research is needed to fill the gap between the various approaches.

INTRODUCTION

Earthquakes occur due to the sudden release of energy within the earth-crust, generally due to slippage at faults. This energy is transmitted to the surface through rock and soil strata, resulting in ground motions at the plant site. These ground motions are the result of wave propagation, reflection, and refraction through a nonhomogeneous medium composed of the various rock and soil strata. Exact analytical models for such a complex phenomenon are not feasible at the present time. In the present design of nuclear power plants, the design motions in terms of response spectra (1) or response spectra consistent motions are specified in the free field. These free field motions represent the earthquake motions at the site in the absence of the power plant structure. When the structure is founded on soil, and to a lesser extent when it is founded on rock, it interacts with its foundation. The forces transmitted to the structure and the feedback to the foundation are complex in nature

---

1. Associate and Assistant Head; 2. Senior Engineering Analyst; 3. Supervisor; and 4. Associate and Head, Structural Analytical Division, Sargent & Lundy, Chicago, Illinois 60603.

and modify the specified free field motions. The soil-structure interaction analysis quantifies these effects. In view of the complexity of the problem, it is not surprising that there is a diversity of views on its solution.

Methods of soil-structure interaction analysis have been published by a number of writers. These methods can be divided into two categories: the direct methods using the finite element approach (2 through 8) and the impedance function approach (9 through 16). The finite element approach can be further divided into the modal synthesis (3,4) approach and the complex frequency response method using a direct solution (8) or a substructure solution (2). The impedance function approach can be divided into the frequency-independent approach (9,10,16) frequency-dependent impedance functions (11,12), numerical impedance functions (13,14), and the substructure method (15).

In the absence of a number of plants undergoing seismic excitation, it is difficult to judge the accuracy of these methods. Reference 19 presents the soil-structure interaction effects on the Humboldt plant; the results of this study were used to show the adequacy of the finite element method. However, corresponding analytical results for the impedance function approach are not available, because the configuration of the plant is such that the compliance function is not suitable. The older studies (17), though useful, can not be extrapolated to determine the applicability of the impedance function approach to seismic environment and embedded structures. Reference 18, though it provides useful experimental and analytical information, does not reduce the uncertainties associated with the soil-structure interaction methods. In addition to the absence of experimental verification of the numerous variations in the soil-structure interaction methods, analytical comparison of the different methods is also difficult because the studies presented in the literature are based on different models and assumptions. In this paper, the various methods of soil-structure interaction are briefly reviewed, and results of analyses for a typical power plant structure using the various methods are compared. An evaluation of the convergence of the numerical scheme used in the computation of the integral equations in the impedance function approach is presented in Appendix B.

VARIOUS METHODS OF ANALYSIS

The direct method using finite elements (2,8) and the impedance function method (9,12) are the two most widely-used methods for power plant soil-structure interaction analysis. The following briefly reviews these methods.

Impedance Function Method - A schematic representation of a soil-structure coupled system for the impedance function analysis is shown in Figure 1. The structure is modeled by a lumped mass-spring system, whereas the soil is represented by two horizontal, one vertical, two rocking, and one torsional impedance function. These impedance functions are complex values and are dependent on the frequency of the excitation.

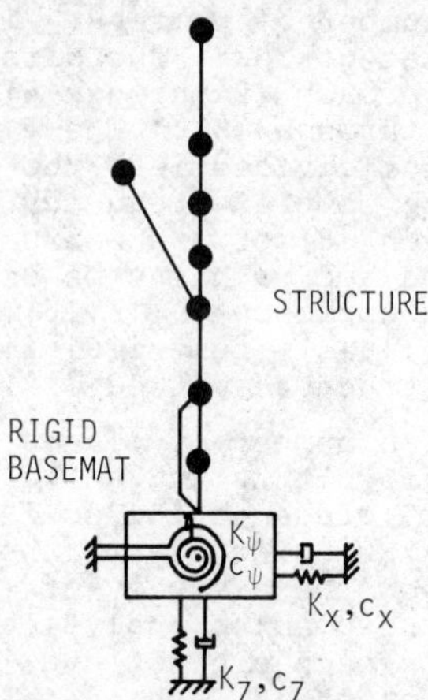

FIGURE 1 SCHEMATIC DETAILS OF A SOIL-STRUCTURE MODEL FOR THE IMPEDANCE FUNCTION METHOD

Analytical methods to compute the soil impedance functions when the rigid circular foundation is at the surface of the soil medium are available when the soil medium is assumed to be an elastic half-space (9), a viscoelastic half-space (20), or a layered viscoelastic half-space (12). For embedded foundations, the embedment effects can be approximated by techniques suggested by Novak (22) or Hall (23). Alternatively, the impedance functions for embedded foundations can be obtained through the finite element method (13,14). In an elastic half space, energy is dissipated through radiation damping, which accounts for the transmission of energy away from the structure foundation. In a viscoelastic half space, the dissipation of energy is due to both the radiation damping and the material damping. The material damping represents the dissipation of energy due to the dynamic motions of the soil medium.

The impedance functions are generally frequency dependent, and the soil-structure system is best solved using the frequency response method (21). However, in many instances, the analysis can be simplified by approximating the soil impedance as constant spring/dash pot systems over the frequency range of interest. In this simplification, the spring stiffness is the real part of the soil impedance at zero frequency, whereas the dash pot constant is the imaginary part of the soil impedance at the fundamental frequency of the soil-structure system. This approximation generally leads to a significant reduction in cost because the soil impedance functions need not be generated for the full range of frequencies. In addition, the response calculations can be performed either by the complex frequency response method or by the direct integration method in the time domain.

Direct Method - In the direct method using finite elements, the coupled soil-structure system is analyzed to the computed motions at the base rock. The base rock motions are obtained through a deconvolution analysis (24) assuming vertically propagating shear waves. The soil and the structure are modeled by two-dimensional plane strain elements. Viscous and transmitting boundaries (8) are used for the soil to simulate the three dimensional effects.

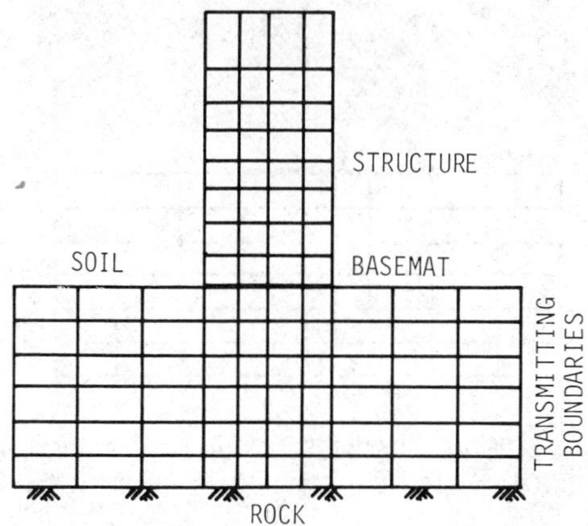

FIGURE 2 SCHEMATIC DETAILS OF THE SOIL-STRUCTURE MODEL FOR A DIRECT FINITE ELEMENT ANALYSIS

In the finite element method proposed by Lysmer et al (8) the soil-structure interaction analysis requires two steps to complete the analysis. In the first step, a simplified two-dimensional structure model is used with the soil model, as shown in Figure 2, to determine the soil-structure interaction effects. In the second step, the response of a detailed three-dimensional structural model (Figure 4) is determined for the interacted base slab motions completed in step one. Approximations and man-time cost are introduced by the simplification and conversion to the two-dimensional structural model required by the first step. Using the finite element soil model, Singh et al (2) presented a substructure formulation for soil-structure interaction analysis which computes the structural responses in one step. The detail of a soil-structure subsystem for this approach is shown in Figure 3. The soil and the structure are treated as separate substructures. The three-dimensional structure is represented by its modal characteristics, i.e., modal shapes, frequencies, and participation factors. The soil is represented by plane strain elements. No modal solution of the soil is performed. The coupled system is analyzed using the complex frequency response analysis method. This substructure formulation has been found to result in significant cost savings since the modal representation of the structure reduces the total degrees of freedom in the coupled soil-structure interaction model and also eliminates the second step of the two-step procedure.

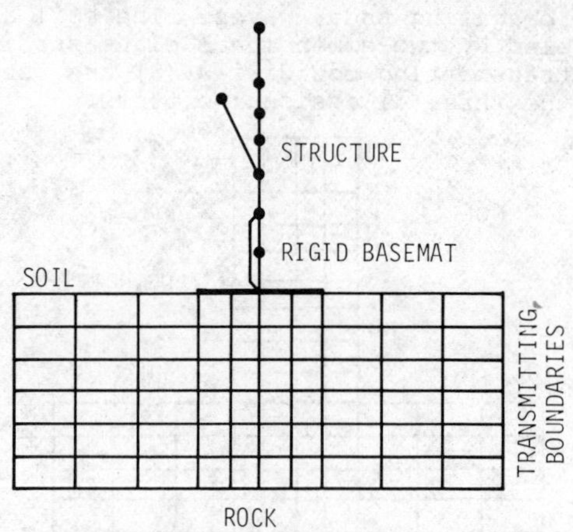

FIGURE 3 SCHEMATIC DETAILS OF THE SOIL-STRUCTURE MODEL FOR A SUBSTRUCTURE FINITE ELEMENT METHOD

NUMERICAL EXAMPLE

In order to compare the responses obtained using the various approaches discussed above, a soil-structure interaction analysis for a typical nuclear power plant was performed using the frequency dependent impedance function, the frequency independent impedance function, the direct (two-step) finite element, and the substructure (one-step) finite element methods. Figure 4 shows the schematic details of the three-dimensional structural model for horizontal excitation. Each floor slab is modeled as a rigid horizontal diaphragm having three in-plane degrees of freedom, two horizontal translations, and one rotation about the vertical axis. For rocking, each slab can move in two additional degrees of freedom (rotation about two horizontal axes). However, since the bending stiffness of the shear wall is large, all slabs are assumed to rock by the same angle.

The mass of the slab and the mass of the equipment and structural components are lumped at the centroid of the included masses. The mass of shear walls is distributed to the connecting floor slabs. Each shear wall and braced

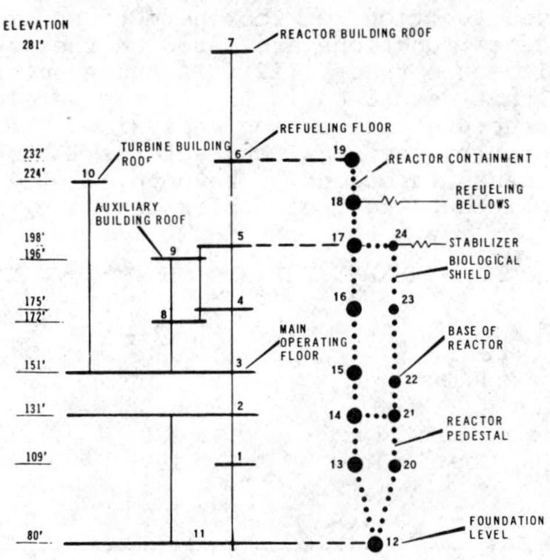

FIGURE 4 SCHEMATIC DETAILS OF THE BUILDING MODEL

vertical frame is modeled as a set of elastic shear springs between floors. To simulate the correct torsional stiffness, these springs in the model are located at the physical location of the shear wall or braced frame in the plant. The containment, the reactor pedestal, and the reactor pressure vessel are modeled by lumped masses connected by flexural members. The flexural and shear stiffnesses of these members correctly simulate the stiffnesses of the components modeled.

Although finite element modeling can account for embedment considerations, the impedance functions for an embedded layered viscoelastic half space are not readily available. In order to make different models comparable, embedment effects are not included, and the structure foundation is assumed to be at the surface of the soil medium. The foundation basemat is 300 ft by 300 ft. The soil depth to bedrock is 65 ft. The properties of the soil profile are listed in Table 1. An acceleration time history, shown in Figure 5, was used as input motion applied at the free field surface. The time history is baseline corrected and scaled to a maximum acceleration of 0.2 g. The response spectrum of the time history closely matches those presented in US NRC Regulatory Guide 1.60 (1). The impedance functions for the sample soil profile and foundation size are presented in Figures 6 through 9 for a frequency range of 0-33 Hz. Figures 6 and 7 present the real and imaginary parts of the impedance function for horizontal motions of the foundation. Figures 8 and 9 present the real and imaginary parts of the impedance function for rocking motions of the foundation. These impedance functions are based on the layered visco-elastic half space theory (12). A substructure technique similar to that described in Ref. 2 was developed to perform the soil-structure interaction analysis. The structure is represented by its modal characteristics and the soil by the impedance functions. The coupled soil-structure system is solved using the complex frequency response method.

TABLE 1 PROPERTIES OF THE SOIL PROFILE

Type:	Soil	Rock
Depth (ft)	65	∞
Shear Modulus (ksf)	2000	140916
Poisson's Ratio	0.38	0.26
Weight Density (pcf)	120	150
Damping Coefficient (%)	10	0

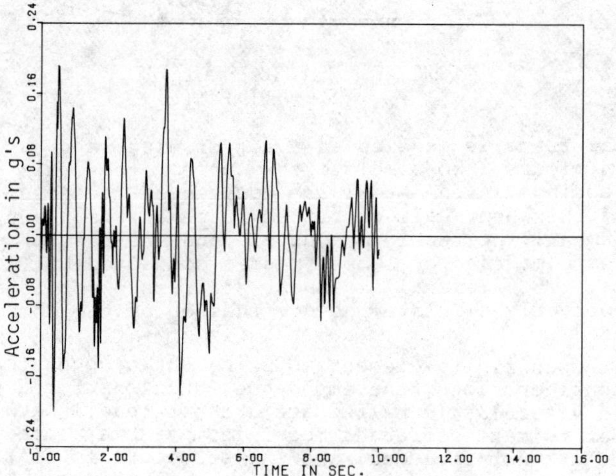

FIGURE 5 ACCELERATION TIME HISTORY

For the frequency independent impedance function solution, the values of the real part of the impedance function at zero frequency were used for the soil spring stiffness, and the values of the imaginary part of the impedance function at the fundamental translational and rocking frequencies were used for the damping constant for the soil spring-dash pot system. These values are listed in Table 2. The soil-structure interaction analysis was performed using the same approach used for the frequency dependent impedance function solution described earlier.

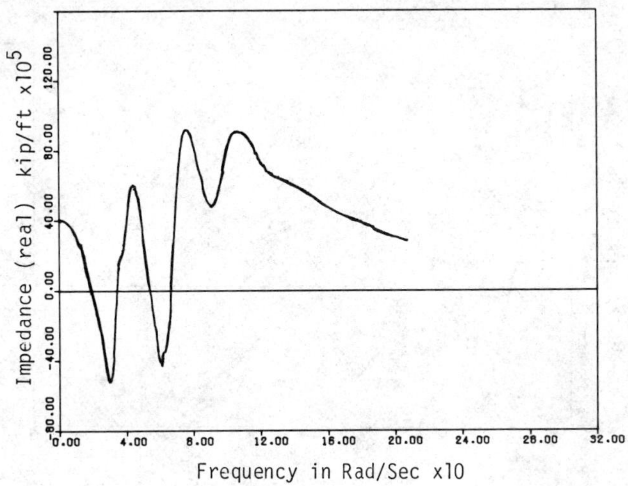

FIGURE 6 REAL PART OF THE HORIZONTAL IMPEDANCE FUNCTION

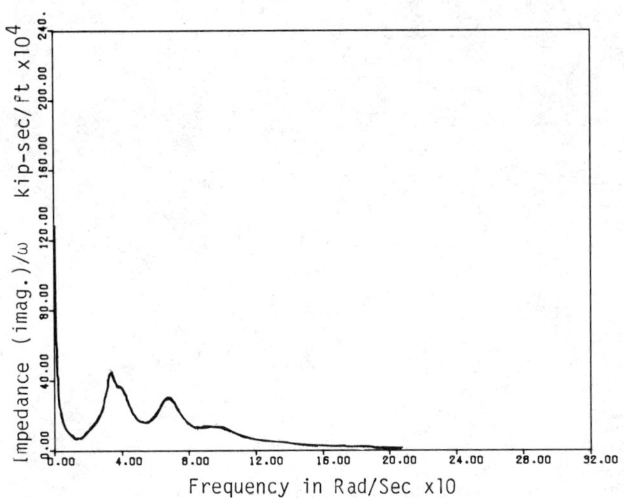

FIGURE 7 IMAGINARY PART OF THE HORIZONTAL IMPEDANCE FUNCTION

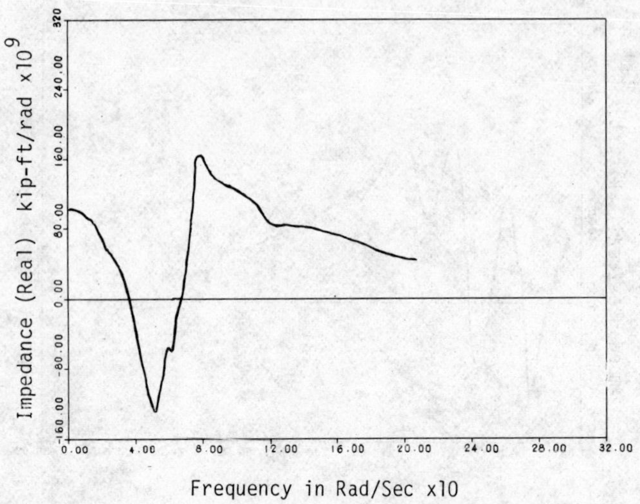

FIGURE 8 REAL PART OF THE ROCKING IMPEDANCE FUNCTION

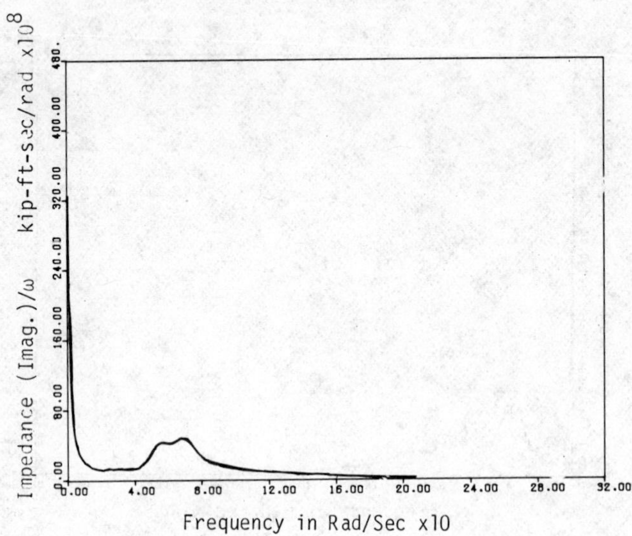

FIGURE 9 IMAGINARY PART OF THE ROCKING IMPEDANCE FUNCTION

TABLE 2 SPRING & DASHPOT CONSTANTS

PARAMETERS	VALUES	UNIT
Translational Stiffness	$.4077x10^{7}$	kip/ft
Rocking Stiffness	$.1028x10^{12}$	kip-ft/rad
Translational Damping	$.3708x10^{6}$	kip-sec/ft
Rocking Damping	$.1315x10^{10}$	kip-ft-sec/rad

In the two-step finite element approach (8), a two-dimensional structural model is needed. This simplified structural model was obtained by matching the shear stiffness and mass distribution of the detailed three-dimensional and the two-dimensional simplified structural models. Rigid horizontal beams were placed at the floor elevations to simulate the floor slab diaphragm action. As shown in Ref. 2, this technique yields a sufficiently accurate two-dimensional model. Viscous and transmitting boundaries were used for the soil elements to simulate the three-dimensional effects. The FLUSH program (8) was used in the first step to determine the interacted basemat motions. The interacted base mat motion was then used, together with the detailed three-dimensional structural model (Figure 4), to obtain the design responses.

In the substructure finite element approach (2), the soil model is the same as that used for the direct approach - i.e., the soil is modeled with two dimensional plane strain elements together with viscous and transmitting boundaries to simulate the three-dimensional effects. The coupled structural model is three-dimensional and is shown in Figure 4. The solution was performed using the substructure technique described in Ref. 2.

In addition to these four alternate soil-structure interaction solutions, a fifth (fixed base) analysis was performed where no soil-structure interaction effects were considered. In this analysis, the structural model shown in Figure 4 was analyzed to the acceleration time history shown in Figure 5. In all five analyses, a 33 Hz frequency cut-off is used.

The response comparisons for the five approaches described above are represented in terms of 2% damping floor response spectra. Figure 10 provides the comparison for the response spectra at the foundation basemat and Figure 11 the comparison for the response spectra at a typical floor slab (Slab 5) within the building. In Figures 10 and 11, the solid line represents the fixed base response, the crossed line the direct finite element approach response, the dashed line the substructure finite element approach response; the dotted line represents the responses obtained using the frequency dependent impedance function approach, and the open circle line represents the responses obtained by the frequency independent impedance function approximation.

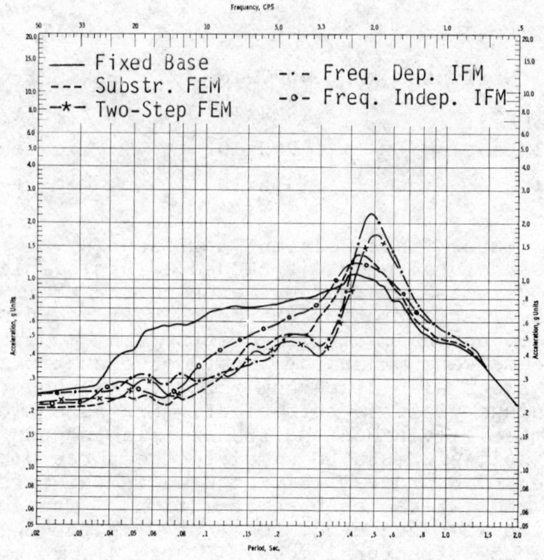

FIGURE 10 RESPONSE SPECTRA COMPARISON AT BASEMAT (SLAB 11)

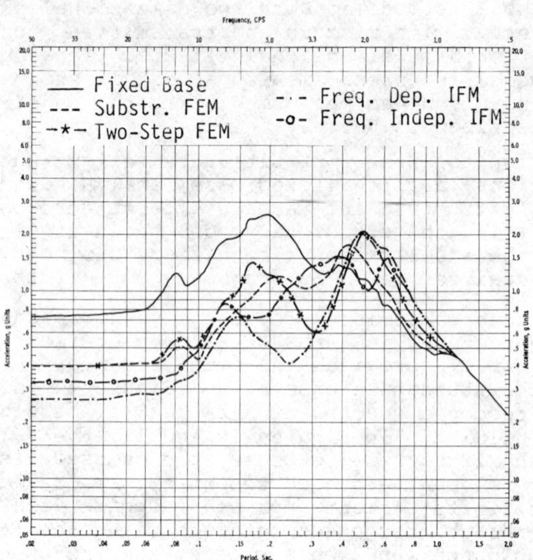

FIGURE 11 RESPONSE SPECTRA COMPARISON AT UPPER FLOOR (SLAB 5)

From Figures 10 and 11, it can be observed that considerable reduction in response due to the soil-structure interaction analysis is anticipated irrespective of the approach used, even though significantly different responses are obtained from the various approaches. It can also be observed that the comparison of responses at the basemat elevation (Figure 10) is better than that at higher elevations (Figure 11) within the structure.

The differences in response between the various approaches may be attributed to the inconsistencies in the basic assumptions required. These assumptions include (i) the treatment of the bedrock as a fixed boundary and the assumption of vertically travelling shear waves in the finite element method; (ii) the absence of rocking in the second step of the two-step finite element method and the approximations introduced by the two-dimensional structural model in step one; (iii) the frequency independent approximation of the impedance functions in the frequency independent impedance function approach; and (iv) the energy dissipation mechanism in terms of material and radiation damping between the finite element and the impedance function approaches.

Table 3 lists the relative computer cost for each of the analyses. It can be observed that the substructure finite element method is the least expensive, while the frequency dependent impedance function approach is the most expensive.

TABLE 3 RELATIVE COMPUTER COST

ANALYSIS	COMPUTER COST
Frequency Dependent Impedance Function	260
Frequency Independent Impedance Function	110
Two-Step Finite Element	240
Substructure Finite Element	100

SUMMARY AND CONCLUSIONS

In this paper, the various methods of soil-structure interaction are briefly reviewed. Soil-structure interaction results and costs for a typical power plant structure are presented using (i) the finite element method using a direct solution; (ii) the finite element method using a substructure solution; (iii) the frequency independent impedance functions; (iv) the frequency dependent impedance function; and (v) when no soil-structure interaction effects are considered. To reduce the uncertainties in the comparison, the assumptions made in the various approaches are identical where possible. An evaluation of the convergence of the numerical scheme used in computation of the integral equations in the impedance function approach is also presented. Based on the comparison, it is concluded that:

1. Considerable reduction in response can be expected due to the soil-structure interaction effects, irrespective of the approach used.

2. The different methods yield significantly different responses at higher elevations within the structure. Thus, the use of the frequency-independent impedance functions or the use of the equivalent two-dimensional structural model in the finite element method should be avoided.

3. The differences in responses can be attributed to the inconsistencies in the assumptions required for each of the basic theoretical derivations. It is felt that further research is needed to fill the gap between the various approaches.

APPENDIX A.- REFERENCES

1. United States Nuclear Regulatory Agency, Regulatory Guide 1.60, Revision 1, December 1973.

2. Singh, A. K., Hsu, T. I., and Holmes, N. A., "Soil-Structure Interaction Using Substructures," presented at the ASCE Specialty Conference on Civil Engineering and Nuclear Power, Knoxville, Tennessee, September 1980.

3. Chu, S. L., Agrawal, P. K., and Singh, S., "Finite Element Treatment of Soil-Structure Interaction Problem for Nuclear Power Plant Under Seismic Excitation," 2nd International Conference on Structural Mechanics in Reactor Technology, Proceedings, Vol. 2, Paper K2/4, September 1973.

4. Singh, S., Srinivasan, R., Agrawal, P. K., and Chu, S. L., "Three-Dimensional Soil-Structure Seismic Analysis Using Finite Elements," Proceedings of the ASCE Specialty Conference on Structural Design of Nuclear Plant Facilities, Chicago, December 1973.

5. Idriss, I. M. and Sadigh, K., "Seismic SSI of Nuclear Power Plant Structures," Journal of the Geotechnical Division, ASCE, July 1976.

6. Seed, H. B. and Lysmer, J., "Soil-Structure Interaction Analysis by Finite Element Methods: State-of-the-Art," Transactions of the 4th SMiRT Conference, San Francisco, California, August 1977.

7. Seed, H. B., Lysmer, J., and Hwang, R., "Soil-Structure Interaction Analysis for Seismic Response," Journal of the Geotechnical Engineering Division, ASCE, May 1975.

8. Lysmer, J., Udaka, T., Tsai, C.-F., and Seed, B. H., "FLUSH, A Computer Program for Approximate 3-D Analysis of Soil-Structure Interaction Problems," EERC Report No. 75-30, University of California, Berkeley, 1975.

9. Richart, R. E. Jr., Hall, J. R. Jr., and Woods, R. D., Vibrations of Soils and Foundations, Prentice-Hall, New Jersey, 1970.

10. Tsai, N. C., Niehoff, I., Swatta, M., and Hadjian, A. H., "The Use of Frequency-Independent Soil-Structure Interaction Parameters," Nuclear Energy and Design, Vol. 31, No. 2, pp. 168-183.

11. Luco, J. E., "Impedance Functions for a Rigid Foundation on a Layered Medium," Nuclear Engineering and Design, Vol. 31, No. 2, December 1975.

12. Luco, J. E., "Vibration of a Rigid Disc on a Layered Viscoelastic Medium," Nuclear Engineering and Design, Vol. 36, No. 3, March 1976.

13. Kausel, E., and Roesset, J. M., "Soil-Structure Interaction Problems for Nuclear Containment Structures," ASCE Power Division Specialty Conference, Denver, Colorado, August 1974.

14. Constantino, C. J., Miller, C. A., and Lufrano, L. A., "Soil-Structure Interaction Parameters from Finite Element Analysis," Nuclear Engineering and Design, Vol. 38, Nos. 2 and 3, August 1975.

15. Gutierrez, J. A. and Chopra, A. K., "A Substructure Method for Earthquake Analysis of Structure-Soil Interaction," Report No. EERCE 76-9, Earthquake Engineering Research Center, University of California, Berkeley, April 1976.

16. Hadjian, A. H., Niehoff, D., and Guss, J., "Simplified Soil-Structure Interaction Analysis with Strain Dependent Soil Properties," Nuclear Engineering and Design, Vol. 31, No. 2, December 1974.

17. Richart, F. E. and Whitman, R. V., "Comparison of Footing Vibration Tests with Theory," Journal of the Soil Mechanics and Foundations Division, ASCE, November 1967.

18. Isenberg, J., Vaughan, D. K., and Sandler, I., "Nonlinear Soil-Structure Interaction," Electric Power Research Institute, Report EPRI NP-945, December 1978.

19. Valera, J. E., et al., "Seismic Soil-Structure Interaction Effects at Humboldt Bay Power Plant," Journal of the Geotechnical Division , ASCE, July 1976.

20. Veletsos, A. S. and Verbic, B., "Vibration of Viscoelastic Foundations," Report 18, Department of Civil Engineering, Rice University, Houston, Texas, April 1973.

21. Hurty, W. C., and Rubinstein, M. F., Dynamics of Structures, Prentice-Hall, Inc., Englewood Cliffs, NJ, 1965.

22. Novak, M., "Effect of Soil on Structural Response to Wind and Earthquake," Earthquake Engineering and Structural Dynamics, Vol. 3, pp. 79-96, 1974.

23. Hall, J. R. and Kissenpfenning, J. F., "Special Topics on Soil-Structure Interaction," Paper U2/2, Proceedings, ELCALAP Seminar, Berlin, September 1975.

24. Schnabel, P. B., Lysmer, J., and Seed, H. B., "SHAKE - A Computer Program for Earthquake Response Analysis of Horizontally Layered Sites," Earthquake Engineering Research Center, Report No. EERC 72-12, University of California, Berkeley, 1972.

APPENDIX B.-NUMERICAL CONVERGENCE OF IMPEDANCE FUNCTIONS

This appendix discusses the numerical convergence of the impedance functions using the method presented by Luco (12) for a rigid circular plate on a layered viscoelastic half space. In Ref. 12, the impedance functions were computed in the lower frequency range where numerical convergence is easily achieved for a wide class of problems. However, for the present paper, it was necessary to compute these functions up to 33 Hz, and it was observed that the accuracy of the computation is a function of the integration step size. It was also observed that the convergence of the numerical scheme also depends on the number of layers in the viscoelastic layered half space and on the relative stiffness between the layers. For several layers with a wide range of relative stiffnesses, convergence of the impedance functions is difficult to achieve, particularly in the high-frequency range. This observation is based on the two numerical example problems presented in this appendix.

For the numerical example, Luco's method (12) was used to compute the impedance functions for a layered foundation medium consisting of N-1 parallel layers resting on a viscoelastic half space. Both the layers and the half space are assumed to be homogeneous and isotropic, with densities ρ_i, shear moduli G_i, and Poisson's ratios σ_i ($i = 1,2,\ldots,N$). In addition, depending on the type of internal friction considered, the relative viscosity coefficient (G_i'/G_1) (for Voigt-type dissipation), or the hysteretic damping coefficient $\xi_i = \omega G_i'/2G_i$ (ω = circular frequency) are assumed for each of the layers forming the soil half space. The geometry of the model and the coordinate systems used are shown in Figure B1.

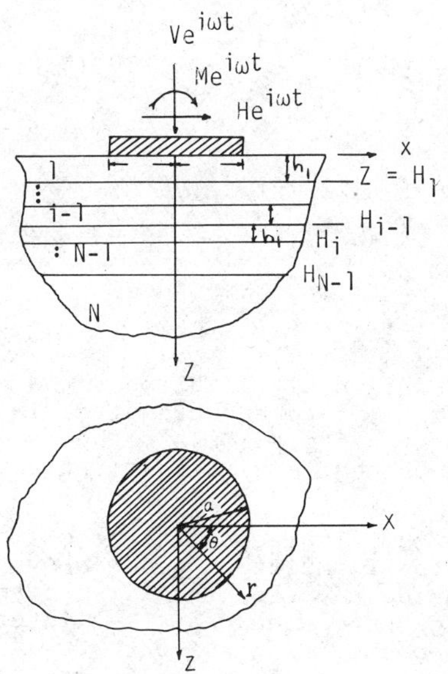

FIGURE B1 DESCRIPTION OF THE MODEL

The integral equations used in computing the impedance functions for vertical vibrations are

$$\phi_v(t) + \int_0^1 K(t,t')\phi_v(t')dt' = 1 \quad (0 \leq t \leq 1) \tag{B1}$$

where

$$K(t,t') = L_1(|t-t'|) + L_1(t+t') \tag{B2}$$

$$L_1(t) = \frac{a_o}{\pi}\int_0^\infty \left[\frac{K\Delta_{22}}{(1-\sigma_1)\Delta_R k_1^2} + 1\right] \cos(a_o K t)\,dK \tag{B3}$$

$$a_o = \omega a/\beta_1; \quad \beta_1^2 = G_1/\rho_1; \quad k_j^2 = G_1\rho_j/G_j^*\rho_1;$$

$$G_j^* = G_j(1+i\omega G_j'/G_j) \quad \text{or} \quad G_j^* = G_j(1+2i\xi_j) \tag{B4}$$

The expressions for Δ_{22} and Δ_R are given in Ref. 12.

The impedance function (I) is given by:

$$I(a_o) = \frac{4G_1 a}{1-\sigma_1}\,[K_{vv}(a_o) + ia_o C_{vv}(a_o)] \tag{B5}$$

where

$$K_{vv}(a_o) = \int_0^1 \mathrm{Re}[\phi_v(t)/k_1^2]dt \tag{B6}$$

$$C_{vv}(a_o) = \frac{1}{a_o}\int_o^1 \mathrm{Im}[\phi_v(t)/k_1^2]dt \tag{B7}$$

For rocking and horizontal vibrations, the equations of motion are of similar type and are given in Ref. 12. Eq. B3 is integrated by using Filon's method and Eqs. B6 and B7 are evaluated using the sixteen-point Gauss quadrature formula.

To study the numerical convergence of the impedance functions using the formulation given above, two example problems are considered. For the first example, the impedance functions are computed at an interval of 0.6136 rad/sec for a frequency range of 0 to 207.4 rad/sec. The radius of the rigid foundation is 150 ft. The geometric and material properties of the two layered half spaces are as follows:

ρ_1 = .003727 kip-sec^2/ft h_1 = 70 ft G_1 = 2000 kip/ft^2

ρ_2 = .004658 kip-sec^2/ft $h_2 = \infty$ G_2 = 140916 kip/ft^2

$\sigma_1 = 0.38$ $\xi_1 = 0.1$

$\sigma_2 = 0.26$ $\xi_2 = 0.1$

To study the convergence, the impedance functions are computed using fine, medium, and coarse integration steps for the infinite integrals using Filon's methods. The sixteen-point Gauss quadrature formula is used to integrate the finite integrals. Integration steps of 0.25 for 0-5 rad/sec and 0.025 for 5-207.4 rad/sec frequency ranges were used in the fine scheme to compute the impedance functions. The corresponding integration steps for the medium scheme were 0.5 and 0.05, and for the coarse scheme 1.0 and 0.1, respectively. The three sets of impedance functions obtained are shown in Figures B2 and B3. The solid line on these figures represents the results based on the fine integration step, the dotted line the results based on the medium integration step, and the dashed line the results based on the coarse integration step. It can be observed that for this example convergence was achieved with the fine integration scheme.

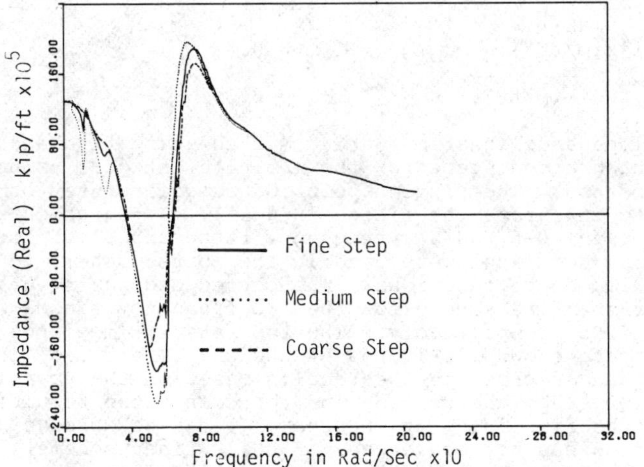

FIGURE B2 REAL PART OF VERTICAL IMPEDANCE FUNCTION

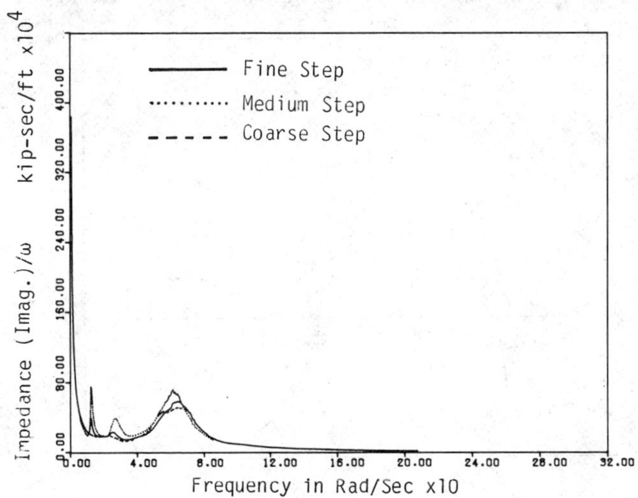

FIGURE B3 IMAGINARY PART OF VERTICAL IMPEDANCE FUNCTION

For the second example, the impedance functions are computed at an interval of 0.6136 rad/sec for a frequency range of 0-207.4 rad/sec. The radius of the rigid foundation is 57 ft. The geometric and material properties of the two-layered half space are as follows:

$\rho_1 = 0.00382$ kip-sec^2/ft^4; $\xi_1 = 0.03$

$\rho_2 = 0.00481$ kip-sec^2/ft^4; $\xi_2 = 0.00$

$G_1 = 1851$ kip/ft^2; $\sigma_1 = 0.42$

$G_2 = 450{,}000$ kip/ft^2; $\sigma_2 = 0.39$

The impedance functions are computed using fine and coarse integration steps for the infinite integral using Filon's method. The sixteen-point Gauss quadrature formula is used to integrate the finite integral. Integration steps of 0.6 for the 0-5 rad/sec frequency range and 0.125 for the 5-207.4 rad/sec range were used in the coarse scheme to compute the impedance functions. The corresponding integration steps for the fine scheme for the two frequency ranges were 0.50 and 0.05, respectively. The two sets of impedance functions are shown in Figures B4 and B5. The solid line on these Figures represents the results based on the coarse integration step; the dotted line represents the results based on the fine integration step. It can be observed that the solution has not converged. A still finer scheme with an integration step of 0.025 for the entire frequency range was tried to achieve convergence; however, no solution beyond 30 Hz could be obtained, due to numerical instability.

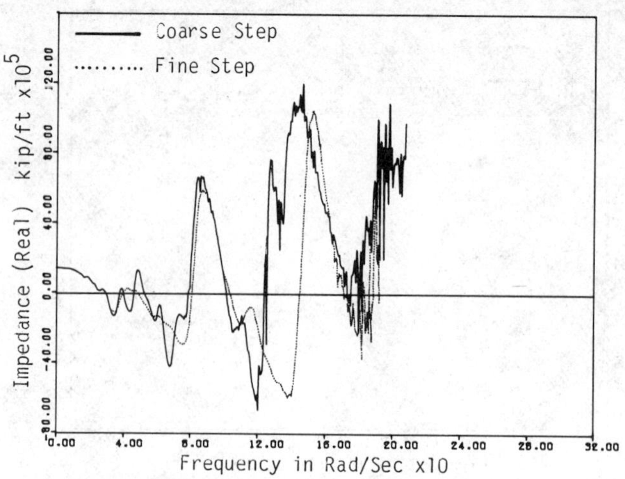

FIGURE B4 REAL PART OF VERTICAL IMPEDANCE FUNCTION

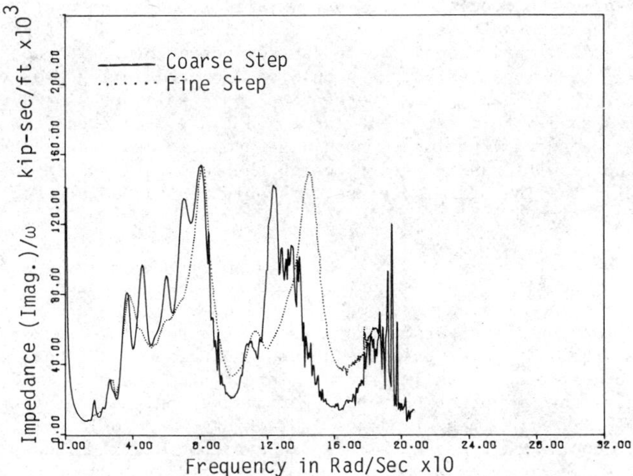

FIGURE B5 IMAGINARY PART OF VERTICAL IMPEDANCE FUNCTION

ISSUES RELATED TO SOIL-STRUCTURE DYNAMIC ANALYSIS

A. H. Hadjian[1], M. ASCE

ABSTRACT

Although the capability to the mathematical solution of the soil-structure interaction problem has been developed, there remain several related issues that could impact on the validity of the computational results. These issues are the subject of this paper. The proper definition of ground motion needs to be further researched. The usefulness of the two-dimensional solutions should be reviewed in light of recent developments. The importance of the non-linear soil behavior during earthquakes should be first established and then, if needed, the proper soil model chosen. Experiments should be properly devised and executed so that the results can be used to check the computational capabilities. And finally, the indiscriminate use of sophisticated computer codes should be cautioned against.

INTRODUCTION

Even though the significant amount of research that has been ongoing for more than a decade has answered most of the questions regarding the impact of the important parameters of soil-structure interaction, it seems that the very same research in turn has raised additional questions. Unquestionably, soil-structure interaction has been studied more intensively than any other branch of earthquake engineering; yet definitive verification of these findings is lacking which could well be the reason for continued research. This paper not only will discuss generally recognized concerns but would also attempt to recommend actions to resolve some of these issues.

GROUND MOTION

It is becoming increasingly important that certain aspects of ground motion be discussed in relation to the soil-structure interaction phenomenon. At the present, for purposes of soil-structure interaction analysis, it is generally assumed that ground motion at a site results from vertically propagating waves. It is felt that this is justified both by seismological considerations and by the extreme simplification this assumption brings to the analysis. However, very little hard data is available at this time to substantiate the assumptions made. In order to resolve this question, data is needed from dense arrays of strong motion instruments. These instruments would need to be deployed at depth as well as on ground surface. At present, some simple arrays are installed both in Japan and California; however, no dense three-dimensional arrays are in operation anywhere. Plans for the deployment

---

[1] Principal Engineer, Bechtel Power Corporation, Los Angeles Power Division, Norwalk, California 90650.

of such arrays were recently formulated by a workshop convened by the International Association for Earthquake Engineering (6). Implementation of these plans should not be delayed. Occasionally, horizontal propagation of waves underneath a foundation is also considered. However, the most general case of oblique incidence is not yet commonly used even though some analytical solutions are now available (7). These solutions resolve only part of the spatial variation of ground motion, namely the phasing of the input motion across the foundation. The question of the amplitude variation is not yet incorporated in these solutions. It is generally accepted that ground motion, even at close distances, of the order of foundation dimensions of structures, exhibits differences in amplitude and frequency content.

One of the significant factors of non-vertically propagating waves is that they give rise to both rocking and twisting input motions to the foundation. Simultaneously, however, there is a reduction in the translational acceleration of the foundation due to finite size effects of the foundation. Present research is directed mainly at the solution of the response problem given the characteristics of the incident waves. The important problem of defining the constituent parts of ground motion, however (i.e., type of wave, angle of incidence, phasing and amplitude) is not being pursued vigorously. It appears that seismology has not yet made its contribution to this important problem, either because investigators are unaware of the problem or it is beyond their immediate interests. Obviously, the overriding question is whether these variables have an important impact on the final response results. Idealized assumptions indicate that these may be significant; however, a more random environment, characteristic of real earthquakes, may not result in similar responses. Nevertheless, the idealized solutions are necessary for our understanding of the underlying phenomena, and they should be continued. Sweeping conclusions from these studies, however, must be made with caution.

Implicit in the above discussion is the question of ground motion variation with depth. A knowledge of ground motion variation with depth is required for the analysis of embedded structures. Obviously, the resultant ground motion recorded by an accelerograph is the aggregate of incoming waves at all possible orientations. Therefore, the expectation that the recorded motion can be deconvolved in a one-dimensional model to provide different ground motions at any specified depth should be viewed only as a first approximation. Both analytical and experimental (blast loadings) work can be carried out to verify if ground motion at depth can be obtained from motion recorded on the ground surface. Since significant reductions of motion with depth are possible, this issue should be addressed on a priority basis.

Another important issue with regard to ground motion is whether averaging of ground motions from earthquakes of different types, magnitudes and epicentral distances is appropriate in establishing design spectra. This concern arises because of (or leads to) the associated problem of site specific motion. Research in determing ground motion at a given site due to a specific fault rupture on a known fault has been ongoing for the last decade and attempts have been made to generate site specific time-histories for actual plant locations. The usefulness of this approach at the present time, in a quantitative sense, is difficult to assess. It is too early in the understanding of the source mechanism and wave propagation processes to justify quantitative results. Qualitatively, though, a series of parametric solutions might be used to help make the engineering decisions to modify standard spectra. The usefulness of site specific motion will depend on how greatly it would affect the ultimate design of structures and equipment. Given the relatively rigid structures of nuclear facilities, it does not seem that this would have a significant impact on the end product.

THREE DIMENSIONALITY OF THE PROBLEM

Cost and possibly accuracy considerations tend to restrict the use of finte-elements to solve soil-structure interaction problems to two-dimensions. The inevitable question regarding the relative merits of two-dimensional finite-element and three-dimensional impedance analyses is still with us. Despite the fact that apparently conflicting results are still being generated by the use of both methods, it is generally agreed that with proper modeling considerations the two solution techniques should provide similar results (4). Nevertheless, the inappropriateness of using 2D solution techniques for building-building interaction and torsional response calculations is obvious. Not so obvious is the issue of damping. Depending on the criteria by which a two-dimensional model is generated for an intrinsically three-dimensional problem, the damping can be over or under-estimated leading to different recommendations by different investigators. The two-dimensional approximation is usually and unnecessarily compounded by ad hoc attempts to revive the three dimensionality of the problem. Fig. 1 from Ref. 4 shows that it is better to leave the results of the two-dimensional solution as is rather than to attempt to resurrect the lost information. In this figure, Curve 4 is the result of an attempt to recover the three dimensional response as given by Curve 1. To resolve all future arguments regarding this basic issue of the dimensions of the problem, the need for truly three-dimensional benchmark analyses still exists. These benchmarks should be devoid of other issues such as non-linear soil behavior, embedment effects and non-vertical incidence of ground motion.

NON-LINEAR SOIL BEHAVIOR

Soil is a non-linear material. Under dynamic loads, soil behavior is characterized by densification/dilatancy, strain hardening/softening, hysteresis and creep. There are two related aspects to the non-linear behavior of soil: constitutive laws and parameters of the model. Early work in soil plasticity was one dimensional and several models have been suggested and used (Fig. 2). Recently, the emphasis has shifted to multi-dimensional state of stress and strain by modifying the basic concepts developed for the study of the plasticity of metals. The resulting "cap" models have been used for the prediction of blast effects and are beginning to be used for soil-structure interaction analysis (10). The cap model is a continuum material model which is based on the classical incremental theory of plasticity (8). More recently, the endochronic theory of plasticity has been advanced that does not rely on the concept of a yield surface as the cap models do (9). No alternative model for non-linear soil behavior is the particulate mechanics approach. Unlike metals, soil is made up of discrete particles having varying shapes, sizes and orientations. Forces are transmitted through particle contact and pore pressure. The random assembly of these irregular particles calls for stochastic methods (5).

The more difficult problem in the study of the non-linear behavior of soils is the definition of the parameters of the models. Laboratory tests are beset with difficulties due to sample disturbance and boundary conditions. Even the applicability of the one-dimensional model parameters to the earthquake environment are being questioned as a result of the study of the response of earth dams to real earthquakes (1).

Given the above discussion, it seems premature to expect definitive results of non-linear soil behavior during earthquakes. The problem is further compounded in soil-structure interaction studies because no distinction is made at the present between the strains due to soil-structure interaction and strains due to the passage of seismic waves. As depicted in Fig. 3, if strain rate is a significant parameter of soil response, then the results from earthquakes are expected to be different not only from free vibration tests but also from blasts. Experiments have shown that rocks continue to behave linearly for strains up to 3% when loaded at strain rates comparable to those from explosives (2), and, therefore, it is expected that soils too should be loading rate sensitive (11). There is evidence to indicate that apparent wave speeds due to earthquakes are generally determined by the geologic structure at depth which could, therefore, be an order of magnitude faster than those due to blasts at the surface. Thus, it becomes necessary to re-evaluate the whole question of significant non-linear response of the overburden during earthquakes. This contention becomes plausible in light of the fact that after earthquakes structures founded on good material seem to remain plumb irrespective of whether soil-structure interaction effects were significant. There is a lack of field evidence that the passage of seismic waves during earthquakes plastically deforms even pre-loaded soil strata.

On the one hand, sophisticated soil models are being developed, and on the other hand, the non-linear behavior of the overburden (as against the non-linear response of an earth structure, such as a dam) during earthquakes is being questioned. However, at the present time, equivalent linear models are generally used for soil-structure interaction analyses. These models are not capable of reproducing correct non-linear behavior since they lack "memory." Hence, residual or permanent displacements cannot be obtained from these analyses. Studies using updated equivalent linearization techniques show that soil-structure interaction strains may have a relatively small effect on the overall response of the structure. However, since these linearized models do not predict local stress concentrations and progressive failure of the foundation materials, these findings may be inconclusive. Geometric non-linearities, such as uplift and side soil separation, are not routinely incorporated in the presently used analysis techniques. However, limited research in this area indicates that, for nuclear plant structures, these non-linearities may not significantly affect the overall results obtained from linearized analyses that explicitly consider these effects in an approximate fashion.

EXPERIMENTAL EVIDENCE

Obviously the best experimental data would be obtained during earthquakes. However, studies based on response records obtained during earthquakes are often incomplete since either "free field" motions are lacking or serious assumptions have to be made regarding those that exist. Motions recorded in basements of structures may not be adequate as they already include soil-structure interaction effects. On the other hand, forced vibration tests can be conducted under controlled conditions and all required data adequately recorded. Although many buildings are regularly vibration tested, soil-structure interaction effects are not given adequate consideration, usually because these effects are small for the tested structures. It is only rarely that vibration testing is carried out for structures where interaction effects are significant. One major drawback of vibration testing with respect to the study of soil-structure interaction effects is the fact that ground motion effects are absent. Thus, the results must be extrapolated with caution to the earthquake environment.

Blast loading has been used as a substitute for actual earthquakes. Blast loading may be either of the conventional chemical type or an underground nuclear explosion. For the latter case, the prototype must obviously be constructed at the test site. Even for prototype structures, conventional blast loading as a means of studying non-linear soil effects should be carefully evaluated as the rate of loading may be significantly less than during earthquakes. During blasting the stiffness characteristics of the surface materials will generally determine the wave passage velocities. During earthquakes apparent wave passage velocities are generally determined by the geologic structure at depth irrespective of the stiffness properties of the surface materials (3). Apparent wave velocities could be about an order of magnitude larger during earthquakes compared to those generated by surface blasts. It is not surprising then that after earthquakes structures founded on good material seem to remain plumb, and models subjected to blast loading from explosives tend to deflect permanently. Extrapolation of the response of models subjected to extremely large strains at relatively slow rates to the real earthquake situation should be made with extreme caution. A related problem is the use of wave velocities in the design of buried long structures mainly because, in the simplified design approach for such structures, stresses due to wave passage are inversely proportional to the velocity of wave propagation.

An important concern in experimental studies is the generation of test data from scaled models. Model testing may not properly consider similitude (buildings and soils cannot be scaled properly in the field) and therefore extrapolation should be made very carefully. In a recent blast-excited series of scaled tests (10), the soil experienced very high ground motions exceeding 8 g's in an attempt at "scaling" the expected accelerations in the prototype. Obviously, the soil did not know about the intent of the experimenters and behaved as it would have under very large accelerations. Since the basic intent of the study was soil-structure interaction response, the severe deformations imposed on the soil by the extremely high g's should invalidate any conclusions regarding the phenomenon of soil-structure interaction. The results of the experiment, though, can be used as a severe test of the capability of non-linear computer codes. Furthermore, the scaled models, erected on non-scaled soils, were de facto ascribed extremely small impedances, in particular for the rocking component of motion (stiffness characteristics vary as R^3 and radiation damping as R^4). Thus, the models responded in severe rocking, much unlike a real structure would respond in a real earthquake environment. In fact, severe uplift and complete side separation of the models occurred during the tests. Obviously, under these adverse circumstances any linear model would be completely inadequate to predict the response of the scaled models. Therefore, to conclude that linear models are totally inappropriate to conduct soil-structure interaction studies is inappropriate since adequate similitude was not achieved in these tests.

Recent advances in centrifuge testing where important similitude effects can be taken into account should be used more often. For example, gravity effects require that the gravitational field be increased linearly with the decrease of the model characteristic length. A 1/100 scale model, therefore requires a 100g gravitational field if gravity effects are to be correctly modeled. On the other hand, the small scale models required for centrifuge testing would have difficulties with boundary conditions. A useful alternative to testing would be the acquisition of data from existing nuclear facilities. This requires a concerted industry-wide effort to augment the instrumentation at existing nuclear facilities in high seismic areas to directly obtain existaction data should a major earthquake occur. Even data from microtremors can be used to validate linear models. These actions should lead to increased public confidence in the seismic design of nuclear power plant facilities.

USE OF COMPUTER CODES

The general availability of advanced computational capabilities to the average user has, overall, hindered the proper solution of the soil-structure interaction problem. It is not unusual that the average user of sophisticated computer codes is not aware of all the explicit and implicit assumptions on which the codes are based and the conditions to which they apply. Early in the development of the soil-structure interaction solution capabilities, conflicting data were generated by the use of different codes presumably solving the same problem. The reason lies in the rapidity by which these codes were being generated. Similar difficulties in computer structural analysis were avoided simply because the structural codes were initially used to solve simple beam problems. The road to sophisticated finite-element capabilities was orderly and at each step of the way known classical problems were solved to check the adequacy of the codes.

Tremendous technological advances, both in software and hardware, are being made. Today we have the ability to perform tasks in hours which just a few years ago may have taken weeks. These capabilities have created a false sense that one is solving the ultimate problem simply because the tool being used is sophisticated. Good engineering is being endangered by a cadre of technicians that are more interested in the arithmetic accuracy of the solution than in the solution of the engineering problem.

Computer codes are rigid. They do what the original developer thought must be done. And, therefore, the danger always exists that the problem will be fitted to the available capability. The question is not whether we can compute or not, but rather whether we know what to compute. For example, it is common to solve the same problem with two different codes. There are obviously two possible outcomes: the results are similar or the results are not similar. If the results do not match, which code is better? If the results do match, does it mean that the correct physics is captured? Proper experimental evidence becomes mandatory to settle these issues.

CONCLUSIONS

Although the capability to the mathematical solution of the soil-structure interaction problem has been developed, there remain several related issues that could impact on the validity of the computational results. The review of these related issues indicates that the need for experimental evidence is great and that these experiments must be properly designed, executed and evaluated.

REFERENCES

1. Abdel-Ghaffar, A. M. and Scott, R. F., (1979) "Shear Moduli and Damping Factors of Earth Dams," Journal of the Geotechnical Engineering Division, ASCE, Vol. 105, No. GT12, Proc. Paper 15034, December, pp. 1406-1426.

2. Grady, D. E., (1977) "Stress Wave Propagation in Rock," Sandia Laboratories, Albuquerque, New Mexico.

3. Hadjian, A. H. and Hadley, D. M., (1981) "Studies of Apparent Seismic Wave Velocity," Proc. International Conference on Recent Advances in Geotechnical Earthquake Engineering and Soil Dynamics, St. Louis, April 26-May 2.

4. Hadjian, A. H., Luco, J. E., Wong, H. L., (1981) "On the Reduction of the 3D Soil-Structure Interaction Problems to 2D Models," Proc. 2nd ASCE-EMD Specialty Conference on Dynamic Response of Structures, Atlanta, Georgia, January 15-16.

5. Hart, M. E., (1977) Mechanics of Particulate Media - A Probabilistic Approach, McGray-Hill, New York.

6. Iwan, W. D., (1979) "The Deployment of Strong Motion Earthquake Instrument Arrays," International Journal of Earthquake Engineering and Structural Dynamics, Vol. 7, pp. 413-426.

7. Luco, J. E. and Wong, H. L., (1979) "Response of Structures to Non-Vertically Incident Seismic Waves," Department of Applied Mechanics and Engineering Science, University of California, San Diego, La Jolla, California, October.

8. Sandler, I. S., DiMaggio, F. L. and Baladi, G. Y., (1976) "Generalized Cap Model for Geological Material," Journal of the Geotechnical Engineering Division, ASCE, Vol. 102, No. GT7, July.

9. Systems, Science and Software, (1980) "New Endochronic Plasticity Model for Soils," EPRI Report NP-1388, April.

10. Weidlinger Associates, (1978) "Non-Linear Soil-Structure Interaction," EPRI Report NP-945, December.

11. Whitman, R. V., (1970) "The Response of Soils to Dynamic Loading," Contract No. DA-22-079-eng-224, U. S. Army Engineers Waterways Experiment Station, Vicksburg, Mississippi, May.

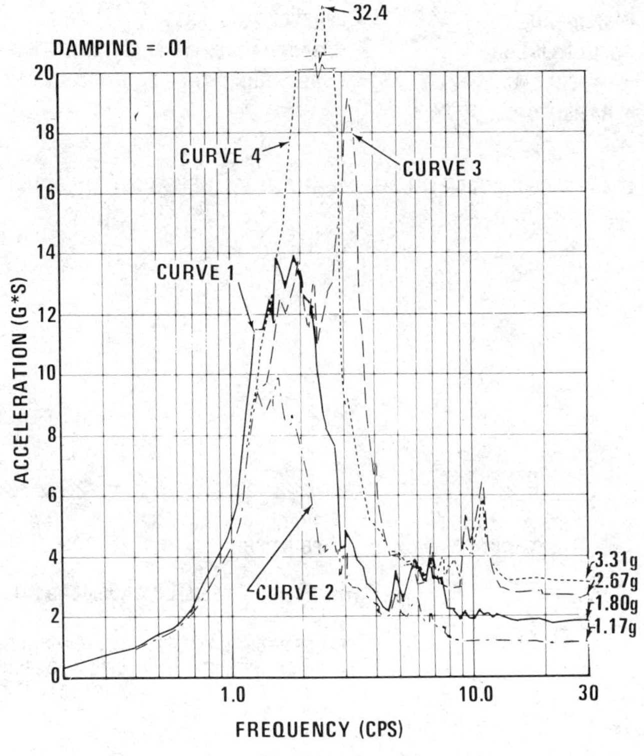

Figure 1. EXAMPLE OF A COMPARISON OF ACCELERATION FLOOR SPECTRA AT TOP OF STRUCTURE (REF. 4)

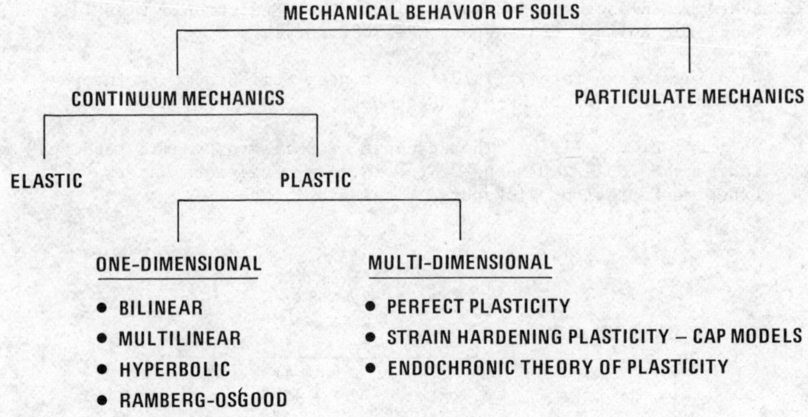

Figure 2. MODELS FOR THE MECHANICAL BEHAVIOR OF SOILS

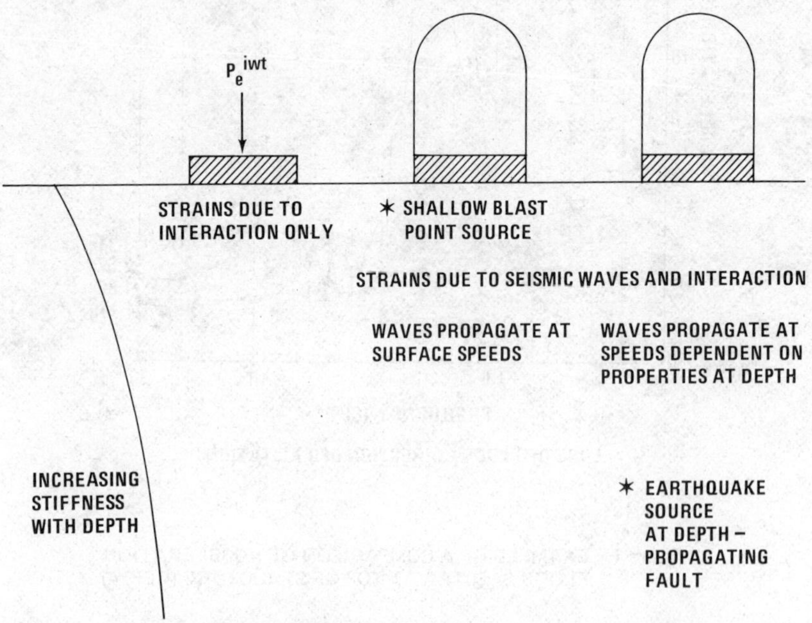

Figure 3. DISTINCTION OF STRAINS DUE TO THREE TYPES OF EXCITATION

INTERACTION EFFECTS OF REACTOR SYSTEM AND ITS SUPPORTS

By

C. -W. Lin,[1] M. ASCE
D. T. Tang,[2] A.M. ASCE
F. Loceff,[3] M. ASCE

ABSTRACT

Simple equations have been developed to evaluate the interaction effects of a coupled system such as the reactor coolant system supported by its concrete structures. The development of the equations are based on solutions from a coupled two-degree-of-freedom system model and results of a statistical study of past plants.

INTRODUCTION

Significant interaction effects in the dynamic response of a typical reactor coolant system have been demonstrated by analysis to exist between the system and its supporting concrete structures when subjected to earthquake type of motions. However, the significance of these interaction effects remains to be substantiated by actual response motions observed during an earthquake. Since dynamic tests conducted on a prototype design are all at extremely low amplitudes, these results usually did not show meaningful interaction effects and their extrapolation to realistic interaction levels is uncertain due to the nonlinear system support features. However, because the system is supported by heavy concrete structures, the system response is not only a function of the system damping, but also significantly affected by the interaction effect between the system and its supporting concrete structures. To determine the interaction effect at a realistic vibration magnitude, a scaled model testing program is utilized. Because model distortion is unavoidable in terms of mass and stiffness similitude requirements, the determination of the extent of the interaction effect is therefore important in the test model design and in the subsequent correlation analysis.

---

1. Advisory Engineer, Structural Technology, Westinghouse Electric Corporation.

2. Senior Engineer, Structural Technology, Westinghouse Electric Corporation.

3. Manager, Piping and Structural Site Engineering, Westinghouse Electric Corporation.

Presented in this paper is a method to numerically identifying the interaction effect for a coupled system such as the reactor coolant system and its supporting concrete structures.

INTERACTION EFFECT

As a simple representation of a coupled system model, Figure 1 shows a two-degree-of-freedom system represented by a primary system with mass m_1 and stiffness k_1, and a secondary system with mass m_2 and stiffness k_2.

When the coupled system is subjected to a random wide band input, the mean square acceleration for m_2 has been shown in Reference 1 as the following:

$$E\,[\ddot{X}_2^{\,2}] = \frac{2\pi\omega_1^{\,2}\,\omega_2^{\,2}\,S_o}{D}\,\{\xi_1\omega_1(\omega_1^{\,2} + R_m\omega_2^{\,2}) + \xi_2\omega_2\,(R_m\omega_1^{\,2} + \mu_e^{\,2}\,\omega_2^{\,2})$$

$$+ 4\,[\xi_1^{\,3}\,\omega_1\omega_2^{\,2} + (\xi_1^{\,2}\,\xi_2\omega_2 + \xi_1\xi_2^{\,2}\,\omega_1)\,\lambda + \xi_2^{\,3}\,\mu_e\omega_1^{\,3}\,\omega_2]$$

$$+ 16\,\xi_1^{\,2}\,\xi_2^{\,2}\,\omega_1\omega_2\,(\xi_1\omega_2 + \xi_2\omega_1)\} \tag{1}$$

where,

$$D = 4\omega_1\omega_2\,\{R_m\omega_1\omega_2\,(\xi_1\omega_2 + \xi_2\omega_1)^2 + \xi_1\xi_2\,(\omega_1^{\,2} - \mu_e\omega_2^{\,2})^2$$

$$+ 4\xi_1\xi_2\omega_1\omega_2\,[(\omega_1\omega_2\,(\xi_1^{\,2} + \mu_e\xi_2^{\,2}) + \xi_1\xi_2\,(\omega_1^{\,2} + \mu_e\omega_2^{\,2})]]\} \tag{2}$$

and

$$\mu_e = 1 + R_m \tag{3}$$

$$\lambda = \omega_1^{\,2} + \mu_e\omega_2^{\,2} \tag{4}$$

In equations (1) through (4), ξ_1 and ξ_2 are damping factors associated with the first mass and the second mass, respectively, and S_o is the spectrum density value and R_m is the mass ratio of m_2 over m_1.

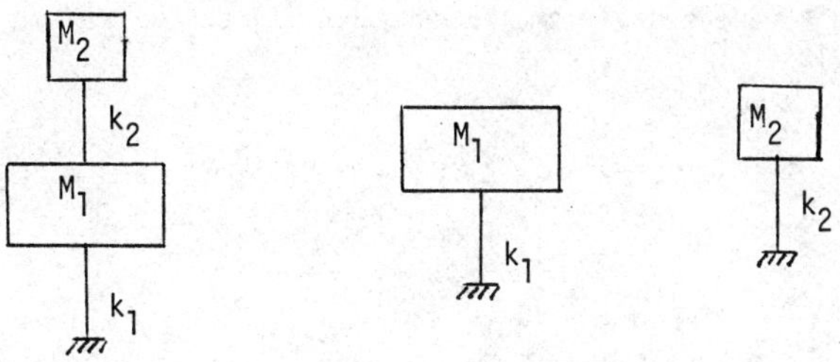

FIGURE 1: DYNAMIC SYSTEMS

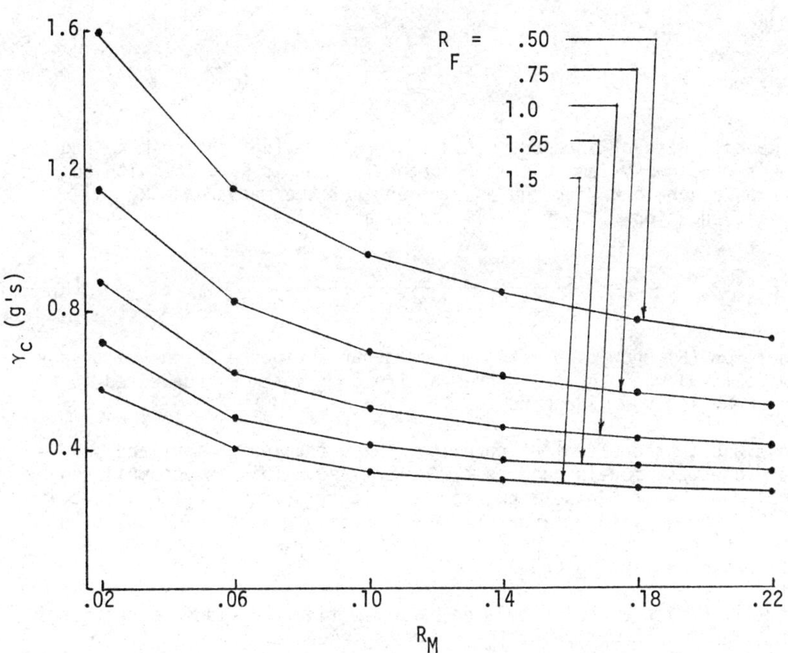

FIGURE 2: RESPONSE OF A SECONDARY SYSTEM - COUPLED MODEL

Since ξ_1 and ξ_2 are usually small, equation (1) can be simplified to yield:

$$E\,[\ddot{X}_2^{\,2}] = \frac{\pi\, S_o\, R_f}{D_1}\ \omega_1[\xi_1\,(1 + R_m R_f^{\,2}) + \xi_2 R_f\,(R_m + \mu_e^{\,2}\, R_f^{\,2})] \tag{5}$$

where

$$D_1 = 2\,\{R_m R_f\,(\xi_1 R_f + \xi_2)^2 + \xi_1\xi_2\,(1 - \mu_e R_f^{\,2})^2$$

$$+ 4\xi_1\xi_2\, R_f\,[R_f\,(\xi_1^{\,2} + \mu_e \xi_2^{\,2})$$

$$+ \xi_1\xi_2\,(1 + \mu_e R_f^{\,2})]\} \tag{6}$$

and

$$R_f = \omega_2/\omega_1 \tag{7}$$

A special case of interest would be to assume that the primary and the secondary systems are in resonance. That is $R_f = 1$. With some manipulation, and with the assumption that the mass ratio R_m is small, equation (5) may be simplified as follows:

$$E\,[\ddot{X}_2^{\,2}] = \frac{\pi\omega_1\, S_o}{2}\ \frac{1 + 3\,R_m}{(R_m + 4\xi_1\,\xi_2)\,(\xi_1 + \xi_2)} \tag{8}$$

Equation (8) shows the maximum mean square response of the secondary system (m_2) when the two-degree-of-freedom system is subjected to a white noise type wide band input.

To evaluate the effect of coupling, the uncoupled primary and secondary one mass models need to the determined. This is accomplished by applying a white noise input to the primary system and the output at m_1 is then applied as input to the secondary system. The solution for the mean square response at m_2 is found the same as Equation (1) while letting R_m equal to 0.

Simplification can also be made by neglecting the higher order terms of ξ_1 and ξ_2 to arrive at the following equation for $R_f = 1$ (i.e., the primary system and the secondary system are in resonance):

$$E\,[\ddot{X}_2^{\,2}] = \frac{\pi\omega_1\, S_o}{8}\ \frac{1}{(\xi_1 + \xi_2)\,\xi_1\xi_2} \tag{9}$$

On the root mean square basis, the effect of coupling can then be evaluated by dividing the square root of Equation (8) with the square root of Equation (9). This results in

$$\frac{(E\,[\ddot{X}_2{}^2])^{1/2}}{(E\,[\ddot{X}_2{}^2])^{1/2}} = 2\left[\frac{(1 + 3\,R_m)\,\xi_1\xi_2}{R_m + 4\xi_1\xi_2}\right]^{1/2} \qquad (10)$$

Equation (10) shows that when $R_m = 0$, the ratio is 1.0, as it should.

COUPLED SYSTEM RESPONSE

The influences of Equation (10) is beyond doubt. However, it would be even more helpful if a simple equation can be found to allow the computation of the peak response values at the secondary mass (m_2) when interaction is considered. To this end, use may be made of the statistical data compiled for the past plants (Reference 2).

In Reference 2, it was found, after a statistical study of all Westinghouse PWR plants with data available at the time of study, that the most probable amplification factor for concrete structure can be represented by

$$A_B = \left(\frac{1}{2\xi_1}\right)^{1/2} \qquad (11)$$

where A_B is the building amplification factor. Equation (11) is the same as the root mean square response of a single-degree-of-freedom system subjected to a wide band white noise input.

In the same study, the probable maximum amplification factor for a single degree of freedom system supported by a concrete structure is found to be

$$A_E = \left(\frac{1}{2\xi_2}\right)^{2/3} \qquad (12)$$

The two thirds power shows in Equation (12) indicating that the equipment response value lies in between the amplification factor for a single-degree-of-freedom system with random input and with sinusoidal input. The equipment response (r) is then computed by multiplying the equipment amplification factor A_E (Equation 12) with the building amplification factor A_B (Equation 11) and the peak design ground acceleration G. That is,

$$r = \left(\frac{1}{2\xi_1}\right)^{1/2}\left(\frac{1}{2\xi_2}\right)^{2/3} G \qquad (13)$$

Equation (13) predicts the peak equipment response on an uncoupled basis. That is, no interaction effect has been considered between the equipment and its supporting concrete structure.

To include the interaction effect into account, one may combine Equation (10) with Equation (5) to result in

$$r_c = \frac{G}{(2\xi_2)^{1/6}} \left[\frac{1 + 3\ R_m}{R_m + 4\xi_1\xi_2} \right]^{1/2} \tag{14}$$

Equation (14) reduces to Equation (13) when $R_m = 0$.

Equation (14) can be used to predict the maximum acceleration response of a secondary system such as the reactor coolant system supported by a concrete structure with dynamic interaction effects included.

It should be noted here that Equation (14) is developed based on the assumption that the secondary system is in resonance with the primary system. In the event that the more accurate information is required where natural frequencies of both secondary and primary systems are known, the necessary response equation can be developed by first taking square root of Equation (5) and then divide it by the same term but with $R_m = 0$. The resulting expression is then multiplied by r of Equation (13) which results in the following, when damping is assumed to be small,

$$r_c = \frac{G}{(2\xi_2)^{1/6}} [\xi_1\ (1 + 3\ R_m R_f^{\ 2}) + \xi_2 R_f^{\ 3}\ (1 + 3\ R_m)$$

$$+ \xi_2 R_f R_m\ (1 - R_f^{\ 2})]^{1/2} \times (\xi_1 + \xi_2)/\{(\xi_1 + \xi_2 R_f^{\ 3})^{1/2}$$

$$\times [R_m R_f\ (\xi_1 R_f + \xi_2)^2 + 4\xi_1\xi_2 R_f^{\ 2}\ (\xi_1 + \xi_2)^2$$

$$+ 4\xi_1^{\ 2}\xi_2^{\ 2}\ (1 - R_f^{\ 2})]^{1/2}\ \} \tag{15}$$

Equation (15) reduces to Equation (14) if it is assumed that the primary system and the secondary system are in resonance, or $R_f = 1$.

NUMERICAL EXAMPLE

As a numerical example to illustrate the application of Equations (14) and (15), consider a secondary system with a typical equipment damping value of 4 percent and a primary system with a typical concrete damping value of 7 percent. Figure 2 shows the predicted peak equipment response for a 0.1g peak ground acceleration for various mass ratios. When $R_m = 0$, the system is essentially uncoupled. As can be seen from this figure, for a system with a moderate mass ratio larger than 0.1, significant reduction in the response acceleration can be expected from the coupled model. This reduction translates into more realistic model design considerations for the combined test and analysis program currently being developed. Application of the equations developed herein to other systems and components could provide important cost savings once their usefulness in predicting response reduction is verified by tests.

CONCLUDING REMARKS

Simple equations have been developed to assess the interaction effect of the reactor coolant system with its supporting concrete structures. The development of the equations is based on a simplified two-degrees-of-freedom model (one for the secondary mass and one for the primary mass) subjected to a wide band white noise input. The results of the two degrees of freedom system are then combined with two equations obtained from a statistical study of data from past Westinghouse pressurized water reactor plants. Numerical results provided show that large response reductions can be expected for the secondary system and the primary system with a moderate mass ratio and when the two systems are close in resonance.

REFERENCES

1. Crandall, S. H., Mark, W. D., "Random Vibrations in Mechanical Systems," Academic Press, 1963.

2. Lin, C. -W., Bohm, G. J., "Building and Equipment Seismic Response for PWR Plants," ASME paper 76-WA-PVP-14.

Computer Controlled Seismic Performance Testing

by Stephen A. Mahin,[1] A. M. ASCE and Mary E. Williams[2]

ABSTRACT

A key question in performing seismic performance tests in the laboratory is whether the loads or displacements imposed on the test specimen are representative of those that would be developed during severe earthquake ground shaking. Currently available test methods suffer from a variety of limitations. This paper examines a relatively new experimental method which may overcome some of these limitations. In this method a computer is used on-line to monitor and control a test specimen so that the *quasi-statically* imposed displacements closely resemble those that would be developed if the specimen were tested dynamically. Experimental information obtained regarding the nonlinear restoring force characteristics of the test specimen at a particular time during the test is used by the computer, along with numerically prescribed data regarding the system s damping and inertial characteristics, to determine the changes in deformed shape that should be imposed on the specimen for a numerically specified ground motion. Following a brief review of previous applications of on-line control methods, the theoretical basis for the method is examined. The method's assumptions and limitations are discussed quantatively and results of a numerical simulation are presented to quantatively assess their significance. The scope of an analytical and experimental research program being initiated at Berkeley on on-line computer control (pseudo-dynamic) methods is then presented.

INTRODUCTION

Buildings and other structures are generally designed to dissipate some of the energy input during severe earthquake ground motions through inelastic deformations. Experimental research remains the most efficient and reliable means for assessing the ability of various types of structural systems to develop these inelastic deformations and for devising structural details to improve the seismic performance of critical regions. Possibly the most realistic method for assessing the nonlinear dynamic response of a particular structural system is to place it on a shaking table and subject it to appropriately selected ground motion time histories. However, currently available shaking tables have severe limitations on the size, weight and strength of structures that they are able to test [1,2]. These limitations often necessitate testing structures at significantly reduced scales, raising the problem

---

1. Asst. Prof. of Civ. Engrg., Univ. of California, Berkeley, CA

2. Engr., Structural Software Development, Berkeley, CA: Formerly Grad. Student, Univ. of California, Berkeley, CA.

of dynamic similitude [1,2]. Such tests also require specialized and costly test equipment, data acquisition systems and personnel.

Forced vibration tests of full-scale buildings into the inelastic range have also been proposed and one such test has been successfully completed [3]. It is difficult with such methods, however, to realistically simulate the distribution and history of forces developed during seismic excitations. While these and other innovative dynamic test methods, such as pulse generators [4] and blast induced ground motions [5], show considerable promise, the high costs involved would likely limit their use to special structures or complex problems involving soil-foundation-building interaction.

The most economical and thus the most common method for obtaining information on inelastic structural behavior is through quasistatic tests in which prescribed histories of load or displacement are imposed on small structural systems or basic subassemblages of larger structural systems. Such tests utilize relatively simple equipment that is available in most structural laboratories and allow for detailed observation of specimens during tests. However, the inelastic behavior of most types of structural systems is very sensitive to the loading history. Thus, considerable care must be exercised in selecting loading histories that are intended to reflect seismic loading conditions.

One method for determining quasistatic loading histories is to perform inelastic dynamic computer analyses of the structure in question and the use the structural deformations computed at appropriate locations to control the experiments [6]. However, since existing analytical idealizations are generally much simpler than actual structural behavior, the computed loading history is not likely to be realistic. Thus, it is more common to assume *a priori* a highly idealized deformation history representative of the cyclic nature of seismic response. Such prescribed displacement histories can be particularly valuable in: assessing the effect of structural details on the inelastic behavior of structures by subjecting different specimens to identical deformation histories; and studying the basic mechanisms affecting the inelastic response of a particular structure by varying the amplitude, rate or pattern of the applied deformation histories. However, such displacement histories do not account for the effect on seismic response of the dynamic and mechanical characteristics of the structural system or the intensity, duration and frequency content of the anticipated seismic excitations. Consequently, it is not possible to directly relate the energy dissipation capacity of a specimen measured in this type of test with that required for seismic safety.

Another difficulty with quasistatic seismic testing is that it is not generally possible to achieve realistic distributions of internal forces in multiple degree-of-freedom systems by imposing a predefined pattern of displacements or forces (such as those corresponding to a first mode shape). Inelastic action can significantly alter the deformed shape of a structure during seismic response (e.g., the formation of a soft story) and "higher mode" effects can control inelastic deformational demands in critical regions of certain types of structures [7].

To facilitate formulation of more rational and reliable seismic-resistant design methods, it is desirable to develop more realistic methods for prescribing displacement or loading histories for quasi-static seismic tests. One potential method for doing this would be on-line computer-actuator control (pseudo-dynamic) methods.

Basic Approach.--The response of structural systems to severe earthquake ground motions can be expressed in terms of a family of nonlinear second-order differential equations. The main source of non-linearity in most cases is the nonlinear relationship between restoring force, R, and displacement, d. Numerical methods exist for solving these equations of motion when the restoring force characteristics can be mathematically idealized. In the on-line computer control method, feedback obtained regarding the nonlinear mechanical characteristics of the test specimen **at** a particular instant in time is used in a computer program, along with numerically prescribed data regarding the system's inertial and damping characteristics, to determine the changes in the deflected shape that should be imposed at that time for a numerically specified ground motion accelerogram. The principal difference between the proposed experimental method and well-established nonlinear dynamic analysis methods is that the restoring force characteristics of the structure are experimentally measured, rather than based on idealized mathematical models, and the computed displacements are used to control displacements in actuators attached to the test specimen, as indicated in Figure 1. Since the numerical procedures explicitly account for the dynamic aspects of the response, the tests are performed quasistatically permitting the use of conventional loading apparatus and instrumentation.

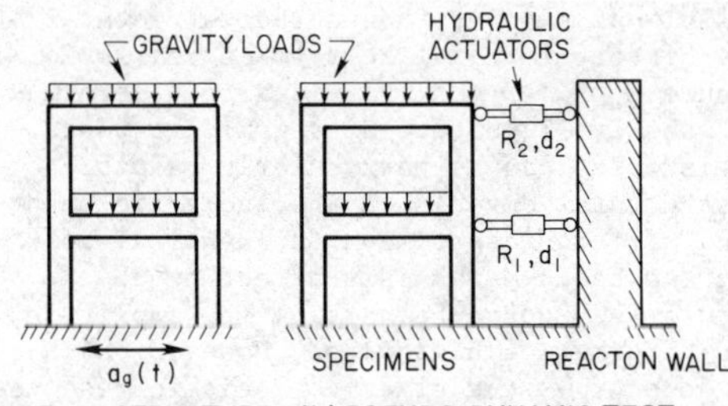

Fig. 1 Pseudo-dynamic Test Idealization

As envisioned the on-line control computer would be most appropriate for structures that would likely sustain large inelastic deformations during severe ground shaking so that uncertainties in idealizing inertial, and especially, viscous damping characteristics would not significantly affect the response. Also, in order to minimize the complexity of the loading apparatus, it is desirable that structures have mass concentrated at discrete points and that as few as possible degrees-of-freedom be needed to describe the overall response. Thus, an ideal structural system would be a multistory structure, with mass concentrated at floor levels, subjected to one or two horizontal components of ground motion (Fig. 1). Structures with significant distributed masses, constructed from materials that have properties that are highly sensitive to loading rate and in which viscous damping is likely to have a significant effect may not be appropriate for this method. Reduced scale specimens may be used but rules of similitude must be followed.

Objectives and Scope.--The main purpose of this paper is to review the general basis, capabilities and limitations of on-line computer control methods for seismic performance testing. The paper is divided into four parts. The first part presents a brief summary of applications of pseudo-dynamic test methods in the U.S. and Japan. The theoretical basis of the method is reviewed in the second part with emphasis on the applicability of various implicit and explicit numerical operators for solving the equations of motion. The third part of the paper qualitatively and quantitatively examines the assumptions and limitations of the method. Sources of error in the force monitoring systems, displacement control and measurement systems, and hydraulic systems are also discussed. Results of numerical simulations are presented for simple inelastic systems to evaluate the reliability of the method. The fourth part of the paper describes an analytical and experimental program being initiated to further assess the accuracy, reliability and practicability of the method.

PREVIOUS APPLICATION OF ON-LINE COMPUTER CONTROL METHODS

Some development work has been carried out in this field in the United States, notably at the University of Michigan, Ann Arbor [8] and the University of California, Berkeley [9]. However, considerable favorable experiences with the method have been gained by researchers in Japan at the Institute of Industrial Science of the University of Tokyo and the Building Research Institute (BRI) of the Ministry of Construction. Following initial experiments on one-degree of freedom reinforced concrete systems [10, 11], tests have also been performed on two-story reinforced concrete [12], steel [13, 14] and braced steel[15] frames. Tests of three-story reinforced concrete frames are planned by BRI before the end of 1980. In addition, work has been done in Japan on single story concrete and steel systems subjected to two horizontal components of ground motion [16, 17] and on equipment supported on floors in multistory buildings [18].

The feasibility of on-line computer control methods is also being evaluated as part of the U.S.-Japan Cooperative Earthquake Research Program [15]. It is proposed to test a full-scale six story reinforced concrete structure using the BRI Large Scale Testing Facility in Tsukuba, Japan. Verification tests will be performed in the U.S. as well as in Japan.

THEORETICAL FORMULATION

Dynamic equilibrium of a viscously damped, multiple degree-of-freedom system which exhibits nonlinear restoring force characteristics during severe earthquake ground motions requires at time $n\Delta t$ that

$$[M]\{a\}_n + [C]\{v\}_n + \{R\}_n = \{P\}_n \tag{1}$$

in which [M] is the effective inertial mass matrix of the system; [C] is its viscous damping matrix; $\{R\}_n$ is the restoring force vector; $\{a\}_n$ and $\{v\}_n$ are vectors of acceleration and velocity, respectively, relative to the ground; and $\{P\}_n$ is the effective load vector on the system. In the case of a ground excited planar structure like that shown in Fig. 1, $\{P\}_n$ equals $-[M]\{1\}a_{gn}$ where a_{gn} is the horizontal component ground accelerator in the plane of the frame. If the beams can be considered inextensible in this structure and mass can be considered lumped at the floor levels, there is only a single lateral degree-of-freedom per floor.

Equation 1 is satisfied only at discrete times. For convenience, time histories are subdivided into time steps of equal duration Δt so that the time at which Eq. 1 is to be solved is equal to Δt times the number of steps n since the beginning of the record. Thus, in the psuedo-dynamic method point n in Eq. 1 corresponds to time $n\Delta t$ in the actual dynamic response, and values of [M], [C] and a_{gn} are numerically specified but $\{R\}_n$ is measured experimentally at that point.

Solution of the equations of motion results in the vector of floor level displacements, $\{d\}_{n+1}$, that are to be imposed on the specimen using hydraulic actuators. This process is repeated for all subsequent time steps necessary to define the response of the test specimen to an entire ground motion time history.

In solving the equations of motion there are a number of considerations that must be taken into account. For example, it is apparent that iterative solutions are not possible in implementing the psuedo-dynamic method due to the path dependency of the restoring force of the test specimen in the inelastic range. Thus, numerical operators which permit calculation of the displacements at time step $(n+1)\Delta t$ based only on previously calculated or measured quantities are essential. In addition, dynamic equilibrium is generally satisfied only at a discrete time during a step. For convenience operators may be classified as being either implicit or explicit depending on when equilibrium is satisfied. Implicit operators are characterized as satisfying dynamic equilibrium at time $(n+1)\Delta t$ whereas explicit operators do so at time $n\Delta t$.

Implicit Integration Operators.--These types of operators are the most commonly used ones in dynamic response analyses and were the ones initially used in implementing on-line computer controlled tests [10]. The basic problems with this approach can be illustrated using the well known Newmark Beta Method [19]. In this method the displacement and velocity of a degree-of-freedom are assumed to vary according to the interpolation functions:

$$\{d\}_{n+1} = \{d\}_n + \Delta t\{v\}_n + \Delta t^2[(0.5-\beta)\{a\}_n + \beta\{a\}_{n+1}] \tag{2}$$

$$\{v\}_{n+1} = \{v\}_n + \Delta t[(1-\gamma)\{a\}_n + \gamma\{a\}_{n+1}] \tag{3}$$

in which β and γ determine how much of the acceleration at the end of the step enters into the expressions for velocity and displacement. To avoid introduction of numerical damping, γ is usually set to 1/2. The value of β can be interpreted in some cases to represent the variation in acceleration over the internal [19]. For example, β equals 1/6 corresponds to a linear variation, β equals 1/4 corresponds to a constant average acceleration, and β equals zero corresponds to impulses concentrated at the beginning and end of the step. By substituting

Eqs. 2 and 3 into Eq. 1 (for point n+1) it may be possible to solve for the state of the system at step n+1, if $\{a\}_n$, $\{v\}_n$ and $\{d\}_n$ are known. For $\beta>0$, there are not sufficient equations for the number of unknowns. However, a solution can be obtained by assuming that the restoring force is related to the displacement by the tangent stiffness during the step, $[K_t]_n$; i.e.,:

$$\{R\}_{n+1} = \{R\}_n + [K_t]_n (\{d\}_{n+1} - \{d\}_n) \tag{4}$$

In an experimental situation a number of problems arise in methods requiring knowledge of the tangent stiffness. First, $[K_t]_n$ would usually have to be extrapolated from data obtained during previous steps. However, in a highly nonlinear system substantial changes in tangent stiffness would be expected from one step to the next. Since interative methods can not be used to correct for any resulting equilibrium violations, end of step corrections would have to be made to the load vector [7]. Second, the tangent stiffness is approximated based on measured forces and displacements and may be sensitive to errors in measuring these quantities. Third, experimental determination of the tangent lateral stiffness for a complex system is difficult because an N degree-of-freedom system will generally have N(N+1)/2 stiffness coefficients to be determined. However, there are only N equations to solve them. Exact solution methods are possible for small systems (N<3), for structures that can be idealized as shear buildings and for other special structures [9]. Approximate least squares estimation procedures or system identification methods would be required for more complex cases, but the reliability of such methods has not been established.

Such problems can be partially avoiked by using β equal to zero. In this case, $[K_t]$ is not needed to solve Eqs. 1 through 3 for the state of the system at point n+1. A flow chart illustrating the application of this method to on-line computer control is presented in Fig. 2. Nonlinearities in the restoring force of the system are automatically taken into account, thereby avoiding the need for iterative solutions to satisfy equilibrium at the end of the step. However, displacements at point n+1 are not affected by changes in acceleration that occur during

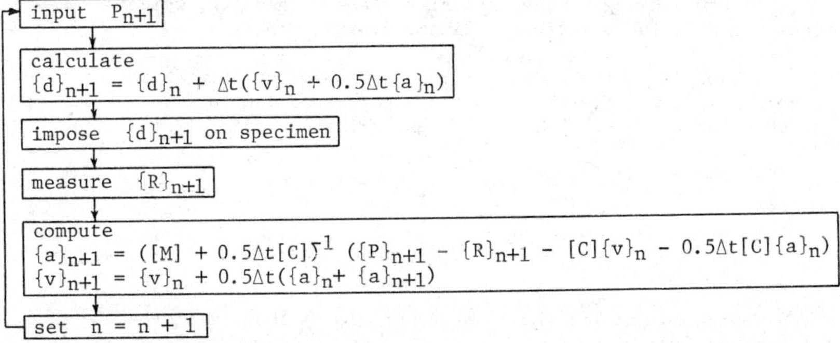

Fig. 2: Flow Chart for Newmark Beta Method; $\beta = 0$, and $\gamma = 1/2$

the previous step. Thus, the method is then explicit and will, as is the case with all numerical integration operators, only give approximate results. Problems regarding the accuracy and stability of this operator will be considered later.

When implementing on-line computer control methods, experimental results have indicated that stress relaxation may occur when a test is stopped between steps to allow for data acquisition and control computations. For small test specimens and high speed computers these interruptions may be nearly imperceptable and cause insignificant changes in the restoring force. However, it may be possible to make the tests continuous by modifying the Newmark method using a concept suggested by Wilson for cases where β equals 1/6. In the Wilson θ method [20] the response during a step is based on equilibrium at a fictitious point beyond the end of the step, i.e., at point $n+\theta$ where θ is greater than unity. The results obtained at the end of the step (point $n+1$), however, are still used as the initial conditions for the next step. Since Eqs. 1 and 2 uncouple when $\beta = 0$, the only change in the on-line procedure indicated in Fig. 2 relates to the initial displacement imposed. In this case the displacement imposed at point $n+\theta$ is computed by:

$$\{d\}_{n+\theta} = \{d\}_n + \theta(\Delta t)\{v\}_n + \theta^2(\Delta t)^2\{a\}_n \tag{5}$$

Because displacements vary linearly with time, the displacement at point n+1 is still given in Fig. 2 and the restoring force for use in the next step is measured when the displacements reach the values given in Fig. 2. The basic advantage of this formulation is that the displacements being imposed on the test specimen continue to increase during the interval between points $n+1$ and $n+\theta$. Computations for the next time step can be can be performed in this interval so that the specimen can always be kept in motion.

Explicit Numerical Operators.--A number of explicit operators exist, including various finite difference methods and the Newmark method with β equal to zero, which can be used to solve the equations of motion without iteration and without knowledge of the tangent stiffness matrix. Recent Japanese on-line computer controlled tests have been performed using the central difference bethod [21] which has been widely used in the solution of linear dynamic problems.

In explicit methods, equlibrium is imposed at point n rather than at point n+1 and difference equations corresponding to a quadratic variation of displacement between n-1 and n+1 are used:

$$\{v\}_n = (\{d\}_{n+1} - \{d\}_{n-1})/2\Delta t \tag{6}$$

$$\{a\}_n = (\{d\}_{n-1} - 2\{d\}_n + \{d\}_{n+1})/\Delta t^2 \tag{7}$$

these can be solved for the displacement to be imposed at the end of the time step:

$$\{d\}_{n+1} = ([M] + \frac{\Delta t}{2}[C])^{-1}(\Delta t^2(\{P\}_n - \{R\}_n) + (\frac{\Delta t}{2}[C] - [M])\{d\}_{n-1} + 2[M]\{d\}_n) \qquad (8)$$

This equation can be applied recursively for each time step. It should be noted that $\{d\}_{n+1}$ does not account for any changes that might occur in the acceleration of the real system between points n-1 and n+1.

A number of algebraically equivalent formulations are possible for Eq. 8 which may have certain numerical advantages [9]. Higher order difference methods have also been suggested for on-line control methods [8].

Numerical Stability of Integration Operators.--The numerical operator used to solve the equations of motion must not only be relatively accurate and efficient, it must also be stable and not likely to propagate errors. The explicit operators examined are relatively simple so that they should not require excessive computational effort.

An integration operator is considered unconditionally stable if the solution for any initial conditions does not grow without bounds for any time step size used. A method is conditionally stable if this requirement can be met so long as the ratio of time step size to the natural period of the system ($\Delta t/T$) is smaller than some limiting value called the stability limit. It can be shown that the basic Newmark method (with β equal to zero and γ equal to 1/2) and the central difference formulation are only conditionally stable for elastic systems and the corresponding stablility limit for both is $1/\pi$[19-21]. Hence, for stability Δt should not exceed $1/\pi$ times the shortest natural period of the structure tested. Stability criteria for inelastic systems are not well understood. Time steps smaller than those required for stability may be necessary to achieve desired accuracy. This could result in impracticable time steps for stiff multiple degree-of-freedom systems.

Accuracy of Integration Operators.--The solution of the equations of motion obtained using the integration operators discussed is only approximate. To assess the accuracy of explicit operators a numerical simulation was performed for single degree-of-freedom systems with linearly elastic and elasto-perfectly plastic restoring force characteristics and with natural periods, T, ranging from 0.05 to 2.5 secs. The N-S component of the 1940 El Centro record was used as the excitation in all of the studies. Viscous damping ratios of 0 and 5 percent of critical were used. Explicit numerical operators corresponding to the Newmark method with $\beta = 0$ (Fig. 1) and the central difference method (Eq. 8) were considered.

The time step size assumed for the analyses were varied as a function of the stability limit. That is, nominal time steps equalling 1/2 and 1/5 of those corresponding to the system's stability limit (i.e. Δt of $T/2\pi$ and $T/10\pi$) were considered in the analyses. However, to prevent misinterpretation of the results Δt was not permitted to exceed the digitization interval of the ground motion (0.02 sec.). Thus, Δt was taken as 0.02 for all systems with a period greater than 0.126 sec. when the nominal step size was $T/2\pi$ and for all systems with T exceeding 0.63 sec when the nominal step size was T/10π.

Linear Elastic Systems.--Although it is not expected that the pseudo-dynamic method would be generally used for elastic systems, the initial simulations were made assuming elastic response. This was done to distinguish more clearly the effect that the choice of numerical operator and viscous damping has on accuracy. To evaluate the results, an index, ε , of the numerical error was taken as the ratio of the maximum displacement predicted by the explicit operator divided by that obtained using an exact solution of the Duhamel integral [22]. In a comprehensive evaluation, other response parameters (e.g., the frequency content) and other ground motions should be considered.

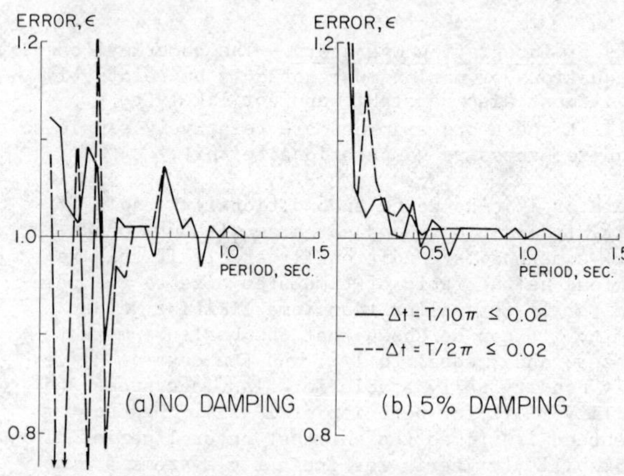

Fig. 3 Error Spectra for Newmark Method (β=0); Elastic Systems; El Centro Record.

Results of the simulations for the Newmark method (β = 0) are shown in the form of error spectra in Fig. 3. It can also be seen that the inclusion of damping generally, but not always, improves the accuracy of this method. From these curves it appears that error tends to diminish considerably with increasing period. The implication of the relatively high errors in the short period range on the higher mode contributions to stiff multiple degree-of-freedom systems remains to be assessed. In the range where the actual time steps used in the analyses are not the same, significantly improved accuracy results with use of the shorter step size, especially for the undamped cases.

Similar trends can be seen in Fig. 4 for the results obtained using the central differnece operator. The principal difference is that errors are not confined only to the short period range.

Inelastic Systems.--Systems with a number of different types of nonlinear restoring force characteristics have been investigated. Results for ideal elasto-perfectly plastic systems will be presented herein. The strengths of the systems, R_y, were selected such that $\eta = R_y/Ma_{gmax}$ remained a constant. In these analyses, η was taken as either 0.5 or 1.0. When systems are designed in this fashion, required displacement ductilities tend to increase with decreasing values of η and period [1]. The error index, , was taken as the predicted ductility divided by that obtained in an analysis based on a linear acceleration method with very small time steps [9]. The time step in all cases was taken as T/10π but not greater than 0.02 sec.

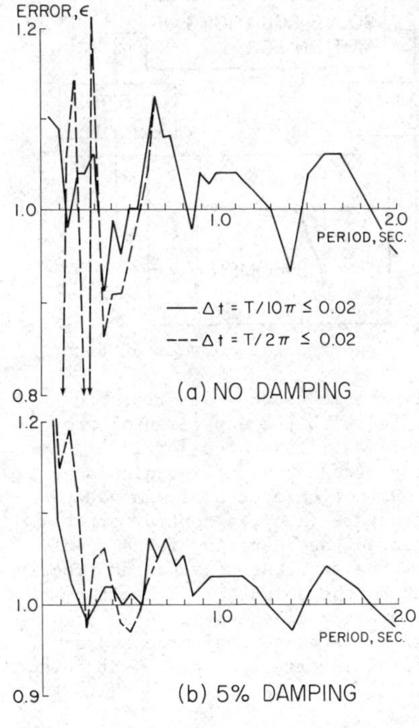

Fig. 4 Error Spectra for Central Difference Method; Elastic Systems; El Centro Record

As shown in Fig. 5a, errors associated with the Newmark β method again tend to diminish with increasing period. Errors also tend to be smaller for smaller values of η (which correspond to systems requiring larger values of ductility). In the case of the central difference method (Fig 5b), however, significant errors occur throughout the entire period range and errors tend to increase with decreasing η . For both methods, errors are reduced with decreasing η for this ground motion in the short period range. Damping has a smaller effect on error for nonlinear systems than it does on elastic systems as can be seen by comparing Figs. 3 and 6 (for α = 0).

EXPERIMENTAL SOURCES OF ERROR

The preceeding information is not by itself sufficient to assess the potential reliability of the on-line computer control method. In addition to the numerical errors examined in the previous section, there are other sources of error associated with implementing the method in an experimental environment. Some of these relate to the reliability of the loading and support apparatus, data acquisition systems, hydraulic systems, and control systems. Additional problems related to idealizing a structure using Eq. 1 will be discussed in the

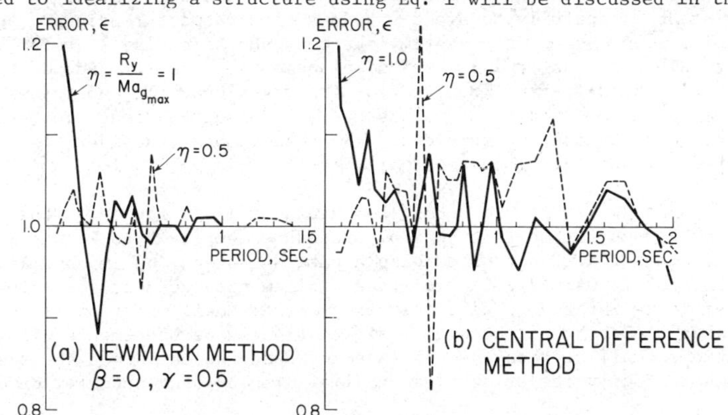

Fig. 5 Error Spectra for Elasto-Perfectly Plastic Systems for El Centro Record; 5% Damping; $\Delta t = T/10\pi$

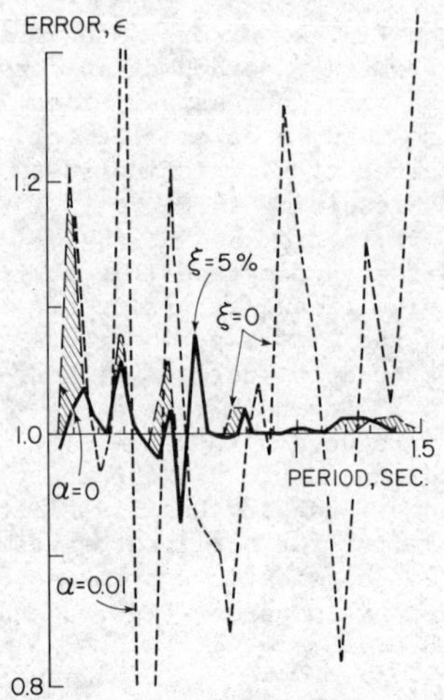

Fig. 6 Effect of Damping and Induced Error on ε ; Newmark Method; El Centro; η=0.5; Δt=T/10π

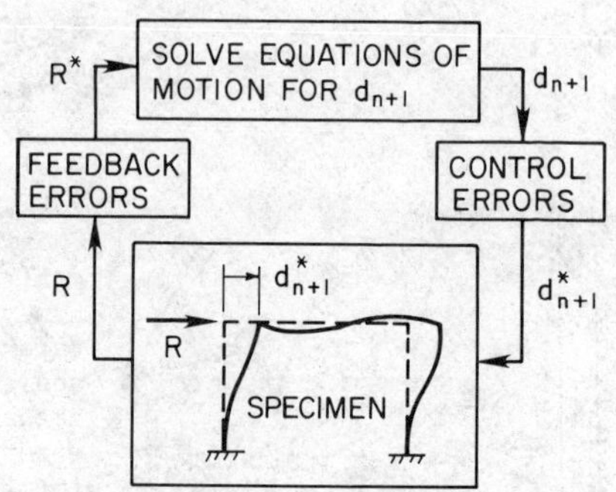

Fig. 7 Experimental Sources of Error

Two basic sources of error are associated with the experimental procedures, as illustrated in Fig. 7. During a step the specified displacements may not be accurately imposed on the structure (i.e., a control error). In addition the restoring force developed by the structure may be incorrectly fedback to the computer. These two errors are not independent as an incorrect displacement will result in an erroneous restoring force and an error in restoring force will result in an incorrect displacement being computed and imposed in the next step.

Sources of Control Errors.--As displacements are used to control the specimen, special care must be taken to avoid unintentional deformation of the specimen and to accurately control and monitor the imposed displacements. Thus, the specimen must be supported in such a way as to avoid undesired movement (slippage) at support and control points. In addition, the selection of the load (and displacement control) points must be done carefully to avoid spurious displacement readings. For example, floor beams in a system like that shown in Fig. 1 are not truly inextensible so that the lateral displacement on the left end of the floor will be different from that on the right end. In some cases rotation of actuator attachment points can also result in inappropriate lateral displacemnet measurements. Some additional considerations related to the control system follow.

1. *Resolution Errors*.--Most computers or microprocessors that would likely be used for on-line control procedures employ 8 to 14 bit words. In the case of a 12 bit word, displacements can only be imposed in increments of 1/2048 ($2/2^{12}$) times the maximum displacement for which the system is calibrated. While this is a rather small fractional error, the absolute error can become significant when the specimen's displacements are small or when the displacement used to calibrate the system is substantially larger than the maximum displacement actually developed.

2. *D-A Conversion Error*.--The digital form of the specific displace-

ment must be converted to an analog form for use by the actuator control system. Due to noise and other electronic problems a small additional error may be introduced.

3. *Instrumentation Errors*.--Displacement transducers are likely to exhibit both nonlinearity and hysteresis under cyclic excursions. Transducers should be selected to minimize these problems and their calibration must be done consistently with the desired level of accuracy.

4. *Actuator Control System Errors*.--A number of problems may arise related to the closed-loop feedback systems used to control the hydraulic actuators. First, the specified (command) displacement may not actually be imposed due to system errors or lack of sensitivity. Second, the system may not impose the desired variation of displacement within a step due to system insensitivity or lack of servovalve capacity. These problems may become accute where sudden changes in specimen stiffness or strength occur; and/or when large differences in the forces or displacements develop in adjacent actuators in multiple degree-of-freedom systems. Recent Japanese tests have indicated the need for carefully monitoring and correcting displacements during a step [14, 17]. Third, the interaction of errors associated with the actuator control feedback system with those associated with the overall experimental system (Fig 7) may lead to instability in some cases. Fourth, the data acquisition system must be coordinated with the actuator control system so that restoring forces are measured when the displacements first reach the specified values. Delay in measurement can result in significant force relaxation as discussed previously.

Sources of Feedback Error.--The restoring force of the structure must be carefully measured for accurate solution of Eq. 1. Some considerations regarding the force feedback system follow.

1. *Load Cell Errors*.--As with the displacement transducers, load cells used to measure actuator loads may be nonlinear and are subject to calibration errors.

2. *Noise*.--Due to high frequency vibration of the specimen associated with testing and the nature of the electronic signal conditioning equipment, noise may introduce uncertainty in the force readings.

3. *A-D Conversion Error*.--Two types of error are introduced by converting the analog force readings to the digital form required by the computer. First, there is usually a small error associated with the converter itself. Second, the truncated digital form of the data results in resolution errors similar to those discussed previously for the control system.

Simulation of System Errors.--Several numerical simulations have been performed to assess the implications of the control and feedback errors. To identify the critical points at which these errors must be limited, algorithms have been formulated to simulate each source of error. However, the general effect of error on the accuracy of the pseudo-dynamic method may be more easily assessed by considering the introduction of a random error into the control and feedback branches of Fig. 7. To do this a psuedo-Gaussian error is assumed with a standard deviation equal to α_d times the maximum displacement computed using an "exact" operator. Similarly the standard deviation of the error in the force feedback signal is taken as α_R times the maximum force developed by the system. It must be noted that many of the sources of error in the actual pseudo-dynamic system are systematic and deterministic rather than random as assumed here.

To illustrate the consequences of these errors, consider the relatively accurate Newmark integration operator (β=0) discussed previously. For an undamped elastic system, selecting $\alpha_d = \alpha_R = \alpha = 0.01$ results in significant additional errors as can be seen in Fig. 6. For nonlinear systems, however, the introduction of random errors seems to have appreciably less effect in the long period range as can be seen in Fig. 8. It is clear that for accurate simulation of dynamic response, sources of control and feedback error must be carefully considered and controlled.

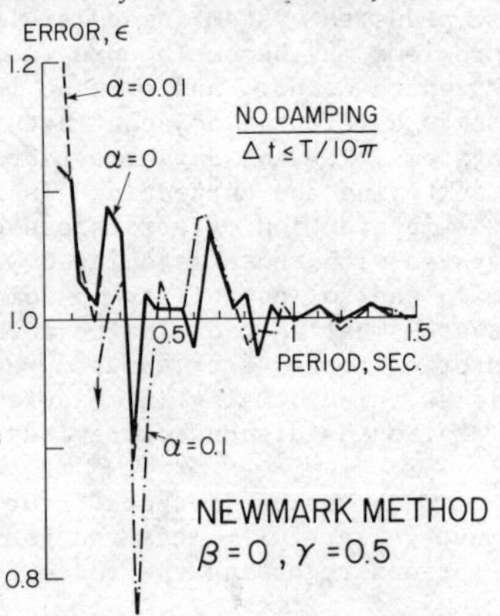

Fig. 8 Effect of Pseudo-Gaussian Error on ε; Elastic Systems; El Centro

<u>Errors Associated with Structural Idealization</u>.--Even if numerical, control and feedback errors can be eliminated completely, pseudo-dynamic test results can differ from those that would result in an actual dynamic test. This is because it is not possible to model the the response of all structures adequately using Eq. 1. For example, many structures of interest (e.g. reinforced concrete shear wall buildings) may have considerable portions of their mass distributed between assumed degrees-of-freedom. In these cases the lumped mass idealization of Eq. 1 may be inappropriate. Moreover representation of the distributed inertial loads on a floor by a concentrated point load may lead to incorrect distributions of member internal forces.

It is reasonable to expect that the effects of viscous damping would be a function of the damage suffered by a structure and would change throughout its dynamic response. However, in a multiple degree-of-freedom system it is difficult to measure experimentally what the damping matrix should be during a test. Although some analytical variation in damping characteristics can be assumed and it is expected that inelastic deformations would be the primary source of energy dissipation in most pseudo-dynamic tests, this effect will introduce some additional error. While Coulomb damping is directly accounted for in the method, this type of damping may be sensitive to rate of loading or high frequency vibrations occuring in the specimens. As indicated previously dynamic properties of certain materials may also be different than those exhibited during quasistatic tests so that the restoring force used in solving Eq. 1 may not be the same as that which would occur during dynamic response.

VERIFICATION TESTS

Because of the uncertainties associated with idealizing structures using Eq. 1 and with the numerical, control and feedback errors discussed previously, it is desirable to perform experiments to verify the accuracy, reliability and practicability of the pseudo-dynamic test method. As part of the U.S. - Japan Cooperative Earthquake Research Program a series of coordinated analytical investigations will be carried out by the BRI in Japan, the University of California, Berkeley and the University of Michigan, Ann Arbor. Research at Berkeley will focus on an assessment of the control and feedback system requirements and on the adequacy of the discrete form of Eq. 1 in representing the dynamic response of structures. Psuedo-dynamic tests will be performed on several representative multiple degree-of-freedom systems and shaking table tests will be performed on matching structures so that the results can be compared to assess (and improve) the pseudo-dynamic method.

CONCLUSIONS

A relatively new method for seismic performance testing has been examined. While the method shows considerable promise for improving the realism of loading histories imposed during quasistatic tests of complete structural systems, there are a number of potential sources of error that must be experimentally evaluated before the method can be used with confidence. Theoretical work to extend the method to tests of of subassemblages of structural systems, based on substructuring concepts, should be explored.

ACKNOWLEDGEMENTS

The financial support provided by the Committee on Research, University of California, Berkeley for conducting this research and of NSF in preparing this paper (Grant PFR-8008584) is greatly appreciated.

REFERENCES

1. Bertero, V. V. et al, "Use of Earthquake Simulators and Large-Scale Loading Facilities," Proceedings, Workshop on ERCBC, Univ. of Calif., Berkeley, 1978.

2. Karwinkler, H., "Experimental Research Needs," Proceedings, Workshop on ERCBC, Univ. of Calif., Berkeley, 1978.

3. Galambos, T. V. and Mayes, R. L., "Dynamic Tests of a Reinforced Concrete Building," Report No. 51, Washington Univ., St Louis, 1978.

4. Safford, B., "Development of Force Pulse Train Generators for the Study of Dynamic Response," ASCE-EMD Conference on Dynamic Response Atlanta, 1981.

5. Lindberg, H., et al, "Simulation of Strong Earthquake Motion with Contained Explosion Line Source Arrays," ASCE-EMD Conference on Dynamic Response, Atlanta, 1981.

6. Townsend, W. H. and Anderson, J. C., "Analytical/Experimental Modeling of Exterior Beam-Column Connections," ACI Convention, Nov. 1978.

7. Mahin, S. A., and Bertero, V. V., "An Evaluation of Some Methods for Predicting Seismic Behavior of Reinforced Concrete Buildings," Report No. EERC 75-5, Univ. of Calif., Berkeley, 1975.

8. Powers, W. F., "Preliminary Analysis of Computer-Actuator

On-Line System Difference Equations," 4th U.S.-Japan Coop. Earthquake Research Planning Group Meeting, Berkeley, 1979.

9. Williams, M. E., "Preliminary Evaluation of Numerical Algorthms Used for On-Line Computer Control Methods for Seismic Performance Testing," Grad. Stud. Report, Univ. of Calif., Berkeley, 1979.

10. Takanashi, K., et al, "Nonlinear Earthquake Response Analysis of Structures By a Computer-Actuator On-Line System," Bull. of the Eq. Resistant Str. Research Center, Inst. of Indust. Sci. Univ. of Tokyo, No. 8, 1974.

11. Okada, T., "The Experimental Inverstigation on ERCBC with Emphasis on the Use of Earthquake Response Simulators in Japan," Proceedings, Workshop on ERCBC, Univ. of Calif., Berkeley, 1978.

12. Okada, T. and Seki, M., "A Simulation of Earthquake Response of Reinforced Concrete Buildings," Proceedings, 6th WCEE, 1977.

13. Takanashi, K., et al, "A Simulation of Earthquake REsponse of Steel Buildings," Proceedings, 6th WCEE, 1977.

14. Takanashi, K., et al, "Nonlinear Earthquake Response Analysis of Structures by a Computer-Actuator On-Line System," Trans. AIJ, Feb. 1980.

15. "Recommendations for a U.S.-Japan Cooperative Research Program Utilizing Large-Scale Testing Facilities," Report UCB/EERC-79/26, Univ. of Calif., Berkeley, 1979.

16. Okada, T., et al, "Response of R/C Columns to Bi-Directional Horiz. Force and Const. Axial Force," Bull. of Eq. Resist. Str. Research Center, No. 10, Inst. of Indust. Sci., Univ. of Tokyo, 1976.

17. Takanashi, K., et al, "Inelastic Response of H-Shaped Cols. to Two-Diminsional Eq. Motions," Bull. ERS Research Center, No. 13, Inst. of Indust. Sci,. Univ. of Tokyo, 1980.

18. Okada, T., et al, "Nonlinear Eq. Response of Equipment Anchored on R/C Building Floor," Bull ERS Research Center, No. 13, Inst. of Industrial Science, Univ. of Tokyo, 1980.

19. Newmark, N. M., "A Method of Computation for Structural Dynamics," Journal of the Eng. Mechanics Div., ASCE, No. EM3, Vol, 85, July 1959.

20. Wilson, E. L., et al, "Nonlinear Dynamic Analysis of Complex Structures," Journal of EESD, Vol 1, 1973.

21. Leech, J. et al, "Stability of Finite-Difference Method for Solving Matrix Equations," AIAA Journal, Vol. 3 No. II, 1965.

22. "Analysis of Strong Motion Earthquake Accelerograms," Report No. EERL 73-81, Calif. Inst. of Tech., 1973.

CYCLIC RESPONSE OF STEEL-STUD FRAMED WALL PANELS

by Thomas S. Tarpy, Jr.[1], A.M., ASCE

Introduction

An experimental investigation is currently under way for determining the in-plane shear resistance of framed wall panels constructed with structural steel-studs and different sheathing or diaphragm materials. The framed wall assembly is shown in Figure 1a. This investigation is a continuation of earlier research for determining the effect of aspect ratio (length/height) on the shear resistance of framed steel-stud wall panels (1).

The earliest known extensive research project involving steel-stud shear wall partitions was conducted by John A. Blume and Associates beginning in the mid-sixties. They developed and conducted a test program for nonstructural wall panels subjected to racking loads (2,3,4,5). Several 8 ft. high by 8 ft. long wall panels with both wood studs and nonstructural metal studs with a base metal thickness of 0.0179 inches were tested. The majority of the panels were constructed with gypsum wallboard, but plaster, plywood, concrete block and combination plywood and gypsum wallboards were also tested. Pop-rivets and friction connections (i.e. the stud is not attached to the track) were used. Both static loading and cyclic load reversals were considered.

Raggett and Ellis (6) tested six wall panels common to low-rise structures. The test program was similar to the test program mentioned above. Since low-rise structures are combinations of shear panels, the results of these tests could be used to predict the behavior of a complete low-rise structure. It was observed that the diagonally braced panels were about 70 percent stiffer than unbraced panels.

Freeman and Honda (7,8) in a study of the response of two almost identical structures involved in the San Fernando earthquake determined the contribution of the nonstructural partitions to the structural performance. As a structure is laterally displaced, the main members behave in an elastic manner and the nonstructural wall partition elements which are attached to the main members distort by some relative displacement. If these wall partitions behave elastically, they add stiffness and damping to the building. By determining the contribution of the nonstructural partitions, the stiffness and damping from the partitions could be subtracted from the stiffness and damping required for the total structure and the main members designed accordingly.

---

[1]Research Associate Professor, Civil Engineering, Vanderbilt University, Nashville, Tennessee; Structural Engineer, Stanley D. Lindsey & Associates, Ltd., Nashville, Tennessee.

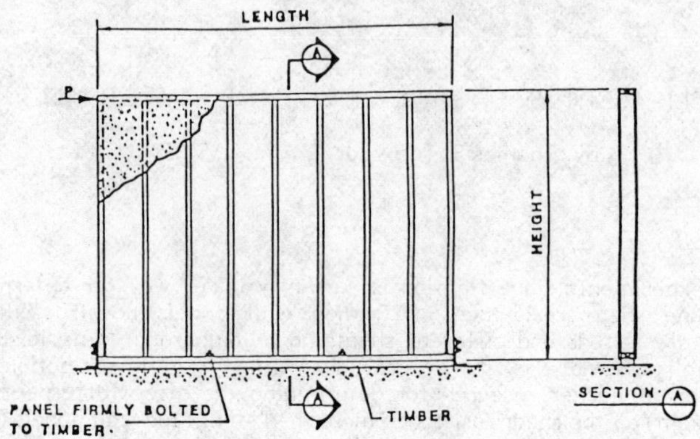

a. Wall Assembly

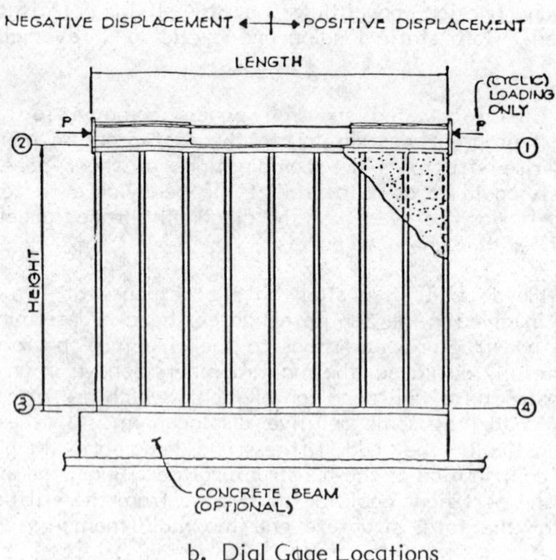

b. Dial Gage Locations

Figure 1. ASTM E564 Racking Load Assembly

The static and dynamic characteristics of wood-stud framed shear walls were investigated both theoretically and experimentally by Medearis (9,10). Solutions were obtained for the natural frequencies, mode shapes, and response quantities for a mathematical model consisting of a distributed mass shear panel having a concentrated roof load at the top. An experimental program was conducted with both cycle and vibration generator excited harmonic-dynamic shear loading. Using dynamic theory, experimental results and a digital computer, typical shear wall responses for harmonic loading were determined. Based on the limited statistical study, Medearis felt additional test results are required for precise conclusions.

Several design codes (11,12,13) currently permit the use of wood-framed shear walls with various types of sheathing or plaster to resist horizontal loads parallel to and in the plane of the walls. These codes specify maximum height-length ratios and allowable shear values per lineal foot for a range of different types of wall construction. The maximum height-length ratio and allowable shear values for a given wall construction are a function of the various types of sheathing or plaster, the sheathing thicknesses, and the nail spacing used to attach the sheathing material to the wood-stud frame. The available shear values for plywood sheathed wood-stud walls are fairly extensive (14). The allowable shear value is a function of nail spacing and placement of the plywood which may be applied directly to the framing studs or over 1/2 inch gypsum wallboard nailed to the studs. The allowable lateral in-plane deflection for wood-framed shear walls is not accurately defined and is controlled by the maximum height-length limitations. Reference (11) also specifies typical tie-down construction details to be used to resist wall uplift or overturning forces produced by forces parallel to the shear wall.

Vertical steel-stud shear walls with sheathing or plaster are not currently permissible by design codes for resisting horizontal in-plane forces unless some form of lateral bracing is used. According to Reference 11, this bracing usually consists of 0.125 inch x 1 inch steel straps used as x-bracing with a maximum angle of 60 degrees to the horizontal. The maximum allowable horizontal load which can be resisted is 1,000 pounds for each brace. The steel-studs are further specified to be a minimum of 16 gage (base metal thickness of 0.0598 inches) and located at a maximum stud spacing of 16 inches on centers.

This paper presents the results of a portion of the experimental test program for determining the shear resistance of framed steel-stud wall panels without the use of the diagonal x-bracing. Wall panels were constructed with structural steel studs using similar types of construction as for wood-stud walls. The purpose of the test program was to define similar allowable shear values and deflection limits for steel-stud walls as currently exist for wood-stud shear walls. Both static and cyclic load reversals are considered.

Test Program

The overall test program consisted of sixteen different types of wall panel construction and anchorage techniques (wall Types A-N; P-R) using static loading procedures. Four of the same wall types (wall Type B, D, E, and H) were also tested using cyclic loading procedures. The number of actual tests included in each wall type was a function of the requirements of

ASTM E 564 - 76 (15). ASTM E564 - 76 is a static test method for determining the shear resistance of framed walls for buildings. Basically, this method requires that if the results of two different tests for a given wall type construction differ by more than 10%, a third test is run and the shear resistance of the wall type is the mean of the lower two values obtained from the three test results.

The actual wall construction and anchorage details for each wall type used to determine the effect of static and cyclic loading conditions and, thus, the need for diagonal x-bracing which effect this comparison by wall types are:

(1) the effect of static versus cyclic loading conditions - wall Types B, D, E and H

(2) the effect of wall panel anchorage details - wall Types A, B, E and H

These parameters of construction were considered to have the most significant influence on the wall performance based on previous research results (1,16). The wall panel elevation and construction details for each wall type are shown in Figures 2 thru 6.

The wall panels were constructed of 3-1/2 inch web by 1-1/2 inch flange by 1/2 inch lip structural steel "C" studs with a base metal thickness of 0.0359 inches (nominal 20 gage). The steel-studs were spaced 24 inches on centers and attached to 3-5/8 inch web by 1-1/2 inch flange structural steel-runner track with a base metal thickness of 0.0359 inches (nominal 20 gage) with #10 x 1/2 inch Low Profile Head Screws. The measured yield strength of the studs for three coupons cut longitudinally from the web ranged from 42 ksi to 51 ksi with a mean value of 48 ksi (33 ksi minimum yield).

The diaphragm material was attached to both sides of the stud frame as noted in Table 1. The gypsum wallboard seams were caulked and taped to complete the construction of the wall panel. The panel caulking was allowed to cure at least 24 hours before the wall panel was moved.

Clip angles were used to attach the wall panel to the test frame at the base at 48 inches on centers for wall Types A and D. Wall Type E was selected to determine if the same shear resistance could be obtained without the added cost associated with clip angles at the base. For this case, 5/8 inch diameter x 3 inch long zinc plated Hex Head Bolts and 1 inch outside diameter zinc plated washers located at four feet centers were used to attach the wall panel to the test frame. Wall Type B construction was the same as wall Type A except that clip angles were included only at the corners of the wall to resist uplift at the end vertical studs. The remaining wall panel attachment was the same as for wall Type E.

Wall Type H was included to determine if the in-plane shear resistance of the wall panels would be affected by using powder-actuated fasteners. A reinforced concrete beam was used as a shim to simulate the concrete floor in actual construction. The concrete beam was anchored to the test frame by 5/8 inch diameter expansion bolts at four feet centers along the base. This detail is shown in Figure 5.

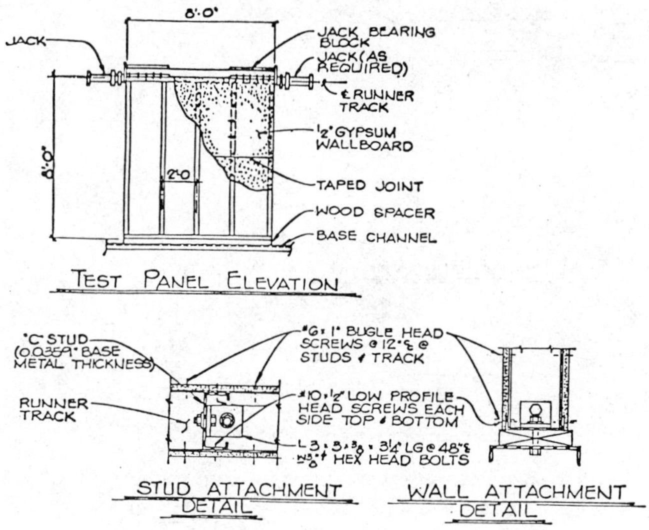

Figure 2. Wall Type A Elevation and Details

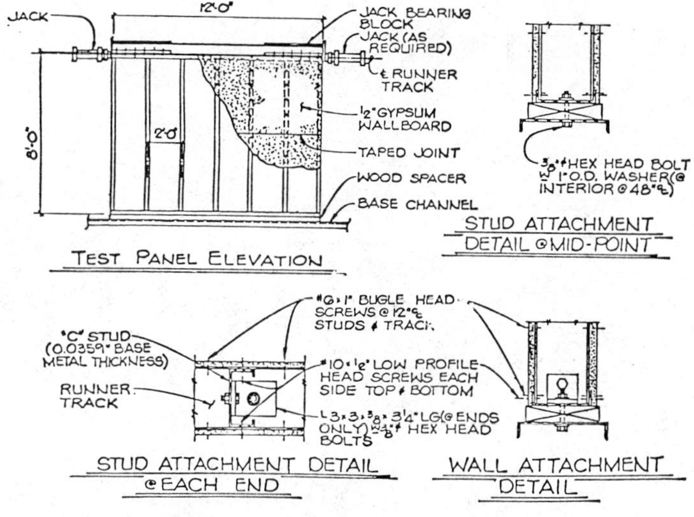

Figure 3. Wall Type B Elevation and Details

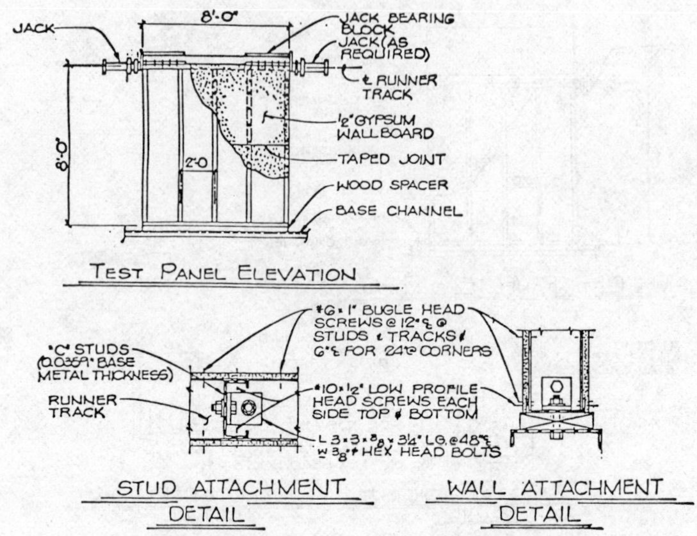

Figure 4. Wall Type D Elevation and Details

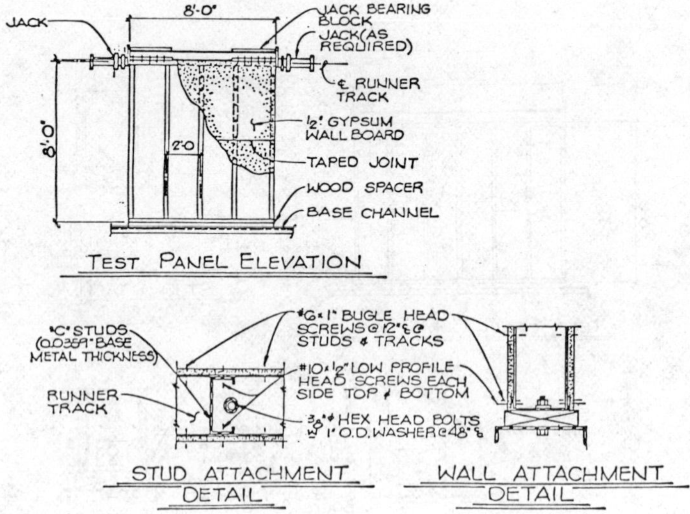

Figure 5. Wall Type E Elevation and Details

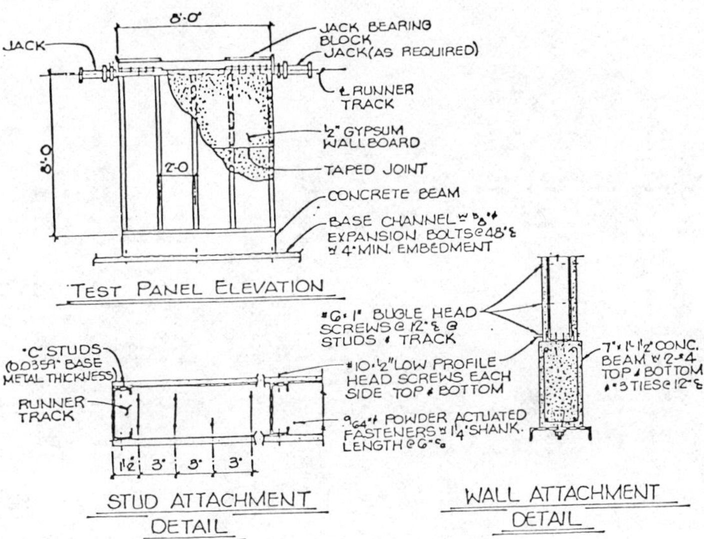

Figure 6. Wall Type H Elevation and Details

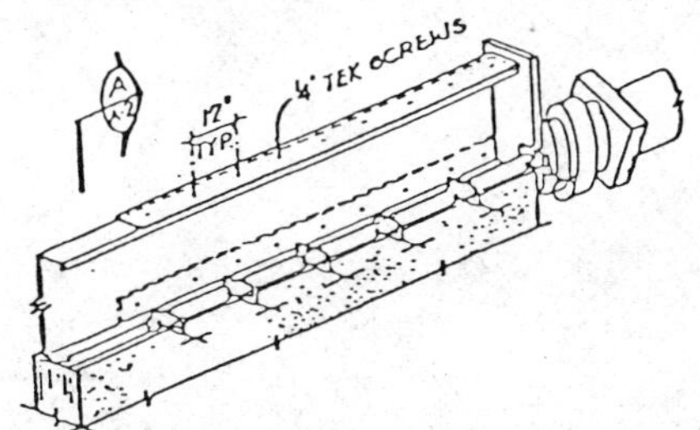

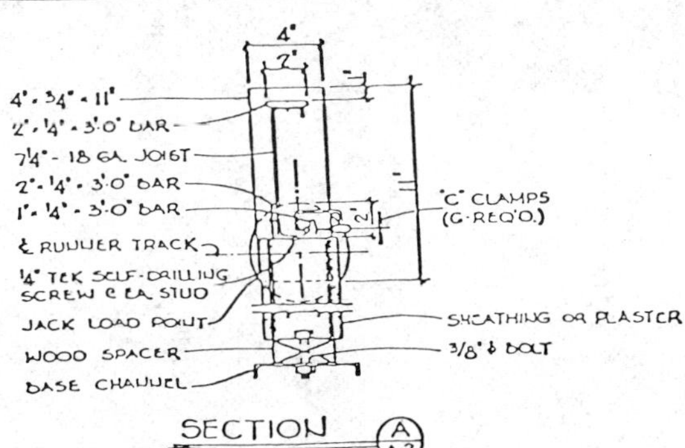

Figure 7. Test Assembly Details

TABLE 1

WALL TYPE CONSTRUCTION

Wall Type	Loading Condition	Wall Height	Wall Length	Type & Thickness of Diaphragm Material	Wall Construction	Diaphragm Attachment	Stud Spacing	Stud Attachment	Wall Anchorage
A	Static	8'-0	8'-0	1/2" Gypsum Wallboard	Single-Ply Each Face	#6x1" Bugle Head Screws @ 12" o.c. Studs & Track	20 gage C-stud @ 24" o.c.	#10x1/2" Low Profile Head Screws Each Side Top and Bottom	L3x3x3/8x3-1/4" LG @ 48" o.c. with 3/8" Hex Head Bolts
B	Static Cyclic	8'-0	12'-0	1/2" Gypsum Wallboard	Single-Ply Each Face	#6x1" Bugle Head Screws @ 12" o.c. Studs and Track	20 gage C-stud @ 24" o.c.	#10x1/2" Low Profile Head Screws Each Side Top and Bottom	L3x3x3/8x3-1/4" LG @ Ends with 3/8 Hex Head Bolts 3/8" Hex Head Bolt with 1" O.D. Washer @ Mid-point
D	Static Cyclic	8'-0	8'-0	1/2" Gypsum Wallboard	Single-Ply Each Face	#6x1" Bugle Head Screws @ 12" o.c. Studs and Track & 6" o.c. for 24" at corners	20 gage C-stud @ 24" o.c.	#10x1/2" Low Profile Head Screws Each Side Top and Bottom	L3x3x3/8x3-1/4" LG @ 48" o.c. with 3/8" Hex Head Bolts
E	Static Cyclic	8'-0	8'-0	1/2" Gypsum Wallboard	Single-Ply Each Face	#6x1"Bugle Head Screws @ 12" o.c. Studs and Track	20 gage C-stud @ 24" o.c.	#10x1/2" Low Profile Head Screws Each Side Top and Bottom	3/8" Hex Head Bolts with 1" Washer @ 48" o.c.
H	Static Cyclic	8'-0	8'-0	1/2" Gypsum Wallboard	Single-Ply Each Face	#6x1" Bugle Head Screws @ 12" o.c. Studs and Track	20 gage C-stud @ 24" o.c.	#10x1/2" Low profile Head Screws Each Side Top and Bottom	9/64" x 1-1/4" Powder Actuated Fasteners @ 6" o.c.

TABLE 2

SUMMARY OF TEST RESULTS

Wall Type	Loading Condition	Wall Height (ft)	Wall Length (ft)	Ultimate Load (lb)	Ult. Shear Strength (lb/ft)	Max. Total Deflection (in)	Total Shear Stiffness (lb/in)	Initial Cracking Load Level (lb)	Initial Cracking Shear Strength (lb/ft)	Initial Cracking Total Deflection (in)
A	Static	8'-0	8'-0	3300	413	0.98	9,700	2400	300	0.40
B	Static	8'-0	12'-0	4500	375	0.58	12,700	3800	317	0.40
	Cyclic	8'-0	12'-0	4500	375	0.22	31,400	3600	300	0.13
D	Static	8'-0	8'-0	4200	525	1.19	10,600	3300	413	0.50
	Cyclic	8'-0	8'-0	2800	350	0.41	20,900	2400	300	0.20
E	Static	8'-0	8'-0	2500	313	1.20	4,100	2000	250	0.70
	Cyclic	8'-0	8'-0	2300	287	0.74	6,200	1800	225	0.30
H	Static	8'-0	8'-0	3100	388	1.48	4,700	2500	313	0.94
	Cyclic	8'-0	8'-0	2500	313	0.52	8,100	1700	213	0.22

Test Set-up

A load bearing block and structural steel-joist member was attached along the top of the wall panels at the point of loading to uniformly distribute the load along the wall and to prevent localized failure of the panel at the point of loading. This detail is shown in Figure 7. It was felt that by attaching the steel joist to the wall panels in this manner the laboratory conditions would represent as closely as possible actual field installation and loading conditions.

Prior to starting a test, displacement indicating dials were mounted on the test frame at locations shown in Figure 1b. The dial gage at the lower right, Gage 4, and the dial gage at the lower left, Gage 3, measured the slippage of the wall panel in the test frame. Gage 2, shown at the point of loading on the upper left of the figure was used as a backup for Gage 1 on the static loading conditions, and for measuring the panel deformation opposite the load point on the right for cyclic loading.

Test Procedure

a) Static Loading

The loading sequence consisted of applying an initial load to the top of the wall panel of approximately ten-percent of the estimated ultimate load carrying capacity of the wall panel using a hydraulic jack/load cell/digital strain indicator combination. This load was held for two-minutes to set the wall panel connections and was then removed. The wall panel was allowed to fully recover and the dial gages set to zero to begin the test at this zero load-deflection condition. The load was then applied incrementally to the wall panel, and displacement measurements recorded at each interval following a two-minute hold period. At load levels of approximately one-third and two-thirds of the estimated ultimate load carrying capacity of the wall panel, the load was fully removed, and the wall panel recovery was recorded after a five-minute hold period. The load was then re-applied to the next higher increment above the back-off load. Loading continued in this manner until the wall panel was no longer capable of holding additional load. The last load, held for two minutes with displacement measurements recorded, was defined as the ultimate load. The loading condition is shown in Figure 8a.

b) Cyclic Loading

The cyclic loading sequence consisted of applying an initial load to the wall panel of approximately ten-percent of the estimated ultimate load carrying capacity of the wall panel in one direction (positive direction) using a hydraulic jack/load cell/digital strain indicator combination. This load was held for two minutes to set the wall panel connections and was then removed. The wall panel was allowed to fully recover before beginning the test at this zero load-deflection condition. The load was then applied incrementally to the right or positive direction, in the direction of preload, to a previously determined load interval.

Displacement measurements were recorded immediately upon reaching the interval load value. After recording the displacements, the load was

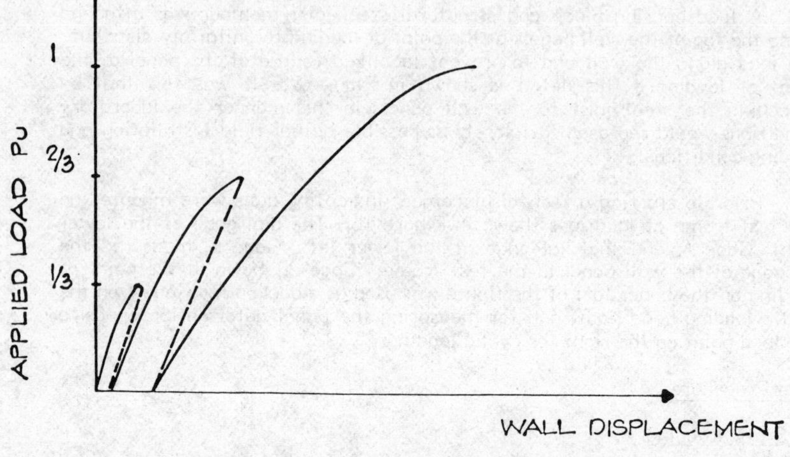

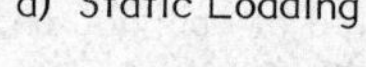
a) Static Loading

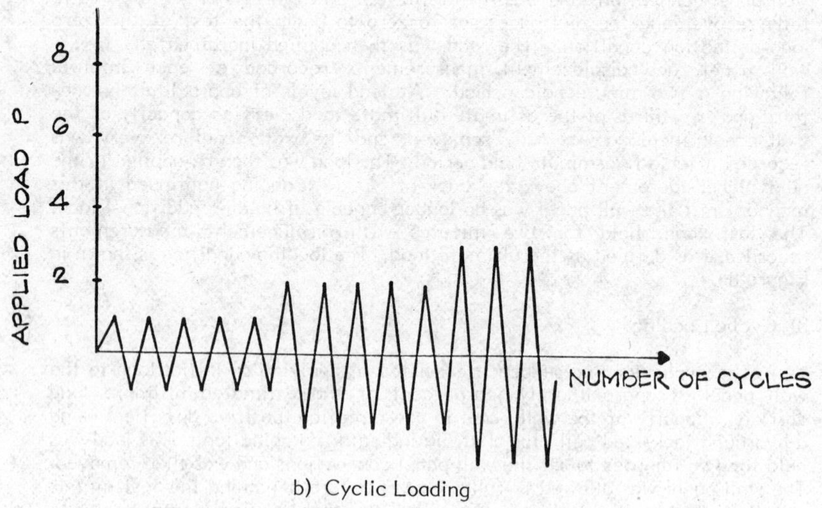

b) Cyclic Loading

Figure 8. Loading Sequence

released and the wall allowed to fully recover at which time another set of displacement readings were obtained. The load was then applied to the left, in the opposite or negative direction from before, until the same load level was obtained. Displacement measurements were recorded and the load fully released. At this zero load, additional displacement measurements were obtained. This process completed one full cycle of the test for a given load interval. The loading sequence continued as before for four additional cycles at the same load interval; the only difference was that the zero load-displacement measurements were not recorded. Upon completion of the fifth cycle, the load was applied in the positive direction to the next higher load interval and the previously described loading sequence repeated for five full cycles. Loading continued in this manner for increasing load intervals until the wall panel was no longer capable of holding additional load. The ultimate load was the last load for which at least one complete cycle was obtained with complete displacement measurements. The loading condition is shown in Figure 8b. Since the time duration of loading was about ten seconds for each load interval, the cyclic loading sequence was considered as low cycle fatigue in nature.

Analysis of Test Results

The information obtained from the test data are load-deflection curves, ultimate shear strength, shear stiffness, and damage threshold load level. The load-deflection curves are plots of the applied load versus the measured total panel deflection.

The total panel deflection, Δ_T is defined as either:

$$\Delta_T = \Delta_{TR} = \Delta_1 - \Delta_4$$

or (1)

$$\Delta_T = \Delta_{TL} = \Delta_2 - \Delta_3$$

Where Δ_1, Δ_2, Δ_3 and Δ_4 are measured deflections (in) at dial gages 1, 2, 3, and 4 as shown in Figure 1b. The subscript TR and TL indicates the direction of loading from the zero load-deflection condition in the positive or negative direction, respectively, at noted in Figure 1b.

The ultimate shear strength, S_U, of the wall panel is defined as:

$$S_U = P_U/b \qquad (2)$$

where P_U is the ultimate load carrying capacity of the wall panel (lb) (i.e. the largest load held for two minutes and gage measurements recorded for the static loading condition or for which one full cycle was completed and gage measurements recorded for the cyclic loading condition), and b is the length of the wall panel (ft.).

The total shear stiffness, G'_T, is determined from the load-deflection curve at a value less than the proportional limit. A suggested reference load level by ASTM is 0.33 P_u. If the selected load level is beyond the proportional limit, a reduced value is chosen. The total shear stiffness for deflection to the right and left is defined as:

$$G'_{TR} = \frac{a}{b} \left(\frac{P}{\Delta_{TR}}\right)$$

and (3)

$$G'_{TL} = \frac{a}{b} \left(\frac{P}{\Delta_{TL}}\right)$$

where P is the load (lb); and Δ_{TR} and Δ_{TL} are the corresponding total deflections (in) at one-third P_u for displacement in the positive and negative direction, respectively, a is the height of the wall panel (ft) and b is the wall panel length (ft).

The damage threshold load level, P', is a visual observation and is defined as the load level at which damage to the sheathing material occurred. As such, the values are based on the general observations of several individuals involved in the testing.

Discussion of Results

The experimental results for wall Types B, D, E and H for a particular wall type construction and loading condition are summarized in Table 2. The corresponding load deflection curves are plotted in Figures 9 thru 12. For the cyclic loading condition, the load deflection curves represent the mean values of two individual tests as discussed previously. For a detailed discussion of the individual test results refer to Reference 17.

All wall types tested experienced the same basic type of failure. The initial sign of distress was the wall base runner tracks deforming around the anchorage device (either clip angle, powder actuated fastener, or washers) at the tension or uplift corner of the wall identified at Gage 3 for positive and Gage 4 for negative displacement. As the load was increased, cracking of the gypsum wallboard occurred at the same locations from the corner fasteners to the edge of the wallboard. This process continued with increased track deformation and increased tearing of the wallboard until the wall panel was no longer able to carry additional load.

Wall Type A is used as the base reference in the following discussion of the effect of various parameters on the shear resistance of the wall panel where possible. This reference is due to the extensive amount of data on wall Type A with variable aspect ratios (1).

a) Effect of Static Versus Cyclic Loading

The effect of the type of loading on the wall panels is seen by comparing wall Types B, D, E and H. In all cases the ultimate shear strength was less than or equal to the value obtained for static loading. This decrease was 33% for Type D, 8% for Type E, and 19% for Type H. The shear strength was independent of loading for Type B. The corresponding effect on shear stiffness was a 147% increase for Type B, 97% for Type D, 51% for Type E and 72% for Type H. This increase was apparently due to the smaller total deflections recorded for the cyclic loading condition than for the static loading condition. The stiffening effect of the corner angles was still apparent for Type B independent of loading.

Cyclic loading had a weakening effect on the wall panels in all cases. This resulted in a decrease in the damage threshold load level. The decreased load levels ranged from 32% for Type H to only 5% for Type B. The range of load levels appears to be related to the degree of corner anchorage rigidity.

b) Effect of Wall Panel Anchorage

The wall panel anchorage effect on the shear strength is seen by comparing wall Types A, B, E and H. The elimination of the clip angles at the interior locations (Type B) had little effect on the shear strength or stiffness. This was due to the stiffening effect the corner angles furnished to the runner track and end vertical stud against local bending and shear deformations. A 24% decrease in shear strength resulted with the substitution of bolt and washers (Type E) in place of the corner angles. The use of powder actuated fasteners (Type H) had a similar restraining effect as the angle for Types A and B because of the spread of the fasteners to as close to the edge of the wall as possible, thus, eliminating the track bending around the anchoring devices. This restraining effect existed as long as the fastener embedment was sufficient against pullout. The type of interior anchorage had little effect on the shear resistance.

The shear stiffness appears to be highly dependent upon the corner anchorage of the wall. The use of corner angles for wall Types A and B resulted in essentially the same value for shear stiffness. The elimination of the angles, however, resulted in a 58% decrease for Type E and a 52% decrease for Type H. This was because of the larger wall panel rotation that occurs when the corner angles are removed.

The influence of corner anchorage is also apparent in the damage threshold load level. The bolt and washer anchorage resulted in a 17% decrease in load level. The use of powder actuated fasteners or corner clip angles resulted in a negligible increase in load level.

Conclusions

The results obtained from this investigation indicate that any of the wall panels, framed with "C" shaped steel studs and gypsum wallboard material, and constructed and anchored as reported herein are a feasible wall of resisting lateral in-plane shear loads when used as vertical diaphragms in buildings. However, it is the professional opinion of the writer that certain design and construction recommendations should be followed. These recommendations are:

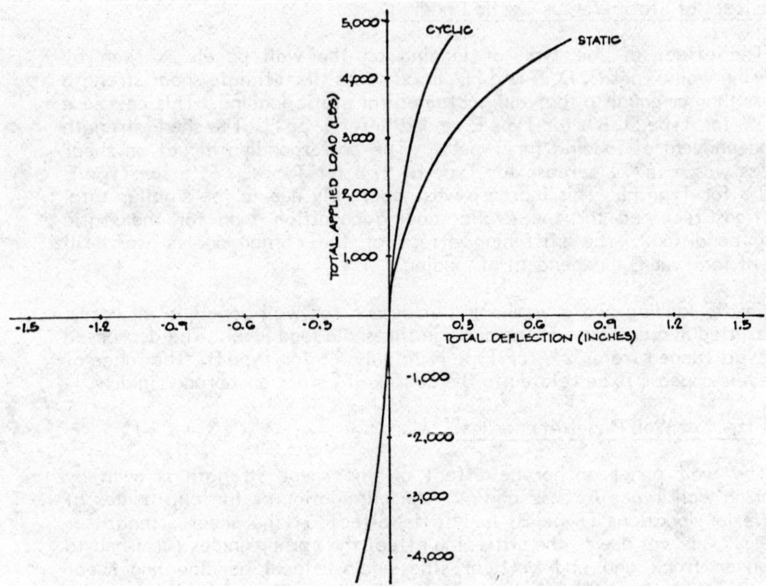

Figure 9. Load Deflection Curves - Wall Type B

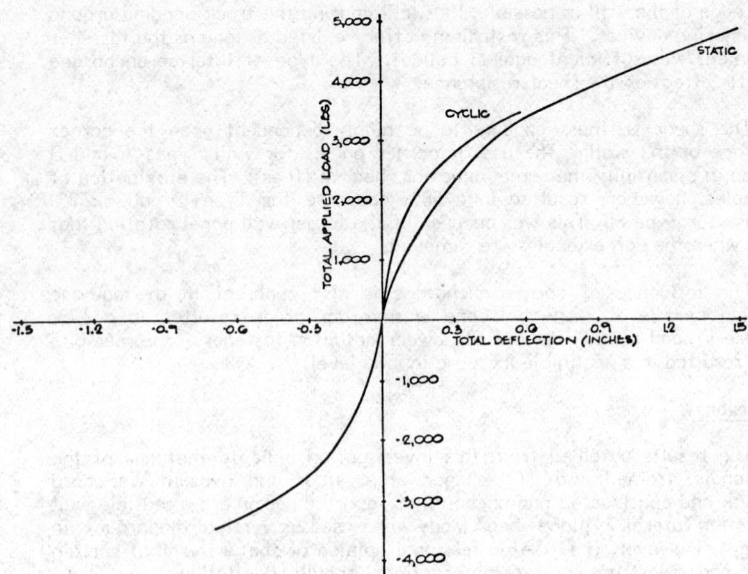

Figure 10. Load Deflection Curves - Wall Type D

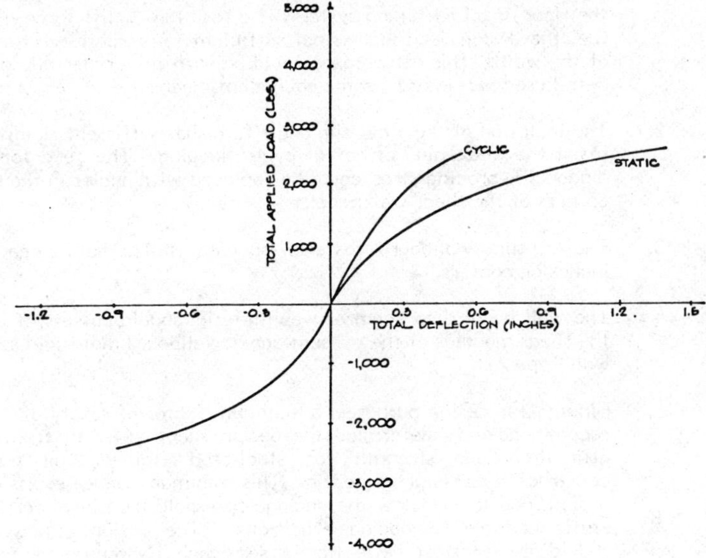

Figure 11. Load Deflection Curves - Wall Type E

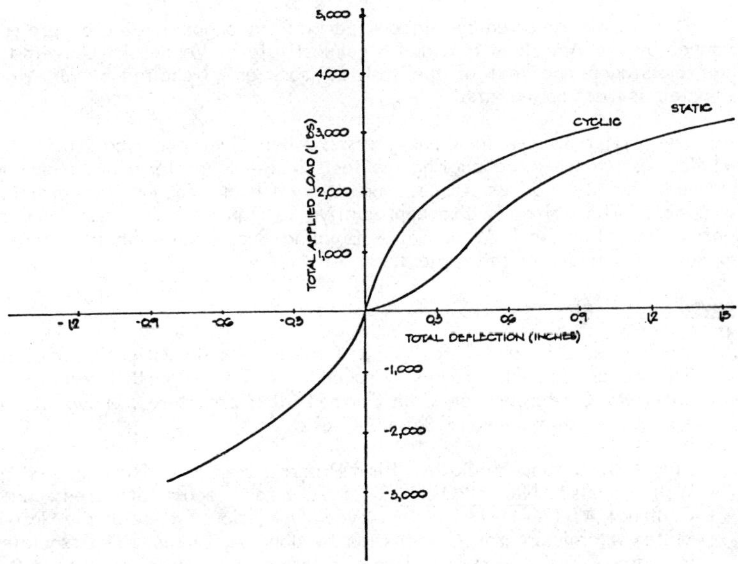

Figure 12. Load Deflection Curves - Wall Type H

1. An attachment should be designed to connect the wall panel to the floor or roof framing systems if a resultant uplift force exists (i.e. the design dead load is not sufficient to prohibit overturning of the wall). This attachment could be with the corner clip angle detail used herein or by some equivalent means.

2. The inclusion of the steel clip angle furnishes sufficient rigidity to resist the weakening effect of cyclic loading. The need for the diagonal x-bracing strap could be replaced with angles in the base corners of the shear wall panels.

3. The gypsum wallboard fastener spacing should not exceed 12 inches on centers.

4. The wall panel diaphragm or web material should possess at least the shear modulus of the gypsum-paper/wallboard material used in wall Type A.

5. Finally, for design purposes, a minimum factor of safety of 2.0 is recommended to determine the design shear strength from the ultimate shear strength for steel-stud framed wall panels constructed as reported herein. This minimum value results in a design load level below the damage threshold load level for both static and cyclic loading conditions. The designer, however, should also consider deflection or serviceability requirements for a particular application.

Acknowlegement

The results reported herein are part of an ongoing research project financed by the American Iron and Steel Institute at Vanderbilt University. Their assistance and that of the task advisory group chaired by Mr. Karl Klippstein is most appreciated.

The writer wishes to thank Messrs. Allen Dyer and Ron Moore for technical assistance in conducting the tests in the laboratory, Ms. Marcelle McDaniel for typing assistance, and Mr. Chuck Vaughn for drafting assistance. The writer is also appreciative to Ramset Fasteners; Rodgers Construction, Inc.; and A. J. Smith Building Supplies for their generous support of materials for this project.

Appendix I. -- References

1. McCreless, C.S., and Tarpy, T.S., "Experimental Investigation of Steel Stud Shear Wall Diaphragms," Proceedings of the Fourth International Specialty Conference on Cold Formed Steel Structures, University of Missouri - Rolla, June, 1978, pp. 647-672.

2. Blume, J.A., and Associates, "First Progress Report on Racking Test of Wall Panels," No. NVO-99-15, a research report prepared under Contract AT (26-1)-99 for the Nevada Operations Office of the United States Atomic Energy Commission by John A. Blume and Associates Research Division, San Francisco, California, July, 1968.

3. Blume, J.A., and Associates, "Second Progress Report on Racking Test of Wall Panels," JAB-99-35, a research report prepared under Contract AT (26-1)-99 for the Nevada Operations Office of the United States Atomic Energy Commission by John A. Blume and Associates, Research Division, San Francisco, California, December, 1971.

4. Freeman, S.A., "Fourth Progress Report on Racking Test of Wall Panels," JAB-99-55, a research report prepared under Contract AT (26-1)-00 for the Nevada Operations Office of the United States Atomic Energy Commission by John A. Blume and Associates, Research Division, San Francisco, California, September, 1974.

5. Freeman, S.A., "Fourth Progress Report on Racking Test of Wall Panels," JAB-99-55, a research report prepared under Contract AT (26-1)-00 for the Nevada Operations Office of the United States Atomic Energy Commission by John A. Blume and Associates, Research Division, San Francisco, California, September, 1974.

6. Raggett, J.D., and Ellis, K.N., "Report on Racking Tests of Low-Rise Wall Panels," John A. Blume and Associates, JAB-99-59, San Francisco, September, 1973.

7. Freeman, S.A. and Honda, K.K., "Response of Two Identical 7-Story Structures to the San Fernando Earthquake of February 9, 1971," John A. Blume and Associates, JAB-99-98, San Francisco, October, 1973.

8. Freeman, S.A., "Racking Tests of High-Rise Building Partitions," Proceedings ASCE, Preprint No. 2462, New Orleans, Louisiana, April 14-18, 1975.

9. Medearis, K., "Static and Dynamic Characteristics of Composite Structures," Bulletin of the Seismological Society of America, Vol., 57, No. 5, October, 1967.

10. Medearis, K., "An Investigation of the Structural Damping Characteristics of Composite Wood Structures Subjected to Cyclic Loading," dissertation presented to Stanford University, Stanford, California, in 1962, in partial fulfillment of the requirements for the degree of Doctor of Philosophy.

11. "Seismic Design for Buildings", U.S. Army Technical Manual 5-809-10, Department of Army, Washington, D.C., April, 1973, Chapter 13.

12. State of California Administrative Code, Title 24, Building Standards, 1974 edition, Dept. of General Services, Sacramento, California, pp. (T-617) - (T-618).

13. Uniform Building Code, 1976 edition including 1978 supplement, International Conference of Building Officials, Whittier, California, Part X, Chapter 47, pp. 603-626.

14. Plywood Diaphragm Construction, 1970 Edition, American Plywood Association, Tacoma, Washington.

15. ASTM E564-76, "Standard Method of Static Load Test for Shear Resistance of Framed Walls for Buildings," American Society for Testing and Materials, Vol. II, 1976 edition, pp 1471-1475.

16. Tarpy, T.S., and Hauenstein, S.F., "Effect of Construction Details on Shear Resistance of Steel-Stud Wall Panels", CE - AISI-1, a research report prepared under Contract No. 1201-412 for the American Iron and Steel Institute by the Civil Engineering Department, Vanderbilt University, Nashville, Tennessee, June, 1978.

17. Siddharth, M.P., "Effect of Construction Techniques and Loading on the Shear Resistance of Steel-Stud Wall Panels," CE-80-80, thesis presented to Vanderbilt University, Nashville, Tennessee, in 1980, in partial fulfillment of the requirements for the degree of Master of Science.

Appendix II. -- Notation

A = Aspect Ratio (length/height)

a = Height of the wall panel (ft)

b = Length of the wall panel (ft)

G'_T = Shear stiffness based on total deflection (lb,in)

G'_{TL} = Shear stiffness based on total deflection in the negative direction (lb/in)

G'_{TR} = Shear stiffness based on total deflection in the positive direction (lb/in)

P_u = Ultimate load (lb)

P' = Damage threshold load level at initial cracking (lb)

S_u = Ultimate shear strength (lb/ft)

Δ_i = Deflection at gage i (in)

Δ_T = Total deflection (in)

Δ_{TL} = Total deflection in the negative direction (in)

Δ_{TR} = Total deflection in the positive direction (in)

SYSTEM IDENTIFICATION IN EARTHQUAKE ENGINEERING

by

James T. P. Yao[1], F. ASCE

and

Anshel J. Schiff[2], M. ASCE

INTRODUCTION

As structural engineers pursuing earthquake engineering research, our motivation and objective to apply system identification techniques are different from those of traditional control engineers because we are mainly concerned with the safety and reliability of certain given structures. Moreover, the situations for such application are different because of the inherent complexities in existing structural systems. In the following, an attempt is made to present a structural engineering viewpoint of this important subject area with an emphasis on the damage evaluation of existing structures.

In the classical problem (and narrow-sense definition) of system identification (e.g., see Sage and Melsa (24)[3] or Eykhoff (10)), a noise-corrupted system state vector, z(t), is observed along with input signal u(t), and input noise w(t). In general,

$$z(t) = h[x(t),\ u(t),\ w(t),\ p(t),\ v(t),\ t]$$

where
$x(t)$ = state vector
$p(t)$ = unknown parameters of the system
$v(t)$ = observation noise
t = time

Moreover, the state vector x(t) is assumed to be governed by a (deterministic or stochastic, linear or nonlinear) differential or integral equation. Solutions of the general system identification problem consist of (a) the determination of the form of the differential or integral equation, and (b) the estimation of unknown parameter vector p(t),

---

[1]Professor of Civil Engineering, Purdue University, W. Lafayette, IN
[2]Professor of Mechanical Engineering, Purdue University, W. Lafayette, IN
[3]Numerals in parentheses refer to corresponding items in Appendix I -- References

which may consist of coefficients of the system differential or integral equation as well as mean and variance values of the system noise w(t) and observation noise v(t). During the past fifteen years, such techniques have been successfully applied to obtain mathematical representations for various civil engineering structures. These resulting mathematical models are certainly more realistic than those without test data in making further analyses of the particular existing structure being studied. Several reviews of the available literature in this subject area are available (9,16,19,23,26,27).

Tests which are conducted for system identification purposes are usually performed at small response amplitudes to avoid the exceedence of any serviceability or safety limit states (18). Consequently, the applicability of the resulting mathematical model is restricted to the linear or slightly nonlinear range of the structural behavior. Although such mathematical representations are more realistic for linear analysis of the structure under consideration, they are not generally applicable for making safety analyses involving extreme loading conditions such as tornadoes or strong-motion earthquakes. In addition, it is well known that nonlinear structural behavior is a function of the previous load-history. Therefore, simple mathematical relationships to simulate the nonlinear, load-dependent, and time-variant behavior of complex structures subject to natural hazards are inadequate while more complex models allow a better representation of the real structural behavior, they require more data. One source of such data is to assess structural damage following each catastrophic event, and then use the results of such an assessment to modify the corresponding mathematical representation (28,32).

The ultimate objective of using methods of structural identification in the field of earthquake engineering is to improve the safety and reduce earthquake damage to engineered structures in a cost-effective manner. In this paper, the use of the expanded concept of structural identification in the following three situations is considered: namely, (a) the analysis of test data, (b) the analysis of strong motion response data, and (c) the analysis of post earthquake test data. Problems common to each of the above applications are reviewed. In addition, an approach to one of the main problems, uncertainty in the analysis, is suggested.

STRUCTURAL IDENTIFICATION

While needs exist for further structural applications of the existing methodology for system identification, it is believed that the identification of other characteristics such as the damage state and some reliability measure, which are of more direct interest to structural engineers, should be studied. In 1978, Liu and Yao (21) presented a comprehensive literature review of damage functions and discussed the general problem of structural identification. Recently, Gorman (15) considered the undesirable consequence of structural damage as a measure of risk. Nevertheless, it is still difficult to clearly define the degree of damage of a prototype complex structural system which exists in the real world. Thus, it is important for structural engineers to estimate the damage state at the time of the test or inspection as well as the use of classical system identification methods.

Almost all civil engineering structures are massive, stiff, and individually designed and constructed. Therefore, it is much more costly to conduct full-scale tests than other types of structures such as airplanes. Nevertheless, many such structures have been tested but usually under small-amplitude dynamic loading conditions (18,19). Recently, destructive and dynamic full-scale tests were performed on an 11-story reinforced concrete building (14) and a 3-span steel highway bridge (3). Experimental data from such full-scale destructive tests are considered to be very important in the development of a rational approach to damage assessment of existing structures (29). In addition, the application of structural identification can provide the designer-engineer with feedback information on the actual behavior of the structural system in the real world.

STRUCTURAL TESTING

While various methods of system identification can be used successfully in the laboratory, the application of these methods to real structures holds the potential for addressing some of the fundamental questions facing the profession. Modern computer methods provide a means of analyzing the basic system of most structures. However, the connectivity and the effect of non-structural members has a significant influence on system response. Testing, as used in this section, refers to the analysis of data obtained from low-level forced vibrations and ambient response data. Test data in this application can be characterized by long time series from several transducers. While the complexity and cost of performing full-scale tests is high, its application to the following areas holds great potential.

Two recent developments provide an increased incentive to use structural identification for locating transducers for strong-motion recording. First, the approach to instrumenting structures to determine its earthquake response is changing. Instead of placing a cluster of three transducers at each of a very limited number of transducers at prescribed locations (basement, mid-level and top) of a structure, more independent transducers are being distributed so that they will measure the response better. A logical extension of this approach is to use structural identification to aid in the process of locating transducers. The second development is that with increased efforts in earthquake prediction there will probably be a time when at least some earthquakes will be predicted. This provides the opportunity to concentrate instrumentation to measure structural response and creates a need to determine pre-earthquake response and transducer deployment.

Fundamental to the response of a system is its ability to dissipate energy. While dissipation during low-amplitude and high-amplitude response will in general be quite different, and understanding of how design and detailing influences dissipation will aid in the ability of its being able to be designed into the system.

In the two applications which have been suggested, it must be realized that low-level test results are being used to obtain future high-amplitude response data. Thus, there is a high degree of uncertainty in the mathematical model used to represent the structural system.

ANALYSIS OF EARTHQUAKE RESPONSE DATA

A significant benefit can be obtained using structural identification is in the evaluation of real earthquake data. This provides the potential of finding out how systems and materials behave at large response levels. Several distinct applications can be identified although they overlap each other to a degree. In general, earthquake response data will be characterized by records of short duration from a limited number of transducers. It should be emphasized that the analysis of response data is a supplement to rather than a substitution for field study of damage by structural engineers. In the following, three specific applications are discussed:

(a) It seems to be desirable to identify the character and location of nonlinear elements within the system. The objective here would be to determine which elements contribute to dissipating energy so that they can be incorporated into designs.

(b) It is always difficult to determine the sequence of events and distinguish the primary causes of damage from secondary failures. Thus, structural identification applied to earthquake data may provide a means of identifying failure mechanisms.

(c) The analysis of system response records using structural identification may provide a means to assess structural damage. Clearly these methods can be used to indicate changes in the response.

There are several major problems to these above mentioned types of analyses. The most interesting data will be those from structures responding in their nonlinear range. Thus, methods must be applicable to nonlinear analysis or be robust enough to be applied to nonlinear response. Piecewise linear methods have been used. Two of the most severe restrictions in evaluating earthquake data is that the excitations will be random in character and the records will be relatively short. Thus, from a statistical point of view, these analyses will have large uncertainties. Finally, there will be uncertainties associated with the system models since the structure will exhibit many different types of nonlinearity.

POST EARTHQUAKE TESTING

Ambient and forced vibration test and other specialized tests after an earthquake can serve several purposes. They are reviewed as follows:

(a) In the aftermath of a severe earthquake, there exists

a need to determine whether a given structure can be occupied prior to detailed inspection and possible repair. Analysis of ambient vibration tests in conjunction with pre-earthquake data and cursory inspection may provide the means for determining if the structure can be occupied.

(b) An extension of structural integrety evaluation is the assessment of damage. Analysis could be used to indicate if restoration should be attempted and could indicate the location and type of damage which is present. For these evaluations, specialized methods, e.g., those associated with nondestructive testing, should be used.

The above use of structural identification will again yield data which exhibit a high degree of uncertainty as to its interpretation.

APPROACHES TO UNCERTAINTY

An element which has been common in the above uses of structural identification is the uncertainty of models and the interpretation of results. First, it is vital to recognize the large uncertainties that are involved. At the present time most authors state the means of results obtained using structural identification but give no indication of their variability. While formal methods of analysis of variance may not be available, frequently simulation is used without assessing the variability of results. The large variability of results suggest the need for special methods of analysis.

Fu and Yao (12) considered the problem of damage assessment in terms of pattern recognition. In the theory of pattern recognition (1,22), data are collected from a physical systems such as an existing building structure. These data would include traditional response data from accelerometers as well as other observations and inspection data such as size, number, and location of cracks. A feature space is then extracted. Finally, a decision function or classifier is applied to obtain the classification, which in our case is the damage state. As an example, the reduction in natural frequency which can be determined with the application of classical system identification techniques can be used as a feature (4,5,7,8,11,17). As another example, the size, number, and location of cracks can be used as another feature (2). Recently, Ishizuka et al (25) discussed the possible application of a rule-inference method for damage assessment of existing structures.

According to Zadeh (34,35), as the complexity of a system increases, our ability of making precise and yet significant statements concerning its behavior diminishes. Consequently, the closer one looks at a real-world problem, which is usually complex, the fuzzier its solution becomes. The application of fuzzy sets in solving civil engineering problems was reviewed recently (30,31). Most civil engineering structures are indeed complex systems, the behavior of which can not be easily and clearly described.

In a recent paper (31), fundamental elements of the theory of fuzzy sets are given by Zadeh (35) and Kaufmann (20) are summarized along with several structural engineering examples from Yao (30) and a simplified version of an example on structural reliability from Brown (6). An attempt was then made to apply the theory of fuzzy sets to the complex problem of damage assessment of existing structures. Recently, Fu and Yao (13) discussed the application of the theory of fuzzy sets to several earthquake engineering problems.

CONCLUDING REMARKS

There exists a need for researchers in this subject area to include in their results an estimation of the variability involved. The large variability of these results is not generally acknowledged by many investigators at present.

One commonly accepted measure of uncertainty has been the probability during these past three decades. Recently, Blockley (private communication in April 1980) pointed out a paradox and suggested that the probability should not be used to measure the plausibility, credibility, or dependability of any assertion.

The available techniques for system identification has now grown to such an extent that several methods are applicable to any given situation. The computer implementation of most system identification methods is a major undertaking, and often includes "tricks" which may not be known to users other than the particular investigators who developed such programs. Thus, it is very important to properly document the software for each practical method of system identification. In this manner, the robustness of various methods can be evaluated through appropriate comparisons of these methods against one another in real-world situation.

Results of system identification studies are considered as a part of the input to the process of structural identification as defined herein. Generally, necessary data can be generated from various testing and inspection procedures. The processes of data reduction and decision making for the purpose of damage assessment remain to be studied further. In this regard, the application of pattern recognition and fuzzy sets shows promise and requires further study.

Although the merits of using active control in earthquake engineering are still being debated (33), such applications would require structural identification techniques. Especially when the control is applied for safety purposes, the reliability of system identification methods as well as that of the control system become paramount.

ACKNOWLEDGMENT

We wish to thank the National Science Foundation for their continued support, without which this paper could not have been prepared.

Vicki Gascho capably typed this manuscript.

APPENDIX I -- REFERENCES

1. Andrews, H. C., Introduction to Mathematical Techniques in Pattern Recognition, Wiley-Interscience, 1972.

2. Aristizabal-Ochoa, J. D., and Sozen, M. A., Behavior of a Ten-Story Reinforced Concrete Walls Subjected to Earthquake Motions, SRS No. 4, Department of Civil Engineering, University of Illinois, Urbana, IL, October 1976.

3. Baldwin, J. W., Jr., Salane, H. J., and Duffield, R. C., Fatigue Test of a Three-Span Composite Highway Bridge, Study 73-1, Department of Civil Engineering, University of Missouri-Columbia, June 1978.

4. Beck, J. L., "Determining Models of Structures from Earthquake Records", Report No. EERL 78-01, California Institute of Technology, Pasadena, CA, June 1978.

5. Beck, J. L., and Jennings, P. C., "Structural Identification Using Linear Models and Earthquake Records", Private Communications, 1978.

6. Brown, C. B., "A Fuzzy Safety Meausre", Journal of the Engineering Mechanics Division, v. 105, n. EM5, October 1979, pp. 885-872.

7. Chen, S. J. Hong, and Yao, J. T. P., "Damage Assessment of Existing Structures", Proceedings, Third ASCE EMD Specialty Conference, University of Texas, Austin, TX, 17-19 September 1979, pp. 661-664.

8. Chen, S. J. H., Yao, J. T. P., and Ting, E. C., "Reliability Evaluation of Structural Damage Using Measurable Data", presented at the ASCE Specialty Conference on Probabilistic Mechanics and Structural Reliability, Tucson, AZ, 10-12 January 1979.

9. Collins, J. D., Young, J. P., and Keifling, L. A. "Methods and Applications of System Identification in Shock and Vibration", System Identification of Vibrating Structures, Edited by W. Pilky and R. Cohen, ASME, New York, 1972, pp. 45-72.

10. Eykhoff, P., System Identification - Parameter and State Estimation, John Wiley & Sons, 1974.

11. Foutch, D. A., Housner, G. W. and Jennings, P. C., Dynamic Responses of Six Multistory Buildings During the San Fernando Earthquake, Report No. EERL 75-02, California Institute of Technology, Pasadena, CA, 1975.

12. Fu, K. S., and Yao, J. T. P., "Pattern Recognition and Damage Assessment", Proceedings, Third ASCE EMD Specialty Conference, University of Texas, Austin, TX, 17-19 September 1979, pp. 344-347.

13. Fu, K. S., and Yao, J. T. P., "Application of Fuzzy Sets in Earthquake Engineering", to be presented at the International Congress on Applied Systems Research and Cybernetics, Acapulco, Mexico, 12-15 December 1980.

14. Galambos, T. V., and Mayes, R. L., Dynamic Tests of a Reinforced Concrete Building, Research Report No. 51, Department of Civil Engineering, Washington University, St. Louis, MO, June 1978.

15. Gorman, M. R., Reliability of Structural Systems, Report No. 79-2, Department of Civil Engineering, Case Western Reserve University, Cleveland, OH, May 1979.

16. Hart, G. C., and Yao, J. T. P., "System Identification in Structural Dynamics, Journal of the Engineering Mechanics Division, ASCE, v. 103, n. EM6, December 1977, pp. 1089-1104.

17. Hidalgo, P., and Clough, R. W., Earthquake Simulator Study of a Reinforced Concrete Frame, Report No. EERC 74-13, Earthquake Engineering Research Center, University of California, Berkeley, California, December 1974.

18. Hudson, D. E., "Dynamic Tests of Full-Scale Structures", Journal of the Engineering Mechanics Division, ASCE, v. 103, n. EM6, Dec. 1977, pp. 1141-1157.

19. Ibanez, P., et al, Review of Analytical and Experimental Techniques for Improving Structural Dynamic Models, Bulletin 249, Welding Research Council, New York, June 1979, 44 pages.

20. Kaufmann, A., Introduction to the Theory of Fuzzy Subsets, Translated by D. L. Swanson, Academic Press, 1975.

21. Liu, S. C. and Yao, J. T. P., "Structural Identification Concept", Journal of the Structural Division, ASCE, v. 104, n. ST12, December 1978, pp. 1845-1858.

22. Mendel, J. M. and Fu, K. S., Editors, Adaptive, Learning and Pattern Recognition Systems, Academic Press, 1970.

23. Rodeman, R., and Yao, J. T. P., Structural Identification - Literature Review, Technical Report No. CE-STR-73-3, School of Civil Engineering, Purdue University, December 1973, 36 pages.

24. Sage, A. P., and Melsa, J. L., System Identification, Academic Press, 1971.

25. Ishizuka, M., Fu, K. S., and Yao, J. T. P., A Rule - Influence Method for Damage Assessment of Existing Structures, Technical Report No. CE-STR-80-8, School of Civil Engineering, Purdue University, West Lafayette, IN, July 1980.

26. Schiff, A. J., "Identification of Large Structures Using Data from Ambient and Low Level Excitation", System Identification of Vibrating Structures, Edited by W. D. Pilkey and R. Cohen, ASME, 1972, pp. 87-120.

27. Ting, E. C., Chen, S. J. Hong, and Yao, J. T. P., System Identification Damage Assessment and Reliability Evaluation of Structures, Technical Report No. CE-STR-78-1, School of Civil Engineering, Purdue University, February 1978, 62 pages.

28. Yao, J. T. P., "Assessment of Seismic Damage in Existing Structures", Proceedings, U.S.-S.E. Asia Symposium on Engineering for Natural Hazard Protection, Edited by A. H-S. Ang, Manila, Philippines, pp. 388-399.

29. Yao, J. T. P., "Damage Assessment and Reliability Evaluation of Existing Structures", Engineering Structures, England, v. 1, October 1979, pp. 245-251.

30. Yao, J. T. P., "Application of Fuzzy Sets in Fatigue and Fracture Reliability", presented at the ASCE/STD/EMD Specialty Conference on Probabilistic Mechanics and Structural Reliability, Tucson, AZ, 10-12 January 1979.

31. Yao, J. T. P., "Damage Assessment of Existing Structures", Journal of the Engineering Mechanics, ASCE, v. 106, No. EM3, August 1980.

32. Yao, J. T. P., and Anderson, C. A., "Reliability Analysis and Assessment of Existing Structural Systems", presented at the Fourth International Conference on Structural Mechanics in Reactor Technology, San Franciso, CA, August 15-19, 1979.

33. Yao, J. T. P., "Identification and Control of Structural Damage", Structural Control, Edited by H. H. E. Leipholz, North-Holland Publishing Company, 1980, pp. 757-777.

34. Zadeh, L. A., "Fuzzy Sets", Information and Control, v. 8, 1965, pp. 338-353.

35. Zadeh, L. A., "Outline of a New Approach to the Analysis of Complex Systems and Decision Processes", IEEE Transactions on Systems, Man and Cybernetics, v. SMC-3, n. 1, January 1973, pp. 28-44.

DYNAMIC STRESSES FROM EXPERIMENTAL AND MODAL ANALYSES

M. A. TUCCHIO[1], J. F. CARNEY III[2], M.ASCE,
H. I. EPSTEIN[3], M.ASCE

ABSTRACT

This paper presents a methodology to determine the dynamic stresses in a structure from a few measured accelerations. A suitable experimental stress analysis technique which uses accelerometers has not been available due to (1) the difficulty in the removal of rigid body modes from accelerometer data and (2) the error involved in the determination of displacements from accelerometer data. The method described herein takes advantage of recent developments in the processing of accelerometer data in conjunction with a scheme which relates the measured accelerations to the dynamic stresses. The theory of the method is developed and checked both analytically and experimentally. A closed form solution for a simply supported beam with suddenly applied end moment is used to provide the analytical support. Base excitation for a cantilever beam is used in the experimental verification.

INTRODUCTION

The object of this paper is to show that the dynamic stresses in a structure can be obtained from measured acceleration data in conjunction with the calculation of the normal modes of the structure. The driving force in performance of this analysis has been the absence of a method that can be used to determine the dynamic stresses in an entire large structure without resorting to hundreds of strain gages. This can be a significant problem when it is not possible to pass the many wires for the strain gages through such structures as the pressure hull of a submarine, the fuselage of an aircraft or the body of a missile silo. It is shown that the proposed technique works well, but is subject to problems associated with the current state-of-the-art in gathering displacement data from accelerometers.

Techniques exist that can be used to determine stresses from measured displacement data but they must use some form of the strain displacement relationship. These techniques are subject to the error

[1] Team Leader, U.S. Naval Underwater Systems Center, New London, CT
[2] Professor of Civil Engineering, University of Connecticut, Storrs, CT
[3] Associate Professor of Engineering, University of Connecticut, Storrs, CT

involved in taking spatial derivatives of experimental data and can be applied only to small areas within the field of measurement.

Review of the literature has not revealed any attempts at the technique described herein. Basically, there are two reasons for this, both of an experimental nature. The first reason is the difficulty arising from methods used in the past for removal of rigid body modes from a structure. This was treated recently by Padagaonkar (1) wherein he avoids the numerical problem associated with accumulative measurement error. The second reason is the difficulty of determining displacements from acceleration data. Several different procedures for this have been proposed over the years. Trifunac (2) and Pecknold (3) attempt to produce corrected accelerograms by giving estimates of initial ground velocity, ground displacement and ground acceleration. According to Pecknold (3), "Unfortunately, there is no completely satisfactory way to avoid the problem at present." In this paper, the parabolic type correction, as established by Trifunac (2) is used with partial success.

The theory of the method is developed from a modal solution of the general equations of motion of an elastic structure in matrix form. A generalized inverse is used to solve for the stresses based on a rectangular eigenvector matrix. This allows the number of accelerometers to be less than the number of modes. The generalized inverse provides a least squares approximation to the overspecified set of equations. A closed form solution for a simply supported beam with suddenly applied end moment is used to provide the analytical support. Considerations of experimental limitations are investigated and discussed in the demonstration problems. The accuracy of measurements as related to the spacial distribution of accelerometers and number of modes is shown to be of major importance.

In the solution of the equations of motion, the application of a generalized inverse is shown to increase the accuracy without increasing the number of measurements. However, the positioning of measurement points is demonstrated to be extremely important. The method gives excellent results within the limits of application delineated in the paper.

MODAL ANALYSIS

The required modal analysis is based on the method developed by Hurty and Rubinstein (4) and Beliveau (5). It will be assumed that the equations of motion can be represented by

$$[M]\ \{\ddot{x}\} + [C]\ \{\dot{x}\} + [K]\ \{x\} = \{P(t)\}, \tag{1}$$

Where $[M]$, $[C]$ and $[K]$ are the mass, damping and stiffness matrices, respectively, the force, $P(t)$, is a function of time, and x is the displacement vector.

For large amounts of damping, $[C]$, it is not possible to uncouple Eq. 1 directly by the separation of variables technique, $\{x\} = [\phi]\{q\}$, where $[\phi]$ is the eigenvector matrix, and $\{q\}$ is the generalized coordinate.

The damping force will be assumed to be linear. The damping need not be proportional, e.g., for any component of (ϕ^r) for an undamped system, or for one in which damping is proportional, (ϕ^r) is distinguished from other components by amplitude only, the phases being equal or 180^0 apart as determined by sign.

By utilizing the identity

$$(M)\{\dot{x}\} - (M)\{\dot{x}\} = 0, \tag{2}$$

the n Eqs. 1 can be expanded to 2n Eqs. 3

$$\left[\begin{array}{c|c} C & M \\ \hline M & 0 \end{array}\right]\left\{\begin{array}{c} \dot{x} \\ \hline \ddot{x} \end{array}\right\} + \left[\begin{array}{c|c} K & 0 \\ \hline 0 & -M \end{array}\right]\left\{\begin{array}{c} x \\ \hline \dot{x} \end{array}\right\} = \left\{\begin{array}{c} P \\ \hline 0 \end{array}\right\} \tag{3}$$

or, if Eq. 3 is written more compactly

$$(B)\ \{\dot{y}\} + (Q)\ \{y\} = \{Y\} \tag{4}$$

where $(B) = \left[\begin{array}{c|c} C & M \\ \hline M & 0 \end{array}\right]$; $(Q) = \left[\begin{array}{c|c} K & 0 \\ \hline 0 & -M \end{array}\right]$; and $\{y\} = \left\{\begin{array}{c} x \\ \dot{x} \end{array}\right\}$; $\{Y\} = \left\{\begin{array}{c} P \\ \hline 0 \end{array}\right\}$ (5)

In order to uncouple Eq. 4 and use the separation of variables technique, y must be of the form:

$$\{y\} = (\Delta)\ \{z\} \tag{6}$$

The transformation matrix (Δ) is constructed column by column using the 2n eigenvectors $\left\{y^{(1)}\right\}, \left\{\bar{y}^{(1)}\right\}, \left\{y^{(2)}\right\}, \left\{\bar{y}^{(2)}\right\}, \ldots, \left\{y^{(n)}\right\}, \left\{\bar{y}^{(n)}\right\}$.

(The bar signifies complex conjugate.) The generalized coordinate, z, is also complex.

Investigation of the availability of measurement techniques to determine {y}, where it is necessary to determine phase as well as amplitude, leads to the conclusion that accurate phase information cannot be obtained with today's technology unless the application is limited to narrow frequency bands where special application equipment for a particular frequency response can be designed. Because of this and the inaccuracy of the determination of displacement from accelerometer measurements, this study is limited to cases where the C matrix is small. The modal analysis, therefore, is unaffected by the presence of damping. This has application in many areas including steel structures where it is common to experience less than 6 percent critical damping. With this restriction, the solution of Eq. 1 is, then, simply expressed as

$$\{x\} = (\phi)\ \{q\} \tag{7}$$

The acceleration amplitude is the only measurement needed to calculate the displacement {x} of Eq. 7. The eigenvector can be calculated using any suitable technique.

Generalized Inverse

In Eq. 7, it is necessary to assume the number of modes to be considered. This is common in all modal analyses. In many structural applications, only the first few modes are of interest; this paper will be restricted to consideration of such problems.

In anticipation of problems associated with the accuracy of {x}, let it be assumed that fewer measurement points are available than the number of modes to be considered. That is, let Eq. 7 be written as

$$\{x\}_{NAC} = (\phi)_{NAC,NMODES}\,\{q\}_{NMODES} \tag{8}$$

in which NAC = number of accelerometers and NMODES = the number of modes, where NMODES $\geq$ NAC.

Now (ϕ) is rectangular and, in effect, {x} is over specified. To maximize the effectiveness of the number of measurements, so that as many modes as possible can be considered, it is assumed that the number of modes, NMODES, is equal to or greater than the number of measurement points or accelerometers, NAC. The determination of the generalized coordinate, {q}, from Eq. 8 involves the pseudoinverse of (ϕ). This can be found by using a least squares polynomial fit (Goldstein (6)) to an NACxNMODES matrix. If the earlier complex eigenvector matrix were retained, it would still be possible to determine a pseudoinverse, as shown by Goldstein (7).

Removal of Rigid Body Modes

Removal of the rigid body modes is necessary so that (K) will not be singular. This is done in a manner that leads to a rigid body solution in which cumulative numerical integration errors are eliminated. Padgaonkar (1) suggests a nine accelerometer scheme which does not use information from a previous time step. For planar motion, only four accelerometers are required.

Normalized Stress Matrix

At this point in the development, the generalized coordinate has been determined from Eq. 8. A stress matrix is now constructed that is compatible with the form of the generalized coordinate.

The stress analysis is performed on a modal basis so that a stress matrix is formed, with columns that represent the stress in the structure due to a displacement equal to the eigenvector for that mode. Theoretically the number of positions considered is not limited. For example, let X_n represent a normal function for a beam; the flexural stresses can be obtained from Eq. 9.

$$\sigma_n = -E_c \frac{d^2X_n}{dx^2} \tag{9}$$

Where σ_n represents the stresses for the nth mode. A stress matrix, (ψ), for a number of modes equal to NMODES and for a number of positions equal to N can be constructed in the form

$$(\psi) = \begin{bmatrix} \sigma_{11} & \sigma_{12} & ------ & \sigma_{NMODE\ 1} \\ \sigma_{21} & & & \\ & & & \\ \sigma_{1\ N} & & ------ & \sigma_{NMODE\ N} \end{bmatrix} \tag{10}$$

The stress matrix can be used in conjunction with the generalized coordinates to determine the required stress distribution,

$$\{\sigma\} = (\psi)\{q\} \tag{11}$$

In Eq. 11, $\{q\}$ is the generalized coordinate determined by experiment and the unsubscripted $\{\sigma\}$ is the desired stress.

It should be noted that unless one knows a-priori the direction of the input displacement,the corresponding calculated stresses can be in phase or 180° out of phase with the measured results. Since the (ϕ) matrix is constructed based on an assumed eigenvector positive sign convention, the actual displacement might very well be of opposite sign.

ANALYTICAL SOLUTION

To test the method, the stresses in a simply supported beam are determined. The loading consists of an instantaneously applied oscillating end moment.

The exact solution for this beam can be obtained in a series form using the methods of Mindlin and Goodman (8) and can be solved for the specific case of a steel beam with the dimensions shown in Fig. 1.

The deflections, w, are calculated for three points on the beam and these are termed "measurements" as if they represent actual measured information.

For a fixed number of "measurement points", which in this case is three, the experimenter has only two variables on which he can exercise judgement: (1) placement of the measurement points and, (2) the number of modes he can consider in the eigenvector problem. It is assumed that the maximum number of modes that can be considered is six. (This is due to the limitations of the solution which neglects shear deformations and rotary inertia.) Frequency content of the applied loading normally is not under the control of the experimenter, so values of ω_f are chosen arbitrarily from 100 to 3600 Hz. These represent frequencies from less than the first resonance to slightly higher than the fourth natural frequency. The placement of the three "measurement points" is varied along the length of the beam according to two coordinate groups

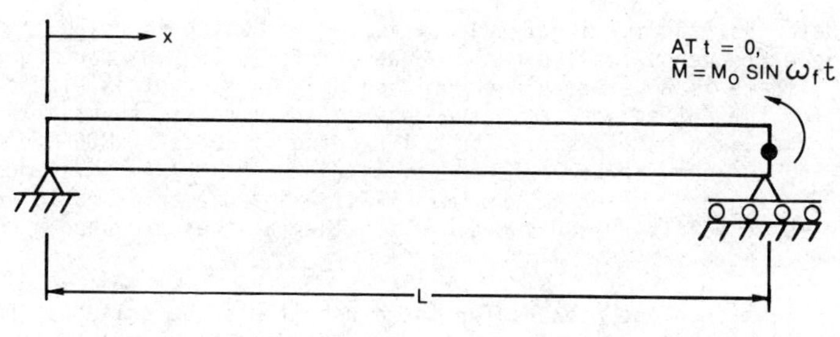

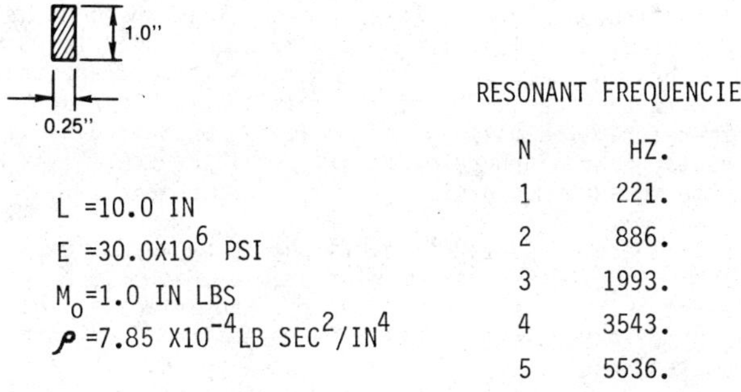

L =10.0 IN
E =$30.0\text{X}10^6$ PSI
M_O=1.0 IN LBS
ρ =7.85 $\text{X}10^{-4}$LB SEC^2/IN^4

RESONANT FREQUENCIES

N	HZ.
1	221.
2	886.
3	1993.
4	3543.
5	5536.

FIGURE 1

SIMPLY SUPPORTED BEAM WITH AN INSTANTANEOUSLY APPLIED END MOMENT

(1) x = 1.5, 5.0, and 7.5 inches,

(2) x = 4.8, 5.0, and 5.2 inches.

This represents a coarse and fine spatial sampling of the data. The exact stress is calculated at the midpoint of the beam (8). The eigenvector and normalized stress matrices are calculated in the usual manner for a simply supported beam.

From Fig. 2 it can be seen that for a σ matrix size of 3 x 3, (i.e., NAC = 3 and NMODES = 3), both groups of measurement point locations give excellent correlation between approximate and exact stresses.

Note that the forcing frequency is 100 Hz, which is below the first resonance frequency for the beam. However, for a 3600 Hz forcing frequency, there is considerable error when using a σ matrix size of 3 x 3 for the coarse spacing. The results are improved substantially by using a 3 x 6 matrix size, i.e., including six modes (NMODES = 6) and using the same three measured points (NAC = 3). The difference between using a 3 x 3 or a 3 x 6 matrix size is not as dramatic when the measured points are spaced close together as they are when a coarse spacing is used.

Insight into the reasons for these results can be gained by referring to the normal modes of a simply supported beam. Curve fitting the first mode with any three points is relatively straightforward. However, the coarse spacing falls on the nodal points for the fourth mode. (The driving frequency was purposely made close to the fourth mode.) A reasonable description of a sine wave requires at least nine equally spaced points on the curve. The forced fitting of any mode higher than one with the coarse distribution is not feasible or appropriate. Therefore, the maximum mode that is considered and included in the rectangular eigenvector matrix should be compatible with the spacing of the measured points.

Theoretically these measurement points should be very close together but, from the experimenter's point of view, the close spacing requires a high degree of measurement accuracy. For example, the displacement value of any point, at best, is determined from integrated accelerometer results to two significant figures. The actual displacement of points close together on a structure could show little relative displacement in a measuring apparatus and give erroneous stress results.

EXPERIMENTAL RESULTS

A 1.0 x 0.25 x 12.1 in. piece of flat steel stock is welded to a 6 x 6 x 0.5 in. steel angle and bolted to the carriage, as shown in Fig. 3. Strain gages are bonded to the beam at locations designated 1 through 4.

The free vibrations solution for the cantilever beam is done using Timoshenko beam theory to determine (ϕ) and (ψ).

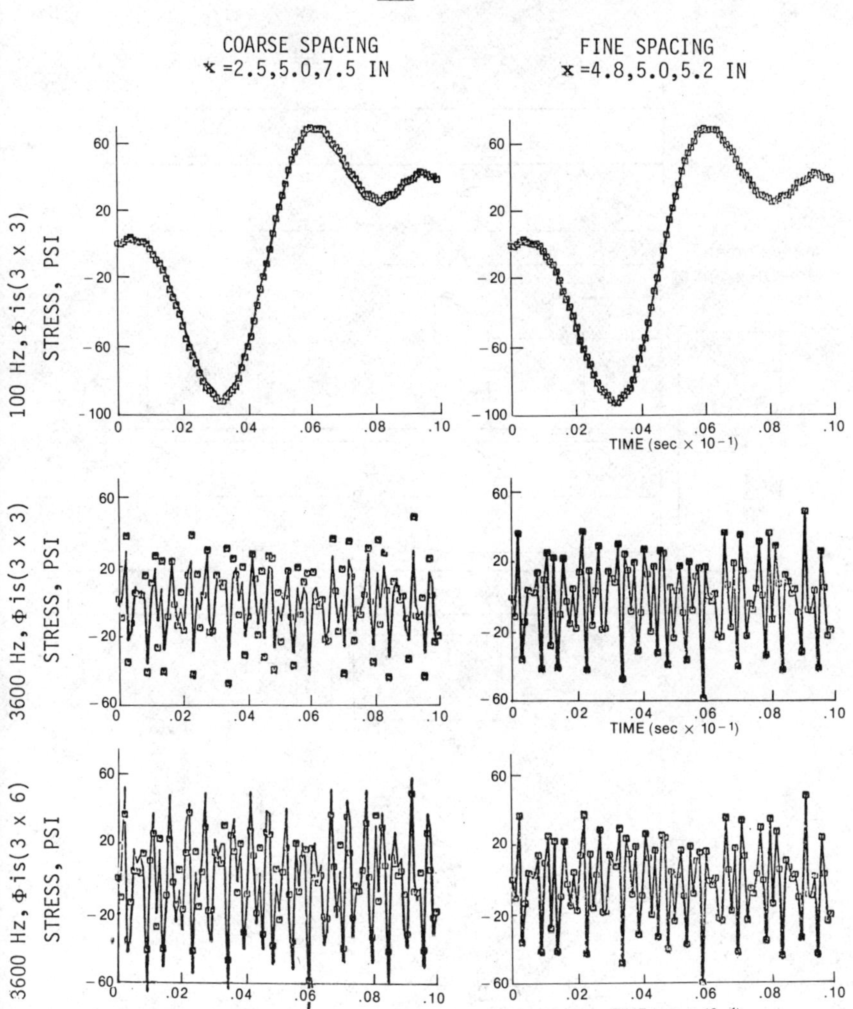

FIGURE 2

APPROXIMATE VS. EXACT SOLUTION

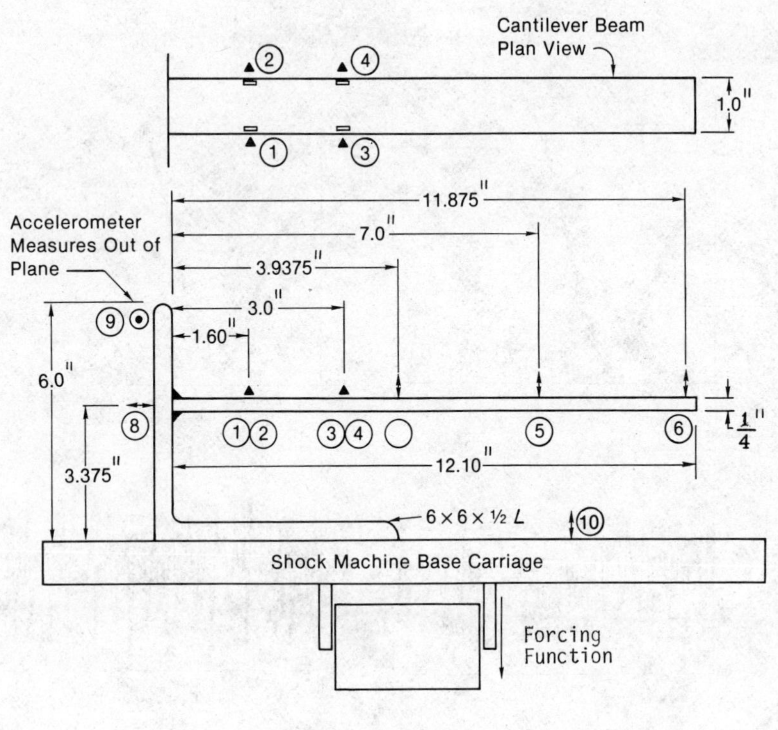

FIGURE 3
CANTILEVER BEAM TEST SET-UP

Integration of Accelerometer Data

Integration of the accelerations to obtain displacement is done using Simpson's rule. However, this does not give satisfactory results for the purposes of this investigation. A review of the recent literature, (2), (3) and discussions with the measurement community show that this problem was by no means unique. Noise and slight low frequency initial conditions can seriously affect the measured response. In this work, it was decided to attempt to solve the integration problem by using a simple boundary value approach. A constant initial acceleration is assumed and the velocity or displacement results are corrected based on this initial condition.

Physically, the carriage is driven down an arbitrary amount and is decelerated by striking lead pads between the moving carriage and the stationary anvil. The velocity of the carriage should be zero after a short period of time. This zero velocity condition is imposed on the first integration of acceleration as a boundary value. This results in a corrected velocity which,upon integration, gives the proper displacement.

In like manner, the relative displacement of the beam with respect to the carriage must be zero after a period of time. Channels 5 and 6 are corrected by forcing this boundary value on the second integration. The drift, or accumulated error, results from an initial small value of acceleration and is assumed to be constant. The velocity correction is linear and the displacement correction is parabolic.

This correction technique works very well for channel 5 and fairly well for channel 6 as can be seen by inspection of Figs. 4 and 5. This observation is based on the attainment of steady state for channel 5 and the continued erratic behavior of channel 6. As will be shown, processing of channel 5 data is sufficient to determine the stresses accurately. Unfortunately it is difficult to select the proper displacement formulation for correction and only by engineering inspection, considering the physical realities, can the quality of the result be judged.

Comparison with Calculated Stresses

In the NAC X NMODES eigenvector array, NAC takes on values of 1 or 2 for channels 5 and/or 6 data and NMODES varies from 1 to 6. It is possible to order the processing of the data in two ways (1) letting channel 5 = x_1, or (2) letting channel 6 = x_1.

Thus, in the calculation $\{x\} = (\phi)\ \{q\}$, the displacement vector may be ordered as

$$\{x\} = \begin{Bmatrix} x \text{ CHANNEL 5} \\ x \text{ CHANNEL 6} \end{Bmatrix} = [\phi_{\text{NAC, NMODES}}] \begin{Bmatrix} q_{\text{NMODES}} \end{Bmatrix} \qquad (12)$$

or

$$\begin{Bmatrix} x \text{ CHANNEL 6} \\ x \text{ CHANNEL 5} \end{Bmatrix} = [\phi_{\text{NAC, NMODES}}] \begin{Bmatrix} q\text{NMODES} \end{Bmatrix}, \qquad (13)$$

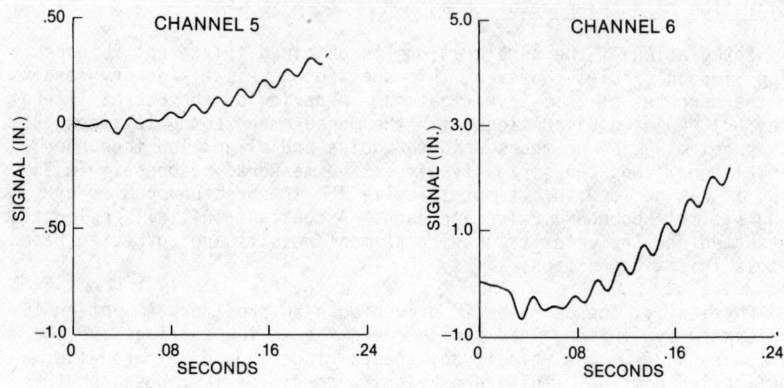

FIGURE 4

DISPLACEMENT OF CHANNELS 5,6 BEFORE CORRECTION

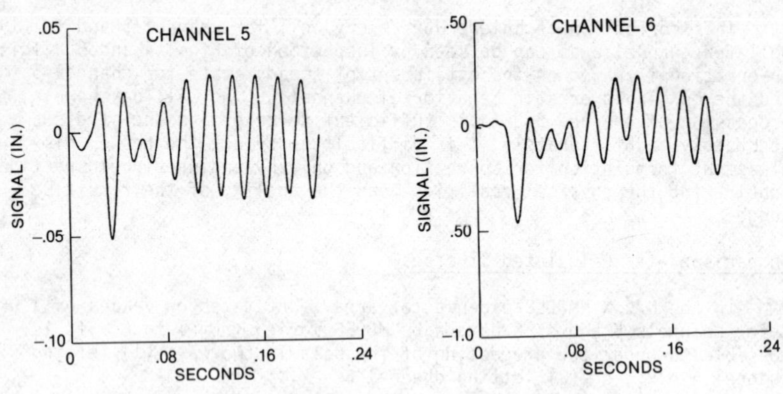

FIGURE 5

DISPLACEMENT OF CHANNELS 5,6 AFTER CORRECTION USING PARABOLIC ERROR ASSUMPTION

Theoretically, it should make no difference whether Eq. 12 or 13 is used. However, because the quality of the integration is not consistent, the actual results vary significantly.

Figure 6 shows the calculated results for one mode with channel 5 data only for the two corresponding strain measurement positions, channels 1 and 3, and calculated positions 1 and 3. There is excellent agreement for position 1 between measured stresses in Fig. 7 and the calculated results. Except for the initial transient, there is also good agreement at position 3 after 0.04 seconds.

When the number of modes is increased to two, so that the eigenvector matrix is now 1 x 2, calculated stresses are higher than measured stresses for positions 1 and 3. Similarly, when using three modes, the calculated stresses are slightly higher than measured values, especially in the neighborhood of the initial transient. Up to six modes are included and the calculated results are consistently higher than the measured values. However, in all cases, the frequency content of the calculated values is precisely the same as that for the measured results.

The number of measurement points is now increased to two so that the eigenvector matrix becomes 2 x 2, 2 x 3, ... 2 x 6. The calculated results are much higher than the measured values for all cases, but the frequency content is preserved. Processing two displacements to calculate the stresses definitely requires much more precision than is currently available.

Instead of using channel 5 in the first position of Eq. 12 , channel 6 is substituted, as in Eq. 13 , with the corresponding exchange of the rows in the eigenvector matrix. The results are not as good, especially in the initial transient.

Because the quality of the integrated data is much better for channel 5 than channel 6, the calculated stresses correspond more closely to measured values when using Eq. 12 with NAC = 1.

SUMMARY AND CONCLUSIONS

The purpose of this investigation was to develop a scheme to process accelerometer data, together with a modal analysis of a structure, to determine the dynamic stresses in the structure. In the experiment several of the comparisons were very good. The analytical demonstration showed excellent agreement.

Unfortunately, the use of displacement data derived from accelerometers is not of sufficient quality to reliably predict the stresses in a structure. The state-of-the-art in deriving displacements from integrated accelerometer data has not progressed to the point where the scheme derived in this paper can be applied consistently. This is borne out by recent literature as well as the work presented here. However, the scheme itself is viable as shown by the sample analytical problem.

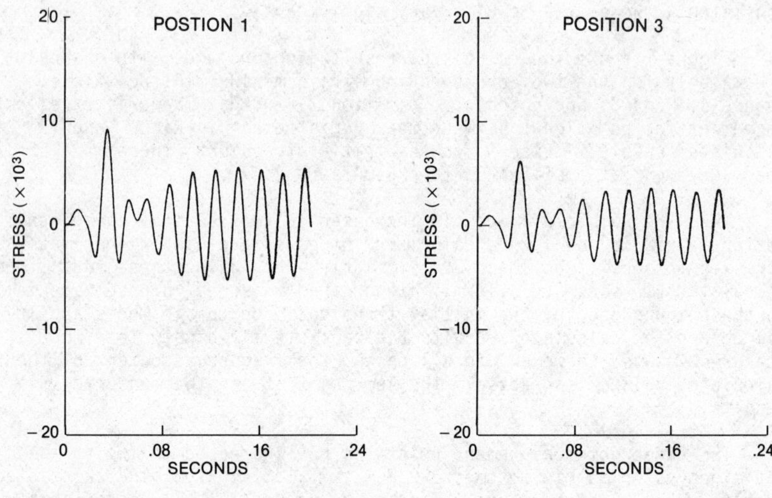

FIGURE 6

CALCULATED STRESSES AT POSITIONS 1 AND 3

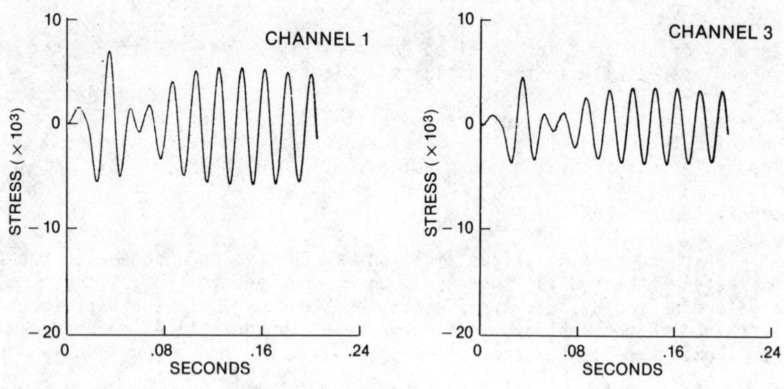

FIGURE 7

MEASURED STRESSES AT POSITIONS 1 AND 3

The placement of the accelerometers on the structure must be chosen carefully and should be based on the eigenvector solution. It is possible to locate the measurement positions in a manner such that a nearly singular eigenvector matrix is obtained. The resulting column of very small numbers in the eigenvector matrix significantly affects the quality of the inverse, and the magnitude of the resulting generalized coordinates may be very inaccurate. This problem arises because of the number of significant figures obtainable from the accelerometer integration in conjunction with the small numbers in the eigenvector column. It would first appear that placing additional accelerometers at antinodes to keep at least one member in the eigenvector column large, could provide a solution. Analytically, this is proved to be a very good technique. In the experiments, however, the small number of significant figures obtainable with the integrated accelerations precluded this strategy. The futility of accurately measuring relative displacements between a large number of positions is the basic problem.

The processing order of the displacements was also shown to affect the accuracy of the calculated stresses. This occurs when two or more accelerometers are used to obtain the stresses and the quality of the integrated acceleration data is not the same. Processing the least accurate displacement first gives erroneous calculated stresses.

REFERENCES

(1) Padagaonkar, A. J., et. al., "Measurement of Angular Acceleration of a Rigid Body Using Linear Accelerometers", Journal of Applied Mechanics, September 1975.

(2) Trifunac, M. D., "Low Frequency Digitization Errors and a New Method of Zero Baseline Correction of Strong-Motion Accelerograms", Earthquake Engineering Research Laboratory, EERL 70-07, California Institute of Technology, Pasadena, CA, 1970.

(3) Pecknold, D.A. and Riddell, R., "Effect of Initial Base Motion on Spectra", Journal of the Engineering Mechanics Division, Proceeding of the American Society of Civil Engineers, Vol. 104, No. EM2, April 1978.

(4) Hurty, W. C. and Rubinstein, M. F., Dynamics of Structures , Prentice Hall, Englewood Cliffs, NJ.,1964

(5) Beliveau, J., "Identification of Viscous Damping in Structures from Modal Information", Journal of Applied Mechanics, June 1976.

(6) Goldstein, M. J., "Linear Least Squares Estimation Using the Generalized Inverse", UNIVAC Scientific Exchange Meeting, Atlanta, Georgia, 15 October, 1970.

(7) Goldstein, M. J., "Reduction of the Pseudo-Inverse of a Hermitian Persymmetric Matrix", Journal of Mathematics of Computation, Vol. 28, 1974.

(8) Mindlin, R. D. and Goodman, L. E., "Beam Vibrations With Time-Dependent Boundary Conditions", Journal of Applied Mechanics, Vol. 17, 1950.

Identification of Wind Force Spectrum

Stepan S. Simonian*
J. H. Wiggins Company

Gary C. Hart**
Member of ASCE

ABSTRACT

A problem of fundamental importance in wind engineering is the quantification of wind gust forces acting on structures. The direct measurement of these forces on full scale structures are very difficult, hence we seek an alternate path to recover these forces from measured response quantities. This leads to an inverse problem, namely, the system identification problem in wind engineering. To test the feasibility of this approach, computer simulated and artificially noise polluted data is utilized as the response measurements. The response of a multistory building undergoing dynamic effects, due to wind gusts, is formulated next using the theory of stationary stochastic processes. The "measured" acceleration response at a floor level is related to the auto power spectral density function of the wind gusts, which are acting on the structure. Two different models for the wind gusts are considered. Next, the wind gust parameters are identified in a recursive fashion employing a novel continuous dynamic programming filter (DPF). This filter is operated continuously in the frequency domain for the first time. Further, Monte Carlo ideas are utilized to uncouple the resulting multiparameter system identification problem into a series of scalar problems, thus achieving considerable computational efficiency.

INTRODUCTION

A problem of fundamental importance in the design of wind-sensitive structures is the quantification of the wind gust forces acting on the structure. Two possible approaches exist to address this problem (a) the direct measurement of the wind gust forces on full scale structures, (b) the estimation of the wind gust forces from measured response quantities (typically velocity or acceleration). The first approach solves the problem directly, whereas the second approach leads to an inverse problem. In this paper, we address the second approach. The basic motivation in attacking the inverse problem to recover the inputs, rather than the direct measurement of these forces is twofold. On the

* Member of the Professional Staff, J.H. Wiggins Company, 1650 South Pacific Coast Highway, Redondo Beach, California 90277.

**Department of Mechanics & Structures, University of California, Los Angeles, California 90024.

one hand a large number of modern structures are permanently equipped with motion sensing instruments (accelerometers, seismometers, etc.); these sensors may be directly utilized to record wind induced motions without significant additional cost. On the other hand, the inherent difficulties in measuring pressures on the full scale structures are well known. These difficulties are of technical, economical, legal and of environmental origin [1].

In order to maximize their usefullness, it is desireable to express or model the wind gust forces in parametric form. The construction of a single universal gust force model to account for all situations is at present not feasible. The next best alternative is the construction of a set of simplified models applicable to particular cases. To be useful, the parameters of the wind gust models must at best have physical meaning and typically depend on local topographical conditions of the terrain, the geometry of the structure and its surroundings and parameters of meteorological origin. Many simplified parametric models of gust forces have appeared in the literature, in this respect the reader may consult the recent book by Simiu and Scanlan [2] for a review and further references.

A sensitivity analysis of these models quickly reveals that the response may be sensitive to the particular numerical values some of these parameters take. Hence, in order to confidently use these models the parameters appearing in these models must be known reasonably well. In the present paper, we develop a methodology to estimate some of these parameters of interest from wind response measurements. We next extend the methodology for the more general case of estimating the force spectrum where no a priori parametric models exist. For both applications, the building parameters may be considered either deterministic or random. The feasibility of the proposed inverse approach is demonstrated on an example of a multistory building structure utilizing computer simulated noisy data.

DIRECT PROBLEM - RESPONSE EQUATIONS

Employing various modeling techniques and invoking physical laws of motion, a set of differential equations result which describe the evolution of the building masses as a function of time. Using standard arguments and procedures from the theory of stationary random vibrations [3], the power spectral density (PSD) matrix of the acceleration (also velocity or displacement) response in the along-wind direction can be written as [3,4,5]

$$\underline{S}^{\ddot{x}}(n) = \underline{\Phi}\, \underline{H}_a^*(n)\, \underline{\Phi}^T\, \underline{S}^f(n)\, \underline{\Phi}\, \underline{H}_a(n)\, \underline{\Phi}^T \tag{1}$$

where $\underline{S}^{\ddot{x}}(n)$ is the NxN PSD matrix of physical accelerations of the building masses, N is the number of degrees of freedom of motion, the matrix $\underline{\Phi}$ is the $N \times \ell$ normalized (with respect to the mass matrix) modal matrix where $\ell \leq N$, the $\ell \times \ell$ matrix $H_a(n)$ is the diagonal complex receptance matrix with respect to accelerations, $\underline{S}^f(n)$ is the NxN PSD matrix of gust forces acting on the structure and n is the frequency in Hz. The super T denotes matrix transposition and the symbol * denotes the complex conjugate.

Given the quantities on the right hand side (rhs) of equation (1), we can calculate the response PSD matrix $\underline{S}^x(n)$ due to gust forces. This is commonly referred to as the direct problem.

Various models for the PSD matrix $\underline{S}^f(n)$ can be constructed assuming different parametric models for the spatial and frequency distribution of the gust forces [2,7]. Since the origin of gust forces is due to fluid flow, it must be clear that the construction of parametric models for $\underline{S}^f$ involves various concepts from experimental and theoretical fluid dynamics as well as local meteorological considerations and various idealizations [2]. As a consequence, it is expected that the models will depend on meteorological parameters, aerodynamic parameters due to the geometry of the structure, as well as on roughness parameters of the local terrain surrounding the structure. The merits of the parameterization of $\underline{S}^f(n)$ must now be obvious: The construction of $\underline{S}^f(n)$ is reduced to the accurate specification of a few fundamental parameters of a more meaningful nature. The direct problem can now be solved by specifying the accurate numerical values of these parameters as well as the true values of the building parameters - i.e., mode shapes $(\underline{\Phi})$, modal damping (ξ_i) and resonant frequencies (n_i).

From the various models for $\underline{S}^f(n)$ proposed in the literature, we choose for the purposes of this paper, a simple model proposed by Davenport [8,9,10]. Davenport's model has been extensively used and incorporated into various building codes. The model can be written as

$$\underline{S}^f(n) = A\, C^2(n)\, S_\upsilon(n)\, \underline{A}^E \tag{2}$$

where

$$\left.\begin{aligned}
A &= \left(\frac{\rho h b C_p V(h)}{1+\alpha}\right)^2 \\
1/C^2(n) &= (1+\alpha)^2\,(1+C_1 n)\,(1+C_2 n) \\
S_\upsilon(n) &= \frac{4\kappa\, V^2(h_o)}{n}\,\frac{x^2}{(1+x^2)^\beta} \\
V\,(Z) &= V_G\left(\frac{Z}{Z_G}\right)^\alpha \\
C_1 &= \frac{8}{3}\,\frac{h}{V(h)},\quad C_2 = 10\,\frac{b}{V(h)} \\
x &= L_s n \,/\, V(h_o)
\end{aligned}\right\} \tag{3}$$

$$A^E_{ir} = \left[\left(\frac{Z_i}{h}\right)^{\alpha+1} - \left(\frac{Z_{i-1}}{h}\right)^{\alpha+1}\right]\left[\left(\frac{Z_r}{h}\right)^{\alpha+1} - \left(\frac{Z_{r-1}}{h}\right)^{\alpha+1}\right] \tag{4}$$

The various parameters in expressions (3) and (4) are defined in Reference [4]:i.e., ρ is the air density, h is the building height, b is the building width normal to the response direction, C_p is the pressure coefficient for the building, V(h) is the along wind component of the mean wind velocity at height h, z is a vertical spatial coordinate measured from ground level, V_G and Z_G are the gradient velocity and length, α and κ are dimensionless parameters which are site dependent and are called power law and roughness coefficients, respectively; L_s is a turbulance scale length parameter and β is an emperically determined parameter; h_o is a reference height usually taken at h_o = 30 ft. Davenport has fixed parameters L_s and β to 4000 ft. and 4/3, respectively; finally Z_j j=0,1,2...., N are the tributory heights of each floor measured from ground level (See Figure 1).

The building properties (rectangular building) are completely specified by the parameters: $\underline{\Phi}$, $\underline{H}_a(\xi_j, n_j, n)$, ξ_j, n_j $(j=1,2,\ldots, \ell \leq N)$, h, b, Z_j. Similarly, the forcing function is completely given by specifying the parameters: κ, V_G, Z_G, α, L_s, h_o, β. Combining equations (1) and (2), we can rewrite equation (1) as

$$\underline{S}^{\ddot{x}}(n) = \underline{\Phi}\ \underline{H}_a^*\ \underline{\Phi}^T\ \underline{A}^E\ \underline{\Phi}\ \underline{H}_a\ \underline{\Phi}^T\ R(n) \tag{5}$$

where

$$R(n) \triangleq A\ C^2(n)\ S_\upsilon(n) \tag{6}$$

To emphasize the dependence of the response on building and force parameters, we write the j-kth component of the matrix equation (5) as

$$S^{\ddot{x}}_{jk} = S^{\ddot{x}}_{jk}(\underline{\Phi},\xi_1,\xi_2,\ldots,\xi_\ell,n_1,\ldots,n_\ell,b,h,Z_j,\kappa,V_G,Z_G,\alpha,L_s,\beta,n) \tag{7}$$

INVERSE PROBLEM

We now pose the inverse problem: Given measured values of $S^{\ddot{x}}_{jk}(n)$ denoted by $S^m_{jk}(n)$ for some j and k, find some of the uncertain parameters in the argument list of equation (7). Expressed in this fashion, we have embedded the inverse problem into a parameter estimation problem. The identification of the uncertain parameters will be carried out using a novel nonlinear recursive filter in the frequency domain.

FILTERING-BACKGROUND AND MOTIVATION

A vast body of literature exists in the theory and applications of linear, and to a lesser degree, nonlinear filters. To a large extent

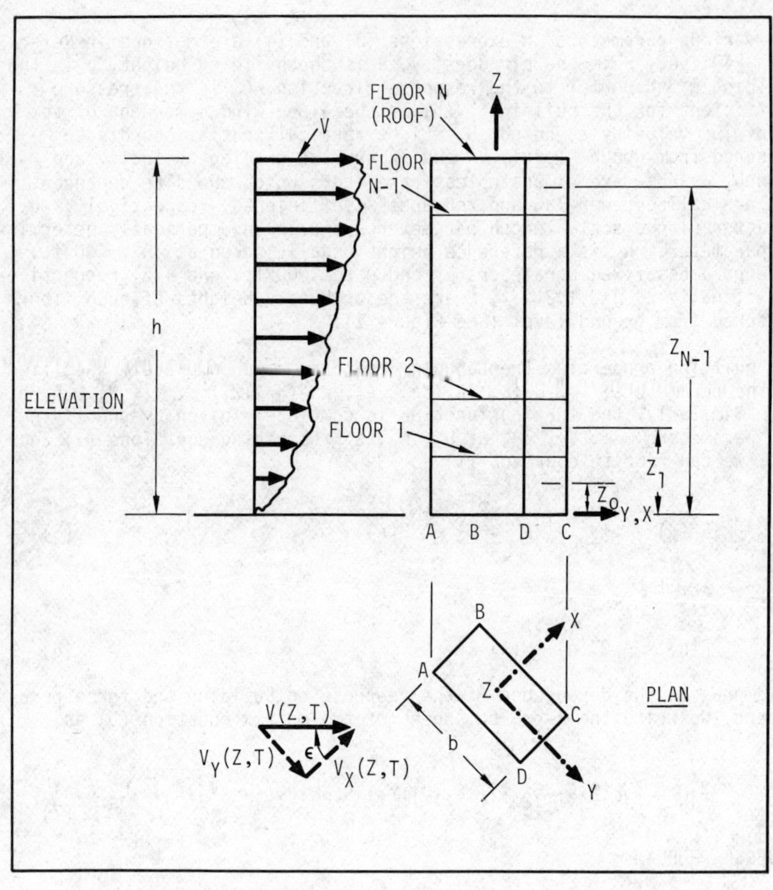

Figure 1. Geometry of Loading

the people responsible for its widespread use have been engineers in the controls, electrical and aerospace fields. Recently filters have been utilized in the area of parameter and state estimation problems in linear and nonlinear structural dynamics [11 to 17]. Although filtering has proved itself to be a powerful tool to the model builder and is applicable to both linear and nonlinear problems, it suffers due to its moderate to large computational requirements. This difficulty is especially acute in structural dynamic application, where it is not uncommon to model structures with hundreds to thousands degrees of freedom. As a consequence, for realistic applications (i.e., greater than ten degrees of freedom) in the realm of parameter identification problems in linear structural dynamics, time domain filtering approaches in their present form are very inefficient. To circumvent this large computational problem the filter has been reformulated for the first time in Simonian [4] in other domains, such as the frequency domain. This reformulation in effect eliminates the velocity and displacement states only retaining as states the parameters of interest, thus achieving a dramatic computational advantage over the time domain formulation. For further details of this reformulation the reader may consult the above mentioned references. A review of alternate possible approaches to reduce the computational burden of parameter identification problems will be the topic of a future communication; nevertheless, we will discuss one such approach in this paper, namely a Monte Carlo extension of the DPF (dynamic programming filter).

THE FILTER

Let the $N_S \times 1$ vector $\underset{\sim}{x}(t)$ denote the state we wish to estimate in the interval $[t_o, t_f]$ where t is an independent variable. The evolution of the state $\underset{\sim}{x}$ is governed by a system of first order differential equations given by

$$\dot{\underset{\sim}{x}}(t) = \underset{\sim}{g}\,[\underset{\sim}{x}(t),\ \underset{\sim}{q},\ t] + \underset{\sim}{z}(t),\quad \underset{\sim}{x}(t_o) = \underset{\sim}{x}^o \tag{8}$$

the overdot on $\underset{\sim}{x}$ denotes differentiation with respect to (wrt) the independent variable t, $\underset{\sim}{g}$ is a given vector valued function of its arguments, the $N_q \times 1$ vector $\underset{\sim}{q}$ denotes a constant random vector with known means and variances with a given distribution. The unknown vector $\underset{\sim}{z}(t)$ is introduced to account for modeling uncertainties, and finally $\underset{\sim}{x}^o$ denotes the initial state of $\underset{\sim}{x}$. Along with equation (8), we have a measurement equation given by

$$\underset{\sim}{a}(t) = \underset{\sim}{h}\,(\underset{\sim}{x}(t),\ \underset{\sim}{q},\ t) + \underline{\Gamma}\ \underset{\sim}{z}(t) + \underset{\sim}{\eta}(t) \tag{9}$$

the $N_m \times 1$ vector $\underset{\sim}{a}(t)$ is an experimentally measured quantity available in the interval $[t_o, t_f]$, $\underset{\sim}{h}$ is a given vector valued function of its arguments, $\underline{\Gamma}$ is a known $N_m \times N_S$ constant matrix and $\underset{\sim}{\eta}(t)$ denotes modeling and measurement errors. Next we define an error functional given by

$$J = \int_{t_o}^{t_f} \left(\underset{\sim}{\eta}^T(t)\ \underline{W}(t)\ \underset{\sim}{\eta}(t) + \underset{\sim}{z}^T(t)\ \underline{K}(t)\ \underset{\sim}{z}(t)\right)dt + \left(\underset{\sim}{x}(t_o)-\hat{\underset{\sim}{x}}^o\right)^T \underline{\Lambda}\left(\underset{\sim}{x}(t_o)-\hat{\underset{\sim}{x}}^o\right) \tag{10}$$

where $\underline{W}$, $\underline{K}$ and $\underline{\Lambda}$ are positive definite specified weighting matrices denoting our confidence in the measurement errors, process modeling errors and initial estimates, respectively; $\underset{\sim}{x}^o$ denotes our best a priori estimate of $\underset{\sim}{x}^o$.

Performing the minimization of J above wrt $\underset{\sim}{z}$ and then wrt $\underset{\sim}{x}$ and denoting the optimal estimate of $\underset{\sim}{x}(t)$ by $\underset{\sim}{e}(t)$, we are led to a system of differential equations to calculate the vector $\underset{\sim}{e}(t)$ [4],

$$\dot{\underset{\sim}{e}}(t) = \underset{\sim}{g}[\underset{\sim}{e}(t),\underset{\sim}{q},t] + [\underline{Q}(t)\underline{h}_e^T\overline{W}+\overline{K}^{-1}\underline{\Gamma}^T\underline{W}]\ (\underset{\sim}{a}(t)-\underset{\sim}{h}(\underset{\sim}{e},\underset{\sim}{q},t))$$

$$\dot{\underline{Q}}(t) = \underline{Q}(t)\underline{V}\ (\underset{\sim}{e},\underset{\sim}{q},t)\underline{Q}(t) + \underline{D}^T(\underset{\sim}{e},\underset{\sim}{q},t)\underline{Q}(t) + \underline{Q}(t)\underline{D}\ (\underset{\sim}{e},\underset{\sim}{q},t) + \overline{K}^{-1}(t)$$

$$\underset{\sim}{e}(t) = \hat{\underset{\sim}{x}}^o,\ \underline{Q}(t_o) = \underline{\Lambda}^{-1} \tag{11}$$

All the appropriate matrices appearing in the above equations have been completely defined in References [4] [18]. Below we present a particular version of equation (11) relevant to the present work.

MONTE CARLO DPF

Let us consider a particular estimation problem for which we have a special case of equation (8), i.e., $\underset{\sim}{g} \equiv \underset{\sim}{o}$; hence, equation (8) reduces to

$$\dot{\underset{\sim}{x}}(t) = \underset{\sim}{z}(t),\ \underset{\sim}{x}(t_o) = \underset{\sim}{x}^o \tag{12}$$

let us also assume that we have a measurement model equation (9) with $\underline{\Gamma} \equiv \underline{o}$, for this case equation (9) reduces to

$$\underset{\sim}{a}(t) = \underset{\sim}{h}\ (\underset{\sim}{x}(t),\ \underset{\sim}{q},\ t) + \underset{\sim}{\eta}(t) \tag{13}$$

In particular let us consider the case where $\underset{\sim}{x}(t)$ and $\underset{\sim}{a}(t)$ are scalars, hence equations (12) and (13) will now be scalars, i.e.,

$$\dot{x}(t) = z(t),\ x(t_o) = x^o \tag{14}$$

$$a(t) = h\ (x(t),\ \underset{\sim}{q},\ t) + \eta(t) \tag{15}$$

the estimation equations for equations (14) and (15) can now be written as

$$\dot{e}(t) = Q(t)\ (\partial h/\partial e)\ W\ [a(t) - h\ (e(t),\ \underset{\sim}{q},\ t)]$$

$$\dot{Q}(t) = Q^2(t)\ V\ (e,\ \underset{\sim}{q},\ t) + K^{-1}$$

$$e(t_o) = \hat{x}^o, \; Q(t_o) = \Lambda^{-1} \tag{16}$$

where

$$V = W\left[(\partial^2 h/\partial e^2)\;(a-h + (\partial h/\partial e)e - (\partial^2 h/\partial e^2)e^2) - (\partial h/\partial e)^2\right] \tag{17}$$

the quantities, W, K and Λ are the scalar versions of the weighting matrices defined in equation (10). For all subsequent numerical work, the second derivative term $\partial^2 h/\partial e^2$ has been dropped out from equation (17), thus reducing the second order filter into first order.

Let us recall that function $h(e, \underset{\sim}{q}, t)$ depends on the uncertain or random $N \times 1$ vector $\underset{\sim}{q}$ whose distribution and first two moments are assumed to be known. Using Monte Carlo (MC) ideas we performn M estimations of x, each time we use a particular random $\underset{\sim}{q}$ (computer generated) denoted by $\underset{\sim}{q}^{(1)}, \underset{\sim}{q}^{(2)}, \ldots, \underset{\sim}{q}^{(M)}$ and get the estimates $\underset{\sim}{e}^{(1)}, e^{(2)}, \ldots, e^{(M)}$ of $\tilde{x}$. For the numerical examples worked out using simulated data, it is observed that the mean of the estimates $e^{(j)}$ converge to the true value of x.

EXAMPLES

In this paper, we present two different numerical experiments. In the first example, we identify the mean wind velocity V(h) at roofheight h. In the second example, we identify the frequency dependent function R(n) defined by equation (6), assuming that no a priori information exists about the form of R(n). All the data used in the numerical experiments have been artificially generated by the computer by adding pseudo-random noise to the known responses.

The identification algorithm for the two examples are given by (See Equations (14) and (15)),

(A): n (frequency in Hz) $\equiv$ t, n is the independent variable
$x = V(h)$
$a(n)$ = measured auto-power spectral density (PSD) of roof acceleration
$h(x(n), \underset{\sim}{q}, n)$ = mathematical model for PSD

(B) $n \equiv t$
$x = R(n)$ (See Equation (5))
$a(n)$ and $h(x(n), \underset{\sim}{q}, n)$, same response quantities as defined above.

(A) Identification of x=V(h), A Monte Carlo Approach

In this example we identify the wind velocity at roof level h, i.e., V(h). Here we directly attack the uncertain parameter problem. In each of the eight runs studied we take $e(n_o) = V^o(h) = \pm\,(1.10)\,V(h) = (1.10)(305.567) = 336.12$ in/sec., $Q(n_o) = \Lambda^{-1} = 100.$, $K^{-1} = 1500.$, $W = 10^7$, with $\pm$ 5% amplitude dependent measurement noise. For all the runs we assume that the random vector is given by $\underset{\sim}{q} = (\xi_1 \;\beta\; C_p \;\kappa\; L_s \;\alpha)^T$. In the beginning of each run we choose a random vector $\underset{\sim}{q}$ whose elements are chosen to lie between ± 10% of their true value (or their best a priori estimates). The ± 10% variability of the elements of q is picked up randomly with a uniform distribution from a random number $\tilde{g}$enerator using subroutine GGUB from the IMSL software package. Two representative results of the numerical experiments are presented in Figures 2 and 3. An interesting result can be seen when we take the average of the eight runs. The average is computed to be $\bar{e} = \hat{V}(h) = 312.39$ in sec., while the true value is 305.57 in/sec., an overestimation of <u>only</u> + 2.23%. This result encourages us to treat the identification of α and η_1 in a similar fashion, the results of which appear in References [4] [19].

B. <u>Identification of x=R(n)</u>

Here we attempt to identify the frequency-dependent gust PSD function R(n) defined in equation (5). Rather than postulating a parametric structure to R(n) we assume here that no a priori information exists on the form of R(n), thus, leaving the burden of extracting this information from the "measured" data.

The simulated "measurements" are generated by contaminating the assumed roof acceleration PSD with noise of the form, $\eta(n) = r(n) A_m$, where r(n) is a zero mean uniformly distributed random variable taking values in the range $-1 < r(n) < 1$; the value chosen for A_m is 0 and .019 $(\text{in/sec}^2)^2$/Hz. We also use the following filter parameters: $e(n_o) = \hat{R}(n_o) = .75$, $Q(n_o) = \Lambda^{-1} = .01$, $K^{-1}=10$, W=10. In Figure 4, we display the identification results for the noise free case (i.e., $A_m=0$) and in Figure 5 the noise polluted case with the above quoted A_m.

The preliminary results in this example in attempting to identify unknown functions, without a priori information, and with very simple models, appear very encouraging. Methods to increase the accuracy of the results, especially in the initial interval where the function initial conditions are unknown, are reported in Reference [4] with additional examples. Other possible approaches to deal with related classes of problems will be the subject of future communications.

CONCLUSIONS

In this paper a new class of inverse problems is identified, namely, <u>the system identification problem in wind engineering</u>. The results reported show that two basic approaches may be feasible to attack the problem. The first approach assumes parametric models are available for the input forces with uncertain parameters, hence the force identification problem is solved as a parameter estimation problem. The second approach attempts to solve the inverse problem in its more fundamental setting; i.e., requiring no a priori assumptions. Results obtained so far show that this second approach is promising, hence should be pursued further, since it provides a very powerful tool to the researcher interested in modeling and identification of complicated phenomena.

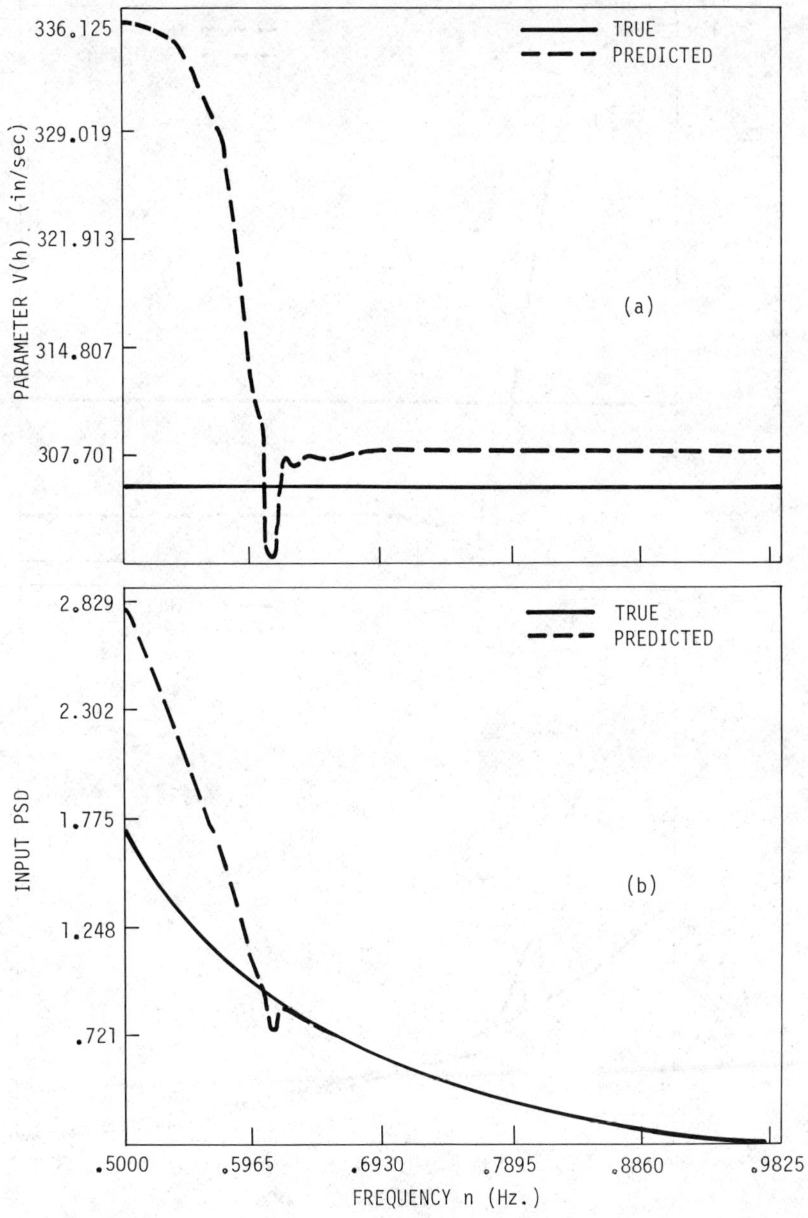

Figure 2. Tracking of (a) Parameter V(h) (b) Input PSD

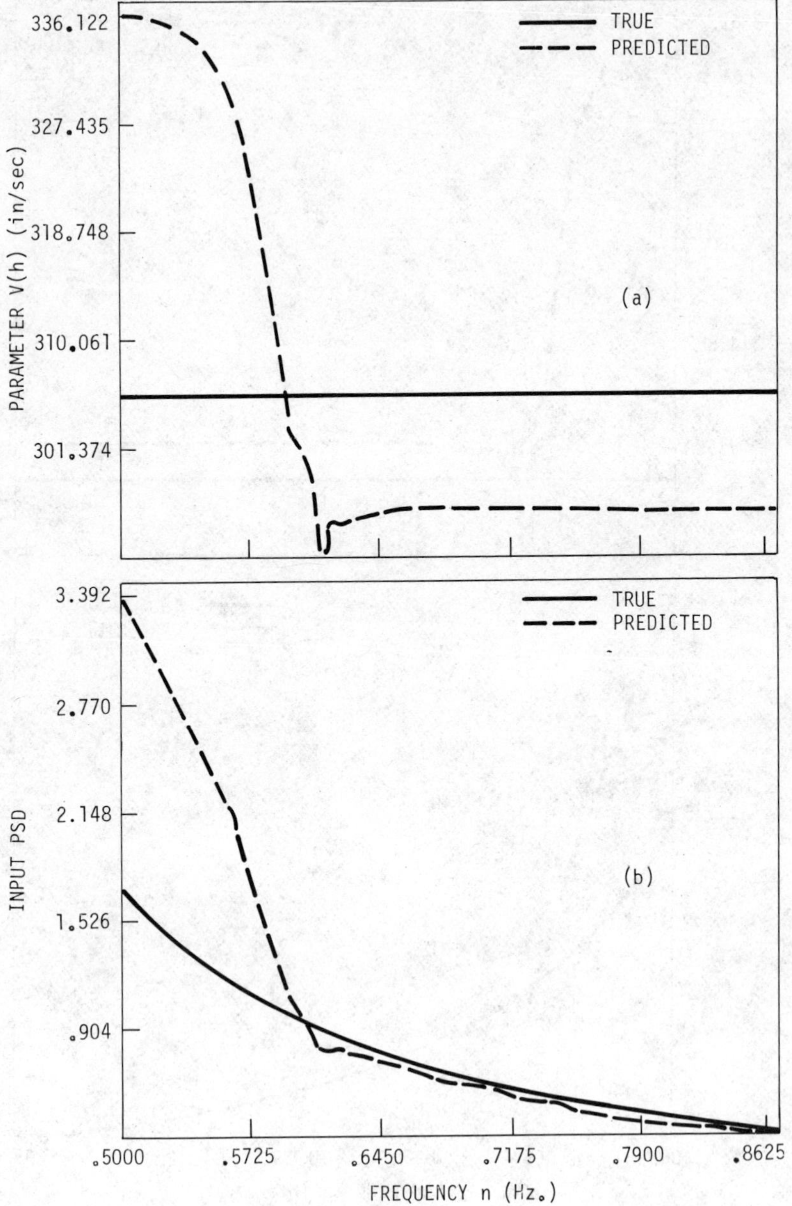

Figure 3. Tracking of (a) Parameter V(h) (b) Input PSD

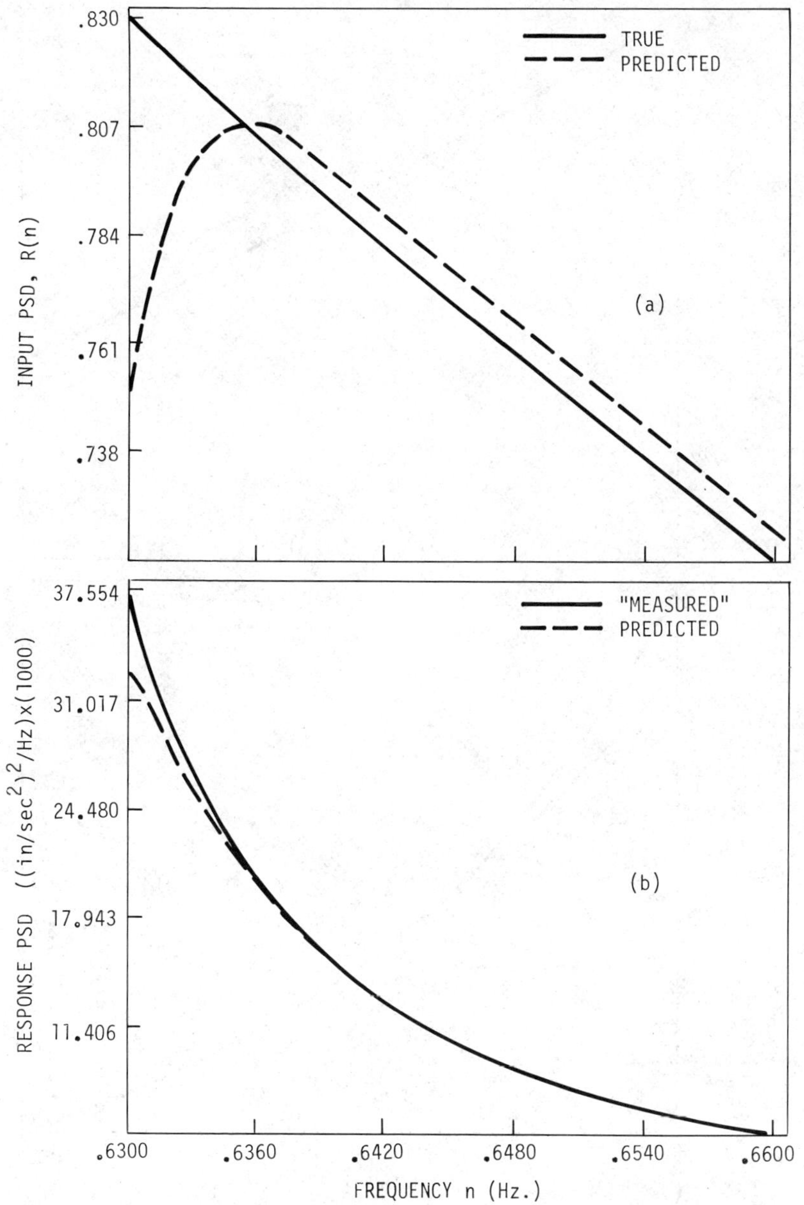

Figure 4. Tracking of (a) Input PSD (b) Response PSD

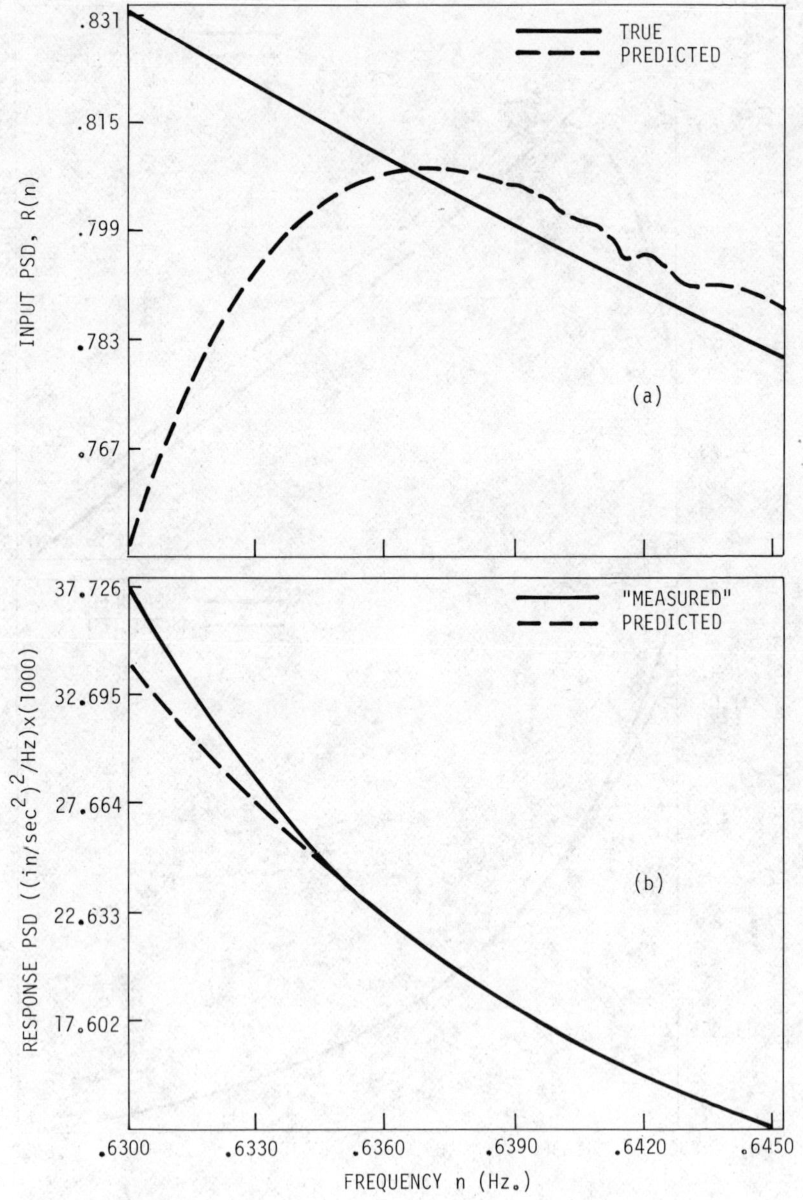

Figure 5. Tracking of (a) Input PSD (b) Response PSD

REFERENCES

1. Simonian, S.S., "System Identification of Wind Forces in Structural Dynamics," Presented for the oral qualifying examination as partial requirement for the Ph.D. degree, University of California, Los Angeles, August 1978.

2. Simiu, E. and Scanlan, R., Wind Effects on Structures, Wiley, New York, N.Y., 1978.

3. Robson, J.D., An Introduction to Random Vibration, Elsevier, Edinburgh University Press, United Kingdom, 1963.

4. Simonian, S.S., "System Identification in Structural Dynamics: An Application to Wind Force Estimation," Ph.D. Dissertation, School of Engineering and Applied Science, University of California, Los Angeles, 1979.

5. Clough, R. and Penzien, J., Dynamics of Structures, McGraw-Hill, New York, N.Y., 1975.

6. Hart, G., "Building Dynamics Due to Stochastic Wind Forces," Journal of Structural Division, ASCE, Vol. 96, No. ST3, March 1973.

7. Ghiocel, D. and D. Lungu, Wind, Snow and Temperature Effects on Structures Based on Probability, Abacus Press, England, 1975.

8. Davenport, A.G., "Gust Loading Factors," Journal of Structural Division, ASCE, Vol. 93, No. ST3, June 1967.

9. Davenport, A.G., "Rationale for Determining Design Wind Velocities," Journal of Structural Division, ASCE, Vol. 86, No. ST5, May 1960.

10. Davenport, A.G., "The Spectrum of Horizontal Gustiness Near the Ground in High Winds," Quarterly Journal of the Royal Meteorological Society, London, pp. 194-211, April 1961.

11. Wells, W.R. and Schwanz, R.C., "Application of Nonlinear Estimation Theory to Parameter Identification of Rigid and Elastic Aircraft," Proc. Fourth Symposium on Nonlinear Estimation Theory and Its Applications, San Diego, 1974.

12. Distefano, N. and Rath, A., "Modeling and Identification in Nonlinear Structural Dynamics-I One Degree of Freedom Models," Earthquake Engr. Research Center, Tech. Report No. EERC 74-15, University of California, Berkeley, December 1974.

13. Simonian, S.S. and Distefano, N., "Derivation of Second Order Nonlinear Filters: Application to Identification Problems in Nonlinear Structural Dynamics," Unpublished Tech. Report, University of California, Berkeley, 1975.

14. Distefano, N. and Pena-Pardo, B., "System Identification of Frames Under Seismic Loads," Journal of Engineering Mechanics Division, ASCE, Vol. 102, No. EM2, April 1976.

15. Thomson, T., Hart, G. and Watson M., "System Identification Methods in Earthquake Engineering," ASCE Fall Convention and Exhibit, Preprint 3098, San Francisco, October 1977.

16. Yun, C.B. and Shinozuka, M., "Identification of Nonlinear Structural Systems," Tech Report No. NSF-ENG-76-12257-3, Columbia University, New York, June 1978.

17. Beck, J.L., "Determining Models of Structures from Earthquake Records," Ph.D. Dissertation, School of Engineering, California Institute of Technology, Pasadena, 1979.

18. Simonian, S.S., "Inverse Problems in Structural Dynamics-I Theory," Inter. J. Numer. Meths. in Engr., in press.

19. Simonian, S.S., "Inverse Problems in Structural Dynamics-II Applications," Inter. J. Numer. Meths. in Engr., in press.

MATHEMATICAL MODELLING OF INDETERMINATE TRUSSES

by

Vernon Charles Matzen*, M. ASCE
and
Joseph Earlton Hardee, Jr.+

Abstract

Parameters representing axial stiffnesses are found for a statically loaded indeterminate truss by minimizing the squared error between the computed displacements and a set of numerically simulated measured displacements. The minimization is carried out with a modified Gauss-Newton alagorithm. Questions of model uniqueness are answered using eigenvalues and eigenvectors of an approximation to the Hessian matrix of the squared error function, and by using a parameter perturbation technique. Mathematical modeling is shown to be an accurate method for determining both the location and magnitude of damage to truss members.

1. Introduction

System identification is the procedure of formulating mathematical models (or improving existing models) of structures using measured excitation and response data. It has been applied successfully to a variety of problems. Examples of these applications can be found in other papers given at this conference and at the previous ASCE-EMD speciality conference on the dynamic response of structures.[1#] There are, however, several questions about system identification which remain unanswered. These pertain to the relationship between the uniqueness of the model on the one hand and the type of loading and number and placement of measurement devices on the other.

In our study of these questions, we have chosen a linearly elastic indeterminate truss loaded both statically and dynamically. This structure would, it seemed to us, be simple enough so that the modeling details could be clearly illustrated and, at the same time, be complex enough so that the answers to the questions would be meaningful.

---

*Assist. Prof. of Civil Engineering, North Carolina State Univ. at Raleigh.

+Research Assist., Dept. of Civil Engineering, North Carolina State Univ. at Raleigh.

#Superscripts refer to entries in the List of References.

After considering these questions, we then investigate one potential application of system identification, that of damage assessment. In a prototype application, we envision damage assessment working in this manner: excitation and response information will be collected both before and after a severe excitation, and then, using these data, "before" and "after" mathematical models will be formulated. Then differences detected between the two mathematical models will be used to provide specific information about structural damage - both the magnitude and location of damage if any occurred.

The uniqueness and damage assessment topics will be investigated separately for the static and dynamic loading cases. In each case, the procedure will be as follows: develop a computer program to formulate the model; verify the program and make preliminary investigations using numerical experiments; and then, using data from laboratory experiments, make the final investigations.

The research program described here is being conducted at NCSU under NSF Grant No. PFR-8007389. In this paper we present the results of our work through the numerical experiments for the static loading case

2. System Identification

The stystem identification process has been divided by Bekey[2] into the following three steps:

> "(a) Determination of the form of the model and isolation [selection] of the unknown parameters.
>
> (b) Selection of a criterion function by means of which the 'goodness of fit' of the model responses to actual system responses can be evaluated, when both model and system are forced by the same inputs.
>
> (c) Selection of an algorithm or strategy for adjustment of the parameters in such a way that the difference between model and system responses, as measured by the criterion of (b) above, are minimized."

Each of these steps is discussed separately in the context of our structure.

2.1 Form of Model and Selection of Parameters

The mathematical model for indeterminate trusses subjected to static loading in the linear range is most often based on the stiffness method of analysis, and it is this method which we use. The model takes the form $\underline{\underline{S}}(\underline{k})\ \underline{\underline{V}} = \underline{\underline{P}}$ (1)

where $(\underline{\underline{.}})$ indicates a two-dimensional array,

$(\underline{.})$ indicates a vector

$\underline{\underline{P}}$ is an NDOF x NLC loading array,

$\underline{\underline{S}}$ is an NDOF x NDOF stiffness matrix,

$\underline{k}$ is an NM x 1 vector of member axial stiffness with the ith element having the form $k_i = A_i E_i / L_i$,

A_i is the corss-sectional area,

E_i is the modulus of elasticity,

L_i is the length.

$\underline{\underline{V}}$ is an NDOF x NLC displacement array,

NDOF is the number of degrees of freedom,

NLC is the number of loading cases, and

NM is the number of members in the truss.

The model is based on the following assumptions:

a) all members are connected at joints with frictionless pins,
b) loads are applied only at joints,
c) the centroidal axes of all members framing into a joint intersect at a common point,
d) all member deformations are axial and within the linear region, and
e) deformations are so small that the initial geometry can be used in all calculations.

If a structure does conform to these assumptions, then eqn. (1) should yield displacements which agree very closely with measured displacements from a laboratory model or prototype structure on which the mathematical model is based. We can now ask a most important question, one on which the entire concept of system identification is founded: "What can we do to the mathematical model if the agreement between computed and measured displacements is not satisfactory?" If the assumptions and boundary conditions are satisfied, and if the measured displacements and loads are accurately known, then there must be errors in the member stiffnesses. The errors might arise from discrepancies in either the cross-sectional areas or elastic modulii. We do not consider these possible errors separately, but rather as combined in the individual stiffnesses. It is therefore the set of member stiffnesses which are consider to be the unknown parameters and which we will adjust until the computed and measured displacements do agree satisfactorily.

2.2 Criterion Function

The scalar criterion function can be based on either of two error vectors - an equation error vector

$$\underline{EQER}\ (\underline{k})^j = \underline{P}^j - \underline{\underline{S}}(\underline{k})\ \underline{V}^j \qquad (2)$$

or a response error vector

$$\underline{RESPER}\ (\underline{k})^j = \underline{\underline{S}}(\underline{k})^{-1}\underline{P}^j - \underline{V}^j \qquad (3)$$

where the superscript j refers to the jth loading condition. Since $\underline{P}^j$ and $\underline{V}^j$ are measured quantities, and $\underline{\underline{S}}(\underline{k})^{-1}\underline{P}^j$ represents the computed displacements, we can rewrite eqn. (3) as

$$\underline{RESPER}\ (\underline{k})^j = \underline{Vc}\,(\underline{k})^j - \underline{Vm}^j \qquad (4)$$

where c and m refer to computed and measured quantities.

It has been shown by Hardee[3] that the equation error vector, while linear in the parameters (the k's), is in practice of little value because each element in this vector requires response information at every degree of freedom (DOF). The response error vector, on the other hand, does not suffer from this limitation, and so it is this vector which we use in the remainder of our work described here.

There are several ways to form the criterion function. We use the square of the Euclidean norm in NDOF - space. This function, often called a squared error function, is given in indical notation by

$$\text{SQERFN}\ (k) = \sum_{j=1}^{NLC} \sum_{i=1}^{NS} [Vc(\underline{k})_i^j - Vm_i^j]^2 \qquad (5)$$

where NS refers to the number of sensors, and NS $<$ NDOF. We use NS instead of NDOF in this equation to emphasize that we may not be able to have sensors at each DOF.

An interpretation of eqn. (5) which will be useful in our description of the parameter adjustment algorithm is that of an NM-dimensional surface (recall that NM is the number of members) in an (NM+1)-dimensional space. The extra dimension is the error function itself. As the values of the k_i's are changed, eqn. (5) maps out different points on this "surface". When the computed and measured displacements match exactly, the elevation of the surface will be zero.

2.3 Parameter Adjustment Algorithm

Equation (5) is highly nonlinear in the member stiffnesses and thus requires, except for trivial examples, an iterative algorithm for its minimization. Such algorithms take the form

$$\underline{k}_{new} = \underline{k}_{old} + \alpha\ \underline{d} \qquad (6)$$

where $\underline{d}$ is an NM x 1 search direction vector and α is a positive scalar. We use a modified Gauss-Newton method for the minimization (see Bekey[2] and Matzen and McNiven[4] for background information and Hardee[3] for details). In this method, the direction vector is found for an initial estimate of the parameters using

$$\underline{d} = \underline{\underline{AH}}(\underline{k})^{-1}\ \underline{G}(\underline{k}) \qquad (7)$$

where $\underline{\underline{AH}}$ is an approximation to the Hessian matrix for eqn. (5),
$\underline{G}$ is the gradient vector also for eqn. (5).

The error surface is then searched in this direction by varying α until the lowest point (to within some tolerance) along this particular error profile is found. At this point, a new search direction is computed. The process is continued until the lowest point on the error surface is reached (to within a given tolerance).

We need to make just a brief comment about constraints. In any one of the line searches, one or more of the stiffnesses may tend to become negative. This is physically impossible and may lead to mathematical difficulties. For these reasons we modify both the direction and line search parts of our algorithm so that the stiffnesses always remain non-negative.

The algorithm is coded in FORTRAN and implemented on an IBM 370/165. Double precision (approximately 16 significant figures) is used for all real variables.

3. Numerical Experiments

To verify the correctness of the computer program and to provide some preliminary information on the uniqueness and damage assessment questions, we performed several numerical experiments. In these experiments, measured data were numerically simulated by assigning values to all of the parameters and then solving eqn. (1) for a specified set of loads. The initial set of parameters used to start the minimization program was then chosen to be somewhat different from the assigned set.

3.1 Sample Problem

We performed numerous experiments with a variety of truss configurations and loadings. The results given here are from just one of those experiments, but are typical of the others. The structure, shown in fig. (1), was a two-story, planar, indeterminate truss subjected to the loading shown. Measured displacement were used at every DOF (i.e. NS=NDOF). The results of this computer run are shown in fig. (2).

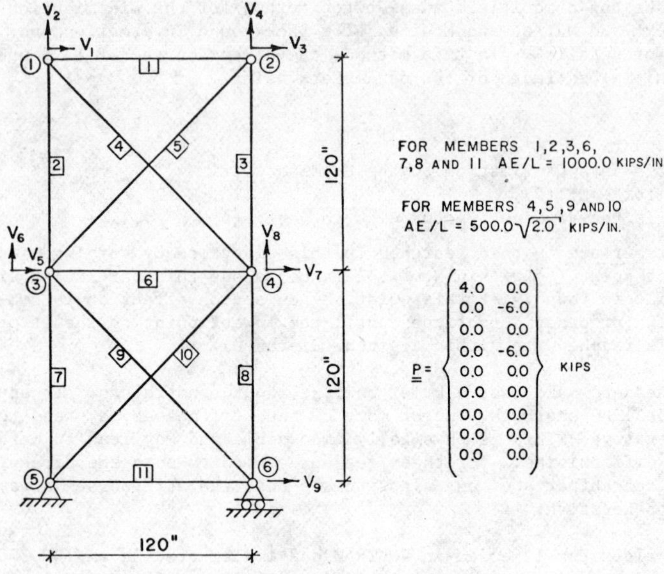

Figure 1

3.2 Questions and Preliminary Answers on Uniqueness

Examples like the one just given demonstrate that, under ideal conditions, the procedure does work. There remain, however, several unanswered questions. Three of the more important questions are discussed below in the context of our numerical experiments. We emphasize that our work is not based on laboratory or prototype experiments. Thus our results must be considered valid only for structures which satisfy all of the assumptions exactly.

Question 1: Is the final set of parameters unique?

A few mathematical definitions* will prove useful in this discussion. A relative (or local) minimum point on a surface is one for which, in a neighborhood of this point, all points are either higher than or equal to it (i.e. there may be a valley of equally low error

*See any book on nonlinear programing (e.g. Luenberger[5]) for more precise forms of these definitions.

INITIAL ELEMENT STIFFNESSES

```
K( 1)=  900.0000  K( 2)=  900.0000  K( 3)=  900.0000  K( 4)=  900.0000
K( 5)=  900.0000  K( 6)=  900.0000  K( 7)=  900.0000  K( 8)=  900.0000
K( 9)=  900.0000  K(10)=  900.0000  K(11)=  900.0000
```

MINIMIZATION TOLERANCES

LINE SLOPE = 1.00D-10 GRADIENT = 1.00D-20 ERROR = 1.00D-15

SUMMARY OF MINIMIZATION

SEARCH DIRECTION NUMBER	NO. OF FUNCT. EVALU.	ROOT-SUM-SQUARE OF GRADIENT	LINE SLOPE	ERROR	ALPHA
1	1	3.528570D-08	-9.060492D-06	4.541822D-06	0.000000D-01
	2	8.767811D-08	1.388284D-05	2.889927D-06	1.000000D 00
	3	4.018720D-08	-6.663319D-07	9.163555D-07	6.397831D-01
	4	4.245584D-08	6.192827D-08	9.075018D-07	6.686471D-01
	5	4.226038D-08	-2.331465D-11	9.074285D-07	6.662740D-01
2	1	4.226038D-08	-1.814779D-06	9.074285D-07	0.000000D-01
	2	4.595029D-09	-1.051324D-07	8.417739D-09	1.000000D 00
	3	2.317675D-09	2.103471D-09	4.323851D-09	1.079965D 00
	4	2.359175D-09	-1.904921D-13	4.322172D-09	1.078368D 00
3	1	2.359175D-09	-8.644339D-09	4.322172D-09	0.000000D-01
	2	2.978875D-11	-7.112099D-11	3.920784D-13	1.000000D 00
4	1	2.978875D-11	-7.841568D-13	3.920784D-13	0.000000D-01
	2	8.303671D-16	-2.220607D-17	3.155026D-22	1.000000D 00

TERMINATION OCCURRED DUE TO THE ERROR BEING LESS THAN OR EQUAL TO THE ERROR TOLERANCE

ERROR = 3.15503D-22

FINAL GRADIENT

```
GR( 1)=  4.6444D-19  GR( 2)= -5.1446D-17  GR( 3)= -4.3320D-17  GR( 4)= -2.0024D-16
GR( 5)= -1.5080D-16  GR( 6)= -1.7341D-19  GR( 7)= -4.2891D-16  GR( 8)= -5.2214D-16
GR( 9)= -2.2206D-16  GR(10)= -3.3655D-16  GR(11)= -5.4606D-17
```

FINAL ELEMENT STIFFNESSES

```
K( 1)= 1000.0000  K( 2)= 1000.0000  K( 3)= 1000.0000  K( 4)=  707.1068
K( 5)=  707.1068  K( 6)= 1000.0000  K( 7)= 1000.0000  K( 8)= 1000.0000
K( 9)=  707.1068  K(10)=  707.1068  K(11)= 1000.0000
```

Figure 2

passing through this point). A strict relative minimum point is one for which all points in the neighborhood are higher. The global minimum is the lowest relative minimum.

We can now restate the question in a more precise form - "Does the set of parameters found define a strict <u>global</u> minimum?" This is a very difficult question to answer. Instead of answering it directly, let us first ask if it defines a strict <u>local</u> minimum. We have used two ways to find out. One is by perturbing the final set of parameters and restarting the program. If it converges to the same point again (to within a specified tolerance, of course) we can be quite certain that the point is a strict local minimum. The other way is to compute the eigenvalues of the approximation to the Hessian matrix. Since the Hessian matrix contains second derivative information about the surface, we can relate the eigenvalues to the principal curvatures of the surface. Thus, at a point where the error and the gradient vector go to <u>zero</u> and eigenvalues of the approximate Hessian are all greater than zero, we know that the point is not only a relative minimum, but a strict relative minimum. In our example, both methods indicated that the final point was a strict relative minimum.

And so we ask again, is the point found a strict global minimum? Mathematically the only way we could tell would be to know the configuration of the <u>entire</u> surface. (Above we used only <u>local</u> surface properties). Since we do not have this information, we cannot say on this basis whether or not the minimum is global. Instead, our answer to this question is based on two observations about the results of our numerical experiments. The first is that in every case in which the program converged to a strict local minimum, the parameters were the same as the assigned set. The second observation is that, for a given structure, the program always converged to the same strict relative minimum regardless of the initial values of the parameters. These observations lead us to the conclusion that the minimum is global and hence that the parameters are unique.

Question 2: What are the restrictions, if any, on the loading?

We know of two restrictions. One is that, for a given set of loading conditions, there must be no truss members which have zero internal forces. The other is that the number of loading conditions times the number of sensors must be greater than or equal to the number of unknown parameters. We have found that if either of these conditions is violated, the approximate Hessian will have at least one zero eigenvalue. The topographical interpretation of this result is informative.

For example, when there is a zero-force member, the eigenvector corresponding to the zero eigenvalue will have only one nonzero element - the one corresponding to the zero force member. Thus the curvature of the surface at the point where $\underline{\underline{AH}}$ is computed (and everywhere else, too, as it turns out) is zero in this one direction and positive in every other principle direction (given by the other eigenvectors). Therefore the surface has a higher order semi-infinite valley of zero error running parallel to the axis of the zero-force member. (Well, at least we think this is informative!) The practical effect of $\underline{\underline{AH}}$ having zero eigenvalues is that it cannot then be legally inverted*. The best remedy is to improve the loading or increase the number of loading conditions until all of the eigenvalues become positive. In the case of a zero-force member, there is another solution - remove the parameter for this member from the parameter list. It then should be possible to find the remaining parameters uniquely.

Question 3: Must there be sensors at every DOF? If not, what is the minimum number needed and where should they be placed?

If in the static case displacements are measured at every DOF, then the data are said to be complete. Then the question is "Do we need complete data?" The answer is no. We arrived at this conclusion by using incomplete data and observing that the program still converged to the correct parameters. For indeterminate trusses, the minimum amount of information required is apparently one displacement for each joint that has a DOF. For example, for the structure described in section 3.1, we have successfully used only horizontal displacements at each joint. If we now also eliminate the horizontal sensor at joint 2, we will find one zero eigenvalue in $\underline{\underline{AH}}$ when it is evaluated at the initial set of parameters. The corresponding eigenvector will tell us which parameters cannot be found uniquely. The solution to the problem of insufficient data is to do either one of two things: increase the number of sensors or remove from the parameter list those which cannot be found uniquely.

3.3 Damage Assessment

Data for damaged structures were simulated by generating sets of "measured" displacements from an originally indeterminate truss which had had one member stiffness decreased in several increments and then finally set to zero. In each case, starting from initial stiffnesses which we would have used for the original structure, the program converged to the correct member stiffnesses, including the reduced or zero stiffness. Thus, in our numerical experiments, it did prove to be possible to determine specifically **where** and to what extent damage occurred.

4. Conclusions

Having completed the numerical experiments for the static loading case, it is worthwhile to look back over this work to see what insight

---

*In certain cases, apparently because of round-off errors, $\underline{\underline{AH}}$ can still be inverted, and the minimization procedure will still converge. Proceeding with known zero or near zero eigenvalues, however, is not recommended.

it provides into the system identification procedure. The first result of interest was that indeterminate trusses are no more difficult to work with than are trusses which are determinate. In fact in some respects, such as selecting loading cases so that there are no zero-forces and selecting the minimum number of sensors, indeterminate trusses are easier to work with.

Our most surprising result, and the one which is the most significant, came from our investigation of the various questions associated with uniqueness. We found that the approximation to the Hessian matrix contains a great deal of infromation which can be used to answer these questions. The eigenvalues and eigenvectors of this matrix tell us precisely which parameters can be identified uniquely and which cannot. The surprise was not that the matrix contained this information at the minimum point on the surface, but rather that the information was contained at every point we tried which was remote from the minimum. The significance of this result is that it now becomes possible to determine at the initial set of parameters what deficiencies, if any, exist in the loading a sensor placement. By remedying these deficiencies the remainder of the minimization procedure should procede smoothly.

Another surprise was the ease with which zero-stiffness members could be located. We found that they were no more difficult to determine than paramers with any other value.

Future work will include static loading of a small laboratory structure. Data from these tests will be used to make a final assessment of the system identification procedure, the questions of uniqueness, and the application of the method to damage assessment. We expect that the conclusions from this work will be similar to those already obtained. Difficultities inherent to laboratory experiments will undoubtedly have some effect, but we expect it to be small.

Following these static tests on our laboratory model, we will proceed with the dynamic loading. We will of course obtain different answers to the questions of uniqueness, but we expect that the answers to these questions will still come from a careful examination of the approximation to the Hessian matrix as well as the application of the parameter perturbation technique. The completion of the work associated with dynamic loading will finish our investigation of model indeterminate trusses and pave the way for the application of the system identification procedure to prototype truss structures.

5. List of References

1. [Preprints of the] ASCE/EMD Specialty Conference on the Dynamic Response of Structures: Instrumentation, Testing Methods and System Identification, ed. by G. C. Hart, UCLA Extension, Los Angeles, March 30-31, 1976.

2. Bekey, G. A., "System Identification - An Introduction and a Survey," Simulation, Vol. 5, No. 4, October 1970, pp. 151-166.

3. Hardee, J. E. Jr., "On the Development of Two System Identification Techniques for the Formulation of Mathematical Models of Statically Loaded Linear Elastic Planar Trusses," MS Thesis, North Carolina State University, Raleigh, 1980.

4. Matzen, V. C., and McNiven, H. D., Investigation of the Inelastic Characteristics of a Single Story Steel Structure Using System Identification and Shaking Table Experiments, Report No. EERC 76-20, Earthquake Engineering Research Center, University of California, Berkeley, August 1976.

5. Luenberger, D. G., Introduction to Linear and Nonlinear Programming, Addison-Wesley Publishing Co., Reading, Mass., 1973.

EXPERIMENTAL DYNAMICS OF HIGHWAY BRIDGES

by

Bruce M. Douglas,[1] M.ASCE
Charles D. Drown,[2] A.M.ASCE and Monty L. Gordon[3]

ABSTRACT

A reinforced concrete and a composite girder bridge were subjected to extensive dynamic field tests. Transverse experimental dynamic excitations were produced by using the quick-release pullback method with two D-8 crawler tractors. Vertical motions were produced by truck traffic or by dropping a loaded sand truck off small ramps. Experimental mode shapes, natural frequencies, and modal damping ratios were determined by analyzing the FM tape recorded field data.

Analytical models were developed by using the SAP IV structural analysis program. System identification was carried out by using multivariable linear regression techniques to fit the analytical models to the observed data.

Some of the preliminary conclusions of the study are that these techniques are effective in identifying (1) the in situ stiffnesses of the pier pile foundations and abutment foundations, (2) the effective moments of inertia of the reinforced concrete bridge superstructure members, (3) and the shear deformation factors. In addition, the test data for a complicated composite girder bridge have been valuable for identifying a relatively simple structural space frame model which will adequately account for the composite girder live load behavior.

INTRODUCTION

The body of experimental knowledge regarding the transverse response of full scale highway bridges from both the static and dynamic point of view is in its infancy. A number of tests on full scale bridges have been performed to measure the vertical static and dynamic response of highway bridges from a number of different points of view [15]. However, only a few have been conducted for purposes of identifying the dynamic mode shapes, natural frequencies, and modal damping ratios [1,3,11,14].

---

[1] Civil Engineering Department, University of Nevada Reno, Reno, NV 89557
[2] Structural Division, State of Nevada, Department of Transportation, 1263 Stewart Street, Carson City, NV 89701
[3] City of Sparks, 431 Prater Way, Sparks, NV 89431.

The inverse problem of using measured dynamic properties of full scale structures to indirectly identify their important structural characteristics is called system identification. The literature of this subject has been comprehensively reviewed in two recent papers [6,9]. Another valuable survey paper dealing with the whole subject of dynamic tests of full scale structures has been prepared by Hudson [7]. Other important experimental work dealing with the dynamic behavior of complex bridges has been conducted by using scale models to predict the behavior of full scale prototypes. [5,16]. These studies have also included the nonlinear behavior of elements of the system.

The purpose of this paper is to summarize the results obtained to date of full scale dynamic tests which were conducted on two Nevada highway bridges. The first bridge is a reinforced concrete box girder interchange at Rose Creek near Winnemucca, and the second is a variable depth skewed composite girder bridge at Deeth near Wells, Nevada. Complete vertical and transverse experimental dynamic data were obtained for the concrete bridge while only limited vertical data were obtained for the composite girder structure.

EXPERIMENTAL METHODS

The experimental methodology used for identifying the mode shapes, natural frequencies, and damping of the structures was essentially identical to that reported for previous dynamic tests [3] of full scale highway bridges. Transverse dynamic excitations were produced by pulling on the bridges with two D-8 crawler tractors and simultaneously quick releasing the cable tensions. Mechanical quick-release hooks were developed with electrical solenoid triggers to achieve simultaneous quick release of the cables. Twenty five thousand pound cable tensions were used during the course of these experiments.

Vertical motions at Rose Creek were caused by running a loaded sand truck off small ramps. At Deeth, the normal vehicular truck traffic was used to cause experimental motions. Vehicular traffic has been used to successfully determine the mode-shapes and frequencies [3] of composite girder bridges.

Accelerometer vibration data were FM tape recorded for analysis. Mode shapes and natural frequencies were obtained by using the Fourier transform of the recorded field acceleration data. Damping estimates were calculated by using a moving window Fourier transform.

ROSE CREEK INTERCHANGE

Description: This symmetrical bridge (Figs. 1 and 2) is a 400-foot long reinforced concrete box girder bridge supported by four single column piers. The piers and abutments are supported on pile foundations because the soil profile consists, in general, of a layer of soft clay over a layer of stiff clay on dense sand and gravel. The deck is supported on the abutment foundations by five elastomeric bearing pads. At the abutments, the only mechanism which is effective in transmitting lateral and longitudinal loads from the deck to the abutments consists of the elastomeric pads. During the course of construction of the

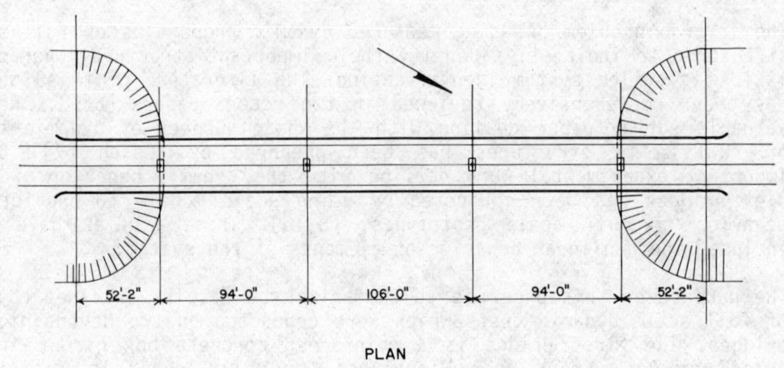

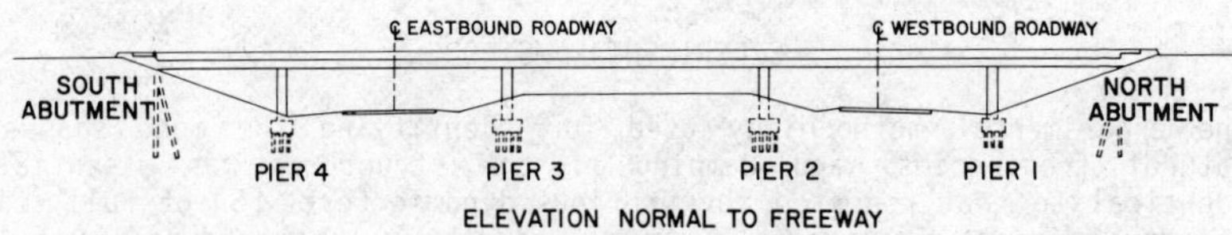

Fig. 1 Rose Creek Plan and Elevation

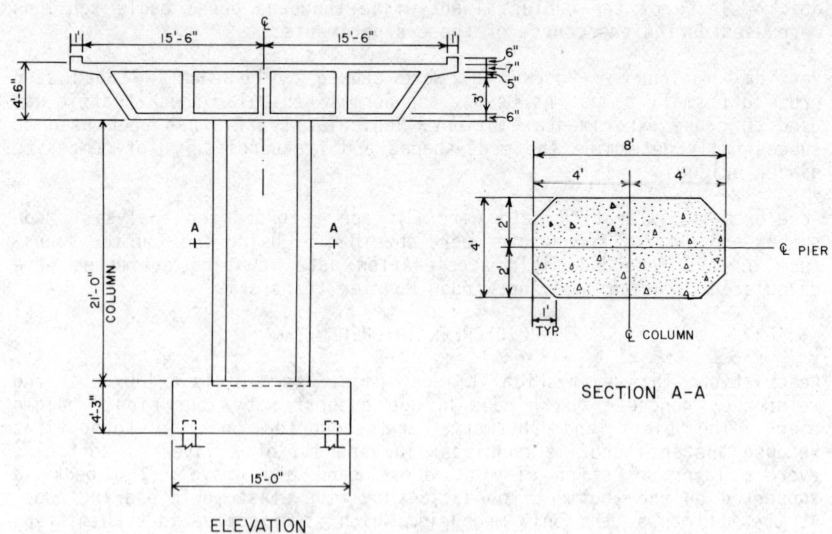

Fig. 2 Rose Creek Typical Pier

superstructure, 51 standard concrete cylinder tests were run. Based on these tests, the average compressive strength of the superstructure concrete is 3550 psi. The modulus of elasticity of the concrete used in all analytical modeling of this structure was indexed to this value through the standard formula.

Experimental Results. From the sand truck vertical motion data the first third, and fourth mode shapes were identified and the first five natural frequencies were obtained (Table 1). Damping estimates in vertical modes 1, 3, 4 and 5 were also made (Table 1).

TABLE 1. FREQUENCIES AND DAMPING RATIOS

Direction	Mode	Expt. Freq. HZ	Th. Freq. HZ	Damping Range	Avg. Damping	No. of Damping Estimates
TRANSVERSE	1	2.7	2.8	1.8 - 2.4	2.1	8
	2	3.8	3.7	4	4	2
	3	5.5	5.7	1.0 - 2.5	1.9	6
	4	8.7	8.4	3.4 - 5.0	4.3	4
	5	11.5	11.7	-	-	-
VERTICAL	1	2.3-2.9	2.7	4.8 - 9	6.5	12
	2	3.5	3.5	-	-	-
	3	4	3.9	1.2 - 2.7	1.8	8
	4	7.6	7.8	1.2 - 2.3	1.7	8
	5	8.1	8.2	1.0 - 2.3	1.5	8

From the quick-release pullback data the first four mode shapes were identified. These are presented in Figs. 3 and 4. The black dots represent the experimental values and the vertical bars represent uncertainties in the measurements. Missing data points represent cases where it was not possible to make estimates from the suite of field data. These represent cases which involved unforseen experimental difficulties, such as broken connectors, etc. The first five transverse natural frequencies were also found along with estimates of the modal damping ratios (Table 1). Other experimental data in Table 1 consist of the number of separate damping estimates which were made, and the maximum and minimum modal damping ratios which were observed.

Analytical Model. A lumped mass space frame model was developed by breaking the bridge deck into elements five feet long and the columns into five elements. The SAP IV program was used for carrying out dynamic response calculations for the 106 member space frame. Preliminary section properties were calculated for each member based on the gross concrete section. The interior girders (Fig. 2) were thicker over the piers which resulted in the bridge deck members being slightly nonprismatic. The foundation compliances for each of the pier foundations were approximated by introducing a transverse linear spring boundary element, and the elastomeric pads at the abutments were modeled with a transverse linear spring and the appropriate moment resisting springs. As indicated previously, the modulus of elasticity of concrete was indexed to the average compressive strength of the

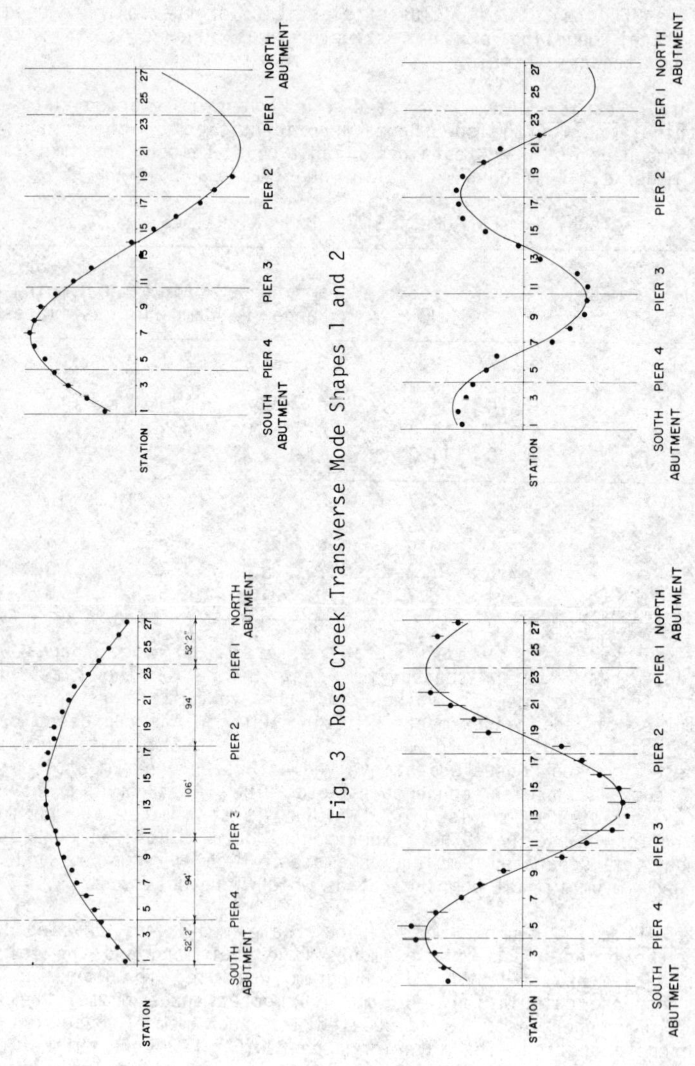

Fig. 3 Rose Creek Transverse Mode Shapes 1 and 2

Fig. 4 Rose Creek Transverse Mode Shapes 3 and 4

superstructure concrete.

System Identification. Identification of the important structural parameters for this bridge was carried out by fitting the lumped mass analytical model to all the observed natural frequencies and mode shapes. This fitting was accomplished by selecting n pertinent structural model paramenters and using n+1 computer solutions to represent each of the measured modal amplitudes and natural frequencies as approximate linear functions of the selected parameters. These approximate linear functions were then used to estimate the values of the n parameters by using multivariable linear regression.

For vertical motions the only significant uncertainties in the analytical model was the extent to which each span of the deck was cracked. Because the bridge was symmetrical only three parameters were needed to characterize the effective moment inertia of the entire bridge deck by individual spans. Within each span, the fraction of the gross concrete section moment of inertia was chosen as the appropriate structural parameter. By using the regression procedure outlined above, the experimental effective moments of inertia were found for each span. They are listed in Table 2 along with the calculated values of effective moment of inertia. The calculated values of effective moment of inertia were determined by obtaining the moment envelope [2] produced by all the design live and dead loads for the deck and using the ACI code formula for the effective moment of inertia. The values listed in column two represent the average of the values calculated at midspan and at the ends of the span in the negative moment region.

TABLE 2. VERTICAL MODEL PARAMETERS

Model Parameter	Parameter Value Obtained By Fitting Model to Data	Calculated Value
Effective Moment of Inertia at Span 1	.44 Ig	.74 Ig
Effective Moment of Inertia of Span 2	.44 Ig	.45 Ig
Effective Moment of Inertia of Span 3	.42 Ig	.44 Ig

The agreement between the experimental and calculated effective moments of inertia in the interior two spans is remarkable. The calculated moment of inertia in the end spans was higher because the effective moment of inertia at the middle of the end span, based on the design moment envelope, was equal to the gross moment of inertia. The experimental results indicate that the end span is cracked, and suggests that this bridge has been overloaded, during its ten year life. The occasion for overloading exists since the bridge is in a relatively remote location, and its principal heavy traffic consists of trucks coming from a nearby gravel pit.

The vertical natural frequencies calculated from the analytical model having the inertia properties listed in Table 2 are given in Table 1. They compare very well with the experimentally determined values.

The analytical model in the transverse sense of motion involves a larger number of unknown, or less well understood parameters. The parameters shown in Table 3 represent a relatively simple model of the structure. A more comprehensive characterization of this structure will be developed as this study continues. The five parameters chosen are listed in Table 3. They are: (1) the effective stiffness of the elastomeric bearing pads under dynamic loads, (2 and 3) the in situ lateral stiffness of the pile foundations, (4) the effective transverse moment of inertia of the deck, (5) the effective shear deformation factor of the transverse girder given as a fraction of the gross area of the concrete section. In this model symmetry was preserved, and uniform shear and inertia factors were used for the entire deck. Partial justification for this is related to the fact that in the vertical sense the effective moment of inertia of the section was experimentally observed to be nearly uniform (Table 2). As will be subsequently shown, the effective transverse moment of inertia of the deck at this level of dynamic excitation is related to the extent of cracking in the vertical sense.

TABLE 3. TRANSVERSE MODEL PARAMETERS

Model Parameter	Parameter Value Obtained From Fitting Model To Experimental Data	Probable Parameter Value Which Would Be Assigned By Bridge Design Profession
Abutment Stiffness (Kip/in)	2900 k/in	Range from free to fixed
Transverse Stiffness Pile Cap at Piers 1 and 4 (kip/in)	500 k/in	Fixed
Transverse Stiffness of Pile Cap at Piers 2 and 3 (kip/in)	1600 k/in	Fixed
Effective Transverse Moment of Inertia of Deck	0.77 I_{gross}	I_{gross}
Effective Shear Area of Deck	0.31 A_{gross}	A_{gross} or Ignore Shear Deformation

The results obtained by fitting the analytical model to all the observed natural frequencies and mode shapes along with their probable values which would be assigned by the bridge design profession are given in Table 3. The "fitted" analytical model having the properties listed in Table 3 was used to calculate the theoretical transverse natural frequencies listed in Table 1. Agreement between the experimental frequencies and calculated frequencies is excellent considering the small number of parameters used in the model. In Figs. 3 and 4 the experimental mode shapes are given as the solid black dots. The solid lines represent the mode shapes obtained from the analytical model having the properties listed in Table 3. Agreement is good. Mode 1 clearly is not perfectly symmetrical, while the analytical model was initially forced to be symmetrical. Differences in the soils from pier to pier is the most probable explanation for the lack of symmetry. A careful examination of the first mode indicates that the torsional stiffness of the piers may be reflected in the relative deformation of the deck at the piers. The mode shapes also do not close (Figs. 3 and 4) at the ends of the bridge because the elastomeric bearing pads allow relative deformation. In summary even this simple analytical model gives results which are in acceptable agreement with the experimental values.

A most interesting result in Table 3 is the magnitude of the abutment stiffness. Based on the elastomeric pads and the stiffness of the specified thickness of expansion material, the abutment stiffness would have been about one tenth of the observed value. Based on information regarding the shear modulus of the elastomeric pads obtained from duPont [4], the calculated transverse stiffness of the pads alone would only be about one sixtieth of the observed dynamic result shown in the table. Because of its magnitude, very careful sensitivity studies were conducted to be certain that the magnitude of the experimental abutment stiffness was correct. This turned out to be one of the least sensitive parameters in the model. In addition, a special field investigation was undertaken to clarify the mechanism of lateral load transfer from the end of the bridge deck to the abutment. The field investigation indicated that the only significant lateral load transfer device at the abutments consisted of the elastomeric pads. A transverse load resisting shear key was present, but the contractor had included three inches of expansion joint material between the edge of the deck and the shear key rather than the one half inch expansion material called for on the plans. Even one-half inch of expansion material resists little compressive load, but three inches of the material resists essentially nothing when compared to the elastomeric pads. In short, the only significant load resisting device at the abutments consists of the elastomeric bearings, which appear to be many times stiffer under small strain dynamic loading than would be expected based on the usual definition of shear modulus for elastomeric pads.

This result triggered a search for the dynamic strain amplitude dependent properties of elastomeric bridge bearings. Very little data exist. Nachtrab and Davidson [12] have presented data that indicates the neoprene elastomeric pads behave nonlinearly, and that the shear modulus is higher at small strain amplitudes. In addition, they present a very limited amount of data to suggest that the modulus increases at higher unidirectional strain rates. Long [10] indicates that the elastomeric pads stiffen under dynamic loading, but no relative numerical values are given. Based on the results suggested by these authors, and the results of this study it would seem appropriate to conduct a comprehensive series of dynamic tests on elastomeric bridge pads.

The in situ stiffness of the pier foundations is an extremely important parameter in determining the lateral load distribution between the piers and the abutments. Based on these observations of pier stiffnesses listed in Table 3, the deflection response of the top of piers 1 and 4 to a unit load is 71 percent due to the transverse deformation of the foundation, 27 percent due to bending of the pier and 2 percent due to shear deformation. At piers 2 and 3, 45 percent of the pier deflection is due to the foundation flexibility, 51 percent to bending and 4 percent due to shear deflection. As should be evident, the soil-structure interaction effect has a dominant effect even at the amplitudes of deformation associated with this testing program.

For comparison with the observed pile group stiffnesses in Table 3, a colleague in the geotechnical area [13] was asked to estimate the stiffnesses of the pile groups based on an analysis of the average soil properties at the site. He estimated that the range of stiffness for the pier foundations 1 and 4 would be 300 to 600 kips/inch and the range for piers 2 and 3 would be 500 to 1000 kips/inch. The observed value for piers 1 and 4 falls within the estimated range but the observed value at piers 2 and 3 is higher than estimated. This probably occurred because no account was taken of the partial burial of piers 2 and 3 when estimating the pile group stiffnesses.

The effective transverse moment of inertia of the deck was found to be 77 percent of the gross moment of inertia. This is consistent with the extent of cracking observed to be produced by the vertical loads. Fig. 5 illustrates the deck section cracked on the lower side by vertical loads. In Fig. 5, H locates the cracked transformed section neutral

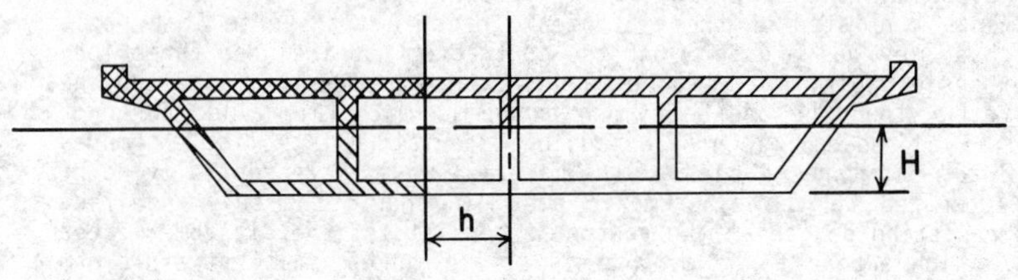

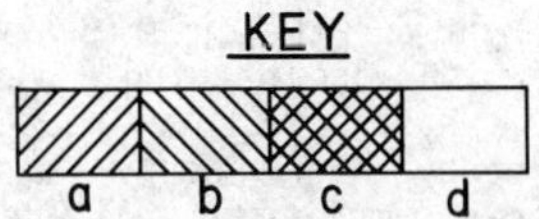

a) DL Compression and TL Tension

b) DL Tension and TL Compression

c) DL and TL Compression

d) DL and TL Tension

DL= Dead Load TL=Experimental Transverse Loads

Fig. 5 Rose Creek girder section showing effect of simultaneous vertical and transverse loads.

axis caused by vertical loads only. The horizontal transformed section neutral axis is indicated by h and was calculated by excluding the concrete in the tension zone caused by both vertical and horizontal loads simultaneously. By calculating the resulting transformed section moment of inertia and averaging over the total deck girder an effective moment of inertia of 66 percent of the gross section value was obtained. This is in reasonable agreement with the observed result. Finally the shear deformation factor of 31 percent of the gross area was examined. If one calculates mean value of the top and bottom deck slab one finds that this is about 28 percent of the gross area of the section. Due to cracking, caused by either positive or negative vertical load moments, the experimental result in Table 3 indicates approximately that only one or the other of the uncracked deck slabs is effective in resisting shear.

DEETH RAILROAD OVERPASS

Description. This four hundred foot bridge is a three span variable depth composite girder bridge. It is skewed 54°, with all girders resting on steel expansion bearings at the piers. Thermal expansion takes place relative to pier 2. The steel girders and the deck framing plan are shown in Fig. 6. The deck section is shown in Fig. 7 and the typical transverse bracing system between girders is shown in Fig. 8. The transverse bracing locations are indicated in the framing plan by the elements acting perpendicular to the girders.

Experimental Results. Vertical motion data caused by truck traffic was taken on the left bridge railing shown in Fig. 7. Experimental mode shapes were calculated from this data. Modes 1, 2, 4 and 6 are indicated by the black dots in Figs. 9 and 10. It should be pointed out that the mode shapes shown only represent the relative modal displacements along the barrier rail. Insufficient data was obtained to allow a characterization of the complete two dimensional modal displacement surface.

The observed natural frequencies in these modes are shown in Table 4.

TABLE 4. NATURAL FREQUENCIES

Mode	1	2	4	6
Observed Frequency, HZ	2.2	2.8	3.4	4.2
Theoretical, HZ	1.85	2.70	3.40	4.50

Analytical Model. The substantial skew and the stiff transverse bracing system (Fig. 8) make this structure behave as a space structure which cannot be idealized as a single girder.

In previous work [3] dealing with a narrow unskewed composite girder bridge, it was shown that an excellent vertical dynamic model of the system could be constructed by treating an individual girder as a beam which was continuously composite over the entire bridge. The only modification necessary, was that an increased modular ratio was required. In the present problem each girder was broken into elements five

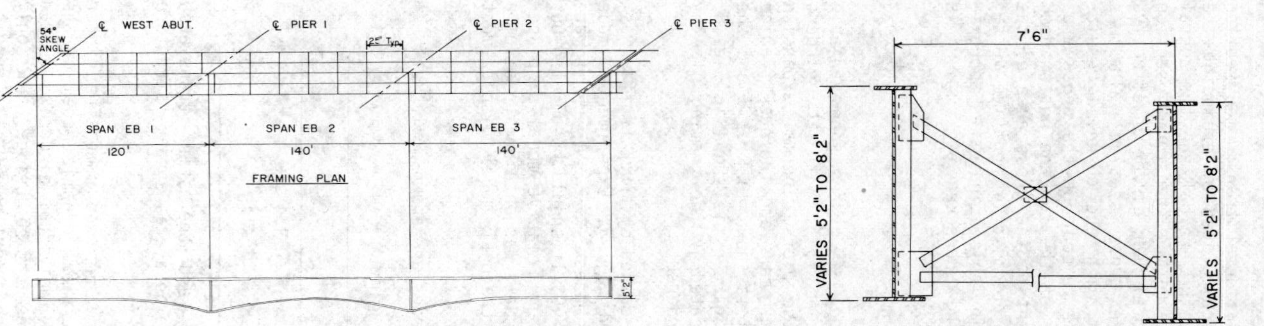

Fig. 6 Deeth Framing Plan and Typical Girder

Fig. 8 Deeth Typical Transverse Brace

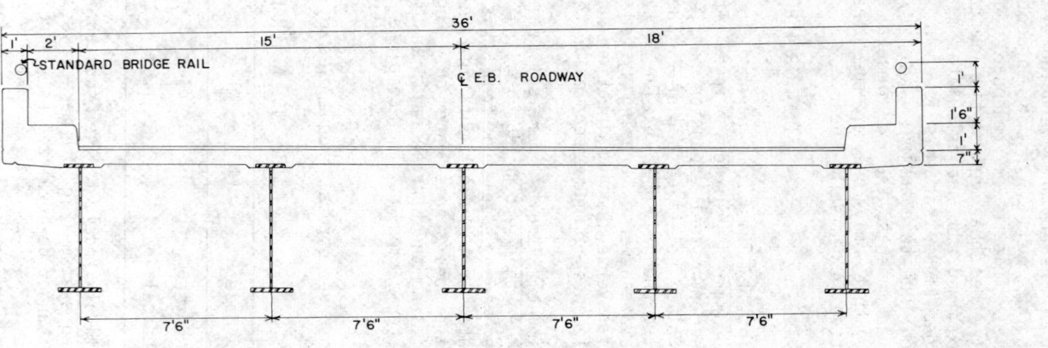

Fig. 7 Deeth Typical Section Normal to Roadway

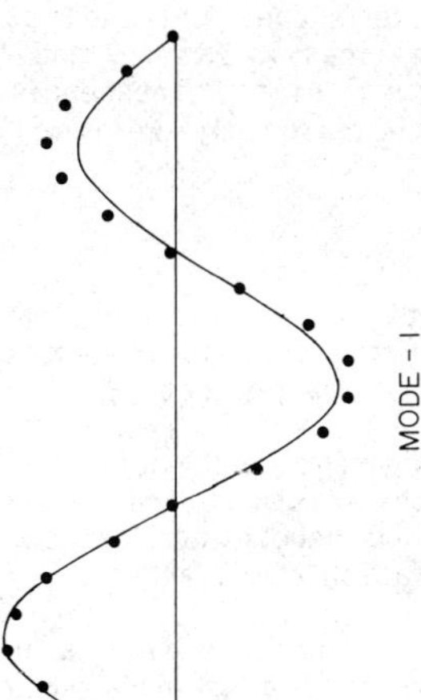

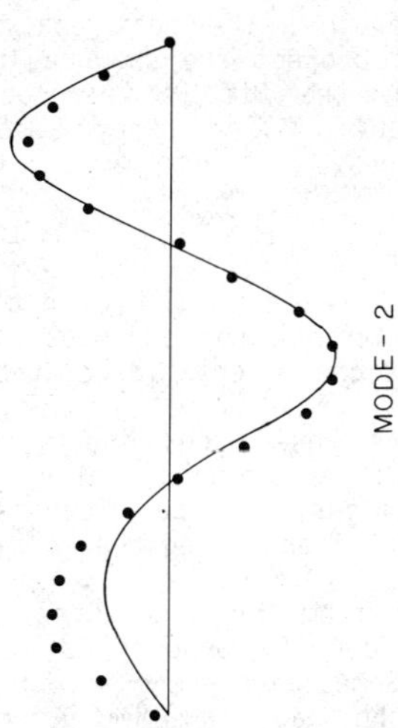

Fig. 9 Deeth Modes 1 and 2

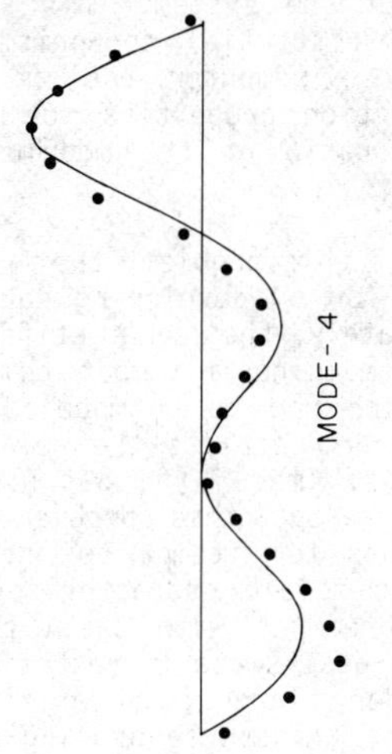

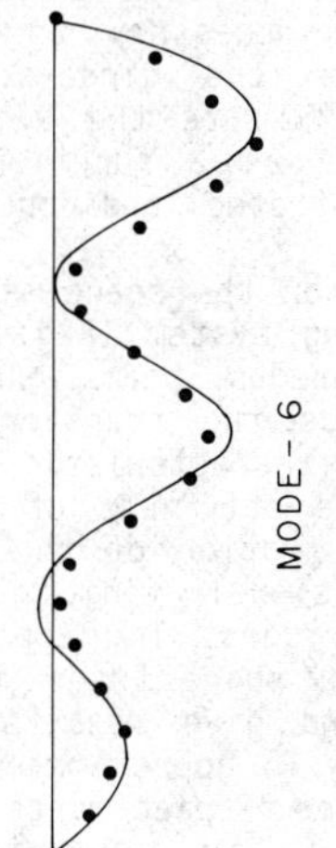

Fig. 10 Deeth Modes 4 and 6

feet long to produce a lumped mass model of the system. This small spacing was necessary to account for the substantial non-prismatic character of the girders. A special purpose computer program was written to compute the average composite section properties for each beam element as a function of the modular ratio of the modulus of elasticity of steel and concrete.

To account for the space structure behavior of this problem, the transverse bracing system (Fig. 8) was accounted for by developing equivalent homogeneous beams which had approximately the same stiffness properties as the transverse braces. The remaining and most complex space frame characteristic of the bridge which had to be modeled was the in plane stiffness of the deck slabs which restricts the relative longitudinal motion of the top girder flanges. Imbsen [8] has used a three dimensional finite element model to treat this problem for composite bridges. The approach taken here was to attempt to develop the simplest space frame model possible which could characterize the structure and give results in reasonable agreement with the experimental data. To approximate the effect of the deck slabs in restraining the relative displacements of the top girder flanges, an artifical torsional rigidity was assigned to each of the transverse bracing members. In the approximate preliminary results shown below, a single value was used for all braces.

Model Results. The fitting procedures described above for Rose Creek were used to fit the analytical lumped mass model to the data. Initially, the dynamic data for this bridge were assumed to be a function of only three parameters. They were: (1) the modular ratio for each composite grider, (2) the beam bending stiffness of the equivalent transverse braces and (3) the artificial torsional stiffness of the transverse braces. The best analytical mode shapes obtained at this stage of the project are shown as the solid lines in Figs. 9 and 10. As can be seen, the fit is reasonable. The calculated frequencies are shown in Table 4, and they also compare reasonably well with the observed values.

CONCLUSIONS

1. The quick release pullback dynamic tests and vertical motion tests were very effective for purposes of identifying the significant structural dynamic characteristics of these two highway bridges.

2. From the Rose Creek study, it is apparent that soil structure interaction effects can greatly affect the distribution of seismic loads in bridges, and the methods described herein are effective in experimentally identifying the in situ foundation compliances.

3. Results from the Rose Creek study also indicate that a testing program to identify properties of elastomeric bridge pads should be initiated. Some parameters which should be included are the axial stress on the pad, maximum dynamic shear strain, frequency of the assigned dynamic shear strain, and initial shear strain about which the cyclic shears are assigned.

4. The experimental techniques were effective in identifying the

structural properties of the reinforced concrete members.

5. The Rose Creek experimental results suggest that these methods may be useful in indicating whether concrete bridges have been overloaded during their lifetime.

6. Preliminary results from the Deeth study indicate that the attempt to develop a relatively simple space frame model to describe the live load behavior of skewed composite girders will be successful.

ACKNOWLEDGEMENTS

We wish to thank Hugh Brinson, Chief Bridge Engineer of the State of Nevada Department of Transportation, and the other members of the Bridge Division for their helpful comments and advice. Thanks are also due Dr. Mehdi Saiidi for his discussions and help in characterizing the interaction behavior of the concrete piers. We also want to thank Roy Imbsen of Engineering Computer Corportion in Sacramento for his ideas, and for accompanying us on a field inspection of the Rose Creek structure. Finally, we wish to thank Wayne Reid for all his efforts while working on this project. This study was supported by the National Science Foundation, Grant Number PFR 77-27596.

REFERENCES

1. Abdel-Ghaffer, Ahmed M., and George W. Housner, "Ambient Vibration Tests of Suspension Bridge," Journal of the Engineering Mechanics Division, ASCE, Vol. 104, No. EM 5, Proc. Paper 14065, October, 1978, pp 983-999.

2. Brinson, Hugh, Chief Bridge Engineer, Nevada Dept. of Transportation, Personal Communication, Mar. 1980.

3. Douglas, Bruce M., "Quick Release Pullback Testing and Analytical Seismic Analysis of Six Span Composite Girder Bridge," Report No. FHWA-RD-76-173, Federal Highway Administration, Offices of Research and Development, Washington, D.C., 73 pp.

4. E.I. duPont de Nemours and Co., "Design of Neoprene Bearing Pads," 1959, pp 1-20.

5. Godden, William G., and Mohammad Asalm, "Dynamic Model Studies of Ruck-A-Chucky Bridge," Journal of the Structural Division, ASCE, Vol. 104, No. ST 12, Proc.Paper 14206, December, 1978, pp 1827-1844.

6. Hart, Gary, C., and Yao, James, T.P., "System Identification in Structural Dynamics," Journal of the Engineering Mechanics Division, ASCE, Vol. 103, No. EM6, Proc. Paper 13443, December, 1977, pp 1089-1104.

7. Hudson, Donald E., "Dynamic Tests of Full-Scale Structures," Journal of the Engineering Mechanics Division, ASCE, Vol. 103, No. EM6, Proc. Paper 13446, December, 1977, pp 1141-1157.

8. Imbsen, Roy A. and Richard V. Nutt, "Load Distribution Study on Highway Bridges Using STRUDL Finite Element Analysis Capabilities," Proceedings, Conference on Computing in Civil Engineering, Atlanta, Georgia, June 26-29, 1978, pp 639-654.

9. Liu, Shih-Chi and James T.P. Yao, "Structural Identification Concept, "Journal of the Structural Division, ASCE, Vol. 104, No. ST 12, Proc. Paper 14207, December, 1978, pp 1845-1858.

10. Long, J.E., Bearings in Structural Engineering, Halstead Press-John Wiley and Sons, New York, 1974, pp 19-21.

11. McLamore, Vernon R., Gary C. Hart, and Ian R. Stubbs, "Ambient Vibration of Two Suspension Bridges," Journal of the Structural Division, ASCE, Vol. 97, No. ST 10, Proc. Paper 8454, October, 1971, pp 2567-2582.

12. Nachtrab, William B. and Robert L. Davidson, "Behavior of Elastomeric Bearing Pads Under Simultaneous Compression and Shear Loads," Highway Research Record, No. 76, Design-Bridges and Structures, 1965, pp 83-101.

13. Norris, Gary M., Assistant Professor, Civil Engineering Department, University of Nevada-Reno, Personal Communication, April 1980.

14. Shepherd, R. and G. K. Sidwell, "Investigation of the Dynamic Properties of Five Concrete Bridges," Fourth Australian Conference on the Mechanics of Structures and Materials," University of Queensland, Brisbane, 1973, p 261-268.

15. Varney, R.F. and C.F. Galambos, "Field Dynamic Loading Studies of Highway Bridges in the U.S., 1948-1965. Highway Research Record, No. 76, Design-Bridges and Structures, 1965, pp 285-305.

16. Williams, David and William G. Godden, "Seismic Response of a Curved Highway Bridge Model," Transportation Research Record 579, Earthquake-Induced Dynamic Response of Bridges and Bridge Measurements, 1976, pp 16-26.

RESISTANCE OF HARDENED BOX-TYPE STRUCTURES

Theodor Krauthammer[(1)], A.M. ASCE

Abstract

Membrane action enhancement in reinforced concrete slabs is a well known and documented structural phenomenon. Here, the contribution of membrane action to the resistance of reinforced concrete roof slabs is evaluated, and the application is demonstrated for the design of hardened box-type structures.

Introduction

The analysis of the resistance of shallow-buried box-type structures under static and dynamic loading conditions, as reflected in recent publications [3, 5, 6, 7, 8], concentrates on the response of the roof. The resistance of the roof was computed therein based on the conventional formulation for reinforced concrete one-way, or two-way slabs. However, the contribution to the resistance of the roof from in-plane forces which are generated by the membrane action of the slab was not considered in those studies.

The present study was initiated for providing a different analytical approach for the prediction of the expected resistance which could be provided by the roof of a hardened box-type structure. The present procedure includes the effects of membrane action as described in the literature [1, 4, 10, 11, 12, 13], and the preliminary results obtained herein may indicate that membrane action enhancement contributes significantly to the overall resistance of such structures.

Membrane Action in Slabs

The resistance of reinforced concrete slabs, as reported in the literature [1, 4, 9, 11, 12, 13] and illustrated in Fig. 1, can be described as follows. Initially, the slab resistance is provided by the conventional one-way, or two-way slab mechanism. However, even at relatively small central deflections the resistance is provided by a compression membrane mechanism, as a result of restraining the outward movement of the slab edges. The peak resistance at B could be between 2 to 8 times higher than the resistance predicted by the yield line theory [13]. The central deflection associated with the peak resistance

---

(1) Visiting Assistant Professor of Civil Engineering, Department of Civil Engineering, University of Illinois at Urbana-Champaign, Urbana, IL.

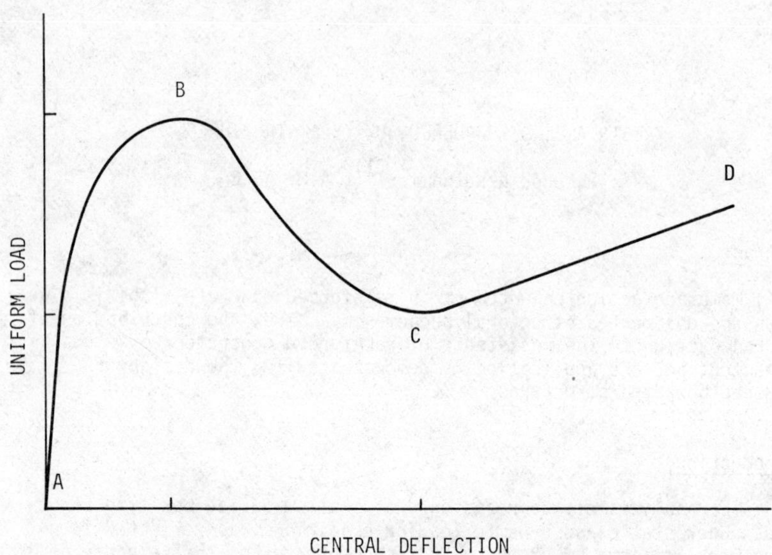

Fig. 1 - Resistance Curve for Reinforced Concrete Rectangular Slabs

is in the range between 0.5d to 0.5t (where d and t are the slab effective and total depths, respectively). After the peak has been reached, a gradual decrease in resistance can be noticed, and eventually the membrane forces in the slab central region are transformed from compression to tension. At that stage (point C) the central deflection is in the range of 2d to 2t (d and t have been defined earlier), and the slab boundaries provide restraint against inward movement of the edges. Extensive tensile cracking is associated with the transition into a tensile membrane, and further deflection of the central zone is associated with an increased resistance which is provided by the steel reinforcement. Failure may be controlled either by shear along the edges (punch-through), or by the fracture of the steel bars in tension. The slab resistance in the compression membrane mode can be computed from the formulation proposed by Park [10]. However, the analysis of hardened box-type structures has to consider the presence of externally applied thrust forces, and their influence on the slab resistance. Therefore, the formulation proposed by Park [10] was modified herein to include the effects of thrust, as follows.

$$w L_x^2 (3 L_y/L_x - 1)/24 = k_1 k_3 f'_c t^2 [(L_y/L_x)(0.188 - 0.281 k_2) + (0.479 - 0.490 k_2)] + 2 [(L_y/L_x) T_x (d_{1x} - d_{2x}) + T_y (d_{1y} - d_{2y})] + [(L_y/L_x) N_x + N_y] \quad (1)$$

where:

w = Uniformly distributed load on the slab (force per unit area).

L_x, L_y = Short and long dimensions of a rectangular slab, respectively.

k_1, k_2, k_3 = Parameters for describing the compressive stress distribution in concrete, and related to the concrete compressive strength.

f'_c = Uniaxial compressive strength of concrete.

t = Slab total thickness.

T_x, T_y = Tensile forces in steel bars along the x and y directions, respectively.

d_{1x}, d_{1y} = Slab effective depth along the x and y directions, respectively.

d_{2x}, d_{2y} = Distance from slab compressive face to the compressive reinforcement along the x and y directions, respectively.

N_x, N_y = Thrust in the x and y directions, respectively.

It is clear from Eq. (1) that increasing the externally applied thrust has a positive effect on the load carrying capacity of the slab under consideration. However, the lateral stability of the slab, and the compressive strength of reinforced concrete may limit that contribution.

The resistance provided by the tensile membrane can be computed from another expression proposed by Park [11]. Nevertheless, a generalized formulation has been derived herein, and is based on the mechanism illustrated in Fig. 2. The tensile membrane resistance is formulated as follows:

$$w = 8\,(M_n + T\,\Delta)/L^2 \qquad (2)$$

where:

w = Uniformly distributed load on membrane (force per unit length).

M_n = Bending moment in the slab central region.

T = Tensile force in reinforcement.

Δ = Central deflection.

L = Membrane span length.

Equation (2) describes the load capacity of the slab before, during, and after the formation of a hinge in the central region of the slab. After the hinge is formed Eq. (2) will transform into the solution provided by Park [11]. When a square slab is considered one may employ Eq. (2) after introducing an appropriate expression for M_n. Another formulation was proposed in Ref. [1], as follows.

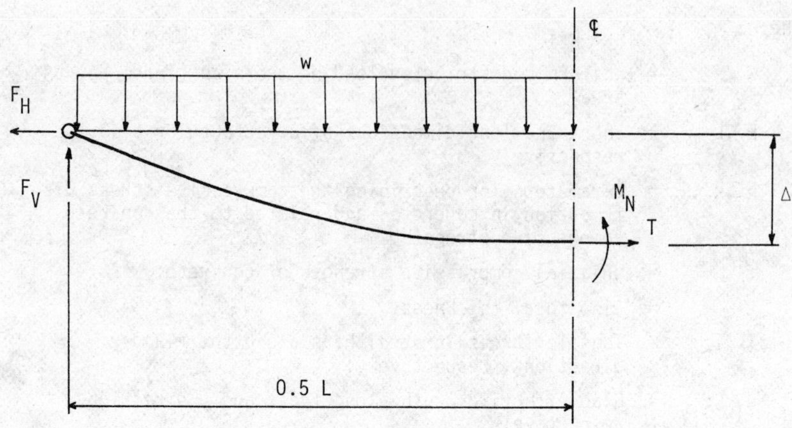

Fig. 2 - Tensile Membrane Mechanism

$$w = 16 \rho d f_s \Delta/L^2 \quad (3)$$

where:

- w = Uniformly distributed load on membrane (force per unit area).
- ρ = Reinforcement ratio.
- d = Slab effective depth.
- Δ = Central deflection.
- L = Membrane span length.
- f_s = Stress in reinforcement

There exists a basic difference between the present approach for describing the tensile membrane behavior, as formulated by Eq. (2), and the method which is represented by Eq. (3), as follows. Eq. (3) defines a straight line passing through the origin of the resistance curve while Eq. (2) defines a straight line which does not pass through the origin. Experimental results obtained by other investigators [1] indicated that normally the straight line segments which represent the tensile membrane resistance do not pass through the origin. Therefore, it was concluded in this study to employ Eq. (2) for describing the tensile membrane resistance.

Formation of Load-Deflection Relationships for Slabs

The analysis of reinforced concrete slabs, both in the static and dynamic modes depends on the resistance curve for such members. Based on the previous discussion one can compute the resistance of the slabs under consideration.

The structures analyzed herein are similar to the reinforced concrete box-type structures that have been studied previously by other investigators [3, 5, 7]. The resistance relationships for a given structure can be derived for the static, or dynamic conditions in which the structure is expected to function. For the static resistance relationships the static material properties are normally employed, while for the dynamic cases the static material properties are usually magnified by up to 30 %, as discussed in Ref. [9].

The structure under consideration is a reinforced concrete box, 4 feet by 4 feet (1.2 meter by 1.2 meter) internal dimensions, 16 feet (4.9 meters) long, and a wall thickness of 5.6 inches (0.142 meters). The longitudinal reinforcement ratio, both in tension and compression was varied between 0.005 to 0.013, and the shear reinforcement ratio was 0.015 for all cases. It was assumed that the concrete compressive strength was 5000 psi (34.5 MPa) and the steel yield strength was 60,000 psi (413.7 MPa). The load function was assumed to be bi-linear as illustrated in Fig. 3, where P_0 = 2000 psi (13.8 MPa) associated with a time t_0, and P_1 = 500 psi (3.4 MPa) associated with a time t_1. The load function can be varied for studying the response of the roof slab, as will be discussed later.

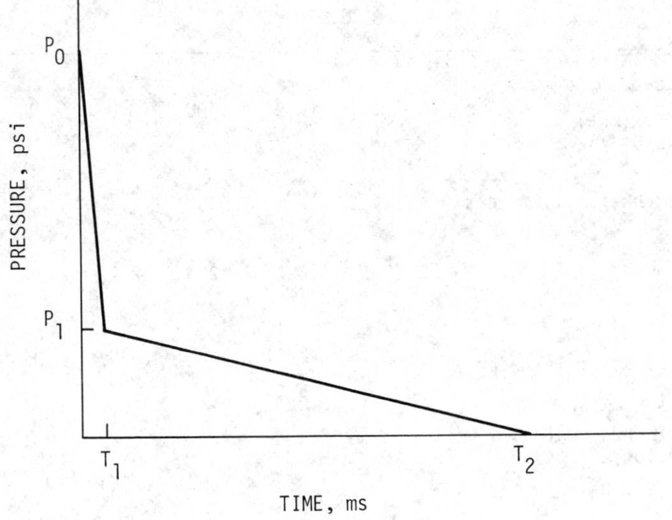

Fig. 3 - Approximated Surface Pressure Pulse

For the given concrete strength the parameters describing the compressive stress distribution over the cross-section are the following; k_1 k_3 = 0.68, and k_2 = 0.43. The initial slope of the resistance curves can be determined from the stiffness of a slab fixed against rotation as $K = \alpha EI/L^5$, and EI is varied to represent the propagation of tensile cracks in the cross-section (E = Modulus of elasticity for concrete; I = Moment of inertia for cross-section; L = span length of roof slab; α is a constant which depends on the L_y/L_x ratio and may be obtained from Ref. [2]).

The resistance of the roof slab was computed herein by employing Eqs. (1) and (2), and the assumed properties of the slab resistance curve as discussed previously. The characteristic points on the resistance curve are presented in Table 1, and the curves are illustrated in Fig. 4. A traditional resistance curve for the roof slab, which is based on the assumption of an elasto-plastic behavior without considering the membrane enhancement, as obtained in Ref. [3] is provided in Fig. 4 to demonstrate the membrane enhancement.

Table 1 -- Characteristic Points for Slab Resistance Curve

	Reinforcement Ratio, ρ in %		
	0.5	1.0	1.3
Peak Load, w_p in psi (MPa)	560 (3.9)	430 (3.0)	390 (2.7)
Deflection, Δ_p in inches (millimeters)	2.6 (66)	2.6 (66)	2.6 (66)
Minimum Load, w_m in psi (MPa)	405 (2.8)	289 (2.0)	190 (1.3)
Deflection, Δ_m in inches (millimeters)	9.2 (234)	8.8 (224)	11.0 (279)
Tensile Membrane Load at 12 inches central deflection, w_T in psi (MPa)	475 (3.3)	357 (2.5)	198 (1.4)

On the Dynamic Analysis

Hardened structures designed for resisting blast loads are normally classified in two principal categories, as follows. The first, includes structures for which a dynamic analysis may not be required, and normally the response can be computed from an equivalent static

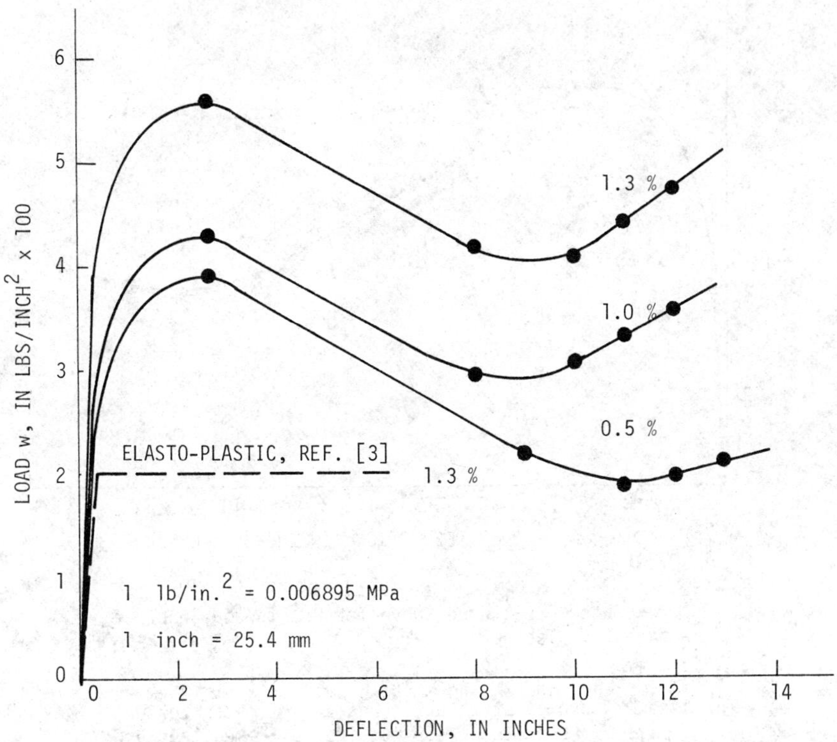

Fig. 4 - Analytical Resistance Curves for Reinforced Concrete Rectangular Slabs

investigation. A dynamic analysis may not be required if the duration of the positive phase of the loading function is more than about five times the natural period of the structural element under consideration. In the event that the loading results from a nuclear detonation in the megaton range of yields, an equivalent static analysis could be employed, as discussed in Ref. [9]. For such cases the resistance curve for the structural element is derived based on static principles, while assigning magnified material properties to the steel and concrete. The resistance curves obtained herein, and illustrated in Fig. 4 have been derived from the equivalent static approach. Therefore, if the structure is designed to perform under long duration loads, as defined previously, the behavior is expected to be illustrated by the resistance curves of Fig. 4. One may also employ a piecewise linear approximation for the resistance curve, as illustrated in Fig. 5.

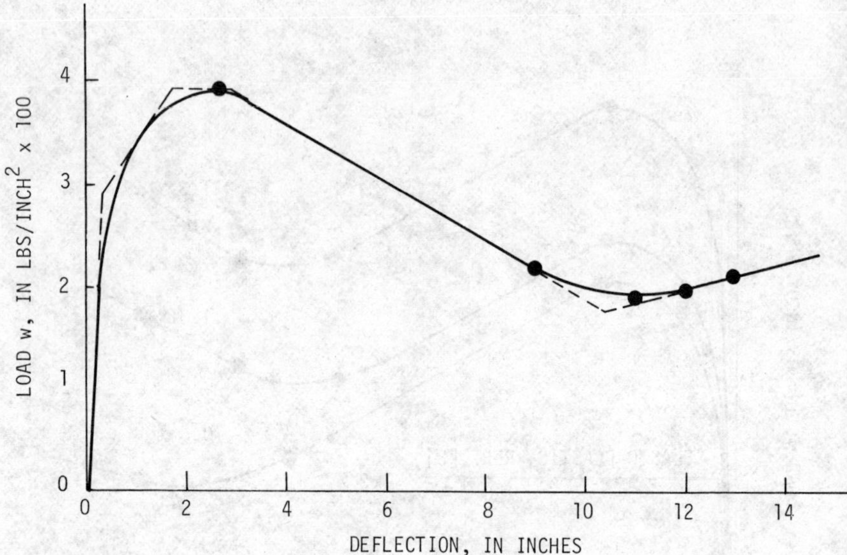

Fig. 5 - Approximated Resistance Curve for Reinforced Concrete Rectangular Slab

The second group of structures consists of cases for which a dynamic analysis is recommended, or required. These structures are subjected to short duration loads which result from a nuclear detonation in the kiloton range of yields, or conventional high explosives. For such cases the duration of the positive phase of the load function is normally less than five times the natural period of the structural element under consideration. A dynamic analysis which can be employed for such cases is by idealization of the responding system as a simple single-degree-of-freedom system, as described in Ref. [3]. The resistance function which is required for such studies was traditionally simulated by an elasto-plastic behavior which did not consider membrane enhancement. Therefore, the resistance curves derived in the present study, as illustrated in Figs. 4 and 5 may provide a more rational approach for studying the response of hardened shallow-buried reinforced concrete box-type structures.

Conclusions and Recommendations

The resistance of the roof slab of a hardened box-type structure was derived herein, and the following conclusions are drawn.

1. Membrane enhancement may provide a major contribution to the resistance of hardened structures, as compared to the results from previous studies.

2. The enhanced resistance, as computed herein conforms with the predicted resistance which is between 2 to 8 times the resistance derived from a yield-line analysis.

3. The analytical procedure for deriving the enhanced structural resistance is relatively simple, and can be employed for the static or dynamic investigation of hardened structures.

4. The design of hardened box-type structures shall include provisions for preventing a premature failure of the slabs, and ensure the membrane action enhancement.

5. The analysis and design of hardened structures is normally regarded as a confidential subject, and most experimental results are therefore classified. The available information is not sufficient for providing the experimental data with which the results of the present study could be compared, and therefore no such attempt was made herein.

Acknowledgement

The helpful suggestions and comments of W. L. Gamble, W. J. Hall, and J. D. Haltiwanger, Professors of Civil Engineering of the University of Illinois at Urbana-Champaign are gratefully acknowledged.

References

1. Brotchie, J. F., and Holley, M. J., "Membrane Action in Slabs," ACI Publication SP-30, 1971.

2. Crawford, E. R., Higgins, C. J., and Bultman, E. H., "The Air Force Manual for Design and Analysis of Hardened Structures," AFWL-TR-74-102, October 1974.

3. Haltiwanger, J. D., and Hall, W. J., "Dynamic Response of Buried Box-Type Structures," ASCE Convention, Atlanta, October 1979.

4. Hopkins, D. C., and Park, R., "Test on a Reinforced Concrete Slab and Beam Floor Designed with Allowance for Membrane Action," ACI Publication SP-30, 1971.

5. Kiger, S. A., and Balsara, J. P., "Response of Shallow-Buried Structures to Blast Loads," Army Science Conference, June 1978.

6. Kiger, S. A., "Static Test of a Shallow-Buried Structure," Technical Report N-78-7, Weapons Effect Laboratory, U.S. Army Engineer Waterways Experiment Station, Vicksburg, Mississippi, October 1978.

7. Kiger, S. A., and Balsara, J. P., "Result of Recent Hardened Structures Research," 100th Symposium on Weapons Effects on Protective Structures, Mannheim, Germany, October 1978.

8. Flathau, W., Balsara, J., and Keiger, S. A., "Response of Shallow-Buried Box Structures," ASCE Convention, Atlanta, October 1979.

9. Krauthammer, T., and Hall, W. J., "Resistance of Reinforced Concrete Structures Under High Intensity Loads," SRS 463, Department of Civil Engineering, University of Illinois at Urbana-Champaign, May 1979.

10. Park, R., "Ultimate Strength of Rectangular Concrete Slabs Under Short-Term Uniform Loading with Edges Restrained Against Lateral Movement," Proceedings, Institution of Civil Engineers (London), V. 28, June 1964, pp. 125-150.

11. Park, R., "Tensile Membrane Behaviour of Uniformly Loaded Rectangular Reinforced Concrete Slabs with Fully Restrained Edges," Magazine of Concrete Research, Vol. 16, No. 46, March 1964, pp. 39-44.

12. Tong, P. Y., and Batchelor, B. de V., "Compressive Membrane Enhancement in Two-Way Bridge Slabs," ACI Publication SP-30, 1971.

13. Wood, R. H., "How Slab Design has Developed in the Past, and What the Indications are for Future Development," ACI Publication SP-30, 1971.

TRANSIENT ANALYSIS OF LIGHT EQUIPMENT IN STRUCTURES

J. L. Sackman[1]

Member, ASCE

J. M. Kelly[2]

ABSTRACT

An analytical method is developed whereby the maximum dynamic response of light equipment attached to a structure subjected to ground motion can be simply estimated. The natural frequency of the equipment, modeled as a single-degree-of-freedom system, is considered to be close or equal to one of the natural frequencies of the N-degree-of-freedom structure. This estimate provides a convenient, rational basis for the structural design of the equipment and its installation.

The approach is based on the transient analysis of lightly damped tuned or slightly nontuned equipment-structure systems in which the mass of the equipment is much smaller than that of the stucture. It is assumed that the information available to the designer is a design spectrum for the ground motion, fixed-base modal properties of the structure, and fixed-base properties of the equipment. The results obtained are simple estimates of the maximum acceleration and displacement of the equipment.

This analytical method is also applied to nontuned equipment-structure systems for which the conventional floor spectrum method is mathematically valid. A closed-form solution is obtained which permits the maximum equipment response to be estimated without the need to compute time histories as is required when the conventional floor spectrum method is used.

INTRODUCTION

In this paper we present a response spectrum approach to the design of relatively light equipment in structures subjected to ground shock. The analysis uses as a model an N-degree-of-freedom structure to which is attached a single-degree-of-freedom component. Significant interaction effects occur when the frequency of such a component is close or equal to one of the natural frequencies of the structure. This situation is referred to as tuning. If tuning is not present, the conventional floor spectrum method, which neglects interaction, is valid.

In previous research on the undamped equipment-structure problem for tuned systems [1] and slightly nontuned systems [2], it was shown that a design ground response spectrum together with fixed-base properties of the structure alone and of the equipment alone can be used to estimate the peak response of the equipment. The analysis took advantage of the mathematical structure of the equations and of perturbation and asymptotic methods made possible by the smallness of the equipment mass in comparison with the mass of the structure to obtain simple results for tuned and nearly tuned systems. In this paper we extend the results to damped tuned, nearly tuned, and completely nontuned systems. For light equipment and small

1. Professor of Engineering Science, Department of Civil Engineering, University of California, Berkeley.
2. Professor of Civil Engineering, Department of Civil Engineering, University of California, Berkeley.

damping, the results obtained can be easily and efficiently implemented by a designer. The most important aspect of the analysis is its extreme simplicity; namely, if the response spectrum for the ground motion is available, the response spectrum for the equipment can be calculated merely by multiplying the former by an amplification factor.

This approach is in contrast to several earlier analyses of equipment response, such as the floor spectrum method in which the equipment is treated as a single-degree-of-freedom system subjected to a base motion that is taken to be that which the structure would experience at the attachment point in the absence of equipment. The floor spectrum method neglects interaction and further requires that an expensive time-history analysis of the structure be performed in order to determine a base motion [3]. While it is possible to account for interaction by considering the system to be an $N+1$-degree-of-freedom model which is then treated by a modal approach and spectral methods or by direct time-history analysis using an ensemble of spectrum-consistent earthquake records, neither approach is ideal. If equipment-structure interaction is important, two closely spaced modes will appear, the contributions of which must be summed when a modal approach is used. There exists, however, no consensus as to an appropriate summation procedure. A disadvantage of the second approach—time-history analysis—is that it is very expensive and for a given item of equipment must be carried out for a wide range of earthquake motion. If the equipment is light, the use of the $N+1$-degree-of-freedom model with standard structural dynamics computer codes may mask significant response.

The method developed here allows the design engineer to use directly the given ground motion design spectrum to determine the response of the equipment. The expense of time-history analyses and the problems of closely spaced modes are thus avoided.

ANALYSIS

The equations of motion of the general N-degree-of-freedom structure to which the single-degree-of-freedom oscillator representing the equipment is attached, as shown in Figure 1, are reduced by standard techniques to modal form (in terms of the N natural modes of the structure alone), and solved by the use of the Laplace transform, and combined with the transformed equation governing the motion of the equipment. The Laplace transform of the equipment acceleration $\bar{\ddot{u}}$ can then be expressed in terms of the transformed ground acceleration $\bar{\ddot{u}}_g$ by

$$\bar{\ddot{u}}\left[(p^2+2\beta\omega p+\omega^2)+p^2\sum_{k=1}^{N}\frac{m\Phi_r^{k2}(2\beta\omega p+\omega^2)}{M_k(p^2+2B_k\Omega_k p+\Omega_k^2)}\right]$$

$$=\sum_{k=1}^{N}\frac{\Phi_r^k\sum_{i=1}^{N}\Phi_i^k\sum_{l=1}^{N}M_{il}R_l(2B_k\Omega_k p+\Omega_k^2)(2\beta\omega p+\omega^2)}{M_k(p^2+2B_k\Omega_k p+\Omega_k^2)}\,\bar{\ddot{u}}_g \tag{1}$$

where M_{ij} is the structural mass matrix, Ω_k, Φ_i^k are the kth natural frequency and mode shape of the structure, $M_k=\sum_{i=1}^{N}\sum_{j=1}^{N}\Phi_i^k\Phi_j^kM_{ij}$ is the generalized mass of the kth mode, B_k is the fraction of critical damping in the kth mode, m, ω, and β are the mass, natural frequency, and damping factor for the equipment, and R_i is a vector of influence coefficients that couple the ground motion to the structure. The solution of eq. (1) will be obtained by residue theory. The poles of the transfer function relating $\bar{\ddot{u}}$ to $\bar{\ddot{u}}_g$ are simple poles and determined by the zeroes of the expression multiplying $\bar{\ddot{u}}$ on the left-hand side of eq. (1). Two different situations arise, one when the equipment is tuned or nearly tuned and the other when the natural frequency of the equipment is well away from any of the structural frequencies (i.e., when the system is grossly nontuned). For light equipment, the poles appear near the natural frequencies of the structure alone and of the equipment alone. For tuned or nearly tuned equipment, two closely spaced poles (tuning poles) occur, located near the equipment frequency and the frequency of the structure to which the equipment is nearly tuned, one below those frequencies and one above them. For a perfectly tuned, undamped system, these two poles coalesce into a double pole as the equipment mass approaches zero. This gives rise to a response in the time domain which grows without bound and is proportional to time regardless of the time history of the input. Thus, the contribution to the sum of the residues at all poles is dominated by the residues at the tuning poles.

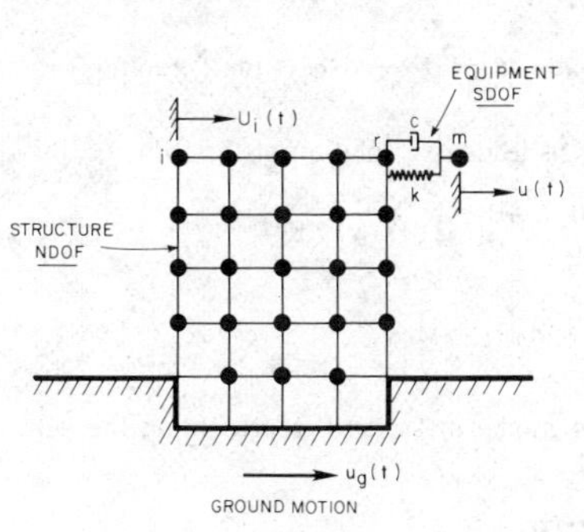

FIGURE 1
EQUIPMENT-STRUCTURE SYSTEM

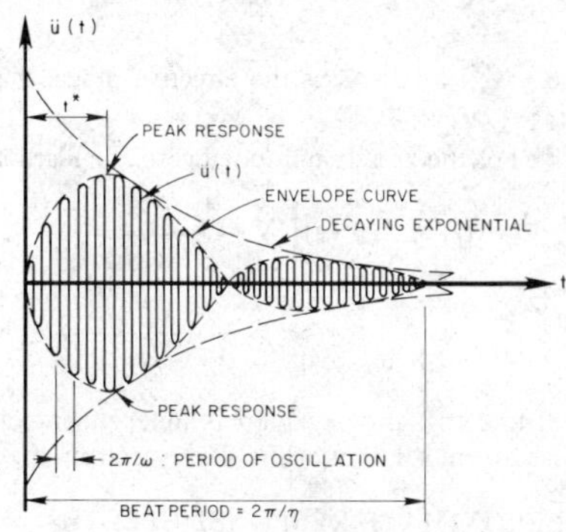

FIGURE 2
EQUIPMENT RESPONSE HISTORY—UNDERDAMPED BEATS

While the contribution to the response from the nontuned poles is quite standard, that from the tuned poles must be obtained by a special approach. Perturbation methods that take advantage of the relative lightness of the equipment and the smallness of the damping factors have been used in references 1, 2, and 4 to obtain rational forms of this result. The complete solution for the response of the equipment in the tuned or nearly tuned cases takes, to the order of the approximation, the form

$$\ddot{u}(t) = \int_0^t \ddot{u}_g(\tau) \left\{ \sum_{\substack{m=1 \\ m \neq n}}^{N} \frac{C_r^m}{1-(\Omega_m/\omega)^2} \Omega_m e^{-B_m \Omega_m (t-\tau)} \sin \Omega_m (t-\tau) \right.$$

$$\left. + \sum_{\substack{m=1 \\ m \neq n}}^{N} \frac{C_r^m}{1-(\omega/\Omega_m)^2} \omega e^{-\beta\omega(t-\tau)} \sin\omega(t-\tau) + C_r^n \ddot{u}_G(t-\tau) \right\} d\tau \tag{2}$$

where r is the structural degree of freedom to which the equipment is attached, n is the structural mode to which the equipment is tuned, and $C_r^n = \Phi_r^n \sum_{i=1}^N \sum_{j=1}^N \Phi_i^n M_{ij} R_j / M_n$ is an effective participation factor for the nth mode at the rth degree of freedom. In the above, $\ddot{u}_G$ is the contribution to the Green's function for the equipment acceleration provided by the residues at the tuning poles. This takes the general form, to dominant terms,

$$\begin{aligned} \ddot{u}_G(t) = \frac{\omega}{\lambda^2+\mu^2} e^{-(\beta+B)\omega t/2} [& \lambda \sinh \frac{\mu}{2}\omega t \cos \frac{\lambda}{2}\omega t \sin(1+\frac{\xi}{2})\omega t \\ & - \lambda \cosh \frac{\mu}{2}\omega t \sin \frac{\lambda}{2}\omega t \cos(1+\frac{\xi}{2})\omega t \\ & - \mu \sinh \frac{\mu}{2}\omega t \cos \frac{\lambda}{2}\omega t \cos(1+\frac{\xi}{2})\omega t \\ & - \mu \cosh \frac{\mu}{2}\omega t \sin \frac{\lambda}{2}\omega t \sin(1+\frac{\xi}{2})\omega t] \end{aligned} \tag{3}$$

where

$$\lambda = \frac{1}{\sqrt{2}}\left(\{[\gamma+\xi^2-(\beta-B_n)^2]^2+4\xi^2(\beta-B_n)^2\}^{1/2}+[\gamma+\xi^2-(\beta-B_n)^2]\right)^{1/2}$$

$$\mu = \frac{1}{\sqrt{2}}\left(\{[\gamma+\xi^2-(\beta-B_n)^2]^2+4\xi^2(\beta-B_n)^2\}^{1/2}-[\gamma+\xi^2-(\beta-B_n)^2]\right)^{1/2}$$

and $\gamma^{eff} = m\Phi_r^{n^2}/M_n$ is the effective modal mass ratio; $\xi = (\Omega_n - \omega)/\omega$ is the detuning parameter.

For the grossly nontuned case, standard analysis leads to a solution of the form

$$\ddot{u}(t) = \int_0^t \ddot{u}_g(\tau)\left\{\sum_{m=1}^{N}\frac{C_r^m}{1-(\Omega_m/\omega)^2}\Omega_m e^{-B_m\Omega_m(t-\tau)}\sin\Omega_m(t-\tau)\right.$$

$$\left. + \sum_{m=1}^{N}\frac{C_r^m}{1-(\omega/\Omega_m)^2}\omega e^{-\beta\omega(t-\tau)}\sin\omega(t-\tau)\right\}d\tau \qquad (4)$$

We note that this equation is independent of the modal mass ratio, representing the general closed-form solution of the floor spectrum method.

APPROXIMATE RESPONSE SPECTRA METHOD

The results given in the previous section can be applied to the design of equipment and equipment mounting. Typically the information available to the designer is a design spectrum. It was shown in references 1 and 2 that for the case of the undamped system, both tuned and nearly tuned, the peak acceleration in the equipment can be related to the response spectrum of the ground motion. The present paper is concerned with the generalization of these results to the damped tuned or nontuned cases.

For the grossly nontuned system where the equipment frequency is far away from all structural frequencies and these are well separated, conventional summation methods, e.g., the square root of the sum of the squares procedure applied to eq. (4), leads to an estimate for $|\ddot{u}|_{max}$ in the form

$$|\ddot{u}|_{max} = \left\{\left[\sum_{m=1}^{N}\frac{C_r^m}{1-(\Omega_m/\omega)^2}S_A(\Omega_m,B_m)\right]^2+\left[\sum_{m=1}^{N}\frac{C_r^m}{1-(\omega/\Omega_m)^2}\right]^2 S_A^2(\omega,\beta)\right\}^{1/2} \qquad (5)$$

where $S_A(\Omega, B)$ is the acceleration design spectrum for the ground motion $\ddot{u}_g$ evaluated at frequency Ω and damping factor B. If maximum displacements are required, then S_D, the displacement design spectrum, may be substituted.

In the case of a tuned or nearly tuned system, the contribution to the peak acceleration of the equipment from the nontuned modes and of the same order from the tuning poles is similarly obtained. Equation (2) is used in conjunction with the square root of the sum of the squares procedure; the appropriate estimate of the contributions to the maximum acceleration $|\ddot{u}|_{max}$ is then given by

$$|\ddot{u}|_{max} = \left\{\left[\sum_{\substack{m=1\\m\neq n}}^{N}\frac{C_r^m}{1-(\Omega_m/\omega)^2}S_A(\Omega_m,B_m)\right]^2+\left[\sum_{\substack{m=1\\m\neq n}}^{N}\frac{C_r^m}{1-(\omega/\Omega_m)^2}\right]^2 S_A^2(\omega,\beta)\right\}^{1/2} \qquad (6)$$

when the equipment is tuned to the nth structural frequency. The contribution to the maximum acceleration from the tuning poles is special and must be treated separately. The procedure is described in detail in reference 4. The approach is based on the recognition that the response predicted by eq. (3) involves a classic beat phenomenon. In some cases the beats are underdamped as shown in Figure 2, but may be overdamped or critically damped. In each case the method used is to locate the peak value of the envelope curve. For light, tuned equipment this peak occurs late, after many cycles of motion at the equipment frequency. If this peak occurs well after the strong portion of the ground motion, then simple approximations can be made which result in an estimate of the tuning contribution to $|\ddot{u}|_{max}$ in the form

$$|\ddot{u}|_{max} = \frac{|C_r^n|e^{-\kappa}}{(\gamma+\xi^2+4\beta B_n)^{1/2}} S_A\left(\frac{\omega+\Omega_n}{2}, \frac{\beta+B_n}{2}\right) \tag{7}$$

where

$$\kappa = (arctan\zeta)/\zeta \qquad \zeta = [\gamma+\xi^2-(\beta-B_n)^2]^{1/2}/(\beta+B_n) \tag{8}$$

Should this peak occur during the significant portion of the ground motion, the estimate given above may not be sufficiently accurate; it will tend to underestimate the peak response of the equipment. A sharper estimate can be obtained from eq. (2) by the use of random vibration theory, a modification that will be the subject of a subsequent paper. On the other hand, if the ground shock is of extremely short duration (i.e., the duration of the significant portion of the ground shock is much smaller than the period of vibration of the equipment), then analysis indicates that the estimate given by eq. (7) must be modified by replacing the term $S_A(\frac{\omega+\Omega_n}{2}, \frac{\beta+B_n}{2})$ by the term $S_A(\frac{\omega+\Omega_n}{2}, 0)$.

The peak value given by eq. (6) occurs early after only a few cycles of equipment motion and that given by eq. (7) occurs late. Although these values should not in principle be taken together, we have recently completed extensive numerical and physical experimentation, the results of which suggest that the estimates can be conveniently applied to equipment mounting design. The physical experiments were carried out at the Earthquake Simulator Laboratory of the Earthquake Engineering Research Center, University of California, Berkeley, and involved the use of a three-story steel frame model structure around one-third full scale. The response of the steel frame model is roughly that of a shear building with natural frequencies at around 2 Hz, 8 Hz, and 15 Hz. The total weight of the model was 39,500 lbs and its height was about 20 feet. Three single-degree-of-freedom oscillators were attached to the concrete blocks at the second and third floors to simulate equipment in a primary structure. The test structure was instrumented to measure displacements and accelerations at each floor and acceleration of the oscillators. The mechanical oscillators, each weighing about 20 lbs, were constructed to correspond to the first three natural frequencies of the model structure.

The first three natural frequencies of the frame were determined accurately and the oscillators were tuned to the structural frequencies. They were then bolted to the floors of the model structure. The Pacoima Dam (1971) and Taft (1950) earthquake signals as well as time-scaled Pacoima Dam, Taft, and El Centro (1940) signals and some approximate square waves were used as input to the shaking table. Figure 3 shows the time history of acceleration in a typical run. The top trace shows the input table acceleration, the second the response of the third floor of the structure, and the next three the acceleration of the three tuned oscillators. The very large magnification of acceleration in the oscillators is immediately obvious. The beat phenomenon is clear. The peak acceleration in the first two oscillators is achieved considerably after the peak acceleration in the input. These two oscillators are obviously responding at the coupled frequencies governed by equipment-structure interaction.

The response of the third oscillator is quite different; a highly irregular pattern appears and local maxima occur during as well as after the excitation. Fourier transforms of the acceleration time histories of the three oscillators have been taken to clarify this response (Figure 4). It is clear from these transforms that the first mode oscillator responded predominantly at the first mode frequency, around 2 Hz; the second mode oscillator responded predominantly at the second mode frequency, around 8 Hz, with a small contribution from the first mode. The dominant contribution to the response of the third mode oscillator was from frequencies around the third mode frequency, 15 Hz, but significant contributions also appeared from the lower modes. Thus the pure damped response predicted by the theory was obscured by contributions from the lower modes.

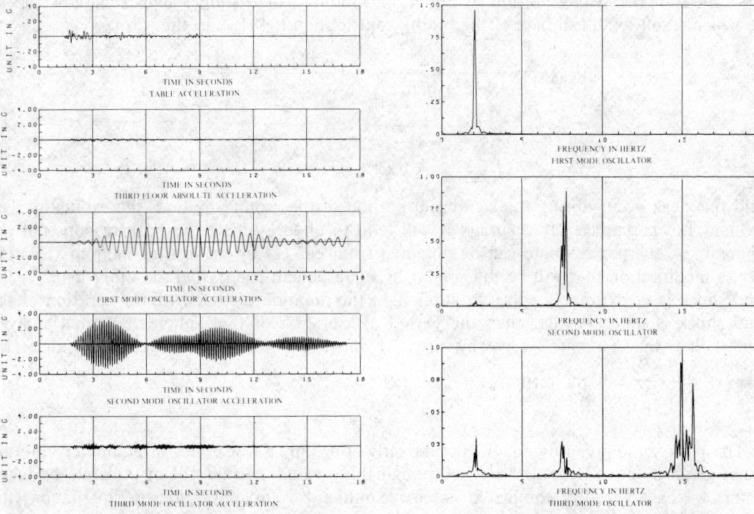

FIGURE 3
ACCELERATION HISTORY FOR SHAKING TABLE, THIRD FLOOR & OSCILLATORS

FIGURE 4
FOURIER TRANSFORMS OF OSCILLATOR ACCELERATION HISTORY

These observations are supported by the numerical experiments and have implications for the application of the theoretical results to design. First, only the structural modes up to and around that of the equipment under consideration need be considered and, second, the late peak acceleration given by eq. (7) should be summed with the early peak acceleration, eq. (6), by an appropriate summation rule. For instance, if the conventional square root of the sum of the squares procedure is used, the estimate is

$$|\ddot{u}|_{\max} = \left\{ \left[\sum_{\substack{m=1\\ m\neq n}}^{N} \frac{C_r^m}{1-(\Omega_m/\omega)^2} S_A(\Omega_m, B_m)\right]^2 + \left[\sum_{\substack{m=1\\ m\neq n}}^{N} \frac{C_r^m}{1-(\omega/\Omega_m)^2}\right]^2 S_A^2(\omega,\beta) + \left[\frac{|C_r^n| e^{-\kappa}}{(\gamma+\xi^2+4\beta B_n)^{1/2}} S_A\left[\frac{\omega+\Omega_n}{2}, \frac{\beta+B_n}{2}\right]\right]^2 \right\}^{1/2} \qquad (9)$$

If the late peak occurs well after the peak of the excitation (as in the first mode oscillator) it will be the dominant term and the result will be nearly the same as if it alone were considered. If on the other hand the late peak occurs during the excitation (as in the third mode oscillator), it should be superposed with the other modal contributions that may occur around the same time and this will be effected by the formula.

In addition to the physical experiments, we have performed numerical experiments using a standard structural analysis program, TABS [5], to compute by a variety of methods the response of a light appendage in a building. The structure analyzed was a ten-story reinforced concrete frame building described and used in reference 6; the appendage was a single-degree-of-freedom oscillator attached to the top floor and which had a mass that gave a mass ratio of 0.001 compared to the modal mass of the first mode of the building. The response of the appendage was calculated using TABS to evaluate the eleven-degree-of-freedom system which consisted of the structure and the appendage. The program computes natural frequencies and mode shapes and then can either compute a time history in each mode (by a method which is exact for piecewise-linear acceleration records) and sum these for the resultant response, or determine the maximum response by the square root of the sum of the squares procedure.

The response of the appendage as a function of appendage frequency (maintaining the same mass ratio) was calculated by both methods for a variety of earthquakes, including the El Centro, Pacoima Dam, and Taft records. Two cases of damping were considered: undamped and 2% of critical damping in each mode. Typical results for the Taft earthquake for both methods of computation are shown in Figure 5 for the undamped case and in Figure 6 for the damped case. The TABS time-history results, which should be exact, are shown by solid dots and the TABS spectrum (SRSS) results are indicated by triangles. The tuned system estimate based on eq. (9) is shown as a solid curve and that for the nontuned system based on eq. (5) by squares.

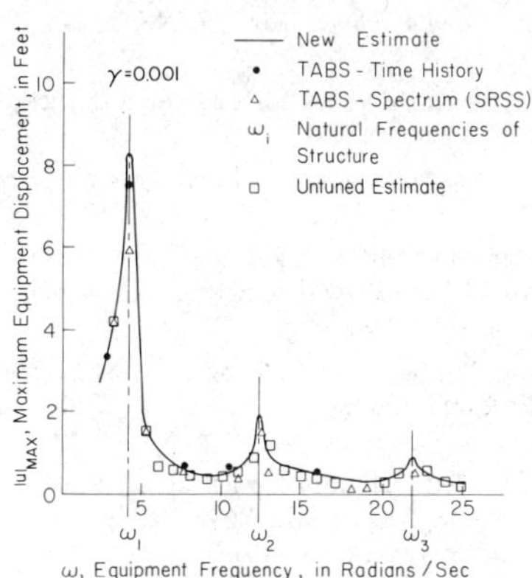

FIGURE 5
EQUIPMENT RESPONSE—TAFT EARTHQUAKE—0% DAMPING

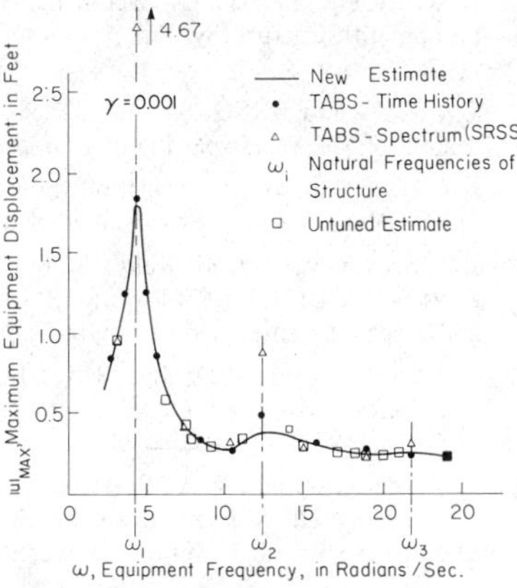

FIGURE 6
EQUIPMENT RESPONSE—TAFT EARTHQUAKE—2% DAMPING

The tuned estimate compares favorably with the nontuned estimate in the regions of gross nontuning examined in this study for both the damped and undamped cases. (This unexpected result is probably due to the even spacing of the natural frequencies of the structure in this example.) Further, the tuned estimate compares favorably with the time-history calculation over the range of appendage frequencies considered, which encompasses the first three modes of the structure. This is surprising in view of the fact that the spectra used in these numerical experiments are jagged spectra for specific earthquakes and not the smoothed design spectra for which the approach was developed. A further result is that the TABS spectrum (SRSS) estimate is very poor near tuning for all modes considered, but is very good for grossly nontuned cases. This is not surprising since it is well known that SRSS is not accurate for closely spaced modes which occur naturally when a light appendage or equipment item has a frequency near one of the frequencies of a structure.

The advantages of this approach are its simplicity and adaptability for practical application. A great deal of computational effort is avoided since time-history analyses need not be performed. The equipment and structure need not be analyzed as an $N+1$-degree-of-freedom system either by modal analysis or matrix-time-marching methods, and errors in estimates of peak response due to the possible unreliability of numerical time integration schemes, or to uncertainty as to the appropriate procedure for summing the contributions of the two closely spaced modes, are thereby avoided. For tuned and nearly tuned systems the method accounts for the important effect of equipment-structure interaction, which effect is completely neglected in the floor spectrum method. The method advanced here does not require that new information be generated. Data available from the building design alone (M_{ij}, R_j, Ω_m, Φ_i^m, B_m), the equipment alone (m, ω, β), and the seismic response spectra (S_D or S_V or S_A) are used. The estimates of peak response have been obtained by rational analysis and are easily evaluated and conveniently used during the design process.

ACKNOWLEDGMENT

This research was partially supported by the National Science Foundation, which support is gratefully acknowledged.

REFERENCES

[1] J. M. Kelly and J. L. Sackman, "Response Spectra Design Methods for Tuned Equipment-Structure Systems," *Journal of Sound and Vibration,* Vol. 59, No. 2, 171-179 (1978).

[2] J.M. Kelly and J. L. Sackman, "Shock Spectra Design Methods for Equipment-Structure Systems," *The Shock and Vibration Bulletin,* No. 49, Part 2 171-176 (1979).

[3] N-C. Tsai, "Spectrum-Compatible Motions for Design Purpose," *Journal of the Engineering Mechanics Division,* Proceedings of the ASCE, Vol. 98, EM2, 345-356 (1972).

[4] J. L. Sackman and J. M. Kelly, "Rational Design Methods for Light Equipment in Structures Subjected to Ground Motion," *Report No. UCB/EERC-78/19,* Earthquake Engineering Research Center, University of California, Berkeley (1978).

[5] E. L. Wilson and H. Dovey, "Three-Dimensional Analysis of Building Systems," *Report No. EERC 72-8,* Earthquake Engineering Research Center, University of California, Berkeley (1972).

[6] S. W. Zagajeski and V. V. Bertero, "Computer-Aided Optimum Design of Ductile Reinforced Concrete Moment-Resisting Frames," *Report No. UCB/EERC-77/16,* Earthquake Engineering Research Center, University of California, Berkeley (1977).

DYNAMIC ANALYSIS OE EQUIPMENT-STRUCTURE SYSTEMS

By

Jerome L. Sackman[1], M. ASCE
Armen Der Kiureghian[1], A.M. ASCE

ABSTRACT

Perturbation methods are employed to determine the dynamic properties of a combined system composed of a multi-degree-of-freedom structure to which is attached a light, single-degree-of-freedom equipment item. These properties are used in conjunction with a response spectrum method to obtain the peak response of the equipment to a dynamic excitation of the structure. This is conveniently given in terms of the modal properties of the structure alone and the equipment alone. The analysis includes the effect of equipment-structure interaction, which is particularly important when the equipment is nearly or perfectly tuned to a modal frequency of the structure.

INTRODUCTION

The dynamic analysis of light equipment in structures is of wide engineering interest. Control equipment, pressure vessels, pumps, etc., in power plants, communication and control items in transportation vehicles, and vital equipment in certain lifeline systems are common examples. In theory, the equipment can be incorporated in a dynamic model of the combined system as a part of the structure and analyzed by a conventional method. However, this approach in practice leads to two difficulties. First, for light equipment, the mass and stiffness matrices of the combined system will have elements with vastly different magnitudes, resulting in numerical difficulties in the dynamic analysis. Second, in design situations, it is often required to study many equipment items attached to a structure or several alternative attachment configurations or locations of a single item. Clearly, such an approach can be prohibitively costly because a multitude of dynamic models of the combined system must be constructed and analyzed. These difficulties can be overcome if, instead of using a conventional approach, an alternative procedure is followed taking advantage of the mismatch between the properties of the light equipment and the structure.

In this paper, closed-form expressions are derived for the dynamic properties of the combined equipment-structure system in terms of those of the structure alone and the equipment alone. These are obtained using a single-degree-of-freedom model for the equipment and a multi-degree-of-freedom one for the structure. This development is based on a perturbation method that uses the intuitively obvious fact that, for light equipment, the dynamic properties of the combined system are not too different from those of the structure. The effect of interaction between the equipment and the structure, which is especially significant when the equipment frequency coincides with one of the structure frequencies, is included in this analysis. By this approach, the numerical difficulties alluded to above are avoided and the computational effort becomes trivial.

[1] Department of Civil Engineering, University of California, Berkeley, California.

Once the properties of the combined equipment-structure system are determined, any dynamic analysis procedure, e.g., time history, response spectrum, or random vibration, can be used to evaluate the response of the equipment to a prescribed deterministic or probabilistic input excitation. When a modal combination method is used in the response spectrum procedure, or in some random vibration approaches, a problem arises when the equipment is light and has a frequency close to one of the structure frequencies. In such cases, closely spaced modes occur in the combined system for which special care must be exercised to account for the correlation between responses in these modes. A procedure for dealing with this problem is presented in another paper in this proceedings (4). Based on this approach, a more thorough development of the response of equipment-structure systems to random excitation will be presented in a forthcoming publication. Here, the main results for the response spectrum approach, which is of primary interest in practical applications, is summarized. By this procedure, the peak response of the equipment is obtained conveniently in terms of information readily available to the designer; namely, the dynamic properties of the structure alone, the dynamic properties of the equipment alone, and the design response spectrum. This formulation includes the equipment-structure interaction which is neglected in the conventional floor spectrum method.

DYNAMIC PROPERTIES OF EQUIPMENT-STRUCTURE SYSTEM

Consider an n-degree-of-freedom structure with mass, damping, and stiffness matrices **M**, **C**, and **K** having elements m_{ij}, c_{ij}, and k_{ij}, $i,j = 1, 2, \ldots, n$, respectively. Assume the structure has classical modes with natural frequencies ω_i, modal vectors $\mathbf{\Phi}_i = \left[\phi_{1i}\ \phi_{2i}\ \ldots\ \phi_{ni}\right]^T$, and modal damping coefficients ζ_i, where a superposed T denotes a transpose. Consider an equipment item, modeled as a single-degree-of-freedom oscillator of mass m_e, natural frequency ω_e, and damping coefficient ζ_e, attached to the structure such that it is only affected by motion in the k-th structure degree of freedom. The combined system, see Fig. 1, is an n+1-degree system with n+1 modes. In general, the equipment frequency can be close or equal , i.e. nearly

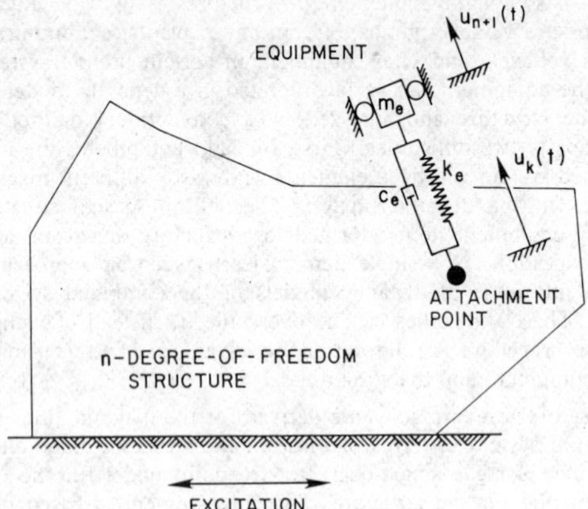

FIG. 1.- Equipment-Structure System

or perfectly tuned , to a natural frequency of the structure, say ω_l. Then, two closely spaced modes will occur in the combined system which require special treatment. It is the objective here to determine the dynamic properties of the combined system, including the case where tuning occurs, in terms of the properties of the two subsystems when the equipment mass is small compared to that of the structure. It is subsequently shown that for such conditions the modes of the combined system are closely related to those of the structure alone. Each structure mode, with contributions from other modes, produces a corresponding mode in the combined system. In addition, all the structure modes together with the equipment generate a "new" mode of the combined system. For convenience, the "new" mode will be identified by the subscript zero with the other modes retaining their original numbers. This process is conceptually illustrated in Fig. 2.

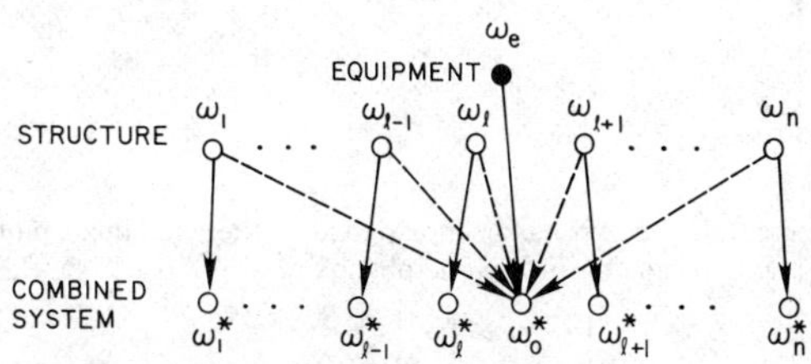

FIG. 2.- Schematic Representation of Modal Frequencies

The system of equations which determine the i-th ($i=0,1,...,n$) frequency and mode shape of the combined system may be obtained from the known properties of the two subsystems as

$$\mathbf{K}^*\mathbf{\Phi}_i^* = \omega_i^{*2}\mathbf{M}^*\mathbf{\Phi}_i^* \tag{1}$$

where a superscript asterisk has been used to denote values associated with the combined system. In the above, $\mathbf{K}^*$ and $\mathbf{M}^*$ are the stiffness and mass matrices for the combined system, and ω_i^* and $\mathbf{\Phi}_i^*$ are the i-th natural frequency and mode shape, given by

$$\mathbf{K}^* = \left[\begin{array}{ccccc|c} k_{11} & \cdots & k_{1k} & \cdots & k_{1n} & 0 \\ \vdots & & \vdots & & \vdots & \vdots \\ k_{k1} & \cdots & k_{kk}+m_e\omega_e^2 & \cdots & k_{kn} & -m_e\omega_e^2 \\ \vdots & & \vdots & & \vdots & \vdots \\ k_{n1} & \cdots & k_{nk} & \cdots & k_{nn} & 0 \\ \hline 0 & \cdots & -m_e\omega_e^2 & \cdots & 0 & m_e\omega_e^2 \end{array}\right]$$

$$\mathbf{M}^* = \left[\begin{array}{ccccc|c} m_{11} & \cdots & m_{1k} & \cdots & m_{1n} & 0 \\ \vdots & & \vdots & & \vdots & \vdots \\ m_{k1} & \cdots & m_{kk} & \cdots & m_{kn} & 0 \\ \vdots & & \vdots & & \vdots & \vdots \\ m_{n1} & \cdots & m_{nk} & \cdots & m_{nn} & 0 \\ \hline 0 & \cdots & 0 & \cdots & 0 & m_e \end{array}\right]$$

and

$$\mathbf{\Phi}_i^* = \left\{\begin{array}{c} \phi_{1i}^* \\ \cdot \\ \cdot \\ \phi_{ki}^* \\ \cdot \\ \cdot \\ \cdot \\ \phi_{ni}^* \\ \hline \phi_{n+1,i}^* \end{array}\right\}$$

where the n+1-th degree of freedom has been assigned to the equipment motion. Partitioning the matrices as shown and expanding Eq. 1, one obtains

$$\mathbf{K}\left\{\begin{array}{c} \phi_{1i}^* \\ \cdot \\ \cdot \\ \phi_{ki}^* \\ \cdot \\ \cdot \\ \phi_{ni}^* \end{array}\right\} + \left\{\begin{array}{c} 0 \\ \cdot \\ \cdot \\ m_e\omega_e^2(\phi_{ki}^* - \phi_{n+1,i}^*) \\ \cdot \\ \cdot \\ 0 \end{array}\right\} = \omega_i^{*2}\mathbf{M}\left\{\begin{array}{c} \phi_{1i}^* \\ \cdot \\ \cdot \\ \phi_{ki}^* \\ \cdot \\ \cdot \\ \phi_{ni}^* \end{array}\right\} \tag{2a}$$

and

$$-m_e\omega_e^2(\phi_{ki}^* - \phi_{n+1,i}^*) = \omega_i^{*2} m_e \phi_{n+1,i}^* \tag{2b}$$

From Eq. 2b,

$$\phi_{n+1,i}^* = -\frac{\omega_e^2}{\omega_i^{*2} - \omega_e^2}\phi_{ki}^* = \alpha_i \phi_{ki}^* \tag{3}$$

where $\alpha_i = \phi_{n+1,i}^*/\phi_{ki}^*$ is the modal amplification factor of the equipment relative to the attachment point. Substituting this in Eq. 2a, yields

$$\mathbf{K}\left\{\begin{array}{c} \phi_{1i}^* \\ \cdot \\ \cdot \\ \phi_{ki}^* \\ \cdot \\ \cdot \\ \phi_{ni}^* \end{array}\right\} + \left\{\begin{array}{c} 0 \\ \cdot \\ \cdot \\ m_e\phi_{ki}^*\omega_i^{*2}\omega_e^2/(\omega_i^{*2} - \omega_e^2) \\ \cdot \\ \cdot \\ 0 \end{array}\right\} = \omega_i^{*2}\mathbf{M}\left\{\begin{array}{c} \phi_{1i}^* \\ \cdot \\ \cdot \\ \phi_{ki}^* \\ \cdot \\ \cdot \\ \phi_{ni}^* \end{array}\right\} \tag{4}$$

The single nonzero term in the second vector in the left hand side is small for light equipment and, therefore, will only slightly modify the frequencies and mode shapes of the original structure. The well known Rayliegh's quotient establishes that first-order errors in mode shapes result in second-order errors in frequencies (6). Thus, as a first approximation, it is assumed here that the portions of modal vectors corresponding to structural degrees of freedom retain their shapes after the equipment is attached; i.e. it is assumed that $\phi_{mi}^* = \phi_{mi}$, for $m = 1, 2, ..., n$ and for $i \neq 0$. This will lead to a second-order approximation in the frequencies which is desirable for the computation of the response, since it can be shown that accuracy in frequencies is more critical than that in mode shapes.

Premultiplying Eq. 4 by Φ_i^T, one obtains

$$\omega_i^2 M_i + m_e \phi_{ki}^2 \frac{\omega_i^{*2} \omega_e^2}{\omega_i^{*2} - \omega_e^2} = \omega_i^{*2} M_i \tag{5}$$

where $M_i = \mathbf{\Phi}_i^T \mathbf{M} \mathbf{\Phi}_i = \omega_i^{-2} \mathbf{\Phi}_i^T \mathbf{K} \mathbf{\Phi}_i$ is the i-th modal mass of the structure. This yields

$$(1+\beta_i)\left(\frac{\omega_i^*}{\omega_i}\right)^4 - 2\left(1+\frac{\beta_i+\gamma_i}{2}\right)\left(\frac{\omega_i^*}{\omega_i}\right)^2 + 1 = 0 \tag{6}$$

where $\beta_i = (\omega_i^2 - \omega_e^2)/\omega_e^2$ and $\gamma_i = m_e/(M_i/\phi_{ki}^2)$ are the *detuning parameter* and the *effective mass ratio*, respectively, for mode i. These two parameters play important roles in the subsequent analysis. The solution of Eq. 6 is

$$\left(\frac{\omega_i^*}{\omega_i}\right)^2 = \begin{cases} \dfrac{1+\dfrac{\beta_i+\gamma_i}{2} - \left[\left(1+\dfrac{\beta_i+\gamma_i}{2}\right)^2 - (1+\beta_i)\right]^{1/2}}{1+\beta_i}, & \beta_i < 0 \\[2ex] \dfrac{1+\dfrac{\beta_i+\gamma_i}{2} + \left[\left(1+\dfrac{\beta_i+\gamma_i}{2}\right)^2 - (1+\beta_i)\right]^{1/2}}{1+\beta_i}, & \beta_i \geqslant 0 \end{cases} \tag{7}$$

where the proper root has been chosen according to the sign of β_i so as to produce ω_i^* near ω_i. (The other root, which yields ω_i^* near ω_e, is the contribution to the "new" mode. This contribution is subsequently retrieved by imposition of orthogonality between the modes.) For $\beta_i = 0$, either root is acceptable; here, the second one is chosen for definiteness. Substituting Eq. 7 in Eq. 3, the modal amplification factor is obtained as

$$\alpha_i = \begin{cases} -\dfrac{1}{\dfrac{\beta_i+\gamma_i}{2} - \left[\left(1+\dfrac{\beta_i+\gamma_i}{2}\right)^2 - (1+\beta_i)\right]^{1/2}}, & \beta_i < 0 \\[2ex] -\dfrac{1}{\dfrac{\beta_i+\gamma_i}{2} + \left[\left(1+\dfrac{\beta_i+\gamma_i}{2}\right)^2 - (1+\beta_i)\right]^{1/2}}, & \beta_i \geqslant 0 \end{cases} \tag{8}$$

It is noted that the magnitude of α_i can be very large for small β_i which occurs in the case of near or perfect tuning.

The new mode, denoted by $\mathbf{\Phi}_0^*$, is obtained through the requirement of orthogonality with the previous modes. Let

$$\mathbf{\Phi}_0^* = \begin{Bmatrix} \phi_{10}^* \\ \cdot \\ \cdot \\ \cdot \\ \phi_{n0}^* \\ 1 \end{Bmatrix} = \begin{Bmatrix} \mathbf{\Phi}_0 \\ 1 \end{Bmatrix} \tag{9}$$

be the new mode shape, scaled so that $\phi_{n+1,0}^* = 1$. Orthogonality requires that $\mathbf{\Phi}_i^{*T} \mathbf{M}^* \mathbf{\Phi}_0^* = 0$, for $i \neq 0$. Using $\mathbf{M}^*$, $\mathbf{\Phi}_i^*$, and $\mathbf{\Phi}_0^*$ as described above, one then obtains

$$\mathbf{\Phi}_i^{*T} \mathbf{M}^* \mathbf{\Phi}_0^* = \mathbf{\Phi}_i^T \mathbf{M} \mathbf{\Phi}_0 + m_e \alpha_i \phi_{ki} = 0, \quad i = 1, 2, \ldots, n \tag{10}$$

which gives

$$\mathbf{\Phi}_i^T \mathbf{M} \mathbf{\Phi}_0 = -m_e \alpha_i \phi_{ki}, \quad i = 1, 2, \ldots, n \tag{11}$$

Collecting all such equations in a matrix form,

$$\mathbf{\Phi}^T\mathbf{M}\mathbf{\Phi}_0 = -\begin{Bmatrix} m_e\alpha_1\phi_{k1} \\ \cdot \\ \cdot \\ \cdot \\ m_e\alpha_n\phi_{kn} \end{Bmatrix} \tag{12}$$

where $\mathbf{\Phi} = \left[\mathbf{\Phi}_1 \ \mathbf{\Phi}_2 \ldots \mathbf{\Phi}_n\right]$ is the modal matrix of the structure alone. It is easy to show that $\mathbf{\Phi}\mathbf{D}^{-1}$ is the inverse of $\mathbf{\Phi}^T\mathbf{M}$, where $\mathbf{D}$ is a diagonal matrix with elements M_i. Premultiplying Eq. 12 by $\mathbf{\Phi}\mathbf{D}^{-1}$, therefore, yields

$$\mathbf{\Phi}_0 = -\mathbf{\Phi}\mathbf{D}^{-1}\begin{Bmatrix} m_e\alpha_1\phi_{k1} \\ \cdot \\ \cdot \\ \cdot \\ m_e\alpha_n\phi_{kn} \end{Bmatrix} = -\begin{Bmatrix} \sum_i \alpha_i\gamma_i\phi_{1i}/\phi_{ki} \\ \cdot \\ \cdot \\ \cdot \\ \sum_i \alpha_i\gamma_i\phi_{ni}/\phi_{ki} \end{Bmatrix} \tag{13}$$

Note that all structure modes contribute to the generation of the new mode. However, observe that generally the shape of the new mode will be dominated by that structure mode which has a frequency closest to the equipment frequency, i.e. the l-th mode, because its modal amplification factor, α_l, is the largest. To determine the frequency of the new mode, Eqs. 9 and 13 are used in conjunction with Eq. 2(b) giving

$$\omega_0^* = \left(1+\sum_i \alpha_i\gamma_i\right)^{1/2}\omega_e \tag{14}$$

Observe that the frequency of the new mode is only slightly removed from the equipment frequency, with the difference being contributed by all structure modes. Again, the largest contribution to this difference comes from the l-th mode.

It is assumed herein that the damping in the equipment can be modeled by a viscous damper connecting the equipment to the attachment point, see Fig. 1. Since the structure is assumed to have modal damping, intuitively it would appear that for light damping the combined system will also very nearly have modal damping. (It is pointed out that this assumption may be critical under certain conditions when the equipment is nearly tuned to a structure mode, as will be discussed later.) The preceding perturbation scheme can be used to determine the damping coefficients for the combined system in terms of those of the two subsystems. To this end, the modal damping relation

$$2\zeta_i^*\omega_i^*M_i^* = \mathbf{\Phi}_i^{*T}\mathbf{C}^*\mathbf{\Phi}_i^* \tag{15}$$

is used, where $\mathbf{C}^*$, the damping matrix of the combined system, is given by

$$\mathbf{C}^* = \left[\begin{array}{ccccc|c} c_{11} & \cdots & c_{1k} & \cdots & c_{1n} & 0 \\ \cdot & & \cdot & & \cdot & \cdot \\ \cdot & & \cdot & & \cdot & \cdot \\ \cdot & & \cdot & & \cdot & \cdot \\ c_{k1} & \cdots & c_{kk}+2\zeta_e\omega_e m_e & \cdots & c_{kn} & -2\zeta_e\omega_e m_e \\ \cdot & & \cdot & & \cdot & \cdot \\ \cdot & & \cdot & & \cdot & \cdot \\ \cdot & & \cdot & & \cdot & \cdot \\ c_{n1} & \cdots & c_{nk} & \cdots & c_{nn} & 0 \\ \hline 0 & \cdots & -2\zeta_e\omega_e m_e & \cdots & 0 & 2\zeta_e\omega_e m_e \end{array}\right] \tag{16}$$

Using the preceding in conjunction with the expressions of β_i and γ_i as well as the identity $2\zeta_i\omega_i M_i = \mathbf{\Phi}_i^T\mathbf{C}\mathbf{\Phi}_i$, Eq. 15 gives

$$\zeta_i^* = \begin{cases} \dfrac{\sqrt{1+\beta_i}\,\zeta_i + (1-\alpha_i)^2\gamma_i\zeta_e}{\sqrt{1+\beta_i}(1+\alpha_i^2\gamma_i)}\,\dfrac{\omega_i}{\omega_i^*}, & i=1,2,...,n \\[2ex] \dfrac{\sum\limits_j\sqrt{1+\beta_j}\alpha_j^2\gamma_j\zeta_j + \left(1+\sum\limits_j\alpha_j\gamma_j\right)^2\zeta_e}{1+\sum\limits_j\alpha_j^2\gamma_j}\,\dfrac{\omega_e}{\omega_i^*}, & i=0 \end{cases} \tag{17}$$

Note that the ratios of frequencies in the preceding relations are close to unity and can be discarded.

To investigate the nature of the error in the preceding analysis, a norm of the second vector on the left hand side of Eq. 4 is examined in comparison with the corresponding norms of the two remaining vectors. This is accomplished by premultiplying Eq. 4 by $\mathbf{\Phi}_i^T$. For light equipment, the scalars thus obtained from the two remaining vectors are of order $\omega_i^2 M_i$, whereas that from the former vector is of order $\omega_i^2 M_i\gamma_i\omega_e^2/(\omega_i^{*2}-\omega_e^2)$. The ratio of these norms, r, which is a measure of the modification of frequencies and mode shapes of the structure resulting from the addition of the light equipment, can be written as

$$r = \frac{\gamma_i}{\left(\dfrac{\omega_i^*}{\omega_i}\right)^2(1+\beta_i)-1} \tag{18}$$

where the ratio of frequencies is given in terms of β_i and γ_i in Eqs. 7. It is easy to show that in the extreme cases of gross detuning, i.e. when $|\beta_i|$ is large, and of perfect tuning, i.e. when β_i is zero, the modification factor r is of order γ_i and $\sqrt{\gamma_i}$, respectively. A plot of $|r/\sqrt{\gamma_i}|$ versus β_i is shown in Fig. 3 for $\gamma_i = 0.001$, 0.01 and 0.05. Observe that for small γ_i, the

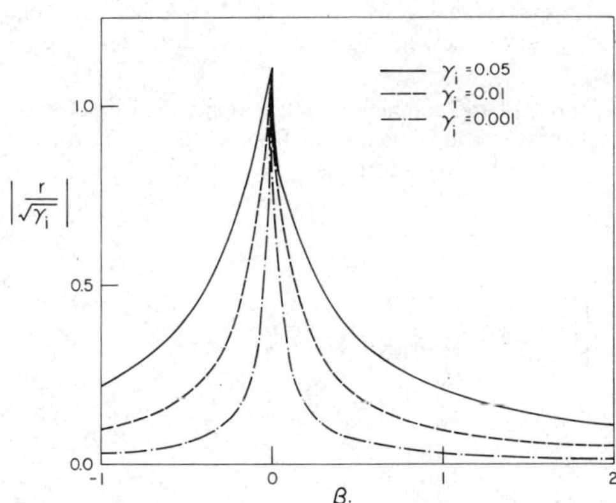

FIG. 3.- Measure of Modification of Modal Properties

modification term rapidly increases very near tuning, i.e. when $\beta_i \rightarrow 0$. This leads to a formal definition of gross detuning as values of β_i for which $|r/\sqrt{\gamma_i}| << 1$. From Eq. 18, this definition can be expressed as $|\beta_i| >> \sqrt{\gamma_i}$.

It is observed from Eq. 18 that errors in the estimated mode shapes, frequencies, and damping coefficients are largest for modes with frequencies closest to the equipment frequency. To achieve a uniform accuracy, it is necessary to improve the results for modes having frequencies near the equipment frequency. For simplicity, it is assumed herein that all but the l-th mode are well spaced from the equipment frequency. Therefore, a refinement of only mode l need be considered. However, the methodology developed here is not restricted to this case and can be extended to the case of multiple tunings.

To construct a refined shape for the l-th mode, the same procedure is used as for the new mode above, i.e. the refined l-th mode is obtained by imposition of orthogonality with all other modes, including the new mode. This results in

$$\mathbf{\Phi}_l^* = \begin{Bmatrix} \sum_{i \neq l} \alpha_i \gamma_i \phi_{1i}/\phi_{ki} - \frac{1}{\alpha_l}\phi_{1l}/\phi_{kl} \\ \cdot \\ \cdot \\ \cdot \\ \sum_{i \neq l} \alpha_i \gamma_i \phi_{ni}/\phi_{ki} - \frac{1}{\alpha_l}\phi_{nl}/\phi_{kl} \\ -1 \end{Bmatrix} \tag{19}$$

where, for the sake of uniformity with $\mathbf{\Phi}_0^*$, this mode has been scaled such that $\phi_{n+1,l}^* = -1$. In this derivation, only the dominant terms in $\mathbf{\Phi}_0^*$, i.e. those arising from the l-th mode, have been included. This is consistent with the order of accuracy in other modes.

The refined shape of the l-th mode could be used to obtain more refined estimates of ω_l^* and ζ_l^* from Eqs. 2(b) and 15, respectively. However, it is clear from Rayliegh's quotient that the second-order correction in the mode shape will result in only a third-order correction in the frequency. Furthermore, by dividing both sides of Eq. 15 by M_i^*, it is observed that $2\zeta_i^*\omega_i^*$ is determined from an expression analogous to Rayliegh's quotient. It follows then that the correction on ζ_l^* is also of third order. In this study, these corrections are neglected for uniformity in approximation.

To obtain the response of the combined system to specified input excitations, it is necessary to compute modal masses and participation factors. Using the mode shapes previously obtained, these are determined in closed form as

$$M_i^* = \mathbf{\Phi}_i^{*T}\mathbf{M}^*\mathbf{\Phi}_i^*$$
$$= \begin{cases} (1+\alpha_i^2\gamma_i)M_i, & i \neq l, 0 \\ \left[1+\alpha_l^2\gamma_l\left\{1+\sum_{j\neq l}\alpha_j^2\gamma_j\right\}\right]\dfrac{M_l}{(\alpha_l\phi_{kl})^2}, & i = l \\ \left\{1+\sum_j \alpha_j^2\gamma_j\right\}m_e, & i = 0 \end{cases} \tag{20}$$

and

$$\Gamma_i^* = \frac{\mathbf{\Phi}_i^{*T}\mathbf{M}^*\mathbf{R}^*}{M_i^*}$$

$$= \begin{cases} \dfrac{\Gamma_i + \alpha_i\gamma_i r_{n+1}^*/\phi_{ki}}{1+\alpha_i^2\gamma_i}, & i \neq l, 0 \\ -\dfrac{\alpha_l\phi_{kl}\Gamma_l - \alpha_l^2\gamma_l\left(\sum\limits_{j\neq l}\alpha_j\phi_{kj}\Gamma_j - r_{n+1}^*\right)}{1+\alpha_l^2\gamma_l\left(1+\sum\limits_{j\neq l}\alpha_j^2\gamma_j\right)}, & i = l \\ -\dfrac{\sum\limits_j \alpha_j\phi_{kj}\Gamma_j - r_{n+1}^*}{1+\sum\limits_j \alpha_j^2\gamma_j}, & i = 0 \end{cases} \tag{21}$$

where $\Gamma_i = \mathbf{\Phi}_i^T\mathbf{MR}/M_i$ is the participation factor associated with the i-th mode of the structure, in which $\mathbf{R} = [r_1 \; r_2 \; \cdots \; r_n]^T$ and $\mathbf{R}^* = [r_1 \; r_2 \; \cdots \; r_n \; r_{n+1}^*]^T$ are the conventional influence vectors coupling the input to the degrees of freedom of the structure and of the combined system, respectively.

PERFECTLY TUNED EQUIPMENT

It is instructive to examine the preceding results in the special case when the equipment is perfectly tuned to the l-th mode and grossly detuned with respect to all other modes. Using Eqs. 7 and 14, the modal frequencies in this case reduce to

$$\omega_i^* \approx \begin{cases} \omega_i, & i \neq l, 0 \\ \left(1+\dfrac{\sqrt{\gamma_l}}{2}\right)\omega_l, & i = l \\ \left(1-\dfrac{\sqrt{\gamma_l}}{2}\right)\omega_e, & i = 0 \end{cases} \tag{22}$$

where $\omega_l = \omega_e$ and terms only up to order $\sqrt{\gamma_l}$ have been retained. The corresponding results for modal amplification factors, obtained from Eqs. 8, are

$$\alpha_i \approx \begin{cases} -\dfrac{1}{\beta_i}, & i \neq l, 0 \\ -\dfrac{1}{\sqrt{\gamma_l}}, & i = l \end{cases} \tag{23}$$

These factors together with γ_i determine the mode shapes as

$$\mathbf{\Phi}_i^* \approx \begin{Bmatrix} \phi_{1i} \\ \cdot \\ \cdot \\ \cdot \\ \phi_{ni} \\ -\dfrac{\phi_{ki}}{\beta_i} \end{Bmatrix}, \quad i \neq l, 0; \quad \mathbf{\Phi}_l^* \approx \begin{Bmatrix} \sqrt{\gamma_l}\dfrac{\phi_{1l}}{\phi_{kl}} - \sum\limits_{i\neq l}\dfrac{\gamma_i}{\beta_i}\dfrac{\phi_{1i}}{\phi_{ki}} \\ \cdot \\ \cdot \\ \cdot \\ \sqrt{\gamma_l}\dfrac{\phi_{nl}}{\phi_{kl}} - \sum\limits_{i\neq l}\dfrac{\gamma_i}{\beta_i}\dfrac{\phi_{ni}}{\phi_{ki}} \\ -1 \end{Bmatrix}; \quad \mathbf{\Phi}_0^* \approx \begin{Bmatrix} \sqrt{\gamma_l}\dfrac{\phi_{1l}}{\phi_{kl}} + \sum\limits_{i\neq l}\dfrac{\gamma_i}{\beta_i}\dfrac{\phi_{1i}}{\phi_{ki}} \\ \cdot \\ \cdot \\ \cdot \\ \sqrt{\gamma_l}\dfrac{\phi_{nl}}{\phi_{kl}} + \sum\limits_{i\neq l}\dfrac{\gamma_i}{\beta_i}\dfrac{\phi_{ni}}{\phi_{ki}} \\ +1 \end{Bmatrix} \tag{24}$$

The corresponding expressions for damping coefficients from Eqs. 17 are

$$\zeta_i^* \approx \begin{cases} \zeta_i, & i \neq l, 0 \\ \dfrac{\zeta_l + \zeta_e}{2}, & i = l, 0 \end{cases} \tag{25}$$

The modal masses from Eqs. 20 reduce to

$$M_i^* \approx \begin{cases} M_i, & i \neq l, 0 \\ 2m_e, & i = l, 0 \end{cases} \tag{26}$$

Note that the small modal masses associated with the l-th and 0-th modes result from the scaling used for these modes, i.e. from the scaling of their last components to 1 and -1, respectively. The expressions for the participation factors from Eqs. 21 are

$$\Gamma_i^* \approx \begin{cases} \Gamma_i, & i \neq l, 0 \\ \dfrac{1}{2}\left(\dfrac{\Gamma_l}{\sqrt{\gamma_l}}\phi_{kl} - \sum\limits_{j \neq l} \dfrac{\Gamma_j}{\beta_j}\phi_{kj} - r_{n+1}^*\right), & i = l \\ \dfrac{1}{2}\left(\dfrac{\Gamma_l}{\sqrt{\gamma_l}}\phi_{kl} + \sum\limits_{j \neq l} \dfrac{\Gamma_j}{\beta_j}\phi_{kj} + r_{n+1}^*\right), & i = 0 \end{cases} \tag{27}$$

Observe that the participation factors for the l-th and 0-th modes are much larger than those for other modes as a consequence of the scaling mentioned above.

The preceding results for perfect tuning exhibit a remarkable symmetry between the l-th and the 0-th modes. Observe that their frequencies are equally spaced on each side of the tuning frequency and their damping coefficients and modal masses are equal. Furthermore, note that the expressions for their modal shapes and participation factors are symmetrical in form. It is also seen that the frequencies, damping coefficients, modal masses, and participation factors of the grossly detuned modes are unaffected by the addition of the light equipment.

RESPONSE OF EQUIPMENT-STRUCTURE SYSTEM

It is well known (e.g., see Ref. 6) that any response of a linear MDOF system to a prescribed input excitation can be obtained as a superposition of modal contributions. In particular, any response quantity of the equipment in the combined system under consideration can be written as

$$R(t) = \sum_{i=0}^{n} R_i(t) = \sum_{i=0}^{n} \Psi_i^* S_i^*(t) \tag{28}$$

where $R_i(t) = \Psi_i^* S_i^*(t)$ is the contribution of mode i to the response, Ψ_i^* is the i-th effective participation factor, and $S_i^*(t)$ is the i-th normal coordinate representing the response of an oscillator of frequency ω_i^* and damping coefficient ζ_i^* to the given excitation. The effective participation factor, Ψ_i^*, depends on the particular response sought and in general is given as a product of the i-th participation factor, Γ_i^*, and a linear combination of the components of the i-th modal vector. For example, for the displacement response of the equipment, it is simply $\Gamma_i^* \phi_{n+1,i}^*$, and for its displacement relative to the attachment point, it is $\Gamma_i^*(\phi_{n+1,i}^* - \phi_{ki}^*)$.

It is interesting to examine the values of the effective participation factors for the equipment displacement in the case of perfect tuning. Using Eqs. 27, these become

$$\Psi_i^* \approx \begin{cases} -\dfrac{\Psi_i}{\beta_i}, & i \neq l, 0 \\ -\dfrac{1}{2}\left(\dfrac{\Psi_l}{\sqrt{\gamma_l}} - \sum_{j \neq l}\dfrac{\Psi_j}{\beta_j} - r_{n+1}^*\right), & i = l \\ +\dfrac{1}{2}\left(\dfrac{\Psi_l}{\sqrt{\gamma_l}} + \sum_{j \neq l}\dfrac{\Psi_j}{\beta_j} + r_{n+1}^*\right), & i = 0 \end{cases} \tag{29}$$

where $\Psi_i = \Gamma_i \phi_{ki}$, $i = 1, 2, ..., n$, is the effective participation factor for the i-th mode for the displacement of the attachment point in the structure alone. It is important to observe the presence of $\sqrt{\gamma_l}$ in the denominators of the leading terms of Ψ_l^* and Ψ_0^*. Because of this term, the effective participation factors for these modes are two large numbers of opposite signs and almost equal magnitudes. This fact has an important bearing on the manner in which these two modes combine to produce the equipment response, as will be shown subsequently.

Depending on the type of input excitation, a deterministic or a probabilistic method is required to evaluate Eq. 28. For a deterministic input time history, any suitable time-integration method could be used with no problems anticipated. On the other hand, if the excitation is stochastic, particular care must be exercised to account for the correlation between normal coordinates, $S_i^*(t)$, which can be significant for modes with closely spaced frequencies, as would occur for the case of tuned or nearly tuned equipment. The problem of closely spaced modes in using the mode-superposition procedure in random vibrations has been addressed in Ref. 2. Interpretations of these results for a general structure in terms of response spectra have been developed in Ref. 3. Details of the application of these developments to the particular problem of equipment-structure systems will be presented in a forthcoming publication. Here, the major results of the response spectrum approach which are of most practical interest to the designer are summarized.

FORMULATION IN TERMS OF RESPONSE SPECTRUM

It has been shown (3) that when the input excitation is wide-band Gaussian and stationary over a duration of time which is long in comparison to the fundamental period of the structure, then, it is possible to obtain statistical measures of the response in terms of the ordinates of the mean response spectrum associated with the input excitation. The mean response spectrum is defined as a function, $\bar{S}_\tau(\omega, \zeta)$, representing the mean peak response over a duration τ of a single-degree-of-freedom oscillator of frequency ω and damping coefficient ζ to a prescribed ensemble of excitations of the type described above. This definition is consistent with that of a "smooth" response spectrum commonly employed in earthquake engineering, with an additional refinement here to account for the duration of excitation. From an engineering point of view, the most critical measure of interest is, $\bar{R}_\tau$, the mean peak response over the duration of excitation. A simple approximate expression for this quantity is (3)

$$\bar{R}_\tau = \left[\sum_{i=0}^{n}\sum_{j=0}^{n}\rho_{0,ij}\bar{R}_{i\tau}\bar{R}_{j\tau}\right]^{1/2} \tag{30}$$

where $\bar{R}_{i\tau} = \Psi_i^* \bar{S}_i^*(\omega_i^*, \zeta_i^*)$ is the mean peak value of the contribution from mode i, and $\rho_{0,ij}$ is the correlation coefficient between responses in modes i and j given by (2)

$$\rho_{0,ij} = \frac{2\sqrt{\zeta_i^*\zeta_j^*}\left[(\omega_i^*+\omega_j^*)^2(\zeta_i^*+\zeta_j^*) + (\omega_i^{*2}-\omega_j^{*2})(\zeta_i^*-\zeta_j^*)\right]}{4(\omega_i^*-\omega_j^{*})^2 + (\zeta_i^*+\zeta_j^*)^2(\omega_i^*+\omega_j^*)^2} \tag{31}$$

As shown in Fig. 4, $\rho_{0,ij}$ rapidly diminishes as ω_i^* and ω_j^* become widely separated. Thus, cross terms in Eq. 30 can be neglected if the frequencies of the combined equipment-structure system are well spaced. In this case Eq. 30 reduces to

$$\overline{R}_\tau = \left[\sum_{i=0}^{n} \overline{R}_{i\tau}^2\right]^{1/2} \tag{32}$$

which is the well known square-root-of-sum-of-squares (SRSS) rule for modal combination. It is important to realize that in most equipment-structure systems closely spaced modes do occur.

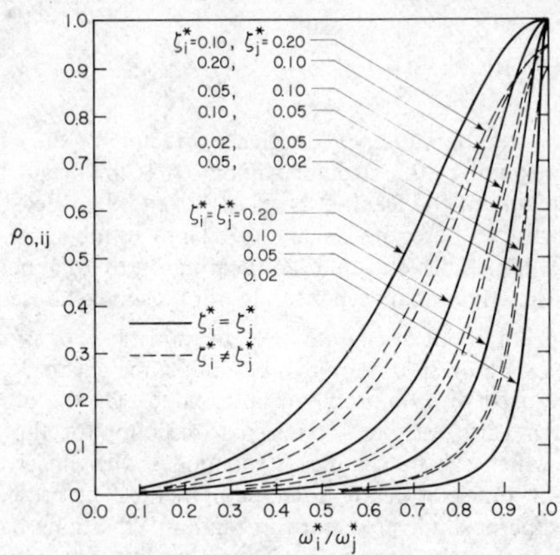

FIG. 4.- Correlation Coefficient Between Modal Responses

Thus, the use of the SRSS method for equipment-structure systems is in general inappropriate and should be avoided. In particular, for tuned or nearly tuned equipment, the effective participation factors Ψ_l^* and Ψ_0^*, associated with the resulting two closely spaced modes, are large numbers of opposite sign and nearly equal magnitude. Because of this, the cross terms associated with these modes are negative and of the same magnitude as the squared terms associated with the individual modes. Hence, it is clear that the SRSS rule, which neglects the cross terms, has the potential for severly overestimating the equipment response. This has been demonstrated in several numerical studies which have indicated that the overestimation can be as much as a factor of 10 or more (5).

It is useful to examine Eq. 30 in the special case of perfect tuning of the equipment to the l-th mode of the structure. In this case, to dominant terms, Eq. 31 for the two closely spaced modes reduces to

$$\rho_{0,l0} = \frac{4\zeta_a^2}{\gamma_l + 4\zeta_a^2} \tag{33}$$

where $\zeta_a = (\zeta_l + \zeta_e)/2$ is the average of the damping coefficients of the equipment and of the structure mode to which it is tuned. Using this expression in conjunction with Eq. 29 in Eq. 30, and assuming that all the other modes are well separated, the mean peak response is obtained as

$$\overline{R}_\tau = \left\{\sum_{i\neq l}\left(\frac{\Psi_i}{\beta_i}\right)^2 \overline{S}_\tau^2(\omega_i,\zeta_i) + \left[\left(\frac{\Psi_l}{\sqrt{2(\gamma_l+4\zeta_a^2)}}\right)^2 + \left(1-\frac{\gamma_l}{2(\gamma_l+4\zeta_a^2)}\right)\left(\sum_{i\neq l}\frac{\Psi_i}{\beta_i}+r_{n+1}^*\right)^2 + \frac{\Psi_l}{2/3}\left(\sum_{i\neq l}\frac{\Psi_i}{\beta_i}+r_{n+1}^*\right)\right]\overline{S}_\tau^2(\omega_e,\zeta_a)\right\}^{1/2} \tag{34}$$

Observe that this result is directly given in terms of the known properties of the structure alone and of the equipment alone.

It is useful to check the preceding result with whatever exact results are available. In Ref. 1, exact results for the response of a single-degree-of-freedom appendage on a single-degree-of-freedom structure subjected to white noise are given. If attention is restricted to the case of perfect tuning with light appendage and small damping, to dominant terms the expression for the response of the appendage can be reduced to

$$\bar{R}_\tau = \frac{1}{\sqrt{2(\gamma+4\zeta_s\zeta_e)}}\bar{S}_\tau(\omega_e,\zeta_a) \tag{35}$$

where γ is the ratio of appendage mass to the structure mass and ζ_s is the damping coefficient of the structure. For such a system, Eq. 34 reduces to

$$\bar{R}_\tau = \frac{1}{\sqrt{2(\gamma+4\zeta_a^2)}}\bar{S}_\tau(\omega_e,\zeta_a) \tag{36}$$

As seen, the latter differs from the exact result only in one term in the denominator of the amplification factor, namely, that the geometric mean of the damping coefficients, $\sqrt{\zeta_s\zeta_e}$, has been replaced by the arithmetic mean, $\zeta_a=(\zeta_s+\zeta_e)/2$. This discrepancy is a consequence of assuming modal damping for the combined system in the present approach. In fact, it can be shown that in general the combined system will not have modal damping when the structure has, except in special cases such as perfect tuning with $\zeta_l=\zeta_e$. It should be realized that the disagreement between the two formulations would only be significant when the mass ratio, γ_l, is smaller or of the same order as the damping term, and at the same time ζ_l and ζ_e are very different.

It is also useful to examine Eq. 30 in the special case of gross detuning, i.e., when the equipment frequency is well spaced from all the structure frequencies. In this case, the cross terms in Eq. 30 can be neglected, yielding

$$\bar{R}_\tau = \left\{\sum_i\left(\frac{\Psi_i}{\beta_i}\right)^2\bar{S}_\tau^2(\omega_i,\zeta_i) + \left(\sum_i\frac{\Psi_i}{\beta_i}+r_{n+1}^*\right)^2\bar{S}_\tau^2(\omega_e,\zeta_e)\right\}^{1/2} \tag{37}$$

This result, which is a valid application of the SRSS rule, agrees with previous results obtained in Refs. 7 and 8.

With the present method, the conventional floor spectrum associated with the k-th degree of freedom (i.e., k-th floor) of a structure can easily be generated. This is done simply by letting all $\gamma_i=0$. Eq. 30 then yields the ordinate of the mean floor spectrum for the particular values of ω_e and ζ_e chosen. This procedure, however, leads to an indefinite form for Eq. 30 when $\omega_e=\omega_l$, because in that case $\Psi_0^*=-\Psi_l^*=\infty$, as may be seen from Eq. 29. By letting $\gamma_l=0$ in Eq. 34, the proper limit for this special case is obtained as

$$\bar{R}_\tau = \left\{\sum_{i\neq l}\left(\frac{\Psi_i}{\beta_i}\right)^2\bar{S}_\tau^2(\omega_i,\zeta_i) + \left[\left(\frac{\Psi_l}{2\sqrt{2}\zeta_a}\right)^2 + \left(\sum_{i\neq l}\frac{\Psi_i}{\beta_i}+r_{n+1}^*\right)^2 + \frac{\Psi_l}{2/3}\left(\sum_{i\neq l}\frac{\Psi_i}{\beta_i}+r_{n+1}^*\right)\right]\bar{S}_\tau^2(\omega_e,\zeta_a)\right\}^{1/2} \tag{38}$$

In the special case of gross detuning, Eq. 37 is the appropriate expression for the floor spectrum. In light of the results developed in this study, there is no point in using the conventional floor spectrum method for a single-degree-of-freedom equipment, since it ignores equipment-structure interaction and is computationally inefficient. Nevertheless, the floor spectrum might still be useful for light multi-degree-of-freedom equipment that is attached to a single floor and in cases where the interaction can be neglected. In that instance, the floor spectrum may be used as the input into the multi-degree-of-freedom equipment in the same way as described above. However, at this time, it is not clear whether this is an accurate procedure, since the floor response of the structure, which is the input into the equipment, is generally not a wide-band process and, therefore, the response spectrum method may not be applicable. This problem will be considered in a future investigation.

ACKNOWLEDGEMENT

This research work was supported by the U.S. National Science Foundation under Grants No. ENG-7905906 and CME- 7923651. The support is gratefully acknowledged.

REFERENCES

[1] Crandall, S.H., and Mark, W.D., *Random Vibration of Mechanical Systems*, Academic Press, New York, N.Y., 1963.

[2] Der Kiureghian, A., "On Response of Structures to Stationary Excitation," *Report No. UCB/EERC-79/32*, Earthquake Engineering Research Center, University of California, Berkeley, CA., December, 1979.

[3] Der Kiureghian, A., "A Response Spectrum Method for Random Vibrations," *Report No. UCB/EERC-80/15*, Earthquake Engineering Research Center, University of California, Berkeley, CA., June, 1980.

[4] Der Kiureghian, A., "Stationary Response of Structures with Closely Spaced Modes," *Proceedings*, Second ASCE/EMD Specialty Conference on Dynamics, Atlanta, GA., January, 1981.

[5] Kelly, J.M., and Sackman, J.L., "Conservatism in Summation Rules for Closely Spaced Modes," *Earthquake Engineering and Structural Dynamics*, Vol. 8, No. 1, January-February, 1980, pp. 63-74.

[6] Meirovitch, L., *Analytical Methods in Vibrations*, Macmillan Co., New York, N.Y., 1967.

[7] Peters, K.A., Schmitz, D., and Wagner, U., "Determination of Floor Response Spectra on the Basis of the Response Spectrum Method," *Nuclear Engineering and Design*, Vol. 44, 1977, pp. 255-262.

[8] Sackman, J.L., and Kelly, J.M., "Seismic Analysis of Internal Equipment and Components in Structures," *Engineering Structures*, Vol. 1, No. 4, July, 1979, pp. 179-190.

SEISMIC-RESISTANT DESIGN OF INDUSTRIAL STORAGE RACKS

C. K. Chen,[I] M. ASCE, Roger E. Scholl,[II] M. ASCE,
and John A. Blume,[III] Hon. M. ASCE

ABSTRACT

Seismic-resistant characteristics of industrial steel storage racks have been studied in order to develop seismic design criteria compatible with seismic design criteria for buildings. The study included portal and cantilever tests of subassemblies and static-cyclic tests, shaking table tests, and linear and nonlinear dynamic mathematical model analyses of typical full-scale steel storage racks. The results of shaking table tests and the correlation of responses predicted mathematically with the experimentally measured responses are summarized. Analysis results from the response spectrum method and a simplified equivalent static analysis method, the *UBC* method, are compared. The lateral force provisions recommended in the 1976 *UBC* (Standard No. 27-11) appear generally to provide adequate seismic resistance in racks similar to those studied except that the load factor (modifier) of 1.25 recommended in the *UBC* for all members in braced frames may not be adequate. A larger load factor or modifications to the rack fabrication are needed to preclude early nonductile damage during strong earthquake shaking. The *UBC*'s simple formulas for estimating periods are inappropriate for racks; designs using these formulas can be erroneous.

INTRODUCTION

Industrial steel storage racks are an important class of structures because about 40% of all goods are stored on racks at some time during the manufacture-to-consumption cycle. The seismic response behavior of these structures is therefore of significant economic importance and is also important to health and safety in connection with the storage of food and medical supplies. Traditionally, criteria for design and construction of industrial racks have been developed by their manufacturers and have been directed primarily at gravity loading, with little attention given to earthquake loading.

This paper is based on a study,[2] by URS/John A. Blume & Associates, Engineers, the overall objective of which was to develop seismic design criteria for industrial steel storage racks through correlation and evaluation of various test results and analytical parameter variation studies. The study included static-cyclic tests of rack subassemblies and full-scale racks; structural performance full-scale shaking-table tests, including testing to determine the effects of loose merchandise

---

(I) Project Engineer, (II) Vice President, (III) President, URS/John A. Blume & Associates, Engineers, San Francisco, California

on rack response to shaking; and engineering analysis reconciliations. This paper briefly summarizes the results of shaking-table tests of typical full-scale racks and correlates these results with results predicted theoretically from mathematical models. Results obtained using response spectrum analysis and the 1976 *Uniform Building Code* (*UBC*)[3] method are compared in order to evaluate the applicability of the *UBC* seismic criteria (Standard No. 27-11) to storage racks.

TEST STRUCTURES AND INSTRUMENTATION

The standard pallet rack selected for testing is currently the most common rack used for merchandise storage. Figure 1 is a photograph of the standard pallet rack assembly on the 20-ft-square shaking table at the University of California, Berkeley.[5] Connection details are also shown in the figure. The standard pallet rack modular assembly consists of prefabricated uprights in the transverse direction and horizontal beams spanning between successive uprights in the longitudinal direction. The uprights have two posts 43 in. apart (outside dimensions) that are connected by 96-in.-long horizontal members spaced 5 ft vertically. The upright posts have bearing plates at the bottom that have a single hole through which floor anchors were installed for the tests considered here. Connections of the upright frame members are button welded. The beam end connections (shelf connectors) are of the clip-in type, and the upright posts are slotted along their full height to allow variations in beam vertical spacings.

In the drive-in storage rack (Figure 2), storage pallets are supported by rail members spanning between support arms that cantilever from the columns rather than by beams spanning the bay width, as in the standard pallet rack. The drive-in rack is accessible from one side, but forklifts cannot pass all the way through. Upright frame (and anchor frame) assemblies are similar in construction to those described for the standard pallet racks. The frames are connected by a continuous rail that supports the pallets. In the direction parallel to the aisle (longitudinal), the upright frames are connected at the top by continuous tie members (overhead tie beams). For the anchor frames, ties (anchor beams) are provided at each story level. The horizontal-load-carrying system for the drive-in rack typically consists of bracing in the direction perpendicular to the aisle (transverse) and frame action in the longitudinal direction.

Stacker racks (Figure 3) are part of an industrial storage system that generally uses floor-running stacker cranes for storage and retrieval of goods in large distribution centers. Stacker cranes are usually remote-controlled and can operate in narrow aisles so that material storage density can be maximized. With computerized controls, stacker racks can provide an efficient, inventory-controlled material-handling system. Stacker rack frame assemblies resemble the drive-in racks previously described, but they are usually more complex structures because they are larger. Horizontal-load-carrying systems generally consist of bracing in the transverse direction and frame action combined with supplemental bracing in the longitudinal direction.

All racks mentioned above were anchored to the table except for one rack (the back-to-back pallet rack), which was left unanchored in order

FIG. 1.--Standard Pallet Rack and Joint Details

FIG. 2.--Drive-In Rack

FIG. 3.--Stacker Rack

to evaluate the difference in seismic performance between the anchored and unanchored conditions. The back-to-back pallet rack assembly is essentially the same as the standard pallet rack except that rigid row spacers are provided to tie two identical pallet racks together.

All members were made of cold-formed steel. The minimum yield stress specified by the manufacturers of the standard pallet and stacker racks, for all members except the diagonal rods of the stacker rack, was 45 ksi. The minimum yield stress of the diagonal rods, and of all members of the drive-in rack, was 36 ksi. The gravity live load was simulated by concrete blocks, weighing 1,000 lb each, tied to the beams or pallet rails.

The data-acquisition system sampled each response channel 50 times per second and stored the data by computer on a magnetic disc; the records were then transferred to magnetic tape for permanent storage. The response quantities measured included accelerations and displacements of each floor and deformations of the columns and the bracing members.

TEST RESULTS

For each rack configuration, tests were conducted in both the longitudinal direction (moment-resisting-frame system) and the transverse direction (braced-frame system). The ground motion was simulated by accelerograms recorded during the 1940 El Centro N-S earthquake (EC) and the 1966 Parkfield earthquake (PF). The designations 1/4 EC and 1/2 EC represent tests performed with the maximum amplitude about 1/4 and 1/2 that of the actual EL Centro record, respectively. Figures 4 and 5 illustrate the 1/2 EC and 1/2 PF input table signals, respectively. For each rack tested, the intensities of the table motions were increased progressively from very slight motions causing only elastic response to severe earthquakes causing material yielding and structural damage.

Table 1 summarizes the building code demand (requirement) and rack capacity of all rack configurations considered here along with their actual performance in the shaking-table tests. The seismic resistance capacities shown were determined using the 1976 *UBC* Zone 4 seismic requirements assuming the best site conditions (i.e., minimum S) and Section 3.6.1 of the American Iron and Steel Institute specifications (AISI 3.6.1)[1] for member capacity calculation. Torsional-flexural behavior was ignored in calculating the M-P stress ratio because there is no reliable method for estimating it. A brief summary of test results follows.

- The fundamental periods of vibration ranged from 2 sec to 3 sec for the loaded standard pallet and drive-in racks in the longitudinal direction (moment-resisting-frame system) and 0.5 sec to 1.0 sec for the loaded standard pallet, drive-in, and stacker racks in the transverse direction (braced-frame system). The first-mode damping values were much larger in the longitudinal direction (ranging from 3% to 9% of critical) than in the transverse direction (ranging from 0.5% to 3% of critical).

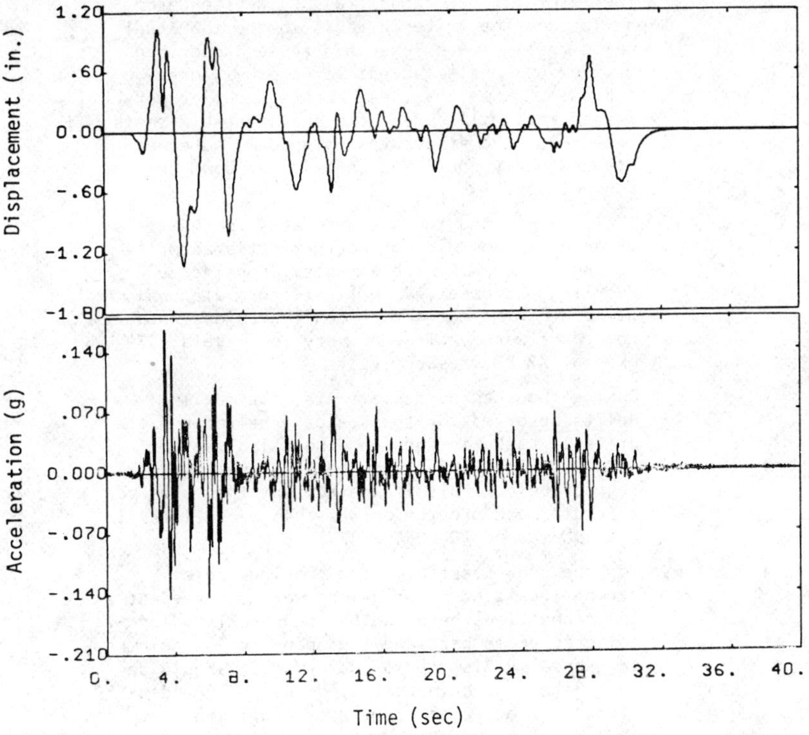

FIG. 4.--Table Motions - 1/2 EC

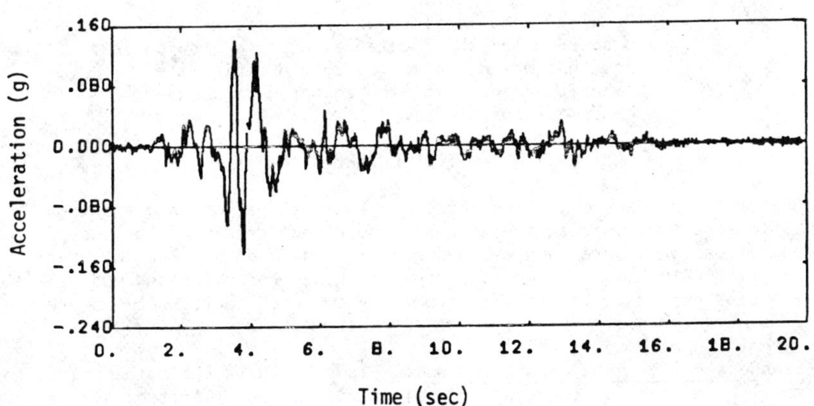

FIG. 5.--Table Motions - 1/2 PF

- The lateral force capacities of the standard pallet and the drive-in racks in the longitudinal direction were less than those prescribed by the *UBC* Zone 4 lateral force provisions Table 1). However, these rack assemblies performed very well during the shaking-table tests because of the structures' high damping capacity and early nonlinear behavior at beam-column connections.
- The drive-in and the stacker racks in the transverse direction did not perform satisfactorily during the shaking-table tests. Considerable buckling was observed in first-story diagonal members of these two rack configurations when the racks were excited at very low levels (1/4 PF and 1/2 PF, respectively).
- During the shaking-table tests, the maximum drifts observed for the standard pallet and the drive-in racks in the longitudinal direction were 0.07 and 0.03 times the story height (H), respectively. This indicates that the racks can tolerate much greater drifts than are allowed in the *UBC* method ($0.005H \times 3/K$).
- For the racks tested on the shaking table, strong localized deformations were observed at the connections between the open-section diagonal bracing members and the open-section columns. In general, this type of deformation should be considered in making detailed response predictions in the braced-frame rack configuration.
- The base plates for all racks that were anchored to the table provided a significant fixity against rotation, which, in turn, reduced the moment at the first-level columns.
- The global and local response amplitudes mea sured from the shaking-table tests for the pallet rack that was not anchored to the table were higher than those for the anchored rack under the same input signal.

ANALYTICAL CORRELATION

One of the primary objectives of the structural performance shaking-table tests was to test the adequacy and effectiveness of various analytical procedures and assumed mathematical models. The computer program DRAIN-2D[4] was used to correlate these procedures and models with the test results. The results from only two rack configurations are presented in this paper.

Standard Pallet Rack - Longitudinal. Figure 6 shows the mathematical model developed for the standard pallet rack in the longitudinal direction. Because the response for the two frames was symmetric, an

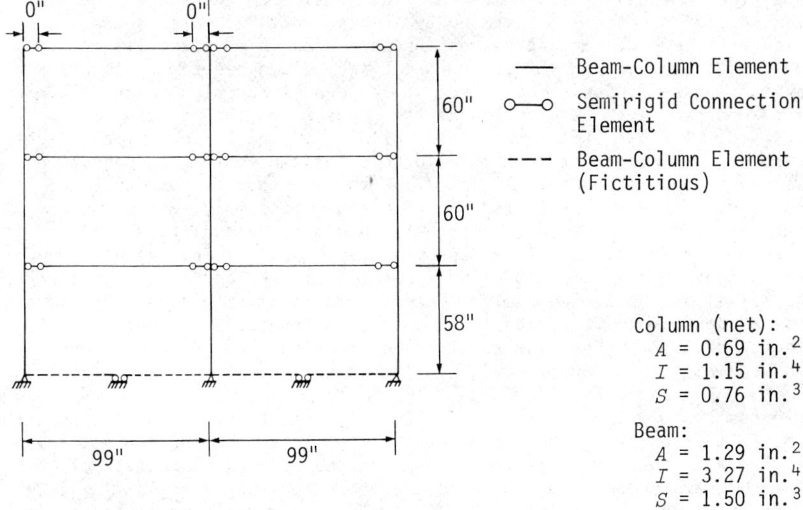

FIG. 6.--Mathematical Model - Standard Pallet Rack, Longitudinal Direction

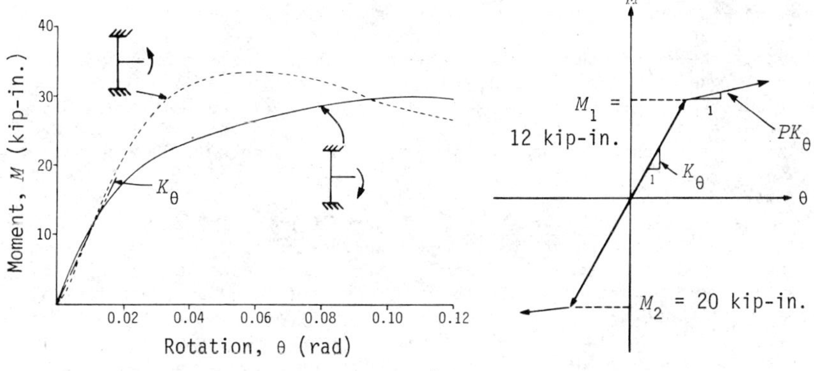

FIG. 7.--Moment-Rotation Relationship for Semirigid Connections

analytical model for a single frame was considered adequate. The minimum section properties, supplied by the manufacturer (also shown in Figure 6), and centerline dimensions were used in modeling all members. Semirigid beam-column joints and semifixed column bases were assumed in evaluating stiffness. The masses of the dead loads plus concrete blocks and wooden pallets were lumped at the nodes where the pallets were located. The p-δ effect was also considered. The bilinear yield mechanism of the moment-rotation relationship for the semirigid connection was idealized as shown in Figure 7. With reference to the experimentally determined M-θ relationship from the subassembly tests,[2] the parameters M_1, M_2, K_θ, and p were assigned and appropriately adjusted in the correlation of calculated and measured results. The calculated yield moments of beams and columns were 65 kip-in. and 34 kip-in., respectively. The average column compression force at yield was estimated to be approximately 22 kips in accordance with AISI 3.6.1 and a safety factor of 1.92. (The assumed yield interaction surface for columns is not shown in this paper but is similar to that shown in Figure 11b.)

Two cases are presented here. For Case 1, the structure was loaded with the full live load (3,000 lb/pallet) and subjected to the input signal of 1/4 PF. Mass-proportional damping corresponding to about 3% of the first-mode viscous damping was assumed for the model. An initial joint rotational spring (K_θ) of 1.4 x 10^6 lb-in./rad and a moment of inertia of the fictitious floor beam (I_f) of 0.2 in.^{4} were assumed. There was good correlation (Figure 8). The response for this case was linear, and no material yielding was detected, either from the analytical or the experimental observations.

For Case 2, the model was also loaded with the full live load and subjected to the input signal of 1/2 EC. This test run was the first instance in which material yielding at the critical column member was observed from the shaking-table tests (the estimated column rotation at yield, θ_y = 1.73 x 10^{-3} rad). Because the data-acquisition system failed at about 12 sec, only response records of 10 sec are included in this presentation. Mass-proportional damping corresponding to about 4.5% of the first-mode viscous damping was prescribed. The parameter K_θ was assigned a value of 10^6 lb-in./rad. Figure 9 shows good agreement between the predicted and measured results. It is apparent that a single basic mathematical model can be used to predict both linear and nonlinear response of the rack structure by varying only damping and joint rotational spring.

Standard Pallet Rack - Transverse. Figure 10 shows the mathematical model and section properties in the transverse direction. Because symmetric response for the three upright frames was found during testing, an analytical model for a single frame was considered adequate for this rack configuration. In modeling this braced-frame system, local deformation at the connections between the braces and column members was considered. The total deformation of the bracing members was taken to be the deformation due to the bracing members and the localized deformation at the connections between the bracing members and column members. Because of this, it was assumed that the composite axial bracing member consisted of two parts, and its stiffness was reduced as shown in Figure 11a. The value of k was assumed and appropriately adjusted in the

Displacement (in.)
2.00
1.00
0.00
-1.00
-2.00
-3.00
3rd Level
0. 2. 4. 6. 8. 10. 12. 14. 16. 18. 20.
Measured
Computed
Time (sec)

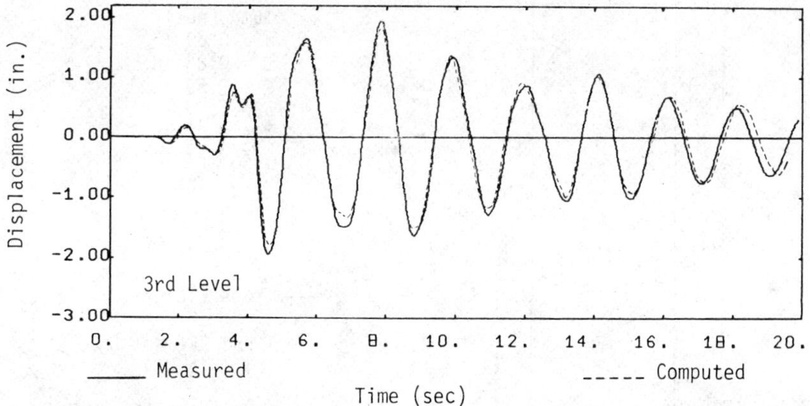

FIG. 8.--Measured and Computed Results - Standard Pallet Rack, Longitudinal Direction, Case 1

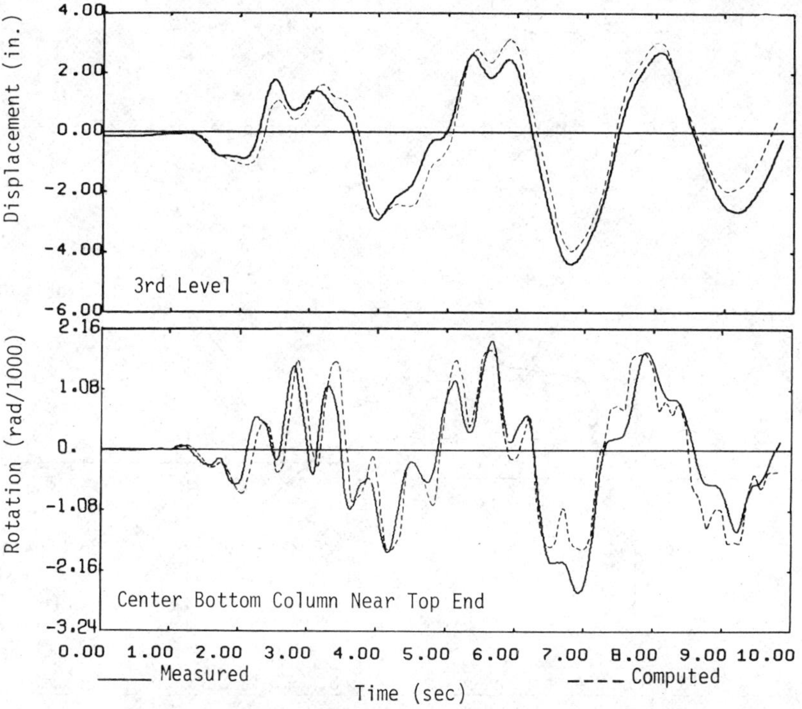

Displacement (in.)
4.00
2.00
0.00
-2.00
-4.00
-6.00
3rd Level
Rotation (rad/1000)
2.16
1.08
0.
-1.08
-2.16
-3.24
Center Bottom Column Near Top End
0.00 1.00 2.00 3.00 4.00 5.00 6.00 7.00 8.00 9.00 10.00
Measured
Computed
Time (sec)

FIG. 9.--Measured and Computed Results - Standard Pallet Rack, Longitudinal Direction, Case 2

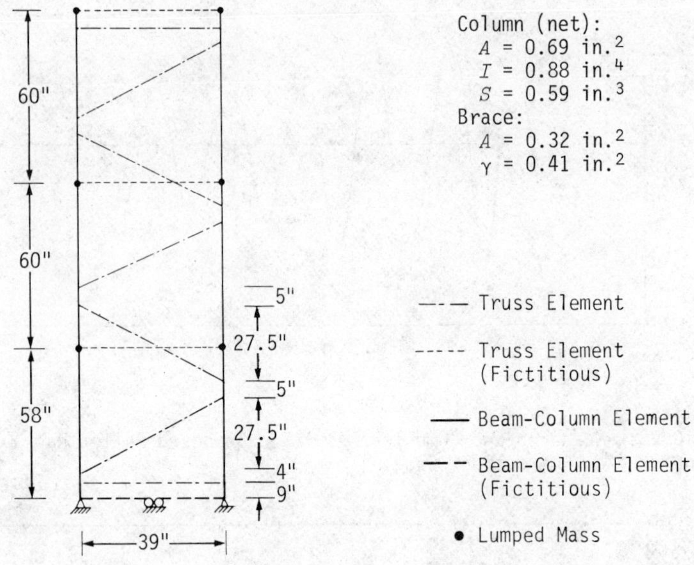

FIG. 10.--Mathematical Model - Standard Pallet Rack Transverse Direction

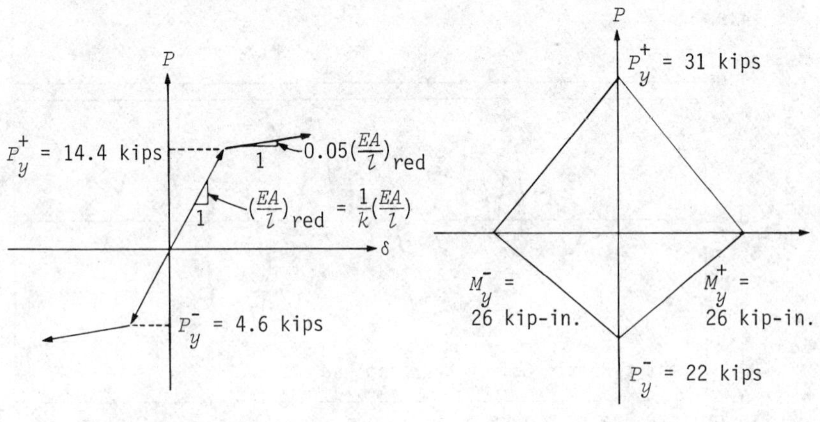

FIG. 11.--Yield Mechanism and Interaction Surface - Standard Pallet Rack, Transverse Direction

correlation of measured and predicted results. The yield interaction surface of the column elements was assumed to be as indicated in Figure 11b.

Three case runs are presented here. The Case 1 model was simulated with 2/3 live load (2,000 lb/pallet) and subjected to the input signal of 1/4 EC. Mass-proportional damping corresponding to about 1.5% of the first-mode critical viscous damping was prescribed. The parameters k and I_f were assumed to be 12 and 0.2 in.^{4}, respectively. Figure 12 shows the measured and predicted third-floor relative story displacements, which correlate well, both in phase and magnitude.

For Case 2, the model was loaded with the full live load, and an input earthquake signal of 1/4 PF was applied. All model parameters and member properties used in the previous case remained unchanged, with the exception of story mass and input signal. Figure 13 shows measured and predicted results. The correlation was considered good except during the latter stage of response, which was essentially free-decay response. The predicted amplitudes were slightly higher than those observed during the test. These higher results could have been reduced by using a higher damping value in the model. The correlation of analytical results with measured results for these two cases again indicates that a single basic mathematical model can be used to predict seismic responses with different storage weights and input earthquake records.

As in the shaking-table test results, significant torsional response and material yielding were observed for Case 3, which was simulated with the full live load subjected to 1/2 EC. A rather brittle fracture occurred at the weld connecting a corner column to the base plate, and noticeable buckling was observed near the base of all except the northeast corner column. Because of the unsymmetric response, theoretical prediction using the two-dimensional model became unrealistic. In modeling this test case, k was assumed to be 14 rather than 12, as was assumed in Cases 1 and 2. The correlation was good during the first few significant cycles until the weld fracture occurred and a significant torsional response took place (results not shown).

Other Rack Configurations. Theoretical predicted results for the drive-in and stacker racks in both principal directions also showed good correction with the experimental results[2] (results not shown in this paper).

EVALUATION OF SEISMIC DESIGN CRITERIA

To achieve the study's overall objective of developing rational seismic design criteria for racks, the *UBC* method, which is simple and widely used for seismic design, was used to assess the adequacy and validity of equivalent static lateral force criteria in rack design. Table 2 summarizes the base shears for each rack considered using the 1976 *UBC* Zone 4 seismic requirements assuming the best site conditions (i.e., minimum S). The base shears from the response spectrum method are also shown in the table. The periods of vibration and the mode shapes were determined from the frequency analysis of the best-fit mathematical models developed for each rack configuration.[2] The *UBC* base shears for the moment-resisting-frame system (the standard pallet and drive-in racks in the longitudinal direction) were multiplied by 1.28 (1.7/1.33) to equate

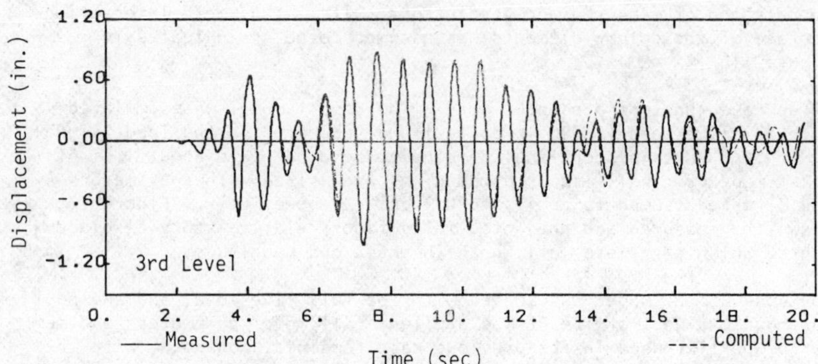

FIG. 12.--Measured and Computed Results - Standard Pallet Rack, Transverse Direction, Case 1

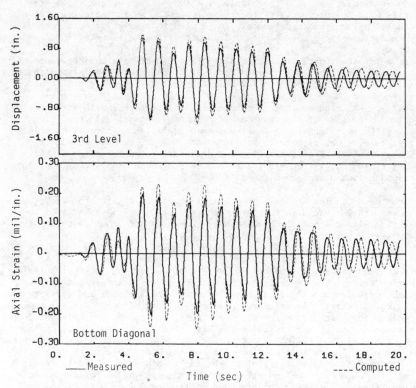

FIG. 13.--Measured and Computed Results - Standard Pallet Rack, Transverse Direction, Case 2

TABLE 1.--Summary of Seismic Performance of Rack Structures*

Rack	Direction	*UBC* Demand versus Rack Capacity†			Behavior During Shaking-Table Tests	
		Critical Element	Stress Ratio		Input Signal	Mode of Damage
			$\frac{P}{P_a}$	$\frac{P}{P_a} + \frac{M}{M_a}$		
Standard Pallet	longi-tudinal	1st floor center column	--	1.32	1-1/3 EC plus vertical	Minor local distress at top of both 1st-floor columns near connectors
"	trans-verse	column near base plates	--	1.32	5/8 EC plus vertical	Noticeable buckling of all columns near bases. Welds of a column base broke.
Drive-In	longi-tudinal	1st floor center column - anchor frame	--	1.37	5/8 EC plus vertical	No visible damage
"	trans-verse	bottom diagonal braces - upright frame	1.20	--	1/4 PF	All bottom diagonal braces of the upright frames buckled.
Stacker	longi-tudinal	first interior columns near bases	--	0.83	3/4 EC	Buckling of interior columns between the bottom and middle rod supports.
"	trans-verse	bottom dia-gonal braces	1.26	--	1/2 PF	All interior bottom diagonals buckled. All interior bottom columns buckled near the base plates.

*The full live load was used in all cases except that of the drive-in rack in the transverse direction, in which 2/3 live load was used.

†The one-third increase permitted for allowable stresses resulting from earthquake forces was included. The periods of vibration were determined from the best-fit mathematical model developed for each rack configuration.

TABLE 2.--Summary of Base Shears for Ultimate-Strength Design: *UBC* Method and Response Spectrum Method

Rack	Direction	Base Shear (lb)				
		UBC Method*	Response Spectrum Method†			
			1/4 EC	1/4 PF	1/2 EC	1/2 PF
Standard Pallet	longitudinal	1,135	564 (3%)	774 (3%)	1,030 (5%)	1,290 (5%)
Standard Pallet	transverse	2,428	1,708 (1.5%)	1,140 (1.5%)	2,799 (3%)	--
Drive-In	longitudinal	3,162	1,892 (4%)	1,494 (4%)	3,485 (6%)	2,888 (6%)
Drive-In	transverse	2,964	2,316 (2%)	2,560 (2%)	4,266 (3%)	--
Stacker	longitudinal	17,040	7,969 (3%)	6,743 (3%)	12,260 (5%)	12,260 (5%)
Stacker	transverse	1,859	1,235 (3%)	1,372 (3%)	1,921 (5%)	2,401 (5%)

*The minimum value of S, for the best site condition, is used.

†Damping values assumed are shown in parentheses.

working-stress design to ultimate-strength design. The base shears for the braced-frame system (the standard pallet, drive-in, and stacker racks in the transverse direction) were multiplied by 1.6 (1.7/1.33 x 1.25). The factor of K in the *UBC* method was assigned values of 1.0 and 1.33 for the moment-resisting-frame system and the braced-frame system, respectively. In the longitudinal direction of the stacker rack, a value of K of 1.33 was assigned because the rack assembly cannot qualify as a dual bracing system consisting of a braced frame and moment-resisting frames.

As shown in Table 2, in the longitudinal direction of the standard pallet and the drive-in racks, the lateral forces determined by the *UBC* method were roughly equivalent to those using the response spectrum method with intensity levels of about 1/2 EC or 1/2 PF. However, in the transverse direction, the *UBC* lateral forces are approximately equivalent to 1/4 to 1/2 the El Centro and Parkfield records. It is evident from the results of the shaking-table tests (Table 1) that the standard pallet and drive-in racks can undergo sizable amounts of inelastic deformation in the longitudinal direction without suffering major damage (in the case of the standard pallet rack, about 1.33 EC in the horizontal direction, with added peak vertical excitation of 0.21g, before observable minor distress occurred). However, in the transverse direction of the drive-in and stacker racks, considerable buckling was observed in the first-story diagonal members when the racks were excited at very low levels (1/4 PF and 1/2 PF, respectively). Thus, it may be concluded that the lateral force provisions recommended in the 1976 *UBC* (Standard No. 27-11) appear generally to provide adequate seismic resistance in racks similar to those studied in this report except that the load factor (modifier) of 1.25 recommended in the *UBC* for all members in braced frames may not be adequate. A larger load factor or some modifications to the rack fabrication are needed to preclude early nonductile damage during strong earthquake shaking.

CONCLUSIONS

The following conclusions and recommendations can be drawn from this study.

- The *UBC* method generally provides adequate earthquake resistance except that a larger load factor or some design modifications to braced-frame systems are needed to preclude early nonductile damage during a strong earthquake. The *UBC* formulas for determining the fundamental periods of vibration, such as $T = 0.05h_n/\sqrt{D}$ and $T = 0.1N$, are not applicable to rack structures. The Rayleigh method (Equation 12-3 in the *UBC*) or a frequency analysis using an appropriate mathematical model (computer-analysis method) are more desirable.
- In general, the responses predicted theoretically for all racks studied in this paper were in good agreement with the experimental results. To develop appropriate mathematical models, rack storage levels are assumed to be sufficiently rigid, and two-dimensional models are considered to be

adequate for practical purposes. Fictitious restraining floor beams can be added to simulate the actual column base fixity condition. Minimum member properties and centerline dimensions are used.

- Further studies to improve the structural performance of the braced-frame system are needed. In addition, experimental investigations are also needed to define the parameter k (localized deformation at connections between the open-section column and open-section bracing members) and the parameter I_f (semifixed column base). Biaxial shaking-table tests, i.e., tests in which the racks are placed at different angles with respect to a principal direction, are also desirable.

ACKNOWLEDGMENTS

The work described in this paper was supported by the National Science Foundation and by the Rack Manufacturers Institute and Automated Storage and Retrieval Systems, sections of the Material Handling Institute. The assistance of R. W. Clough, who directed the shaking table tests at the University of California, Berkeley, and of H. W. Krawinkler, who directed the subassembly tests at Stanford University, is greatly appreciated.

REFERENCES

1. American Iron and Steel Institute, *Specification for the Design of Cold-Formed Steel Structural Members*, New York, 1968.

2. Chen, C. K., R. E. Scholl, and J. A. Blume, *Seismic Study of Industrial Steel Storage Racks*, URS/John A. Blume & Associates, Engineers, San Francisco, California, June 1980.

3. International Conference of Building Officials, *Uniform Building Code*, Whittier, California, 1976.

4. Kanaan, A., and G. H. Powell, *General Purpose Computer Program for Inelastic Dynamic Response of Plane Structures*, EERC 73-5, Earthquake Engineering Research Center, University of California, Berkeley, April 1973 (revised September 1973 and August 1975).

5. Rea, D., and J. Penzien, "Structural Research Using an Earthquake Simulator," *Proceedings*, Structural Engineers Association of California Conference, Monterey, California, 1972.

DYNAMICS OF LOW-PROFILE SHIP LOADING CONTAINER CRANES

Dr.Robert T. Ratay[1] and Tibor Szegezdy[2], Members, ASCE

ABSTRACT

Accelerometer measurements and dynamic analyses were carried out to study the dynamic characteristics and dynamic loadings of container handling shore cranes of low-profile configuration.

The large dynamic loads and the many millions of stress cycles during the lives of these cranes create serious fatigue problems.

Vibrations were recorded during all phases of normal crane work: gantry travel, boom shuttling, container handling and in the stowed-in-wind position.

Previously unaccounted for dynamic loads and oscillations were measured.

The entire structure was modeled for frequency, mode-shape and forced vibration analysis which was performed with the aid of the STRUDL computer code.

Mode shapes and natural frequencies were explored and quantified.

INTRODUCTION

Containerized shipping has been a fast growing mode of material transport throughout the world and the industry has created new and improved equipment for more efficient operations. One of the important links along the route of a container as it is transported, from warehouse to port to ship to port and to final destination, is the crane that handles it between shore and ship.

---

[1] Consultant and Associate Partner, SEVERUD-PERRONE-SZEGEZDY-STURM, Consulting Engineers, New York, N.Y.

[2] Principal Partner, SEVERUD-PERRONE-SZEGEZDY-STURM, Consulting Engineers, New York, N.Y.

The "standard" ship loading container cranes are tall A-frame and similar shaped, rather rigid structures. Height limitations imposed on many ports by air traffic near airports dictated the development of "low-profile" crane frames. By virtue of their configuration these low-profile cranes are more flexible and have been observed to have different dynamic characteristics than the A-frame types with which the industry had long experience.

These cranes have traditionally been designed for equivalent static loads specified by codes, such as those of the AISE (1), FEM (2), BSI (3) and others, or developed by crane designers and manufacturers. The dynamic load amplifications and the millions of stress cycles during the life of a crane create serious fatigue problems which are considered in a design sometime only by the use of impact factors and reduced allowable stresses (fatigue allowables) under an estimated number of repeated loads. Because of the rapid load pick-up and acceleration of the load, large dynamic forces are created throughout the crane, and they experience nearly continuous and occasionally violent vibrations. These vibrations are generated primarily by boom shuttling, container handling and trolley travel. There is also a continuous wind-induced ambient oscillation. Once disturbed, the amplitudes decrease extremely slowly because of the minimal damping in the structure. It appears, however, that no published information is available on the real dynamic behavior of these cranes. A good understanding of the dynamics of them would help in the specification of the real dynamic forces and the actual number and magnitude of repeated loads during the various phases of operation.

Field measurements and numerical analyses were undertaken to learn in an organized, scientific and practical manner about the dynamics of a type of low-profile container cranes in operation. This work was part of an extensive program of structural review and analysis of this type of cranes, performed by Severud-Perrone-Szegezdy-Sturm, Consulting Engineers.

It was believed at the commencement, and confirmed during the progress of the work, that careful observation and sound understanding of the vibrations would shed light on some unanticipated and apparently premature fatigue problems that are experienced by some cranes of this type.

This paper describes and evaluates a program of accelerometer measurements and computer analyses that was carried out to study the dynamic characteristics and dynamic loadings of container handling shore cranes of low-profile configuration.

DESCRIPTION OF THE CRANES

A schematic drawing of the type of cranes investigated is

shown in Figure 1 with the identification of the major parts.

The cranes are of the shuttling boom and travelling trolley type for the specific purpose of handling truck trailer-size shipping containers between ship and shore. They consist of a 60 feet wide x 100 feet long x 120 feet high rail-mounted gantry frame that travels along the dock, and a 25 feet wide x 15 feet tall x 300 feet long three-dimensional boom truss that shuttles in the direction of its long axis between its operating and stowed (non-operating) positions. The boom lower chords are supported on rollers which are attached to four hangers which in turn are suspended from the boom support beams. The container pick-up head block hangs on an intricate system of cables which in turn are hung off a trolley which travels with the operator's cab between ship and shore on rails attached to the lower truss chords of the boom. The entire crane is of welded steel construction.

Figure 2 illustrates the container handling operation. Figures 3 through 6 are photographs of various views: Figure 3 shows three cranes with the booms in fully extended operating position, a container being lifted from a truck trailer chassis (to be deposited in a stack of containers to be ready for loading into a ship at a later time) by the first crane; and a ship tied up in loading-unloading position under the third crane. Figure 4 shows two cranes with the booms in fully retracted, or stowed, position.

Figure 5 is intended to show the gantry's structural configuration looking from the landside (the horizontal member near the lower edge of the picture is a power rail, not a structural member of the crane). Figure 6 shows the boom extended over the water with a container being moved.

The approximate shapes and sizes of the gantry and boom members are the following:

Gantry	Legs	48" x 48" box column
	Boom support beams	48" x 72" box girder
	Leg tie girders	50" x 60" box girder
	Stowage girders	50" x 72" box girder
	Sill beams	48" x 98" box girder
Boom	Upper chords	14" diameter
	Lower chords	30" wide-flange
	Truss verticals	14" wide-flange
	Truss diagonals	8" diameter pipe
	Internal bracing	4" and 8" square tube
	Machinery house	1/4" plate with stiffeners

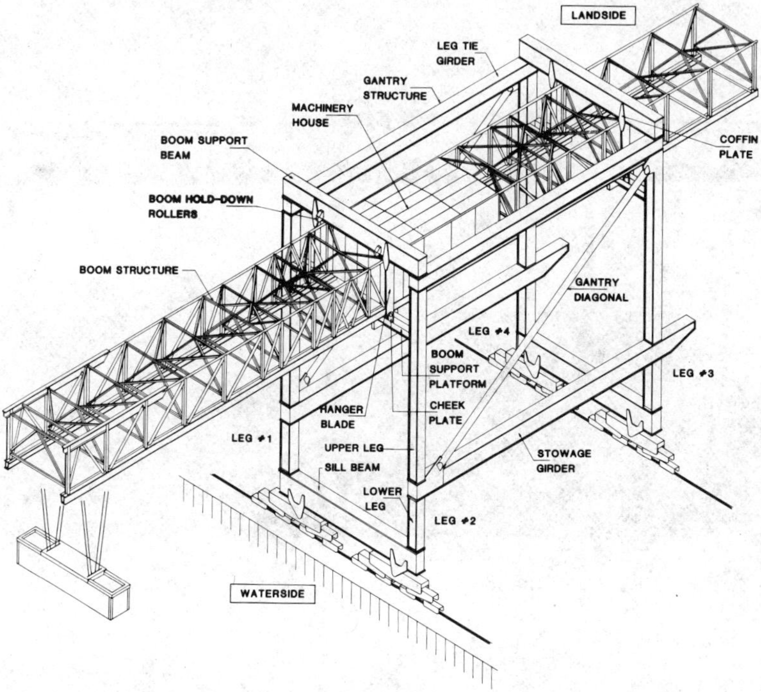

LANDSIDE
LEG TIE GIRDER
GANTRY STRUCTURE
MACHINERY HOUSE
BOOM SUPPORT BEAM
COFFIN PLATE
BOOM HOLD-DOWN ROLLERS
BOOM STRUCTURE
GANTRY DIAGONAL
LEG #4
BOOM SUPPORT PLATFORM
LEG #3
HANGER BLADE
CHEEK PLATE
LEG #1
UPPER LEG
STOWAGE GIRDER
SILL BEAM
LOWER LEG
LEG #2
WATERSIDE

FIGURE 1.- Low-Profile Container Crane

BACKREACH 43'
MAXIMUM OUTREACH 115'
OPERATOR'S LINE OF SIGHT
HATCH COVER STOWAGE (PIGGY BACK ARRANGEMENT)
OPERATOR'S CAB
STACKING FRAMES STOWAGE
WING BUTTRESSES
CONTAINER
STACKING FRAME RACK
HATCH COVER RACK
COLLECTOR SYSTEM
VESSEL
STANDARD CONTAINERS
CAR CARRIER
OPTIONAL
ALTERNATE
ACTIVE
TRAFFIC LANES

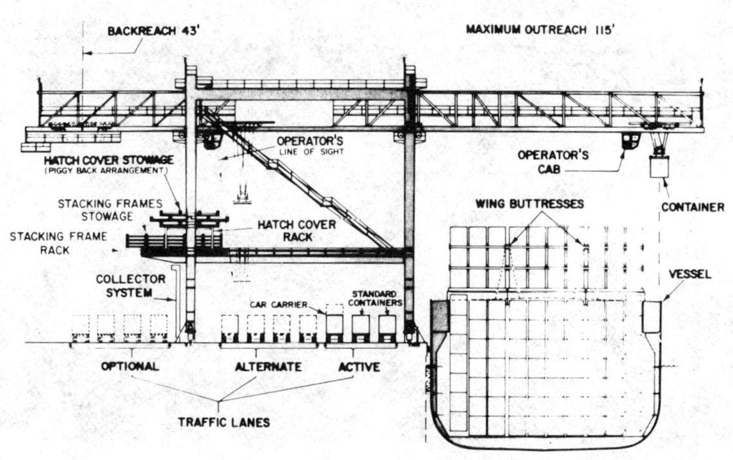

FIGURE 2.- Crane Operation

FIGURE 3.- Crane Booms in Operating Position

FIGURE 4.- Crane Booms in Stowed Position

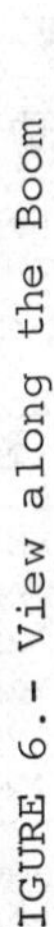

FIGURE 6.- View along the Boom

FIGURE 5.- View from the Land Side

FIELD INSTRUMENTATION

The S-5 Automatic Vibration Monitor System was used for field measurements of vibrations. This system consists of two triaxial detector packages (three-component seismometers) connected by shielded cable to a seven channel permanent recording oscillograph.

This equipment can be operated in either manual or automatic mode; in the automatic mode the instrument continuously monitors the vibrations. The recorder is capable of processing the generated electrical signals directly into acceleration, velocity, or displacement at the operator's option. A permanent record of each vibration event is produced on light-sensitive paper which includes the date and time of the event. Time pulses are automatically produced on each record in order that the period of the vibration can be determined. Two instrument locations can be recorded simultaneously, with each location monitoring vibration levels in three mutually perpendicular directions: longitudinal, transverse and vertical. Deviations of the recorded lines from their positions of rest amount to the actual movement of the monitored location expressed in terms of acceleration, velocity, or displacement.

Pairs of seismometers were placed at various locations on one of the cranes: at each end of the boom and on the boom support platforms. Cyclic displacements and accelerations were recorded continuously and simultaneously in three orthogonal directions. The instruments were monitored during all phases of normal crane work: gantry travel, boom shuttling, container handling and stowed-in-wind.

The purpose of the field measurements was three-fold: to get a better feel and understanding of the vibrations at various locations; to demonstrate the presence of vibrations in the various operating and non-operating conditions; and to confirm certain results of the computer analysis.

Figures 7, 8 and 9 are selected portions of the vibration recordings, illustrating the nature and form of the data.

Figure 7 is a copy of the displacement recordings on a waterside boom support platform and at the waterside end of the fully extended boom, made during and after picking up a container from the ship. Observe the large vertical amplitudes at the end of the boom, and the slow decay of oscillations. Also observe the significant transverse vibrations at both the support platform and the end of the boom; in fact, these transverse displacements appear to be building up with time, probably as a result of transverse swaying of the suspended container and the traveling of the trolley.

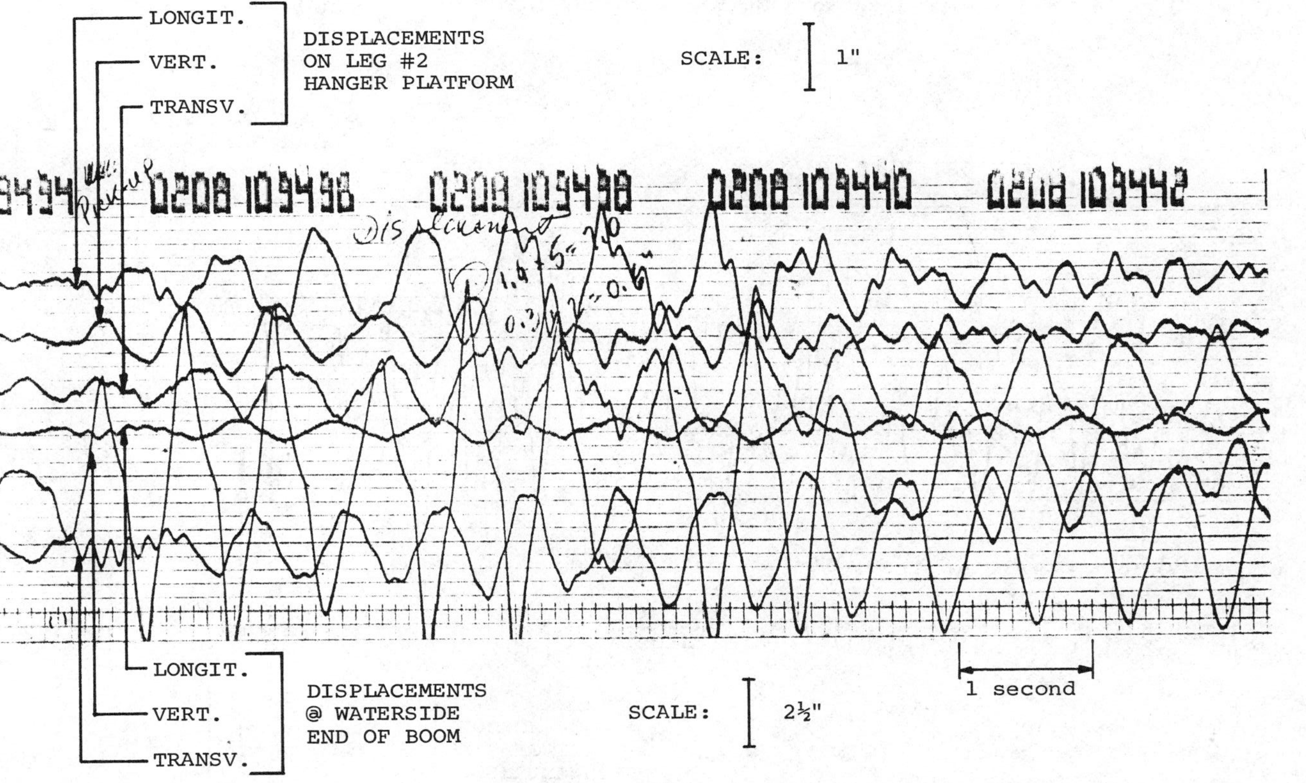

FIGURE 7.- Vibration Records Following Container Pick-up From Ship

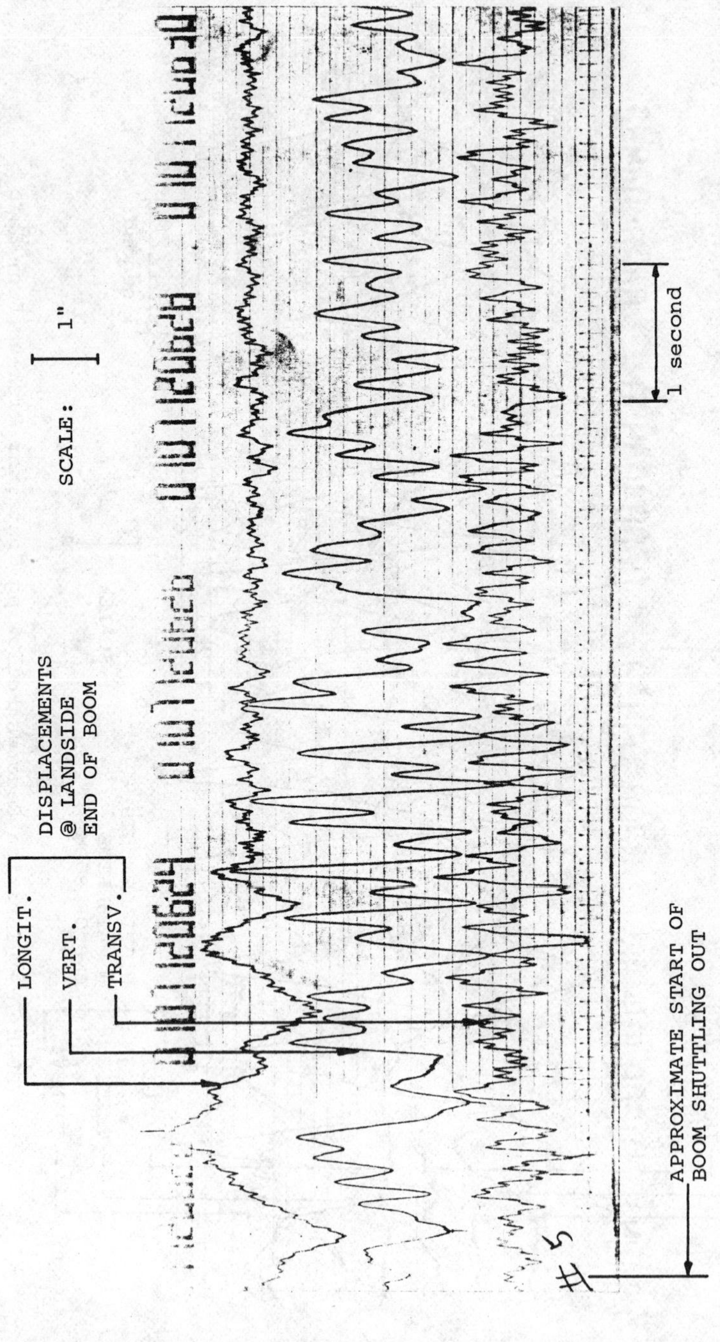

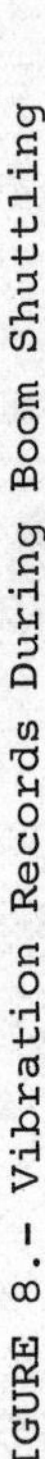

FIGURE 8.- Vibration Records During Boom Shuttling

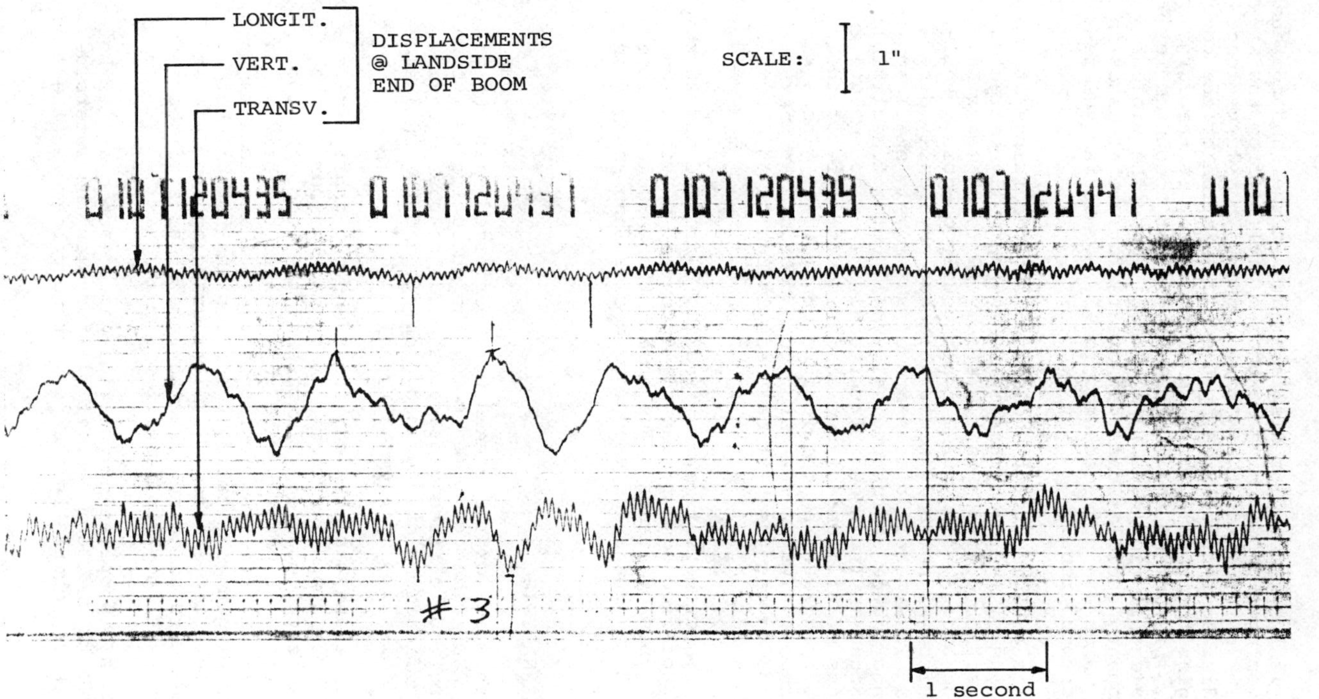

FIGURE 9.- Vibration Records in Stowed Position in Wind.

Figure 8 is the tracing of displacements at the landside end of the boom at the beginning of boom shuttling from the stowed position out towards the water. Observe the large longitudinal displacement created by the acceleration of the boom as it begins to travel. Of much significance is the large vertical vibration of the boom, which is not accounted for by equivalent static loads in a design. Note the violent high frequency oscillations that create fatigue stresses in the boom which are not accounted for by normal fatigue design load conditions.

Figure 9 is a segment of the displacement recordings at the landside end of the boom when it is in the fully retracted stowed position during a fairly steady 20 to 25 mph wind across the boom. There was a continuously on-going oscillation which was not only measurable but physically discernible. These oscillations are of relatively small amplitude, hence induce correspondingly small but not negligible stress ranges in the boom. Importantly, they accumulate into hundreds of millions of cycles during the life of the crane, thus aggrevating the fatigue conditions. These cycles are not always considered in a design. An unexpected finding was the vertical oscillation of the boom in the cross-wind.

The seismometer measurements revealed, clarified and to some extent quantified some previously unconsidered vibrations of these types of cranes. The observations indicate more severe fatigue loads and many more cycles than assumed in routine designs.

COMPUTER ANALYSIS

Dynamic analyses of the cranes were performed with the aid of the STRUDL DYNAL computer code. The objective was to determine periods of free vibrations, mode shapes, dynamic load magnifications due to load pick-up, and the influence of the flexibility of the gantry.

The entire crane was modeled, including the gantry frame, boom and boom support system. Every main member of the gantry frame was modeled. In case of the boom, the two top and two bottom chords were represented by their actual cross-sectional areas as four separate longitudinal members; the truss diagonals, verticals and top and bottom chord bracings were represented by weightless diaphragm plates. The plates were sized to have the same shear stiffnesses as the trusses they replaced.

The boom in its operating position was attached at its bottom chords to the boom support platforms on the waterside, and at its top chords to the boom support beams on the landside.

Masses of the gantry members were input as distributed weights along the members. Masses of the boom were input as lumped masses at several top and bottom chord nodes along the length of the boom. The weights of the trolley and Lifted Load were lumped at the 115-foot outreach.

Undamped periods, mode-shapes and participation factors were computed.

The significant modes and their periods are listed in Table 1; the first five significant normalized mode shapes are drawn in Figures 10.a through e.

Table 1. - Mode Shapes and Periods
(Flexible gantry, flexible boom)

Mode Number	Mode Shape	Period (seconds)
1	Transversse sway of gantry	2.6
2	Transverse twist of gantry	2.1
3	Vertical flexure and tilting of boom	0.84
6	Vertical flexure of boom	0.34
7	Longitudinal sway of gantry	0.31
7	Vertical flexure of boom	0.31
11	Vertical flexure of boom	0.2
14	Transverse flexure of boom	0.17

It is noted that the fundamental and the second modes are in transverse sways of the gantry frame at periods of 2.6 and 2.1 seconds respectively; and the lowest vertical flexural mode shape of the boom is the third mode at a period of 0.84 second. This explains the large transverse vibrations that are observed on these cranes.

A transient forced-vibration analysis was also performed on the same computer model. (STRUDL DYNAL applies the technique of modal superposition). The forcing function was applied at the maximum outreach of the boom (115 feet), uniformly increasing from zero to the total weight of the Lifted Load (120 kips) during a 0.2 second time interval. The time interval was established from the speed of the hoisting engine and the stretch of the hoisting cables.

The time-history of response at 1% viscous damping was calculated through several response cycles.

The calculated transient response in the vertical plane was at a dominant period of 0.84 second; the Dynamic Load Factor was 1.85. This Dynamic Load Factor is significantly higher than the Impact Factors recommended by design codes (1, 2, 3).

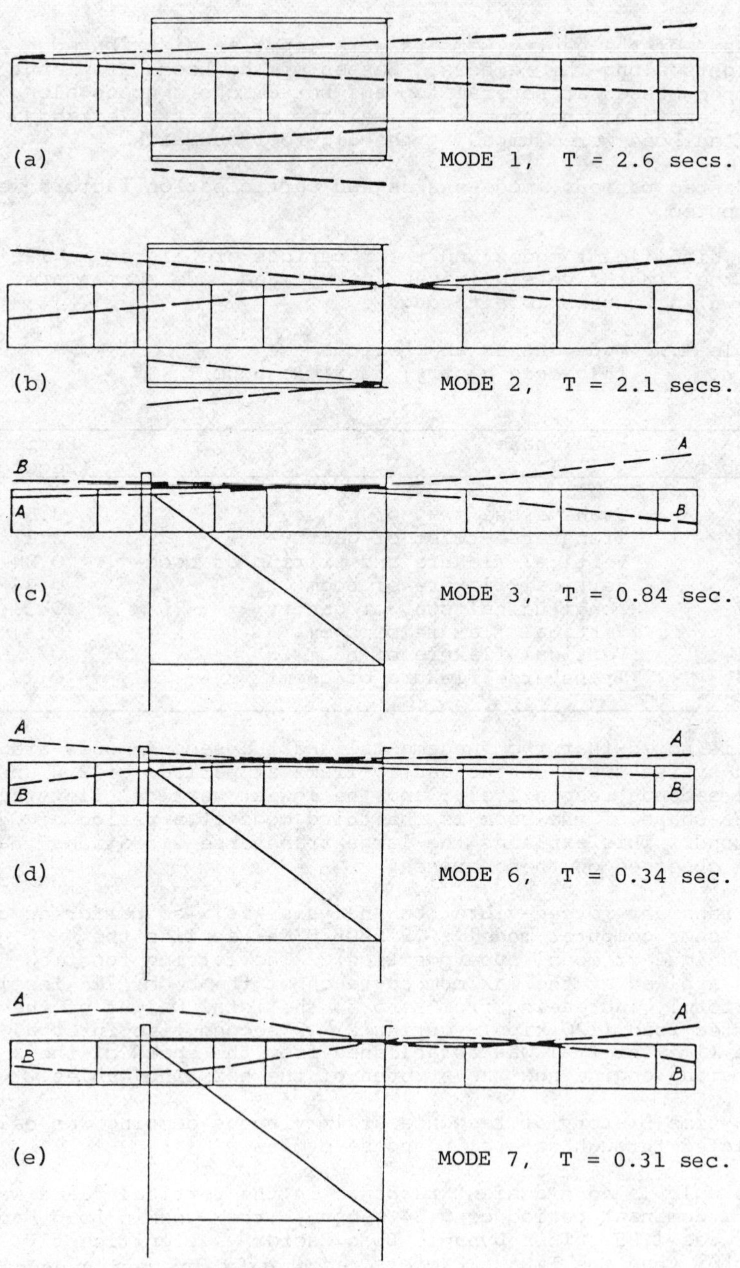

FIGURE 10.- Mode Shapes and Natural Periods.

The foregoing analyses were repeated with the one difference that the gantry frame was made infinitely stiff. The purpose was to isolate the dynamic characteristics of the boom alone, and to explore the effect of the flexibility of the gantry frame.

The significant modes of the boom on rigid supports and their periods are listed in Table 2.

TABLE 2. - Mode Shapes and Periods (Rigid gantry, flexible boom)

Mode Number	Mode Shape	Period (seconds)
1	Vertical flexure of boom	0.72
2	Transverse flexure of boom	0.47
3	Vertical flexure of boom	0.13
4	Vertical flexure of boom	0.01
4	Transverse flexure of boom	0.01

It is noted that on the rigid gantry the first and second modes of the boom are in vertical and transverse bending at periods of 0.72 and 0.47 seconds, respectively. A comparison of these results to those obtained with the flexible gantry frame shows the strong effect of the flexibility of the gantry in the transverse direction. The calculated Dynamic Load Factor was 1.81, which is essentially the same as that with the flexible gantry.

CONCLUSIONS

The seismometer measurements and dynamic analyses indicated strong vibrations in the type of low-profile ship loading container cranes that were investigated. The results showed large dynamic load magnifications and very large numbers of load cycles that result in severe fatigue load conditions for these cranes. Persistent oscillations exist after container pick-up, during boom shuttling, and even in the stationary stowed position in wind.

The flexibility of the gantry frame in the transverse direction strongly contributes to the modes and dominant frequencies of vibrations of the cranes.

This work was preliminary and exploratory in nature and should be evaluated as such. More complete field instrumentation and more refined forced vibration analyses are advisable for firmer conclusions.

This first step of field instrumentation and dynamic analysis is expected to be followed by additional measurements and analyses which are hoped to improve design specifications, codes and standards for container handling shore cranes.

ACKNOWLEDGEMENT

Algis Ratas and John Baranello of SEVERUD-PERRONE-SZEGEZDY-STURM are thanked for their valuable work in all phases of the project; Fred Severud performed the computer analyses.

REFERENCES

1. "Specification for Electric Overhead Traveling Cranes for Steel Mill Service," AISE Standard No. 6, Association of Iron and Steel Engineers, May 1, 1969.

2. "Rules for the Design of Hoisting Appliances," Section I, Heavy Lifting Equipment, Federation Europeenne de la Manutention, December, 1970.

3. "Specification for Permissible Stresses in Cranes and Design Rules," BS 2573, Part 1, Structures, British Standards Institution, 1977.

PARAMETER ESTIMATION FROM FULL-SCALE CYCLIC TESTING

Jean-Guy Béliveau[1], Member ASCE and Michel Favillier[2]

ABSTRACT

In this paper, a method is presented for estimating both stiffness and damping parameters of actual building structures based on acceleration measurements of each story undergoing sinusoidal excitation. Both the acceleration amplitude and phase angle relative to the exciters located on the roof are considered. The optimization is a modified Newton-Raphson scheme which minimizes an objective function based on Bayesian inference. The technique is applied to a five-story steel frame in which the range of frequency excitation encompasses the first two natural frequencies in each of the three lateral motions of a particular story, i.e. the two orthogonal lateral displacements and torsion about the center of geometry.

INTRODUCTION

The correct modelling of a structure is not always evident from the structural plans and whatever assumptions are made regarding its stiffness often does not adequately portray the measured results of the structure as built. The most obvious measurements for dynamic considerations are the natural frequencies and the associated mode shapes. Though the easiest experimental setup required would involve observation during ambient motion, say in a strong wind, consideration is given here to forced excitation where both amplitude and phase of the response with the exciter are considered at a number of frequencies. Time and exciter limitations precluded testing all frequencies, and thus, the frequency range studied encompasses only two natural frequencies in each of the two lateral displacements and in torsion of the floor, herein supposed as a rigid diaphram. Secondly, though the five story steel frame studied has obvious coupling due to both eccentric loading on the roof and unsymmetric location of structural elements, the results recorded here are limited to uncoupled motions and to a shear building representations. A significant variation with previous

1) Associate Professor, Department of Civil Engineering, University of Sherbrooke, Sherbrooke, Québec.

2) Graduate Student, Department of Civil Engineering, University of Sherbrooke, Sherbrooke, Québec.

tests of this type is that both amplitude and phase angle measurements are used in identifying the building parameters.

A description of the building and experimental setup is followed by a description of the building model. A procedure for initial estimates, parameter estimation and sensitivity of the measurement with respect to the parameters is next developped. Finally, the results for a five story rigid steel frame are given.

DESCRIPTION OF BUILDING AND EXPERIMENT

The building tested is a five-story office building on the University of California Campus in Los Angeles. It is built on top of a one-story concrete building and is separated from Boelter Hall and the Math Science Building by two six-inch seismic joints at the North and South end of each story as shown in Fig. 1. There are two bays in the North-South direction and three bays in the East-West direction. Due to equipment restrictions there is no column at the intersection of column lines I and 3 as shown on the typical floor plan, Fig. 2. All lateral loads are resisted by the rigid steel frame composed of WF sections, there being no stairwells or other resisting elements. The column schedule is given in Fig. 3 and the beams for the roof, level 8000, are shown in Fig. 4.

Two centrifugal exciters each consisting of two counter rotating buckets plus additional weights were mounted on the roof. These generated sinusoidal loads of variable frequency in the N-S and E-W direction separately as well as in torsion. Further description and characteristics of the exciters may be found in (Hudson, 1961).

The columns are encased in concrete and a brick facing exists on the east and west exterior walls. The first floor, level 4000, has no floor, and a 26000 pound cooling tower is located on the roof six feet east and four feet north of the geometric center. Levels 5000, 6000 and 7000 have 2" concrete slabs and the roof has a $2\frac{1}{2}$" concrete slab. The story masses and mass moments of inertia are given in Table 1.

This building has been tested and studied on a number of occasions. First a series of dynamic tests were performed at various stages of construction to monitor how the dynamic characteristics were changing in time (Benya, 1967). The completed building was tested at a later time, (Shanman, 1969) where the intrinsic nonlinearity in natural frequency and damping ratio as a function of the force level was observed. Two and three dimensional models of the building were calculated from the member sections, (Bunce, 1970) and a comparison of measured and calculated response to an underground blast was made (Scott, 1973).

The tests reported herein were performed in March 1976, after the February 9, 1971 San Fernándo Earthquake which affected the building. The natural frequencies measured were somewhat smaller, corresponding to a softening of the building, than those reported before, as shown in Table 2. Unlike previous tests, phase angle as well as response amplitude measurements were made, requiring that the data

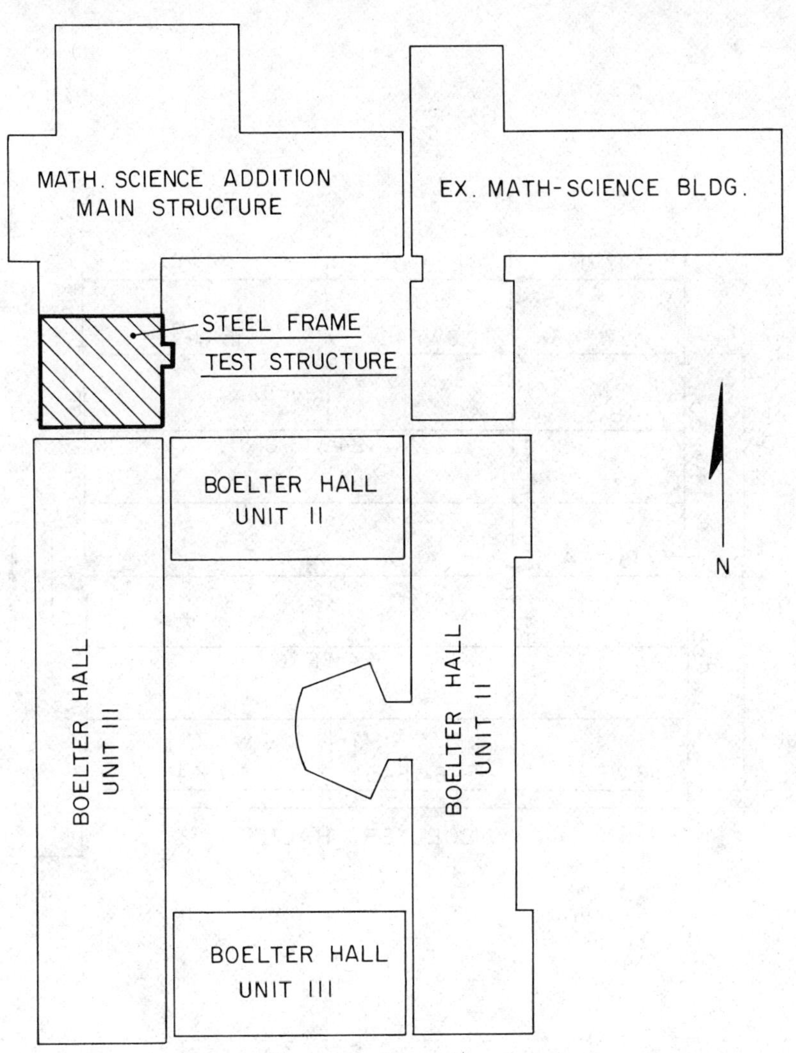

FIG. 1 _ LOCATION PLAN.

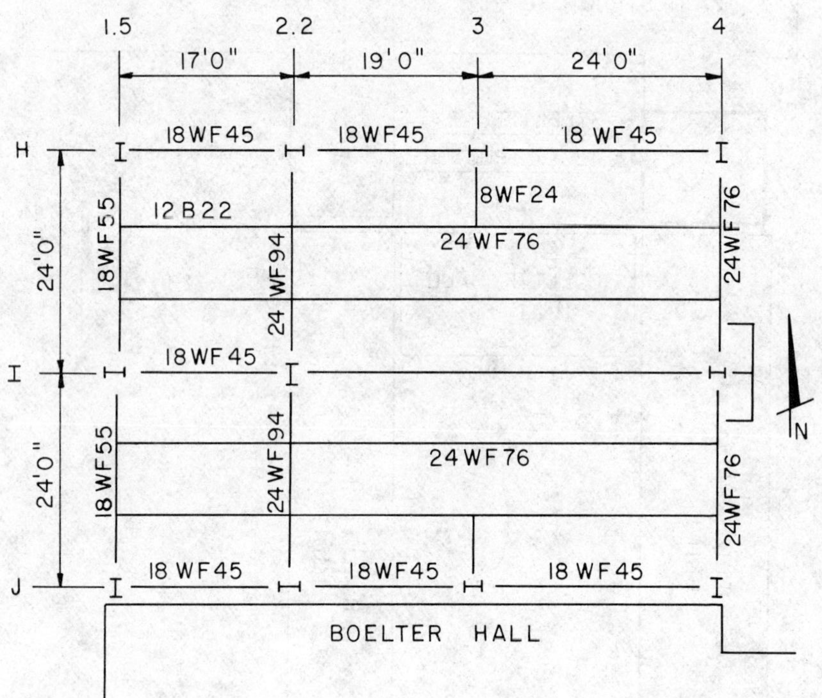

FIG. 2_ TYPICAL FLOOR PLAN.

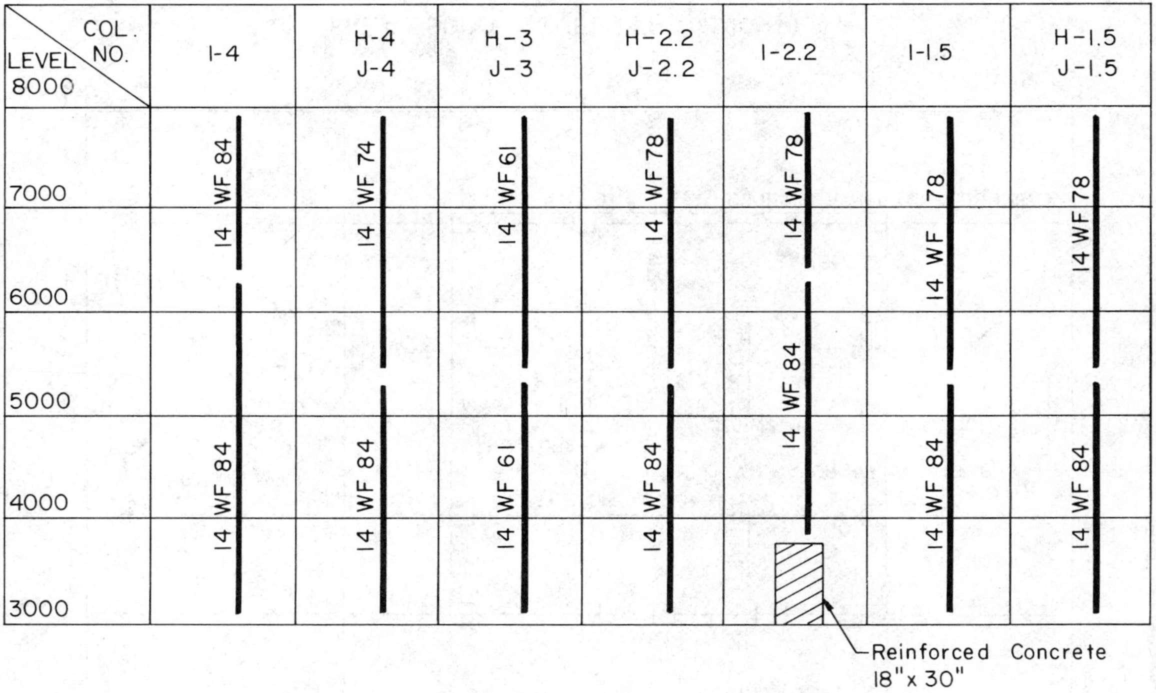

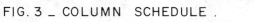

LEVEL \ COL. NO.	I-4	H-4 J-4	H-3 J-3	H-2.2 J-2.2	I-2.2	I-1.5	H-1.5 J-1.5
8000							
7000	14 WF 84	14 WF 74	14 WF 61	14 WF 78	14 WF 78	14 WF 78	14 WF 78
6000					14 WF 84		
5000	14 WF 84	14 WF 84	14 WF 61	14 WF 84		14 WF 84	14 WF 84
4000					Reinforced Concrete 18"x 30"		
3000							

FIG. 3 _ COLUMN SCHEDULE .

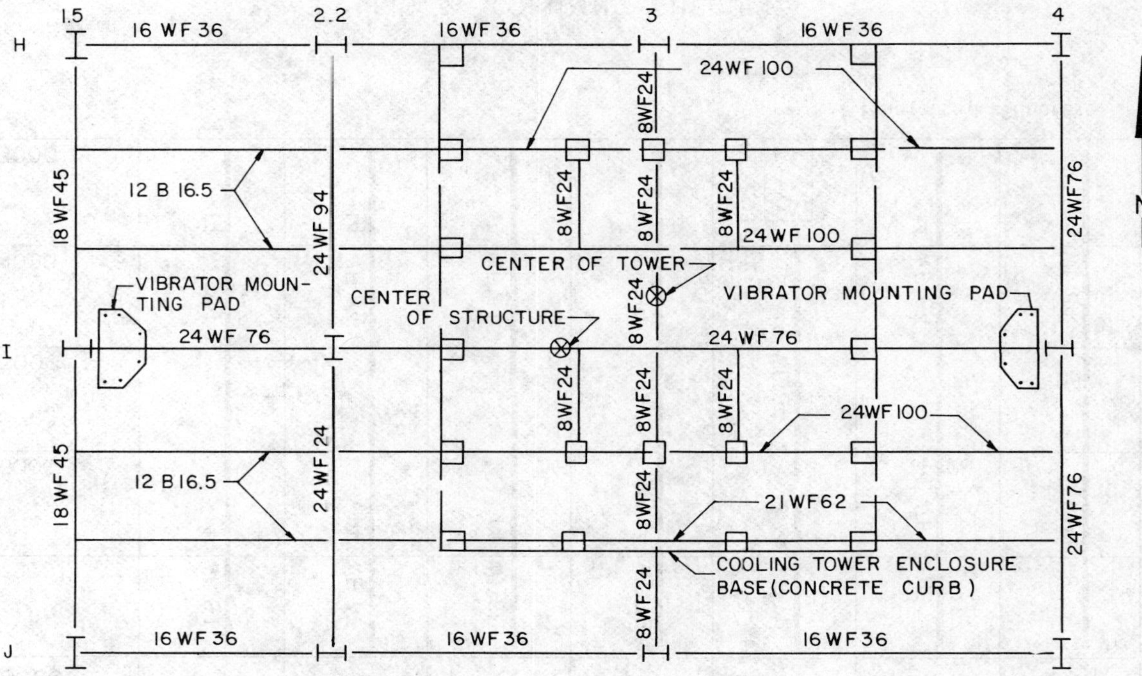

FIG. 4-8000 LEVEL FRAMING PLAN.

TABLE 1 - MASS AND MASS MOMENT OF INERTIA

Story	Level	Mass (kip-sec^2/ft)	Mass Moment of Inertia (kip-sec^2-ft)
1	4000	2.098	1036.
2	5000	8.205	4036.
3	6000	8.300	4079.
4	7000	8.470	4166.
5	8000	8.765	4317.

TABLE 2 - MEASURED AND ESTIMATES OF NATURAL FREQUENCIES (radians/sec)

Mode	Measured (Shanman, 1969)	Measured	Initial Estimate	Best Fit
EW 1	9.55	8.29	8.36	8.61
NS 1	12.50	11.31	10.24	11.31
T 1	14.51	12.57	10.93	12.38
EW 2	27.71	24.13	23.94	25.13
NS 2	35.75	30.98	31.10	31.73
T 2	42.73	37.00	33.18	35.63

be digitized for accuracy in the phase angle. Tests were done on the three degrees of excitation and measurements were taken at all five floors at a number of different excitation frequencies incorporating the two first natural frequencies in each of the three motions. Two levels of loads were used for the lowest and second frequency for a total of four hundred data points, each having four measurements, one being the force, the others acceleration, two of which were parallel, the third being perpendicular to these. Ten seconds were recorded on Sanborn recorders and a HP-39-60A tape recorder. These were later digitized by means of a Kinemetric Digital Datal System at every one hundredth of a second. These signals were finally transformed to lateral accelerations of the center of geometry of each floor and its rotational acceleration as well, both in amplitude and in phase relatioship with the exciter signal, at each frequency of excitation. Further details of the method used to arrive at these data, the experimental aspects, and characteristics of the building, are given elsewhere (Favillier, 1979).

BUILDING MODEL

For purposes of parameter estimation, the equations of structutal dynamics for the lateral loads and rotations are assumed to be uncoupled and to satisfy linear ordinary differential equations of motion.

$$[M]\{\ddot{x}\} + [C]\{\dot{x}\} + [K]\{x\} = \{f\} \tag{1}$$

in which [M], [C], [K] are the mass, damping and stiffness matrices, {f} is the vector of loads and x, $\dot{x}$, $\ddot{x}$ are the lateral displacements, or rotation and the corresponding first and second order time derivatives.

For sinusoidal loads and displacements

$$\{f\} = \tfrac{1}{2}\{F\}e^{i\omega t} + \tfrac{1}{2}\{F^*\}e^{-i\omega t} \tag{2}$$

$$\{x\} = \tfrac{1}{2}\{X\}e^{i\omega t} + \tfrac{1}{2}\{X^*\}e^{-i\omega t} \tag{3}$$

in which ω is the excitation frequency in radians per second and F^* and X^* represent the complex conjugates of the complex quantities, F and X respectively.

When these are substituted into the equations of motion, Eqn 1, there results

$$\{X\} = [H]\{F\} \tag{4}$$

in which, [H], the transfer matrix is the inverse of the impedance matrix [Z]

$$[Z] = [K] - \omega^2[M] + i\omega[C] \tag{5}$$

For purposes of this study, the mass matrix is supposed diagonal, the stiffness matrix is supposed to be closely coupled as in a shear building, and the viscous damping matrix is assured to yield normal modes

$$[M] = \begin{bmatrix} m_1 & 0 & 0 & 0 & 0 \\ 0 & m_2 & 0 & 0 & 0 \\ 0 & 0 & m_3 & 0 & 0 \\ 0 & 0 & 0 & m_4 & 0 \\ 0 & 0 & 0 & 0 & m_5 \end{bmatrix} \qquad (6)$$

$$[K] = \begin{bmatrix} k_1+k_2 & -k_2 & 0 & 0 & 0 \\ -k_2 & k_2+k_3 & -k_3 & 0 & 0 \\ 0 & -k_3 & k_3+k_4 & -k_4 & 0 \\ 0 & 0 & -k_4 & k_4+k_5 & -k_5 \\ 0 & 0 & 0 & -k_5 & k_5 \end{bmatrix} \qquad (7)$$

$$[C] = 2\,[M]\,[\phi]\,[\zeta]\,[\Omega]\,[\phi]'\,[M] \qquad (8)$$

in which ' represents the transpose and $[\zeta]$ are the modal damping ratios (assumed to be zero if not measured). $[\Omega]$ are the undamped natural frequencies and $[\phi]$ is the matrix of normal modes. $[\zeta]$ and $[\Omega]$ are diagonal matrices. The modes and frequencies satisfy the following eigenvalue formulation

$$\left[[K] - \Omega_j^2\,[M]\right]\{\phi_j\} = 0 \qquad j = 1, 2 \ldots 5 \qquad (9)$$

and may be normalized such that

$$[\phi]'\,[M]\,[\phi] = [I] \qquad (10)$$

$$[\phi]'\,[K]\,[\phi] = [\Omega]\,[\Omega] \qquad (11)$$

$$[\phi]'\,[C]\,[\phi] = 2\,[\zeta]\,[\Omega] \qquad (12)$$

$\{\phi_j\}$, the j -th column of $[\phi]$, is the normal mode associated with Ω_j, which are normalized with respect to the mass matrix, [M].

Since only the top story is excited by an inertial exciter the force vector is proportional to the square of the frequency for a constant mass in the buckets

$$\{f\} = c_1\,\omega^2 \begin{Bmatrix} 0 \\ 0 \\ 0 \\ 0 \\ 1 \end{Bmatrix} \qquad (13)$$

where the constant c_1 depends on the mass in the buckets and whether the excitation is lateral or in torsion.

For small damping and with the exciting frequency close to a natural frequency, the response to this load is approximately proportional to the corresponding mode, and the acceleration then would be proportional to the fourth power of frequency times this vector.

$$\{\ddot{X}\} = c_2 \, \omega_j^4 \, \{\phi_j\} \tag{14}$$

where c_2 is a constant.

INITIAL ESTIMATES

Although these are only approximate normal mode shapes, Eqn 9 may be reorganized into a set of equations for solving for the unknown parameters $k_1, \ldots k_5$ (Nielsen, 1966)

$$[B] \;\; \{x\} = [b] \tag{15}$$

When there are two frequencies and mode shapes measured and five degrees of freedom, there results

$$\{x\} = \begin{Bmatrix} k_1 \\ k_2 \\ k_3 \\ k_4 \\ k_5 \end{Bmatrix} \qquad [B] = \begin{bmatrix} [B_1] \\ \hline [B_2] \end{bmatrix} \tag{16, 17}$$

$$\{b\} = \begin{Bmatrix} \Omega_1^2 \; [M] \; \{\phi_1\} \\ \hline \Omega_2^2 \; [M] \; \{\phi_2\} \end{Bmatrix} \tag{18}$$

in which

$$[B_i] = \begin{bmatrix} \phi_{1i} & \phi_{1i} - \phi_{2i} & 0 & 0 & 0 \\ 0 & \phi_{2i} - \phi_{1i} & \phi_{2i} - \phi_{3i} & 0 & 0 \\ 0 & 0 & \phi_{3i} - \phi_{2i} & \phi_{3i} - \phi_{4i} & 0 \\ 0 & 0 & 0 & \phi_{4i} - \phi_{3i} & \phi_{4i} - \phi_{5i} \\ 0 & 0 & 0 & & \phi_{5i} - \phi_{4i} \end{bmatrix} \tag{19}$$

An estimate of the stiffness parameters would then be obtained by premultiplying both sides of Eqn 15 by the transpose of B, corresponding to a least squares estimate.

The masses are assumed known in this development. A similar relationship would hold for damping quantities if a tridiagonal form had been chosen. But, for modal damping ratios, a good initial estimate is available by using the half-power bandwidth of the displacement response.

$$\zeta_j = \frac{\Delta\Omega_j}{2\Omega_j} \tag{20}$$

in which ζ_j is the appropriate modal damping ratio associated with Ω_j and $\Delta\Omega_j$ is the half-power bandwidth. Since only two frequencies were measured in tests reported here, the other three damping ratios are set equal to zero in Eqn 8. Thus only seven parameters are to be estimated, k_1, k_2, k_3, k_4, k_5, ζ_1, ζ_2, with the object being to match the response amplitude and phase angle of the response at each story.

PARAMETER ESTIMATION

The procedure for initial estimates yields parameter values which are approximate only. The values depend on the number of modes considered, do not in general diagonalize the mass or stiffness matrix, and when substituted into the eigenvalue relation, Eqn 9, yield natural frequencies and mode shapes different from the values used in arriving at the estimates. They do, however, provide estimates which at least yield values close to the measured results at the resonant frequencies.

In order to consider all the data available, i.e. amplitude and phase angle at many different frequencies and stories, the objective function used should consider all these measurements including, where possible, their respective weights. In addition, a quantitative evaluation of the initial estimates should be incorporated, in which the weighting matrix can be obtained from the relationship used in arriving at the initial estimates (Jenkins, 1969).

For nonlinear least squares, in which both the measurements and a priori estimates are assumed to satisfy a normal distribution, the objective function to be minimized is (Béliveau, 1976)

$$S\ (\delta p) = \{\varepsilon^{\ell}\}\ [W^d]\ \{\varepsilon^{\ell}\} + \{e^{\ell}\}'\ [W^p]\ \{e^{\ell}\} \tag{21}$$

in which $[W^d]$ and $[W^p]$ are weighting natrices for the data and parameters respectively. $\{e\}$ is related to the corrections to the parameters at ℓ-th iteration and $\{\varepsilon\}$ is the vector of errors between the quantities predicted by the parameters and those measured, using a first order Taylor series

$$\{e^{\ell}\} = \{p^o\} - \{p^{\ell}\} - \{\delta^p\} \tag{22}$$

$$\{\varepsilon^{\ell}\} = \{y\} - \{g\} - \left[\frac{\partial g}{\partial p}\right]'\ \{\delta p\} \tag{23}$$

in which $\{y\}$ and $\{g\}$ are the measured and predicted values respectively. The parameter variations are given by (Béliveau, 1976)

$$\{\delta p\} = [A]^{-1}\ \{r\} \tag{24}$$

in which

$$[A] = \left[\frac{\partial g}{\partial p}\right]' [W^d] \left[\frac{\partial g}{\partial p}\right] + [W^p] \tag{25}$$

$$\{r\} = \left[\frac{\partial g}{\partial p}\right]' [W^d] \{y - g\} + [W^p] \{p^o - p^\ell\} \tag{26}$$

The sensitivity matrix for this modified Newton-Raphson scheme is based solely on first order derivatives of the measured quantities, with respect to the unknown parameter {p} which reduces the numerical effort relative to a Newton-Raphson scheme involving a second order expansion of the objective function S.

DATA SENSITIVITY

The measured quantities are the amplitude, $|x|$, and phase angles, ψ, of the accelerations at each story relative to the exciter load and correspond to the complex quantity, x

$$x = y + i z \tag{27}$$

$$|x| = \sqrt{y^2 + z^2} \tag{28}$$

$$\psi = \tan^{-1}(z/y) \tag{29}$$

The sensitivity of $|x|$ and ψ with respect to the parameters is given by (Béliveau, 1976)

$$\frac{\partial |x|}{\partial p} = \frac{1}{|x|} \mathcal{R} \left(x^* \frac{\partial x}{\partial p}\right) \tag{30}$$

$$\frac{\partial \psi}{\partial p} = \frac{1}{|x|^2} \mathcal{I} \left(x^* \frac{\partial x}{\partial p}\right) \tag{31}$$

in which $\mathcal{R}$ and $\mathcal{I}$ represent the real and imaginary portions, respectively. $|x|$ and ψ for each measurement then form the vector {y} each element of which is associated with the corresponding predicted value, {g}.

The sensitivity of {x} with respect to the parameters {p} is given by

$$\left\{\frac{\partial X}{\partial p}\right\} = - [H] \left[\frac{\partial Z}{\partial p}\right] [H] \{F\} \tag{32}$$

in which [H] is the inverse of [Z] given in Eqn 5 and

$$\left[\frac{\partial Z}{\partial p}\right] = \left[\frac{\partial K}{\partial p}\right] + i \omega \left[\frac{\partial C}{\partial p}\right] - \omega^2 \left[\frac{\partial M}{\partial p}\right] \tag{33}$$

For the model supposed here the last term in Eqn 33 is zero, since the mass elements are considered as known, and the sensitivity of [C] from Eqn 8, is a function of the eigenvalue and eigenvector sensitivity. These are determined from relations dependent only on the associated frequency and mode shape (Nelson, 1976).

NUMERICAL RESULTS

The mass and mass moment of inertia used for the building tested are summarized in Table 1 and correspond to values used before for this structure (Scott, 1973). The frequencies measured once the building was complete, together with those measured more recently are summarized in Table 2 along with the values based on initial estimates and based on estimates obtained iteratively by the method outlined.

The measured natural frequencies are lower than those obtained right after the structure was completed and correspond to a softening effect on the building, part of which is probably due to earthquakes in the interim. The ratio between first and second frequencies is changed very slightly for the EW and torsional motions. The best-fit values are all higher than those calculated based on the initial estimates.

The measured and calculated mode shapes, based on initial estimates and best fit, are shown in Fig. 5. A typical best fit and initial fit for amplitude in levels of gravity ($1\ g = 32.178\ ft/sec^2$) is given in Fig. 6 with corresponding phase angle in Fig. 7 for the acceleration response at the top story in the East-West direction in the neighborhood of the first frequency.

Typically, all the data at each story was fit in each optimization. Thus three different passes were made, one for each direction of motion. The weighting matrix for both the initial estimates and data was assumed to be a diagonal matrix with diagonal equal to one over the square of the standard deviation given in Table 3. These were chosen arbitrarily. Also shown in the Table are the initial values of the parameters (in appropriate units), the final estimates and their standard deviations equal to the square root of the diagonal element of the variance-covariance matrix. The final objective function S^f is expressed as a fraction of the initial value S^o. The number of iterations is also indicated.

Estimates for closely coupled damping matrice were also obtained, but some of the damping constants were negative. Attempts could have been made to determine more general stiffness representations, i.e. not limited to shear building, and to consider strong masses as parameters as well, but were not pursued. A number of attempts were made to model the structure in three dimensions, i.e. incorporating all the data at once, and to estimate parameters for a shear building, with twenty-five stiffness elements, ten of which were coupling terms, and six damping ratios. This did not converge, however, based on initial estimates obtained previously, or from a three-dimensional initial estimate similar in formulation to that given previously.

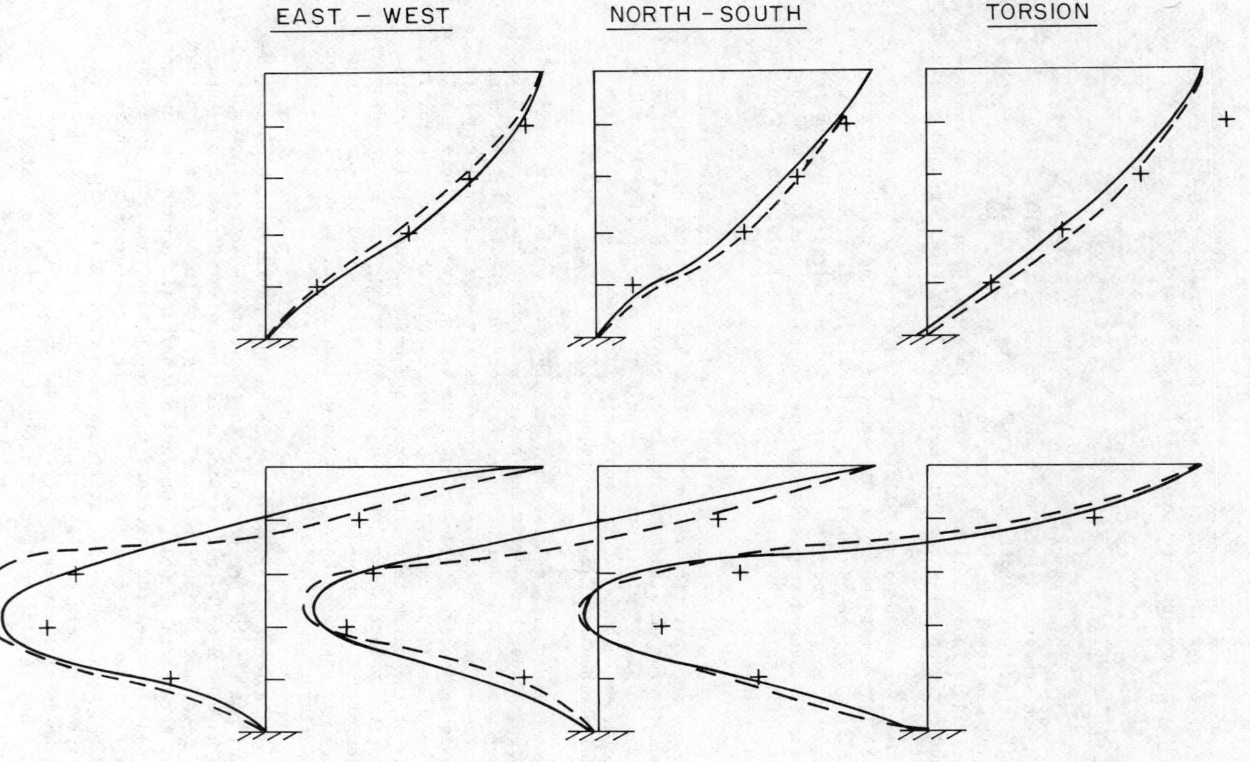

FIG.5_ MODE SHAPES (+ DATA, ---INITIAL ESTIMATE, —— BEST FIT).

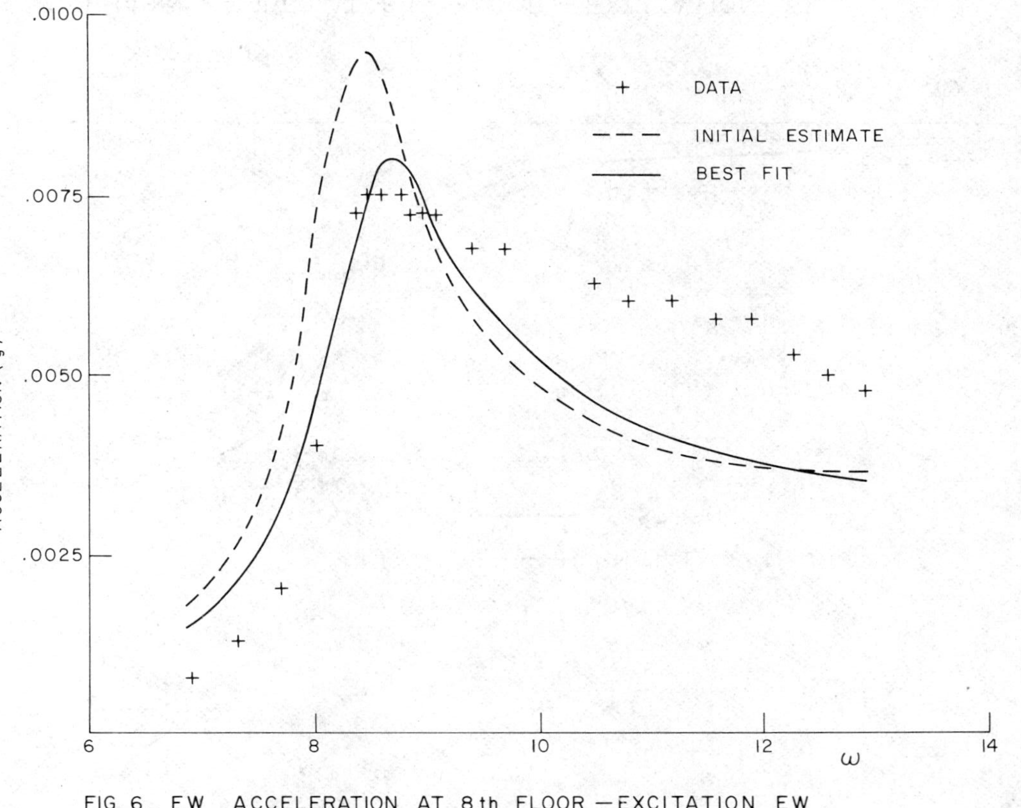

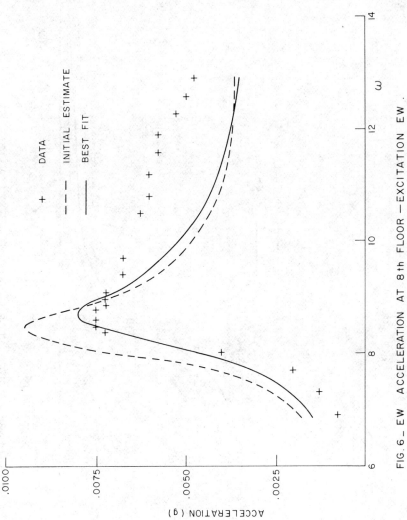

FIG. 6 – EW ACCELERATION AT 8th FLOOR – EXCITATION EW.

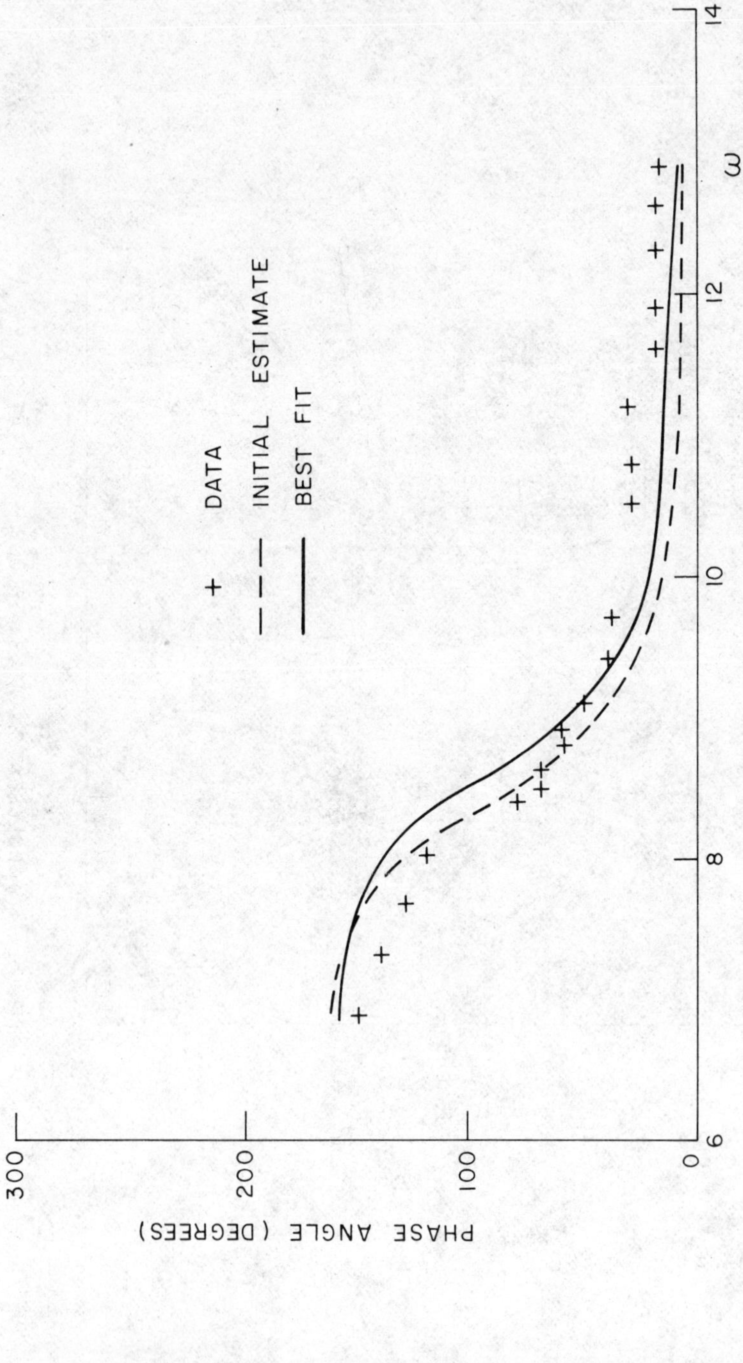

FIG. 7 – PHASE ANGLE AT 8th FLOOR – EXCITATION EW.

TABLE 3 - RESULTS OF PARAMETER ESTIMATION

Parameter	East-West (kip/ft)				North-South (kip/ft)				Torsion ($x10^5$ kip-ft)			
	p^o	σ^o	p^f	σ^f	p^o	σ^o	p^f	σ^f	p^o	σ^o	p^f	σ^f
k_1	9058.	500.	9193.	1.58	19514.	100.	17042	24.	54.47	1.43	86.15	.50
k_2	6448.	500.	6509.	.82	6814.	100.	10608.	11.	59.14	1.43	84.88	.49
k_3	8469.	500.	8332.	1.72	14452.	100.	17696.	32.	88.22	1.43	77.88	.49
k_4	5036.	500.	7248.	.48	8750.	100.	9758.	3.5	47.85	1.43	51.98	.09
k_5	6351.	500.	5510.	.36	12737.	100.	9696.	6.4	85.28	1.43	76.95	.44
ζ_1	.06	.001	.073	$5. \times 10^{-6}$	.062	.001	.064	1×10^{-5}	.058	.001	.112	$.25 \times 10^{-3}$
ζ_2	.051	.001	.063	$7. \times 10^{-6}$	.073	.002	.065	4×10^{-5}	.088	.002	.074	$.29 \times 10^{-3}$
σ_d		2×10^{-6} g				2×10^{-5} g				5×10^{-6} g/ft		
σ_ψ		1×10^{-3} rad				1×10^{-2} rad				1×10^{-2} rad		
No. of iterations		5				8				9		
S^f/S^o		.41				.36				.18		

CONCLUSION

A method is developped to estimate dynamic parameters of buildings from frequency response data incorporating both the amplitude of acceleration at each floor and the phase angle relative to the excitation load on the roof. Initial estimates to the stiffness were based on the approximate mode shapes at the first two natural frequencies for a shear-building representation of an assumed uncoupled building. Story masses are assumed as known, and initial estimates to the damping ratios are obtained from the half-power method. The procedure is based on a modified Newton-Raphson method for modifying the parameters with the sensitivity of the data with respect to the parameters determined analytically.

Caution is in order, however, as the parameters obtained are based on the actual data, and extrapolation to frequencies outside the range tested is not recommended. A more complicated model of the building could be used, but again, caution is needed depending on the number of modes observed, and since good initial estimates are required for convergence. Finally, measurements in all three degrees of freedom are required for a three-dimensional representation.

ACKNOWLEDGEMENTS

This research is sponsored by the National Sciences and Engineering Research Council of Canada, Grant No. A9306. Professor G.C. Hart, of the University of California, Los Angeles, was instrumental in the realization of the experimental aspects.

REFERENCES

[1] Hudson, D.E., "A New Vibration Exciter for Dynamic Tests of Full-Scale Structures", Earthquake Engineering Research Laboratory, California Institute of Technology, Pasadena, California, 1961.

[2] Benya, J.S., "A Forced Vibration Study of a Framed Building", M.S. Mechanics and Structures Department, University of California, Los Angeles, 1967.

[3] Shanman, R.D., "A Forced Vibration Study of a Completed Five Story Frame Building", M.S. Mechanics and Structures Department, University of California, Los Angeles, 1969.

[4] Bunce, B.T., "Two and Three Dimensional Modeling and Dynamic Analysis of the Five Story Math-Science Addition", M.S. Mechanics and Structures Department, University of California, Los Angeles, 1970.

[5] Scott, J.P., "Three-Dimensional Dynamic Response of a Multistory Building", M.S. Structures and Dynamics, Department University of California, Los Angeles, California, 1973.

[6] Favillier, M., "Identification des paramètres dynamiques d'un bâtiment soumis à des charges cycliques", mémoire de maîtrise, Département de génie civil, Université de Sherbrooke, Sherbrooke, Québec, 1979.

[7] Nielsen, N.N., "Vibration Tests of a Nine-Story Steel Frame Building", *Journal of the Engineering Mechanics Division*, American Society of Civil Engineers, No. EM1, 1966, pp. 81-110.

[8] Béliveau, J.-G., "Identification of Viscous Damping in Structures from Modal Information", *Journal of Applied Mechanics*, Vol. 98, No. 2, June 1976, pp. 335-339.

[9] Jenkins, G.M. and Watts, D.G., "Spectral Analysis and its Applications", Holden-Day, San Francisco, 1969.

[10] Nelson, R.B., "Simplified Calculation of Eigenvector Relations", *AIAA Journal*, Vol. 14, No. 9, September 1976, pp. 1201-1205.

IDENTIFICATION OF A CLASS OF NONLINEAR DYNAMIC SYSTEMS

by

Sami F. Masri[1], George A. Bekey[1], Hassan Sassi[2], Thomas K. Caughey[3]

Abstract

A method is presented for the nonparametric identification of chain-like nonlinear multidegree-of-freedom dynamic systems that are characteristic of many practical engineering problems. The method requires information regarding the system response to determine, by means of regression techniques involving the use of two dimensional orthogonal functions, an approximate expression for the surface of each of the system state variables. The method is shown to be simple to apply, computationally efficient, accurate in identifying virtually any type of structural nonlinearity with almost any form of probing signals, and quite tolerant to measurement noise pollution.

Introduction

The identification and modeling of multidegree-of-freedom dynamic systems through the use of experimental data is a problem of considerable importance in the area of structural dynamics (see, e.g., [1-11].*) Indicative of the wide range of applicability of this subject is a recent survey article [6] which contains over 130 references related to system and parameter identification.

System identification techniques can be classified on the basis of their search space: (a) parametric methods that search in parameter space, and (b) nonparametric methods that search in function space. Parametric methods are effective when adequate information about the structure of the model is available and only the parameter magnitudes are to be determined. Nonparametric methods, on the other hand, produce the "best" functional representation of the system without a priori assumptions about the system model.

In many practical dynamic problems, such as mechanical equipment, the structure of the model is by no means clear. As a result, an increasing amount of attention has recently been devoted to nonparametric identification methods such as the ones that use the Volterra-series [12] or Wiener-kernel [13,14] approach.

---

[1] Professor, School of Engineering, University of Southern California.

[2] Graduate Student, School of Engineering, U.S.C.

[3] Professor, Division of Engineering, California Institute of Technology.

* Numbers in brackets designate items in the reference list.

However, the traditional nonparametric identification techniques have their own problems. Some of these include restrictions on the nature of the dynamic systems to be identified and on the input signal that can be used. Furthermore, when dealing with systems that incorporate commonly encountered nonlinearities, such as polynomial nonlinearities, the evaluation of higher-order terms requires a prohibitive amount of computational effort, coupled with very demanding storage requirements.

This paper presents a relatively simple and straightforward approach to the nonparametric identification and modeling of a broad class of nonlinear multidegree-of-freedom dynamic systems. Such a class of models encompasses many practical engineering problems including certain equipment systems, tall buildings, and other similar structures, all of which are governed by equations of motion that are characteristic of chain-like systems.

Identification Procedure

Consider the multidegree-of-freedom (MDOF) chain-like structure similar to the one shown in Fig. 1 which consists of n lumped masses each of magnitude m_i. The structure may be subjected to base excitation $S(t)$ and/or directly applied forces $F_i(t)$. The absolute displacement of m_i is measured by $x_i(t)$; the corresponding relative displacement with respect to the moving support is given by $y_i(t) = x_i(t) - S(t)$; and the interstory relative motion is specified by $z_i(t) = x_i(t) - x_{i-1}(t)$ for $i > 1$, and $z_1(t) = x_1(t) - S(t)$. Differentiation with respect to time is indicated by a dot.

The arbitrary nonlinear elements interposed between the masses are represented by functions $G^{(i)}$ which are dependent on the relative displacement and velocity across the terminals of each element. In addition, each $G^{(i)}$ depends on a vector $\underset{\sim}{p}^{(i)}$ whose parameters characterize the nature of the nonlinear element.

The equations of motion for the chain-like systems under discussion can be expressed as

$$\left.\begin{aligned} G^{(n)}(\underset{\sim}{p}^{(n)}, z_n, \dot{z}_n) &= F_n(t) - m_n\ddot{x}_n \\ &\vdots \\ G^{(2)}(\underset{\sim}{p}^{(2)}, z_2, \dot{z}_2) &= F_2(t) - m_2\ddot{x}_2 + G^{(3)}(\underset{\sim}{p}^{(3)}, z_3, \dot{z}_3) \\ G^{(1)}(\underset{\sim}{p}^{(1)}, z_1, \dot{z}_1) &= F_1(t) - m_1\ddot{x}_1 + G^{(2)}(\underset{\sim}{p}^{(2)}, z_2, \dot{z}_2) \end{aligned}\right\} \quad (1)$$

In order to proceed with this method, the time histories of the excitation as well as the system response are needed. In addition, the magnitude of the concentrated masses must be known, which usually poses no problems in practical cases. However, a more accurate estimate of the mass distribution may be obtained by invoking the impulse-momentum relationship [9].

It is seen that since all of the terms appearing on the right-hand

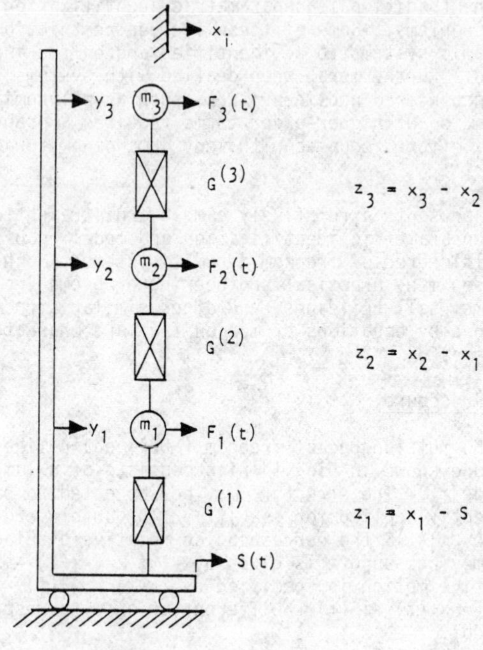

(a) System model

$$G^{(i)}(z, \dot{z}) = p_1^{(i)} z + p_2^{(i)} \dot{z} + p_3^{(i)} z^3$$

ELEMENT CHARACTERISTICS

$G^{(i)}$

i	1	2	3
m_i	12.0	12.0	12.0
p(1, i)	7214.0	14900.0	14900.0
p(2, i)	21.0	60.0	60.0
p(3, i)	72.14	0.0	-223.0

(b) System characteristics

FIGURE 1. EXAMPLE OF NONLINEAR 3 DOF SYSTEM

side of Eq. (1) are available from measurement or can be readily estimated, the time history of the various restoring functions $G^{(i)}$ can be computed. Then, by digitizing the analog measurements at an adequate sampling rate Δt, the magnitude of $G^{(i)}$ corresponding to discrete values of its associated state-variables z_i and $\dot{z}_i$, can be determined.

As outlined in [10], the main idea behind the present method is to estimate each of the real restoring forces $G^{(i)}(\underset{\sim}{p}^{(i)}, z_i, \dot{z}_i)$ by an approximate function $\hat{G}^{(i)}(z_i, \dot{z}_i)$ expressed in terms of two-dimensional orthogonal polynomials. Thus,

$$G(\underset{\sim}{p}, z, \dot{z}) \approx \hat{G}(z,\dot{z}) = \sum_{i=0}^{m2} \sum_{j=0}^{n2} C_{ij}T_i(z')T_j(\dot{z}') \tag{2}$$

where the T's are Chebyshev polynomials and the prime symbol indicates normalized parameters.

In the special case where the nonlinearity of a given element does not involve cross-product terms, function $\hat{G}(z,\dot{z})$ can be expressed as the sum of two separate one-dimensional orthogonal polynomials:

$$G(z,\dot{z}) = g(z) + h(\dot{z}) \tag{3}$$

Functions g and h can be estimated by [10]

$$g(z) \approx \hat{g}(z) = \sum_{i=0}^{m1} a_i T_i(z') \tag{4}$$

and
$$h(\dot{z}) \approx h(\dot{z}) = \sum_{j=0}^{n1} b_j T_j(\dot{z}') \tag{5}$$

Applications

As an illustration, consider the nonlinear 3DOF systems shown in Fig. 1. Note that the nonlinearity used here is assumed to be of the polynomial type (3rd order without cross-product terms) that is defined by three coefficients, with $p_1^{(i)}$ corresponding to the elastic stiffness of element i, $p_2^{(i)}$ being the viscous damping term for element i, and $p_3^{(i)}$ corresponding to the coefficient of the nonlinear cubic displacement term. Thus, depending on the sign of $p_3^{(i)}$, the form of $G^{(i)}$ shown in Fig. 1(b) can be made to represent restoring forces with hardening or softening nonlinearities -- a commonly encountered type of nonlinearity in physical systems.

The numerical values given in the table shown in Fig. 1(b) for the mass and linear stiffness properties are first-order approximations of the properties of a three-story steel frame structure that has been extensively analyzed, both analytically and experimentally [15,16] at the University of California, Berkeley (UCB). Since this model has been subjected to numerous tests on the large testing machine at UCB whereby arbitrary base excitation can be applied, it is assumed in the following examples that F_1, F_2, and F_3 are zero and that S(t) is some prescribed excitation. Also, for the sake of illustration, the simulated "experiment" will be conducted on two hypothetical versions of the UCB structure shown in Fig. 1: (a) the first version (henceforth referred to as UCB-1) will consist of a completely linear model (i.e., only linear terms involving $p_1^{(i)}$ and $p_2^{(i)}$ appear in $G^{(i)}$), and (b) a nonlinear version (UCB-2) that includes the cubic nonlinearities in the stiffness terms. The numerical values for the restoring function parameters for both model versions are shown in Fig. 1(b).

Applying a swept-sine excitation of the form $S(t) = S_0 \sin(at^2 + bt)$ to the two example structures discussed above results in the time histories of the restoring functions shown in Fig. 2. Plots of the restoring functions versus their corresponding interstory relative displacement are given in Fig. 3.

Processing the structural response [9] gives the sample identification results shown in Figs. 4 and 5. Three dimensional plots for the surfaces corresponding to the estimated restoring functions, as a function of their state variables, are given in Fig. 6. A comparison between the exact and approximate restoring functions' time histories is shown in Fig. 7.

The form of nonlinearity discussed so far involved polynomial types without cross-product terms. On the other hand, hysteretic-type restoring functions, which are widely encountered in all areas of structural mechanics, not only involve cross-product terms of displacement and velocity, but are not even expressible in polynomial forms. The wide applicability of the present method is illustrated in Fig. 8 by the identification of a modified version of the nonlinear model UCB-2 which includes a hysteretic element.

A further illustration of the usefulness of the method in detecting the essential features of nonlinear components is presented in Fig. 9 where the response characteristics of a nonlinear restoring function with a gap nonlinearity is shown.

Sample results for the nonparametric identification of a nonlinear 3DOF system using the Wiener-kernel approach [9] are shown in Figs. 10 and 11.

Acknowledgement

This study was supported in part by a grant from the National Science Foundation and by a contract with the Nuclear Regulatory Commission. The authors appreciate the assistance of D. Oba in the preparation of the manuscript.

References

1. Andronikou, A.M., and Bekey, G.A., "Identifiability of Hysteretic Systems," Proc. of the 18th IEEE Conf. on Decision and Control, Dec. 1979, pp. 1072-1073.

2. Bekey, G.A., "System Identification, an Introduction and a Survey," Simulation, Oct. 1970, pp. 151-166.

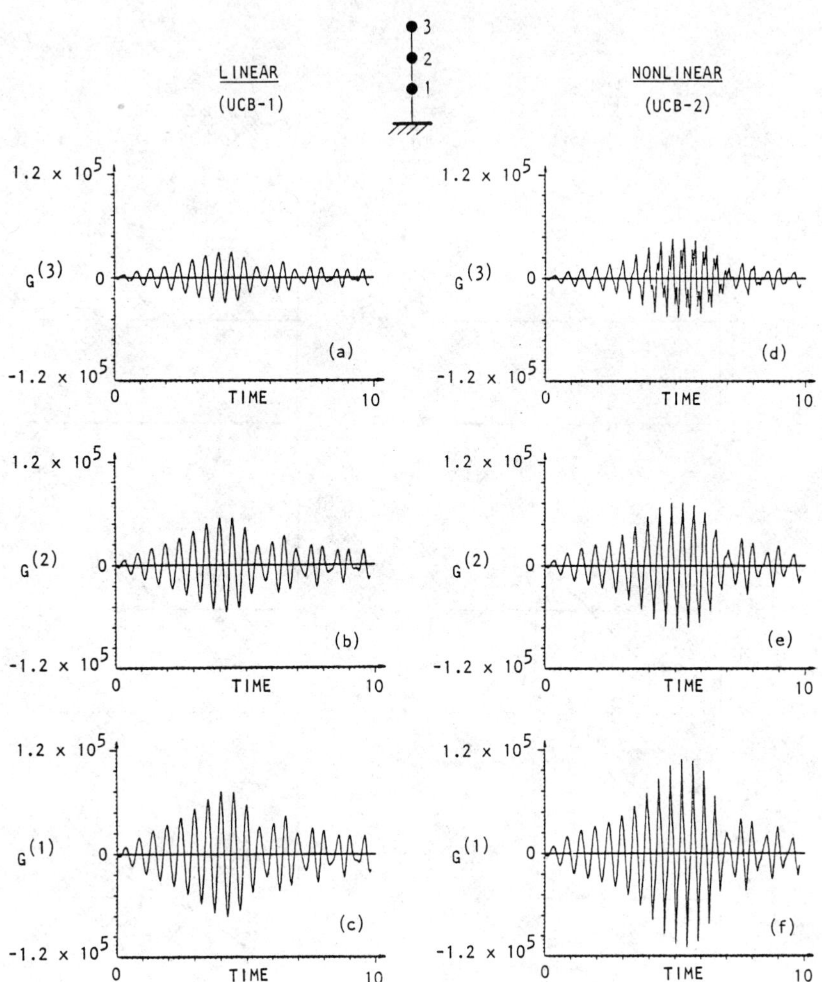

FIGURE 2. RESTORING FORCE

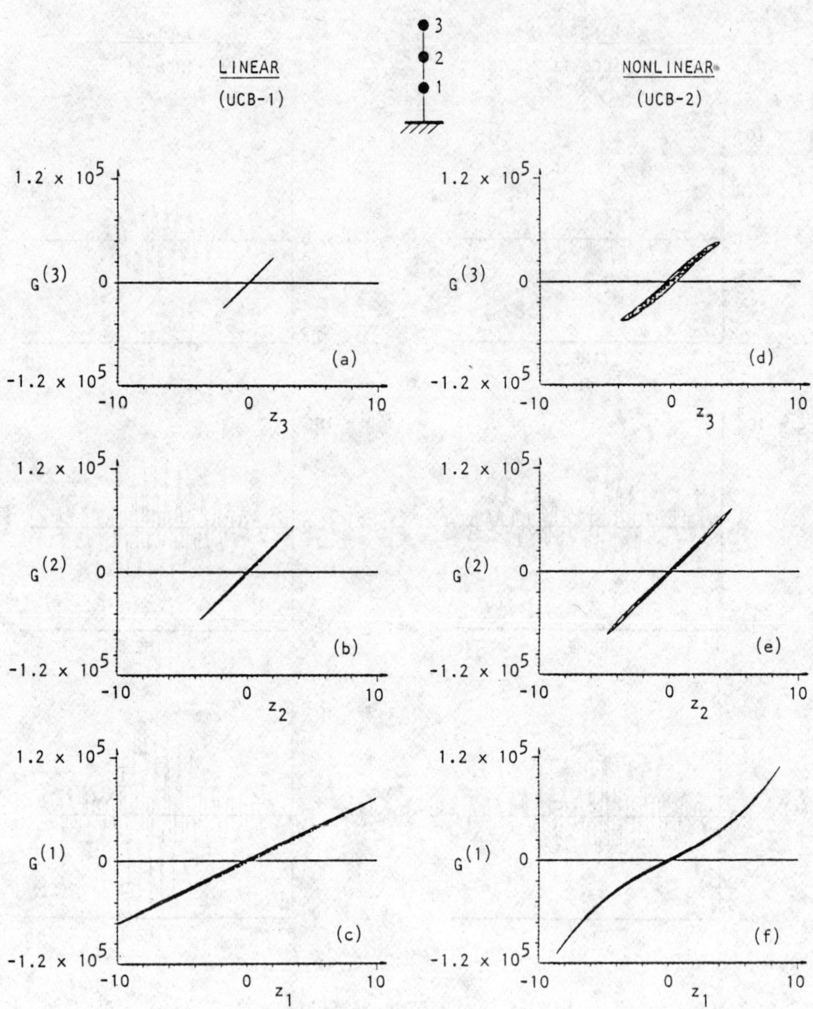

FIGURE 3. STATE VARIABLES: Z VS. FORCE

Actual System: LINEAR (Element 1)

EXACT $G(z,\dot{z})$: $7214.0z + 21.0\dot{z}$

Identification Excitation: $S(t) = S_o \sin(at^2 + bt)$

$S_o = 1.0$

Motion Range: $z_{min} = -10.09$ $\quad z_{max} = 9.93$

$\dot{z}_{min} = -1.29 \times 10^2$ $\quad \dot{z}_{max} = 1.30 \times 10^2$

1D Fit: $G(z,\dot{z}) = g(z) + h(\dot{z})$

$g(z) = -6.51 \times 10^3 T_o(z') + 7.23 \times 10^5 T_1(z')$

$h(\dot{z}) = 1.05 \times 10^3 T_o(\dot{z}') + 6.85 \times 10^5 T_1(\dot{z}')$

Coefficient of 2D Chebyshev Series:

$$G(z,\dot{z}) = \sum_{i=0}^{m2} \sum_{j=0}^{n2} C_{ij} T_i(z') T_j(\dot{z}')$$

i \ j	$T_o(z')$	$T_1(\dot{z}')$
$T_o(z')$	-1.85×10^3	5.21×10^3
$T_o(z')$	7.22×10^4	-4.98

Polynomial Series:

1D Fit: $g(z) = -5.86 \times 10^2 + 7.22 \times 10^4 z'$

$h(\dot{z}) = -1.33 \times 10^3 + 5.29 \times 10^3 \dot{z}'$

2D Fit: $$G(z,\dot{z}) = \sum_{\ell=0}^{m2} \sum_{m=0}^{n2} A_{\ell m} z'^{\ell} \dot{z}'^{m}$$

ℓ \ m	1	$\dot{z}'$
1	-1.28×10^3	40.25
z'	7.22×10^3	-3.77×10^{-3}

FIGURE 4

Actual System: NONLINEAR (Element 1)

EXACT $G(z,\dot{z})$: $7214.0z + 21.0\dot{z} + 72.14z^3$

Identification Excitation: $S(t) = S_o \sin(at^2 + bt)$

$S_o = 1.0$

Motion Range: $z_{min} = -8.66$ $\quad z_{max} = 8.61$

$\dot{z}_{min} = -1.39 \times 10^2$ $\quad \dot{z}_{max} = 1.39 \times 10^2$

1D Fit: $\hat{G}(z,z) = g(z) + h(\dot{z})$

$$\hat{g}(z) = 1.23 \times 10^5 T_o(z') + 2.34 \times 10^6 T_1(z') + 1.22 \times 10^5 T_2(z') + 7.64 \times 10^6 T_3(z')$$

$$\hat{h}(\dot{z}) = -5.35 \times 10^2 T_o(\dot{z}') + 4.13 \times 10^5 T_1(\dot{z}')$$

Coefficient of 2D Chebyshev series:

$$G(z,\dot{z}) = \sum_{i=0}^{m2} \sum_{j=0}^{n2} C_{ij} T_i(z') T_j(\dot{z}')$$

i \ j	$T_o(\dot{z}')$	$T_o(\dot{z}')$
$T_o(z')$	-9.31×10^2	2.97×10^3
$T_1(z')$	9.77×10^4	6.27
$T_2(z')$	-1.97×10^2	1.27
$T_3(z')$	1.19×10^2	-11.11

Polynomial series:

1D Fit: $g(z) = -2.46 \times 10^2 + 6.22 \times 10^4 z' - 3.56z'^3 + 4.74z'^3$

$h(\dot{z}) = -5.35 \times 10^2 + 3.00 \times 10^3 \dot{z}'$

2D Fit:
$$G(z,\dot{z}) = \sum_{\ell=0}^{m2} \sum_{m=0}^{n2} A_{\ell m} z'^{\ell} \dot{z}'^{m}$$

ℓ \ m	1	$\dot{z}'$
1	-1.12×10^3	21.41
z'	7.20×10^3	3.30×10^{-2}
z'^2	-10.39	2.83×10^{-4}
z'^3	73.56	4.97×10^{-4}

FIGURE 5

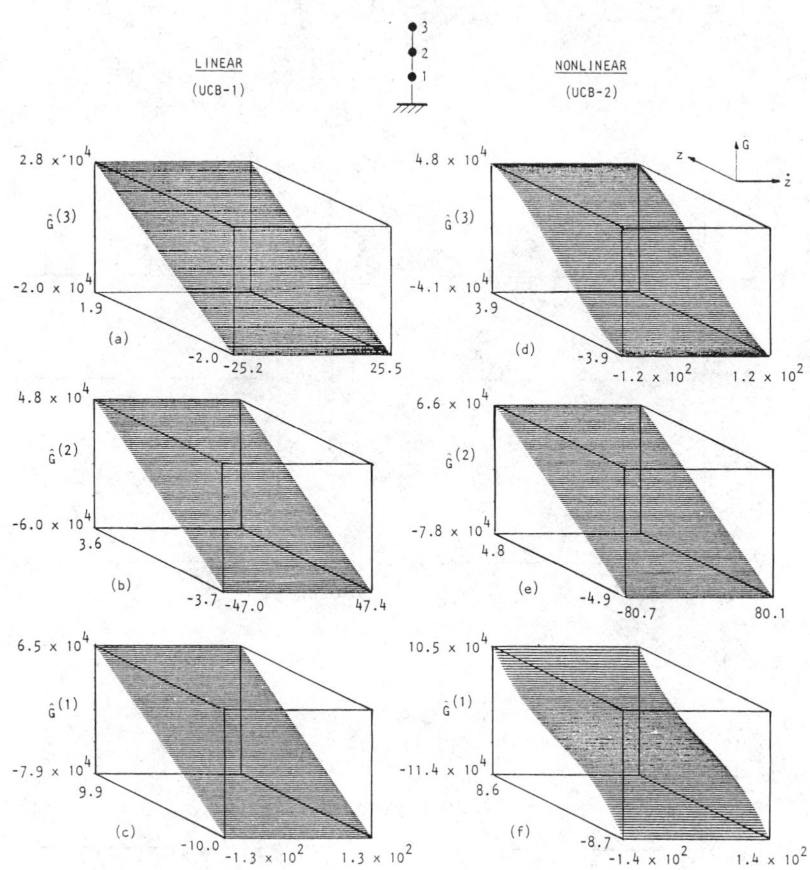

FIGURE 6. RESTORING FUNCTIONS

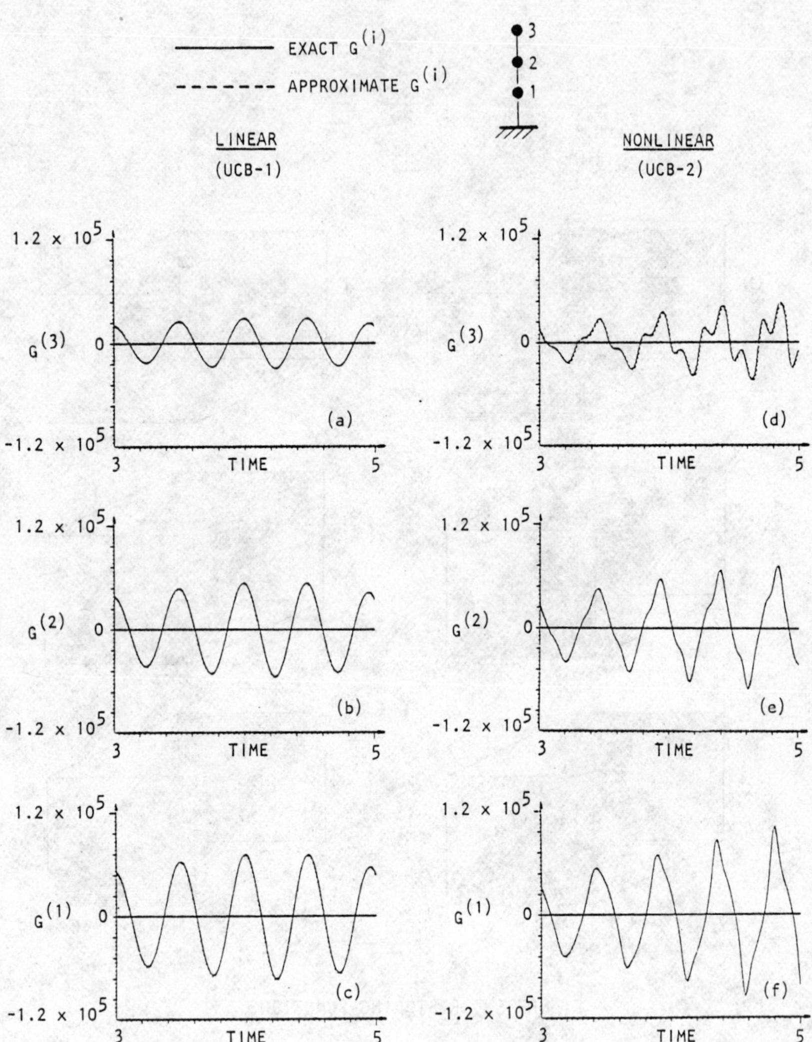

FIGURE 7. COMPARISON BETWEEN EXACT AND APPROXIMATE RESTORING FUNCTIONS

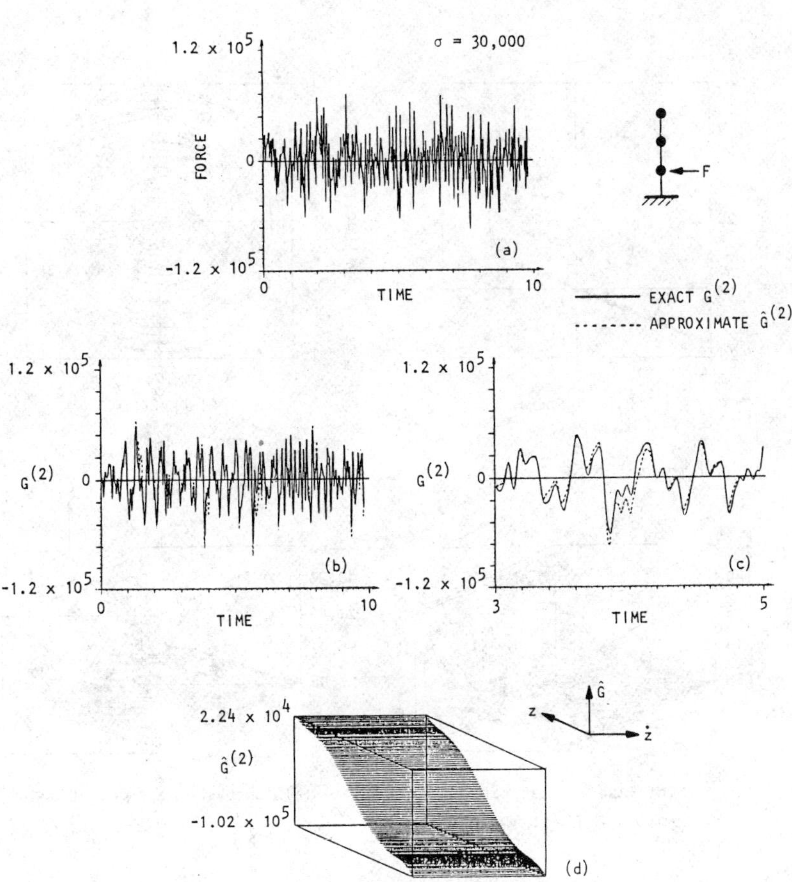

FIGURE 8. EXCITATION AND RESTORING FUNCTION OF HYSTERETIC ELEMENT NO. 2 IN MODEL UCB-3

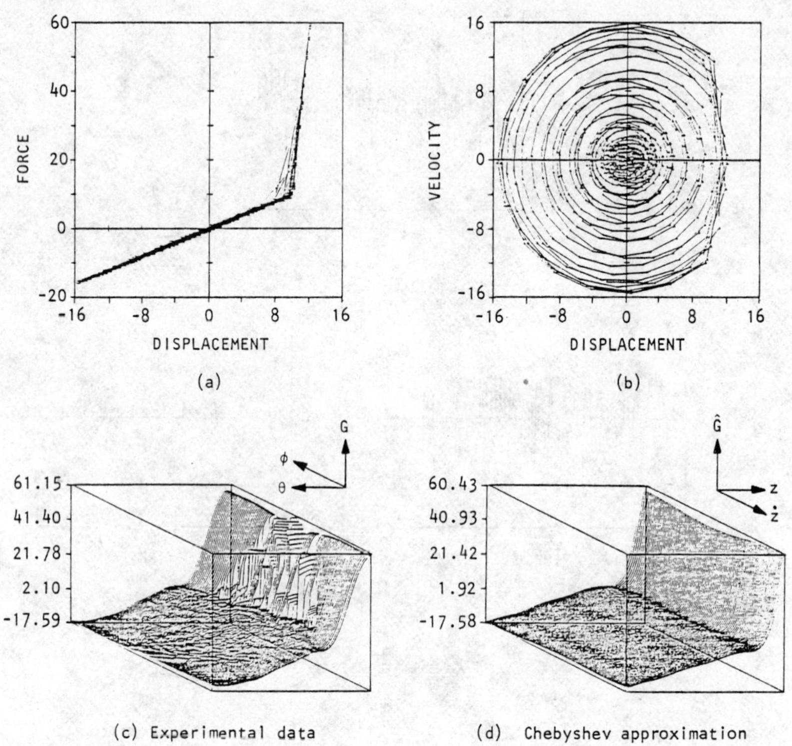

(c) Experimental data

(d) Chebyshev approximation

FIGURE 9. CHARACTERISTICS OF A NONLINEAR ELEMENT WITH A GAP

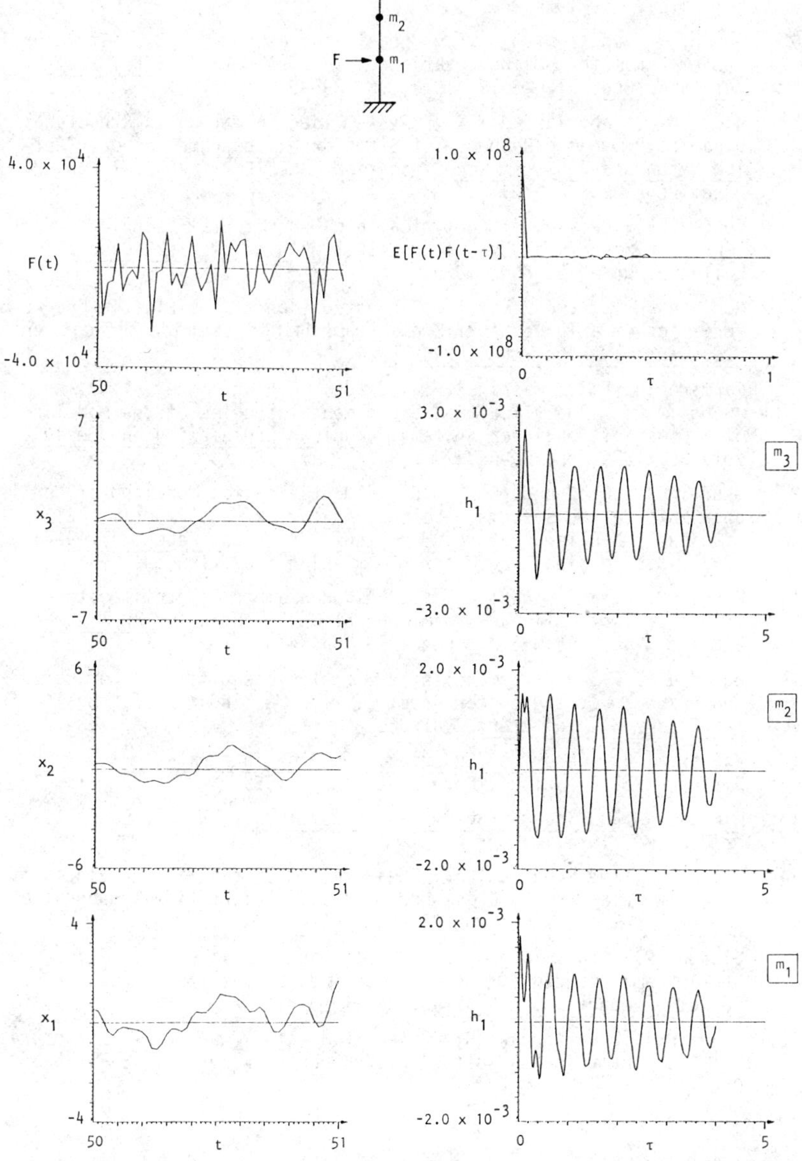

FIGURE 10. RESPONSE OF A NONLINEAR 3 DOF SYSTEM TO RANDOM EXCITATION

FIGURE 11. WIENER KERNELS OF A NONLINEAR SYSTEM

3. Eshleman, R.L., "System Modeling," The Shock and Vibration Digest, Vol. 4(5), May 1972, p. 1.

4. Gersh, W., and Foutch, D.A., "Least-Squares Estimates of Structural System Parameters Using Covariance Function Data," IEEE Trans. Automatic Control, AC-19, 1974, pp. 898-903.

5. Hart, G.C., and Yao, J.T.P., "System Identification in Structural Dynamics," Dynamic Response of Structures: Instrumentation, Testing Methods and System Identification, ASCE/EMD Technical Conference, UCLA, Mar. 1976.

6. Ibañez, P., "Review of Analytical and Experimental Techniques for Improving Structural Dynamic Models," Welding Research Council Bulletin, No. 249, June 1979.

7. McNiven, H.D., and Kaya, I., "Investigation of the Elastic Characteristics of a Three Story Steel Frame Using System Identification," U.C. Berkeley Report No. EERC 78-24, Nov. 1978.

8. Marmarelis, V.Z., Masri, S.F., Udwadia, F.E., Caughey, T.K., and Jeong, J.D., "Analytical and Experimental Studies of the Modeling of a Class of Nonlinear Systems," Nuclear Engineering and Design, Vol. 55, 1979, pp. 59-68.

9. Masri, S.F., and Anderson, J.C., "Identification/Modeling of Nonlinear Multidegree Systems," Vol. 3, Analytical and Experimental Studies of Nonlinear System Modeling - A Progress Report AT(49-24-0262), U.S. Nuclear Regulatory Commission, Apr. 1980.

10. Masri, S.F., and Caughey, T.K., "A Nonparametric Identification Technique for Nonlinear Dynamic Problems," Journal of Applied Mechanics, Vol. 46, No. 2, June 1979, pp. 433-447.

11. Udwadia, F.E., and Marmarelis, P.Z., "The Identification of Building Systems - I. The Linear Case; II. The Nonlinear Case," Bull. of the Seism. Soc. of Amer., Vol. 66, No. 1, Feb. 1976, pp. 153-171.

12. Volterra, V., Theory of Functionals and of Integral and Integro-Differential Equations, Dover, New York, 1959.

13. Wiener, N., Nonlinear Problems in Random Theory, M.I.T. Press, Cambridge, MA, 1958.

14. Lee, Y.W., and Schetzen, M., "Measurement of the Wiener Kernels of a Nonlinear System by Cross Correlation," International Journal of Control, Vol. 2, 1965, pp. 237-254.

15. Clough, R.W., and Tang, D.T., "Earthquake Simulated Study of a Steel Frame Structure, Vol. I: Experimental Results," U.C. Berkeley Report No. EERC 75-6, April 1975.

16. Tang, D.T., "Earthquake Simulated Study of a Steel Frame Structure, Vol. II: Analytical Results," U.C. Berkeley Report No. EERC 75-36, Oct. 1975.

FRICTION DEVICES TO CONTROL SEISMIC RESPONSE

[1] Cedric Marsh, M.ASCE; [2] Avtar S. Pall, M.ASCE

INTRODUCTION

Of all the methods available to extract kinetic energy from a moving body, the most widely adopted is undoubtedly the simple friction brake. Predictability, reliability and repeatable performance are taken for granted. The essential components are cheap and widely available.

A sliding friction joint, with a given pre-load, using flat brake lining pads, will slip at the same force after long periods of inactivity. Furthermore, it possesses a rectangular hysteresis loop with negligible fade over many more cycles of reversal than are encountered in successive earthquakes.

The capability of engineering friction devices to satisfy the requirements of predictable slip load and uniform hysteretic behaviour makes them ideal for dissipating seismic energy. Much greater quantities of energy can be disposed of than by any method that involves such damaging processes as yielding steel or cracking concrete, and the device is always ready to do its job, regardless of how many times it has performed.

A friction joint limits the maximum force, and thus acts as a fuse. When compelled to slide, it acts as a damper.

How anti-seismic friction devices can be incorporated into buildings is the subject of this paper.

GENERAL

It is recognized that well engineered modern buildings of any size can be designed to resist the most severe earthquakes. During major earthquakes, a large amount of kinetic energy is fed into the structure. Except for the case of special structures, such as nuclear reactors, it is economically not feasible to reconcile the energy elastically. Since elastic capacity of the structure is limited, the balance energy must be dissipated to avoid the total collapse of the structure. Generally, energy is mainly consumed in the building as damage to both structural and non-structural members. Although the building will remain standing to safeguard life, damage, both primary and secondary, may be as econo-

[1] Professor; [2] Research Associate, Centre for Building Studies, Concordia University, Montréal

mically significant as collapse. It is in the reduction of this damage that the use of friction devices will find its principal appeal.

Proposals, for the means of dissipating energy by friction, fall into the following main areas :

1. Low rise buildings, with a high natural frequency, which move with the ground motion without significant elastic oscillations (height/base width < 3).

2. Concrete shear wall buildings of more than 10 storeys.

3. Steel :

   3.1 Rigid frames.
   3.2 Rigid frames with diagonal bracing.

4. Curtain walls and infill panels.

LOW RISE RIGID BUILDINGS

The ideal way to reduce seismic effects in a building is to isolate the building from the forcing ground motion allowing the ground to move without exerting any force on the building. To achieve this, it is proposed to support the building (Fig. 1) on sliding surfaces. A friction force is required, sufficient to react the wind, but under seismic action the magnitude of the lateral force that the building can experience will be limited to this value for any direction of force. There must be sufficient clearance, in all directions, to allow the building to travel, relative to the ground, a distance equal to the maximum final ground displacement plus the maximum amplitude of the ground motion. This arrangement was modelled by Newmark (2) and represents a simplification of the problem, studied by Becker (1), of the sliding of horizontal joints in panelized concrete buildings.

To avoid any uplift which would result in vertical impact, the effective coefficient of friction will not exceed b/3h, where b and h are the building breadth and height respectively. This value may be lower than that provided by brake linings, in which case some of the supports will be on reduced friction surfaces, so proportioned that the total friction force available provides the desired value. A need for a higher coefficient of friction than brake linings possess does not arise.

In practice, the building superstructure will rest on the foundation, or on basement walls below grade, with the contact surfaces large enough to permit the anticipated movement. Suitable protection from weather or moisture will be required to ensure no change in the slip load.

Such a building would be perhaps one of the few that could survive an earthquake even if situated directly on the fault line.

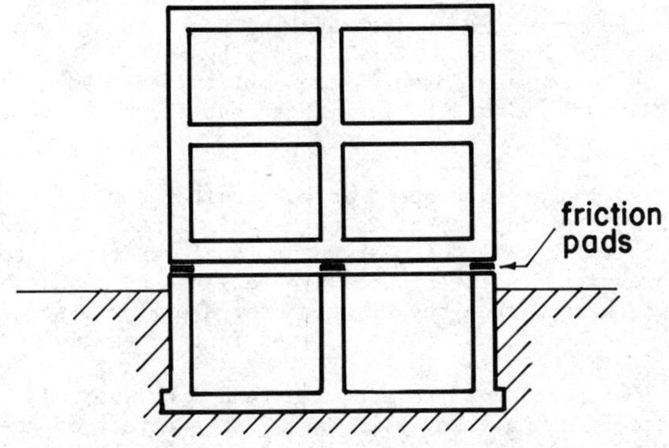

FIG. 1 - SLIDING SUPPORTS FOR LOW RISE BUILDINGS

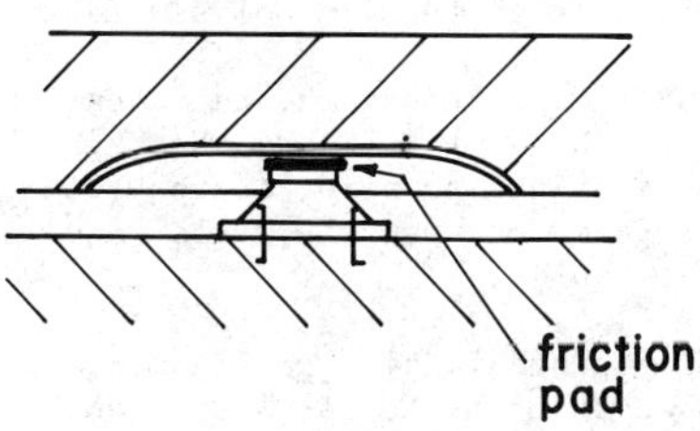

FIG. 2 - DISHED SLIDING SUPPORT

Detailing of the supports and the transfer of loads from the building to the foundations is standard engineering.

In order to provide some restraint on the total movement, the surfaces on which the friction pads slide may be dished (Fig. 2). This will create an increasing force as the amplitude of the ground motion increases, but it will always be less than that exerted on a rigidly founded building.

This force isolation concept is most suitable for low rise buildings in which the overturning moment is not significant. The overturning moment on the building will not change the total reactive force provided by friction; however, if uplift forces are resisted by anchor bolts or pre-tensioning wires, then the friction force will be increased by the overturning moment and must be considered in the design.

If vertical accelerations are of a random nature independent of the horizontal acceleration, they will have little effect on the general behaviour, but should the vertical and horizontal acceleration be in phase there will be a tendency to move the building in one direction, and allowance will need to be made for this.

MULTISTOREY CONCRETE SHEAR WALL BUILDING

A study of shear walls coupled by vertical sliding friction joints is given (Ref. 3, to 6) (Fig. 3). In this case, the shear walls are considered to be anchored to the foundation, and to act as independent vertical cantilevers joined by friction devices, somewhat in the style of a leaf spring.

The sliding joint is comprised of steel plates anchored into the concrete to which a plate with slotted holes is bolted, trapping a flat brake lining pad between the plates. Pre-load on the pad is controlled by the yielding of the high tension bolts used in the joint.

Under wind loads no slip occurs, but under the higher lateral inertial forces created by an earthquake the shear force in the vertical interface causes the joints to slip, thereby limiting the magnitude of the vertical shear force between the walls, and, as the oscillations continue, absorbing energy.

For a simple shear wall, divided equally into two, a joint can be provided such that the total elastic energy at maximum amplitude is dissipated in one quarter cycle.

To obtain this in a coupled wall, each half of which is of breadth, b, thickness, t, and height, h, with a limiting stress, f, (based on the concrete area only), the slip load, q, per unit height, is given approximately by :

$$q = btf/5h$$

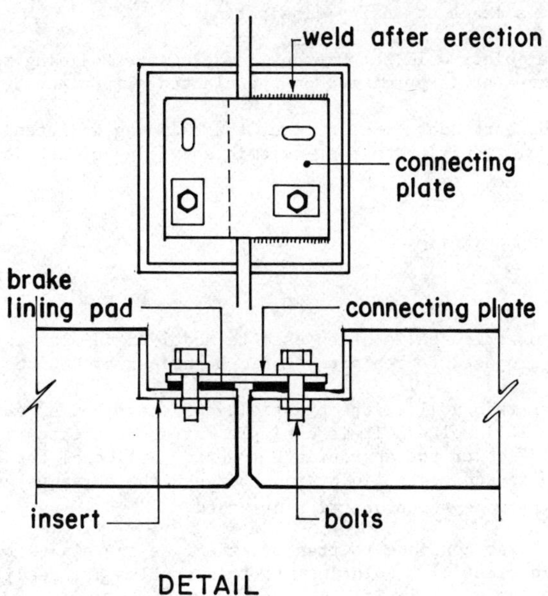

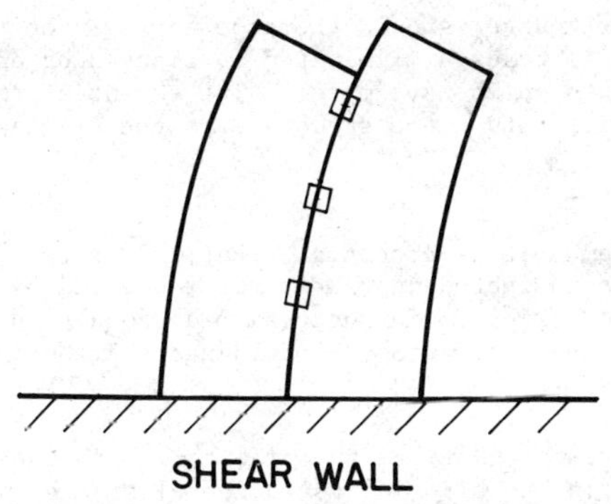

FIG. 3 - FRICTION JOINTS FOR COUPLED SHEAR WALLS

Any set of vertical walls which meet to form an L, an I, a box, or other form, can be connected by a vertical line of friction devices to provide a means of energy dissipation.

These joints will act without damaging the building fabric so long as free movement is permitted on the elected slip planes.

As the earthquake motion dies, the building will tend to its original position, requiring no post-seismic servicing of the joints.

STEEL FRAMED BUILDINGS

Rigid Frames

Coulomb friction is often used to model rigid-plastic behaviour. It is now proposed to replace plastic behaviour by Coulomb friction.

Rather than utilize the plasticity of steel beams, and the permanent damage it involves, it is a simple matter to create real hinges that will "yield" under the appropriate moment. Splice plates with high tension bolts and brake lining pads between the surfaces have been shown (3, 4) to provide the properties demanded.

Beams must continue to transmit shear as the hinges rotate (columns will remain elastic). Solutions to this problem are readily devised and some ideas are shown in Fig. 4.

Analysis for steel frames which incorporate these joints will follow the existing procedures for plastic analysis.

After an earthquake, should there be any permanent distortion of the building, it is readily eliminated by slackening and retightening the bolts in joints which have moved. The essential elasticity of the building frame will return the structure to the original configuration.

Braced Frames

In braced frames it is essential, that, after the diagonal has slipped, there is an increasing shear force carried by some part of the structure to force slippage in successive diagonals in a vertical panel of the building, and to return the building to its original position as the earthquake passes.

A second vertical panel of the bracing, which does not slip, will limit the travel in the slip joints to the elastic strains in the diagonals of the second panel. Theoretically, this can provide a useful source of energy dissipation, but it is of limited value.

If there is a single, slipping, braced plane in a rigid framed building, the deformation in the rigid frame after the diagonals slip is large relative to the elastic elongation of the diagonals and the slip travel can be sufficient to provide critical damping. The rigid frame

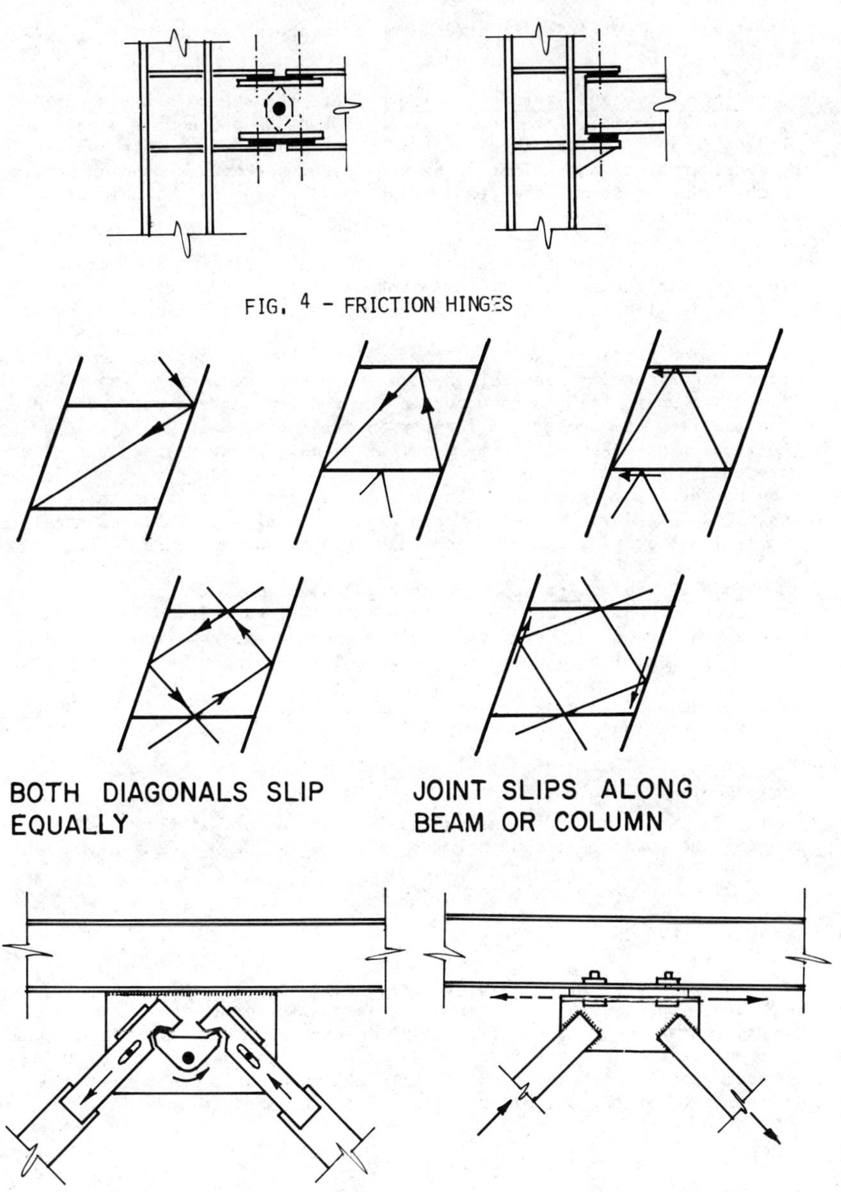

FIG. 4 - FRICTION HINGES

FIG. 5 - SLIDING BRACING, TENSION AND COMPRESSION

itself will be sufficiently strong to carry approximately half the maximum shear force.

Compressive and Tension Bracing. Where the shear force is carried by diagonal bracing which can resist both tension and compression, the sliding devices will be incorporated in the joints of the diagonals. However, in order not to introduce unnecessary secondary bending moments in the frame, both the compression and tension diagonals should slip simultaneously.

In Fig. 5 two approaches are shown. The sliding motion occurs either along the diagonals or along the beams or columns. How this can be achieved is illustrated in Fig. 5 for warren, K, and diamond bracing.

Tension Bracing. To avoid the use of heavy compression members, cross braced tension members, sometimes pre-stressed, are often used. If sliding under load is to occur in the tension members only, some device to allow the members to recover in the second half cycle is required. The problem has been solved by a number of special arrangements that ensure that only the tension bar carries load, independently of the direction of the shear force, or that the compression force is as small as desired. Some devices are shown in Fig. 6. Analysis of braced steel frames incorporating these devices has been conducted by Pall (7).

Coupled Braced Frames. The contiguous columns of a pair of independent braced bays may be connected together by vertically sliding joints in the manner used for coupled concrete shear wall. This will rely on the axial strains in the inner columns to provide the slip travel.

For an optimum arrangement, the slip load, q per unit height, will be approximately :

$$q = R/2h$$

where R is the axial loads, due to lateral forces, in the exterior columns of the braced pairs of columns, and h is the building height.

CURTAIN WALLS AND INFILL PANELS

Any rigid frame, which provides an elastic restoring force, can be provided with exterior or interior rigid panels for the vertical surfaces, connected by means of friction joints. The behaviour can be made to simulate that of energy absorbing diagonal bracing, or may be independent panels providing damping while making a large contribution to the strength of the building. Figure 7 illustrates these modes.

CONCLUSIONS

A number of techniques have been described by which friction might be used to control the response of buildings to earthquakes. Friction

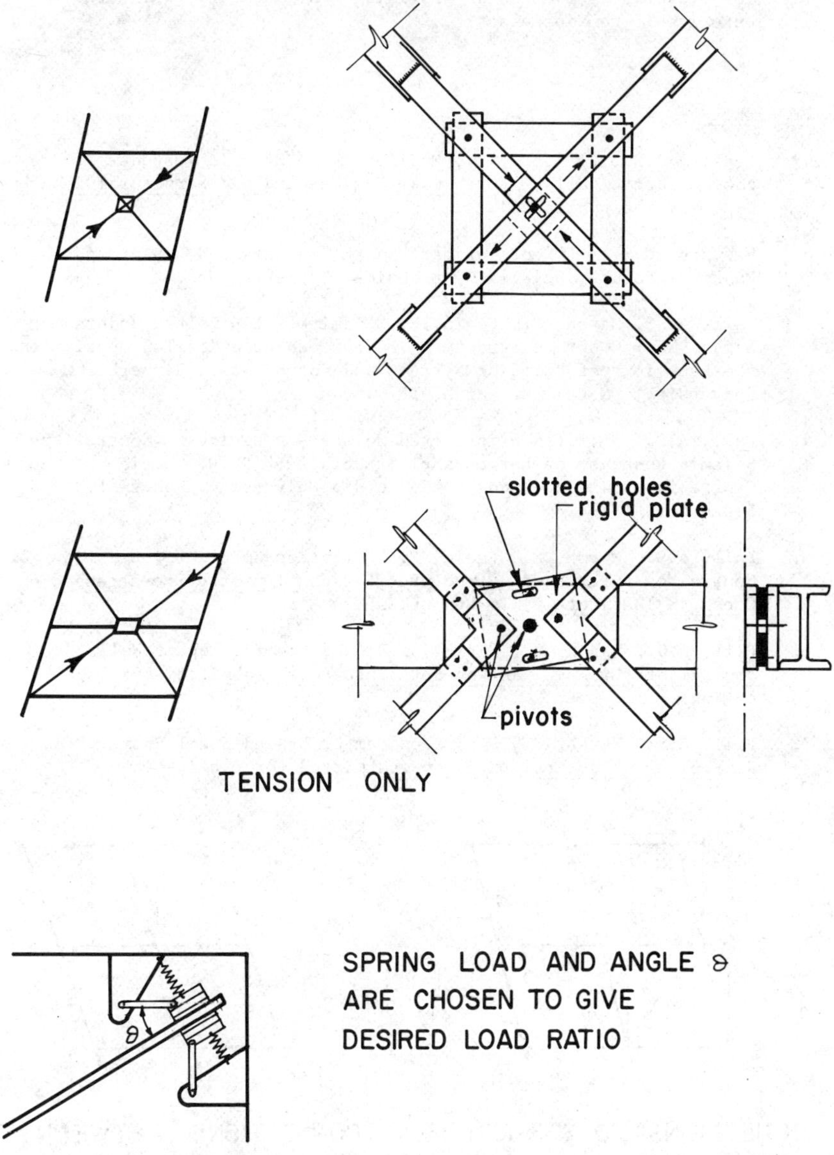

slotted holes
rigid plate
pivots

TENSION ONLY

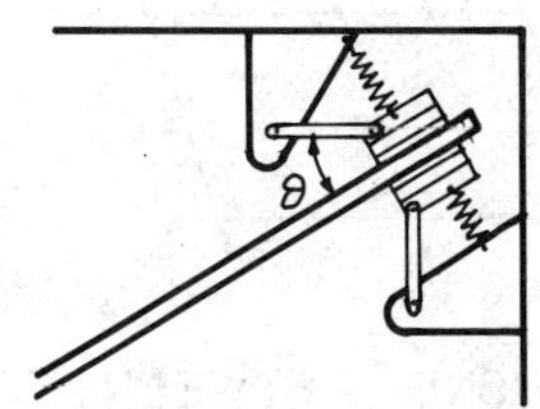

SPRING LOAD AND ANGLE θ
ARE CHOSEN TO GIVE
DESIRED LOAD RATIO

HIGH TENSION, LOW COMPRESSION

FIG. 6 - SLIDING BRACING, TENSION ONLY

replaces yielding as a means of providing non-linearity, while the structure itself remains elastic.

REFERENCES

1. Becker, J.M., Llorente, C., Mueller, P., "Seismic Response of Precase Concrete Walls," Earthquake Engineering and Structural Dynamics. (To be published).

2. Newmark, N.M., "Effects of Earthquakes on Dams and Embankments," Rankine Lecture, British Geotechnical Society, 1965.

3. Pall, A.S., Marsh, C., Fazio, P., "Limited Slip Bolted Joints for Large Panel Concrete Structures," Proc. International Symposium on the Behaviour of Building Systems and Components. Vanderbilt University, Nashville, March, 1979.

4. Pall, A.S., "Limited Slip Bolted Joints - A Device to Control the Seismic Response of Large Panel Structures," Ph.D. Thesis, Centre for Building Studies, Concordia University, Montréal, September, 1979.

5. Pall, A.S., Marsh, C., Fazio, P., "Friction Joints for Seismic Control of Large Panel Structures," Journal of the Pre-stressed Concrete Institute. (To be published).

6. Pall, A.S., Marsh, C., "Friction Damped Concrete Shear Walls," Journal of the American Concrete Institute (submitted for publication).

7. Pall, A.S., Marsh, C., "Friction Damped Braced Steel Frames," ASCE Journal of Structural Division (submitted for publication).

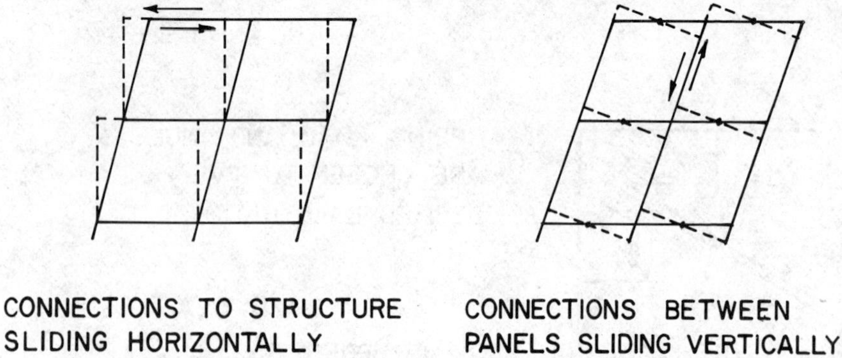

FIG. 7 - SLIDING PANELS

DAMPING OF AN OFFSHORE PLATFORM MODEL BY RANDOM DEC METHOD

by

J. C. S. Yang[I], M. S. Aggour[II], N. Dagalakis[III], F. Miller[IV]

ABSTRACT

The purpose of this investigation is to examine the feasibility of using a new analysis technique called Random Decrement to accurately extract frequencies, and damping coefficients for an offshore platform structure supported by different foundation conditions.

The structure tested was a model of an oil drilling platform. It was a welded-steel space frame with four primary legs supported on a steel plate, and braced with horizontal and diagonal members. A special air cushion supported base was constructed and was connected to a Calidyne 5000 lb. shaker.

Three different foundation cases were investigated: (1) fixed (bolted) base, (2) free base, (3) embedded about 2 inches in soil.

In all cases tests were ran to measure the damping of the structure. Sinusoidal and random input excitation was applied in a single horizontal direction. The response of the structure was monitored with accelerometers placed at different positions on the structure.

The damping coefficient was first calculated from the frequency response curve (half power point) and the logarithmic decrement of the free response curve. Then the damping coefficient was calculated from the random decrement analysis of the response of a random input excitation and it was compared with the coefficients calculated from the first two classical analysis techniques.

INTRODUCTION

It is generally accepted that in the design of structures to safely withstand ground motions from earthquakes, more precise methods of analysis are needed. This is especially critical for the design of

---

I Professor, Mechanical Engineering Dept., University of Maryland, College Park, MD.

II Associate Professor, Civil Engineering Dept., University of Maryland, College Park, MD.

III Assistant Professor, Mechanical Engineering Dept., University of Maryland, College Park, MD.

IV Graduate Student, Civil Engineering Dept., University of Maryland, College Park, MD.

important structures such as multi-story buildings, nuclear power plants, offshore platforms, etc. Sophisticated analytical techniques such as the finite-element method are being developed to satisfy these needs. The results of analyses using these techniques clearly show that there is a strong interaction between structures and their soil foundations, and accurate analysis of structural response during earthquakes are strongly dependent on the damping values used in the analysis (1,2,3,4,5,6,7).

Various attempts have been made to provide means for obtaining structural damping information used in the design of structures with foundations and used in the monitoring of the response to the applied forces. However, most of these techniques are only suitable for use under controlled laboratory conditions and are of little use for structures in service.

A simple, direct, and precise method is needed for translating the response time history into a form meaningful to the observer. Spectral power density has been considered, with damping measured by the half-power bandwidth method, but this was found to have a large measurement variance, especially when the bandwidth was small. In addition, when two modes are close this method cannot be applied.

A new technique called "Random Decrement" has been developed and explored initially by H. Cole (8) and continued by J. C. S. Yang (9) which advanced the state-of-the-art in the measurement of damping in structures subjected to random excitation when only response data is available. The method was originally developed as a technique for determining damping characteristics of models being tested in wind tunnels (8). The method is currently used in the aerospace industry to measure damping in tunnel models and aircraft in flight (10,11). Subsequently the method is used to measure damping in plates, beams, mobile homes, piping systems, etc. (9,12), and to detect structural failure in piping systems, offshore platforms, etc. (13,14).

The basic concept of the "random decrement signature" is based on the fact that a random response of a structure due to a random input is composed of two parts: a) a deterministic part (impulse and/or step function), and b) a random part. By averaging enough samples of the same random response, the random part will average out, leaving the deterministic part. It can be shown that by proper digital processing, the deterministic part that remains is the free decay response; from which the damping can be measured. Hence the Randomdec technique uses the free decay responses of a structure under random load to identify its vibration parameters, namely frequencies, damping and modal vectors. This method offers the obvious advantage that the input function can be any type of random loading which occurs during its service life, such as seismic, wind, ocean waves, traffic, etc. and the random decrement signal is independent of the input to the structure and represents that particular structure tested.

A brief, rather intuitive explanation of the principles of Random Decrement Technique is given in Appendix A.

EXPERIMENTAL SET-UP

Test Model

A model oil platform was constructed. It is a welded-steel space frame with four primary legs and an upper and a lower plate, braced with horizontal and diagonal members. The primary legs of the model had a diameter of 25 mm (0.984 inch), and the horizontal and diagonal members had diameter of 15 mm (0.591 inch). The structure, its design and the dimensions of all the components are shown in Figure 1.

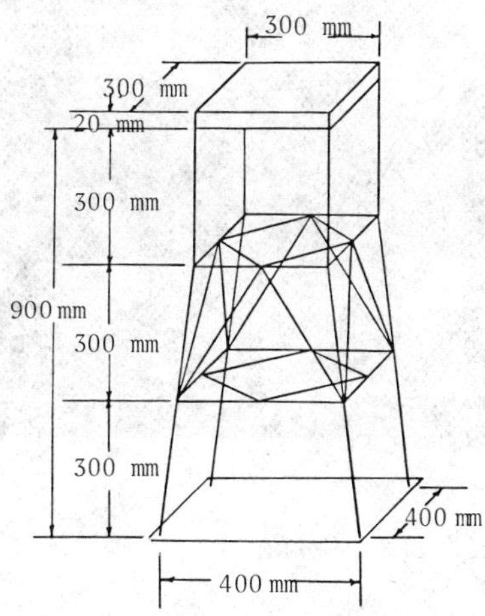

Fig.1 Offshore Platform Model

Shake Table

The platform model was placed on a special air cushion supported base connected at either ends to a Calidyne 5000 lb. shaker, or an MB Electronic, PM 50, 30 lb. shaker. The shake table consisted of a 4'x6' sheet of 3/4" plywood with 1"x8" pine wood sides supported on a cast iron grating. The grating is sitting on two rubber inner tubes. The two shakers are attached to the wooden sides of the table with their axis of motion in the longitudinal direction. An accelerometer is placed on the base of the platform in the direction of motion. A picture of the shake Table with the Test Model is given in Figure 2.

Soil Used

A fine sand was used to represent the foundation for the platform. Results from sieve tests shown in Fig. 3 performed according to ASTM Standard C-136 indicated that the sand to be fine and poorly graded and can be classified as SP according to the Unified Soil Classification System. Oven Dry Unit weight of the sand was 97.5 lb/ft^3.

From ASTM Standard C-128, the Bulk Specific Gravity of the sand was 2.633. The relative density of the sand in the loose state, determined according to ASTM Standard D-2049, was 0.895, and in the dense state it was 0.574. The average thickness of the soil in the shake table was 5 inch.

Fig. 2 Shake Table and Test Model

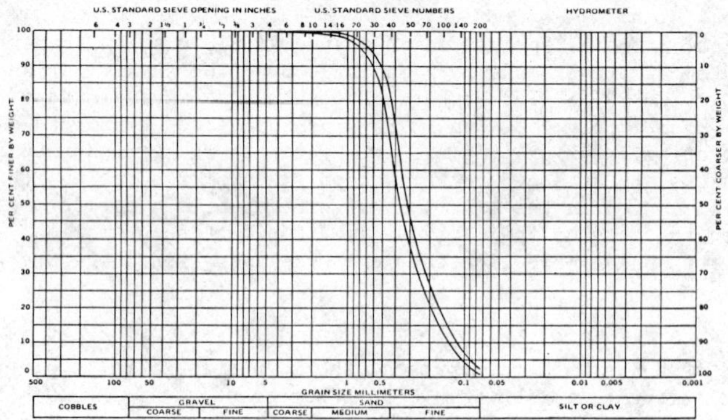

Fig. 3 Limits of Sieve Tests on Soil Used

DAMPING DETERMINATION

A laboratory experiment was conducted to measure the damping of the structure with three different foundation conditions: (1) fixed (bolted) base; (2) free base; and (3) embedded (2 inches) in soil. Sinusoidal and random input excitation was applied in a single horizontal direction. The response of the structure was monitored with accelerometers placed at different positions on the structure.

Sine Sweep

With the 30 lb. shaker bolted to the end of the shake table, sinusoidal vibration was inputted into the shake table. The output of the accelerometer was monitored on a Tektronix 5A18N Trace Amplifier Oscilloscope and the signal strength, peak-to-peak, was recorded along with the frequency of the accelerometer signal at that signal strength: the frequency was found by using a 10 second count off, a Tektronix DC 504 Counter/timer and dividing the output by 10.

The points were then plotted on an arithmetic scale graph and damping calculated using the half-power point technique.

Free Response

The Free Response of the system was arrived at by exciting the table at the system resonant frequency and suddenly removing the electrical cable running from the shaker amplifier to the vibrator, allowing the system to lose energy through damping. The signal of the accelerometer during the loss of energy was stored on the oscilloscope screen and transferred onto tracing paper. The logarithmic decrement of the signal was then calculated.

Frequency Spectrum

The frequency spectrum of the system was found through computer analysis of a recorded signal from an accelerometer (Columbia 504-3) on the platform while the Shake Table was undergoing random excitation.

The random input is produced by a Random Noise Generator (General Radio Company 1390-B), connected to a power amplifier (Electrochyne, N-100) whose signal was the input into the 30 lb shaker (MB Electronics, PM 50).

The platform response is picked up by the accelerometer glued to a support near the top of the platform. The output signal is amplified by a charge amplifier (Kistler, 504) and then recorded on magnetic tape with an AC/DC tape recorder (B & K, 7003).

The recorded response signal is first passed through a signal amplifier then through a band pass filter (I claco, P.11) before it is fed into a programmable amplifier, which is tied to a minicomputer (Cromenco, 32-K bit memory).

The computer program is set up such that it automatically amplifies the input signal to the full range capacity of ± 8 volts of the following analog-to-digital converter (8 bits). The digitized signal is then stored into the memory of the computer and the power spectral density function is evaluated. The execution of the analysis is controlled by a terminal (TEKTRONIX, 4066-1) which can also be used to display the resulting power spectral density. If a hard copy is desired it can be obtained from a digital plotter. (HOUSTON INSTRUMENTS, HIPLOT TM) A schematic diagram of the computer set-up is shown in Figure 4.

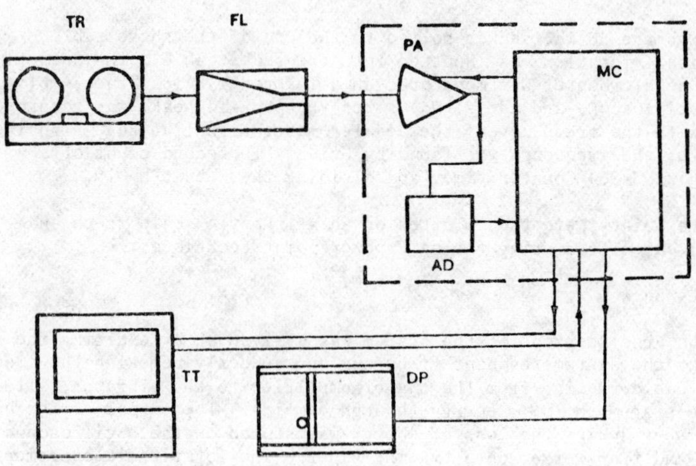

Fig. 4 Analysis of the recorded response signal, with TR: tape recorder, FL: filter, PA: programmable amplifier, AD: digitizer, MC: minicomputer, TT: tektronix terminal, DP: digital plotter

Random Decrement Technique

For the Random Decrement Technique the same procedure as was used for the power spectrum was used. However, for the fixed and free conditions the signal was first passed through a 3rd Octave filter (B & K 2112 Audio-Frequency Spectrometer) and then into the minicomputer. Center frequencies of the filter were chosen after analyzing the power spectrum. When smaller bandwidths were needed for close spaced modes, a resonant filter contained within the minicomputer was used. The signal was then analyzed by the Random Decrement computer program within the minicomputer. For the platform in the sand condition the signal was passed through the signal amplifier then into the minicomputer. After the signals were analyzed the output in all cases was recorded using the digital plotter.

Numerical Model

A finite element space frame model of the structure was developed using the GIFTS and the NASTRAN computer programs. The horizontal and the diagonal braces were modeled by beam elements, the top plate was modeled by plate elements which allow bending and membrane flexibility. From the computer model program, the frequency of the first natural mode of oscillation of the model platform was found to be 59.8 cycles/sec.

RESULTS

The computer generated power spectral density curves of the response for the different cases tested are shown in Figure 5.

Sine sweep test results are shown on Figure 6. Damping values were obtained, for the different cases tested, from half power point measurements.

From the power spectral density curves appropriate narrow band filters were selected to isolate individual modes for random decrement analysis. Typical random decrement free vibration decay curves are given in Fig. 7 for each case tested. Damping coefficients are calculated from the logarithmic decrement of the random decrement free response curve.

A summary of the damping measurements from all the techniques are given in Table 1.

TABLE 1: Summary of Results

Test Condition	Conventional Techniques				Random Decrement	
	Sine Sweep		Free Response			
	Freq.	Damping	Freq.	Damping	Freq.	Damping
Fixed	15.1 66.0	.045 -.060 .0025-.0045	15.1 66.0	.024 .0046	14.2 63.8	.015 .0030
Free	64.9	.004 -.006	64.9	.0045	62.8	.0030
Platform Embedded 2 inches in Sand	6.9 59.5	.065 -.085 .02 -.04	6.9 59.5	.070 .029	8.1 60.5	.065 .017

DISCUSSION AND CONCLUSIONS

The limitations imposed by the digital filter characteristics at low frequencies considerably affected the calculation of damping ratio. However the results from Table 1 show that for the comparisons which were made, the damping ratios differed by less than a factor of 1.4 except for the case where the structure is fixed to the support.

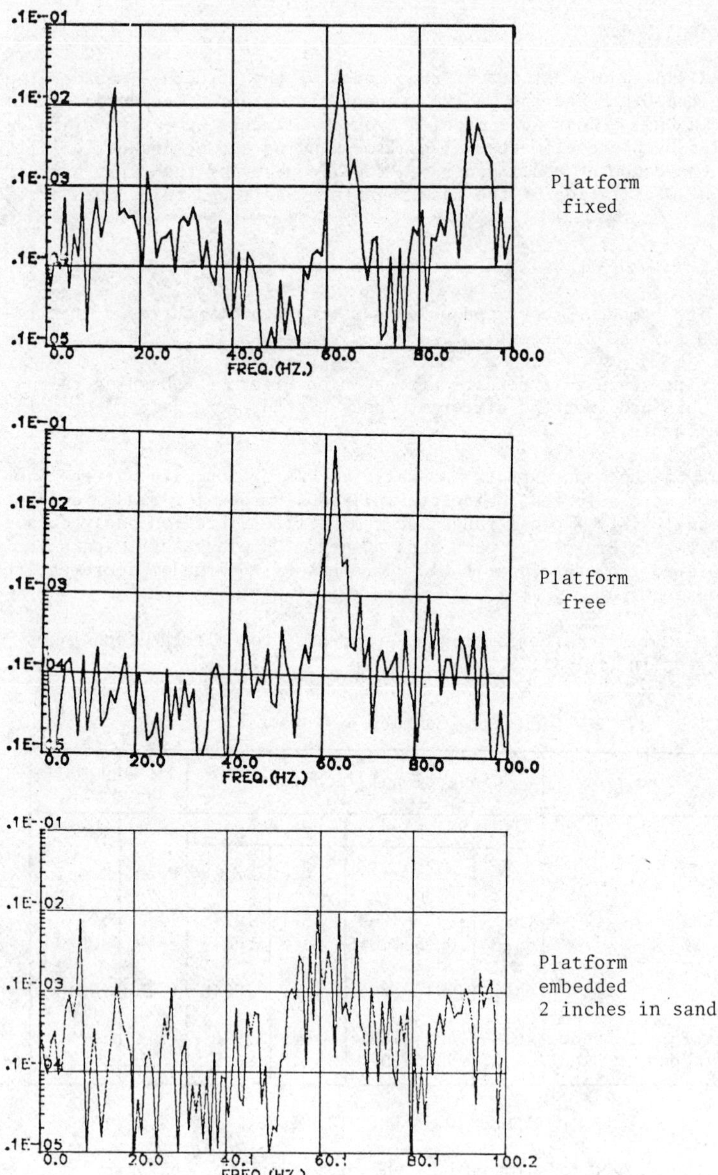

Fig. 5 Power Spectral Density Curves

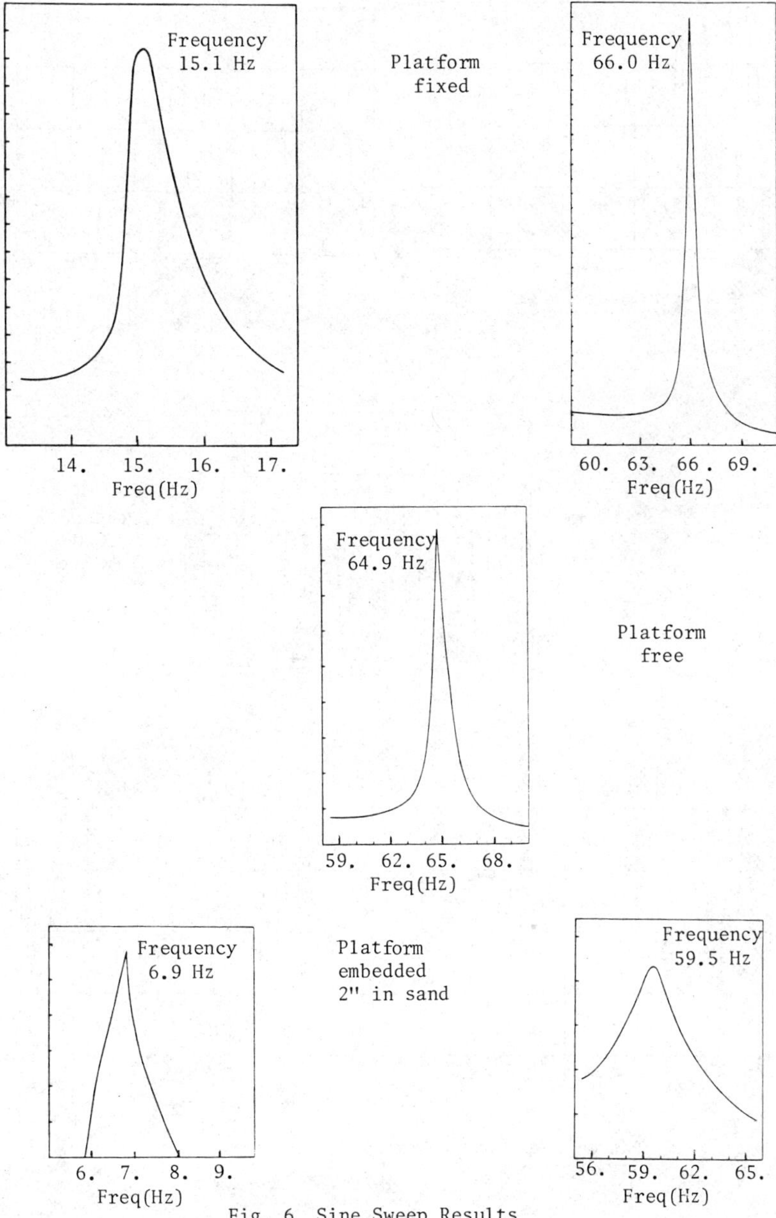

Fig. 6 Sine Sweep Results

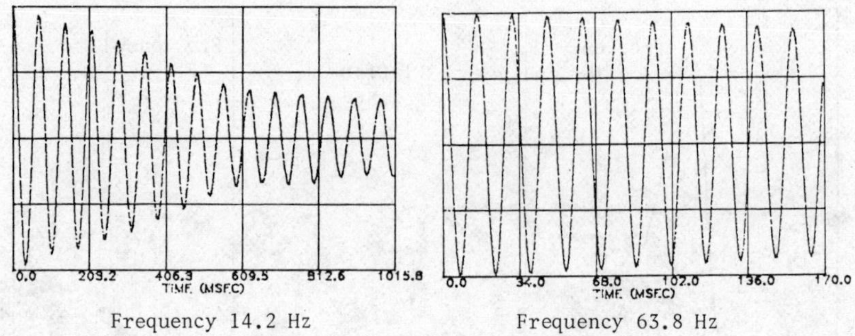

Frequency 14.2 Hz

Frequency 63.8 Hz

Platform fixed

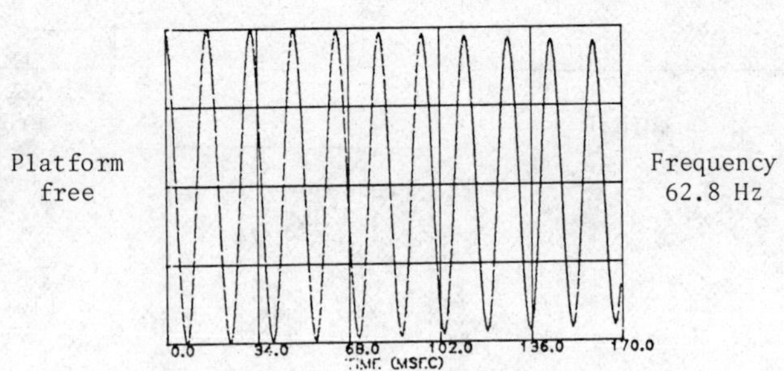

Platform free

Frequency 62.8 Hz

Frequency 8.1 Hz

Frequency 60.5 Hz

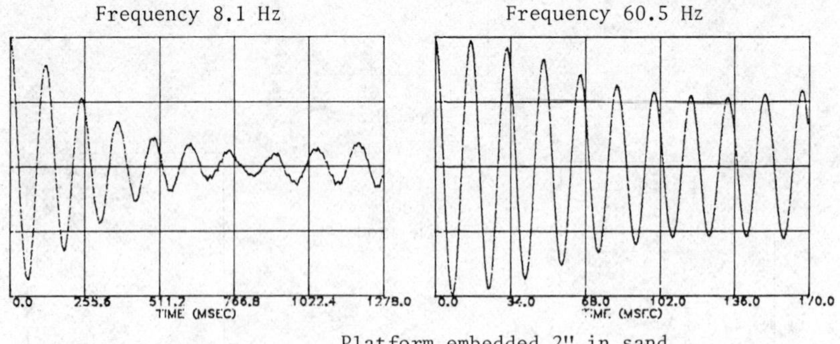

Platform embedded 2" in sand

Fig. 7 Random Dec. Free Vibration Decay Curves

The damping ratio at the fundamental frequency differs by a factor of 3.

A comparison of values of the resonant frequencies obtained from the transfer function and sine sweep plots with those calculated from the periods of the random decrement signatures show that all differed by less than 6 percent, except for the fundamental mode of the structure embedded in soil. The difference is about 14%. Considering that the random decrement signatures were calculated from only response data while it was necessary to also use the excitation data for the calculation of the transfer function, these results are very good.

The satisfactory results obtained from the experimental comparisons indicate that the random decrement technique is a valid method of measuring damping ratios.

ACKNOWLEDGMENT

We wish to thank the Civil and Mechanical Engineering Departments of the University of Maryland and the National Science Foundation for their support of this research.

REFERENCES

1. Degenkolb, H.J., "Earthquake Forces on Tall Structures", Bethlehem Steel.

2. Housner, George W., "Vibration of Structures Induced by Seismic Waves", Shock and Vibration Handbook, Vol. 3, 1961 Edition.

3. Ashok K. Vaish, et al., "Earthquake Analysis of Structure-Foundation Systems, Univ. of California, Berkeley, Calif., May 1973.

4. Tsui, N.C., 1974, "Model Damping for Soil-Structure Interaction", Journal of the Engineering Mechanics Division, ASCE, Vol. 100, No. EM2, April.

5. Roesset, J.M., R.V. Whitman, and R. Dobry, 1973, "Model Analysis for Structures with Foundation Interaction", Journal of the Structural Division, ASCE, Vol. 99, No. 573, March.

6. Novak, M., 1973, "The Effect of Embedment on Vibration of Footings and Structures", Proceedings, 5th World Conference in Earthquake Engineering, Vol. 2, pp. 2658-2661.

7. D'Appolonia Consulting Engineers, 1979, Seismic Input and Soil-Structure Interaction, a report to the U.S. Nuclear Regulatory Commission NUREG/CR-0693, February.

8. Cole, H.A., Jr.: Method and Apparatus for Measuring the Damping Characteristics of a Structure. United States Patent No.3,620,069, Nov. 1971.

9. Yang, J.C.S. and Caldwell, D., "Measurement of Damping and the Detection of Damages in Structures by the Random Decrement Technique", 46th Shock and Vibration Bulletin, August 1976, pp.129-136.

10. Cole, H.A., Jr.: On-Line Failure Detection and Damping Measurements of Aerospace Structures by Random Decrement Signatures. NASA CR-2205, 1972.

11. Brignac, W.J., et al.: The Random Decrement Technique Applied to the YF-16 Flight Flutter Tests. AIAA/ASME/SAE 16th Structures Conference, Denver, CO, May 1975.

12. Yang, J.C.S., "The Measurement of Damping in Mobile Homes", ATR Report No.00-76-5, National Bureau of Standards, 1976.

13. Yang, J.C.S. and Caldwell, D., "A Method for Detecting Structural Deterioration in Piping Systems", ASME Probabilistic Analysis and Design of Nuclear Power Plant Structures Manual, PVB-PB-030, Dec., 1978, pp. 97-117.

14. Yang, J.C.S., "Detection of Incipient Structural Failure by the Random Decrement Method", USGS Research and Development Program for Outer Continental Shelf Oil and Gas Operations Report No. 78-902, 1978, pp. 16-20.

APPENDIX A

RANDOM DECREMENT TECHNIQUE

The response $x(t)$ of a linear system is governed by the following basic equation:

$$m\,\ddot{x}(t) + c\,\dot{x}(t) + k\,x(t) = f(t) \tag{1}$$

The solution of this differential equation depends on its initial conditions and the excitation $f(t)$. Since, for linear systems the superposition law applies, the response can be decomposed into three parts: response due to initial displacement $x_d(t)$, response due to initial velocity $x_v(t)$ and finally the response due to the forcing function $x_f(t)$.

The Randomdec analysis consists of averaging N segments of the length τ_1 of the system response in the following manner: the starting time t_i of each segment is selected such that $x_i(t_i)=x_s$=constant and the slope $\dot{x}_i(t_i)$ is alternating positive and negative. This process can be represented in mathematical form:

$$\delta(\tau) = \frac{1}{N} \sum_{i=1} x_i(t_i+\tau) \tag{2}$$

where $x_i(t_i) = x_s$ $\quad i = 1, 2, 3 \ldots.$

$\dot{x}_i(t_i) = \geq 0$ $\quad i = 1, 3, 5 \ldots.$

$\dot{x}_i(t_i) = \leq 0$ $\quad i = 2, 4, 6 \ldots.$

The function $\delta(\tau)$ is called the Randomdec signature and is only defined in the time interval $0 \leq \tau \leq \tau_1$. The meaning of the Randomdec signature can now be determined. If the parts due to initial velocity are averaged together, they cancel out because alternately parts with positive and negative initial slopes are taken and their distribution is random. Furthermore, if the parts due to the excitation are averaged they also vanish because, by definition, the excitation is random. Finally only the parts due to initial displacement are left and their average is the Randomdec signature representing the free vibration decay curve of the system due to an initial displacement, which corresponds to the bias level x_s. (Fig 1)

In reality the Randomdec computer converts each segment into digital form and adds it to the previous segments (Fig 2); the average is then stored in the memory and can be displayed on a screen. The number of segments to be averaged for the Randomdec signature depends on the signal shape, usually 400 to 500 averages are sufficient to produce a repeatable signature.

One particularly interesting characteristic of Randomdec technique should be mentioned: it requires no knowledge of the excitation f(t) as long as it is random. Neither the type nor the intensity of the input affect the signature.

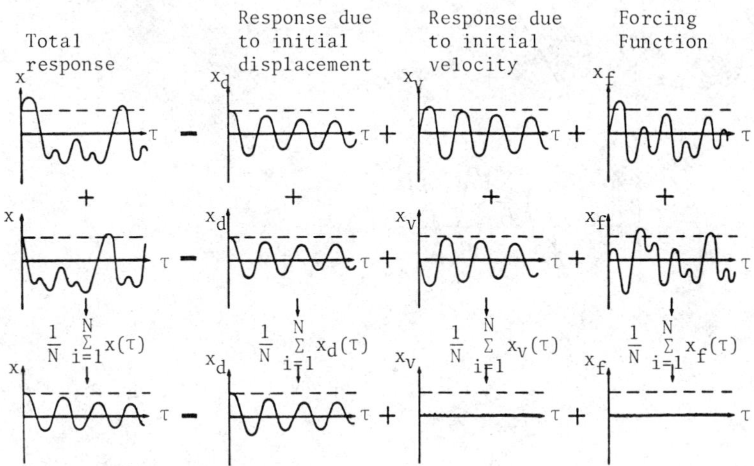

Fig. 1 Principles of Random Decrement Technique

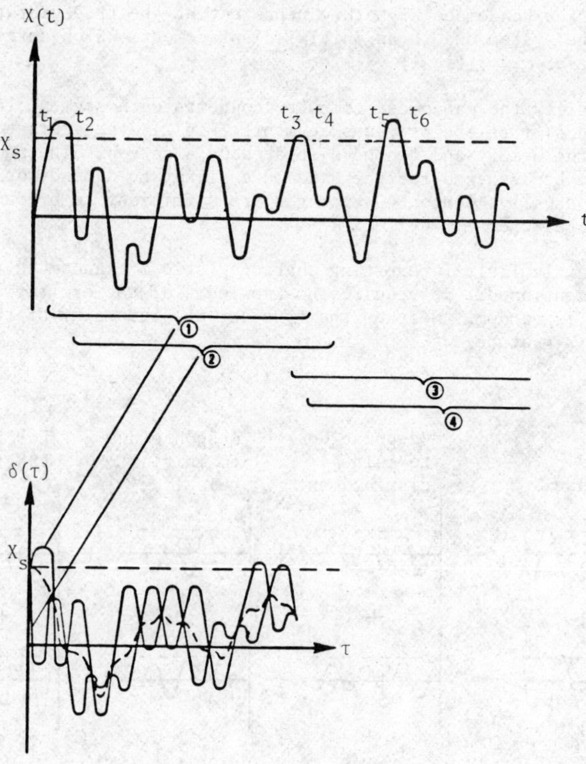

Fig. 2 Extraction of the Random Decrement Signature

Structural Damping In The Time Domain

Allen J. Bronowicki *

Abstract

The structural damping model is widely recognized as being superior to the viscous damping model for representing the energy dissipation characteristics of actual structures and soils. Previous authors have stated, however, that structural damping may not be used in transient applications. It is shown that structural damping produces oscillations which decay in the positive frequency and which grow in the negative frequency, i.e. the roots of the equations are not complex conjugates. A new definition of the structural damping restoring force is proposed which allows computation of real-valued, decaying transient response. An efficient method of complex modal response is derived which makes use of the complex eigenvalues while maintaining decaying motion.

1.0 Introduction

A credible model of structural damping for use in transient response analyses would be of great value. For most structural materials energy dissipation bears a stronger correlation to the magnitude of deformation than to the rate of deformation [1, 2]. The mechanisms involved in damping include: magnetoelasticity, dislocation motions, interface slip and impact, and anelasticity or internal friction. Most if not all of these mechanisms are better modeled using rate-independent structural damping. Viscous damping is a rate-dependent mechanism which is most easily visualized as arising from the shear forces in fluid flow. Viscous forces evince themselves in structural applications only in specialized viscoelastic materials such as solid rocket propellants. In addition to its use in representing linear dissipation effects, the structural damping mechanism is also most useful in the development of equivalent linear models for non-linear hysteretic structures. The value of an equivalent linear damper is based upon the area within the hysteresis loop of the particular structural element [3]. This area represents the amount of energy dissipated in one cycle of motion, and is a function of the amplitude of motion and not of the rate of motion. An equivalent structural damper would produce the same energy dissipation at all frequencies of motion, whereas an equivalent viscous damper would not.

Previous authors [4, 5] have stated that structural damping may be used only in steady state applications. This conclusion is certainly justified on the basis of the previous, and in certain respects erroneous, definition of structural damping. This definition originated with Soroka [6], who chose to ignore the mathematics and postulated that a damping force of the form $i\ h\ k\ x$ produces only positive damping forces. The misconception was propagated by Scanlon and Rosenbaum [7] who rearranged their equations to enable

*Member of the Technical Staff, TRW (DSSG), Redondo Beach, California

response to decay for both positive and negative frequencies. In this paper a new definition of the structural damping restoring force, with a frequency dependence of sorts, is proposed. This definition predicts positively damped motions in both frequency directions. It also allows the calculation of real-valued transient response. It is then pointed out that this predicted response violates the widely perceived notion that no earthly object should possess the powers of precognition. Despite this drawback it is felt that the damping model put forth here offers a more realistic measure of peak response, the response quantity of most importance in the design of a structure, than does the equivalent viscous damping model.

2. Eigenvalues of Damped Single Degree of Freedom Systems

The implications of the standard assumption in modeling structural damping are best seen by contrasting solutions for a simple system using both structural and viscous dissipation functions. The homogeneous equation of motion for a viscously damped oscillator is:

$$m\ddot{x} + c\dot{x} + kx = 0 \qquad (1)$$

where the dot denotes time derivative.

Assuming the existence of a non-trivial solution of the form e^{pt} the following characteristic equation is obtained:

$$mp^2 + cp + k\,p = 0 \qquad (2)$$

which may be solved to yield the complex conjugate pair of roots:

$$p_{1,2} = -\zeta\omega_o \pm i\omega_o\sqrt{1-\zeta^2} \equiv -\zeta\omega_o \pm i\omega \qquad (3)$$

where ω_o is the undamped natural frequency $\sqrt{k/m}$, ω is the damped frequency and ζ is the damping ratio, $C/(2m\,\omega_o)$. The solution will assume the form

$$x(t) = e^{-\zeta\omega_o t}(Z_1 e^{i\omega t} + Z_2 e^{-i\omega t}) \qquad (4)$$

For a given set of initial conditions the complex coefficients Z_1, and Z_2 may be evaluated to supply the actual solution.

Given the initial conditions $x(o) = x_o$ and $\dot{x}(o) = v_o$ the solution becomes:

$$x = e^{-\zeta\omega_o t}\left[x_o \cos\omega t + \frac{v_o + \zeta\omega_o x_o}{\omega}\sin\omega t\right] \qquad (5)$$

which is everywhere real-valued and, for positive damping, decays with increasing time.

The assumption of structural damping requires that the restoring force on a particle be applied at the time of maximum velocity but in proportion to displacement. This is commonly achieved by multiplying the elastic restoring force by i and the structural damping ratio h to give the homogeneous equation of motion [8].

$$m\ddot{x} + (1 + ih)\,k\,x = 0 \qquad (6)$$

Again assuming a solution of the form e^{pt} the characteristic equation is obtained as:

$$mp^2 + (1 + ih)k = 0 \qquad (7)$$

which yields the complex roots:

$$p_{1,2} = (-\zeta\omega_o + i\omega) \quad (8)$$

where the damping ratio, ζ, is now defined as $(1 + h^2)^{1/4}$ sin h/2 and the damped frequency, ω, is $\omega_o(1 + h^2)^{1/4}\cos h/2$. For small values of h the damped frequency is now greater than the undamped frequency. This is due to the increase in the magnitude of the stiffness in the complex plane by the addition of structural damping, which may be described more accurately as an imaginary stiffness. The roots are not complex conjugates, and may be interpreted as providing positive damping in the positive frequency and negative damping in the negative frequency. Imposition of real initial conditions on the general solution:

$$\chi(t) = Z_1 e^{-\zeta\omega_o t} e^{i\omega t} + Z_2 e^{+\zeta\omega_o t} e^{-i\omega t} \quad (9)$$

will not produce a real-valued solution. In fact the solution obtained will be complex and will diverge with increasing time. This behavior underlies the inability of the standard structural damping model to produce real, well-behaved transient solutions. In fact, any attempt to construct a real-valued transient solution by integrating the complex-valued equations of motion through time cannot succeed due to the presence of the imaginary terms in the differential equation.

3. Equivalent Viscous Damping

A common and effective approach for the treatment of structural damping in multi-degree of freedom models is the construction of an equivalent viscous damping matrix. For time harmonic motions at a single frequency the two damping models can be made to be completely equivalent, but for structures that must respond across a spectrum or in a transient manner the equivalence is not complete.

The amount of energy dissipated by a single degree of freedom oscillator in one cycle is:

$$\Delta E = hk\pi\chi^2 \text{; Structural} \quad (10a)$$

$$\Delta E = \Omega c\pi\chi^2 \text{; Viscous} \quad (10b)$$

where χ is the amplitude of motion and Ω is the circular frequency. By equating the two at a chosen frequency Ω_o, one may obtain an equivalent viscous damper, $C_{eq} = hk/\Omega_o$. This equivalent viscous damper will provide a greater amount of energy dissipation than the structural damper at response frequencies Ω greater than Ω_o, and vice versa for lower frequencies. In the case of multi-degree of freedom systems it is common to assume structural damping matrices in proportion to the stiffness matrix, giving a homogeneous equation of motion of the form:

$$[m]\{\ddot{\chi}\} + (1 + ih)[k]\{\chi\} = \{0\} \quad (12)$$

An equivalent viscous damping matrix can be constructed as $[C_{eq}] = h[k]/\Omega_o$ for a given frequency Ω_o. It is common to choose the reference frequency Ω_o as the mean power frequency of response [9]. As this frequency is not known apriori the analysis must proceed iteratively in order to converge upon its value.

In the proportional damping assumption one forms the viscous damping matrix as a linear combination of the mass and stiffness matrices, i.e.

$$[c] = \alpha[m] + \beta[k] \tag{13}$$

The equation of motion:

$$[m]\{\ddot{x}\} + [c]\{\dot{x}\} + [k]\{x\} = 0 \tag{14}$$

for real, symmetric mass and stiffness matrices may then be diagonalized using the eigenvectors, $\{\psi_i\}$, of the undamped problem. These eigenvectors are usually normalized such that:

$$\begin{aligned} \{\psi_i\}^T [m] \{\psi_i\} &= 1 \\ \{\psi_i\}^T [k] \{\psi_i\} &= \omega_i^2 \end{aligned} \tag{15}$$

where ω_i is the modal natural frequency. The damping matrix diagonalizes as:

$$\{\psi_i\}^T [c] \{\psi_i\} = \alpha + \beta\omega_i \equiv 2\zeta_i \omega_i \tag{16}$$

The modal damping ratio, ζ_i, is found to be:

$$\zeta_i = 1/2(\alpha/\omega_i + \beta\omega_i) \tag{17}$$

In the case of equivalent structural damping we choose $\alpha = 0$ and $\beta = h/\Omega_o$, and hence the modal damping ratios become:

$$\zeta_i = h\omega_i/(2\Omega_o) \tag{18}$$

The modal damping ratios are thus seen to increase with increasing modal natural frequency, rather than staying constant as one would prefer for true structural damping. This approach has the additional disadvantage of increasing energy dissipation with increasing excitation frequency, Ω.

The drawbacks of the equivalent viscous damping method arise from its attempt to model a displacement dependent force (independent of frequency) with a velocity dependent force (linear in frequency). A true structural damping model would avoid these difficulties.

4. A Rational Definition of Structural Damping

The rationale behind the construction of a complex stiffness matrix is to produce, through multiplication by i, a restoring force one quarter cycle ahead of the displacement. This makes sense only when the motion is time-harmonic, since a cycle is not well-defined otherwise. A plot in the complex plane of the displacement, x, and the restoring force, ihkx, for time depedence $e^{-i\omega t}$ is shown in Figure 1. For motion having time dependence of the form $e^{-i\omega t}$ it is apparent, as may be seen in Figure 1, that the displacement, $\bar{x}$, must be multiplied by -i in order to produce a restoring force having a 90° lead. For motions at zero frequency, i.e. for static problems, it is desirable that the displacement be real and hence that there should be no imaginary restoring or damping force. Taking into account these three considerations the structural damping restoring force, F_R may be defined

$$F_R \equiv i \operatorname{sgn}(\Omega)\, c_s\, \chi \tag{19}$$

where the function sgn (Ω) is defined as:

$$\operatorname{sgn}(\Omega) \begin{Bmatrix} -1; & \Omega < 0 \\ 0; & \Omega = 0 \\ 1; & \Omega > 0 \end{Bmatrix} \tag{20}$$

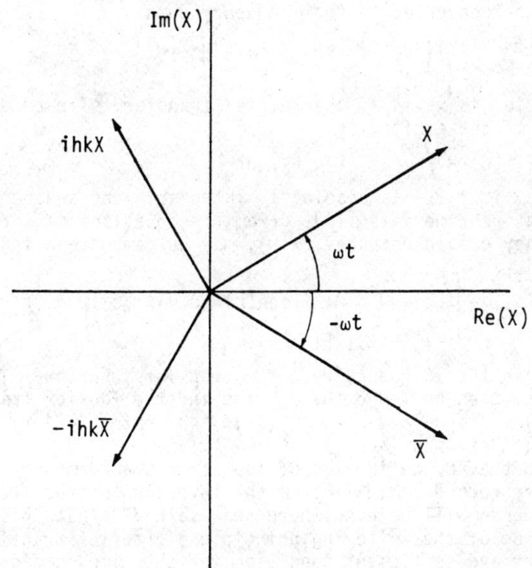

Displacement and Structural Damping Restoring Force in the Complex Plane.

Figure 1

and the constant c_s is the structural damping value, often expressed as a fraction of the stiffness k, i.e. $c_s = hk$.

5.0 Construction of a Transient Solution

Using the definition (19) of the structural damping force the equations of motion of a damped multi-degree of freedom oscillator become:

$$[m]\left\{\ddot{\chi}\right\} + [c]\left\{\dot{\chi}\right\} + ([k] + i\,\mathrm{sgn}(\Omega)\,[c_s])\left\{\chi\right\} = \left\{f(t)\right\} \tag{21}$$

Taking the Fourier Transform of both sides of (21) the equations of motion can be converted to the frequency domain.

$$\left[(K) - \Omega^2\,[m] \quad + i\left(\mathrm{sgn}(\Omega)\,[c_s] + \Omega\,[c]\right)\right]\left\{X(\Omega)\right\} = \left\{F(\Omega)\right\} \tag{22}$$

This operation is valid if the Fourier transform of each of the components of the excitation:

$$F_j\,(\Omega) = \int_{-\infty}^{\infty} f_j(t)e^{-i\Omega t}\,dt \tag{23}$$

exists, i.e. if $f_j(t)$ is absolutely integrable and square integrable over the entire time axis [10]. From the equations of motion (22) the frequency domain response, $X(\Omega)$, may be computed using a matrix inversion as

$$\left\{X(\Omega)\right\} = \left[\left([k] - \Omega^2[m]\right) + i(\mathrm{sgn}(\Omega)\,[c_s] + \Omega\,[c])\right]^{-1}\left\{F(\Omega)\right\} \equiv [H(\Omega)]\{F(\Omega)\} \tag{24}$$

where $H(\Omega)$ is the frequency response transfer function. Time domain response may then be found through the inverse Fourier transform

$$\left\{\chi(t)\right\} = \frac{1}{2\pi}\int_{-\infty}^{\infty} [\,H(\Omega)\,]\left\{F(\Omega)\right\} e^{i\Omega t}\,d\Omega \tag{25}$$

Due to the new definition of the structural damping force (19) the response record obtained from the inversion of the frequency domain response will be everywhere real-valued. That this is true is a consequence of the following principle of Fourier transforms: "A necessary and sufficient condition for the time domain component of a Fourier transform pair to be real is that the frequency domain component have a real part which is even in frequency and an imaginary part which is odd in frequency." Stated more succinctly, given a frequency domain function $X(\Omega)$, the corresponding time domain function, $\chi(t)$, will be real-valued if and only if $X(\Omega) = \overline{X}(-\Omega)$, where $\overline{}$ denotes the complex conjugate. Certainly, since f(t) is real, then $F(\Omega)$ will have its real part even in Ω and its imaginary part odd in Ω. Examination of (24) will reveal that $H(\Omega)$ also possesses this property and hence the product of the excitation and the frequency response function, $X(\Omega)$, will also share this property and thus lead to a real temporal response. Use of the prior definition of the structural damping force would result in a frequency response function without the benign symmetry properties in Ω. In a sense this explains the inability of the prior structural damping model to be applied in transient solutions.

An interesting implication of the use of complex stiffness for temporal solutions is that the structural model exhibits precognizance and will actually respond before the first time of application of any force to the structure. By way of explanation one could say that the 90° phase lead of damping force over displacement in the frequency domain translates into a time lead, or precognizance, for this force in the time domain. That a solution possessing precognizance could be constructed owes to the existence in the Fourier transform $F(\Omega)$, of the information content of the entire temporal excitation, $f(t)$. The transfer function $H(\Omega)$ is then able to extract this information,looking, in essence into the future. Of course a time history integration approach, having no information about the excitation at future times, would never be able to do this.

6.0 Complex Eigenvectors as a Basis for Response

The method of computing response outlined in the previous section is simple, but not computationally efficient for systems having more than a few degrees of freedom. Computation of the frequency response function requires a matrix inversion at each frequency. This becomes costly given the large number of frequencies necessary to represent a transient solution using a discrete Fourier transform [11]. A more efficient approach would be to first diagonalize the equations of motion using the complex system eigenvectors. Response may then be calculated using only matrix multiplications and simple algebra.

The equations of motion (21) may be expressed in first order form as [8]:

$$\left[\begin{array}{c|c} c & m \\ \hline m & o \end{array}\right] \left\{\begin{array}{c} \dot{x} \\ \hline \ddot{x} \end{array}\right\} + \left[\begin{array}{c|c} k+i\,\mathrm{sgn}(\Omega)c_s & o \\ \hline o & -m \end{array}\right] \left\{\begin{array}{c} x \\ \hline \dot{x} \end{array}\right\} = \left\{\begin{array}{c} f(t) \\ \hline o \end{array}\right\} \qquad (26)$$

or, defining a new vector q containing both displacement and velocity information, they may be stated equivalently as:

$$[A]\left\{\dot{q}\right\} + [B(\Omega)]\left\{q\right\} = \left\{f_q(t)\right\} \qquad (27)$$

or:

$$\left\{\dot{q}\right\} - [D(\Omega)]\left\{q\right\} = [A]^{-1}\left\{f_q\right\} \qquad (28)$$

where we have defined the dynamic matrix $D(\Omega)$ such that:

$$[D(\Omega)] = -[A]^{-1}[B(\Omega)] \qquad (29)$$

For a given value of Ω equation (28) with its right hand side set to zero will define an eigenproblem. Ignoring the case $\Omega=0$ for the time being, we may assign two values to the dynamic matrix in accordance with the definition (20) of $\mathrm{sgn}(\Omega)$:

$$\begin{aligned} [D^+] &= [D(\Omega)] \quad ; \quad \Omega > 0 \\ [D^-] &= [D(\Omega)] \quad ; \quad \Omega < 0 \end{aligned} \qquad (30)$$

It may be seen that D^+ is the complex conjugate of D^-. Two eigenproblems are thus defined in the form:

$$[D^{+,-}]\left\{\psi_i\right\} = \lambda_i\left\{\psi_i\right\} \qquad (31)$$

Denoting the eigenvectors of D^+ and D^- as ψ^+ and ψ^-, respectively, and the eigenvalues as λ_i^+ and λ_i^-, respectively, it may be shown that:

$$\left\{\psi_i^+\right\} = \left\{\overline{\psi_i^-}\right\} \qquad \lambda_i^+ = \overline{\lambda_i^-} \tag{32}$$

where again the ——— denotes complex conjugate.

The two sets of eigenvectors obtained may each be used as a basis for the response vector q. This relationship may be expressed:

$$\left\{q\right\} = [\Psi^+]\{Z^+\} = [\Psi^-]\{Z^-\} \tag{33}$$

where Ψ is the modal matrix, whose columns are the eigenvectors ψ_i, and Z is a vector of modal responses in the time domain. Insertion of either of transformations (33) into the equations of motion (28) and premultiplication by the inverse of the modal matrix produces the diagonalized equations of motion in modal coordinates:

$$\left\{\dot{Z}\right\} + [\Lambda]\left\{Z\right\} = [\Psi]^{-1}[A]^{-1}\left\{f_q\right\} \tag{34}$$

where Λ is a diagonal matrix of eigenvalues.
Transformation to frequency domain allows calculation of modal frequency response as:

$$\left\{Z(\Omega)\right\} = \left(i\Omega[I] - [\Lambda]\right)^{-1} [\Psi]^{-1} [A]^{-1} \left\{F_q(\Omega)\right\} \tag{35}$$

Physical response in the frequency domain, $Q(\Omega)$, is obtained through the pre-multiplication of (35) by the modal matrix, giving:

$$\left\{Q(\Omega)\right\} = [H_q(\Omega)] \left\{F_q(\Omega)\right\} \tag{36}$$

where the frequency response transfer function, $H_q(\Omega)$, is given by:

$$[H_q(\Omega)] = [\Psi]\left(i\,[I] - [\Lambda]\right)^{-1} [\Psi]^{-1} [A]^{-1} \tag{37}$$

In order for the transient response, q(t), to be real it is necessary that $H_q(\Omega)$ should possess the beneficial symmetry properties described in Section 5, i.e. that:

$$[H_q(\Omega)] = \left[\bar{H}_q(-\Omega)\right] \tag{38}$$

This may be accomplished by requiring that Ψ^+ and Λ^+ be used in the calculation of H_q (Ω) for Ω greater than zero and that Ψ^- and Λ^- be used for Ω less than zero. For Ω equal to zero the static solution, with zero complex stiffness in accordance with definition (20), must be used.

Transient response is obtained by taking the inverse Fourier transform of the frequency domain response, as in (25). Since it is known that $F(\Omega) = \bar{F}(-\Omega)$ and that $H_q(\Omega) = \bar{H}_q(-\Omega)$, it must also be true that $Q(\Omega) = \bar{Q}(-\Omega)$. Employing this symmetry property in the inverse Fourier transform it can be shown that the transient response is given by:

$$\left\{q(t)\right\} = 2\mathrm{Re}\left[\frac{1}{2\pi}\int_0^\infty [H_q(\Omega)] \left\{F_q(\Omega)\right\} e^{i\Omega t} d\Omega\right] \tag{39}$$

Thus it shown that the time domain response is real and that computations only for the zero and positive frequencies are required. Calculation of the second set of eigenvectors ψ^- and Λ^- is not required. Calculation of a matrix inverse at each frequency is also not necessary, as it was in the method of Section 5.

7.0 Results

The major influence of damping on the response of a single degree of freedom oscillator occurs at the resonant frequency, where response is inversely porportional to the amount of damping in the system. As was shown in Section 3, an equivalent viscous damping model may be found to emulate the structural damping model at one given frequency, and if this frequency is chosen as the resonant frequency the response predicted by the two models will be quite close across the entire spectrum. However, for multi-degree of freedom models this fortuitous circumstance no longer holds. The equivalent viscous model will filter out the response of the higher modes to a much greater degree than is consistent with the assumption of a damping force proportional to displacement. Hence the discrepancies between the two models will be most evident for multi-degree of freedom systems, especially those in which the contribution to response by the higher modes is significant.

In order to contrast the performance of the two damping models, consider the four degree of freedom oscillator shown in Figure 2. The system consists of four masses connected by four identical springs of stiffness $k = 1$ and by four identical dampers. The structural damping ratio, h, is chosen as 0.2, giving $C_s = hk = 0.2$. This produces modal damping ratios of approximately ten percent. The equivalent viscous damping model is evaluated at the undamped natural frequency of the first mode, .0745 cycles per second, giving a value of $c = hk/\omega_1 = .4273$. The damped and undamped natural frequencies and the modal damping ratios for the two models are given in Table 1. Note that the modal damping ratios of the viscous damping model increase with modal frequency, while the modal structural damping ratios remain constant.

The transient response of the top mass, x_4, due to an impulsive force applied at x_4 is shown in Figure 3 for both damping models. The force is a triangular pulse, commencing at t = 50 seconds, rising linearly to 1 at t = 50.6 seconds and falling linearly to zero at t = 51.2 seconds. This force was chosen to excite all the modes of the structure. Response of the viscous damping model may be seen to be predominantly first mode, whereas the structural damping model exhibits considerable second mode motion in addition. The peak responses are .510 and -.258 for the structural damping model and .383 and -.218 for the viscous model. Thus peak response is 25% higher on the positive side and 16% higher on the negative side for the structural versus the viscous damping model. The precognizance effect is barely visible in the response of the structural damping model as the top mass moves down only slightly prior to the application of the load.

8.0 Conclusions

A structural damping model has been constructed which may be applied in the transient analysis of linear structures. The values of peak response predicted by this model will generally be considerably higher than those predicted by the equivalent viscous damping model which filters out an excessive amount of the higher mode participation. The structural damping model will produce response time histories which exhibit precognizance, an interesting phenomenon and in violation of fundamental physical principles but probably not of great concern since the primary objective in structural analysis is

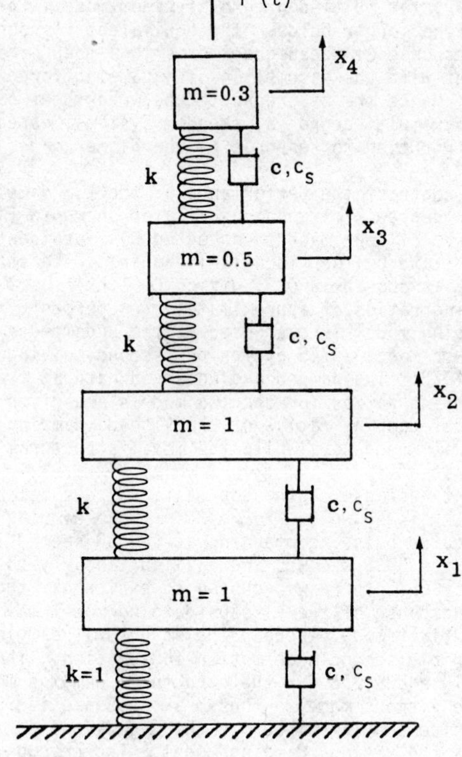

Four Degree of Freedom Damped Oscillator (c and c_s represent viscous and Structural Damping, respectively).

Figure 2

Natural Frequencies and Modal Damping Ratios
of the Four Degree of Freedom Oscillator

Mode	1	2	3	4
Undamped Frequency (Hertz)	.0745	.1967	.2778	.4070
Damped Frequency (Structural)	.0749	.1977	.2791	.4091
Damped Frequency (Viscous)	.0741	.1897	.2577	.3409
Modal Damping Ratio (Structural)	.0985	.0985	.0985	.0985
Modal Damping Ratio (Viscous)	.1000	.2640	.3729	.5464

Table 1

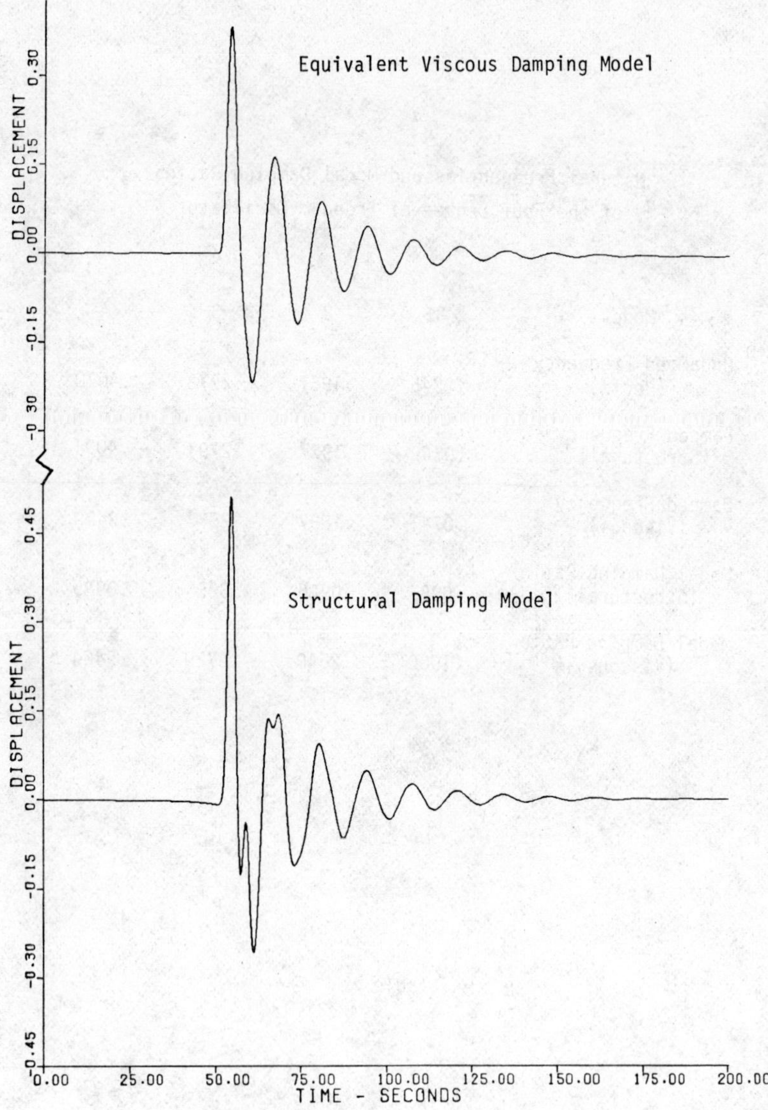

Transient Response of Tip Mass of 4 DOF Oscillator to Triangular Pulse

Figure 3

the prediction of peak response. Since structural damping is generally conceded to be the most realistic mechanism for energy dissipation in solids [1, 2, 4-7] it is felt that the proposed model is the most accurate for the prediction of response in linear and also in non-linear hysteretic structures. Confirmation could come in the form of transient tests using various structural materials, or, where the actual shape of the hysteresis loop is known sufficiently well, by comparison with non-linear transient analyses.

9.0 References

1. Lazan, B. J., Damping of Materials and Members in Structural Mechanics, Pergamon, Oxford, 1968.
2. Goodman, L. E., "Material Damping and Slip Damping," Shock and Vibration Handbook Ch. 36, Harris, C. M. and Crede, C. E. editors, McGraw-Hill, N.Y., 1976.
3. Caughey, T.K., "Equivalent Linearization Techniques," Journal of the Acoustical Society of America, Vol. 35, November, 1963, pp. 1706-1711.
4. Fung, Y.C., The Theory of Aeroelasticity, Wiley, N.Y., 1955.
5. Meirovitch, L., Analytical Methods in Vibrations, MacMillan, London, 1967, pp. 400-405.
6. Soroka, W.W., "Note on the Relations Between Viscous and Structural Damping Coefficients," Journal of the Aeronautical Sciences, Vol. 16, July 1949, pp. 409-410, 448.
7. Scanlon and Rosenbaum, Introduction to the Study of Aircraft Vibration and Flutter, MacMillan, N.Y., 1954, pp. 85-88.
8. Hurty, W.C. and Rubinstein, M.F., Dynamics of Structures, Prentice-Hall, Englewood Cliffs, N.J., 1964.
9. Hasselman, T.K., Bronowicki, A.J. and Chrostowski, J.C., "Probabilistic Response of Offshore Platforms to Seismic Excitation," Offshore Technology Conference, Paper No. OTC 2353, Houston, 1975.
10. Weinberger, H.F., A First Course in Partial Differential Equations, Ginn, Waltham, Mass., 1965.
11. Brigham, E.O., The Fast Fourier Transform, Prentice-Hall, Englewood Cliffs, N.J., 1974.

SYSTEM IDENTIFICATION: REVIEW AND EXAMPLES

by

Ken Blakely, Student Member, ASCE,
Paul Ibáñez, Shelley Griffith, Bob Cooper,
and John Stoessel*

ABSTRACT

This paper presents an example of system identification, in which the parameters of an analytical model are systematically and rationally adjusted to equate the model's response with that of the actual structure. The entire system identification process is presented for a simple structure: formulation of a pre-test analytical model, acquisition and reduction of data taken from static and dynamic tests of the structure, eigenparameter identification, and post-test model refinement via Bayesian parameter identification techniques. Data from full-scale structures are also presented.

INTRODUCTION

The objective of this paper is to illustrate the range of techniques available to investigate the system dynamics of a single structure. Testing and analysis methods are presented to correlate an analytical model with test data; this is known as system identification.

In this paper a simple test structure is used, upon which the various aspects of the system identification process are demonstrated: formulation of a pre-test analytical model, static and dynamic testing, eigenparameter identification, and post-test model refinement using Bayesian parameter identification.

The entire process was performed using a minicomputer based, field portable vibration analysis system [1]**. Hardware consists of a 32K word Data General NOVA-3 minicomputer with two 2.5 Mbyte disks, 9 track magnetic tape drive, incremental X-Y plotter, CRT input/output, and a line printer. Digitization of up to 64 channels is provided by a Data Products RTP device and 64 channels of anti-aliasing filters. Up to 10,000 samples/second can be digitized.

This system has been programmed to gather and reduce both transient and sinusoidal data, present the results graphically, and perform eigenparameter identification, Bayesian model modification, finite element modeling, and spectral analysis.

---

*ANCO Engineers, Inc., Santa Monica, California

**Numbers in brackets designate references listed at the end of the paper.

While this paper focuses primarily on a single test structure, data obtained with the vibration analysis system from previous full scale tests are also shown. In addition, experience in applying system identification techniques to large, complex structures is outlined.

PRE-TEST ANALYSES

Prior to dynamically testing a complex structure, such as a nuclear power plant containment, a piping system, an offshore oil platform, or an arch dam, analyses are performed to determine eigenparameters (resonance frequencies and mode shapes) and structural response (displacements and stresses) to the applied excitation. These analyses aid in judicious placement of the shakers to best excite all modes in the frequency range of interest. In some cases a pre-test safety analysis must be performed to insure that the dynamic testing will not damage the structure or adversely affect its operation.

The test specimen is a seven foot high, three-dimensional structure made of angles and threaded pipe, joined by couplings and welded joints (Figure 1). A steel brace is C-clamped to the structure to create nonsymmetrical structural response. During testing the steel foot pads were bolted to the floor.

A linear elastic finite element model of the structure was formulated. Figure 2 shows a plot of the model geometry, which is stored in the minicomputer. Pipe and beam elements were used to discretize the structure. Each pipe was modeled as being continuous, since it was felt that shear and moment continuity existed at the couplings. The pipes are modeled as having an outside diameter of 1.315 inches and a wall thickness of .133 inches. The angles are 2.5 x 2.5 x 3/16 inches. An eccentric-mass shaker was clamped to the structure at node 19 for the dynamic tests; this was modeled as a lumped weight of 25 pounds.

Each of the two static tests were simulated. For the first, a unit load was applied to the analytical model at node 12, in the X direction, and nodal displacements were calculated. In the second test, nodal displacements were calculated for a unit load applied to node 5, in the Y direction. Analytical and test results are compared in the parameter identification section of this paper.

Eigenparameters were then calculated for the model. The first three resonance frequencies were determined to be 10.69 Hz (bending in the X direction), 13.22 Hz (Y bending), and 22.30 Hz (X,Y bending).

TEST PROCEDURE AND DATA PRESENTATION

Static loads were applied to the test specimen via a hand-operated winch. A force gauge was used to measure the applied force and displacement gauges were used to measure resulting nodal displacements. Loads were applied to node 12, in the X direction, for the first static test. Slight softening under increasing load is observed, as shown in Figure 3. Much greater softening is observed for the second static test (force at node 5, Y direction), shown in Figure 4. It was felt that the brace was loosening at the larger force levels; therefore, it was decided to use the first data point for each test (50 and 80 lbs., respectively) for identification purposes.

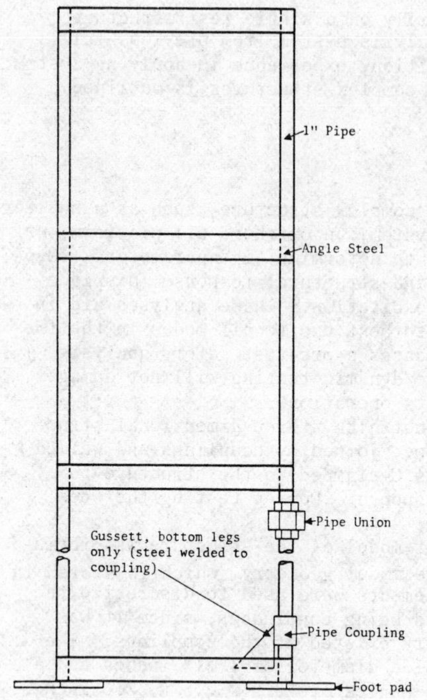

FIGURE 1: SCHEMATIC OF THE TEST STRUCTURE

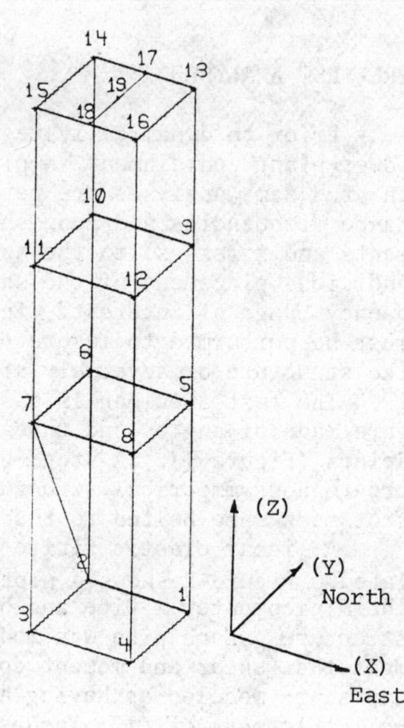

FIGURE 2: GEOMETRY OF THE TEST STRUCTURE

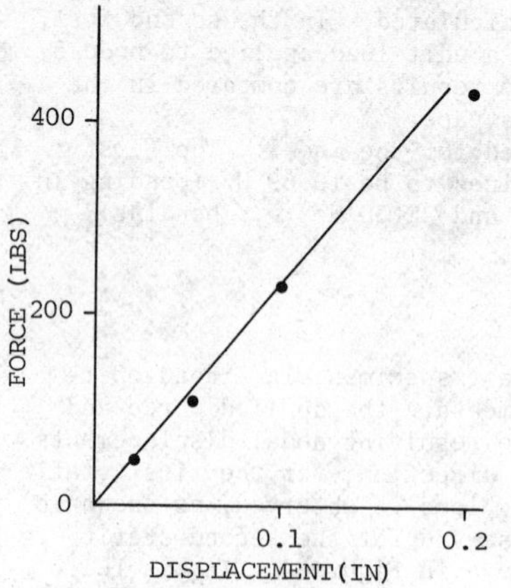

FIGURE 3: NODE 14(X) DATA FOR STATIC TEST 1

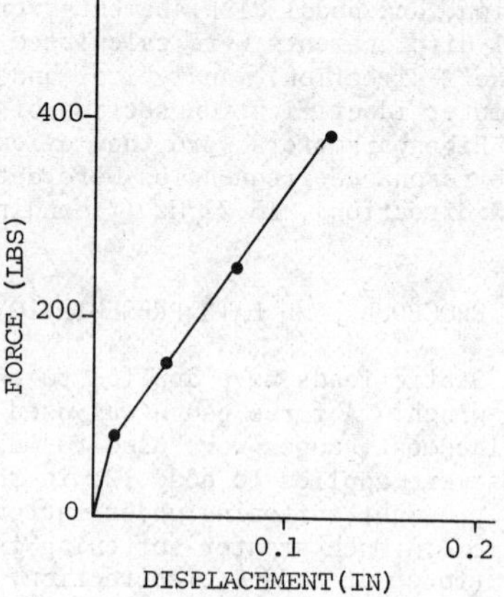

FIGURE 4: NODE 7(Y) DATA FOR STATIC TEST 2

Two dynamic forced vibration tests were then performed with the shaker at node 19, first in the ±Y direction, then in the ±X direction. Accelerometers were located on the east and north faces of the test specimen, two in each direction at each floor level. The shaker was controlled by the minicomputer to step through the frequency range of interest (5 to 25 Hz). Analog signals from the accelerometers are fed through an analog-to-digital converter; magnitude and phase (relative to the forcing function) for each accelerometer are then calculated via the computer software. Once data are taken on all channels at a particular frequency, the shaker is automatically set by the minicomputer to the next frequency point. Shaker control, raw data acquisition (in the form of a digitized time history), and data analysis (magnitude and phase) are all performed by the computer program FSINE. Figure 5 shows an example of forced vibration data taken on the test specimen. The program XSHAPE plots response shape data, examples of which are shown in Figure 6.

Snapback tests were performed by pulling on the top of the specimen with a cable and then relieving the force via a quick-release mechanism. This technique is especially useful in exciting the lowest few modes of a structure. Figure 7 shows a resulting acceleration time history for a snap test on the test specimen. Typical piping system results for a snapback test are shown in Figure 8 for both the time history and corresponding FFT.

As a final illustration of testing methods, tap tests were conducted on the test structure by lightly striking it with a rubber hammer. Two accelerometers were used, moved about the structure for each test; data were taken on a two-channel spectrum analyzer. The first X bending mode was found to be 8.8 Hz; first Y bending was at 11.0 Hz, and X,Y bending at 17.2 Hz.

Note that frequencies obtained via tap tests are higher than those measured during the forced vibration tests. This is attributed to several of the couplings working loose during the shaker tests, considerably altering the test structure; therefore, the tap test results were used for model verification.

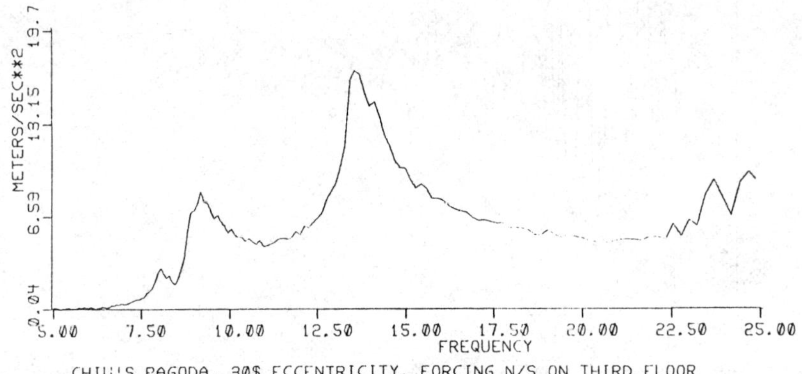

FIGURE 5: ACCELERATION AMPLITUDE VERSUS FREQUENCY FOR FORCED VIBRATION TEST

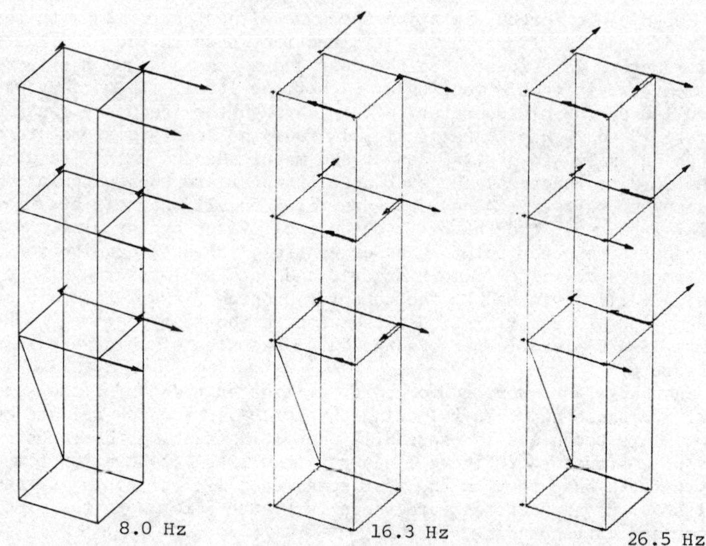

FIGURE 6: FORCED VIBRATION RESPONSE SHAPE DATA

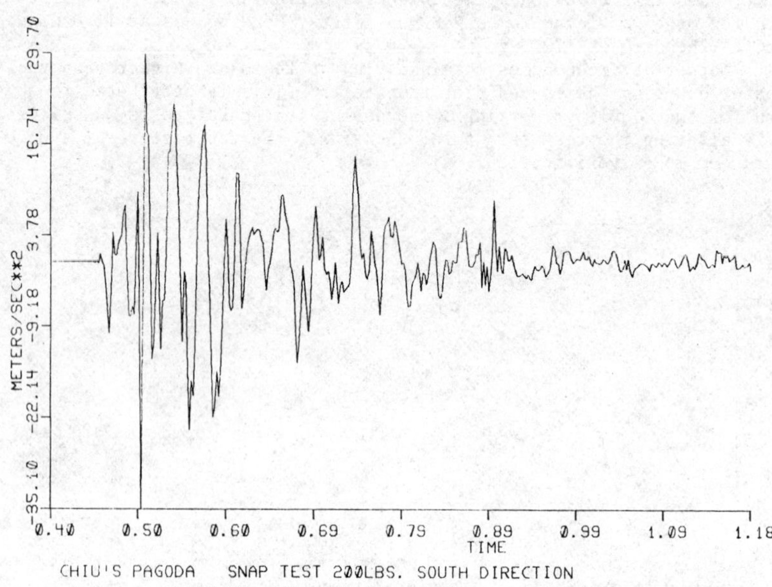

FIGURE 7: TYPICAL SHAPBACK TIME HISTORY OF TEST SPECIMEN

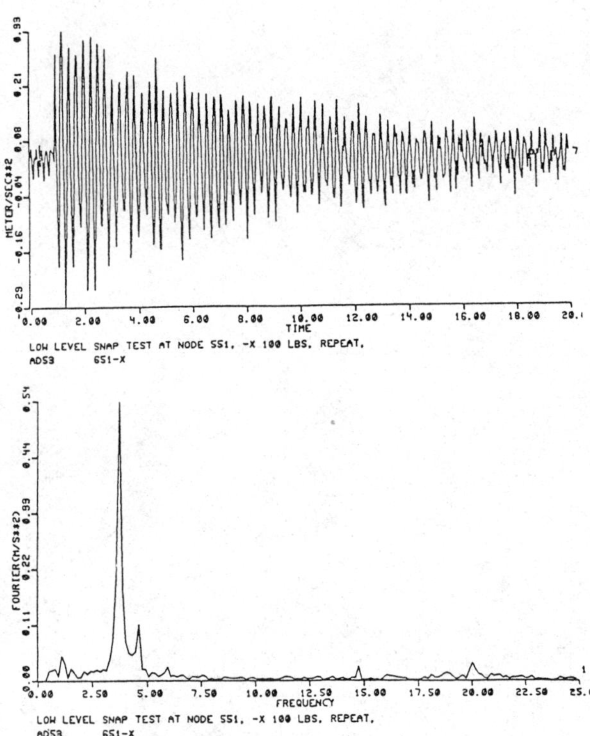

FIGURE 8: TIME HISTORY (TOP) AND FFT (BOTTOM) OF TYPICAL PIPING SYSTEM SNAPBACK TEST DATA

PARAMETER IDENTIFICATION

Eigenparameter identification involves determining mode shapes and resonance frequencies from forced vibration data, which is not pure modal response due to the use of one or two shakers (note that M shakers need be used for purely exciting a single mode of an M degree-of-freedom system). The computer program ANSPI [2] iterates on initial estimates of eigenparameters and the forced vibration test data to produce eigenparameters that best fit the test data. Near pure modal response is assumed. ANSPI was not used in the study presented herein, but was used on some recent tests. Figure 9 shows some forced vibration data taken on a three-story building plus results after running ANSPI.

After gathering and reducing data from a vibration test the measured eigenparameters and response shapes are compared to those derived from the analytical model. More often than not, test and analytical results are found to differ. If the purpose of the test is to verify an analytical model (as opposed to performing a "proof test" for

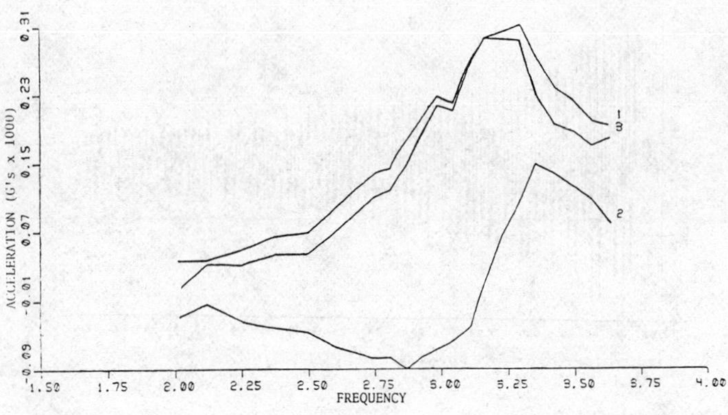

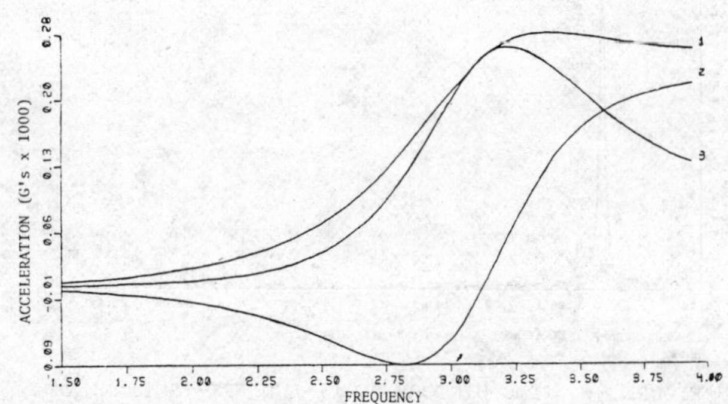

FIGURE 9: FORCED VIBRATION RESPONSE OF A MULTI-STORY BUILDING: RAW DATA (TOP), SMOOTHED DATA VIA ANSPI (BOTTOM)

structural integrity), then the model must be revised to reflect the test data.

System identification in its most rudimentary form is performed in a trial and error manner: parameters are varied until the model matches the test data. The most obvious drawback of this approach is the expense of engineering and computing time necessary to achieve an accurate model. Traditional system identification involves the rational calculation of model parameters to match test data within some prescribed error bounds. One such method is Bayesian parameter identification [3,4], which incorporates both experimental and analytical uncertainties. This method is used herein to determine necessary changes in model parameters.

Bayesian identification is used to determine model parameters that minimize, in a least mean square sense, an error function associated with differences between analytical and test results and original and updated model parameters. Updated responses are written in a Taylor series expansion as

$$\{\bar{R}\} = \{R\} + [S]\{\Delta C\} \tag{1}$$

in which $\{\bar{R}\}$ is the vector of updated responses, $\{R\}$ is the vector of prior responses, $[S]$ is a rectangular partial derivative matrix (the i,j term is the partial derivative of the i response with respect to the j parameter), and $\{\Delta C\}$ is the vector of differences between prior and revised parameters.

Equation (1) is used in the derivation of the iterative equation to determine updated model parameters [5,6]:

$$\{\hat{C}\} = \{C^i\} + ([S]^t[W_R][S] + [W_c])^{-1}([W_c]\{C^0-C^i\} + [S]^t[W_R]\{R^0-R^i\}) \tag{2}$$

in which $\{\hat{C}\}$ is a vector of updated parameters, $\{C^i\}$ is the vector of parameters from the previous iteration, $[W_R]$ is the inverse of the covariance matrix (usually diagonal) of responses, $[W_C]$ is the inverse of the parameter covariance matrix (again, usually diagonal), $\{C^0\}$ is a vector of initial model parameters, $\{R^0\}$ is a vector of "target values" (usually test data), and $\{R^i\}$ is a vector of model responses using $\{C^i\}$.

Equation (2) is used iteratively to update model parameters. After each step, $\{\hat{C}\}$ becomes $\{C^i\}$ and a new $\{R^i\}$ is calculated. The sensitivity matrix $[S]$ is recomputed each time. For the study presented in this paper, $[S]$ was calculated via a finite difference approach: a small change was made in each of the $\{C^i\}$ and the model was rerun. Greater computing efficiency is achieved by calculating $[S]$ in closed form; work is currently in progress to do this. The matrices $[W_R]$ and $[W_C]$ remain constant throughout each of the iterations in equation (2).

The five parameters to be identified in the static model are pipe outside diameter (relative uncertainty of ±5%) and wall thickness (±10%), area (±5%) and moment of inertia (±5%) of the angles, and area (±10%) of the brace. Each displacement gauge measurement had an absolute uncertainty of ±.001 inch, which converted to relative uncertainties on the order of ten percent for each of the measured displacements. Results are presented in Table 1 for test data, displacements calculated with the pre-test model, and resulting analytical displacements after updating the parameters from each iteration of the Bayesian estimation and rerunning the finite element model. After the third iteration the pipe outside diameter had decreased by 1.5%, pipe wall thickness decreased by 8.4%, there was no change in area of the angles, moment of inertia of the angles decreased by 3.6%, and brace area decreased .4%. The last column in the table shows predicted displacements $(\bar{R})$ using equation (1), where $\{R\}$ are the analytical displacements, $[S]$ is the sensitivity matrix, and $\{\Delta C\}$ are the changes in model parameters, each from the second Bayesian iteration. The predicted responses, $\bar{R}$, as calculated by the Taylor series expansion, are seen to agree well with finite element results from the third iteration.

Table 1 shows good agreement between measured displacements and those calculated during the third iteration, especially at nodes for which the pre-test model and test data are within fifteen percent of each other. The Bayesian algorithm does little where measured and initial analytical responses are markedly different. For the study presented in this paper, changing the properties of each beam and pipe element individually might have helped, at the expense of greater

TABLE 1

COMPARISON OF MEASURED AND ANALYTICAL STATIC DISPLACEMENTS

		ANALYTICAL DISPLACEMENTS				
Node[1]	Measured Displ[2]	Pre-Test Model	First Iter.	Second Iter.	Third Iter.	$\overline{R}$
6X	.011	.01530	.01709	.01764	.01523	.01696
7X	.025	.02649	.02964	.03054	.02950	.02937
11X	.045	.03774	.04209	.04337	.04191	.04175
14X	.022	.02110	.02309	.02386	.02307	.02298
15X	.045	.03861	.04286	.04420	.04272	.04256
11Y	-.006	-.00280	-.00316	-.00323	-.00320	-.00311
7Y	.010	.00051	.00049	.00055	.00054	.00052
8Y	.038	.03545	.03970	.04094	.03953	.03937
11Y	.007	.00119	.00119	.00128	.00125	.00125
14Y	.006	.00143	.00140	.00152	.00149	.00149
15Y	.008	.00143	.00140	.00152	.00149	.00149
15X	.020	.01780	.01989	.02053	.01981	.01974

[1]The first six nodes are for the X direction test; the second six nodes are for the Y direction test.

[2]All units are inches.

computing cost. It is up to the analyst to choose reasonable parameters to be varied, keeping in mind modeling uncertainties, accuracy to which the model matches test data, and cost.

The pre-test model was updated to match the first three resonance frequencies measured during the tap tests of the structure. Parameters varied were pipe outside diameter (relative uncertainty of ±5%) and wall thickness (±10%), shaker weight (±2%), and the weight density of the pipes and beams (±5%). Table 2 shows measured and analytical results from the Bayesian iterations. Frequencies calculated using [S], {R}, and {ΔC} from the second iteration are denoted $\overline{R}$; these are seen to equal frequencies calculated by rerunning the finite element model in the third iteration (indeed, the problem converged, as there were no changes in the frequencies between the second and third iterations).

TABLE 2

COMPARISON OF MEASURED AND ANALYTICAL RESONANCE FREQUENCIES

		ANALYTICAL FREQUENCIES				
Mode	Measured Frequency[1]	Pre-Test Model	First Iter.	Second Iter.	Third Iter.	$\overline{R}$
X bending	8.8	10.69	8.86	8.83	8.83	8.83
Y bending	11.0	13.22	10.98	10.94	10.94	10.94
X,Y bending	17.2	22.30	18.73	18.67	18.67	18.67

[1]Units are Hertz.

Convergence of the Bayesian algorithm can be defined in several ways, including no change between iterations in computed response (as for the dynamic case with the test structure), no change between iterations in the model parameters, no change between iterations in the magnitude of the error function, or simply when the analyst feels he should stop (as for the static case with the test structure).

Changing the covariance matrices, $[W_R]$ and $[W_C]$, can cause the Bayesian estimation algorithm to give different models. This is illustrated in Table 3. It has been noted that the algorithm can produce *different* models which match target values exactly. The first author [7] once created three different stiffness matrices which, when used with the same mass matrix, yielded identical results for the lowest three eigenvalues; however, static loadings produced differing displacements between the three models. This shows that non-unique models can result when using the algorithm. When performing vibration tests for model verification, care is taken to insure that all parameters of interest, usually resonance frequencies and mode shapes, are measured directly; just because a model correctly predicts first and second resonance frequencies is no guarantee that it will correctly predict the third. The higher the mode, the greater the influence of local variations in mass and stiffness upon the eigenparameters. This has been seen by the authors when dealing with offshore oil platform dynamics.

System identification as performed via Bayesian estimation could perhaps be better termed *parameter* identification; that is, the *system* has already been defined (topology, connectivity, and governing equations of motion) and the system *parameters* which best fit the data, within prescribed error bounds, are determined. Sometimes the selected

TABLE 3

INFLUENCE OF COVARIANCE MATRICES ON PARAMETER CHANGES

Parameter/ Measurement	W^0(%)*	W^1(%)**	W^0(%)	W^1(%)	W^0(%)	W^1(%)
pipe o.d.	±5	-10.8	±5	-10.1	±5	-11.4
pipe t	±10	- 7.4	±10	- 7.3	±10	- 8.2
shaker wt.	±2	.1	±2	.1	±2	.0
wt. density	±5	3.6	±5	3.1	±5	6.0
first mode	±2		±2		±2	
second mode	±2		±5		±2	
third mode	±5		±10		±2	
Frequencies given by Bayesian est. ($\overline{R}$)						
first mode		8.82		8.86		8.62
second mode		10.94		10.99		10.68
third mode		18.70		18.78		18.27

*W^0 are the initial uncertainties of the four model parameters and three measured frequencies.

**W^1 are the percent changes in the model parameters after the first iteration.

parameters are unable to yield a truly close fit to the data (as in the third resonance frequency of the test specimen). Then, the parameter choice must be viewed critically and the *system* examined. A test performed on a containment structure [8] showed significant discrepancies between analytical and measured eigenparameters. No amount of *parameter* estimation could yield the correct results until it was discovered that there was a styrofoam liner between the inner and outer containments, which had not been modeled. Once the *system* was changed to incorporate the coupling of the containments via the styrofoam, *parameter* estimation yielded an acceptable model.

FUTURE OF SYSTEM IDENTIFICATION

Research is in progress to generalize Bayesian estimation. Two major extensions of the algorithm have been developed and implemented thus far. In the first, linear constraint relations are written between model parameters [9]. The vector of parameter values is partitioned into dependent and independent parameters, linked by the constraint relation $\{Y\} = [A]\{X\} + \{B\}$. This also has the same effect as using off diagonal terms in $[W_C]$. Related to this is the use of inequality constraints, which avoids physically impossible parameter changes (negative stiffnesses, for example); this would incorporate many aspects of nonlinear programming. The other extension of the Bayesian algorithm makes use of time dependent data [10]. The error function includes a term that is the difference (as a function of time) between an analytical and measured response, integrated over time. This is necessary when identifying nonlinear models.

The complete system identification process--pre-test analysis, data acquisition and reduction, eigenparameter estimation, and model updating--has been applied by the authors to several complex structures, including containment structures, piping systems, offshore oil platforms, arch dams, and multistory buildings. The sensitivity matrix, [S], is being implemented in closed form into a finite element program. The effects of varying $[W_R]$ and $[W_C]$ are being investigated for real structures.

Much still remains to be done, however. Force identification techniques (given a verified model and its response, predict the forces acting upon it) and methods of identifying nonlinear models are two such areas. System identification can be applied to any type of analytical model which does not reproduce measured results. Using system identification in such diverse fields as economics, geophysics, medicine, and meteorology, as well as applying it to the wide variety of civil structures to increase their structural integrity and cost effectiveness, is a challenge for the years to come.

CONCLUSIONS

This paper has reviewed the various aspects of system identification: formulation of a pre-test analytical model, structural testing, data acquisition and reduction, eigenparameter estimation, and post-test model refinement. The process was performed on a simple structure so as to clarify the presentation. Application of system identification to large, complex structures was also discussed in the paper.

Bayesian estimation was used for model refinement. The pre-test analytical model of the test structure was updated to match measured data taken from static and dynamic tests. For the static model, modal displacements that were within about fifteen percent of measured displacements matched test data much more closely after model refinement; pre-test displacements differing significantly from measured displacements were changed little during Bayesian estimation. The dynamic model was updated using three different covariance matrices for response; the greater the uncertainty in resonance frequency, the less closely the updated model matched it.

In all cases, static and dynamic, Bayesian estimation yielded post-test models which matched the measured data significantly better than did the pre-test model.

ACKNOWLEDGEMENTS

The authors wish to acknowledge several of our colleagues for aiding us with this paper. Our appreciation goes to Michael W. Dobbs, Rebecca Ghadaifchian, William E. Gundy, Peter E. Rentz, and William B. Walton for their significant contributions.

REFERENCES

1. Ibáñez, P., and Spencer, R.B., "Experience With a Field Computerized Vibration Analysis System," SAE paper 791074, December, 1979.

2. Ibáñez, P., "Identification of Dynamic Parameters of Linear and Nonlinear Structural Models from Experimental Data," *Nuclear Engineering and Design*, Vol. 25, No. 1, 1973.

3. Hart, G.C., and Collins, J.D., "The Treatment of Randomness in Finite Element Modeling," SAE paper 700842, October, 1970.

4. "Review of Analytical and Experimental Techniques for Improving Structural Dynamic Models," prepared for the Pressure Vessel Research Committee by ANCO Engineers, Inc., January, 1977.

5. Isenberg, J., "Progressing from Least Squares to Bayesian Estimation," presented at the 1979 ASME Winter Annual Meeting in New York City, Dec. 2-7, 1979.

6. Collins, J.D., Hart, G.C., Hasselman, T.K., and Kennedy, B., "Statistical Identification of Structures," *AIAA Journal*, Vol. 12, No. 2, February, 1974.

7. Blakely, K.D., "Application of Bayesian Identification," term project for UCLA Engineering Extension course X413-6, June, 1978.

8. Gundy, W.E., et al, "A Comparison of Vibration Tests and Analysis on Nuclear Power Plant Structures and Piping," Transactions of the 4th International Conference on Structural Mechanics in Reactor Technology, held in San Francisco, Aug. 15-19, 1977.

9. Walton, W.B., "Bayesian Identification With Linear Constraint Relations Among the Parameters," ANCO Engineers, Inc. internal research and development project, May, 1978.

10. Walton, W.B., "Iterative Bayesian Identification--Integral Form of Criterion Function," ANCO Engineers, Inc. internal research and development project, May, 1980.

DAMPING OF A FLOOR SAMPLE

By Johann H. Rainer,[1] Affil. ASCE, and Gerald Pernica[1]

ABSTRACT

The dynamic behavior of floors was investigated by constructing a 28 ft × 29 ft (8534 × 8839 mm) single bay floor sample, consisting of a composite concrete slab and open-web steel joists. This paper is primarily concerned with an examination of the effect of four different test methods on the modal damping ratios determined for the fundamental mode of vibration of the floor sample. It was found that the commonly used heel impact test gave consistently higher values of modal damping than a new shaker impact test, white noise or steady state shaker excitations. The higher damping values were a result of the additional damping contributed to the floor sample by the person performing the heel impact. The effect on the fundamental damping ratio of the following structural and architectural modifications to the floor sample were also investigated: varying the edge conditions of the floor sample, adding a suspended ceiling or a gypsum board ceiling, and, bolting steel-neoprene damping strips to the bottom chord of the joists.

INTRODUCTION

Dynamic properties of floors have received considerable attention lately because many floors have exhibited unsatisfactory vibration behavior. More efficient designs and higher strength materials tend to make these components softer and lighter and thus more sensitive to vibration sources such as persons walking, mechanical equipment, and road and rail traffic. To arrive at acceptable designs or appropriate remedial measures, various criteria for floor vibrations in residential buildings have been proposed. A number of these have been summarized and compared by Murray. (6)

Damping is one of the most important parameters that affects floor vibrations. Highly damped floors are rarely cause for complaint; those with low damping are likely to be troublesome. In the former, vibrations caused by footsteps die out quickly, but in the latter they persist and can build up and propagate from the source to adjacent regions of a floor system. The amount of damping in a floor has to be determined by measurements since it cannot as yet be reliably computed or predicted. It is therefore important that test methods give consistent and reproducible results. In this investigation it was found that the values of damping determined from measurements can differ

---

[1]Research Officers, Noise and Vibration Section, Division of Building Research, National Research Council of Canada, Ottawa, Ontario K1A 0R6

greatly depending on the experimental procedure. Reasons and explanations for these differences are presented as well as damping ratios for several floor and ceiling configurations. The majority of the results only deal with the fundamental mode of the floor sample.

DESCRIPTION OF FLOOR SAMPLE AND INSTRUMENTATION

The experimental floor sample consisted of six 18-in. (457-mm) deep open web steel joists forming a composite construction with a 3-in. (76-mm) concrete slab. The sample had a total width of 29 ft (8840 mm) and a span of 28 ft (8534 mm). The shoes of the joists rested on the top flange of the end girders which in turn were supported by four corner steel columns. The columns were joined by edge beams parallel to the joists (Fig. 1). After record* No. 147, a 6- × 9-in. (152 × 229-mm) concrete stiffening beam was added on the underside of the slab normal to the joists, as indicated by the dotted lines in Fig. 1. After record No. 166, a 1-in. (25-mm) concrete topping was applied to the concrete slab. The joists were instrumented with strain gages and accelerometers were attached to the top of the slab. The accelerometer stations are shown in Fig. 2, labelled 1 to 28; not all were used for every test. Transducer signals were fed into signal conditioners and three 7-channel FM magnetic tape recorders. A portable spectrum analyzer, strip chart recorder, and oscilloscope proved invaluable in assessing the quality of the signals and in providing partial answers as the test progressed. An electrodynamic shaker** was the exciter. An external, band-limited white noise source, sinusoidal signal oscillator and a variable d-c supply provided the desired inputs and offset positions for the shaker armature. The locations where the floor sample was excited are shown in Fig. 2, labelled A to E.

METHODS OF DYNAMIC TESTING

Two impulse methods and two continuous excitations were employed to test the floor sample. Heel impact and shaker impact were the impulse methods; continuous excitations were white noise and steady-state shaker tests.

Heel Impact. - In the heel impact test, a person weighing approximately 170 lb. (77 kg) shifts his weight onto the balls of his feet (thus raising his heels approximately 2 in. (50 mm) off the floor) and then suddenly shifts the body weight onto his heels and impacts the floor. This test has been widely used in floor vibration studies and was incorporated in floor vibration criteria (1, 4) because it is simple, relatively reproducible, and readily available. The impulse has a roughly triangular shape and lasts 50 ms. (5) The initial peak acceleration amplitude is obtained from the resulting free vibrations of the floor and the damping value is calculated from the rate of decay. Filtering is usually required to eliminate vibration components other than those for the mode of vibration that is of interest. For the

---

* A sequential numbering scheme was used during testing.

** Type "Electro-seis" Model 113, manufactured by Acoustic Power Systems, Inc., Anaheim, California.

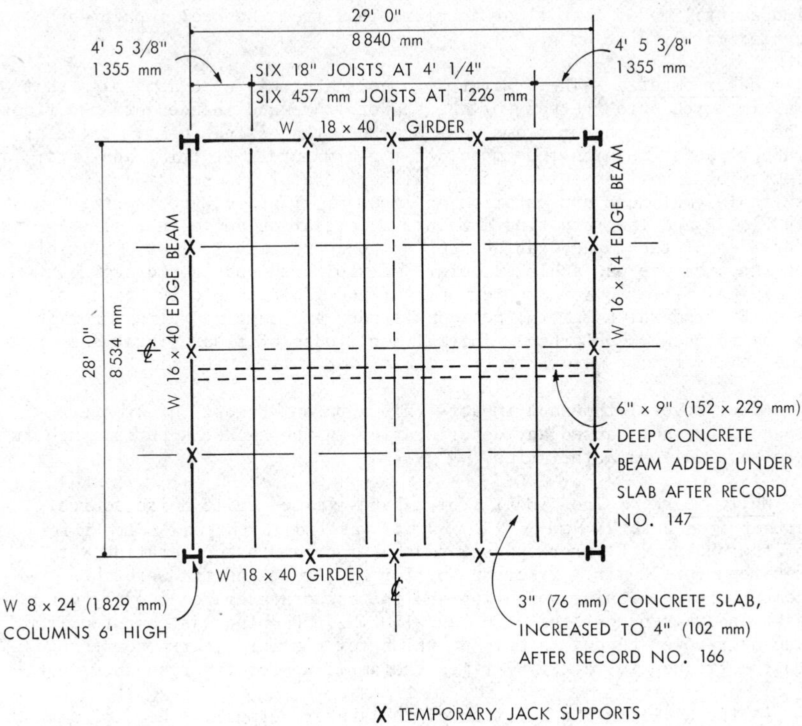

29' 0"
8 840 mm
4' 5 3/8"
1 355 mm
SIX 18" JOISTS AT 4' 1/4"
SIX 457 mm JOISTS AT 1 226 mm
4' 5 3/8"
1 355 mm
W 18 x 40 GIRDER
W 16 x 40 EDGE BEAM
W 16 x 14 EDGE BEAM
28' 0"
8 534 mm
℄
6" x 9" (152 x 229 mm)
DEEP CONCRETE
BEAM ADDED UNDER
SLAB AFTER RECORD
NO. 147
W 18 x 40 GIRDER
℄
W 8 x 24 (1 829 mm)
COLUMNS 6' HIGH
3" (76 mm) CONCRETE SLAB,
INCREASED TO 4" (102 mm)
AFTER RECORD NO. 166

X TEMPORARY JACK SUPPORTS

FIG. 1—Layout of floor sample

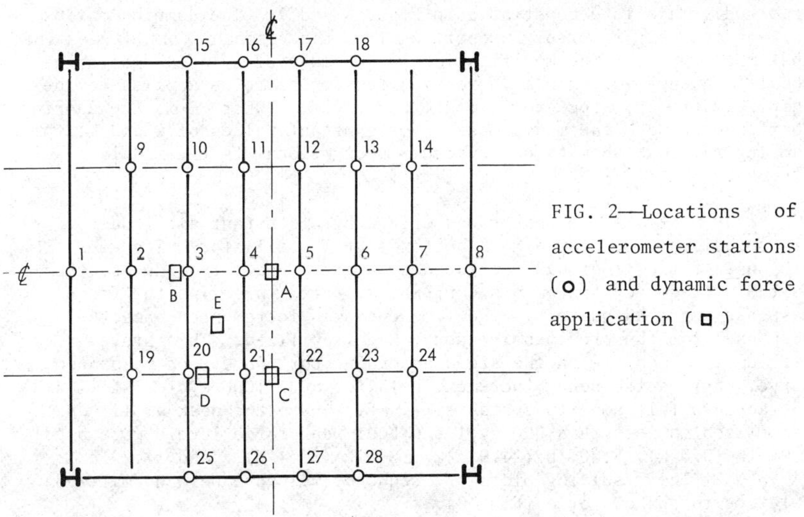

FIG. 2—Locations of accelerometer stations (○) and dynamic force application (□)

fundamental mode, peak floor accelerations from the heel impact were approximately 3 to 4% g.

Shaker Impact. - The armature of the shaker was used as a mass that was impacted onto the base of the shaker frame and thence onto the floor. The base of the shaker frame was covered with a ¼-in. (6-mm) cellular rubber pad to cushion the impact of the armature. By trial-and-error, a rectangular voltage pulse from a pulse generator was selected to give suitable amplitude and duration of impulse. (The voltage input is shown in Fig. 3 and the resulting armature acceleration pulse in Fig. 4.) The duration of the impact can be seen to be approximately 7 ms. Lift-off of the armature was achieved before multiple rebounds could occur. A typical filtered impulse response of the floor sample is shown in Fig. 5, from which initial peak amplitudes and damping ratios from the decay rates were obtained. Initial amplitudes of floor vibrations from the shaker impact were 1 to 1.5% g for the fundamental mode.

After several thousand impacts this apparent "abuse" of armature impacts has not caused any deterioration in the shaker's performance or its physical characteristics.

White Noise. - The signal from a band-limited white noise source, either from 0 to 25 Hz, or 0 to 50 Hz, was fed to the power amplifier of the shaker. Floor response and shaker armature acceleration were recorded for 20 min. Discrete Fourier amplitude spectra were then computed with Hanning weighting and the spectra averaged over 10 min. with analysis bandwidths of 60 and 150 mHz. Damping was computed from the half-power bandwidth method, which for Fourier spectra corresponds to the frequencies at 0.707 times the amplitude of the resonance peak.

As the acceleration spectrum of the shaker output was held constant over a wide frequency range, relative spectral peak amplitudes also provide a measure of change in damping as floor modifications are examined. This is demonstrated in Figs. 6 and 7. The damping ratios for any mode are inversely proportional to the resonance amplitudes for that mode. But if the modification is accompanied by a change in the modal frequency, ω, then a correction factor has to be applied to the damping ratio, β, since for a mode, $\beta = C/(2M\omega)$, where C is the damping coefficient and M the modal mass. For constant values of C and M, the damping ratio is seen to be inversely proportional to the modal frequency.

Steady State. - The shaker was driven by a constant amplitude sinusoidal voltage from a push-button signal oscillator. The vibration response of the floor was monitored on a strip chart recorder so that suitable frequency increments could be selected to define the frequency response curve adequately. The frequency was incremented once the response from the previous frequency had stabilized. This proved difficult to achieve on the steep ascending portion of the resonance curve, where a frequency increment held for over 1000 cycles still did not achieve full amplitude stability. Elsewhere the peak amplitudes stabilized in less than 300 cycles. Four peak force levels were used: 2.02 lb (9.0 N), 5.60 lb (24.9 N), 8.84 lb (39.4 N), and 11.2 lb (49.8 N). The resulting plots of frequency response were normalized relative to shaker force (Fig. 8).

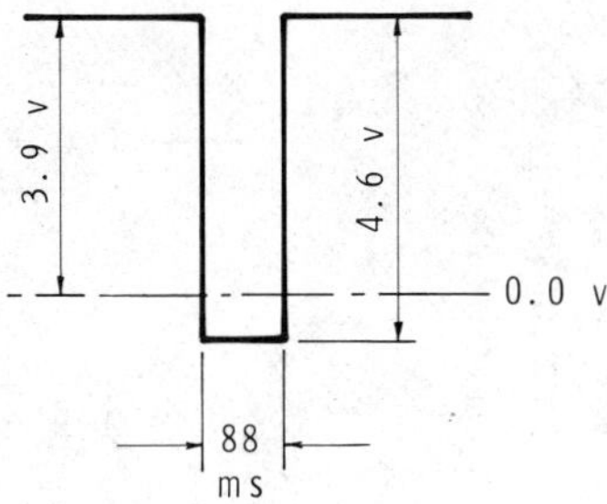

FIG. 3.—Voltage pulse to shaker

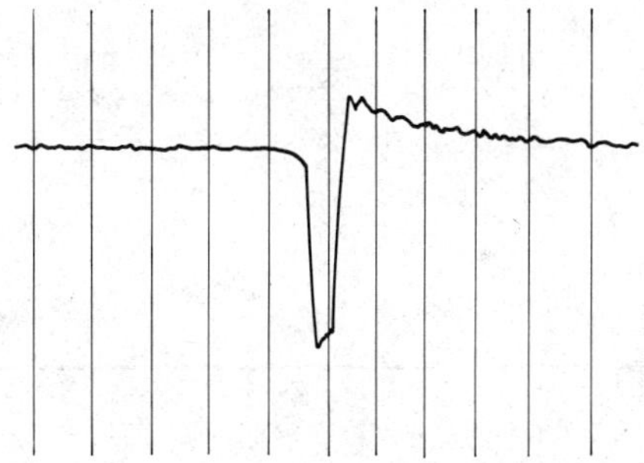

FIG. 4.—Measured acceleration of shaker armature (0.01 s/division)

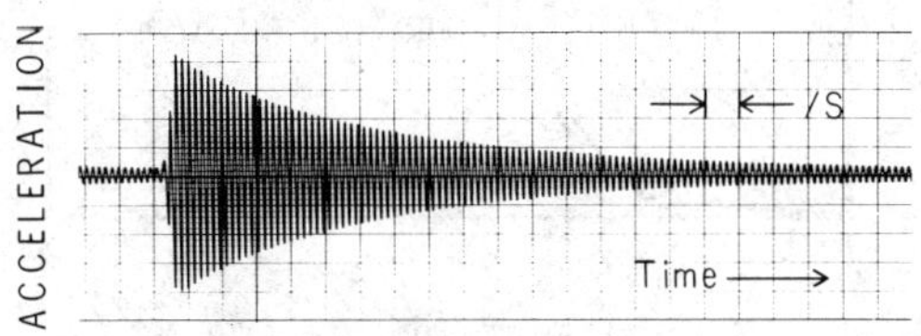

FIG. 5.—Typical impulse decay curve (Station 3, shaker impact at B, record 177)

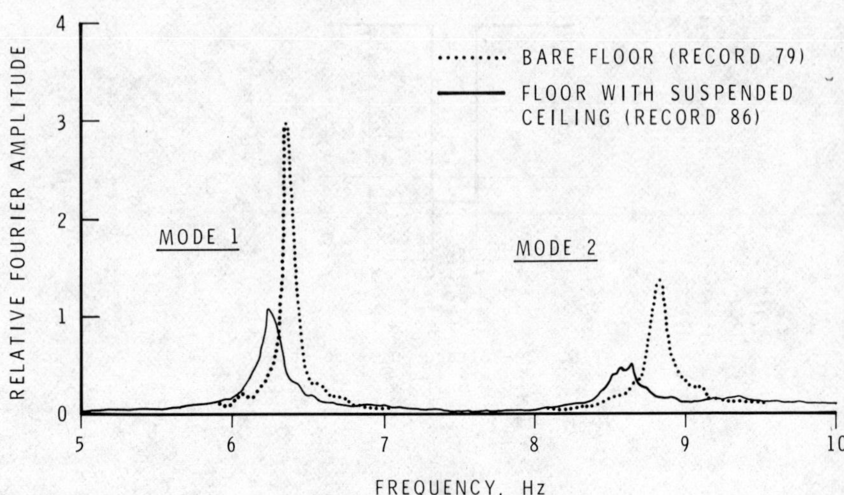

FIG. 6.—Effect of suspended ceiling on Fourier amplitude spectra of floor response (Station 3, white noise shaker excitation at E, end girders fully supported)

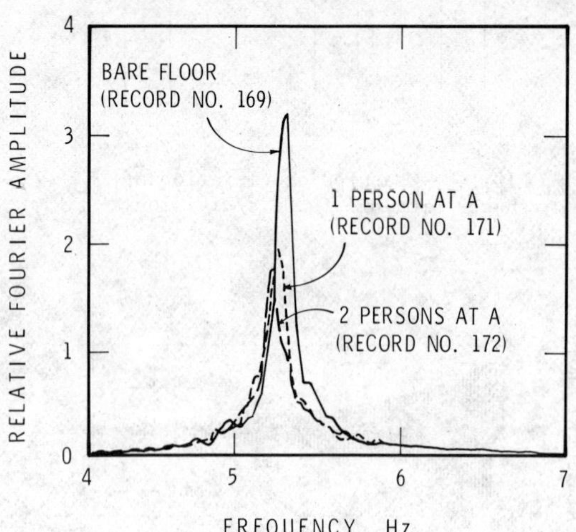

FIG. 7.—Effect of people on Fourier amplitude spectra of floor response (Station 3, white noise shaker excitation at (A)

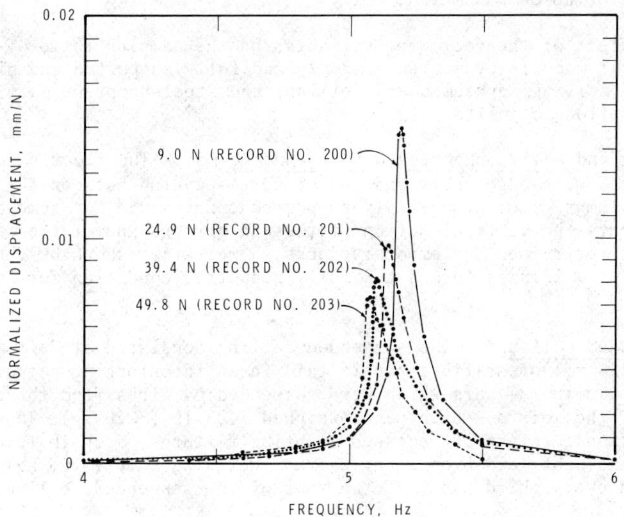

FIG. 8.—Normalized frequency-response curves for increasing force levels

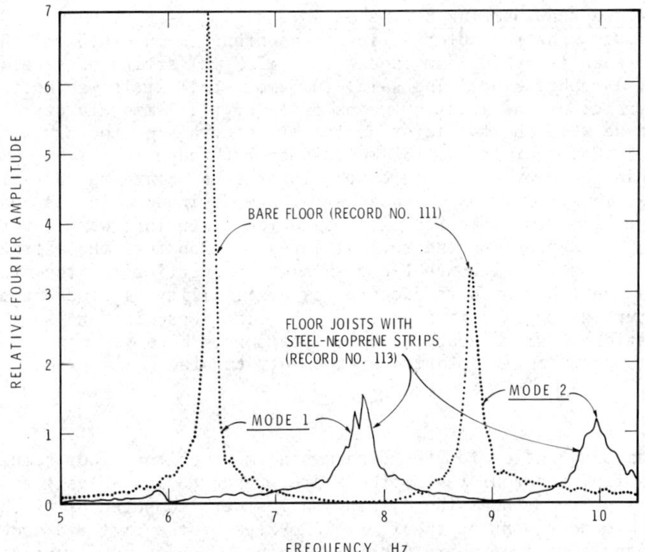

FIG. 9.—Effect of steel-neoprene strips on Fourier amplitude spectra of floor response (Station 3, white noise shaker excitation at E, girders fully supported

VARIATION OF FLOOR PARAMETERS

The effect of the following variables on the damping ratio of the fundamental mode is presented: simply and fully supported end girders; suspended ceiling; gypsum board ceiling; and steel-neoprene damping strips applied to joists.

Simply and Fully Supported Girders and Beams. - The floor sample was constructed with edge beams and end girders spanning between the four corner columns, i.e., the simply supported condition. For some of the measurements the vertical motion of the end girders and/or the edge beams was restrained by temporary jacks, 3 per side, shown by the symbol X in Fig. 1. This is termed the fully supported condition for beams or girders.

Suspended Ceiling. - The suspended ceiling consisted of 2 ft × 4 ft (610 × 1220 mm) press-fibre panels laid in an interlocking network of inverted aluminum T bars which were suspended by wires from the lower chords of the joists. Each panel weighed 7.25 lb (3.29 kg). To simulate the additional weight of recessed lighting fixtures, a 20-lb (9.07-kg) bag of lead shot laid over 2 angle-irons weighing 8 lb (3.63 kg) was placed on every third panel of two rows of the suspended ceiling.

Gypsum Board Ceiling. - Sheet metal channels were wired tightly onto the underside of the joists and sheets of ½-in. (12.7-mm) gypsum board 4 ft × 8 ft (1220 × 2440 mm) were screwed onto the channels. The ceiling terminated at the beams and girders.

Steel-neoprene Damping Strips on Joists. - A 3-in. (76-mm) wide sandwich was constructed of strips of neoprene, ¼ in. (6.3 mm) thick, glued between ¼-in. (6.3-mm) and ½-in. (12.7-mm) strips of steel by means of a rubber glue. The ¼-in. (6.3-mm) steel strip was bolted onto the underside of the joist by means of ½-in. (12.7-mm) diameter bolts inserted between the two joist angles at intervals of roughly 2 ft (610 mm). Holes of 1¼-in. (32-mm) diameter through the lower steel strip and the neoprene permitted the lower steel plate to restrain the neoprene. Thus longitudinal shear strains were induced in the neoprene core as the joists deflected. The sandwich dimensions were determined according to a previous theoretical investigation into the action of damping strips (2) although the mechanical properties for the present sandwich construction were dictated by availability of stock items. The manufacture of these strips proved to be time consuming and some local delamination occurred near the bolt locations, where the upper steel plate was permanently deformed by the bolt tension.

RESULTS

The damping ratios for the fundamental mode of the floor sample as obtained from the four test methods are presented in Tables 1 and 2. For the impulse methods the damping ratios were computed using the logarithmic decrement of the first 20 cycles of the filtered accelerometer signals, except where otherwise noted. For the white noise and steady state results the half-power bandwidth method was employed. The numbers used to identify the various diagrams and sets of damping values in the tables are the record numbers associated with the measurement

TABLE 1.—Variation of Damping Ratios with Number of People and with Experimental Method

Description of Floor and Loading	Natural Frequency, Mode 1 Hz	Damping ratio, δ, % of critical							
		From impulse decay, $\beta = \frac{1}{2\pi n} \ell n \frac{Xo}{Xn}$				From half-power bandwidth, $\beta = \frac{\Delta f}{2f}$			
		Shaker Impact		Heel Impact		White Noise		Steady State	
		Record Number, Location of Force	δ_S %	Record Number, Location of Force	δ_H %	Record Number, Location of Force	δ_W %	Record Number, Location of Force	δ_{SS} %
1. 4" slab, 6 × 9" beam, No temporary supports									
a) bare floor	5.3	179, A	0.60			169, A	0.46	200, A 201, A 202, A 203, A	0.53 0.63 0.69 0.89
b) 1 person at A		179, A	0.93	166, A	0.88	171, A	0.60		
c) 2 persons at A		179, A	1.18	166, A	1.32	172, A	0.81		
2. 3" slab, 6 × 9" beam, No temporary supports									
a) bare floor	5.4	164, A	0.58			150, B 147, E	0.27 0.40		
b) 1 person at A		164, A	1.10	165, A	0.99				
c) 2 persons at A		164, A	1.38	165, A	1.75				
3. 3" slab, 6 × 9" beam, girders and beams fully supported									
a) bare floor	6.7	159, A	0.49						
b) 1 person at A		160, A	1.48	155, A	1.42				
c) 2 persons at A		160, A	2.92*						
d) 3 persons at A		160, A	3.76*						

* 10 cycles

TABLE 2.—Damping Ratios for Various Floor Modifications, 3 in. (76.2 mm) Slab

Description of Floor and Modifications	Natural Frequency, Mode 1	Damping ratio, δ, % of critical					
		Shaker Impact		Heel Impact		White Noise, from half-power bandwidth*	
	Hz	Record Number, Location of Force	δ_S %	Record Number, Location of Force	δ_H %	Record Number, Location of Force	δ_W %
A. Simply supported beams and girders							
a) floor without modifications	5.41		(0.4)**	135, A	0.91	146, E	0.37
b) suspended ceiling with "lights"	5.30	99, E	0.62	98, A B C D	1.10 0.90 0.88 0.85	97, E	0.47
c) gypsum board ceiling	5.38	104, E	0.44	103, A B C D	0.98 0.97 0.88 0.81	100, E	0.38
d) steel-neoprene damping strips	5.8 } double 6.2 } peak	121, E	0.76	119, A	0.97	118, E	0.65 } 0.77 }
B. End girders fully supported, except as noted							
a) floor without modifications							
i) end girders and edge beams fully supported	6.4	76, E	0.37	77, A B C D	1.19 0.85 0.82 0.72		
ii) only end girders fully supported	6.38		(0.4)**	112, A B C D	1.25 0.95 0.86 0.65	111, E	0.20
b) suspended ceiling with "lights"	6.20	92, E	0.71	93, A B C D	1.67 1.35 1.17 1.06	94, E	0.58
c) gypsum board ceiling	6.36	109, E	0.44	108, A B C D	1.30 1.18 1.07 0.96	106, E	0.40
d) steel-neoprene damping strips	7.8	115, E	1.0	114, A C	1.74 1.65	113, E	0.95

* Analysis bandwidth = 0.060 Hz

** Extrapolated from comparisons between other cases of shaker impacts and heel impact tests

procedure; the letter symbols refer to the location of dynamic force application (Fig. 2).

<u>Damping for Simply and Fully Supported Girders.</u> - From the results given in Table 2A for the simply supported girders, and Table 2B for the fully supported girders, it can be seen that except for one white noise comparison (record No. 111 in Table 2B vs. record No. 146 in Table 2A) damping ratios for the simply supported case are lower than for the fully supported one. This is contrary to what one might expect on a cursory examination of damping in a single-degree-of-freedom equivalent system. Since $\beta = C/(2M\omega)$, one might expect β to decrease as the frequency, ω, is increased by supporting the girders. However, the increase in floor frequency is also accompanied by changes in M and C. The equivalent mass, M, is reduced due to changes in the mode shape and the damping coefficient C is changed since damping mechanisms that are associated with the girder motions are modified. Prediction of damping ratios corresponding to changes in structural configurations should therefore be approached cautiously. In most cases prediction of damping ratios may not be possible since changes in the damping coefficient C are not measurable or quantifiable.

<u>Comparisons of Damping Ratios from Heel Impact, Shaker Impact and White Noise Excitation.</u> - Damping ratios from shaker impact and heel impact tests differed substantially for the same floor configuration (Table 1). This was recognized to be the result of the additional damping contributed to the floor sample by the person performing the heel impact. This is illustrated by comparing the damping ratios from the shaker impacts, first on the empty floor, then with 1 person, 2 persons, and sometimes 3 persons present. For the simply supported case the percentage of critical damping increased by approximately 0.3 per person for the 4-in. (102-mm) slab, 0.4 per person for the 3-in. (76-mm) slab, and by 1.0 per person for the 3-in. (76-mm) slab with supported girders. This pattern of increased damping with increasing number of persons is evident from the heel impact tests. With the same number of people on the floor during shaker impact and heel impact tests damping ratios were comparable. A similar difference in damping values for the heel impact and steady state shaker tests was found in a field test of a floor (7).

Another variable that affects the damping ratio as obtained from the heel impact is the location of the person performing the test. Tables 2A and 2B show that the damping ratios from the heel impact decrease considerably as the impactor moves from locations A through D. This can be ascribed to a decreased effectiveness of the human "damper" as the modal amplitude of vibration decreases. It should be emphasized that this is not a phenomenon of the amplitude dependent damping mechanism of the floor, but rather one of amplitude dependence of the human "damper" as it changes location relative to the modal deflection pattern of the floor. The influence of people on the damping level of floors has also been noted and described by Lenzen. (4)

The effect of humans on the damping level of the floor is also illustrated by the white noise results where a decrease in the resonance peak indicates a proportional increase in the damping ratio. In Fig. 7 the ratio of decrease of the peaks is 1:0.64:0.58 showing a 56% increase in damping for 1 person and a further 10% increase for the second person. The increase in damping ratio for the second person is smaller for white noise than for the impact tests. One can attribute this to the fact that a person finds it difficult to remain motionless over an extended period of time. Consequently some floor vibration will be induced by involuntary movements of the subjects on the floor.

Steady State. - The results from the steady state tests presented in Table 1 and in the frequency response curves in Fig. 8 indicate a trend of increasing damping ratios with increasing amplitude of vibration. The frequency response curves in Fig. 8 also show a reduction in resonance frequency of 0.107 Hz as the vibration amplitudes at resonance increase from 0.0054 to 0.0148 in. (0.137 to 0.375 mm). The floor thus exhibits the characteristics of a softening spring system. Such a system can exhibit stability problems in the ascending portion of the resonance curve, giving rapid increases in amplitude for very small increases in frequency. (3) The descending portion of the frequency response curve is relatively higher than the ascending portion, also a typical characteristic of softening spring systems. Given such non-linear behavior, the half-power bandwidth method for calculating damping ratios is not strictly applicable. For small non-linearities, however, this should not introduce large errors and in Table 1 the damping ratios from the frequency response curves are seen to be comparable to those obtained from the shaker impact method. Whether this non-linear behavior is present in the other floor configurations is not certain since only this one set of steady state tests was carried out.

White Noise Excitation. - Band-limited white noise, when applied to the shaker, produced a random-type excitation with nearly constant force output over the selected frequency range. Natural frequencies and mode shapes were readily determined from the Fourier spectra of the resulting response signals. Unfortunately, it was not as easy to obtain reliable damping ratios from the half-power bandwidth method for this lightly damped system.

As seen in Tables 1 and 2, damping ratios from white noise shaker tests are not fully consistent with the shaker impact or heel impact tests. The overall trends of an increase in modal damping with an increase in the number of people on the floor, and an increase in damping with the steel-neoprene strips are present, but the degree of consistency is not sufficient to permit their use in quantitative comparisons of test methods or structural configurations.

For consistent results, the excitation needs to be stationary, which means a relatively long recording time. The analysis should then be carried out to sufficient resolution so that the bandwidth of the Fourier transform process is a small fraction of the half-power bandwidth used to calculate the damping ratio. A high resolution analysis may, however, accentuate small local nonuniformities in spectral content of the excitation and thereby introduce a distortion in the Fourier spectrum of the response.

Damping for Different Ceiling Types. - Damping ratios for the floor sample with two different types of ceiling are presented in Table 2A for the simply supported end girders and in Table 2B for the fully supported end girders. For both support conditions the suspended ceiling produced increases of 0.2 to 0.3 in the percentage of critical damping compared to that of the bare floor. The gypsum board ceiling, on the other hand, showed practically no increase in damping. Two possible causes suggest themselves. First, the large number of loose panels inserted into the T-frame of the suspended ceiling provides a horizontal source of energy dissipation along the edges of the T-bars as the floor vibrates up and down. Second, the suspended ceiling may act like a tuned damper and thereby reduce the floor motion. The more rigidly installed gypsum boards do not appear to offer the same opportuities for energy dissipation. The small increase in fundamental frequency of the floor sample indicates that the gypsum board actually contributes to the floor stiffness and as a consequence relatively less energy is dissipated by friction. The relative effectiveness of these two types of ceiling is opposite to that reported by Lenzen. (4) This would indicate that details of installation and construction of ceilings play a role in the mobilization of damping.

Steel-neoprene Damping Strips. - The steel-neoprene sandwich strips bolted to the lower chord of the steel joists increased the damping ratio of the floor sample by about 0.2% of critical for simply supported end girders and by about 0.5% of critical for the fully supported ones (Tables 2A and 2B). In addition, the steel-neoprene strips increased the fundamental frequency by about 20% (Tables 1 and 2). However, these changes in characteristics of the floor sample were still not sufficient to result in acceptable vibration behavior under footstep excitation. The effectiveness of these particular steel-neoprene sandwich strips has thus been somewhat disappointing.

For the white noise tests, Fig. 9 shows that the strips reduced the root-mean-square acceleration by a factor of 4.7 for the first mode, and 2.0 for the second mode.

DISCUSSION

This floor sample was constructed to investigate the effect of various parameters on the dynamic characteristics of floors. Because the physical dimensions of the sample were limited, its behavior is not representative of multi-bay floor systems in buildings where adjacent panels and different support conditions yield additional sources of damping. The low level of damping of this sample, however, provided an opportunity to detect relatively small changes in damping with changes in floor configurations and experimental methods.

Even though the heel impact test does not yield the damping characteristics of an unoccupied structural floor system, it is the simplest and probably most realistic test method to be used in tests that involve the reaction of people to floor vibrations, since at least one person has to be present on the floor to perceive the vibrations. However, for the application of floor vibration data to situations other than human perceptibility, such as placement of sensitive instrumentation or machinery, the heel impact test may overestimate the amount of damping present in lightly damped floors and thus give an erroneous assessment.

The non-linear stiffness and damping characteristics of the floor sample were most distinctly portrayed by the steady-state tests. Non-linear effects were not discernible from the Fourier spectra obtained from the shaker impact, heel impact, or white noise tests. It should be noted, however, that for these three methods no special attempt was made to investigate non-linear behavior by varying the amplitude of floor excitation.

The non-linear behavior in this floor sample may in part be ascribed to the structural modifications made to the original floor, namely the addition of the transverse 6 in. × 9 in. (152 × 229 mm) concrete beam and the 1-in. (25-mm) concrete topping. Both beam and topping showed some cracks which would contribute to non-linear vibration behavior.

CONCLUSIONS

In a lightly damped system the damping ratios obtained from the heel impact test are larger than those from a shaker impact test or other shaker tests due to the additional damping contributed by the person performing the heel impact test. Increasing the number of persons present on the floor (up to 3 investigated here) results in a proportional increase in the measured damping ratios.

The damping ratio obtained from the heel impact test varies with the location of the impactor, being largest when the floor was impacted at the point of largest modal amplitude and decreasing as the modal amplitudes corresponding to the impact location decreased.

For the floor sample with structural modifications, the steady-state frequency response curves indicated an increase in damping ratios with increasing amplitudes of floor vibration. The same conclusion was reached when the damping ratio was evaluated by the half-power bandwidth method. The frequency response curves also showed a reduction in resonance frequency with increasing amplitude of floor vibration and, together with the shape of the frequency response curves, indicated a softening spring system.

With this floor sample, a suspended ceiling was found to provide more damping in the fundamental mode than a gypsum board ceiling. Higher levels of damping were achieved with steel-neoprene sandwich strips but neither the ceilings nor the sandwich strips provided what could be considered large increases of damping over those exhibited by the original bare floor.

ACKNOWLEDGEMENTS

This paper describes results obtained from a cooperative testing program involving Public Works Canada and the Division of Building Research of the National Research Council of Canada. The authors wish to thank all contributors for their help in this study.

APPENDIX I. - REFERENCES

1. Allen, D.E., and Rainer, J.H., "Vibration Criteria for Long-Span Floors," Canadian Journal of Civil Engineering, Vol. 3, No. 2, June 1976, pp. 165-173.

2. Farah, A., Ibrahim, I.M., and Green, R., "Damping of Floor Vibrations by Constrained Viscoelastic Layers," Canadian Journal of Civil Engineering, Vol. 4, No. 4, Dec., 1977, pp. 405-411.

3. Jacobsen, L.S., and Ayre, R.S., Engineering Vibrations, McGraw-Hill Book Co., Inc., New York, 1958, p. 286.

4. Lenzen, K.H., "Vibration of Steel Joist-Concrete Slab Floors," AISC Engineering Journal, Vol. 3, No. 3, July 1966, pp. 133-136.

5. Lenzen, K.H., and Murray, T.M., "Vibration of Steel Beam Concrete Slab Floor Systems," Report No. 29, University of Kansas, Lawrence, KA, 1969.

6. Murray, T.M., "Acceptability Criterion for Occupant-Induced Floor Vibrations," Sound and Vibration, Vol. 13, No. 11, Nov., 1979, pp. 24-30.

7. Rainer, J.H., "Dynamic Tests on a Steel-Joist Concrete-Slab Floor," Canadian Journal of Civil Engineering, Vol. 7, No. 2, June 1980, pp. 213-224.

ASEISMIC DESIGN OF A 31-STORY FRAME-WALL BUILDING

by Mark Fintel[1], M. ASCE, and S. K. Ghosh[2], M. ASCE

SUMMARY

Earthquake-resistant design of a 31-story reinforced concrete building is carried out on the basis of inelastic response history analyses under carefully selected input motions. The design approach makes it possible to predetermine the sequence of plastification, provide ductility details only where required, and balance the strength and ductility requirements of individual members. Efficiency, economy and desired structural performance are achieved as a result.

INTRODUCTION

For economic reasons, it is not practical to design ordinary building structures with so much strength that they will remain elastic throughout the strongest earthquakes to which they may be subjected. Structures may be required to undergo considerable amounts of inelastic deformation in earthquakes of moderate to intermediate intensity. Modern building codes prescribe designs based on this anticipated inelastic behavior. However, structural members are designed to resist internal forces which are determined by elastically analyzing a structure under prescribed static loads applied to the various mass locations. The forces so determined may be quite different from those resulting from an actual inelastic earthquake response of the structure. Also, the distribution and magnitude of inelastic deformations in individual structural members cannot be determined through elastic analysis under code-specified static loads. As a result, ductility or inelastic deformability must be provided throughout the entire structure, although inelasticity may actually be confined to certain levels and locations.

It has recently become possible to perform a realistic analysis of the earthquake response of multistory concrete and steel structures at a reasonable cost, using two-dimensional dynamic inelastic (response history) analysis computer programs which incorporate proper hysteretic characteristics of reinforced concrete and steel

---

[1]Director, [2]Principal Structural Engineer, Advanced Engineering Services, Portland Cement Association, Skokie, IL 60077.

members. Designs based on such analyses make it possible to provide ductility details only where required, and to strike the most desirable balance between the strength and ductility requirements of individual members. The sequence in which the various structural members become inelastic can also be predetermined. For example, the beams may be made to yield before the columns.

The inelastic design approach mentioned above represents a significant departure from the empirical code approach. The suggested procedure uses carefully selected earthquake accelerograms as loading, and dynamic inelastic response history analyses to determine member forces and deformations. The procedure is applied in this paper to a 31-story reinforced concrete frame-shear wall building located in an area of considerable seismicity. The application of the inelastic approach to the analysis and design of the structure results in efficiency, economy and desired structural performance.

STRUCTURE, MODELING AND ANALYSIS

The Building

The building considered is circular in plan, with a diameter of approximately 110 ft (33.5 m), and an area of approximately 9,500 sq ft (885.2 m^2) per floor (Fig. 1a). The building has a total height of 440 ft (134 m) over 31 stories (Fig. 1b). It is located in an area with a seismicity equivalent to that of Zone 3 of the Uniform Building Code[1] (UBC) four-zone seismic risk classification.

The Lateral Load Resisting System

The lateral load resisting system in the E-W direction (the uncoupled direction) consists of the core walls bending about their major axes; the walls are connected at every floor level to the columns on each side through the main beams. In the orthogonal (N-S) direction, the two C-segments of the core walls are connected through two coupling beams at every floor level, and form a box which is connected on both sides to the peripheral columns through the main beams. This orthogonal lateral load resisting system consists of the coupled wall segments bending about their minor axes in interaction with the peripheral columns. The analysis of the building in the uncoupled (E-W) direction only is considered in this paper.

Modeling of Structure for Static and Dynamic Analyses

The two columns on each side of the central core are lumped into a single column, and the two rows of main beams are lumped into a single row. The resulting model has three column lines. The inner line represents the central core; the two outer column lines represent the lumped columns at the two ends. The links between the outer column lines and the inner line are the lumped main beams.

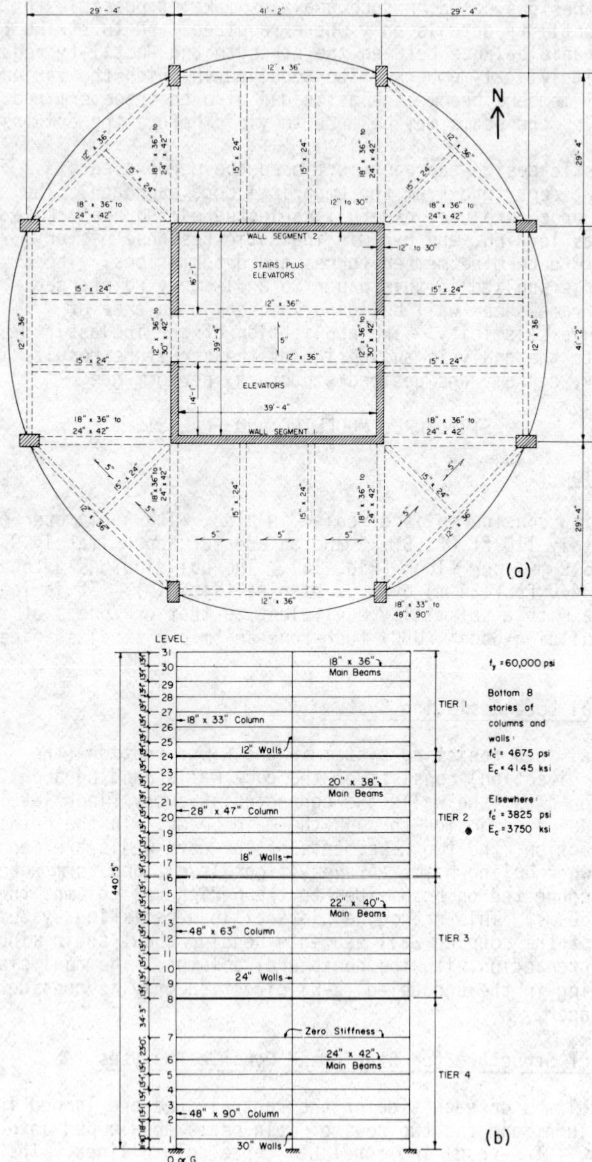

Fig. 1: (a) Floor plan,
(b) Sectional and analytical model of building studied

In the dynamic analyses the masses were concentrated at every floor level. Each node had three degrees of freedom--horizontal translation (all the nodes on the same floor were constrained to undergo the same horizontal translation), vertical translation and rotation.

Static Analyses, Periods, and Mode Shapes

Elastic static analysis of the building under UBC-79[1] 20 psf Zone (20 psf = 0.96 kPa) wind forces was carried out using a standard computer program. The computed wind deflection at the top of the building was 1.42 in. (36 mm) in the uncoupled direction. This results in a wind drift ratio of 1/3730, which indicates that the structure is quite rigid.

Elastic static analysis of the structure under UBC[1] Zone 3 equivalent static seismic forces (with K = 1.0) was also carried out. The results are presented later in this paper.

The natural undamped periods and mode shapes of the analytical model of the structure were determined using a standard computer program. The structure in the direction of analysis was found to have first and second mode periods of 1.995 and 0.391 seconds, respectively.

Dynamic Analysis

Dynamic inelastic response history analysis, by the computer program DRAIN-2D[2], was used to determine the amount and distribution of inelastic deformations in the various structural members. Simplified dynamic analysis by modal superposition, used in conjunction with elastic analysis, cannot provide the needed information.

DRAIN-2D is a general purpose program for the dynamic analysis of plane inelastic structures. The dynamic response is determined using step-by-step integration, assuming a constant response acceleration during each time step.

Program DRAIN-2D accounts for inelastic effects by allowing the formation of concentrated "point hinges" at the ends of elements where the moments equal or exceed the specified yield moments. The moment vs. end rotation characteristics of elements are defined in terms of a basic bilinear relationship which develops into a hysteretic loop with unloading and reloading stiffnesses decreasing in loading cycles subsequent to yielding. The modified Takeda model[3] (Fig. 2a), developed for reinforced concrete, has been utilized in the program to represent the above characteristics.

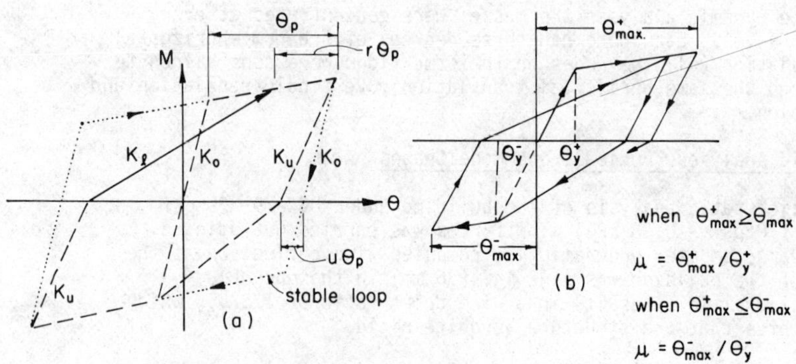

Fig. 2: (a) Takeda model
(b) Definition of ductility

Definition of Ductility

The ductility discussed in this paper is based on rotations over the hinging regions of structural members. The hinging region is the length of a member over which the bending moments, while on a loading cycle, exceed the specified yield moments. Rotational ductility is defined as:

$$\mu = \theta_{max}/\theta_y$$

where θ_{max} is the maximum rotation, as the hinging region goes through many cycles of response; and θ_y is the rotation corresponding to yielding, while deforming in the same direction in which θ_{max} occurs (Fig. 2b). Yielding is defined as corresponding to the intersection between the initial elastic and the post-yield branches of the moment-rotation diagram of the hinging region.

In the case of wide columns and walls, the length of the hinging region is roughly equal to the width of the structural element, and may extend over one full story or even several stories. In the case of beams, the hinging length is usually equal to the depth of the beam, and is considerably less than the span length, except in the case of very short and deep beams. It should be noted that in beams connecting walls and/or columns subjected to lateral movement, two hinges rotating in opposite directions may form at the two ends of a beam.

INPUT MOTION

The principal ground motion characteristics affecting dynamic structural response are intensity, duration and frequency content. Intensity provides a characteristic measure of the amplitude of the acceleration pulses in a record. Duration refers to the length of the record during which relatively large amplitude pulses occur, with due allowance made for a reasonable build-up time. The frequency characteristics of a given ground motion have to do with the relative energy content of the different component waves (having different frequencies) which make up the motion.

A geotechnical investigation of the site indicated the design earthquake intensity* to be two-thirds that of the 1940 El Centro N-S record.

Earlier studies[5] have established that a 10-second duration of strong ground motion is long enough to determine the envelope of response quantities for design purposes.

To select the frequency content of the input motion to be used in dynamic analysis, an analytical investigation had to be carried out. The structure considered in this study was found by computation to have an initial fundamental period of about 2 seconds in the uncoupled direction, as indicated earlier. Four input motions were selected as being potentially critical for structures in this period range. Two of these accelerograms - El Centro, 1940, N-S and E-W components - have broad-band response spectra ascending beyond the period value of interest (Fig. 3). The other two

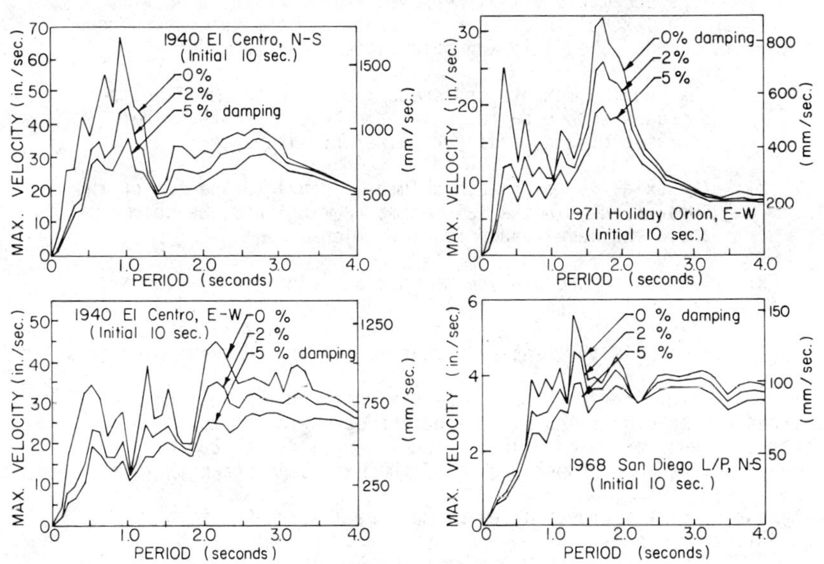

Fig. 3: Relative velocity response spectra of input motions considered

* Measured in terms of spectrum intensity[4] or the area under the 5%-damped relative velocity response spectrum between periods of 0.1 and 3.0 seconds.

accelerograms - Holiday Inn, Orion, 1971, E-W component and San Diego Light and Power Building, 1968, N-S component - have their spectra peaking close to the above period value (Fig. 3).

Inelastic dynamic analyses of the structure using DRAIN-2D[2] were carried out under all four of the input motions mentioned above, each normalized to the design earthquake intensity. In these analyses, the columns and walls were kept elastic throughout their seismic response. The yield levels for the main beams were chosen close to the maximum factored elastic moments caused in these beams by UBC Zone 3 static seismic forces. Some of these beams yielded slightly under one or more of the input motions. The results of these analyses are presented in Fig. 4. The figure is self-explanatory, and clearly indicates that El Centro, 1940, E-W component is the critical input motion for the structure in the uncoupled direction. The first 10 seconds of this motion, normalized to the intensity given above, is referred to as the design earthquake in the remainder of this paper.

EFFECTS OF MEMBER STRENGTH OPTIMIZATION ON INTERNAL FORCES

Figure 5 is a schematic visual representation of the internal forces due to lateral loads in a structural system such as the one being discussed. The following observations can be made:

1. The bending moments and shear forces in the main beams decrease with a decrease in main beam strength. The corresponding ductility requirements increase.

2. The axial force in each column is equal to the sum of the shear forces in the main beams framing into the column above the level under consideration.

3. The sum of the bending moments at column and wall bases is equal to the external bending moment minus the couple $T_c \times L_c$, where $T_c = C_c$ is the column axial force, and L_c is the center-to-center distance between columns.

It is apparent from the above observations that the internal forces can be controlled by varying the strengths of the main beams. A desired combination of the internal forces can be achieved through a proper choice of the main beam strength levels.

The choice of the proper level of main beam strength is an optimization process to accomplish the following:

1. Reduce the bending moments in the columns and walls to avoid inefficiency and construction difficulties;

2. Limit or avoid net tension in the columns;

3. Limit axial compression in the columns to levels below the desired maximum;

MAX. BENDING MOMENTS IN MAIN BEAMS

LEVEL
HOLIDAY ORION, 1971, E-W
SAN DIEGO L/P, 1968, N-S
EL CENTRO, 1940, N-S
EL CENTRO, 1940, E-W
20,000 40,000 60,000 in.-k
2,000 4,000 6,000 kN-m

MAX. BENDING MOMENTS IN CORE WALL

LEVEL
HOLIDAY ORION, 1971, E-W
EL CENTRO, 1940, N-S
EL CENTRO, 1940, E-W
SAN DIEGO L/P, 1968, N-S
2,000,000 4,000,000 6,000,000 8,000,000 10,000,000 12,000,000 14,000,000 16,000,000 in.-k
20,000 40,000 60,000 80,000 100,000 120,000 140,000 kN-m

MAX AXIAL FORCES IN COLUMNS

LEVEL
SAN DIEGO L/P, 1968, N-S
HOLIDAY ORION, 1971, E-W
EL CENTRO, 1940, E-W
EL CENTRO, 1940, N-S
2,000 4,000 6,000 k
10,000 20,000 30,000 kN

MAX. DISPLACEMENTS

LEVEL
EL CENTRO, 1940, E-W
EL CENTRO, 1940, N-S
HOLIDAY ORION, 1971, E-W
SAN DIEGO L/P, 1968, N-S
UNCOUPLED DIRECTION
DESIGN EARTHQUAK INTENSITY
1.0 2.0 3.0 4.0 5.0 6.0 7.0 8.0 9.0 10.0 11.0 12.0 in.
100 200 300 mm

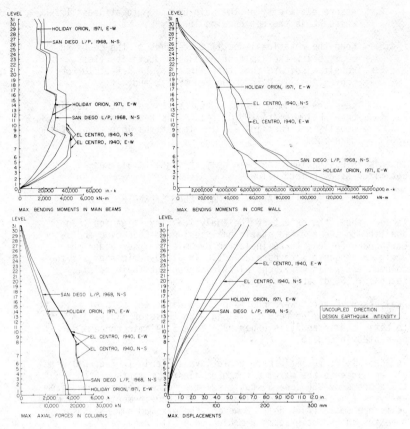

Fig. 4: Selection of critical input motion

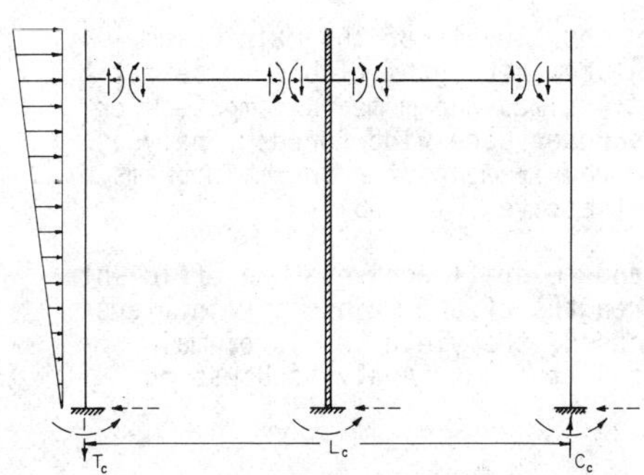

Fig. 5:

Schematic representation of internal forces due to lateral loads in structure considered

4. Assure elasticity of the main beams up to at least 1.4WL, where WL is the design wind load;

5. Keep the rotational ductility demands of the coupling beams below the limit of available ductility (test results indicate that this limit may be set at 10 for conventionally reinforced beams[6,7]); and

6. Keep the maximum nominal shear stresses in the coupling beams below $6\sqrt{f'_c}$ $(0.50\sqrt{f'_c})$. This requirement is usually not restrictive, in view of the long spans of the main beams.

In this investigation, the main beam strength levels were optimized as described below.

OPTIMIZATION OF MAIN BEAM STRENGTH

Table 1 shows the extent of the analytical investigation. In addition to the elastic static analyses under Code wind and earthquake forces, three inelastic dynamic analyses under the design earthquake were carried out. In all dynamic analyses the columns and walls were kept elastic throughout their seismic response. 5% of critical damping was assumed.

Analyses B, C and D were run with three different sets of main beam strengths. The results, along with those of Analysis A (elastic static analysis under UBC[1] Zone 3 seismic forces), are presented in Fig. 6.

In Analysis B, the main beam yield levels (Table 1, Fig. 6a) were chosen close to the maximum factored elastic moments caused in these beams by UBC Zone 3 static seismic forces (as computed from Analysis A). The corresponding ductility requirements in the main beams are small (Fig. 6b). However, the maximum nominal shear stresses in the main beams are high enough to require shear reinforcement (Fig. 6c); the axial forces in the columns are also high (Fig. 6d).

In an effort to optimize the strength levels of the main beams, Analysis C was run with the main beam strengths (Table 1) set at the minimum level of just over 1.4 times the moments computed for these beams from static analysis under Code wind forces (Analysis W). The corresponding ductility requirements for the main beams in the upper two tiers are now excessive (Fig. 6b).

To remedy the above situation, and to still arrive at an efficient and economical solution, the strengths of the main beams in Tiers 1 and 2 were substantially increased, the yield levels of main beams in Tier 3 were augmented a little, and Analysis D was run.

Table 1: Summary of Analytical Investigations

#	Analysis	Tier	Main Beams (Lumped) Computed Max. Moments, Factored, in.-k (kN-m)	
W	Elastic Static Analysis under UBC 20 psf Zone Equivalent Wind Forces	1	4,235 (478)	
		2	6,858 (775)	
		3	9,714 (1,098)	
		4	11,502 (1,300)	
A	Elastic Static Analysis under UBC Zone 3 Equivalent Seismic Forces with K = 1.0	1	22,366 (2,527)	
		2	33,192 (3,750)	
		3	43,766 (4,945)	
		4	46,655 (5,271)	
			Chosen Yield Moment, in.-k (kN-m)	**Ductility**
B	Inelastic Dynamic Analysis Under Design Earthquake	1	21,000 (2,373)	2.62
		2	31,500 (3,559)	2.19
		3	38,500 (4,350)	1.65
		4	42,000 (4,745)	1.61
C	Inelastic Dynamic Analysis Under Design Earthquake	1	5,600 (633)	23.55
		2	6,900 (780)	18.12
		3	9,800 (1,107)	10.25
		4	11,600 (1,311)	5.51
D	Inelastic Dynamic Analysis Under Design Earthquake	1	10,800 (1,220)	10.28
		2	10,800 (1,220)	10.05
		3	10,800 (1,220)	8.35
		4	11,600 (1,311)	5.12

The results indicate that the ductility requirements in the main beams are substantially reduced (the largest value is just over 10 - Fig. 6b). Should a further reduction in ductility requirements be desired, a proportional increase in beam strength would be necessary. The shear capacity requirements in the main beams are now low; so are the axial forces in the columns. The other response quantities do not suffer any adverse effect in going from Analysis C to Analysis D (Fig. 6e). The yield levels chosen for the main beams in Analysis D are thus satisfactory. It may be noted that these yield levels are 48% (Tier 1), 33% (Tier 2) and 25% (Tiers 3, 4) of the maximum factored static moments caused in these beams by UBC Zone 3 earthquake forces.

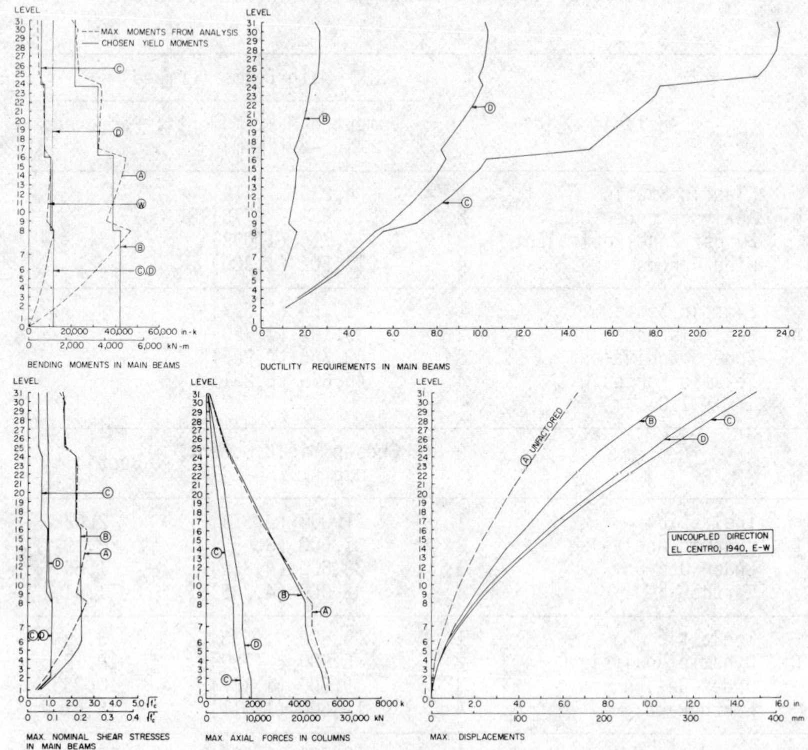

Fig. 6: Choice of main beam strength - selected results of dynamic analysis

CONCLUSION

An inelastic dynamic approach to the earthquake-resistant design of reinforced concrete structures is illustrated in this paper. The approach uses earthquake accelerograms as loading, and dynamic inelastic response history analysis to determine member forces and deformations.

A predetermined sequence of energy dissipating mechanisms is established by reducing the strength of a selected group of members. A desired response is thereby imposed on the structure. Detailing for ductility is required only in the predetermined hinging regions.

Reducing the main beam strengths below the levels indicated by an elastic analysis, while making sure that they can accommodate the increased ductility demands, will:

1. Advance the onset of yielding during an earthquake, thus enhancing the utilization of energy dissipation capacity of selected members.

2. Reduce moment input into columns from beams, thus protecting columns from flexural hinging.

3. Reduce shear in beams, materially improving their ductility.

4. Decrease the congestion of reinforcement at the joints, as well as shear in the joints.

5. Reduce the seismic axial forces in the columns.

An optimization procedure is employed in the determination of appropriate strength levels for the beams. Through a number of analyses, it is usually possible to design into members a desirable balance between flexural strength, shear capacity and ductility. This promotes economy and improves structural performance.

REFERENCES

1. International Conference of Building Officials, "Uniform Building Code," 1979 Edition, Whittier, California.

2. Kanaan, A.E., and Powell, G.H., "A General Purpose Computer Program for Inelastic Dynamic Response of Plane Structures," Report No. EERC 73-22, University of California, Berkeley, August 1975.

3. Takeda, T., Sozen, M.A., and Nielsen, N.N., "Reinforced Concrete Response to Simulated Earthquake," Journal of the Structural Division, ASCE, Proc. Vol. 96, NO. ST12, December 1970, pp. 2557-2573.

4. Derecho, A.T., Ghosh, S.K., Iqbal, M., Freskakis, G.N., and Fintel, M., "Structural Walls in Earthquake-Resistant Buildings - Analytical Investigation, Dynamic Analysis of Isolated Structural Walls - PARAMETRIC STUDIES," Final Report to the National Science Foundation, Portland Cement Association, Skokie, IL, April 1978.

5. Housner, G.W., "Behavior of Structures during Earthquakes," Journal of the Engineering Mechanics Division, ASCE, Proceedings Vol. 85, No. EM4, October 1959, pp. 109-129.

6. Wang, T.Y., Bertero, V.V., and Popov, E.P., "Hysteretic Behavior of Reinforced Concrete Framed Walls," Report No. EERC 75-23, University of California, Berkeley, December 1975.

7. Ma, S.-Y. M., Popov, E.P., and Bertero, V.V., "Experimental and Analytical Studies on the Hysteretic Behavior of Reinforced Concrete Rectangular and T-Beams," Report No. EERC 76-2, University of California, Berkeley, 1976.

8. Bertero, V.V., and Popov, E.P., "Hysteretic Behavior of Ductile Moment-Resisting Reinforced Concrete Frame Components," Report No. EERC 75-16, University of California, Berkeley, April 1975.

ELASTIC-PLASTIC DEFLECTIONS DUE TO PULSE LOADING

P. S. Symonds*, F. ASCE

ABSTRACT

The paper describes and illustrates the application of an approach to estimating deflections (peak transient and permanent) of structures subjected to pulse loading. Mode form rigid-plastic solutions furnish the main plastic response. These are introduced and succeeded by elastic motions, so that "first order" influences of elastic effects are included.

1 INTRODUCTION AND BACKGROUND

The paper summarizes the main ideas of a simple approximate method for the major deflection magnitude and response duration of a structure subjected to pulse loading such that plastic deformation occurs. The approach makes use of some concepts used in previously published methods (e.g. Martin and Symonds [1], Kaliszky [2], Forrestal and Wesenberg [3], among others). However, the main emphasis of the new approach is the removal of limitations and restrictions of previous approximate methods, while retaining simplicity and directness. It is applicable over the range from wholly elastic to mainly plastic, and from small deflections to magnitudes such that geometrical nonlinearities have strong influences.

Comparisons with experimental results will be illustrated over this wide range of response conditions. The present examples assume "impulsive" loading (vanishingly short duration pulses). This idealization is suitable for the tests considered, which were made with close range explosive loading. However, the general method is applicable to finite pulses of arbitrary shape and duration.

We consider structures of ductile metals subject to standard yield conditions and flow rules, usually with strain rate dependence. Fracture and other localized deformations are excluded. Strain hardening is not treated explicitly other than by choice of yield stress. Strain rate sensitivity is accounted for by use of a dynamic yield stress which is explicitly determined in the method.

The applications to date have been published for frames [4] and beams with full end constraints [5, 6]. Applications to other one-dimensional structures are expected to be straightforward, as are those to transverse loading of plates and shells such that initial response is mainly flexural. Situations such as axial impact on a long column where deformation must be analyzed as a propagation phenomenon, as well as buckling, are

*Brown University

excluded.

The proposed approach is frankly empirical and approximate, but it appears to contain all the essential elements of structural dynamics and compatibility and of elastic-plastic material behavior. Reasonable agreement with test results is shown over a fairly wide range of conditions. However comparisons with tests have been made to date only for rather simple structures of small scale, with loading approaching impulsive conditions. Further test comparisons are needed, as well as further comparisons with numerical solutions.

The approach depends strongly on concepts of the "mode approximation" technique [1]. This was proposed for impulsive loading specified by a field of initial velocities over the structure, which is then free of external loading. The response of the structure is written in "mode" (separated variable) form. Full solutions of this form exist for a structure with fixed displacement constraints that is subjected to constant external loads (here zero), provided that rigid-perfectly plastic behavior is assumed, and linear field equations of dynamics and kinematices are used. General mode fields of velocity and strain rate are written as

$$\dot{u}_i(\underline{x}, t) = \dot{w}_*(t)\phi_i(\underline{x}) \; ; \quad \dot{q}_j(\underline{x}, t) = \dot{w}_*(t)k_j(\underline{x}) \qquad (1a, b)$$

where $\phi_i(\underline{x})$ is a vector-valued "shape function" of space variables $\underline{x}$, with $i = 1, 2, 3$, normalized so that the scalar amplitude $\dot{w}_*(t)$ is the velocity of a point of main interest, and $w_*(t)$ is its deflection; $\dot{q}_j$ are generalized strain rates, with $j = 1, 2 .. r$. Such a form cannot generally give an exact solution for a specified initial velocity field. However a particular choice [1] of the initial velocity amplitude $\dot{w}_*^{o} = \dot{w}_*(0)$ minimizes the "difference" Δ_o between the specified initial velocities $\dot{u}_i^{o}(\underline{x})$ and those of the approximating mode field $\dot{w}_*^{o}\phi_i$, defining

$$\Delta_o = \frac{1}{2}\int\rho(\dot{u}_i^{o} - \dot{w}_*^{o}\phi_i)(\dot{u}_i^{o} - \dot{w}_*^{o}\phi_i)\,dV \qquad (2)$$

where ρ is specific mass and the integration is over the structure; the repeated suffix implies summation. Minimizing with respect to $\dot{w}_*^{o}$ furnishes

$$\dot{w}_*^{o} = \int\rho\phi_i\dot{u}_i^{o}dV/\int\rho\phi_i\phi_i dV \qquad (3a)$$

$$\Delta_o^{min} = \frac{1}{2}\int\rho\dot{u}_i^{o}\dot{u}_i^{o}dV - \frac{1}{2}(\dot{w}_*^{o})^2\int\rho\phi_i\phi_i dV = K_o - K_o^* \qquad (3b)$$

where K_o and K_o^* are the kinetic energies respectively of the specified field and of the ("closest") mode field. Equation (3a) will be referred to as the "minimum Δ_o" formula. Once the initial velocity is determined, approximate final deflections and response times are readily obtained.

This technique avoids the difficulties usually met in determining the initial phase of a rigid-plastic response, in which travelling hinges and changing boundaries of plastic regions must be found. The "minimum Δ_o" expression Eq. (3a) provides a rational choice of initial mode velocity amplitude, in the sense of minimizing a non-increasing "error" functional [1]. Further comments on this are made at a later point.

The elementary rigid-plastic theory, to which the mode approximation technique belongs, enables much progress to be made in understanding basic response phenomena, but the assumptions of rigid-perfectly plastic material behavior, linear (small deflection) field equations, and impulsive loading are not satisfied in many practical problems. Extensions to include strain rate sensitivity [7 - 9] together with nonlinear effects due to finite deflections have been made, e.g. [7,10,11]. The basic theory has been studied, e.g. [12-14] . The extensions to meet practical conditions have been successful, but in most cases at the price of adding considerable complexity. The approach summarized in this paper emphasizes simplicity, and in addition to including rate sensitivity and finite deflections also allows the assumption of rigid-plastic behavior to be relaxed by including elastic response in a "first-order" manner. At the same time, it facilitates treatment of more general finite pulses.

2 PROPOSED GENERAL APPROACH

The basic elements of the approach are listed as follows:

(a) The response is taken to occur in three distinct stages, namely: an elastic initial response; a subsequent rigid-perfectly plastic stage; and a final elastic vibration. Thus, while both elastic and plastic behaviors are treated, they are artificially separated; the response is treated as either entirely elastic or entirely plastic.

(b) In the rigid-plastic stage the behavior is taken as perfectly plastic, and the solution is taken initially in a mode form. If the material is rate sensitive, a raised yield stress is used, obtained as described below (f).

(c) In general the deflections reach magnitudes such that geometry changes must be included. Two situations may be distinguished: (i) When the structural action remains essentially flexural, as in a simple beam, frame, ring, etc., a modification of the initial mode deformation pattern can be found which satisfies the geometrical constraints. This will be a one-degree of freedom extension of the original mode solution, and if not also a full solution, should be close to one. (ii) In cases like a fully clamped beam or plate (with all edge motions prevented), as deflections increase the structural action changes from flexural to membrane. In these cases the finite deflection part of the rigid-plastic stage (when deflections exceed a certain magnitude), may be treated by considering membrane action only, using an appropriate mode solution. The rigid-plastic stage in these cases thus will consist of two distinct consecutive modes.

(d) Each mode form motion will have its velocity amplitude determined from the velocity field at the end of the preceding phase by applying the "minimum Δ_o" formula Eq. (3a), treating the preceding velocity field as the specified initial velocity field $\dot{u}_i^o$.

(e) A "global yield condition" is stipulated in order to terminate the initial elastic stage. Thus, when stresses computed from the elastic response determine an "effective stress" such that this condition is satisfied, at time t_e, that instant is taken as the end of the elastic stage, and the velocity field at this time used to determine the velocity amplitude at the start of the subsequent rigid-plastic stage. In a frame type structure the global yield condition may be taken as an average (root mean square or of absolute values) of bending moments at the points where plastic hinges occur in the succeeding rigid-plastic mode response.

(f) When the material is strain rate sensitive, the strain rates of the elastic stage are used in a constitutive law of visco-plastic behavior in order to determine a strain rate dependent global yield condition. The time t_e of ending the elastic stage is the time when the computed stresses furnish a global "effective stress" which equals that required by the strain rate dependent yield condition. The terminating condition may be visualized (Fig. 5c) as the intersection of a rising curve of effective stress computed from elastic stresses and a falling curve of limiting effective stress computed from elastic strain rates in a stress-strain rate law of viscoplastic behavior [3]. The critical effective stress so determined is used in the subsequent rigid-plastic stage as the appropriate dynamic yield stress.

(g) The solution in the initial elastic stage may be obtained by a hand calculation, using a discrete model with one or a few degrees of freedom, and assuming an idealized load pulse: for example, a delta-function (impulsive) or a rectangular form. Alternatively, a computer code may be used to solve the elastic problem. This would enable details of the shape and duration of the pulse to be used, so that general pulses, as well as impact or ground shock, may be dealt with. Typically the initial elastic stage involves only linear elastic behaviour at small strains and deflections. An exceptional case is considered in [6] and commented on briefly later in this paper.

(h) The rigid-plastic stage, first at small and then typically at finite deflections, dissipates energy and brings the structure to rest at time t_m. At this instant the structure reaches its peak deflection. Subsequently an elastic vibration takes place. In the present method its estimated initial amplitude is subtracted from the transient peak deflection to give the permanent plastic deflection. This residual elastic motion can be determined only roughly. It results from initial conditions of zero velocity and limiting elastic strains in the plastic zones of the rigid-plastic response; strain rate effects are absent.

The main novelties in this scheme are the inclusion of elastic deflections as well as plastic deformations, but artificially separated from them; and the use of the "minimum Δ_o" formula to determine the starting velocity amplitude of each stage from the closing field of the previous stage, continuity being imposed only with respect to the amplitude $w_*(t)$. Some further comments are offered on these features. Regarding the relation of <u>elastic</u> behavior of dynamic plastic structural response, we can distinguish three possibilities: (A) Elastic effects are <u>minor</u>;

plastic flow takes place in regions such that large general plastic motion of the structure occurs, as in a kinematic mechanism. (B) Elastic effects predominate; plastic deformation may occur perhaps in one highly stressed zone, but general plastic motion does not occur. (C) Intermediate between (A) and (B): both elastic and plastic effects are important; plastic zones do appear so that general motion of the structure is possible, but the energy and momentum available are such that plastic deflections are moderate.

In case (A) the present method may agree closely with a rigid-plastic treatment but offers advantages. In case (B) the present method predicts no plastic deformation, in contrast to a rigid-plastic treatment which would clearly be inappropriate. In case (C) the present method should be more realistic than a rigid-plastic treatment; a "first order" accounting for elastic effects is provided.

In essence, the present method uses rigid-plastic analysis in its simplest form, but requires no prior assumption for its validity: where a rigid-plastic analysis fails, the present method gives an approximate but reasonable answer.

Rigid-plastic analyses are valid, in the sense of agreeing with an elastic-plastic treatment, in certain circumstances. Unfortunately, there is no general criterion for this. An "energy ratio" criterion has often been cited: a rigid-plastic solution should be valid if the ratio R of plastic work to maximum elastic strain energy capacity of the structure greatly exceeds unity. This is useful if the loading is impulsive, but cannot serve as a general criterion for pulses with finite duration τ. For these, additional criteria must play a role, such as the ratio of τ to the fundamental period T of vibration of the structure. This is illustrated in Fig. 1 for a mass-spring model with elastic-perfectly plastic behavior, subjected to a force pulse of rectangular shape, duration τ, and amplitude P_o. The error in final deflection given by a rigid-plastic solution relative to that of an elastic-plastic analysis is shown as a function both of energy ratio R and of time ratio τ/T. It is positive for impulsive loading ($\tau = 0$), but becomes negative as the pulse duration increases. Evidently to have an error of less than 25 per cent, for example, it is not sufficient merely to specify a lower limit on R.

From another point of view, in the proposed method the initial elastic stage takes the place of the initial phase of a rigid-plastic solution in which travelling plastic hinges are present; typically, to satisfy initial conditions rigid-plastic analysis requires such a transient phase, which ultimately converges to a mode form motion. Regardless of the ultimate validity of a rigid-plastic analysis, the motion in its initial travelling hinge stage is basically a fiction: the initial response must be elastic. The yield condition at the most highly stressed region is reached only after a finite time interval, and a longer interval passes before yielding occurs simultaneously in several regions so that general motion of the structure can occur as a plastic mechanism. The end of this interval corresponds to reaching the global yield condition used in the present elastic-plastic approach. In the present examples the elastic interval of this approach has roughly the same duration as the initial travelling hinge phase in the rigid-plastic analysis of the same problem.

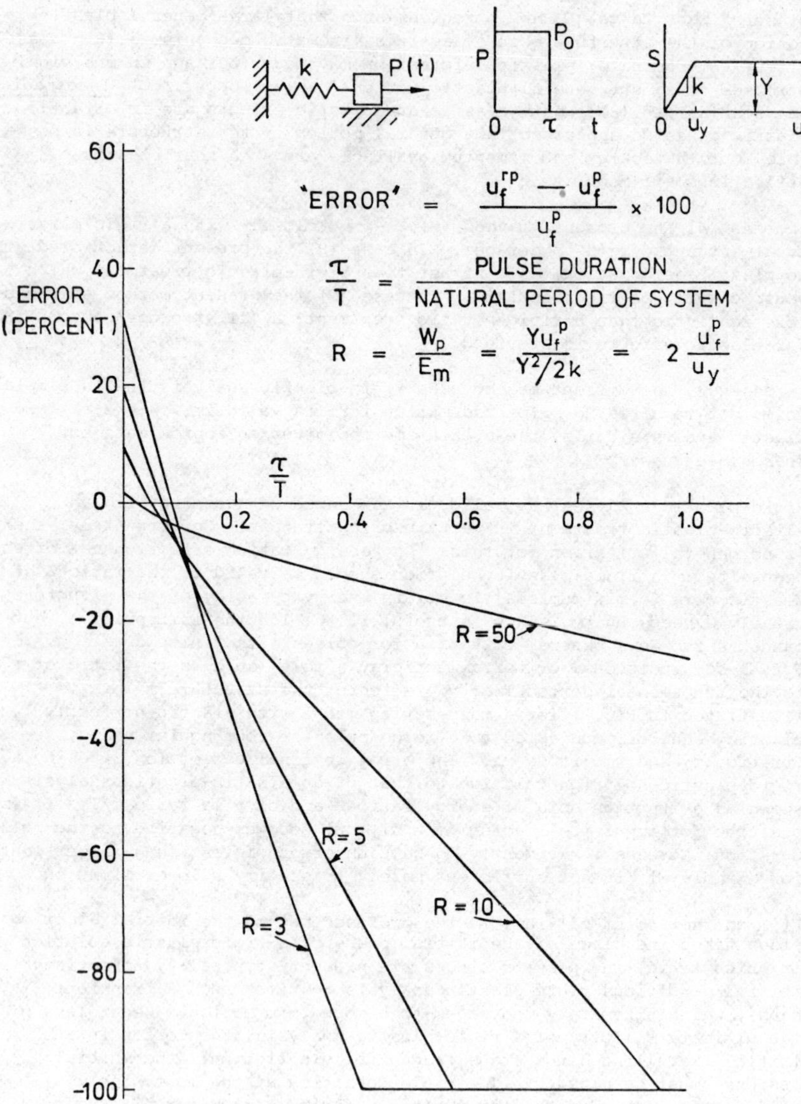

FIGURE 1. One-degree of freedom elastic-perfectly plastic system: relative error of rigid-plastic model as function of energy ratio R and of ratio of pulse duration to natural period of elastic system.

Why is the "minimum Δ_o" formula Eq. (3a) preferred for determining the amplitude of each new mode-form velocity field from the preceding field? Originally [1] this formula was proposed as the condition for minimizing the difference (in the mean-square, mass-weighted sense) between the actual velocity field and the approximating mode field; the function Δ_o is the initial value of a non-increasing functional of the difference between the two velocity fields. It can be interpreted in terms of momentum conservation, and derived on that basis [1]; thus it is the limiting case for short pulses of Kaliszky's theorem [2] for general pulses. In a broader form for elastic or elastic-plastic behavior, it can be used in an initial velocity problem of an elastic structure to derive the "best" amplitudes of a finite number of superimposed normal modes [12]; the results are identical with those of the classical elastic theory.

In the present context it appears the best available device for matching a mode velocity field with arbitrary amplitude to the preceding field; superposition of modes is not permitted here, of course. Among obvious alternatives, such as matching kinetic energy or the velocity amplitude itself, these take no account of the distribution over the structure of the specified velocity field. In contrast, the minimum Δ_o formula is sensitive to this distribution.

3 APPLICATIONS AND COMPARISONS

The method has been applied to two types of structure: rigid-jointed frames and clamped beams. These illustrate the two types of influence of finite deflections mentioned in (c). Test results and some numerical solutions are available for comparison. Most of the tests were done with parameters such that a rigid-plastic analysis (using the mode approximation technique and accounting for effects of finite deflections and rate sensitivity) turns out to give good estimates of final deflections; however, one test series [19] on higher strength alloys had conditions such that a rigid-plastic treatment is useless. Here we show only some representative examples for this case and others.

(I) The rigid-jointed frames for which applications have been published to date are shown in Fig. 2, where the impulsive loading is indicated as concentrated on the type (a) frames and uniformly distributed on the type (b) frames. Reference [4] describes comparisons with tests on frames of mild steel [15,16], alpha-titanium [15], and aluminum [17]. Here we show only some comparisons for mild steel frames of type (a); they are typical.

The rigid-plastic mode form solution is illustrated in Fig. 3. For small deflections the plastic hinges are at midpoint C and at the corners B; for finite deflections it is sufficient to consider the pattern shown, with plastic hinges at H in the columns. They move down at the same speed as the midpoint C.

If an "exact" rigid-plastic solution were sought, for the initial velocity fields as shown for the two frame types, an initial response phase involving travelling plastic hinges would have to be solved. In the type (a) frames travelling hinges move out from the central block, while for the frames of type (b) they move in from the corners. The

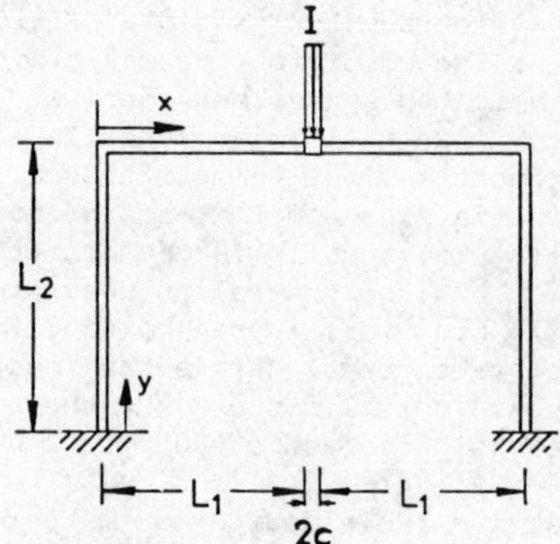

FIGURE 2a. Frame type (a). Impulsive load applied to steel block on horizontal member.

FIGURE 2b. Frame type (b). Uniform impulsive load applied to horizontal member.

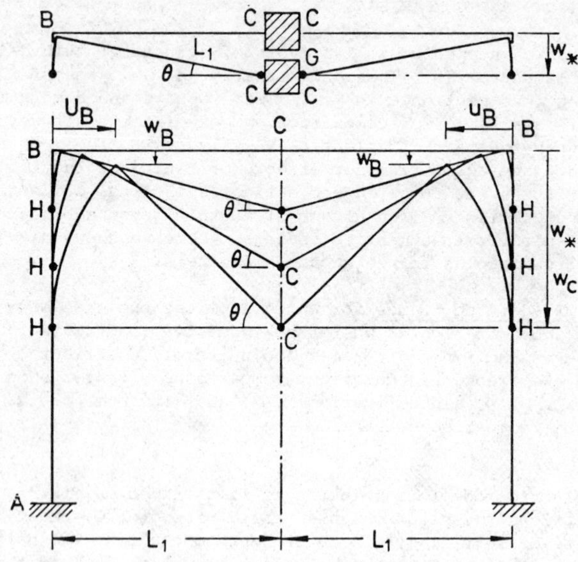

FIGURE 3. Patterns of finite deflections. Sketch shows successive positions of plastic hinges at C and H, corresponding to rotation angle θ.

need to consider these travelling hinges is obviated in the mode approximation technique [1] as proposed originally for rigid-perfectly plastic behavior. It is eliminated also in the present three-stage approach for elastic-plastic behavior.

In [4] the initial elastic stage is described using a simple two-DoF model derived from standard matrices of stiffness and consistent mass. Since the subsequent rigid-plastic stage has plastic hinges at C and at B in the columns, the condition for terminating the elastic stage is written in terms of bending moments and curvature rates at these sections. Root-mean-square averages were used to define a global "effective moment" M_e and a global "dynamic yield moment" M_v. The usual power-law relation [15] was used to describe the rate sensitive behavior of steel and titanium. The intersection time t_e such that

$$\frac{M_e(t_e)}{M_o} = \frac{M_v(t_e)}{M_o} = \mu \tag{4}$$

where M_o is the static fully plastic moment, furnishes the dynamic yield factor μ used in the subsequent rigid-elastic stage. The velocity field computed at time t_e is taken as the "specified" velocity field for determining the starting velocity amplitude of the rigid-plastic small deflection field by means of the minimum Δ_o formula Eq. (3a). The deflection at C w_C^e at this instant is the initial displacement of the rigid-plastic stage.

Motion in the rigid-plastic pattern of Fig. 3 continues until the velocity is reduced to zero, the mode deflection then reaching its peak amplitude. The subsequent elastic vibration is determined from starting conditions of zero velocity and displacements corresponding to reaching the static fully plastic moment at C and B in the columns. The amplitude w_C^r of this vibration is subtracted from the peak displacement to estimate the permanent plastic deflection. Thus the peak and final displacements, respectively, are given by

$$w_C^m = w_C^e + w_C^{RP} \; ; \quad w_C^f = w_C^e + w_C^{RP} - w_C^r \tag{5a, b}$$

where w_C^{RP} is the displacement amplitude of the rigid-plastic stage with initial velocity determined by the elastic stage in the manner indicated.

Figure 4 shows the resulting peak and final deflections as function of the impulse, with test points from [15] and curves computed by the rigid-plastic mode approximation and for wholly elastic behaviour. It is seen that the present method and the rigid-plastic mode approximation are fairly close, and both agree well with the deflections found in the tests. It should be pointed out however that the strain rate factor μ used for the rigid-plastic calculation was actually that obtained in the present elastic-plastic method, Eq. (4). This factor is not obtainable solely from a rigid-plastic analysis involving flexure; such an analysis is concerned with rotation rates and cannot furnish strain rates without additional information.

It should be mentioned also that the calculations reported in [4] were somewhat cruder in that it was there assumed that the elastic recovery deflection was the same as the deflection at the end of the initial

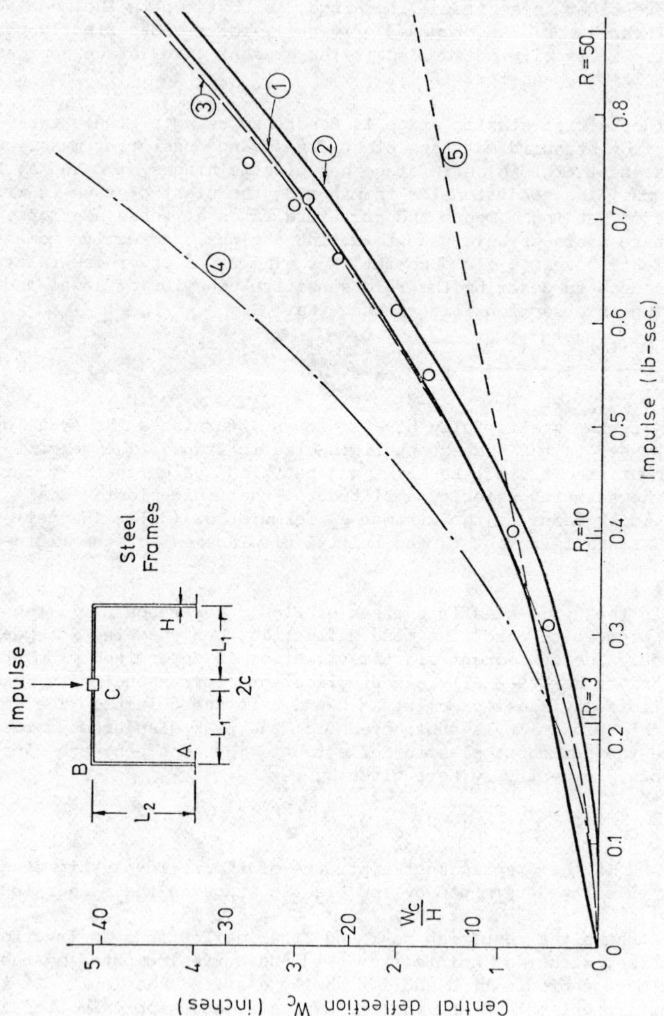

FIGURE 4. Circles show final deflections at midpoint measured in tests [15]. Mild steel frames had L_1 = 5.63 in. (143 mm), L_2 = 8.00 in. (203 mm), H = 0.123 in. (3.1 mm), width b = 0.75 in. (19.1 mm), σ_o = 32.1 ksi (228 N/mm^2). Curves (1) and (2): maximum and final deflections by present method. Curves (3) and (4): deflections by rigid-plastic mode approximation; (3) with strain rate correction factor μ taken from present elastic-plastic results, curve (4) with no strain rate correction. Curve (5) shows wholly elastic deflection amplitude.

elastic response. In some cases a discontinuous dependence on the impulse is therefore predicted. This discontinuity in the estimated final displacement is almost completely eliminated in the present calculation.

(II) We briefly note applications to fully clamped beams, accounting for the change from flexural to membrane behavior, and with deflections up to several multiples of the thickness. Applications to mild steel beams as used in tests [18] are discussed in [5], while reference [6] considers tests on titanium 6Al-4V and aluminum 7075-T6 beams [19] and on aluminum 6061-T651 beams [20].

Figures 5 - 7 depict the typical steps in the calculation, Fig. 5b indicating the initial elastic stage and Fig. 5c its terminating condition. In general the rigid-plastic stage is described by two mode form solutions. The three-hinge flexural mode of Fig. 6 applies for deflections smaller than the thickness. For subsequent larger deflections the membrane mode sketched in Fig. 7 is appropriate. The minimum Δ_o formula Eq. (3a) was used to determine the initial velocity amplitudes of both these mode fields, from the elastic field in the first case and from the preceding flexural rigid-plastic mode field in the second case.

The treatment of steel beams in [5] is straightforward and probably typical. Deflections in the elastic range are small, as are the differences between peak and final deflections (like those shown in Fig. 4). Estimates obtained assuming rigid-plastic behavior, using the two rigid-plastic modes and the minimum Δ_o formula to determine their amplitudes, are found in good agreement with results of the tests and the present method. (Again, the important strain rate corrections were taken from the present elastic-plastic approach.)

The experiments on higher strength beams [19] provide a more severe test of the present method. Figure 8 shows comparisons for beams of titanium 6Al-4V; similar results are shown in [6] for aluminum 7075-T6. For both these cases the same type of rigid-plastic approximation is unrealistic. The application of the present approach is complicated by the fact that the elastic stage itself must be treated nonlinearly, since the deflections reach several thicknesses in it. The end of the elastic stage was taken to be governed by the approximate relation

$$\frac{|M|}{M_o'} + \left(\frac{N}{N_o'}\right)^2 = 1 \tag{6}$$

where M_o', N_o' are dynamic fully plastic moment and axial force, as computed from elastic strain rates. The elastic recovery was similarly analyzed, but substituting static values M_o, N_o. Despite these complications, as shown in Fig. 8 the estimated peak and final displacements agree remarkably well with those of the tests. This may well be fortuitous; such close agreement is of course not demanded of an approximation technique. It does suggest that the present approach may be on the right track. Useful applications to other structures and to more general pulse loading may be expected.

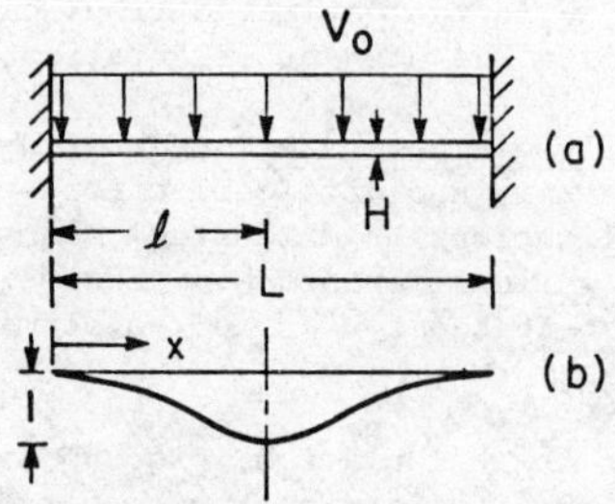

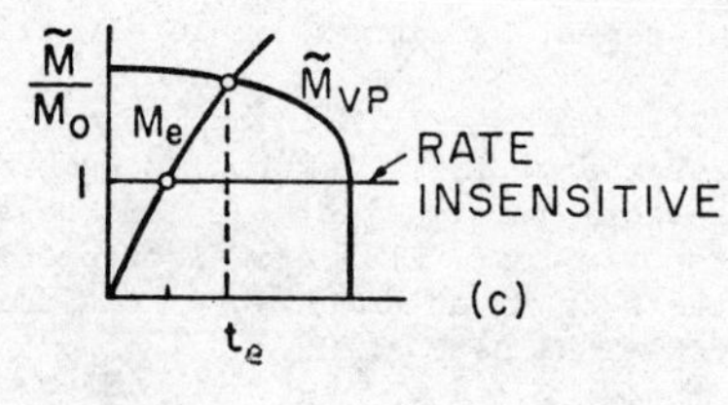

FIGURE 5a. Fully fixed beam.

FIGURE 5b. First stage elastic response.

FIGURE 5c. Condition for end of first stage.

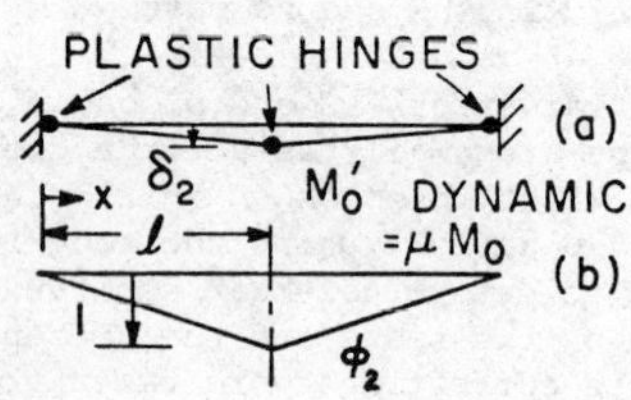

FIGURE 6a. Second stage rigid-plastic mode for small deflections.

FIGURE 6b. Mode shape function.

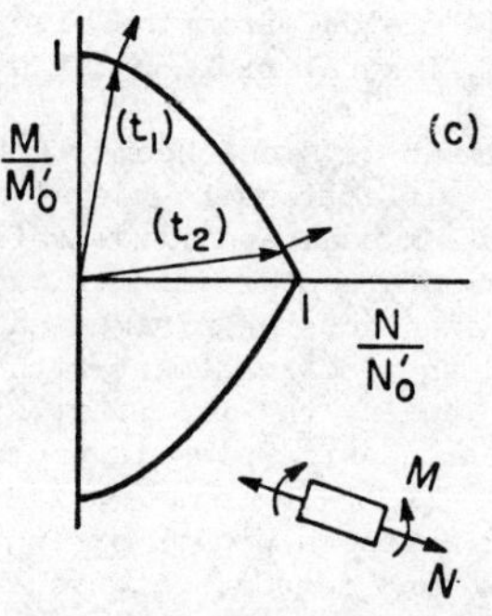

FIGURE 6c. Relation between fully plastic moment M and axial force N.

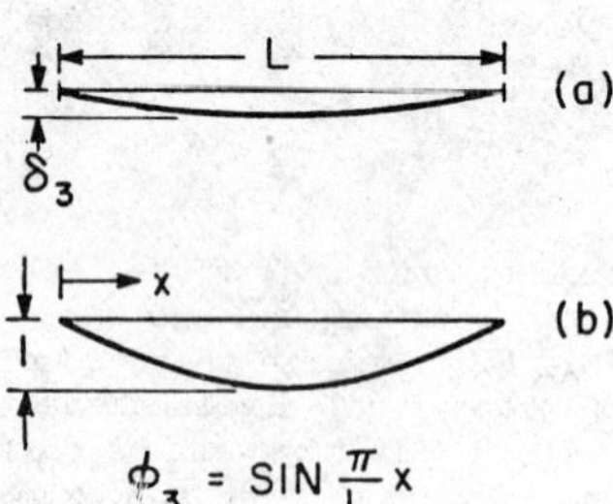

FIGURE 7a. Second stage rigid-plastic solution for finite deflections (membrane).

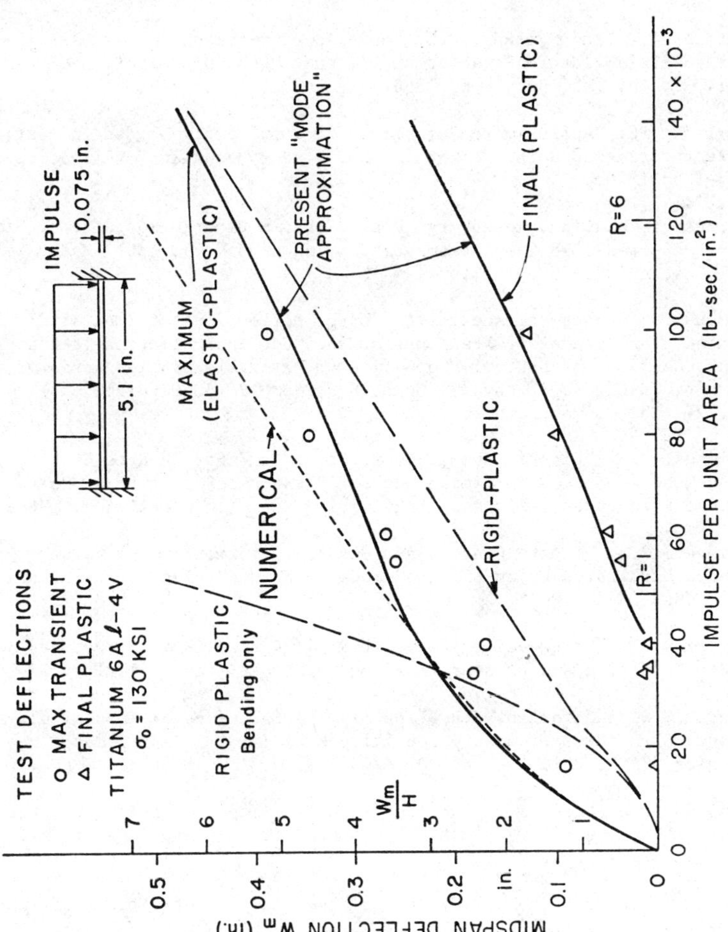

FIGURE 7b. Mid-span deflections of fully fixed titanium alloy beams. Test points from [19] (explosive loading). Heavy curves by present elastic-plastic method. Dashed curves by rigid-plastic mode approximation, using either two modes or bending mode only; strain rate correction by present elastic-plastic results. Numerical solution for maximum displacement by finite difference DEPROSS program [19].

REFERENCES

1. Martin, J.B. and Symonds, P.S. "Mode Approximations for Impulsively Loaded Rigid-Plastic Structures", J. Eng. Mech. Div. Proc. ASCE Vol. 92, No. EM5, pp.43-66, 1966.

2. Kaliszky, S. "Approximate Solutions for Impulsively Loaded Inelastic Structures and Continua", Int. J. Non-linear Mechanics, Vol. 5, pp. 143-158, 1970.

3. Forrestal, M.J. and Wesenberg, D.L. "Elastic-Plastic Response of Simply-Supported 1018 Steel Beams to Impulse Loads", J. Appl. Mech. Vol. 44, pp.779-780, 1977.

4. Symonds, P.S. and Raphanel, J.L. "Large Deflections of Impulsively Loaded Plane Frames - Extensions of the Mode Approximation Technique", Proc. Conf. on Mechanical Properties of Materials at High Rates of Strain, Institute of Physics Conf. Ser. No. 47 (J. Harding, Ed.) pp. 277-287, London and Bristol, 1979.

5. Symonds, P.S. "Elastic, Finite Deflection, and Strain Rate Effects in a Mode Approximation Technique for Plastic Deformation of Pulse Loaded Structures", J. Mech. Engineering Sci. (in press, Aug. 1980).

6. Symonds, P.S. "Finite Elastic and Plastic Deformations of Pulse-Loaded Structures by an Extended Mode Technique", Int. J. Mech. Sci. (in press, Aug. 1980).

7. Kaliszky, S. "Large Deformations of Rigid-Viscoplastic Structures under Impulsive and Pressure Loading" J.Struct.Mech. Vol. 1,p.295,1973.

8. Lee, L.S.S. and Martin, J.B. "Approximate Solutions of Impulsively Loaded Structures of a Rate Sensitive Material", J. Appl. Math. Physics (ZAMP) Vol. 21, pp. 1011-1032, 1970.

9. Symonds, P.S. "Approximation Techniques for Impulsively Loaded Structures of Rate Sensitive Plastic Behavior", SIAM J. Appl. Math. Vol. 25, pp. 462-473, 1973.

10. Symonds, P.S. and Chon, C.T. "Large Viscoplastic Deflections of Impulsively Loaded Plane Frames", Int. J. Solids Structures Vol.15, pp. 15-31, 1979.

11. Symonds, P.S. and Chon, C.T. "Finite Viscoplastic Deflections of an Impulsively Loaded Plate by the Mode Approximation Technique", J. Mech. Phys. Solids Vol. 27, pp. 115-133, 1979.

12. Martin, J.B. and Lee, L.S.S. "Approximate Solutions for Impulsively Loaded Elastic-Plastic Beams" J. Appl. Mech. Vol. 35, pp.803-809,1968.

13. Martin, J.B. "Extremum Principles for a Class of Dynamic Rigid-Plastic Problems", Int. J. Solids Structures Vol. 8, pp.1185-1204, 1972.

14. Symonds, P.S. and Chon, C.T. "On Dynamic Plastic Mode Form Solutions" J. Mech. Phys. Solids Vol. 26, pp. 21-35, 1978.

15. Bodner, S.R. and Symonds, P.S. "Experiments on Dynamic Plastic Loading of Frames", Int. J. Solids Structures Vol. 15, pp. 1-13, 1979.

16. Hanan, D. "The Permanent Deformation of Steel Portal Frames Subjected to Impulsive Loading", M.Sc. (Engineering) thesis, University of Cape Town, Department of Civil Engineering, 1977.

17. Hashmi, S.J. and Al-Hassani, S.T.S. "Large Deflexion Response of Square Frames to Distributed Impulsive Loads", Int. J. Mech. Sci. Vol. 17, pp. 513-523, 1975.

18. Symonds, P.S. and Jones, Norman "Impulsive Loading of Fully Clamped Beams with Finite Plastic Deflections and Strain Rate Sensitivity", Int. J. Mech. Sci. Vol. 14, pp. 49-69, 1972.

19. Lindholm, U.S. and Bessey, R.L. "Elastic-Viscoplastic Response of Clamped Beams under Uniformly Distributed Impulse", Technical Report AFML-TR-68-396, Jan. 1969, of Southwest Research Institute to Air Force Materials Laboratory, Wright-Patterson A.F.B., Ohio.

20. Witmer, E.A., Merlis, F. and Spilker, R.L. "Experimental Transient and Permanent Deformation Studies of Steel Sphere-Impacted or Impulsively Loaded Aluminum Beams with Clamped Ends", NASA CR 123922, M.I.T. ASRL TR 154-11, NASA Grant NGR 22-009-339, October 1975, pp. 1-102.

ACKNOWLEDGEMENT

Support of research by the U.S. Army Research Office under Grant DAAG-78-G-0085 is gratefully acknowledged.

DYNAMIC TORSIONAL RESPONSE OF SYMMETRIC STRUCTURES

by

Pradip K. Syamal[1] and O.A. Pekau[2], Members ASCE

ABSTRACT

A method for investigating induced torsional oscillations of nonlinear symmetric structures subjected to only translational excitation is presented. Nonlinear elastic resisting elements are assumed. Two mathematical relations, one for the upper bound and the other for the lower bound, establish the zone of instability and are derived from a set of damped coupled Mathieu-Hill variational equations. The latter originate from perturbation of the original equations of motion. Stability curves are presented which can usefully be employed in evaluating the susceptibility of a symmetric or nominally symmetric structure to possible torsional motion. Also, a relationship between torsional damping and the system parameters for stable torsional response is derived. It can be used to determine the minimum torsional damping required to prevent the initiation of a torsional component of response in symmetric structures. Finally, the influence of various system parameters on the initiation of torsional oscillation is discussed [viz., uncoupled torsional and translational frequencies, damping coefficients, building aspect ratio].

INTRODUCTION

Recent studies (6,10,11) of symmetric structures subjected to ground motion indicate that unexpected torsional oscillations may accompany the translational response under certain circumstances. Newmark (6) observed that such torsional vibrations can be induced in symmetric structures by the rotational component of ground motion which accompanies the two horizontal components. On the other hand, Tso and Asmis (10,11) showed that nonlinearity in the resisting elements can induce torsional response in a nominally symmetric structure subjected only to translational excitation. Asmis (1) also investigated parametric resonance in nonlinear mechanical systems and observed that nonlinear coupling between modes can cause an interaction effect between external parametric and internal resonances. He also noted that the instability zone of the external parametric resonance is a function of the ratio of the damping coefficients. Finally, he concluded that the susceptibility to resonance and parametric excitation of a nonlinear system subjected

---

[1]Graduate Student in Civ. Engrg., Concordia University, Montreal, P.Q., Canada; also Senior Engineer, Canatom Inc., Montreal, P.Q., Canada.
[2]Associate Professor, Dept. of Civ. Engrg., Concordia University, Montreal, P.Q., Canada.

to external periodic excitation is dependent upon the relationship between fundamental frequencies of the system and the external exciting frequency. Other studies (3,4), conducted for axisymmetric elastic bodies exhibiting nonlinearity of the softening type, established the extent of instability zones and the influence of frequency ratios.

Nigam and Housner (7,8) investigated the inelastic response of a symmetric simple frame subjected to both earthquake as well as sinusoidal base excitation. After comparing the response obtained for elastic, elastoplastic, and two-directional elastoplastic behaviour with interaction, they concluded that the interaction has a significant influence on response.

This paper presents the results of a detailed investigation of the torsional response induced in symmetric structures due to the nonlinearity of resisting elements. Attention is directed toward identifying the basic system parameters governing the initiation of such behaviour. Results are presented in the form of stability curves showing the effects of various system parameters, such as building aspect ratio and torsional and translational damping coefficients among others.

GOVERNING EQUATIONS AND METHOD OF ANALYSIS

The model selected for study is an idealized single-storey structure; the floor diaphragm is assumed to be rigid and the resisting elements are nonlinear elastic, having eccentric centre of mass with respect to the centre of resistance (Fig. 1). The force-displacement relationship of the lateral resisting elements is assumed to be of the softening type with cubic nonlinearity expressed as

$$R_x(\delta_j) = k_x\delta_j[1 - \lambda(\frac{\delta_j}{\delta_0})^2], \; j = 1,2 \tag{1}$$

$$R_y(\delta_j) = k_y\delta_j[1 - \lambda(\frac{\delta_j}{\delta_0})^2], \; j = 3,4 \tag{2}$$

in which k_x, k_y are the linear stiffnesses of resisting elements in the x- and y- directions, respectively; δ_0 is some convenient reference displacement; λ is a measure of nonlinearity of the force-displacement relation with $\lambda \ll 1$; $\delta_j (j = 1,\ldots,4)$ are the displacements of the lateral resisting elements (Fig. 1); and $R_x(\delta_j)$, $R_y(\delta_j)$ represent the restoring forces of element 'j'. The nondimensional equations of motion for such a system, with degrees-of-freedom u, v, θ and subjected to sinusoidal ground acceleration in the x-direction of magnitude

$$\ddot{u}_g = U \cos \omega t \tag{3}$$

have been derived elsewhere (9) and can be written as

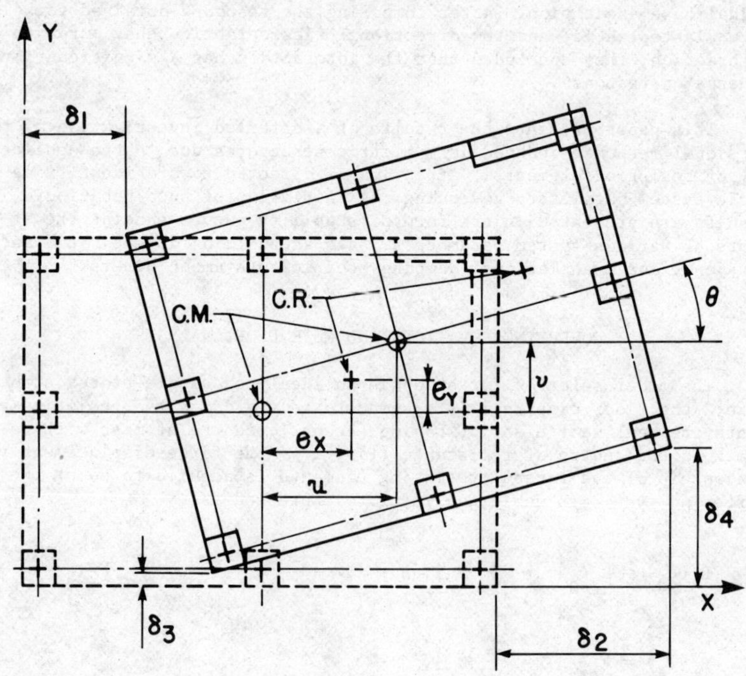

Fig. 1 Structural Geometry and Building Displacements

$$\ddot{\Lambda}_x + f_1\dot{\Lambda}_x + f_2\Lambda_x - (f_3\Lambda_x^3 + f_4\Lambda_x^2\Lambda_\theta + f_5\Lambda_x\Lambda_\theta^2 + f_6\Lambda_\theta + f_7\Lambda_\theta^3) = -\cos\Omega\tau \quad (4)$$

$$\ddot{\Lambda}_\theta + h_1\dot{\Lambda}_\theta + h_2\Lambda_\theta - (h_4\Lambda_\theta^3 + h_9\Lambda_\theta^2\Lambda_x + h_3\Lambda_\theta\Lambda_x^2 + h_8\Lambda_x^3) - (h_6\Lambda_\theta^3 + h_{12}\Lambda_\theta^2\Lambda_y + h_5\Lambda_\theta\Lambda_y^2 + h_{11}\Lambda_y^3) - (h_7\Lambda_x + h_{10}\Lambda_y) = 0 \quad (5)$$

$$\ddot{\Lambda}_y + g_1\dot{\Lambda}_y + g_2\Lambda_y - (g_3\Lambda_y^3 + g_4\Lambda_y^2\Lambda_\theta + g_5\Lambda_y\Lambda_\theta^2 + g_6\Lambda_\theta + g_7\Lambda_\theta^3) = 0 \quad (6)$$

in which f_i, g_i, h_i are nondimensional constants that may be expressed as functions of the following variables: nonlinearity parameter λ; damping coefficients ζ_x, ζ_y, ζ_θ; fundamental frequencies of the uncoupled system ω_x, ω_y, ω_θ; building aspect ratio r (= a/b); mass eccentricities with respect to the centre of resistance e_x, e_y; and Γ, the mass radius of gyration about the vertical axis through the centre of mass. Also, in the above set of equations of motion, variables defined as follows have been employed:

$$\omega_x^2 = \frac{K_x}{M} \; ; \quad \omega_y^2 = \frac{K_y}{M} \; ; \quad \omega_\theta^2 = \frac{K_{\theta m}}{M\Gamma^2} \quad (7)$$

$$\tau = \omega_x t \quad (8)$$

$$\delta_x = \frac{U}{\omega_x^2} \; ; \quad \delta_y = \frac{U}{\omega_y^2} \; ; \quad \delta_\theta = \frac{U}{\Gamma\omega_\theta^2} \quad (9)$$

$$\Lambda_x(\tau) = \frac{u(t)}{\delta_x} \; ; \quad \Lambda_y(\tau) = \frac{v(t)}{\delta_y} \; ; \quad \Lambda_\theta(\tau) = \frac{\theta(t)}{\delta_\theta} \quad (10)$$

$$\Omega_y = \frac{\omega_y}{\omega_x} \; ; \quad \Omega_\theta = \frac{\omega_\theta}{\omega_x} \; ; \quad \Omega = \frac{\omega}{\omega_x} \quad (11)$$

where τ represents nondimensional time; Λ_x, Λ_y, Λ_θ represent nondimensional response of the system; K_x, K_y are total translational stiffness in x- and y- directions, respectively; $K_{\theta m}$ is the torsional stiffness of the structure with respect to the centre of mass; and M is the total mass of the structure.

Eqs. 4 to 6 can be solved by applying the method of averaging, i.e., method of slowly varying amplitude popularly known as the Kryloff-Bogoliuboff method (5). This averaging procedure leads to a set of six nonlinear simultaneous algebraic equations with approximate solution given by

$$\Lambda_x(\tau) = \bar{P}\cos(\Omega\tau + \bar{\Phi}) \tag{12a}$$

$$\Lambda_\theta(\tau) = \bar{R}\cos(\Omega\tau + \bar{\chi}) \tag{12b}$$

$$\Lambda_y(\tau) = \bar{Q}\cos(\Omega\tau + \bar{\Psi}) \tag{12c}$$

where $\bar{P}$, $\bar{Q}$, $\bar{R}$ are the average amplitudes of response and $\bar{\Phi}$, $\bar{\chi}$, $\bar{\Psi}$ are average phase angles.

To investigate the stability of the solution, the perturbation technique (3,4) is employed. Imposing small perturbations $\xi_x(\tau)$, $\xi_\theta(\tau)$ and $\xi_y(\tau)$ on the solutions for $\Lambda_x(\tau)$, $\Lambda_\theta(\tau)$ and $\Lambda_y(\tau)$ yields the following set of damped coupled Mathieu-Hill variational equations (9):

$$\underset{\sim}{C}\,\ddot{\underset{\sim}{\xi}} + 2\,\underset{\sim}{C}\,\underset{\sim}{\varepsilon}\,\dot{\underset{\sim}{\xi}} + [\underset{\sim}{E} - \tfrac{1}{2}\underset{\sim}{A} - \tfrac{1}{2}\cos 2\Omega\tau\,\underset{\sim}{B}]\underset{\sim}{\xi} = \underset{\sim}{0} \tag{13}$$

where

$$\underset{\sim}{C} = \begin{bmatrix} \frac{1}{f_2} & 0 & 0 \\ 0 & \frac{1}{h_2} & 0 \\ 0 & 0 & \frac{1}{g_2} \end{bmatrix} \;;\quad \underset{\sim}{\varepsilon} = \frac{1}{2}\begin{bmatrix} f_1 & 0 & 0 \\ 0 & h_1 & 0 \\ 0 & 0 & g_1 \end{bmatrix} \tag{14}$$

In the above, $\underset{\sim}{E}$ is a 3 x 3 identity matrix and the elements of $\underset{\sim}{A}$ and $\underset{\sim}{B}$ are functions of f_i, g_i, h_i, $\bar{P}$, $\bar{Q}$ and $\bar{R}$. The principal region of instability of Eq. 3 can be approximated by (2)

$$\begin{vmatrix} \underset{\sim}{E} - \frac{1}{2}\underset{\sim}{A} + \frac{1}{4}\underset{\sim}{B} - \Omega^2\underset{\sim}{C} & -2\,\Omega\,\underset{\sim}{C}\,\underset{\sim}{\varepsilon} \\ 2\,\Omega\,\underset{\sim}{C}\,\underset{\sim}{\varepsilon} & \underset{\sim}{E} - \frac{1}{2}\underset{\sim}{A} - \frac{1}{4}\underset{\sim}{B} - \Omega^2\underset{\sim}{C} \end{vmatrix} = 0 \tag{15}$$

which, upon expansion, becomes a sixth-order algebraic equation in Ω^2.

Eq. 15 involves the following variables: geometric parameters $\underset{\sim}{e}_x$, e_y, Γ and aspect ratio r; damping coefficients ζ_x, ζ_y, ζ_θ; fundamental frequencies ω_x, ω_y, ω_θ; nonlinearity parameter λ; and the input frequency ω. Hence, the influence of these parameters on the region of instability can easily be established. Eq. 15 is expanded and solved for input frequency ratio Ω yielding two expressions, one representing an upper bound and the other a lower bound, for the zone of torsional instability of an unsymmetric system.

APPLICATION OF ANALYSIS TO SYMMETRIC STRUCTURES

The typical symmetric single-story structure shown in Fig. 2 is considered, in which the elements of lateral resistance are assumed to be distributed along the perimeter of the structure. The geometric parameters, damping coefficients, fundamental uncoupled frequencies and the nonlinearity parameter give rise to the following values of

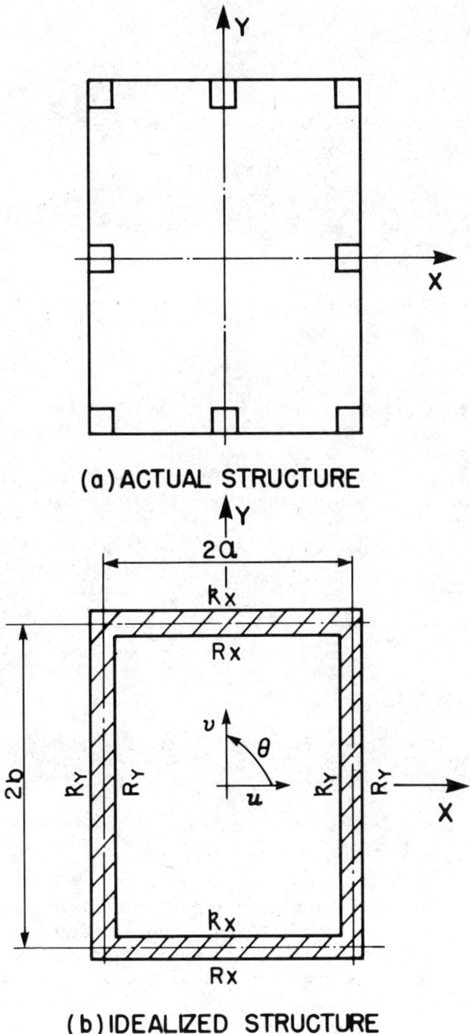

Fig. 2 Plan of Symmetric Structure

f_i, g_i and h_i for such a system:

$$f_1 = 2\zeta_x \; ; \quad g_1 = 2\zeta_y\Omega_y \; ; \quad h_1 = 2\zeta_\theta\Omega_\theta \; ; \quad f_2 = 1 \; ; \quad g_2 = \Omega_y^2 \; ; \tag{16a}$$

$$h_2 = \Omega_\theta^2 \; ; \quad f_3 = \lambda \; ; \quad g_3 = \frac{\lambda}{\Omega_y^2} \; ; \quad h_3 = 3\lambda\gamma_1^2 \; ; \quad f_4 = f_6 = f_7 = 0 \; ; \tag{16b}$$

$$g_4 = g_6 = g_7 = 0 \; ; \quad h_4 = \lambda\left(\frac{\gamma_1}{\Omega_\theta}\right)^4 \; ; \quad f_5 = 3\lambda\left(\frac{\gamma_1}{\Omega_\theta^2}\right)^2 \; ; \quad g_5 = 3\lambda\left(\frac{\beta_1\Omega_y}{\Omega_\theta^2}\right)^2 ; \tag{16c}$$

$$h_5 = 3\lambda\left(\frac{\beta_1}{\Omega_y}\right)^2 \; ; \quad h_7 = h_8 = h_9 = h_{10} = h_{11} = h_{12} = 0 \; ; \tag{16d}$$

$$h_6 = \lambda\Omega_y^2\left(\frac{\beta_1}{\Omega_\theta}\right)^4 \; ; \quad r = \frac{a}{b} \; ; \quad \beta_1 = \left(\frac{3r^2}{1 + r^2}\right)^{\frac{1}{2}} \; ; \quad \gamma_1 = \left(\frac{3}{1 + r^2}\right)^{\frac{1}{2}} \tag{16e}$$

Substituting Eq. 16 into Eq. 15, neglecting torsional damping, and expanding lead to the following upper and lower bounds:

$$\Omega_\theta^2 = \Omega^2 + \tfrac{3}{4}(3h_4\bar{R}^2 + h_3\bar{P}^2 + 3h_6\bar{R}^2 + h_5\bar{Q}^2)$$
$$+ \left(\frac{9g_5h_5\bar{Q}^2\bar{R}^2}{4g_2 - 9g_3\bar{Q}^2 - 3g_5\bar{R}^2 - 4\Omega_\theta^2}\right) + \left(\frac{9f_5h_3\bar{P}^2\bar{R}^2}{4f_2 - 9f_3\bar{P}^2 - 3f_5\bar{R}^2 - 4\Omega_\theta^2}\right) \tag{17}$$

and

$$\Omega_\theta^2 = \Omega^2 + \tfrac{1}{4}(3h_4\bar{R}^2 + h_3\bar{P}^2 + 3h_6\bar{R}^2 + h_5\bar{Q}^2)$$
$$+ \left(\frac{g_5h_5\bar{Q}^2\bar{R}^2}{4g_2 - 3g_3\bar{Q}^2 - g_5\bar{R}^2 - 4\Omega_\theta^2}\right) + \left(\frac{f_5h_3\bar{P}^2\bar{R}^2}{4f_2 - 3f_3\bar{P}^2 - f_5\bar{R}^2 - 4\Omega_\theta^2}\right) \tag{18}$$

With the further approximation that translational response in only the x-direction predominates (compared to y- and rotational directions) when the excitation is in the x-direction, the values of $\bar{Q}$ and $\bar{R}$ in Eqs. 17 and 18 are set to zero, whereas the magnitude of $\bar{P}$ is taken to be

$$\bar{P}^2 = [(1 - \Omega^2)^2 + (2\zeta_x\Omega)^2]^{-1} = D \tag{19}$$

Substituting Eq. 19 into Eqs. 17 and 18 and neglecting terms involving $\bar{Q}$ and $\bar{R}$, the following upper and lower bound instability curves result, respectively:

$$\Omega_\theta^2 = \Omega^2 + \frac{27\lambda D}{4(1 + r^2)} \tag{20}$$

$$\Omega_\theta^2 = \Omega^2 + \frac{9\lambda D}{4(1 + r^2)} \tag{21}$$

In order to determine the influence of torsional damping on the stability, the condition for Ω to be real (Eq. 15) is investigated. To avoid complex values of boundary frequencies, the largest value of torsional damping for possible dynamic instability is expressed as

$$2\Omega_\theta h_1 = \frac{1}{2}(3h_4\bar{R}^2 + h_3\bar{P}^2 + 3h_6\bar{R}^2 + h_5\bar{Q}^2)$$

$$+ \left(\frac{9g_5h_5\bar{Q}^2\bar{R}^2}{4g_2 - 9g_3\bar{Q}^2 - 3g_5\bar{R}^2 - 4\Omega_\theta^2}\right) + \left(\frac{9f_5h_3\bar{P}^2\bar{R}^2}{4f_2 - 9f_3\bar{P}^2 - 3f_5\bar{R}^2 - 4\Omega_\theta^2}\right)$$

$$- \left(\frac{g_5h_5\bar{Q}^2\bar{R}^2}{4g_2 - 3g_3\bar{Q}^2 - g_5\bar{R}^2 - 4\Omega_\theta^2}\right) - \left(\frac{f_5h_3\bar{P}^2\bar{R}^2}{4f_2 - 3f_3\bar{P}^2 - f_5\bar{R}^2 - 4\Omega_\theta^2}\right) \quad (22)$$

Substituting Eq. 19 into Eq. 22 and again neglecting terms involving $\bar{Q}$ and $\bar{R}$, the influence of torsional damping on the stability is established as

$$\zeta_\theta = \frac{9\lambda D}{8\Omega_\theta^2(1 + r^2)} \quad (23)$$

Eq. 23 represents the largest value of torsional damping ζ_θ for which dynamic instability is possible in a symmetric building.

RESULTS AND DISCUSSION

To demonstrate the influence of aspect ratio, fundamental frequency and damping, Figs. 3 and 4 show the regions of instability in Ω_θ - Ω parameter space. It is noted that the region of instability becomes more pronounced for Ω values between 0.8 and 1.2. It is also interesting to note that torsional frequency ratio Ω_θ need not be equal to input frequency ratio Ω to create an unstable condition for torsional oscillation of the structure.

For an unsymmetric system, torsional stability is defined as a bound on the magnitude of the rotational displacement of the system over time. On the other hand, torsional stability in a symmetric system means having no torsional response when the input excitation is purely translational. It is seen from Figs. 3 and 4 that the torsional frequency ratio must be greater than the excitation frequency ratio in order for torsional response to occur. At a value of Ω equal to unity, the likelihood of torsional instability is most pronounced; on either side of this point (i.e., $\Omega < 0.8$ or $\Omega > 1.2$) the unstable region narrows and approaches the $\Omega_\theta = \Omega$ line. Hence, away from $\Omega = 1$, the unstable region becomes sufficiently narrow that coupled torsional oscillation is unlikely.

As shown in Fig. 3, the unstable region increases with decrease in aspect ratio r of the structure and shifts away from the $\Omega_\theta = \Omega$ line. Similarly, the unstable region decreases, particularly in the neighbourhood of $\Omega = 1$, and shifts towards the $\Omega_\theta = \Omega$ line with increase in the

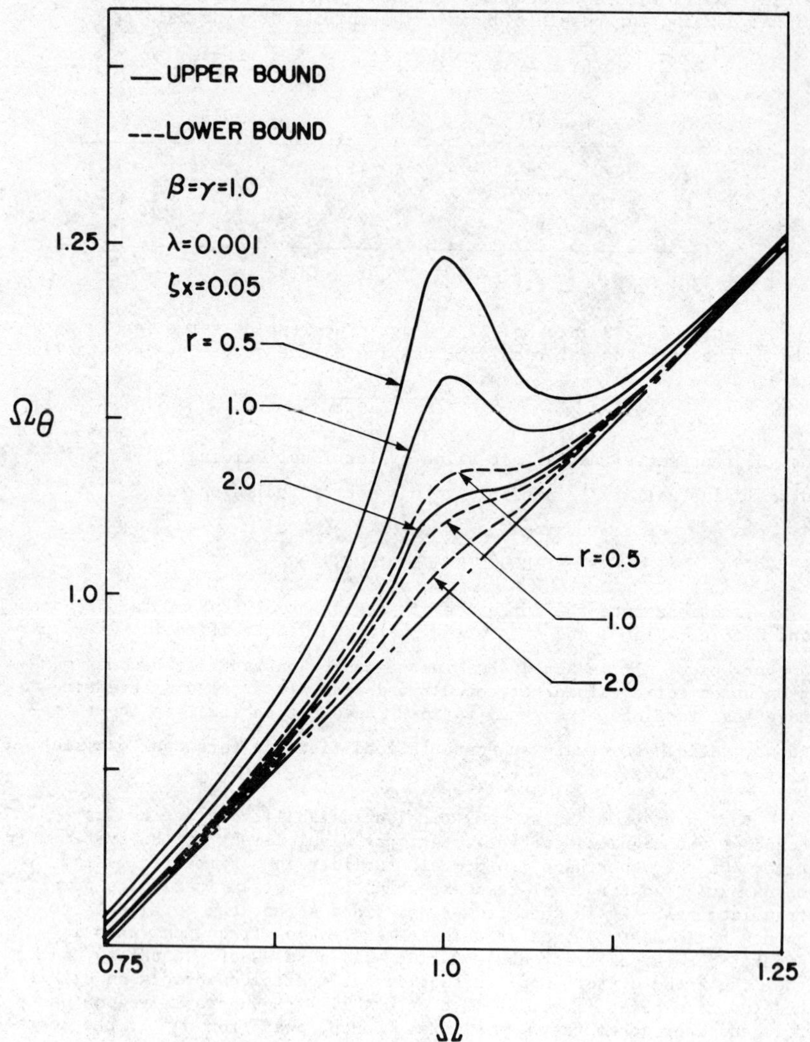

Fig. 3 Effect of Building Aspect Ratio (r) on the Region of Induced Torsional Response

value of translational damping ratio ζ_x.

From the above observations it is evident that uncoupled torsional frequency ratio Ω_θ is an important parameter influencing the torsional stability of a symmetric system. Tso and Asmis (11) have indicated that the critical values for Ω_θ is restricted to the range 1.11 - 1.67. However, it can be observed in Figs. 3 and 4 that the critical range for Ω_θ is strongly influenced by aspect ratio r as well as by the translational damping coefficient ζ_x.

Figs. 5 and 6 show plots of the torsional damping ratio ζ_θ required to prevent initiation of torsional oscillation. In Fig. 5 curves are plotted in ζ_θ - Ω_θ parameter space for various input frequency ratios. For each curve shown, the area below the curve represents an unstable zone and the area above it denotes stable translational response. The ζ_θ label of the curves is the magnitude of minimum torsional damping necessary in the system to prevent the occurrence of torsional oscillation. It is noted that, in the neighbourhood of $\Omega = 1$, higher torsional damping is necessary to ensure stability compared to that required for other values of Ω, for given system frequency ratio Ω_θ. It is also observed from these curves that, at a particular value of Ω, torsional response may be avoided at lower torsional damping when the torsional frequency of the structure increases.

Similarly, Fig. 6 shows the effect of aspect ratio r on the minimum torsional damping for stability. Again, the area under each curve is defined as the unstable zone, which is seen to decrease with increase in aspect ratio.

CONCLUSIONS

In general, the results obtained in this study can usefully be employed in evaluating the susceptibility of a proposed or existing building structure to possible torsional motions, even though the structure is symmetric and excitation is purely translational.

Based on the data presented, the following conclusions are noted:

1. In a symmetric structure torsional motion may be induced due to nonlinear coupling between torsional and translational displacements, caused by nonlinear force-displacement characteristics of the resisting elements.
2. Torsional instability depends mainly on torsional uncoupled frequency ratio Ω_θ. It has been shown that torsional oscillations are induced only when Ω_θ is greater than input frequency ratio Ω.
3. Translational damping helps to decrease the instability zone of torsional response, i.e., the smaller the translational damping, the greater the likelihood of significant torsional response.
4. Building plan aspect ratio influences the initiation of

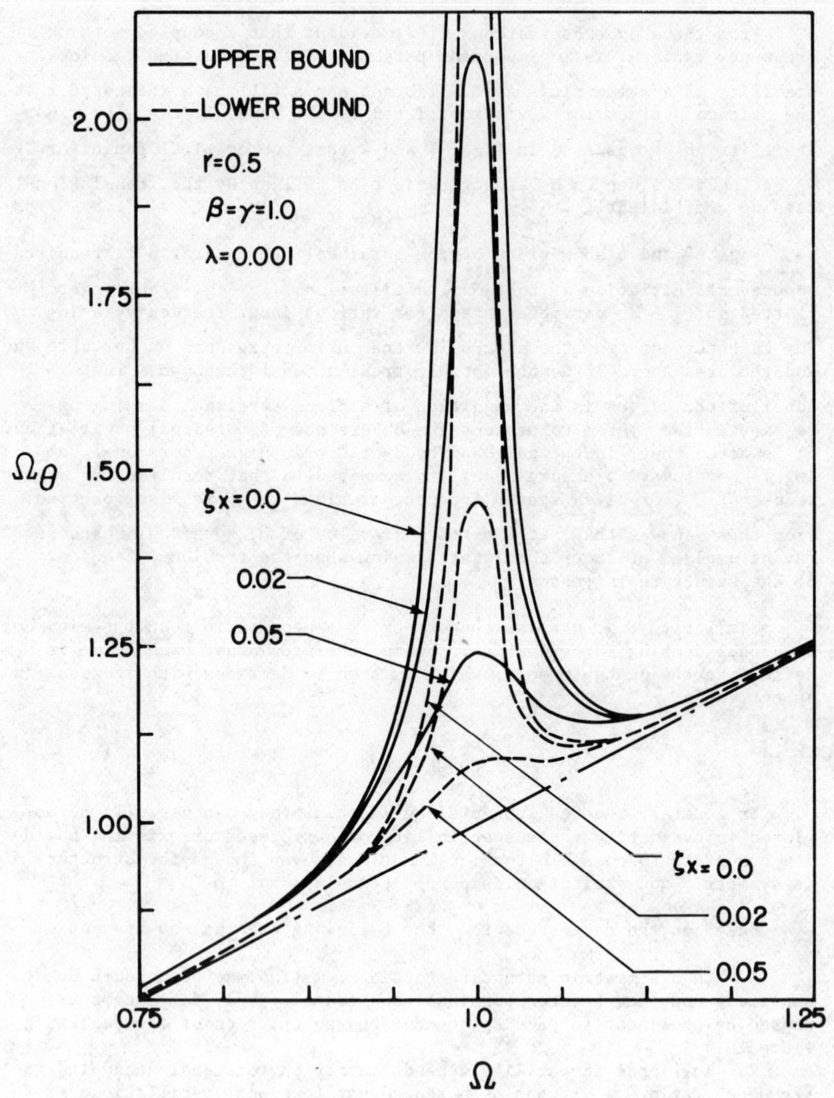

Fig. 4 Effect of Translational Damping Coefficient (ζ_x) on the Region of Induced Torsional Response

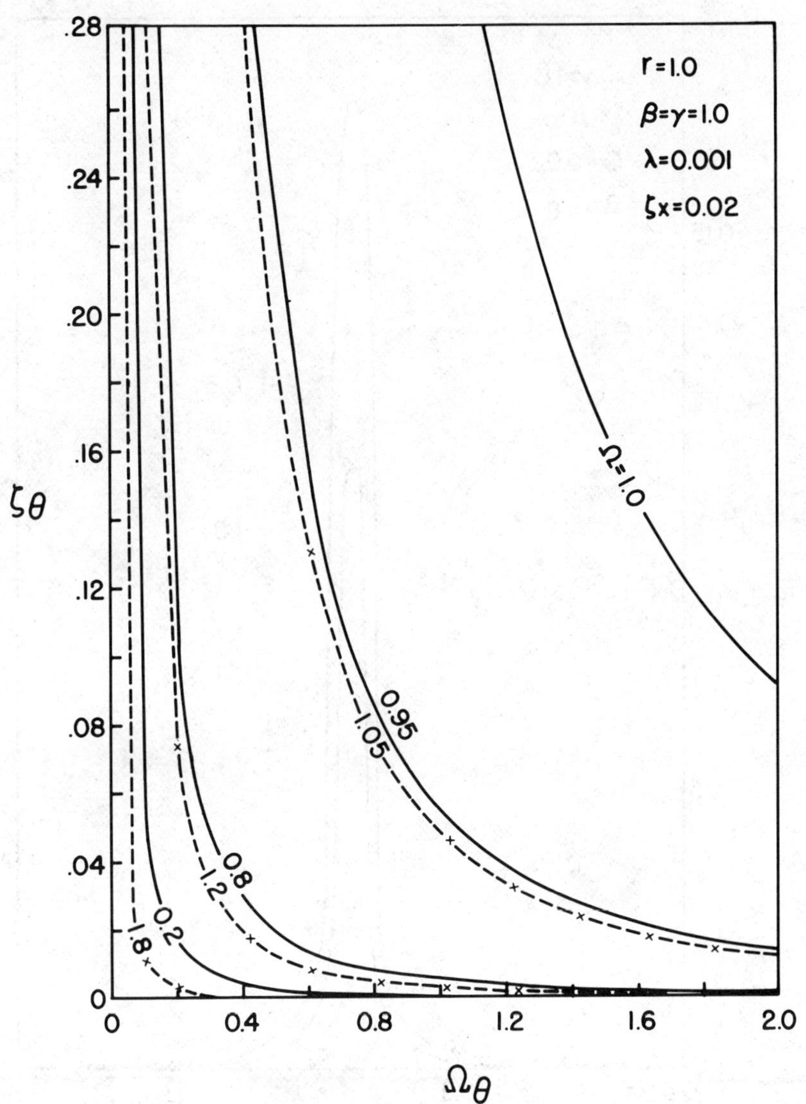

Fig. 5 Effect of Input Frequency Ratio (Ω) on the Minimum Torsional Damping Preventing Induced Torsional Oscillation

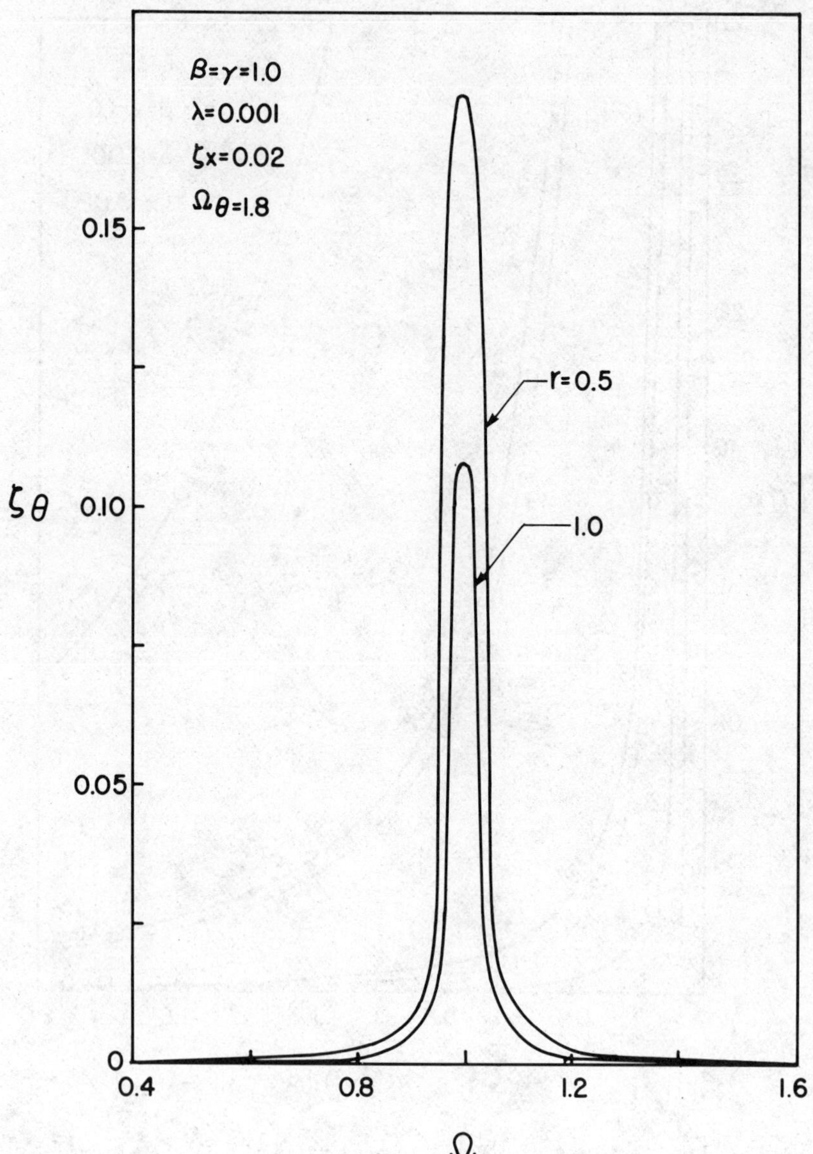

Fig. 6 Effect of Building Aspect Ratio (r) on Minimum Torsional Damping

torsional oscillations in a symmetric structure. Torsional response is more likely if the ground motion is in the direction parallel to the short overall dimension of a symmetric or nominally symmetric building.

5. Torsional damping is also an important parameter in controlling the initiation of torsional oscillations. A system with high torsional frequency requires a smaller value of torsional damping to prevent torsional oscillations for given input frequency. Additionally, larger torsional damping is necessary for structures with low aspect ratios.

APPENDIX I.-REFERENCES

1. Asmis, K.G., "Parametric Resonances in Nonlinear Mechanical Systems," thesis presented to McMaster University, at Hamilton, Ontario, in 1971 in partial fulfilment of the requirements for the degree of Doctor of Philosophy.
2. Bolotin, V.V., The Dynamic Stability of Elastic Systems, Holden-Day Inc., San Francisco, 1964.
3. Evensen, D.A., "Nonlinear Flexural Vibrations of Thin Circular Rings," Journal of Applied Mechanics, ASME Transactions, Vol. 33, 1966, pp. 553-560.
4. Evensen, D.A., "Nonlinear Flexural Vibrations of Thin Circular Rings," thesis presented to the California Institute of Technology, at Pasadena, California, in 1964 in partial fulfilment of the requirements for the degree of Doctor of Philosophy.
5. Kryloff, N., and Bogoliuboff, N., Introduction to Nonlinear Mechanics, Princeton University Press, New Jersey, 1947.
6. Newmark, N.M., "Torsion in Symmetric Buildings," Proceedings of the Fourth World Conference on Earthquake Engineering, Santiago, Chile, Vol. 2, 1969, pp. A3-19 to A3-32.
7. Nigam, N.C., and Housner, G.W., "Elastic and Inelastic Response of Framed Structures During Earthquakes," Proceedings of the Fourth World Conference on Earthquake Engineering, Santiago, Chile, Vol. 2, 1969, pp. A4-89 to A4-104.
8. Nigam, N.C., "Inelastic Interactions in the Dynamic Response of Structures," thesis presented to the California Institute of Technology, at Pasadena, California, in 1967 in partial fulfilment of the requirements for the degree of Doctor of Philosophy.
9. Syamal, P.K., and Pekau, O.A., "Coupling in the Dynamic Response of Nonlinear Unsymmetric Structures," presented at the October 6-8, 1980, Symposium on Computational Methods in Nonlinear Structural and Solid Mechanics, held in Washington, D.C.
10. Tso, W.K., "Induced Torsional Oscillations in Symmetrical Structures," International Journal of Earthquake Engineering and Structural Dynamics, Vol. 3, No. 4, April-June 1975, pp. 337-346.
11. Tso, W.K., and Asmis, K.G., "Torsional Vibration of Symmetrical Structures," Proceedings of the First Canadian Conference on Earthquake Engineering, Vancouver, Canada, 1971, pp. 178-186.

APPENDIX II.-NOTATION

The following symbols are used in the paper:

a,b = plan dimensions of building;

D	=	square of nondimensional amplitude of a single-degree-of freedom system;
e_x, e_y	=	static eccentricities;
$\bar{P}, \bar{Q}, \bar{R}$	=	average nondimensional response amplitudes in x-, y- and θ- directions, respectively;
r	=	building plan aspect ratio (a/b);
t	=	time;
U	=	amplitude of sinusoidal ground acceleration;
$u(t), v(t)$	=	lateral displacements of mass centre;
x, y	=	principal coordinate directions;
Γ	=	mass radius of gyration;
δ_j	=	in-plane displacements of resisting elements;
$\zeta_x, \zeta_y, \zeta_\theta$	=	critical damping ratios;
$\theta(t)$	=	rotational displacement;
$\Lambda_x(\tau), \Lambda_y(\tau), \Lambda_\theta(\tau)$	=	nondimensional response;
λ	=	coefficient of nonlinearity;
ξ_x, ξ_y, ξ_θ	=	nondimensional displacement perturbations;
τ	=	nondimensional time;
Ω_y	=	translational fundamental frequency ratio (ω_y/ω_x);
Ω_θ	=	torsional fundamental frequency ratio (ω_θ/ω_x);
Ω	=	sinusoidal ground excitation frequency ratio (ω/ω_x);
$\omega_x, \omega_y, \omega_\theta$	=	uncoupled fundamental frequencies of the structure;
ω	=	frequency of sinusoidal ground excitation.

3-D NONLINEAR DYNAMIC ANALYSIS OF STACKED BLOCKS

BY KIYOSHI MUTO[1], F.ASCE HIROMOTO TAKASE[2]
KIYOMI HORIKOSHI[2] HIROMICHI UENO[2]

INTRODUCTION

Since 1974 the aurhors have developed several computer codes of COLLAN-1, COLLAN-2V and COLLAN-2H, in order to simulate series of vibration tests of HTGR (High Temperature Gas Cooled Reactor) core. This paper describes first the 3-dimensional analysis COLLAN-3 of stacked blocks as the extended version of the 2-dimensional COLLAN-2V code, and second some numerical examples by this code and the COLLAN-2V. The block element is treated as a rigid body and the stacked blocks are connected at one corner of the adjacent (upper or lower) blocks. Since the top and bottom of each block are constrained by dowel pins, the block is restricted not to slip or twist on plane, but allowed to rotate about a horizontal axis. For formulating the equation of motion the principle of conservation of impulse and momentum is applied for each block motion, and the collision theory based on coefficient of restitution is applied for the horizontal impact between adjacent blocks or between blocks and boundary wall. By using the assumptions, mentioned above, a computer code called COLLAN-3 has been developed and 3-dimensional seismic excitation is available for two horizontal and one vertical directions.

As to the application of the analytical method several numerical examples are described in this paper regarding the tipping behaviors of 3 kinds of blocks and the overturn of book-shelves placed in a tall building subjected to earthquake ground motion. In addition, the simulation analysis of the forces vibration experiment of stacked blocks surrounded with rigid walls [1] is described comparing with experimental results conducted by Ikushima et al.

THEORETICAL DEVELOPMENT

The coordinate system is chosen as shown in Fig. 1. Using the

---

[1] Prof. Emeritus, Tokyo Univ., Director, Muto Inst., Tokyo, Japan
[2] System Engr., Kajima Corp., Research Engr., Muto Inst., Tokyo Japan

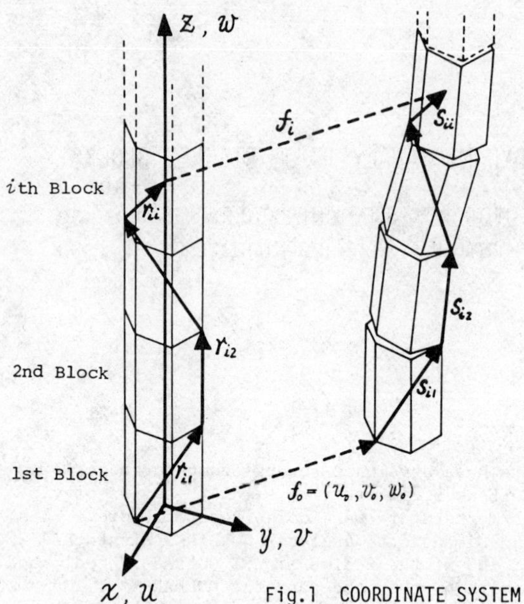

Fig.1 COORDINATE SYSTEM

ground displacement; $f_o = (u_o, v_o, w_o)$, the displacement of the i-th block; $f_i = (u_i, v_i, w_i)$, is expressed as follows:

$$f_i = \sum_{j=1}^{N} S_{ij} + f_o - \sum_{j=1}^{N} r_{ij} \tag{1}$$

where r_{ij} and S_{ij} are pendulum vectors before and after movement respectively, and also mean the j-th block components (r_{ij} S_{ij} are assumed to be zero if $j>i$) of the zigzaged route from the contact point of the column bottom to the center of gravity G of the i-th block.

Since the block is allowed only to rotate about horizontal axes, S_{ij} can be written by r_{ij} as a form of linear combination:

$$S_{ij} = T(\theta_j) \cdot r_{ij} \tag{2}$$

where $\theta_j = (\theta_j^x, \theta_j^y)$ is the rotational displacement of the j-th block and $T(\theta_j)$ is a 3x3 transformation matrix as shown below:

$$T(\theta_j) = \begin{bmatrix} 1 & 0 & \theta_j^y \\ 0 & 1 & -\theta_j^x \\ -\theta_j^y & \theta_j^x & 1 \end{bmatrix} \tag{3}$$

and then, for the velocity

$$\dot{S}_{ij} = K(\dot{\theta}_j) r_{ij} \tag{4}$$

where the matrix $K(\dot{\theta}_j)$ is the derivative of $T(\theta_j)$.

Let $P_i = (Q_i^x, Q_i^y, N)$ as shown in Fig. 2 be the internal force vector of the i-th block acting upon the $(i-1)$-th block. The equilibriums of force and moment about two horizontal axes through point A are expressed as follows:

$$P_{i+1} - P_i - m_i(\ddot{f}_i + g) = 0 \tag{5}$$

$$[S_{i+1i} \cdot P_{i+1}] - [S_{ii} \cdot m_i(\ddot{f}_i + g)] - I_i \ddot{\theta}_i = 0 \tag{6}$$

where g is the acceleration vector of gravity and m_i and I_i are the mass matrix and the rotational inertia matrix of the i-th block, respectively, and notation [•] indicates vector multiplication. Since $P_{N+1} = 0$ at the top block, P_i can be expressed by Eq.(6) as follows:

$$P_i = -\sum_{j=i}^{N} m_j (\ddot{f}_j + g) \tag{7}$$

By substituting P_i into Eq.(6), the following equation of motion can be obtained.

$$I_i \ddot{\theta}_i - \sum_{j=1}^{N} [S_{ji} \cdot m_j (\ddot{f}_j + g)] = 0 \tag{8}$$

Expressing Eq.(8) as each component, the equation of motion can be obtained regarding the rotational displacement as follows:

$$I_i^x \ddot{\theta}_i^x + \sum_{k=1}^{N} (\Phi_{ik}^{yy} + \Phi_{ik}^{zz}) \ddot{\theta}_k^x - \sum_{k=1}^{N} \Phi_{ik}^{yx} \ddot{\theta}_k^y = -\Psi_i^y (\ddot{w}_0 + g) + \Psi_i^z \ddot{v}_0 \tag{9a}$$

$$-\sum_{k=1}^{N} \Phi_{ik}^{xy} \ddot{\theta}_k^x + I_i^y \ddot{\theta}_i^y + \sum_{k=1}^{N} (\Phi_{ik}^{xx} + \Phi_{ik}^{zz}) \ddot{\theta}_k^y = \Psi_i^x (\ddot{w}_0 + g) - \Psi_i^z \ddot{u}_0 \tag{9b}$$

where $\Phi_{ik}^{yx} = \sum_{j=1}^{N} m_j S_{ji}^y r_{jk}^x$ and $\Psi_i^x = \sum_{j=1}^{N} m_j S_{ji}^x$.

Eq.(9) is also expressed in terms of kinetic and potential energy of a block.

Let $T = (I^x \dot{\theta}^{x2} + I^y \dot{\theta}^{y2})/2 + m(\dot{u}^2 + \dot{v}^2 + \dot{w}^2)/2$ and $V = mg(-\theta^y r_x + \theta^x r_y)$

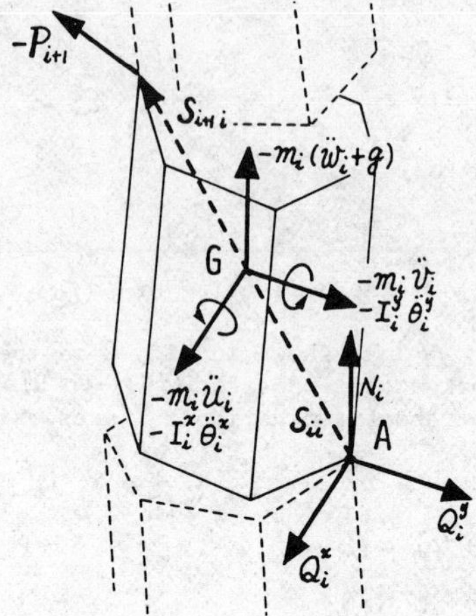

Fig.2 EQUILIBRIUM FORCE

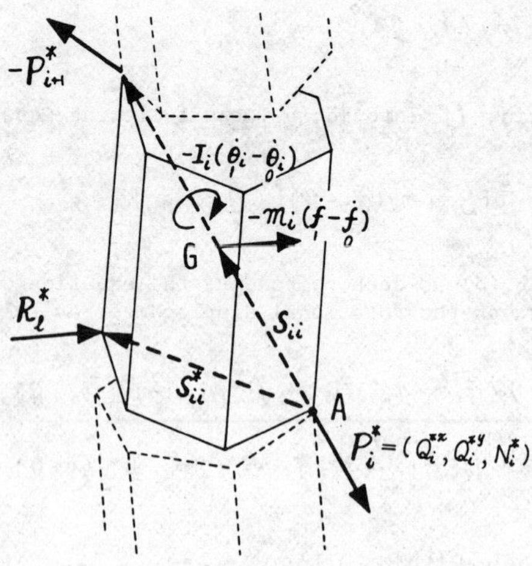

Fig.3 EQUILIBRIUM OF IMPULSE AND MOMENTUM

be kinetic energy and potential energy of a block, respectively. Substituting these T, V into Lagrange's equation, the following equation can be obtained.

$$\begin{bmatrix} I^x + m(r_z^2 + r_y^2) & -m r_z r_y \\ -m r_z r_y & I^y + m(r_z^2 + r_x^2) \end{bmatrix} \begin{Bmatrix} \ddot{\theta}^x \\ \ddot{\theta}^y \end{Bmatrix} = m \begin{Bmatrix} -(\ddot{w}_0 + g) r_y + \ddot{v}_0 r_z \\ (\ddot{w}_0 + g) r_x - \ddot{u}_0 r_z \end{Bmatrix} \tag{10}$$

When collisions occur between upper and lower blocks (rocking collision) and between horizontal adjacent blocks and/or wall, the velocities of all blocks are changed. As general equilibriums of impulse and momentum, the case that the impulse R_ℓ^* exerts on the l-th block as shown in Fig.3 is considered. In the same manner as described previously, let $P_i^* = (Q_i^{*x}, Q_i^{*y}, N_i^*)$ be the internal impulse vector of the i-th block exerting on the $(i-1)$-th block. The equilibriums of impulse and momentum of the i-th block can be represented as follows:

$$P_{i+1}^* - P_i^* - m_i(\underset{1}{\dot{f}_i} - \underset{0}{\dot{f}_i}) - \delta_{i\ell} R_\ell^* = 0 \tag{11}$$

$$[S_{i+1\,i} \cdot P_{i+1}^*] - [S_{ii} \cdot m_i(\underset{1}{\dot{f}_i} - \underset{0}{\dot{f}_i})] - I_i(\underset{1}{\dot{\theta}_i} - \underset{0}{\dot{\theta}_i}) + \delta_{i\ell}[S_{\ell\ell}^* \cdot R_\ell^*] = 0 \tag{12}$$

where suffixes 0 and 1 mean the conditions just before and after impact respectively, and δ_{ij} is the Kronecker's symbol, that is $\delta_{i\ell}=1$ if $i=\ell$ and $\delta_{i\ell}=0$ if $i\neq\ell$. By eliminating P_i^* in Eq.(11), the generalized equation of equilibrium of impulse and momentum is derived in the following form:

$$I_i(\underset{1}{\dot{\theta}_i} - \underset{0}{\dot{\theta}_i}) + \sum_{j=1}^{N}[S_{ji} \cdot m_j(\underset{1}{\dot{f}_j} - \underset{0}{\dot{f}_j})] - [S_{\ell j} \cdot R_\ell^*] = 0 \tag{13}$$

The occurrence of rocking collision between upper and lower blocks is judged by the relative vertical displacement of these two blocks. Since the external impulse $R_\ell^* = 0$, Eq.(13) becomes as follows:

$$I_i \underset{1}{\dot{\theta}_i} + \sum_{j=1}^{N}[S_{ji} \cdot m_j \underset{1}{\dot{f}_j}] = I_i \underset{0}{\dot{\theta}_i} + \sum_{j=1}^{N}[S_{ji} \cdot m_j \underset{0}{\dot{f}_j}] \tag{14}$$

In case that a block collides with an adjacent block or with a boundary wall, the external impulse R_ℓ^* in Eq.(13) is unknown. When the impulse at the m-th column is $-R_\ell^*$ and the impulse at $(m+1)$-th column is $+R_\ell^*$, R_ℓ^* can be expressed to be $R_\ell^* = \alpha t^*$ $(|t^*|=1)$ where t^* is the direction of collision given by the situation of the two blocks and α is an unknown scalar. The equation of equilibrium (14) is rewritten with respect to m-th and $(m+1)$-th columns as follows, respectively:

$$ {}^{m+1}I_i({}^{m+1}\underset{1}{\dot{\theta}}_i - {}^{m+1}\underset{0}{\dot{\theta}}_i) + \sum_{j=1}^{N}[{}^{m+1}S_{ji}\cdot {}^{m+1}m_j({}^{m+1}\underset{1j}{\dot{f}} - {}^{m+1}\underset{0j}{\dot{f}})] - \alpha[{}^{m+1}S^*_{\ell j}\cdot t^*] = 0 \tag{15a} $$

$$ {}^{m}I_i({}^{m}\underset{1}{\dot{\theta}}_i - {}^{m}\underset{0}{\dot{\theta}}_i) + \sum_{j=1}^{N}[{}^{m}S_{ji}\cdot {}^{m}m_j({}^{m}\underset{1}{\dot{f}}_j - {}^{m}\underset{0}{\dot{f}}_j)] + \alpha[{}^{m}S^*_{\ell j}\cdot t^*] = 0 \tag{15b} $$

On the other hand, the collision theory is expressed as follows:

$$ {}^{m}\underset{1}{\dot{u}}{}^*_\ell - {}^{m+1}\underset{1}{\dot{u}}{}^*_\ell = -e^*({}^{m}\underset{0}{\dot{u}}{}^*_\ell - {}^{m+1}\underset{0}{\dot{u}}{}^*_\ell) \tag{16} $$

where e^* is the coefficient of restitution, $\dot{u}^*_\ell$ is the velocity of the block in the direction of collision expressed as follows:

$$ \dot{u}^*_\ell = \pm(\dot{f}^*_\ell \cdot t^*) \tag{17} $$

The rotational velocity and impulse can be obtained by Eqs.(15) and (16).

As for the collision between a block and a boundary wall, the equations can also be derived in the same manner described above.

NUMERICAL EXAMPLES

Numerical examples in case of the tipping behavior of the blocks are shown in Fig.4.

The first numerical example represents the free vibration of a block having the initial condition; $\theta^y=-0.1$ rad. and $\dot{\theta}^x=0.8$ rad/sec. The time histories of the rotational displacements are shown in Fig.5 and the loci of the center of gravity projected on horizontal plane are shown in Fig.6. It can be found from Figs.5 and 6 that the motion of rectangular block decreases more rapidly than the others.

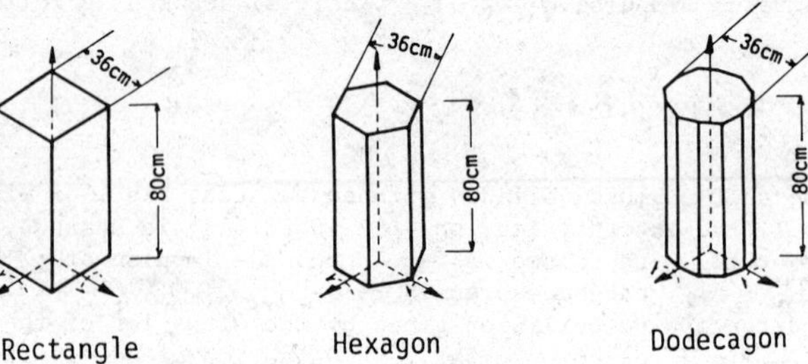

Fig.4 BLOCKS FOR NUMERICAL EXAMPLES

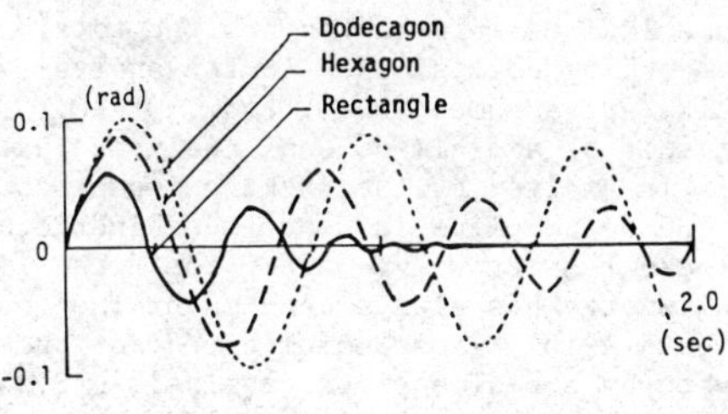

Time history of θ^x

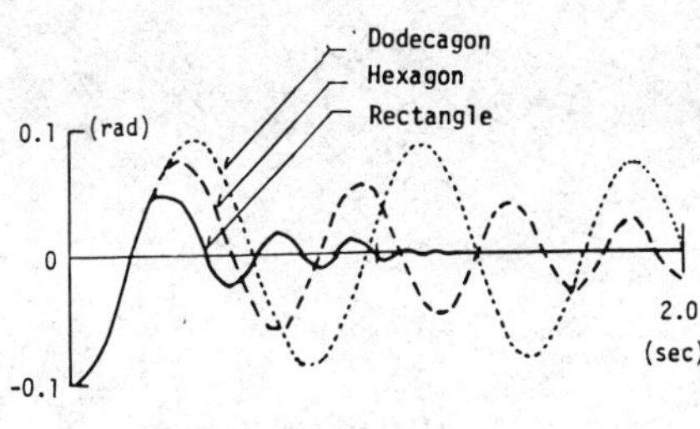

Time history of θ^y

Fig.5 COMPARISON OF DIFFERENT TYPE OF BLOCK ROCKING

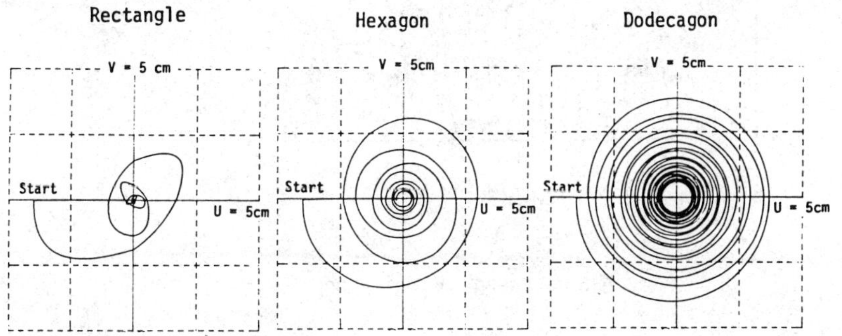

Fig.6 HORIZONTAL LOCI OF CENTER OF GRAVITY

The second numerical example represents the total energy of the hexagonal block by rocking collisions. In the analysis its initial position and velocity are assumed; $\Theta^x=0.$ rad., $\Theta^y=-0.1$ rad., $\dot{\Theta}^x=\dot{\Theta}^y=0$. It is the unique result of a 3-dimensional analysis that the rocking motion about x-axis is excited by the rocking impact about y-axis. In Fig.7 broken line and dotted line indicate the kinetic and the potential energies of the block, respectively. After the first impact, a certain degree of potential energy has always been maintained due to the rocking interaction about both axis. These complicated motions of the block could not be analysed by a 2-dimensional analysis. Symbol ● indicates the result of experiment. The analysis simulates them fairly well.

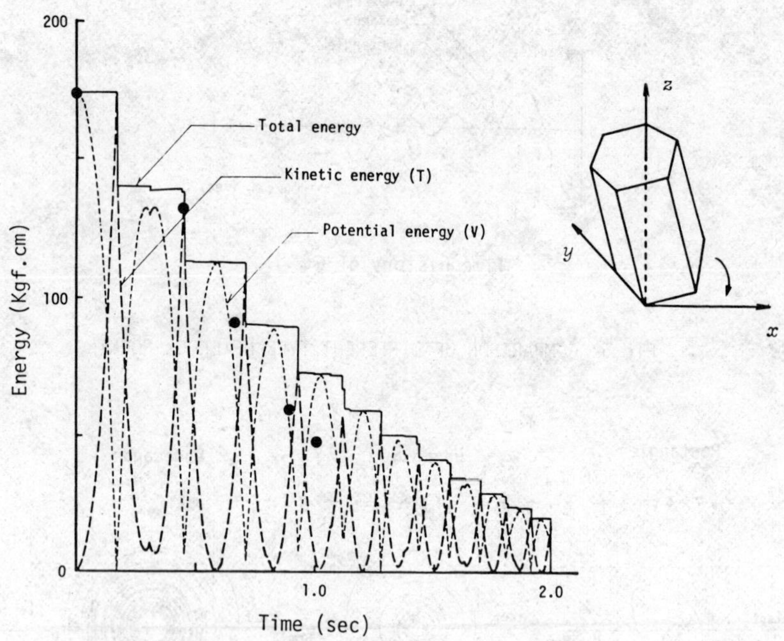

Fig.7 ENERGY OF ONE BLOCK

2-DIMENSIONAL DYNAMIC ANALYSIS OF BOOK-SHELF

Fig.8 shows the results of 2-dimensional dynamic analysis of book-shelves (aspect ratio B/H=0.135) placed on every floor of a 20 storied steel building subjected to EL CENTRO earthquake ground motions with different intensities.

In the analyses the maximum accelerations of these earthquake waves are assumed to be from 100 gal to maximum 500 gal with each increment of 10 gal. The fundamental vibration period of the building is about 2.0 seconds. In Fig.8, solid line indicates the maximum response accelerations of floors in case of 200 gal ground motion. The analyses show that the overturning phenomena of shelves depend not only on the magnitudes of floor accelerations well, but also on the periodical characteristics of these waves.

It suggests a special characteristics of non-linear response that in the middle floors of the building NOT FALL zones can be seen in FALL zone in spite of large floor accelerations.

Symbols o, ● indicate NOT ROCK & MOVE and FALL in the experimental results of the forced vibration test conducted by Kajima Construction Institute under same conditions with the analysis. The analytical results are in fairly good agreement with these experimental ones.

Fig.9 shows the response of two kinds of book-shelves under two different conditions. The first and the second ones are 1-tier and 3-tier (placed in the center of the 1st floor). The others are placed on the same floor against walls with 0.5cm gap. In these analyses it is ascertained that 3-tier shelves are stable compared with 1-tier ones subjected to earthquake vibrations, and that placing them against walls are very effective for the shelves to resist overturning.
The latter is understood, judging from the experimental results and many experiences about actual earthquake damages of furnitures in the past.

FORCED VIBRATION OF STACKED BLOCKS

The analyses for seismic safety of HTGR core consisting of stacked graphite blocks are important problems which have already been investigated by many researchers. Some useful computer programs have been developed, but almost all of them use spring-dashpot models. In order to confirm the reliability of program code COLLAN-3 as mentioned above, several collision problems are analyzed by using this code. Fig.10 shows the 13 stacked hexagonal blocks which are self supporting and provided with gaps between surrounding rigid walls. A numerical model of the column is effectively idealized into 7 stacked blocks thus saving computer cost. The coefficient of restitution is assumed to be constantly 0.6 and the integral time interval is 1/500 sec.

Figs.11 and 12 show the maximum responses subjected to sinusoidal excitations in the flat-to-flat direction (x-direction) and the

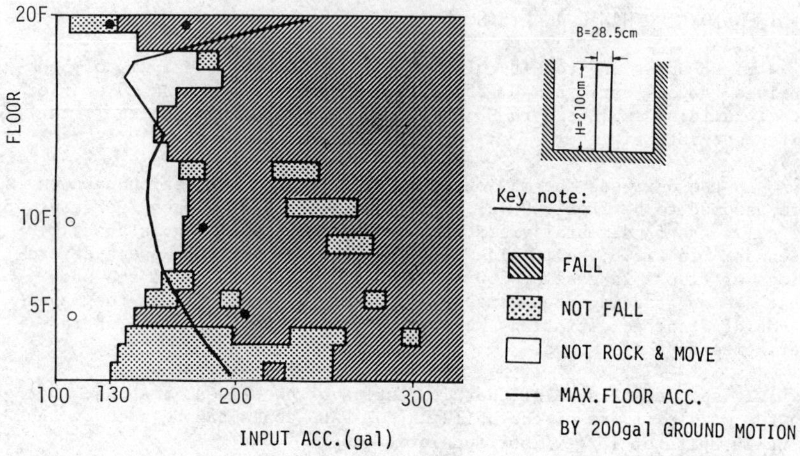

Fig.8 FALL OF BOOK-SHELF IN EACH FLOOR OF A 20 STORIED BUILDING

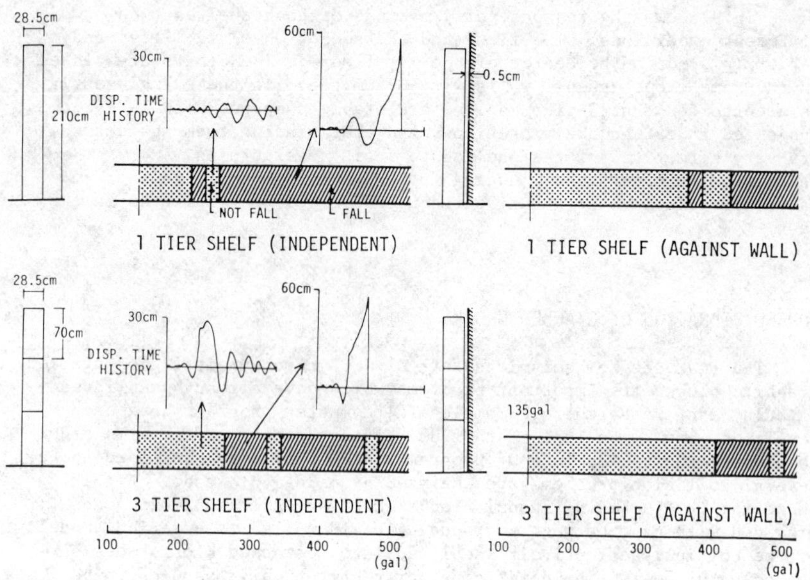

Fig.9 COMPARISON OF FALL OF TIER SHELF

corner-to-corner direction (y-direction) of the hexagonal block. Symbols MV*, VEL, and R* represent the maximum momentum and absolute velocity at center of gravity, and external impulse at block edge by impact of block and wall, respectively. Compared with responses of both directions, the response velocities are almost the same, but the momenta and impulses in the y-direction are smaller by about 0.4 times of those in the x-direction because of the interaction in both directions. The perpendicular response is prominent in case of y-directional excitation, which corresponds with the experiment.

Fig.13 shows the maximum responses subjected to EL CENTRO earthquake. The responses are very small compared with the former sinusoidal excitation in spite of the same intensity level of ground acceleration. The amplifications of velocity are small, and the perpendicular collision between blocks and wall does not occur. The momenta MV* in the y-direction indicated by broken line is generated only by the rocking collisions.

Fig.14 shows the resonance curves of impact velocities subjected to 15 cycles of sinusoidal excitations with the acceleration amplitude of 500 gal. In cases of 5mm and 15mm gap, these computational results are almost same as the experimental ones. However in case of 25mm small difference can be seen due to the numerical reduction in modelling.

Fig.15 shows the resonance curves of impact velocities subjected to sinusoidal excitations with different intensities. With increase of exciting amplitude, the resonance peak gradually shifts to high frequency region, as a characteristics phenomenon of non-linear dynamic response. The same property of response was found in the experiment.

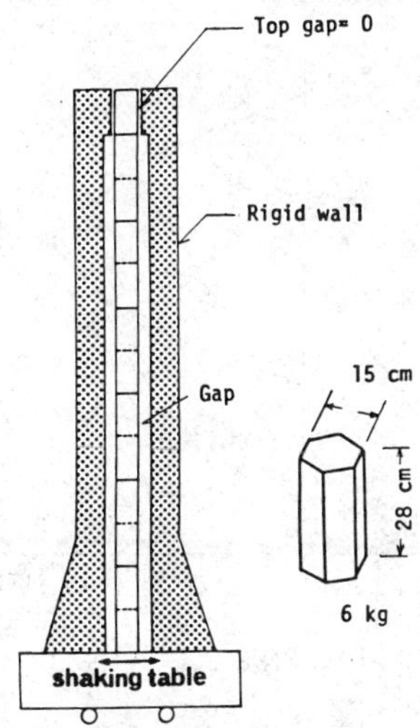

Fig.10 NUMERICAL MODEL

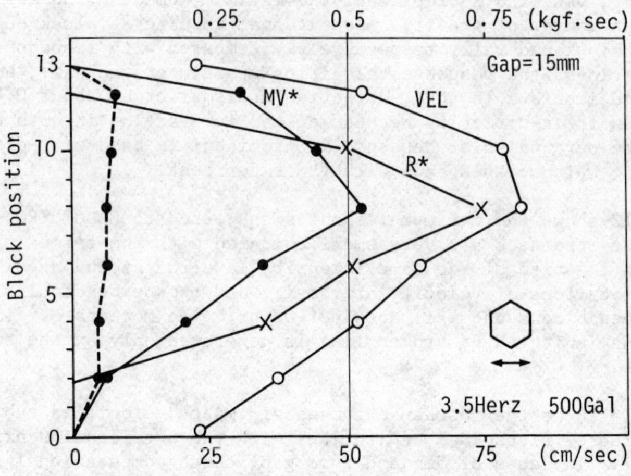

Fig.11 EFFECT OF EXCITING DIRECTION (X)

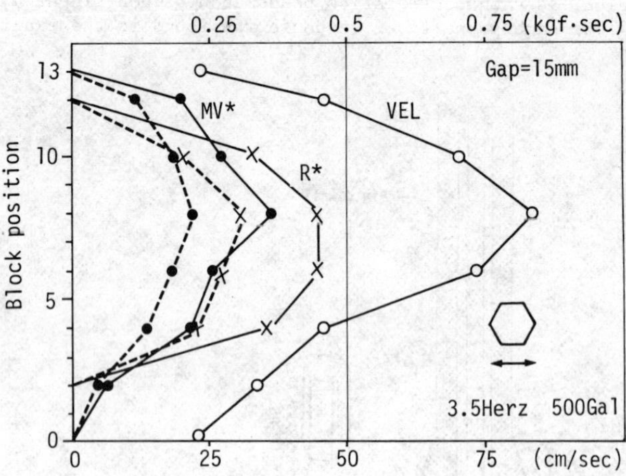

Fig.12 EFFECT OF EXCITING DIRECTION (Y)

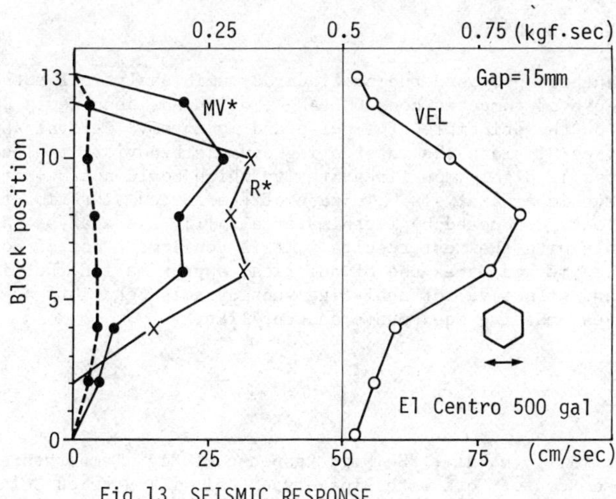

Fig.13 SEISMIC RESPONSE

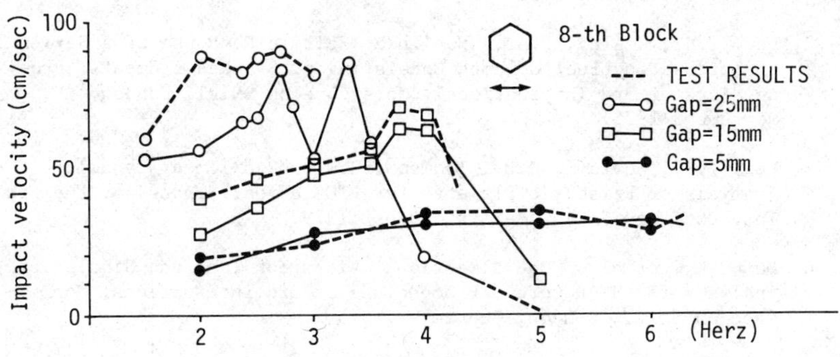

Fig.14 EFFECT OF GAP WIDTH

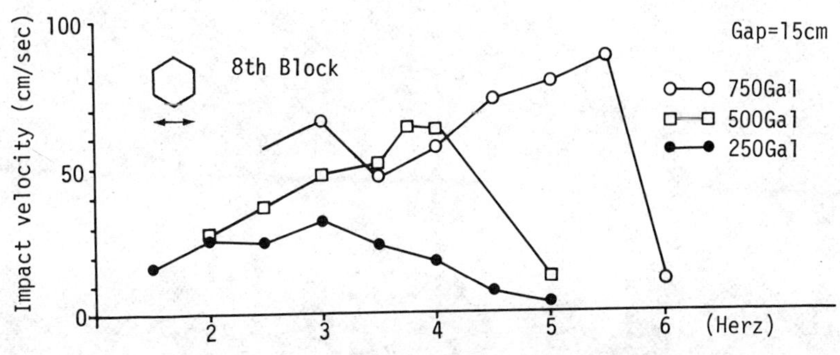

Fig.15 EFFECT OF INPUT ACCELERATION LEVEL

SUMMARY AND CONCLUSIONS

The authors proposed the nonlinear dynamic analytical method of blocks by use of computer code COLLAN-3, which has been newly developed, based on the principle of impulse and momentum. Several application analyses simulate the interesting dynamic behavior of blocks and book-shelves in highly non-linear region which could not be obtained by a 2-dimensional analysis. In case of the forced vibration test of stacked blocks conducted by Ikushima et al. [1], the analyses agree considerably with the test results. It is confirmed that of course the proposed method makes use of numerical approximation, but it is suitable and effective for analyzing such dynamic behaviors as blocks, book-shelves, massive equipments and the like.

REFERENCES

1. IKUSHIMA,T., et al., "Seismic Response of High Temperature Gas-Cooled Reactor Core with Block-Type Fuel, (I) Stacked Column Vibration Experiment," Journal of the Atomic Energy Society of Japan, Vol.22 No.1, JAN. 1980.

2. Lee, T.H., Wesley, D.A., "Nonlinear Seismic Response of a Series of Interacting Fuel Columns Consisting of Stacked Elements," Proceedings of 3rd International Conference on SMiRT, London, September 1975.

3. Lee, T.H., Shatoff, H.D., Thomson, R.W., "Non-linear Dynamic Analysis of Prismatic Elements for HTGR cores," Trans. of the 5th Int. Conf. on SMiRT, Berlin, August 1979.

4. Muto, K., et al., "One Dimensional Vibration Test and Simulation Analysis for HTGR Core," Proceedings of 3rd International Conference on SMiRT, London, September 1975.

5. Muto, K., et al., "Two-dimensional Vibration Test and Simulation Analysis for a Vertical Slice Model of HTGR Core," Proceedings of 4th International Conference on SMiRT, San Francisco, August 1977.

Subject Index

Author Index